Universitext

Universitext

Universitext is a series of textbooks that presents material from a wide variety of mathematical disciplines at master's level and beyond. The books, often well class-tested by their author, may have an informal, personal even experimental approach to their subject matter. Some of the most successful and established books in the series have evolved through several editions, always following the evolution of teaching curricula, into very polished texts.

Thus as research topics trickle down into graduate-level teaching, first textbooks written for new, cutting-edge courses may make their way into *Universitext*.

More information about this series at http://www.springer.com/series/223

Dorina Mitrea

Distributions, Partial Differential Equations, and Harmonic Analysis

Second Edition

Springer

Dorina Mitrea
Department of Mathematics
University of Missouri
Columbia, MO, USA

ISSN 0172-5939   ISSN 2191-6675 (electronic)
Universitext
ISBN 978-3-030-03295-1   ISBN 978-3-030-03296-8 (eBook)
https://doi.org/10.1007/978-3-030-03296-8

Library of Congress Control Number: 2018960188

Mathematics Subject Classification (2010): 35A08, 35A09, 35A20, 35B05, 35B53, 35B65, 35C15, 35D30, 35E05, 35G05, 35G35, 35H10, 35J05, 35J30, 35J47, 42B20, 42B37, 46E35

This Springer imprint is published by the registered company Springer Nature Switzerland AG
The registered company address is: Gewerbestrasse 11, 6330 Cham, Switzerland

Dedicată cu drag lui Diana și Adrian,
Love, Mom

Preface to the Second Edition

The main additions in the current edition pertain to fundamental solutions and Sobolev spaces. The list of fundamental solutions from Chapter 7 has been expanded to include the Helmholtz operator (§7.6), the perturbed Dirac operator (§7.10), and their iterations (§7.7 and §7.11). Understanding quantitative and qualitative features of the said fundamental solutions is of paramount importance in scattering theory, where Helmholtz and perturbed Dirac operators play a prominent role. Special emphasis is placed on elucidating the nature of their singularity at the origin and asymptotic behavior at infinity. In this vein, a number of useful results concerning the behavior of Hankel functions are summarized in §14.10.

The new material concerning Sobolev spaces is contained in Chapter 12. Our approach builds the theory from ground up and is completely self-contained. Presently, we limit ourselves to L^2-based Sobolev spaces, where the connection with the Fourier analysis developed earlier in the monograph is most apparent. Such an approach is also natural from a pedagogical point of view. This being said, many key results have been given proofs which canonically adapt to more general situations (such as L^p-based Sobolev spaces, weighted Sobolev spaces, Sobolev spaces with vanishing traces). The starting point is the treatment of global Sobolev spaces $H^s(\mathbb{R}^n)$ in $\mathbb{R}^n$ of arbitrary smoothness $s \in \mathbb{R}$, via the Fourier transform (cf. §12.1). The theory is natural and elegant since the Fourier transform is an isometry on $L^2(\mathbb{R}^n)$. The next step is the consideration of Sobolev spaces in arbitrary open subsets Ω of $\mathbb{R}^n$. There are two natural venues to define the latter brand of spaces. One approach, yielding the scale $H^s(\Omega)$ with $s \in \mathbb{R}$, proceeds via restriction from the corresponding spaces in $\mathbb{R}^n$ (cf. §12.2). For integer amounts of smoothness $m \in \mathbb{N}$, one may also introduce Sobolev spaces $\mathcal{H}^m(\Omega)$ in an intrinsic fashion, demanding that distributional derivatives up to order m are square-integrable in Ω (cf. §12.3).

In relation to Sobolev spaces in an open set $\Omega \subseteq \mathbb{R}^n$, two basic theorems are proved in the case when Ω is a bounded Lipschitz domain. The first is the density of restrictions to Ω of smooth compactly supported functions from $\mathbb{R}^n$ in the intrinsic Sobolev space $\mathcal{H}^m(\Omega)$. The second is the construction of Calderón's extension operator from $\mathcal{H}^m(\Omega)$ to $\mathcal{H}^m(\mathbb{R}^n)$. Among other things, these results allow the identification of the intrinsic Sobolev space $\mathcal{H}^m(\Omega)$ with the restriction Sobolev space $H^m(\Omega)$ whenever $m \in \mathbb{N}$ and Ω is a bounded Lipschitz domain.

In §12.4, we treat L^2-based Sobolev spaces $H^{1/2}$, of fractional order 1/2, on boundaries of Lipschitz domains. These are defined as spaces of square-integrable functions satisfying a finiteness condition involving a suitable Gagliardo–Slobodeckij semi-norm. In the setting of bounded Lipschitz domains, this study ties up with the earlier theory via extension and trace results, as explained in §12.5. Specifically, having first established the density of Lipschitz functions on $\partial\Omega$, where Ω is a bounded Lipschitz domain, in the space $H^{1/2}(\partial\Omega)$, we then prove the existence of a linear and bounded trace operator from $H^1(\Omega)$ into $H^{1/2}(\partial\Omega)$, and of a linear and bounded extension operator from $H^{1/2}(\partial\Omega)$ into $H^1(\Omega)$.

The work on this project has been supported in part by the Simons Foundation grants # 426669 and # 200750 and by a University of Missouri Research Leave grant. The author wishes to express her gratitude to these institutions.

Columbia, MO, USA   Dorina Mitrea
July 2018

Preface to the First Edition

This book has been written from the personal perspective of a mathematician working at the interface between partial differential equations and harmonic analysis. Its aim is to offer, in a concise, rigorous, and largely self-contained form, a rapid introduction to the theory of distributions and its applications to partial differential equations and harmonic analysis. This is done in a format suitable for a graduate course spanning either over one semester, when the focus is primarily on the foundational aspects, or over a two-semester period that allows for the proper amount of time to cover all intended applications as well.

Throughout, a special effort has been made to develop the theory of distributions not as an abstract edifice but rather give the reader a chance to see the rationale behind various seemingly technical definitions, as well as the opportunity to apply the newly developed tools (in the natural build-up of the theory) to concrete problems in partial differential equations and harmonic analysis, at the earliest opportunity.

In addition to being suitable as a textbook for a graduate course, the monograph has been designed so that it may also be used for independent study since the presentation is reader-friendly, mostly self-sufficient (for example, all auxiliary results originating outside the scope of the present monograph have been carefully collected and presented in the appendix), and a large number of the suggested exercises have complete solutions.

Columbia, MO, USA   Dorina Mitrea
March 2013

Contents

Introduction

It has long been recognized that there is a large overlap and intricate interplay among Distribution Theory (DT), Partial Differential Equations (PDE), and Harmonic Analysis (HA). The purpose of this monograph is to guide a reader with a background in basic real analysis through the journey taking her/him to the stage when such connections become self-evident.

Another goal of the present book is to convince the reader that traditional distinctions made among these branches of mathematics are largely artificial and are often simply a matter of choice in focus. Indeed, given the manner in which they complement, motivate, and draw inspiration from one another, it is not necessarily a stretch to attempt to pursue their development virtually simultaneously.

Concerning the triumvirate DT, PDE, HA, while there exist a number of good reference texts available on the market, they are by and large conceived in such a way that they either emphasize more one of these topics, typically at the detriment of the others, or are simply not particularly well suited for a nonspecialist. By way of contrast, not only is the present text written in a way that brings together and blends the aforementioned topics in a unified, coherent body of results, but the resulting exposition is also sufficiently detailed and reader-friendly so that it may be read independently, outside the formal classroom setting. Indeed, the book is essentially self-contained, presents a balanced treatment of the topics involved, and contains a large number of exercises (upwards of two hundred, more than half of which are accompanied by solutions), which have been carefully chosen to amplify the effect, and substantiate the power and scope, of the theory discussed here.

While the topics treated are classical, the material is not entirely standard since a number of results are new even for a seasoned practitioner, and the overall architectural design of the monograph (including the way in which certain topics are covered) is original.

Regarding its inception, the present monograph is an expanded version of the notes I prepared for a course on distribution theory I taught in the Spring of 2007 and the Spring of 2011, at the University of Missouri. My intention was to present the theory of distributions not as an abstract edifice but rather give the student a chance to instantaneously see the justification and practical benefits of the multitude of seemingly technical definitions and results, as well as give her/him the opportunity to immediately see how the newly introduced concepts (in the natural build-up of the theory) apply to concrete problems in partial differential equations and harmonic analysis.

Special care has been paid to the pedagogical aspect of the presentation of the material in the book. For example, a notable feature of the present monograph is the fact that fundamental solutions for some of the most basic differential operators in mathematical physics and engineering, including Laplace, heat, wave, polyharmonic, Dirac, Lamé, Stokes, and Schrödinger, are systematically deduced starting from first principles. This stands in contrast with the more common practice in the literature in which one starts with a certain distribution (whose origins are fairly obscure) and simply checks that the distribution in question is a fundamental solution for a given differential operator. Another feature is the emphasis placed on the interrelations between topics. For example, a clear picture is presented as to how DT vastly facilitates the computation of fundamental solutions, and the development of singular integral operators, tools which, in turn, are used to solve PDE as well as represent and estimate solutions of PDE.

The presentation is also conceived in such a way as to avoid having to confront heavy duty topology/functional analysis up front, in the main narrative. For example, the jargon associated with the multitude of topologies on various spaces of test functions and distributions is minimized by deferring to an appendix the technical details while retaining in the main body of the monograph only those consequences that are most directly relevant to the fluency of the exposition.

While the core material I had in mind deals primarily with the theory of distributions, the monograph is ultimately devised in such a way as to make the present material a solid launching pad for a number of subsequent courses, dealing with allied topics, including:

- Harmonic Analysis
- Partial Differential Equations
- Boundary Integral Methods
- Sobolev Spaces
- Pseudodifferential Operators

For example, the theory of singular integral operators of convolution type in $L^2(\mathbb{R}^n)$ is essentially developed here, in full detail, up to the point where more specialized tools from harmonic analysis (such as the Hardy–Littlewood maximal operator and the Calderón–Zygmund lemma) are typically involved in order to further extend this theory (via a weak-$(1, 1)$ estimate, interpolation with L^2, and then duality) to L^p spaces with $p \in (1, \infty)$, as in [8], [11], [19], [26], [52], [68], [69], [70], among others.

Regarding connections with partial differential equations, the Poisson problem in the whole space,

$$Lu = f \quad in \quad \mathcal{D}'(\mathbb{R}^n), \tag{1}$$

is systematically treated here for a variety of differential operators L, including the Laplacian, the biharmonic operator, and the Lamé system. While Sobolev spaces are not explicitly considered (the interested reader may consult in this regard [2], [3], [15], [43], [47], [48], [82], to cite just a small fraction of a large body of literature on this topic), they lurk just beneath the surface as their presence is implicit in estimates of the form

$$\sum_{|\alpha|=m} \|\partial^\alpha u\|_{L^p(\mathbb{R}^n)} \leq C(L,p)\|f\|_{L^p(\mathbb{R}^n)}, \qquad 1 < p < \infty, \tag{2}$$

where m denotes the order of the elliptic differential operator L.

Such estimates are deduced from an integral representation formula for u, involving a fundamental solution of L, and estimates for singular integral operators of convolution type in $\mathbb{R}^n$. In particular, this justifies two features of the present monograph: (1) the emphasis placed on finding explicit formulas for fundamental solutions for a large number of operators of basic importance in mathematics, physics, and engineering; (2) the focus on the theory of singular integral operators of convolution type, developed alongside the distribution theory. In addition, the analysis of the Cauchy problem formulated and studied for the heat and wave operators once more underscores the significance of the fundamental solutions for the named operators. As a whole, this material is designed to initiate the reader into the field of partial differential equations. At the same time, it complements, and works well in tandem with, the treatment of this subject in [5], [14], [24], [33], [34], [35], [50], [74], [75], [81].

Whenever circumstances permit it, other types of problems are brought into play, such as the Dirichlet and Neumann problem in the upper half-space for the Laplacian, as well as more general second order systems. In turn, the latter genre of boundary value problems motivates introducing and developing boundary integral methods, and serves as an opportunity to highlight the basic role that layer potential operators play in this context. References dealing with the latter topic include [37], [39], [50], [53], [80].

The analysis of the structure of the boundary layer potential operators naturally intervening in this context also points to the possibility of considering larger classes of operators where the latter may be composed, inverted, etc., in a stable fashion. This serves as an excellent motivation for the introduction of such algebras of operators as pseudodifferential and Fourier integral operators, a direction which the interested reader may then pursue in, e.g., [27], [73], [75], [81], to name a few sources.

A brief description of this book's contents is as follows.

Chapters 1–2 are devoted to the development of the most basic aspects of the theory of distributions. Starting from the discussion of the Cauchy problem for a vibrating infinite string as a motivational example, the notion of weak derivative is introduced as a mean of extending the notion of solution to a more general setting, where the functions involved may lack standard pointwise differentiability properties. After touching upon classes of test functions, the space of distributions is then introduced and studied from the perspective of a topological vector space with various other additional features (such as the concept of support, and a partially defined convolution product). Chapter 3 contains material pertaining to the Schwartz space of functions rapidly decaying at infinity and the Fourier transform in such a setting.

In Chapter 4 the action of the Fourier transform is further extended to the setting of tempered distributions, and several distinguished subclasses of tempered distributions are introduced and studied (including homogeneous and principal value distributions). The foundational material developed up to this point already has significant applications to harmonic analysis and partial differential equations. For example, a general, higher dimensional jump-formula is deduced in this chapter for a certain class of tempered distributions (that includes the classical harmonic Poisson kernel) which is later used as the main tool in deriving information about the boundary behavior of layer potential operators associated with various partial differential operators and systems. Also, one witnesses here how singular integral operators of central importance to harmonic analysis (such as the Riesz transforms) naturally arise as an extension to L^2 of the convolution product of tempered distributions of principal value type with Schwartz functions.

The first explicit encounter with the notion of fundamental solution takes place in Chapter 5, where the classical Malgrange–Ehrenpreis theorem is presented. Subsequently, in Chapter 6, the concept of hypoelliptic operator is introduced and studied. In particular, here a classical result, due to L. Schwartz, is proved to the effect that a necessary and sufficient condition for a linear, constant coefficient differential operator to be hypoelliptic in the entire ambient space is that the named operator possesses a fundamental solution with singular support consisting of the origin alone. In Chapter 6 we also prove an integral representation formula and interior estimates for a subclass of hypoelliptic operators, which are subsequently used to show that null solutions of these operators are real-analytic.

One of the main goals in Chapter 7 is identifying (starting from first principles) all fundamental solutions that are tempered distributions for scalar elliptic operators. While the natural starting point is the Laplacian, this study encompasses a variety of related operators, such as the bi-Laplacian, the poly-harmonic operator, the Cauchy–Riemann operator, the Dirac operator, as well as general second-order constant coefficient strongly elliptic operators. Having accomplished this task then makes it possible to prove the well-posedness of the Poisson problem (1) (equipped with a boundary condition at infinity), and derive qualitative/quantitative properties for the solution such as (2). Along the way, Cauchy-like integral operators are also introduced and their connections with Hardy spaces is brought to light in the setting of both complex and Clifford analysis.

Chapter 8 has a twofold aim: determine all fundamental solutions that are tempered distributions for the heat operator and related versions (including the Schrödinger operator), then use this as a toll in the solution of the generalized

Cauchy problem for the heat operator. The same type of program is then carried out in Chapter 9, this time in connection with the wave operator.

While the analysis up to this point has been largely confined to scalar operators, the final two chapters in the monograph are devoted to studying systems of differential operators. The material in Chapter 10 is centered around two such basic systems: the Lamé operator arising in the theory of elasticity, and the Stokes operator arising in hydrodynamics. Among other things, all their fundamental solutions that are tempered distributions are identified, and the well-posedness of the Poisson problem for the Lamé system is established. The former issue is then revisited in the first part of Chapter 11 from a different perspective, and subsequently generalized to the case of (homogeneous) constant coefficient systems of arbitrary order. In Chapter 11 we also show that integral representation formulas and interior estimates hold for null solutions of homogeneous systems with nonvanishing full symbol. As a consequence, we prove that such null solutions are real-analytic and satisfy reverse Hölder estimates. The final topic addressed in Chapter 11 pertains to layer potentials associated with arbitrary constant coefficient second-order systems in the upper half-space, and the relevance of these operators vis-a-vis to the solvability of boundary value problems for such systems in this setting.

For completeness, a summary of topological and functional analysis results in reference to the description of the topology and equivalent characterizations of convergence in spaces of test functions and in spaces of distributions is included in the appendix (which also contains a variety of foundational results from calculus, measure theory, and special functions originating outside the scope of this book). One aspect worth noting in this regard is that the exposition in the main body of the book may be followed even without being fully familiar with all these details by alternatively taking, as the starting point, the characterization of convergence in the various topologies considered here (summarized in the main text under the heading **Fact**) as definitions. Such an approach makes the topics covered in the present monograph accessible to a larger audience while, at the same time, provides a full treatment of the topological and functional analysis background accompanying the theory of distributions for the reader interested in a more in-depth treatment.

Finally, each book chapter ends with bibliographical references tailored to its respective contents under the heading **Further Notes**, as well as with a number of additional exercises, selectively solved in Chapter 13.

Common Notational Conventions

Throughout this book the set of natural numbers will be denoted by $\mathbb{N}$, that is $\mathbb{N} := \{1, 2, \ldots\}$, while $\mathbb{N}_0 := \mathbb{N} \cup \{0\}$. For each $k \in \mathbb{N}$ set $k! := 1 \cdot 2 \cdots (k-1) \cdot k$, and make the convention that $0! := 1$. The letter $\mathbb{C}$ will denote the set of complex numbers, and $\bar{z}$ denotes the complex conjugate of $z \in \mathbb{C}$. Also the real and imaginary parts of a complex number z are denoted by $\mathrm{Re}\, z$ and $\mathrm{Im}\, z$, respectively. The symbol i is reserved for the complex imaginary unit $\sqrt{-1} \in \mathbb{C}$. The letter $\mathbb{R}$ will denote the set of real numbers and its n-fold Cartesian product of $\mathbb{R}$ with itself (where $n \in \mathbb{N}$) is denoted by $\mathbb{R}^n$. That is,

$$\mathbb{R}^n := \{x = (x_1, \ldots, x_n) : x_1, \ldots, x_n \in \mathbb{R}\} \tag{3}$$

considered with the usual vector space and inner product structure, i.e.,

$$x + y := (x_1 + y_1, \ldots, x_n + y_n), \quad cx := (cx_1, \ldots, cx_n), \quad x \cdot y := \sum_{j=1}^{n} x_j y_j,$$
$$\forall x = (x_1, \ldots, x_n) \in \mathbb{R}^n, \ \forall y = (y_1, \ldots, y_n) \in \mathbb{R}^n, \ \forall c \in \mathbb{R}. \tag{4}$$

The standard orthonormal basis of vectors in $\mathbb{R}^n$ is denoted by $\{e_j\}_{1 \leq j \leq n}$, where we have set $e_j := (0, \ldots, 0, 1, 0, \ldots, 0) \in \mathbb{R}^n$ with the only nonzero component on the j-th slot. We shall also consider the two canonical (open) half-spaces of $\mathbb{R}^n$, denoted by $\mathbb{R}_\pm^n := \{x = (x_1, \ldots, x_n) \in \mathbb{R}^n : \pm x_n > 0\}$. Hence, $\mathbb{R}_+^n = \mathbb{R}^{n-1} \times (0, \infty)$ and $\mathbb{R}_-^n = \mathbb{R}^{n-1} \times (-\infty, 0)$.

Given a multi-index $\alpha = (\alpha_1, \ldots, \alpha_n) \in \mathbb{N}_0^n$, we set

$$\mathrm{supp}\, \alpha := \{j \in \{1, \ldots, n\} : \alpha_j \neq 0\}, \tag{5}$$

$$\alpha! := \alpha_1! \alpha_2! \cdots \alpha_n! \quad \text{and} \quad |\alpha| := \sum_{j=1}^{n} \alpha_j, \tag{6}$$

$$\partial^\alpha := \prod_{j \in \text{supp } \alpha} \partial_j^{\alpha_j} \quad \text{where} \quad \partial_j := \frac{\partial}{\partial x_j} \quad \text{for } j = 1, \ldots, n, \tag{7}$$

$$x^\alpha := \prod_{j \in \text{supp } \alpha} x_j^{\alpha_j} \quad \text{for every } x = (x_1, \ldots, x_n) \in \mathbb{C}^n, \tag{8}$$

with the convention that $\partial^{(0,\ldots,0)}$ is the identity operator and $x^{(0,\ldots,0)} := 1$. Also if $\beta = (\beta_1, \ldots, \beta_n) \in \mathbb{N}_0^n$ is another multi-index we shall write $\beta \leq \alpha$ provided $\beta_j \leq \alpha_j$ for each $j \in \{1, \ldots, n\}$, in which case we set $\alpha - \beta := (\alpha_1 - \beta_1, \ldots, \alpha_n - \beta_n)$. We shall also say that $\beta < \alpha$ if $\beta \leq \alpha$ and $\beta \neq \alpha$. Recall that the Kronecker symbol is defined by $\delta_{jk} := 1$ if $j = k$ and $\delta_{jk} := 0$ if $j \neq k$.

All functions in this monograph are assumed to be complex-valued unless otherwise indicated. Derivatives of a function f defined on the real line are going to be denoted using f', f'', etc., or $f^{(k)}$, or $\frac{d^k f}{dx^k}$.

Throughout the book, Ω denotes an arbitrary open subset of $\mathbb{R}^n$. If A is an arbitrary subset of $\mathbb{R}^n$, then $\mathring{A}$, $\overline{A}$, and ∂A denote its interior, its closure, and its boundary, respectively. In addition, if B is another arbitrary subset of $\mathbb{R}^n$, then their set theoretic difference is denoted by $A \backslash B := \{x \in A : x \notin B\}$. In particular, the complement of A is $A^c := \mathbb{R}^n \backslash A$. For any $E \subset \mathbb{R}^n$ we let χ_E stand for the characteristic function of the set E (i.e., $\chi_E(x) = 1$ if $x \in E$ and $\chi_E(x) = 0$ if $x \in \mathbb{R}^n \backslash E$).

For $k \in \mathbb{N}_0 \cup \{\infty\}$, we will work with the following classes of functions that are vector spaces over $\mathbb{C}$:

$$C^k(\Omega) := \{\varphi : \Omega \to \mathbb{C} : \partial^\alpha \varphi \text{ continuous } \forall \alpha \in \mathbb{N}_0^n, |\alpha| \leq k\}, \tag{9}$$

$$C^k(\overline{\Omega}) := \{\varphi|_{\overline{\Omega}} : \varphi \in C^k(U), \ U \subseteq \mathbb{R}^n \text{ open set containing } \overline{\Omega}\}, \tag{10}$$

$$C_0^k(\Omega) := \{\varphi \in C^k(\Omega) : \text{supp } \varphi \text{ compact subset of } \Omega\}. \tag{11}$$

As usual, for any Lebesgue-measurable (complex-valued) function f defined on a Lebesgue-measurable set $E \subseteq \mathbb{R}^n$ and any $p \in [1, \infty]$ we write

$$\|f\|_{L^p(E)} := \begin{cases} \left(\int_E |f|^p \, dx\right)^{1/p} & \text{if } 1 \leq p < \infty, \\ \text{ess-sup}_E |f| & \text{if } p = \infty, \end{cases} \tag{12}$$

and denote by $L^p(E)$ the Banach space of (equivalence classes of) Lebesgue-measurable functions f on E satisfying $\|f\|_{L^p(E)} < \infty$. Also, we will work with locally integrable functions and with compactly supported integrable functions. For $p \in [1, \infty]$ these are defined as

$$L^p_{loc}(\Omega) := \{f : \Omega \to \mathbb{C} : f \text{ Lebesgue measurable such that}$$
$$\|f\|_{L^p(K)} < \infty, \quad \forall K \subset \Omega \text{ compact set}\}, \tag{13}$$

and, respectively, as

$$L^p_{comp}(\Omega) := \{f \in L^p(\Omega) : \text{supp} f \text{ compact subset of } \Omega\}, \tag{14}$$

where supp f is defined in (2.5.12).

Given a measure space (X, μ), a measurable set $A \subseteq X$ with $0 < \mu(A) < \infty$, and a function $f \in L^1(A, \mu)$, we define the integral average of f over A by

$$\fint_A f \, d\mu := \frac{1}{\mu(A)} \int_A f \, d\mu. \tag{15}$$

If E is a Lebesgue-measurable subset of $\mathbb{R}^n$, the Lebesgue measure of E is denoted by $|E|$. If $x \in \mathbb{R}^n$ and radius $r > 0$, set $B(x, r) := \{y \in \mathbb{R}^n : |y - x| < r\}$ for the ball of center x and radius R and its boundary is denoted by $\partial B(0, r)$. The unit sphere in $\mathbb{R}^n$ centered at zero is $S^{n-1} := \{x \in \mathbb{R}^n : |x| = 1\} = \partial B(0, 1)$ and its surface measure is denoted by ω_{n-1}.

For $n, m \in \mathbb{N}$ and $\mathfrak{R}$ an arbitrary commutative ring with multiplicative unit we denote by $\mathcal{M}_{n \times m}(\mathfrak{R})$ the collection of all $n \times m$ matrices with entries from $\mathfrak{R}$. If $B \in \mathcal{M}_{n \times m}(\mathfrak{R})$, then B^T denotes its transpose and if $n = m$, then $\det B$ denotes the determinant of the matrix B, while $I_{n \times n}$ denotes the identity matrix.

Regarding semi-orthodox notational conventions, $A := B$ stands for "A is defined as being B", while $A =: B$ stands for "B is defined as being A". Also, the letter C when used as a multiplicative constant in various inequalities, is allowed to vary from line to line. Whenever necessary, its dependence on the other parameters $a, b, \ldots$ implicit in the estimate in question is stressed by writing $C(a, b, \ldots)$ or $C_{a,b,\ldots}$ in place of just C.

Chapter 1
Weak Derivatives

Abstract Starting from the discussion of the Cauchy problem for a vibrating infinite string as a motivational example, the notion of a weak derivative is introduced as a mean of extending the notion of solution to a more general setting, where the functions involved may lack standard pointwise differentiability properties. Here two classes of test functions are also defined and discussed.

1.1 The Cauchy Problem for a Vibrating Infinite String

The partial differential equation

$$\partial_1^2 u - \partial_2^2 u = 0 \quad \text{in} \quad \mathbb{R}^2 \tag{1.1.1}$$

was derived by Jean d'Alembert in 1747 to describe the displacement $u(x_1, x_2)$ of a violin string as a function of time and distance along the string. Assuming that the string is infinite and that at time $x_2 = 0$ the displacement is given by some function $\varphi_0 \in C^2(\mathbb{R})$ leads to the following global Cauchy problem

$$
\begin{cases}
u \in C^2(\mathbb{R}^2), \\
\partial_1^2 u - \partial_2^2 u = 0 \quad \text{in } \mathbb{R}^2, \\
u(\cdot, 0) = \varphi_0 \quad \text{in } \mathbb{R}, \\
(\partial_2 u)(\cdot, 0) = 0 \quad \text{in } \mathbb{R}.
\end{cases}
\tag{1.1.2}
$$

Thanks to the regularity assumption on φ_0, it may be checked without difficulty that the function

$$u(x_1, x_2) := \tfrac{1}{2}[\varphi_0(x_1 + x_2) + \varphi_0(x_1 - x_2)], \quad \text{for every } (x_1, x_2) \in \mathbb{R}^2, \tag{1.1.3}$$

© Springer Nature Switzerland AG 2018
D. Mitrea, *Distributions, Partial Differential Equations, and Harmonic Analysis*,
Universitext, https://doi.org/10.1007/978-3-030-03296-8_1

1

is a solution of (1.1.2). This being said, the expression of u in (1.1.3) continues to be meaningful under much less restrictive assumptions on φ_0. For example, u is a well-defined continuous function in $\mathbb{R}^2$ whenever $\varphi_0 \in C^0(\mathbb{R})$. While in this case expression (1.1.3) is no longer a classical solution of (1.1.2), it is natural to ask whether it is possible to identify a new (and possibly weaker) sense in which (1.1.3) would continue to satisfy $\partial_1^2 u - \partial_2^2 u = 0$.

To answer this question, fix a function $u \in C^2(\mathbb{R}^2)$ satisfying $\partial_1^2 u - \partial_2^2 u = 0$ pointwise in $\mathbb{R}^2$. If $\varphi \in C_0^\infty(\mathbb{R}^2)$ is an arbitrary function and $R \in (0, \infty)$ is a number such that $\operatorname{supp} \varphi \subseteq (-R, R) \times (-R, R)$, then integration by parts gives

$$0 = \iint_{\mathbb{R}^2} (\partial_1^2 u - \partial_2^2 u)\varphi \, dx$$

$$= \int_{-R}^{R} \int_{-R}^{R} \partial_1^2 u(x_1, x_2)\varphi(x_1, x_2) \, dx_1 \, dx_2$$

$$- \int_{-R}^{R} \int_{-R}^{R} \partial_2^2 u(x_1, x_2)\varphi(x_1, x_2) \, dx_1 \, dx_2$$

$$= \iint_{\mathbb{R}^2} (\partial_1^2 \varphi - \partial_2^2 \varphi)u \, dx. \tag{1.1.4}$$

Note that the condition $\iint_{\mathbb{R}^2}(\partial_1^2\varphi - \partial_2^2\varphi)u \, dx = 0$ for all $\varphi \in C_0^\infty(\mathbb{R}^2)$ is meaningful even if $u \in C^0(\mathbb{R}^2)$, which suggests the following definition.

Definition 1.1. A function $u \in C^0(\mathbb{R}^2)$ is called a weak (generalized) solution of the equation $\partial_1^2 u - \partial_2^2 u = 0$ in $\mathbb{R}^2$ if

$$\iint_{\mathbb{R}^2} (\partial_1^2 \varphi - \partial_2^2 \varphi)u \, dx = 0 \quad \text{for all } \varphi \in C_0^\infty(\mathbb{R}^2). \tag{1.1.5}$$

Returning to (1.1.3), let us now check that, under the assumption $\varphi_0 \in C^0(\mathbb{R})$, the function u defined in (1.1.3) is a generalized solution of $\partial_1^2 u - \partial_2^2 u = 0$ in $\mathbb{R}^2$. Concretely, fix $\varphi \in C_0^\infty(\mathbb{R}^2)$ and write

$$\iint_{\mathbb{R}^2} (\partial_1^2 \varphi - \partial_2^2 \varphi)u \, dx \tag{1.1.6}$$

$$= \frac{1}{2} \int_{\mathbb{R}} \int_{\mathbb{R}} [\partial_1^2\varphi(x_1, x_2) - \partial_2^2\varphi(x_1, x_2)](\varphi_0(x_1 + x_2) + \varphi_0(x_1 - x_2)) \, dx_1 \, dx_2$$

$$= \frac{1}{4} \int_{\mathbb{R}} \int_{\mathbb{R}} (\partial_1^2\varphi)\left(\frac{y_1 + y_2}{2}, \frac{y_1 - y_2}{2}\right)(\varphi_0(y_1) + \varphi_0(y_2)) \, dy_1 \, dy_2$$

$$- \frac{1}{4} \int_{\mathbb{R}} \int_{\mathbb{R}} (\partial_2^2\varphi)\left(\frac{y_1 + y_2}{2}, \frac{y_1 - y_2}{2}\right)(\varphi_0(y_1) + \varphi_0(y_2)) \, dy_1 \, dy_2,$$

where for the last equality in (1.1.6) we have made the following change of variables: $y_1 = x_1 + x_2$ and $y_2 = x_1 - x_2$. If we now let $\psi(y_1, y_2) := \varphi\left(\frac{y_1+y_2}{2}, \frac{y_1-y_2}{2}\right)$ for

$(y_1, y_2) \in \mathbb{R}^2$, then

$$\partial_1 \psi(y_1, y_2) \tag{1.1.7}$$
$$= \frac{1}{2}(\partial_1 \varphi)\left(\frac{y_1 + y_2}{2}, \frac{y_1 - y_2}{2}\right) + \frac{1}{2}(\partial_2 \varphi)\left(\frac{y_1 + y_2}{2}, \frac{y_1 - y_2}{2}\right)$$

and

$$\partial_2 \partial_1 \psi(y_1, y_2) \tag{1.1.8}$$
$$= \frac{1}{4}(\partial_1^2 \varphi)\left(\frac{y_1 + y_2}{2}, \frac{y_1 - y_2}{2}\right) - \frac{1}{4}(\partial_2 \partial_1 \varphi)\left(\frac{y_1 + y_2}{2}, \frac{y_1 - y_2}{2}\right)$$
$$+ \frac{1}{4}(\partial_1 \partial_2 \varphi)\left(\frac{y_1 + y_2}{2}, \frac{y_1 - y_2}{2}\right) - \frac{1}{4}(\partial_2^2 \varphi)\left(\frac{y_1 + y_2}{2}, \frac{y_1 - y_2}{2}\right)$$

which, when used in (1.1.6), give

$$\iint_{\mathbb{R}^2} (\partial_1^2 \varphi - \partial_2^2 \varphi) u \, dx = \int_{\mathbb{R}} \int_{\mathbb{R}} \partial_1 \partial_2 \psi(y_1, y_2)[\varphi_0(y_1) + \varphi_0(y_2)] \, dy_1 \, dy_2. \tag{1.1.9}$$

Let $R \in (0, \infty)$ be such that $\operatorname{supp} \varphi \subset (-R, R) \times (-R, R)$. Then the support of ψ is contained in the set of points $(y_1, y_2) \in \mathbb{R}^2$ satisfying $-2R \leq y_1 + y_2 \leq 2R$ and $-2R \leq y_1 - y_2 \leq 2R$. Hence, if $R' > 2R$ we have $\operatorname{supp} \psi \subset (-R', R') \times (-R', R')$ and integration by parts yields

$$\iint_{\mathbb{R}^2} (\partial_1^2 \varphi - \partial_2^2 \varphi) u \, dx$$

$$= \int_{-R'}^{R'} \varphi_0(y_1) \int_{-R'}^{R'} \partial_2 \partial_1 \psi(y_1, y_2) \, dy_2 \, dy_1$$

$$+ \int_{-R'}^{R'} \varphi_0(y_2) \int_{-R'}^{R'} \partial_1 \partial_2 \psi(y_1, y_2) \, dy_1 \, dy_2$$

$$= \int_{-R'}^{R'} \varphi_0(y_1)[\partial_1 \psi(y_1, R') - \partial_1 \psi(y_1, -R')] \, dy_1$$

$$+ \int_{-R'}^{R'} \varphi_0(y_2)[\partial_2 \psi(R', y_2) - \partial_2 \psi(-R', y_2)] \, dy_2 = 0. \tag{1.1.10}$$

In summary, this proves that

for every $\varphi_0 \in C^0(\mathbb{R})$ the function u defined as in (1.1.3)
is a weak solution of the equation $\partial_1^2 u - \partial_2^2 u = 0$ in $\mathbb{R}^2$. $\qquad$ (1.1.11)

We emphasize that in general, there is no reason to expect that u has any pointwise differentiability properties if φ_0 is merely continuous.

1.2 Weak Derivatives

Convention. Unless otherwise specified, Ω denotes an arbitrary open subset of $\mathbb{R}^n$.

Before proceeding with the definition of weak (Sobolev) derivatives we discuss a phenomenon that serves as motivation for the definition. Let $f \in C^m(\Omega)$, where $m \in \mathbb{N}_0$. Given an arbitrary $\varphi \in C_0^\infty(\Omega)$, consider a function $F \in C_0^m(\mathbb{R}^n)$ that agrees with f in a neighborhood of $\operatorname{supp}\varphi$. For example, we may take $F = \psi\widetilde{f}$, where tilde denotes the extension by zero outside Ω and $\psi \in C_0^\infty(\mathbb{R}^n)$ is identically one in a neighborhood of $\operatorname{supp}\varphi$ (see Proposition 14.34 for the construction of such a function). Also, pick $R > 0$ large enough so that $\operatorname{supp}\varphi \subset B(0, R)$. Then for each $\alpha \in \mathbb{N}_0^n$ with $|\alpha| \leq m$, integration by parts (cf. Theorem 14.60 for a precise formulation) and support considerations yield

$$\int_\Omega (\partial^\alpha f)\varphi \, dx = \int_{B(0,R)} (\partial^\alpha F)\varphi \, dx$$

$$= (-1)^{|\alpha|} \int_{B(0,R)} F \, \partial^\alpha \varphi \, dx = (-1)^{|\alpha|} \int_\Omega f \, \partial^\alpha \varphi \, dx. \qquad (1.2.1)$$

This computation suggests the following definition.

Definition 1.2. If $f \in L_{loc}^1(\Omega)$ and $\alpha \in \mathbb{N}_0^n$, we say that $\partial^\alpha f$ belongs to $L_{loc}^1(\Omega)$ in a weak (Sobolev) sense provided there exists some $g \in L_{loc}^1(\Omega)$ with the property that

$$\int_\Omega g\varphi \, dx = (-1)^{|\alpha|} \int_\Omega f \, \partial^\alpha \varphi \, dx \qquad \text{for every} \quad \varphi \in C_0^\infty(\Omega). \qquad (1.2.2)$$

Whenever this happens, we shall write $\partial^\alpha f = g$ and call g the weak derivative of order α of f.

The fact that the concept of weak derivative is unambiguously defined is then ensured by the next theorem.

Theorem 1.3. *If $g \in L_{loc}^1(\Omega)$ and $\int_\Omega g\varphi \, dx = 0$ for each $\varphi \in C_0^\infty(\Omega)$ then $g = 0$ almost everywhere on Ω.*

Proof. Consider a function ϕ satisfying (see (14.3.3) for a concrete example)

$$\phi \in C_0^\infty(\mathbb{R}^n), \quad \phi \geq 0, \quad \operatorname{supp}\phi \subseteq \overline{B(0,1)},$$

$$\text{and} \quad \int_{\mathbb{R}^n} \phi(x) \, dx = 1, \qquad (1.2.3)$$

and for each $\varepsilon > 0$ define

$$\phi_\varepsilon(x) := \frac{1}{\varepsilon^n}\phi\left(\frac{x}{\varepsilon}\right) \qquad \text{for each } x \in \mathbb{R}^n. \qquad (1.2.4)$$

Then for each $\varepsilon > 0$ we have

$$\phi_\varepsilon \in C_0^\infty(\mathbb{R}^n), \quad \phi_\varepsilon \geq 0, \quad \operatorname{supp}\phi_\varepsilon \subseteq \overline{B(0,\varepsilon)},$$

$$\text{and} \quad \int_{\mathbb{R}^n} \phi_\varepsilon(x)\,dx = 1. \tag{1.2.5}$$

Fix now $x \in \Omega$ and $\varepsilon \in (0, \operatorname{dist}(x, \partial\Omega))$. Then $B(x,\varepsilon) \subseteq \Omega$ which, in light of (1.2.5), implies $\phi_\varepsilon(x - \cdot)\big|_\Omega \in C_0^\infty(\Omega)$. Consequently, under the current assumptions on g we have

$$\int_\Omega g(y)\phi_\varepsilon(x - y)\,dy = 0. \tag{1.2.6}$$

In particular, if we also assume that x is a Lebesgue point for g (i.e., the limit (14.2.11) holds for this x with f replaced by g), then (1.2.6) combined with the properties of ϕ_ε allow us to write

$$|g(x)| = \left| \int_\Omega g(x)\phi_\varepsilon(x - y)\,dy - \int_\Omega g(y)\phi_\varepsilon(x - y)\,dy \right|$$

$$\leq \frac{1}{\varepsilon^n} \int_{B(x,\varepsilon)} |g(x) - g(y)| \phi\left(\frac{y}{\varepsilon}\right) dy$$

$$\leq \frac{c}{|B(x,\varepsilon)|} \int_{B(x,\varepsilon)} |g(x) - g(y)|\,dy \xrightarrow[\varepsilon \to 0^+]{} 0, \tag{1.2.7}$$

where $c := \frac{\omega_{n-1}}{n}\|\phi\|_{L^\infty(\mathbb{R}^n)}$. The convergence in (1.2.7) is due to Lebesgue's Differentiation Theorem (cf. Theorem 14.14). This proves that $g(x) = 0$ for every $x \in \Omega$ that is a Lebesgue point for g, hence $g = 0$ almost everywhere in Ω. $\qquad\square$

Next we present a few examples related to the notion of weak (Sobolev) derivative.

Example 1.4. Consider the function

$$f : \mathbb{R} \longrightarrow \mathbb{R}, \quad f(x) := \begin{cases} x, & x > 0, \\ 0, & x \leq 0, \end{cases} \quad \forall\, x \in \mathbb{R}. \tag{1.2.8}$$

Note that f is continuous on $\mathbb{R}$ but not differentiable at 0. Nonetheless, f has a weak derivative of order one that is equal to the Heaviside function

$$H : \mathbb{R} \longrightarrow \mathbb{R}, \quad H(x) := \begin{cases} 1, & x > 0, \\ 0, & x \leq 0, \end{cases} \quad \forall\, x \in \mathbb{R}. \tag{1.2.9}$$

Indeed, if $\varphi \in C_0^\infty(\mathbb{R})$, then integration by parts yields

$$-\int_{-\infty}^\infty f(x)\varphi'(x)\,dx = -\int_0^\infty x\varphi'(x)\,dx = -x\varphi(x)\Big|_0^\infty + \int_0^\infty \varphi(x)\,dx$$

$$= \int_{-\infty}^\infty H(x)\varphi(x)\,dx, \tag{1.2.10}$$

which shows that

$$\int_{\mathbb{R}} H\varphi\,dx = -\int_{\mathbb{R}} f\varphi'\,dx, \qquad \forall\,\varphi \in C_0^\infty(\mathbb{R}). \tag{1.2.11}$$

Note that $H \in L_{loc}^1(\mathbb{R})$ hence H is the weak (or Sobolev) derivative of order one of the function f in $\mathbb{R}$.

Example 1.5. Does there exist a function $g \in L_{loc}^1(\mathbb{R})$ such that g is a weak derivative (of order one) of the Heaviside function? To answer this question first observe that for each $\varphi \in C_0^\infty(\mathbb{R})$ we have

$$-\int_{-\infty}^{\infty} H(x)\varphi'(x)\,dx = -\int_0^{\infty} \varphi'(x)\,dx = \varphi(0). \tag{1.2.12}$$

Suppose that H has a weak derivative of order one, and call this $g \in L_{loc}^1(\mathbb{R})$. Then, by Definition 1.2 and (1.2.12), we have $\int_{\mathbb{R}} g\varphi\,dx = \varphi(0)$ for all $\varphi \in C_0^\infty(\mathbb{R})$. This forces $\int_0^{\infty} g\varphi\,dx = 0$ for all $\varphi \in C_0^\infty(\mathbb{R} \setminus \{0\})$. In concert with Theorem 1.3, the latter yields $g = 0$ almost everywhere on $\mathbb{R} \setminus \{0\}$. When combined with (1.2.12), this gives that $0 = \int_{\mathbb{R}} g\varphi\,dx = \varphi(0)$ for all $\varphi \in C_0^\infty(\mathbb{R})$, leading to a contradiction (as there are functions $\varphi \in C_0^\infty(\mathbb{R})$ with $\varphi(0) \neq 0$). Thus, a weak (Sobolev) derivative of order one of H does not exist.

Having defined the notion of weak (Sobolev) derivatives for locally integrable functions, we return to the notion of weak solution considered in Definition 1.1 in a particular case, and extend this to more general partial differential equations. To set the stage, let $P(x, \partial)$ be a linear partial differential operator of order $m \in \mathbb{N}$ of the form

$$P(x, \partial) := \sum_{|\alpha| \leq m} a_\alpha(x)\partial^\alpha, \quad a_\alpha \in C^{|\alpha|}(\Omega), \quad \alpha \in \mathbb{N}_0^n, \quad |\alpha| \leq m. \tag{1.2.13}$$

Also, suppose $f \in C^0(\Omega)$ is a given function and that u is a classical solution of the partial differential equation $P(x, \partial)u = f$ in Ω. That is, assume $u \in C^m(\Omega)$ and the equation holds pointwise in Ω. Then, for each $\varphi \in C_0^\infty(\Omega)$, integration by parts gives

$$\int_\Omega f\varphi\,dx = \sum_{|\alpha| \leq m} \int_\Omega \varphi a_\alpha(\partial^\alpha u)\,dx = \int_\Omega \Big[\sum_{|\alpha| \leq m} (-1)^{|\alpha|}\partial^\alpha(a_\alpha\varphi)\Big]u\,dx. \tag{1.2.14}$$

Hence, if we define

$$P^\top(x, \partial)\varphi := \sum_{|\alpha| \leq m} (-1)^{|\alpha|}\partial^\alpha(a_\alpha\varphi), \tag{1.2.15}$$

and call it the $\texttt{transpose}$ of the operator $P(x, \partial)$, the resulting equation from (1.2.14) becomes

$$\int_\Omega f\varphi\,dx = \int_\Omega [P^\top(x, \partial)\varphi]u\,dx, \qquad \forall\,\varphi \in C_0^\infty(\mathbb{R}^n). \tag{1.2.16}$$

Thus, any classical solution u of $P(x, \partial)u = f$ in Ω satisfies (1.2.16). On the other hand, there might exist functions $u \in L^1_{loc}(\Omega)$ that satisfy (1.2.16) but are not classical solutions of the given equation. Such a scenario has been already encountered in (1.1.11) (cf. also the subsequent comment). This motivates the following general definition (compare with Definition 1.1 corresponding to $P(x, \partial) = \partial_1^2 - \partial_2^2$).

Definition 1.6. Let $u, f \in L^1_{loc}(\Omega)$ be given and assume that $P(x, \partial)$ is as in (1.2.13). Then $P(x, \partial)u = f$ is said to hold in the weak (or Sobolev) sense if

$$\int_\Omega f\varphi \, \mathrm{d}x = \int_\Omega [P^\top(x, \partial)\varphi]u \, \mathrm{d}x, \qquad \forall \varphi \in C_0^\infty(\Omega). \tag{1.2.17}$$

From the comments in the preamble to Definition 1.6 we know that if u is a classical solution of $P(x, \partial)u = f$ in Ω for some $f \in C^0(\Omega)$, then u is also a weak (Sobolev) solution of the same equation. Conversely, if $u \in C^m(\Omega)$ is a weak solution of the partial differential equation $P(x, \partial)u = f$ for some given function $f \in C^0(\Omega)$, then by Definition 1.6 and integration by parts we obtain

$$\int_\Omega f\varphi \, \mathrm{d}x = \int_\Omega [P^\top(x, \partial)\varphi]u \, \mathrm{d}x = \int_\Omega \varphi \, P(x, \partial)u \, \mathrm{d}x. \tag{1.2.18}$$

Since $\varphi \in C_0^\infty(\Omega)$ is arbitrary, Theorem 1.3 then forces $f = P(x, \partial)u$ almost everywhere in Ω, hence ultimately everywhere in Ω, since the functions in question are continuous.

In summary, the above discussion shows that the notion of weak solution of a partial differential equation is a natural, unambiguous, and genuine generalization of the concept of classical solution, in the following precise sense:

- any classical solution is a weak solution,

- any sufficiently regular weak solution is classical,

- weak solutions may exist even in the absence of classical ones.

1.3 The Spaces $\mathcal{E}(\Omega)$ and $\mathcal{D}(\Omega)$

A major drawback of Definition 1.2 is that while the right-hand side of (1.2.2) is always meaningful, it cannot always be written it in the form given by the left-hand side of (1.2.2). In addition, it might be the case that some locally integrable function in Ω may admit weak (Sobolev) derivatives of a certain order and not of some intermediate lower order (see the example in Exercise 1.29). The remedy is to focus on the portion of (1.2.2) that always makes sense. Specifically, given $f \in L^1_{loc}(\Omega)$ and $\alpha \in \mathbb{N}_0^n$, define the mapping

$$g_\alpha : C_0^\infty(\Omega) \to \mathbb{C}, \quad g_\alpha(\varphi) := (-1)^{|\alpha|} \int_\Omega f(\partial^\alpha \varphi) \, \mathrm{d}x, \quad \forall \varphi \in C_0^\infty(\Omega). \tag{1.3.1}$$

The functional in (1.3.1) has the following properties.

1. g_α is linear, i.e., $g_\alpha(\lambda_1\varphi_1 + \lambda_2\varphi_2) = \lambda_1 g_\alpha(\varphi_1) + \lambda_2 g_\alpha(\varphi_2)$ for every $\lambda_1, \lambda_2 \in \mathbb{C}$, and every $\varphi_1, \varphi_2 \in C_0^\infty(\Omega)$.
2. For each $\varphi \in C_0^\infty(\Omega)$ we may estimate

$$|g_\alpha(\varphi)| \leq \int_\Omega |f| |\partial^\alpha \varphi| \, dx \leq \left(\int_{\text{supp}\,\varphi} |f| \, dx \right) \sup_{x \in \text{supp}\,\varphi} |\partial^\alpha \varphi(x)|. \qquad (1.3.2)$$

The fact that the term $\int_{\text{supp}\,\varphi} |f| \, dx$ in (1.3.2) depends on φ is inconvenient if we want to consider the continuity of g_α in some sense. Nonetheless, if a priori a compact set $K \subset \Omega$ is fixed and the requirement $\text{supp}\,\varphi \subseteq K$ is imposed, then (1.3.2) becomes

$$|g_\alpha(\varphi)| \leq \left(\int_K |f| \, dx \right) \sup_{x \in K} |\partial^\alpha \varphi(x)| \qquad (1.3.3)$$

and, this time, $\int_K |f| \, dx$ is a constant independent of φ. This observation motivates considering an appropriate topology τ on $C^\infty(\Omega)$. For the exact definition of this topology see Section 14.1. We will not elaborate here more on this subject other than highlighting those key features of τ that are particularly important for our future investigations. To record the precise statements of these features, introduce

$$\mathcal{E}(\Omega) := (C^\infty(\Omega), \tau), \qquad (1.3.4)$$

a notation which emphasizes that $\mathcal{E}(\Omega)$ is the vector space $C^\infty(\Omega)$ equipped with the topology τ. We then have:

Fact 1.7 *A sequence $\{\varphi_j\}_{j \in \mathbb{N}} \subset C^\infty(\Omega)$ converges in $\mathcal{E}(\Omega)$ to some $\varphi \in C^\infty(\Omega)$ as $j \to \infty$ if and only if*

$$\forall K \subset \Omega \text{ compact}, \ \forall \alpha \in \mathbb{N}_0^n, \quad \text{we have} \quad \lim_{j \to \infty} \sup_{x \in K} \left| \partial^\alpha(\varphi_j - \varphi)(x) \right| = 0, \quad (1.3.5)$$

in which case we use the notation $\varphi_j \xrightarrow[j \to \infty]{\mathcal{E}(\Omega)} \varphi$.

Fact 1.8 *$\mathcal{E}(\Omega)$ is a locally convex, metrizable, and complete topological vector space over $\mathbb{C}$.*

It is easy to see that as a consequence of Fact 1.7 we have the following result.

Remark 1.9. A sequence $\{\varphi_j\}_{j \in \mathbb{N}} \subset C^\infty(\Omega)$ converges in $\mathcal{E}(\Omega)$ to a function $\varphi \in C^\infty(\Omega)$ as $j \to \infty$, if and only if for any compact set $K \subset \Omega$ and any $m \in \mathbb{N}_0$ one has

$$\lim_{j \to \infty} \sup_{\alpha \in \mathbb{N}_0^n, |\alpha| \leq m} \sup_{x \in K} \left| \partial^\alpha(\varphi_j - \varphi)(x) \right| = 0. \qquad (1.3.6)$$

Exercise 1.10. Prove that if $\varphi_j \xrightarrow[j \to \infty]{\mathcal{E}(\Omega)} \varphi$ then the following also hold:

(1) $\partial^\alpha \varphi_j \xrightarrow[j\to\infty]{\mathcal{E}(\Omega)} \partial^\alpha \varphi$ for each $\alpha \in \mathbb{N}_0^n$;

(2) $a\varphi_j \xrightarrow[j\to\infty]{\mathcal{E}(\Omega)} a\varphi$ for each $a \in C_0^\infty(\Omega)$.

A standard way of constructing a sequence of smooth functions in $\mathbb{R}^n$ that converges in $\mathcal{E}(\mathbb{R}^n)$ to a given $f \in C^\infty(\mathbb{R}^n)$ is by taking the convolution of f with dilations of a function as in (1.2.3). This construction is discussed in detail next.

Example 1.11. Let $f \in C^\infty(\mathbb{R}^n)$ be given. Then a sequence of functions from $C^\infty(\mathbb{R}^n)$ that converges to f in $\mathcal{E}(\mathbb{R}^n)$ may be constructed as follows. Recall ϕ from (1.2.3) and define

$$\phi_j(x) := j^n \phi(jx) \qquad \text{for } x \in \mathbb{R}^n \text{ and each } j \in \mathbb{N}. \tag{1.3.7}$$

Clearly, for each $j \in \mathbb{N}$ we have

$$\phi_j \in C_0^\infty(\mathbb{R}^n), \quad \operatorname{supp} \phi_j \subseteq \overline{B(0, 1/j)}, \quad \text{and} \quad \int_{\mathbb{R}^n} \phi_j \, dx = 1. \tag{1.3.8}$$

Now if we further set for each $j \in \mathbb{N}$

$$f_j(x) := \int_{\mathbb{R}^n} f(x-y)\phi_j(y) \, dy$$

$$= \int_{B(0,1)} f(x - z/j)\phi(z) \, dz \quad \text{for each } x \in \mathbb{R}^n, \tag{1.3.9}$$

then $f_j \in C^\infty(\mathbb{R}^n)$. Also, if K is an arbitrary compact set in $\mathbb{R}^n$ and $\alpha \in \mathbb{N}_0^n$, then

$$|\partial^\alpha f_j(x) - \partial^\alpha f(x)| \le \int_{B(0,1)} |\partial^\alpha f(x - z/j) - \partial^\alpha f(x)| \, \phi(z) \, dz$$

$$\le \frac{1}{j} \max_{|\beta|=|\alpha|+1} \|\partial^\beta f\|_{L^\infty(\widetilde{K})} \qquad \forall \, x \in K, \tag{1.3.10}$$

where $\widetilde{K} := \{x \in \mathbb{R}^n : \operatorname{dist}(x, K) \le 1\}$. Hence $f_j \xrightarrow[j\to\infty]{\mathcal{E}(\mathbb{R}^n)} f$, as desired.

The previous approximation result may be further strengthened as indicated in the next two exercises.

Exercise 1.12. Prove that $C_0^\infty(\mathbb{R}^n)$ is sequentially dense in $\mathcal{E}(\mathbb{R}^n)$. That is, show that for every $f \in C^\infty(\mathbb{R}^n)$ there exists a sequence of functions $\{f_j\}_{j\in\mathbb{N}}$ from $C_0^\infty(\mathbb{R}^n)$ with the property that $f_j \xrightarrow[j\to\infty]{\mathcal{E}(\mathbb{R}^n)} f$.

Hint: Let $\psi \in C_0^\infty(\mathbb{R}^n)$ be such that $\psi(x) = 1$ whenever $|x| < 1$. Then given $f \in C^\infty(\mathbb{R}^n)$ define $f_j(x) := \psi(x/j)f(x)$, for every $x \in \mathbb{R}^n$ and every $j \in \mathbb{N}$.

Exercise 1.13. Prove that $C_0^\infty(\Omega)$ is sequentially dense in $\mathcal{E}(\Omega)$.

Hint: Consider the sequence of compacts

$$K_j := \{x \in \Omega : \ \mathrm{dist}(x, \partial\Omega) \geq \tfrac{1}{j}\} \cap \overline{B(0, j)}, \quad \forall j \in \mathbb{N}. \tag{1.3.11}$$

Then $\bigcup_{j\in\mathbb{N}} K_j = \Omega$ and $K_j \subset \mathring{K}_{j+1}$ for every $j \in \mathbb{N}$. For each $j \in \mathbb{N}$ pick a function $\psi_j \in C_0^\infty(\Omega)$ with $\psi_j \equiv 1$ in a neighborhood of K_j and $\mathrm{supp}\,\psi_j \subseteq K_{j+1}$ (cf. Proposition 14.34). If $f \in C^\infty(\Omega)$, define $f_j := \psi_j f$ for every $j \in \mathbb{N}$.

Moving on, we focus on defining a topology on $C_0^\infty(\Omega)$ that suits the purposes we have in mind. Since $C_0^\infty(\Omega) \subset \mathcal{E}(\Omega)$, one option would be to consider the topology induced by this larger ambient on $C_0^\infty(\Omega)$. However, this topology has the distinct drawback of not preserving the property of being compactly supported under convergence. Here is an example to that effect.

Example 1.14. Consider the function

$$\varphi(x) := \begin{cases} e^{\frac{1}{\left|x - \frac{1}{2}\right|^2 - \frac{1}{4}}} & \text{if } 0 < x < 1, \\ 0 & \text{if } x \leq 0 \text{ or } x > 1, \end{cases} \qquad \text{for each } x \in \mathbb{R}. \tag{1.3.12}$$

Note that $\varphi \in C^\infty(\mathbb{R})$, $\mathrm{supp}\,\varphi = [0, 1]$, and $\varphi > 0$ in $(0, 1)$. For each $j \in \mathbb{N}$ define

$$\varphi_j(x) := \varphi(x - 1) + \tfrac{1}{2}\varphi(x - 2) + \cdots + \tfrac{1}{j}\varphi(x - j), \quad \forall x \in \mathbb{R}. \tag{1.3.13}$$

Then $\varphi_j \in C^\infty(\mathbb{R})$, $\mathrm{supp}\,\varphi_j = [1, j+1]$, and $\varphi_j \xrightarrow[j\to\infty]{\mathcal{E}(\mathbb{R})} \varphi$ where

$$\varphi(x) := \sum_{j=1}^{\infty} \tfrac{1}{j}\varphi(x - j) \quad \text{for each } x \in \mathbb{R}. \tag{1.3.14}$$

Clearly this limit function does not have compact support.

The flaw just highlighted is remedied by introducing a different topology on $C_0^\infty(\Omega)$ that is finer than the one inherited from $\mathcal{E}(\Omega)$. First, for each $K \subset \Omega$ compact, denote by $\mathcal{D}_K(\Omega)$ the vector space consisting of functions from $C^\infty(\Omega)$ supported in K endowed with the topology induced by $\mathcal{E}(\Omega)$. Second, consider on $C_0^\infty(\Omega)$ the inductive limit topology of the spaces $\left\{\mathcal{D}_K(\Omega)\right\}_{\substack{K \subset \Omega \\ \text{compact}}}$ and denote the resulting topological vector space by $\mathcal{D}(\Omega)$. For precise definitions see Section 14.1. The topology induced on $\{\varphi \in C_0^\infty(\Omega) : \ \mathrm{supp}\,\varphi \subseteq K\}$ by this inductive limit topology coincides with the topology on $\mathcal{D}_K(\Omega)$. Two features that are going to be particularly important for our analysis are singled out below (see Section 14.1.0.4 in this regard).

Fact 1.15 $\mathcal{D}(\Omega)$ *is a locally convex and complete topological vector space over* $\mathbb{C}$.

Fact 1.16 *A sequence* $\{\varphi_j\}_{j\in\mathbb{N}} \subset C_0^\infty(\Omega)$ *converges in* $\mathcal{D}(\Omega)$ *to some* $\varphi \in C_0^\infty(\Omega)$ *as* $j \to \infty$ *if and only if the following two conditions are satisfied:*

(1) there exists a compact set $K \subset \Omega$ such that $\operatorname{supp} \varphi_j \subseteq K$ for all $j \in \mathbb{N}$ and $\operatorname{supp} \varphi \subseteq K$;

(2) for any $\alpha \in \mathbb{N}_0^n$ we have $\lim_{j \to \infty} \sup_{x \in K} |\partial^\alpha (\varphi_j - \varphi)(x)| = 0$.

We abbreviate (1)–(2) by simply writing $\varphi_j \xrightarrow[j \to \infty]{\mathcal{D}(\Omega)} \varphi$.

In view of Fact 1.7 one obtains the following consequence of Fact 1.16.

Remark 1.17. $\varphi_j \xrightarrow[j \to \infty]{\mathcal{D}(\Omega)} \varphi$ *if and only if*

(1) there exists a compact set $K \subset \Omega$ such that $\operatorname{supp} \varphi_j \subseteq K$ for all $j \in \mathbb{N}$, and

(2) $\varphi_j \xrightarrow[j \to \infty]{\mathcal{E}(\Omega)} \varphi$.

If one now considers the identity map from $\mathcal{D}(\Omega)$ into $\mathcal{E}(\Omega)$, a combination of Remark 1.17, and Theorem 14.6 yields that this map is continuous. Hence, if we also take into account Exercise 1.12, it follows that

$$\mathcal{D}(\Omega) \text{ is continuously and densely embedded into } \mathcal{E}(\Omega). \qquad (1.3.15)$$

Exercise 1.18. Suppose ω is an open subset of Ω and consider the map

$$\iota : C_0^\infty(\omega) \to C_0^\infty(\Omega), \qquad \iota(\varphi) := \begin{cases} \varphi & \text{on } \omega, \\ 0 & \text{on } \Omega \setminus \omega, \end{cases} \qquad \forall \, \varphi \in C_0^\infty(\omega). \quad (1.3.16)$$

Prove that if $\varphi_j \xrightarrow[j \to \infty]{\mathcal{D}(\omega)} \varphi$ then $\iota(\varphi_j) \xrightarrow[j \to \infty]{\mathcal{D}(\Omega)} \iota(\varphi)$. Use Theorem 14.6 to conclude that $\iota : \mathcal{D}(\omega) \to \mathcal{D}(\Omega)$ is continuous.

Exercise 1.19. Let $x_0 \in \mathbb{R}^n$ and consider the translation by x_0 map defined as

$$t_{x_0} : \mathcal{D}(\mathbb{R}^n) \longrightarrow \mathcal{D}(\mathbb{R}^n)$$
$$t_{x_0}(\varphi) := \varphi(\cdot - x_0), \quad \forall \, \varphi \in C_0^\infty(\mathbb{R}^n). \qquad (1.3.17)$$

Prove that t_{x_0} is linear and continuous.

Hint: Use Theorem 14.6.

Exercise 1.20. Prove that if $\varphi_j \xrightarrow[j \to \infty]{\mathcal{D}(\Omega)} \varphi$ then the following also hold:

(1) $\partial^\alpha \varphi_j \xrightarrow[j \to \infty]{\mathcal{D}(\Omega)} \partial^\alpha \varphi$ for each $\alpha \in \mathbb{N}_0^n$;

(2) $a \varphi_j \xrightarrow[j \to \infty]{\mathcal{D}(\Omega)} a \varphi$ for each $a \in C^\infty(\Omega)$.

Exercise 1.21. Prove that the map $\mathcal{D}(\Omega) \ni \varphi \mapsto a\varphi \in \mathcal{D}(\Omega)$ is linear and continuous for every $a \in C^\infty(\Omega)$.

Hint: Use Exercise 1.20 and Theorem 14.6.

Exercise 1.22. Prove that if $\varphi_j \xrightarrow[j\to\infty]{\mathcal{E}(\Omega)} \varphi$ then $a\varphi_j \xrightarrow[j\to\infty]{\mathcal{D}(\Omega)} a\varphi$ for each $a \in C_0^\infty(\Omega)$.

Also show that if $\varphi_j \xrightarrow[j\to\infty]{\mathcal{E}(\Omega)} \varphi$ and $a \in C_0^\infty(\omega)$ for some open subset ω of Ω, then $a\varphi_j \xrightarrow[j\to\infty]{\mathcal{D}(\omega)} a\varphi$.

As a consequence of Remark 1.17, we see that the topology $\mathcal{D}(\Omega)$ is finer than the topology $C_0^\infty(\Omega)$ inherits from $\mathcal{E}(\Omega)$, and an example of a sequence of smooth, compactly supported functions in Ω convergent in $\mathcal{E}(\Omega)$ to a limit which does not belong to $\mathcal{D}(\Omega)$ has been given in Example 1.14. The example below shows that even if the limit function is in $\mathcal{D}(\Omega)$, one should still not expect that convergence in $\mathcal{E}(\Omega)$ of a sequence of smooth, compactly supported functions in Ω implies convergence in $\mathcal{D}(\Omega)$.

Example 1.23. Let φ be as in (1.3.12) and for each $j \in \mathbb{N}$ set $\varphi_j(x) := \varphi(x-j)$, $x \in \mathbb{R}$. Clearly, $\varphi_j \in C^\infty(\mathbb{R})$ and $\operatorname{supp}\varphi_j = [j, j+1]$ for all $j \in \mathbb{N}$. If $K \subset \mathbb{R}$ is compact, then there exists $j_0 \in \mathbb{N}$ such that $K \subseteq [-j_0, j_0]$. Consequently, $\operatorname{supp}\varphi_j \cap K = \varnothing$ for $j \geq j_0$. Thus, trivially, $\sup_{x\in K}\left|\varphi_j^{(k)}(x)\right| = 0$ if $j \geq j_0$ which shows that $\varphi_j \xrightarrow[j\to\infty]{\mathcal{E}(\mathbb{R})} 0$. Consider next the issue whether $\{\varphi_j\}_{j\in\mathbb{N}}$ converge in $\mathcal{D}(\mathbb{R})$. If this were to be the case, there would exist $r \in (0, \infty)$ such that $\operatorname{supp}\varphi_j \subseteq [-r, r]$ for every j. However, $\bigcup_{j=1}^\infty \operatorname{supp}\varphi_j = [1, \infty)$ which leads to a contradiction. Thus, $\{\varphi_j\}_{j\in\mathbb{N}}$ does not converge in $\mathcal{D}(\mathbb{R})$.

For $n, m \in \mathbb{N}$, denote by $\mathcal{M}_{n\times m}(\mathbb{R})$ the collection of all $n \times m$ matrices with entries in $\mathbb{R}$. Recall that a map $L : \mathbb{R}^m \to \mathbb{R}^n$ is linear if and only if there exists a matrix $A \in \mathcal{M}_{n\times m}(\mathbb{R})$ such that $L(x) = Ax$ for every $x \in \mathbb{R}^m$, where Ax denotes the multiplication of the matrix A with the vector x viewed as an element in $\mathcal{M}_{m\times 1}(\mathbb{R})$. Moreover, such a matrix is unique. In the sequel, we follow the standard practice of denoting by A the linear map associated with a matrix A. It is well-known that a linear mapping $A \in \mathcal{M}_{n\times n}(\mathbb{R})$ is invertible if and only if it is open. The following lemma shows that the composition to the right with an invertible matrix defines a linear and continuous mapping from $\mathcal{D}(\mathbb{R}^n)$ into itself.

Lemma 1.24. *Suppose $A \in \mathcal{M}_{n\times n}(\mathbb{R})$ is such that $\det A \neq 0$. Then the composition mapping*

$$\mathcal{D}(\mathbb{R}^n) \ni \varphi \mapsto \varphi \circ A \in \mathcal{D}(\mathbb{R}^n)$$

(1.3.18)

is well defined, linear and continuous.

Proof. Let $\varphi \in C_0^\infty(\mathbb{R}^n)$. By the Chain Rule we have $\varphi \circ A \in C^\infty(\mathbb{R}^n)$. We claim that

$$\operatorname{supp}(\varphi \circ A) = \{x \in \mathbb{R}^n : Ax \in \operatorname{supp}\varphi\}.$$

(1.3.19)

Indeed, if $x \in \mathbb{R}^n$ and $x \notin \operatorname{supp}(\varphi \circ A)$ then there exists $r > 0$ such that $\varphi \circ A = 0$ on $B(x, r)$. Hence, the open set $O := A(B(x, r))$ contains Ax and $\varphi = 0$ on O, which implies $O \cap \operatorname{supp} \varphi = \varnothing$. In particular x does not belong to the set in the right-hand side of (1.3.19). Conversely, if $x \in \mathbb{R}^n$ is such that $Ax \notin \operatorname{supp} \varphi$, then there exists $r > 0$ such that $\varphi = 0$ on $B(Ax, r)$. Since $A^{-1}(B(Ax, r))$ is open, contains x, and $\varphi \circ A$ vanishes identically on it, we conclude that x does not belong to the set in the left-hand side of (1.3.19). The proof of the claim is finished. Moreover, $\{x \in \mathbb{R}^n : Ax \in \operatorname{supp} \varphi\} = A^{-1}(\operatorname{supp} \varphi)$, so (1.3.19) may be ultimately recast as

$$\operatorname{supp}(\varphi \circ A) = A^{-1}(\operatorname{supp} \varphi), \qquad \forall \varphi \in C_0^\infty(\mathbb{R}^n). \tag{1.3.20}$$

Given that φ has compact support and A^{-1} is continuous, from (1.3.20) we see that $\varphi \circ A$ has compact support. This proves that the map in (1.3.18) is well defined, while its linearity is clear. To show that this map is also continuous, by Fact 1.15 and Theorem 14.6, matters are reduced to proving sequential continuity at 0. The latter is now a consequence of Fact 1.16, (1.3.20), the continuity of A^{-1}, and the Chain Rule. $\qquad\qquad\square$

Convention. In what follows, we will often identify a function $f \in C_0^\infty(\Omega)$ with its extension by zero outside its support, which makes such an extension belong to $C_0^\infty(\mathbb{R}^n)$.

Exercise 1.25. Suppose Ω_1, Ω_2 are open sets in $\mathbb{R}^n$ and let $F : \Omega_1 \to \Omega_2$ be a C^∞ diffeomorphism.

(1) Prove that $\varphi \circ F \in C_0^\infty(\Omega_1)$ for all $\varphi \in C_0^\infty(\Omega_2)$.
(2) Prove that if $\varphi_j \xrightarrow[j \to \infty]{\mathcal{D}(\Omega_2)} \varphi$ then $\varphi_j \circ F \xrightarrow[j \to \infty]{\mathcal{D}(\Omega_1)} \varphi \circ F$.
(3) Prove that the map $\mathcal{D}(\Omega_2) \ni \varphi \longmapsto \varphi \circ F \in \mathcal{D}(\Omega_1)$ is linear and continuous.
(4) Prove that the map $\mathcal{D}(\Omega_2) \ni \varphi \longmapsto |\det(DF)| \varphi \circ F \in \mathcal{D}(\Omega_1)$ is linear and continuous, where DF denotes the Jacobian matrix of F.

Hint: For *(1)* show that $\operatorname{supp}(\varphi \circ F) = F^{-1}(\operatorname{supp} \varphi)$, for *(2)* you may use Fact 1.16 and the Chain Rule. Given Fact 1.15, *(2)*, and Exercise 1.20, in order to prove *(3)* and *(4)* you may apply Theorem 14.6.

Further Notes for Chapter 1. The concept of weak derivative goes back to the pioneering work of the Soviet mathematician Sergei Lvovich Sobolev (1908–1989). Although we shall later extend the scope of taking derivatives in a generalized sense to the larger class of distributions, a significant portion of partial differential equations may be developed solely based on the notion of weak derivative. For example, this is the approach adopted in [14], where distributions are avoided altogether. A good reference to the topological aspects that are most pertinent to the spaces of test functions considered here is [76], though there are many other monographs dealing with these issues. The interested reader may consult [12], [65], [77], and the references therein.

1.4 Additional Exercises for Chapter 1

Exercise 1.26. Given $\varphi_0 \in C^2(\mathbb{R})$, $\varphi_1 \in C^1(\mathbb{R})$, and $F \in C^1(\mathbb{R}^2)$, show that the function $u : \mathbb{R}^2 \to \mathbb{R}$ defined by

$$u(x_1, x_2) := \tfrac{1}{2}[(\varphi_0(x_1 + x_2) + \varphi_0(x_1 - x_2)] + \tfrac{1}{2} \int_{x_1-x_2}^{x_1+x_2} \varphi_1(t)\,dt$$

$$- \tfrac{1}{2} \int_0^{x_2} \Big(\int_{x_1-(x_2-t)}^{x_1+(x_2-t)} F(\xi, t)\,d\xi \Big) dt, \qquad \forall\, (x_1, x_2) \in \mathbb{R}^2, \qquad (1.4.1)$$

is a classical solution of the problem

$$\begin{cases} u \in C^2(\mathbb{R}^2), \\ \partial_1^2 u - \partial_2^2 u = F \ \text{ in } \ \mathbb{R}^2, \\ u(\cdot, 0) = \varphi_0 \ \text{ in } \ \mathbb{R}, \\ (\partial_2 u)(\cdot, 0) = \varphi_1 \ \text{ in } \ \mathbb{R}. \end{cases} \qquad (1.4.2)$$

Exercise 1.27. Determine the values of $a \in \mathbb{R}$ for which the function $f : \mathbb{R} \to \mathbb{R}$ defined by $f(x) := \begin{cases} x, \text{ if } x \geq a, \\ 0, \text{ if } x < a, \end{cases}$ for each $x \in \mathbb{R}$, has a weak derivative.

Exercise 1.28. Consider $f : \mathbb{R} \to \mathbb{R}$ defined by $f(x) := \begin{cases} 1, \text{ if } x \geq 2, \\ 0, \text{ if } x < 2, \end{cases}$ for each $x \in \mathbb{R}$. Does the function f have a weak derivative?

Exercise 1.29. Let $f : \mathbb{R}^2 \to \mathbb{R}$ be defined by $f(x, y) := H(x) + H(y)$ for each $(x, y) \in \mathbb{R}^2$ (where H is the Heaviside function from (1.2.9)). Prove that for $\alpha = (1, 1)$ and $\beta = (1, 0)$ the weak derivatives $\partial^\alpha f$ and $\partial^{\alpha+\beta} f$ exist, while the weak derivative $\partial^\beta f$ does not exist.

Exercise 1.30. Compute the weak derivative of order one of $f : (-1, 1) \to \mathbb{R}$ defined by $f(x) := \text{sgn}(x)\sqrt{|x|}$ for every $x \in (-1, 1)$, where

$$\text{sgn}(x) := \begin{cases} 1 & \text{if } x > 0, \\ 0 & \text{if } x = 0, \\ -1 & \text{if } x < 0. \end{cases} \qquad (1.4.3)$$

Exercise 1.31. Let $f : \mathbb{R}^2 \to \mathbb{R}$ be defined by $f(x, y) := x|y|$ for each $(x, y) \in \mathbb{R}^2$. Prove that the weak derivative $\partial_1^2 \partial_2 f$ exists, while the weak derivative $\partial_1 \partial_2^2 f$ does not.

Exercise 1.32. Let $f : \mathbb{R}^2 \to \mathbb{R}$ be defined by $f(x, y) := H(x) - \text{sgn}(y)$ for each $(x, y) \in \mathbb{R}^2$, and let $\alpha = (\alpha_1, \alpha_2) \in \mathbb{N}_0^2$. Prove that $\partial^\alpha f$ exists in the weak sense if and only if $\alpha_1 \geq 1$ and $\alpha_2 \geq 1$.

Exercise 1.33. Let $f : \mathbb{R} \to \mathbb{R}$ be defined by $f(x) := \sin|x|$ for every $x \in \mathbb{R}$. Does f' exist in the weak sense? How about f''?

Exercise 1.34. Let ω, Ω be open subsets of $\mathbb{R}^n$ with $\omega \subseteq \Omega$. Suppose $f \in L^1_{loc}(\Omega)$ is such that $\partial^\alpha f$ exists in the weak sense in Ω for some $\alpha \in \mathbb{N}_0^n$. Show that the weak derivative $\partial^\alpha(f|_\omega)$ exists and equals $(\partial^\alpha f)|_\omega$ almost everywhere in ω.

Exercise 1.35. Suppose $n \in \mathbb{N}$, $n \geq 2$ and $a \in (0, n)$. Let $f(x) := \frac{1}{|x|^a}$ for each $x \in \mathbb{R}^n \setminus \{0\}$ and note that $f \in L^1_{loc}(\mathbb{R}^n)$. Prove that $\partial_j f$, $j \in \{1, \ldots, n\}$, exists in the weak sense if and only if $a < n - 1$.

Exercise 1.36. Let Ω be an open subset of $\mathbb{R}^n$ and let $\alpha, \beta \in \mathbb{N}_0$. Suppose f belongs to $L^1_{loc}(\Omega)$ and is such that the weak derivatives $\partial^\alpha f$ and $\partial^\beta(\partial^\alpha f)$ exist. Prove that $\partial^{\alpha+\beta} f$ exists and equals the weak derivative $\partial^\beta(\partial^\alpha f)$.

Exercise 1.37. Let $\varepsilon \in (0, 1)$ and consider the function $f : \mathbb{R}^n \to \mathbb{R}$ defined by

$$f(x) := \begin{cases} |x|^{-\varepsilon} & \text{if } x \in \mathbb{R}^n \setminus \{0\}, \\ 1 & \text{if } x = 0, \end{cases} \qquad \forall x \in \mathbb{R}^n.$$

Prove that $\partial_j f$ exists in the weak sense for each $j \in \{1, \ldots, n\}$ if and only if $n \geq 2$. Also compute the weak derivatives $\partial_j f$, $j \in \{1, \ldots, n\}$, in the case when $n \geq 2$.

Exercise 1.38. Assume that $a, b \in \mathbb{R}$ are such that $a < b$.

(a) Prove that if $f \in L^1_{loc}((a, b))$ is such that the weak derivative f' exists and is equal to zero almost everywhere on (a, b), then there exists some complex number c such that $f = c$ almost everywhere on (a, b).

Hint: Fix $\varphi_0 \in C_0^\infty((a, b))$ with $\int_a^b \varphi_0(t)\, dt = 1$. Then every $\varphi \in C_0^\infty((a, b))$ is of the form $\varphi = \varphi_0 \int_a^b \varphi(t)\, dt + \psi'$, for some $\psi \in C_0^\infty((a, b))$.

(b) Assume that $g \in L^1_{loc}((a, b))$ and $x_0 \in (a, b)$. Prove that the function defined by $f(x) := \int_{x_0}^x g(t)\, dt$, for $x \in (a, b)$, belongs to $L^1_{loc}((a, b))$ and has a weak derivative that is equal to g almost everywhere on (a, b).

(c) Let $f \in L^1_{loc}((a, b))$ be such that the weak derivative $f^{(k)}$ exists for some $k \in \mathbb{N}$. Prove that all the weak derivatives $f^{(j)}$ exist for each $j \in \mathbb{N}$ with $j < k$.

Hint: Prove that if $g(x) := \int_{x_0}^x h(t)\, dt$ where $x_0 \in (a, b)$ is a fixed point and $h := f^{(k)}$, and if φ_0 is as in the hint to *(a)*, then

$$f^{(k-1)} = g - \int_a^b g(t)\varphi_0(t)\, dt + (-1)^{k-1} \int_a^b f(t)\varphi_0^{(k-1)}(t)\, dt.$$

(d) Let $f \in L^1_{loc}((a, b))$ such that $f^{(k)} = 0$ for some $k \in \mathbb{N}$. Prove that there exist $a_0, a_1, \ldots, a_{k-1} \in \mathbb{C}$ such that $f(x) = \sum_{j=0}^{k-1} a_j x^j$ for almost every $x \in (a, b)$.

(e) If Ω is an open set in $\mathbb{R}^n$ for $n \geq 2$ and $f \in L_{loc}^1(\Omega)$ is such that the weak derivative $\partial^\alpha f$ exists for some $\alpha \in \mathbb{N}_0^n$, does it follow that $\partial^\beta f$ exists in a weak sense for all $\beta \leq \alpha$?

Exercise 1.39. Let $\theta \in C_0^\infty(\mathbb{R}^n)$ and $m \in \mathbb{N}$. Prove that the sequence

$$\varphi_j(x) := e^{-j} j^m \theta(jx), \qquad \forall x \in \mathbb{R}^n, \quad j \in \mathbb{N},$$

converges in $\mathcal{D}(\mathbb{R}^n)$.

Exercise 1.40. Let $\theta \in C_0^\infty(\mathbb{R}^n)$, $h \in \mathbb{R}^n \setminus \{0\}$, and set

$$\varphi_j(x) := \theta(x + jh), \qquad \forall x \in \mathbb{R}^n, \quad j \in \mathbb{N}.$$

Prove that $\varphi_j \xrightarrow[j \to \infty]{\mathcal{E}(\mathbb{R}^n)} 0$. Is it true that $\varphi_j \xrightarrow[j \to \infty]{\mathcal{D}(\mathbb{R}^n)} 0$?

Exercise 1.41. Let $\theta \in C_0^\infty(\mathbb{R}^n)$ be not identically zero, and for each $j \in \mathbb{N}$ define

$$\varphi_j(x) := \frac{1}{j} \theta(jx), \qquad \forall x \in \mathbb{R}^n.$$

Prove that the sequence $\{\varphi_j\}_{j \in \mathbb{N}}$ does not converge in $\mathcal{D}(\mathbb{R}^n)$.

Exercise 1.42. Let $\theta \in C_0^\infty(\mathbb{R}^n)$, $h \in \mathbb{R}^n \setminus \{0\}$. Prove that the sequence

$$\varphi_j(x) := e^{-j} \theta(j(x - jh)), \qquad \forall x \in \mathbb{R}^n, \quad j \in \mathbb{N},$$

converges in $\mathcal{D}(\mathbb{R}^n)$ if and only if $\theta(x) = 0$ for all $x \in \mathbb{R}^n$.

Exercise 1.43. Consider $\theta \in C_0^\infty(\mathbb{R}^n)$ not identically zero, and for each $j \in \mathbb{N}$ define

$$\varphi_j(x) := \frac{1}{j} \theta\left(\frac{x}{j}\right), \qquad \forall x \in \mathbb{R}^n.$$

Does $\{\varphi_j\}_{j \in \mathbb{N}}$ converge in $\mathcal{D}(\mathbb{R}^n)$? How about in $\mathcal{E}(\mathbb{R}^n)$?

Exercise 1.44. Suppose that $\{\varphi_j\}_{j \in \mathbb{N}}$ is a sequence of functions in $C_0^\infty(\Omega)$ with the property that $\lim_{j \to \infty} \int_{\mathbb{R}^n} f(x)\varphi_j(x)\, dx = 0$ for every $f \in L_{loc}^1(\Omega)$. Is it true that $\varphi_j \xrightarrow[j \to \infty]{\mathcal{D}(\mathbb{R}^n)} 0$?

Chapter 2
The Space $\mathcal{D}'(\Omega)$ of Distributions

Abstract In this chapter the space of distributions is introduced and studied from the perspective of a topological vector space with various other additional features, such as the concept of support, multiplication with a smooth function, distributional derivatives, tensor product, and a partially defined convolution product. Here the nature of distributions with higher order gradients continuous or bounded is also discussed.

2.1 The Definition of Distributions

Building on the idea emerging in (1.3.1), we now make the following definition that is central to all subsequent considerations.

Definition 2.1. $u : \mathcal{D}(\Omega) \to \mathbb{C}$ is called a distribution on Ω if u is linear and continuous.

By design, distributions are simply elements of the dual space of the topological vector space $\mathcal{D}(\Omega)$. Given a functional $u : \mathcal{D}(\Omega) \to \mathbb{C}$ and a function $\varphi \in C_0^\infty(\Omega)$, we use the traditional notation $\langle u, \varphi \rangle$ in place of $u(\varphi)$ (in particular, $\langle u, \varphi \rangle$ is a complex number).

While any linear and continuous functional is sequentially continuous, the converse is not always true. Nonetheless, for linear functionals on $\mathcal{D}(\Omega)$, continuity is equivalent with sequential continuity. This remarkable property, itself a consequence of Theorem 14.6, is formally recorded below.

Fact 2.2 *Let* $u : \mathcal{D}(\Omega) \to \mathbb{C}$ *be a linear map. Then u is a distribution on Ω if and only if for every sequence $\{\varphi_j\}_{j \in \mathbb{N}}$ contained in $C_0^\infty(\Omega)$ with the property that $\varphi_j \xrightarrow[j \to \infty]{\mathcal{D}(\Omega)} \varphi$ for some function $\varphi \in C_0^\infty(\Omega)$, we have $\lim_{j \to \infty} \langle u, \varphi_j \rangle = \langle u, \varphi \rangle$ (where the latter limit is considered in $\mathbb{C}$).*

© Springer Nature Switzerland AG 2018   17
D. Mitrea, *Distributions, Partial Differential Equations, and Harmonic Analysis*,
Universitext, https://doi.org/10.1007/978-3-030-03296-8_2

Remark 2.3. In general, if X, Y are topological vector spaces and $\Lambda : X \to Y$ is a linear map, then Λ is sequentially continuous on X if and only if Λ is sequentially continuous at the zero vector $0 \in X$. When combined with Fact 2.2, this shows that a linear map $u : \mathcal{D}(\Omega) \to \mathbb{C}$ is a distribution on Ω if and only if $\lim_{j \to \infty} \langle u, \varphi_j \rangle = 0$ for every sequence $\{\varphi_j\}_{j \in \mathbb{N}} \subset C_0^\infty(\Omega)$ with $\varphi_j \xrightarrow[j \to \infty]{\mathcal{D}(\Omega)} 0$.

Another important characterization of continuity of complex-valued linear functionals u defined on $\mathcal{D}(\Omega)$ is given by the following proposition.

Proposition 2.4. *Let $u : \mathcal{D}(\Omega) \to \mathbb{C}$ be a linear map. Then u is a distribution if and only if for each compact set $K \subset \Omega$ there exist $k \in \mathbb{N}_0$ and $C \in (0, \infty)$ such that*

$$|\langle u, \varphi \rangle| \le C \sup_{\substack{x \in K \\ |\alpha| \le k}} |\partial^\alpha \varphi(x)| \quad \text{for all } \varphi \in C_0^\infty(\Omega) \text{ with } \operatorname{supp} \varphi \subseteq K. \tag{2.1.1}$$

Proof. Fix $u : \mathcal{D}(\Omega) \to \mathbb{C}$ linear and suppose that for each compact set $K \subset \Omega$ there exist $k \in \mathbb{N}_0$ and $C \in (0, \infty)$ satisfying (2.1.1). To show that u is a distribution, let $\varphi_j \xrightarrow[j \to \infty]{\mathcal{D}(\Omega)} 0$. Then, there exists a compact set $K \subseteq \Omega$ such that $\operatorname{supp} \varphi_j \subseteq K$ for all $j \in \mathbb{N}$, and $\partial^\alpha \varphi_j \xrightarrow[j \to \infty]{} 0$ uniformly on K for any $\alpha \in \mathbb{N}_0^n$. For this compact set K, by our hypotheses, there exist $C > 0$ and $k \in \mathbb{N}_0$ such that (2.1.1) holds, and hence,

$$|\langle u, \varphi_j \rangle| \le C \sup_{\substack{x \in K \\ |\alpha| \le k}} |\partial^\alpha \varphi_j(x)| \xrightarrow[j \to \infty]{} 0, \tag{2.1.2}$$

which implies that $\langle u, \varphi_j \rangle \xrightarrow[j \to \infty]{} 0$. From this and Remark 2.3 it follows that u is a distribution in Ω.

To prove the converse implication we reason by contradiction. Suppose that there exists a compact set $K \subseteq \Omega$ such that for every $j \in \mathbb{N}$, there exists a function $\varphi_j \in C_0^\infty(\Omega)$ with $\operatorname{supp} \varphi_j \subseteq K$ and

$$|\langle u, \varphi_j \rangle| > j \sup_{\substack{x \in K \\ |\alpha| \le j}} |\partial^\alpha \varphi_j(x)|. \tag{2.1.3}$$

Define $\psi_j := \frac{1}{\langle u, \varphi_j \rangle} \varphi_j$. Then for each $j \in \mathbb{N}$ we have $\psi_j \in C^\infty(\Omega)$, $\operatorname{supp} \psi_j \subseteq K$, and $\langle u, \psi_j \rangle = 1$. On the other hand, from (2.1.3) we see that

$$\sup_{\substack{x \in K \\ |\alpha| \le j}} |\partial^\alpha \psi_j(x)| < \frac{1}{j} \quad \forall\, j \in \mathbb{N}. \tag{2.1.4}$$

Now let $\alpha \in \mathbb{N}_0^n$ be arbitrary. Then (2.1.4) implies that

$$\sup_{\substack{x \in K \\ |\alpha| \le j}} |\partial^\alpha \psi_j| < \frac{1}{j} \quad \text{whenever} \quad j \ge |\alpha|, \tag{2.1.5}$$

thus $\psi_j \xrightarrow[j\to\infty]{\mathcal{D}(\Omega)} 0$. Since u is a distribution in Ω, the latter implies $\lim_{j\to\infty} \langle u, \psi_j \rangle = 0$, contradicting the fact that $\langle u, \psi_j \rangle = 1$ for each $j \in \mathbb{N}$. This completes the proof of the proposition. $\qquad\square$

Remark 2.5. Recall that for each compact set $K \subset \Omega$ we denote by $\mathcal{D}_K(\Omega)$ the vector space of functions in $C^\infty(\Omega)$ with support contained in K endowed with the topology inherited from $\mathcal{E}(\Omega)$.

A closer look at the topology in $\mathcal{D}_K(\Omega)$ reveals that Proposition 2.4 may be rephrased as saying that a linear map $u : \mathcal{D}(\Omega) \to \mathbb{C}$ is a distribution in Ω if and only if $u\big|_{\mathcal{D}_K(\Omega)}$ is continuous for each compact set $K \subset \Omega$. In fact, the topology on $\mathcal{D}(\Omega)$ is the smallest topology on $C_0^\infty(\Omega)$ with this property.

Definition 2.6. Let u be a distribution in Ω. If the nonnegative integer k intervening in (2.1.1) may be taken to be independent of K, then u is called a `distribution of finite order`. If u is a distribution of finite order, then the `order of` u is by definition the smallest $k \in \mathbb{N}_0$ satisfying condition (2.1.1) for every compact set $K \subset \Omega$.

Here are a few important examples of distributions.

Example 2.7. For each $f \in L^1_{loc}(\Omega)$ define the functional $u_f : \mathcal{D}(\Omega) \to \mathbb{C}$ by

$$u_f(\varphi) := \int_\Omega f(x)\varphi(x)\,dx, \qquad \forall\,\varphi \in C_0^\infty(\Omega). \tag{2.1.6}$$

Then clearly u_f is linear and, if K is an arbitrary compact set contained in Ω, then

$$|u_f(\varphi)| \le \sup_{x\in K} |\varphi(x)| \int_K |f|\,dx = C \sup_{x\in K} |\varphi(x)|, \tag{2.1.7}$$
$$\text{for all } \varphi \in C_0^\infty(\Omega) \text{ with } \operatorname{supp}\varphi \subseteq K.$$

Hence, by Proposition 2.4, u_f is a distribution in Ω. Moreover, (2.1.7) also shows that u_f is a distribution of order 0.

Remark 2.8. A distribution whose action is defined as in (2.1.6) will be referred to as a `distribution of function type`.

To simplify notation, if $f \in L^1_{loc}(\Omega)$ we will often simply use f (in place of u_f) for the distribution of function type defined as in (2.1.6). This is justified by the fact that the linear map

$$\iota : L^1_{loc}(\Omega) \to \{u : \mathcal{D}(\Omega) \to \mathbb{C} : u \text{ linear and continuous}\}$$
$$\iota(f) := u_f \quad \text{for each } f \in L^1_{loc}(\Omega), \tag{2.1.8}$$

is one-to-one. Indeed, if $\iota(f) = 0$ for some $f \in L^1_{loc}(\Omega)$, then $\int_\Omega f\varphi\,dx = 0$ for all functions $\varphi \in C_0^\infty(\Omega)$, which in turn, based on Theorem 1.3, implies that $f = 0$ almost everywhere in Ω. Since ι is also linear, the desired conclusion follows.

Example 2.9. We have that $\ln|x| \in L^1_{loc}(\mathbb{R}^n)$, thus $\ln|x|$ is a distribution in $\mathbb{R}^n$.

To see that indeed $\ln|x|$ is locally integrable in $\mathbb{R}^n$, observe first that

$$\left|\ln t\right| \leq \frac{1}{\varepsilon}\, \max\,\{t^\varepsilon, t^{-\varepsilon}\} \quad \text{for all } t > 0 \text{ and } \varepsilon > 0. \tag{2.1.9}$$

This is justified by starting with the elementary inequality $\ln t \leq t$ for all $t > 0$, then replacing t by $t^{\pm\varepsilon}$. In turn, for every $R > 0$ and $\varepsilon \in (0, 1)$, estimate (2.1.9) gives

$$\int_{B(0,R)} \left|\ln|x|\right| dx \leq \varepsilon^{-1} \int_{B(0,R)} \max\,\{|x|^\varepsilon, |x|^{-\varepsilon}\}\, dx < \infty. \tag{2.1.10}$$

Let us now revisit the functional from (1.3.1).

Example 2.10. For a given $f \in L^1_{loc}(\Omega)$ and multi-index $\alpha \in \mathbb{N}_0^n$, consider the functional $g_\alpha : \mathcal{D}(\Omega) \to \mathbb{C}$ defined by

$$g_\alpha(\varphi) := (-1)^{|\alpha|} \int_\Omega f\, \partial^\alpha \varphi\, dx \quad \text{for each } \varphi \in C_0^\infty(\Omega). \tag{2.1.11}$$

Clearly this is a linear mapping. Moreover, if $\varphi_j \xrightarrow[j\to\infty]{\mathcal{D}(\Omega)} 0$, then there exists K compact subset of Ω such that $\operatorname{supp}\varphi_j \subseteq K$ for every $j \in \mathbb{N}$, hence

$$|g_\alpha(\varphi_j)| \leq \int_K |f| |\partial^\alpha \varphi_j|\, dx$$

$$\leq \|\partial^\alpha \varphi_j\|_{L^\infty(K)} \|f\|_{L^1(K)} \to 0 \quad \text{as } j \to \infty. \tag{2.1.12}$$

By invoking Remark 2.3, it follows that g_α is a distribution in Ω.

Next we consider a set of examples of distributions that are not of function type. As a preamble, observe that $f(x) := \frac{1}{x}$ for $x \neq 0$, is not locally integrable on $\mathbb{R}$. Nonetheless, it is possible to associate to this function a certain distribution, not as in (2.1.6), but in the specific manner described below.

Example 2.11. Consider the mapping P.V. $\frac{1}{x} : \mathcal{D}(\mathbb{R}) \to \mathbb{C}$ defined by

$$\left(\text{P.V.}\,\frac{1}{x}\right)(\varphi) := \lim_{\varepsilon \to 0^+} \int_{|x| \geq \varepsilon} \frac{\varphi(x)}{x}\, dx, \qquad \forall\, \varphi \in C_0^\infty(\mathbb{R}). \tag{2.1.13}$$

We claim that P.V. $\frac{1}{x}$ is a distribution of order one in $\mathbb{R}$.

First, we prove that the mapping (2.1.13) is well defined. Let $\varphi \in C_0^\infty(\mathbb{R})$ and suppose $R > 0$ is such that $\operatorname{supp}\varphi \subset (-R, R)$. Fix $\varepsilon \in (0, R)$ and observe that since $\frac{1}{x}$ is odd on $\mathbb{R} \setminus \{0\}$, we have

$$\int_{|x| \geq \varepsilon} \frac{\varphi(x)}{x}\, dx = \int_{\varepsilon \leq |x| \leq R} \frac{\varphi(x)}{x}\, dx = \int_{\varepsilon \leq |x| \leq R} \frac{\varphi(x) - \varphi(0)}{x}\, dx. \tag{2.1.14}$$

In addition, $\left|\frac{\varphi(x)-\varphi(0)}{x}\right| \leq \sup_{|y| \leq R} |\varphi'(y)|$ for each $x \in \mathbb{R} \setminus \{0\}$. Thus, by Lebesgue's Dominated Convergence Theorem the limit $\lim_{\varepsilon \to 0^+} \int_{|x| \geq \varepsilon} \frac{\varphi(x)}{x} \, dx$ exists and is equal to $\int_{|x| \leq R} \frac{\varphi(x)-\varphi(0)}{x} \, dx$. This shows that the mapping P.V. $\frac{1}{x}$ is well defined, and from definition it is clear that P.V. $\frac{1}{x}$ is linear. Furthermore, it is implicit in the argument above that

$$\left|\left(\text{P.V.} \frac{1}{x}\right)(\varphi)\right| \leq 2R \sup_{|x| \leq R} |\varphi'(x)|, \qquad \forall \, \varphi \in C_0^\infty((-R, R)). \tag{2.1.15}$$

In concert with Proposition 2.4 this shows that P.V. $\frac{1}{x}$ is a distribution in $\mathbb{R}$ of order at most one. We are left with showing that P.V. $\frac{1}{x}$ does not have order 0. Consider the compact $K = [0, 1]$ and for each $j \in \mathbb{N}$ let $\varphi_j \in C_0^\infty((0, 1))$ be such that $0 \leq \varphi_j \leq 1$ and $\varphi_j \equiv 1$ on $\left[\frac{1}{j+2}, 1 - \frac{1}{j+2}\right]$. Then from the very definition of P.V. $\frac{1}{x}$ and the fact that φ_j vanishes near zero,

$$\left|\left\langle \text{P.V.} \frac{1}{x}, \varphi_j \right\rangle\right| = \int_0^1 \frac{\varphi_j(x)}{x} \, dx \geq \int_{\frac{1}{j+2}}^{1-\frac{1}{j+2}} \frac{1}{x} \, dx = \ln(j + 1) \tag{2.1.16}$$

for each $j \in \mathbb{N}$. Since $\sup_{x \in K} |\varphi_j(x)| \leq 1$ and $\lim_{j \to \infty} \ln(j + 1) = \infty$, the inequality in (2.1.16) shows that there is no constant $C \in (0, \infty)$ with the property that

$$\left|\left\langle \text{P.V.} \frac{1}{x}, \varphi \right\rangle\right| \leq C \sup_{x \in K} |\varphi(x)| \text{ for all } \varphi \in C_0^\infty(\mathbb{R}) \text{ with } \operatorname{supp} \varphi \subseteq K. \tag{2.1.17}$$

This proves that P.V. $\frac{1}{x}$ does not have order 0.

Remark 2.12. An inspection of the proof in Example 2.11 shows that

$$\left\langle \text{P.V.} \frac{1}{x}, \varphi \right\rangle = \int_{|x| \leq 1} \frac{\varphi(x) - \varphi(0)}{x} \, dx + \int_{|x| > 1} \frac{\varphi(x)}{x} \, dx \tag{2.1.18}$$

for each $\varphi \in C_0^\infty(\mathbb{R})$.

Example 2.13. An important distribution is the Dirac distribution δ defined by

$$\delta(\varphi) := \varphi(0), \qquad \forall \, \varphi \in C_0^\infty(\mathbb{R}^n). \tag{2.1.19}$$

It is not difficult to check that δ is a distribution in $\mathbb{R}^n$ of order 0. A natural question to ask is whether δ is a distribution of function type. To answer this question, suppose there exists $f \in L^1_{loc}(\mathbb{R}^n)$ such that

$$\varphi(0) = \langle \delta, \varphi \rangle = \int_{\mathbb{R}^n} f\varphi \, dx \text{ for all } \varphi \in C_0^\infty(\mathbb{R}^n). \tag{2.1.20}$$

This implies that $\int_{\mathbb{R}^n} f\varphi \, dx = 0$ for every $\varphi \in C_0^\infty(\mathbb{R}^n)$ with the property that $0 \notin \operatorname{supp} \varphi$. Hence, by Theorem 1.3 we have $f = 0$ almost everywhere on $\mathbb{R}^n \setminus \{0\}$, thus $f = 0$ almost everywhere in $\mathbb{R}^n$. Consequently,

$$\varphi(0) = \langle \delta, \varphi \rangle = \int_{\mathbb{R}^n} f\varphi \, dx = 0, \qquad \forall \varphi \in C_0^\infty(\mathbb{R}^n), \tag{2.1.21}$$

which is false. This proves that the Dirac distribution is not of function type.

Example 2.14. The Dirac distribution δ is sometimes referred to as having "mass at zero" since, for each $x_0 \in \mathbb{R}^n$, we may similarly $\delta_{x_0} : C_0^\infty(\mathbb{R}^n) \to \mathbb{C}$ by setting $\delta_{x_0}(\varphi) := \varphi(x_0)$. Then δ_{x_0} is a distribution in $\mathbb{R}^n$ (called the Dirac distribution with mass at x_0) and the convention we make is to drop the subscript x_0 if $x_0 = 0 \in \mathbb{R}^n$.

Example 2.15. Let μ be either a complex Borel measure on Ω, or a Borel positive measure on Ω that is locally finite (i.e., satisfies $\mu(K) < \infty$ for every compact $K \subset \Omega$). Consider

$$\mu : \mathcal{D}(\Omega) \to \mathbb{C}, \quad \mu(\varphi) := \int_\Omega \varphi \, d\mu, \quad \forall \varphi \in C_0^\infty(\Omega). \tag{2.1.22}$$

The mapping in (2.1.22) is well defined, linear, and if K is an arbitrary compact set in Ω, then

$$|\mu(\varphi)| \le |\mu|(K) \sup_{x \in K} |\varphi(x)|, \qquad \forall \varphi \in C^\infty(\Omega), \ \ \mathrm{supp}\,\varphi \subset K. \tag{2.1.23}$$

By Proposition 2.4 we have that μ induces a distribution in Ω. The estimate in (2.1.23) also shows that μ is a distribution of order zero.

Next we discuss the validity of a converse implication to the implication in Example 2.15.

Proposition 2.16. *Let u be a distribution in Ω of order zero. Then the distribution u extends uniquely to a linear map $\Lambda_u : C_0^0(\Omega) \to \mathbb{C}$ that is locally bounded, in the following sense: for each compact set $K \subset \Omega$ there exists $C_K \in (0, \infty)$ with the property that*

$$|\Lambda_u(\varphi)| \le C_K \sup_{x \in K} |\varphi(x)|, \qquad \forall \varphi \in C_0^0(\Omega) \ \text{with} \ \mathrm{supp}\,\varphi \subseteq K. \tag{2.1.24}$$

In addition, the functional Λ_u satisfies the following properties.

(i) Let $\{K_j\}_{j \in \mathbb{N}}$ be a sequence of compact subsets of Ω satisfying $K_j \subset \mathring{K}_{j+1}$ for $j \in \mathbb{N}$ and $\Omega = \bigcup_{j=1}^\infty K_j$. Then there exists a sequence of complex regular Borel measures μ_j on K_j, $j \in \mathbb{N}$, with the following properties:

(a) $\mu_j(E) = \mu_\ell(E)$ for every $\ell \in \mathbb{N}$, every Borel set $E \subset \mathring{K}_\ell$, and every $j \ge \ell$;
(b) for each $j \in \mathbb{N}$ one has

$$\Lambda_u(\varphi) = \int_{K_j} \varphi \, d\mu_j, \quad \forall \varphi \in C_0^0(\Omega) \ \text{with} \ \mathrm{supp}\,\varphi \subset K_j. \tag{2.1.25}$$

(ii) There exist two Radon measures μ_1, μ_2, taking Borel sets from Ω into $[0, \infty]$ (i.e., measures satisfying the regularity properties (ii)–(iv) in Theorem 14.25), such

that

$$\mathrm{Re}\,[\Lambda_u(\varphi)] = \int_{\Omega} \varphi\,d\mu_1 - \int_{\Omega} \varphi\,d\mu_2, \quad \forall\,\varphi \in C_0^0(\Omega) \ \textit{real-valued.} \tag{2.1.26}$$

Furthermore, a similar conclusion is valid for $\mathrm{Im}\,\Lambda_u$.

The following approximation result is useful in the proof of Proposition 2.16.

Lemma 2.17. *Let* $k \in \mathbb{N}_0$ *and suppose K is a compact subset of Ω. Define the compact set*

$$K_0 := \{x \in \Omega : \ \mathrm{dist}\,(x, K) \le \tfrac{1}{2}\,\mathrm{dist}\,(K, \partial\Omega)\}. \tag{2.1.27}$$

Then for every $\varphi \in C_0^k(\Omega)$ *with* $\mathrm{supp}\,\varphi \subseteq K$ *there exists a sequence* $\{\varphi_j\}_{j \in \mathbb{N}}$ *of functions in* $C_0^{\infty}(\Omega)$ *with* $\mathrm{supp}\,\varphi_j \subseteq K_0$ *and*

$$\begin{aligned} &\forall\,K_1 \subset \Omega \text{ compact and } \forall\,\alpha \in \mathbb{N}_0^n \text{ with } |\alpha| \le k, \text{ we}\\ &\text{have } \lim_{j \to \infty} \sup_{x \in K_1} |\partial^{\alpha}\varphi_j(x) - \partial^{\alpha}\varphi(x)| = 0. \end{aligned} \tag{2.1.28}$$

Proof. Recall the sequence of functions $\{\phi_{1/j}\}_{j \in \mathbb{N}}$ defined in (1.2.3)–(1.2.5) corresponding to $\varepsilon := \frac{1}{j}$, $j \in \mathbb{N}$. Pick $j_0 \in \mathbb{N}$ such that $\frac{1}{j_0} < \frac{1}{2}\,\mathrm{dist}\,(K, \partial\Omega)$ and then set

$$\begin{aligned} \varphi_j(x) :&= \int_{\mathbb{R}^n} \varphi(y)\phi_{1/j}(x - y)\,dy\\ &= \int_{\mathbb{R}^n} \varphi(x - y)\phi_{1/j}(y)\,dy, \qquad \forall\,x \in \Omega, \quad j \ge j_0. \end{aligned} \tag{2.1.29}$$

In light of (1.2.5) and (2.1.29) we have that

$$\varphi_j \in C_0^{\infty}(\Omega) \ \text{ and } \ \mathrm{supp}\,\varphi_j \subseteq K_0, \ \text{ for all } \ j \ge j_0. \tag{2.1.30}$$

Pick some $j_1 \in \mathbb{N}$ such that $\frac{1}{j_1} < \frac{1}{4}\,\mathrm{dist}(K, \partial\Omega)$ (note that $j_1 \ge j_0$). Then the set $\widetilde{K} := K_0 + B(0, 1/j_1)$ is a compact subset of Ω.

Fix $\varepsilon > 0$ arbitrary. Since φ is continuous in Ω, it is uniformly continuous on the compact $\widetilde{K}$, hence there exists $\delta > 0$ such that $|\varphi(x_1) - \varphi(x_2)| \le \varepsilon$ for every $x_1, x_2 \in \widetilde{K}$ satisfying $|x_1 - x_2| \le \delta$. Furthermore, choose $j_2 \in \mathbb{N}$ satisfying $\frac{1}{j_2} \le \delta$ and $j_2 \ge j_1$. At this point, for each $x \in K_0$, and each $j \ge j_2$, we may write

$$\begin{aligned} |\varphi_j(x) - \varphi(x)| &= \left| \int_{B(0,1/j)} [\varphi(x - y) - \varphi(x)]\phi_{1/j}(y)\,dy \right|\\ &= \int_{B(0,1/j)} |\varphi(x - y) - \varphi(x)|\phi_{1/j}(y)\,dy\\ &\le \varepsilon \int_{B(0,1/j)} \phi_{1/j}(y)\,dy = \varepsilon. \end{aligned} \tag{2.1.31}$$

In summary, we have proved that for each $\varepsilon > 0$ there exists $j_2 \in \mathbb{N}$ such that $|\varphi_j(x) - \varphi(x)| \leq \varepsilon$ for every $x \in K_0$. This shows that the sequence defined in (2.1.29) satisfies

$$\lim_{j \to \infty} \sup_{x \in K_0} |\varphi_j(x) - \varphi(x)| = 0. \qquad (2.1.32)$$

Next, observe that since φ is of class C^k, from (2.1.29) we also have

$$\partial^\alpha \varphi_j(x) = \int_{\mathbb{R}^n} (\partial^\alpha \varphi)(y) \phi_{1/j}(x - y) \, dy, \qquad \forall\, x \in \Omega, \qquad (2.1.33)$$

for every $\alpha \in \mathbb{N}_0$, $|\alpha| \leq k$, and all $j \geq j_0$. This and an argument similar to that used for the proof of (2.1.32) imply

$$\lim_{j \to \infty} \sup_{x \in K_0} |\partial^\alpha \varphi_j(x) - \partial^\alpha \varphi(x)| = 0, \quad \forall\, \alpha \in \mathbb{N}_0, \ |\alpha| \leq k. \qquad (2.1.34)$$

Now the fact that the sequence $\{\varphi_j\}_{j \geq j_0}$ satisfies (2.1.28) follows from the support condition (2.1.30) and (2.1.34). This finishes the proof of the lemma. $\qquad \square$

We are now ready to present the proof of Proposition 2.16.

Proof of Proposition 2.16. Let u be a distribution in Ω of order zero and let K be a compact set contained in Ω. Fix $\varphi \in C_0^0(\Omega)$ such that $\mathrm{supp}\, \varphi \subseteq K$ and apply Lemma 2.17 with $k := 0$. Hence, with K_0 as in (2.1.27), there exists a sequence $\{\varphi_j\}_{j \in \mathbb{N}}$ of functions in $C_0^\infty(\Omega)$ with $\mathrm{supp}\, \varphi_j \subseteq K_0$ satisfying (2.1.28).

The fact that u is a distribution of order zero, implies the existence of some $C = C(K_0) \in (0, \infty)$ such that (2.1.1) holds with $k = 0$. The latter combined with (2.1.28) implies

$$|\langle u, \varphi_j \rangle - \langle u, \varphi_k \rangle| \leq C \sup_{x \in K_0} |\varphi_j(x) - \varphi_k(x)| \xrightarrow[j,k \to \infty]{} 0. \qquad (2.1.35)$$

Hence, the sequence of complex numbers $\{\langle u, \varphi_j \rangle\}_{j \in \mathbb{N}}$ is Cauchy, thus convergent in $\mathbb{C}$, which allows us to define

$$\Lambda_u(\varphi) := \lim_{j \to \infty} \langle u, \varphi_j \rangle. \qquad (2.1.36)$$

Proving that this definition is independent of the selection of $\{\varphi_j\}_{j \in \mathbb{N}}$ is done by interlacing sequences. Specifically, if $\{\varphi_j'\}_{j \in \mathbb{N}}$ is another sequence of functions in $C_0^\infty(\Omega)$, supported in K_0 and satisfying (2.1.28), then by considering the sequence $\{\psi_j\}_{j \geq 2}$ defined by $\psi_{2j+1} := \varphi_j$ and $\psi_{2j} := \varphi_j'$ for every $j \geq 2$, we obtain that $\{\langle u, \psi_j \rangle\}_{j \geq 2}$ is convergent, thus its subsequences $\{\langle u, \varphi_j \rangle\}_{j \in \mathbb{N}}$ and $\{\langle u, \varphi_j' \rangle\}_{j \in \mathbb{N}}$ are also convergent and have the same limit. In turn, the independence of the definition of $\Lambda_u(\varphi)$ on the approximating sequence $\{\varphi_j\}_{j \in \mathbb{N}}$ readily implies that $\Lambda_u : C_0^0(\Omega) \to \mathbb{C}$ is a linear mapping.

In addition, the fact that u is a distribution of order zero implies the existence of a finite constant $C = C(K_0) > 0$ such that

$$|\langle u, \varphi_j \rangle| \leq C \sup_{x \in K_0} |\varphi_j(x)| \quad \text{for each } j \in \mathbb{N}. \qquad (2.1.37)$$

Taking the limit as $j \to \infty$ in (2.1.37) gives $|\Lambda_u(\varphi)| \le C \sup_{x \in K} |\varphi(x)|$ on account of (2.1.36). Moreover, since K_0 has an explicit construction in terms of K (recall (2.1.27)), the constant C above is ultimately dependent on K, thus $C = C(K)$ as wanted. This proves the local boundedness of Λ_u in the sense of (2.1.24). To show that this linear extension Λ_u of u to $C_0^0(\Omega)$ is unique in the class of linear and locally bounded mappings, it suffices to prove that if $\Lambda : C_0^0(\Omega) \to \mathbb{C}$ is a linear locally bounded mapping that vanishes on $C_0^\infty(\Omega)$ then Λ is identically zero on $C_0^0(\Omega)$. To this end, pick an arbitrary $\varphi \in C_0^\infty(\Omega)$, set $K := \operatorname{supp}\varphi$ and, as before, apply Lemma 2.17 to obtain a sequence of functions $\varphi_j \in C_0^\infty(\Omega)$, $j \in \mathbb{N}$, supported in the fixed compact neighborhood K_0 of K such that (2.1.28) holds with $k = 0$. Then

$$|\Lambda(\varphi)| = |\Lambda(\varphi - \varphi_j)| \le C \sup_{x \in K_0} |\varphi(x) - \varphi_j(x)| \xrightarrow[j \to \infty]{} 0, \tag{2.1.38}$$

proving that $\Lambda(\varphi) = 0$, as wanted.

Moving on, consider a sequence of compact sets satisfying the hypothesis of (i). For example the sequence of compacts $\{K_j\}_{j \in \mathbb{N}}$ defined by (1.3.11) will do. Based on what we have proved so far and Riesz's representation theorem for complex measures (see Theorem 14.26), it follows that there exists a sequence of complex regular Borel measures μ_j on K_j, $j \in \mathbb{N}$, such that (2.1.25) holds.

Now fix $\ell \in \mathbb{N}$ and a Borel set $E \subset \mathring{K}_\ell$. If $j \ge \ell$, since the measures μ_j, μ_ℓ are regular, in order to prove that $\mu_j(E) = \mu_\ell(E)$ it suffices to show that $\mu_j(F) = \mu_\ell(F)$ for every compact set $F \subseteq E$. Fix such a compact set and choose a sequence of compact sets $\{F_k\}_{k \in \mathbb{N}}$ contained in $\mathring{K}_\ell$, such that $F_{k+1} \subset \mathring{F}_k$ for each $k \in \mathbb{N}$ and $\bigcap_{k=1}^\infty F_k = F$. Applying Uryshon's lemma (see Proposition 14.28) we obtain a sequence $\varphi_k \in C_0^0(\Omega)$, $0 \le \varphi_k \le 1$, $\operatorname{supp}\varphi_k \subseteq F_k$ and $\varphi_k \equiv 1$ on F, for each $k \in \mathbb{N}$. Then $\lim_{k \to \infty} \varphi_k(x) = \chi_F(x)$ for $x \in \Omega$ and we may write

$$\mu_j(F) = \int_{K_j} \chi_F \, \mathrm{d}\mu_j = \lim_{k \to \infty} \int_{K_j} \varphi_k \, \mathrm{d}\mu_j = \lim_{k \to \infty} \langle u, \varphi_k \rangle = \lim_{k \to \infty} \int_{K_\ell} \varphi_k \, \mathrm{d}\mu_\ell$$

$$= \int_{K_\ell} \chi_F \, \mathrm{d}\mu_\ell = \mu_\ell(F). \tag{2.1.39}$$

This completes the proof of (i). Finally, the claim in (ii) follows by invoking Riesz's representation theorem for locally bounded functionals (cf. Theorem 14.27). □

Among other things, Proposition 2.16 is a useful ingredient in the following representation theorem for positive distributions.

Theorem 2.18. *Let u be a distribution in Ω such that $\langle u, \varphi \rangle \ge 0$ for every nonnegative function $\varphi \in C_0^\infty(\Omega)$. Then there exists a unique positive Borel regular measure μ on Ω such that*

$$\langle u, \varphi \rangle = \int_\Omega \varphi \, \mathrm{d}\mu, \quad \forall \varphi \in C_0^\infty(\Omega). \tag{2.1.40}$$

Proof. First we prove that u has order zero. To do so, let K be a compact set in Ω and fix $\psi \in C_0^\infty(\Omega)$, $\psi \ge 0$ and satisfying $\psi \equiv 1$ on K. Then, if $\varphi \in C_0^\infty(\Omega)$ has

$\operatorname{supp}\varphi \subseteq K$ and is real valued, it follows that

$$(\sup_{x\in K} |\varphi(x)|).\psi \pm \varphi \geq 0, \tag{2.1.41}$$

hence by the positivity of u, we have

$$(\sup_{x\in K} |\varphi(x)|)\langle u,\psi\rangle \pm \langle u,\varphi\rangle \geq 0.$$

Thus,

$$|\langle u,\varphi\rangle| \leq \langle u,\psi\rangle \sup_{x\in K} |\varphi(x)|,$$

$$\text{for all real valued } \varphi \in C_0^0(\Omega) \text{ with } \operatorname{supp}\varphi \subseteq K. \tag{2.1.42}$$

If now $\varphi \in C_0^\infty(\Omega)$ with $\operatorname{supp}\varphi \subseteq K$ is complex valued, say $\varphi = \varphi_1 + i\varphi_2$ and φ_1, φ_2, are real valued, then by using (2.1.42) we may write

$$|\langle u,\varphi\rangle| = \sqrt{|\langle u,\varphi_1\rangle|^2 + |\langle u,\varphi_2\rangle|^2} \leq \langle u,\psi\rangle[\,\sup_{x\in K} |\varphi_1(x)| + \sup_{x\in K} |\varphi_2(x)|\,]$$

$$\leq 2\langle u,\psi\rangle \sup_{x\in K} |\varphi(x)|. \tag{2.1.43}$$

The fact that the distribution u has order zero now follows from (2.1.43). Having established this, we may apply Proposition 2.16 to conclude that u may be uniquely extended to a linear map $\Lambda_u : C_0^0(\Omega) \to \mathbb{C}$ that is locally bounded.

We next propose to show that this linear map is positive. In this respect, we note that if $\varphi \geq 0$ then the functions $\{\Phi_j\}_{j\in\mathbb{N}}$ used in the proof of Proposition 2.16 (which are constructed by convolving φ with a nonnegative mollifier) may also be taken to be nonnegative. When combined with (2.1.36) and the fact that u is positive, this shows that the extension Λ_u of u to $C_0^0(\Omega)$ as defined in (2.1.36) is a positive functional. Consequently, Riesz's representation theorem for positive functionals (see Theorem 14.25) may be invoked to conclude that there exists a unique positive Borel regular measure μ on Ω such that

$$\Lambda_u(\varphi) = \int_\Omega \varphi \, d\mu, \quad \forall\, \varphi \in C_0^0(\Omega). \tag{2.1.44}$$

Now (2.1.40) follows by specializing (2.1.44) to $\varphi \in C_0^\infty(\Omega)$. This finishes the proof of the theorem.   □

The first part in the statement of Proposition 2.16 has a natural generalization corresponding to distributions of any finite order.

Proposition 2.19. *Let u be a distribution in Ω of order $k \in \mathbb{N}_0$. Then there exists a linear map $\Lambda_u : C_0^k(\Omega) \to \mathbb{C}$ satisfying $\Lambda_u\big|_{C_0^\infty(\Omega)} = u$ and with the property that*

for each compact set $K \subset \Omega$ there exists $C \in (0,\infty)$ such that

$$|\Lambda_u(\varphi)| \leq C \sup_{\substack{x\in K \\ |\alpha|\leq k}} |\partial^\alpha \varphi(x)|, \quad \forall\, \varphi \in C_0^k(\Omega) \text{ with } \operatorname{supp}\varphi \subseteq K. \tag{2.1.45}$$

Moreover, Λ_u is unique satisfying these properties.

Proof. Let K be an arbitrary compact subset of Ω and let $\varphi \in C_0^k(\Omega)$ be such that $\operatorname{supp}\varphi \subseteq K$. Apply Lemma 2.17 to obtain a sequence $\{\varphi_j\}_{j\in\mathbb{N}} \subset C_0^\infty(\Omega)$ of functions supported in the compact K_0 defined in relation to K as in (2.1.27), and satisfying (2.1.28). Using the fact that u is a distribution of order k and the properties of the sequence $\{\varphi_j\}_{j\in\mathbb{N}}$ we may run an argument similar to that from the first part of the proof Proposition 2.16 to define a mapping $\Lambda_u : C_0^k(\Omega) \to \mathbb{C}$ as in (2.1.36). That this mapping is well defined, linear, and satisfies the desired properties is proved much as is the corresponding result in Proposition 2.16 (with the obvious adjustments due to the fact that u has finite order k, rather than order zero). $\qquad\square$

Example 2.20. For each multi-index $\alpha \in \mathbb{N}_0^n$ the distribution $\partial^\alpha \delta$ is a distribution of order $|\alpha|$ in $\mathbb{R}^n$. By Proposition 2.19 (and formula (2.1.36)) this distribution extends uniquely to the linear map

$$\Lambda_{\partial^\alpha\delta} : C_0^{|\alpha|}(\mathbb{R}^n) \to \mathbb{C} \quad \text{defined by}$$

$$\Lambda_{\partial^\alpha\delta}(\varphi) := (-1)^{|\alpha|}(\partial^\alpha\varphi)(0) \ \text{ for each } \ \varphi \in C_0^{|\alpha|}(\mathbb{R}^n),$$

(2.1.46)

which satisfies (2.1.45) with $k := |\alpha|$.

Remark 2.21. Let $k \in \mathbb{N}_0$.

(1) If $w : C_0^k(\Omega) \to \mathbb{C}$ is a linear map, then for every $\alpha \in \mathbb{N}_0^n$ we may define the mapping $\partial^\alpha w : C_0^{k+|\alpha|}(\Omega) \to \mathbb{C}$ by $\partial^\alpha w(\varphi) := (-1)^{|\alpha|}w(\partial^\alpha\varphi)$ for every $\varphi \in C_0^{k+|\alpha|}(\Omega)$.

(2) If u is a distribution in Ω of order k, then for each $\alpha \in \mathbb{N}_0^n$ the distribution $\partial^\alpha u$ has finite order $k + |\alpha|$ and the extension $\Lambda_{\partial^\alpha u}$ to $C_0^{k+|\alpha|}(\Omega)$ given by Proposition 2.16 is equal to $\partial^\alpha \Lambda_u$ on $C_0^{k+|\alpha|}(\Omega)$, the latter being the mapping defined in part *(1)* in relation to Λ_u in place of w. Indeed, since both $\Lambda_{\partial^\alpha u}$ and $\partial^\alpha\Lambda_u$ are linear functionals on $C_0^{k+|\alpha|}(\Omega)$ extending the distribution $\partial^\alpha u$ and satisfying (2.1.45), the uniqueness portion of Proposition 2.16 guarantees that they must be equal.

2.2 The Topological Vector Space $\mathcal{D}'(\Omega)$

The space of distributions in Ω endowed with the natural addition and scalar multiplication of linear mappings becomes a vector space over $\mathbb{C}$. Indeed, if u_1, u_2, u are distributions in Ω and $\lambda \in \mathbb{C}$, we define $u_1 + u_2 : \mathcal{D}(\Omega) \to \mathbb{C}$ and $\lambda u : \mathcal{D}(\Omega) \to \mathbb{C}$ by setting

$$(u_1 + u_2)(\varphi) := \langle u_1, \varphi \rangle + \langle u_2, \varphi \rangle \ \text{ and } \ (\lambda u)(\varphi) := \lambda\langle u, \varphi \rangle,$$

$$\text{for each test function } \ \varphi \in C_0^\infty(\Omega).$$

(2.2.1)

It is not difficult to check that $u_1 + u_2$ and λu are distributions in Ω.

The topology we consider on the vector space of distributions in Ω is the weak$*$-topology induced by $\mathcal{D}(\Omega)$ (for details see Section 14.1.0.5), which makes it a topological vector space over $\mathbb{C}$ and we denote this topological vector space by $\mathcal{D}'(\Omega)$. As a byproduct of this definition we have the following important properties of $\mathcal{D}'(\Omega)$.

Fact 2.22 $\mathcal{D}'(\Omega)$ *is a locally convex topological vector space over* $\mathbb{C}$. *In addition, a sequence* $\{u_j\}_{j\in\mathbb{N}}$ *in* $\mathcal{D}'(\Omega)$ *converges to some* $u \in \mathcal{D}'(\Omega)$ *as* $j \to \infty$ *in* $\mathcal{D}'(\Omega)$ *if and only if* $\langle u_j, \varphi \rangle \xrightarrow[j\to\infty]{} \langle u, \varphi \rangle$ *for every* $\varphi \in C_0^\infty(\Omega)$, *in which case we use the notation*

$$u_j \xrightarrow[j\to\infty]{\mathcal{D}'(\Omega)} u.$$

Moreover, the topological space $\mathcal{D}'(\Omega)$ *is complete, in the following sense. If the sequence* $\{u_j\}_{j\in\mathbb{N}} \subset \mathcal{D}'(\Omega)$ *is such that* $\lim_{j\to\infty}\langle u_j, \varphi \rangle$ *exists (in* $\mathbb{C}$*) for every* $\varphi \in C_0^\infty(\Omega)$ *then the functional* $u : \mathcal{D}(\Omega) \to \mathbb{C}$ *defined by* $u(\varphi) := \lim_{j\to\infty}\langle u_j, \varphi \rangle$ *for every* $\varphi \in C_0^\infty(\Omega)$ *is a distribution in* Ω.

Note that from Fact 2.22 it is easy to see that if a sequence $\{u_j\}_{j\in\mathbb{N}}$ in $\mathcal{D}'(\Omega)$ is convergent then its limit is unique. Indeed, if such a sequence would have two limits, say $u, v \in \mathcal{D}'(\mathbb{R}^n)$, then it would follow that for each $\varphi \in C_0^\infty(\mathbb{R}^n)$, the sequence of numbers $\{\langle u_j, \varphi \rangle\}_{j\in\mathbb{N}}$ would converge to both $\langle u, \varphi \rangle$ and $\langle v, \varphi \rangle$, thus $\langle u, \varphi \rangle = \langle v, \varphi \rangle$. Hence, $u = v$.

Remark 2.23. Assume that we are given $u \in \mathcal{D}'(\Omega)$ and a sequence $u_\varepsilon \in \mathcal{D}'(\Omega)$, $\varepsilon \in (0, \infty)$. We make the convention that $u_\varepsilon \xrightarrow[\varepsilon\to 0^+]{\mathcal{D}'(\Omega)} u$ is understood in the sense that for every sequence of positive numbers $\{\varepsilon_j\}_{j\in\mathbb{N}}$ satisfying $\lim_{j\to\infty} \varepsilon_j = 0$ we have

$$u_{\varepsilon_j} \xrightarrow[j\to\infty]{\mathcal{D}'(\Omega)} u.$$

Example 2.24. Let ϕ be as in (1.2.3) and recall the sequence of functions $\{\phi_j\}_{j\in\mathbb{N}}$ from (1.3.7). Interpreting each $\phi_j \in L^1_{loc}(\mathbb{R}^n)$ as distribution in $\mathbb{R}^n$, for each function $\varphi \in C_0^\infty(\mathbb{R}^n)$ we have

$$\langle \phi_j, \varphi \rangle = \int_{\mathbb{R}^n} \phi_j(x)\varphi(x)\,dx = \int_{\mathbb{R}^n} \phi(y)\varphi(y/j)\,dy, \qquad \forall\, j \in \mathbb{N}. \tag{2.2.2}$$

Thus, by Lebesgue's Dominated Convergence Theorem (cf. Theorem 14.15),

$$\langle \phi_j, \varphi \rangle = \int_{\mathbb{R}^n} \phi(y)\varphi(y/j)\,dy \xrightarrow[j\to\infty]{} \varphi(0)\int_{\mathbb{R}^n} \phi(y)\,dy = \varphi(0) = \langle \delta, \varphi \rangle. \tag{2.2.3}$$

This proves that $\phi_j \xrightarrow[j\to\infty]{\mathcal{D}'(\mathbb{R}^n)} \delta$.

Exercise 2.25. Prove that if $p \in [1, \infty]$ and if $\{f_j\}_{j\in\mathbb{N}}$ is a sequence of functions in $L^p(\Omega)$ which converges in $L^p(\Omega)$ to some $f \in L^p(\Omega)$ then $f_j \xrightarrow[j\to\infty]{\mathcal{D}'(\Omega)} f$.

Hint: Use Hölder's inequality.

Exercise 2.26. (a) Assume that $f \in L^1(\mathbb{R}^n)$ and for each $\varepsilon > 0$ define the function $f_\varepsilon(x) := \varepsilon^{-n} f(x/\varepsilon)$ for each $x \in \mathbb{R}^n$. Prove that for each $g \in L^\infty(\mathbb{R}^n)$ that is continuous at $0 \in \mathbb{R}^n$ we have

$$\int_{\mathbb{R}^n} f_\varepsilon(x) g(x)\, dx \xrightarrow[\varepsilon \to 0^+]{} \Big(\int_{\mathbb{R}^n} f(x)\, dx \Big) g(0). \qquad (2.2.4)$$

(b) Use part *(a)* to prove that if $f \in L^1(\mathbb{R}^n)$ is given and for every $j \in \mathbb{N}$ we define $f_j(x) := j^n f(jx)$ for each $x \in \mathbb{R}^n$, then each f_j belongs to $L^1(\mathbb{R}^n)$ (hence $f_j \in \mathcal{D}'(\mathbb{R}^n)$) and

$$f_j \xrightarrow[j \to \infty]{\mathcal{D}'(\mathbb{R}^n)} c\,\delta \quad \text{where} \quad c := \int_{\mathbb{R}^n} f(x)\, dx. \qquad (2.2.5)$$

Hint: To justify the claim in part *(a)*, make a change of variables to write the integral $\int_{\mathbb{R}^n} f_\varepsilon(x) g(x)\, dx$ as $\int_{\mathbb{R}^n} f(y) g(\varepsilon y)\, dy$, then use Lebesgue's Dominated Convergence Theorem.

In the last part of this section we discuss the composition with invertible linear maps of distributions in $\mathbb{R}^n$. Specifically, let $A \in \mathcal{M}_{n \times n}(\mathbb{R})$ be such that $\det A \neq 0$. Then for every $f \in L^1_{loc}(\mathbb{R}^n)$ one has $f \circ A \in L^1_{loc}(\mathbb{R}^n)$. By Example 2.7 we have $f, f \circ A \in \mathcal{D}'(\mathbb{R}^n)$. In addition,

$$\langle f \circ A, \varphi \rangle = \int_{\mathbb{R}^n} f(Ax) \varphi(x)\, dx = |\det A|^{-1} \int_{\mathbb{R}^n} f(y) \varphi(A^{-1}y)\, dy$$

$$= |\det A|^{-1} \langle f, \varphi \circ A^{-1} \rangle, \qquad \forall \varphi \in C_0^\infty(\mathbb{R}^n). \qquad (2.2.6)$$

This and Exercise 1.24 justify extending the operator of composition with linear maps to $\mathcal{D}'(\mathbb{R}^n)$ as follows.

Proposition 2.27. *Let $A \in \mathcal{M}_{n \times n}(\mathbb{R})$ be such that $\det A \neq 0$. For each distribution $u \in \mathcal{D}'(\mathbb{R}^n)$, define the mapping $u \circ A : \mathcal{D}(\mathbb{R}^n) \to \mathbb{C}$ by setting*

$$(u \circ A)(\varphi) := |\det A|^{-1} \langle u, \varphi \circ A^{-1} \rangle, \qquad \forall \varphi \in \mathcal{D}(\mathbb{R}^n). \qquad (2.2.7)$$

Then $u \circ A \in \mathcal{D}'(\mathbb{R}^n)$.

Proof. This is an immediate consequence of (1.3.18). $\qquad \square$

Exercise 2.28. Let $A, B \in \mathcal{M}_{n \times n}(\mathbb{R})$ be such that $\det A \neq 0$ and $\det B \neq 0$. Then the following identities hold in $\mathcal{D}'(\mathbb{R}^n)$:

(1) $(u \circ A) \circ B = u \circ (AB)$ for every $u \in \mathcal{D}'(\mathbb{R}^n)$;
(2) $u \circ (\lambda A) = \lambda u \circ A$ for every $u \in \mathcal{D}'(\mathbb{R}^n)$ and every $\lambda \in \mathbb{R}$;
(3) $(u + v) \circ A = u \circ A + v \circ A$ for every $u, v \in \mathcal{D}'(\mathbb{R}^n)$.

2.3 Multiplication of a Distribution with a C^{∞} Function

The issue we discuss in this section is the definition of the multiplication of a distribution $u \in \mathcal{D}'(\Omega)$ with a smooth function $a \in C^{\infty}(\Omega)$. First we consider the case when u is of function type, i.e., $u = u_f$ for some $f \in L^1_{loc}(\Omega)$. In this particular case, $af \in L^1_{loc}(\Omega)$ thus it defines a distribution u_{af} on Ω and

$$\langle u_{af}, \varphi \rangle = \int_{\Omega} (af)\varphi \, dx = \int_{\Omega} f(a\varphi) \, dx = \langle f, a\varphi \rangle, \qquad \forall \varphi \in C_0^{\infty}(\Omega). \qquad (2.3.1)$$

In the general case, this suggests defining au as in the following proposition.

Proposition 2.29. *Let $u \in \mathcal{D}'(\Omega)$ and $a \in C^{\infty}(\Omega)$. Then the mapping*

$$au : \mathcal{D}(\Omega) \to \mathbb{C} \quad \text{defined by} \quad (au)(\varphi) := \langle u, a\varphi \rangle, \quad \forall \varphi \in C_0^{\infty}(\Omega), \qquad (2.3.2)$$

is linear and continuous, hence a distribution on Ω.

Proof. The fact that au is linear is immediate. To show that au is also continuous we make use of Remark 2.3. To this end, consider a sequence $\varphi_j \xrightarrow[j\to\infty]{\mathcal{D}(\Omega)} 0$. By *(2)* in Exercise 1.20 we have $a\varphi_j \xrightarrow[j\to\infty]{\mathcal{D}(\Omega)} 0$. Since u is a distribution on Ω, the latter convergence implies $\lim_{j\to\infty}\langle u, a\varphi_j \rangle = 0$. Moreover, from (2.3.2) we have $(au)(\varphi_j) = \langle u, a\varphi_j \rangle$ for each $j \in \mathbb{N}$. Hence, $\lim_{j\to\infty}(au)(\varphi_j) = 0$ proving that au is continuous. $\qquad \square$

Exercise 2.30. Let $f \in L^1_{loc}(\Omega)$ and $a \in C^{\infty}(\Omega)$. With the notation from (2.1.6), prove that $au_f = u_{af}$ in $\mathcal{D}'(\Omega)$.

Remark 2.31.

(1) When more information about $u \in \mathcal{D}'(\Omega)$ is available, (2.3.2) may continue to yield a distribution under weaker regularity demands on the function a than the current assumption that $a \in C^{\infty}(\Omega)$. In general, however, the condition $a \in C^{\infty}(\Omega)$ may not be weakened if (2.3.2) is to yield a distribution for arbitrary $u \in \mathcal{D}'(\Omega)$.

(2) As observed later (see Remark 2.36), one may not define the product of two arbitrary distributions in a way that ensures associativity.

(3) Based on (2.3.2) and (2.2.1), it follows that if u, u_1, $u_2 \in \mathcal{D}'(\Omega)$, and if a, a_1, $a_2 \in C^{\infty}(\Omega)$, then $a(u_1 + u_2) = au_1 + au_2$, $(a_1 + a_2)u = a_1u + a_2u$, and $a_1(a_2u) = (a_1a_2)u$, where the equalities are considered in $\mathcal{D}'(\Omega)$.

Exercise 2.32. The following properties hold.

(1) If $u \in \mathcal{D}'(\Omega)$ and $a_j \xrightarrow[j\to\infty]{\mathcal{E}(\Omega)} a$, then $a_j u \xrightarrow[j\to\infty]{\mathcal{D}'(\Omega)} au$.

(2) If $a \in C^{\infty}(\Omega)$ and $u_j \xrightarrow[j\to\infty]{\mathcal{D}'(\Omega)} u$, then $au_j \xrightarrow[j\to\infty]{\mathcal{D}'(\Omega)} au$.

(3) For each $a \in C^\infty(\Omega)$ the mapping $\mathcal{D}'(\Omega) \ni u \mapsto au \in \mathcal{D}'(\Omega)$ is linear and continuous.

Hint: Observe that the map in *(3)* is the transpose (recall the definition from (14.1.10)) of the linear and continuous map in Exercise 1.21, hence Proposition 14.2 applies.

Example 2.33. Recall the Dirac distribution defined in (2.1.19) and assume that some function $a \in C^\infty(\Omega)$ has been given. Then, for every $\varphi \in C_0^\infty(\mathbb{R}^n)$ we may write

$$\langle a\delta, \varphi \rangle = \langle \delta, a\varphi \rangle = (a\varphi)(0) = a(0)\varphi(0) = \langle a(0)\delta, \varphi \rangle.$$

This shows that

$$a\delta = a(0)\delta \quad \text{in} \quad \mathcal{D}'(\mathbb{R}^n) \quad \text{for every } a \in C^\infty(\Omega). \tag{2.3.3}$$

As a consequence,

$$x^m \delta = 0 \quad \text{in } \mathcal{D}'(\mathbb{R}), \qquad \text{for every } m \in \mathbb{N}. \tag{2.3.4}$$

Example 2.34. The goal is to solve the equation

$$xu = 1 \quad \text{in } \mathcal{D}'(\mathbb{R}). \tag{2.3.5}$$

Clearly, this equation does not have a solution u of function type. Recall the distribution defined in Example 2.11. Then for every $\varphi \in C_0^\infty(\mathbb{R})$ we may write

$$\left\langle x\left(\text{P.V.} \frac{1}{x}\right), \varphi \right\rangle = \left\langle \text{P.V.} \frac{1}{x}, x\varphi(x) \right\rangle = \lim_{\varepsilon \to 0^+} \int_{|x| \geq \varepsilon} \frac{x\varphi(x)}{x} \, \mathrm{d}x$$

$$= \int_{\mathbb{R}} \varphi(x) \, \mathrm{d}x = \langle 1, \varphi \rangle. \tag{2.3.6}$$

Thus,

$$x\left(\text{P.V.} \frac{1}{x}\right) = 1 \quad \text{in} \quad \mathcal{D}'(\mathbb{R}). \tag{2.3.7}$$

Given (2.3.4), it follows that $u := \text{P.V.} \frac{1}{x} + c\,\delta$ will also be a solution of (2.3.5) for any $c \in \mathbb{C}$. We will see later (c.f. Remark 2.78) that in fact any solution of (2.3.5) is of the form $\text{P.V.} \frac{1}{x} + c\,\delta$, where $c \in \mathbb{C}$.

Exercise 2.35. Let $\psi \in C^\infty(\mathbb{R})$. Determine a solution $u \in \mathcal{D}'(\mathbb{R})$ of the equation $xu = \psi$ in $\mathcal{D}'(\mathbb{R})$.

Remark 2.36. Suppose one could define the product of distributions as an associative operation, in a manner compatible with the multiplication by a smooth function. Considering then δ, x, and $\text{P.V.} \frac{1}{x} \in \mathcal{D}'(\mathbb{R})$, one would then necessarily have

$$0 = (\delta \cdot x)\left(\text{P.V.} \frac{1}{x}\right) = \delta\left[x \cdot \left(\text{P.V.} \frac{1}{x}\right)\right] = \delta \cdot 1 = \delta \quad \text{in } \mathcal{D}'(\mathbb{R}), \tag{2.3.8}$$

thanks to (2.3.4) and (2.3.7), leading to the false conclusion that $\delta = 0$ in $\mathcal{D}'(\mathbb{R})$.

2.4 Distributional Derivatives

We are now ready to define derivatives of distributions. One of the most basic attributes of the class of distributions, compared with other classes of locally integrable functions, is that distributions may be differentiated unrestrictedly within this environment (with the resulting objects being still distributions), and that the operation of distributional differentiation retains some of the most basic properties as in the case of ordinary differentiable functions (such as a suitable product formula, symmetry of mixed derivatives, etc.). In addition, the differentiation of distributions turns out to be compatible with the pointwise differentiation in the case when the distribution in question is of function type, given by a sufficiently regular function.

To develop some sort of intuition, we shall start our investigation by looking first at a distribution of function type, and try to generalize the notion of weak (Sobolev) derivative from Definition 1.2. As noted earlier, if $f \in L^1_{loc}(\Omega)$, the mapping defined in (1.3.1) is a distribution on Ω. This suggests making the following definition.

Definition 2.37. If $\alpha \in \mathbb{N}_0^n$ and $u \in \mathcal{D}'(\Omega)$, the distributional derivative (or the derivative in the sense of distributions) of order α of the distribution u is the mapping $\partial^\alpha u : \mathcal{D}(\Omega) \to \mathbb{C}$ defined by

$$\partial^\alpha u(\varphi) := (-1)^{|\alpha|}\langle u, \partial^\alpha \varphi \rangle, \qquad \forall\, \varphi \in C_0^\infty(\Omega). \tag{2.4.1}$$

Remark 2.38. Note that if $u \in \mathcal{D}'(\Omega)$ is of function type, say $u = u_f$ for some $f \in L^1_{loc}(\Omega)$, and if the weak derivative $\partial^\alpha f$ exists, i.e., one can find $g \in L^1_{loc}(\Omega)$ such that (1.2.2) holds, then according to Definition 2.37 we have that the distributional derivative $\partial^\alpha u$ is equal to the distribution u_g in $\mathcal{D}'(\Omega)$. In short, $\partial^\alpha u_f = u_{\partial^\alpha f}$ in this case. Thus, Definition 2.37 generalizes Definition 1.2.

That the class of distributions in Ω is stable under taking distributional derivatives is proved in the next proposition.

Proposition 2.39. *For each $\alpha \in \mathbb{N}_0^n$ and each $u \in \mathcal{D}'(\Omega)$ we have $\partial^\alpha u \in \mathcal{D}'(\Omega)$.*

Proof. Fix $\alpha \in \mathbb{N}_0^n$ and $u \in \mathcal{D}'(\Omega)$. That $\partial^\alpha u : \mathcal{D}(\Omega) \to \mathbb{C}$ is a linear map is easy to see. To prove that it is also continuous, let $\varphi_j \xrightarrow[j\to\infty]{\mathcal{D}(\Omega)} 0$. Since by item (1) in Exercise 1.20 we have $\partial^\alpha \varphi_j \xrightarrow[j\to\infty]{\mathcal{D}(\Omega)} 0$ and $u \in \mathcal{D}'(\Omega)$, we may write

$$\partial^\alpha u(\varphi_j) = (-1)^{|\alpha|}\langle u, \partial^\alpha \varphi_j \rangle \xrightarrow[j\to\infty]{} 0. \tag{2.4.2}$$

Remark 2.3 then shows that $\partial^\alpha u \in \mathcal{D}'(\Omega)$. $\square$

Exercise 2.40. Suppose that $m \in \mathbb{N}$ and $f \in C^m(\Omega)$. Prove that for any $\alpha \in \mathbb{N}_0^n$ satisfying $|\alpha| \leq m$, the distributional derivative of order α of u_f is the distribution of function type given by the derivative, in the classical sense, of order α of f, that is, $\partial^\alpha(u_f) = u_{\partial^\alpha f}$ in $\mathcal{D}'(\Omega)$.

Proposition 2.41. *Let $m \in \mathbb{N}_0$ and assume that*

$$P(x,\partial) := \sum_{|\alpha| \leq m} a_\alpha(x)\partial^\alpha, \quad a_\alpha \in C^\infty(\Omega), \quad \alpha \in \mathbb{N}_0^n, \quad |\alpha| \leq m. \tag{2.4.3}$$

Also, suppose that $u \in C^m(\Omega)$. Then $P(x,\partial)u$, computed in $\mathcal{D}'(\Omega)$, coincides as a distribution with the distribution induced by $P(x,\partial)u$, computed pointwise in Ω.

Proof. This follows from (2.4.3), Exercise 2.40, and Exercise 2.30. □

Remark 2.42. Recall that, given a set $E \subseteq \mathbb{R}^n$, a function $f : E \to \mathbb{C}$ is called Lipschitz provided there exists $M \in [0, \infty)$ such that

$$|f(x) - f(y)| \leq M|x - y| \quad \forall\, x, y \in E. \tag{2.4.4}$$

The number $C := \inf \{M$ for which (2.4.4) holds$\}$ is referred to as the Lipschitz constant of f. We agree to denote by $Lip(E)$ the collection of all Lipschitz functions on E. A classical useful result (due to E. McShane) concerning this space is that for any set $E \subseteq \mathbb{R}^n$ we have

$$Lip(E) = \{F|_E : F \in Lip(\mathbb{R}^n)\}. \tag{2.4.5}$$

Indeed given any $f \in Lip(E)$ with Lipschitz constant M one may check that the function $F(x) := \inf\{f(y) + M|x - y| : y \in E\}$ for all $x \in \mathbb{R}^n$ belongs to $Lip(\mathbb{R}^n)$, has Lipschitz constant M, and satisfies $f = F|_E$.

We will also prove (see Theorem 2.114) that if $\Omega \subseteq \mathbb{R}^n$ is an arbitrary open set and $f : \Omega \to \mathbb{C}$ is Lipschitz then the distributional derivatives $\partial_k f$, $k = 1, \ldots, n$, belong to $L^\infty(\Omega)$. Consequently,

$$f : \Omega \to \mathbb{R} \text{ Lipschitz} \implies \partial_k(u_f) = u_{\partial_k f} \text{ in } \mathcal{D}'(\Omega) \text{ for } k = 1, \ldots, n. \tag{2.4.6}$$

Some basic properties of differentiation in the distributional sense are summarized below.

Proposition 2.43. *The following properties of distributional differentiation hold.*

(1) Any distribution is infinitely differentiable (i.e., $\mathcal{D}'(\Omega)$ is stable under the action of ∂^α for any $\alpha \in \mathbb{N}_0$).

(2) If $u \in \mathcal{D}'(\Omega)$ and $k, \ell \in \{1, ..., n\}$ then $\partial_k \partial_\ell u = \partial_\ell \partial_k u$ in $\mathcal{D}'(\Omega)$.

(3) If $u_j \xrightarrow[j\to\infty]{\mathcal{D}'(\Omega)} u$ and $\alpha \in \mathbb{N}_0^n$, then $\partial^\alpha u_j \xrightarrow[j\to\infty]{\mathcal{D}'(\Omega)} \partial^\alpha u$.

(4) For any $u \in \mathcal{D}'(\Omega)$ and any $a \in C^\infty(\Omega)$ we have $\partial_j(au) = (\partial_j a)u + a(\partial_j u)$ in $\mathcal{D}'(\Omega)$.

Proof. The first property follows immediately from the definition of distributional derivatives. To prove the remaining properties, fix an arbitrary $\varphi \in C_0^\infty(\Omega)$. Then, using (2.4.1) repeatedly and the symmetry of mixed partial derivatives for smooth functions (Schwarz's theorem), we have

$$\langle \partial_k \partial_\ell u, \varphi \rangle = -\langle \partial_\ell u, \partial_k \varphi \rangle = \langle u, \partial_\ell \partial_k \varphi \rangle = \langle u, \partial_k \partial_\ell \varphi \rangle$$
$$= -\langle \partial_k u, \partial_\ell \varphi \rangle = \langle \partial_\ell \partial_k u, \varphi \rangle, \quad \forall k, \ell \in \{1, \dots, n\},$$

which implies *(2)*. Let now $\{u_j\}_{j \in \mathbb{N}}$, u, and α satisfy the hypotheses in *(3)*. Based on (2.4.1) and Fact 2.22 we may write

$$\langle \partial^\alpha u_j, \varphi \rangle = (-1)^{|\alpha|} \langle u_j, \partial^\alpha \varphi \rangle \xrightarrow[j \to \infty]{} (-1)^{|\alpha|} \langle u, \partial^\alpha \varphi \rangle = \langle \partial^\alpha u, \varphi \rangle,$$

hence $\partial^\alpha u_j$ converges to $\partial^\alpha u$ in $\mathcal{D}'(\Omega)$ as $j \to \infty$ and the proof of *(3)* is complete. Finally, for $u \in \mathcal{D}'(\Omega)$ and $a \in C^\infty(\Omega)$, using (2.4.1) and Leibniz's product formula for derivatives of smooth functions we can write

$$\langle \partial_j(au), \varphi \rangle = -\langle au, \partial_j \varphi \rangle = -\langle u, a(\partial_j \varphi) \rangle = -\langle u, \partial_j(a\varphi) \rangle + \langle u, (\partial_j a)\varphi \rangle$$
$$= \langle \partial_j u, a\varphi \rangle + \langle (\partial_j a)u, \varphi \rangle = \langle a(\partial_j u) + (\partial_j a)u, \varphi \rangle, \quad (2.4.7)$$

from which *(4)* follows.   □

Example 2.44. Recall the Heaviside function H from (1.2.9). This is a locally integrable function thus it defines a distribution on $\mathbb{R}$ that we denote also by H. Then, the computation in (1.2.12) implies

$$\langle H', \varphi \rangle = -\langle H, \varphi' \rangle = \langle \delta, \varphi \rangle, \qquad \forall \varphi \in C_0^\infty(\mathbb{R}). \quad (2.4.8)$$

Hence,

$$H' = \delta \quad \text{in } \mathcal{D}'(\mathbb{R}). \quad (2.4.9)$$

Exercise 2.45. Prove that for every function $a \in C^\infty(\Omega)$ and every $\alpha \in \mathbb{N}_0^n$ we have

$$a(\partial^\alpha \delta) = \sum_{\beta \leq \alpha} \frac{\alpha!}{\beta!(\alpha - \beta)!} (-1)^{|\beta|} (\partial^\beta a)(0) \partial^{\alpha-\beta} \delta \quad \text{in } \mathcal{D}'(\Omega). \quad (2.4.10)$$

Hint: Use formula (14.2.6) when computing $\partial^\alpha(a\varphi)$ for $\varphi \in C_0^\infty(\Omega)$.

Exercise 2.46. Prove that for every $c \in \mathbb{R}$ one has

$$(e^{-c|x|})' = -c\, e^{-cx} H(x) + c\, e^{cx} H(-x) \quad \text{in } \mathcal{D}'(\mathbb{R}). \quad (2.4.11)$$

Hint: Show $e^{-c|x|} = e^{-cx} H(x) + e^{cx} H(-x)$ in $\mathcal{D}'(\mathbb{R})$ and then use part *(4)* in Proposition 2.43 and (2.4.9).

Next, we look at the issue of existence of antiderivatives for distributions on open intervals.

Proposition 2.47. *Let I be an open interval in $\mathbb{R}$ and suppose $u_0 \in \mathcal{D}'(I)$.*

(1) The equation $u' = u_0$ in $\mathcal{D}'(I)$ admits at least one solution.

(2) If $u_1, u_2 \in \mathcal{D}'(I)$ are such that $(u_1)' = u_0$ in $\mathcal{D}'(I)$ and $(u_2)' = u_0$ in $\mathcal{D}'(I)$, then there exists $c \in \mathbb{C}$ such that $u_1 - u_2 = c$ in $\mathcal{D}'(I)$.

Proof. Suppose $I = (a, b)$, where $a \in \mathbb{R} \cup \{-\infty\}$ and $b \in \mathbb{R} \cup \{+\infty\}$, and define the set $\mathcal{A}(I) := \{\varphi' : \varphi \in C_0^\infty(I)\}$. We claim that

$$\text{if } \varphi \in C_0^\infty(I), \text{ then } \varphi \in \mathcal{A}(I) \iff \int_I \varphi(x)\, dx = 0. \tag{2.4.12}$$

The left-to-right implication in (2.4.12) is clear by the fundamental theorem of calculus. To prove the converse implication, suppose $\varphi \in C_0^\infty(I)$ is such that $\int_I \varphi(x)\, dx = 0$. Then the function $\psi(x) := \int_a^x \varphi(t)\, dt$, $x \in I$, satisfies $\psi \in C^\infty(I)$, $\operatorname{supp} \psi \subseteq \operatorname{supp} \varphi$, and $\psi' = \varphi$ on I, thus $\varphi \in \mathcal{A}(I)$. This finishes the justification of (2.4.12).

Next, fix $\varphi_0 \in C_0^\infty(I)$ with the property that $\int_I \varphi_0(x)\, dx = 1$ and consider the map $\Theta : \mathcal{D}(I) \to \mathcal{D}(I)$ defined by

$$\Theta(\varphi) := \theta_\varphi \text{ for each } \varphi \in C_0^\infty(I), \text{ where}$$

$$\theta_\varphi(x) := \int_a^x \left[\varphi(t) - \left(\int_I \varphi(y)\, dy \right) \varphi_0(t) \right] dt, \quad \forall\, x \in I. \tag{2.4.13}$$

Since the integral of $\varphi - \left(\int_I \varphi(x)\, dx \right) \varphi_0$ over I is zero, our earlier discussion shows that Θ is well defined. Since Θ is linear, Theorem 14.6 and Fact 1.15 imply that Θ is continuous if and only if it is sequentially continuous. The latter property may be verified from definitions. Finally, from (2.4.13) and the fundamental theorem of calculus we have

$$\Theta(\varphi') = \varphi \quad \text{for every} \quad \varphi \in C_0^\infty(I). \tag{2.4.14}$$

Fix an arbitrary distribution u_0 on I and define $u := -u_0 \circ \Theta$. Thanks to the properties of Θ, we have that u is a distribution on I. In concert with (2.4.14), the definition of u implies that $\langle u, \varphi' \rangle = -\langle u_0, \varphi \rangle$ for every $\varphi \in C_0^\infty(I)$, proving that $u' = u_0$ in $\mathcal{D}'(I)$. This finishes the proof of the statement in *(1)*.

Moving on, suppose $u \in \mathcal{D}'(I)$ is such that $u' = 0$ in $\mathcal{D}'(I)$. Then if φ_0 is as earlier in the proof, for any $\varphi \in C_0^\infty(I)$ we may write

$$\langle u, \varphi \rangle = \left\langle u, \varphi - \left(\int_I \varphi(x)\, dx \right) \varphi_0 \right\rangle + \left(\int_I \varphi(x)\, dx \right) \langle u, \varphi_0 \rangle$$

$$= \langle u, (\theta_\varphi)' \rangle + \langle \langle u, \varphi_0 \rangle, \varphi \rangle = \langle c, \varphi \rangle, \tag{2.4.15}$$

where $c := \langle u, \varphi_0 \rangle \in \mathbb{C}$. Hence, $u = c$ in $\mathcal{D}(I)$. By linearity, this readily implies the statement in *(2)*. The proof of the proposition is now complete. $\qquad\square$

Proposition 2.48. *Let I be an open interval in $\mathbb{R}$, $g \in C^\infty(I)$, and $f \in C^k(I)$ for some $k \in \mathbb{N}_0$. If $u \in \mathcal{D}'(I)$ satisfies $u' + gu = f$ in $\mathcal{D}'(I)$, then $u \in C^{k+1}(I)$.*

Proof. Fix $a \in I$ and define $F(x) := e^{\int_a^x g_0(t)\,dt}$ for $x \in I$. Then $F \in C^\infty(I)$ and we may use *(4)* in Proposition 2.43 and the equation satisfied by u to write (keeping in mind that f is continuous)

$$\left(Fu - \int_a^x F(t)f(t)\,dt \right)' = F'u + Fu' - Ff = 0 \quad \text{in} \quad \mathcal{D}'(I). \tag{2.4.16}$$

By Proposition 2.47, $Fu = \int_a^x F(t)f(t)\,dt + c$ in $\mathcal{D}'(I)$ for some constant $c \in \mathbb{C}$. Since $\int_a^x F(t)f(t)\,dt \in C^{k+1}(I)$ and $\frac{1}{F} \in C^\infty(I)$, we conclude that

$$u = \frac{1}{F(x)} \int_a^x F(t)f(t)\,dt + \frac{c}{F(x)} \in C^{k+1}(I), \tag{2.4.17}$$

as desired. $\square$

We close this section by presenting a higher degree version of the product formula for differentiation from part *(4)* in Proposition 2.43.

Proposition 2.49 (Generalized Leibniz Formula). *Suppose* $f \in C^\infty(\Omega)$ *and let* $u \in \mathcal{D}'(\Omega)$. *Then for every* $\alpha \in \mathbb{N}_0^n$ *one has*

$$\partial^\alpha(fu) = \sum_{\beta \leq \alpha} \frac{\alpha!}{\beta!(\alpha - \beta)!} (\partial^\beta f)(\partial^{\alpha-\beta} u) \quad \text{in} \quad \mathcal{D}'(\Omega). \tag{2.4.18}$$

Proof. The first step is to observe that for each $j \in \{1,\dots,n\}$ and each $k \in \mathbb{N}_0$ we have

$$\partial_j^k(fu) = \sum_{0 \leq \ell \leq k} \frac{k!}{\ell!(k-\ell)!} (\partial_j^\ell f)(\partial_j^{k-\ell} u) \quad \text{in} \quad \mathcal{D}'(\Omega), \tag{2.4.19}$$

which is proved by induction on k making use of part *(4)* in Proposition 2.43. Hence, given any $\alpha = (\alpha_1, \alpha_2, \dots, \alpha_n) \in \mathbb{N}_0^n$, via repeated applications of (2.4.19) we obtain

$$\partial_1^{\alpha_1}(fu) = \sum_{0 \leq \beta_1 \leq \alpha_1} \frac{\alpha_1!}{\beta_1!(\alpha_1 - \beta_1)!} (\partial_1^{\beta_1} f)(\partial_1^{\alpha_1-\beta_1} u) \tag{2.4.20}$$

and

$$\partial_1^{\alpha_1} \partial_2^{\alpha_2}(fu) \tag{2.4.21}$$

$$= \sum_{0 \leq \beta_1 \leq \alpha_1} \sum_{0 \leq \beta_2 \leq \alpha_2} \frac{\alpha_1!}{\beta_1!(\alpha_1 - \beta_1)!} \frac{\alpha_2!}{\beta_2!(\alpha_2 - \beta_2)!} (\partial_1^{\beta_1} \partial_2^{\beta_2} f)(\partial_1^{\alpha_1-\beta_1} \partial_2^{\alpha_2-\beta_2} u).$$

By induction, we may then infer

$$\partial^\alpha(fu) = \sum_{0 \leq \beta_1 \leq \alpha_1} \cdots \sum_{0 \leq \beta_n \leq \alpha_n} \frac{\alpha_1!}{\beta_1!(\alpha_1 - \beta_1)!} \cdots \frac{\alpha_n!}{\beta_n!(\alpha_n - \beta_n)!} \times$$

$$\times (\partial_1^{\beta_1} \cdots \partial_n^{\beta_n} f)(\partial_1^{\alpha_1 - \beta_1} \cdots \partial_n^{\alpha_n - \beta_n} u)$$

$$= \sum_{0 \leq \beta \leq \alpha} \frac{\alpha!}{\beta!(\alpha - \beta)!} (\partial^\beta f)(\partial^{\alpha - \beta} u), \tag{2.4.22}$$

as claimed. $\qquad\qquad\qquad\qquad\qquad\qquad\qquad\qquad\qquad\qquad\qquad\qquad\qquad\qquad\qquad\square$

2.5 The Support of a Distribution

In preparation to discussing the notion of support of a distribution, we first define the restriction of a distribution to an open subset of the Euclidean domain on which the distribution is considered. Necessarily, such a definition should generalize restrictions at the level of locally integrable functions. We start from the observation that if $f \in L^1_{loc}(\Omega)$ and ω is a non-empty open subset of Ω, then $f|_\omega \in L^1_{loc}(\omega)$. Thus,

$$\langle f|_\omega, \varphi \rangle = \int_\omega f\varphi \, dx = \int_\Omega f \iota(\varphi) \, dx \quad \text{for each} \ \varphi \in C_0^\infty(\omega), \tag{2.5.1}$$

where ι is the map from (1.3.16).

Proposition 2.50. *Let Ω be a non-empty open subset of $\mathbb{R}^n$ and suppose ω is a non-empty open subset of Ω. Also, recall the map ι from (1.3.16). Then for every $u \in \mathcal{D}'(\Omega)$, the mapping arising as the* restriction *of the distribution u to ω, i.e.,*

$$u|_\omega : \mathcal{D}(\omega) \to \mathbb{C} \quad \text{defined by} \quad (u|_\omega)(\varphi) := \langle u, \iota(\varphi) \rangle, \quad \forall \varphi \in C_0^\infty(\omega), \tag{2.5.2}$$

is linear and continuous. Hence, $u|_\omega \in \mathcal{D}'(\omega)$.

Proof. It is immediate that the map in (2.5.2) is well defined and linear. To see that it is also continuous we use Proposition 2.4. Let K be a compact set contained in ω. Then $K \subset \Omega$ and, since $u \in \mathcal{D}'(\Omega)$, Proposition 2.4 applies and gives $k \in \mathbb{N}_0$ and $C \in (0, \infty)$ such that (2.1.1) holds. In particular, for each $\varphi \in C_0^\infty(\omega)$ with $\operatorname{supp}\varphi \subseteq K$,

$$\left| (u|_\omega)(\varphi) \right| = \left| \langle u, \iota(\varphi) \rangle \right| \leq C \sup_{\substack{x \in K \\ |\alpha| \leq k}} |\partial^\alpha \varphi(x)|. \tag{2.5.3}$$

The conclusion that $u|_\omega \in \mathcal{D}'(\omega)$ now follows. For an alternative proof of the continuity of $u|_\omega$ one may use Fact 2.2 and Exercise 1.18. $\qquad\qquad\qquad\qquad\qquad\square$

Exercise 2.51.

(1) Prove that the definition of the restriction of a distribution from (2.5.2) generalizes the usual restriction of functions. More specifically, using the notation in-

troduced in (2.1.6), show that if ω is an open subset of Ω and $f \in L^1_{loc}(\Omega)$, then $(u_f)\big|_\omega = u_{f|_\omega}$ in $\mathcal{D}'(\omega)$.

(2) Prove that the operation of differentiation of a distribution commutes with the operation of restriction of a distribution to open sets, that is, if ω is an open subset of Ω, then

$$\partial^\alpha(u\big|_\omega) = (\partial^\alpha u)\big|_\omega, \qquad \forall\, u \in \mathcal{D}'(\Omega), \quad \forall\, \alpha \in \mathbb{N}_0^n. \tag{2.5.4}$$

(3) Prove that the operation of multiplication of a distribution by a smooth function behaves naturally relative to restriction to open subsets. Specifically, show that if ω is an open subset of Ω then

$$(\varphi u)\big|_\omega = (\varphi\big|_\omega)(u\big|_\omega), \qquad \forall\, u \in \mathcal{D}'(\Omega), \quad \forall\, \varphi \in C^\infty(\Omega). \tag{2.5.5}$$

The next proposition shows that a distribution is uniquely determined by its local behavior.

Proposition 2.52. *If $u_1, u_2 \in \mathcal{D}'(\Omega)$ are such that for each $x_0 \in \Omega$ there exists an open subset ω of Ω with $x_0 \in \omega$ and satisfying $u_1\big|_\omega = u_2\big|_\omega$ in $\mathcal{D}'(\omega)$, then $u_1 = u_2$ in $\mathcal{D}'(\Omega)$.*

Proof. Observe that this proposition may be viewed as a reconstruction problem, thus it is meaningful to try to use a partition of unity. Let $\varphi \in C_0^\infty(\Omega)$ be arbitrary, fixed and set $K := \operatorname{supp}\varphi$. The goal is to prove that $\langle u_1, \varphi \rangle = \langle u_2, \varphi \rangle$. From hypotheses it follows that for each $x \in K$ there exists an open neighborhood $\omega_x \subset \Omega$ of x with the property that $u_1\big|_{\omega_x} = u_2\big|_{\omega_x}$. Based on the fact that K is compact, the cover $\{\omega_x\}_{x \in K}$ of K may be refined to a finite one, consisting of, say $\omega_1, \dots, \omega_N$. These are open subsets of Ω and satisfy

$$K \subset \bigcup_{j=1}^N \omega_j \quad\text{and}\quad u_1\big|_{\omega_j} = u_2\big|_{\omega_j} \quad \text{for } j = 1, \dots, N. \tag{2.5.6}$$

Consider a partition of unity $\{\psi_j : j = 1, \dots, N\}$ subordinate to the cover $\{\omega_j\}_{j=1}^N$ of K, as given by Theorem 14.37. In particular, for each $j = 1, \dots, N$, the function $\psi_j \in C_0^\infty(\Omega)$ satisfies $\operatorname{supp}\psi_j \subset \omega_j$. Moreover, $\sum_{j=1}^N \psi_j = 1$ on K. Consequently, using the linearity of distributions and (2.5.6), we obtain

$$\langle u_1, \varphi \rangle = \Big\langle u_1, \varphi \sum_{j=1}^N \psi_j \Big\rangle = \sum_{j=1}^N \langle u_1, \varphi\psi_j \rangle = \sum_{j=1}^N \langle u_2, \varphi\psi_j \rangle$$

$$= \Big\langle u_2, \sum_{j=1}^N \varphi\psi_j \Big\rangle = \langle u_2, \varphi \rangle. \tag{2.5.7}$$

Thus, $\langle u_1, \varphi \rangle = \langle u_2, \varphi \rangle$ and the proof of the proposition is complete. $\qquad\square$

Exercise 2.53. Let $k \in \mathbb{N}_0 \cup \{\infty\}$ and suppose $u \in \mathcal{D}'(\Omega)$ is such that for each $x \in \Omega$ there exists a number $r_x > 0$ and a function $f_x \in C^k(B(x, r_x))$ such that $B(x, r_x) \subset \Omega$ and $u\big|_{B(x, r_x)} = f_x$ in $\mathcal{D}'(B(x, r_x))$. Prove that $u \in C^k(\Omega)$.

Hint: Use Theorem 14.42 to obtain a partition of unity $\{\psi_j\}_{j \in J}$ subordinate to the cover $\{B(x, r_x)\}_{x \in \Omega}$ of Ω, then show that $f := \sum\limits_{j \in J} \psi_j f_j$ is a function in $C^k(\Omega)$ satisfying $u = f$ in $\mathcal{D}'(\Omega)$.

Now we are ready to define the notion of support of a distribution. Recall that if $f \in C^0(\Omega)$ then its support is defined to be the closure relative to Ω of the set $\{x \in \Omega : f(x) \neq 0\}$. However, the value of an arbitrary distribution at a point is not meaningful. The fact that $\Omega \setminus \operatorname{supp} f$ is the largest open set contained in Ω on which $f = 0$ suggests the introduction of the following definition.

Definition 2.54. The support of a distribution $u \in \mathcal{D}'(\Omega)$ is defined as

$$\operatorname{supp} u \qquad\qquad (2.5.8)$$

$$:= \{x \in \Omega : \text{ there is no } \omega \text{ open such that } x \in \omega \subseteq \Omega \text{ and } u\big|_\omega = 0\}.$$

Based on (2.5.8), it follows that

$$\Omega \setminus \operatorname{supp} u = \{x \in \Omega : \exists \omega \text{ open set such that } x \in \omega \subseteq \Omega \text{ and } u\big|_\omega = 0\}, \quad (2.5.9)$$

which is an open set. Hence, $\operatorname{supp} u$ is relatively closed in Ω. Moreover, if we apply Proposition 2.52 to the distributions u and $0 \in \mathcal{D}'(\Omega \setminus \operatorname{supp} u)$ we obtain that

$$u\big|_{\Omega \setminus \operatorname{supp} u} = 0, \qquad \forall u \in \mathcal{D}'(\Omega). \qquad (2.5.10)$$

In other words, $\Omega \setminus \operatorname{supp} u$ is the largest open subset of Ω on which the restriction of u is zero.

Example 2.55. Recall the Dirac distribution δ from (2.1.19). We claim that $\operatorname{supp} \delta = \{0\}$. Indeed, if $\varphi \in C_0^\infty(\mathbb{R}^n \setminus \{0\})$ it follows that $\langle \delta, \varphi \rangle = \varphi(0) = 0$. By Proposition 2.52, $\delta\big|_{\mathbb{R}^n \setminus \{0\}} = 0$, thus $\operatorname{supp} \delta \subseteq \{0\}$. To prove the opposite inclusion, consider an arbitrary open subset ω of Ω such that $0 \in \omega$. Then there exists $\varphi \in C_0^\infty(\omega)$ such that $\varphi(0) = 1$, and hence, $\langle \delta, \varphi \rangle = 1 \neq 0$, which in turn implies that $\delta\big|_\omega \neq 0$. Consequently, $0 \in \operatorname{supp} \delta$ as desired. Similarly, if $x_0 \in \mathbb{R}^n$, then $\operatorname{supp} \delta_{x_0} = \{x_0\}$, where δ_{x_0} is as in Example 2.14.

Example 2.56. If $f \in C^0(\Omega)$ then $\operatorname{supp} u_f = \operatorname{supp} f$, where u_f is the distribution from (2.1.6). Indeed, since $f = 0$ in $\Omega \setminus \operatorname{supp} f$, we have $\int_\Omega f(x) \varphi(x) \, dx = 0$ for every $\varphi \in C_0^\infty(\Omega \setminus \operatorname{supp} f)$, hence $\operatorname{supp} u_f \subseteq \operatorname{supp} f$. Also, if $x \in \Omega \setminus \operatorname{supp} u_f$ then there exists an open neighborhood ω of x with $\omega \subseteq \Omega$ and such that $u_f\big|_\omega = 0$. Thus, for every $\varphi \in C_0^\infty(\omega)$ one has $0 = \langle u_f, \varphi \rangle = \int_\omega f(x) \varphi(x) \, dx$. Invoking Theorem 1.3 we arrive at the conclusion that $f = 0$ almost everywhere in ω hence, ultimately, $f = 0$ in ω (since f is continuous in ω). Consequently, $x \notin \operatorname{supp} f$ and this proves that $\operatorname{supp} f \subseteq \operatorname{supp} u_f$.

Exercise 2.57. Let $u, v \in \mathcal{D}'(\Omega)$ be such that $\operatorname{supp} u \cap \operatorname{supp} v = \varnothing$ and $u + v = 0$ in $\mathcal{D}'(\Omega)$. Prove that $u = 0$ and $v = 0$ in $\mathcal{D}'(\Omega)$.

Hint: Note that $v\big|_{\Omega \setminus \operatorname{supp} u} = (u + v)\big|_{\Omega \setminus \operatorname{supp} u} = 0$ and $v\big|_{\Omega \setminus \operatorname{supp} v} = 0$. Combine these with the fact that $(\Omega \setminus \operatorname{supp} u) \cup (\Omega \setminus \operatorname{supp} v) = \Omega$ and Proposition 2.52 to deduce that $v = 0$.

Exercise 2.58. Prove that

$$\operatorname{supp}(\partial^\alpha u) \subseteq \operatorname{supp} u, \qquad \forall u \in \mathcal{D}'(\Omega), \ \forall \alpha \in \mathbb{N}_0^n. \tag{2.5.11}$$

Hint: Use (2.5.10) and (2.5.4).

We propose to extend the scope of the discussion in Example 2.56 as to make it applicable to functions that are merely locally integrable (instead of continuous). This requires defining a suitable notion of support for functions that lack continuity, and we briefly address this issue first.

Given an arbitrary set $E \subseteq \mathbb{R}^n$ and an arbitrary function $f : E \to \mathbb{C}$, we define the support of f as

$$\operatorname{supp} f := \{x \in E : \not\exists r > 0 \text{ such that } f = 0 \text{ a.e. in } B(x, r) \cap E\}. \tag{2.5.12}$$

From this definition one may check without difficulty that

$$E \setminus \operatorname{supp} f = E \cap \bigcup_{x \in E \setminus \operatorname{supp} f} B(x, r_x) \tag{2.5.13}$$

where for each $x \in E \setminus \operatorname{supp} f$ the number $r_x > 0$ is such that $f = 0$ a.e. in $B(x, r_x) \cap E$. Moreover, since $\mathbb{R}^n$ has the Lindelöf property, the above union can be refined to a countable one. Based on these observations, the following basic properties of the support may be deduced:

$$\operatorname{supp} f \text{ is a relatively closed subset of } E, \tag{2.5.14}$$

$$f = 0 \text{ a.e. in } E \setminus \operatorname{supp} f, \tag{2.5.15}$$

$$\operatorname{supp} f \subseteq F \text{ if } F \text{ relatively closed subset of } E, f = 0 \text{ a.e. on } E \setminus F, \tag{2.5.16}$$

$$\operatorname{supp} f = \operatorname{supp} g \text{ if } g : E \to \mathbb{C} \text{ is such that } f = g \text{ a.e. on } E. \tag{2.5.17}$$

In addition, if the set $E \subseteq \mathbb{R}^n$ is open and the function $f : E \to \mathbb{C}$ is continuous, then $\operatorname{supp} f$ may be described as the closure in E of the set $\{x \in E : f(x) \neq 0\}$, which is precisely our earlier notion of support in this context.

Exercise 2.59. If $f \in L^1_{loc}(\Omega)$ then $\operatorname{supp} u_f = \operatorname{supp} f$, where u_f is the distribution from (2.1.6).

Hint: Use (2.5.12), (2.5.9), part *(1)* in Exercise 2.51, and the fact that the injection in (2.1.8) is one-to-one.

2.6 Compactly Supported Distributions and the Space $\mathcal{E}'(\Omega)$

Next we discuss the issue of extending the action of a distribution $u \in \mathcal{D}'(\Omega)$ to a subclass of $C^\infty(\Omega)$ that is possibly larger than $C_0^\infty(\Omega)$. Observe that if f belongs to $L^1_{loc}(\Omega)$, the expression $\int_\Omega f\varphi\,dx$ is meaningful for functions $\varphi \in C^\infty(\Omega)$ with the property that $\operatorname{supp} f \cap \operatorname{supp}\varphi$ is a compact subset of Ω. A particular case is when $\operatorname{supp}\varphi \cap \operatorname{supp} f = \varnothing$ in which scenario $\int_\Omega f\varphi\,dx = 0$. This observation is the motivation behind the following theorem.

Theorem 2.60. *Let $u \in \mathcal{D}'(\Omega)$ and consider a relatively closed subset F of Ω satisfying $\operatorname{supp} u \subseteq F$. Set*

$$M_F := \{\varphi \in C^\infty(\Omega) : \operatorname{supp}\varphi \cap F \text{ is a compact set in } \mathbb{R}^n\}. \tag{2.6.1}$$

Then there exists a unique linear map $\widetilde{u} : M_F \to \mathbb{C}$ satisfying the following conditions:

(i) $\langle \widetilde{u}, \varphi \rangle = \langle u, \varphi \rangle$ for every $\varphi \in C_0^\infty(\Omega)$, and
(ii) $\langle \widetilde{u}, \varphi \rangle = 0$ for every $\varphi \in C^\infty(\Omega)$ with $\operatorname{supp}\varphi \cap F = \varnothing$,

where, if $\psi \in M_F$ then $\langle \widetilde{u}, \psi \rangle$ denotes $\widetilde{u}(\psi)$.

    Moreover, extensions of u constructed with respect to different choices of F act in a compatible fashion. More precisely, if $F_1, F_2 \subseteq \Omega$ are two relatively closed sets in Ω with the property that $\operatorname{supp} u \subseteq F_j$, $j = 1, 2$, then $\langle \widetilde{u}_1, \varphi \rangle = \langle \widetilde{u}_2, \varphi \rangle$ for every $\varphi \in M_{F_1} \cap M_{F_2}$, where $\widetilde{u}_1, \widetilde{u}_2$, are the extensions of u constructed as above relative to the sets F_1 and F_2, respectively.

Before presenting the proof of this theorem, a few comments are in order.

Remark 2.61. Retain the context of Theorem 2.60.

(a) One has $C_0^\infty(\Omega) \subseteq M_F$ and $\{\varphi \in C^\infty(\Omega) : \operatorname{supp}\varphi \cap F = \varnothing\} \subseteq M_F$.
(b) M_F is a vector subspace of $C^\infty(\Omega)$, albeit not a topological subspace of $\mathcal{E}(\Omega)$.
(c) If $F = \operatorname{supp} u$ we are in the setting discussed prior to the statement of Theorem 2.60. Also, if $F_1 \subseteq F_2$ then $M_{F_2} \subseteq M_{F_1}$. In particular, the largest M_F corresponds to the case when $F = \operatorname{supp} u$.
(d) If $\operatorname{supp} u$ is compact and we take $F = \operatorname{supp} u$ then $M_F = C^\infty(\Omega)$. In such a scenario, Theorem 2.60 gives an extension of u, originally defined as linear functional on $C_0^\infty(\Omega)$, to a linear functional defined on the larger space $C^\infty(\Omega)$. From a topological point of view, this extension turns out to be a continuous mapping of $\mathcal{E}(\Omega)$ into $\mathbb{C}$ (as we will see later, in Theorem 2.67).

Proof of Theorem 2.60. Fix a relatively closed subset F of Ω satisfying $\operatorname{supp} u \subseteq F$. First we prove the uniqueness statement in the first part of the theorem. Suppose $\widetilde{u}_1$, $\widetilde{u}_2 : M_F \to \mathbb{C}$ satisfy (i) and (ii). Fix $\varphi \in M_F$ and consider a function $\psi \in C_0^\infty(\Omega)$ such that $\psi \equiv 1$ in an open neighborhood W of $F \cap \operatorname{supp}\varphi$. That such a function ψ exists is guaranteed by Proposition 14.34. Decompose $\varphi = \varphi_0 + \varphi_1$ where $\varphi_0 := \psi\varphi \in C_0^\infty(\Omega)$ and $\varphi_1 := (1 - \psi)\varphi \in C^\infty(\Omega)$.

In general, if $A \subseteq \mathbb{R}^n$ and $f \in C^0(\mathbb{R}^n)$, it may be readily verified that $f = 0$ on A if and only if $\operatorname{supp} f \subseteq \overline{(A^c)}$. Making use of this observation we obtain that $\operatorname{supp}(1 - \psi) \subseteq \overline{W^c} = (\overset{\circ}{W})^c = W^c$. It follows that $\operatorname{supp} \varphi_1 \subseteq W^c \cap \operatorname{supp} \varphi$, hence $\operatorname{supp} \varphi_1 \cap F = \varnothing$. Thus, by (i) and (ii) written for $\widetilde{u}_1$ and $\widetilde{u}_2$, we have

$$\langle \widetilde{u}_1, \varphi \rangle = \langle \widetilde{u}_1, \varphi_0 \rangle + \langle \widetilde{u}_1, \varphi_1 \rangle = \langle u, \varphi_0 \rangle + 0 = \langle \widetilde{u}_2, \varphi_0 \rangle$$

$$= \langle \widetilde{u}_2, \varphi_0 \rangle + \langle \widetilde{u}_2, \varphi_1 \rangle = \langle \widetilde{u}_2, \varphi \rangle, \tag{2.6.2}$$

which implies that $\widetilde{u}_1 = \widetilde{u}_2$.

To prove the existence of an extension satisfying properties (i) and (ii), we make use of the decomposition of φ already employed in the proof of uniqueness. The apparent problem is that such a decomposition is not unique. However that is not the case. Suppose $\varphi \in M_F$ is such that $\varphi = \varphi_0' + \varphi_1' = \varphi_0'' + \varphi_1''$ for some functions $\varphi_0', \varphi_0'' \in C_0^\infty(\Omega)$ and $\varphi_1', \varphi_1'' \in C^\infty(\Omega)$ satisfying

$$\operatorname{supp} \varphi_1' \cap F = \varnothing = \operatorname{supp} \varphi_1'' \cap F.$$

Then, $\varphi_0' - \varphi_0'' = \varphi_1'' - \varphi_1'$, and since $\operatorname{supp}(\varphi_1'' - \varphi_1') \cap F = \varnothing$, we also have

$$\operatorname{supp}(\varphi_0' - \varphi_0'') \cap F = \varnothing,$$

which in turn implies $\operatorname{supp}(\varphi_0' - \varphi_0'') \subseteq \Omega \setminus \operatorname{supp} u$. The latter condition entails

$$0 = \langle u, \varphi_0' - \varphi_0'' \rangle = \langle u, \varphi_0' \rangle - \langle u, \varphi_0'' \rangle.$$

This suggests defining the extension

$$\widetilde{u} : M_F \longrightarrow \mathbb{C}, \quad \langle \widetilde{u}, \varphi \rangle := \langle u, \psi\varphi \rangle \quad \text{for each } \varphi \in M_F \text{ and}$$

each $\psi \in C_0^\infty(\Omega)$ with $\psi \equiv 1$ in a neighborhood of $\operatorname{supp} \varphi \cap F$. $\tag{2.6.3}$

Clearly $\widetilde{u}$ as in (2.6.3) is linear and, based on the previous reasoning, independent of the choice of ψ, thus well defined. We claim that this extension also satisfies (i) and (ii). Indeed, if $\varphi \in C_0^\infty(\Omega)$, we choose $\psi \equiv 1$ on $\operatorname{supp} \varphi$. Then necessarily $\langle \widetilde{u}, \varphi \rangle = \langle u, \varphi \rangle$, so the extension in (2.6.3) satisfies (i). Also, if $\varphi \in C^\infty(\Omega)$ is such that $\operatorname{supp} \varphi \cap F = \varnothing$, we may choose $\psi \in C_0^\infty(\Omega)$ such that $\operatorname{supp} \psi \cap F = \varnothing$ which forces $\langle \widetilde{u}, \varphi \rangle = \langle u, \psi\varphi \rangle = 0$, hence our extension satisfies (ii) as well. This proves the claim.

We are left with proving the compatibility of extensions. Let $F_1, F_2 \subseteq \Omega$ be relatively closed sets in Ω each containing $\operatorname{supp} u$. Denote by $\widetilde{u}_1$ and $\widetilde{u}_2$ the linear extensions of u to M_{F_1} and M_{F_2}, respectively, constructed as above relative to the sets F_1 and F_2. For $\varphi \in M_{F_1} \cap M_{F_2}$ let $\psi \in C_0^\infty(\Omega)$ be such that $\psi \equiv 1$ on an open neighborhood of the set $\operatorname{supp} \varphi \cap F_1$ and on an open neighborhood of the set $\operatorname{supp} \varphi \cap F_2$. Then by (2.6.3), $\langle \widetilde{u}_1, \varphi \rangle = \langle u, \psi\varphi \rangle = \langle \widetilde{u}_2, \varphi \rangle$. The proof of the theorem is now complete. $\square$

Remark 2.62 In the context of Theorem 2.60 consider $u \in \mathcal{D}'(\Omega)$, $a \in C^\infty(\Omega)$, and $\alpha \in \mathbb{N}_0^n$. Then the extension given in Theorem 2.60 satisfies the following properties:

(1) $\langle \widetilde{au}, \varphi \rangle = \langle \widetilde{u}, a\varphi \rangle$ for every $\varphi \in M_F$;
(2) $\langle \widetilde{\partial^\alpha u}, \varphi \rangle = (-1)^{|\alpha|} \langle \widetilde{u}, \partial^\alpha \varphi \rangle$ for every $\varphi \in M_F$.

Indeed, since by Theorem 2.60 an extension with properties (*i*) and (*ii*) is unique, the statement in *(1)* above will follow if one proves that the actions of the linear functionals considered in the left- and right-hand sides of the equality in *(1)* coincide on $C_0^\infty(\Omega)$ and on $C^\infty(\Omega)$ functions with supports outside F, which are immediate from (2.6.3) and properties of distributions. A similar approach works for the proof of *(2)*.

We introduce the following notation

$$\mathcal{D}'_c(\Omega) := \{ u \in \mathcal{D}'(\Omega) : \text{supp } u \text{ is a compact subset of } \Omega \}. \qquad (2.6.4)$$

By applying Theorem 2.60 to $u \in \mathcal{D}'_c(\Omega)$ and $F := \text{supp } u$, in which case $M_F = C^\infty(\Omega)$, it follows that there exists a linear map $\widetilde{u} : C^\infty(\Omega) \to \mathbb{C}$ satisfying (*i*) and (*ii*) in the statement of this theorem. In fact, this extension turns out to be continuous with respect to the topology $\mathcal{E}(\Omega)$, an issue that we will address shortly.

The dual of $\mathcal{E}(\Omega)$ is the space

$$\{ v : \mathcal{E}(\Omega) \to \mathbb{C} : v \text{ linear and continuous} \}. \qquad (2.6.5)$$

Whenever $v : \mathcal{E}(\Omega) \to \mathbb{C}$ is linear and continuous, and whenever $\varphi \in C^\infty(\Omega)$, we use the notation $\langle v, \varphi \rangle$ in place of $v(\varphi)$. The following is an equivalent characterization of continuity for linear functionals on $\mathcal{E}(\Omega)$ (see Section 14.1.0.2 for more details).

Fact 2.63 *A linear functional $v : \mathcal{E}(\Omega) \to \mathbb{C}$ is continuous (for details see (14.1.22)) if and only if there exist a compact $K \subset \Omega$, a number $m \in \mathbb{N}_0$, and a constant $C \in (0, \infty)$, such that*

$$|v(\varphi)| \leq C \sup_{\alpha \in \mathbb{N}_0^n, |\alpha| \leq m} \sup_{x \in K} |\partial^\alpha \varphi(x)|, \qquad \forall \, \varphi \in C^\infty(\Omega). \qquad (2.6.6)$$

In the current setting, functionals on $\mathcal{E}(\Omega)$ are continuous if and only if they are sequentially continuous. This can be seen by combining the general result presented in Theorem 14.1 with Fact 1.8. A direct proof, applicable to the specific case of linear functionals on $\mathcal{E}(\Omega)$, is given in the next proposition.

Proposition 2.64. *Let $v : \mathcal{E}(\Omega) \to \mathbb{C}$ be a linear map. Then v is continuous if and only if v is sequentially continuous.*

Proof. The general fact that any linear and continuous functional on topological vector spaces is sequentially continuous gives the left-to-right implication. To prove the converse implication, it suffices to check continuity at zero. This is done reasoning by contradiction. Assume that

$$v(\varphi_j) \xrightarrow[j\to\infty]{} 0 \quad \text{whenever} \quad \varphi_j \xrightarrow[j\to\infty]{\mathcal{E}(\Omega)} 0, \tag{2.6.7}$$

but that v is not continuous at $0 \in \mathcal{E}(\Omega)$. Then for each compact subset K of Ω and every $j \in \mathbb{N}$, there exists $\varphi_j \in \mathcal{E}(\Omega)$ such that

$$|v(\varphi_j)| > j \sup_{\substack{x\in K \\ |\alpha|\le j}} |\partial^\alpha \varphi_j(x)|. \tag{2.6.8}$$

Consider now a nested sequence of compact sets $\{K_j\}_{j\in\mathbb{N}}$ such that $\bigcup_{j=1}^{\infty} K_j = \Omega$. For each $j \in \mathbb{N}$, let φ_j be as given by (2.6.8) corresponding to $K := K_j$ and define the function $\psi_j := \frac{\varphi_j}{v(\varphi_j)}$ which belongs to $\mathcal{E}(\Omega)$. Then

$$v(\psi_j) = 1 \quad \text{and} \quad \sup_{x\in K_j,\, |\alpha|\le j} |\partial^\alpha \psi_j(x)| \le \frac{1}{j} \quad \text{for every} \ \ j \in \mathbb{N}. \tag{2.6.9}$$

Thus, for each fixed $\alpha \in \mathbb{N}_0^n$ and every compact subset K of Ω there exists some $j_0 \ge |\alpha|$ with the property that that $K \subset K_{j_0}$ and $\sup_{x\in K} |\partial^\alpha \psi_j(x)| < \frac{1}{j}$ for all $j \ge j_0$. The latter implies $\psi_j \xrightarrow[j\to\infty]{\mathcal{E}(\Omega)} 0$ which, in light of (2.6.7), further implies $v(\psi_j) \xrightarrow[j\to\infty]{} 0$. Since this contradicts the fact that $v(\psi_j) = 1$ for every $j \in \mathbb{N}$, the proof is finished. $\square$

The topology we consider on the dual of $\mathcal{E}(\Omega)$ is the weak$*$-topology, and we denote the resulting topological vector space by $\mathcal{E}'(\Omega)$ (see Section 14.1.0.2 for more details). A significant byproduct of this set up is singled out next.

Fact 2.65 $\mathcal{E}'(\Omega)$ *is a locally convex topological vector space over* $\mathbb{C}$, *which is not metrizable, but is complete.*

In addition, we have the following important characterization of continuity in $\mathcal{E}'(\Omega)$.

Fact 2.66 *A sequence* $\{u_j\}_{j\in\mathbb{N}} \subset \mathcal{E}'(\Omega)$ *converges to* $u \in \mathcal{E}'(\Omega)$ *as* $j \to \infty$ *in* $\mathcal{E}'(\Omega)$, *something we will indicate by writing* $u_j \xrightarrow[j\to\infty]{\mathcal{E}'(\Omega)} u$, *if and only if* $\langle u_j, \varphi \rangle \xrightarrow[j\to\infty]{} \langle u, \varphi \rangle$ *for every* $\varphi \in \mathcal{E}(\Omega)$.

We are now ready to state and prove a result that gives a complete characterization of the class of functionals that are extensions as in Theorem 2.60 of distributions $u \in \mathcal{D}'(\Omega)$ with compact support.

Theorem 2.67. *The spaces* $\mathcal{D}'_c(\Omega)$ *and* $\mathcal{E}'(\Omega)$ *are algebraically isomorphic.*

Proof. Consider the mapping $\iota : \mathcal{D}'_c(\Omega) \to \mathcal{E}'(\Omega)$, $\iota(u) := \widetilde{u}$, where $\widetilde{u}$ is the extension of u given by Theorem 2.60 corresponding to $F := \operatorname{supp} u$. Then $\mathcal{M}_{\operatorname{supp} u} = C^\infty(\Omega)$ and, to conclude that ι is well defined, there remains to show that the functional $\widetilde{u}$ is continuous on $\mathcal{E}(\Omega)$. With this goal in mind, note that while in general the function $\psi \in C_0^\infty(\Omega)$ used in the construction of $\widetilde{u}$ (as in (2.6.3)) depends on φ, given that we are currently assuming that $\operatorname{supp} u$ is compact, we may take $\psi \equiv 1$

on a neighborhood of supp u (originally we only needed $\psi \equiv 1$ on a neighborhood of supp $u \cap$ supp $\varphi \subseteq$ supp u). Let $K_0 := \text{supp}\,\psi$. Then, for each $\varphi \in C^\infty(\Omega)$ we have that $\varphi\psi$ belongs to $C_0^\infty(\Omega)$, satisfies the support condition supp $(\varphi\psi) \subseteq K_0$, and $\langle \widetilde{u}, \varphi \rangle = \langle u, \psi\varphi \rangle$. Fix $\varphi \in C^\infty(\Omega)$. Since $u \in \mathcal{D}'(\Omega)$, corresponding to the compact set K_0 there exist $k_0 \in \mathbb{N}_0$ and a finite constant $C \geq 0$ such that

$$|\langle \widetilde{u}, \varphi \rangle| = |\langle u, \psi\varphi \rangle| \leq C \sup_{\substack{x \in K_0 \\ |\alpha| \leq k_0}} |\partial^\alpha(\psi\varphi)|. \tag{2.6.10}$$

Starting with Leibniz's formula (14.2.6) applied to $\psi\varphi$, we estimate

$$|\partial^\alpha(\psi\varphi)| = \left| \sum_{\beta \leq \alpha} \frac{\alpha!}{\beta!(\alpha-\beta)!} \partial^{\alpha-\beta}\psi \partial^\beta\varphi \right| \leq C' \sup_{\substack{x \in K_0 \\ |\beta| \leq k_0}} |\partial^\beta\varphi(x)|, \tag{2.6.11}$$

for some finite constant $C' = C'(\alpha, \psi) > 0$. Combining (2.6.10) and (2.6.11), we obtain

$$|\langle \widetilde{u}, \varphi \rangle| \leq C \cdot C' \sup_{\substack{x \in K_0 \\ |\beta| \leq k_0}} |(\partial^\beta\varphi)(x)|, \tag{2.6.12}$$

hence $\widetilde{u} \in \mathcal{E}'(\Omega)$, proving that ι is well defined.

Moving on, it is clear that ι is linear, hence to conclude that it is injective, it suffices to show that if $\iota(u) = 0$ for some $u \in \mathcal{D}'_c(\Omega)$, then $u = 0$. Consider $u \in \mathcal{D}'_c(\Omega)$ such that $\iota(u) = 0$. Then by (i) in Theorem 2.60, $u = \iota(u)\big|_{C_0^\infty(\Omega)} = 0$, as desired.

Consider now the task of proving that ι is surjective. To get started, pick an arbitrary $v \in \mathcal{E}'(\Omega)$ and set $u := v\big|_{C_0^\infty(\Omega)}$. Clearly $u : \mathcal{D}(\Omega) \to \mathbb{C}$ is linear. Since $v \in \mathcal{E}'(\Omega)$, Fact 2.63 ensures the existence of a compact set $K \subset \Omega$, nonnegative integer k, and finite constant $C > 0$, such that

$$|\langle v, \varphi \rangle| \leq C \sup_{\substack{x \in K \\ |\alpha| \leq k}} |\partial^\alpha\varphi(x)|, \qquad \forall\, \varphi \in C^\infty(\Omega). \tag{2.6.13}$$

Then, for each compact subset A of Ω and $\varphi \in C_0^\infty(\Omega)$ with supp $\varphi \subseteq A$, by regarding φ as being in $\mathcal{E}(\Omega)$ we may use (2.6.13) to write

$$|\langle u, \varphi \rangle| \leq C \sup_{\substack{x \in K \\ |\alpha| \leq k}} |\partial^\alpha\varphi(x)| = C \sup_{\substack{x \in K \cap A \\ |\alpha| \leq k}} |\partial^\alpha\varphi(x)| \leq C \sup_{\substack{x \in A \\ |\alpha| \leq k}} |\partial^\alpha\varphi(x)|. \tag{2.6.14}$$

From (2.6.14) we may now conclude (invoking Proposition 2.4) that $u \in \mathcal{D}'(\Omega)$.

Next, we claim that supp $u \subseteq K$. Indeed, if $\varphi \in C_0^\infty(\Omega)$ is a test function with supp $\varphi \cap K = \varnothing$ then from (2.6.13) we obtain $|\langle u, \varphi \rangle| = 0$, thus $u = 0$ on $\Omega \setminus K$. Hence, the claim is proved which, in turn, shows that $u \in \mathcal{D}'_c(\Omega)$.

To finish the proof of the surjectivity of ι, it suffices to show that $\iota(u) = v$. Denote by $\widetilde{u}_K$ the extension of u given by Theorem 2.60 with $F := K$. Then reasoning as in the proof of the fact that ι is well defined, we obtain $\widetilde{u}_K \in \mathcal{E}'(\Omega)$. By (i) in Theorem 2.60 it follows that $\widetilde{u}_K\big|_{C_0^\infty(\Omega)} = u$. Also, if $\varphi \in C^\infty(\Omega)$ satisfies supp $\varphi \cap$

$K = \varnothing$, then by (ii) in Theorem 2.60 we have $\langle \widetilde{u}_K, \varphi \rangle = 0$, while (2.6.13) implies $\langle v, \varphi \rangle = 0$. Hence, the uniqueness result in Theorem 2.60 yields $\widetilde{u}_K = v$. On the other hand, since K and supp u are compact, we have $M_K = C^\infty(\Omega) = M_{\text{supp}\,u}$. Now the last conclusion in Theorem 2.60 gives $\widetilde{u}_K = \iota(u)$. Consequently, $\iota(u) = v$, and the surjectivity of ι is proved. This finishes the proof of the theorem. $\square$

In light of the significance of $\mathcal{D}'_c(\Omega)$, Theorem 2.67 provides a natural algebraic identification

$$\mathcal{E}'(\Omega) = \{u \in \mathcal{D}'(\Omega) : \text{supp}\,u \text{ is a compact subset of } \Omega\}. \qquad (2.6.15)$$

Remark 2.68. The spaces $\mathcal{D}'_c(\Omega)$ and $\mathcal{E}'(\Omega)$ are not topologically isomorphic since there exist sequences of distributions with compact support that converge in $\mathcal{D}'(\Omega)$ but not in $\mathcal{E}'(\Omega)$. For example, take the sequence $\{\delta_j\}_{j\in\mathbb{N}} \subset \mathcal{D}'(\mathbb{R})$ of Dirac distributions with mass at $j \in \mathbb{N}$, that have been defined in Example 2.14. Then it is easy to check that the sequence $\{\delta_j\}_j$ converges to 0 in $\mathcal{D}'(\mathbb{R})$ but not in $\mathcal{E}'(\mathbb{R})$.

Theorem 2.67 nonetheless proves that the identity mapping is well defined from $\mathcal{E}'(\Omega)$ into $\mathcal{D}'(\Omega)$. Keeping this in mind and relying on (1.3.15) and Proposition 14.4, we see that

$$\mathcal{E}'(\Omega) \text{ is continuously embedded into } \mathcal{D}'(\Omega). \qquad (2.6.16)$$

This corresponds to the dual version of (1.3.15). In particular, the operation of restriction to an open subset ω of Ω is a well-defined linear mapping

$$\mathcal{E}'(\Omega) \ni u \mapsto u\big|_\omega \in \mathcal{D}'(\omega). \qquad (2.6.17)$$

Moreover, $u\big|_\omega \in \mathcal{E}'(\omega)$ whenever the support of $u \in \mathcal{E}'(\Omega)$ is contained in ω.

Proposition 2.69. *Let ω and Ω be open subsets of $\mathbb{R}^n$ such that $\omega \subseteq \Omega$. Then every $u \in \mathcal{E}'(\omega)$ extends to a functional $\widetilde{u} \in \mathcal{E}'(\Omega)$ by setting*

$$\widetilde{u} : \mathcal{E}(\Omega) \to \mathbb{C}, \quad \langle \widetilde{u}, \varphi \rangle := \langle u, \psi\varphi \rangle, \quad \forall\, \varphi \in C^\infty(\Omega), \qquad (2.6.18)$$

where $\psi \in C_0^\infty(\omega)$ is such that $\psi \equiv 1$ in a neighborhood of supp u.

Proof. We first claim that the mapping in (2.6.18) is well defined. To see why this is the case, suppose $\psi_j \in C_0^\infty(\omega)$, $\psi_j \equiv 1$ in a neighborhood of supp u, for $j = 1, 2$. Then for each function $\varphi \in C^\infty(\Omega)$ we have $(\psi_1 - \psi_2)\varphi \in C_0^\infty(\omega)$ and in addition supp $u \cap \text{supp}\,[(\psi_1 - \psi_2)\varphi] = \varnothing$. These further imply that necessarily $\langle u, (\psi_1 - \psi_2)\varphi \rangle = 0$. This proves that the definition of $\widetilde{u}$ is independent of the choice of ψ with the given properties. The functional defined in (2.6.18) is also linear while its continuity is a consequence of Proposition 2.64 and Exercise 1.22. In addition, if $\varphi \in C^\infty(\omega)$ is given and if $\psi \in C_0^\infty(\omega)$ is such that $\psi \equiv 1$ in a neighborhood of supp u, then supp $[(1 - \psi)\varphi] \cap \text{supp}\,u = \varnothing$. Hence

$$\langle \widetilde{u}, \varphi \rangle = \langle u, \psi\varphi \rangle = \langle u, (1 - \psi)\varphi \rangle + \langle u, \varphi \rangle = \langle u, \varphi \rangle, \qquad (2.6.19)$$

proving that $\widetilde{u}$ is an extension of u. $\square$

Exercise 2.70. In the context of Proposition 2.69 prove that

$$\partial^\alpha \widetilde{u} = \widetilde{\partial^\alpha u} \quad \text{in } \mathcal{D}'(\Omega) \tag{2.6.20}$$

for every $u \in \mathcal{E}'(\omega)$ and every $\alpha \in \mathbb{N}_0^n$.

Remark 2.71.

(1) In the sequel, we will often drop $\widetilde{}$ from the notation of the extension (as defined in the proof of Theorem 2.67 or Proposition 2.69) of a compactly supported distribution. More precisely, if $u \in \mathcal{D}'_c(\Omega)$ we will simply use u for the extension of u to a functional in $\mathcal{E}'(\Omega)$, as well as for its extension to a functional in $\mathcal{E}'(O)$, where O is an open subset of $\mathbb{R}^n$ containing Ω.

(2) Whenever necessary, if $u \in \mathcal{E}'(\Omega)$, $\varphi \in C_0^\infty(\Omega)$, and $\psi \in C^\infty(\Omega)$, we will use the notation ${}_{\mathcal{D}'}\langle u, \varphi \rangle_{\mathcal{D}}$ for the action of u on φ as a functional in $\mathcal{D}'(\Omega)$, and the notation ${}_{\mathcal{E}'}\langle u, \psi \rangle_{\mathcal{E}}$ for the action of u on ψ as a functional in $\mathcal{E}'(\Omega)$.

Proposition 2.72. *Let $u \in \mathcal{D}'(\Omega)$ and $\psi \in C_0^\infty(\Omega)$. Then $\psi u \in \mathcal{E}'(\Omega)$ and*

$${}_{\mathcal{E}'}\langle \psi u, \varphi \rangle_{\mathcal{E}} = {}_{\mathcal{D}'}\langle u, \psi\varphi \rangle_{\mathcal{D}}, \quad \forall \varphi \in \mathcal{E}(\Omega). \tag{2.6.21}$$

Proof. Since $\psi u \in \mathcal{D}'(\Omega)$ and $\operatorname{supp} \psi u \subseteq \operatorname{supp} \psi$, we have $\psi u \in \mathcal{D}'_c(\Omega)$, thus $\psi u \in \mathcal{E}'(\Omega)$ (i.e., ψu extends as an element in $\mathcal{E}'(\Omega)$). Let $\phi \in C_0^\infty(\Omega)$ be such that $\phi \equiv 1$ on a neighborhood of $\operatorname{supp} \psi$. Then for every $\varphi \in \mathcal{E}(\Omega)$,

$${}_{\mathcal{E}'}\langle \psi u, \varphi \rangle_{\mathcal{E}} = {}_{\mathcal{D}'}\langle \psi u, \phi\varphi \rangle_{\mathcal{D}} = {}_{\mathcal{D}'}\langle u, \psi\phi\varphi \rangle_{\mathcal{D}}$$

$$= {}_{\mathcal{D}'}\langle u, \psi\varphi \rangle_{\mathcal{D}} \tag{2.6.22}$$

proving (2.6.21). $\qquad\square$

Exercise 2.73. Let $u_j \xrightarrow[j\to\infty]{\mathcal{D}'(\Omega)} u$ be such that there exists a compact K in $\mathbb{R}^n$ that is contained in Ω and with the property that $\operatorname{supp} u_j \subseteq K$ for every $j \in \mathbb{N}$. Prove that $\operatorname{supp} u \subseteq K$ and $u_j \xrightarrow[j\to\infty]{\mathcal{E}'(\mathbb{R}^n)} u$. Consequently, we also have $u_j \xrightarrow[j\to\infty]{\mathcal{E}'(\Omega)} u$.

Exercise 2.74. Let $k \in \mathbb{N}_0$ and assume that $c_\alpha \in \mathbb{C}$ for $\alpha \in \mathbb{N}_0^n$ with $|\alpha| \leq k$. Prove that

$$\sum_{|\alpha| \leq k} c_\alpha \partial^\alpha \delta = 0 \quad \text{in } \mathcal{D}'(\mathbb{R}^n) \iff \text{each } c_\alpha = 0. \tag{2.6.23}$$

Hint: Use (14.2.7).

Exercise 2.75. Prove that if $u \in \mathcal{D}'(\mathbb{R}^n)$ and $\operatorname{supp} u \subseteq \{a\}$ for some $a \in \mathbb{R}^n$, then u has a unique representation of the form

$$u = \sum_{|\alpha| \leq k} c_\alpha \partial^\alpha \delta_a, \tag{2.6.24}$$

for some $k \in \mathbb{N}_0$ and coefficients $c_\alpha \in \mathbb{C}$.

Sketch of proof:

  (I) Via a translation, reduce matters to the case $a = 0$.

 (II) Use Fact 2.63 to determine $k \in \mathbb{N}_0$.

(III) Fix $\psi \in C_0^\infty(B(0, 1))$ such that $\psi \equiv 1$ on $B(0, \frac{1}{2})$ and for $\varepsilon > 0$ define the function $\psi_\varepsilon(x) := \psi(\frac{x}{\varepsilon})$ for every $x \in \mathbb{R}^n$. Prove that $u = \psi_\varepsilon u$ in $\mathcal{D}'(\mathbb{R}^n)$.

(IV) For $\varphi \in C_0^\infty(\mathbb{R}^n)$ consider the k-th order Taylor polynomial for φ at 0, i.e.,

$$\varphi_k(x) := \sum_{|\beta| \le k} \frac{1}{\beta!} \partial^\beta \varphi(0)\, x^\beta, \qquad \forall\, x \in \mathbb{R}^n. \tag{2.6.25}$$

Prove that for $\alpha \in \mathbb{N}_0^n$ satisfying $|\alpha| \le k$ one has

$$\partial^\alpha(\varphi - \varphi_k) = \partial^\alpha \varphi - (\partial^\alpha \varphi)_{k-|\alpha|}.$$

 (V) Show that for each $\varphi \in C_0^\infty(\mathbb{R}^n)$ there exists some constant $c \in (0, \infty)$ such that $|\langle u, (\varphi - \varphi_k)\psi_\varepsilon \rangle| \le c\,\varepsilon$.

(VI) Combine all the above to obtain that

$$\langle u, \varphi \rangle = \Big\langle \sum_{|\alpha| \le k} \Big[\frac{(-1)^{|\alpha|}}{\alpha!} \langle u, x^\alpha \rangle \Big] \partial^\alpha \delta,\, \varphi \Big\rangle \qquad \forall\, \varphi \in C_0^\infty(\mathbb{R}^n). \tag{2.6.26}$$

(VII) Prove that the representation in (VI) is unique.

Example 2.76. Let $m \in \mathbb{N}$. We are interested in solving the equation

$$x^m u = 0 \quad \text{in} \quad \mathcal{D}'(\mathbb{R}). \tag{2.6.27}$$

In this regard, assume that $u \in \mathcal{D}'(\mathbb{R})$ solves (2.6.27) and note that if φ belongs to $C_0^\infty(\mathbb{R} \setminus \{0\})$, then $\frac{1}{x^m}\varphi \in C_0^\infty(\mathbb{R} \setminus \{0\})$. This observation permits us to write

$$\langle u, \varphi \rangle = \Big\langle x^m u, \frac{1}{x^m}\varphi \Big\rangle = 0, \qquad \forall\, \varphi \in C_0^\infty(\mathbb{R} \setminus \{0\}), \tag{2.6.28}$$

which proves that $\operatorname{supp} u \subseteq \{0\}$. In particular, $u \in \mathcal{E}'(\mathbb{R})$. Applying Exercise 2.75 we conclude that there exists $N \in \mathbb{N}_0$ such that $u = \sum_{k=0}^{N} c_k \delta^{(k)}$ in $\mathcal{D}'(\mathbb{R})$, for some $c_k \in \mathbb{C}$, $k = 0, 1, 2, \ldots, N$. We claim that

$$c_\ell = 0 \quad \text{whenever} \quad m \le \ell \le N. \tag{2.6.29}$$

To see why this is true, observe that since $u \in \mathcal{E}'(\mathbb{R})$ it makes sense to apply u to any function in $C^\infty(\mathbb{R})$. In particular, it is meaningful to apply u to any polynomial. Concerning (2.6.29), if $N \le m - 1$ there is nothing to prove, while in the case when $N \ge m$ for each $\ell \in \{m, \ldots, N\}$ we may write

$$0 = \langle x^m u, x^{\ell-m} \rangle = \langle u, x^\ell \rangle = \sum_{k=0}^{N} c_k \langle \delta^{(k)}, x^\ell \rangle$$

$$= \sum_{k=0}^{N} (-1)^k c_k \Big[\frac{d^k}{dx^k}(x^\ell) \Big]\Big|_{x=0} = (-1)^\ell \ell! \, c_\ell. \tag{2.6.30}$$

This proves (2.6.29) which, in turn, forces u to have the form

$$u = \sum_{k=0}^{m-1} c_k \delta^{(k)} \quad \text{for some} \quad c_k \in \mathbb{C}, \ k = 0, 1, \ldots, m-1. \tag{2.6.31}$$

Conversely, one may readily verify that any distribution u as in (2.6.31) solves (2.6.27). In conclusion, any solution u of (2.6.27) is as in (2.6.31).

Exercise 2.77. Let $m \in \mathbb{N}$, $a \in \mathbb{R}$, and $u \in \mathcal{D}'(\mathbb{R})$. Prove u is a solution of the equation $(x - a)^m u = 0$ in $\mathcal{D}'(\mathbb{R})$ if and only if u is of the form $u = \sum_{k=0}^{m-1} c_k \delta_a^{(k)}$ for some $c_k \in \mathbb{C}$, $k = 0, 1, \ldots, m-1$.

Remark 2.78. You have seen in Example 2.34 that P.V. $\frac{1}{x}$ is a solution of the equation $xu = 1$ in $\mathcal{D}'(\mathbb{R})$. Hence, if $v \in \mathcal{D}'(\mathbb{R})$ is another solution of this equation, then the distribution $v - \text{P.V.} \frac{1}{x}$ is a solution of the equation $xu = 0$ in $\mathcal{D}'(\mathbb{R})$. By Example 2.76, it follows that $v - \text{P.V.} \frac{1}{x} = c\,\delta$, where $c \in \mathbb{C}$. Thus, the general solution of the equation $xu = 1$ in $\mathcal{D}'(\mathbb{R})$ is $u = \text{P.V.} \frac{1}{x} + c\,\delta$, for $c \in \mathbb{C}$.

Example 2.79. Let $N \in \mathbb{N}$ and $a_j \in \mathbb{R}^n$, $j \in \{1, \ldots, N\}$, be a finite family of distinct points. If $u \in \mathcal{D}'(\mathbb{R}^n)$ is such that $\operatorname{supp} u \subseteq \{a_1, a_2, \ldots, a_N\}$, then u has a unique representation of the form

$$u = \sum_{j=1}^{N} \sum_{|\alpha| \le k_j} c_{\alpha, j} \partial^\alpha \delta_{a_j}, \qquad k_j \in \mathbb{N}_0, \quad c_{\alpha, j} \in \mathbb{C}. \tag{2.6.32}$$

To justify formula (2.6.32), fix a family of pairwise disjoint balls $B_j := B(a_j, r_j)$, $j \in \{1, \ldots, N\}$, and for each j select a function $\psi_j \in C_0^\infty(B_j)$ satisfying $\psi_j \equiv 1$ in a neighborhood of $\overline{B(a_j, r_j/2)}$. Then for each $j \in \{1, \ldots, N\}$ we have that $\psi_j u$ is a compactly supported distribution in $\mathbb{R}^n$ with $\operatorname{supp}(\psi_j u) \subseteq \{a_j\}$. We may now apply Exercise 2.75 to obtain that there exist $k_j \in \mathbb{N}_0$ and $c_{\alpha, j} \in \mathbb{C}$ such that $\psi_j u = \sum_{|\alpha| \le k_j} c_{\alpha, j} \partial^\alpha \delta_{a_j}$ in $\mathcal{D}'(\mathbb{R}^n)$. In addition, since $\sum_{j=1}^{N} \psi_j \equiv 1$ in a neighborhood of $\operatorname{supp} u$, we have

$$\Big\langle \sum_{j=1}^{N} u\psi_j, \varphi \Big\rangle = \sum_{j=1}^{N} \langle \psi_j u, \varphi \rangle = \sum_{j=1}^{N} \langle u, \psi_j \varphi \rangle = \Big\langle u, \varphi \sum_{j=1}^{N} \psi_j \Big\rangle$$

$$= \langle u, \varphi \rangle \quad \text{for each} \ \varphi \in C_0^\infty(\mathbb{R}^n). \tag{2.6.33}$$

Hence, $u = \sum_{j=1}^{N} \psi_j u$ in $\mathcal{D}'(\mathbb{R}^n)$ which, given (2.6.33), proves (2.6.32).

Example 2.80. Let $a, b \in \mathbb{R}$ be such that $a \neq b$. We are interested in solving the equation

$$(x - a)(x - b)u = 0 \quad \text{in} \quad \mathcal{D}'(\mathbb{R}). \tag{2.6.34}$$

The first observation is that any solution u of this equation satisfies the support condition $\mathrm{supp}\, u \subseteq \{a, b\}$. Indeed, if we take an arbitrary $\varphi \in C_0^\infty(\mathbb{R} \setminus \{a, b\})$, then $\frac{1}{(x-a)(x-b)}\varphi$ belongs to the space $C_0^\infty(\mathbb{R} \setminus \{a, b\})$ and

$$\langle u, \varphi \rangle = \left\langle (x - a)(x - b)u, \tfrac{1}{(x-a)(x-b)} \varphi \right\rangle = 0. \tag{2.6.35}$$

Hence, we may apply Example 2.79 to conclude that

$$u = \sum_{j=0}^{N_1} c_j \delta_a^{(j)} + \sum_{j=0}^{N_2} d_j \delta_b^{(j)} \quad \text{in} \quad \mathcal{D}'(\mathbb{R}), \tag{2.6.36}$$

where $N_1, N_2 \in \mathbb{N}$, $\{c_j\}_{0 \leq j \leq N_1} \subset \mathbb{C}$ and $\{d_j\}_{0 \leq j \leq N_2} \subset \mathbb{C}$. Moreover, by dropping terms with zero coefficients, there is no loss of generality in assuming that

$$c_{N_1} \neq 0 \quad \text{and} \quad d_{N_2} \neq 0. \tag{2.6.37}$$

In this scenario, we make the claim that $N_1 = N_2 = 0$. To prove this claim, suppose first that $N_1 \geq 1$. Then, using (2.6.36) and the hypotheses on u, we obtain

$$0 = \langle (x - a)(x - b)u, (x - a)^{N_1-1}(x - b)^{N_2} \rangle = \langle u, (x - a)^{N_1}(x - b)^{N_2+1} \rangle$$

$$= \sum_{j=0}^{N_1} c_j \langle \delta_a^{(j)}, (x - a)^{N_1}(x - b)^{N_2+1} \rangle + \sum_{j=0}^{N_2} d_j \langle \delta_b^{(j)}, (x - a)^{N_1}(x - b)^{N_2+1} \rangle$$

$$= \sum_{j=0}^{N_1} (-1)^j c_j \Big[\frac{d^j}{dx^j}\big((x - a)^{N_1}(x - b)^{N_2+1}\big) \Big]\Big|_{x=a}$$

$$+ \sum_{j=0}^{N_2} (-1)^j d_j \Big[\frac{d^j}{dx^j}\big((x - a)^{N_1}(x - b)^{N_2+1}\big) \Big]\Big|_{x=b}$$

$$= (-1)^{N_1} c_{N_1} N_1! (a - b)^{N_2+1}. \tag{2.6.38}$$

Since by assumption $a \neq b$, from (2.6.38) we obtain $c_{N_1} = 0$. This contradicts (2.6.37) and shows that necessarily $N_1 = 0$. Similarly, we obtain that $N_2 = 0$, hence any solution of (2.6.34) has the form

$$u = c\,\delta_a + d\,\delta_b \quad \text{in} \quad \mathcal{D}'(\mathbb{R}), \quad c, d \in \mathbb{C}. \tag{2.6.39}$$

Conversely, it is clear that any distribution as in (2.6.39) solves (2.6.34). To sum up, (2.6.39) describes all solutions of (2.6.34).

2.7 Tensor Product of Distributions

Let $m, n \in \mathbb{N}$, U be an open subset of $\mathbb{R}^m$, V be an open subset of $\mathbb{R}^n$, and consider two complex-valued functions $f \in L^1_{loc}(U)$ and $g \in L^1_{loc}(V)$. Then the tensor product of the functions f and g is defined as

$$f \otimes g : U \times V \to \mathbb{C}, \quad (f \otimes g)(x, y) := f(x)g(y) \quad \text{for each} \quad (x, y) \in U \times V. \quad (2.7.1)$$

In particular, it follows from (2.7.1) that $f \otimes g \in L^1_{loc}(U \times V)$. When f, g, and $f \otimes g$ are regarded as distributions, for each $\varphi \in C_0^\infty(U \times V)$ we obtain that

$$\langle f \otimes g, \varphi \rangle = \iint_{U \times V} f(x)g(y)\varphi(x, y)\, dx\, dy = \int_U f(x) \left(\int_V g(y)\varphi(x, y)\, dy \right) dx$$

$$= \int_V g(y) \left(\int_U f(x)\varphi(x, y)\, dx \right) dy \qquad (2.7.2)$$

or, concisely,

$$\langle f \otimes g, \varphi \rangle = \langle f(x), \langle g(y), \varphi(x, y) \rangle \rangle = \langle g(y), \langle f(x), \varphi(x, y) \rangle \rangle. \qquad (2.7.3)$$

If, in addition, the test function φ has the form $\varphi_1 \otimes \varphi_2$, for some $\varphi_1 \in C_0^\infty(U)$ and $\varphi_2 \in C_0^\infty(V)$, then (2.7.3) becomes

$$\langle f \otimes g, \varphi_1 \otimes \varphi_2 \rangle = \langle f, \varphi_1 \rangle \langle g, \varphi_2 \rangle. \qquad (2.7.4)$$

This suggests a natural way to define tensor products of general distributions granted the availability of the following density result for $\mathcal{D}(U \times V)$.

Proposition 2.81. *Let $m, n \in \mathbb{N}$, U be an open subset of $\mathbb{R}^m$, and V be an open subset of $\mathbb{R}^n$. Then the set*

$$C_0^\infty(U) \otimes C_0^\infty(V) := \left\{ \sum_{j=1}^N \varphi_j \otimes \psi_j : \varphi_j \in C_0^\infty(U), \psi_j \in C_0^\infty(V), N \in \mathbb{N} \right\} \quad (2.7.5)$$

is sequentially dense in $\mathcal{D}(U \times V)$.

Before proceeding with the proof of Proposition 2.81 we state and prove two lemmas.

Lemma 2.82. *Suppose that the sequence $\{f_j\}_{j \in \mathbb{N}} \subset \mathcal{E}(\mathbb{R}^n)$ and $f \in \mathcal{E}(\mathbb{R}^n)$ are such that*

$$\|\partial^\alpha f_j - \partial^\alpha f\|_{L^\infty(B(0,j))} < \frac{1}{j} \quad \text{for all } \alpha \in \mathbb{N}_0^n \text{ satisfying } |\alpha| \leq j. \tag{2.7.6}$$

Then $f_j \xrightarrow[j\to\infty]{\mathcal{E}(\mathbb{R}^n)} f$.

Proof. Suppose $\{f_j\}_{j\in\mathbb{N}}$ and f satisfy the current hypotheses. Pick an arbitrary $\varepsilon > 0$, a multi-index $\alpha \in \mathbb{N}_0^n$, and fix a compact subset K of $\mathbb{R}^n$. Then there exists $j_0 \in \mathbb{N}$ such that $K \subset B(0, j_0)$. If we now fix $j^* \in \mathbb{N}$ with the property that $j^* > \max\{\frac{1}{\varepsilon}, |\alpha|, j_0\}$, it follows that for each $j \geq j^*$ we have $|\alpha| \leq j^* \leq j$ and

$$\|\partial^\alpha f_j - \partial^\alpha f\|_{L^\infty(K)} \leq \|\partial^\alpha f_j - \partial^\alpha f\|_{L^\infty(B(0,j))} < \frac{1}{j} < \frac{1}{j^*} < \varepsilon. \tag{2.7.7}$$

Hence, $\partial^\alpha f_j$ converges uniformly on K to $\partial^\alpha f$. Since α and K are arbitrary, we conclude that $f_j \xrightarrow[j\to\infty]{\mathcal{E}(\mathbb{R}^n)} f$. $\square$

Lemma 2.83. *For every $f \in C_0^\infty(\mathbb{R}^n)$ there exists a sequence $\{P_j\}_{j\in\mathbb{N}}$ of polynomials in $\mathbb{R}^n$ such that $P_j \xrightarrow[j\to\infty]{\mathcal{E}(\mathbb{R}^n)} f$.*

Proof. For each $t > 0$ define the function

$$f_t(x) := (4\pi t)^{-\frac{n}{2}} \int_{\mathbb{R}^n} e^{-\frac{|x-y|^2}{4t}} f(y)\,\mathrm{d}y, \qquad \forall x \in \mathbb{R}^n. \tag{2.7.8}$$

The first goal is to prove that

$$f_t \xrightarrow[t\to 0^+]{\mathcal{E}(\mathbb{R}^n)} f. \tag{2.7.9}$$

To this end, consider the function u defined by

$$u(x,t) := \begin{cases} f_t(x), & x \in \mathbb{R}^n,\ t > 0, \\ f(x), & x \in \mathbb{R}^n,\ t = 0, \end{cases} \qquad \forall (x,t) \in \mathbb{R}^n \times [0,\infty). \tag{2.7.10}$$

From definition it is clear that u is continuous on $\mathbb{R}^n \times (0, \infty)$. We claim that, in fact, u is continuous on $\mathbb{R}^n \times [0, \infty)$. Indeed, by making use of the change of variables $x - y = 2\sqrt{t}z$, we may write

$$u(x,t) = \pi^{-\frac{n}{2}} \int_{\mathbb{R}^n} e^{-|z|^2} f(x - 2\sqrt{t}z)\,\mathrm{d}z, \qquad \forall x \in \mathbb{R}^n,\ \forall t > 0. \tag{2.7.11}$$

Hence, for each $x_* \in \mathbb{R}^n$, Lebesgue Dominated Convergence Theorem (cf. Theorem 14.15) gives

$$\lim_{\substack{x\to x_* \\ t\to 0^+}} u(x,t) = f(x_*)\pi^{-\frac{n}{2}} \int_{\mathbb{R}^n} e^{-|z|^2}\,\mathrm{d}z = f(x_*) = u(x_*, 0), \tag{2.7.12}$$

proving that u is continuous at points of the form $(x_*, 0)$.

Being continuous on $\mathbb{R}^n \times [0, \infty)$, u is uniformly continuous on every compact subset of $\mathbb{R}^n \times [0, \infty)$, thus uniformly continuous on sets of the form $K \times [0, 1]$, where $K \subset \mathbb{R}^n$ is compact. Fix such a compact K and fix $\varepsilon \in (0, 1)$ arbitrary. Then, there exists $\delta > 0$ such that if $(x_1, t_1), (x_2, t_2) \in K \times [0, 1]$ satisfy $|(x_1, t_1) - (x_2, t_2)| \leq \delta$ then we necessarily have $|u(x_1, t_1) - u(x_2, t_2)| \leq \varepsilon$. In particular, if $x \in K$ and $t \in (0, \delta)$, then $|u(x, t) - u(x, 0)| \leq \varepsilon$, that is $\|f_t - f\|_{L^\infty(K)} \leq \varepsilon$ for all $0 < t < \delta$. This proves that $\lim_{t \to 0^+} f_t(x) = f(x)$ uniformly on compact sets in $\mathbb{R}^n$.

The derivatives of f_t enjoy the same type of properties as f_t. More precisely, by a direct computation (involving also the change of variables $x - y = 2\sqrt{tz}$) we see that for each $t > 0$ and each $\alpha \in \mathbb{N}_0^n$ we have

$$\partial_x^\alpha f_t(x) = (4\pi t)^{-\frac{n}{2}} \int_{\mathbb{R}^n} e^{-\frac{|x-y|^2}{4t}} (\partial^\alpha f)(y) \, dy, \qquad \forall \, x \in \mathbb{R}^n. \tag{2.7.13}$$

In addition, as before, we obtain that $\lim_{t \to 0^+} (\partial_x^\alpha f_t) = \partial^\alpha f$ uniformly on compact sets in $\mathbb{R}^n$. This completes the proof of (2.7.9).

Next, recall that the Taylor expansion of the function e^s, $s \in \mathbb{R}$, about the origin is $e^s = \sum_{j=0}^{\infty} \frac{s^j}{j!}$ with the series converging uniformly on compact subsets of $\mathbb{R}$. In addition, for each $N \in \mathbb{N}$, the remainder $R_N(s) := e^s - \sum_{j=0}^{N} \frac{s^j}{j!}$ satisfies $|R_N(s)| \leq e^C \frac{C^{N+1}}{(N+1)!}$ whenever $|s| \leq C$. Fix $t > 0$ and a compact subset K of $\mathbb{R}^n$. Then there exists $C > 0$ such that $\frac{|x-y|^2}{4t} \leq C$ for every $x \in K$ and every $y \in \text{supp} f$, so

$$\lim_{N \to \infty} \left| R_N\left(-\frac{|x-y|^2}{4t} \right) \right| \leq \lim_{N \to \infty} e^C \frac{C^{N+1}}{(N+1)!} = 0. \tag{2.7.14}$$

Consequently,

$$\partial_x^\alpha f_t(x) = (4\pi t)^{-\frac{n}{2}} \sum_{j=0}^{\infty} \frac{1}{j!} \int_{\mathbb{R}^n} \left(-\frac{|x-y|^2}{4t} \right)^j \partial^\alpha f(y) \, dy, \tag{2.7.15}$$

for each $\alpha \in \mathbb{N}_0^n$ and each $t > 0$, and the series in (2.7.15) converges uniformly for x in a compact set in $\mathbb{R}^n$. In addition, integrating by parts, we may write

$$\partial_x^\alpha f_t(x) = (4\pi t)^{-\frac{n}{2}} \sum_{j=0}^{\infty} \frac{(-1)^{|\alpha|}}{j} \int_{\mathbb{R}^n} \partial_y^\alpha \left[\left(-\frac{|x-y|^2}{4t} \right)^j \right] f(y) \, dy, \tag{2.7.16}$$

where, for each $t > 0$ fixed, the series in (2.7.16) converges uniformly on compact sets in $\mathbb{R}^n$. Hence, if for each $t > 0$ we define the sequence of polynomials

$$P_{t,k}(x) := (4\pi t)^{-\frac{n}{2}} \sum_{j=0}^{k} \frac{1}{j!} \int_{\mathbb{R}^n} \left(-\frac{|x-y|^2}{4t} \right)^j f(y) \, dy, \qquad \forall \, x \in \mathbb{R}^n, \ \forall \, k \in \mathbb{N}, \tag{2.7.17}$$

then the above proof implies that for each $t > 0$ we have $P_{t,k} \xrightarrow[k \to \infty]{\mathscr{E}(\mathbb{R}^n)} f_t$.

Next, we claim that there exists a sequence of positive numbers $\{t_j\}_{j \in \mathbb{N}}$ with the property that for each $j \in \mathbb{N}$ we have

$$\|\partial^\alpha f_{t_j} - \partial^\alpha f\|_{L^\infty(B(0,j))} < \tfrac{1}{2j} \qquad \text{for every } \alpha \in \mathbb{N}_0^n \text{ with } |\alpha| \le j. \tag{2.7.18}$$

To construct a sequence $\{t_j\}_{j \in \mathbb{N}}$ satisfying (2.7.18) we proceed by induction. First, consider the compact set $\overline{B(0,1)}$. For each $\alpha \in \mathbb{N}_0^n$ satisfying $|\alpha| \le 1$, based on (2.7.9), there exists $\ell_\alpha^1 \in \mathbb{N}$ with the property that

$$\left\|\partial^\alpha f_t - \partial^\alpha f\right\|_{L^\infty(B(0,1))} < \tfrac{1}{2} \quad \text{for all } t \in (0, 1/\ell_\alpha^1]. \tag{2.7.19}$$

Define $t_1 := \min\left\{\dfrac{1}{\ell_\alpha^1} : \alpha \in \mathbb{N}_0^n, \ |\alpha| \le 1\right\}$.

Suppose that, for some $j \ge 2$, we have already selected $t_1, \ldots, t_{j-1}$ satisfying (2.7.18). Let $\alpha \in \mathbb{N}_0^n$ be such that $|\alpha| \le j$. Based on (2.7.9), there exists $\ell_\alpha^j \in \mathbb{N}$ with the property that $\ell_\alpha^j \ge \ell_\alpha^{j-1}$ whenever $|\alpha| \le j - 1$, and such that

$$\left\|\partial^\alpha f_t - \partial^\alpha f\right\|_{L^\infty(B(0,j))} < \tfrac{1}{2j} \quad \text{for all } t \in (0, 1/\ell_\alpha^j]. \tag{2.7.20}$$

Now define $t_j := \min\left\{\dfrac{1}{\ell_\alpha^j} : \alpha \in \mathbb{N}_0^n, \ |\alpha| \le j\right\}$. In particular, this choice ensures that $t_j \le t_{j-1}$. Proceeding by induction it follows that the sequence $\{t_j\}_{j \in \mathbb{N}}$ constructed in this manner satisfies (2.7.18).

Our next claim is that for each $t > 0$ and each $j \in \mathbb{N}$ there exists $k_{t,j} \in \mathbb{N}$ such that

$$\|\partial^\alpha P_{t,k_{t,j}} - \partial^\alpha f_t\|_{L^\infty(B(0,j))} < \tfrac{1}{2j} \quad \text{for every } \alpha \in \mathbb{N}_0^n \text{ with } |\alpha| \le j. \tag{2.7.21}$$

To prove this, fix $t > 0$ and $j \in \mathbb{N}$. Since $P_{t,k} \xrightarrow[k \to \infty]{\mathscr{E}(\mathbb{R}^n)} f_t$ and $\overline{B(0,j)}$ is a compact subset of $\mathbb{R}^n$, it follows that for each $\alpha \in \mathbb{N}_0^n$ satisfying $|\alpha| \le j$ there exists $k_\alpha^* \in \mathbb{N}$ such that

$$\|\partial^\alpha P_{t,k} - \partial^\alpha f_t\|_{L^\alpha(B(0,j))} < \tfrac{1}{2j}, \qquad \text{for } k \ge k_\alpha^*. \tag{2.7.22}$$

If we now define $k_{t,j} := \max\{k_\alpha^* : \alpha \in \mathbb{N}_0^n, \ |\alpha| \le j\}$, then estimate (2.7.21) holds for this $k_{t,j}$. This completes the proof the claim.

Here is the endgame in the proof of the lemma. For each $j \in \mathbb{N}$, let $t_j > 0$ be as constructed above so that (2.7.18) holds, for this t_j let $k_{t_j,j}$ be as defined above so that (2.7.21) holds, and set $P_j := P_{t_j,k_{t_j,j}}$. Hence, for each $j \in \mathbb{N}$ and every $\alpha \in \mathbb{N}_0^n$ satisfying $|\alpha| \le j$ we have

$$\|\partial^\alpha P_j - \partial^\alpha f\|_{L^\infty(B(0,j))} = \|\partial^\alpha P_{t_j,k_{t_j,j}} - \partial^\alpha f\|_{L^\infty(B(0,j))}$$
$$\le \tfrac{1}{2j} + \tfrac{1}{2j} = \tfrac{1}{j}. \tag{2.7.23}$$

The fact that $P_j \xrightarrow[j\to\infty]{\mathcal{E}(\mathbb{R}^n)} f$ now follows from (2.7.23) by invoking Lemma 2.82.   □

Before turning to the proof of Proposition 2.81 we introduce some notation. For $m, n \in \mathbb{N}$, if U is an open subset of $\mathbb{R}^m$, V is an open subset of $\mathbb{R}^n$, and $A \subseteq U \times V$, the projections of A on U and V, respectively, are

$$\pi_U(A) := \{x \in U : \exists y \in V \text{ such that } (x, y) \in A\},$$
$$\pi_V(A) := \{y \in V : \exists x \in U \text{ such that } (x, y) \in A\}. \tag{2.7.24}$$

We are ready to present the proof of the density result stated at the beginning of this section.

Proof of Proposition 2.81. Let $\varphi \in C_0^\infty(U \times V)$. By Lemma 2.83, there exists a sequence of polynomials $\{P_j\}_{j\in\mathbb{N}}$ in $\mathbb{R}^{n+m}$ with the property that $P_j \xrightarrow[j\to\infty]{\mathcal{E}(\mathbb{R}^{n+m})} \varphi$. Set $K := \operatorname{supp}\varphi$, $K_1 := \pi_U(K)$, and $K_2 := \pi_V(K)$. Then K_1 and K_2 are compact sets in $\mathbb{R}^m$ and $\mathbb{R}^n$, respectively. Fix a compact set $L_1 \subset U$ such that $K_1 \subset \mathring{L}_1$ and a compact set $L_2 \subset V$ such that $K_2 \subset \mathring{L}_2$. Then there exists a function $\varphi_1 \in C_0^\infty(U)$ with $\operatorname{supp}\varphi_1 \subseteq L_1$, $\varphi_1 \equiv 1$ in a neighborhood of K_1, and a function $\varphi_2 \in C_0^\infty(V)$ satisfying $\operatorname{supp}\varphi_2 \subseteq L_2$ and $\varphi_2 \equiv 1$ in a neighborhood of K_2. Consequently,

$$\varphi_1 \otimes \varphi_2 \in C_0^\infty(\mathbb{R}^{n+m}) \quad \text{and} \quad \operatorname{supp}(\varphi_1 \otimes \varphi_2) \subseteq L_1 \times L_2. \tag{2.7.25}$$

By *(2)* in Exercise 1.10 it follows that

$$(\varphi_1 \otimes \varphi_2)P_j \xrightarrow[j\to\infty]{\mathcal{E}(\mathbb{R}^{n+m})} (\varphi_1 \otimes \varphi_2)\varphi. \tag{2.7.26}$$

Hence, since
$$\operatorname{supp}[(\varphi_1 \otimes \varphi_2)P_j] \subseteq L_1 \times L_2 \text{ for every } j \in \mathbb{N}$$

and since $(\varphi_1 \otimes \varphi_2)\varphi = \varphi$, we obtain

$$(\varphi_1 \otimes \varphi_2)P_j \xrightarrow[j\to\infty]{\mathcal{D}(U\times V)} \varphi. \tag{2.7.27}$$

Upon observing that $(\varphi_1 \otimes \varphi_2)P_j \in C_0^\infty(U) \otimes C_0^\infty(V)$ for every $j \in \mathbb{N}$, the desired conclusion follows.   □

The next proposition is another important ingredient used to define the tensor product of two distributions.

Proposition 2.84. *Fix $m, n \in \mathbb{N}$, let U be an open subset of $\mathbb{R}^m$, and let V be an open subset of $\mathbb{R}^n$. Then for each distribution $u \in \mathcal{D}'(U)$ the following properties hold.*

(a) If for each $\varphi \in C_0^\infty(U \times V)$ we define the mapping

$$\psi : V \to \mathbb{C}, \quad \psi(y) := \langle u(x), \varphi(x, y)\rangle \text{ for all } y \in V, \tag{2.7.28}$$

then $\psi \in C_0^\infty(V)$.

(b) The mapping $\mathcal{D}(U \times V) \ni \varphi \mapsto \psi \in \mathcal{D}(V)$, with ψ as defined in (a), is linear and continuous.

Prior to presenting the proof of this result we make the following remark.

Remark 2.85. In the definition of ψ in part *(a)* of Proposition 2.84, the use of the notation $u(x)$ does NOT mean that the distributions u is evaluated at x since the latter is not meaningful. The notation $\langle u(x), \varphi(x, y) \rangle$ should be understood in the following sense: for each $y \in V$ fixed, the distribution u acts on the function $\varphi(\cdot, y)$.

We now turn to the task of presenting the proof of Proposition 2.84.

Proof of Proposition 2.84. Fix $\varphi \in C_0^\infty(U \times V)$ and let $K := \operatorname{supp} \varphi$ that is a compact subset of $U \times V$. Also, consider ψ as in (2.7.28) and recall the projections π_U, π_V from (2.7.24). Then clearly $\operatorname{supp} \psi \subseteq \pi_V(K)$, thus ψ has compact support. Next we prove that ψ is continuous on V. Let $\{y_j\}_{j \in \mathbb{N}}$ be a sequence in V such that $\lim\limits_{j \to \infty} y_j = y_0$ for some $y_0 \in V$. Since $u \in \mathcal{D}'(U)$, based on the definition of ψ and Fact 2.2, in order to conclude that $\lim\limits_{j \to \infty} \psi(y_j) = \psi(y_0)$ it suffices to show that $\varphi(\cdot, y_j) \xrightarrow[j \to \infty]{\mathcal{D}(U)} \varphi(\cdot, y_0)$. It is clear that for every $j \in \mathbb{N}$ we have $\varphi(\cdot, y_j) \in C_0^\infty(U)$ and $\operatorname{supp} \varphi(\cdot, y_j) \subseteq \pi_U(K)$. Moreover, since $\varphi \in C^\infty(U \times V)$ it follows that $\partial_x^\alpha \varphi$ is continuous on K for every $\alpha \in \mathbb{N}_0^m$, thus uniformly continuous on K. Consequently,

$$(\partial_x^\alpha \varphi)(\cdot, y_j) \xrightarrow[j \to \infty]{} (\partial_x^\alpha \varphi)(\cdot, y_0) \quad \text{uniformly on } \pi_U(K).$$

This completes the proof of the fact that ψ is continuous on V.

To continue, we claim that ψ is of class C^1 on V. Fix $y \in V$ and j in $\{1, ..., n\}$. Recall that $\mathbf{e}_j$ is the unit vector in $\mathbb{R}^n$ with the j-th component equal to 1, and let $h \in \mathbb{R} \setminus \{0\}$. Since V is open, there exists $\varepsilon_0 > 0$ such that if $|h| < \varepsilon_0$ then $y + h\mathbf{e}_j \in V$. Make the standing assumption that $|h| < \varepsilon_0$ and set

$$R_h(x, y) := \frac{\varphi(x, y + h\mathbf{e}_j) - \varphi(x, y)}{h} - \frac{\partial \varphi}{\partial y_j}(x, y), \qquad \forall\, x \in U. \tag{2.7.29}$$

Then

$$\frac{\psi(y + h\mathbf{e}_j) - \psi(y)}{h} - \left\langle u(x), \frac{\partial \varphi}{\partial y_j}(x, y) \right\rangle = \langle u(x), R_h(x, y) \rangle, \qquad \forall\, x \in U. \tag{2.7.30}$$

Suppose

$$\lim_{h \to 0} R_h(\cdot, y) = 0 \quad \text{in} \quad \mathcal{D}(U). \tag{2.7.31}$$

Then $\lim\limits_{h \to 0} \langle u, R_h(\cdot, y) \rangle = 0$, which in view of (2.7.30) implies

$$\partial_j \psi(y) = \left\langle u, \frac{\partial \varphi}{\partial y_j}(\cdot, y) \right\rangle. \tag{2.7.32}$$

Moreover, since $\frac{\partial \varphi}{\partial y_j} \in C_0^\infty(U \times V)$ by reasoning as in the proof of the continuity of ψ on V, we also obtain that $\partial_j \psi$ is continuous on V. Hence, since $j \in \{1, ..., n\}$ is arbitrary, to complete the proof of the claim, we are left with showing (2.7.31).

Clearly supp $[R_h(\cdot, y)] \subseteq \pi_U(K)$. Applying Taylor's formula to φ in the variable y for each fixed $x \in U$ we obtain

$$\varphi(x, y + he_j) = \varphi(x, y) + h\frac{\partial \varphi}{\partial y_j}(x, y) + h^2 \int_0^1 (1 - t)\frac{\partial^2 \varphi}{\partial y_j^2}(x, y + the_j)\, dt. \quad (2.7.33)$$

Hence, (2.7.29) and (2.7.33) imply

$$R_h(x, y) = h \int_0^1 (1 - t)\frac{\partial^2 \varphi}{\partial y_j^2}(x, y + the_j)\, dt. \quad (2.7.34)$$

Consequently, for every $\beta \in \mathbb{N}_0^m$, we have

$$\partial_x^\beta R_h(x, y) = h \int_0^1 (1 - t)\partial_x^\beta \frac{\partial^2 \varphi}{\partial y_j^2}(x, y + the_j)\, dt, \qquad \forall\, x \in U. \quad (2.7.35)$$

Since the integral in the right-hand side of (2.7.35) is bounded by a constant independent of h, x and y, it follows that $\lim_{h \to 0} \partial_x^\beta R_h(\cdot, y) = 0$ uniformly on $\pi_U(K)$. Combined with the support information on $R_h(\cdot, y)$, this implies (2.7.31) and completes the proof of the claim that $\psi \in C^1(V)$. By induction, we obtain $\psi \in C^\infty(V)$, completing the proof of the statement in part (a) of the proposition.

The linearity of the mapping in part (b) is immediate since u is a linear mapping. To show that the mapping in (b) is also continuous, since $\mathcal{D}(V)$ is locally convex, by Theorem 14.6 it suffices to prove that it is sequentially continuous. To this end, let $\varphi_j \xrightarrow[j \to \infty]{\mathcal{D}(U \times V)} \varphi$. In particular, there exists a compact subset K of $U \times V$ such that supp $\varphi_j \subseteq K$ for all $j \in \mathbb{N}$ and

$$\partial^\alpha \varphi_j \xrightarrow[j \to \infty]{} \partial^\alpha \varphi \quad \text{uniformly on } K, \text{ for every } \alpha \in \mathbb{N}_0^{m+n}. \quad (2.7.36)$$

To proceed, for every $y \in V$ set $\psi(y) := \langle u, \varphi(\cdot, y)\rangle$ and $\psi_j(y) := \langle u, \varphi_j(\cdot, y)\rangle$ for each $j \in \mathbb{N}$. The goal is to prove that $\psi_j \xrightarrow[j \to \infty]{\mathcal{D}(V)} \psi$. Applying Proposition 2.4 to the distribution u and compact $\pi_U(K)$ yields $k \in \mathbb{N}_0$ and $C > 0$ for which (2.1.1) holds with K replaced by $\pi_U(K)$. Then, for every $\beta \in \mathbb{N}_0^n$, we have

$$\sup_{y\in\pi_V(K)} \left|\partial^\beta\psi_j(y) - \partial^\beta\psi(y)\right| = \sup_{y\in\pi_V(K)} \left|\langle u(x), \partial^\beta_y\varphi_j(\cdot,y) - \partial^\beta_y\varphi(\cdot,y)\rangle\right|$$

$$\leq \sup_{\substack{y\in\pi_V(K)\\|\gamma|\leq k}} \sup_{x\in\pi_U(K)} \left|\partial^\gamma_x\partial^\beta_y\varphi_j(x,y) - \partial^\gamma_x\partial^\beta_y\varphi(x,y)\right|$$

$$= C \sup_{\substack{(x,y)\in K\\|\gamma|\leq k}} \left|\partial^\gamma_x\partial^\beta_y\varphi_j(x,y) - \partial^\gamma_x\partial^\beta_y\varphi(x,y)\right| \xrightarrow[j\to\infty]{} 0, \qquad (2.7.37)$$

where $\gamma \in \mathbb{N}_0^m$ and the convergence to zero in (2.7.37) is due to (2.7.36) applied for $\alpha := (\gamma, \beta)$. Thus, $\psi_j \xrightarrow[j\to\infty]{\mathcal{D}(V)} \psi$ and the proof of the statement in part *(b)* is complete.

$\square$

Remark 2.86. Let $m, n \in \mathbb{N}$, and consider an open subset U of $\mathbb{R}^m$ along with an open subset V of $\mathbb{R}^n$.

(1) For each $\varphi \in C^\infty(V \times U)$ define the function

$$\varphi^\top : U \times V \to \mathbb{C}, \quad \varphi^\top(x,y) := \varphi(y,x), \quad \forall x \in U, \forall y \in V. \qquad (2.7.38)$$

Then $\varphi^\top \in C^\infty(U \times V)$. In addition, if φ has compact support in $V \times U$ then $\varphi^\top$ also has compact support in $U \times V$. Moreover, by invoking Fact 1.15, Theorem 14.6, and Fact 1.16, we obtain that the mapping

$$\mathcal{D}(V \times U) \ni \varphi \mapsto \varphi^\top \in \mathcal{D}(U \times V) \qquad (2.7.39)$$

is linear and continuous.

(2) Let $v \in \mathcal{D}'(V)$ and for each $\varphi \in C_0^\infty(U \times V)$ define the mapping

$$\eta : U \to \mathbb{C}, \quad \eta(x) := \langle v(y), \varphi(x,y) \rangle \text{ for all } x \in U. \qquad (2.7.40)$$

Then $\eta \in C_0^\infty(U)$ and the mapping $\mathcal{D}(U \times V) \ni \varphi \mapsto \eta \in \mathcal{D}(U)$ is linear and continuous. Indeed, this follows by applying Proposition 2.84 with the current $V, U, v, \varphi^\top, x, y$ in place of U, V, u, φ, y, x.

(3) The fact that the map in (2.7.39) is linear and continuous yields the following result:

if $w \in \mathcal{D}'(U \times V)$, then the map $w^\top : \mathcal{D}(V \times U) \to \mathbb{C}$ defined by $w^\top(\varphi) := \langle w, \varphi^\top \rangle$ for each $\varphi \in \mathcal{D}(V \times U)$ is actually a distribution on $V \times U$, i.e., $w^\top \in \mathcal{D}'(V \times U)$.

$$(2.7.41)$$

We are now ready to define the tensor product of distributions.

Theorem 2.87. *Let* $m, n \in \mathbb{N}$, *U be an open subset of $\mathbb{R}^m$, and V be an open subset of $\mathbb{R}^n$. Consider $u \in \mathcal{D}'(U)$ and $v \in \mathcal{D}'(V)$. Then the following statements are true.*

(i) There exists a unique distribution $u \otimes v \in \mathcal{D}'(U \times V)$, called the tensor product *of u and v, with the property that*

$$\langle u \otimes v, \varphi_1 \otimes \varphi_2 \rangle = \langle u, \varphi_1 \rangle \langle v, \varphi_2 \rangle$$

$$\forall \varphi_1 \in C_0^\infty(U), \ \forall \varphi_2 \in C_0^\infty(V). \tag{2.7.42}$$

(ii) The action of the distributions $u \otimes v \in \mathcal{D}'(U \times V)$ is given by

$$\langle u \otimes v, \varphi \rangle = \langle v(y), \langle u(x), \varphi(x,y) \rangle \rangle \tag{2.7.43}$$

$$= \langle u(x), \langle v(y), \varphi(x,y) \rangle \rangle \ \text{for each} \ \varphi \in C_0^\infty(U \times V).$$

(iii) The tensor product just defined satisfies $u \otimes v = (v \otimes u)^\top$ in $\mathcal{D}'(U \times V)$, where $(v \otimes u)^\top$ is the distribution defined in (2.7.41) corresponding to u replaced by $v \otimes u$.

Proof. For each $\varphi \in C_0^\infty(U \times V)$ consider the function

$$\psi(y) := \langle u(x), \varphi(x,y) \rangle \quad \text{for} \quad y \in V.$$

By Proposition 2.84, we have $\psi \in C_0^\infty(V)$ and the mapping

$$\mathcal{D}(U \times V) \ni \varphi \mapsto \psi \in \mathcal{D}(V) \quad \text{is linear and continuous.} \tag{2.7.44}$$

Hence, $\langle v, \psi \rangle$ is meaningful and we may define

$$u \otimes v : \mathcal{D}(U \times V) \longrightarrow \mathbb{C}$$

$$\langle u \otimes v, \varphi \rangle := \langle v(y), \langle u(x), \varphi(x,y) \rangle \rangle \quad \text{for every} \quad \varphi \in C_0^\infty(U \times V). \tag{2.7.45}$$

As defined, the mapping $u \otimes v$ is the composition of two linear and continuous mappings, hence it is linear and continuous. Also, if $\varphi_1 \in C_0^\infty(U)$ and $\varphi_2 \in C_0^\infty(V)$, then

$$\langle u \otimes v, \varphi_1 \otimes \varphi_2 \rangle = \langle v(y), \langle u(x), \varphi_1(x)\varphi_2(y) \rangle \rangle$$

$$= \langle v(y), \varphi_2(y)\langle u(x), \varphi_1(x) \rangle \rangle$$

$$= \langle v(y), \varphi_2(y) \rangle \langle u(x), \varphi_1(x) \rangle, \tag{2.7.46}$$

thus the mapping $u \otimes v$ defined in (2.7.45) satisfies (2.7.42).

To prove the uniqueness statement in part *(i)*, suppose $w_1, w_2 \in \mathcal{D}'(U \times V)$ are such that

$$\langle w_j, \varphi_1 \otimes \varphi_2 \rangle = \langle u, \varphi_1 \rangle \langle v, \varphi_2 \rangle, \quad j = 1, 2,$$

$$\text{for every} \quad \varphi_1 \in C_0^\infty(U), \ \varphi_2 \in C_0^\infty(V). \tag{2.7.47}$$

Then it follows that $\langle w_1, \varphi \rangle = \langle w_2, \varphi \rangle$ for every $\varphi \in C_0^\infty(U) \otimes C_0^\infty(V)$, which in concert with Proposition 2.81 and the continuity of w_1 and w_2 implies $w_1 = w_2$ in $\mathcal{D}'(U \times V)$. This completes the proof of the statement in *(i)*.

The reasoning used to prove the statement in *(i)*, this time relying on *(2)* in Remark 2.86 in place of Proposition 2.84, also yields that the mapping

$$w : \mathcal{D}(U \times V) \longrightarrow \mathbb{C}$$

$$\langle w, \varphi \rangle := \big\langle u(x), \langle v(y), \varphi(x, y)\rangle \big\rangle \quad \text{for every} \quad \varphi \in C_0^\infty(U \times V) \tag{2.7.48}$$

is linear and continuous and satisfies

$$\langle w, \varphi_1 \otimes \varphi_2 \rangle = \langle v(y), \varphi_2(y)\rangle \langle u(x), \varphi_1(x)\rangle, \tag{2.7.49}$$

for every $\varphi_1 \in C_0^\infty(U)$ and every $\varphi_2 \in C_0^\infty(V)$. The uniqueness result proved in *(i)* then gives $u \otimes v = w$. Hence, (2.7.43) holds as wanted.

As for the statement in *(iii)*, observe that based on (2.7.45) we have

$$v \otimes u : \mathcal{D}(V \times U) \to \mathbb{C} \quad \text{and}$$

$$\langle v \otimes u, \psi \rangle = \langle u(x), \langle v(y), \psi(y, x)\rangle\rangle \quad \text{for every} \quad \psi \in C_0^\infty(V \times U). \tag{2.7.50}$$

Hence, $(v \otimes u)^\top : \mathcal{D}(U \times V) \to \mathbb{C}$ and

$$\langle (v \otimes u)^\top, \varphi \rangle = \langle v \otimes u, \varphi^\top \rangle = \langle u(x), \langle v(y), \varphi^\top(y, x)\rangle\rangle$$

$$= \langle u(x), \langle v(y), \varphi(x, y)\rangle\rangle \quad \text{for every} \quad \varphi \in C_0^\infty(U \times V). \tag{2.7.51}$$

In particular, for every $\varphi_1 \in C_0^\infty(U)$ and $\varphi_2 \in C_0^\infty(U)$ we have

$$\langle (v \otimes u)^\top, \varphi_1 \otimes \varphi_2 \rangle = \langle u, \varphi_1\rangle\langle v, \varphi_2\rangle = \langle u \otimes v, \varphi_1 \otimes \varphi_2\rangle. \tag{2.7.52}$$

The uniqueness result from part *(i)* now implies $u \otimes v = (v \otimes u)^\top$ in $\mathcal{D}'(U \times V)$.   □

Remark 2.88. If $u \in \mathcal{D}'(U)$ and $v \in L_{loc}^1(V)$, then the statement in part *(iii)* in Theorem 2.87 becomes

$$\Big\langle u(x), \int_V v(y)\varphi(x, y)\,dy \Big\rangle = \int_V v(y)\langle u(x), \varphi(x, y)\rangle\,dy, \quad \forall \varphi \in C_0^\infty(U \times V). \tag{2.7.53}$$

The interpretation of (2.7.53) is that the distribution u commutes with the integral.

We next establish a number of basic properties for the tensor products of distributions.

Theorem 2.89. *Let* $m, n \in \mathbb{N}$, *U be an open subset of* $\mathbb{R}^m$, *and V be an open subset of* $\mathbb{R}^n$. *Assume that* $u \in \mathcal{D}'(U)$ *and* $v \in \mathcal{D}'(V)$. *Then the following properties hold.*

(a) $supp\, u \otimes v = supp\, u \times supp\, v$.
(b) $\partial_x^\alpha \partial_y^\beta(u \otimes v) = (\partial_x^\alpha u) \otimes (\partial_y^\beta v)$ *for every* $\alpha \in \mathbb{N}_0^m$ *and every* $\beta \in \mathbb{N}_0^n$.
(c) $(f \otimes g) \cdot (u \otimes v) = (fu) \otimes (gv)$ *for every* $f \in C^\infty(U)$ *and every* $g \in C^\infty(V)$.
(d) *The mapping* $\mathcal{D}'(U) \times \mathcal{D}'(V) \ni (u, v) \mapsto u \otimes v \in \mathcal{D}'(U \times V)$ *is bilinear and separately sequentially continuous.*

(e) The tensor product of distributions is associative.

Proof. We start by proving the set theoretic equality from *(a)*. For the right-to-left inclusion, fix $(x_0, y_0) \in \operatorname{supp} u \times \operatorname{supp} v$. If $C \subseteq U \times V$ is an open neighborhood of (x_0, y_0), then there exists an open set $A \subseteq U$ containing x_0 and an open set $B \subseteq V$ containing y_0 such that $A \times B \subset C$. In particular, since $x_0 \in \operatorname{supp} u$ and $y_0 \in \operatorname{supp} v$, there exist $\varphi_1 \in C_0^\infty(A)$ and $\varphi_2 \in C_0^\infty(B)$ with the property that $\langle u, \varphi_1 \rangle \neq 0$ and $\langle v, \varphi_2 \rangle \neq 0$. If we now set $\varphi := \varphi_1 \otimes \varphi_2$, then $\varphi \in C_0^\infty(C)$ and $\langle u \otimes v, \varphi \rangle = \langle u, \varphi_1 \rangle \langle v, \varphi_2 \rangle \neq 0$. Hence $(x_0, y_0) \in \operatorname{supp}(u \otimes v)$, finishing the proof of the right-to-left inclusion in *(a)*.

To prove the opposite inclusion, observe that $\operatorname{supp}(u \otimes v) \subseteq \operatorname{supp} u \times \operatorname{supp} v$ is equivalent to

$$(U \times V) \setminus (\operatorname{supp} u \times \operatorname{supp} v) \subseteq (U \times V) \setminus \operatorname{supp}(u \otimes v). \tag{2.7.54}$$

Write the left-hand side of (2.7.54) as $D_1 \cup D_2$, where $D_1 := (U \setminus \operatorname{supp} u) \times V$ and $D_2 := U \times (V \setminus \operatorname{supp} v)$. Note that D_1 and D_2 are open sets in $\mathbb{R}^m \times \mathbb{R}^n$. Since the support of a distribution is the smallest relatively closed set outside of which the distribution vanishes, for (2.7.54) to hold it suffices to show that $\langle u \otimes v, \varphi \rangle = 0$ for every $\varphi \in C_0^\infty(D_1 \cup D_2)$. Fix such a function φ, set $K := \operatorname{supp} \varphi$, and consider a partition of unity subordinate to the covering $\{D_1, D_2\}$ of K, say

$$\psi_j \in C_0^\infty(D_j), \ j \in \{1, 2\}, \quad \psi_1 + \psi_2 = 1 \text{ in a neighborhood of } K. \tag{2.7.55}$$

Then $\varphi \psi_1 \in C_0^\infty(D_1)$, $\varphi \psi_2 \in C_0^\infty(D_2)$ (with the understanding that ψ_1 and ψ_2 have been extended by zero outside their supports), and $\varphi = \varphi \psi_1 + \varphi \psi_2$ on $U \times V$. Since $\pi_U(D_1) \cap \operatorname{supp} u = \varnothing$ and $\pi_V(D_2) \cap \operatorname{supp} v = \varnothing$, we may write

$$\langle u \otimes v, \varphi \rangle = \langle v(y), \langle u(x), \varphi(x, y) \psi_1(x, y) \rangle \rangle$$

$$+ \langle u(x), \langle v(y), \varphi(x, y) \psi_2(x, y) \rangle \rangle = 0. \tag{2.7.56}$$

This completes the proof of the equality of sets from part *(a)*.

To prove the identity in *(b)*, fix $\alpha \in \mathbb{N}_0^m, \beta \in \mathbb{N}_0^n$, and let $\varphi_1 \in C_0^\infty(U), \varphi_2 \in C_0^\infty(V)$. Then starting with the definition of distributional derivatives and then using (2.7.42) we may write

$$\langle \partial_x^\alpha \partial_y^\beta (u \otimes v), \varphi_1 \otimes \varphi_2 \rangle = (-1)^{|\alpha| + |\beta|} \langle u \otimes v, \partial_x^\alpha \partial_y^\beta (\varphi_1 \otimes \varphi_2) \rangle$$

$$= (-1)^{|\alpha| + |\beta|} \langle u \otimes v, \partial_x^\alpha \varphi_1 \otimes \partial_y^\beta \varphi_2 \rangle$$

$$= (-1)^{|\alpha| + |\beta|} \langle u, \partial_x^\alpha \varphi_1 \rangle \langle v, \partial_y^\beta \varphi_2 \rangle$$

$$= \langle \partial_x^\alpha u, \varphi_1 \rangle \langle \partial_y^\beta u, \varphi_2 \rangle$$

$$= \langle (\partial_x^\alpha u) \otimes (\partial_y^\beta v), \varphi_1 \otimes \varphi_2 \rangle. \tag{2.7.57}$$

By the uniqueness statement in part *(i)* of Theorem 2.87 we deduce from (2.7.57) that $(\partial_x^\alpha u) \otimes (\partial_y^\beta v) = \partial_x^\alpha \partial_y^\beta (u \otimes v)$, completing the proof of the identity in *(b)*.

Moving on to the proof of the statement in *(c)*, note that if $f \in C^\infty(U)$ and $g \in C^\infty(V)$, then $f \otimes g \in C^\infty(U \times V)$. The latter, combined with the definition of multiplication of a distribution with a smooth function and (2.7.42), permits us to write

$$\langle (f \otimes g) \cdot (u \otimes v), \varphi_1 \otimes \varphi_2 \rangle = \langle u \otimes v, (f \otimes g) \cdot (\varphi_1 \otimes \varphi_2) \rangle \tag{2.7.58}$$

$$= \langle u \otimes v, (f\varphi_1) \otimes (g\varphi_2) \rangle = \langle u, f\varphi_1 \rangle \langle v, g\varphi_2 \rangle$$

$$= \langle fu, \varphi_1 \rangle \langle gv, \varphi_2 \rangle = \langle (fu) \otimes (gv), \varphi_1 \otimes \varphi_2 \rangle,$$

for every $\varphi_1 \in C_0^\infty(U)$, $\varphi_2 \in C_0^\infty(V)$. The identity in *(c)* now follows from (2.7.58) by once again invoking the uniqueness result from part *(i)* in Theorem 2.87.

The bilinearity of the mapping $(u, v) \mapsto u \otimes v$ is a consequence of the definition of $u \otimes v$ and (2.2.1). To prove that this mapping is also separately sequentially continuous, let $u_j \xrightarrow[j \to \infty]{\mathcal{D}'(U)} u$ and fix $v \in \mathcal{D}'(V)$. If $\varphi \in C_0^\infty(U \times V)$, then $\langle v(y), \varphi(x, y) \rangle \in C_0^\infty(U)$ by Proposition 2.84, and we may use Fact 2.22 to write

$$\langle u_j \otimes v, \varphi \rangle = \langle u_j(x), \langle v(y), \varphi(x, y) \rangle \rangle \xrightarrow[j \to \infty]{} \langle u(x), \langle v(y), \varphi(x, y) \rangle \rangle = \langle u \otimes v, \varphi \rangle. \tag{2.7.59}$$

Similarly, if $u \in \mathcal{D}'(U)$ is fixed and $v_j \xrightarrow[j \to \infty]{\mathcal{D}'(V)} v$, then $u \otimes v_j \xrightarrow[j \to \infty]{\mathcal{D}'(U \times V)} u \otimes v$.

Finally, we are left with proving the associativity of the tensor product of distributions. To this end, let $k \in \mathbb{N}$, W be an open subset of $\mathbb{R}^k$, and $w \in \mathcal{D}'(W)$ be arbitrary. By Theorem 2.87, $u \otimes v \in \mathcal{D}'(U \times V)$, $v \otimes w \in \mathcal{D}'(V \times W)$ and, furthermore,

$$(u \otimes v) \otimes w \in \mathcal{D}'(U \times V \times W), \qquad u \otimes (v \otimes w) \in \mathcal{D}'(U \times V \times W). \tag{2.7.60}$$

The goal is to prove that

$$(u \otimes v) \otimes w = u \otimes (v \otimes w) \quad \text{in} \quad \mathcal{D}'(U \times V \times W). \tag{2.7.61}$$

In this regard we first note that for each $\varphi \in C_0^\infty(U)$, $\psi \in C_0^\infty(V)$, and $\eta \in C_0^\infty(W)$, we may write

$$\langle (u \otimes v) \otimes w, (\varphi \otimes \psi) \otimes \eta \rangle = \langle u, \varphi \rangle \langle v, \psi \rangle \langle w, \eta \rangle$$

$$= \langle u \otimes (v \otimes w), \varphi \otimes (\varphi \otimes \eta) \rangle. \tag{2.7.62}$$

Define $C_0^\infty(U) \otimes C_0^\infty(V) \otimes C_0^\infty(W)$ as

$$\Big\{ \sum_{j=1}^N \varphi_j \otimes \psi_j \otimes \eta_j : \varphi_j \in C_0^\infty(U), \psi_j \in C_0^\infty(V), \eta_j \in C_0^\infty(W), N \in \mathbb{N} \Big\}, \tag{2.7.63}$$

and note that this set is sequentially dense in $\mathcal{D}(U \times V \times W)$ (which can be proved by reasoning as in the proof of Proposition 2.81). Granted this, (2.7.61) is implied by (2.7.62), completing the proof of the theorem. $\qquad\square$

Exercise 2.90. Let $n, m \in \mathbb{N}$ and pick $x_0 \in \mathbb{R}^n$ and $y_0 \in \mathbb{R}^m$ arbitrary. Prove that $\delta_{x_0} \otimes \delta_{y_0} = \delta_{(x_0, y_0)}$ in $\mathcal{D}'(\mathbb{R}^{n+m})$.

We close this section by revisiting the result proved in Proposition 2.84 and establishing a related version that is going to be useful later on.

Proposition 2.91. *Let $m, n \in \mathbb{N}$, U be an open subset of $\mathbb{R}^m$, and V be an open subset of $\mathbb{R}^n$. Assume that $u \in \mathcal{E}'(U)$, $\varphi \in C^\infty(U \times V)$, and define the function*

$$\psi : V \to \mathbb{C}, \qquad \psi(y) := \langle u(x), \varphi(x, y) \rangle, \qquad \forall y \in V. \qquad (2.7.64)$$

Then $\psi \in C^\infty(V)$ and for every $\alpha \in \mathbb{N}_0^n$ we have

$$\partial^\alpha \psi(y) = \langle u(x), \partial_y^\alpha \varphi(x, y) \rangle, \qquad \forall y \in V. \qquad (2.7.65)$$

Proof. Fix some $\eta \in C_0^\infty(U)$ that satisfies $\eta \equiv 1$ in a neighborhood of $\operatorname{supp} u$. Then for each $\theta \in C_0^\infty(V)$ we may write

$$(\theta\psi)(y) = \langle (\eta u)(x), \theta(y)\varphi(x, y) \rangle = \langle u(x), (\eta \otimes \theta)(x, y)\varphi(x, y) \rangle$$

$$= \langle u(x), [(\eta \otimes \theta)\varphi](x, y) \rangle, \qquad \forall y \in V. \qquad (2.7.66)$$

Given that $(\eta \otimes \theta)\varphi \in C_0^\infty(U \times V)$, Proposition 2.84 applies and gives that the rightmost side of (2.7.66) depends in a C^∞ manner on the variable $y \in V$. Hence, $\theta\psi \in C^\infty(V)$ and since $\theta \in C_0^\infty(V)$ has been arbitrarily chosen we deduce that $\psi \in C^\infty(V)$. This takes care of the first claim in the statement of the proposition.

Moving on, observe that it suffices to prove (2.7.65) when $|\alpha| = 1$, since the general case then follows by iteration. With this in mind, fix some $j \in \{1, \ldots, n\}$ and pick an arbitrary point $y^* \in V$. Also, select $\theta \in C_0^\infty(V)$ such that $\theta \equiv 1$ near y^*. These properties of θ permit us to compute

$$\partial_{y_j}[(\eta \otimes \theta)\varphi](x, y^*) = \eta(x)\{(\partial_j\theta)(y^*)\varphi(x, y^*) + \theta(y^*)(\partial_{y_j}\varphi)(x, y^*)\}$$

$$= \eta(x)(\partial_{y_j}\varphi)(x, y^*), \qquad \forall x \in U. \qquad (2.7.67)$$

Making use of (2.7.66), (2.7.32), and (2.7.67), we may then write

$$\partial_j\psi(y^*) = \partial_j(\theta\psi)(y^*) = \langle u(x), \partial_{y_j}[(\eta \otimes \theta)\varphi](x, y^*) \rangle$$

$$= \langle u(x), \eta(x)(\partial_{y_j}\varphi)(x, y^*) \rangle$$

$$= \langle u(x), (\partial_{y_j}\varphi)(x, y^*) \rangle. \qquad (2.7.68)$$

This corresponds precisely to formula (2.7.65) written at the point $y = y^*$ and for the multi-index $\alpha = (0, \ldots, 0, 1, 0, \ldots, 0) \in \mathbb{N}_0^n$ with the nonzero component on the j-th slot. As remarked earlier, this suffices to finish the proof. $\qquad\square$

2.8 The Convolution of Distributions in $\mathbb{R}^n$

Recall that, as a consequence of Fubini's Theorem, given any $f, g \in L^1(\mathbb{R}^n)$, the function $h : \mathbb{R}^n \times \mathbb{R}^n \to \mathbb{C}$ defined by $h(x, y) := f(x - y)g(y)$ for every point $(x, y) \in \mathbb{R}^n \times \mathbb{R}^n$, is absolutely integrable on $\mathbb{R}^n \times \mathbb{R}^n$ and

$$
\iint_{\mathbb{R}^n \times \mathbb{R}^n} |h(x, y)| \, dx \, dy = \int_{\mathbb{R}^n} \int_{\mathbb{R}^n} |f(x - y)g(y)| \, dx \, dy
$$
$$
= \int_{\mathbb{R}^n} |g(y)| \Big(\int_{\mathbb{R}^n} |f(x - y)| \, dx \Big) dy
$$
$$
= \|f\|_{L^1(\mathbb{R}^n)} \|g\|_{L^1(\mathbb{R}^n)}. \tag{2.8.1}
$$

Hence, the `convolution` of f and g defined as

$$
f * g : \mathbb{R}^n \to \mathbb{C}, \quad (f * g)(x) := \int_{\mathbb{R}^n} f(x - y)g(y) \, dy \quad \text{for each } x \in \mathbb{R}^n, \tag{2.8.2}
$$

satisfies $f * g \in L^1(\mathbb{R}^n)$ (and a natural estimate). We would like to extend this definition to functions that are not necessarily in $L^1(\mathbb{R}^n)$. Specifically, assume that $f, g \in L^1_{loc}(\mathbb{R}^n)$ have the property that

$$
M_{B(0,r)} := \{(x, y) \in \operatorname{supp} f \times \operatorname{supp} g : x + y \in \overline{B(0, r)}\} \tag{2.8.3}
$$
$$
\text{is a compact set in } \mathbb{R}^n \times \mathbb{R}^n \text{ for every } r \in (0, \infty).
$$

In this scenario, consider the function $G : \mathbb{R}^n \to [0, \infty]$ defined by

$$
G(x) := \int_{\mathbb{R}^n} |f(x - y)| \, |g(y)| \, dy \quad \text{for each} \quad x \in \mathbb{R}^n. \tag{2.8.4}
$$

Note that for every $r \in (0, \infty)$, by making a natural change of variables and using the fact that $f \otimes g \in L^1_{loc}(\mathbb{R}^n \times \mathbb{R}^n)$, we obtain

$$
\int_{|x| \leq r} G(x) \, dx = \iint_{M_{B(0,r)}} |f(z)| |g(y)| \, dy \, dz < \infty. \tag{2.8.5}
$$

Thus the function G is locally integrable, hence finite almost everywhere in $\mathbb{R}^n$.

In conclusion, whenever $f, g \in L^1_{loc}(\mathbb{R}^n)$ satisfy (2.8.3), for almost every $x \in \mathbb{R}^n$ the integral $(f * g)(x) := \int_{\mathbb{R}^n} f(x - y)g(y) \, dy$ is absolutely convergent and we have $f * g \in L^1_{loc}(\mathbb{R}^n)$. Furthermore, having fixed an arbitrary $\varphi \in C_0^\infty(\mathbb{R}^n)$ we may write

$$
\langle f * g, \varphi \rangle = \int_{\mathbb{R}^n} \int_{\mathbb{R}^n} f(x - y)g(y)\varphi(x) \, dy \, dx
$$
$$
= \int_{\mathbb{R}^n} \int_{\mathbb{R}^n} f(z)g(y)\varphi(z + y) \, dy \, dz. \tag{2.8.6}
$$

To proceed, observe that the function φ^Δ defined by

$$\varphi^\Delta : \mathbb{R}^n \times \mathbb{R}^n \to \mathbb{C}, \qquad \varphi^\Delta(x, y) := \varphi(x + y) \quad \text{for every} \quad x, y \in \mathbb{R}^n, \qquad (2.8.7)$$

satisfies $\varphi^\Delta \in C^\infty(\mathbb{R}^n \times \mathbb{R}^n)$ though, in general, the support of φ^Δ is not compact. Formally, the last double integral in (2.8.6) has the same expression as $\langle f \otimes g, \varphi^\Delta \rangle$. However, under the current assumptions on φ, f, and g, it is not clear that this pairing may be interpreted in the standard distributional sense. Indeed, even though $f \otimes g$ is a well-defined distribution in $\mathcal{D}'(\mathbb{R}^n \times \mathbb{R}^n)$ (cf. Theorem 2.87), the function φ^Δ does not belong to $C_0^\infty(\mathbb{R}^n \times \mathbb{R}^n)$, as it lacks the compact support property. Nonetheless, (2.8.3) implies

$$\operatorname{supp} \varphi^\Delta \cap \operatorname{supp} (f \otimes g) \quad \text{is a compact set in } \mathbb{R}^n \times \mathbb{R}^n. \qquad (2.8.8)$$

Theorem 2.60 applies with $F := \operatorname{supp}(f \otimes g)$ and allows us to uniquely extend the action of the distribution $f \otimes g$ to the set of functions $\psi \in C^\infty(\mathbb{R}^n \times \mathbb{R}^n)$ satisfying the property that $\operatorname{supp} \psi \cap \operatorname{supp}(f \otimes g)$ is a compact set in $\mathbb{R}^n \times \mathbb{R}^n$. Denote this unique extension by $\widetilde{f \otimes g}$. Then $\langle \widetilde{f \otimes g}, \varphi^\Delta \rangle$ is well defined, and it is meaningful to set $\langle f * g, \varphi \rangle := \langle \widetilde{f \otimes g}, \varphi^\Delta \rangle$. This discussion justifies making the following definition.

Definition 2.92. Suppose u, $v \in \mathcal{D}'(\mathbb{R}^n)$ are such that

for every compact subset K of $\mathbb{R}^n$ the set
$$M_K := \{(x, y) \in \operatorname{supp} u \times \operatorname{supp} v : x + y \in K\} \text{ is compact in } \mathbb{R}^n \times \mathbb{R}^n. \qquad (2.8.9)$$

Granted this, define the convolution of the distributions u and v as the functional $u * v : \mathcal{D}(\mathbb{R}^n) \to \mathbb{C}$ whose action on each $\varphi \in C_0^\infty(\mathbb{R}^n)$ is given by

$$\langle u * v, \varphi \rangle := \langle \widetilde{u \otimes v}, \varphi^\Delta \rangle \qquad (2.8.10)$$

where $\varphi^\Delta(x, y) := \varphi(x + y)$ for every $x, y \in \mathbb{R}^n$, and $\widetilde{u \otimes v}$ is the unique extension of $u \otimes v$ obtained by applying Theorem 2.60 with $F := \operatorname{supp}(u \otimes v)$.

Remark 2.93. Retain the context of Definition 2.92.

(1) If $\varphi \in C_0^\infty(\mathbb{R}^n)$ and $\psi \in C_0^\infty(\mathbb{R}^n \times \mathbb{R}^n)$ is such that $\psi \equiv 1$ in a neighborhood of $M_{\operatorname{supp}\varphi}$ then
$$\langle u * v, \varphi \rangle = \langle u \otimes v, \psi \varphi^\Delta \rangle. \qquad (2.8.11)$$

(2) If (2.8.9) holds for the compacts $K_j = \overline{B(0, j)}$, $j \in \mathbb{N}$, then (2.8.9) holds for arbitrary compact sets $K \subset \mathbb{R}^n$.

(3) Condition (2.8.9) is always satisfied if either u or v is compactly supported.

The issue of continuity of the convolution map introduced in Definition 2.92 is discussed next.

Theorem 2.94. *If u, $v \in \mathcal{D}'(\mathbb{R}^n)$ are such that (2.8.9) holds, then $u * v$ belongs to $\mathcal{D}'(\mathbb{R}^n)$. In particular, the convolution between two distributions in $\mathbb{R}^n$, one of which is compactly supported, is always well defined and is a distribution in $\mathbb{R}^n$.*

Proof. Let $u, v \in \mathscr{D}'(\mathbb{R}^n)$ be such that (2.8.9) is satisfied. From Theorem 2.60, we have that $\widetilde{u \otimes v}$ is linear, hence $u*v$ is linear as well. Let $\varphi_j \xrightarrow[j\to\infty]{\mathscr{D}(\mathbb{R}^n)} 0$. Then there exists a compact subset K of $\mathbb{R}^n$ such that $\operatorname{supp}\varphi_j \subseteq K$ for each $j \in \mathbb{N}$, and $\lim\limits_{j\to\infty} \partial^\alpha \varphi_j = 0$ uniformly on K, for every $\alpha \in \mathbb{N}_0^n$. In particular, $M_{\operatorname{supp}\varphi_j} \subseteq M_K$ for every $j \in \mathbb{N}$. Hence, if we fix $\psi \in C_0^\infty(\mathbb{R}^n \times \mathbb{R}^n)$ such that $\psi \equiv 1$ in a neighborhood of M_K, then part *(1)* in Remark 2.93 gives

$$\langle u * v, \varphi_j \rangle = \langle u \otimes v, \psi \varphi_j^{\Delta} \rangle, \qquad \forall\, j \in \mathbb{N}. \tag{2.8.12}$$

Moreover, we claim that

$$\psi \varphi_j^{\Delta} \xrightarrow[j\to\infty]{\mathscr{D}(\mathbb{R}^n \times \mathbb{R}^n)} 0. \tag{2.8.13}$$

To prove this claim, note that $\operatorname{supp}(\psi\,\varphi_j^{\Delta}) \subseteq \operatorname{supp}\psi$ for every $j \in \mathbb{N}$ and for every $\alpha_1, \alpha_2, \beta_1, \beta_2 \in \mathbb{N}_0^n$ we have

$$\sup_{(x,y)\in\operatorname{supp}\psi} \left| (\partial_x^{\alpha_1} \partial_y^{\beta_1} \psi)(x,y)(\partial_x^{\alpha_2} \partial_y^{\beta_2} \varphi_j^{\Delta})(x,y) \right|$$

$$\leq \sup_{(x,y)\in\operatorname{supp}\psi} \left| \partial_x^{\alpha_1} \partial_y^{\beta_1} \psi(x,y) \right| \| \partial^{\alpha_2 + \beta_2} \varphi_j \|_{L^\infty(\mathbb{R}^n)}$$

$$\leq \| \partial_x^{\alpha_1} \partial_y^{\beta_1} \psi \|_{L^\infty(\operatorname{supp}\psi)} \| \partial^{\alpha_2+\beta_2} \varphi_j \|_{L^\infty(K)} \xrightarrow[j\to\infty]{} 0. \tag{2.8.14}$$

Hence, (2.8.13) follows by combining (2.8.14) with Leibniz's formula (14.2.6). Since $u \otimes v \in \mathscr{D}'(\mathbb{R}^n \times \mathbb{R}^n)$, from (2.8.12) and (2.8.13) we deduce

$$\langle u * v, \varphi_j \rangle = \langle u \otimes v, \psi \varphi_j^{\Delta} \rangle \xrightarrow[j\to\infty]{} 0. \tag{2.8.15}$$

On account of Remark 2.3, this proves that $u * v \in \mathscr{D}'(\mathbb{R}^n)$. Finally, the last statement of the theorem is a consequence of what we proved so far and part *(3)* in Remark 2.93. $\qquad\square$

Remark 2.95. Combined, Theorem 2.94 and the discussion regarding (2.8.6) yield the following result. If $f, g \in L_{loc}^1(\mathbb{R}^n)$ are such that for each compact $K \subset \mathbb{R}^n$ the set $\{(x,y) : x \in \operatorname{supp} f,\ y \in \operatorname{supp} g,\ x + y \in K\}$ is compact, then $u_f * u_g = u_{f*g}$ in $\mathscr{D}'(\mathbb{R}^n)$. That is, $f * g \in L_{loc}^1(\mathbb{R}^n)$, the convolution between the distributions u_f and u_g (recall (2.1.6)) is well defined, and $u_f * u_g$ is a distribution of function type that is equal to the distribution u_{f*g}. In particular,

$$\left.\begin{array}{c} f \in L_{loc}^1(\mathbb{R}^n),\ \ g \in L_{comp}^1(\mathbb{R}^n) \\ \text{or} \\ f \in L_{comp}^1(\mathbb{R}^n),\ \ g \in L_{loc}^1(\mathbb{R}^n) \end{array}\right\} \implies u_f * u_g = u_{f*g}. \tag{2.8.16}$$

The main properties of the convolution of distributions, whenever meaningfully defined, are stated and proved in the next theorem. Recall that for any $A, B \subseteq \mathbb{R}^n$ the set $A \pm B$ is defined as $\{x \pm y : x \in A, \ y \in B\}$.

Theorem 2.96. *The following statements are true.*

(a) If $u, v \in \mathcal{D}'(\mathbb{R}^n)$ are two distributions with the property that (2.8.9) is satisfied, then $supp\,(u * v) \subseteq supp\,u + supp\,v$.
(b) If $u, v \in \mathcal{D}'(\mathbb{R}^n)$ are such that (2.8.9) is satisfied, then $u * v = v * u$.
(c) If $u, v, w \in \mathcal{D}'(\mathbb{R}^n)$ are such that

$$\begin{cases} \textit{for every compact subset } K \textit{ of } \mathbb{R}^n \textit{ the set} \\ M'_K := \{(x, y, z) \in supp\,u \times supp\,v \times supp\,w : x + y + z \in K\} \\ \textit{is compact in } \mathbb{R}^n \times \mathbb{R}^n \times \mathbb{R}^n, \end{cases} \qquad (2.8.17)$$

then $(u * v) * w$ and $u * (v * w)$ are well defined, belong to $\mathcal{D}'(\mathbb{R}^n)$, and are equal.
(d) Let $u \in \mathcal{D}'(\mathbb{R}^n)$, $\alpha \in \mathbb{N}_0^n$. Then $(\partial^\alpha u) * \delta = \partial^\alpha u = u * \partial^\alpha \delta$. In particular, $u * \delta = u$.
(e) If the distributions $u, v \in \mathcal{D}'(\mathbb{R}^n)$ are such that (2.8.9) is satisfied and $\alpha \in \mathbb{N}_0^n$, then $\partial^\alpha(u * v) = (\partial^\alpha u) * v = u * (\partial^\alpha v)$.

Proof. Let $u, v \in \mathcal{D}'(\mathbb{R}^n)$ be such that (2.8.9) holds. Since $supp\,u + supp\,v$ is closed, the inclusion in *(a)* will follow as soon as we show that

$$u * v \big|_{\mathbb{R}^n \setminus (supp\,u + supp\,v)} = 0. \qquad (2.8.18)$$

Pick an arbitrary function $\varphi \in \mathcal{D}(\mathbb{R}^n \setminus (supp\,u + supp\,v))$. Then

$$supp\,\varphi^\Delta \cap supp\,(u \otimes v) = \varnothing, \qquad (2.8.19)$$

hence $\langle \widetilde{u \otimes v}, \varphi^\Delta \rangle = 0$, which implies $\langle u * v, \varphi \rangle = 0$, as wanted.

Next we show the statement in *(b)*. Take $u, v \in \mathcal{D}'(\mathbb{R}^n)$ for which (2.8.9) holds and let $\varphi \in C_0^\infty(\mathbb{R}^n)$. Choose some $\psi \in C_0^\infty(\mathbb{R}^n \times \mathbb{R}^n)$ with the property that $\psi \equiv 1$ in a neighborhood of

$$M_1 = \{(x, y) \in supp\,u \times supp\,v : x + y \in supp\,\varphi\}. \qquad (2.8.20)$$

By Remark 2.93 and *(iii)* in Theorem 2.87 we have

$$\langle u * v, \varphi \rangle = \langle u \otimes v, \psi \varphi^\Delta \rangle = \langle (v \otimes u)^\top, \psi \varphi^\Delta \rangle$$
$$= \langle v \otimes u, \psi^\top (\varphi^\Delta)^\top \rangle. \qquad (2.8.21)$$

Definition (2.7.38) in our current setting implies $\psi^\top \in C_0^\infty(\mathbb{R}^n \times \mathbb{R}^n)$ and $\psi^\top \equiv 1$ in a neighborhood of

$$M_2 = \{(y, x) \in supp\,v \times supp\,u : x + y \in supp\,\varphi\}. \qquad (2.8.22)$$

Also, it is immediate that $(\varphi^\Delta)^\top = \varphi^\Delta$. Hence, by Remark 2.93, we see that

$$\langle v \otimes u, \psi^\top (\varphi^\mathcal{A})^\top \rangle = \langle v \otimes u, \psi^\top \varphi^\mathcal{A} \rangle = \langle v * u, \varphi \rangle. \tag{2.8.23}$$

From (2.8.21) and (2.8.23) it follows that $\langle u * v, \varphi \rangle = \langle v * u, \varphi \rangle$, finishing the proof of the statement in *(b)*.

To prove the statement in *(c)*, suppose $u, v, w \in \mathcal{D}'(\mathbb{R}^n)$ satisfy (2.8.17). Define the functional $u * v * w : \mathcal{D}(\mathbb{R}^n) \to \mathbb{C}$ by setting

$$\langle u * v * w, \varphi \rangle := \langle \widetilde{u \otimes v \otimes} w, \varphi^\mathcal{A} \rangle, \qquad \text{for every } \varphi \in C_0^\infty(\mathbb{R}^n), \tag{2.8.24}$$

where $\varphi^\mathcal{A}(x, y, z) := \varphi(x + y + z)$ for each $x, y, x \in \mathbb{R}^n$, and where we have denoted by $\widetilde{u \otimes v \otimes} w$ the unique extension of $u \otimes v \otimes w$ obtained by applying Theorem 2.60 for $F := \operatorname{supp}(u \otimes v \otimes w)$. The mapping in (2.8.24) is well defined since if φ is an arbitrary test function in $C_0^\infty(\mathbb{R}^n)$ then $\varphi^\mathcal{A} \in C^\infty(\mathbb{R}^n \times \mathbb{R}^n \times \mathbb{R}^n)$ and, based on (2.8.17), the set

$$M'_{\operatorname{supp}\varphi} = \operatorname{supp}(u \otimes v \otimes w) \cap \operatorname{supp}\varphi^\mathcal{A} \quad \text{is compact in } \mathbb{R}^n \times \mathbb{R}^n \times \mathbb{R}^n. \tag{2.8.25}$$

Reasoning as in the proof of Theorem 2.94, it follows that $u * v * w$ belongs to $\mathcal{D}'(\mathbb{R}^n)$ and

$$\langle u * v * w, \varphi \rangle = \langle u \otimes v \otimes w, \psi \varphi^\mathcal{A} \rangle \quad \text{for } \varphi \in C_0^\infty(\mathbb{R}^n) \text{ and for each}$$
$$\psi \in C_0^\infty(\mathbb{R}^n \times \mathbb{R}^n \times \mathbb{R}^n) \text{ with } \psi \equiv 1 \text{ in a neighborhood of } M'_{\operatorname{supp}\varphi}. \tag{2.8.26}$$

Given the freedom in selecting ψ as in (2.8.26), we choose to take ψ as follows. Let $\pi_j : \mathbb{R}^n \times \mathbb{R}^n \times \mathbb{R}^n \to \mathbb{R}^n$, $j = 1, 2, 3$, be the projections defined for each $x, y, z \in \mathbb{R}^n$ by $\pi_1((x, y, z)) := x$, $\pi_2((x, y, z)) := y$, and $\pi_3((x, y, z)) := z$. Given a function $\varphi \in C_0^\infty(\mathbb{R}^n)$, fix

$$\psi_j \in C_0^\infty(\mathbb{R}^n) \text{ with } \psi_j \equiv 1 \text{ near } \pi_j(M'_{\operatorname{supp}\varphi}), \quad j = 1, 2, 3, \tag{2.8.27}$$

then choose

$$\psi := \psi_1 \otimes \psi_2 \otimes \psi_3 \in C_0^\infty(\mathbb{R}^n \times \mathbb{R}^n \times \mathbb{R}^n). \tag{2.8.28}$$

The next two claims are designed to prove that $u * (v * w)$ exists.

Claim 1. *For every compact K in $\mathbb{R}^n$ the set*

$$N_K := \{(y, z) \in \operatorname{supp} v \times \operatorname{supp} w : y + z \in K\}$$

is compact in $\mathbb{R}^n \times \mathbb{R}^n$.

To see why this is true, start by observing that, for every $x_0 \in \operatorname{supp} u$, the set $K + x_0$ is compact in $\mathbb{R}^n$ and

$$B := \{(x, y, z) \in \{x_0\} \times \operatorname{supp} v \times \operatorname{supp} w : x + y + z \in K + x_0\} \tag{2.8.29}$$

is closed and contained in M'_{K+x_0}. Since (2.8.17) ensures that M'_{K+x_0} is compact in $\mathbb{R}^n \times \mathbb{R}^n \times \mathbb{R}^n$, it follows that B is compact in $\mathbb{R}^n \times \mathbb{R}^n \times \mathbb{R}^n$. In addition, the mapping

$\theta : \mathbb{R}^n \times \mathbb{R}^n \times \mathbb{R}^n \to \mathbb{R}^n \times \mathbb{R}^n$, defined by $\theta(x, y, z) := (y, z)$ for each $x, y, z \in \mathbb{R}^n$, is continuous and $\theta(B) = N_K$. Therefore, N_K must be compact, and Claim 1 is proved. The latter ensures that $v * w$ exists.

Claim 2. *For every $K \subset \mathbb{R}^n$ compact, the set*

$$P_K := \{(x, z) \in \text{supp}\, u \times \text{supp}\,(v * w) : x + z \in K\}$$

is compact in $\mathbb{R}^n \times \mathbb{R}^n$.

By part *(a)* in the theorem, we have $\text{supp}\,(v * w) \subseteq \text{supp}\, v + \text{sup}\, w$. Thus

$$P_K \subseteq \{(x, z) \in \text{supp}\, u \times (\text{supp}\, v + \text{supp}\, w) : x + z \in K\}$$

$$= \{(x, y + t) : x \in \text{supp}\, u, \ y \in \text{supp}\, v, \ t \in \text{supp}\, w, \ x + y + t \in K\} \quad (2.8.30)$$

and the last set in (2.8.30) is closed in $\mathbb{R}^n \times \mathbb{R}^n$. If we now set

$$\sigma : \mathbb{R}^n \times \mathbb{R}^n \times \mathbb{R}^n \to \mathbb{R}^n \times \mathbb{R}^n,$$
$$\sigma(x, y, t) := (x, y + t) \quad \text{for every } x, y, t \in \mathbb{R}^n, \quad (2.8.31)$$

then σ is continuous and $P_K \subseteq \sigma(M'_K)$. By (2.8.17), we have that M'_K is compact in $\mathbb{R}^n \times \mathbb{R}^n \times \mathbb{R}^n$, hence $\sigma(M'_K)$ is compact in $\mathbb{R}^n \times \mathbb{R}^n$. The set P_K being closed in $\mathbb{R}^n \times \mathbb{R}^n$, we may conclude that P_K is compact in $\mathbb{R}^n \times \mathbb{R}^n$. This proves Claim 2 and, as a consequence, the fact that $u * (v * w)$ exists.

With an eye toward proving $u * (v * w) = u * v * w$, we dispense of two more claims.

Claim 3. *Let K be an arbitrary compact subset of $\mathbb{R}^n$ and for u, v, w as before introduce the set $A := (\text{supp}\, v + \text{supp}\, w) \cap (K - \text{supp}\, u)$. The A is also compact in $\mathbb{R}^n$.*

Rewrite A as

$$A = \{t \in \text{supp}\, v + \text{supp}\, w : t = \omega - x, \ \text{for some } \omega \in K, \ x \in \text{supp}\, u\}.$$

Then if $t \in A$, it follows that there exist $y \in \text{supp}\, v$, $z \in \text{supp}\, w$, $\omega \in K$ and $x \in \text{supp}\, u$ such that $t = y + z = \omega - x$. Hence, $(x, y, z) \in \text{supp}\, u \times \text{supp}\, v \times \text{supp}\, w$ and $x + y + z = \omega$, which implies that $(x, y, z) \in M'_K$ and $A \subseteq \nu(M'_K)$, where

$$\nu : \mathbb{R}^n \times \mathbb{R}^n \times \mathbb{R}^n \to \mathbb{R}^n,$$
$$\nu(x_1, x_2, x_3) := x_2 + x_3 \quad \text{for every } x_1, x_2, x_3 \in \mathbb{R}^n. \quad (2.8.32)$$

We may now conclude that A is compact since the map ν is continuous, M'_K is compact and A is closed. This proves Claim 3.

Claim 4. *Fix $\varphi \in C_0^\infty(\mathbb{R}^n)$ and set $K := \text{supp}\, \varphi$. Also, let A be as in Claim 3 corresponding to this K, and suppose $\eta \in C_0^\infty(\mathbb{R}^n)$ is such that $\eta \equiv 1$ in a neighborhood* *of A. Then, with ψ_1 as in* (2.8.27), *we have*

$$\psi_1 \otimes \eta = 1 \quad \text{on} \quad (\operatorname{supp} u \times \operatorname{supp}(v * w)) \cap \operatorname{supp} \varphi^\Delta \tag{2.8.33}$$

where, as before, $\varphi^\Delta(x, y) = \varphi(x + y)$ *for* $x, y \in \mathbb{R}^n$.

To prove this claim, since $\psi_1 \equiv 1$ on $\pi_1(M_K')$ and $\eta \equiv 1$ on A, it suffices to show that

$$(\operatorname{supp} u \times \operatorname{supp}(v * w)) \cap \operatorname{supp} \varphi^\Delta \subseteq \pi_1(M_K') \times A. \tag{2.8.34}$$

To justify the latter inclusion, let $x \in \operatorname{supp} u$, $y \in \operatorname{supp} v$, and $z \in \operatorname{supp} w$, be such that $(x, y + z) \in \operatorname{supp} \varphi^\Delta$. Then $x + y + z \in K$ which forces $x \in \pi_1(M_K')$ as well as $y + z \in K - \operatorname{supp} u \subseteq A$. Thus, $(x, y + z) \in \pi_1(M_K') \times A$ and this completes the proof of Claim 4.

Consider now an arbitrary function $\varphi \in C_0^\infty(\mathbb{R}^n)$ and set $K := \operatorname{supp} \varphi$. Also, assume that ψ_1, ψ_2, ψ_3 are as in (2.8.27), and let η be as in Claim 4. Making use of the definition of the convolution and tensor products, and keeping in mind (2.8.33), we may write

$$\langle u * (v * w), \varphi \rangle = \big\langle u(x) \otimes (v * w)(t), \psi_1(x)\eta(t)\varphi(x + t) \big\rangle \tag{2.8.35}$$

$$= \big\langle u(x), \langle (v * w)(t), \psi_1(x)\eta(t)\varphi(x + t) \rangle \big\rangle$$

$$= \big\langle u(x), \langle v(y) \otimes w(z), \psi_1(x)\eta(y + z)\varphi(x + y + z)\psi_2(y)\psi_3(z) \rangle \big\rangle.$$

A few words explaining the origin of the last equality are in order. According to the definition of the convolution, passing from $v * w$ to $v \otimes w$ requires that we consider the set

$$C := (\operatorname{supp} v \times \operatorname{supp} w) \cap \operatorname{supp} \eta^\Delta \cap \operatorname{supp}[\varphi(x + \cdot)]^\Delta. \tag{2.8.36}$$

Since C is closed and satisfies $C \subset \pi_2(M_K') \times \pi_3(M_K')$, it follows that C is compact. Now, the fact that $\psi_2 \otimes \psi_3 \equiv 1$ in a neighborhood of C justifies the presence of $\psi_2 \otimes \psi_3$ in the last term in (2.8.35).

Going further, since

$$\psi_1(x)\psi_2(y)\psi_3(z) = \psi_1(x)\psi_2(y)\psi_3(z)\eta(y + z) \quad \text{for } (x, y, z) \text{ near } M_K', \tag{2.8.37}$$

referring to (2.8.26) and (2.8.28) allows us to rewrite (2.8.35) in the form

$$\langle u * (v * w), \varphi \rangle = \big\langle u(x), \langle v(y) \otimes w(z), \psi_1(x)\psi_2(y)\psi_3(z)\varphi(x + y + z) \rangle \big\rangle$$

$$= \big\langle u \otimes v \otimes w, (\psi_1 \otimes \psi_2 \otimes \psi_3)\varphi^\Delta \big\rangle = \langle u * v * w, \varphi \rangle. \tag{2.8.38}$$

Since $\varphi \in C_0^\infty(\mathbb{R}^n)$ was arbitrary, it follows that $u * (v * w) = u * v * w$. Similarly, it can be seen that $(u * v) * w = u * v * w$ and this completes the proof of the statement in *(c)*.

Moving on to *(d)*, fix $u \in \mathcal{D}'(\mathbb{R}^n)$ and $\alpha \in \mathbb{N}_0^n$. Since the Dirac distribution δ has compact support, by Theorem 2.94 it follows that both $\delta * u$ and $(\partial^\alpha \delta) * u$ are well defined and belong to $\mathcal{D}'(\mathbb{R}^n)$. Let $\varphi \in C_0^\infty(\mathbb{R}^n)$ and consider a function $\psi \in C_0^\infty(\mathbb{R}^n \times \mathbb{R}^n)$ with $\psi = 1$ on a neighborhood of the set $(\{0\} \times \text{supp}\, u) \cap \text{supp}\, \varphi^\Delta$. Starting with Definition 2.92, then using (2.4.1) combined with Proposition 2.39, then Leibniz's formula (14.2.6), then (2.1.19), the support condition for ψ and then (2.4.1) again, we may write

$$\langle \partial^\alpha \delta * u, \varphi \rangle = \langle (\partial_x^\alpha \delta) \otimes u(y), \psi(x,y)\varphi(x+y) \rangle$$

$$= \Big\langle u(y), \langle \partial^\alpha \delta(x), \psi(x,y)\varphi(x+y) \rangle \Big\rangle$$

$$= (-1)^{|\alpha|} \Big\langle u(y), \langle \delta(x), \partial_x^\alpha(\psi(x,y)\varphi(x+y)) \rangle \Big\rangle$$

$$= (-1)^{|\alpha|} \Big\langle u(y), \Big\langle \delta(x), \sum_{\beta \leq \alpha} \frac{\alpha!}{\beta!(\alpha-\beta)!} \partial_x^\beta \psi(x,y)(\partial_x^{\alpha-\beta}\varphi)(x+y) \Big\rangle \Big\rangle$$

$$= (-1)^{|\alpha|} \Big\langle u(y), \sum_{\beta \leq \alpha} \frac{\alpha!}{\beta!(\alpha-\beta)!} (\partial_x^\beta \psi)(0,y)(\partial_x^{\alpha-\beta}\varphi)(y) \Big\rangle$$

$$= (-1)^{|\alpha|} \langle u(y), \psi(0,y)\partial^\alpha\varphi(y) \rangle$$

$$= (-1)^{|\alpha|} \langle u, \partial^\alpha\varphi \rangle = \langle \partial^\alpha u, \varphi \rangle. \tag{2.8.39}$$

In particular, if $|\alpha| = 0$, the above implies $\delta * u = u$. When combined with *(b)*, this finishes the proof of the statement in *(d)*.

Finally, by making use of the results from *(d)* and *(c)* we have

$$\partial^\alpha(u * v) = \partial^\alpha(\delta * (u * v)) = \partial^\alpha \delta * (u * v) = (\partial^\alpha \delta * u) * v$$

$$= (\partial^\alpha u) * v. \tag{2.8.40}$$

A similar argument also shows that $\partial^\alpha(u * v) = u * (\partial^\alpha v)$. The proof of the theorem is now complete. $\qquad\square$

Exercise 2.97. Prove that for a distribution $u \in \mathcal{D}'(\mathbb{R}^n)$,

$$u = \delta \iff u * f = f \quad \text{for each} \quad f \in C_0^\infty(\mathbb{R}^n). \tag{2.8.41}$$

Hint: For the right-to-left implication use $f = \phi_j$, where ϕ_j is as in Example 2.24, and let $j \to \infty$.

Next, we extend the translation map (1.3.17) to distributions.

Proposition 2.98. *For each $x_0 \in \mathbb{R}^n$ and each $u \in \mathcal{D}'(\mathbb{R}^n)$ fixed, the translation mapping $\mathcal{D}(\mathbb{R}^n) \ni \varphi \mapsto \langle u, t_{-x_0}(\varphi) \rangle \in \mathbb{C}$ is linear and continuous.*

Denoting this map by $t_{x_0} u$ thus yields a distribution in $\mathbb{R}^n$ that satisfies

$$\langle t_{x_0} u, \varphi \rangle = \langle u, t_{-x_0}(\varphi) \rangle, \qquad \forall \varphi \in C_0^\infty(\mathbb{R}^n). \tag{2.8.42}$$

Proof. This follows by observing that the mapping in question is the composition $u \circ t_{-x_0}$ where the latter translation operator is consider in the sense of Exercise 1.19.
□

Exercise 2.99. Fix $x_0 \in \mathbb{R}^n$ and recall the distribution δ_{x_0} from Example 2.14. Prove that $\delta_{x_0} = t_{x_0} \delta$ in $\mathcal{D}'(\mathbb{R}^n)$. Also show that $\delta_{x_0} * u = t_{x_0} u$ for every $u \in \mathcal{D}'(\mathbb{R}^n)$. In particular, if $x_1 \in \mathbb{R}^n$ is arbitrary, then $\delta_{x_0} * \delta_{x_1} = \delta_{x_0+x_1}$ in $\mathcal{D}'(\mathbb{R}^n)$.

Remark 2.100.
(1) If $u, v \in \mathcal{E}'(\mathbb{R}^n)$ then $u * v \in \mathcal{E}'(\mathbb{R}^n)$. This is an immediate consequence of part *(a)* in Theorem 2.96.
(2) There exists a sequence $\{u_j\}_{j \in \mathbb{N}}$ of compactly supported distributions in $\mathbb{R}$ that converges to some $u \in \mathcal{D}'(\mathbb{R})$ and such that $u_j * v$ does not necessarily converge to $u * v$ in $\mathcal{D}'(\mathbb{R})$ for every $v \in \mathcal{D}'(\mathbb{R})$.

To see an example in this regard, consider the sequence of compactly supported distributions $\{\delta_j\}_{j \in \mathbb{N}}$ that satisfies $\delta_j \xrightarrow[j \to \infty]{\mathcal{D}'(\mathbb{R})} 0$. Then if 1 denotes the distribution on $\mathbb{R}$ given by the constant function 1, Exercise 2.99 gives $\delta_j * 1 = 1$ for each j, and the constant sequence $\{1\}_{j \in \mathbb{N}} \subset \mathcal{D}'(\mathbb{R})$ does not converge in $\mathcal{D}'(\mathbb{R})$ to $0 * 1 = 0$. This shows that sequential continuity for convolution of distributions cannot be expected in general.
(3) Condition (2.8.17) is necessary for the operation of convolution of distributions to be associative. To see this, consider the distributions $1, \delta'$, and H on $\mathbb{R}$. Then we have $\operatorname{supp} \delta' = \{0\}$, $\operatorname{supp} 1 = \mathbb{R}$, $\operatorname{supp} H = [0, \infty)$. If K is a compact set in $\mathbb{R}$, the set

$$M_K' = \{(x, 0, z) : x \in \mathbb{R},\ z \in [0, \infty),\ x + z \in K\} \tag{2.8.43}$$

is not compact in $\mathbb{R} \times \mathbb{R} \times \mathbb{R}$, thus (2.8.17) does not hold. Furthermore, $1 * \delta' = 1' * \delta = 0 * \delta = 0$ so $(1 * \delta') * H = 0$, while $1 * (\delta' * H) = 1 * (\delta * H') = 1 * (\delta * \delta) = 1 * \delta = 1$ and clearly $(1 * \delta') * H \neq 1 * (\delta' * H)$ in $\mathcal{D}'(\mathbb{R})$.

Proposition 2.101. *The following statements are true.*

(1) If $u \in \mathcal{E}'(\mathbb{R}^n)$ and $v_j \xrightarrow[j \to \infty]{\mathcal{D}'(\mathbb{R}^n)} v$, then $u * v_j \xrightarrow[j \to \infty]{\mathcal{D}'(\mathbb{R}^n)} u * v$.

(2) If $u_j \xrightarrow[j \to \infty]{\mathcal{D}'(\mathbb{R}^n)} u$ and there exists $K \subset \mathbb{R}^n$ compact with $\operatorname{supp} u_j \subseteq K$ for every

$j \in \mathbb{N}$, then $u_j * v \xrightarrow[j \to \infty]{\mathcal{D}'(\mathbb{R}^n)} u * v$ for every $v \in \mathcal{D}'(\mathbb{R}^n)$.

Proof. To see why *(1)* is true, fix $\varphi \in C_0^\infty(\mathbb{R}^n)$. Then by definition, for each $j \in \mathbb{N}$ we have $\langle u * v_j, \varphi \rangle = \langle u \otimes v_j, \psi_j \varphi^\Delta \rangle$ for any smooth compactly supported function ψ_j with $\psi_j = 1$ on a neighborhood of $(\operatorname{supp} u \times \operatorname{supp} v_j) \cap \operatorname{supp} \varphi^\Delta$. Note that

$$(\operatorname{supp} u \times \operatorname{supp} v_j) \cap \operatorname{supp} \varphi^\Delta \subseteq \operatorname{supp} u \times (\operatorname{supp} \varphi - \operatorname{supp} u) \tag{2.8.44}$$

and $\operatorname{supp} u \times (\operatorname{supp} \varphi - \operatorname{supp} u)$ is compact since both $\operatorname{supp} u$ and $\operatorname{supp} \varphi$ are compact. Hence, if we fix $\psi \in C_0^\infty(\mathbb{R}^n \times \mathbb{R}^n)$ such that $\psi = 1$ in a neighborhood of the set

$\operatorname{supp} u \times (\operatorname{supp} \varphi - \operatorname{supp} u)$, then $\langle u * v_j, \varphi \rangle = \langle u \otimes v_j, \psi \varphi^{\Delta} \rangle$ for every $j \in \mathbb{N}$. Based on (d) in Theorem 2.89, we may write

$$\langle u * v_j, \varphi \rangle = \langle u \otimes v_j, \psi \varphi^{\Delta} \rangle \xrightarrow[j \to \infty]{} \langle u \otimes v, \psi \varphi^{\Delta} \rangle = \langle u * v, \varphi \rangle, \tag{2.8.45}$$

and the desired conclusion follows.

Assume next the hypotheses in part (2) of the proposition. In particular, these entail $\operatorname{supp} u \subseteq K$. Let $\varphi \in C_0^{\infty}(\mathbb{R}^n)$ and note that

$$(\operatorname{supp} u_j \times \operatorname{supp} v) \cap \operatorname{supp} \varphi^{\Delta} \subseteq K \times (\operatorname{supp} \varphi - K), \quad \forall j \in \mathbb{N}. \tag{2.8.46}$$

Then if $\psi \in C_0^{\infty}(\mathbb{R}^n \times \mathbb{R}^n)$ is a function with the property that $\psi = 1$ in a neighborhood of $K \times (\operatorname{supp} \varphi - K)$, we have $\langle u_j * v, \varphi \rangle = \langle u_j \otimes v, \psi \varphi^{\Delta} \rangle$ for every $j \in \mathbb{N}$. Hence,

$$\langle u_j * v, \varphi \rangle = \langle u_j \otimes v, \psi \varphi^{\Delta} \rangle \xrightarrow[j \to \infty]{} \langle u \otimes v, \psi \varphi^{\Delta} \rangle = \langle u * v, \varphi \rangle, \tag{2.8.47}$$

where for the convergence in (2.8.47) we used part (d) in Theorem 2.89. $\qquad \square$

When convolving an arbitrary distribution with a distribution of function type given by a compactly supported smooth function, the resulting distribution is of function type. This fact is particularly useful in applications and we prove it next.

Proposition 2.102. *If $u \in \mathcal{D}'(\mathbb{R}^n)$ and $g \in C_0^{\infty}(\mathbb{R}^n)$, then the distribution $u * g$ is of function type given by the function*

$$f : \mathbb{R}^n \to \mathbb{C}, \qquad f(x) := \langle u(y), g(x - y) \rangle, \qquad \forall x \in \mathbb{R}^n, \tag{2.8.48}$$

that satisfies $f \in C^{\infty}(\mathbb{R}^n)$. Moreover, if u is compactly supported then so is f. In short,

$$\mathcal{D}'(\mathbb{R}^n) * C_0^{\infty}(\mathbb{R}^n) \subseteq C^{\infty}(\mathbb{R}^n) \quad \text{and} \quad \mathcal{E}'(\mathbb{R}^n) * C_0^{\infty}(\mathbb{R}^n) \subseteq C_0^{\infty}(\mathbb{R}^n). \tag{2.8.49}$$

Proof. Let $\phi : \mathbb{R}^n \times \mathbb{R}^n \to \mathbb{C}$ be defined by $\phi(x, y) := g(x - y)$ for each point $(x, y) \in \mathbb{R}^n \times \mathbb{R}^n$. Then $\phi \in C^{\infty}(\mathbb{R}^n \times \mathbb{R}^n)$ and $\phi(x, \cdot) \in C_0^{\infty}(\mathbb{R}^n)$ for each $x \in \mathbb{R}^n$. This shows that the function f in (2.8.48) is well defined.

To prove that f is of class C^{∞} in $\mathbb{R}^n$, pick an arbitrary point $x^* \in \mathbb{R}^n$ and pick a function $\psi \in C_0^{\infty}(\mathbb{R}^n)$ with the property that $\psi = 1$ in a neighborhood of $\overline{B(x^*, 1)}$. Then, for every $x \in B(x^*, 1)$,

$$\left(f \big|_{B(x^*, 1)} \right)(x) = \langle u(y), g(x - y) \rangle = \langle u(y), \psi(x) g(x - y) \rangle. \tag{2.8.50}$$

Since the function $\mathbb{R}^n \times \mathbb{R}^n \ni (x, y) \mapsto \psi(x) g(x - y) \in \mathbb{C}$ is of class C_0^{∞}, we may invoke Proposition 2.84 to conclude that $f \big|_{B(x^*, 1)} \in C^{\infty}(B(x^*, 1))$. Given that $x^* \in \mathbb{R}^n$ has been arbitrarily chosen, it follows that $f \in C^{\infty}(\mathbb{R}^n)$.

We now turn to the task of showing that the distribution $u * g$ is of function type and is given by f. To this end, fix $\varphi \in C_0^{\infty}(\mathbb{R}^n)$ and consider a function

$\psi \in C_0^\infty(\mathbb{R}^n \times \mathbb{R}^n)$ such that $\psi = 1$ in a neighborhood of

the set $\{(x, y) : x \in \operatorname{supp} u,\ y \in \operatorname{supp} g,\ x + y \in \operatorname{supp} \varphi\}$. $\qquad$ (2.8.51)

Then, starting with Definition 2.92 we write

$$\langle u * g, \varphi \rangle = \langle u \otimes g, \psi \varphi^\Delta \rangle$$

$$= \big\langle u(x), \langle g(y), \psi(x, y) \varphi(x + y) \rangle \big\rangle$$

$$= \Big\langle u(x), \int_{\mathbb{R}^n} g(y) \psi(x, y) \varphi(x + y)\, dy \Big\rangle$$

$$= \Big\langle u(x), \int_{\mathbb{R}^n} g(y) \varphi(x + y)\, dy \Big\rangle$$

$$= \Big\langle u(x), \int_{\mathbb{R}^n} g(z - x) \varphi(z)\, dz \Big\rangle$$

$$= \int_{\mathbb{R}^n} \langle u(x), g(z - x) \rangle \varphi(z)\, dz$$

$$= \langle f, \varphi \rangle. \qquad (2.8.52)$$

For the third equality in (2.8.52) we have used the fact that g is a distribution of function type, for the fourth equality we used condition (2.8.51), the fifth equality is based on a change of variables, the sixth equality follows from (2.7.53), while the last equality is a consequence of the definition of f.

To complete the proof of the proposition there remains to notice that, by part (a) in Theorem 2.96, we have $\operatorname{supp} f \subseteq \operatorname{supp} u + \operatorname{supp} g$. In particular, if u is compactly supported then so is f. $\qquad \square$

Exercise 2.103. Prove that if $u \in \mathcal{E}'(\mathbb{R}^n)$ and $g \in C^\infty(\mathbb{R}^n)$ then the distribution $u * g$ is of function type given by the function

$$f : \mathbb{R}^n \to \mathbb{C}, \qquad f(x) := \langle u(y), g(x - y) \rangle, \qquad \forall\, x \in \mathbb{R}^n, \qquad (2.8.53)$$

that satisfies $f \in C^\infty(\mathbb{R}^n)$. In short, $\mathcal{E}'(\mathbb{R}^n) * C^\infty(\mathbb{R}^n) \subseteq C^\infty(\mathbb{R}^n)$.

Hint: Use Proposition 2.91 to show that $f \in C^\infty(\mathbb{R}^n)$, then reason as in Proposition 2.102 to take care of the remaining claims.

Exercise 2.104. Let Ω be a bounded open set in $\mathbb{R}^n$. Suppose $u \in L^1(\Omega)$ and define $\widetilde{u}$ to be the extension by zero outside Ω of u. Also assume that $g \in L^1_{loc}(\mathbb{R}^n)$ is given. Prove that $\widetilde{u} \in \mathcal{D}'(\mathbb{R}^n)$, it satisfies $\operatorname{supp} u \subseteq \overline{\Omega}$, hence $\widetilde{u} * g$ is well defined in $\mathcal{D}'(\mathbb{R}^n)$, and that the distribution $\widetilde{u} * g$ is of function type given by the function

$$(\widetilde{u} * g)(x) = \int_\Omega g(x - y) u(y)\, dy, \qquad \forall\, x \in \mathbb{R}^n. \qquad (2.8.54)$$

Hint: Apply (2.8.16) with $f = \widetilde{u}$.

Exercise 2.105. Prove that if $u \in \mathcal{D}'(\mathbb{R}^n)$ is arbitrary and $\varphi_j \xrightarrow[j \to \infty]{\mathcal{D}(\mathbb{R}^n)} \varphi$, then one has $u * \varphi_j \xrightarrow[j \to \infty]{\mathcal{E}(\mathbb{R}^n)} u * \varphi$.

Next, we prove that distributions of function type given by smooth, compactly supported functions are dense in the class of distributions. First we treat the simpler case when the distributions are considered in $\mathbb{R}^n$. The case when the distributions are considered on arbitrary open subsets of $\mathbb{R}^n$ needs to be handled with a bit more care.

Theorem 2.106. *The set $C_0^\infty(\mathbb{R}^n)$ is sequentially dense in $\mathcal{D}'(\mathbb{R}^n)$.*

Proof. First we will show that $C^\infty(\mathbb{R}^n)$ is sequentially dense in $\mathcal{D}'(\mathbb{R}^n)$. Let ϕ be as in (1.2.3) and recall the sequence of functions $\{\phi_j\}_{j \in \mathbb{N}}$ from (1.3.7). In particular, we have that $\operatorname{supp} \phi_j \subseteq \overline{B(0,1)}$ for every $j \in \mathbb{N}$. Recall from Example 2.24 that $\phi_j \xrightarrow[j \to \infty]{\mathcal{D}'(\mathbb{R}^n)} \delta$.

Let $u \in \mathcal{D}'(\mathbb{R}^n)$ be arbitrary and define

$$u_j := u * \phi_j \quad \text{in } \mathcal{D}'(\mathbb{R}^n), \qquad \forall j \in \mathbb{N}. \tag{2.8.55}$$

By Proposition 2.102 we have $u_j \in C^\infty(\mathbb{R}^n)$ for all $j \in \mathbb{N}$. Also, by part *(2)* in Proposition 2.101 and part *(d)* in Theorem 2.96 we obtain

$$u_j = u * \phi_j \xrightarrow[j \to \infty]{\mathcal{D}'(\mathbb{R}^n)} u * \delta = u. \tag{2.8.56}$$

This completes the proof of the fact that $C^\infty(\mathbb{R}^n)$ is sequentially dense in $\mathcal{D}'(\mathbb{R}^n)$.

Moving on, and keeping the notation introduced so far, fix $\psi \in C_0^\infty(\mathbb{R}^n)$ satisfying $\psi(x) = 1$ if $|x| < 1$ and set

$$w_j(x) := \psi(x/j)(u * \phi_j)(x) \qquad \forall x \in \mathbb{R}^n, \quad \forall j \in \mathbb{N}. \tag{2.8.57}$$

It is immediate that $w_j \in C_0^\infty(\mathbb{R}^n)$. Moreover, if $\varphi \in C_0^\infty(\mathbb{R}^n)$ is given, then there exists $j_0 \in \mathbb{N}$ with the property that $\operatorname{supp} \varphi \subset B(0, j_0)$. Therefore, for all $j \geq j_0$ we may write

$$\langle w_j, \varphi \rangle = \int_{\mathbb{R}^n} \psi\left(\frac{x}{j}\right)(u * \phi_j)(x) \varphi(x) \, dx = \int_{\mathbb{R}^n} (u * \phi_j)(x) \varphi(x) \, dx$$

$$= \langle u * \phi_j, \varphi \rangle \xrightarrow[j \to \infty]{} \langle u, \varphi \rangle. \tag{2.8.58}$$

This shows that $w_j \xrightarrow[j \to \infty]{\mathcal{D}'(\mathbb{R}^n)} u$. Hence, ultimately, $C_0^\infty(\mathbb{R}^n)$ is sequentially dense in $\mathcal{D}'(\mathbb{R}^n)$, finishing the proof of the theorem. $\qquad \square$

The same type of result as in Theorem 2.106 actually holds in arbitrary open subsets of the Euclidean ambient.

Theorem 2.107. *The set $C_0^\infty(\Omega)$ is sequentially dense in $\mathcal{D}'(\Omega)$.*

Proof. Fix $u \in \mathcal{D}'(\Omega)$ and recall the sequence of compact sets introduced in (1.3.11). Then $\bigcup_{j \in \mathbb{N}} K_j = \Omega$ and $K_j \subset \mathring{K}_{j+1}$ for every $j \in \mathbb{N}$. For each $j \geq 2$ consider a function

$$\psi_j \in C_0^\infty(\Omega), \quad \psi_j = 1 \text{ on a neighborhood of } K_{j-1}, \quad \operatorname{supp} \psi_j \subseteq K_j, \qquad (2.8.59)$$

and define $u_j := \psi_j u \in \mathcal{D}'(\Omega)$. Since $\operatorname{supp} u_j \subseteq K_j$, Proposition 2.69 gives that each u_j may be extended to a distribution in $\mathcal{E}'(\mathbb{R}^n)$, which we continue to denote by u_j. If we now set

$$\varepsilon_j := \frac{1}{4} \operatorname{dist}(K_j, \partial K_{j+1}) > 0, \qquad \forall\, j \in \mathbb{N}, \qquad (2.8.60)$$

then the definition of the compacts in (1.3.11) forces $\varepsilon_j \searrow 0$ as $j \to \infty$. Having fixed some $\phi \in C_0^\infty(\mathbb{R}^n)$ satisfying $\operatorname{supp} \phi \subseteq \overline{B(0,1)}$ and $\int_{\mathbb{R}^n} \phi(x)\, dx = 1$, define

$$\phi_j(x) := \varepsilon_j^{-n} \phi(x/\varepsilon_j), \qquad \forall\, x \in \mathbb{R}^n, \quad \forall\, j \in \mathbb{N}. \qquad (2.8.61)$$

Thus,

$$\phi_j \in C_0^\infty(\mathbb{R}^n), \quad \operatorname{supp} \phi_j \subseteq \overline{B(0,\varepsilon_j)} \subseteq \overline{B(0,1)}, \qquad \forall\, j \in \mathbb{N}, \qquad (2.8.62)$$

and reasoning as in Example 2.24 we see that

$$\phi_j \xrightarrow[j \to \infty]{\mathcal{D}'(\mathbb{R}^n)} \delta. \qquad (2.8.63)$$

Fix $j \in \mathbb{N}$ with $j \geq 2$ and introduce $w_j := u_j * \phi_j$. By combining part *(1)* in Remark 2.100 with Proposition 2.102 we obtain that $w_j \in C_0^\infty(\mathbb{R}^n)$. In addition, *(a)* in Theorem 2.96 and (2.8.60) imply that $\operatorname{supp} w_j \subseteq K_j + \overline{B(0,\varepsilon_j)} \subset K_{j+1}$, so in fact $w_j \in C_0^\infty(\Omega)$.

To complete the proof of the theorem it suffices to show that $w_j \xrightarrow[j \to \infty]{\mathcal{D}'(\Omega)} u$. To this end, fix $\varphi \in C_0^\infty(\Omega)$ and observe that based on (1.3.11) there exists some $j_0 \in \mathbb{N}$ such that $\operatorname{supp} \varphi \subseteq K_{j_0}$. Then for $j > j_0 + 1$ we may write

$$\langle w_j, \varphi \rangle = \langle u_j * \phi_j, \varphi \rangle = \int_\Omega (u_j * \phi_j)(x)\varphi(x)\, dx$$

$$= \int_\Omega \langle u_j(y), \phi_j(x-y) \rangle \varphi(x)\, dx = \int_\Omega \langle u(y), \psi_j(y)\phi_j(x-y) \rangle \varphi(x)\, dx$$

$$= \int_\Omega \langle u(y), \psi_{j_0+2}(y)\phi_j(x-y) \rangle \varphi(x)\, dx = \langle u_{j_0+2} * \phi_j, \varphi \rangle. \qquad (2.8.64)$$

For the second equality in (2.8.64) we used the fact that $u_j * \phi_j \in C_0^\infty(\mathbb{R}^n)$ for $j \geq 2$, for the third equality we used Proposition 2.102, while for the fourth the fact that $u_j = \psi_j u$ for every $j \in \mathbb{N}$. These observations also give the last equality in (2.8.64). As for the penultimate equality in (2.8.64), observe that if $j > j_0 + 1$, $x \in \mathrm{supp}\,\varphi \subseteq K_{j_0}$ and $x - y \in \mathrm{supp}\,\phi_j \subseteq \overline{B(0, \varepsilon_j)}$, then

$$y \in K_{j_0} - \overline{B(0, \varepsilon_j)} \subseteq K_{j_0} - \overline{B(0, \varepsilon_{j_0})} \subset K_{j_0+1}, \qquad (2.8.65)$$

thus $\psi_j(y) = 1 = \psi_{j_0+2}(y)$.

If we now combine (2.8.63) with (2.8.62) and part (2) in Proposition 2.101, it follows that $u_{j_0+2} * \phi_j \xrightarrow[j\to\infty]{\mathcal{D}'(\mathbb{R}^n)} u_{j_0+2} * \delta = u_{j_0+2}$. Hence, if we return with this in (2.8.64), we may write

$$\lim_{j\to\infty} \langle w_j, \varphi \rangle = \lim_{j\to\infty} \langle u_{j_0+2} * \phi_j, \varphi \rangle = \langle u_{j_0+2}, \varphi \rangle = \langle \psi_{j_0+2} u, \varphi \rangle$$

$$= \langle u, \varphi \rangle, \qquad (2.8.66)$$

since $\psi_{j_0+2} = 1$ in a neighborhood of the support of φ. This finishes the proof of the theorem. $\qquad \square$

Proposition 2.108. *Suppose* Ω_1, Ω_2 *are open sets in* $\mathbb{R}^n$ *and let* $F : \Omega_1 \to \Omega_2$ *be a* C^∞ *diffeomorphism. For each* $u \in \mathcal{D}'(\Omega_2)$, *define the mapping* $u \circ F : \mathcal{D}(\Omega_1) \to \mathbb{C}$ *by setting*

$$(u \circ F)(\varphi) := \left\langle u, |\det(DF^{-1})|(\varphi \circ F^{-1}) \right\rangle, \quad \forall \varphi \in \mathcal{D}(\Omega_1), \qquad (2.8.67)$$

where DF^{-1} *denotes the Jacobian matrix of* F^{-1}. *Then the following are true.*

(1) $u \circ F \in \mathcal{D}'(\Omega_1)$ *and the mapping*

$$\mathcal{D}'(\Omega_2) \ni u \mapsto u \circ F \in \mathcal{D}'(\Omega_1) \ \text{ is sequentially continuous.} \qquad (2.8.68)$$

(2) If $u = u_f$ *for some* $f \in L^1_{loc}(\Omega_2)$, *then* $u \circ F = u_{f \circ F}$ *in* $\mathcal{D}'(\Omega_1)$.
(3) If $F = (F_1, \ldots, F_n)$, *then the following generalized Chain Rule formula holds*

$$\partial_j(u \circ F) = \sum_{k=1}^{n} (\partial_j F_k)[(\partial_k u) \circ F] \quad \text{in } \mathcal{D}'(\Omega_1) \qquad (2.8.69)$$

for all $j \in \{1, \ldots, n\}$ *and all* $u \in \mathcal{D}'(\Omega_2)$.

Proof. Fix $u \in \mathcal{D}'(\Omega_2)$. The fact that $u \circ F \in \mathcal{D}'(\Omega_1)$ is an immediate consequence of Exercise 1.25. Also, if $u_j \xrightarrow[j\to\infty]{\mathcal{D}'(\Omega_2)} u$ and $\varphi \in C_0^\infty(\Omega_1)$, since Exercise 1.25 gives $|\det(DF^{-1})|(\varphi \circ F^{-1}) \in C_0^\infty(\Omega_1)$, in view of definition (2.8.67) and Fact 2.22 we have $u_j \circ F \xrightarrow[j\to\infty]{\mathcal{D}'(\Omega_1)} u \circ F$.

Next, suppose $u = u_f$ for some $f \in L^1_{loc}(\Omega_2)$. Then for each $\varphi \in C_0^\infty(\Omega_1)$ we may write

$$\langle u \circ F, \varphi \rangle = \langle u, |\det(DF^{-1})|(\varphi \circ F^{-1}) \rangle$$

$$= \int_{\Omega_2} f(y)(\varphi \circ F^{-1})(y)|\det(DF^{-1}(y))|\,\mathrm{d}y$$

$$= \int_{\Omega_1} (f \circ F)(x)\varphi(x)\,\mathrm{d}x = \langle u_{f \circ F}, \varphi \rangle. \qquad (2.8.70)$$

The first equality in (2.8.70) is based on (2.8.67), the second uses the fact that $u = u_f$, the third uses the change of variables $y = F(x)$, and the last one uses the fact that $f \circ F \in L^1_{loc}(\Omega_1)$. This proves the statement in item *(2)*.

There remains to prove the Chain Rule formula (2.8.69). Suppose first that u is a distribution of function type given by a function $\psi \in C_0^\infty(\Omega_2)$, that is $u = u_\psi$. Then by item *(2)* we have $u \circ F = u_{\psi \circ F}$ in $\mathcal{D}'(\Omega_1)$. Hence, invoking Exercise 2.40 and the Chain Rule for pointwise differentiable functions we have

$$\partial_j(u \circ F) = u_{\partial_j(\psi \circ F)} = u_{\sum_{k=1}^n (\partial_j F_k)(x)[(\partial_k \psi) \circ F]}$$

$$= \sum_{k=1}^n u_{(\partial_j F_k)(x)[(\partial_k \psi) \circ F]}$$

$$= \sum_{k=1}^n (\partial_j F_k)[(\partial_k u) \circ F] \quad \text{in } \mathcal{D}'(\Omega_1). \qquad (2.8.71)$$

This proves (2.8.69) in the case when $u = u_\psi$ for some $\psi \in C_0^\infty(\Omega_2)$. Recall that by Theorem 2.107 the set $C_0^\infty(\Omega_2)$ is sequentially dense in $\mathcal{D}'(\Omega_2)$. Since the following operations with distributions are sequentially continuous: differentiation (item *(3)* in Proposition 2.43), composition with C^∞ diffeomorphisms (item *(1)* in the current proposition), and multiplication with C^∞ functions ((*4*) in Exercise 2.32), we may conclude that (2.8.69) holds for arbitrary $u \in \mathcal{D}'(\Omega_2)$. $\qquad\square$

2.9 Distributions with Higher Order Gradients Continuous or Bounded

We have seen that if $u \in C^m(\mathbb{R}^n)$ for some $m \in \mathbb{N}$, then its distributional derivatives up to order m are distributions of function type, each given by the corresponding pointwise derivative of u. A more subtle question pertains to the possibility of deducing regularity results for distributions whose distributional derivatives of a certain order are of function type, and these functions exhibit a certain amount of smoothness. In this section we prove two main results in this regard. In the first one (see Theorem 2.112), we show that if a distribution $u \in \mathcal{D}'(\Omega)$ has all distributional derivatives of order $m \in \mathbb{N}$ continuous, then in fact $u \in C^m(\Omega)$. In the second main result (see Theorem 2.114) we prove that a distribution in $\mathcal{D}'(\mathbb{R}^n)$ is of function type given by a Lipschitz function if and only if all its first-order distributional derivatives are bounded functions in $\mathbb{R}^n$.

We start by proving a weaker version of Theorem 2.112.

Proposition 2.109. *If $u \in \mathcal{D}'(\mathbb{R}^n)$ and there exists some $m \in \mathbb{N}_0$ such that for each $\alpha \in \mathbb{N}_0^n$ satisfying $|\alpha| \leq m$ the distributional derivative $\partial^\alpha u$ belongs to $C^0(\mathbb{R}^n)$, then $u \in C^m(\mathbb{R}^n)$.*

Proof. Recall the sequence of distributions $\{\phi_j\}_{j\in\mathbb{N}}$ from Example 2.24 and for u satisfying the hypothesis of the proposition let $\{u_j\}_{j\in\mathbb{N}}$ be as in (2.8.55). In particular, (2.8.56) holds, thus the distributional and classical derivatives of each u_j coincide.

Next we make the following claim:

$$\lim_{j\to\infty} \partial^\alpha u_j = \partial^\alpha u \quad \text{uniformly on compact sets in } \mathbb{R}^n$$

$$\text{for every multi-index } \alpha \in \mathbb{N}_0^n \text{ satisfying } |\alpha| \leq m. \tag{2.9.1}$$

To prove this claim, observe that by part *(e)* in Theorem 2.96, for each $j \in \mathbb{N}$ we have $\partial^\alpha u_j = (\partial^\alpha u) * \phi_j$ for every $\alpha \in \mathbb{N}_0^n$. Fix $\alpha \in \mathbb{N}_0^n$ satisfying $|\alpha| \leq m$. Since by the current hypotheses the distributional derivative $\partial^\alpha u$ is continuous, by invoking Proposition 2.102 we may write

$$\partial^\alpha u_j(x) = ((\partial^\alpha u) * \phi_j)(x) = \langle \partial^\alpha u(y), \phi_j(x-y) \rangle$$

$$= \int_{\mathbb{R}^n} (\partial^\alpha u)(y)\phi_j(x-y)\,dy = \int_{\mathbb{R}^n} (\partial^\alpha u)(x-z)\phi_j(z)\,dz$$

$$= \int_{\mathbb{R}^n} (\partial^\alpha u)(x-y/j)\phi(y)\,dy, \qquad \forall\, x \in \mathbb{R}^n. \tag{2.9.2}$$

Fix a compact set K in $\mathbb{R}^n$. Making use of (2.9.2) and the properties of ϕ (recall (1.2.3)) we estimate

$$\sup_{x\in K} |\partial^\alpha u_j(x) - \partial^\alpha u(x)| = \sup_{x\in K} \left| \int_{\mathbb{R}^n} [(\partial^\alpha u)(x-y/j) - (\partial^\alpha u)(x)]\phi(y)\,dy \right|$$

$$\leq \sup_{\substack{x\in K \\ y\in B(0,1)}} \left| (\partial^\alpha u)(x-y/j) - (\partial^\alpha u)(x) \right|. \tag{2.9.3}$$

Since $\partial^\alpha u$ is continuous in $\mathbb{R}^n$ it follows that $\partial^\alpha u$ is uniformly continuous on compact subsets of $\mathbb{R}^n$, thus

$$\lim_{j\to\infty} \sup_{x\in K} |\partial^\alpha u_j(x) - \partial^\alpha u(x)| \tag{2.9.4}$$

$$\leq \lim_{j\to\infty} \sup_{\substack{x\in K \\ y\in B(0,1)}} \left| (\partial^\alpha u)(x-z/j) - (\partial^\alpha u)(x) \right| = 0,$$

completing the proof of the claim.

With (2.9.1) in hand, we may invoke Lemma 2.110 below and proceed by induction on $|\alpha|$ to conclude that, as desired, $u \in C^m(\mathbb{R}^n)$. $\square$

Lemma 2.110. *Suppose the functions* $\{u_j\}_{j \in \mathbb{N}}$ *and* u *are such that:*

(i) $u_j \in C^1(\Omega)$ *for every* $j \in \mathbb{N}$,

(ii) $\lim\limits_{j \to \infty} u_j = u$ *uniformly on compact subsets of* $\mathbb{R}^n$ *contained in* Ω, *and*

(iii) for each $k \in \{1, \dots, n\}$ *there exists a function* $v_k \in C^0(\Omega)$ *with the property that* $\lim\limits_{j \to \infty} \partial_k u_j = v_k$ *uniformly on compact subsets of* $\mathbb{R}^n$ *contained in* Ω.

Then $u \in C^1(\Omega)$ *and* $\partial_k u = v_k$ *for each* $k \in \{1, \dots, n\}$.

Proof. From the start, since uniform convergence on compact sets preserves continuity, we have that $u \in C^0(\Omega)$. Fix $x \in \Omega$ and $k \in \{1, 2, \dots, n\}$ and let $t_0 > 0$ be such that $x + t\mathbf{e}_k \in \Omega$ whenever $t \in [-t_0, t_0]$ where, as before, $\mathbf{e}_k$ is the unit vector in $\mathbb{R}^n$ whose k-th component is equal to 1. Then, for each $t \in [-t_0, t_0]$, we may write

$$u(x + t\mathbf{e}_k) - u(x) = \lim_{j \to \infty} [u_j(x + t\mathbf{e}_k) - u_j(x)] \qquad (2.9.5)$$

$$= \lim_{j \to \infty} \int_0^t \frac{d}{ds}[u_j(x + s\,\mathbf{e}_k)]\, ds$$

$$= \lim_{j \to \infty} \int_0^t (\partial_k u_j)(x + s\,\mathbf{e}_k)\, ds = \int_0^t v_k(x + s\,\mathbf{e}_k)\, ds.$$

The first equality in (2.9.5) is based on *(ii)* while the last one is a consequence of *(iii)*. Next, since v_k is continuous on Ω, by the Mean Value Theorem for integrals it follows that there exists some s_t, belonging to the interval with end-points 0 and t, such that $\int_0^t v_k(x + s\,\mathbf{e}_k)\, ds = t v_k(x + s_t\mathbf{e}_k)$. Hence,

$$\lim_{t \to 0} \frac{u(x + t\mathbf{e}_k) - u(x)}{t} = \lim_{t \to 0} v_k(x + s_t\mathbf{e}_k) = v_k(x), \qquad (2.9.6)$$

which proves that $(\partial_k u)(x)$ exists and is equal to $v_k(x)$. Thus, $\partial_k u = v_k \in C^0(\Omega)$ for every k, which shows that $u \in C^1(\Omega)$. $\square$

Lemma 2.111. *Let* $u \in \mathcal{D}'(\Omega)$ *be such that for each* $j \in \{1, \dots, n\}$, *the distributional derivatives* $\partial_j u$ *are of function type and belong to* $C^0(\Omega)$. *Then* $u \in C^0(\Omega)$.

Proof. Since $\nabla u \in [C^0(\Omega)]^n$ the function $v(x) := \int_0^1 (\nabla u)(tx) \cdot x\, dt$ for $x \in \Omega$ (where "$\cdot$" denotes the dot product of vectors) is well defined and belongs to $C^0(\Omega)$. Given the current goals, by Exercise 2.53 it suffices to show that for each $x_0 \in \Omega$ there exists an open set $\omega \subset \Omega$ such that $x_0 \in \omega$ and $u\big|_\omega = v\big|_\omega$ in $\mathcal{D}'(\omega)$.

To this end, fix $x_0 \in \Omega$ and let $r \in (0, \mathrm{dist}(x_0, \partial\Omega))$. Consequently, we have $B(x_0, r) \subset \Omega$. Without loss of generality in what follows we may assume that $x_0 = 0$ (since translations interact favorably, in a reversible manner, with both hypotheses and conclusion). Let $\varphi \in C_0^\infty(\Omega)$ be such that $\mathrm{supp}\, \varphi \subset B(0, r)$ and fix $j \in \{1, \dots, n\}$. Then we have

$$\langle \partial_j v, \varphi \rangle = -\langle v, \partial_j \varphi \rangle = - \int_\Omega \Big(\int_0^1 (\nabla u)(tx) \cdot x \, \partial_j \varphi(x) \, dt \Big) \, dx$$

$$= - \int_0^1 \Big(\int_\Omega (\nabla u)(tx) \cdot x \, \partial_j \varphi(x) \, dx \Big) \, dt$$

$$= - \lim_{\varepsilon \to 0^+} \int_\varepsilon^1 \Big(\int_\Omega (\nabla u)(tx) \cdot x \, \partial_j \varphi(x) \, dx \Big) \, dt$$

$$= - \lim_{\varepsilon \to 0^+} \int_\varepsilon^1 \Big(\int_\Omega \nabla u(y) \cdot \frac{y}{t^{n+1}} (\partial_j \varphi)(y/t) \, dy \Big) \, dt$$

$$= - \lim_{\varepsilon \to 0^+} \int_\Omega \sum_{k=1}^n \partial_k u(y) \Big(\int_\varepsilon^1 \frac{y_k}{t^{n+1}} (\partial_j \varphi)(y/t) \, dt \Big) \, dy$$

$$= \lim_{\varepsilon \to 0^+} \Big\langle u(y), \sum_{k=1}^n \int_\varepsilon^1 \partial_{y_k} \Big[\frac{y_k}{t^{n+1}} (\partial_j \varphi)(y/t) \Big] \, dt \Big\rangle. \tag{2.9.7}$$

For the fourth equality in (2.9.7) we have used Lebesgue's Dominated Convergence Theorem (cf. Theorem 14.15) and for the fifth one the change of variables $tx = y$ (note that $\operatorname{supp} \varphi(\cdot/t) \subset B(0, r) \subset \Omega$ since $\operatorname{supp} \varphi \in B(0, r)$ and $t \in (0, 1]$). Furthermore, for each $\varepsilon > 0$, differentiating with respect to y and then integrating by parts in t, gives

$$\sum_{k=1}^n \int_\varepsilon^1 \partial_{y_k} \Big[\frac{y_k}{t^{n+1}} (\partial_j \varphi)(y/t) \Big] \, dt$$

$$= \int_\varepsilon^1 \sum_{k=1}^n \Big[\frac{1}{t^{n+1}} (\partial_j \varphi)(y/t) + \frac{y_k}{t^{n+2}} (\partial_k \partial_j \varphi)(y/t) \Big] \, dt$$

$$= \int_\varepsilon^1 \Big[\frac{n}{t^{n+1}} (\partial_j \varphi)(y/t) - \frac{1}{t^n} \frac{d}{dt} [(\partial_j \varphi)(y/t)] \Big] \, dt$$

$$= - \frac{1}{t^n} (\partial_j \varphi)(y/t) \Big|_{t=\varepsilon}^{t=1} = -\partial_j \varphi(y) + \frac{1}{\varepsilon^n} (\partial_j \varphi)(y/\varepsilon). \tag{2.9.8}$$

By combining (2.9.7) and (2.9.8) we obtain

$$\langle \partial_j v, \varphi \rangle = -\langle u, \partial_j \varphi \rangle + \lim_{\varepsilon \to 0^+} \Big\langle u(y), \frac{1}{\varepsilon^{n-1}} \partial_{y_j} [\varphi(y/\varepsilon)] \Big\rangle$$

$$= \langle \partial_j u, \varphi \rangle + \lim_{\varepsilon \to 0^+} \Big\langle \partial_j u(y), \frac{1}{\varepsilon^{n-1}} \varphi(y/\varepsilon) \Big\rangle. \tag{2.9.9}$$

Since $\partial_j u \in C^0(\Omega)$, the pairing under the limit in the rightmost term in (2.9.9) is given by an integral in which we make the change of variables $x = y/\varepsilon$ to further compute

$$\lim_{\varepsilon \to 0^+} \left\langle \partial_j u(y), \frac{1}{\varepsilon^{n-1}} \varphi(y/\varepsilon) \right\rangle = \lim_{\varepsilon \to 0^+} \frac{1}{\varepsilon^{n-1}} \int_\Omega (\partial_j u)(y) \varphi(y/\varepsilon) \, dy$$

$$= \lim_{\varepsilon \to 0^+} \varepsilon \int_\Omega (\partial_j u)(\varepsilon x) \varphi(x) \, dx = 0 \qquad (2.9.10)$$

given that, by the continuity of $\partial_j u$ at $0 \in \Omega$,

$$\lim_{\varepsilon \to 0^+} \int_\Omega (\partial_j u)(\varepsilon x) \varphi(x) \, dx = \partial_j u(0) \int_\Omega \varphi(x) \, dx. \qquad (2.9.11)$$

In summary, from (2.9.9), (2.9.10), (2.9.11), the fact that φ is an arbitrary element in $\mathcal{D}(B(0, r))$, and that j is arbitrary in $\{1, \ldots, n\}$, we conclude that $\nabla v\big|_{B(0,r)} = \nabla u\big|_{B(0,r)}$ in $\mathcal{D}'(B(0, r))$. By Exercise 2.158 there exists $c \in \mathbb{C}$ such that $u\big|_{B(0,r)} - v\big|_{B(0,r)} = c$ in $\mathcal{D}'(B(0, r))$. Since $v \in C^0(\Omega)$, the latter implies $u\big|_{B(0,r)} \in C^0(B(0, r))$ as desired. This completes the proof of the lemma. $\square$

After these preparations, we are ready to state and prove our first main result.

Theorem 2.112. *Let $u \in \mathcal{D}'(\Omega)$ and suppose that there exists $m \in \mathbb{N}_0$ such that for each $\alpha \in \mathbb{N}_0^n$ satisfying $|\alpha| = m$ the distributional derivative $\partial^\alpha u$ is continuous on Ω. Then $u \in C^m(\Omega)$.*

Proof. We prove the theorem by induction on m. For $m = 0$ there is nothing to prove. Suppose $m = 1$. Applying Lemma 2.111, we obtain that $u \in C^0(\Omega)$. To prove that $u \in C^1(\Omega)$, it suffices to show that u is of class C^1 in a neighborhood of any point $x_0 \in \Omega$. Fix $x_0 \in \Omega$, pick a number $r > 0$ with the property that $B(x_0, r) \subset \Omega$ and a function $\psi \in C_0^\infty(\mathbb{R}^n)$ such that $\mathrm{supp}\, \psi \subset B(x_0, r)$ and $\psi \equiv 1$ on $B(x_0, r/2)$. Then by Proposition 2.69 the distribution ψu extends to $v := \widetilde{\psi u} \in \mathcal{E}'(\mathbb{R}^n)$. Also, $v \in C^0(\mathbb{R}^n)$ since $u \in C^0(\Omega)$ and ψ is compactly supported in Ω. Using (2.5.4) and part *(4)* in Proposition 2.43 we obtain

$$\partial_j v = \widetilde{\partial_j(\psi u)} = \widetilde{(\partial_j \psi) u} + \widetilde{\psi \partial_j u} \in C^0(\mathbb{R}^n). \qquad (2.9.12)$$

Applying Proposition 2.109 with $m = 1$ then gives $v \in C^1(\mathbb{R}^n)$. In addition, since by design $v\big|_{B(x_0, r/2)} = u\big|_{B(x_0, r/2)}$, we conclude that $u\big|_{B(x_0, r/2)} \in C^1(B(x_0, r/2))$, as wanted.

Assume now that the theorem is true for all nonnegative integers up to, and including, some $m \in \mathbb{N}$. Take $u \in \mathcal{D}'(\Omega)$ with the property that $\partial^\alpha u \in C^0(\Omega)$ for all $\alpha \in \mathbb{N}_0^n$ satisfying $|\alpha| = m + 1$ and fix $j \in \{1, \ldots, n\}$. Then $\partial_j u \in \mathcal{D}'(\Omega)$ satisfies $\partial^\alpha(\partial_j u) \in C^0(\Omega)$ for all $\alpha \in \mathbb{N}_0^n$ with $|\alpha| = m$. By the induction hypothesis, it follows that $\partial_j u \in C^m(\Omega)$. In particular, $\partial_j u \in C^0(\Omega)$. This being true for all $j \in \{1, \ldots, n\}$, what we already proved for $m = 1$ implies $u \in C^1(\Omega)$. Thus, u is a C^1 function in Ω whose first-order partial derivatives are of class C^m in Ω. Then necessarily $u \in C^{m+1}(\Omega)$ as wanted. The proof of the theorem is finished. $\square$

Exercise 2.113. Let $u \in \mathcal{D}'(\mathbb{R}^n)$ be such that for some $N \in \mathbb{N}_0$ we have $\partial^\alpha u = 0$ in $\mathcal{D}'(\mathbb{R}^n)$ for each $\alpha \in \mathbb{N}_0^n$ with $|\alpha| > N$. Prove that u is a polynomial of degree less than or equal to N.

Hint: Use Theorem 2.112 to conclude that $u \in C^\infty(\mathbb{R}^n)$ then invoke Taylor's formula (14.2.9).

Moving on to the second issue discussed at the beginning of this section we recall (cf. Remark 2.42) that a function $f : \Omega \to \mathbb{C}$ is called Lipschitz (in Ω) provided there exists some constant $M \in [0, \infty)$ such that

$$|f(x) - f(y)| \le M|x - y|, \qquad \forall\, x, y \in \Omega, \tag{2.9.13}$$

and that the Lipschitz constant of f is the smallest M for which (2.9.13) holds. Our next task is to prove the following theorem, which provides a distributional characterization of Lipschitzianity.

Theorem 2.114. *For $f \in \mathcal{D}'(\mathbb{R}^n)$ and a number $M \in [0, \infty)$ the following two statements are equivalent:*

(i) f is given by a Lipschitz function in $\mathbb{R}^n$ with Lipschitz constant less than or equal to M;
(ii) for each $k \in \{1, \dots, n\}$, the distributional derivative $\partial_k f$ belongs to $L^\infty(\mathbb{R}^n)$ and $\|\partial_k f\|_{L^\infty(\mathbb{R}^n)} \le M$.

As a consequence of this distributional characterization of Lipschitzianity and the extension result recorded in (2.4.5) we have that

$$\text{if } \Omega \subseteq \mathbb{R}^n \text{ is an open set and } f : \Omega \to \mathbb{C} \text{ is a Lipschitz function then} \atop \partial_j f \in L^\infty(\Omega) \text{ for each } j = 1, \dots, n. \tag{2.9.14}$$

Proof. Fix a distribution $f \in \mathcal{D}'(\mathbb{R}^n)$ and let ϕ be as in (1.2.3). Consider the sequence of functions $\{\phi_j\}_{j\in\mathbb{N}}$ from (1.3.7), that satisfies the properties listed in (1.3.8). Furthermore, set

$$f_j := f * \phi_j \quad \text{in } \mathcal{D}'(\mathbb{R}^n), \qquad \forall\, j \in \mathbb{N}. \tag{2.9.15}$$

By Proposition 2.102 we have

$$f_j \in C^\infty(\mathbb{R}^n) \quad \text{and} \quad f_j(x) = \langle f, \phi_j(x - \cdot)\rangle, \quad \forall\, j \in \mathbb{N}. \tag{2.9.16}$$

Also, since (as proved in Example 2.24) one has $\phi_j \xrightarrow[j\to\infty]{\mathcal{D}'(\mathbb{R}^n)} \delta$, by part *(2)* in Proposition 2.101 and part *(d)* in Theorem 2.96 one obtains

$$f_j = f * \phi_j \xrightarrow[j\to\infty]{\mathcal{D}'(\mathbb{R}^n)} f * \delta = f. \tag{2.9.17}$$

Next we proceed with the proof of *(i)* $\Longrightarrow$ *(ii)*. Suppose f is Lipschitz with Lipschitz constant $\le M$. In particular, the formula in (2.9.16) becomes

$$f_j(x) = \int_{\mathbb{R}^n} f(y)\phi_j(x - y)\,\mathrm{d}y = \int_{\mathbb{R}^n} f(x - y)\phi_j(y)\,\mathrm{d}y, \tag{2.9.18}$$
$$\text{for all } x \in \mathbb{R}^n \text{ and all } j \in \mathbb{N}.$$

We claim that for each $j \in \mathbb{N}$ one has

$$|f_j(x) - f_j(y)| \le M|x - y|, \quad \forall\, x, y \in \mathbb{R}^n \quad \text{and} \quad \|\nabla f_j\|_{L^\infty(\mathbb{R}^n)} \le M. \tag{2.9.19}$$

Indeed, if $j \in \mathbb{N}$ is fixed, then from (2.9.18) we obtain

$$|f_j(x) - f_j(y)| \le \int_{\mathbb{R}^n} |f(x - z) - f(y - z)|\phi_j(z)\, dz \le M|x - y|, \tag{2.9.20}$$

for every $x, y \in \mathbb{R}^n$. In turn, since f_j is smooth, we have

$$(\partial_k f_j)(x) = \lim_{h \to 0} \frac{f_j(x + h\mathbf{e}_k) - f_j(x)}{h}, \quad \forall\, x \in \mathbb{R}^n, \ \forall\, k \in \{1, \dots, n\}. \tag{2.9.21}$$

In combination with (2.9.20) this implies $\|\nabla f_j\|_{L^\infty(\mathbb{R}^n)} \le M$, completing the proof of the claims made in (2.9.19).

Next, fix $k \in \{1, \dots, n\}$ and $\varphi \in C_0^\infty(\mathbb{R}^n)$. Then based on (2.9.17) we may write

$$\langle \partial_k f, \varphi \rangle = -\langle f, \partial_k \varphi \rangle = -\lim_{j \to \infty} \langle f_j, \partial_k \varphi \rangle = \lim_{j \to \infty} \langle \partial_k f_j, \varphi \rangle$$

$$= \lim_{j \to \infty} \int_{\mathbb{R}^n} (\partial_k f_j)(x)\varphi(x)\, dx. \tag{2.9.22}$$

Using the second estimate in (2.9.19) we obtain

$$\left| \int_{\mathbb{R}^n} (\partial_k f_j)(x)\varphi(x)\, dx \right| \le \|\nabla f_j\|_{L^\infty(\mathbb{R}^n)} \|\varphi\|_{L^1(\mathbb{R}^n)} \le M\|\varphi\|_{L^1(\mathbb{R}^n)}. \tag{2.9.23}$$

Hence, from (2.9.23) and (2.9.22) it follows that

$$\left| \langle \partial_k f, \varphi \rangle \right| \le \limsup_{j \to \infty} \left| \int_{\mathbb{R}^n} (\partial_k f_j)(x)\varphi(x)\, dx \right| \le M\|\varphi\|_{L^1(\mathbb{R}^n)}. \tag{2.9.24}$$

Consequently, the linear assignment

$$C_0^\infty(\mathbb{R}^n) \ni \varphi \mapsto \langle \partial_k f, \varphi \rangle \in \mathbb{R} \tag{2.9.25}$$

is continuous in the L^1-norm and has norm less than or equal to M. Since $C_0^\infty(\mathbb{R}^n)$ is dense in $L^1(\mathbb{R}^n)$, the linear functional in (2.9.25) extends to a linear, bounded functional $\Lambda_k : L^1(\mathbb{R}^n) \to \mathbb{C}$ with norm less or equal to M. Thus, $\Lambda_k \in (L^1(\mathbb{R}^n))^* = L^\infty(\mathbb{R}^n)$ has norm less than or equal to M, which implies that there exists a unique

$$g_k \in L^\infty(\mathbb{R}^n) \quad \text{with} \quad \|g_k\|_{L^\infty(\mathbb{R}^n)} \le M \tag{2.9.26}$$

and such that

$$\Lambda_k(h) = \int_{\mathbb{R}^n} g_k(x)h(x)\, dx, \qquad \forall\, h \in L^1(\mathbb{R}^n). \tag{2.9.27}$$

Granted (2.9.27) and keeping in mind that Λ_k is an extension of the linear assignment in (2.9.25), we arrive at the conclusion that

$$\langle \partial_k f, \varphi \rangle = \int_{\mathbb{R}^n} g_k(x) \varphi(x) \, dx, \qquad \forall \varphi \in C_0^\infty(\mathbb{R}^n). \tag{2.9.28}$$

The identity in (2.9.28) yields $\partial_k f = g_k$ in $\mathcal{D}'(\mathbb{R}^n)$, which proves *(ii)* in view of (2.9.26).

Conversely, suppose *(ii)* is true. Fix $j \in \mathbb{N}$ and note that from (2.9.15) and part *(e)* in Theorem 2.96, for every $k \in \{1, \dots, n\}$ one has $\partial_k f_j = (\partial_k f) * \phi_j$ in $\mathcal{D}'(\mathbb{R}^n)$. Thus, using Proposition 2.102, the current assumptions on f, and the properties of ϕ_j, we have

$$|(\partial_k f_j)(x)| = |\langle \partial_k f, \phi_j(x - \cdot)\rangle| = \left| \int_{\mathbb{R}^n} (\partial_k f)(y) \phi_j(x - y) \, dy \right|$$

$$\leq \|\partial_k f\|_{L^\infty(\mathbb{R}^n)} \int_{\mathbb{R}^n} \phi_j(x - y) \, dy \leq M, \qquad \forall x \in \mathbb{R}^n, \tag{2.9.29}$$

for each $k \in \{1, \dots, n\}$. Now fix $x_0 \in \mathbb{R}^n$ and consider the sequence of functions $\{g_j\}_{j \in \mathbb{N}}$ given by

$$g_j(x) := f_j(x) - f_j(x_0), \quad \text{for} \quad x \in \mathbb{R}^n. \tag{2.9.30}$$

Then $g_j \in C^\infty(\mathbb{R}^n)$ and based on the Mean Value Theorem, (2.9.30), and (2.9.29) we also obtain

$$|g_j(x) - g_j(y)| \leq M|x - y|, \quad \forall x, y \in \mathbb{R}^n, \quad \text{and} \tag{2.9.31}$$

$$|g_j(x)| \leq M|x - x_0|, \quad \forall x \in \mathbb{R}^n. \tag{2.9.32}$$

By a corollary of the classical Arzelá–Ascoli theorem (see Theorem 14.24), there exists a subsequence $\{g_{j_\ell}\}_{\ell \in \mathbb{N}}$ that converges uniformly on any compact subset of $\mathbb{R}^n$ to some function $g \in C^0(\mathbb{R}^n)$. As such, for every $\varphi \in C_0^\infty(\mathbb{R}^n)$ we may write

$$\lim_{\ell \to \infty} \langle g_{j_\ell}, \varphi \rangle = \lim_{\ell \to \infty} \int_{\operatorname{supp} \varphi} g_{j_\ell}(x) \varphi(x) \, dx$$

$$= \int_{\operatorname{supp} \varphi} g(x) \varphi(x) \, dx = \langle g, \varphi \rangle \tag{2.9.33}$$

which goes to show that $g_{j_\ell} \xrightarrow[\ell \to \infty]{\mathcal{D}'(\mathbb{R}^n)} g$. The latter, (2.9.17), and (2.9.30), give that whenever $\psi \in C_0^\infty(\mathbb{R}^n)$ is such that $\int_{\mathbb{R}^n} \psi(x) \, dx = 0$ we have

$$\langle g, \psi \rangle = \lim_{\ell \to \infty} \langle g_{j_\ell}, \psi \rangle = \lim_{\ell \to \infty} \int_{\mathbb{R}^n} g_{j_\ell}(x) \psi(x) \, dx$$

$$= \lim_{\ell \to \infty} \int_{\mathbb{R}^n} f_{j_\ell}(x) \psi(x) \, dx = \lim_{\ell \to \infty} \langle f_{j_\ell}, \psi \rangle = \langle f, \psi \rangle. \qquad (2.9.34)$$

Thus, we may apply Exercise 2.157 to conclude that there exists some constant $c \in \mathbb{C}$ with the property that $f = g + c$ in $\mathcal{D}'(\mathbb{R}^n)$. This proves that the distribution f is of function type and is given by the function $g + c$. Moreover, writing the estimate in (2.9.31) with j replaced by j_ℓ and then taking the limit as $\ell \to \infty$ (recall that $\{g_{j_\ell}\}_\ell$ converges pointwise to g) leads to

$$|g(x) - g(y)| \le M|x - y|, \quad \forall \, x, y \in \mathbb{R}^n, \qquad (2.9.35)$$

which in concert with the fact that $f(x) - f(y) = g(x) - g(y)$ proves that f is a Lipschitz function with Lipschitz constant $\le M$. Now the proof of *(ii)* $\Longrightarrow$ *(i)* is finished. $\square$

In connection with Theorem 2.114 we recall Rademacher's theorem which gives that Lipschitz functions are pointwise differentiable at almost every point.

Theorem 2.115. *Any Lipschitz function* $f : \mathbb{R}^n \to \mathbb{R}$ *is differentiable almost everywhere.*

Exercise 2.116. Assume $f : \mathbb{R}^n \to \mathbb{R}$ is a Lipschitz function. On the one hand, according to Rademacher's theorem for each $k \in \{1, \dots, n\}$ the pointwise partial derivative with respect to the k-th variable, temporarily denoted by $\partial_k^{pw} f$, exists a.e. in $\mathbb{R}^n$. On the other hand, Theorem 2.114 gives that for each $k \in \{1, \dots, n\}$ the partial derivative $\partial_k f$ computed in the sense of distributions belongs to $L^\infty(\mathbb{R}^n)$. Follow the outline below to show that the two brands of derivatives mentioned above coincide a.e. in $\mathbb{R}^n$. As a consequence, the pointwise partial derivatives of order one of f are bounded.

Step I. Rademacher's theorem ensures that

$$(\partial_k^{pw} f)(x) = \lim_{j \to \infty} g_j(x) \quad \text{for a.e. } x \in \mathbb{R}^n, \qquad (2.9.36)$$

where for each $j \in \mathbb{N}$ we have set $g_j(x) := \dfrac{f(x + j^{-1} \mathbf{e}_k) - f(x)}{j^{-1}}$ for all $x \in \mathbb{R}^n$. Since each g_j is continuous, conclude that $\partial_k^{pw} f$ is measurable in $\mathbb{R}^n$.

Step II. If $M := \sup\limits_{x, y \in \mathbb{R}^n, \, x \ne y} \dfrac{|f(x) - f(y)|}{|x - y|} \in [0, \infty)$ then for every $j \in \mathbb{N}$ we have $|g_j(x)| \le M$ for all $x \in \mathbb{R}^n$. Use this and Step I to conclude that $\partial_k^{pw} f \in L^\infty(\mathbb{R}^n)$.

Step III. Fix a test function $\varphi \in C_0^\infty(\mathbb{R}^n)$ and use Lebesgue's Dominated Convergence Theorem to write

$$\int_{\mathbb{R}^n} (\partial_k^{pw} f)(x)\varphi(x)\,dx = \lim_{j\to\infty} \int_{\mathbb{R}^n} g_j(x)\varphi(x)\,dx$$

$$= \lim_{j\to\infty} \left\{ \int_{\mathbb{R}^n} jf(x + j^{-1}\mathbf{e}_k)\varphi(x)\,dx - \int_{\mathbb{R}^n} jf(x)\varphi(x)\,dx \right\}$$

$$= \lim_{j\to\infty} \left\{ \int_{\mathbb{R}^n} jf(y)\varphi(y - j^{-1}\mathbf{e}_k)\,dy - \int_{\mathbb{R}^n} jf(y)\varphi(y)\,dy \right\}$$

$$= \lim_{j\to\infty} \int_{\mathbb{R}^n} f(y)\frac{\varphi(y - j^{-1}\mathbf{e}_k) - \varphi(y)}{j^{-1}}\,dy$$

$$= -\int_{\mathbb{R}^n} f(y)(\partial_k\varphi)(y)\,dy. \tag{2.9.37}$$

Use this to derive the desired conclusion.

Exercise 2.117. Let Ω be an arbitrary open subset of $\mathbb{R}^n$ and fix $f \in Lip(\Omega)$. Then for each $k \in \{1, \ldots, n\}$ the following properties hold.

(i) The pointwise partial derivative with respect to the k-th variable, denoted by $\partial_k^{pw} f$, exists a.e. in Ω. Moreover, as a function, $\partial_k^{pw} f$ belongs to $L^\infty(\Omega)$.
(ii) The partial derivative $\partial_k f$ computed in the sense of distributions in Ω coincides with the locally integrable function $\partial_k^{pw} f$.

Hint: Use (2.4.5) (with $E := \Omega$) and Exercise 2.116.

We close this section by a discussion regarding locally Lipschitz functions. Specifically, given an open set Ω in $\mathbb{R}^n$ we define the set of locally Lipschitz functions in Ω by

$$Lip_{loc}(\Omega) := \{f : \Omega \to \mathbb{C} : f \in Lip(K) \quad \forall K \subset \Omega \text{ compact set}\}, \tag{2.9.38}$$

There are other useful characterizations of local Lipschitzianity, as discussed next.

Exercise 2.118. Let Ω be an open subset of $\mathbb{R}^n$ and suppose $f : \Omega \to \mathbb{C}$ is a given function. Then the following are equivalent.

(1) $f \in Lip_{loc}(\Omega)$;
(2) for each $x \in \Omega$ there exists $r_x \in (0, \text{dist}(x, \partial\Omega))$ such that the restriction $f|_{B(x,r_x)}$ belongs to $Lip(B(x, r_x))$.

Hint: The implication *(1) ⇒ (2)* is immediate, while the reverse implication may be established with the help of the Lebesgue Number Theorem (cf. Theorem 14.44).

Exercise 2.119. If Ω is an arbitrary open subset of $\mathbb{R}^n$ then $C^1(\Omega) \subseteq Lip_{loc}(\Omega)$.

Hint: Use Exercise 2.118 and the Mean Value Theorem.

Exercise 2.120. Let $\Omega \subseteq \mathbb{R}^n$, $O \subseteq \mathbb{R}^m$ be two arbitrary open sets, and assume that $\Psi : O \to \Omega$ has Lipschitz components. Then if $f \in Lip_{loc}(\Omega)$ it follows that $f \circ \Psi \in Lip_{loc}(O)$.

Hint: For any given compact set $K \subseteq O$ it follows that $W := \Psi(K)$ is a compact subset of Ω, and $(f \circ \Psi)\big|_K = (f\big|_W) \circ (\Psi\big|_K) \in Lip(K)$.

Proposition 2.121. *Let Ω be an open subset in $\mathbb{R}^n$ and let $f \in Lip_{loc}(\Omega)$. Then for each $j \in \{1, \dots, n\}$ the pointwise partial derivative $\partial_j^{pw} f$ exists at almost every point in Ω and belongs to $L_{loc}^\infty(\Omega)$. Moreover, as a locally integrable function, $\partial_j^{pw} f$ equals the distributional derivative $\partial_j f \in \mathcal{D}'(\Omega)$. In other words,*

$$\int_\Omega \partial_j^{pw} f \varphi \, dx = - \int_\Omega f \partial_j \varphi \, dx, \qquad \forall \varphi \in C_0^\infty(\Omega). \tag{2.9.39}$$

Proof. If for each $k \in \mathbb{N}$ we introduce

$$\Omega_k := \{ x \in \Omega : \text{dist}(x, \partial\Omega) > 1/k \text{ and } |x| < k \}, \tag{2.9.40}$$

then

each Ω_k is an open, relatively compact, subset of Ω,

$$\overline{\Omega_k} \subseteq \Omega_{k+1}, \text{ and } \bigcup_{k \in \mathbb{N}} \Omega_k = \Omega. \tag{2.9.41}$$

Since $f \in Lip_{loc}(\Omega)$, it follows that for each $k \in \mathbb{N}$ we have $f\big|_{\Omega_k} \in Lip(\Omega_k)$. As such, Exercise 2.117 ensures that

for each index $j \in \{1, \dots, n\}$ the pointwise derivative $\partial_j^{pw}(f\big|_{\Omega_k})$ exists
a.e. in Ω_k, belongs to $L^\infty(\Omega_k)$, and $\partial_j^{pw}(f\big|_{\Omega_k}) = \partial_j(f\big|_{\Omega_k})$ in $\mathcal{D}'(\Omega_k)$. $\tag{2.9.42}$

Granted these, it is then immediate from (2.9.41) that for each $j \in \{1, \dots, n\}$ the pointwise partial derivative $\partial_j^{pw} f$ exists at almost every point in Ω and belongs to $L_{loc}^\infty(\Omega)$. There remains to prove (2.9.39). To this end, fix an arbitrary $\varphi \in C_0^\infty(\Omega)$ and pick $k \in \mathbb{N}$ such that $\text{supp}\,\varphi \subseteq \Omega_k$. Then, having fixed some j in $\{1, \dots, n\}$, we may write

$$\int_\Omega f \partial_j \varphi \, dx = \int_{\Omega_k} f \partial_j \varphi \, dx = \langle f\big|_{\Omega_k}, \partial_j(\varphi\big|_{\Omega_k}) \rangle$$

$$= -\langle \partial_j(f\big|_{\Omega_k}), \varphi\big|_{\Omega_k} \rangle = - \int_{\Omega_k} \partial_j^{pw}(f\big|_{\Omega_k}) \varphi \, dx$$

$$= - \int_\Omega \partial_j^{pw} f \varphi \, dx. \tag{2.9.43}$$

Above, the first and last equalities are consequences of the support condition on φ, the second equality is the interpretation of the integral as the distributional pairing over the open set Ω_k, the third equality is based on the definition of the distributional derivative (considered in $\mathcal{D}'(\Omega_k)$), while the fourth equality is a consequence of the last property in (2.9.42). $\qquad\qquad\qquad\qquad\qquad\qquad\qquad\qquad\qquad\qquad$ $\square$

Further Notes for Chapter 2. The material in this chapter is at the very core of the theory of distributions since it provides a versatile calculus for distributions that naturally extends the scope of the standard calculus for ordinary functions. The definition of distributions used here is essentially that of the French mathematician Laurent Schwartz (1915–2002), cf. [66], though nowadays there are many books dealing at length with the classical topics discussed here. These include [10], [17], [21], [20], [22], [23], [27], [28], [33], [34], [65], [71], [74], [76], [77], [78], and the reader is referred to these sources for other angles of exposition. In particular, in [34], [27], [74], distributions are defined on smooth manifolds, while in [54] the notion of distributions is adapted to rough settings.

2.10 Additional Exercises for Chapter 2

Exercise 2.122. Consider the mapping $u : \mathcal{D}(\mathbb{R}) \to \mathbb{C}$ by setting

$$u(\varphi) := \sum_{j=1}^{\infty}\left[\varphi\left(\frac{1}{j^2}\right) - \varphi(0)\right], \qquad \forall\,\varphi \in C_0^{\infty}(\mathbb{R}). \tag{2.10.1}$$

Prove that u is well defined. Is u a distribution? If yes, what is the order of u?

Exercise 2.123. Prove that there exists $u \in \mathcal{D}'(\Omega)$ for which the following statement is false:

For each compact K contained in Ω there exist $C > 0$ and $k \in \mathbb{N}_0$ such that

$$|\langle u, \varphi \rangle| \leq C \sup_{x \in K, |\alpha| \leq k} |\partial^{\alpha}\varphi(x)|, \qquad \forall\,\varphi \in C_0^{\infty}(\Omega).$$

$$\tag{2.10.2}$$

Exercise 2.124. Prove that $|x|^N \ln|x| \in L_{loc}^1(\mathbb{R}^n)$ whenever N is a real number satisfying $N > -n$. Thus, in particular, $|x|^N \ln|x| \in \mathcal{D}'(\mathbb{R}^n)$ when $N > -n$.

Exercise 2.125. Suppose $n \geq 2$ and given $\xi \in S^{n-1}$ define $f(x) := \ln|x \cdot \xi|$ for each $x \in \mathbb{R}^n \setminus \{0\}$. Prove that $f \in L_{loc}^1(\mathbb{R}^n)$. In particular, $\ln|x \cdot \xi| \in \mathcal{D}'(\mathbb{R}^n)$.

Exercise 2.126. Prove that $(\ln|x|)' = \text{P.V.}\,\frac{1}{x}$ in $\mathcal{D}'(\mathbb{R})$, where $\text{P.V.}\,\frac{1}{x}$ is the distribution defined in (2.1.13).

Exercise 2.127. Let $f : \mathbb{R} \to \mathbb{R}$ be defined by $f(x) = x\ln|x| - x$ if $x \in \mathbb{R} \setminus \{0\}$ and $f(0) = 0$. Prove that $f \in C^0(\mathbb{R})$ and the distributional derivative of f in $\mathcal{D}'(\mathbb{R})$ equals $\ln|x|$.

Exercise 2.128. Suppose $n \geq 2$. Prove that $\partial_j(\ln|x|) = \frac{x_j}{|x|^2}$ in $\mathcal{D}'(\mathbb{R}^n)$ for every $j \in \{1, \ldots, n\}$.

Exercise 2.129. Suppose $n \geq 3$. Prove that $\partial_j(\frac{1}{|x|^{n-2}}) = (2-n)\frac{x_j}{|x|^n}$ in $\mathcal{D}'(\mathbb{R}^n)$ for every $j \in \{1, \ldots, n\}$.

Exercise 2.130. Let $\theta \in C_0^\infty(\mathbb{R})$ be supported in the interval $(-1, 1)$ and such that $\int_\mathbb{R} \theta(t)\,dt = 1$. For each $j \in \mathbb{N}$, define

$$\psi_j(x) := \int_{1/j}^{j} j\,\theta(jx - jt)\,dt, \qquad \forall\, x \in \mathbb{R}.$$

Prove that $\psi_j \xrightarrow[j \to \infty]{\mathcal{D}'(\mathbb{R})} H$, where H is the Heaviside function.

Exercise 2.131. Let $u : \mathcal{D}(\mathbb{R}) \to \mathbb{C}$ defined by $u(\varphi) := \sum\limits_{j=1}^{\infty} \varphi^{(j)}(j)$ for each function $\varphi \in C_0^\infty(\mathbb{R})$. Prove that $u \in \mathcal{D}'(\mathbb{R})$ and that this distribution does not have finite order.

Exercise 2.132. For each $j \in \mathbb{N}$ define

$$f_j(x) := \frac{1}{j|x|^{n-\frac{1}{j}}} \qquad \text{for } x \in \mathbb{R}^n \setminus \{0\}.$$

Prove that $f_j \xrightarrow[j \to \infty]{\mathcal{D}'(\mathbb{R}^n)} \omega_{n-1}\delta$, where ω_{n-1} is the area of the unit sphere in $\mathbb{R}^n$.

Exercise 2.133. For each $\varepsilon > 0$ define

$$f_\varepsilon(x) := \frac{1}{\pi}\frac{\varepsilon}{x^2 + \varepsilon^2} \qquad \text{for } x \in \mathbb{R}.$$

Prove that $f_\varepsilon \xrightarrow[\varepsilon \to 0^+]{\mathcal{D}'(\mathbb{R})} \delta$.

Exercise 2.134. For each $\varepsilon > 0$ define

$$f_\varepsilon(x) := (4\pi\varepsilon)^{-\frac{n}{2}}e^{-\frac{|x|^2}{4\varepsilon}} \qquad \text{for } x \in \mathbb{R}^n.$$

Prove that $f_\varepsilon \xrightarrow[\varepsilon \to 0^+]{\mathcal{D}'(\mathbb{R}^n)} \delta$.

Exercise 2.135. Recall that $i = \sqrt{-1} \in \mathbb{C}$ and, for each $\varepsilon \in (0, \infty)$, define

$$f_\varepsilon^\pm(x) := \frac{1}{x \pm i\varepsilon} \qquad \text{for } x \in \mathbb{R}.$$

Also, recall the distribution from (2.1.13). Prove that

$$\frac{1}{x \pm i\varepsilon} \longrightarrow \mp i\pi\delta + \text{P.V.}\,\frac{1}{x} \qquad \text{as} \quad \varepsilon \to 0^+ \text{ in } \mathcal{D}'(\mathbb{R}). \tag{2.10.3}$$

This is the so-called Sokhotsky's formula.

Exercise 2.136. Prove that the sequence $\left\{\frac{\sin(jx)}{\pi x}\right\}_{j \in \mathbb{N}}$ converges to δ in $\mathscr{D}'(\mathbb{R})$ as $j \to \infty$.

Exercise 2.137. In each case determine if the given sequence of distributions in $\mathscr{D}'(\mathbb{R})$ indexed over $j \in \mathbb{N}$ converges and determine its limit whenever convergent. Below $m \in \mathbb{N}$ is fixed and $i = \sqrt{-1}$.

(a) $f_j(x) := \dfrac{\sqrt{j}}{\sqrt{\pi}} e^{-jx^2}$ for $x \in \mathbb{R}$;

(b) $f_j(x) ::= j^m \cos(jx)$ for $x \in \mathbb{R}$;

(c) $f_j(x) = \dfrac{2j^3 x^2}{\pi(1 + j^2 x^2)^2}$ for $x \in \mathbb{R}$;

(d) $u_j := (-1)^j \delta_{\frac{1}{j}}$;

(e) $u_j := \dfrac{j}{2}\left[\delta_{\frac{1}{j}} - \delta_{-\frac{1}{j}}\right]$;

(f) $f_j(x) := \dfrac{1}{x} \chi_{|x| \geq \frac{1}{j}}$ for $x \in \mathbb{R} \setminus \{0\}$;

(g) $f_j(x) := \dfrac{1}{j\pi} \dfrac{\sin^2(jx)}{x^2}$ for $x \in \mathbb{R} \setminus \{0\}$;

(h) $f_j(x) := j^m e^{ijx}$ for $x \in \mathbb{R}$;

(j) $f_j(x) := \begin{cases} je^{ijx} & \text{if } x > 0, \\ 0 & \text{if } x \leq 0, \end{cases}$ for $x \in \mathbb{R}$.

Exercise 2.138. Let $a \in \mathbb{R}$. Compute $(H(\cdot - a))'$ in $\mathscr{D}'(\mathbb{R})$.

Exercise 2.139. Consider the function

$$f : \mathbb{R} \longrightarrow \mathbb{R} \quad \text{defined by} \quad f(x) := \begin{cases} x & \text{if } x > a, \\ 0 & \text{if } x \leq a, \end{cases} \quad \forall\, x \in \mathbb{R}, \qquad (2.10.4)$$

where $a \in \mathbb{R}$ is fixed. Compute $(u_f)'$ in $\mathscr{D}'(\mathbb{R})$, where u_f is defined as in (2.1.6).

Exercise 2.140. Let $f : \mathbb{R} \to \mathbb{R}$ be defined by $f(x) := \sin|x|$ for every $x \in \mathbb{R}$. Compute $(u_f)'$ and $(u_f)''$ in $\mathscr{D}'(\mathbb{R})$.

Exercise 2.141. Let $I \subseteq \mathbb{R}$ be an open interval, $x_0 \in I$, and $f \in C^1(I \setminus \{x_0\})$ be such that $f' \in L^1_{loc}(I)$ (here f' is the pointwise derivative of f in $I \setminus \{x_0\}$). Prove that the one-sided limits $\lim\limits_{x \to x_0^+} f(x)$, $\lim\limits_{x \to x_0^-} f(x)$ exist and are finite, that $f \in L^1_{loc}(I)$, and that

$$(u_f)' = u_{f'} + \left[\lim_{x \to x_0^+} f(x) - \lim_{x \to x_0^-} f(x) \right] \delta_{x_0} \quad \text{in} \quad \mathscr{D}'(I).$$

Remark 2.142. Prove that there exist pointwise differentiable functions on $\mathbb{R}$ for which the distributional derivative does not coincide with the classical derivative. You may consider the function f defined by $f(x) := x^2 \cos(\frac{1}{x^2})$ for $x \neq 0$ and $f(0) := 0$, and show that $f \in C^1(\mathbb{R} \setminus \{0\})$ and also f is differentiable at the origin, while $f' \notin L^1_{loc}(\mathbb{R})$.

Exercise 2.143. Let $I \subseteq \mathbb{R}$ be an open interval, $x_0 \in I$, and let $m \in \mathbb{N}$. Suppose that $f \in C^\infty(I \setminus \{x_0\})$ is such that the pointwise derivatives $f', f'', \dots, f^{(m)}$, belong to $L^1_{loc}(I)$. Prove that for every $k \in \{0, , 1, \dots, m-1\}$ the limits $\lim_{x \to x_0^+} f^{(k)}(x)$ and $\lim_{x \to x_0^-} f^{(k)}(x)$ exist, are finite, and that

$$
u_f^{(m)} = u_{f^{(m)}} + \Big[\lim_{x \to x_0^+} f(x) - \lim_{x \to x_0^-} f(x) \Big] \delta_{x_0}^{(m-1)}
$$

$$
+ \Big[\lim_{x \to x_0^+} f'(x) - \lim_{x \to x_0^-} f'(x) \Big] \delta_{x_0}^{(m-2)} + \cdots
$$

$$
+ \Big[\lim_{x \to x_0^+} f^{(m-1)}(x) - \lim_{x \to x_0^-} f^{(m-1)}(x) \Big] \delta_{x_0} \quad \text{in } \mathcal{D}'(I).
$$

Exercise 2.144. Let $I \subseteq \mathbb{R}$ be an open interval and consider a sequence $\{x_k\}_{k \in \mathbb{N}}$ of points in I with no accumulation point in I. Suppose we are given a function $f \in C^1(I \setminus \{x_k : k \in \mathbb{N}\})$ such that its pointwise derivative f' belongs to $L^1_{loc}(I)$. Prove that for each $k \in \mathbb{N}$ the limits $\lim_{x \to x_k^\pm} f(x)$ exist and are finite, f belongs to $L^1_{loc}(I)$, and

$$
(u_f)' = u_{f'} + \sum_{k=1}^{\infty} \Big[\lim_{x \to x_k^+} f(x) - \lim_{x \to x_k^-} f(x) \Big] \delta_{x_k} \quad \text{in } \mathcal{D}'(I).
$$

Exercise 2.145. Let $f : \mathbb{R} \to \mathbb{R}$ be the function defined by $f(x) := \lfloor x \rfloor$ for each $x \in \mathbb{R}$, where $\lfloor x \rfloor$ is the integer part of x. Determine $(u_f)'$.

Exercise 2.146. Let $\Sigma \subset \mathbb{R}^n$ be a surface of class C^1 as in Definition 14.45, and denote by σ the surface measure on Σ. Define the mapping $\delta_\Sigma : C_0^\infty(\mathbb{R}^n) \to \mathbb{C}$ by $\delta_\Sigma(\varphi) := \int_\Sigma \varphi(x) \, d\sigma(x)$ for each $\varphi \in C_0^\infty(\mathbb{R}^n)$. Prove that $\delta_\Sigma \in \mathcal{D}'(\mathbb{R}^n)$, it has order zero, and $\text{supp} \, \delta_\Sigma = \Sigma$. Also show that if $g \in L^\infty(K \cap \Sigma)$ for each compact set K in $\mathbb{R}^n$, and if we define

$$
(g\delta_\Sigma)(\varphi) := \int_\Sigma g(x)\varphi(x) \, d\sigma(x), \qquad \forall \, \varphi \in C_0^\infty(\mathbb{R}^n), \tag{2.10.5}
$$

then $g\delta_\Sigma \in \mathcal{D}'(\mathbb{R}^n)$.

Exercise 2.147. Let $\Omega \subset \mathbb{R}^n$ be a domain of class C^1 (recall Definition 14.59) and denote by $\nu = (\nu_1, \dots, \nu_n)$ its outward unit normal. Denote by $\delta_{\partial\Omega}$ the distribution defined as in Exercise 2.146 corresponding to $\Sigma := \partial\Omega$.

Set $\Omega_+ := \Omega$ and $\Omega_- := \mathbb{R}^n \setminus \overline{\Omega}$. Suppose $f \in L^1_{loc}(\mathbb{R}^n)$ has the property that, for each $k \in \{1, 2\dots, n\}$, its distributional derivative $\partial_k f$ belongs to $L^1_{loc}(\mathbb{R}^n)$. In addition, assume that the restrictions $f_\pm := f|_{\Omega_\pm}$ satisfy $f_\pm \in C^1(\Omega_\pm)$ and that they may be extended to ensure that $f_\pm \in C^0(\overline{\Omega_\pm})$. Prove that for each k in $\{1, 2\dots, n\}$ the following equality holds:

$$
\partial_k u_f = u_{\partial_k f} + s_{\partial\Omega}(f)\nu_k \delta_{\partial\Omega} \quad \text{in } \mathcal{D}'(\mathbb{R}^n),
$$

where $s_{\partial\Omega}(f) : \partial\Omega \to \mathbb{C}$ is defined by

$$s_{\partial\Omega}(f)(x) := f_-(x) - f_+(x)$$

$$= \lim_{\mathbb{R}^n\setminus\overline{\Omega}\ni y\to x} f(y) - \lim_{\Omega\ni y\to x} f(y) \quad \text{for every } x \in \partial\Omega. \tag{2.10.6}$$

Exercise 2.148. Let $\Omega \subset \mathbb{R}^n$ be a bounded domain of class C^1 with outward unit normal $\nu = (\nu_1, \ldots, \nu_n)$. Prove that $\partial_k \chi_\Omega = -\nu_k \delta_{\partial\Omega}$ in $\mathcal{D}'(\mathbb{R}^n)$ for each $k \in \{1, 2 \ldots, n\}$.

Exercise 2.149. Suppose $R \in (0, \infty)$ and let $u \in \mathcal{D}'(\mathbb{R}^n)$ be such that

$$(|x|^2 - R^2)u = 0 \quad \text{in } \mathcal{D}'(\mathbb{R}^n). \tag{2.10.7}$$

Prove that u has compact support. Give an example of a distribution u satisfying condition (2.10.7).

Exercise 2.150. Let $f \in C^0(\Omega)$ be such that $u_f \in \mathcal{E}'(\Omega)$. Prove that f has compact support and $\operatorname{supp} u_f = \operatorname{supp} f$.

Exercise 2.151. Compute the derivatives of order $m \in \mathbb{N}$ of each distribution on $\mathbb{R}$ given below.

(a) $|x|$ (b) $\operatorname{sgn} x$ (c) $\cos x\, H$ (d) $\sin x\, H$ (e) $x^2 \chi_{[-1,1]}$

Exercise 2.152. Consider the set $A := \{(x, y) \in \mathbb{R}^2 : |x - 2| + |y - 1| < 1\} \subset \mathbb{R}^2$. Compute $(\partial_1^2 - \partial_2^2)\chi_A$ in $\mathcal{D}'(\mathbb{R}^2)$.

Exercise 2.153. Let $f : \mathbb{R}^2 \to \mathbb{R}$ be defined by $f(x, y) := \chi_{[0,1]}(x - y)$ for $x, y \in \mathbb{R}$. Compute $\partial_1(u_f), \partial_2(u_f)$ in $\mathcal{D}'(\mathbb{R}^2)$. Prove that $\partial_1^2(u_f) - \partial_2^2(u_f) = 0$ in $\mathcal{D}'(\mathbb{R}^2)$.

Exercise 2.154. Let $\psi \in C^\infty(\Omega)$ be such that $\psi(x) \neq 0$ for every $x \in \Omega$. Prove that for each $v \in \mathcal{D}'(\Omega)$ there exists a unique solution $u \in \mathcal{D}'(\Omega)$ of the equation $\psi u = v$ in $\mathcal{D}'(\Omega)$.

Exercise 2.155. Let $\psi \in C^\infty(\Omega)$ and suppose $u_1, u_2 \in \mathcal{D}'(\Omega)$ are such that $u_1 \neq u_2$ and $\psi u_1 = \psi u_2$ in $\mathcal{D}'(\Omega)$. Prove that the set $\{x \in \Omega : \psi(x) = 0\}$ is not empty.

Exercise 2.156. Suppose $\{\Omega_j\}_{j\in I}$ is an open cover of the open set $\Omega \subseteq \mathbb{R}^n$ and there exists a family of distributions $\{u_j\}_{j\in I}$ such that $u_j \in \mathcal{D}'(\Omega_j)$ for each $j \in I$ and $u_j|_{\Omega_j\cap\Omega_k} = u_k|_{\Omega_j\cap\Omega_k}$ in $\mathcal{D}'(\Omega_j \cap \Omega_k)$ for every $j, k \in I$ such that $\Omega_j \cap \Omega_k \neq \varnothing$. Prove that there exists a unique distribution $u \in \mathcal{D}'(\Omega)$ with the property that $u|_{\Omega_j} = u_j$ in $\mathcal{D}'(\Omega_j)$ for every $j \in I$.

Exercise 2.157. Let $u \in \mathcal{D}'(\mathbb{R}^n)$ be such that $\langle u, \varphi \rangle = 0$ for every $\varphi \in C_0^\infty(\mathbb{R}^n)$ satisfying $\int_{\mathbb{R}^n} \varphi(x)\, dx = 0$. Prove that there exists $c \in \mathbb{C}$ such that $u = c$ in $\mathcal{D}'(\mathbb{R}^n)$.

Exercise 2.158. Let $\Omega \subseteq \mathbb{R}^n$ be open and connected and let $u \in \mathcal{D}'(\Omega)$ be such that $\partial_1 u = \partial_2 u = \cdots = \partial_n u = 0$ in $\mathcal{D}'(\Omega)$. Prove that there exists $c \in \mathbb{C}$ such that $u = c$ in $\mathcal{D}'(\Omega)$.

Exercise 2.159. Let $u \in \mathcal{D}'(\mathbb{R}^n)$ be such that $x_n u = 0$ in $\mathcal{D}'(\mathbb{R}^n)$. Prove that there exists $v \in \mathcal{D}'(\mathbb{R}^{n-1})$ such that $u(x', x_n) = v(x') \otimes \delta(x_n)$ in $\mathcal{D}'(\mathbb{R}^n)$.

Exercise 2.160. Let $u \in \mathcal{D}'(\mathbb{R}^n)$ be such that $x_1 u = \cdots = x_n u = 0$ in $\mathcal{D}'(\mathbb{R}^n)$. Determine the expression for u.

Exercise 2.161. Let $u \in \mathcal{D}'(\mathbb{R}^n)$ be such that $\partial_n u = 0$ in $\mathcal{D}'(\mathbb{R}^n)$. Prove that there exists $v \in \mathcal{D}'(\mathbb{R}^{n-1})$ such that $u(x', x_n) = v(x') \otimes 1$ in $\mathcal{D}'(\mathbb{R}^n)$, where 1 denotes the constant function (equal to 1) in $\mathbb{R}$.

Exercise 2.162. Let $v, w \in \mathcal{D}'(\mathbb{R})$ and define the distribution

$$u(x_1, x_2) := 1 \otimes v(x_2) + w(x_1) \otimes 1 \quad \text{in} \quad \mathcal{D}'(\mathbb{R}^2),$$

where 1 denotes the constant function (equal to 1) in $\mathbb{R}$. Prove that $\partial_1 \partial_2 u = 0$ in $\mathcal{D}'(\mathbb{R}^2)$.

Exercise 2.163. Let $u(x_1, x_2, x_3) := H(x_1) \otimes \delta(x_2) \otimes \delta(x_3)$ in $\mathcal{D}'(\mathbb{R}^3)$, where H is the Heaviside function on the real line. Compute $\partial_1 u, \partial_2 u, \partial_3 u$ in $\mathcal{D}'(\mathbb{R}^3)$.

Exercise 2.164. Consider the sequence

$$f_j(x) := (2\pi)^{-n} \int_{[-j,j] \times \cdots \times [-j,j]} e^{ix \cdot \xi} \, d\xi, \qquad \forall \, x \in \mathbb{R}^n, \ \forall \, j \in \mathbb{N}. \qquad (2.10.8)$$

Prove that $f_j \xrightarrow[j \to \infty]{\mathcal{D}'(\mathbb{R}^n)} \delta$.

Exercise 2.165. Solve each equation in $\mathcal{D}'(\mathbb{R})$ for u.

(1) $(x - 1)u = \delta$;
(2) $xu = a$, where $a \in C^\infty(\mathbb{R})$;
(3) $xu = v$, where $v \in \mathcal{D}'(\mathbb{R})$.

Exercise 2.166. Prove that the given convolutions exist and then compute them.

(a) $H * H$

(b) $H(-x) * H(-x)$

(c) $x^2 H * (\sin x \, H)$

(d) $\chi_{[0,1]} * (xH)$

(e) $|x|^2 * \delta_{\partial B(0,r)}$ where $r > 0$ and $\delta_{\partial B(0,r)}$ is as defined in Exercise 2.146 corresponding to the surface $\Sigma := \partial B(0, r)$.

Exercise 2.167. Let $a \in \mathbb{R}^n \setminus \{0\}$, $u_j := \delta_{ja}$, $v_j := \delta_{-ja}$, for each $j \in \mathbb{N}$. Determine $\lim_{j \to \infty} u_j$, $\lim_{j \to \infty} v_j$, $\lim_{j \to \infty}(u_j * v_j)$, in $\mathcal{D}'(\mathbb{R}^n)$.

Exercise 2.168. For each $j \in \mathbb{N}$, consider the functions $f_j(x) := \frac{(-1)^j j}{2} \chi_{\left[-\frac{1}{j}, \frac{1}{j}\right]}(x)$ and $g_j(x) := (-1)^j$, for every $x \in \mathbb{R}$. Determine if the given limits exist in $\mathcal{D}'(\mathbb{R})$.

(a) $\lim\limits_{j \to \infty} f_j$

(b) $\lim\limits_{j \to \infty} g_j$

(c) $\lim\limits_{j \to \infty} (f_j * g_j)$

Exercise 2.169. Let $u \in \mathcal{D}'(\mathbb{R}^n)$ and consider the map $\Lambda : \mathcal{D}(\mathbb{R}^n) \to \mathcal{E}(\mathbb{R}^n)$ given by $\Lambda(\varphi) := u * \varphi$, for every $\varphi \in C_0^\infty(\mathbb{R}^n)$. Prove that Λ is a well-defined, linear, and continuous map. Also prove that Λ commutes with translations, that is, if $x_0 \in \mathbb{R}^n$ and $\varphi \in C_0^\infty(\mathbb{R}^n)$, then $t_{x_0}(\Lambda(\varphi)) = \Lambda(t_{x_0}\varphi)$, where t_{x_0} is the map from (1.3.17).

Exercise 2.170. Suppose $\Lambda : \mathcal{D}(\mathbb{R}^n) \to \mathcal{E}(\mathbb{R}^n)$ is a linear, continuous map that commutes with translations (in the sense explained in Exercise 2.169). Prove that there exists a unique $u \in \mathcal{D}'(\mathbb{R}^n)$ such that $\Lambda(\varphi) = u * \varphi$ for every $\varphi \in C_0^\infty(\mathbb{R}^n)$.

Exercise 2.171. Let $u \in \mathcal{E}'(\mathbb{R}^n)$ be such that $\langle u, x^\alpha \rangle = 0$ for every $\alpha \in \mathbb{N}^n$. Prove that $u = 0$ in $\mathcal{E}'(\mathbb{R}^n)$.

Exercise 2.172. Let $u : \mathcal{E}(\mathbb{R}) \to \mathbb{C}$ be the functional defined by

$$u(\varphi) := \lim_{k \to \infty} \left[\left(\sum_{j=1}^{k} \varphi(\tfrac{1}{j}) \right) - k\varphi(0) - \varphi'(0) \ln k \right], \qquad \forall \, \varphi \in C^\infty(\mathbb{R}).$$

Prove that $u \in \mathcal{E}'(\mathbb{R})$ and determine $\operatorname{supp} u$.

Exercise 2.173. For each $j \in \mathbb{N}$ consider the function $f_j : \mathbb{R} \to \mathbb{R}$ defined by $f_j(x) := \frac{j}{2}$ if $|x| \le \frac{1}{j}$ and $f_j(x) := 0$ if $|x| > \frac{1}{j}$. Prove that $f_j \xrightarrow[j \to \infty]{\mathcal{E}'(\mathbb{R})} \delta$.

Exercise 2.174. For each $j \in \mathbb{N}$ consider the function $f_j : \mathbb{R} \to \mathbb{R}$ defined by $f_j(x) := \frac{1}{j}$ if $|x| \le j$ and $f_j(x) := 0$ if $|x| > j$. Prove that the sequence $\{f_j\}_{j \in \mathbb{N}}$ converges in $\mathcal{D}'(\mathbb{R})$ but not in $\mathcal{E}'(\mathbb{R})$.

Exercise 2.175. Let $\psi \in C_0^\infty(\mathbb{R}^n)$ be such that $\int_{\mathbb{R}^n} \psi(x) \, dx = 1$ and for each $j \in \mathbb{N}$ define $f_j : \mathbb{R}^n \to \mathbb{C}$ by $f_j(x) := j^n \psi(jx)$ for each $x \in \mathbb{R}^n$. Prove that $f_j \xrightarrow[j \to \infty]{\mathcal{E}'(\mathbb{R}^n)} \delta$.

Exercise 2.176. Let $\{x_j\}_{j \in \mathbb{N}}$ be a sequence of points in $\mathbb{R}^n$. Prove that $\{x_j\}_{j \in \mathbb{N}}$ is convergent in $\mathbb{R}^n$ if and only if $\{\delta_{x_j}\}_{j \in \mathbb{N}}$ is convergent in $\mathcal{E}'(\mathbb{R}^n)$.

Exercise 2.177. Let $a \in \mathbb{R}$ and $k \in \mathbb{N}_0$ be given. Prove that

$$(x + a)\delta_a^{(k)} = 2a \, \delta_a^{(k)} + k\delta_a^{(k-1)} \quad \text{in} \quad \mathcal{D}'(\mathbb{R}), \tag{2.10.9}$$

$$(x^2 - a^2)\delta_a^{(k)} = -2k \, a \, \delta_a^{(k-1)} + k(k-1)\delta_a^{(k-2)} \quad \text{in} \quad \mathcal{D}'(\mathbb{R}), \tag{2.10.10}$$

with the convention that $\delta_a^{(-m)} := 0 \in \mathcal{D}'(\mathbb{R})$ for each $m \in \mathbb{N}$.

Chapter 3
The Schwartz Space and the Fourier Transform

Abstract This chapter contains material pertaining to the Schwartz space of functions rapidly decaying at infinity and the Fourier transform in such a setting.

3.1 The Schwartz Space of Rapidly Decreasing Functions

Recall that if $f \in L^1(\mathbb{R}^n)$ then the Fourier transform of f is the mapping $\widehat{f} : \mathbb{R}^n \to \mathbb{C}$ defined by

$$\widehat{f}(\xi) := \int_{\mathbb{R}^n} e^{-ix\cdot\xi} f(x)\, dx \quad \text{for each} \quad \xi \in \mathbb{R}^n, \tag{3.1.1}$$

where $i := \sqrt{-1} \in \mathbb{C}$. Note that under the current assumptions the integral in (3.1.1) is absolutely convergent (which means that $\widehat{f}$ is well-defined pointwise in $\mathbb{R}^n$) and one has

$$\sup_{\xi\in\mathbb{R}^n} |\widehat{f}(\xi)| \le \|f\|_{L^1(\mathbb{R}^n)} \quad \text{and} \quad \widehat{f} \in C^0(\mathbb{R}^n). \tag{3.1.2}$$

where the second condition is seen by applying Lebesgue's Dominated Convergence Theorem. Hence, the mapping

$$\mathcal{F} : L^1(\mathbb{R}^n) \to \{g \in C^0(\mathbb{R}^n) : g \text{ is bounded}\}, \quad \mathcal{F}f := \widehat{f}, \quad \forall f \in L^1(\mathbb{R}^n), \tag{3.1.3}$$

called the Fourier transform, is well defined. Besides being continuous, functions belonging to the image of $\mathcal{F}$ also vanish at infinity. This property is proved next.

Proposition 3.1. *If $f \in L^1(\mathbb{R}^n)$, then* $\lim\limits_{|\xi|\to\infty} \widehat{f}(\xi) = 0$.

Proof. First, consider the case when $f \in C_0^\infty(\mathbb{R}^n)$. In this scenario, integrating by parts gives $|\xi|^2 \widehat{f}(\xi) = -\widehat{\Delta f}(\xi)$ for every $\xi \in \mathbb{R}^n \setminus \{0\}$, where $\Delta f := \sum\limits_{j=1}^{n} \partial_j^2 f$. Hence,

© Springer Nature Switzerland AG 2018
D. Mitrea, *Distributions, Partial Differential Equations, and Harmonic Analysis,*
Universitext, https://doi.org/10.1007/978-3-030-03296-8_3

$$|\widehat{f}(\xi)| \leq \frac{|\widehat{\Delta f}(\xi)|}{|\xi|^2} \leq \frac{\|\Delta f\|_{L^1(\mathbb{R}^n)}}{|\xi|^2}, \qquad \forall \, \xi \in \mathbb{R}^n \setminus \{0\}, \tag{3.1.4}$$

from which it is clear that $\lim_{|\xi| \to \infty} \widehat{f}(\xi) = 0$ in this case.

Consider now the case when f is an arbitrary function in $L^1(\mathbb{R}^n)$. Since $C_0^\infty(\mathbb{R}^n)$ is dense in the latter space, for each $\varepsilon > 0$ fixed there exists $g \in C_0^\infty(\mathbb{R}^n)$ such that $\|f - g\|_{L^1(\mathbb{R}^n)} \leq \frac{\varepsilon}{2}$. Also, from what we proved so far, there exists $R > 0$ with the property that $|\widehat{g}(\xi)| \leq \frac{\varepsilon}{2}$ whenever $|\xi| > R$. Keeping this in mind and using the linearity of the Fourier transform as well as the estimate in (3.1.2) we may write

$$|\widehat{f}(\xi)| \leq |\widehat{(f-g)}(\xi)| + |\widehat{g}(\xi)| \leq \|f - g\|_{L^1(\mathbb{R}^n)} + \frac{\varepsilon}{2} \leq \varepsilon, \quad \text{if } |\xi| > R. \tag{3.1.5}$$

From this, the desired conclusion follows. The proof of the proposition is therefore complete. $\qquad\qquad\Box$

We are very much interested in the possibility of extending the action of the Fourier transform from functions to distributions, though this is going to be accomplished later. For now, we note the following consequence of Fubini's theorem:

$$\int_{\mathbb{R}^n} \widehat{f}(\xi) g(\xi) \, d\xi = \int_{\mathbb{R}^n} f(x) \widehat{g}(x) \, dx, \qquad \forall \, f, g \in L^1(\mathbb{R}^n). \tag{3.1.6}$$

Identity (3.1.6) might suggest defining the Fourier transform of a distribution based on duality. However, there is a serious impediment in doing so. Specifically, while for every $\varphi \in C_0^\infty(\mathbb{R}^n)$ we have $\widehat{\varphi} \in C^\infty(\mathbb{R}^n)$ (as may be seen directly from (3.1.1)) and $\widehat{\varphi}$ decays at infinity (as proved in Proposition 3.1), we nonetheless have

$$\mathcal{F}(C_0^\infty(\mathbb{R}^n)) \not\subset C_0^\infty(\mathbb{R}^n). \tag{3.1.7}$$

In fact, we claim that

$$\varphi \in C_0^\infty(\mathbb{R}^n) \text{ and } \widehat{\varphi} \in C_0^\infty(\mathbb{R}^n) \implies \varphi = 0. \tag{3.1.8}$$

To see that this is the case, suppose $\varphi \in C_0^\infty(\mathbb{R}^n)$ is such that $\widehat{\varphi}$ has compact support in $\mathbb{R}^n$, and pick an arbitrary point $x^* = (x_1^*, \ldots, x_n^*) \in \mathbb{R}^n$. Define the function $\Phi : \mathbb{C} \to \mathbb{C}$ by setting

$$\Phi(z) := \int_{\mathbb{R}^n} e^{-izx_1 + \sum_{j=2}^n x_j^* x_j} \varphi(x_1, \ldots, x_n) \, dx_1 \cdots dx_n, \quad \text{for } z \in \mathbb{C}. \tag{3.1.9}$$

Then Φ is analytic in $\mathbb{C}$ and $\Phi(t) = \widehat{\varphi}(t, x_2^*, \ldots, x_n^*)$ for every $t \in \mathbb{R}$. Given that $\widehat{\varphi}$ has compact support, this implies that Φ vanishes on $\mathbb{R} \setminus [-R, R]$ if $R > 0$ is suitably large. The identity theorem for ordinary analytic functions of one complex variable then forces $\Phi = 0$ everywhere in $\mathbb{C}$. In particular, $\widehat{\varphi}(x^*) = \Phi(x_1^*) = 0$. Since $x^* \in \mathbb{R}^n$ has been chosen arbitrarily, we conclude that $\widehat{\varphi} = 0$ in $\mathbb{R}^n$. However, as we will see in the sequel, the Fourier transform is injective on $C_0^\infty(\mathbb{R}^n)$, so (3.1.8) follows.

To overcome the difficulty highlighted in (3.1.7), we introduce a new (topological vector) space of functions, that contains $C_0^\infty(\mathbb{R}^n)$, is invariant under $\mathcal{F}$, and whose dual is a subspace of $\mathcal{D}'(\mathbb{R}^n)$. This is the space of Schwartz functions, named after the French mathematician Laurent–Moïse Schwartz (1915–2002) who pioneered the theory of distributions and first considered this space in connection with the Fourier transform.

Before presenting the definition of Schwartz functions, we introduce some notation, motivated by the observation that each time a partial derivative of $\widehat{\varphi}$ is taken, the exponential introduces i as a multiplicative factor. To adjust for this factor, it is therefore natural to re-normalize the ordinary partial differentiation operators as follows:

$$D_j := \tfrac{1}{i}\partial_j, \quad j = 1,\ldots,n, \quad D := (D_1,\ldots,D_n),$$
$$D^\alpha := D_1^{\alpha_1}\cdots D_n^{\alpha_n}, \quad \forall\, \alpha = (\alpha_1,\ldots,\alpha_n) \in \mathbb{N}_0^n. \tag{3.1.10}$$

At times, we will also use subscripts to specify the variable with respect to which the differentiation is taken. For example, D_x^α stands for D^α with the additional specification that the differentiation is taken with respect to the variable $x \in \mathbb{R}^n$.

Fix now $\alpha, \beta \in \mathbb{N}_0^n$ and observe that for each $\varphi \in C_0^\infty(\mathbb{R}^n)$ integration by parts implies

$$\xi^\beta D_\xi^\alpha \widehat{\varphi}(\xi) = \int_{\mathbb{R}^n} \xi^\beta e^{-ix\cdot\xi}(-x)^\alpha \varphi(x)\,dx = \int_{\mathbb{R}^n} \left[(-D_x)^\beta(e^{-ix\cdot\xi})\right](-x)^\alpha \varphi(x)\,dx$$

$$= \int_{\mathbb{R}^n} e^{-ix\cdot\xi} D_x^\beta\left[(-x)^\alpha \varphi(x)\right] dx. \tag{3.1.11}$$

Hence,

$$\sup_{\xi \in \mathbb{R}^n} \left|\xi^\beta D_\xi^\alpha \widehat{\varphi}(\xi)\right| \leq \int_{\mathbb{R}^n} |D_x^\beta(x^\alpha \varphi(x))|\,dx < \infty. \tag{3.1.12}$$

The conclusion from (3.1.12) is that derivatives of any order of $\widehat{\varphi}$ decrease at ∞ faster than any polynomial. This suggests making the following definition.

Definition 3.2. The `Schwartz class` of rapidly decreasing functions is defined as

$$\mathcal{S}(\mathbb{R}^n) := \{\varphi \in C^\infty(\mathbb{R}^n) : \sup_{x \in \mathbb{R}^n} |x^\beta \partial^\alpha \varphi(x)| < \infty, \ \forall\, \alpha, \beta \in \mathbb{N}_0^n\}. \tag{3.1.13}$$

We shall simply say that φ is a `Schwartz function` if $\varphi \in \mathcal{S}(\mathbb{R}^n)$. Obviously,

$$C_0^\infty(\mathbb{R}^n) \subset \mathcal{S}(\mathbb{R}^n) \subset C^\infty(\mathbb{R}^n) \tag{3.1.14}$$

though both inclusions are strict. An example of a Schwartz function that is not compactly supported is provided below.

Exercise 3.3. Prove that for each fixed number $a \in (0, \infty)$, the function f, defined by $f(x) := e^{-a|x|^2}$ for each $x \in \mathbb{R}^n$, belongs to $\mathcal{S}(\mathbb{R}^n)$ and has the property that $\operatorname{supp} f = \mathbb{R}^n$.

Other elementary observations pertaining to the Schwartz class from Definition 3.2 are recorded below.

Remark 3.4. One has

$$S(\mathbb{R}^n) = \{\varphi \in C^\infty(\mathbb{R}^n) : \sup_{x \in \mathbb{R}^n} |x^\beta D^\alpha \varphi(x)| < \infty, \ \forall \alpha, \beta \in \mathbb{N}_0^n\}, \tag{3.1.15}$$

and if $\varphi \in C^\infty(\mathbb{R}^n)$, then $\varphi \in S(\mathbb{R}^n)$ if and only if

$$\sup_{x \in \mathbb{R}^n} [(1 + |x|)^m |\partial^\alpha \varphi(x)|] < \infty, \qquad \forall m \in \mathbb{N}_0, \ \forall \alpha \in \mathbb{N}_0^n, \ |\alpha| \leq m. \tag{3.1.16}$$

Indeed, (3.1.15) is immediate from Definition 3.2. Also, the second claim in the remark readily follows from the observation that for each $m \in \mathbb{N}$ there exists a constant $C \in [1, \infty)$ with the property that

$$C^{-1} |x|^m \leq \sum_{|\gamma|=m} |x^\gamma| \leq C |x|^m, \qquad \forall x \in \mathbb{R}^n. \tag{3.1.17}$$

In turn, the second inequality in (3.1.17) is seen by noting that for each $\alpha \in \mathbb{N}_0^n$ with nonempty support and each $x = (x_1, \dots, x_n) \in \mathbb{R}^n$ we have (recall (0.0.5), (0.0.8))

$$|x^\alpha| = \prod_{j \in \mathrm{supp}\,\alpha} |x_j|^{\alpha_j} \leq \prod_{j \in \mathrm{supp}\,\alpha} |x|^{\alpha_j} = |x|^{|\alpha|}. \tag{3.1.18}$$

To justify the first inequality in (3.1.17), consider the function $g(x) := \sum_{|\gamma|=m} |x^\gamma|$ for $x \in \mathbb{R}^n$. Then its restriction to S^{n-1} attains a nonzero minimum, and the desired inequality follows by rescaling.

Exercise 3.5. Prove that if $f \in S(\mathbb{R}^n)$ then for every $\alpha, \beta \in \mathbb{N}_0^n$ and $N \in \mathbb{N}$ there exists $C = C_{f,N,\alpha,\beta} \in (0, \infty)$ such that

$$\left| x^\alpha \partial^\beta f(x) \right| \leq \frac{C}{(1 + |x|)^N} \qquad \text{for every } x \in \mathbb{R}^n. \tag{3.1.19}$$

Use this to deduce that $S(\mathbb{R}^n) \subset L^p(\mathbb{R}^n)$ for every $p \in [1, \infty]$.

In particular, $S(\mathbb{R}^n) \subset L^1(\mathbb{R}^n)$ which, in concert with (3.1.3), allows us to consider the Fourier transform on $S(\mathbb{R}^n)$. Also, $\mathcal{F}(C_0^\infty(\mathbb{R}^n)) \subseteq S(\mathbb{R}^n)$ as seen from the computation in (3.1.12).

Clearly, $S(\mathbb{R}^n)$ is a vector space when endowed with the canonical operations of addition of functions and multiplication by complex numbers. For a detailed discussion regarding the topology we consider on $S(\mathbb{R}^n)$ see Section 14.1.0.6. We continue to denote by $S(\mathbb{R}^n)$ the respective topological vector space and we single out here a few important facts that are useful for our analysis.

Fact 3.6. $S(\mathbb{R}^n)$ *is a Frechét space, i.e.,* $S(\mathbb{R}^n)$ *is a locally convex, metrizable, complete, topological vector space over* $\mathbb{C}$.

Fact 3.7. *A sequence $\{\varphi_j\}_{j\in\mathbb{N}} \subset \mathcal{S}(\mathbb{R}^n)$ converges in $\mathcal{S}(\mathbb{R}^n)$ to some $\varphi \in \mathcal{S}(\mathbb{R}^n)$ provided*

$$\sup_{x\in\mathbb{R}^n} \left| x^\beta \partial^\alpha [\varphi_j(x) - \varphi(x)] \right| \xrightarrow[j\to\infty]{} 0, \qquad \forall\, \alpha, \beta \in \mathbb{N}_0^n, \qquad (3.1.20)$$

in which case we use the notation $\varphi_j \xrightarrow[j\to\infty]{\mathcal{S}(\mathbb{R}^n)} \varphi$.

Exercise 3.8. Use (3.1.17) to prove that $\varphi_j \xrightarrow[j\to\infty]{\mathcal{S}(\mathbb{R}^n)} \varphi$ if and only if

$$\sup_{\substack{x\in\mathbb{R}^n \\ \alpha\in\mathbb{N}_0^n, |\alpha|\le m}} \left[(1+|x|)^m \left| \partial^\alpha [\varphi_j(x) - \varphi(x)] \right| \right] \xrightarrow[j\to\infty]{} 0, \qquad \forall\, m \in \mathbb{N}_0. \qquad (3.1.21)$$

It is useful to note that the Schwartz class embeds continuously into Lebesgue spaces.

Exercise 3.9. Prove that if $p \in [1, \infty]$ and a sequence of functions $\{\varphi_j\}_{j\in\mathbb{N}}$ in $\mathcal{S}(\mathbb{R}^n)$ converges in $\mathcal{S}(\mathbb{R}^n)$ to some $\varphi \in \mathcal{S}(\mathbb{R}^n)$ then $\{\varphi_j\}_{j\in\mathbb{N}}$ also converges in $L^p(\mathbb{R}^n)$ to φ.

Hint: Use Exercise 3.8.

For further reference we also single out the analogue of (2.7.39) for the class of Schwartz functions.

Remark 3.10. Let $m, n \in \mathbb{N}$. Given $\varphi \in C^\infty(\mathbb{R}^n \times \mathbb{R}^m)$ recall the definition of the function $\varphi^\top \in C^\infty(\mathbb{R}^m \times \mathbb{R}^n)$ from (2.7.38) (presently used with $V := \mathbb{R}^n$ and $U := \mathbb{R}^m$). Then a combination of Fact 3.6, Theorem 14.1, Fact 3.7, and (2.7.38), implies that the mapping

$$\mathcal{S}(\mathbb{R}^n \times \mathbb{R}^m) \ni \varphi \mapsto \varphi^\top \in \mathcal{S}(\mathbb{R}^m \times \mathbb{R}^n) \qquad (3.1.22)$$

is linear and continuous.

Definition 3.11. The space of slowly increasing functions in $\mathbb{R}^n$ is defined as

$$\mathcal{L}(\mathbb{R}^n) := \{a \in C^\infty(\mathbb{R}^n) : \forall\, \alpha \in \mathbb{N}_0^n \ \exists\, k \in \mathbb{N}_0 \ \text{ such that}$$

$$\sup_{x\in\mathbb{R}^n} \left[(1+|x|)^{-k} |\partial^\alpha a(x)| \right] < \infty \}. \qquad (3.1.23)$$

Note that an immediate consequence of Definition 3.11 is that $\mathcal{L}(\mathbb{R}^n)$ is stable under differentiation (i.e., if $a \in \mathcal{L}(\mathbb{R}^n)$ then $\partial^\alpha a \in \mathcal{L}(\mathbb{R}^n)$ for every $\alpha \in \mathbb{N}_0^n$). Also,

$$\mathcal{S}(\mathbb{R}^n) \subset \mathcal{L}(\mathbb{R}^n), \qquad (3.1.24)$$

though $\mathcal{L}(\mathbb{R}^n)$ contains many additional functions that lack decay as, for example, the class of polynomials (other examples are contained in the two exercises below).

Exercise 3.12. Prove that the function $f(x) := e^{i|x|^2}$, $x \in \mathbb{R}^n$, belongs to $\mathcal{L}(\mathbb{R}^n)$.

Exercise 3.13. Prove that for each $s \in \mathbb{R}$ the function $f(x) := (1 + |x|^2)^s$, $x \in \mathbb{R}^n$, belongs to $\mathcal{L}(\mathbb{R}^n)$.

Some other basic properties of the Schwartz class are collected in the next theorem.

Theorem 3.14. *The following statements are true.*

(a) *For each $a \in \mathcal{L}(\mathbb{R}^n)$, the mapping $S(\mathbb{R}^n) \ni \varphi \mapsto a\varphi \in S(\mathbb{R}^n)$ is well defined, linear, and continuous.*

(b) *For every $\alpha \in \mathbb{N}_0^n$, the mapping $S(\mathbb{R}^n) \ni \varphi \mapsto \partial^\alpha \varphi \in S(\mathbb{R}^n)$ is well defined, linear, and continuous.*

(c) *$\mathcal{D}(\mathbb{R}^n) \hookrightarrow S(\mathbb{R}^n) \hookrightarrow \mathcal{E}(\mathbb{R}^n)$ and the embeddings are continuous.*

(d) *$C_0^\infty(\mathbb{R}^n)$ is sequentially dense in $S(\mathbb{R}^n)$. Also, the Schwartz class $S(\mathbb{R}^n)$ is sequentially dense in $\mathcal{E}(\mathbb{R}^n)$.*

(e) *If $m, n \in \mathbb{N}$ and $f \in S(\mathbb{R}^m)$, $g \in S(\mathbb{R}^n)$, then $f \otimes g \in S(\mathbb{R}^m \times \mathbb{R}^n)$ and the mapping*

$$S(\mathbb{R}^m) \times S(\mathbb{R}^n) \ni (f, g) \mapsto f \otimes g \in S(\mathbb{R}^m \times \mathbb{R}^n) \tag{3.1.25}$$

is bilinear and continuous.

(f) *If $f, g \in S(\mathbb{R}^n)$ then $f * g \in S(\mathbb{R}^n)$ and the mapping*

$$S(\mathbb{R}^n) \times S(\mathbb{R}^n) \ni (f, g) \mapsto f * g \in S(\mathbb{R}^n) \tag{3.1.26}$$

is bilinear and continuous.

Proof. Clearly, the mappings in *(a)* and *(b)* are linear. By Fact 3.6 and Theorem 14.1, their continuity is equivalent with sequential continuity at 0, something that can be easily checked using Fact 3.7. Moving on to the statement in *(c)*, we first prove that $\mathcal{D}(\mathbb{R}^n)$ embeds continuously into $S(\mathbb{R}^n)$. Consider the mapping $\iota : \mathcal{D}(\mathbb{R}^n) \to S(\mathbb{R}^n)$ defined by $\iota(\varphi) := \varphi$ for each $\varphi \in C_0^\infty(\mathbb{R}^n)$. From (3.1.14) this is a well-defined and linear mapping. To see that ι is sequentially continuous at $0 \in \mathcal{D}(\mathbb{R}^n)$, consider $\varphi_j \xrightarrow[j\to\infty]{\mathcal{D}(\mathbb{R}^n)} 0$. Then there exists a compact set $K \subset \mathbb{R}^n$ with the property that $\operatorname{supp}\varphi_j \subseteq K$ for every $j \in \mathbb{N}$, and $\lim_{j\to\infty} \sup_{x\in K} |\partial^\alpha \varphi_j| = 0$ for every $\alpha \in \mathbb{N}_0^n$. Hence, for any $\alpha, \beta \in \mathbb{N}_0^n$,

$$\sup_{x\in\mathbb{R}^n} \left| x^\beta \partial^\alpha \varphi_j(x) \right| = \sup_{x\in K} \left| x^\beta \partial^\alpha \varphi_j(x) \right| \le C \sup_{x\in K} \left| \partial^\alpha \varphi_j(x) \right| \xrightarrow[j\to\infty]{} 0, \tag{3.1.27}$$

proving that ι is sequentially continuous at the origin. Recalling now Fact 3.6 and Theorem 14.6, we conclude that ι is continuous.

Our next goal is to show that $S(\mathbb{R}^n)$ embeds continuously in $\mathcal{E}(\mathbb{R}^n)$. From (3.1.14) we have that the identity $\iota : S(\mathbb{R}^n) \to \mathcal{E}(\mathbb{R}^n)$ given by $\iota(f) := f$, for each $f \in S(\mathbb{R}^n)$, is a well-defined linear map. By Fact 3.6, Fact 1.8, and Theorem 14.1, ι is continuous if and only if it is sequentially continuous at zero. However, if $f_j \xrightarrow[j\to\infty]{S(\mathbb{R}^n)} 0$ then for any compact set $K \subset \mathbb{R}^n$ and any $\alpha \in \mathbb{N}_0^n$,

$$\lim_{j\to\infty}\sup_{x\in K}|\partial^\alpha f_j(x)| \le \lim_{j\to\infty}\sup_{x\in\mathbb{R}^n}|\partial^\alpha f_j(x)| = 0. \tag{3.1.28}$$

This shows that ι is sequentially continuous at zero, finishing the proof (c).

Next, we prove the statement in (d). Let $f \in \mathcal{S}(\mathbb{R}^n)$ be arbitrary and, for some fixed $\psi \in C_0^\infty(\mathbb{R}^n)$ satisfying $\psi \equiv 1$ in a neighborhood of $\overline{B(0,1)}$, define the sequence of functions $f_j : \mathbb{R}^n \to \mathbb{C}$ by setting $f_j(x) := \psi(\frac{x}{j})f(x)$ for every $x \in \mathbb{R}^n$ and every $j \in \mathbb{N}$. Then $f_j \in C_0^\infty(\mathbb{R}^n)$ and $f_j = f$ on $\overline{B(0,j)}$ for each $j \in \mathbb{N}$. We claim that

$$f_j \xrightarrow[j\to\infty]{\mathcal{S}(\mathbb{R}^n)} f. \tag{3.1.29}$$

To see that this is the case, if $\alpha,\beta \in \mathbb{N}_0^n$ are arbitrary, by making use of Leibniz's formula (14.2.6) and the fact that $\psi(\frac{x}{j}) = 1$ for each $x \in \overline{B(0,j)}$, we may write

$$\sup_{x\in\mathbb{R}^n}\left|x^\beta\partial^\alpha(f_j(x) - f(x))\right| = \sup_{x\in\mathbb{R}^n}\left|x^\beta \sum_{\gamma\le\alpha}\frac{\alpha!}{\gamma!(\alpha-\gamma)!}\partial^\gamma f(x)\partial^{\alpha-\gamma}\left[\psi\left(\frac{x}{j}\right) - 1\right]\right|$$

$$\le \sup_{|x|\ge j}\left|x^\beta \sum_{\gamma<\alpha}\frac{\alpha!}{\gamma!(\alpha-\gamma)!}\partial^\gamma f(x)\partial^{\alpha-\gamma}\left[\psi\left(\frac{x}{j}\right)\right]\right|$$

$$+ \sup_{|x|\ge j}\left|x^\beta\partial^\alpha f(x)\left[\psi\left(\frac{x}{j}\right) - 1\right]\right|. \tag{3.1.30}$$

Since $\psi \in C_0^\infty(\mathbb{R}^n)$, it follows that there exists a finite constant $C > 0$, depending only on ψ and α, such that

$$\sup_{|x|\ge j}\left|\partial^{\alpha-\gamma}\left[\psi\left(\frac{x}{j}\right)\right]\right| \le \frac{C}{j}, \qquad \forall\gamma\in\mathbb{N}_0^n,\ \gamma<\alpha,\ \forall j\in\mathbb{N}. \tag{3.1.31}$$

Also, since $f \in \mathcal{S}(\mathbb{R}^n)$, we may invoke (3.1.19) to conclude that there exists some $C = C_{f,\alpha,\beta} \in (0,\infty)$ such that

$$\sup_{|x|\ge j}\left|x^\beta\partial^\alpha f(x)\left[\psi\left(\frac{x}{j}\right) - 1\right]\right| \le \frac{C}{j}[1 + \|\psi\|_{L^\infty(\mathbb{R}^n)}]. \tag{3.1.32}$$

Combining (3.1.30), (3.1.31), (3.1.32), and keeping in mind that $f \in \mathcal{S}(\mathbb{R}^n)$, we obtain

$$\sup_{x\in\mathbb{R}^n}|x^\beta\partial^\alpha(f_j(x) - f(x))| \tag{3.1.33}$$

$$\le \frac{C}{j}\sup_{x\in\mathbb{R}^n}\left|x^\beta \sum_{\gamma\le\alpha}\frac{\alpha!}{\gamma!(\alpha-\gamma)!}\partial^\gamma f(x)\right| + \frac{C}{j} \xrightarrow[j\to\infty]{} 0.$$

This shows that $f_j \xrightarrow[j\to\infty]{S(\mathbb{R}^n)} f$ and completes the proof of the fact that $C_0^\infty(\mathbb{R}^n)$ is sequentially dense in $S(\mathbb{R}^n)$. The sequential continuity of $S(\mathbb{R}^n)$ in $\mathcal{E}(\mathbb{R}^n)$ is a consequence of Exercise 1.13 and (3.1.14). This completes the proof of *(d)*.

The claims in part *(e)* follow using the observation that

$$\left|(x,y)^{(\alpha,\beta)} \partial_x^\gamma \partial_y^\mu (f \otimes g)(x,y)\right| = \left|x^\alpha \partial^\gamma f(x)\right|\left|y^\beta \partial^\mu g(y)\right|, \tag{3.1.34}$$

for every $(x,y) \in \mathbb{R}^m \times \mathbb{R}^n$, for every $f \in S(\mathbb{R}^m)$, $g \in S(\mathbb{R}^n)$, and every $\alpha, \gamma \in \mathbb{N}_0^m$, $\beta, \mu \in \mathbb{N}_0^n$.

Consider now the statement in *(f)*. Since $S(\mathbb{R}^n) \subset L^1(\mathbb{R}^n)$ (cf. Exercise 3.5) the convolution between two functions in $S(\mathbb{R}^n)$ is meaningfully defined. To see that $S(\mathbb{R}^n) * S(\mathbb{R}^n) \subset S(\mathbb{R}^n)$, fix some arbitrary $f, g \in S(\mathbb{R}^n)$ and $\alpha, \beta \in \mathbb{N}_0^n$. Then, making use of the binomial theorem (cf. Theorem 14.9) as well as Exercise 3.5, we may estimate

$$\sup_{x\in\mathbb{R}^n} \left|x^\beta \partial_x^\alpha (f * g)(x)\right| = \sup_{x\in\mathbb{R}^n} \left|\int_{\mathbb{R}^n} ((x-y)+y)^\beta \partial_x^\alpha f(x-y)g(y)\,dy\right|$$

$$\leq \sup_{x\in\mathbb{R}^n} \sum_{\gamma\leq\beta} \frac{\beta!}{\gamma!(\beta-\gamma)!} \int_{\mathbb{R}^n} |(x-y)^{\beta-\gamma}(\partial^\alpha f)(x-y)||y^\gamma g(y)|\,dy \tag{3.1.35}$$

$$\leq C_{\alpha,\beta} \sup_{z\in\mathbb{R}^n} \left[(1+|z|)^{|\beta|}|\partial^\alpha f(z)|\right] \int_{\mathbb{R}^n} (1+|y|)^{|\beta|}|g(y)|\,dy$$

$$\leq C_{\alpha,\beta}\left(\sup_{z\in\mathbb{R}^n} \left[(1+|z|)^{|\beta|}|\partial^\alpha f(z)|\right]\right)\left(\sup_{y\in\mathbb{R}^n} \left[(1+|y|)^{|\beta|+n+1}|g(y)|\right]\right) < \infty.$$

This implies $f * g \in S(\mathbb{R}^n)$. The fact that the mapping in *(e)* is bilinear is immediate from definitions. As regards its continuity, we may invoke again Theorem 14.1 and Fact 3.6 to reduce matters to proving sequential continuity instead. However, the latter is apparent from the estimate in (3.1.35). This finishes the proof of the theorem. $\square$

Exercise 3.15. Assume that $\psi \in S(\mathbb{R}^n)$ is given and, for each $j \in \mathbb{N}$, define the function $\psi_j(x) := \psi(\frac{x}{j})$ for every $x \in \mathbb{R}^n$. Prove that

$$\psi_j f \xrightarrow[j\to\infty]{S(\mathbb{R}^n)} \psi(0)f \qquad \text{for every } f \in S(\mathbb{R}^n). \tag{3.1.36}$$

Hint: Adapt estimates (3.1.30)–(3.1.31) to the current setting and, in place of (3.1.32), this time use the Mean Value Theorem for the term $\psi(x/j) - \psi(0)$ to get a decay factor of the order $1/j$.

Proposition 3.16. *Let $m, n \in \mathbb{N}$. Then $C_0^\infty(\mathbb{R}^m) \otimes C_0^\infty(\mathbb{R}^n)$ is sequentially dense in $S(\mathbb{R}^m \times \mathbb{R}^n)$.*

Proof. Since the topology on $S(\mathbb{R}^m \times \mathbb{R}^n)$ is metrizable (recall Fact 3.6), there exists a distance function $d : S(\mathbb{R}^m \times \mathbb{R}^n) \times S(\mathbb{R}^m \times \mathbb{R}^n) \to [0, \infty)$ that induces its topology.

Hence,

$$f_j \xrightarrow[j\to\infty]{S(\mathbb{R}^m\times\mathbb{R}^n)} f \quad \text{if and only if} \quad \lim_{j\to\infty} d(f_j, f) = 0. \qquad (3.1.37)$$

Now let $f \in S(\mathbb{R}^m \times \mathbb{R}^n)$. Then by part *(d)* in Theorem 3.14, there exists a sequence $\{f_j\}_{j\in\mathbb{N}} \subset C_0^\infty(\mathbb{R}^m \times \mathbb{R}^n)$ with the property that $d(f_j, f) < \frac{1}{j}$ for every $j \in \mathbb{N}$. Furthermore, by Proposition 2.81, for each fixed number $j \in \mathbb{N}$, there exists a sequence $\{g_{j_k}\}_{k\in\mathbb{N}} \subset C_0^\infty(\mathbb{R}^m) \otimes C_0^\infty(\mathbb{R}^n)$ such that $g_{j_k} \xrightarrow[k\to\infty]{\mathcal{D}(\mathbb{R}^m\times\mathbb{R}^n)} f_j$. In particular, by *(c)* in Theorem 3.14,

$$g_{j_k} \xrightarrow[k\to\infty]{S(\mathbb{R}^m\times\mathbb{R}^n)} f_j \quad \text{for each } j \in \mathbb{N}, \qquad (3.1.38)$$

thus

$$\lim_{k\to\infty} d(g_{j_k}, f_j) = 0 \quad \text{for each } j \in \mathbb{N}. \qquad (3.1.39)$$

Condition (3.1.39) implies that for each $j \in \mathbb{N}$ there exists $k_j \in \mathbb{N}$ with the property that $d(g_{j_{k_j}}, f_j) < \frac{1}{j}$. Now the sequence $\{g_{j_{k_j}}\}_{j\in\mathbb{N}} \subset C_0^\infty(\mathbb{R}^m) \otimes C_0^\infty(\mathbb{R}^n)$ satisfies

$$d(g_{j_{k_j}}, f) \le d(g_{j_{k_j}}, f_j) + d(f_j, f) < \frac{2}{j} \quad \text{for every} \quad j \in \mathbb{N}. \qquad (3.1.40)$$

In turn, this forces $g_{j_{k_j}} \xrightarrow[j\to\infty]{S(\mathbb{R}^m\times\mathbb{R}^n)} f$, from which the desired conclusion follows. $\qquad\square$

The analogue of Lemma 1.24 corresponding to the Schwartz class is stated next.

Exercise 3.17. Suppose $A \in M_{n\times n}(\mathbb{R})$ is such that $\det A \neq 0$. Prove that the composition mapping

$$S(\mathbb{R}^n) \ni \varphi \mapsto \varphi \circ A \in S(\mathbb{R}^n) \quad \text{is well defined, linear, and continuous.} \qquad (3.1.41)$$

Hint: To prove continuity you may use the linearity of the map in (3.1.41), Fact 3.6, and Theorem 14.1, to reduce matters to proving sequential continuity at 0.

We conclude this section by proving that $\mathcal{L}(\mathbb{R}^n) * S(\mathbb{R}^n) \subseteq C^\infty(\mathbb{R}^n)$.

Proposition 3.18. *For every function $f \in \mathcal{L}(\mathbb{R}^n)$ and every function g in $S(\mathbb{R}^n)$ one has $\int_{\mathbb{R}^n} |f(x-y)||g(y)|\, dy < \infty$ for each $x \in \mathbb{R}^n$, and the convolution $f * g$ defined by*

$$(f * g)(x) := \int_{\mathbb{R}^n} f(x-y)g(y)\, dy \quad \text{for each} \quad x \in \mathbb{R}^n, \qquad (3.1.42)$$

has the property that $f * g \in C^\infty(\mathbb{R}^n)$.

Proof. If f, g are as in the statement, then from (3.1.23) and Exercise 3.5 it follows that there exists $M \in \mathbb{N}$ such that for every $N \in \mathbb{N}$ there exists $C \in (0, \infty)$ such that

$$\int_{\mathbb{R}^n} |f(x-y)||g(y)|\, dy \le C \int_{\mathbb{R}^n} (1+|x-y|)^M (1+|y|)^{-N}\, dy \qquad (3.1.43)$$

for every $x \in \mathbb{R}^n$. For each fixed point $x \in \mathbb{R}^n$ choose now $N \in \mathbb{N}$ such that $N > M+n$ and note that this ensures

$$\int_{\mathbb{R}^n} (1 + |x - y|)^M (1 + |y|)^{-N} \, dy < \infty, \tag{3.1.44}$$

proving the first claim in the statement. The fact that $f * g \in C^\infty(\mathbb{R}^n)$ is now seen in a similar fashion given that $\partial^\alpha f$ continues to be in $\mathcal{L}(\mathbb{R}^n)$ for every $\alpha \in \mathbb{N}_0^n$. □

Exercise 3.19. Prove that $L^p(\mathbb{R}^n) * S(\mathbb{R}^n) \subseteq C^\infty(\mathbb{R}^n)$ for every $p \in [1, \infty]$.

Hint: Use the blue print as in the proof of Proposition 3.18, using Hölder's inequality in place of estimates for slowly increasing functions, and arrange matters so that all derivatives fall on the Schwartz function.

We conclude this section with an integration by parts formula that will be useful shortly.

Lemma 3.20. *If $f \in \mathcal{L}(\mathbb{R}^n)$ and $g \in S(\mathbb{R}^n)$, then for every $\alpha \in \mathbb{N}_0^n$ the following integration by parts formula holds:*

$$\int_{\mathbb{R}^n} (\partial^\alpha g)(x) f(x) \, dx = (-1)^{|\alpha|} \int_{\mathbb{R}^n} g(x)(\partial^\alpha f)(x) \, dx. \tag{3.1.45}$$

Proof. Fix $f \in \mathcal{L}(\mathbb{R}^n)$ and $g \in S(\mathbb{R}^n)$. Since the classes $\mathcal{L}(\mathbb{R}^n)$ and $S(\mathbb{R}^n)$ are stable under differentiation, it suffices to show that for each $j \in \{1, ..., n\}$ we have

$$\int_{\mathbb{R}^n} (\partial_j g)(x) f(x) \, dx = - \int_{\mathbb{R}^n} g(x)(\partial_j f)(x) \, dx, \tag{3.1.46}$$

since (3.1.45) then follows by iterating (3.1.46). To this end, fix some $j \in \{1, ..., n\}$ along with some arbitrary $R \in (0, \infty)$. The classical integration by parts formula in the bounded, smooth, domain $B(0, R) \subset \mathbb{R}^n$ then reads (cf. (14.8.4))

$$\int_{B(0,R)} (\partial_j g)(x) f(x) \, dx = - \int_{B(0,R)} g(x)(\partial_j f)(x) \, dx$$

$$+ \int_{\partial B(0,R)} g(x) f(x)(x_j/R) \, d\sigma(x). \tag{3.1.47}$$

From part *(a)* in Theorem 3.14 we know that $fg \in S(\mathbb{R}^n)$. Based on this and Exercise 3.5, it follows that

$$\left| \int_{\partial B(0,R)} g(x) f(x)(x_j/R) \, d\sigma(x) \right| \leq \omega_{n-1} R^{n-1} \sup_{|x|=R} |(fg)(x)|$$

$$\leq CR^{-1} \xrightarrow[R \to \infty]{} 0. \tag{3.1.48}$$

On the other hand, $(\partial_j g) f, g \partial_j f \in S(\mathbb{R}^n) \subset L^1(\mathbb{R}^n)$. As such, taking the limit with $R \to \infty$ in (3.1.47) yields (3.1.46) on account of Lebesgue's Dominated Convergence Theorem and (3.1.48). $\qquad\qquad\qquad\qquad\qquad\qquad\qquad\qquad\qquad\qquad\qquad\qquad$ $\square$

3.2 The Action of the Fourier Transform on the Schwartz Class

Originally, we have defined the Fourier transform in (3.1.3), as a mapping acting on functions from $L^1(\mathbb{R}^n)$. Since $S(\mathbb{R}^n)$ is contained in $L^1(\mathbb{R}^n)$, it makes sense to consider the Fourier transform acting on the Schwartz class. In this section, we study the main properties of the Fourier transform in this setting. The reader is advised that we use the symbols $\mathcal{F}$ and $\widehat{\cdot}$ interchangeably to denote this Fourier transform.

To state our first major result pertaining to the Fourier transform in this setting, recall the notation introduced in (3.1.10).

Theorem 3.21. *The following statements are true.*

(a) If $f \in S(\mathbb{R}^n)$ and $\alpha \in \mathbb{N}_0^n$ are arbitrary, then $\widehat{D^\alpha f}(\xi) = \xi^\alpha \widehat{f}(\xi)$ for every $\xi \in \mathbb{R}^n$.

(b) If $f \in S(\mathbb{R}^n)$ and $\alpha \in \mathbb{N}_0^n$ are arbitrary, then $\widehat{x^\alpha f}(\xi) = (-D)^\alpha \widehat{f}(\xi)$ for every $\xi \in \mathbb{R}^n$.

(c) The Fourier transform, originally introduced in the context of (3.1.3), induces a mapping $\mathcal{F} : S(\mathbb{R}^n) \to S(\mathbb{R}^n)$ that is linear and continuous.

(d) If $m, n \in \mathbb{N}$, $f \in S(\mathbb{R}^m)$ and $g \in S(\mathbb{R}^n)$, then $\widehat{f \otimes g} = \widehat{f} \otimes \widehat{g}$.

Proof. Fix $f \in S(\mathbb{R}^n)$ and $\alpha \in \mathbb{N}_0^n$. Then the decay of f (cf. (3.1.19)) ensures that we may differentiate under the integral sign in (3.1.1) and obtain

$$D^\alpha \widehat{f}(\xi) = \int_{\mathbb{R}^n} e^{-ix \cdot \xi} (-x)^\alpha f(x) \, dx$$

$$= \widehat{(-x)^\alpha f}(\xi), \qquad \forall \xi \in \mathbb{R}^n. \tag{3.2.1}$$

From this, the statement in *(b)* readily follows. Also, if $\beta \in \mathbb{N}_0^n$ is arbitrary, then using the first identity in (3.2.1), the fact that $\xi^\beta e^{-ix \cdot \xi} = (-D_x)^\beta (e^{-ix \cdot \xi})$, and the integration by parts formula from Lemma 3.20, we obtain

$$\xi^\beta D^\alpha \widehat{f}(\xi) = \int_{\mathbb{R}^n} (-D_x)^\beta (e^{-ix \cdot \xi}) (-x)^\alpha f(x) \, dx$$

$$= \int_{\mathbb{R}^n} e^{-ix \cdot \xi} D_x^\beta [(-x)^\alpha f(x)] \, dx, \qquad \forall \xi \in \mathbb{R}^n. \tag{3.2.2}$$

The formula in *(a)* follows by specializing (3.2.2) to the case when the multi-index is $\alpha = (0, \ldots, 0) \in \mathbb{N}_0^n$. In addition, starting with (3.2.2) we may estimate

$$\sup_{\xi \in \mathbb{R}^n} |\xi^\beta D^\alpha \widehat{f}(\xi)| \le \left(\int_{\mathbb{R}^n} (1 + |x|^2)^{-n} \, dx \right) \times$$

$$\times \sup_{x \in \mathbb{R}^n} \left\{ (1 + |x|^2)^n \left| D_x^\beta [x^\alpha f(x)] \right| \right\} < \infty, \tag{3.2.3}$$

where the finiteness condition is a consequence of the membership of f to $\mathcal{S}(\mathbb{R}^n)$. Clearly, (3.2.1) also implies that $\widehat{f} \in C^\infty(\mathbb{R}^n)$ which, in combination with (3.2.3), shows that $\widehat{f} \in \mathcal{S}(\mathbb{R}^n)$. Hence, the mapping in *(c)* is well defined. The fact that this mapping is linear is immediate from definition. In addition, if $f_j \xrightarrow[j \to \infty]{\mathcal{S}(\mathbb{R}^n)} 0$, then based on the first inequality in (3.2.3) we have that, for each $m, k \in \mathbb{N}_0$,

$$\sup_{\substack{\xi \in \mathbb{R}^n \\ |\alpha| \le m, |\beta| \le k}} |\xi^\beta \partial^\alpha \widehat{f_j}(\xi)|$$

$$\le C \sup_{\substack{x \in \mathbb{R}^n \\ |\alpha| \le m, |\beta| \le k}} \left[(1 + |x|^2)^n \left| \partial^\beta (x^\alpha f_j(x)) \right| \right] \to 0 \text{ as } j \to \infty. \tag{3.2.4}$$

In view of Exercise 3.8, this proves $\mathcal{F} f_j \xrightarrow[j \to \infty]{\mathcal{S}(\mathbb{R}^n)} 0$. The latter combined with Fact 3.6 and Theorem 14.1 then implies that $\mathcal{F}$ is continuous from $\mathcal{S}(\mathbb{R}^n)$ into $\mathcal{S}(\mathbb{R}^n)$.

At this stage we are left with proving the statement in *(d)*. To this end, fix some $f \in \mathcal{S}(\mathbb{R}^m)$ and some $g \in \mathcal{S}(\mathbb{R}^n)$. Then by *(e)* in Theorem 3.14, we have $f \otimes g \in \mathcal{S}(\mathbb{R}^m \times \mathbb{R}^n)$, so $\mathcal{F}(f \otimes g)$ is well defined. Furthermore, by applying Fubini's theorem, we may write

$$\widehat{f \otimes g}(\xi, \eta) = \int_{\mathbb{R}^m} \int_{\mathbb{R}^n} e^{-ix \cdot \xi - iy \cdot \eta} (f \otimes g)(x, y) \, dy \, dx$$

$$= \int_{\mathbb{R}^m} e^{-ix \cdot \xi} f(x) \, dx \int_{\mathbb{R}^n} e^{-iy \cdot \eta} g(y) \, dy$$

$$= \widehat{f}(\xi) \widehat{g}(\eta) = (\widehat{f} \otimes \widehat{g})(\xi, \eta), \qquad \forall \xi \in \mathbb{R}^m, \ \forall \eta \in \mathbb{R}^n. \tag{3.2.5}$$

This finishes the proof of the theorem. $\square$

Example 3.22. Suppose $\lambda \in \mathbb{C}$ satisfies $\operatorname{Re}(\lambda) > 0$ and if $\lambda = re^{i\theta}$ for $r > 0$ and $-\pi/2 < \theta < \pi/2$, set $\lambda^{\frac{1}{2}} := \sqrt{r} e^{i\theta/2}$. Consider the function $f(x) := e^{-\lambda |x|^2}$ for $x \in \mathbb{R}^n$. Then $f \in \mathcal{S}(\mathbb{R}^n)$ and

$$\widehat{f}(\xi) = \left(\frac{\pi}{\lambda} \right)^{\frac{n}{2}} e^{-\frac{|\xi|^2}{4\lambda}} \quad \text{for each } \xi \in \mathbb{R}^n. \tag{3.2.6}$$

Proof. Fix $\lambda \in \mathbb{C}$ satisfying the given hypotheses. Then Exercise 3.3 ensures that f is a Schwartz function. Also, $f(x) = e^{-\lambda x_1^2} \otimes \cdots \otimes e^{-\lambda x_n^2}$ for each point $x = (x_1 \dots, x_n)$ in $\mathbb{R}^n$. As such, part *(d)* in Theorem 3.21 shows that it suffices to prove (3.2.6) when $n = 1$, in which case $f(x) = e^{-\lambda x^2}$ for every $x \in \mathbb{R}$. Suppose that this is the case and observe that f satisfies $f' + 2\lambda x f = 0$ in $\mathbb{R}$. By taking the Fourier transform of both sides of this differential equation, and using *(a)-(b)* in Theorem 3.21, we arrive at $\xi \widehat{f} + 2\lambda (\widehat{f})' = 0$. The solution to this latter ordinary differential equation is $\widehat{f}(\xi) = \widehat{f}(0)\mathrm{e}^{-\frac{\xi^2}{4\lambda}}$ for $\xi \in \mathbb{R}$. There remains to show that $\widehat{f}(0) = (\frac{\pi}{\lambda})^{\frac{1}{2}}$. Since by definition, $\widehat{f}(0) = \int_{\mathbb{R}} f(x)\,\mathrm{d}x = \int_{\mathbb{R}} \mathrm{e}^{-\lambda x^2}\,\mathrm{d}x$, we are left with showing that

$$\int_{\mathbb{R}} \mathrm{e}^{-\lambda x^2}\,\mathrm{d}x = (\tfrac{\pi}{\lambda})^{\frac{1}{2}} \quad \text{whenever } \lambda \in \mathbb{C} \text{ has } \mathrm{Re}(\lambda) > 0. \tag{3.2.7}$$

Corresponding to the case when $\lambda \in \mathbb{R}_+$, the identity $\int_{\mathbb{R}} \mathrm{e}^{-\lambda x^2}\,\mathrm{d}x = (\frac{\pi}{\lambda})^{\frac{1}{2}}$ is a standard exercise in basic calculus. To extend this to complex λ's observe that the function

$$h(z) := \int_{\mathbb{R}} \mathrm{e}^{-zx^2}\,\mathrm{d}x - (\frac{\pi}{z})^{\frac{1}{2}} \quad \text{for } z \in \mathbb{C} \text{ with } \mathrm{Re}(z) > 0,$$

is analytic and equal to zero for every $z \in \mathbb{R}_+$. This forces $h(z) = 0$ for all z in $\mathbb{C}$ with $\mathrm{Re}(z) > 0$. Thus, $\widehat{f}(0) = (\frac{\pi}{\lambda})^{\frac{1}{2}}$, as desired. $\qquad\qquad\Box$

Exercise 3.23. Let $a \in (0, \infty)$ and $b \in \mathbb{R}$ be fixed. Show that if $x \in \mathbb{R}$ then

$$\mathcal{F}(\mathrm{e}^{-ax^2 + ibx})(\xi) = (\tfrac{\pi}{a})^{\frac{1}{2}} \mathrm{e}^{-\frac{(\xi - b)^2}{4a}} \qquad \text{for every } \xi \in \mathbb{R}. \tag{3.2.8}$$

Hint: First prove that $\mathcal{F}(\mathrm{e}^{-ax^2 + ibx})(\xi) = (\mathcal{F}(\mathrm{e}^{-ax^2}))(\xi - b)$ for every $\xi \in \mathbb{R}$ then use Example 3.22.

Exercise 3.24. Prove that if $A \in M_{n\times n}(\mathbb{R})$ is such that $\det A \neq 0$, then

$$\widehat{\varphi \circ A^{-1}} = |\det A|\, (\widehat{\varphi} \circ A^\top), \qquad \forall\, \varphi \in S(\mathbb{R}^n), \tag{3.2.9}$$

where A^{-1} and $A^\top$ denote, respectively, the inverse and the transpose of the matrix A.

Next we note a consequence of Theorem 3.21 of basic importance. As motivation, suppose $P(D) = \sum\limits_{|\alpha| \leq m} a_\alpha D^\alpha$ is a differential operator of order $m \in \mathbb{N}$, with constant coefficients $a_\alpha \in \mathbb{C}$, for every $\alpha \in \mathbb{N}_0^n$ with $|\alpha| \leq m$. Furthermore, assume that $f \in S(\mathbb{R}^n)$ has been given. Then any solution $u \in S(\mathbb{R}^n)$ of the differential equation $P(D)u = f$ in $\mathbb{R}^n$ also satisfies $P(\xi)\widehat{u}(\xi) = \widehat{f}(\xi)$ for each $\xi \in \mathbb{R}^n$, where we have set $P(\xi) := \sum\limits_{|\alpha| \leq m} a_\alpha \xi^\alpha$. In particular, if $P(\xi)$ has no zeros, then $\widehat{u}(\xi) = \frac{\widehat{f}(\xi)}{P(\xi)}$, for every $\xi \in \mathbb{R}^n$. This gives a formula for the Fourier transform of u. In order to find a formula for u itself, the natural question that arises is whether we can reconstruct u from $\widehat{u}$. The next theorem provides a positive answer to this question in the class of Schwartz functions.

Theorem 3.25. *The mapping $\mathcal{F} : S(\mathbb{R}^n) \to S(\mathbb{R}^n)$ is an algebraic and topologic isomorphism, that is, it is bijective, continuous, and its inverse is also continuous. In addition, its inverse is the operator $\mathcal{F}^{-1} : S(\mathbb{R}^n) \to S(\mathbb{R}^n)$ given by the formula*

$$(\mathcal{F}^{-1}g)(x) = (2\pi)^{-n} \int_{\mathbb{R}^n} e^{ix\cdot\xi} g(\xi) \, d\xi, \qquad \forall \, x \in \mathbb{R}^n, \quad \forall \, g \in \mathcal{S}(\mathbb{R}^n). \qquad (3.2.10)$$

Proof. The proof of the fact that the mapping $\mathcal{F}^{-1} : \mathcal{S}(\mathbb{R}^n) \to \mathcal{S}(\mathbb{R}^n)$ defined as in (3.2.10) is well defined, linear, and continuous is similar to the proof of part *(c)* in Theorem 3.21. There remains to show that $\mathcal{F}^{-1} \circ \mathcal{F} = I = \mathcal{F} \circ \mathcal{F}^{-1}$ on $\mathcal{S}(\mathbb{R}^n)$, where I is the identity operator on $\mathcal{S}(\mathbb{R}^n)$. To proceed, observe that the identity $\mathcal{F}^{-1} \circ \mathcal{F} = I$ is equivalent to

$$(2\pi)^{-n} \int_{\mathbb{R}^n} e^{ix\cdot\xi} \widehat{f}(\xi) \, d\xi = f(x), \qquad \forall \, x \in \mathbb{R}^n, \quad \forall \, f \in \mathcal{S}(\mathbb{R}^n). \qquad (3.2.11)$$

As regards (3.2.11), fix a function $f \in \mathcal{S}(\mathbb{R}^n)$ along with a point $x \in \mathbb{R}^n$. Recall (cf. (3.1.1)) that

$$\widehat{f}(\xi) = \int_{\mathbb{R}^n} e^{-iy\cdot\xi} f(y) \, dy, \quad \forall \, \xi \in \mathbb{R}^n. \qquad (3.2.12)$$

As such, one is tempted to directly replace $\widehat{f}(\xi)$ in (3.2.11) by the right-hand side of (3.2.12) and then use Fubini's theorem to reverse the order of integration in the variables ξ and y. There is, however, a problem in carrying out this approach, since the function $e^{i(x-y)\cdot\xi} f(y)$, considered jointly in the variable $(\xi, y) \in \mathbb{R}^n \times \mathbb{R}^n$, does not belong to $L^1(\mathbb{R}^n \times \mathbb{R}^n)$, hence Fubini's theorem is not necessarily applicable. To remedy this problem, we introduce a "convergence factor" in the form of a suitable family of Schwartz functions ψ^ε, indexed by $\varepsilon > 0$ (to be specified shortly), designed to provide control in the variable ξ thus ensuring the applicability of Fubini's theorem.

The idea is to consider $\int_{\mathbb{R}^n} e^{ix\cdot\xi} \psi^\varepsilon(\xi) \widehat{f}(\xi) \, d\xi$ in place of $\int_{\mathbb{R}^n} e^{ix\cdot\xi} \widehat{f}(\xi) \, d\xi$ and write (granted that $\psi^\varepsilon \in \mathcal{S}(\mathbb{R}^n)$)

$$\int_{\mathbb{R}^n} e^{ix\cdot\xi} \psi^\varepsilon(\xi) \widehat{f}(\xi) \, d\xi = \int_{\mathbb{R}^n} e^{ix\cdot\xi} \psi^\varepsilon(\xi) \int_{\mathbb{R}^n} e^{-iy\cdot\xi} f(y) \, dy \, d\xi$$

$$= \int_{\mathbb{R}^n \times \mathbb{R}^n} e^{-i(y-x)\cdot\xi} \psi^\varepsilon(\xi) f(y) \, dy \, d\xi$$

$$= \int_{\mathbb{R}^n} \left(\int_{\mathbb{R}^n} e^{-i(y-x)\cdot\xi} \psi^\varepsilon(\xi) \, d\xi \right) f(y) \, dy$$

$$= \int_{\mathbb{R}^n} \widehat{\psi^\varepsilon}(y-x) f(y) \, dy = \int_{\mathbb{R}^n} \widehat{\psi^\varepsilon}(z) f(x+z) \, dz. \qquad (3.2.13)$$

Given the goal we have in mind (cf. (3.2.11)), as well as the format of the current identity, we find it convenient to define ψ^ε by setting

$$\psi^\varepsilon(\xi) := \varphi(\varepsilon\xi) \quad \text{for each } \xi \in \mathbb{R}^n \text{ and } \varepsilon > 0, \qquad (3.2.14)$$

where $\varphi \in \mathcal{S}(\mathbb{R}^n)$ is to be specified momentarily. The rationale behind this choice is that, as we will see next, the limits as $\varepsilon \to 0^+$ of the most extreme sides in

(3.2.13) are reasonably easy to compute. This will eventually allow us to deduce (3.2.11) from (3.2.13) by letting $\varepsilon \to 0^+$. Concretely, from (3.2.14) we obtain that $\lim_{\varepsilon \to 0^+} \psi^\varepsilon(\xi) = \varphi(0)$, while from the definition of the Fourier transform it is immediate that

$$\widehat{\psi^\varepsilon}(z) = \varepsilon^{-n}\widehat{\varphi}\left(\frac{z}{\varepsilon}\right) = (\widehat{\varphi})_\varepsilon(z) \quad \text{for every} \quad z \in \mathbb{R}^n. \tag{3.2.15}$$

Keeping this in mind and employing part (a) in Exercise 2.26 we obtain

$$\lim_{\varepsilon \to 0^+} \int_{\mathbb{R}^n} \widehat{\psi^\varepsilon}(z) f(x+z) \, dz = \lim_{\varepsilon \to 0^+} \int_{\mathbb{R}^n} (\widehat{\varphi})_\varepsilon(z) f(x+z) \, dz$$

$$= \left(\int_{\mathbb{R}^n} \widehat{\varphi}(z) \, dz\right) f(x). \tag{3.2.16}$$

Also, on account of (3.2.14) and the fact that $\widehat{f} \in \mathcal{S}(\mathbb{R}^n) \subset L^1(\mathbb{R}^n)$, Lebesgue's Dominated Convergence Theorem gives

$$\lim_{\varepsilon \to 0^+} \int_{\mathbb{R}^n} e^{ix\cdot\xi} \psi^\varepsilon(\xi) \widehat{f}(\xi) \, d\xi = \lim_{\varepsilon \to 0^+} \int_{\mathbb{R}^n} e^{ix\cdot\xi} \varphi(\varepsilon\xi) \widehat{f}(\xi) \, d\xi$$

$$= \varphi(0) \int_{\mathbb{R}^n} e^{ix\cdot\xi} \widehat{f}(\xi) \, d\xi. \tag{3.2.17}$$

In summary, (3.2.13), (3.2.16), and (3.2.17), show that whenever $\varphi \in \mathcal{S}(\mathbb{R}^n)$ is such that $\int_{\mathbb{R}^n} \widehat{\varphi}(z) \, dz \neq 0$ we have

$$C \int_{\mathbb{R}^n} e^{ix\cdot\xi} \widehat{f}(\xi) \, d\xi = f(x), \qquad \forall x \in \mathbb{R}^n, \quad \forall f \in \mathcal{S}(\mathbb{R}^n), \tag{3.2.18}$$

where the normalization constant C is given by

$$C := \frac{\varphi(0)}{\int_{\mathbb{R}^n} \widehat{\varphi}(z) \, dz}. \tag{3.2.19}$$

As such, (3.2.11) will follow as soon as we prove that $C = (2\pi)^{-n}$. For this task, we have the freedom of choosing the function $\varphi \in \mathcal{S}(\mathbb{R}^n)$ and a candidate that springs to mind is the Schwartz function from Example 3.22 (say, in the particular case when $\lambda = 1$). Hence, if $\varphi(x) := e^{-|x|^2}$ for each $x \in \mathbb{R}^n$, formula (3.2.6) gives

$$\widehat{\varphi}(\xi) = \pi^{\frac{n}{2}} e^{-\frac{|\xi|^2}{4}} \quad \text{for each} \quad \xi \in \mathbb{R}^n. \tag{3.2.20}$$

Consequently,

$$\int_{\mathbb{R}^n} \widehat{\varphi}(\xi) \, d\xi = \pi^{\frac{n}{2}} \int_{\mathbb{R}^n} e^{-\frac{|\xi|^2}{4}} \, d\xi = \pi^{\frac{n}{2}} \left(\int_{\mathbb{R}} e^{-\frac{|t|^2}{4}} \, dt\right)^n = \pi^{\frac{n}{2}} (4\pi)^{\frac{n}{2}} = (2\pi)^n. \tag{3.2.21}$$

where the second equality is simply Fubini's theorem, while the third equality is provided by (3.2.7) with $\lambda := 1/4$. Since in this case we also have $\varphi(0) = 1$, it follows that $C = (2\pi)^{-n}$, as wanted. This finishes the justification of the identity $\mathcal{F}^{-1} \circ \mathcal{F} = I$ on $S(\mathbb{R}^n)$. The same approach also works to show $\mathcal{F} \circ \mathcal{F}^{-1} = I$, completing the proof of the theorem.   □

In what follows, for an arbitrary function $f : \mathbb{R}^n \to \mathbb{C}$ we define

$$f^{\vee}(x) := f(-x), \qquad \forall\, x \in \mathbb{R}^n. \tag{3.2.22}$$

Exercise 3.26. Prove that the mapping

$$S(\mathbb{R}^n) \ni f \mapsto f^{\vee} \in S(\mathbb{R}^n) \tag{3.2.23}$$

is well defined, linear, and continuous.

Hint: Use Fact 3.6 and Theorem 14.1.

Recall that $\bar{z}$ denotes the complex conjugate of $z \in \mathbb{C}$.

Exercise 3.27. Let $f \in S(\mathbb{R}^n)$. Then the following formulas hold:

(1) $\widehat{f^{\vee}} = (\widehat{f}\,)^{\vee}$;
(2) $\overline{\widehat{f}} = \widehat{\overline{f}^{\vee}}$;
(3) $\widehat{\widehat{f}} = (2\pi)^n f^{\vee}$;
(4) $\int_{\mathbb{R}^n} f(x)\,\mathrm{d}x = \widehat{f}(0)$ and $\int_{\mathbb{R}^n} \widehat{f}(\xi)\,\mathrm{d}\xi = (2\pi)^n f(0)$.

Proposition 3.28. *Let $f, g \in S(\mathbb{R}^n)$. Then the following identities hold:*

(a) $\int_{\mathbb{R}^n} f(x)\widehat{g}(x)\,\mathrm{d}x = \int_{\mathbb{R}^n} \widehat{f}(\xi)g(\xi)\,\mathrm{d}\xi$;
(b) $\int_{\mathbb{R}^n} f(x)\overline{g(x)}\,\mathrm{d}x = (2\pi)^{-n} \int_{\mathbb{R}^n} \widehat{f}(\xi)\overline{\widehat{g}(\xi)}\,\mathrm{d}\xi$ *an identity referred to in the literature as* Parseval's identity;
(c) $\widehat{f * g} = \widehat{f} \cdot \widehat{g}$;
(d) $\widehat{f \cdot g} = (2\pi)^{-n}\widehat{f} * \widehat{g}$.

Proof. The identity in *(a)* follows via a direct computation using Fubini's theorem. Also, based on Exercise 3.27, we have $\overline{\widehat{\widehat{g}}} = \overline{\widehat{g}^{\vee}} = (2\pi)^n \overline{g}$ which, when combined with *(a)* gives *(b)*. The identity in *(c)* follows using Fubini's theorem. Specifically, for each $\xi \in \mathbb{R}^n$ we may write

$$\widehat{f * g}(\xi) = \int_{\mathbb{R}^n} e^{-ix\cdot\xi}(f * g)(x)\,\mathrm{d}x = \int_{\mathbb{R}^n} e^{-ix\cdot\xi} \int_{\mathbb{R}^n} f(x-y)g(y)\,\mathrm{d}y\,\mathrm{d}x$$

$$= \int_{\mathbb{R}^u} g(y) \int_{\mathbb{R}^n} e^{-ix\cdot\xi} f(x-y)\,\mathrm{d}x\,\mathrm{d}y$$

$$= \int_{\mathbb{R}} e^{-iy\cdot\xi} g(y) \int_{\mathbb{R}^n} e^{-iz\cdot\xi} f(z)\,\mathrm{d}z\,\mathrm{d}y = \widehat{f}(\xi)\widehat{g}(\xi), \tag{3.2.24}$$

as wanted. Next, the identity from *(c)* combines with Exercise 3.27 to yield

$$\widehat{\widetilde{f} * \widehat{g}} = \widehat{\widehat{f}} \cdot \widehat{\widehat{g}} = (2\pi)^{2n} f^\vee \cdot g^\vee = (2\pi)^{2n} (f \cdot g)^\vee. \tag{3.2.25}$$

Applying now the Fourier transform to the most extreme sides of (3.2.25) and once again invoking Exercise 3.27, we obtain

$$(2\pi)^{-n} \widehat{\widehat{f} * \widehat{g}} = (2\pi)^{-2n} \widehat{\widehat{\widehat{f} * \widehat{g}}}^{\vee} = \widehat{(f \cdot g)^\vee}^{\vee} = \widehat{f \cdot g}. \tag{3.2.26}$$

This completes the proof of the proposition. $\qquad\square$

Remark 3.29.

(1) It is not difficult to see via a direct computation that we also have

$$\int_{\mathbb{R}^n} f(x) \widehat{g}(x) \, dx = \int_{\mathbb{R}^n} \widehat{f}(\xi) g(\xi) \, d\xi, \qquad \forall f \in L^1(\mathbb{R}^n), \ \forall g \in \mathcal{S}(\mathbb{R}^n). \tag{3.2.27}$$

(2) Parseval's identity written for $f = g \in \mathcal{S}(\mathbb{R}^n)$ becomes

$$\|\widehat{f}\|_{L^2(\mathbb{R}^n)} = (2\pi)^{\frac{n}{2}} \|f\|_{L^2(\mathbb{R}^n)}. \tag{3.2.28}$$

As a consequence, since $C_0^\infty(\mathbb{R}^n)$ is dense in $L^2(\mathbb{R}^n)$, the Fourier transform $\mathcal{F}$ may be extended to a linear operator from $L^2(\mathbb{R}^n)$ into itself, and the latter identity continues to hold for every $f \in L^2(\mathbb{R}^n)$. In summary, this extension of $\mathcal{F}$, originally considered as in part *(c)* of Theorem 3.21, satisfies

$$\mathcal{F} : L^2(\mathbb{R}^n) \to L^2(\mathbb{R}^n) \text{ is linear and continuous and}$$
$$\|\mathcal{F} f\|_{L^2(\mathbb{R}^n)} = (2\pi)^{\frac{n}{2}} \|f\|_{L^2(\mathbb{R}^n)}, \quad \forall f \in L^2(\mathbb{R}^n). \tag{3.2.29}$$

Based on this, part *(3)* in Exercise 3.27, the continuity of the linear mapping $L^2(\mathbb{R}^n) \ni f \mapsto f^\vee \in L^2(\mathbb{R}^n)$, and the density of Schwartz functions in $L^2(\mathbb{R}^n)$, we further deduce that

$$\mathcal{F}(\mathcal{F} f) = (2\pi)^n f^\vee, \quad \forall f \in L^2(\mathbb{R}^n). \tag{3.2.30}$$

Combined with (3.2.29), this proves that

$$\mathcal{F} : L^2(\mathbb{R}^n) \to L^2(\mathbb{R}^n) \text{ is a linear, continuous, isomorphism,}$$
$$\text{and} \ \ \mathcal{F}^{-1} f = (2\pi)^{-n} (\mathcal{F} f^\vee) = (2\pi)^{-n} (\mathcal{F} f)^\vee, \quad \forall f \in L^2(\mathbb{R}^n). \tag{3.2.31}$$

We will continue to use the notation $\widehat{f}$ for $\mathcal{F} f$ whenever $f \in L^2(\mathbb{R}^n)$. The identity in (3.2.29) is called `Plancherel's identity`. The same type of density argument shows that formula from part *(b)* of Proposition 3.28 extends to

$$\int_{\mathbb{R}^n} f(x) \overline{g(x)} \, dx = (2\pi)^{-n} \int_{\mathbb{R}^n} \widehat{f}(\xi) \overline{\widehat{g}(\xi)} \, d\xi, \qquad \forall f, g \in L^2(\mathbb{R}^n), \tag{3.2.32}$$

to which we continue to refer as Parseval's identity.

(3) An inspection of the computation in (3.2.24) reveals that the identity $\widehat{f * g} = \widehat{f} \cdot \widehat{g}$ remains valid if $f, g \in L^1(\mathbb{R}^n)$.

Exercise 3.30. Prove that $\int_{\mathbb{R}^n} f(x) \widehat{g}(x) \, dx = \int_{\mathbb{R}^n} \widehat{f}(\xi) g(\xi) \, d\xi$ for all $f, g \in L^2(\mathbb{R}^n)$.

Hint: Use part *(a)* in Proposition 3.28, (3.2.29), and the fact that $C_0^\infty(\mathbb{R}^n)$ is sequentially dense in $L^2(\mathbb{R}^n)$ to first prove the desired identity for $f \in L^2(\mathbb{R}^n)$ and $g \in \mathcal{S}(\mathbb{R}^n)$.

Further Notes for Chapter 3. The basic formalism associated with the Fourier transform goes back to the French mathematician and physicist Joseph Fourier (1768–1830) in a more or less precise form. A distinguished attribute of this tool, of fundamental importance in the context of partial differential equations, is the ability to render the action of a constant coefficient differential operator simply as ordinary multiplication by its symbol on the Fourier transform side. As the name suggest, the Schwartz space of rapidly decreasing functions has been formally introduced by Laurent Schwartz who was the first to recognize its significance in the context of the Fourier transform. Much of the elegant theory presented here is due to him.

3.3 Additional Exercises for Chapter 3

Exercise 3.31. Prove that if $f \in L^1_{comp}(\mathbb{R}^n)$ then $\widehat{f} \in C^\infty(\mathbb{R}^n)$.

Exercise 3.32. Prove that if $f \in L^1(\mathbb{R}^n)$ is real-valued and odd, then so is $\widehat{f}$.

Exercise 3.33. Prove that if $f \in \mathcal{S}(\mathbb{R}^n)$ then for every $\alpha, \beta \in \mathbb{N}_0^n$ one has

$$\lim_{R \to \infty} \left[\sup_{|x| \geq R} \left| x^\alpha \partial^\beta f(x) \right| \right] = 0. \tag{3.3.1}$$

Exercise 3.34. Let $\varphi \in C_0^\infty(\mathbb{R}^n)$ be such that $\varphi \neq 0$ and for each $j \in \mathbb{N}$ set

$$\varphi_j(x) := e^{-j} \varphi(x/j), \qquad \forall \, x \in \mathbb{R}^n.$$

Prove that $\varphi_j \xrightarrow[j \to \infty]{\mathcal{S}(\mathbb{R}^n)} 0$ but the sequence $\{\varphi_j\}_{j \in \mathbb{N}}$ does not converge in $\mathcal{D}(\mathbb{R}^n)$.

Exercise 3.35. Let $\varphi \in C_0^\infty(\mathbb{R}^n)$ be such that $\varphi \neq 0$ and for each $j \in \mathbb{N}$ set

$$\varphi_j(x) := \frac{1}{j} \varphi(x/j), \qquad \forall \, x \in \mathbb{R}^n.$$

Prove that $\varphi_j \xrightarrow[j \to \infty]{\mathcal{E}(\mathbb{R}^n)} 0$ but the sequence $\{\varphi_j\}_{j \in \mathbb{N}}$ does not converge in $\mathcal{S}(\mathbb{R}^n)$.

Exercise 3.36. Let $\theta \in C_0^\infty(\mathbb{R})$ be such that $\theta(x) = 1$ for $|x| \leq 1$, and let ψ in $C^\infty(\mathbb{R})$ be such that $\psi(x) = 0$ for $x \leq -1$ and $\psi(x) = e^{-x}$ for $x \geq 0$. For each $j \in \mathbb{N}$ then set

$$\varphi_j(x) := \frac{1}{j}\psi(x)\theta(x/j), \qquad \forall\, x \in \mathbb{R}.$$

Prove that the sequence $\{\varphi_j\}_{j \in \mathbb{N}}$ converges in $\mathcal{S}(\mathbb{R})$.

Exercise 3.37. Determine which of the following functions belongs to $\mathcal{S}(\mathbb{R}^n)$.

(a) $e^{-(x_1 + x_2^2 + \cdots + x_n^2)}$

(b) $(x_1^2 + x_2^2 + \cdots + x_n^2)^{n!} e^{-|x|^2}$

(c) $(1 + |x|^2)^{-2^n}$

(d) $\frac{\sin(e^{-|x|^2})}{1 + |x|^2}$

(e) $\frac{\cos(e^{-|x|^2})}{(1 + |x|^2)^n}$

(f) $e^{-|x|^2} \sin(e^{x_1^2})$

(g) $e^{-(Ax)\cdot x}$, where $A \in \mathcal{M}_{n \times n}(\mathbb{R})$ is symmetric and satisfies $(Ax) \cdot x > 0$ for all $x \in \mathbb{R}^n \setminus \{0\}$ (as before, "$\cdot$" denotes the dot product of vectors in $\mathbb{R}^n$).

Exercise 3.38. Let $A \in \mathcal{M}_{n \times n}(\mathbb{R})$ be symmetric and such that $(Ax) \cdot x > 0$ for every $x \in \mathbb{R}^n \setminus \{0\}$. Prove that if we define $f(x) := e^{-(Ax)\cdot x}$ for $x \in \mathbb{R}^n$, then $\widehat{f}(\xi) = \frac{\pi^{\frac{n}{2}}}{\sqrt{\det A}} e^{-\frac{(A^{-1}\xi)\cdot\xi}{4}}$ for every $\xi \in \mathbb{R}^n$.

Exercise 3.39. Prove that $f : \mathbb{R}^2 \to \mathbb{R}$ defined by $f(x_1, x_2) := e^{-(x_1^2 + x_1 x_2 + x_2^2)}$ for $(x_1, x_2) \in \mathbb{R}^2$ belongs to $\mathcal{S}(\mathbb{R}^2)$, then compute its Fourier transform.

Exercise 3.40. If $P(x)$ is a polynomial in $\mathbb{R}^n$, compute the Fourier transform of the function defined by $f(x) := P(x)e^{-|x|^2}$ for each $x \in \mathbb{R}^n$.

Exercise 3.41. If $a \in (0, \infty)$ and $x_0 \in \mathbb{R}^n$ are fixed, compute the Fourier transform of the function defined by $f(x) := e^{-a|x|^2} \sin(x \cdot x_0)$ for each $x \in \mathbb{R}^n$.

Exercise 3.42. Let $\varphi \in \mathcal{S}(\mathbb{R})$. Prove that the equation $\psi' = \varphi$ has a solution $\psi \in \mathcal{S}(\mathbb{R})$ if and only if $\int_{\mathbb{R}} \varphi(x)\,dx = 0$.

Exercise 3.43. Does the equation $\psi' = e^{-x^2}$ have a solution in $\mathcal{S}(\mathbb{R})$?

Exercise 3.44. Fix $x_0 \in \mathbb{R}^n$. Prove that the translation map t_{x_0} from (1.3.17) extends linearly and continuously as a map from $\mathcal{S}(\mathbb{R}^n)$ into itself. More precisely, show that the translation map $t_{x_0} : \mathcal{S}(\mathbb{R}^n) \to \mathcal{S}(\mathbb{R}^n)$ defined by $t_{x_0}(\varphi) := \varphi(\cdot - x_0)$ for every $\varphi \in \mathcal{S}(\mathbb{R}^n)$, is linear and continuous.

Also, prove that for every $\varphi \in \mathcal{S}(\mathbb{R}^n)$ the following identities hold in $\mathcal{S}(\mathbb{R}^n)$

$$\mathcal{F}(t_{x_0}(\varphi))(\xi) = e^{-ix_0\cdot\xi}\,\widehat{\varphi}(\xi) \quad \text{and} \quad t_{x_0}(\widehat{\varphi}) = \mathcal{F}(e^{ix_0\cdot x}\varphi). \tag{3.3.2}$$

Chapter 4
The Space of Tempered Distributions

Abstract The action of the Fourier transform is extended to the setting of tempered distributions, and several distinguished subclasses of tempered distributions are introduced and studied, including homogeneous and principal value distributions. Significant applications to harmonic analysis and partial differential equations are singled out. For example, a general, higher dimensional jump-formula is deduced in this chapter for a certain class of tempered distributions, which includes the classical harmonic Poisson kernel that is later used as the main tool in deriving information about the boundary behavior of layer potential operators associated with various partial differential operators and systems. Also, one witnesses here how singular integral operators of central importance to harmonic analysis, such as the Riesz transforms, naturally arise as an extension to the space of square integrable functions, of the convolution product of tempered distributions of principal value type with Schwartz functions.

4.1 Definition and Properties of Tempered Distributions

The algebraic dual of $S(\mathbb{R}^n)$ is the vector space

$$\{u : S(\mathbb{R}^n) \to \mathbb{C} : u \text{ is linear and continuous}\}. \tag{4.1.1}$$

Functionals u belonging to this space are called `tempered distributions` (a piece of terminology justified a little later). An important equivalent condition for a linear functional on $S(\mathbb{R}^n)$ to be a tempered distribution is stated next (see (14.1.33)).

Fact 4.1 *A linear functional $u : S(\mathbb{R}^n) \to \mathbb{C}$ is continuous if and only if there exist $m, k \in \mathbb{N}_0$, and a finite constant $C > 0$, such that*

$$|u(\varphi)| \leq C \sup_{\alpha, \beta \in \mathbb{N}_0^n, |\alpha| \leq m, |\beta| \leq k} \sup_{x \in \mathbb{R}^n} \left| x^\beta \partial^\alpha \varphi(x) \right|, \qquad \forall \varphi \in S(\mathbb{R}^n). \tag{4.1.2}$$

© Springer Nature Switzerland AG 2018
D. Mitrea, *Distributions, Partial Differential Equations, and Harmonic Analysis*,
Universitext, https://doi.org/10.1007/978-3-030-03296-8_4

Exercise 4.2. Prove that a linear functional $u : S(\mathbb{R}^n) \to \mathbb{C}$ is continuous if and only if there exist $m, k \in \mathbb{N}_0$, and a finite constant $C > 0$, such that

$$|u(\varphi)| \le C \sup_{\alpha \in \mathbb{N}_0^n, |\alpha| \le m, 0 \le j \le k} \sup_{x \in \mathbb{R}^n} \left| (1 + |x|)^j \partial^\alpha \varphi(x) \right|, \qquad \forall \varphi \in S(\mathbb{R}^n). \qquad (4.1.3)$$

Hint: Use Fact 4.1 and (3.1.17).

From Fact 3.6 and Theorem 14.1 we also know that any $f : S(\mathbb{R}^n) \to \mathbb{C}$ (linear or not) is continuous if and only if it is sequentially continuous. As a consequence, we have the following characterization of tempered distributions.

Proposition 4.3. *Let* $u : S(\mathbb{R}^n) \to \mathbb{C}$ *be linear. Then* u *is a tempered distribution if and only* $u(\varphi_j) \xrightarrow[j \to \infty]{} 0$ *whenever* $\varphi_j \xrightarrow[j \to \infty]{S(\mathbb{R}^n)} 0$.

As discussed in Example 2.7, to any locally integrable function f in $\mathbb{R}^n$ one can associate an "ordinary" distribution $u_f \in \mathcal{D}'(\mathbb{R}^n)$. This being said, it is not always the case that u_f is actually a tempered distribution (for more on this, see Remark 4.16). This is, however, true if the locally integrable function f becomes integrable at infinity after being tempered by a polynomial. We elaborate on this issue in the next example.

Example 4.4. Let $f \in L_{loc}^1(\mathbb{R}^n)$ be such that there exist some $m \in [0, \infty)$ and some $R \in (0, \infty)$ with the property that

$$\int_{|x| \ge R} |x|^{-m} |f(x)| \, dx < \infty. \qquad (4.1.4)$$

We claim that the distribution of function type defined by f is a tempered distribution, that is, the mapping

$$u_f : S(\mathbb{R}^n) \to \mathbb{C}, \qquad u_f(\varphi) := \int_{\mathbb{R}^n} f\varphi \, dx, \qquad \forall \varphi \in S(\mathbb{R}^n), \qquad (4.1.5)$$

is a well-defined tempered distribution. To see that this is the case, pick $N \in \mathbb{N}_0$ such that $N \ge m$ and, for an arbitrary $\varphi \in S(\mathbb{R}^n)$, estimate

$$\int_{\mathbb{R}^n} |f\varphi| \, dx \le \int_{|x|<R} |f(x)\varphi(x)| \, dx + \int_{|x| \ge R} |x|^{-m} |f(x)| |x|^m |\varphi(x)| \, dx$$

$$\le C \sup_{x \in \mathbb{R}^n, 0 \le k \le N} |x|^k |\varphi(x)| \left[\int_{|x|<R} |f(x)| \, dx + \int_{|x| \ge R} |x|^{-m} |f(x)| \, dx \right]$$

$$\le C \sup_{x \in \mathbb{R}^n, 0 \le k \le N} \left[|x|^k |\varphi(x)| \right], \qquad (4.1.6)$$

where we have used the easily checked fact that, since $N \ge m$, we have

$$\sup_{|x| \ge R} \left[|x|^m |\varphi(x)| \right] \le R^{m-N} \sup_{x \in \mathbb{R}^n} \left[|x|^N |\varphi(x)| \right]. \qquad (4.1.7)$$

Estimate (4.1.6) shows that the functional u_f is well defined on $\mathcal{S}(\mathbb{R}^n)$. Clearly u_f is linear, and (4.1.6) also implies (based on Fact 4.1) that u_f is continuous on $\mathcal{S}(\mathbb{R}^n)$. Hence, u_f is a tempered distribution.

In the sequel, we will often use the notation f instead of u_f, whenever f is such that the operator u_f as in (4.1.5) is linear and continuous.

Exercise 4.5. Prove that for each $a \in (-n, \infty)$ the function $|x|^a$ is a tempered distribution in $\mathbb{R}^n$.

Exercise 4.6. Let $f : \mathbb{R}^n \to \mathbb{C}$ be a Lebesgue measurable function with the property that there exists $m \in \mathbb{R}$ such that

$$\int_{\mathbb{R}^n} (1 + |x|)^m |f(x)|\, dx < \infty. \tag{4.1.8}$$

Then the mapping u_f as in (4.1.5) is well defined and is a tempered distribution.

Hint: Use Example 4.4.

Exercise 4.7. Prove that $\mathcal{L}(\mathbb{R}^n)$, the space of slowly increasing functions defined in (3.1.23), is contained in the space of tempered distributions.

Example 4.8. Let $p \in [1, \infty]$ and $f \in L^p(\mathbb{R}^n)$ be arbitrary. We claim that u_f defined in (4.1.5) is a tempered distribution. To prove this claim, since f is measurable, by Exercise 4.6, it suffices to show that f satisfies (4.1.8). If $p = 1$, then (4.1.8) holds for $m = 0$, while if $p = \infty$, (4.1.8) holds for any $m < -n$. If $p \in (1, \infty)$, by applying Hölder's inequality, we have

$$\int_{\mathbb{R}^n} (1 + |x|)^m |f(x)|\, dx \leq \|f\|_{L^p(\mathbb{R}^n)} \left(\int_{\mathbb{R}^n} (1 + |x|)^{\frac{mp}{p-1}}\, dx \right)^{\frac{p-1}{p}} < \infty$$

provided we choose $m < -\frac{n(p-1)}{p}$. In summary, we proved that

$$\begin{gathered} \text{for each } p \in [1, \infty] \text{ the space } L^p(\mathbb{R}^n) \text{ is} \\ \text{a subspace of the space of tempered distributions.} \end{gathered} \tag{4.1.9}$$

As an immediate consequence of Exercise 3.5 and (4.1.9) we therefore obtain

$$\mathcal{S}(\mathbb{R}^n) \text{ is a subspace of the space of tempered distributions.} \tag{4.1.10}$$

Example 4.9. From Example 2.9 we know that $\ln|x| \in L^1_{loc}(\mathbb{R}^n)$ and, hence, defines a distribution in $\mathbb{R}^n$. We claim that the distribution $\ln|x|$ is, in fact, a tempered distribution. Indeed, this follows from Exercise 4.6, since (4.1.8) holds for any $m < -n$ (seen by using estimate (2.1.9) with $0 < \varepsilon < \min\{n, -m - n\}$).

The topology we consider on the space of tempered distributions is the weak$*$-topology and we denote this topological vector space by $\mathcal{S}'(\mathbb{R}^n)$. In particular, from the general discussion in Section 14.1 we have:

Fact 4.10 $S'(\mathbb{R}^n)$ *is a locally convex topological space.*

Fact 4.11 *A sequence* $\{u_j\}_{j \in \mathbb{N}} \subset S'(\mathbb{R}^n)$ *converges to some* $u \in S'(\mathbb{R}^n)$ *in* $S'(\mathbb{R}^n)$ *if and only if* $\langle u_j, \varphi \rangle \xrightarrow[j \to \infty]{} \langle u, \varphi \rangle$ *for every* $\varphi \in S(\mathbb{R}^n)$, *in which case we write*

$$u_j \xrightarrow[j \to \infty]{S'(\mathbb{R}^n)} u.$$

It is useful to note that, for each $p \in [1, \infty]$, the space of p-th power integrable functions in $\mathbb{R}^n$ embeds continuously in the space of tempered distributions.

Exercise 4.12. Prove that if $p \in [1, \infty]$ and a sequence of functions $\{f_j\}_{j \in \mathbb{N}}$ in $L^p(\mathbb{R}^n)$ converges in $L^p(\mathbb{R}^n)$ to some $f \in L^p(\mathbb{R}^n)$ then $f_j \xrightarrow[j \to \infty]{S'(\mathbb{R}^n)} f$.

Hint: Use Hölder's inequality and (4.1.9).

Exercise 4.13. Assume that ϕ is as in (1.2.3) and recall the sequence of functions $\{\phi_j\}_{j \in \mathbb{N}}$ from (1.3.7). Prove that $\phi_j \xrightarrow[j \to \infty]{S'(\mathbb{R}^n)} \delta$.

Hint: Reason as in Example 2.24.

Let $m, n \in \mathbb{N}$ be arbitrary. As a consequence of Remark 3.10 we obtain (compare to (2.7.41)):

> if $w \in S'(\mathbb{R}^m \times \mathbb{R}^n)$, then the map $w^\top : S(\mathbb{R}^n \times \mathbb{R}^m) \to \mathbb{C}$ defined by $w^\top(\varphi) := \langle w, \varphi^\top \rangle$ for each $\varphi \in S(\mathbb{R}^n \times \mathbb{R}^m)$ is a tempered distribution in $\mathbb{R}^n \times \mathbb{R}^m$, i.e., $w^\top \in S'(\mathbb{R}^n \times \mathbb{R}^m)$.
>
> (4.1.11)

Theorem 4.14. *For each* $n, m \in \mathbb{N}$ *the following statements are true:*

(a) $\mathcal{E}'(\mathbb{R}^n) \hookrightarrow S'(\mathbb{R}^n) \hookrightarrow \mathcal{D}'(\mathbb{R}^n)$, *where all the embeddings are injective and continuous.*

(b) *For each* $a \in \mathcal{L}(\mathbb{R}^n)$ *and each* $u \in S'(\mathbb{R}^n)$, *the distribution* $au \in \mathcal{D}'(\mathbb{R}^n)$ *extends uniquely to a tempered distribution, which will be denoted by* au *and its action is given by* $\langle au, \varphi \rangle = \langle u, a\varphi \rangle$ *for every* $\varphi \in S(\mathbb{R}^n)$. *Moreover, the mapping*

$$S'(\mathbb{R}^n) \ni u \mapsto au \in S'(\mathbb{R}^n) \tag{4.1.12}$$

is linear, and continuous.

(c) *For every* $\alpha \in \mathbb{N}_0^n$ *and each* $u \in S'(\mathbb{R}^n)$, *the distribution* $\partial^\alpha u \in \mathcal{D}'(\mathbb{R}^n)$ *extends uniquely to a tempered distribution, that will be denoted by* $\partial^\alpha u$ *and its action is given by*

$$\langle \partial^\alpha u, \varphi \rangle = (-1)^{|\alpha|} \langle u, \partial^\alpha \varphi \rangle \quad \text{for every} \quad \varphi \in S(\mathbb{R}^n). \tag{4.1.13}$$

Moreover, the mapping

$$S'(\mathbb{R}^n) \ni u \mapsto \partial^\alpha u \in S'(\mathbb{R}^n) \tag{4.1.14}$$

is linear and continuous.

(d) If $u \in S'(\mathbb{R}^m)$ and $v \in S'(\mathbb{R}^n)$ then the distribution $u \otimes v$ originally defined in $\mathcal{D}'(\mathbb{R}^m \times \mathbb{R}^n)$ extends uniquely to a tempered distribution, that will be denoted by $u \otimes v$ and its action is given by

$$\langle u \otimes v, \varphi \rangle = \langle u(x), \langle v(y), \varphi(x, y) \rangle \rangle = \langle v(y), \langle u(x), \varphi(x, y) \rangle \rangle,$$
$$\forall \varphi \in S(\mathbb{R}^m \times \mathbb{R}^n). \tag{4.1.15}$$

Hence,

$$\langle u \otimes v, \varphi_1 \otimes \varphi_2 \rangle = \langle u, \varphi_1 \rangle \langle v, \varphi_2 \rangle$$
$$\text{for each } \varphi_1 \in S(\mathbb{R}^m) \text{ and each } \varphi_2 \in S(\mathbb{R}^n). \tag{4.1.16}$$

In addition, $u \otimes v = (v \otimes u)^\top$ in $S'(\mathbb{R}^m \times \mathbb{R}^n)$, for every $u \in S'(\mathbb{R}^m)$ and every $v \in S'(\mathbb{R}^n)$. Moreover, the mapping

$$S'(\mathbb{R}^m) \times S'(\mathbb{R}^n) \ni (u, v) \mapsto u \otimes v \in S'(\mathbb{R}^m \times \mathbb{R}^n) \tag{4.1.17}$$

is bilinear and separately continuous.

Proof. The statement in *(a)* is a consequence of parts *(c)* and *(d)* in Theorem 3.14 and duality (cf. Proposition 14.4). Next, fix $a \in \mathcal{L}(\mathbb{R}^n)$ and let $u \in S'(\mathbb{R}^n)$ be arbitrary. Then based on part *(a)* we have $u \in \mathcal{D}'(\mathbb{R}^n)$. Hence, by Proposition 2.29, au exists and belongs to $\mathcal{D}'(\mathbb{R}^n)$. We will show that au may be extended uniquely to a tempered distribution. Define

$$\widetilde{au} : S(\mathbb{R}^n) \to \mathbb{C}, \qquad \widetilde{au}(\varphi) := \langle u, a\varphi \rangle \quad \forall \varphi \in S(\mathbb{R}^n). \tag{4.1.18}$$

Since $\widetilde{au}$ is the composition of $u \in S'(\mathbb{R}^n)$ with the map in part *(a)* of Theorem 3.14, both of which are linear and continuous, it follows that $\widetilde{au} \in S'(\mathbb{R}^n)$. In addition, $\widetilde{au}\big|_{C_0^\infty(\mathbb{R}^n)} = au$, so if we invoke *(d)* in Theorem 3.14, we obtain that $\widetilde{au}$ is the unique continuous extension of au to $S(\mathbb{R}^n)$. The map in (4.1.12) is also continuous as seen from Theorem 3.14 and the general fact that the transpose of any linear and continuous operator between two topological vector spaces is continuous at the level of dual spaces equipped with weak∗-topologies (cf. Proposition 14.2). Re-denoting $\widetilde{au}$ simply as au now finishes the proof of the statement in *(b)*. The proof of *(c)* is similar to the proof of *(b)*.

Turning our attention to *(d)*, let $u \in S'(\mathbb{R}^m)$ and $v \in S'(\mathbb{R}^n)$. By part *(a)* we have $u \in \mathcal{D}'(\mathbb{R}^m)$ and $v \in \mathcal{D}'(\mathbb{R}^n)$, hence (by Theorem 2.87) $u \otimes v$ belongs to $\mathcal{D}'(\mathbb{R}^m \times \mathbb{R}^n)$. We construct an extension $\widetilde{u \otimes v} : S(\mathbb{R}^{n+m}) \to \mathbb{C}$ by setting

$$\langle \widetilde{u \otimes v}, \varphi \rangle := \langle u(x), \langle v(y), \varphi(x, y) \rangle \rangle, \qquad \forall \varphi \in S(\mathbb{R}^m \times \mathbb{R}^n). \tag{4.1.19}$$

To see that this mapping is in $S'(\mathbb{R}^m \times \mathbb{R}^n)$, fix $\varphi \in S(\mathbb{R}^m \times \mathbb{R}^n)$. Clearly we have $\varphi(x, \cdot) \in S(\mathbb{R}^n)$ for each $x \in \mathbb{R}^m$. We claim that

$$\psi(x) := \langle v(y), \varphi(x, y) \rangle \text{ for each } x \in \mathbb{R}^m \implies \psi \in S(\mathbb{R}^m). \tag{4.1.20}$$

Indeed, by reasoning as in the proof of Proposition 2.84, we obtain that ψ belongs to $C^\infty(\mathbb{R}^m)$. Also, for $\alpha, \beta \in \mathbb{N}_0^m$ we have $x^\beta \partial^\alpha \psi(x) = \langle v(y), x^\beta \partial_x^\alpha \varphi(x, y) \rangle$ and, since $v \in \mathcal{S}'(\mathbb{R}^n)$, there exist $C > 0$, and $\ell, k \in \mathbb{N}_0$, such that v satisfies (4.1.2). Thus, for every $x \in \mathbb{R}^m$, we have

$$\left| x^\beta \partial^\alpha \psi(x) \right| = \left| \langle v(y), x^\beta \partial_x^\alpha \varphi(x, y) \rangle \right| \leq C \sup_{\substack{y \in \mathbb{R}^n \\ |\gamma| \leq \ell, |\delta| \leq k}} \left| y^\gamma x^\beta \partial_x^\alpha \partial_y^\delta \varphi(x, y) \right|. \tag{4.1.21}$$

Therefore, since $\varphi \in \mathcal{S}(\mathbb{R}^m \times \mathbb{R}^n)$, estimate (4.1.21) further yields

$$\sup_{x \in \mathbb{R}^m} \left| x^\beta \partial^\alpha \psi(x) \right| \leq C \sup_{\substack{x \in \mathbb{R}^m, y \in \mathbb{R}^n \\ |\gamma| \leq \ell, |\delta| \leq k}} \left| y^\gamma x^\alpha \partial_x^\beta \partial_y^\delta \varphi(x, y) \right| < \infty. \tag{4.1.22}$$

This shows that $\psi \in \mathcal{S}(\mathbb{R}^m)$, finishing the proof of the claim. Consider next the mapping

$$\mathcal{S}(\mathbb{R}^m \times \mathbb{R}^n) \ni \varphi \mapsto \psi \in \mathcal{S}(\mathbb{R}^m). \tag{4.1.23}$$

This is linear by design, as well as continuous (as seen from (4.1.22), Fact 3.6, and Theorem 14.1). Thus, the composition between u and the map in (4.1.23) gives rise to a linear and continuous map, which proves that $\widetilde{u \otimes v}$ is a tempered distribution in $\mathbb{R}^m \times \mathbb{R}^n$. In addition, (4.1.19) and *(ii)* in Theorem 2.87 imply

$$\widetilde{u \otimes v} \Big|_{C_0^\infty(\mathbb{R}^m \times \mathbb{R}^n)} = u \otimes v, \tag{4.1.24}$$

which in combination with *(d)* in Theorem 3.14, gives that $\widetilde{u \otimes v}$ is the unique continuous extension of $u \otimes v$ to $\mathcal{S}(\mathbb{R}^m \times \mathbb{R}^n)$.

Next, we note that the reasoning above also yields that

$$\eta(y) := \langle u(x), \varphi(x, y) \rangle \quad \text{for each } y \in \mathbb{R}^n \implies \eta \in \mathcal{S}(\mathbb{R}^n) \tag{4.1.25}$$

and that the mapping

$$\mathcal{S}(\mathbb{R}^m \times \mathbb{R}^n) \ni \varphi \mapsto \eta \in \mathcal{S}(\mathbb{R}^n) \quad \text{is linear and continuous.} \tag{4.1.26}$$

Consequently, the composition between v and the map in (4.1.26) gives rise to a linear and continuous map w defined by

$$w(\varphi) := \langle v(y), \langle u(x), \varphi(x, y) \rangle \rangle \quad \text{for each } \varphi \in \mathcal{S}(\mathbb{R}^m \times \mathbb{R}^n). \tag{4.1.27}$$

By *(ii)* in Theorem 2.87, we have that $w \big|_{C_0^\infty(\mathbb{R}^m \times \mathbb{R}^n)} = u \otimes v$. Since we have proved that $\widetilde{u \otimes v}$ is the unique continuous extension of $u \otimes v$ to $\mathcal{S}(\mathbb{R}^m \times \mathbb{R}^n)$, we must have $w = \widetilde{u \otimes v}$. This completes the proof of (4.1.15). Moreover, from (4.1.19) we see that

$$\langle \widetilde{u \otimes v}, \varphi_1 \otimes \varphi_2 \rangle = \langle u, \varphi_1 \rangle \langle v, \varphi_2 \rangle, \quad \forall \varphi_1 \in \mathcal{S}(\mathbb{R}^m), \ \forall \varphi_2 \in \mathcal{S}(\mathbb{R}^n). \tag{4.1.28}$$

Similarly, we define $\widetilde{v \otimes u} \in \mathcal{S}'(\mathbb{R}^n \times \mathbb{R}^m)$, the unique extension of the distribution $v \otimes u \in \mathcal{D}'(\mathbb{R}^m \times \mathbb{R}^n)$ to a tempered distribution. Recall (4.1.11) and consider the tempered distribution $(\widetilde{v \otimes u})^\top$ in $\mathcal{S}'(\mathbb{R}^m \times \mathbb{R}^n)$. Then for each test function $\varphi \in C_0^\infty(\mathbb{R}^m \times \mathbb{R}^n)$ we may write

$$\langle (\widetilde{v \otimes u})^\top, \varphi \rangle = \langle \widetilde{v \otimes u}, \varphi^\top \rangle = \langle v \otimes u, \varphi^\top \rangle = \langle (v \otimes u)^\top, \varphi \rangle$$

$$= \langle u \otimes v, \varphi \rangle = \langle \widetilde{u \otimes v}, \varphi \rangle, \tag{4.1.29}$$

where for the fourth equality in (4.1.29) we have used item *(iii)* in Theorem 2.87. This proves that

$$\widetilde{u \otimes v}\big|_{C_0^\infty(\mathbb{R}^m \times \mathbb{R}^n)} = \left[(\widetilde{v \otimes u})^\top\right]\big|_{C_0^\infty(\mathbb{R}^m \times \mathbb{R}^n)}. \tag{4.1.30}$$

Thus, using *(d)* in Theorem 3.14, we conclude $\widetilde{u \otimes v} = (\widetilde{v \otimes u})^\top$.

Clearly, $\mathcal{S}'(\mathbb{R}^m) \times \mathcal{S}'(\mathbb{R}^n) \ni (u, v) \mapsto \widetilde{u \otimes v} \in \mathcal{S}'(\mathbb{R}^m \times \mathbb{R}^n)$ is bilinear, and our goal is to show that this is also separately continuous. First we will prove that this map is continuous in the first variable. For this, we shall rely on the abstract description of open sets in the weak*-topology from (14.1.9). Specifically, having fixed $v \in \mathcal{S}'(\mathbb{R}^n)$, pick an arbitrary finite set $A \subset \mathcal{S}(\mathbb{R}^m \times \mathbb{R}^n)$ along with some number $\varepsilon \in (0, \infty)$, and introduce

$$O_{A,\varepsilon} := \{ w \in \mathcal{S}'(\mathbb{R}^m \times \mathbb{R}^n) : |\langle w, \psi \rangle| < \varepsilon, \ \forall \psi \in A \}. \tag{4.1.31}$$

If we now define $\widetilde{A} := \{\langle v(y), \psi(\cdot, y)\rangle : \psi \in A\}$, then from what we proved earlier (cf. (4.1.20)) it follows that $\widetilde{A}$ is a subset of $\mathcal{S}'(\mathbb{R}^m)$. Also, $\widetilde{A}$ is finite since A is finite. Hence, if we now set

$$\widetilde{O}_{\widetilde{A},\varepsilon} := \{ u \in \mathcal{S}'(\mathbb{R}^m) : |\langle u, \varphi \rangle| < \varepsilon, \ \forall \varphi \in \widetilde{A} \}, \tag{4.1.32}$$

using (4.1.19) we have $\widetilde{u \otimes v} \in O_{A,\varepsilon}$ for every $u \in \widetilde{O}_{\widetilde{A},\varepsilon}$. In light of (14.1.9), this shows that $u \mapsto \widetilde{u \otimes v}$ is continuous. On account of formula (4.1.15), a similar proof also gives that $v \mapsto \widetilde{u \otimes v}$ is continuous.

Finally, abbreviating matters by simply writing $u \otimes v$ in place of $\widetilde{u \otimes v}$, all claims in part *(d)* of the statement of the theorem now follow. $\square$

Remark 4.15.

(i) In view of *(a)* in Theorem 4.14 and *(d)* in Theorem 3.14 we have:

$$\text{if } u, v \in \mathcal{S}'(\mathbb{R}^n) \text{ and } u = v \text{ in } \mathcal{D}'(\mathbb{R}^n), \text{ then } u = v \text{ in } \mathcal{S}'(\mathbb{R}^n). \tag{4.1.33}$$

(ii) Whenever $u \in \mathcal{S}'(\mathbb{R}^n)$ its support is understood as defined in (2.5.8).

Remark 4.16. The inclusion $\mathcal{S}'(\mathbb{R}^n) \hookrightarrow \mathcal{D}'(\mathbb{R}^n)$ (which goes to show that any tempered distribution is indeed an ordinary distribution) is actually strict. This follows from the observation that, in contrast to the case of ordinary distributions,

$$L_{loc}^1(\mathbb{R}^n) \not\subset \mathcal{S}'(\mathbb{R}^n). \tag{4.1.34}$$

To justify (4.1.34), take $\lambda > 0$ arbitrary, fixed, and consider the function

$$f : \mathbb{R}^n \to \mathbb{C}, \qquad f(x) := e^{|x|^\lambda} \quad \text{for every} \quad x \in \mathbb{R}^n. \tag{4.1.35}$$

Clearly $f \in L^1_{loc}(\mathbb{R}^n)$, hence $f \in \mathcal{D}'(\mathbb{R}^n)$. However, this distribution cannot be extended to a tempered distribution. To see why this is true, suppose there exists $u \in \mathcal{S}'(\mathbb{R}^n)$ such that $u|_{C_0^\infty(\mathbb{R}^n)} = f$ in $\mathcal{D}'(\mathbb{R}^n)$. Since u is a tempered distribution there exist some finite constant $C \geq 0$ and numbers $k, m \in \mathbb{N}_0$ such that

$$|\langle u, \varphi \rangle| \leq C \sup_{x \in \mathbb{R}^n, |\alpha| \leq k, |\beta| \leq m} |x^\beta \partial^\alpha \varphi(x)|, \quad \forall \varphi \in \mathcal{S}(\mathbb{R}^n). \tag{4.1.36}$$

Now fix a function

$$\psi \in C_0^\infty(\mathbb{R}^n), \quad \psi \geq 0, \quad \operatorname{supp} \psi \subseteq \overline{B(0,1)},$$
$$\int_{1/2 \leq |x| \leq 1} \psi(x) \, dx = 1, \tag{4.1.37}$$

and for each $j \in \mathbb{N}$ define $\psi_j(x) := \psi(x/j)$ for every $x \in \mathbb{R}^n$. Then by (4.1.36), for each $j \in \mathbb{N}$, we have

$$|\langle u, \psi_j \rangle| \leq C \sup_{x \in \overline{B(0,j)}, |\alpha| \leq k, |\beta| \leq m} |x^\beta \partial^\alpha \psi_j(x)| \leq C' j^{m-k}, \tag{4.1.38}$$

for some $C' \in [0, \infty)$ independent of j. On the other hand, (4.1.37) forces the lower bound

$$|\langle u, \psi_j \rangle| = \int_{|x| \leq j} e^{|x|^\lambda} \psi_j(x) \, dx \geq \int_{j/2 \leq |x| \leq j} e^{|x|^\lambda} \psi_j(x) \, dx$$

$$\geq e^{(j/2)^\lambda} j^{-n} \int_{1/2 \leq |x| \leq 1} \psi(x) \, dx = e^{(j/2)^\lambda} j^{-n}, \quad \forall j \in \mathbb{N}. \tag{4.1.39}$$

Comparing (4.1.38) and (4.1.39) yields a contradiction choosing j large enough. Hence, f cannot be extended to a tempered distribution.

Remark 4.17. What prevents locally integrable functions of the form (4.1.35) from belonging to $\mathcal{S}'(\mathbb{R}^n)$ is the fact that their growth at infinity is not tempered enough and, in fact, this observation justifies the very name "tempered distribution."

Exercise 4.18. Let $\lambda \in \mathbb{R}$ be such that $\lambda < n - 1$, and fix some j in $\{1, \ldots, n\}$. For these choices, consider the functions f, g defined, respectively, by $f(x) := |x|^{-\lambda}$ and $g(x) := -\lambda x_j |x|^{-\lambda-2}$ for each $x \in \mathbb{R}^n \setminus \{0\}$. Prove that $f, g \in \mathcal{S}'(\mathbb{R}^n)$ and that $\partial_j f = g$ in $\mathcal{S}'(\mathbb{R}^n)$. In short,

$$\partial_j[|x|^{-\lambda}] = -\lambda x_j |x|^{-\lambda-2} \quad \text{in} \quad \mathcal{S}'(\mathbb{R}^n)$$
$$\text{whenever } \lambda < n - 1 \text{ and } j \in \{1, \ldots, n\}. \tag{4.1.40}$$

Hint: To prove (4.1.40) use (4.1.13), and integration by parts coupled with a limiting argument to extricate the singularity at the origin.

Given a tempered distribution, we may consider the convolution between its restriction to $C_0^\infty(\mathbb{R}^n)$ and any compactly supported distribution. A natural question, addressed in the next theorem, is whether the distribution obtained via such a convolution may be extended to a tempered distribution.

Theorem 4.19. *The following statements are true:*

(a) *If $u \in \mathcal{E}'(\mathbb{R}^n)$ and $v \in \mathcal{S}'(\mathbb{R}^n)$, then the distribution $u * v \in \mathcal{D}'(\mathbb{R}^n)$ extends uniquely to a tempered distribution, that will be denoted by $u * v$. Also, the distribution $v * u \in \mathcal{D}'(\mathbb{R}^n)$ extends uniquely to a tempered distribution and $u * v = v * u$ in $\mathcal{S}'(\mathbb{R}^n)$. Moreover, the mapping*

$$\mathcal{E}'(\mathbb{R}^n) \times \mathcal{S}'(\mathbb{R}^n) \ni (u, v) \mapsto u * v \in \mathcal{S}'(\mathbb{R}^n) \tag{4.1.41}$$

is bilinear, and for every $u \in \mathcal{E}'(\mathbb{R}^n)$, $v \in \mathcal{S}'(\mathbb{R}^n)$, we have

$$\partial^\alpha(u * v) = (\partial^\alpha u) * v = u * (\partial^\alpha v) \quad in \quad \mathcal{S}'(\mathbb{R}^n), \quad \forall \alpha \in \mathbb{N}_0^n. \tag{4.1.42}$$

(b) *If $u_j \xrightarrow[j \to \infty]{\mathcal{D}'(\mathbb{R}^n)} u$ and there exists a compact set $K \subset \mathbb{R}^n$ with the property that $\operatorname{supp} u_j \subset K$ for every $j \in \mathbb{N}$, then $u_j * v \xrightarrow[j \to \infty]{\mathcal{S}'(\mathbb{R}^n)} u * v$ for each $v \in \mathcal{S}'(\mathbb{R}^n)$.*

(c) *If $v_j \xrightarrow[j \to \infty]{\mathcal{S}'(\mathbb{R}^n)} v$, then $u * v_j \xrightarrow[j \to \infty]{\mathcal{S}'(\mathbb{R}^n)} u * v$ for each $u \in \mathcal{E}'(\mathbb{R}^n)$.*

(d) *If $u \in \mathcal{S}'(\mathbb{R}^n)$ then $\delta * u = u = u * \delta$ for every $u \in \mathcal{S}'(\mathbb{R}^n)$.*

(e) *Let $a \in \mathcal{S}(\mathbb{R}^n)$ and $u \in \mathcal{S}'(\mathbb{R}^n)$. Then the mapping*

$$a * u : \mathcal{S}(\mathbb{R}^n) \to \mathbb{C}, \quad (a * u)(\varphi) := \langle u, a^\vee * \varphi \rangle \quad \forall \varphi \in \mathcal{S}(\mathbb{R}^n), \tag{4.1.43}$$

is well defined, linear, and continuous, hence $a * u$ belongs to $\mathcal{S}'(\mathbb{R}^n)$. We also define $u * a := a * u$. If $a \in C_0^\infty(\mathbb{R}^n)$, then the map in (4.1.43) is the unique continuous extension of $a * u \in \mathcal{D}'(\mathbb{R}^n)$ to a tempered distribution. In addition, the mapping

$$\mathcal{S}(\mathbb{R}^n) \times \mathcal{S}'(\mathbb{R}^n) \ni (a, u) \mapsto a * u \in \mathcal{S}'(\mathbb{R}^n) \tag{4.1.44}$$

is bilinear and separately sequentially continuous. Also, for every $a \in \mathcal{S}(\mathbb{R}^n)$ and $u \in \mathcal{S}'(\mathbb{R}^n)$ we have

$$\partial^\alpha(a * u) = (\partial^\alpha a) * u = a * (\partial^\alpha u) \quad in \quad \mathcal{S}'(\mathbb{R}^n), \quad \forall \alpha \in \mathbb{N}_0^n. \tag{4.1.45}$$

Proof. Fix $u \in \mathcal{E}'(\mathbb{R}^n)$ and $v \in \mathcal{S}'(\mathbb{R}^n)$. Because of (a) in Theorem 4.14 and Theorem 2.94, it follows that $u * v$ exists as an element in $\mathcal{D}'(\mathbb{R}^n)$. We will prove that $u * v$ extends uniquely to an element in $\mathcal{S}'(\mathbb{R}^n)$. Fix $\psi \in C_0^\infty(\mathbb{R}^n)$ such that $\psi \equiv 1$ in a neighborhood of $\operatorname{supp} u$. Recall the notation introduces in (2.8.7) and let 1 denote the constant function equal to 1 in $\mathbb{R}^n$. We then claim that the map

$$S(\mathbb{R}^n) \ni \varphi \mapsto (\psi \otimes 1)\varphi^{\Delta} \in S(\mathbb{R}^n \times \mathbb{R}^n) \quad \text{is linear and continuous.} \qquad (4.1.46)$$

Indeed, if $\varphi \in S(\mathbb{R}^n)$, then $(\psi \otimes 1)\varphi^{\Delta} \in C^{\infty}(\mathbb{R}^n \times \mathbb{R}^n)$ and if we pick arbitrary $\alpha, \beta, \gamma, \delta \in \mathbb{N}_0^n$, then by Leibniz's formula (cf. Proposition 14.10), the binomial theorem (cf. Theorem 14.9), and the compactness of the support of ψ, we may write

$$\sup_{x \in \mathbb{R}^n, y \in \mathbb{R}^n} \left| x^{\alpha} y^{\beta} \partial_x^{\gamma} [\psi(x) \partial_y^{\delta} \varphi(x + y)] \right|$$

$$= \sup_{x \in \mathrm{supp}\,\psi, y \in \mathbb{R}^n} \left| x^{\alpha} y^{\beta} \partial_x^{\gamma} [\psi(x)(\partial^{\delta} \varphi)(x + y)] \right|$$

$$\leq C \sum_{\mu \leq \gamma} \sup_{x \in \mathrm{supp}\,\psi, y \in \mathbb{R}^n} \left| (y + x - x)^{\beta} (\partial^{\mu} \psi)(x)(\partial^{\gamma - \mu + \delta} \varphi)(x + y) \right|$$

$$\leq C \sum_{\mu \leq \gamma} \sum_{\eta \leq \beta} \sup_{x \in \mathrm{supp}\,\psi, y \in \mathbb{R}^n} \left| (y + x)^{\eta} (\partial^{\gamma - \mu + \delta} \varphi)(x + y) \right|$$

$$\leq C \sum_{\mu \leq \gamma} \sum_{\eta \leq \beta} \sup_{z \in \mathbb{R}^n} \left| z^{\eta} \partial^{\gamma - \mu + \delta} \varphi(z) \right| < \infty, \qquad (4.1.47)$$

where all constants are independent of φ. Hence, $(\psi \otimes 1)\varphi^{\Delta}$ belongs to $S(\mathbb{R}^n \times \mathbb{R}^n)$ which proves that the map in (4.1.46) is well defined. Since this map is also linear, it suffices to check its continuity at zero. This, in turn, follows from Fact 3.6, Theorem 14.1, and the fact that the final estimate in (4.1.47) implies that the map in (4.1.46) is sequentially continuous. If we now define

$$\widetilde{u * v} : S(\mathbb{R}^n) \longrightarrow \mathbb{C},$$
$$(\widetilde{u * v})(\varphi) := \langle u(x) \otimes v(y), \psi(x)\varphi(x + y) \rangle, \quad \forall \varphi \in S(\mathbb{R}^n), \qquad (4.1.48)$$

then (d) in Theorem 4.14 combined with (4.1.46) gives that $\widetilde{u * v}$ is a tempered distribution in $\mathbb{R}^n \times \mathbb{R}^n$. In addition, this definition is independent of the choice of ψ selected as above. Indeed, if $\psi_1, \psi_2 \in C_0^{\infty}(\mathbb{R}^n)$ are such that each equals 1 in some neighborhood of $\mathrm{supp}\,u$ then, for each $\varphi \in S(\mathbb{R}^n)$,

$$\langle u(x) \otimes v(y), (\psi_1(x) - \psi_2(x))\varphi(x + y) \rangle = 0. \qquad (4.1.49)$$

To see that the map in (4.1.48) is an extension of $u * v \in \mathcal{D}'(\mathbb{R}^n)$, pick some $\varphi \in C_0^{\infty}(\mathbb{R}^n)$ along with a function $\eta \in C_0^{\infty}(\mathbb{R}^n)$ with the property that $\eta \equiv 1$ in a neighborhood of $\mathrm{supp}\,\varphi - \mathrm{supp}\,u$. Then the function $\Psi(x, y) := \psi(x)\eta(y)$ for each $x, y \in \mathbb{R}^n$, belongs to $C_0^{\infty}(\mathbb{R}^n \times \mathbb{R}^n)$ and is equal to 1 in a neighborhood of $(\mathrm{supp}\,u \times \mathrm{supp}\,v) \cap \mathrm{supp}\,\varphi^{\Delta}$. Upon recalling (2.8.11), this permits us to write

$$\langle u(x) \otimes v(y), \psi(x)\varphi(x + y) \rangle = \langle u(x) \otimes v(y), \Psi(x, y)\varphi(x + y) \rangle$$

$$= \langle u * v, \varphi \rangle. \qquad (4.1.50)$$

This shows that the mapping in (4.1.48) is an extension of $u * v \in \mathcal{D}'(\mathbb{R}^n)$ which, together with *(d)* in Theorem 3.14, implies that $\widetilde{u * v}$ is the unique extension of $u * v \in \mathcal{D}'(\mathbb{R}^n)$ to a tempered distribution.

A similar construction realizes $\widetilde{v * u}$ as a tempered distribution. Moreover, since $u * v = v * u$ in $\mathcal{D}'(\mathbb{R}^n)$, from (4.1.33) we deduce that $\widetilde{u * v} = \widetilde{v * u}$ in $\mathcal{S}'(\mathbb{R}^n)$. It is also clear from (4.1.48) that the mapping

$$\mathcal{E}'(\mathbb{R}^n) \times \mathcal{S}'(\mathbb{R}^n) \ni (u, v) \mapsto \widetilde{u * v} \in \mathcal{S}'(\mathbb{R}^n) \tag{4.1.51}$$

is bilinear. After dropping the tilde, all claims in statement *(a)* now follow, with the exception of (4.1.42). Regarding the latter, we first note that the equalities in (4.1.42) hold when interpreted in $\mathcal{D}'(\mathbb{R}^n)$, thanks to part *(e)* in Theorem 2.96. Given that all distributions involved are tempered, we may invoke (4.1.33) to conclude that the named equalities also hold in $\mathcal{S}'(\mathbb{R}^n)$.

Moving on, let $u_j \xrightarrow[j \to \infty]{\mathcal{D}'(\mathbb{R}^n)} u$ be such that there exists a compact set $K \subset \mathbb{R}^n$ with the property that $\operatorname{supp} u_j \subseteq K$ for each $j \in \mathbb{N}$. Then, by Exercise 2.73, we have $\operatorname{supp} u \subseteq K$ and $u_j \xrightarrow[j \to \infty]{\mathcal{E}'(\mathbb{R}^n)} u$. The latter combined with *(a)* in Theorem 4.14 implies $u_j \xrightarrow[j \to \infty]{\mathcal{S}'(\mathbb{R}^n)} u$. Fix now an arbitrary $v \in \mathcal{S}'(\mathbb{R}^n)$. From the current part *(a)* it follows that $u * v \in \mathcal{S}'(\mathbb{R}^n)$ and $u_j * v \in \mathcal{S}'(\mathbb{R}^n)$, for every $j \in \mathbb{N}$. If $\varphi \in \mathcal{S}(\mathbb{R}^n)$ then definition (4.1.48) implies that

$$\langle u_j * v, \varphi \rangle = \langle u_j(x) \otimes v(y), \psi(x)\varphi(x + y) \rangle, \qquad \forall j \in \mathbb{N}, \tag{4.1.52}$$

where $\psi \in C_0^\infty(\mathbb{R}^n)$ is a fixed function chosen so that $\psi \equiv 1$ in a neighborhood of K. The proof of (4.1.46) shows that $(\psi \otimes 1)\varphi^\Delta \in \mathcal{S}(\mathbb{R}^n \times \mathbb{R}^n)$. Since the map in (4.1.17) is separately continuous, we may write

$$\langle u_j * v, \varphi \rangle = \langle u_j \otimes v, (\psi \otimes 1)\varphi^\Delta \rangle \xrightarrow[j \to \infty]{} \langle u \otimes v, (\psi \otimes 1)\varphi^\Delta \rangle = \langle u * v, \varphi \rangle. \tag{4.1.53}$$

This proves the statement in *(b)*. The proof of the statement in *(c)* is similar. Also the statement in *(d)* is an immediate consequence of part *(d)* in Theorem 2.96 and (4.1.33).

Turning our attention to the statement in *(e)*, we first note that, as noted in Exercise 3.26, the mapping $\mathcal{S}(\mathbb{R}^n) \ni a \mapsto a^\vee \in \mathcal{S}(\mathbb{R}^n)$ is well defined, linear, and continuous. Based on this and *(f)* in Theorem 3.14 we may then conclude that the map in (4.1.43) is well defined, linear, and continuous, as a composition of linear and continuous maps. Assume next that $a \in C_0^\infty(\mathbb{R}^n)$. Then by Proposition 2.102, for each $u \in \mathcal{D}'(\mathbb{R}^n)$ we have $a * u \in C^\infty(\mathbb{R}^n)$ and $(a * u)(x) = \langle u(y), a(x - y) \rangle$ for every $x \in \mathbb{R}^n$, hence

$$\langle a * u, \varphi \rangle = \int_{\mathbb{R}^n} \langle u(y), a(x - y) \rangle \varphi(x) \, dx \tag{4.1.54}$$

$$= \left\langle u(y), \int_{\mathbb{R}^n} a(x - y) \varphi(x) \, dx \right\rangle = \langle u, a^\vee * \varphi \rangle, \quad \forall \varphi \in C_0^\infty(\mathbb{R}^n).$$

This shows that the map in (4.1.43) is an extension of $a * u \in \mathcal{D}'(\mathbb{R}^n)$ to a tempered distribution. The uniqueness of such an extension is then seen from *(d)* in Theorem 3.14.

Regarding the map in (4.1.44), its bilinearity is immediate. If $a_j \xrightarrow[j \to \infty]{\mathcal{S}(\mathbb{R}^n)} a$ then $a_j^\vee \xrightarrow[j \to \infty]{\mathcal{S}(\mathbb{R}^n)} a^\vee$, so by *(f)* in Theorem 3.14, $a_j^\vee * \varphi \xrightarrow[j \to \infty]{\mathcal{S}(\mathbb{R}^n)} a^\vee * \varphi$ for every φ in $\mathcal{S}(\mathbb{R}^n)$, hence

$$\langle u, a_j^\vee * \varphi \rangle \xrightarrow[j \to \infty]{} \langle u, a^\vee * \varphi \rangle \quad \text{for each } u \in \mathcal{S}'(\mathbb{R}^n),$$

proving that the map in (4.1.44) is sequentially continuous in the first variable. Moreover, if $u_j \xrightarrow[j \to \infty]{\mathcal{S}'(\mathbb{R}^n)} u$ and $a \in \mathcal{S}(\mathbb{R}^n)$, then

$$\langle a * u_j, \varphi \rangle = \langle u_j, a^\vee * \varphi \rangle \xrightarrow[j \to \infty]{} \langle u, a^\vee * \varphi \rangle = \langle a * u, \varphi \rangle \tag{4.1.55}$$

for each $\varphi \in \mathcal{S}(\mathbb{R}^n)$. Thus, $a * u_j \xrightarrow[j \to \infty]{\mathcal{S}'(\mathbb{R}^n)} a * u$. This finishes the proof of the fact that the mapping (4.1.44) is bilinear and separately sequentially continuous. Finally, (4.1.45) is a consequence of (4.1.42), part *(d)* in Theorem 3.14, the separate sequential continuity of (4.1.44), and part *(c)* in Theorem 4.14. This completes the proof of the theorem. $\qquad \square$

In the last part of this section we present a density result.

Proposition 4.20. *The space $C_0^\infty(\mathbb{R}^n)$ is sequentially dense in $\mathcal{S}'(\mathbb{R}^n)$. In particular, the Schwartz class $\mathcal{S}(\mathbb{R}^n)$ is sequentially dense in $\mathcal{S}'(\mathbb{R}^n)$.*

Proof. Pick an arbitrary $u \in \mathcal{S}'(\mathbb{R}^n)$. Fix a function $\psi \in C_0^\infty(\mathbb{R}^n)$ such that $\psi \equiv 1$ on $B(0, 1)$ and, for each $j \in \mathbb{N}$, define $\psi_j(x) := \psi(x/j)$ for every $x \in \mathbb{R}^n$. We claim that

$$\psi_j u \xrightarrow[j \to \infty]{\mathcal{S}'(\mathbb{R}^n)} u. \tag{4.1.56}$$

Indeed, if $\varphi \in \mathcal{S}(\mathbb{R}^n)$ is arbitrary, then by reasoning as in the proof of (3.1.29) we deduce that $\psi_j \varphi \xrightarrow[j \to \infty]{\mathcal{S}(\mathbb{R}^n)} \varphi$. Hence,

$$\langle \psi_j u, \varphi \rangle = \langle u, \psi_j \varphi \rangle \xrightarrow[j \to \infty]{} \langle u, \varphi \rangle, \tag{4.1.57}$$

from which (4.1.56) follows. In light of (4.1.56) it suffices to show that any tempered distribution with compact support is the limit in $\mathcal{S}'(\mathbb{R}^n)$ of a sequence from $C_0^\infty(\mathbb{R}^n)$. To this end, suppose that $u \in \mathcal{S}'(\mathbb{R}^n)$ is compactly supported. Also, choose a function

ϕ as in (1.2.3) and recall the sequence $\{\phi_j\}_{j \in \mathbb{N}} \subset C_0^\infty(\mathbb{R}^n)$ from (1.3.7). Exercise 4.13 then gives $\phi_j \xrightarrow[j \to \infty]{S'(\mathbb{R}^n)} \delta$ which, in concert with parts *(d)–(e)* in Theorem 4.19, implies that $\phi_j * u \xrightarrow[j \to \infty]{S'(\mathbb{R}^n)} u$. At this stage there remains to observe that $\phi_j * u \in C_0^\infty(\mathbb{R}^n)$ for every $j \in \mathbb{N}$, thanks to (2.8.49). $\qquad\square$

4.2 The Fourier Transform Acting on Tempered Distributions

The goal here is to extend the action of the Fourier transform, considered earlier on the Schwartz class of rapidly decreasing functions, to the space of tempered distributions. To get some understanding of how this can be done in a natural fashion, we begin by noting that if $f \in L^1(\mathbb{R}^n)$ is arbitrary then (3.2.27) gives

$$\int_{\mathbb{R}^n} \widehat{f}(\xi)\varphi(\xi)\,\mathrm{d}\xi = \int_{\mathbb{R}^n} f(x)\widehat{\varphi}(x)\,\mathrm{d}x, \qquad \forall\,\varphi \in S(\mathbb{R}^n). \tag{4.2.1}$$

Since $f \in L^1(\mathbb{R}^n)$ and $\widehat{f} \in L^\infty(\mathbb{R}^n)$, based on (4.1.9) we have $f, \widehat{f} \in S'(\mathbb{R}^n)$. Hence, (4.2.1) may be rewritten as $\langle \widehat{f}, \varphi \rangle = \langle f, \widehat{\varphi} \rangle$, for every $\varphi \in S(\mathbb{R}^n)$. This suggests the definition made below for the Fourier transform of a tempered distribution.

Proposition 4.21. *Let* $u \in S'(\mathbb{R}^n)$. *Then the mapping*

$$\widehat{u} : S(\mathbb{R}^n) \to \mathbb{C}, \quad \widehat{u}(\varphi) := \langle u, \widehat{\varphi} \rangle, \qquad \forall\,\varphi \in S(\mathbb{R}^n), \tag{4.2.2}$$

is well defined, linear, and continuous. Hence, $\widehat{u} \in S'(\mathbb{R}^n)$.

Proof. This is an immediate consequence of the fact that $u \in S'(\mathbb{R}^n)$, part *(c)* in Theorem 3.21, and the identity $\widehat{u} = u \circ \mathcal{F}$, where $\mathcal{F}$ is the Fourier transform on $S(\mathbb{R}^n)$. $\qquad\square$

Example 4.22. Since $\delta \in S'(\mathbb{R}^n)$ we may write

$$\langle \widehat{\delta}, \varphi \rangle = \langle \delta, \widehat{\varphi} \rangle = \widehat{\varphi}(0) = \int_{\mathbb{R}^n} \varphi(x)\,\mathrm{d}x = \langle 1, \varphi \rangle, \qquad \forall\,\varphi \in S(\mathbb{R}^n), \tag{4.2.3}$$

thus

$$\widehat{\delta} = 1 \quad \text{in} \quad S'(\mathbb{R}^n). \tag{4.2.4}$$

Also, if $x_0 \in \mathbb{R}^n$, then for each $\varphi \in S(\mathbb{R}^n)$ we may write

$$\langle \widehat{\delta_{x_0}}, \varphi \rangle = \langle \delta_{x_0}, \widehat{\varphi} \rangle = \widehat{\varphi}(x_0) = \int_{\mathbb{R}^n} \mathrm{e}^{-\mathrm{i}x_0 \cdot x}\varphi(x)\,\mathrm{d}x = \langle \mathrm{e}^{-\mathrm{i}x_0 \cdot x}, \varphi \rangle, \tag{4.2.5}$$

proving

$$\widehat{\delta_{x_0}} = \mathrm{e}^{-\mathrm{i}x_0 \cdot x} \quad \text{in} \quad S'(\mathbb{R}^n). \tag{4.2.6}$$

Remark 4.23. The map $S'(\mathbb{R}^n) \ni u \mapsto \widehat{u} \in S'(\mathbb{R}^n)$ (that is further analyzed in part *(a)* of Theorem 4.26) is an extension of the map in (3.1.3). Indeed, if $f \in L^1(\mathbb{R}^n)$ then the map u_f from (4.1.5) is a tempered distribution and we may write

$$\langle \widehat{u_f}, \varphi \rangle = \langle u_f, \widehat{\varphi} \rangle = \int_{\mathbb{R}^n} f(x)\widehat{\varphi}(x)\,\mathrm{d}x = \int_{\mathbb{R}^n} \widehat{f}(\xi)\varphi(\xi)\,\mathrm{d}\xi$$

$$= \langle \widehat{f}, \varphi \rangle \qquad \forall\, \varphi \in S(\mathbb{R}^n), \tag{4.2.7}$$

where $\widehat{f}$ is as in (3.1.1) and for the third equality we used (3.2.27). This shows that

$$\widehat{u_f} = u_{\widehat{f}} \quad \text{in} \quad S'(\mathbb{R}^n), \quad \forall\, f \in L^1(\mathbb{R}^n). \tag{4.2.8}$$

In particular, (4.2.8) also holds if $f \in S(\mathbb{R}^n)$ (since $S(\mathbb{R}^n) \subset L^1(\mathbb{R}^n)$ by Exercise 3.5).

Example 4.24. Suppose $a \in (0, \infty)$ and consider the function $f(x) := \frac{1}{x^2 + a^2}$ for every $x \in \mathbb{R}$. Since $f \in L^1(\mathbb{R})$, by (4.1.9) we may regard f as a tempered distribution. The goal is to compute $\widehat{f}$ in $S'(\mathbb{R})$. From (4.2.8) we know that this is the same as the Fourier transform of f as a function in $L^1(\mathbb{R})$, that is

$$\widehat{f}(\xi) = \int_{\mathbb{R}} \frac{e^{-ix\xi}}{x^2 + a^2}\,\mathrm{d}x, \qquad \forall\, \xi \in \mathbb{R}. \tag{4.2.9}$$

To compute the integral in (4.2.9) we use the calculus of residues applied to the function

$$g(z) := \frac{e^{-iz\xi}}{z^2 + a^2} \quad \text{for} \quad z \in \mathbb{C} \setminus \{ia, -ia\}. \tag{4.2.10}$$

In this regard, denote the residue of g at z by $\mathrm{Res}_g(z)$. We separate the computation in two cases.

Case 1. *Assume $\xi \in (-\infty, 0)$. For each $R \in (a, \infty)$ consider the contour $\Gamma := \Gamma_1 \cup \Gamma_2$, where*

$$\Gamma_1 := \{t : -R \le t \le R\}, \tag{4.2.11}$$

$$\Gamma_2 := \{Re^{i\theta} : 0 \le \theta \le \pi\}. \tag{4.2.12}$$

The function g has one pole $z = ia$ enclosed by Γ and the residue theorem gives

$$\int_{\Gamma_1} g(z)\,\mathrm{d}z + \int_{\Gamma_2} g(z)\,\mathrm{d}z = 2\pi i\, \mathrm{Res}_g(ia). \tag{4.2.13}$$

Let us analyze each term in (4.2.13). First,

$$\mathrm{Res}_g(ia) = \left. \frac{e^{-iz\xi}}{z + ia} \right|_{z=ia} = \frac{e^{a\xi}}{2ia} = \frac{e^{-a|\xi|}}{2ia}. \tag{4.2.14}$$

Second, given that for each fixed $\xi \in \mathbb{R}$ the function $\mathbb{R} \ni t \mapsto \frac{e^{-it\xi}}{t^2+a^2} \in \mathbb{C}$ is absolutely integrable, Lebesgue's Dominated Convergence Theorem gives

$$\int_{\Gamma_1} g(z)\,dz = \int_{-R}^{R} \frac{e^{-it\xi}}{t^2+a^2}\,dt \xrightarrow{R\to\infty} \int_{\mathbb{R}} \frac{e^{-it\xi}}{t^2+a^2}\,dt = \widehat{f}(\xi). \qquad (4.2.15)$$

Third,

$$\left| \int_{\Gamma_2} g(z)\,dz \right| = \left| \int_0^\pi \frac{e^{\xi R \sin\theta - i\xi R\cos\theta}}{R^2 e^{i2\theta}+a^2}\, iRe^{i\theta}\,d\theta \right| \leq \frac{C}{R} \xrightarrow{R\to\infty} 0, \qquad (4.2.16)$$

since $e^{\xi R \sin\theta} \leq 1$ for $\theta \in [0,\pi]$ (recall that we are assuming $\xi < 0$). Hence, letting $R \to \infty$ in (4.2.13) and making use of (4.2.14), (4.2.15), and (4.2.16), we arrive at

$$\widehat{f}(\xi) = 2\pi i\, \frac{e^{-a|\xi|}}{2ia} = \frac{\pi}{a} e^{-a|\xi|} \qquad \text{whenever } \xi \in (0,\infty). \qquad (4.2.17)$$

Case 2. *Assume $\xi \in (0,\infty)$.* This time we consider a contour that encloses the pole $z = -ia$. Specifically, we keep Γ_1 as before (with $R \in (a,\infty)$) and set

$$\Gamma_3 := \{Re^{i\theta} : \pi \leq \theta \leq 2\pi\}. \qquad (4.2.18)$$

Applying the residue theorem for the contour $\Gamma_1 \cup \Gamma_3$ we then obtain

$$-\int_{\Gamma_1} g(z)\,dz + \int_{\Gamma_3} g(z)\,dz = 2\pi i\, \mathrm{Res}_g(-ia) = \frac{e^{-iz\xi}}{z-ia}\bigg|_{z=-ia} = -\frac{e^{-a|\xi|}}{2ia}. \qquad (4.2.19)$$

A computation similar to that in (4.2.16) (note that $e^{\xi R \sin\theta} \leq 1$ for each $\xi > 0$ and each $\theta \in [\pi, 2\pi]$) gives $\lim_{R\to\infty} \left| \int_{\Gamma_2} g(z)\,dz \right| = 0$, which when used in (4.2.19) yields $\widehat{f}(\xi) = \frac{\pi}{a} e^{-a|\xi|}$ for $\xi \in (0,\infty)$.

Case 3. *Assume $\xi = 0$.* It is immediate from (4.2.9) that

$$\widehat{f}(0) = \int_{\mathbb{R}} \frac{1}{x^2+a^2}\,dx = \frac{\pi}{a}. \qquad (4.2.20)$$

In summary, we have proved

$$\mathcal{F}\left(\frac{1}{x^2+a^2}\right)(\xi) = \frac{\pi}{a} e^{-a|\xi|} \qquad \text{for every } \xi \in \mathbb{R}, \text{ as well as in } \mathcal{S}'(\mathbb{R}). \qquad (4.2.21)$$

In the next example we compute the Fourier transform of a tempered distribution induced by a slowly growing function that is not absolutely integrable in $\mathbb{R}^n$.

Example 4.25. Let $a \in (0,\infty)$ and consider the function $f(x) := e^{-ia|x|^2}$ for each $x \in \mathbb{R}^n$. It is not difficult to check that $f \in \mathcal{L}(\mathbb{R}^n)$, thus by Exercise 4.7 the function f may be regarded as a tempered distribution (identifying it with $u_f \in \mathcal{S}'(\mathbb{R}^n)$ associated with f as in (4.1.5)). We claim that

$$\widehat{\mathrm{e}^{-ia|x|^2}} = \left(\frac{\pi}{ia}\right)^{\frac{n}{2}} \mathrm{e}^{i\frac{|\xi|^2}{4a}} \quad \text{in} \quad \mathcal{S}'(\mathbb{R}^n). \tag{4.2.22}$$

To prove (4.2.22), fix $\varphi \in \mathcal{S}(\mathbb{R}^n)$ and starting with the definition of the Fourier transform on $\mathcal{S}'(\mathbb{R}^n)$ write

$$\begin{aligned}
\left\langle \widehat{\mathrm{e}^{-ia|x|^2}}, \varphi \right\rangle &= \left\langle \mathrm{e}^{-ia|x|^2}, \widehat{\varphi} \right\rangle = \int_{\mathbb{R}^n} \mathrm{e}^{-ia|\xi|^2} \widehat{\varphi}(\xi)\, \mathrm{d}\xi \\
&= \lim_{\varepsilon \to 0^+} \int_{\mathbb{R}^n} \mathrm{e}^{-(ia+\varepsilon)|\xi|^2} \widehat{\varphi}(\xi)\, \mathrm{d}\xi = \lim_{\varepsilon \to 0^+} \left\langle \widehat{\varphi}, \mathrm{e}^{-(ia+\varepsilon)|\xi|^2} \right\rangle \\
&= \lim_{\varepsilon \to 0^+} \left\langle \varphi, \mathcal{F}(\mathrm{e}^{-(ia+\varepsilon)|x|^2}) \right\rangle = \lim_{\varepsilon \to 0^+} \left\langle \varphi, \left(\frac{\pi}{ia+\varepsilon}\right)^{\frac{n}{2}} \mathrm{e}^{-\frac{|\xi|^2}{4(ia+\varepsilon)}} \right\rangle \\
&= \lim_{\varepsilon \to 0^+} \left(\frac{\pi}{ia+\varepsilon}\right)^{\frac{n}{2}} \int_{\mathbb{R}^n} \mathrm{e}^{-\frac{|\xi|^2}{4(ia+\varepsilon)}} \varphi(\xi)\, \mathrm{d}\xi \\
&= \left(\frac{\pi}{ia}\right)^{\frac{n}{2}} \int_{\mathbb{R}^n} \mathrm{e}^{-\frac{|\xi|^2}{4ia}} \varphi(\xi)\, \mathrm{d}\xi \\
&= \left\langle \left(\frac{\pi}{ia}\right)^{\frac{n}{2}} \mathrm{e}^{i\frac{|\xi|^2}{4a}}, \varphi \right\rangle. \tag{4.2.23}
\end{aligned}$$

Above, the first equality is based on (4.2.2), while the second equality is a consequence of the way in which the slowly growing function $\mathrm{e}^{-ia|x|^2}$ is regarded as a tempered distribution. The third and second-to-last equalities are based on Lebesgue's Dominated Convergence Theorem. In the fourth equality we interpret the Schwartz function $\widehat{\varphi}$ as being in $\mathcal{S}'(\mathbb{R}^n)$ and rely on the fact that $\mathrm{e}^{-(ia+\varepsilon)|x|^2} \in \mathcal{S}(\mathbb{R}^n)$ for each $\varepsilon > 0$ (as noted in Example 3.22). The fifth equality once uses Remark 4.23 and once again (4.2.2). The sixth equality is a consequence of formula (3.2.6), while the rest is routine.

Theorem 4.26. *The following statements are true:*

(a) *The mapping* $\mathcal{F} : \mathcal{S}'(\mathbb{R}^n) \to \mathcal{S}'(\mathbb{R}^n)$ *defined by* $\mathcal{F}(u) := \widehat{u}$, *for every u in $\mathcal{S}'(\mathbb{R}^n)$, is bijective, continuous, and its inverse is also continuous.*

(b) $\widehat{D^\alpha u} = \xi^\alpha \widehat{u}$ *for all* $\alpha \in \mathbb{N}_0^n$ *and all* $u \in \mathcal{S}'(\mathbb{R}^n)$.

(c) $\widehat{x^\alpha u} = (-D)^\alpha \widehat{u}$ *for all* $\alpha \in \mathbb{N}_0^n$ *and all* $u \in \mathcal{S}'(\mathbb{R}^n)$.

(d) *If* $u \in \mathcal{S}'(\mathbb{R}^m)$ *and* $v \in \mathcal{S}'(\mathbb{R}^n)$, *then* $\widehat{u \otimes v} = \widehat{u} \otimes \widehat{v}$.

Proof. Recall from Theorem 3.25 that the map $\mathcal{F} : \mathcal{S}(\mathbb{R}^n) \to \mathcal{S}(\mathbb{R}^n)$ is linear, continuous, bijective, and its inverse is also continuous. Since the transpose of this map (in the sense of (14.1.10)) is precisely the Fourier transform in the context of part (a) of the current statement, from Propositions 14.2–14.3 it follows that the mapping $\mathcal{F} : \mathcal{S}'(\mathbb{R}^n) \to \mathcal{S}'(\mathbb{R}^n)$ is also well defined, linear, continuous, bijective, and has a continuous inverse.

Consider next $u \in \mathcal{S}'(\mathbb{R}^n)$ and $\alpha \in \mathbb{N}_0^n$. Then for every $\varphi \in \mathcal{S}(\mathbb{R}^n)$, using part (c) in Theorem 4.14, part (a) in Theorem 3.21, (4.2.2), and part (b) in Theorem 4.14, we may write

$$\langle \widehat{D^\alpha u}, \varphi \rangle = \langle D^\alpha u, \widehat{\varphi} \rangle = (-1)^{|\alpha|} \langle u, D^\alpha \widehat{\varphi} \rangle = \langle u, \widehat{\xi^\alpha \varphi} \rangle$$

$$= \langle \widehat{u}, \xi^\alpha \varphi \rangle = \langle \xi^\alpha \widehat{u}, \varphi \rangle, \qquad (4.2.24)$$

and

$$\langle \widehat{x^\alpha u}, \varphi \rangle = \langle x^\alpha u, \widehat{\varphi} \rangle = \langle u, x^\alpha \widehat{\varphi} \rangle = \langle u, \widehat{D^\alpha \varphi} \rangle$$

$$= \langle \widehat{u}, D^\alpha \varphi \rangle = \langle (-D)^\alpha \widehat{u}, \varphi \rangle. \qquad (4.2.25)$$

This proves the claims in *(b)–(c)*.

We are left with the proof of the statement in *(d)*. Fix $u \in \mathcal{S}'(\mathbb{R}^m)$ and some $v \in \mathcal{S}'(\mathbb{R}^n)$. Then based on statement *(d)* in Theorem 4.14 we have that $u \otimes v$ belongs to $\mathcal{S}'(\mathbb{R}^m \times \mathbb{R}^n)$. Hence, starting with (4.2.2), then using *(d)* in Theorem 3.21 and (4.1.16), we may write

$$\langle \widehat{u \otimes v}, \varphi \otimes \psi \rangle = \langle u \otimes v, \widehat{\varphi \otimes \psi} \rangle = \langle u \otimes v, \widehat{\varphi} \otimes \widehat{\psi} \rangle \qquad (4.2.26)$$

$$= \langle u, \widehat{\varphi} \rangle \langle v, \widehat{\psi} \rangle = \langle \widehat{u}, \varphi \rangle \langle \widehat{v}, \psi \rangle$$

$$= \langle \widehat{u} \otimes \widehat{v}, \varphi \otimes \psi \rangle, \quad \forall \varphi \in \mathcal{S}(\mathbb{R}^m), \ \forall \psi \in \mathcal{S}(\mathbb{R}^n).$$

Consequently, $\widehat{u \otimes v}\big|_{C_0^\infty(\mathbb{R}^m) \otimes C_0^\infty(\mathbb{R}^n)} = \widehat{u} \otimes \widehat{v}\big|_{C_0^\infty(\mathbb{R}^m) \otimes C_0^\infty(\mathbb{R}^n)}$, which in combination with Proposition 3.16 proves the statement in *(d)*. $\qquad \square$

Exercise 4.27. Recall (4.1.5) and (4.1.9). Prove that $\widehat{u_f} = u_{\widehat{f}}$ in $\mathcal{S}'(\mathbb{R}^n)$ for each $f \in L^2(\mathbb{R}^n)$ and each $f \in L^1(\mathbb{R}^n)$.

Hint: Use (3.2.27) and Exercise 3.30.

Lemma 4.28 (Riemann–Lebesgue Lemma). *If $f \in L^1(\mathbb{R}^n)$, then the tempered distribution u_f satisfies $\widehat{u_f} \in C^0(\mathbb{R}^n)$.*

Proof This is a consequence of Exercise 4.27 and (3.1.3). $\qquad \square$

Proposition 4.29. *Prove that for every $f, g \in L^2(\mathbb{R}^n)$ one has $\widehat{f * g} = \widehat{f} \cdot \widehat{g}$ in $\mathcal{S}'(\mathbb{R}^n)$.*

Proof. Let $f, g \in L^2(\mathbb{R}^n)$ be arbitrary. Using Young's inequality (cf. Theorem 14.17) we see that $f * g \in L^\infty(\mathbb{R}^n)$, hence $f * g \in \mathcal{S}'(\mathbb{R}^n)$ (recall Example 4.8). Also, from (3.2.29) we have $\widehat{f}, \widehat{g} \in L^2(\mathbb{R}^n)$, which further implies that $\widehat{f} \cdot \widehat{g}$ belongs to $L^1(\mathbb{R}^n) \subseteq \mathcal{S}'(\mathbb{R}^n)$. This shows that the equality $\widehat{f * g} = \widehat{f} \cdot \widehat{g}$ is meaningful in $\mathcal{S}'(\mathbb{R}^n)$.

The proof is now based on a density argument. First, since $C_0^\infty(\mathbb{R}^n)$ is dense in $L^2(\mathbb{R}^n)$, from (3.1.14) and Exercise 3.9 it follows that $\mathcal{S}(\mathbb{R}^n)$ is dense in $L^2(\mathbb{R}^n)$. Consequently there exist sequences $\{f_j\}_{j \in \mathbb{N}}$ and $\{g_j\}_{j \in \mathbb{N}}$ in $\mathcal{S}(\mathbb{R}^n)$ such that

$$f_j \xrightarrow[j \to \infty]{L^2(\mathbb{R}^n)} f \quad \text{and} \quad g_j \xrightarrow[j \to \infty]{L^2(\mathbb{R}^n)} g. \qquad (4.2.27)$$

Invoking (3.2.29) we obtain

$$\widehat{f_j} \xrightarrow[j\to\infty]{L^2(\mathbb{R}^n)} \widehat{f} \quad \text{and} \quad \widehat{g_j} \xrightarrow[j\to\infty]{L^2(\mathbb{R}^n)} \widehat{g}. \tag{4.2.28}$$

Then, for each $j \in \mathbb{N}$, using Hölder's inequality we write

$$\|\widehat{f_j} \cdot \widehat{g_j} - \widehat{f} \cdot \widehat{g}\|_{L^1(\mathbb{R}^n)} \le \|\widehat{f_j}(\widehat{g_j} - \widehat{g})\|_{L^1(\mathbb{R}^n)} + \|(\widehat{f_j} - \widehat{f})\widehat{g}\|_{L^1(\mathbb{R}^n)}$$

$$\le \|\widehat{f_j}\|_{L^2(\mathbb{R}^n)}\|\widehat{g_j} - \widehat{g}\|_{L^2(\mathbb{R}^n)}$$

$$+ \|\widehat{g}\|_{L^2(\mathbb{R}^n)}\|\widehat{f_j} - \widehat{f}\|_{L^2(\mathbb{R}^n)}, \tag{4.2.29}$$

which when combined with (4.2.28) implies $\widehat{f_j} \cdot \widehat{g_j} \xrightarrow[j\to\infty]{L^2(\mathbb{R}^n)} \widehat{f} \cdot \widehat{g}$. The latter and Exercise 4.12 further yield

$$\widehat{f_j} \cdot \widehat{g_j} \xrightarrow[j\to\infty]{\mathcal{S}'(\mathbb{R}^n)} \widehat{f} \cdot \widehat{g}. \tag{4.2.30}$$

On the other hand, from (4.2.27) and Young's inequality (cf. (14.2.16) applied with $p = p' = 2$) we have $f_j * g_j \xrightarrow[j\to\infty]{L^\infty(\mathbb{R}^n)} f * g$, hence $f_j * g_j \xrightarrow[j\to\infty]{\mathcal{S}'(\mathbb{R}^n)} f * g$ according to Exercise 4.12. Since the Fourier transform is continuous on $\mathcal{S}'(\mathbb{R}^n)$, it follows that

$$\widehat{f_j * g_j} \xrightarrow[j\to\infty]{\mathcal{S}'(\mathbb{R}^n)} \widehat{f * g}. \tag{4.2.31}$$

The desired conclusion follows from (4.2.30), (4.2.31), and the formula $\widehat{f_j * g_j} = \widehat{f_j} \cdot \widehat{g_j}$ in $\mathcal{S}(\mathbb{R}^n)$ (see part (c) of Proposition 3.28). $\qquad\square$

Exercise 4.30. Let $\theta \in \mathcal{S}(\mathbb{R}^n)$ be such that $\int_{\mathbb{R}^n} \theta(x)\,dx = 1$. For each $\varepsilon > 0$ set $\theta_\varepsilon(x) := \varepsilon^{-n}\theta(x/\varepsilon)$ for every $x \in \mathbb{R}^n$. Also, let $f \in L^2(\mathbb{R}^n)$ and for each $\varepsilon > 0$ define the function $f_\varepsilon := f * \theta_\varepsilon$, which by Young's inequality belongs to $L^2(\mathbb{R}^n)$. Prove that $f_\varepsilon \xrightarrow[j\to\infty]{L^2(\mathbb{R}^n)} f$.

Hint: Use Plancherel's identity (cf. the identity in (3.2.29)), Proposition 4.29, and the fact that for each $\varepsilon > 0$ we have $\widehat{\theta_\varepsilon} \in L^\infty(\mathbb{R}^n)$ and $\widehat{\theta_\varepsilon}(\xi) = \widehat{\theta}(\varepsilon\xi)$ for every $\xi \in \mathbb{R}^n$.

Example 4.31. Let $a \in (0, \infty)$. We are interested in computing the Fourier transform of the bounded function $\frac{x}{x^2+a^2}$, viewed as a distribution in $\mathcal{S}'(\mathbb{R})$. In this vein, consider the auxiliary function $f(x) := \frac{1}{x^2+a^2}$ for $x \in \mathbb{R}$, and recall from (4.2.21) that $\widehat{f}(\xi) = \frac{\pi}{a}e^{-a|\xi|}$ in $\mathcal{S}'(\mathbb{R})$. This formula, part (c) in Theorem 4.26, and (2.4.11), allow us to write

$$\mathcal{F}\Big(\frac{x}{x^2+a^2}\Big)(\xi) = \mathcal{F}(xf)(\xi) = i\frac{d}{d\xi}[\widehat{f}(\xi)] = i\frac{d}{d\xi}\Big(\frac{\pi}{a}e^{-a|\xi|}\Big)$$

$$= \frac{\pi i}{a}\Big(-a\,e^{-a\xi}H(\xi) + a\,e^{a\xi}H(-\xi)\Big)$$

$$= -\pi i(\operatorname{sgn}\xi)e^{-a|\xi|} \quad \text{in} \quad \mathcal{S}'(\mathbb{R}). \qquad (4.2.32)$$

Proposition 4.32. *For each $u \in \mathcal{S}'(\mathbb{R}^n)$ define the mapping*

$$u^{\vee} : \mathcal{S}(\mathbb{R}^n) \to \mathbb{C}, \qquad u^{\vee}(\varphi) := \langle u, \varphi^{\vee}\rangle, \quad \forall\,\varphi \in \mathcal{S}(\mathbb{R}^n). \qquad (4.2.33)$$

Then $u^{\vee} \in \mathcal{S}'(\mathbb{R}^n)$ and

$$\widehat{\widehat{u}} = (2\pi)^n u^{\vee}. \qquad (4.2.34)$$

Proof. Fix $u \in \mathcal{S}'(\mathbb{R}^n)$. Then (4.2.33) is simply the composition of u with the mapping from Exercise 3.26. Since both are linear and continuous, it follows that $u^{\vee} \in \mathcal{S}'(\mathbb{R}^n)$. Formula (4.2.34) then follows by combining (4.2.2) with the identity in part *(3)* of Exercise 3.27. $\qquad\square$

Exercise 4.33. Show that

$$\widehat{1} = (2\pi)^n\delta \quad \text{in} \quad \mathcal{S}'(\mathbb{R}^n). \qquad (4.2.35)$$

Exercise 4.34. Prove that for any tempered distribution u the following equivalence holds:

$$u^{\vee} = -u \ \text{ in } \mathcal{S}'(\mathbb{R}^n) \quad \Longleftrightarrow \quad (\widehat{u})^{\vee} = -\widehat{u} \ \text{ in } \mathcal{S}'(\mathbb{R}^n). \qquad (4.2.36)$$

One suggestive way of summarizing (4.2.36) is to say that a tempered distribution is odd if and only if its Fourier transform is odd.

Theorem 4.35. *The following statements are true:*

(a) If $a \in \mathcal{S}(\mathbb{R}^n)$ and $u \in \mathcal{S}'(\mathbb{R}^n)$, then $\widehat{u*a} = \widehat{a}\,\widehat{u}$ in $\mathcal{S}'(\mathbb{R}^n)$, where $\widehat{a}$ is viewed as an element from $\mathcal{S}(\mathbb{R}^n)$.

(b) If $u \in \mathcal{E}'(\mathbb{R}^n)$ then the tempered distribution $\widehat{u}$ is of function type given by the formula $\widehat{u}(\xi) = \langle u(x), e^{-ix\cdot\xi}\rangle$ for every $\xi \in \mathbb{R}^n$, and $\widehat{u} \in \mathcal{L}(\mathbb{R}^n)$.

(c) If $u \in \mathcal{E}'(\mathbb{R}^n)$, $v \in \mathcal{S}'(\mathbb{R}^n)$ then $\widehat{u*v} = \widehat{u}\widehat{v}$, where $\widehat{u}$ is viewed as an element in $\mathcal{L}(\mathbb{R}^n)$.

Proof. By *(d)* in Theorem 4.19 we have $u * a \in \mathcal{S}'(\mathbb{R}^n)$, hence $\widehat{u*a}$ exists and belongs to $\mathcal{S}'(\mathbb{R}^n)$. Also, since $\widehat{a} \in \mathcal{S}(\mathbb{R}^n)$, by (3.1.24) and *(b)* in Theorem 4.14 it follows that $\widehat{a}\,\widehat{u} \in \mathcal{S}'(\mathbb{R}^n)$. Then, we may write

$$\langle \widehat{u*a}, \varphi\rangle = \langle u*a, \widehat{\varphi}\rangle = \langle u, a^{\vee}*\widehat{\varphi}\rangle = (2\pi)^{-n}\langle u, \widehat{\widehat{a}}*\widehat{\varphi}\rangle$$

$$= \langle u, \widehat{\widehat{a}\,\varphi}\rangle = \langle \widehat{u}, \widehat{a}\,\varphi\rangle = \langle \widehat{a}\,\widehat{u}, \varphi\rangle, \qquad \forall\,\varphi \in \mathcal{S}(\mathbb{R}^n). \qquad (4.2.37)$$

For the second equality in (4.2.37) we used (4.1.43), for the third we used part *(3)* in Exercise 3.27, while for the fourth we used *(d)* in Proposition 3.28. This proves the statement in *(a)*.

Moving on to the proof of *(b)*, fix some $u \in \mathcal{E}'(\mathbb{R}^n)$ and introduce the function $f(\xi) := \langle u(x), e^{-ix\cdot\xi} \rangle$ for every $\xi \in \mathbb{R}^n$. From Proposition 2.91 it follows that $f \in C^\infty(\mathbb{R}^n)$ and

$$\partial^\alpha f(\xi) = \langle u(x), \partial_\xi^\alpha [e^{-ix\cdot\xi}] \rangle, \quad \forall \xi \in \mathbb{R}^n, \quad \text{for every} \quad \alpha \in \mathbb{N}_0^n. \tag{4.2.38}$$

In addition, since $u \in \mathcal{E}'(\mathbb{R}^n)$, there exist a compact subset K of $\mathbb{R}^n$, along with a constant $C \in (0, \infty)$, and a number $k \in \mathbb{N}_0$, such that u satisfies (2.6.6). Combining all these facts, for each $\alpha \in \mathbb{N}_0^n$, we may estimate

$$\left| \partial^\alpha f(\xi) \right| = \left| \langle u(x), \partial_\xi^\alpha e^{-ix\cdot\xi} \rangle \right| \leq C \sup_{x \in K, |\beta| \leq k} \left| \partial_x^\beta \partial_\xi^\alpha e^{-ix\cdot\xi} \right|$$

$$\leq C \sup_{x \in K, |\beta| \leq k} \left| x^\alpha \xi^\beta \right| \leq C(1 + |\xi|)^k. \tag{4.2.39}$$

From (4.2.39) and the fact that f is smooth we deduce that $f \in \mathcal{L}(\mathbb{R}^n)$. Hence, if we now recall Exercise 4.7, it follows that $f \in \mathcal{S}'(\mathbb{R}^n)$.

We are left with proving that $\widehat{u} = f$ as tempered distributions. To this end, fix $\theta \in C_0^\infty(\mathbb{R}^n)$ such that $\theta \equiv 1$ in a neighborhood of supp u. Then, for every $\varphi \in C_0^\infty(\mathbb{R}^n)$ one has

$$\langle \widehat{u}, \varphi \rangle = \langle u, \widehat{\varphi} \rangle = \langle u(\xi), \theta(\xi)\widehat{\varphi}(\xi) \rangle = \langle u(\xi), \int_{\mathbb{R}^n} e^{-ix\cdot\xi} \theta(\xi)\varphi(x)\, dx \rangle. \tag{4.2.40}$$

At this point recall Remark 2.88 (note that the function $e^{-ix\cdot\xi}\theta(\xi)\varphi(x)$ belongs to $C_0^\infty(\mathbb{R}^n \times \mathbb{R}^n)$ and one may take $\nu = 1$ in (2.7.53)) which allows one to rewrite the last term in (4.2.40) and conclude that

$$\langle \widehat{u}, \varphi \rangle = \int_{\mathbb{R}^n} \langle u(\xi), e^{-ix\cdot\xi} \rangle \varphi(x)\, dx = \langle f, \varphi \rangle, \quad \forall \varphi \in C_0^\infty(\mathbb{R}^n). \tag{4.2.41}$$

Hence the tempered distributions $\widehat{u}$ and f coincide on $C_0^\infty(\mathbb{R}^n)$. By (4.1.33), we therefore have $\widehat{u} = f$ in $\mathcal{S}'(\mathbb{R}^n)$, completing the proof of the statement in *(b)*.

Regarding the formula in part *(c)*, while informally this is similar to the formula proved in part *(a)*, the computation in (4.2.37) through which the latter has been deduced no longer works in the current case, as various ingredients used to justify it break down (given that $\widehat{u}$ is now only known to belong to $\mathcal{L}(\mathbb{R}^n)$ and not necessarily to $\mathcal{S}(\mathbb{R}^n)$). This being said, we may employ what has been established in part *(a)* together with a limiting argument to get the desired result.

Specifically, assume that $u \in \mathcal{E}'(\mathbb{R}^n)$ and $v \in \mathcal{S}'(\mathbb{R}^n)$. Then $u * v$ is a tempered distribution in $\mathbb{R}^n$ (recall item *(a)* in Theorem 4.19) hence, $\widehat{u * v}$ is well defined in $\mathcal{S}'(\mathbb{R}^n)$. Also, from *(b)* we have $\widehat{u} \in \mathcal{L}(\mathbb{R}^n)$, thus $\widehat{u}\widehat{v}$ is meaningful and belongs to $\mathcal{S}'(\mathbb{R}^n)$. To proceed, recall the sequence $\{\phi_j\}_{j \in \mathbb{N}}$ from Example 1.11. In particular, (1.3.8) holds and $\phi_j \xrightarrow[j \to \infty]{\mathcal{D}'(\mathbb{R}^n)} \delta$ (see Example 2.24). Thus, the statement from part *(b)* in Theorem 4.19 applies and gives that $u * \phi_j \xrightarrow[j \to \infty]{\mathcal{S}'(\mathbb{R}^n)} u * \delta = u$. Moreover, since

$u \in \mathcal{E}'(\mathbb{R}^n)$, it follows that $u * \phi_j \in \mathcal{E}'(\mathbb{R}^n)$ (by statement *(a)* in Theorem 2.96) and $\operatorname{supp}(u * \phi_j) \subseteq \operatorname{supp} u + B(0, 1)$ for every $j \in \mathbb{N}$ (by part *(1)* in Remark 2.100). Hence, one may apply *(b)* in Theorem 4.19 to further conclude that the convergence $(u * \phi_j) * v \xrightarrow[j\to\infty]{\mathcal{S}'(\mathbb{R}^n)} u * v$ holds. Given *(a)* in Theorem 4.26, the latter implies (recall Fact 4.11)

$$\lim_{j\to\infty} \langle \mathcal{F}((u * \phi_j) * v), \varphi \rangle = \langle \widehat{u * v}, \varphi \rangle, \qquad \forall \varphi \in \mathcal{S}(\mathbb{R}^n). \tag{4.2.42}$$

Note that (2.8.49) gives $u * \phi_j \in C_0^\infty(\mathbb{R}^n)$ for every $j \in \mathbb{N}$. Hence, based on what we have proved already in part *(a)*, we obtain (keeping in mind that $\widehat{u} \in \mathcal{L}(\mathbb{R}^n)$)

$$\mathcal{F}((u * \phi_j) * v) = \widehat{u * \phi_j} \, \widehat{v} = \widehat{\phi_j} \, \widehat{u} \, \widehat{v} = \widehat{\phi}(\cdot/j) \, \widehat{u} \, \widehat{v} \quad \text{in} \quad \mathcal{S}'(\mathbb{R}^n), \qquad \forall j \in \mathbb{N}. \tag{4.2.43}$$

In concert, (4.2.42) and (4.2.43) yield that for each $\varphi \in \mathcal{S}(\mathbb{R}^n)$

$$\langle \widehat{u * v}, \varphi \rangle = \lim_{j\to\infty} \langle \widehat{u} \, \widehat{v}, \widehat{\phi}(\cdot/j) \, \varphi \rangle = \langle \widehat{u} \, \widehat{v}, \widehat{\phi}(0) \, \varphi \rangle = \langle \widehat{u} \, \widehat{v}, \varphi \rangle \tag{4.2.44}$$

since $\widehat{\phi}(\cdot/j) \, \varphi \xrightarrow[j\to\infty]{\mathcal{S}(\mathbb{R}^n)} \widehat{\phi}(0) \, \varphi$ by Exercise 3.15, and $\widehat{\phi}(0) = 1$. This proves the statement in *(c)* and finishes the proof of the theorem. $\qquad\square$

Example 4.36. If $a \in (0, \infty)$ then $\chi_{[-a,a]}$, the characteristic function of the interval $[-a, a]$, belongs to $\mathcal{E}'(\mathbb{R})$ and by statement *(b)* in Theorem 4.35 we have

$$\widehat{\chi_{[-a,a]}}(\xi) = \int_{-a}^{a} e^{-ix\xi} \, dx = \begin{cases} 2\frac{\sin(a\xi)}{\xi} & \text{for } \xi \in \mathbb{R} \setminus \{0\}, \\ 2a & \text{for } \xi = 0. \end{cases} \tag{4.2.45}$$

Exercise 4.37. Use Exercise 2.75 and statement *(b)* in Theorem 4.35 to prove that if $u \in \mathcal{S}'(\mathbb{R}^n)$ is such that $\operatorname{supp} \widehat{u} \subseteq \{a\}$ for some $a \in \mathbb{R}^n$, then there exist $k \in \mathbb{N}_0$ and constants $c_\alpha \in \mathbb{C}$, for $\alpha \in \mathbb{N}_0^n$ with $|\alpha| \le k$, such that

$$u = \sum_{|\alpha| \le k} c_\alpha x^\alpha e^{ix \cdot a} \quad \text{in} \quad \mathcal{S}'(\mathbb{R}^n). \tag{4.2.46}$$

In particular, if $a = 0$ then u is a polynomial in $\mathbb{R}^n$.

Example 4.38. Let $a \in \mathbb{R}$ and consider the function $f(x) := \sin(ax)$ for $x \in \mathbb{R}$. Then $f \in C^\infty(\mathbb{R}) \cap L^\infty(\mathbb{R})$, hence $f \in \mathcal{S}'(\mathbb{R})$. Suppose $a \ne 0$. We shall compute the Fourier transform of f in $\mathcal{S}'(\mathbb{R})$ by making use of a technique relying on the ordinary differential equation f satisfies. More precisely, since $f'' + a^2 f = 0$ in $\mathbb{R}$, the same equation holds in $\mathcal{S}'(\mathbb{R})$, thus by *(a)* and *(b)* in Theorem 4.26 we have $(\xi^2 - a^2)\widehat{f} = 0$ in $\mathcal{S}'(\mathbb{R})$. By Example 2.80 this implies $\widehat{f} = C_1 \delta_a + C_2 \delta_{-a}$ in $\mathcal{S}'(\mathbb{R})$ for some $C_1, C_2 \in \mathbb{C}$. Applying again the Fourier transform to the last equation, using (4.2.34) and part *(b)* from Theorem 4.35, we have

$$2\pi \sin(-ax) = \widehat{\sin(ax)} = C_1\widehat{\delta_a} + C_2\widehat{\delta_{-a}}$$
$$= C_1\langle \delta_a(\xi), e^{-ix\xi}\rangle + C_2\langle \delta_{-a}(\xi), e^{-ix\xi}\rangle$$
$$= C_1 e^{-iax} + C_2 e^{iax}$$
$$= (C_1 + C_2)\cos(ax) + i(C_1 - C_2)\sin(-ax). \tag{4.2.47}$$

The resulting identity in (4.2.47) forces $C_1 = -i\pi$ and $C_2 = i\pi$. Plugging these constants back in the expression for $\widehat{f}$ yields

$$\widehat{\sin(ax)} = -i\pi\delta_a + i\pi\delta_{-a} \quad \text{in } \mathcal{S}'(\mathbb{R}). \tag{4.2.48}$$

It is immediate that (4.2.48) also holds if $a = 0$.

Example 4.39. Let $a, b \in \mathbb{R}$. Then the function $g(x) := \sin(ax)\sin(bx)$ for $x \in \mathbb{R}$ satisfies $g \in L^\infty(\mathbb{R})$, thus $g \in \mathcal{S}'(\mathbb{R})$ (cf. (4.1.9)). Applying the Fourier transform to the identity in (4.2.48) and using (4.2.34) we obtain

$$\sin(ax) = \frac{i}{2}\widehat{\delta_a} - \frac{i}{2}\widehat{\delta_{-a}} \quad \text{in } \mathcal{S}'(\mathbb{R}). \tag{4.2.49}$$

Also, making use of (4.2.49), part *(c)* in Theorem 4.35, and then Exercise 2.99, we may write

$$\sin(ax)\sin(bx) = \left(\frac{i}{2}\widehat{\delta_a} - \frac{i}{2}\widehat{\delta_{-a}}\right)\left(\frac{i}{2}\widehat{\delta_b} - \frac{i}{2}\widehat{\delta_{-b}}\right)$$
$$= -\frac{1}{4}\mathcal{F}(\delta_{a+b} - \delta_{a-b} - \delta_{b-a} + \delta_{-a-b}) \quad \text{in } \mathcal{S}'(\mathbb{R}). \tag{4.2.50}$$

Hence, another application of the Fourier transform gives (relying on (4.2.34))

$$\mathcal{F}(\sin(ax)\sin(bx)) = -\frac{\pi}{2}(\delta_{a+b} - \delta_{a-b} - \delta_{b-a} + \delta_{-a-b}) \quad \text{in } \mathcal{S}'(\mathbb{R}). \tag{4.2.51}$$

Example 4.40. Let $a \in (0, \infty)$ and consider the function $f(x) := e^{ia|x|^2}$ for $x \in \mathbb{R}^n$. Then $f \in L^\infty(\mathbb{R}^n)$ thus $f \in \mathcal{S}'(\mathbb{R}^n)$ by (4.1.9). The goal is to compute the Fourier transform of f in $\mathcal{S}'(\mathbb{R}^n)$. The starting point is the observation that

$$f(x) = e^{iax_1^2} \otimes e^{iax_2^2} \otimes \cdots \otimes e^{iax_n^2} \quad \forall x = (x_1, \ldots, x_n) \in \mathbb{R}. \tag{4.2.52}$$

Invoking part *(d)* in Theorem 4.26 then reduces matters to computing the Fourier transform of f in the case $n = 1$.

Assume that $n = 1$, in which case $f(x) = e^{iax^2}$ for $x \in \mathbb{R}$. Then f satisfies the differential equation: $f' - 2ia\,xf = 0$ in $\mathcal{S}'(\mathbb{R})$. Taking the Fourier transform in $\mathcal{S}'(\mathbb{R})$ and using the formulas from *(b)–(c)* in Theorem 4.26 we obtain

$$(\widehat{f})' + \frac{i}{2a}\xi\widehat{f} = 0 \quad \text{in } \mathcal{S}'(\mathbb{R}). \tag{4.2.53}$$

The format of (4.2.53) suggests that we consider the ordinary differential equation $y' + \frac{i}{2a}\xi y = 0$ in $\mathbb{R}$, in the unknown $y = y(\xi)$. One particular solution of this o.d.e. is $y(\xi) = e^{-i\frac{\xi^2}{4a}}$. Note that both y and $1/y$ belong to $\mathcal{L}(\mathbb{R})$. In particular, it makes sense to consider the tempered distribution $u := (1/y)\widehat{f}$, whose derivative is

$$u' = (1/y)'\widehat{f} + (1/y)(\widehat{f})' = -(y'/y^2)\widehat{f} + (1/y)(\widehat{f})'$$

$$= \left(\left(\frac{i}{2a}\xi\right)/y\right)\widehat{f} - \frac{i}{2a}\xi(1/y)\widehat{f} = 0 \quad \text{in} \quad \mathcal{S}'(\mathbb{R}). \tag{4.2.54}$$

From this and part *(2)* in Proposition 2.47 we then deduce that there exists some constant $c \in \mathbb{C}$ such that $u = c$ in $\mathcal{S}'(\mathbb{R})$ which, given the significance of u and y, forces $\widehat{f} = c\,e^{-i\frac{\xi^2}{4a}}$ in $\mathcal{S}'(\mathbb{R})$ for some $c \in \mathbb{C}$. We are therefore left with determining the constant c. This may be done by choosing a suitable Schwartz function and then computing the action of $\widehat{f}$ on it. Take $\varphi(\xi) := e^{-\frac{\xi^2}{4a}}$ for $\xi \in \mathbb{R}$. Since $\widehat{\varphi}(x) = \sqrt{4a\pi}e^{-ax^2}$ for $x \in \mathbb{R}$ (recall Example 3.22), we may write

$$c \int_{\mathbb{R}} e^{-i\frac{\xi^2}{4a} - \frac{\xi^2}{4a}} \, d\xi = \langle \widehat{f}, \varphi \rangle = \langle f, \widehat{\varphi} \rangle = \sqrt{4a\pi} \int_{\mathbb{R}} e^{iax^2 - ax^2} \, dx. \tag{4.2.55}$$

The two integrals in (4.2.55) may be computed by applying formula (3.2.7) with $\lambda := \frac{1+i}{4a}$ and $\lambda := a(1-i)$, respectively. After some routine algebra (i.e., computing these integrals, replacing their values in (4.2.55), then solving for c), we find that $c = \sqrt{\frac{\pi}{a}}e^{i\frac{\pi}{4}}$. In summary, this analysis proves that

$$\widehat{e^{ia|x|^2}}(\xi) = \left(\frac{\pi}{a}\right)^{\frac{n}{2}} e^{i\frac{n\pi}{4}} e^{-i\frac{|\xi|^2}{4a}} \quad \text{in} \quad \mathcal{S}'(\mathbb{R}^n). \tag{4.2.56}$$

Partial Fourier Transforms

In the last part of this section we define partial Fourier transforms. To set the stage, fix $m, n \in \mathbb{N}$. We shall denote by x, ξ generic variables in $\mathbb{R}^n$, and by y, η generic variables in $\mathbb{R}^m$. The partial Fourier transform with respect to the variable x of a function $\varphi \in \mathcal{S}(\mathbb{R}^{n+m})$, denoted by $\widehat{\varphi}_x$ or $\mathcal{F}_x\varphi$, is defined by

$$\widehat{\varphi}_x(\xi, y) := \int_{\mathbb{R}^n} e^{-ix\cdot\xi}\varphi(x, y) \, dx, \qquad \forall \xi \in \mathbb{R}^n, \ \forall y \in \mathbb{R}^m. \tag{4.2.57}$$

Reasoning in a similar manner as in the proof of Theorem 3.25, it follows that

$$\mathcal{F}_x : \mathcal{S}(\mathbb{R}^{n+m}) \to \mathcal{S}(\mathbb{R}^{n+m}) \text{ is bijective,}$$

continuous, with continuous inverse

$$\tag{4.2.58}$$

and its inverse is given by

$$(\mathcal{F}_x^{-1}\psi)(x,\eta) := (2\pi)^{-n}\int_{\mathbb{R}^n} e^{ix\cdot\xi}\psi(\xi,\eta)\,d\xi,$$

(4.2.59)

for all $(x,\eta) \in \mathbb{R}^{n+m}$ and all $\psi \in S(\mathbb{R}^{n+m})$.

Furthermore, analogously to Proposition 4.21, the partial Fourier transform $\mathcal{F}_x$ extends to $S'(\mathbb{R}^{n+m})$ as a continuous map by setting

$$\langle \mathcal{F}_x u, \varphi \rangle := \langle u, \mathcal{F}_x\varphi \rangle, \qquad \forall u \in S'(\mathbb{R}^{n+m}), \quad \forall \varphi \in S(\mathbb{R}^{n+m}),$$

(4.2.60)

and this extension is an isomorphism from $S'(\mathbb{R}^{n+m})$ into itself, with continuous inverse denoted by $\mathcal{F}_x^{-1}$. Moreover, the action of $\mathcal{F}_x$ enjoys properties analogous to those established for the "full" Fourier transform in Theorem 3.21, Exercise 3.27, Theorem 4.26, and Proposition 4.32.

Exercise 4.41. Let $\widehat{\ }$ denote the full Fourier transform in $\mathbb{R}^{n+m}$. Prove that for each function $\varphi \in S(\mathbb{R}^n \times \mathbb{R}^m)$ and each $(\xi,\eta) \in \mathbb{R}^n \times \mathbb{R}^m$ we have

$$(\mathcal{F}_x\mathcal{F}_y\varphi(x,y))(\xi,\eta) = (\mathcal{F}_y\mathcal{F}_x\varphi(x,y))(\xi,\eta) = \widehat{\varphi}(\xi,\eta).$$

(4.2.61)

Also, show that

$$\mathcal{F}_x\mathcal{F}_y u = \mathcal{F}_y\mathcal{F}_x u = \widehat{u} \quad \text{in } S'(\mathbb{R}^{n+m}), \quad \forall u \in S'(\mathbb{R}^n \times \mathbb{R}^m).$$

(4.2.62)

Exercise 4.42. Prove that

$$\mathcal{F}_x(\delta(x) \otimes \delta(y)) = 1(\xi) \otimes \delta(y) \quad \text{in } S'(\mathbb{R}^{n+m}).$$

(4.2.63)

4.3 Homogeneous Distributions

Let $A \in M_{n\times n}(\mathbb{R})$ be such that $\det A \neq 0$. Then for every $f \in L^1(\mathbb{R}^n)$ one has $f \circ A \in L^1(\mathbb{R}^n)$, thus $f, f \circ A \in S'(\mathbb{R}^n)$ by (4.1.9). Moreover,

$$\langle f \circ A, \varphi \rangle = \int_{\mathbb{R}^n} f(Ax)\varphi(x)\,dx = |\det A|^{-1}\int_{\mathbb{R}^n} f(y)\varphi(A^{-1}y)\,dy$$

$$= |\det A|^{-1}\langle f, \varphi \circ A^{-1} \rangle, \qquad \forall \varphi \in S(\mathbb{R}^n).$$

(4.3.1)

The resulting identity in (4.3.1) and Exercise 3.17 justifies the following extension (compare with Proposition 2.27).

Proposition 4.43. *Let $A \in M_{n\times n}(\mathbb{R})$ be such that $\det A \neq 0$. For each u in $S'(\mathbb{R}^n)$, define the mapping $u \circ A : S(\mathbb{R}^n) \to \mathbb{C}$ by setting*

$$(u \circ A)(\varphi) := |\det A|^{-1}\langle u, \varphi \circ A^{-1} \rangle, \qquad \forall \varphi \in S(\mathbb{R}^n).$$

(4.3.2)

Then $u \circ A \in S'(\mathbb{R}^n)$.

Proof. This is an immediate consequence of (3.1.41). $\qquad\qquad\qquad\qquad\qquad$ □

The identities in Exercise 2.28 have natural analogues in the current setting.

Exercise 4.44. Let $A, B \in \mathcal{M}_{n \times n}(\mathbb{R})$ be such that $\det A \neq 0$ and $\det B \neq 0$. Then the following identities hold in $\mathcal{S}'(\mathbb{R}^n)$:

(1) $(u \circ A) \circ B = u \circ (AB)$ for every $u \in \mathcal{S}'(\mathbb{R}^n)$;
(2) $u \circ (\lambda A) = \lambda u \circ A$ for every $u \in \mathcal{S}'(\mathbb{R}^n)$ and every $\lambda \in \mathbb{R}$;
(3) $(u + v) \circ A = u \circ A + v \circ A$ for every $u, v \in \mathcal{S}'(\mathbb{R}^n)$.

In the next proposition we study how the Fourier transform interacts with the operator of composition by an invertible matrix. Recall that the transpose of a matrix A is denoted by $A^\top$.

Proposition 4.45. *Assume that $A \in \mathcal{M}_{n \times n}(\mathbb{R})$ is such that $\det A \neq 0$. Then for each $u \in \mathcal{S}'(\mathbb{R}^n)$,*

$$\widehat{u \circ A} = |\det A|^{-1} \widehat{u} \circ (A^\top)^{-1}. \tag{4.3.3}$$

Proof. For each $\varphi \in \mathcal{S}(\mathbb{R}^n)$, based on (4.2.2), (4.3.2), and (3.2.9), we may write

$$\langle \widehat{u \circ A}, \varphi \rangle = \langle u \circ A, \widehat{\varphi} \rangle = |\det A|^{-1} \langle u, \widehat{\varphi} \circ A^{-1} \rangle$$

$$= \langle u, \widehat{\varphi \circ A^\top} \rangle = \langle \widehat{u}, \varphi \circ A^\top \rangle$$

$$= |\det A|^{-1} \langle \widehat{u} \circ (A^\top)^{-1}, \varphi \rangle. \tag{4.3.4}$$

This proves (4.3.3). $\qquad\qquad\qquad\qquad\qquad\qquad\qquad\qquad\qquad\qquad\qquad\qquad$ □

The mappings in (3.1.41) and (4.3.2) corresponding to $A := t I_{n \times n}$, for some number $t \in (0, \infty)$, are called `dilations` and will be denoted by τ_t. More precisely, for each $t \in (0, \infty)$ we have

$$\tau_t : \mathcal{S}(\mathbb{R}^n) \to \mathcal{S}(\mathbb{R}^n), \quad (\tau_t \varphi)(x) := \varphi(tx), \quad \forall \varphi \in \mathcal{S}(\mathbb{R}^n), \ \forall x \in \mathbb{R}^n, \tag{4.3.5}$$

and

$$\tau_t : \mathcal{S}'(\mathbb{R}^n) \to \mathcal{S}'(\mathbb{R}^n),$$
$$\langle \tau_t u, \varphi \rangle := t^{-n} \langle u, \tau_{\frac{1}{t}} \varphi \rangle, \quad \forall u \in \mathcal{S}'(\mathbb{R}^n), \ \forall \varphi \in \mathcal{S}(\mathbb{R}^n). \tag{4.3.6}$$

Exercise 4.46. Prove that for each $t \in (0, \infty)$ the following are true:

$$\mathcal{F}(\tau_t \varphi) = t^{-n} \tau_{\frac{1}{t}} \mathcal{F}(\varphi) \quad \text{in} \quad \mathcal{S}(\mathbb{R}^n), \quad \forall \varphi \in \mathcal{S}(\mathbb{R}^n), \tag{4.3.7}$$

$$\mathcal{F}(\tau_t u) = t^{-n} \tau_{\frac{1}{t}} \mathcal{F}(u) \quad \text{in} \quad \mathcal{S}'(\mathbb{R}^n), \quad \forall u \in \mathcal{S}'(\mathbb{R}^n). \tag{4.3.8}$$

Hint: Use (3.2.9) with $A = \frac{1}{t} I_{n \times n}$ to prove (4.3.7), then use (4.3.7) and (4.3.6) to prove (4.3.8).

To proceed, we make a couple of definitions.

Definition 4.47. A linear transformation $A \in M_{n \times n}(\mathbb{R})$ is called `orthogonal` provided A is invertible and $A^{-1} = A^\top$.

Some of the most basic attributes of an orthogonal matrix A are

$$(A^\top)^{-1} = A, \quad |\det A| = 1, \quad |Ax| = |x| \text{ for every } x \in \mathbb{R}^n. \qquad (4.3.9)$$

Definition 4.48. A tempered distribution $u \in S'(\mathbb{R}^n)$ is called `invariant under orthogonal transformations` provided $u = u \circ A$ in $S'(\mathbb{R}^n)$ for every orthogonal matrix $A \in M_{n \times n}(\mathbb{R})$.

Proposition 4.49. *Let $u \in S'(\mathbb{R}^n)$. Then u is invariant under orthogonal transformations if and only if $\widehat{u}$ is invariant under orthogonal transformations.*

Proof. This is a direct consequence of (4.3.3) and the fact that any orthogonal matrix A satisfies (4.3.9). □

Next we take a look at homogeneous functions to gain some insight into how this notion may be defined in the setting of distributions.

Definition 4.50. (1) A nonempty open set O in $\mathbb{R}^n$ is called a `cone-like region` if $tx \in O$ whenever $x \in O$ and $t \in (0, \infty)$.

(2) Given a cone-like region $O \subseteq \mathbb{R}^n$, call a function $f : O \to \mathbb{C}$ `positive homogeneous of degree` $k \in \mathbb{R}$ if $f(tx) = t^k f(x)$ for every $t > 0$ and every $x \in O$.

Exercise 4.51. Prove that if $O \subseteq \mathbb{R}^n$ is a cone-like region, $N \in \mathbb{N}$, and $f \in C^N(O)$ is positive homogeneous of degree $k \in \mathbb{R}$ on O, then $\partial^\alpha f$ is positive homogeneous of degree $k - |\alpha|$ on O for every $\alpha \in \mathbb{N}_0^n$ with $|\alpha| \le N$.

Exercise 4.52. Prove that if $f \in C^0(\mathbb{R}^n \setminus \{0\})$ is positive homogeneous of degree $1 - n$ on $\mathbb{R}^n \setminus \{0\}$, and $g \in C^0(S^{n-1})$, then

$$\int_{\partial B(0,R)} g(x/R) f(x) \, d\sigma(x) = \int_{S^{n-1}} g(x) f(x) \, d\sigma(x), \quad \forall R \in (0, \infty). \qquad (4.3.10)$$

Exercise 4.53. Prove that if $f \in C^0(\mathbb{R}^n \setminus \{0\})$ is positive homogeneous of degree $k \in \mathbb{R}$, then $|f(x)| \le \|f\|_{L^\infty(S^{n-1})} |x|^k$ for every $x \in \mathbb{R}^n \setminus \{0\}$.

Hint: Write $f(x) = f(\frac{x}{|x|}) |x|^k$ for every $x \in \mathbb{R}^n \setminus \{0\}$.

Exercise 4.54. Show that if $f \in C^0(\mathbb{R}^n \setminus \{0\})$ is positive homogeneous of degree $k \in \mathbb{R}$ with $k > -n$, then $f \in S'(\mathbb{R}^n)$.

Hint: Make use of Exercise 4.53 and the result discussed in Example 4.4.

After this preamble, we are ready to extend the notion of positive homogeneity to tempered distributions.

Definition 4.55. A distribution $u \in \mathcal{S}'(\mathbb{R}^n)$ is called `positive homogeneous` of degree $k \in \mathbb{R}$ provided $\tau_t u = t^k u$ in $\mathcal{S}'(\mathbb{R}^n)$ for every $t > 0$.

Exercise 4.56. Prove that $\delta \in \mathcal{S}'(\mathbb{R}^n)$ is positive homogeneous of degree $-n$.

Exercise 4.57. Let $f \in L^1_{loc}(\mathbb{R}^n)$ be such that the integrability condition (4.1.4) is satisfied for some $m \in [0, \infty)$ and some $R \in (0, \infty)$, and let $k \in \mathbb{R}$. Show that the tempered distribution u_f is positive homogeneous of degree k, if and only if f is positive homogeneous of degree k.

Exercise 4.58. Prove that if $u \in \mathcal{S}'(\mathbb{R}^n)$ is positive homogeneous of degree $k \in \mathbb{R}$ then for every $\alpha \in \mathbb{N}_0^n$ the tempered distribution $\partial^\alpha u$ is positive homogeneous of degree $k - |\alpha|$. Deduce from this that for every $\alpha \in \mathbb{N}_0^n$ the tempered distribution $\partial^\alpha \delta$ is positive homogeneous of degree $-n - |\alpha|$.

Proposition 4.59. *Let $k \in \mathbb{R}$. If $u \in \mathcal{S}'(\mathbb{R}^n)$ is positive homogeneous of degree k, then $\widehat{u}$ is positive homogeneous of degree $-n - k$.*

Proof. Let $u \in \mathcal{S}'(\mathbb{R}^n)$ be positive homogeneous of degree k, and fix $t > 0$. Then (4.3.8) and the assumption on u give

$$\tau_t \widehat{u} = t^{-n} \mathcal{F}(\tau_{\frac{1}{t}} u) = t^{-n} \mathcal{F}(t^{-k} u) = t^{-n-k} \widehat{u} \quad \text{in} \quad \mathcal{S}'(\mathbb{R}^n), \tag{4.3.11}$$

hence $\widehat{u}$ is positive homogeneous of degree $-n - k$. $\qquad\square$

Proposition 4.60. *If $u \in \mathcal{S}'(\mathbb{R}^n)$, $u\big|_{\mathbb{R}^n \setminus \{0\}} \in C^\infty(\mathbb{R}^n \setminus \{0\})$, and $u\big|_{\mathbb{R}^n \setminus \{0\}}$ is positive homogeneous of degree k, for some $k \in \mathbb{R}$, then $\widehat{u}\big|_{\mathbb{R}^n \setminus \{0\}} \in C^\infty(\mathbb{R}^n \setminus \{0\})$.*

Proof. Fix u satisfying the hypotheses of the proposition. By *(c)* in Proposition 4.26, for each $\alpha \in \mathbb{N}_0^n$ one has $D_\xi^\alpha \widehat{u} = \widehat{(-x)^\alpha u}$ in $\mathcal{S}'(\mathbb{R}^n)$. Also, it is not difficult to check that $(-x)^\alpha u \in \mathcal{S}'(\mathbb{R}^n)$ continues to satisfy the hypotheses of the proposition with k replaced by $k + |\alpha|$. Hence, the desired conclusion follows once we prove that $\widehat{u}\big|_{\mathbb{R}^n \setminus \{0\}}$ is continuous on $\mathbb{R}^n \setminus \{0\}$.

To this end, assume first that $k < -n$ and fix $\psi \in C_0^\infty(\mathbb{R}^n)$ such that $\psi \equiv 1$ on $B(0, 1)$. Use this to decompose $u = \psi u + (1 - \psi) u$. Since $\psi u \in \mathcal{E}'(\mathbb{R}^n)$ part *(b)* in Theorem 4.35 gives $\widehat{\psi u} \in C^\infty(\mathbb{R}^n)$. Furthermore, $(1 - \psi) u$ vanishes near the origin while outside $\operatorname{supp} \psi$ becomes $u(x) = u(|x| \frac{x}{|x|}) = |x|^k u(\frac{x}{|x|})$. Given the current assumption on k, this behavior implies $(1 - \psi) u \in L^1(\mathbb{R}^n)$, hence $\widehat{(1 - \psi) u} \in C^0(\mathbb{R}^n)$ by Lemma 4.28. To summarize, this analysis shows that

$$\widehat{u} \in C^0(\mathbb{R}^n) \quad \text{whenever} \quad k + n < 0. \tag{4.3.12}$$

To deal with the case $k + n \geq 0$, pick some multi-index $\alpha \in \mathbb{N}_0^n$ arbitrary and define $v_\alpha := D^\alpha u \in \mathcal{S}'(\mathbb{R}^n)$. Since $u\big|_{\mathbb{R}^n \setminus \{0\}} \in C^\infty(\mathbb{R}^n \setminus \{0\})$ differentiating $u(tx) = t^k u(x)$ yields

$$t^{|\alpha|}(D^\alpha u)(tx) = t^k (D^\alpha u)(x) \quad \text{for} \quad x \in \mathbb{R}^n \setminus \{0\} \quad \text{and} \quad t > 0. \tag{4.3.13}$$

Given that $v_\alpha\big|_{\mathbb{R}^n\setminus\{0\}} \in C^\infty(\mathbb{R}\setminus\{0\})$, the latter translates into

$$v_\alpha(tx) = t^{k-|\alpha|}v_\alpha(x) \quad \text{for} \quad x \in \mathbb{R}^n\setminus\{0\} \quad \text{and} \quad t > 0. \tag{4.3.14}$$

Hence, $v_\alpha\big|_{\mathbb{R}^n\setminus\{0\}}$ is homogeneous of degree $k - |\alpha|$. Based on what we proved earlier (cf. (4.3.12)), it follows that $\widehat{v_\alpha} \in C^0(\mathbb{R}^n)$ whenever $k - |\alpha| < -n$. In terms of the original distribution u this amounts to saying that

$$\xi^\alpha \widehat{u} \in C^0(\mathbb{R}^n), \qquad \forall\, \alpha \in \mathbb{N}_0^n \quad \text{with} \quad |\alpha| > k + n. \tag{4.3.15}$$

The endgame in the proof is then as follows. Given an arbitrary $k \in \mathbb{R}$, pick a natural number N with the property that $2N > k + n$. Writing (cf. (14.2.3)) for each $\xi \in \mathbb{R}^n$

$$|\xi|^{2N} = \sum_{|\alpha|=N} \frac{N!}{\alpha!} \xi^{2\alpha}, \tag{4.3.16}$$

we obtain

$$|\xi|^{2N}\widehat{u} = \sum_{|\alpha|=N} \frac{N!}{\alpha!} \xi^{2\alpha}\widehat{u} \quad \text{in} \quad \mathcal{S}'(\mathbb{R}^n). \tag{4.3.17}$$

Collectively, (4.3.17), (4.3.15), and the assumption on N imply

$$|\xi|^{2N}\widehat{u} \in C^0(\mathbb{R}^n). \tag{4.3.18}$$

Since $\frac{1}{|\xi|^{2N}}\big|_{\mathbb{R}^n\setminus\{0\}} \in C^\infty(\mathbb{R}^n\setminus\{0\})$, the membership in (4.3.18) further implies that $\widehat{u}\big|_{\mathbb{R}^n\setminus\{0\}} \in C^0(\mathbb{R}^n\setminus\{0\})$. This completes the proof of the proposition. $\qquad\square$

An inspection of the proof of Proposition 4.60 shows that several other useful versions could be derived, two of which are recorded below.

Exercise 4.61. If $u \in \mathcal{S}'(\mathbb{R}^n)$, $u\big|_{\mathbb{R}^n\setminus\{0\}} \in C^\infty(\mathbb{R}^n\setminus\{0\})$ and $u\big|_{\mathbb{R}^n\setminus\{0\}}$ is positive homogeneous of degree k, for some $k \in \mathbb{R}$ satisfying $k < -n$, then $\widehat{u} \in C^{k_0}(\mathbb{R}^n)$, where $k_0 := \max\{j \in \mathbb{N}_0,\ j + k < -n\}$.

Exercise 4.62. Assume that $u \in \mathcal{S}'(\mathbb{R}^n)$, $u\big|_{\mathbb{R}^n\setminus\{0\}} \in C^N(\mathbb{R}^n\setminus\{0\})$ where $N \in \mathbb{N}$ is even, and $u\big|_{\mathbb{R}^n\setminus\{0\}}$ is positive homogeneous of degree k, for some $k \in \mathbb{R}$ satisfying $k < N - n$. Then $\widehat{u}\big|_{\mathbb{R}^n\setminus\{0\}} \in C^m(\mathbb{R}^n\setminus\{0\})$ for every $m \in \mathbb{N}_0$ satisfying $m < N - n - k$.

Exercise 4.63. Consider a function $b \in C^0(\mathbb{R}^n\setminus\{0\})$ that is positive homogeneous of degree k, for some $k \in \mathbb{R}$ satisfying $k > -n$. Prove that the following are true.

(1) The mapping $u_b : \mathcal{S}(\mathbb{R}^n) \to \mathbb{C}$ acting on each $\varphi \in \mathcal{S}(\mathbb{R}^n)$ according to

$$u_b(\varphi) := \int_{\mathbb{R}^n} b(x)\varphi(x)\,dx, \tag{4.3.19}$$

is a well-defined (via an absolutely convergent integral) tempered distribution in $\mathbb{R}^n$, i.e., $u_b \in \mathcal{S}'(\mathbb{R}^n)$.

(2) Let $N \in \mathbb{N}$ be even and such that $N \geq 4$. Suppose $b \in C^N(\mathbb{R}^n \setminus \{0\})$ is positive homogeneous of degree k, for some $k \in \mathbb{R}$ satisfying $1 - n < k < N - n - 1$. Fix an index $j \in \{1, \ldots, n\}$ and consider the tempered distribution $u_{\partial_j b}$ defined as in (4.3.19) with $\partial_j b$ in place of b. Then, when restricted to $\mathbb{R}^n \setminus \{0\}$, the distributions $\widehat{u_{\partial_j b}}$ and $\widehat{u_b}$ are given by functions of class C^m, for each integer $0 \leq m < N - n - k - 1$, and satisfy the pointwise equality

$$\left(\widehat{u_{\partial_j b}} \big|_{\mathbb{R}^n \setminus \{0\}} \right)(\xi) = i\xi_j \left(\widehat{u_b} \big|_{\mathbb{R}^n \setminus \{0\}} \right)(\xi), \quad \forall \xi \in \mathbb{R}^n \setminus \{0\}. \tag{4.3.20}$$

Sketch of proof of (2):
Step I. Consider the function $a_j(\xi) := i\xi_j$ for each $\xi \in \mathbb{R}^n$ which belongs to $\mathcal{L}(\mathbb{R}^n)$. The goal here is to show that

$$\widehat{u_{\partial_j b}} = a_j \widehat{u_b} \quad \text{in } \mathcal{S}'(\mathbb{R}^n). \tag{4.3.21}$$

One way to see this is to use Remark 2.38 to obtain that $u_{\partial_j b} = \partial_j u_b$ in $\mathcal{S}'(\mathbb{R}^n)$ and then apply item *(2)* in Theorem 4.26.

A more direct proof of (4.3.21) is as follows. Pick $\varphi \in \mathcal{S}(\mathbb{R}^n)$ and use integration by parts, Lebesgue's Dominated Convergence Theorem, and the fact that for each $x \in \mathbb{R}^n$ we have $\partial_j \widehat{\varphi}(x) = -\widehat{a_j \varphi}(x)$ to write

$$\langle \widehat{u_{\partial_j b}}, \varphi \rangle = \langle u_{\partial_j b}, \widehat{\varphi} \rangle = \int_{\mathbb{R}^n} (\partial_j b)(x) \widehat{\varphi}(x) \, dx$$

$$= \lim_{\varepsilon \to 0^+} \lim_{R \to \infty} \int_{\varepsilon < |x| < R} (\partial_j b)(x) \widehat{\varphi}(x) \, dx$$

$$= \lim_{\varepsilon \to 0^+} \lim_{R \to \infty} \left[-\int_{\varepsilon < |x| < R} b(x) \partial_j \widehat{\varphi}(x) \, dx + \int_{|x|=R} b(x) \widehat{\varphi}(x) \frac{x_j}{R} d\sigma(x) \right.$$

$$\left. - \int_{|x|=\varepsilon} b(x) \widehat{\varphi}(x) \frac{x_j}{\varepsilon} d\sigma(x) \right]$$

$$= \int_{\mathbb{R}^n} b(x) \widehat{a_j \varphi}(x) \, dx = \langle u_b, \widehat{a_j \varphi} \rangle = \langle \widehat{u_b}, a_j \varphi \rangle$$

$$= \langle a_j \widehat{u_b}, \varphi \rangle. \tag{4.3.22}$$

In justifying the fifth equality in (4.3.22) we also need to observe that the limits as $R \to \infty$ and as $\varepsilon \to 0^+$ of the integrals over $|x| = R$ and over $|x| = \varepsilon$, respectively, are equal to zero. Indeed, by Exercise 4.53 we have $|b(x)| \leq \|b\|_{L^\infty(S^{n-1})} |x|^k$ for every $x \in \mathbb{R}^n \setminus \{0\}$. Hence, for each $R > 0$, using also Remark 3.4, we can estimate

$$\left| \int_{|x|=R} b(x) \widehat{\varphi}(x) \frac{x_j}{R} d\sigma(x) \right| \tag{4.3.23}$$

$$\leq \|b\|_{L^\infty(S^{n-1})} \sup_{x \in \mathbb{R}^n} \left[(1 + |x|)^{k+n} |\widehat{\varphi}(x)| \right] \omega_{n-1} \frac{R^{k+n-1}}{(1+R)^{k+n}} \xrightarrow[R \to \infty]{} 0.$$

Similarly,

$$\left| \int_{|x|=\varepsilon} b(x)\widehat{\varphi}(x) \frac{x_j}{\varepsilon} d\sigma(x) \right|$$

$$\leq \|b\|_{L^\infty(S^{n-1})} \|\widehat{\varphi}\|_{L^\infty(\mathbb{R}^n)} \, \omega_{n-1} \varepsilon^{k+n-1} \xrightarrow[\varepsilon \to 0^+]{} 0, \qquad (4.3.24)$$

where for the limit we used the current assumption that $k > 1 - n$.

Step II. By Exercise 4.51 we have that $\partial_j b$ is positive homogeneous of degree $k - 1$ in $\mathbb{R}^n \setminus \{0\}$. Now apply Exercise 4.62 (with $k - 1$ in place of k and $N - 2$ in place of N; note that the smoothness N in Exercise 4.62 should be an even natural number) and use the assumption $k < N - n - 1$ to conclude that $\widehat{u_{\partial_j b}}\big|_{\mathbb{R}^n \setminus \{0\}} \in C^m(\mathbb{R}^n \setminus \{0\})$ for each integer $0 \leq m < N - n - k - 1$. By Exercise 4.62 we also have $\widehat{u_b}\big|_{\mathbb{R}^n \setminus \{0\}} \in C^m(\mathbb{R}^n \setminus \{0\})$ for each integer $0 \leq m < N - n - k$. Now invoke (4.3.21) to finish the proof of (4.3.20).

Next, we take on the task of computing the Fourier transform of certain homogeneous tempered distributions that will be particularly important later in applications. Recall the gamma function Γ from (14.5.1).

Proposition 4.64. *Let* $\lambda \in (0, n)$ *and set* $f_\lambda(x) := |x|^{-\lambda}$, *for each* $x \in \mathbb{R}^n \setminus \{0\}$. *Then* $f_\lambda \in \mathcal{S}'(\mathbb{R}^n)$, $\widehat{f_\lambda}\big|_{\mathbb{R}^n \setminus \{0\}} \in C^\infty(\mathbb{R}^n \setminus \{0\})$, *and*

$$\widehat{f_\lambda}(\xi) = 2^{n-\lambda} \pi^{\frac{n}{2}} \frac{\Gamma\left(\frac{n-\lambda}{2}\right)}{\Gamma\left(\frac{\lambda}{2}\right)} |\xi|^{\lambda-n} \quad \text{for every} \quad \xi \in \mathbb{R}^n \setminus \{0\}. \qquad (4.3.25)$$

Proof. Fix $\lambda \in (0, n)$. Exercise 4.5 then shows that $f_\lambda \in \mathcal{S}'(\mathbb{R}^n)$. Clearly, $|x|^{-\lambda}$ is invariant under orthogonal transformations and is positive homogeneous of degree $-\lambda$. Hence, by Proposition 4.49 and Proposition 4.59, it follows that $\widehat{f_\lambda}$ is invariant under orthogonal transformations and positive homogeneous of degree $-n + \lambda$. In addition, $\widehat{f_\lambda}\big|_{\mathbb{R}^n \setminus \{0\}} \in C^\infty(\mathbb{R}^n \setminus \{0\})$ by Proposition 4.60.

Fix $\xi \in \mathbb{R}^n \setminus \{0\}$ and choose an orthogonal matrix $A \in M_{n \times n}(\mathbb{R})$ with the property that $A\xi = (0, 0, \ldots, 0, |\xi|)$ (such a matrix may be obtained by completing the vector $v_n := \frac{\xi}{|\xi|}$ to an orthonormal basis $\{v_1, \ldots, v_n\}$ in $\mathbb{R}^n$ and then taking A to be the matrix mapping each v_j into $\mathbf{e}_j$ for $j = 1, \ldots, n$). Then

$$\widehat{f_\lambda}(\xi) = \widehat{f_\lambda}(A\xi) = \widehat{f_\lambda}(0, \ldots, 0, |\xi|) = c_\lambda |\xi|^{\lambda-n}, \qquad (4.3.26)$$

where $c_\lambda := \widehat{f_\lambda}(0, \ldots, 0, 1) \in \mathbb{C}$. As such, we are left with determining the value of c_λ. We do so by apply $\widehat{f_\lambda}$ to the particular Schwartz function $\varphi(x) := e^{-\frac{|x|^2}{2}}$ for $x \in \mathbb{R}^n$. From Example 3.22 we know that $\widehat{\varphi}(\xi) = (2\pi)^{\frac{n}{2}} e^{-\frac{|\xi|^2}{2}}$ for every $\xi \in \mathbb{R}^n$. Based on this formula and (4.3.26), the identity $\langle \widehat{f_\lambda}, \varphi \rangle = \langle f_\lambda, \widehat{\varphi} \rangle$ may be rewritten as

$$c_\lambda \int_{\mathbb{R}^n} |\xi|^{\lambda-n} e^{-\frac{|\xi|^2}{2}} \, d\xi = (2\pi)^{\frac{n}{2}} \int_{\mathbb{R}^n} |x|^{-\lambda} e^{-\frac{|x|^2}{2}} \, dx. \qquad (4.3.27)$$

The two integrals in (4.3.27) may be computed simultaneously, by adopting a slightly more general point of view, as follows. For any $p > -n$ use polar coordinates (cf. (14.9.9)) and a natural change of variables to write (for the definition and properties of the gamma function Γ see (14.5.1) and the subsequent comments)

$$
\begin{aligned}
\int_{\mathbb{R}^n} |\xi|^p e^{-\frac{|\xi|^2}{2}} \, d\xi &= \omega_{n-1} \int_0^\infty \rho^{p+n-1} e^{-\frac{\rho^2}{2}} \, d\rho \\
&= \omega_{n-1} 2^{\frac{p+n-2}{2}} \int_0^\infty t^{\frac{p+n-2}{2}} e^{-t} \, dt \\
&= 2^{\frac{p+n-2}{2}} \omega_{n-1} \Gamma\left(\frac{p+n}{2}\right).
\end{aligned}
\tag{4.3.28}
$$

When used with $p := \lambda - n$ and $p := -\lambda$, formula (4.3.28) allows us to rewrite (4.3.27) as

$$
c_\lambda 2^{\frac{\lambda-n+n-2}{2}} \omega_{n-1} \Gamma\left(\frac{\lambda}{2}\right) = (2\pi)^{\frac{n}{2}} 2^{\frac{-\lambda+n-2}{2}} \omega_{n-1} \Gamma\left(\frac{-\lambda+n}{2}\right).
\tag{4.3.29}
$$

This gives $c_\lambda = 2^{n-\lambda} \pi^{\frac{n}{2}} \frac{\Gamma\left(\frac{n-\lambda}{2}\right)}{\Gamma\left(\frac{\lambda}{2}\right)}$, finishing the proof of (4.3.25).   □

A remarkable consequence of Proposition 4.64 is singled out below.

Corollary 4.65. *Assume that $n \in \mathbb{N}$, $n \geq 2$, and fix $\lambda \in [0, n-1)$. Then for each $j \in \{1, \ldots, n\}$, we have*

$$
\mathcal{F}\left(\frac{x_j}{|x|^{\lambda+2}}\right) = -i\, 2^{n-\lambda-1} \pi^{\frac{n}{2}} \frac{\Gamma\left(\frac{n-\lambda}{2}\right)}{\Gamma\left(\frac{\lambda}{2}+1\right)} \frac{\xi_j}{|\xi|^{n-\lambda}} \quad in \quad \mathcal{S}'(\mathbb{R}^n).
\tag{4.3.30}
$$

In particular, corresponding to the case when $\lambda = n - 2$, formula (4.3.30) becomes

$$
\mathcal{F}\left(\frac{x_j}{|x|^n}\right) = -i\, \omega_{n-1} \frac{\xi_j}{|\xi|^2} \quad in \quad \mathcal{S}'(\mathbb{R}^n).
\tag{4.3.31}
$$

Proof. Fix an integer $n \geq 2$, and suppose first that $\lambda \in (0, n-1)$. In this regime, both (4.3.25) and (4.1.40) hold. In concert with part *(b)* in Theorem 4.26, these give

$$
\begin{aligned}
-\lambda \mathcal{F}\left(\frac{x_j}{|x|^{\lambda+2}}\right) &= \mathcal{F}\left(\partial_j f_\lambda\right) = i\xi_j \widehat{f_\lambda}(\xi) \\
&= i\, 2^{n-\lambda} \pi^{\frac{n}{2}} \frac{\Gamma\left(\frac{n-\lambda}{2}\right)}{\Gamma\left(\frac{\lambda}{2}\right)} \xi_j |\xi|^{\lambda-n} \quad in \quad \mathcal{S}'(\mathbb{R}^n).
\end{aligned}
\tag{4.3.32}
$$

Hence, whenever $\lambda \in (0, n-1)$,

$$
\mathcal{F}\left(\frac{x_j}{|x|^{\lambda+2}}\right) = C_\lambda \xi_j |\xi|^{\lambda-n} \quad in \quad \mathcal{S}'(\mathbb{R}^n),
\tag{4.3.33}
$$

where we have set

$$C_\lambda := -\mathrm{i}\, 2^{n-\lambda-1} \pi^{\frac{n}{2}} \frac{\Gamma\left(\frac{n-\lambda}{2}\right)}{\frac{\lambda}{2}\Gamma\left(\frac{\lambda}{2}\right)} = -\mathrm{i}\, 2^{n-\lambda-1} \pi^{\frac{n}{2}} \frac{\Gamma\left(\frac{n-\lambda}{2}\right)}{\Gamma\left(\frac{\lambda}{2}+1\right)} \qquad (4.3.34)$$

and the last equality follows from (14.5.2). This proves formula (4.3.30) in the case when $\lambda \in (0, n-1)$. The case when $\lambda = 0$ then follows from what we have just proved, by passing to limit $\lambda \to 0^+$ in (4.3.30) and observing that all quantities involved depend continuously on λ, in an appropriate sense. Finally, (4.3.31) is a direct consequence of (4.3.30) and (14.5.6).   □

4.4 Principal Value Tempered Distributions

Recall the distribution P.V. $\frac{1}{x} \in \mathcal{D}'(\mathbb{R})$ from Example 2.11. As seen from Exercise 4.107, we have P.V. $\frac{1}{x} \in \mathcal{S}'(\mathbb{R})$. The issue we address in this section is the generalization of this distribution to higher dimensions. The key features of the function $\Theta(x) := \frac{1}{x}$, $x \in \mathbb{R} \setminus \{0\}$, that allowed us to define P.V. $\frac{1}{x}$ as a tempered distribution on the real line are as follows: first, $\Theta \in C^0(\mathbb{R} \setminus \{0\})$, second, Θ is positive homogeneous of degree -1, and third, $\Theta(1) + \Theta(-1) = 0$.

Moving from one dimension to $\mathbb{R}^n$, this suggests considering the class of functions satisfying

$$\Theta \in C^0(\mathbb{R}^n \setminus \{0\}), \quad \text{positive homogeneous of degree } -n,$$
$$\int_{S^{n-1}} \Theta \, \mathrm{d}\sigma = 0. \qquad (4.4.1)$$

It is worth noting that a function Θ as above typically fails to be in $L^1_{loc}(\mathbb{R}^n)$ (though obviously $\Theta \in L^1_{loc}(\mathbb{R}^n \setminus \{0\})$). As such, associating a distribution with Θ necessarily has to be a more elaborate process than the one identifying functions from $L^1_{loc}(\mathbb{R}^n)$ with distributions in $\mathbb{R}^n$. This has already been the case when defining P.V. $\frac{1}{x}$ and here we model the same type of definition in higher dimensions.

Specifically, given Θ as in (4.4.1) consider the linear mapping

$$\text{P.V.}\, \Theta : \mathcal{S}(\mathbb{R}^n) \to \mathbb{C},$$
$$(\text{P.V.}\, \Theta)(\varphi) := \lim_{\varepsilon \to 0^+} \int_{|x| \geq \varepsilon} \Theta(x)\varphi(x)\, \mathrm{d}x, \quad \forall\, \varphi \in \mathcal{S}(\mathbb{R}^n). \qquad (4.4.2)$$

That this definition does the job is proved next.

Proposition 4.66. *Let Θ be a function satisfying* (4.4.1). *Then the map* P.V. Θ *considered in* (4.4.2) *is well defined and is a tempered distribution in $\mathbb{R}^n$. In addition,* $(\text{P.V.}\, \Theta)\big|_{\mathbb{R}^n \setminus \{0\}} = \Theta\big|_{\mathbb{R}^n \setminus \{0\}}$ *in $\mathcal{D}'(\mathbb{R}^n \setminus \{0\})$.*

Before proceeding with the proof of Proposition 4.66 we recall a definition and introduce a class of functions that will be used in the proof.

Definition 4.67. A function $\psi : \mathbb{R}^n \to \mathbb{C}$ is called `radial` if $\psi(x)$ depends only on $|x|$ for every $x \in \mathbb{R}^n$, that is, $\psi(x) = f(|x|)$ for all $x \in \mathbb{R}^n$, where $f : \mathbb{R} \to \mathbb{C}$.

Next, consider the class of functions

$$\begin{cases} \psi : \mathbb{R}^n \to \mathbb{C}, \quad \psi \in C^1(\mathbb{R}^n), \\ \psi \text{ is radial and } \psi(0) = 1, \\ \exists\, \varepsilon_o \in (0, \infty) \text{ such that } \psi \text{ decays like } |x|^{-\varepsilon_o} \text{ at } \infty, \end{cases} \tag{4.4.3}$$

(for example $\psi(x) = e^{-\frac{|x|^2}{2}}$, $x \in \mathbb{R}^n$, satisfies (4.4.3)) and set

$$Q := \{\psi : \psi \text{ satisfies (4.4.3)}\}. \tag{4.4.4}$$

Now we are ready to return to Proposition 4.66.

Proof of Proposition 4.66. Fix an arbitrary ψ satisfying (4.4.3). Then, making use of formula (14.9.9) and the properties of Θ and ψ, for each $\varepsilon \in (0, \infty)$ we have

$$\int_{|x| \geq \varepsilon} \Theta(x)\psi(x)\,\mathrm{d}x = \int_{|x| \geq \varepsilon} \frac{\Theta(\frac{x}{|x|})}{|x|^n}\psi(x)\,\mathrm{d}x$$

$$= \int_\varepsilon^\infty \frac{\psi(\rho)}{\rho} \int_{S^{n-1}} \Theta(\omega)\,\mathrm{d}\sigma(\omega)\,\mathrm{d}\rho = 0. \tag{4.4.5}$$

Hence, $\Theta[\varphi - \varphi(0)\psi] \in L^1(\mathbb{R}^n)$ for every $\varphi \in S(\mathbb{R}^n)$, which when combined with (4.4.5) and Lebesgue Dominated Convergence Theorem 14.15 yields

$$(\text{P.V.}\,\Theta)(\varphi) = \int_{\mathbb{R}^n} \Theta(x)[\varphi(x) - \varphi(0)\psi(x)]\,\mathrm{d}x, \quad \forall\, \varphi \in S(\mathbb{R}^n). \tag{4.4.6}$$

Note that because of (4.4.5), the right-hand side in (4.4.6) is independent of the choice of ψ. Estimating the right-hand side of (4.4.6) (using Exercise 4.53, the decay at infinity of functions from Q and the Schwartz class, and the Mean Value Theorem near the origin) shows that there exists a constant $C \in (0, \infty)$ independent of φ with the property that

$$\left|(\text{P.V.}\,\Theta)(\varphi)\right| \leq C \sup_{x \in \mathbb{R}^n, |\alpha| \leq 1, |\beta| \leq 1} \left|x^\alpha \partial^\beta \varphi(x)\right|, \quad \forall\, \varphi \in S(\mathbb{R}^n). \tag{4.4.7}$$

Since from (4.4.6) we see that P.V. Θ is linear, in light of Fact 4.1 estimate (4.4.7) implies P.V. $\Theta \in S'(\mathbb{R}^n)$ as wanted. The fact that the restriction in the distributional sense of P.V. Θ to $\mathbb{R}^n \setminus \{0\}$ is equal to the restriction of the function Θ to $\mathbb{R}^n \setminus \{0\}$ is immediate from definitions. $\qquad\square$

Remark 4.68. (1) As already alluded to, if $n = 1$ and we take $\Theta(x) := \frac{1}{x}$ for each $x \in \mathbb{R} \setminus \{0\}$, then P.V. $\Theta = $ P.V. $\frac{1}{x}$.

(2) Suppose Θ is as in (4.4.1). Since identity (4.4.6) holds for any $\psi \in Q$, we may select $\psi \in Q$ that also satisfies $\psi \equiv 1$ on $\overline{B(0,1)}$ and observe that for this choice of ψ we have

$$\int_{\mathbb{R}^n} \Theta(x)[\varphi(x) - \varphi(0)\psi(x)] \, dx \tag{4.4.8}$$

$$= \int_{|x| \leq 1} \Theta(x)[\varphi(x) - \varphi(0)] \, dx + \int_{|x| > 1} \Theta(x)\varphi(x) \, dx, \quad \forall \varphi \in S(\mathbb{R}^n).$$

Hence,

$$\langle \text{P.V.} \, \Theta, \varphi \rangle = \int_{|x| \leq 1} \Theta(x)[\varphi(x) - \varphi(0)] \, dx$$

$$+ \int_{|x| > 1} \Theta(x)\varphi(x) \, dx, \quad \forall \varphi \in S(\mathbb{R}^n). \tag{4.4.9}$$

Example 4.69. If $j \in \{1, \ldots, n\}$, the function Θ defined by $\Theta(x) := \frac{x_j}{|x|^{n+1}}$ for each $x \in \mathbb{R}^n \setminus \{0\}$ satisfies (4.4.1). By Proposition 4.66 we have P.V. $\frac{x_j}{|x|^{n+1}}$ belongs to $S'(\mathbb{R}^n)$ and part *(2)* in Remark 4.68 gives that for every $\varphi \in S(\mathbb{R}^n)$

$$\left\langle \text{P.V.} \, \frac{x_j}{|x|^{n+1}}, \varphi \right\rangle = \lim_{\varepsilon \to 0^+} \int_{|x| \geq \varepsilon} \frac{x_j \varphi(x)}{|x|^{n+1}} \, dx \tag{4.4.10}$$

$$= \int_{|x| \leq 1} \frac{x_j(\varphi(x) - \varphi(0))}{|x|^{n+1}} \, dx + \int_{|x| > 1} \frac{x_j \varphi(x)}{|x|^{n+1}} \, dx.$$

The next proposition elaborates on the manner in which principal value tempered distributions convolve with Schwartz functions.

Proposition 4.70. *Let Θ be a function satisfying the conditions in (4.4.1). Then for each $\varphi \in S(\mathbb{R}^n)$ one has that $(\text{P.V.} \, \Theta) * \varphi \in S'(\mathbb{R}^n) \cap C^\infty(\mathbb{R}^n)$ and*

$$\big((\text{P.V.} \, \Theta) * \varphi\big)(x) = \lim_{\varepsilon \to 0^+} \int_{|x-y| \geq \varepsilon} \Theta(x-y)\varphi(y) \, dy, \quad \forall x \in \mathbb{R}^n. \tag{4.4.11}$$

Proof. Fix an arbitrary $\varphi \in S(\mathbb{R}^n)$ and note that since P.V. $\Theta \in S'(\mathbb{R}^n)$ (c.f. Proposition 4.66), part *(e)* in Theorem 4.19 implies $(\text{P.V.} \, \Theta) * \varphi \in S'(\mathbb{R}^n)$. Let $\psi \in C_0^\infty(\mathbb{R}^n)$ be such that $\psi \equiv 1$ near the origin. Then $1 - \psi \in \mathcal{L}(\mathbb{R}^n)$ and it makes sense to consider $(1 - \psi)(\text{P.V.} \, \Theta)$ in $S'(\mathbb{R}^n)$ (cf. part *(b)* in Theorem 4.14). Hence, we may decompose P.V. $\Theta = u + v$ where

$$u := \psi \, \text{P.V.} \, \Theta \in \mathcal{E}'(\mathbb{R}^n) \quad \text{and} \quad v := (1 - \psi)(\text{P.V.} \, \Theta) \in S'(\mathbb{R}^n). \tag{4.4.12}$$

The last part in Proposition 4.66 also permits us to identify $v = (1 - \psi)\Theta$ in $L_{loc}^1(\mathbb{R}^n)$. By Exercise 4.53 we therefore have $v \in L^p(\mathbb{R}^n)$ for every $p \in (1, \infty)$ which, in combination with Exercise 3.19, allows us to conclude that $v * \varphi$ belongs to $C^\infty(\mathbb{R}^n)$

and

$$(v * \varphi)(x) = \int_{\mathbb{R}^n} (1 - \psi(y))\Theta(y)\varphi(x - y)\,dy, \quad \forall x \in \mathbb{R}^n. \tag{4.4.13}$$

Since the above integral is absolutely convergent, by Lebesgue's Dominated Convergence Theorem permits we further express this as

$$(v * \varphi)(x) = \lim_{\varepsilon \to 0^+} \int_{|y| \geq \varepsilon} (1 - \psi(y))\Theta(y)\varphi(x - y)\,dy, \quad \forall x \in \mathbb{R}^n. \tag{4.4.14}$$

Thanks to Exercise 2.103 we also have $u * \varphi \in C^\infty(\mathbb{R}^n)$ and

$$(u * \varphi)(x) = \langle \psi\,\mathrm{P.V.}\,\Theta, \varphi(x - \cdot) \rangle \quad \text{for each } x \in \mathbb{R}^n. \tag{4.4.15}$$

On the other hand, the definition of the principal value gives that for each $x \in \mathbb{R}^n$

$$\langle \psi\,\mathrm{P.V.}\,\Theta, \varphi(x - \cdot) \rangle = \langle \mathrm{P.V.}\,\Theta, \psi(\cdot)\varphi(x - \cdot) \rangle$$

$$= \lim_{\varepsilon \to 0^+} \int_{|y| \geq \varepsilon} \Theta(y)\psi(y)\varphi(x - y)\,dy. \tag{4.4.16}$$

Collectively, these arguments show that, for each $x \in \mathbb{R}^n$,

$$\big((\mathrm{P.V.}\,\Theta) * \varphi\big)(x) = (u * \varphi)(x) + (v * \varphi)(x)$$

$$= \lim_{\varepsilon \to 0^+} \int_{|y| \geq \varepsilon} \Theta(y)\psi(y)\varphi(x - y)\,dy$$

$$+ \lim_{\varepsilon \to 0^+} \int_{|y| \geq \varepsilon} (1 - \psi(y))\Theta(y)\varphi(x - y)\,dy$$

$$= \lim_{\varepsilon \to 0^+} \int_{|x-y| \geq \varepsilon} \Theta(x - y)\varphi(y)\,dy, \tag{4.4.17}$$

proving (4.4.11).   □

The next example discusses a basic class of principal value tempered distributions arising naturally in applications.

Example 4.71. Let $\Phi \in C^1(\mathbb{R}^n \setminus \{0\})$ be positive homogeneous of degree $1 - n$. Then for each $j \in \{1, \ldots, n\}$ it follows that $\partial_j \Phi$ satisfies the conditions in (4.4.1). Consequently, $\mathrm{P.V.}(\partial_j \Phi)$ is a well-defined tempered distribution.

To see why this is true fix $j \in \{1, \ldots, n\}$ and note that $\partial_j \Phi \in C^0(\mathbb{R}^n \setminus \{0\})$ and $\partial_j \Phi$ is positive homogeneous of degree $-n$ (cf. Exercise 4.51). Moreover, using Exercise 4.52, then integrating by parts based on (14.8.4), and then using (14.9.5), we obtain

$$0 = \int_{|x|=2} \Phi(x)\frac{x_j}{2}\,d\sigma(x) - \int_{|x|=1} \Phi(x)x_j\,d\sigma(x) = \int_{1<|x|<2} \partial_j\Phi(x)\,dx$$

$$= \int_1^2 \int_{S^{n-1}} (\partial_j\Phi)(\rho\omega)\rho^{n-1}\,d\sigma(\omega)\,d\rho = \Big(\int_1^2 \frac{d\rho}{\rho}\Big)\int_{S^{n-1}} (\partial_j\Phi)(\omega)\,d\sigma(\omega)$$

$$= (\ln 2)\int_{S^{n-1}} (\partial_j\Phi)(\omega)\,d\sigma(\omega). \tag{4.4.18}$$

This shows that $\int_{S^{n-1}} \partial_j\Phi\,d\sigma = 0$, hence $\partial_j\Phi$ satisfies all conditions in (4.4.1).

Principal value tempered distributions often arise when differentiating certain types of functions exhibiting a point singularity. Specifically, we have the following theorem.

Theorem 4.72. *Let* $\Phi \in C^1(\mathbb{R}^n \setminus \{0\})$ *be a function that is positive homogeneous of degree* $1 - n$. *Then for each* $j \in \{1,\dots,n\}$, *the distributional derivative* $\partial_j\Phi$ *satisfies*

$$\partial_j\Phi = \Big(\int_{S^{n-1}} \Phi(\omega)\omega_j\,d\sigma(\omega)\Big)\delta + \text{P.V.}(\partial_j\Phi) \qquad \text{in } \mathcal{S}'(\mathbb{R}^n). \tag{4.4.19}$$

Proof. From the properties of Φ, Exercise 4.53, Exercise 4.54, and Exercise 4.6 it follows that $\Phi \in \mathcal{S}'(\mathbb{R}^n)$. Fix $j \in \{1,\dots,n\}$. By Example 4.71 we have that $\text{P.V.}(\partial_j\Phi) \in \mathcal{S}'(\mathbb{R}^n)$. Hence, invoking (4.1.33), to conclude (4.4.19) it suffices to prove that the equality in (4.4.19) holds in $\mathcal{D}'(\mathbb{R}^n)$. To this end, fix some $\varphi \in C_0^\infty(\mathbb{R}^n)$ and using Lebesgue Dominated Convergence Theorem 14.15 and integration by parts (based on (14.8.4)) write

$$\langle \partial_j\Phi, \varphi \rangle = -\langle \Phi, \partial_j\varphi \rangle = -\int_{\mathbb{R}^n} \Phi(x)\partial_j\varphi(x)\,dx$$

$$= -\lim_{\varepsilon\to 0^+} \int_{|x|\geq\varepsilon} \Phi(x)\partial_j\varphi(x)\,dx$$

$$= \lim_{\varepsilon\to 0^+} \int_{|x|\geq\varepsilon} \partial_j\Phi(x)\varphi(x)\,dx + \lim_{\varepsilon\to 0^+} \int_{|x|=\varepsilon} \frac{x_j}{\varepsilon}\Phi(x)\varphi(x)\,d\sigma(x)$$

$$= \big\langle \text{P.V.}(\partial_j\Phi), \varphi \big\rangle + \lim_{\varepsilon\to 0^+} \int_{|x|=\varepsilon} \frac{x_j}{\varepsilon}\Phi(x)\varphi(x)\,d\sigma(x). \tag{4.4.20}$$

Also, for each $\varepsilon \in (0, \infty)$ we have

$$\int_{|x|=\varepsilon} \frac{x_j}{\varepsilon} \, \Phi(x)\varphi(x) \, d\sigma(x)$$

$$= \int_{|x|=\varepsilon} \frac{x_j}{\varepsilon} \, \Phi(x)[\varphi(x) - \varphi(0)] \, d\sigma(x) + \varphi(0) \int_{|x|=\varepsilon} \frac{x_j}{\varepsilon} \, \Phi(x) \, d\sigma(x)$$

$$= \int_{|x|=\varepsilon} \frac{x_j}{\varepsilon} \, \Phi(x)[\varphi(x) - \varphi(0)] \, d\sigma(x) + \varphi(0) \int_{S^{n-1}} \omega_j \Phi(\omega) \, d\sigma(\omega), \quad (4.4.21)$$

where for the last equality in (4.4.21) we used Exercise 4.52. In addition, using the fact that $\varphi \in C_0^\infty(\mathbb{R}^n)$ and Exercise 4.53 we may estimate

$$\left| \int_{|x|=\varepsilon} \frac{x_j}{\varepsilon} \, \Phi(x)[\varphi(x) - \varphi(0)] \, d\sigma(x) \right|$$

$$\leq \varepsilon \|\nabla\varphi\|_{L^\infty(\mathbb{R}^n)} \|\Phi\|_{L^\infty(S^{n-1})} \int_{|x|=\varepsilon} \frac{1}{|x|^{n-1}} \, d\sigma(x)$$

$$= \|\nabla\varphi\|_{L^\infty(\mathbb{R}^n)} \|\Phi\|_{L^\infty(S^{n-1})} \omega_{n-1} \varepsilon \xrightarrow[\varepsilon \to 0^+]{} 0. \quad (4.4.22)$$

Combining (4.4.20), (4.4.21), and (4.4.22), we conclude

$$\langle \partial_j \Phi, \varphi \rangle = \langle \text{P.V.}(\partial_j \Phi), \varphi \rangle + \Big(\int_{S^{n-1}} \omega_j \Phi(\omega) \, d\sigma(\omega) \Big) \langle \delta, \varphi \rangle. \quad (4.4.23)$$

This yields that $\partial_j \Phi = \text{P.V.}(\partial_j \Phi) + \Big(\int_{S^{n-1}} \omega_j \Phi(\omega) \, d\sigma(\omega) \Big) \delta$ in $\mathcal{D}'(\mathbb{R}^n)$ and completes the proof of the theorem. $\qquad \Box$

4.5 The Fourier Transform of Principal Value Distributions

From Proposition 4.66 we know that whenever Θ is a function satisfying the conditions in (4.4.1), the principal value distribution P.V. Θ belongs to $\mathcal{S}'(\mathbb{R}^n)$. As such, its Fourier transform makes sense as a tempered distribution. This being said, in many applications (cf. the discussion in Remark 4.95), it is of basic importance to actually identify this distribution. The general aim of this section is to do just that, though as a warm-up, we deal with the following particular (yet relevant) case.

Proposition 4.73. *For each $k \in \{1, \dots, n\}$ consider the function*

$$\Phi_k(x) := \frac{x_k}{|x|^n}, \qquad \forall \, x \in \mathbb{R}^n \setminus \{0\}. \quad (4.5.1)$$

Then for each $j \in \{1, \dots, n\}$ the function $\partial_j \Phi_k$ satisfies the conditions in (4.4.1) and

$$\mathcal{F}(\text{P.V.}(\partial_j \Phi_k)) = \omega_{n-1}\frac{\xi_j\xi_k}{|\xi|^2} - \frac{\omega_{n-1}}{n}\delta_{jk} \quad in \quad \mathcal{S}'(\mathbb{R}^n). \tag{4.5.2}$$

Proof. Fix $j, k \in \{1, \ldots, n\}$. From Example 4.71 it is clear that the function $\partial_j \Phi_k$ is as in (4.4.1). Moreover, Theorem 4.72 gives that

$$\partial_j \Phi_k = \Big(\int_{S^{n-1}} \omega_k\omega_j \, d\sigma(\omega)\Big)\delta + \text{P.V.}(\partial_j \Phi_k) \qquad in \quad \mathcal{S}'(\mathbb{R}^n). \tag{4.5.3}$$

Taking into account (14.9.45), and applying the Fourier transform to both sides, this yields

$$\mathcal{F}(\text{P.V.}(\partial_j \Phi_k)) = \mathcal{F}(\partial_j \Phi_k) - \frac{\omega_{n-1}}{n}\delta_{jk} \quad in \quad \mathcal{S}'(\mathbb{R}^n). \tag{4.5.4}$$

On the other hand, since by part *(b)* in Theorem 4.26 and Corollary 4.65 we have

$$\mathcal{F}(\partial_j \Phi_k) = i\,\xi_j\mathcal{F}(\Phi_k) = \omega_{n-1}\frac{\xi_j\xi_k}{|\xi|^2} \quad in \quad \mathcal{S}'(\mathbb{R}^n), \tag{4.5.5}$$

formula (4.5.2) follows from (4.5.4)–(4.5.5).   □

The next theorem shows that the Fourier transform of principal value distributions P.V. Θ is given by bounded functions. As we shall see in Section 4.9, such a result plays a key role in establishing the L^2 boundedness of singular integral operators.

Theorem 4.74. *Let Θ be a function satisfying the conditions in (4.4.1). Then the function given by the formula*

$$m_\Theta(\xi) := -\int_{S^{n-1}} \Theta(\omega)\log(i(\xi\cdot\omega))\,d\sigma(\omega) \qquad for \ \xi \in \mathbb{R}^n \setminus \{0\}, \tag{4.5.6}$$

is well-defined, positive homogeneous of degree zero, satisfies

$$\int_{S^{n-1}} m_\Theta(\xi)\,d\sigma(\xi) = 0, \tag{4.5.7}$$

$$\big\|m_\Theta\big\|_{L^\infty(\mathbb{R}^n)} \le C_n\|\Theta\|_{L^\infty(S^{n-1})}, \tag{4.5.8}$$

where $C_n \in (0, \infty)$ is defined as

$$C_n := \frac{\pi\omega_{n-1}}{2} + \int_{S^{n-1}} \Big|\ln\Big|\frac{\xi}{|\xi|}\cdot\omega\Big|\Big|\,d\sigma(\omega), \tag{4.5.9}$$

and

$$(m_\Theta)^\vee = m_{\Theta^\vee}. \tag{4.5.10}$$

Moreover, the Fourier transform of the tempered distribution P.V. Θ *is of function type and*

$$\mathcal{F}(\text{P.V.}\,\Theta) = m_\Theta \quad in \quad \mathcal{S}'(\mathbb{R}^n), \tag{4.5.11}$$

and

$$\overline{m_\Theta} = m_{\overline{\Theta}} \quad \text{if } \Theta \text{ is an even function.} \tag{4.5.12}$$

In addition,

$$\text{if } \Theta \in C^k(\mathbb{R}^n \setminus \{0\}) \text{ for some } k \in \mathbb{N}_0 \cup \{\infty\},$$
$$\text{then } m_\Theta\big|_{\mathbb{R}^n \setminus \{0\}} \in C^k(\mathbb{R}^n \setminus \{0\}). \tag{4.5.13}$$

Proof. First, we show that the integral in (4.5.6) is absolutely convergent for each vector $\xi \in \mathbb{R}^n \setminus \{0\}$. To see this, fix an arbitrary $\xi \in \mathbb{R}^n \setminus \{0\}$ and observe that for each $\omega \in S^{n-1}$ we have

$$\log(i(\xi \cdot \omega)) = \ln|\xi \cdot \omega| + i\frac{\pi}{2}\text{sgn}\,(\xi \cdot \omega)$$

$$= \ln|\xi| + \ln\left|\frac{\xi}{|\xi|} \cdot \omega\right| + i\frac{\pi}{2}\,\text{sgn}\,(\xi \cdot \omega). \tag{4.5.14}$$

Applying Proposition 14.65 with $f(t) := \ln|t|$, $t \in \mathbb{R}$, and $v := \frac{\xi}{|\xi|}$, yields

$$\int_{S^{n-1}}\left|\ln\left|\frac{\xi}{|\xi|} \cdot \omega\right|\right| d\sigma(\omega) = 2\omega_{n-2}\int_0^1|\ln s|(\sqrt{1-s^2})^{n-3}\,ds < \infty. \tag{4.5.15}$$

As a by-product, we note that this implies that the constant C_n from (4.5.9) is finite. Next, from (4.5.15) and (4.5.9) we obtain

$$\int_{S^{n-1}}\left|\Theta(\omega)\left(\ln\left|\frac{\xi}{|\xi|} \cdot \omega\right| + i\frac{\pi}{2}\text{sgn}\,(\xi \cdot \omega)\right)\right| d\sigma(\omega)$$

$$\leq \|\Theta\|_{L^\infty(S^{n-1})}\left(\frac{\pi\omega_{n-1}}{2} + \int_{S^{n-1}}\left|\ln\left|\frac{\xi}{|\xi|} \cdot \omega\right|\right| d\sigma(\omega)\right)$$

$$= C_n\|\Theta\|_{L^\infty(S^{n-1})} < \infty. \tag{4.5.16}$$

From (4.5.14) and (4.5.16) it is now clear that the integral in (4.5.6) is absolutely convergent for each fixed $\xi \in \mathbb{R}^n \setminus \{0\}$. This proves that m_Θ is well defined in $\mathbb{R}^n \setminus \{0\}$.

Going further, since $\int_{S^{n-1}} \Theta\,d\sigma = 0$, from (4.5.6) and (4.5.14) we see that, for each $\xi \in \mathbb{R}^n \setminus \{0\}$,

$$m_\Theta(\xi) = -\int_{S^{n-1}}\Theta(\omega)\left(\ln\left|\frac{\xi}{|\xi|} \cdot \omega\right| + i\frac{\pi}{2}\text{sgn}\,(\xi \cdot \omega)\right) d\sigma(\omega). \tag{4.5.17}$$

Having justified this, we see that m_Θ is positive homogeneous of degree zero and that (4.5.8) follows based on (4.5.16). Also, (4.5.10) is obtained directly from (4.5.6) by changing variables $\omega \mapsto -\omega$.

Next, we turn to (4.5.7). Observe that if $\omega \in S^{n-1}$ is arbitrary, then there exists some unitary transformation $\mathcal{R}$ in $\mathbb{R}^n$ such that $\mathcal{R}\omega = \mathbf{e}_1$. This combined with (14.9.11) allows us to write

$$\int_{S^{n-1}} \log(i(\xi \cdot \omega)) \, d\sigma(\xi) = \int_{S^{n-1}} \log(i(\mathcal{R}^{\top}\xi) \cdot \omega) \, d\sigma(\xi)$$

$$= \int_{S^{n-1}} \log(i(\xi \cdot \mathbf{e}_1)) \, d\sigma(\xi) = C \qquad (4.5.18)$$

Hence, (4.5.6), Fubini's Theorem, (4.5.18), and the last condition in (4.4.1), further imply

$$\int_{S^{n-1}} m_{\Theta}(\xi) \, d\sigma(\xi) = -\int_{S^{n-1}} \theta(\omega) \int_{S^{n-1}} \log(i(\xi \cdot \omega)) \, d\sigma(\xi) \, d\sigma(\omega)$$

$$= -C \int_{S^{n-1}} \theta(\omega) \, d\sigma(\omega) = 0. \qquad (4.5.19)$$

This proves (4.5.7).

To show (4.5.11), fix an arbitrary function $\varphi \in S(\mathbb{R}^n)$. Using (4.4.2), Lebesgue's Dominated Convergence Theorem, Fubini's Theorem, and (14.9.7) write

$$\langle \mathcal{F}(\text{P.V.}\,\Theta), \varphi \rangle = \langle \text{P.V.}\,\Theta, \widehat{\varphi} \rangle = \lim_{\varepsilon \to 0^+} \int_{|x| \geq \varepsilon} \Theta(x) \widehat{\varphi}(x) \, dx \qquad (4.5.20)$$

$$= \lim_{\substack{\varepsilon \to 0^+ \\ R \to \infty}} \int_{\varepsilon \leq |x| \leq R} \Theta(x) \widehat{\varphi}(x) \, dx$$

$$= \lim_{\substack{\varepsilon \to 0^+ \\ R \to \infty}} \int_{\mathbb{R}^n} \varphi(\xi) \int_{\varepsilon \leq |x| \leq R} \Theta(x) \, e^{-ix \cdot \xi} \, dx \, d\xi$$

$$= \lim_{\substack{\varepsilon \to 0^+ \\ R \to \infty}} \int_{\mathbb{R}^n} \varphi(\xi) \int_{S^{n-1}} \Theta(\omega) \int_{\varepsilon}^{R} e^{-i(\omega \cdot \xi)\rho} \frac{d\rho}{\rho} \, d\sigma(\omega) \, d\xi$$

$$= \lim_{\substack{\varepsilon \to 0^+ \\ R \to \infty}} \int_{\mathbb{R}^n} \varphi(\xi) \int_{S^{n-1}} \Theta(\omega) \int_{\varepsilon}^{R} [e^{-i(\omega \cdot \xi)\rho} - \cos\rho] \frac{d\rho}{\rho} \, d\sigma(\omega) \, d\xi,$$

where the last equality uses $\int_{S^{n-1}} \Theta(\omega) \left(\int_{\varepsilon}^{R} \frac{\cos\rho}{\rho} d\rho \right) d\sigma(\omega) = 0$ for each $\varepsilon, R > 0$ (itself a consequence of the fact that Θ has mean value zero over S^{n-1}). At this stage, we wish to invoke Lebesgue's Dominated Convergence Theorem in order to absorb the limit inside the integral. To see that this theorem is applicable in the current context, we first note that for each $\xi \in \mathbb{R}^n \setminus \{0\}$ and each $\omega \in S^{n-1}$ such that $\xi \cdot \omega \neq 0$, formulas (4.11.5) and (4.11.6) give

$$\lim_{\substack{\varepsilon \to 0^+ \\ R \to \infty}} \int_\varepsilon^R \left[e^{-i(\omega \cdot \xi)\rho} - \cos\rho \right] \frac{d\rho}{\rho}$$

$$= \lim_{\substack{\varepsilon \to 0^+ \\ R \to \infty}} \left[\int_\varepsilon^R \frac{\cos((\omega \cdot \xi)\rho) - \cos\rho}{\rho} \, d\rho - i \int_\varepsilon^R \frac{\sin((\omega \cdot \xi)\rho)}{\rho} \, dr \right]$$

$$= -\ln|\omega \cdot \xi| - i\frac{\pi}{2} \operatorname{sgn}(\omega \cdot \xi) = -\log(i(\omega \cdot \xi)). \tag{4.5.21}$$

This takes care of the pointwise convergence aspect of Lebesgue's theorem. To verify the uniform domination aspect, based on (4.11.7)–(4.11.8) we first estimate

$$\int_{\mathbb{R}^n} |\varphi(\xi)| \int_{S^{n-1}} |\Theta(\omega)| \left(\sup_{0 < \varepsilon < R} \left| \int_\varepsilon^R \left[e^{-i(\omega \cdot \xi)\rho} - \cos\rho \right] \frac{d\rho}{\rho} \right| \right) d\sigma(\omega) \, d\xi$$

$$\leq \int_{\mathbb{R}^n} |\varphi(\xi)| \int_{S^{n-1}} |\Theta(\omega)| (2|\ln|\omega \cdot \xi|| + 4) \, d\sigma(\omega) \, d\xi$$

$$\leq \|\Theta\|_{L^\infty(S^{n-1})} \times \tag{4.5.22}$$

$$\times \left[4\omega_{n-1} \|\varphi\|_{L^1(\mathbb{R}^n)} + 2 \int_{\mathbb{R}^n} |\varphi(\xi)| \int_{S^{n-1}} |\ln|\omega \cdot \xi|| \, d\sigma(\dot{\omega}) \, d\xi \right],$$

and note that, further,

$$\int_{\mathbb{R}^n} |\varphi(\xi)| \int_{S^{n-1}} |\ln|\omega \cdot \xi|| \, d\sigma(\omega) \, d\xi \leq \int_{\mathbb{R}^n} |\varphi(\xi)| \int_{S^{n-1}} \left| \ln \left| \frac{\xi}{|\xi|} \cdot \omega \right| \right| d\sigma(\omega) \, d\xi$$

$$+ \omega_{n-1} \int_{\mathbb{R}^n} |\varphi(\xi)| |\ln|\xi|| \, d\xi. \tag{4.5.23}$$

From (4.5.22)–(4.5.23), the fact that $\varphi \in S(\mathbb{R}^n)$, and (4.5.15), we may therefore conclude that

$$\int_{\mathbb{R}^n} |\varphi(\xi)| \int_{S^{n-1}} |\Theta(\omega)| \left(\sup_{0 < \varepsilon < R} \left| \int_\varepsilon^R \left[e^{-i(\omega \cdot \xi)\rho} - \cos\rho \right] \frac{d\rho}{\rho} \right| \right) d\sigma(\omega) \, d\xi < \infty. \tag{4.5.24}$$

Having established (4.5.21) and (4.5.24) we may now use Lebesgue's Dominated Convergence Theorem in the context of (4.5.20) to obtain

$$\langle \mathcal{F}(\text{P.V.}\,\Theta), \varphi \rangle = \int_{\mathbb{R}^n} \varphi(\xi) \, m_\Theta(\xi) \, d\xi. \tag{4.5.25}$$

From this, the fact that $m_\Theta \in L^\infty(\mathbb{R}^n)$, and keeping in mind that $\varphi \in S(\mathbb{R}^n)$ was arbitrary, (4.5.11) follows.

In order to show (4.5.12), note that if Θ is assumed to be even, then

$$\int_{S^{n-1}} \overline{\Theta(\omega)} \, \text{sgn} \, (\xi \cdot \omega) \, d\sigma(\omega) = \int_{\substack{\omega \in S^{n-1} \\ \omega \cdot \xi > 0}} \overline{\Theta(\omega)} \, d\sigma(\omega) - \int_{\substack{\omega \in S^{n-1} \\ \omega \cdot \xi < 0}} \overline{\Theta(\omega)} \, d\sigma(\omega)$$

$$= \int_{\substack{\omega \in S^{n-1} \\ \omega \cdot \xi > 0}} \overline{\Theta(\omega)} \, d\sigma(\omega) - \int_{\substack{\omega \in S^{n-1} \\ (-\omega) \cdot \xi < 0}} \overline{\Theta(-\omega)} \, d\sigma(\omega)$$

$$= \int_{\substack{\omega \in S^{n-1} \\ \omega \cdot \xi > 0}} \left[\overline{\Theta(\omega)} - \overline{\Theta(-\omega)} \right] d\sigma(\omega) = 0. \qquad (4.5.26)$$

Consequently,

$$\overline{m_\Theta(\xi)} = - \int_{S^{n-1}} \overline{\Theta(\omega)} \Big(\ln \Big| \tfrac{\xi}{|\xi|} \cdot \omega \Big| - i \tfrac{\pi}{2} \text{sgn} \, (\xi \cdot \omega) \Big) d\sigma(\omega)$$

$$= - \int_{S^{n-1}} \overline{\Theta(\omega)} \Big(\ln \Big| \tfrac{\xi}{|\xi|} \cdot \omega \Big| + i \tfrac{\pi}{2} \text{sgn} \, (\xi \cdot \omega) \Big) d\sigma(\omega)$$

$$+ i\pi \int_{S^{n-1}} \overline{\Theta(\omega)} \, \text{sgn} \, (\xi \cdot \omega) \, d\sigma(\omega)$$

$$= m_{\overline{\Theta}}(\xi), \qquad (4.5.27)$$

proving (4.5.12).

There remains to prove (4.5.13). To this end, suppose $\Theta \in C^k(\mathbb{R}^n \setminus \{0\})$ for some $k \in \mathbb{N}_0 \cup \{\infty\}$. Fix $\ell \in \{1, \dots, n\}$ and let $\mathbf{e}_\ell$ be the unit vector in $\mathbb{R}^n$ with one on the ℓ-th component. Introduce the open set

$$O_\ell := \{\xi = (\xi_1, \dots, \xi_n) \in \mathbb{R}^n : \xi_\ell > 0\}. \qquad (4.5.28)$$

For each $\xi = (\xi_1, \dots, \xi_n) \in O_\ell$ define the linear map $R_{\ell,\xi} : \mathbb{R}^n \to \mathbb{R}^n$ by

$$R_{\ell,\xi}(x) := x - \frac{x \cdot \xi + x_\ell |\xi|}{|\xi| + \xi_\ell} \mathbf{e}_\ell + \frac{x_\ell(|\xi| + 2\xi_\ell) - x \cdot \xi}{|\xi|(|\xi| + \xi_\ell)} \xi, \quad \forall x \in \mathbb{R}^n. \qquad (4.5.29)$$

By Exercise 4.130 (presently used with $\zeta := \mathbf{e}_\ell$ and $\eta := \tfrac{\xi}{|\xi|}$) we have that this is a unitary transformation,

$$\xi \cdot R_{\ell,\xi}(x) = |\xi| x_\ell, \quad \forall \xi \in O_\ell, \quad \forall x \in \mathbb{R}^n, \qquad (4.5.30)$$

and the joint application

$$O_\ell \times \mathbb{R}^n \ni (\xi, x) \mapsto R_{\ell,\xi}(x) \in \mathbb{R}^n \quad \text{is of class } C^\infty. \qquad (4.5.31)$$

Starting with (4.5.17), then using the invariance under unitary transformations of the operation of integration over S^{n-1} (cf. (14.9.11)) and (4.5.30), for each $\xi \in O_\ell$ we may then write

$$m_\Theta(\xi) = -\int_{S^{n-1}} \Theta(\omega)\Big[\ln\Big|\tfrac{\xi}{|\xi|}\cdot\omega\Big| + i\tfrac{\pi}{2}\operatorname{sgn}(\xi\cdot\omega)\Big]d\sigma(\omega)$$

$$= -\int_{S^{n-1}} \Theta(R_{\ell,\xi}(\omega))\Big[\ln\Big|\tfrac{\xi}{|\xi|}\cdot R_{\ell,\xi}(\omega)\Big| + i\tfrac{\pi}{2}\operatorname{sgn}(\xi\cdot R_{\ell,\xi}(\omega))\Big]d\sigma(\omega)$$

$$= -\int_{S^{n-1}} \Theta(R_{\ell,\xi}(\omega))\Big[\ln|\omega_\ell| + i\tfrac{\pi}{2}\operatorname{sgn}(\omega_\ell)\Big]d\sigma(\omega). \tag{4.5.32}$$

In turn, (4.5.32), (4.5.31), and the current assumption $\Theta \in C^k(\mathbb{R}^n \setminus \{0\})$ imply

$$m_\Theta\big|_{O_\ell} \in C^k(O_\ell). \tag{4.5.33}$$

If we also set

$$O_\ell^- := \{x = (x_1,\dots,x_n) \in \mathbb{R}^n : x_\ell < 0\} \tag{4.5.34}$$

and for each given $\xi \in O_\ell^-$ define the linear map $R_{\ell,\xi}^- : \mathbb{R}^n \to \mathbb{R}^n$ by

$$R_{\ell,\xi}^-(x) := x + \frac{x\cdot\xi - x_\ell|\xi|}{|\xi| - \xi_\ell}\,\mathbf{e}_\ell - \frac{x_\ell(|\xi| - 2\xi_\ell) + x\cdot\xi}{|\xi|(|\xi| - \xi_\ell)}\,\xi, \quad \forall x \in \mathbb{R}^n, \tag{4.5.35}$$

running the reasoning that yielded (4.5.33) this time with O_ℓ replaced by O_ℓ^- and $R_{\ell,\xi}$ replaced by $R_{\ell,\xi}^-$ (now invoking Exercise 4.130 with $\zeta := \mathbf{e}_\ell$ and $\eta := -\tfrac{\xi}{|\xi|}$), and having identity (4.5.30) replaced by $\xi\cdot R_{\ell,\xi}^-(x) = -|\xi|x_\ell$ for all $\xi \in O_\ell^-$ and all $x \in \mathbb{R}^n$, ultimately implies

$$m_\Theta\big|_{O_\ell^-} \in C^k(O_\ell^-). \tag{4.5.36}$$

Upon observing that $\mathbb{R}^n \setminus \{0\} = \bigcup_{\ell=1}^{n}(O_\ell \cup O_\ell^-)$, from (4.5.33) and (4.5.36), we conclude that $m_\Theta\big|_{\mathbb{R}^n\setminus\{0\}} \in C^k(\mathbb{R}^n \setminus \{0\})$, as wanted. This finishes the proof of the theorem. $\square$

Exercise 4.75. Complete the following outline aimed at extending the convolution product to the class of principal value distributions in $\mathbb{R}^n$. Assume that Θ_1, Θ_2 are two given functions as in (4.4.1).

Step 1. Pick an arbitrary function $\psi \in C_0^\infty(\mathbb{R}^n)$ with the property that $\psi \equiv 1$ near the origin, and show that the following convolutions are meaningfully defined in $\mathcal{S}'(\mathbb{R}^n)$:

$$u_{00} := \big(\psi\,\mathrm{P.V.}\,\Theta_1\big) * \big(\psi\,\mathrm{P.V.}\,\Theta_2\big)$$

$$u_{01} := \big(\psi\,\mathrm{P.V.}\,\Theta_1\big) * \big((1-\psi)\,\mathrm{P.V.}\,\Theta_2\big)$$

$$u_{10} := \big((1-\psi)\,\mathrm{P.V.}\,\Theta_1\big) * \big(\psi\,\mathrm{P.V.}\,\Theta_2\big)$$

$$u_{11} := \big((1-\psi)\,\mathrm{P.V.}\,\Theta_1\big) * \big((1-\psi)\,\mathrm{P.V.}\,\Theta_2\big). \tag{4.5.37}$$

For u_{00}, u_{01}, and u_{10}, use Proposition 4.66 and part *(a)* of Theorem 4.19. Show that $u_{11} = f_1 * f_2$ where $f_j := (1-\psi)\Theta_j$, $j = 1, 2$, are functions belonging to $L^2(\mathbb{R}^n)$ (here the behavior of Θ_1, Θ_2 at infinity is relevant). Use Young's inequality to conclude that u_{11} is meaningfully defined in $L^\infty(\mathbb{R}^n)$.

Step 2. With the same cutoff function ψ as in Step 1, define

$$\left(\text{P.V.}\,\Theta_1\right) * \left(\text{P.V.}\,\Theta_2\right) := u_{00} + u_{01} + u_{10} + u_{11} \quad \text{in} \quad S'(\mathbb{R}^n), \tag{4.5.38}$$

and use this definition to show that

$$\mathcal{F}\left[\left(\text{P.V.}\,\Theta_1\right) * \left(\text{P.V.}\,\Theta_2\right)\right] = m_{\Theta_1} m_{\Theta_2} \quad \text{in} \quad S'(\mathbb{R}^n), \tag{4.5.39}$$

where $m_{\Theta_1}, m_{\Theta_2}$ are associated with Θ_1, Θ_2 as in (4.5.6). To do so, compute first $\widehat{u_{jk}}$ for $j, k \in \{0, 1\}$, using part *(c)* in Theorem 4.35, Proposition 4.29, and Theorem 4.74.

Step 3. Use (4.5.39) to show that the definition in (4.5.38) is independent of the cutoff function ψ chosen at the beginning.

As an application, show that

$$\left(\text{P.V.}\,\frac{1}{x}\right) * \left(\text{P.V.}\,\frac{1}{x}\right) = -\pi^2 \delta \quad \text{in} \quad S'(\mathbb{R}). \tag{4.5.40}$$

4.6 Tempered Distributions Associated with $|x|^{-n}$

Let us consider the effect of dropping the cancelation condition in (4.4.1). Concretely, given a function $\Psi \in C^0(\mathbb{R}^n \setminus \{0\})$ that is positive homogeneous of degree $-n$, if one sets $C := \int_{S^{n-1}} \Psi \, d\sigma$, then $\Psi(x) = \Psi_0(x) + C|x|^{-n}$ for every x in $\mathbb{R}^n \setminus \{0\}$, where Ψ_0 satisfies (4.4.1).

From Section 4.4 we know that one may associate to Ψ_0 the tempered distribution P.V. Ψ_0. As such, associating a tempered distribution to the original function Ψ hinges on how to meaningfully associate a tempered distribution to $\frac{1}{|x|^n}$. In fact, we shall associate to $\frac{1}{|x|^n}$ a family of tempered distributions in the manner described below.

Recall the class of functions Q from (4.4.4), fix $\psi \in Q$, and define

$$w_\psi(\varphi) := \int_{\mathbb{R}^n} \frac{1}{|x|^n} [\varphi(x) - \varphi(0)\psi(x)] \, dx, \quad \forall \varphi \in S(\mathbb{R}^n). \tag{4.6.1}$$

This gives rise to a well-defined, linear mapping $w_\psi : S(\mathbb{R}^n) \to \mathbb{C}$. In addition, much as in the proof of (4.4.7), there exists a constant $C \in [0, \infty)$ with the property that

$$|w_\psi(\varphi)| \leq C \sup_{x \in \mathbb{R}^n, |\alpha| \leq 1, |\beta| \leq 1} |x^\alpha \partial^\beta \varphi(x)|, \quad \forall \varphi \in S(\mathbb{R}^n). \tag{4.6.2}$$

In light of Fact 4.1 this shows that w_ψ is a tempered distribution. An inspection of (4.6.1) also reveals that $w_\psi\big|_{\mathbb{R}^n \setminus \{0\}} = |x|^{-n}$ in $\mathcal{D}'(\mathbb{R}^n \setminus \{0\})$. This is the sense in which we shall say that we have associated to $|x|^{-n}$ the family of tempered distributions $\{w_\psi\}_{\psi \in Q}$.

Our next goal is to determine a formula for the Fourier transform of the tempered distribution in (4.6.1). As a preamble, note that the class Q is invariant under dilations (recall (4.3.5)); that is, for every $t \in (0, \infty)$ we have

$$\tau_t(\psi) \in Q, \qquad \forall\, t \in (0, \infty), \quad \forall\, \psi \in Q. \tag{4.6.3}$$

Theorem 4.76. *For each $\psi \in Q$, the Fourier transform of the tempered distribution w_ψ defined in (4.6.1) is of function type and*

$$\widehat{w_\psi}(\xi) = -\omega_{n-1} \ln |\xi| + C_\psi, \quad \forall\, \xi \in \mathbb{R}^n \setminus \{0\}, \tag{4.6.4}$$

for some constant $C_\psi \in \mathbb{C}$.

Proof. Fix $\psi \in Q$. As noted earlier, $w_\psi\big|_{\mathbb{R}^n \setminus \{0\}} = \frac{1}{|x|^n}$, hence Proposition 4.60 applies and yields

$$\widehat{w_\psi}\big|_{\mathbb{R}^n \setminus \{0\}} \in C^\infty(\mathbb{R}^n \setminus \{0\}). \tag{4.6.5}$$

To establish (4.6.4) we proceed by proving a series of claims.

Claim 1. $\widehat{w_\psi}$ *is invariant under orthogonal transformations.* let $A \in M_{n \times n}(\mathbb{R})$ be an orthogonal matrix. Then

$$\langle w_\psi \circ A, \varphi \rangle = \langle w_\psi, \varphi \circ A^{-1} \rangle = \int_{\mathbb{R}^n} \frac{1}{|x|^n} [\varphi(A^{-1}x) - \varphi(0)\psi(A^{-1}x)]\, dx$$

$$= \langle w_\psi, \varphi \rangle, \qquad \forall\, \varphi \in \mathcal{S}(\mathbb{R}^n). \tag{4.6.6}$$

This shows that w_ψ is invariant under orthogonal transformations which, together with Proposition 4.49, implies that $\widehat{w_\psi}$ is also invariant under orthogonal transformations.

Claim 2. $\tau_t(w_\psi) = t^{-n} w_{\tau_t \psi}$ *in* $\mathcal{S}'(\mathbb{R}^n)$. To see why this is true, for any $t \in (0, \infty)$ and any $\varphi \in \mathcal{S}(\mathbb{R}^n)$ write

$$\langle \tau_t w_\psi, \varphi \rangle = t^{-n} \langle w_\psi, \tau_{\frac{1}{t}} \varphi \rangle = t^{-n} \int_{\mathbb{R}^n} \frac{1}{|x|^n} \left[\varphi\left(\frac{x}{t}\right) - \varphi(0)\psi(x) \right] dx$$

$$= \int_{\mathbb{R}^n} \frac{1}{t^n |y|^n} [\varphi(y) - \varphi(0)\psi(ty)]\, dy = t^{-n} \langle w_{\tau_t \psi}, \varphi \rangle, \tag{4.6.7}$$

from which the desired identity follows.

Claim 3. *For every $t \in (0, \infty)$ we have*

$$\tau_t w_\psi - t^{-n} w_\psi = \left(t^{-n} \int_{\mathbb{R}^n} \frac{\psi(x) - \psi(tx)}{|x|^n}\, dx \right) \delta \quad \text{in } \mathcal{S}'(\mathbb{R}^n). \tag{4.6.8}$$

To prove the current claim, consider $\psi_1, \psi_2 \in Q$ and write

$$\langle w_{\psi_1} - w_{\psi_2}, \varphi \rangle = \int_{\mathbb{R}^n} \frac{\varphi(0)}{|x|^n} [\psi_2(x) - \psi_1(x)] \, dx$$

$$= \left\langle \left(\int_{\mathbb{R}^n} \frac{\psi_2(x) - \psi_1(x)}{|x|^n} \, dx \right) \delta, \varphi \right\rangle \qquad \forall \varphi \in \mathcal{S}(\mathbb{R}^n). \qquad (4.6.9)$$

As a result,

$$w_{\psi_1} - w_{\psi_2} = \left(\int_{\mathbb{R}^n} \frac{1}{|x|^n} [\psi_2(x) - \psi_1(x)] \, dx \right) \delta \quad \text{in } \mathcal{S}'(\mathbb{R}^n). \qquad (4.6.10)$$

Combining now the identity from Claim 2 and (4.6.10) yields

$$\tau_t w_\psi - t^{-n} w_\psi = t^{-n} [w_{\tau_t \psi} - w_\psi]$$

$$= \left(t^{-n} \int_{\mathbb{R}^n} \frac{\psi(x) - \psi(tx)}{|x|^n} \, dx \right) \delta \quad \text{in } \mathcal{S}'(\mathbb{R}^n) \qquad (4.6.11)$$

for every $t \in (0, \infty)$. This proves Claim 3.

Claim 4. *The following identity is true:*

$$\int_{\mathbb{R}^n} \frac{\psi(x) - \psi(tx)}{|x|^n} \, dx = \omega_{n-1} \ln t \qquad \forall t \in (0, \infty). \qquad (4.6.12)$$

To set the stage for the proof of Claim 4 observe that there exists some C^1 function $\eta : [0, \infty) \to \mathbb{R}$ decaying at ∞, satisfying $\eta(0) = 1$, and $\psi(x) = \eta(|x|)$ for every $x \in \mathbb{R}^n$. In particular, $(\nabla \psi)(x) = \eta'(|x|) \frac{x}{|x|}$ for every $x \in \mathbb{R}^n \setminus \{0\}$. Focusing attention on (4.6.12), note that the intervening integral is absolutely convergent. In turn, this allows us to write for each fixed $t \in (0, \infty)$

$$\int_{\mathbb{R}^n} \frac{\psi(x) - \psi(tx)}{|x|^n} \, dx = \lim_{R \to \infty} \int_{|x| < R} \frac{1}{|x|^n} \left(\int_t^1 \frac{d}{ds} [\psi(sx)] \, ds \right) dx$$

$$= \lim_{R \to \infty} \int_{|x| < R} \frac{1}{|x|^{n-1}} \left(\int_t^1 \eta'(s|x|) \, ds \right) dx$$

$$= \omega_{n-1} \lim_{R \to \infty} \int_0^R \left(\int_t^1 \eta'(s\rho) \, ds \right) d\rho \qquad (4.6.13)$$

$$= \omega_{n-1} \lim_{R \to \infty} \int_t^1 \left(\int_0^R \frac{1}{s} \frac{d}{d\rho} [\eta(s\rho)] \, d\rho \right) ds$$

$$= \omega_{n-1} \lim_{R \to \infty} \int_t^1 \frac{1}{s} [\eta(sR) - \eta(0)] \, ds = \omega_{n-1} \ln t,$$

as wanted.

Claim 5. *The following identity holds:*

$$\tau_t \widehat{w_\psi} = \widehat{w_\psi} - \omega_{n-1} \ln t \quad \text{in } \mathbb{R}^n \setminus \{0\}, \quad \forall t \in (0, \infty). \qquad (4.6.14)$$

To prove (4.6.14), we first combine (4.6.8) and (4.6.12) and obtain

$$\tau_t w_\psi = t^{-n} w_\psi + (\omega_{n-1} t^{-n} \ln t)\delta \quad \text{in} \quad S'(\mathbb{R}^n), \quad \forall t \in (0, \infty). \tag{4.6.15}$$

Hence, for each $t \in (0, \infty)$ and each $\varphi \in S(\mathbb{R}^n)$, we have

$$\begin{aligned}
\langle \tau_t \widehat{w_\psi}, \varphi \rangle &= t^{-n} \left\langle \widehat{w_\psi}, \tau_{\frac{1}{t}} \varphi \right\rangle = \langle w_\psi, \tau_t \widehat{\varphi} \rangle = t^{-n} \langle \tau_{\frac{1}{t}} w_\psi, \widehat{\varphi} \rangle \\
&= t^{-n} \langle t^n w_\psi - (\omega_{n-1} t^n \ln t)\delta, \widehat{\varphi} \rangle = \langle w_\psi - (\omega_{n-1} \ln t)\delta, \widehat{\varphi} \rangle \\
&= \langle \widehat{w_\psi} - \omega_{n-1} \ln t, \varphi \rangle.
\end{aligned} \tag{4.6.16}$$

The first and third equality in (4.6.16) is based on the definition of τ_t acting on $S'(\mathbb{R}^n)$, the second uses (4.3.7), while the fourth makes use of (4.6.15). Now (4.6.14) follows from (4.6.16) and the fact that $\widehat{w_\psi}\big|_{\mathbb{R}^n \setminus \{0\}} \in C^\infty(\mathbb{R}^n \setminus \{0\})$.

We are ready to complete the proof of Theorem 4.76. First, Claim 1 ensures that $\widehat{w_\psi}\big|_{\mathbb{R}^n \setminus \{0\}}$ is constant on S^{n-1}, thus

$$C_\psi := \widehat{w_\psi}\big|_{S^{n-1}} \tag{4.6.17}$$

is a well-defined complex number. Hence, given any $\xi \in \mathbb{R}^n \setminus \{0\}$, taking $t := \frac{1}{|\xi|}$ in (4.6.14) yields

$$C_\psi = \widehat{w_\psi}\left(\frac{\xi}{|\xi|}\right) = \tau_{\frac{1}{|\xi|}} \widehat{w_\psi}(\xi) = \widehat{w_\psi}(\xi) + \omega_{n-1} \ln|\xi|, \tag{4.6.18}$$

from which (4.6.4) follows. $\qquad\square$

Remark 4.77. It is possible to enlarge the scope of our earlier considerations by relaxing the hypotheses on the function ψ. Specifically, in place of (4.4.3) assume now that $\psi : \mathbb{R}^n \to \mathbb{C}$ is such that

$$\begin{aligned}
&\psi \text{ is radial, of class } C^1 \text{ near origin, satisfies } \psi(0) = 1, \\
&\text{and } \exists \varepsilon_o, C \in (0, \infty) \text{ such that } |\psi(x)| \leq C(1 + |x|)^{-\varepsilon_o}.
\end{aligned} \tag{4.6.19}$$

Such a function ψ still induces a tempered distribution as in (4.6.1), and we claim that formula (4.6.4) continues to be valid in this more general case as well.

To justify this claim, observe that the proof of Theorem 4.76 goes through for ψ as in (4.6.19), albeit (4.6.12) requires more care, since we are now dropping the global C^1 requirement on η. However, this may be remedied by working with η_ε in place of η, where η_ε is defined via mollification as follows. If $\{\phi_\varepsilon\}$ is the sequence from (1.2.3)–(1.2.5), then

$$\eta_\varepsilon(x) := \int_{\mathbb{R}^n} \eta(y)\phi_\varepsilon(x - y)\, dy, \quad \forall x \in \mathbb{R}^n. \tag{4.6.20}$$

Now observe that if $\psi_\varepsilon(x) := \eta_\varepsilon(|x|)$ then $\psi_\varepsilon/\psi_\varepsilon(0) \in Q$, hence

$$\int_{\mathbb{R}^n} \frac{\psi_\varepsilon(x) - \psi_\varepsilon(tx)}{|x|^n}\,\mathrm{d}x = \omega_{n-1}\psi_\varepsilon(0)\ln t \qquad \forall\, t \in (0, \infty), \tag{4.6.21}$$

and finally passing to limit $\varepsilon \to 0^+$ in (4.6.21).

Remark 4.77 applies, in particular, to the function $\psi := \chi_{B(0,1)}$. This gives that

$$\left\langle w_{\chi_{B(0,1)}}, \varphi \right\rangle = \int_{|x|>1} \frac{\varphi(x)}{|x|^n}\,\mathrm{d}x + \int_{|x|\le 1} \frac{\varphi(x) - \varphi(0)}{|x|^n}\,\mathrm{d}x, \quad \forall\, \varphi \in \mathcal{S}(\mathbb{R}^n), \tag{4.6.22}$$

is a tempered distribution, $\mathcal{F}(w_{\chi_{B(0,1)}}) \in \mathcal{S}'(\mathbb{R}^n)$ is of function type, and there exists a constant $c \in \mathbb{C}$ such that

$$\mathcal{F}(w_{\chi_{B(0,1)}})(\xi) = -\omega_{n-1}\ln|\xi| + c, \quad \forall\, \xi \in \mathbb{R}^n \setminus \{0\}. \tag{4.6.23}$$

In the last part of this section we compute the constant c from (4.6.23) in the case $n = 1$. Before doing so we recall that Euler's constant γ is defined by

$$\gamma := \lim_{k\to\infty}\Big[\sum_{j=1}^{k}\frac{1}{j} - \ln k\Big]. \tag{4.6.24}$$

The number γ plays an important role in analysis, as it appears prominently in a number of basic formulas. Two such identities which are relevant for us here are as follows:

$$\int_0^\infty \mathrm{e}^{-x}\ln x\,\mathrm{d}x = -\gamma, \qquad \int_0^\infty \mathrm{e}^{-x^2}\ln x\,\mathrm{d}x = -\frac{\sqrt{\pi}}{4}(\gamma + 2\ln 2). \tag{4.6.25}$$

Example 4.78. We claim that

$$\mathcal{F}(w_{\chi_{(-1,1)}})(\xi) = -2\ln|\xi| - 2\gamma \quad \text{in } \mathcal{S}'(\mathbb{R}), \tag{4.6.26}$$

where $w_{\chi_{(-1,1)}}$ is defined in (4.6.22). To see why this is true, note that (4.6.23) written for $n = 1$ (in which case $\omega_{n-1} = 2$) becomes

$$\mathcal{F}(w_{\chi_{(-1,1)}})(x) = -2\ln|x| + c \quad \text{in } \mathcal{S}'(\mathbb{R}). \tag{4.6.27}$$

Consequently, to obtain (4.6.26) it remains to prove that $c = -2\gamma$ when $n = 1$.

The idea for determining the value of c is to apply $\mathcal{F}(w_{\chi_{(-1,1)}})$ to the Schwartz function e^{-x^2}. First, from (4.6.23) we have

$$\left\langle \mathcal{F}(w_{\chi_{(-1,1)}}), \mathrm{e}^{-x^2} \right\rangle = \left\langle -2\ln|x| + c, \mathrm{e}^{-x^2} \right\rangle. \tag{4.6.28}$$

Second, we compute the term in the left-hand side of (4.6.28). Specifically, based on the definition of the Fourier transform of a tempered distribution, (3.2.6), and (4.6.22) we may write

$$\left\langle \mathcal{F}(w_{\chi(-1,1)}), \mathrm{e}^{-x^2} \right\rangle = \left\langle w_{\chi(-1,1)}, \mathcal{F}(\mathrm{e}^{-x^2}) \right\rangle = \left\langle w_{\chi(-1,1)}, \sqrt{\pi}\mathrm{e}^{-x^2/4} \right\rangle$$

$$= \sqrt{\pi} \int_{|x|>1} \frac{\mathrm{e}^{-x^2/4}}{|x|}\, \mathrm{d}x + \sqrt{\pi} \int_{|x|\le 1} \frac{\mathrm{e}^{-x^2/4} - 1}{|x|}\, \mathrm{d}x$$

$$= 2\sqrt{\pi} \int_{1/2}^{\infty} \frac{\mathrm{e}^{-x^2}}{x}\, \mathrm{d}x + 2\sqrt{\pi} \int_{0}^{1/2} \frac{\mathrm{e}^{-x^2} - 1}{x}\, \mathrm{d}x$$

$$=: I + II. \tag{4.6.29}$$

Going further, integrating by parts then changing variables gives

$$I = 2\sqrt{\pi}\mathrm{e}^{-1/4} \ln 2 + 4\sqrt{\pi} \int_{1/2}^{\infty} (\ln x)\mathrm{e}^{-x^2} x\, \mathrm{d}x$$

$$= 2\sqrt{\pi}\mathrm{e}^{-1/4} \ln 2 + \sqrt{\pi} \int_{1/\sqrt{2}}^{\infty} (\ln x)\mathrm{e}^{-x}\, \mathrm{d}x. \tag{4.6.30}$$

Regarding II, for each $\varepsilon > 0$ an integration by parts and then a change of variables yield

$$\int_{\varepsilon}^{1/2} \frac{\mathrm{e}^{-x^2} - 1}{x}\, \mathrm{d}x \tag{4.6.31}$$

$$= -(\mathrm{e}^{-1/4} - 1)\ln 2 - (\mathrm{e}^{-\varepsilon^2} - 1)\ln \varepsilon + 2 \int_{\varepsilon}^{1/2} (\ln x)\mathrm{e}^{-x^2} x\, \mathrm{d}x$$

$$= -(\mathrm{e}^{-1/4} - 1)\ln 2 - (\mathrm{e}^{-\varepsilon^2} - 1)\ln \varepsilon + \frac{1}{2} \int_{\sqrt{\varepsilon}}^{1/\sqrt{2}} (\ln x)\mathrm{e}^{-x}\, \mathrm{d}x.$$

Observe that since

$$\lim_{\varepsilon \to 0^+} (\mathrm{e}^{-\varepsilon^2} - 1)\ln \varepsilon = \frac{1}{2} \lim_{y \to \infty} \frac{\ln(\ln y - \ln(y-1))}{y} = 0, \tag{4.6.32}$$

by taking the limit as $\varepsilon \to 0^+$ in (4.6.31) we may conclude that

$$II = -2\sqrt{\pi}(\mathrm{e}^{-1/4} - 1)\ln 2 + \sqrt{\pi} \int_{0}^{1/\sqrt{2}} (\ln x)\mathrm{e}^{-x}\, \mathrm{d}x. \tag{4.6.33}$$

Hence, from (4.6.30), (4.6.33), and the first identity in (4.6.25), it follows that

$$I + II = 2\sqrt{\pi}\ln 2 + \sqrt{\pi} \int_{0}^{\infty} (\ln x)\mathrm{e}^{-x}\, \mathrm{d}x = 2\sqrt{\pi}\ln 2 - \sqrt{\pi}\gamma. \tag{4.6.34}$$

This takes care of the term in the left-hand side of (4.6.26). To compute the term in the right-hand side of (4.6.26), write

$$\langle -2\ln|\xi| + c, e^{-x^2} \rangle = -2 \int_{\mathbb{R}} (\ln|x|) e^{-x^2} dx + c \int_{\mathbb{R}} e^{-x^2} dx$$

$$= -4 \int_0^\infty (\ln x) e^{-x^2} dx + c \sqrt{\pi}$$

$$= \sqrt{\pi}(\gamma + 2\ln 2) + c \sqrt{\pi}. \tag{4.6.35}$$

For the last equality in (4.6.35) we used the second identity in (4.6.25). A combination of (4.6.28), (4.6.29), (4.6.34), and (4.6.35), then implies $c = -2\gamma$, as desired.

4.7 A General Jump-Formula in the Class of Tempered Distributions

The aim in this section is to prove a very useful formula expressing the limits of certain sequences of tempered distributions $\{\Phi_\varepsilon\}_{\varepsilon \neq 0}$ as the index parameter $\varepsilon \in \mathbb{R} \setminus \{0\}$ approaches the origin from either side. The trademark feature (which also justifies the name) of this formula is the presence of a jump-term of the form $\pm C\delta$ (where the sign is correlated to $\operatorname{sgn} \varepsilon$) in addition to a suitable principal value tempered distribution. A conceptually simple example of this phenomenon has been presented in Exercise 2.135 which our theorem contains as a simple special case (see Remark 4.82).

With the notational convention that for points $x \in \mathbb{R}^n$ we write $x = (x', t)$, where $x' = (x_1, \ldots, x_{n-1}) \in \mathbb{R}^{n-1}$ and $t \in \mathbb{R}$, our main result in this regard reads as follows.

Theorem 4.79. *If $\Phi \in C^4(\mathbb{R}^n \setminus \{0\})$ is odd and positive homogeneous of degree $1 - n$, then*

$$\lim_{\varepsilon \to 0^\pm} \Phi(x', \varepsilon) = \pm \frac{i}{2} \widehat{\Phi}(0', 1) \delta(x') + \text{P.V.}\, \Phi(x', 0) \quad in \ \ \mathcal{S}'(\mathbb{R}^{n-1}). \tag{4.7.1}$$

A few comments before presenting the proof of this result are in order. First, above we have employed the earlier convention of writing $u(x')$ for a distribution u in $\mathbb{R}^{n-1}$ simply to stress that the test functions to which u is applied are considered in the variable $x' \in \mathbb{R}^{n-1}$.

Second, for each fixed $\varepsilon \in \mathbb{R} \setminus \{0\}$, applying Exercise 4.53 yields

$$|\Phi(x', \varepsilon)| \leq \|\Phi\|_{L^\infty(S^{n-1})} (|x'|^2 + \varepsilon^2)^{-(n-1)/2} \quad \text{for each} \quad x' \in \mathbb{R}^{n-1}. \tag{4.7.2}$$

Having observed this, the discussion in Example 4.4 then shows that $\Phi(\cdot, \varepsilon)$ belongs to $\mathcal{S}'(\mathbb{R}^{n-1})$.

Third, it is worth recalling an earlier convention to the effect that given a family of tempered distributions u_ε, indexed by $\varepsilon \in I$, $I = (a, b) \subseteq \mathbb{R}$ open interval, we say that $u_\varepsilon \to u \in \mathcal{D}'(\mathbb{R}^n)$ in $\mathcal{S}'(\mathbb{R}^n)$ as $I \ni \varepsilon \to a^+$ provided $u_{\varepsilon_j} \to u$ in $\mathcal{S}'(\mathbb{R}^n)$ for every sequence $\{\varepsilon_j\}_{j \in \mathbb{N}} \subseteq I$ such that $\varepsilon_j \to a^+$ as $j \to \infty$. In particular, this interpretation is in effect for (4.7.1).

Fourth, from Exercise 4.54 we know that $\Phi \in \mathcal{S}'(\mathbb{R}^n)$. In addition, Exercise 4.62 applied to Φ shows that $\widehat{\Phi}\big|_{\mathbb{R}^n \setminus \{0\}} \in \mathcal{C}^1(\mathbb{R}^n \setminus \{0\})$. In particular, $\widehat{\Phi}(0', 1)$ is meaningfully defined.

Fifth, $\Phi(\cdot, 0)$ (viewed as a function in $\mathbb{R}^{n-1} \setminus \{0'\}$) is continuous and positive homogeneous of degree $-(n-1)$ in $\mathbb{R}^{n-1} \setminus \{0'\}$ and, being odd, satisfies $\int_{S^{n-2}} \Phi(\cdot, 0) \, d\sigma = 0$. As such, the conditions in (4.4.1) are satisfied (with $n - 1$ in place of n); hence, Proposition 4.66 ensures that P.V. $\Phi(\cdot, 0)$ is a well-defined tempered distribution in $\mathbb{R}^{n-1}$.

Proof of Theorem 4.79. Assume $\Phi \in \mathcal{C}^4(\mathbb{R}^n \setminus \{0\})$ is odd and positive homogeneous of degree $1 - n$, and let $\varphi \in \mathcal{S}(\mathbb{R}^{n-1})$. Then, for any fixed $\varepsilon > 0$, write

$$
\lim_{t \to 0^+} \int_{\mathbb{R}^{n-1}} \Phi(x', t) \varphi(x') \, dx' = \lim_{t \to 0^+} \int_{|x'| > \varepsilon} \Phi(x', t) \varphi(x') \, dx'
$$

$$
+ \lim_{t \to 0^+} \int_{|x'| < \varepsilon} \Phi(x', t) [\varphi(x') - \varphi(0')] \, dx'
$$

$$
+ \varphi(0') \lim_{t \to 0^+} \int_{|x'| < \varepsilon} \Phi(x', t) \, dx'
$$

$$
=: I_\varepsilon + II_\varepsilon + III_\varepsilon. \tag{4.7.3}
$$

Note that, making the change of variable $y' := x'/t$, we obtain

$$
III_\varepsilon = \varphi(0') \lim_{t \to 0^+} \int_{|y'| < \varepsilon/t} \Phi(y', 1) \, dy' = \varphi(0') \lim_{r \to \infty} \int_{|y'| < r} \Phi(y', 1) \, dy', \tag{4.7.4}
$$

that is independent of ε. Also, since for every $x' \in \mathbb{R}^{n-1}$ and $t \in \mathbb{R} \setminus \{0\}$,

$$
|\Phi(x', t)| \le \frac{\|\Phi\|_{L^\infty(S^{n-1})}}{|(x', t)|^{n-1}} \le \frac{\|\Phi\|_{L^\infty(S^{n-1})}}{|x'|^{n-1}}, \tag{4.7.5}
$$

$$
|\varphi(x') - \varphi(0')| \le \|\nabla \varphi\|_{L^\infty(\mathbb{R}^{n-1})} |x'|, \tag{4.7.6}
$$

it follows that $|II_\varepsilon| \le C\varepsilon$, hence

$$
\lim_{\varepsilon \to 0^+} II_\varepsilon = 0. \tag{4.7.7}
$$

Finally, it is clear from Lebesgue's Dominated Convergence Theorem (cf. Theorem 14.15) that

$$
I_\varepsilon = \int_{|x'| > \varepsilon} \Phi(x', 0) \varphi(x') \, dx'. \tag{4.7.8}
$$

By passing to the limit as $\varepsilon \to 0^+$ we therefore conclude that for any function $\varphi \in \mathcal{S}(\mathbb{R}^{n-1})$

$$\lim_{t \to 0^+} \int_{\mathbb{R}^{n-1}} \Phi(x', t)\varphi(x')\, dx' = \lim_{\varepsilon \to 0^+} \int_{|x'|>\varepsilon} \Phi(x', 0)\varphi(x')\, dx'$$

$$+ \varphi(0') \lim_{r \to \infty} \int_{|x'|<r} \Phi(x', 1)\, dx'. \qquad (4.7.9)$$

To proceed, make the change of variable $x' \mapsto -x'$ in each integral of (4.7.9) and use the fact that Φ is odd. After re-denoting t by $-t$ and $\varphi^\vee$ by φ this yields the identity

$$\lim_{t \to 0^-} \int_{\mathbb{R}^{n-1}} \Phi(x', t)\varphi(x')\, dx' = \lim_{\varepsilon \to 0^+} \int_{|x'|>\varepsilon} \Phi(x', 0)\varphi(x')\, dx'$$

$$+ \varphi(0') \lim_{r \to \infty} \int_{|x'|<r} \Phi(x', -1)\, dx', \qquad (4.7.10)$$

for any $\varphi \in \mathcal{S}(\mathbb{R}^{n-1})$. Collectively, (4.7.9) and (4.7.10) may then be written as

$$\lim_{t \to 0^\pm} \int_{\mathbb{R}^{n-1}} \Phi(x', t)\varphi(x')\, dx' \qquad (4.7.11)$$

$$= a_\pm \varphi(0') + \lim_{\varepsilon \to 0^+} \int_{|x'|>\varepsilon} \Phi(x', 0)\varphi(x')\, dx', \quad \forall \varphi \in \mathcal{S}(\mathbb{R}^{n-1}),$$

where

$$a_\pm := \lim_{r \to \infty} \int_{|x'|<r} \Phi(x', \pm1)\, dx'. \qquad (4.7.12)$$

Regarding (4.7.12), we note that the limits

$$\lim_{r \to \infty} \int_{|x'|<r} \Phi(x', \pm1)\, dx' \quad \text{exist in } \mathbb{C}. \qquad (4.7.13)$$

To see that this is the case, first observe that by Exercise 4.53,

$$\int_{|x'|<r} |\Phi(x', t)|\, dx' < \infty, \quad \forall t \neq 0, \ \forall r > 0. \qquad (4.7.14)$$

Consider the case of the choice of the sign "plus" in (4.7.13) (the case of the sing "minus" is treated analogously). For this choice of sign we write for each fixed $r > 0$

$$\int_{|x'|<r} \Phi(x', 1)\, dx' = \frac{1}{2} \int_{|x'|<r} \Phi(x', 1)\, dx' + \frac{1}{2} \int_{|x'|<r} \Phi(x', 1)\, dx'$$

$$= \frac{1}{2} \int_{|x'|<r} \Phi(x', 1)\, dx' - \frac{1}{2} \int_{|x'|<r} \Phi(x', -1)\, dx'$$

$$= \int_{|x'|<r} \tfrac{1}{2}[\Phi(x', 1) - \Phi(x', -1)]\, dx', \qquad (4.7.15)$$

where we have used the fact that Φ is odd. Next, the Mean Value Theorem and the fact that $\nabla\Phi$ is positive homogeneous of degree $-n$ allow us to estimate

$$|\Phi(x', 1) - \Phi(x', -1)| \le \frac{2\|\nabla\Phi|_{L^\infty(S^{n-1})}}{|x'|^n} \quad \text{for } |x'| \text{ large,} \qquad (4.7.16)$$

which implies that

$$\int_{\mathbb{R}^{n-1}} |\Phi(x', 1)\, dx' - \Phi(x', -1)|\, dx' < \infty. \qquad (4.7.17)$$

Consequently, by Lebesgue's Dominated Convergence Theorem

$$\lim_{r\to\infty} \int_{|x'|<r} \Phi(x', 1)\, dx' = \int_{\mathbb{R}^{n-1}} \tfrac{1}{2}[\Phi(x', 1)\, dx' - \Phi(x', -1)]\, dx' \in \mathbb{C}, \qquad (4.7.18)$$

proving (4.7.13).

There remains to identify the actual values of $a_\pm$ and we organize the remainder of the proof as a series of claims, starting with:

Claim 1: If $u_0 \in \mathcal{E}'(\mathbb{R}^n)$ and $u_1 \in L^1(\mathbb{R}^n)$, then

$$u := u_0 + u_1 \quad \text{has the property that } \widehat{u} \in C^0(\mathbb{R}^n). \qquad (4.7.19)$$

Proof of Claim 1. First note that $\widehat{u} = \widehat{u_0} + \widehat{u_1}$. Since $\widehat{u_0} \in C^\infty(\mathbb{R}^n)$ by part *(b)* in Theorem 4.35 and $\widehat{u_1} \in C^0(\mathbb{R}^n)$ by Lemma 4.28, it follows that $\widehat{u} \in C^0(\mathbb{R}^n)$.

Claim 2: Assume that $\xi_n \ne 0$ and that $\xi' \in \mathbb{R}^{n-1}$ is such that $|\xi'| \le C'$ for some $C' \in (0, \infty)$. Also, suppose that $\Theta \in C^1(\mathbb{R}^n \setminus \{0\})$ satisfies

$$\Theta(\lambda x) = \lambda^{-1}\Theta(x), \qquad \forall\, x \in \mathbb{R}^n \setminus \{0\}, \quad \forall\, \lambda \in \mathbb{R} \setminus \{0\}. \qquad (4.7.20)$$

Then there exists $C \in (0, \infty)$ independent of ξ_n such that

$$|\Theta(\xi', -\xi_n) + \Theta(0', \xi_n)| \le \frac{C}{|\xi_n|^2}. \qquad (4.7.21)$$

Proof of Claim 2. From (4.7.20) it follows that $(\nabla\Theta)(\lambda x) = \lambda^{-2}(\nabla\Theta)(x)$ for every $x \in \mathbb{R}^n \setminus \{0\}$ and every $\lambda \in \mathbb{R} \setminus \{0\}$. Based on (4.7.20) and this observation, we may estimate

$$|\Theta(\xi', -\xi_n) + \Theta(0', \xi_n)| = \frac{1}{|\xi_n|} |\Theta(-\xi'/\xi_n, 1) - \Theta(0', 1)|$$

$$\leq \frac{1}{|\xi_n|} \cdot \left| \frac{\xi'}{\xi_n} \right| \cdot \sup_{t \in [0,1]} |(\nabla \Theta)(-t\xi'/\xi_n, 1)|$$

$$\leq \frac{1}{|\xi_n|} \cdot \left| \frac{\xi'}{\xi_n} \right| \cdot \sup_{t \in [0,1]} \frac{\|\nabla \Theta\|_{L^\infty(S^{n-1})}}{|(-t\xi'/\xi_n, 1)|^2}$$

$$\leq \frac{C'}{|\xi_n|^2} \cdot \|\nabla \Theta\|_{L^\infty(S^{n-1})}, \tag{4.7.22}$$

where the last line uses the assumption that $|\xi'| \leq C'$. Since $\nabla \Theta$ is continuous, hence bounded, on S^{n-1}, the desired result follows by taking the constant to be $C := C' \|\nabla \Theta\|_{L^\infty(S^{n-1})} < \infty$.

Claim 3: If $\varphi \in S(\mathbb{R}^{n-1})$ is such that $\widehat{\varphi}$ has compact support, then

$$F(\xi_n) := -(2\pi)^{1-n} \int_{\mathbb{R}^{n-1}} [\widehat{\Phi}(\xi', -\xi_n) + \widehat{\Phi}(0', \xi_n)] \widehat{\varphi}(\xi') \, d\xi'. \tag{4.7.23}$$

is well defined and integrable on $\mathbb{R} \setminus [-1, 1]$.

Proof of Claim 3. From Proposition 4.59 it follows that $\widehat{\Phi}$ is a positive homogeneous tempered distribution of degree $-n - (1 - n) = -1$. This readily implies that $\widehat{\Phi}\big|_{\mathbb{R}^n \setminus \{0\}}$, viewed as a continuous function in $\mathbb{R}^n \setminus \{0\}$, is also positive homogeneous of degree -1. Moreover, from (4.2.36) we deduce that $\widehat{\Phi}$ is an odd function in $\mathbb{R}^n \setminus \{0\}$. As a consequence, the function $\Theta := \widehat{\Phi}\big|_{\mathbb{R}^n \setminus \{0\}} \in C^1(\mathbb{R}^n \setminus \{0\})$ satisfies (4.7.20).

Hence, Claim 2 applies to this Θ and, granted the compact support condition on $\widehat{\varphi}$, it follows that there exists $C \in (0, \infty)$ such that

$$|F(\xi_n)| \leq \frac{C}{|\xi_n|^2}, \quad \forall \xi_n \neq 0, \tag{4.7.24}$$

from which the desired conclusion follows.

Claim 4: Assume $\varphi \in S(\mathbb{R}^{n-1})$ is such that $\widehat{\varphi}$ has compact support. Then the function

$$f(t) := \int_{\mathbb{R}^{n-1}} \Phi(x', t) \varphi(x') \, dx' + \frac{1}{2i} (\text{sgn} \, t) \widehat{\Phi}(0', 1) \varphi(0'), \tag{4.7.25}$$

originally defined for $t \in \mathbb{R} \setminus \{0\}$, has a continuous extension to all of $\mathbb{R}$.

Proof of Claim 4. Decompose $f = f_1 + f_2$ where $f_1(t) := \int_{\mathbb{R}^{n-1}} \Phi(x', t) \varphi(x') \, dx'$ and $f_2(t) := \frac{1}{2i} (\text{sgn} \, t) \widehat{\Phi}(0', 1) \varphi(0')$ for each $t \in \mathbb{R} \setminus \{0\}$. From (13.4.18) and (3.2.10) we know that

$$\widehat{\text{sgn}}(\xi_n) = -2i \, \text{P.V.} \left(\frac{1}{\xi_n} \right), \quad \text{and} \quad \varphi(0') = (2\pi)^{1-n} \int_{\mathbb{R}^{n-1}} \widehat{\varphi}(\xi') \, d\xi'. \tag{4.7.26}$$

Also, $\widehat{\Phi}$ is odd and positive homogeneous of degree -1 in $\mathbb{R}^n \setminus \{0\}$, and since $\text{P.V.}\left(\frac{1}{\xi_n}\right)$ restricted to $\mathbb{R} \setminus \{0\}$ is simply $\frac{1}{\xi_n}$, it follows that

$$\text{P.V.}\left(\frac{1}{\xi_n}\right)\widehat{\Phi}(0', 1) = \widehat{\Phi}(0', \xi_n) \quad \text{on } \mathbb{R} \setminus \{0\}.$$

Keeping these in mind we conclude that

$$\widehat{f_2}(\xi_n) = -(2\pi)^{1-n} \int_{\mathbb{R}^{n-1}} \widehat{\Phi}(0', \xi_n)\widehat{\varphi}(\xi') \, d\xi' \quad \text{for} \quad \xi_n \in \mathbb{R} \setminus \{0\}. \tag{4.7.27}$$

As far as f_1 is concerned, observe first that $f_1 \in \mathcal{S}'(\mathbb{R})$. Indeed, for every $\psi \in \mathcal{S}(\mathbb{R})$ we have $\langle f_1, \psi \rangle = \langle \Phi, \varphi \otimes \psi \rangle$. Since $\varphi \otimes \psi \in \mathcal{S}(\mathbb{R}^n)$ and, as already noted, $\Phi \in \mathcal{S}'(\mathbb{R}^n)$, the desired conclusion follows. The next order of business is to compute the Fourier transform of f_1. With this goal in mind, pick an arbitrary $\psi \in \mathcal{S}(\mathbb{R})$ and write (keeping in mind Exercise 3.27)

$$\langle \widehat{f_1}, \psi \rangle = \langle f_1, \widehat{\psi} \rangle = \langle \Phi, \varphi \otimes \widehat{\psi} \rangle$$

$$= (2\pi)^{1-n} \langle \Phi, \mathcal{F}(\widecheck{\varphi} \otimes \psi) \rangle = (2\pi)^{1-n} \langle \widehat{\Phi}, \widecheck{\varphi} \otimes \psi \rangle$$

$$= (2\pi)^{1-n} \big\langle \langle \widehat{\Phi}(\xi', \xi_n), \widehat{\varphi}(-\xi') \rangle, \psi(\xi_n) \big\rangle \tag{4.7.28}$$

which proves that

$$\widehat{f_1}(\xi_n) = (2\pi)^{1-n} \int_{\mathbb{R}^{n-1}} \widehat{\Phi}(\xi', \xi_n)\widehat{\varphi}(-\xi') \, d\xi' \quad \text{for} \quad \xi_n \in \mathbb{R} \setminus \{0\}. \tag{4.7.29}$$

By combining (4.7.27) with (4.7.29) we arrive at the conclusion that, for each $\xi_n \in \mathbb{R} \setminus \{0\}$,

$$\widehat{f}(\xi_n) = (2\pi)^{1-n} \int_{\mathbb{R}^{n-1}} \widehat{\Phi}(\xi', \xi_n)\widehat{\varphi}(-\xi') \, d\xi' - (2\pi)^{1-n} \int_{\mathbb{R}^{n-1}} \widehat{\Phi}(0', \xi_n)\widehat{\varphi}(\xi') \, d\xi'$$

$$= -(2\pi)^{1-n} \int_{\mathbb{R}^{n-1}} \widehat{\Phi}(\xi', -\xi_n)\widehat{\varphi}(\xi') \, d\xi' - (2\pi)^{1-n} \int_{\mathbb{R}^{n-1}} \widehat{\Phi}(0', \xi_n)\widehat{\varphi}(\xi') \, d\xi'$$

$$= F(\xi_n), \tag{4.7.30}$$

where the second equality uses the fact that $\widehat{\Phi}$ is odd in $\mathbb{R}^n \setminus \{0\}$. Hence, $\widehat{f} = F$ on $\mathbb{R} \setminus \{0\}$, where F is defined in Claim 3. Now select $\theta \in C_0^\infty(\mathbb{R})$ with $\theta \equiv 1$ on $[-1, 1]$ and write

$$\widehat{f} = (1 - \theta)\widehat{f} + \theta\widehat{f}. \tag{4.7.31}$$

Since $(1 - \theta)\widehat{f} = (1 - \theta)F \in L^1(\mathbb{R})$ by Claim 3, and $\theta\widehat{f} \in \mathcal{E}'(\mathbb{R})$, we may conclude from Claim 1 that the Fourier transform of $\widehat{f}$ belongs to $C^0(\mathbb{R})$ hence, ultimately, that f itself belongs to $C^0(\mathbb{R})$. This completes the proof of Claim 4.

Claim 5: Assume that $\varphi \in \mathcal{S}(\mathbb{R}^{n-1})$ is such that $\widehat{\varphi}$ has compact support. Then

$$\lim_{t \to 0^+} \int_{\mathbb{R}^{n-1}} \Phi(x', t)\varphi(x')\, dx' - \lim_{t \to 0^-} \int_{\mathbb{R}^{n-1}} \Phi(x', t)\varphi(x')\, dx'$$

$$= i\, \widehat{\Phi}(0', 1)\varphi(0'). \tag{4.7.32}$$

Proof of Claim 5. This follows from Claim 4 by writing for each $t \in \mathbb{R} \setminus \{0\}$

$$\int_{\mathbb{R}^{n-1}} \Phi(x', t)\varphi(x')\, dx' = -\frac{1}{2i}(\operatorname{sgn} t)\widehat{\Phi}(0', 1)\varphi(0') + f(t) \tag{4.7.33}$$

with f continuous on $\mathbb{R}$.

Claim 6: For $a_\pm$ originally defined in (4.7.12) one has

$$a_\pm = \pm\frac{i}{2}\, \widehat{\Phi}(0', 1). \tag{4.7.34}$$

Proof of Claim 6. Let $\psi \in C_0^\infty(\mathbb{R}^{n-1})$ be such that

$$\int_{\mathbb{R}^{n-1}} \psi(x')\, dx' = 1, \tag{4.7.35}$$

and set $\varphi := \widehat{\psi}$. Then $\varphi \in \mathcal{S}(\mathbb{R}^{n-1})$, $\widehat{\varphi} = (2\pi)^{n-1}\psi^\vee$ has compact support, and

$$\varphi(0') = \widehat{\psi}(0') = \int_{\mathbb{R}^{n-1}} \psi(x')\, dx' = 1. \tag{4.7.36}$$

From (4.7.11) and Claim 5, one obtains

$$i\, \widehat{\Phi}(0', 1) = a_+ - a_-. \tag{4.7.37}$$

However, since Φ is odd, from the definition of $a_\pm$ in (4.7.12) we see that $a_- = -a_+$. This forces the equalities in (4.7.34) and finishes the proof of Claim 6.

At this stage, we note that (4.7.1) is a consequence of (4.7.11) and (4.7.34). The proof of Theorem 4.79 is therefore complete. □

The proof just completed offers a bit more and, below, we bring to the forefront one such by-product.

Proposition 4.80. *Assume that $\Phi \in C^4(\mathbb{R}^n \setminus \{0\})$ is odd and positive homogeneous of degree $1 - n$. Also, let $\xi \in S^{n-1}$ be arbitrary and set*

$$H_\xi := \{x \in \mathbb{R}^n : x \cdot \xi = 0\}. \tag{4.7.38}$$

Then

$$\widehat{\Phi}(\xi) = -2i \lim_{r \to \infty} \int_{x \in H_\xi, |x| < r} \Phi(x + \xi)\, d\sigma(x), \tag{4.7.39}$$

where σ denotes the surface measure on H_ξ (viewed as surface in $\mathbb{R}^n$).

Proof. We start by observing that (4.7.12) and (4.7.34) imply that

$$\widehat{\Psi}(\mathbf{e}_n) = -2i \lim_{r \to \infty} \int_{x' \in \mathbb{R}^{n-1}, |x'| < r} \Psi((x', 0) + \mathbf{e}_n) \, dx'$$

for each function Ψ of class C^4, odd, and positive (4.7.40)

homogeneous of degree $1 - n$, in $\mathbb{R}^n \setminus \{0\}$.

To proceed, fix a vector $\xi \in S^{n-1}$ and denote by H_ξ the hyperplane in $\mathbb{R}^n$ orthogonal to ξ. Consider then an orthogonal matrix $A \in \mathcal{M}_{n \times n}(\mathbb{R})$ satisfying

$$A H_\xi = \mathbb{R}^{n-1} \times \{0\} \quad \text{and} \quad A\xi = \mathbf{e}_n. \tag{4.7.41}$$

For example, if $v_1, \ldots, v_{n-1}$ is an orthonormal basis in H_ξ and we set $v_n := \xi$, then the conditions $A v_j = \mathbf{e}_j$ for $j \in \{1, \ldots, n\}$ define an orthogonal matrix A such that the conditions in (4.7.41) hold. If we now introduce $\Psi : \mathbb{R}^n \setminus \{0\} \to \mathbb{C}$ by setting $\Psi := \Phi \circ A^{-1}$ then $\Psi \in C^4(\mathbb{R}^n \setminus \{0\})$ is odd, positive homogeneous of degree $1 - n$, and $\widehat{\Psi} = \widehat{\Phi} \circ A^{-1}$ (as seen from (4.3.3)). Using these observations and (4.7.40), we may write

$$\widehat{\Phi}(\xi) = \widehat{\Psi}(A\xi) = \widehat{\Psi}(\mathbf{e}_n) = -2i \lim_{r \to \infty} \int_{x' \in \mathbb{R}^{n-1}, |x'| < r} \Psi((x', 0) + \mathbf{e}_n) \, dx'. \tag{4.7.42}$$

Given a number $r \in (0, \infty)$, consider now the surface $\Sigma := \{x \in H_\xi : |x| < r\}$ and note that if $O := \{x' \in \mathbb{R}^{n-1} : |x'| < r\}$ then

$$P : O \to \Sigma, \quad P(x') := A^{-1}(x', 0) \text{ for each } x' \in O, \tag{4.7.43}$$

is a global C^∞ parametrization of Σ, satisfying $|\partial_1 P \times \cdots \times \partial_{n-1} P| = 1$ at each point in the set O. Consequently, (14.6.6) gives

$$\int_{x \in H_\xi, |x| < r} \Phi(x + \xi) \, d\sigma(x) = \int_O \Phi(P(x') + \xi) \, dx'$$

$$= \int_O \Phi(A^{-1}(x', 0) + A^{-1}\mathbf{e}_n) \, dx'$$

$$= \int_{x' \in \mathbb{R}^{n-1}, |x'| < r} \Psi((x', 0) + \mathbf{e}_n) \, dx'. \tag{4.7.44}$$

At this stage, (4.7.39) is clear from (4.7.42) and (4.7.44). $\qquad \square$

The power of Theorem 4.79 is most apparent by considering the consequence discussed in Corollary 4.81 below, which sheds light on the boundary behavior $\lim_{t \to 0^+} F^{\pm}(x', t)$ of functions of the following type:

$$F^{\pm}(x', t) := \int_{\mathbb{R}^{n-1}} \Phi(x' - y', \pm t)\varphi(y') \, dy', \qquad \forall (x', t) \in \mathbb{R}^n_+, \qquad (4.7.45)$$

where Φ is as in the statement of Theorem 4.79 and φ is a Schwartz function on $\mathbb{R}^{n-1}$ (which is canonically identified with $\partial \mathbb{R}^n_+$). Functions of the form (4.7.45) play a prominent role in partial differential equations and harmonic analysis as they arise naturally in the treatment of boundary value problems via boundary integral methods. We shall return to this topic in Section 11.6.

Corollary 4.81. *Let the function $\Phi \in C^4(\mathbb{R}^n \setminus \{0\})$ be odd and positive homogeneous of degree $1 - n$, and assume that $\varphi \in \mathcal{S}(\mathbb{R}^{n-1})$. Then for every $x' \in \mathbb{R}^{n-1}$ one has*

$$\lim_{t \to 0^{\pm}} \int_{\mathbb{R}^{n-1}} \Phi(x' - y', t)\varphi(y') \, dy' = \pm \frac{i}{2} \widehat{\Phi}(0', 1)\varphi(x')$$

$$+ \lim_{\varepsilon \to 0^+} \int_{\substack{y' \in \mathbb{R}^{n-1} \\ |x' - y'| > \varepsilon}} \Phi(x' - y', 0)\varphi(y') \, dy'. \qquad (4.7.46)$$

Proof. Given any $\varphi \in \mathcal{S}(\mathbb{R}^{n-1})$ and any $x' \in \mathbb{R}^{n-1}$, write

$$\lim_{t \to 0^{\pm}} \int_{\mathbb{R}^{n-1}} \Phi(x' - y', t)\varphi(y') \, dy' = \lim_{t \to 0^{\pm}} \int_{\mathbb{R}^{n-1}} \Phi(z', t)\varphi(x' - z') \, dz'$$

$$= \lim_{t \to 0^{\pm}} \langle \Phi(\cdot, t), \varphi(x' - \cdot) \rangle \qquad (4.7.47)$$

$$= \pm \frac{i}{2} \widehat{\Phi}(0', 1)\varphi(x') + \langle \text{P.V.} \, \Phi(\cdot, 0), \varphi(x' - \cdot) \rangle$$

then notice that

$$\langle \text{P.V.} \, \Phi(\cdot, 0), \varphi(x' - \cdot) \rangle = \lim_{\varepsilon \to 0^+} \int_{\substack{z' \in \mathbb{R}^{n-1} \\ |z'| > \varepsilon}} \Phi(z', 0)\varphi(x' - z') \, dz'$$

$$= \lim_{\varepsilon \to 0^+} \int_{\substack{y' \in \mathbb{R}^{n-1} \\ |x' - y'| > \varepsilon}} \Phi(x' - y', 0)\varphi(y') \, dy'. \qquad (4.7.48)$$

Now (4.7.46) follows from (4.7.47)–(4.7.48). $\qquad \square$

Remark 4.82. Consider $\Phi : \mathbb{R}^2 \setminus \{(0, 0)\} \to \mathbb{C}$ given by $\Phi(x, y) := \frac{1}{x + iy}$ for all $(x, y) \in \mathbb{R}^2 \setminus \{(0, 0)\}$. Then Φ is odd and homogeneous of degree -1. Moreover, since under the canonical identification $\mathbb{R}^2 \equiv \mathbb{C}$ the function Φ becomes $\Phi(z) = \frac{1}{z}$ for $z \in \mathbb{C} \setminus \{0\}$, it follows that $\widehat{\Phi}(\xi) = \frac{2\pi}{i} \cdot \frac{1}{\xi}$ for all $\xi \in \mathbb{C} \setminus \{0\}$ (for details, see Proposition 7.44). In particular, this yields $\widehat{\Phi}(0, 1) = \frac{2\pi}{i} \cdot \frac{1}{i} = -2\pi$. Consequently, (4.7.1) becomes in this case

$$\lim_{\varepsilon \to 0^+} \frac{1}{x \pm i\varepsilon} = \mp i\pi \delta + \text{P.V.} \frac{1}{x} \quad \text{in } \mathcal{S}'(\mathbb{R}), \tag{4.7.49}$$

which is in agreement with Sokhotsky's formula from (2.10.3).

Theorem 4.79 also suggests a natural procedure for computing the Fourier transform of certain principal value distributions. While the latter topic has been treated in Section 4.5, where the general formula (4.5.11) has been established, such a procedure remains of interest since the integral in (4.5.6) may not always be readily computed from scratch. Specifically, we have the following result.

Corollary 4.83. *Assume that the function $\Phi \in C^4(\mathbb{R}^n \setminus \{0\})$ is odd and positive homogeneous of degree $1 - n$. Then*

$$\mathcal{F}'\big(\text{P.V.}\,\Phi(\cdot, 0)\big) = \mp \frac{i}{2}\,\widehat{\Phi}(0', 1) + \lim_{\varepsilon \to 0^{\pm}} \mathcal{F}'(\Phi(\cdot, \varepsilon)) \quad \text{in } \mathcal{S}'(\mathbb{R}^{n-1}), \tag{4.7.50}$$

where $\mathcal{F}'$ denotes the Fourier transform in $\mathbb{R}^{n-1}$.

Proof. Formula (4.7.50) is a direct consequence of Theorem 4.79, the continuity of $\mathcal{F}'$ on $\mathcal{S}'(\mathbb{R}^{n-1})$, and the fact that $\mathcal{F}'(\delta(x')) = 1$. $\qquad\qquad\square$

Here is an example implementing this procedure in a case that is going to be useful later on, when discussing the Riesz transforms in $\mathbb{R}^n$.

Proposition 4.84. *For each $j \in \{1, \ldots, n\}$ we have (with $\mathcal{F}$ denoting the Fourier transform in $\mathbb{R}^n$)*

$$\mathcal{F}\Big(\text{P.V.} \frac{x_j}{|x|^{n+1}}\Big) = -\frac{i\omega_n}{2} \frac{\xi_j}{|\xi|} \quad \text{in } \mathcal{S}'(\mathbb{R}^n). \tag{4.7.51}$$

Proof. Fix some $j \in \{1, \ldots, n\}$ and consider the function defined by

$$\Phi(x, t) := \frac{x_j}{\big(|x|^2 + t^2\big)^{\frac{n+1}{2}}}, \qquad \forall\, (x, t) \in \mathbb{R}^{n+1} \setminus \{0\}. \tag{4.7.52}$$

Note that the function Φ is C^∞, odd, and positive homogeneous of degree $-n$ in $\mathbb{R}^{n+1} \setminus \{0\}$. Moreover, if hat denotes the Fourier transform in $\mathbb{R}^{n+1}$, from Corollary 4.65 we have

$$\widehat{\Phi}(\xi, \eta) = -i\omega_n \frac{\xi_j}{|\xi|^2 + \eta^2} \quad \text{in } \mathcal{S}'(\mathbb{R}^{n+1}), \tag{4.7.53}$$

from which we see that $\widehat{\Phi}(0, 1) = 0$. Keeping this in mind, formula (4.7.50) (used with $n + 1$ in place on n) then yields

$$\mathcal{F}\Big(\text{P.V.} \frac{x_j}{|x|^{n+1}}\Big) = \lim_{t \to 0^+} \mathcal{F}_x \Phi(x, t) \quad \text{in } \mathcal{S}'(\mathbb{R}^n), \tag{4.7.54}$$

where $\mathcal{F}_x$ denotes the Fourier transform in the variable x in $\mathbb{R}^n$. Next, recall the discussion at the end of Section 4.2 regarding partial Fourier transforms and compute

$$(\mathcal{F}_x \Phi(x,t))(\xi) = \mathcal{F}_\eta^{-1}[\widehat{\Phi}(\xi,\eta)](t) = -i\omega_n \xi_j \mathcal{F}_\eta^{-1}\Big[\frac{1}{|\xi|^2 + \eta^2}\Big](t)$$

$$= -i\omega_n (2\pi)^{-1} \xi_j \mathcal{F}_\eta\Big[\frac{1}{|\xi|^2 + \eta^2}\Big](t)$$

$$= -\frac{i\omega_n}{2} \frac{\xi_j}{|\xi|} e^{-|\xi||t|} \quad \text{in} \quad \mathcal{S}'(\mathbb{R}^n), \tag{4.7.55}$$

where the last equality makes uses of (4.2.21). Since by Lebesgue's Dominated Convergence Theorem

$$\lim_{t \to 0}\Big[\frac{\xi_j}{|\xi|} e^{-|\xi||t|}\Big] = \frac{\xi_j}{|\xi|} \quad \text{in} \quad \mathcal{S}'(\mathbb{R}^n), \tag{4.7.56}$$

formula (4.7.51) now follows from (4.7.54)–(4.7.56). $\qquad\qquad\square$

Remark 4.85. Alternatively, one may prove (4.7.51) by combining (4.3.25) and (4.4.19). Indeed, if $j \in \{1,\dots,n\}$ is fixed, identity (4.4.19) written for the function $\Phi(x) := \frac{1}{1-n}|x|^{-(n-1)}$, $x \in \mathbb{R}^n \setminus \{0\}$, becomes

$$\frac{1}{1-n}\partial_j[|x|^{-(n-1)}] = \text{P.V.}\Big(\frac{x_j}{|x|^{n+1}}\Big) \quad \text{in} \quad \mathcal{S}'(\mathbb{R}^n). \tag{4.7.57}$$

Hence, taking the Fourier transform of (4.7.57), then using *(b)* in Theorem 4.26, then (4.3.25) with $\lambda := n - 1$, and formulas (14.5.2) and (14.5.2), we obtain

$$\mathcal{F}\Big(\text{P.V.}\frac{x_j}{|x|^{n+1}}\Big) = \frac{1}{1-n}\mathcal{F}\big(\partial_j[|x|^{-(n-1)}]\big) = \frac{i}{1-n}\xi_j \mathcal{F}(|x|^{-(n-1)})$$

$$= -\frac{i}{n-1}\frac{2\pi^{\frac{n}{2}}\Gamma\big(\frac{1}{2}\big)}{\Gamma\big(\frac{n-1}{2}\big)}\frac{\xi_j}{|\xi|} = -\frac{i\omega_n}{2}\frac{\xi_j}{|\xi|} \quad \text{in} \quad \mathcal{S}'(\mathbb{R}^n). \tag{4.7.58}$$

Corollary 4.83 may also be combined with Theorem 4.74 to obtain the following result pertaining to certain limits of Fourier transforms of tempered distributions.

Corollary 4.86. *Suppose that* $\Phi \in C^4(\mathbb{R}^n \setminus \{0\})$ *is odd and positive homogeneous of degree* $1 - n$*. Then in* $\mathcal{S}'(\mathbb{R}^{n-1})$*,*

$$\lim_{\varepsilon \to 0^{\pm}} \mathcal{F}'[\Phi(\cdot,\varepsilon)](\xi') = \pm\frac{i}{2}\widehat{\Phi}(0',1) - \int_{S^{n-2}} \Phi(\theta,0)\log(i(\xi'\cdot\theta))\,d\sigma(\theta) \tag{4.7.59}$$

where $\mathcal{F}'$ denotes the Fourier transform in $\mathbb{R}^{n-1}$ and S^{n-2} denotes the unit sphere centered at the origin in $\mathbb{R}^{n-1}$.

Proof. This follows from Corollary 4.83, (4.5.11), and (4.5.6). $\qquad\qquad\square$

4.8 The Harmonic Poisson Kernel

The goal of this section is to introduce and discuss the harmonic Poisson kernel.

Definition 4.87. Define the `harmonic Poisson kernel` $P : \mathbb{R}^n \setminus \{0\} \to \mathbb{R}$ by setting

$$P(x', x_n) := \frac{2}{\omega_{n-1}} \frac{x_n}{|x|^n}, \qquad \forall\, x = (x', x_n) \in \mathbb{R}^n \setminus \{0\}. \tag{4.8.1}$$

Furthermore, for each $x' \in \mathbb{R}^{n-1}$ set $p(x') := P(x', 1)$, i.e., consider

$$p(x') := \frac{2}{\omega_{n-1}} \frac{1}{(1 + |x'|^2)^{\frac{n}{2}}}, \qquad \forall\, x' \in \mathbb{R}^{n-1}. \tag{4.8.2}$$

In our next result we discuss the boundary behavior of a mapping $\mathscr{P}$ defined below, taking Schwartz functions from $\mathbb{R}^{n-1}$ into functions defined in $\mathbb{R}^n_\pm$, which in partial differential equations is referred to as the harmonic double layer potential operator.

Proposition 4.88. *Given any $\varphi \in \mathcal{S}(\mathbb{R}^{n-1})$ and any $x = (x', x_n) \in \mathbb{R}^n$ with $x_n \neq 0$, define (where P denotes the harmonic Poisson kernel)*

$$(\mathscr{P}\varphi)(x) := \int_{\mathbb{R}^{n-1}} P(x' - y', x_n)\varphi(y')\,\mathrm{d}y'$$

$$= \frac{2}{\omega_{n-1}} \int_{\mathbb{R}^{n-1}} \frac{x_n}{(|x' - y'|^2 + x_n^2)^{\frac{n}{2}}}\, \varphi(y')\,\mathrm{d}y'. \tag{4.8.3}$$

Then for each $x' \in \mathbb{R}^{n-1}$ one has

$$\lim_{x_n \to 0^\pm} (\mathscr{P}\varphi)(x', x_n) = \pm\, \varphi(x'). \tag{4.8.4}$$

Proof. The function $P : \mathbb{R}^n \setminus \{0\} \to \mathbb{R}$ from (4.8.1) is C^∞, odd, and positive homogeneous of degree $1 - n$. Moreover, Corollary 4.65 gives

$$\widehat{P}(\xi) = -2\mathrm{i}\frac{\xi_n}{|\xi|^2} \quad \text{in} \quad \mathcal{S}'(\mathbb{R}^n). \tag{4.8.5}$$

Consequently, $\widehat{P}(0', 1) = -2\mathrm{i}$. Also, since $P(x', 0) = 0$ we have P.V. $P(x', 0) = 0$. At this point, Corollary 4.81 applies (with P playing the role of Φ) and yields (4.8.4). $\qquad\square$

Remark 4.89. When specializing Theorem 4.79 to the case $\Phi := P$, with P defined as in (4.8.1), we obtain (making use of (4.8.5))

$$\lim_{x_n \to 0^\pm} P(x', x_n) = \pm\,\delta(x') \quad \text{in} \quad \mathcal{S}'(\mathbb{R}^{n-1}). \tag{4.8.6}$$

This may be regarded as a higher dimensional generalization of the result presented in Exercise 2.133, to which (4.8.6) reduces in the case when $n = 2$ and $x_n = \varepsilon > 0$.

Among other things, the following proposition sheds light on the normalization of the harmonic Poisson kernel introduced in (4.8.1).

Proposition 4.90. *The function p defined in (4.8.2) satisfies the following properties:*

(1) One has $p \in \bigcap\limits_{1 \leq q \leq \infty} L^q(\mathbb{R}^{n-1})$ *and*

$$\int_{\mathbb{R}^{n-1}} p(x')\, dx' = 1. \tag{4.8.7}$$

(2) For each t > 0 set

$$p_t(x') := t^{1-n} p(x'/t) \quad \text{for} \quad x' \in \mathbb{R}^{n-1}. \tag{4.8.8}$$

Then for each $t \in (0, \infty)$ *we have* $p_t \in \bigcap\limits_{1 \leq q \leq \infty} L^q(\mathbb{R}^{n-1})$ *and its Fourier transform is*

$$\widehat{p_t}(\xi') = e^{-t|\xi'|} \quad \text{in} \quad \mathcal{S}'(\mathbb{R}^{n-1}). \tag{4.8.9}$$

(3) The family $\{p_t\}_{t>0}$ *has the semigroup property, that is,*

$$p_{t_1} * p_{t_2} = p_{t_1+t_2} \quad \forall\, t_1, t_2 \in (0, \infty). \tag{4.8.10}$$

Proof. That $p \in L^q(\mathbb{R}^{n-1})$ for any $q \in [1, \infty]$ is immediate from its expression. Fix $t \in (0, \infty)$ and let p_t be as in part *(2)*. Applying Exercise 2.26 (with $n - 1$ in place of n, p in place of f, and t in place of $1/j$), we obtain

$$p_t(x') \xrightarrow[t \to 0^+]{\mathcal{D}'(\mathbb{R}^{n-1})} c\, \delta(x') \quad \text{where} \quad c := \int_{\mathbb{R}^{n-1}} p(x')\, dx'. \tag{4.8.11}$$

On the other hand, as seen from (4.8.2) and (4.8.1),

$$p_t(x') = P(x', t) \qquad \text{for each } t > 0 \text{ and } x' \in \mathbb{R}^{n-1}, \tag{4.8.12}$$

so (4.8.6) gives

$$p_t(x') \to \delta(x') \quad \text{in} \quad \mathcal{S}'(\mathbb{R}^n) \quad \text{as} \quad t \to 0^+. \tag{4.8.13}$$

Now (4.8.7) follows by combining (4.8.13) with (4.8.11).

Moving on to the proof of part *(2)*, observe first that $p_t \in L^q(\mathbb{R}^{n-1})$ for each $t > 0$, if $q \in [1, \infty]$, thus $p_t \in \mathcal{S}'(\mathbb{R}^{n-1})$ (as a consequence of (4.1.9)). In particular, its Fourier transform in $\mathcal{S}'(\mathbb{R}^{n-1})$ is meaningfully defined and equal to its Fourier transform as a function in $L^1(\mathbb{R}^{n-1})$ (by Remark (4.23)) and belongs to $C^0(\mathbb{R}^{n-1})$ (by (3.1.3)).

In what follows we will be using partial Fourier transforms (cf. the discussion at the end of Section 4.2) and the notation $\xi' \in \mathbb{R}^{n-1}$ and $\eta \in \mathbb{R}$. The idea we will pursue is to compute $\mathcal{F}_{x'}[p_t(x')](\xi')$ by making use of (4.8.12) and the fact that we know $\widehat{P}(\xi', \eta)$, the Fourier transform of P in $\mathcal{S}'(\mathbb{R}^n)$ (cf. (4.8.5)). With this in mind, for each $\xi' \in \mathbb{R}^{n-1} \setminus \{0\}$ and $t \in \mathbb{R}$ we write

$$\mathcal{F}_{x'}[P(x', t)](\xi') = \mathcal{F}_\eta^{-1}[\widehat{P}(\xi', \eta)](t) = \mathcal{F}_\eta^{-1}\left[-2i\frac{\eta}{|\xi'|^2 + \eta^2}\right](t)$$

$$= 2i(2\pi)^{-1}\mathcal{F}_\eta\left[\frac{\eta}{|\xi'|^2 + \eta^2}\right](t) = \frac{i}{\pi}(-\pi i)(\text{sgn } t)e^{-|t||\xi'|}$$

$$= (\text{sgn } t)e^{-|t||\xi'|}. \tag{4.8.14}$$

For the second equality in (4.8.14) we have used (4.8.5), for the third the fact that $\mathcal{F}_\eta^{-1}g = (2\pi)^{-1}\mathcal{F}_\eta g^\vee$ for every $g \in \mathcal{S}'(\mathbb{R}^n)$, while for the fourth we have used (4.2.32) (applied with $a := |\xi'|$). In particular, (4.8.14) implies that

$$\mathcal{F}_{x'}[p_t(x')](\xi') = e^{-t|\xi'|} \qquad \text{for } \xi' \in \mathbb{R}^{n-1} \setminus \{0\} \text{ and } t > 0. \tag{4.8.15}$$

Now (4.8.9) follows from (4.8.15) and the fact that $p_t \in C^0(\mathbb{R}^{n-1})$ for every $t > 0$. Regarding (4.8.10), fix $t_1, t_2 \in (0, \infty)$ and by using the property of the Fourier transform singled out in part (3) of Remark 3.29 and (4.8.10) we obtain

$$\mathcal{F}_{x'}(p_{t_1} * p_{t_2})(\xi') = \mathcal{F}_{x'}(p_{t_1})(\xi')\mathcal{F}_{x'}(p_{t_2})(\xi')$$

$$= e^{-t_1|\xi'|}e^{-t_2|\xi'|} = e^{-(t_1+t_2)|\xi'|}$$

$$= \mathcal{F}_{x'}(p_{t_1+t_2}) \qquad \forall \xi' \in \mathbb{R}^{n-1}, \ \forall t > 0. \tag{4.8.16}$$

Now (4.8.10) follows from (4.8.16) by recalling that $\mathcal{F}_{x'}$ is an isomorphism on $\mathcal{S}'(\mathbb{R}^{n-1})$. $\qquad\square$

We wish to note that the harmonic Poisson kernel presented above plays a basic role in the treatment of boundary value problems for the Laplacian $\Delta := \partial_1^2 + \cdots + \partial_n^2$ in the upper-half space. More specifically, for any given function $\varphi \in \mathcal{S}(\mathbb{R}^{n-1})$ (called the boundary datum), the Dirichlet problem

$$\begin{cases} u \in C^\infty(\mathbb{R}^n_+), \\ \Delta u = 0 & \text{in } \mathbb{R}^n_+, \\ u\big|_{\partial\mathbb{R}^n_+}^{ver} = \varphi & \text{on } \mathbb{R}^{n-1} \equiv \partial\mathbb{R}^n_+, \end{cases} \tag{4.8.17}$$

has as a solution the function

$$u := \mathscr{P}\varphi \quad \text{in } \mathbb{R}^n_+, \tag{4.8.18}$$

i.e., (cf. (4.8.3))

$$u(x) = \frac{2}{\omega_{n-1}} \int_{\mathbb{R}^{n-1}} \frac{x_n}{(|x' - y'|^2 + x_n^2)^{\frac{n}{2}}} \varphi(y')\, dy', \qquad x = (x', x_n) \in \mathbb{R}^n_+. \tag{4.8.19}$$

In the last line of (4.8.17), the symbol $u\big|_{\partial\mathbb{R}_+^n}^{ver}$ stands for the "vertical limit" of u to the boundary of the upper-half space, understood at each point $x' \in \mathbb{R}^{n-1}$ as

$$\left(u\big|_{\partial\mathbb{R}_+^n}^{ver}\right)(x') := \lim_{x_n \to 0^+} u(x', x_n). \tag{4.8.20}$$

To see why u as in (4.8.19) is a solution of (4.8.17) note that (4.8.1) implies $u \in C^\infty(\mathbb{R}_+^n)$, that u has the limit to the boundary equal to φ based on (4.8.4), while the fact that $\Delta u = 0$ in $\mathbb{R}_+^n$ follows from (4.8.19) by checking directly that, for each $y' \in \mathbb{R}^{n-1}$ fixed, $\Delta\big[x_n(|x' - y'|^2 + x_n^2)^{-\frac{n}{2}}\big] = 0$ for all $(x', x_n) \in \mathbb{R}_+^n$.

Definition 4.91. The conjugate harmonic Poisson kernels are the mappings $Q_j : \mathbb{R}^n \setminus \{0\} \to \mathbb{R}$, $j \in \{1, \dots, n-1\}$, defined by

$$Q_j(x', x_n) := \frac{2}{\omega_{n-1}} \frac{x_j}{|x|^n}, \qquad \forall\, x = (x', x_n) \in \mathbb{R}^n \setminus \{0\}, \quad j \in \{1, \dots, n-1\}. \tag{4.8.21}$$

Furthermore, for each $x' \in \mathbb{R}^{n-1}$ set $q_j(x') := Q_j(x', 1)$, that is,

$$q_j(x') := \frac{2}{\omega_{n-1}} \frac{x_j}{\left(1 + |x'|^2\right)^{\frac{n}{2}}}, \qquad \forall\, x' \in \mathbb{R}^{n-1}, \quad j \in \{1, \dots, n-1\}. \tag{4.8.22}$$

Proposition 4.92. *The functions defined in (4.8.2) satisfy the following properties:*

(1) For each $j \in \{1, \dots, n-1\}$ and each $p \in (1, \infty)$ one has $q_j \in L^p(\mathbb{R}^{n-1})$.
(2) For each $t > 0$ set

$$(q_j)_t(x') := t^{1-n} q_j(x'/t) \quad \text{for} \quad x' \in \mathbb{R}^{n-1}, \quad j \in \{1, \dots, n-1\}. \tag{4.8.23}$$

Then for each $j \in \{1, \dots, n-1\}$, each $t \in (0, \infty)$, and each $p \in (1, \infty)$, we have $(q_j)_t \in L^p(\mathbb{R}^{n-1})$ and its Fourier transform is

$$\widehat{(q_j)_t}(\xi') = -i\frac{\xi_j}{|\xi'|} e^{-t|\xi'|} \quad \text{in} \quad \mathcal{S}'(\mathbb{R}^{n-1}). \tag{4.8.24}$$

(3) The following identity holds:

$$\frac{d}{dt}[p_t(x')] = -\sum_{j=1}^{n-1} \partial_j\big[(q_j)_t(x')\big] \qquad \forall\, t \in (0, \infty), \ \forall\, x' \in \mathbb{R}^{n-1}, \tag{4.8.25}$$

where p_t is as in (4.8.23).

Proof. The claim in *(1)* is an immediate consequence of (4.8.22). To prove *(2)*, fix $j \in \{1, \dots, n-1\}$. Using (4.8.23) we obtain

$$(q_j)_t(x') = \frac{2}{\omega_{n-1}} \frac{x_j}{(t^2 + |x'|^2)^{\frac{n}{2}}} = Q_j(x', t) \tag{4.8.26}$$

for each $x' \in \mathbb{R}^{n-1}$ and each $t \in (0, \infty)$.

As in the proof of Proposition 4.90, we will be using partial Fourier transforms (cf. the discussion at the end of Section 4.2) and the notation $\xi' \in \mathbb{R}^{n-1}$ and $\eta \in \mathbb{R}$. The idea we will pursue is to compute $\mathcal{F}_{x'}[(q_j)_t(x')](\xi')$ by making use of (4.8.26) and the fact that, by (4.3.31), the Fourier transform of Q_j is

$$\widehat{Q_j}(\xi', \eta) = -2i \frac{\xi_j}{|\xi'|^2 + \eta^2} \quad \text{in} \quad \mathcal{S}'(\mathbb{R}^n). \tag{4.8.27}$$

Hence, for each $t \in (0, \infty)$ we may write

$$\mathcal{F}_{x'}[(q_j)_t(x')](\xi') = \mathcal{F}_\eta^{-1}[\widehat{Q_j}(\xi', \eta)](t) = \mathcal{F}_\eta^{-1}\Big[-2i \frac{\xi_j}{|\xi'|^2 + \eta^2} \Big](t)$$

$$= -2i(2\pi)^{-1} \xi_j \mathcal{F}_\eta\Big[\frac{1}{|\xi'|^2 + \eta^2} \Big](t) = -\frac{i\xi_j}{\pi} \frac{\pi}{|\xi'|} e^{-t|\xi'|}$$

$$= -\frac{i\xi_j}{|\xi'|} e^{-t|\xi'|} \quad \text{in} \quad \mathcal{S}'(\mathbb{R}^{n-1}). \tag{4.8.28}$$

For the third equality in (4.8.28) we used the fact that $\mathcal{F}_\eta^{-1} g = (2\pi)^{-1} \mathcal{F}_\eta g^\vee$ for every $g \in \mathcal{S}'(\mathbb{R}^n)$, while for the fourth we used (4.2.21) (applied with $a := |\xi'|$). This completes the proof of (2). Finally, (4.8.25) follows by a direct computation based on the Chain Rule. $\qquad \qquad \square$

4.9 Singular Integral Operators

In Example 4.69 we have already encountered the principal value tempered distributions P.V. $\frac{x_j}{|x|^{n+1}}$, for $j \in \{1, \ldots, n\}$. The operators R_j, $j \in \{1, ..., n\}$, defined by convolving with these distributions, i.e.,

$$R_j \varphi := \Big(\text{P.V.} \frac{x_j}{|x|^{n+1}} \Big) * \varphi, \qquad \forall \varphi \in \mathcal{S}(\mathbb{R}^n), \tag{4.9.1}$$

are called the Riesz transforms in $\mathbb{R}^n$. In the particular case when $n = 1$ the corresponding operator, i.e.,

$$H \varphi := \Big(\text{P.V.} \frac{1}{x} \Big) * \varphi, \qquad \forall \varphi \in \mathcal{S}(\mathbb{R}), \tag{4.9.2}$$

is called the Hilbert transform. These operators play a fundamental role in harmonic analysis and here the goal is to study a larger class of operators containing the aforementioned examples. We begin by introducing this class.

The format of Proposition 4.70 suggests making the following definition.

Definition 4.93. For each function $\Theta \in C^0(\mathbb{R}^n \setminus \{0\})$ that is positive homogeneous of degree $-n$ and such that $\int_{S^{n-1}} \Theta(\omega) \, d\sigma(\omega) = 0$, define the singular integral operator

$$(T_\Theta \varphi)(x) := \lim_{\varepsilon \to 0^+} \int_{|x-y| \geq \varepsilon} \Theta(x-y)\varphi(y)\, dy, \qquad \forall\, x \in \mathbb{R}^n,\ \forall\, \varphi \in S(\mathbb{R}^n). \quad (4.9.3)$$

Proposition 4.70 ensures that the above definition is meaningful. Moreover, for the class of singular integral operators just defined, the following result holds (further properties are deduced in Theorem 4.100).

Proposition 4.94. *Let Θ be a function satisfying the conditions in* (4.4.1) *and consider the singular integral operator T_Θ associated with Θ as in* (4.9.3)*. Then*

$$T_\Theta : S(\mathbb{R}^n) \longrightarrow S'(\mathbb{R}^n) \quad \text{is linear and sequentially continuous.} \qquad (4.9.4)$$

Moreover, for each $\varphi \in S(\mathbb{R}^n)$ we have

$$T_\Theta \varphi \in C^\infty(\mathbb{R}^n) \quad and \quad T_\Theta \varphi = (\mathrm{P.V.}\,\Theta) * \varphi \quad in \quad S'(\mathbb{R}^n), \qquad (4.9.5)$$

as well as

$$\widehat{T_\Theta \varphi} = \widehat{\varphi}\,\mathcal{F}(\mathrm{P.V.}\,\Theta) = \widehat{\varphi}\, m_\Theta \quad in \quad S'(\mathbb{R}^n), \qquad (4.9.6)$$

where m_Θ is as in (4.5.6)*.*

Proof. All claims are consequences of Definition 4.93, Proposition 4.70, part *(e)* in Theorem 4.19, part *(a)* in Theorem 4.35, and (4.5.11). $\qquad\square$

Remark 4.95. In the harmonic analysis parlance, a mapping

$$T : S(\mathbb{R}^n) \to S'(\mathbb{R}^n) \qquad (4.9.7)$$

with the property that there exists $m \in S'(\mathbb{R}^n)$ such that $\widehat{T\varphi} = \widehat{\varphi}\, m$ in $S'(\mathbb{R}^n)$ for every $\varphi \in S(\mathbb{R}^n)$ is called a `multiplier` (and the tempered distribution m is referred to as the `symbol of the multiplier`). Note that any multiplier T is necessarily a linear and sequentially bounded mapping from $S(\mathbb{R}^n)$ into $S'(\mathbb{R}^n)$. Also, with this piece of terminology, for any function Θ as in (4.4.1) and m_Θ as in (4.5.6) we have

$$\text{the operator } T_\Theta \text{ is a multiplier with symbol } \mathcal{F}(\mathrm{P.V.}\,\Theta) = m_\Theta. \qquad (4.9.8)$$

To set the stage for the subsequent discussion, we recall that given a linear and bounded map (also referred to as a bounded operator) $T : L^2(\mathbb{R}^n) \to L^2(\mathbb{R}^n)$, its adjoint is the unique map $T^* : L^2(\mathbb{R}^n) \to L^2(\mathbb{R}^n)$ with the property that

$$\int_{\mathbb{R}^n} (Tf)(x)\overline{g(x)}\, dx = \int_{\mathbb{R}^n} f(x)\overline{(T^*g)(x)}\, dx, \qquad \forall\, f, g \in L^2(\mathbb{R}^n). \qquad (4.9.9)$$

Exercise 4.96. Let $T : L^2(\mathbb{R}^n) \to L^2(\mathbb{R}^n)$ be a linear and bounded operator which is a multiplier with a bounded symbol, i.e., there exists $m \in L^\infty(\mathbb{R}^n)$ with the property that $\widehat{T\varphi} = \widehat{\varphi}\, m$ in $S'(\mathbb{R}^n)$ (hence also in $L^2(\mathbb{R}^n)$) for every $\varphi \in S(\mathbb{R}^n)$. Prove that its adjoint T^* is also a multiplier with symbol $\overline{m}$.

Hint: Use (4.9.9) and Parseval's identity (3.2.32).

Some of the properties of the symbol $\mathcal{F}(\text{P.V. } \Theta)$ are discussed in Theorem 4.74. An explicit computation of this Fourier transform has been done in the case corresponding to $\Theta(x) := \frac{x_j}{|x|^{n+1}}$ for some $j \in \{1, \ldots, n\}$ (see (4.7.51)). The fact that such functions appear in the definition of the Riesz transforms in $\mathbb{R}^n$ (cf. (4.9.1)), warrants revisiting these operators.

Theorem 4.97. *Consider the Riesz transforms, originally introduced as in* (4.9.1). *Then for each* $j \in \{1, \ldots, n\}$ *the following properties hold:*

(a) The operator

$$R_j : S(\mathbb{R}^n) \to S'(\mathbb{R}^n) \quad \text{is well defined,}$$

$$\text{linear, and sequentially continuous.} \tag{4.9.10}$$

(b) For each $\varphi \in S(\mathbb{R}^n)$ *we have* $R_j\varphi \in C^\infty(\mathbb{R}^n)$ *and* R_j *may be expressed as the singular integral operator*

$$(R_j\varphi)(x) = \lim_{\varepsilon \to 0^+} \int_{|x-y| \geq \varepsilon} \frac{x_j - y_j}{|x - y|^{n+1}} \varphi(y) \, dy, \qquad \forall \, x \in \mathbb{R}^n. \tag{4.9.11}$$

(c) The j-*th Riesz transform is a multiplier with symbol given by the function* $m_j(\xi) := -\frac{i\omega_n}{2} \frac{\xi_j}{|\xi|} \in L^\infty(\mathbb{R}^n)$. *That is, for each* $\varphi \in S(\mathbb{R}^n)$,

$$\widehat{R_j\varphi}(\xi) = -\frac{i\omega_n}{2} \frac{\xi_j}{|\xi|} \widehat{\varphi}(\xi) \quad \text{in} \quad S'(\mathbb{R}^n). \tag{4.9.12}$$

(d) For each $\varphi \in S(\mathbb{R}^n)$, *we have that* $R_j\varphi$ *originally viewed as a tempered distribution belongs to the subspace* $L^2(\mathbb{R}^n)$ *of* $S'(\mathbb{R}^n)$, *and*

$$\|R_j\varphi\|_{L^2(\mathbb{R}^n)} \leq (\omega_n/2)\|\varphi\|_{L^2(\mathbb{R}^n)}. \tag{4.9.13}$$

(e) The j-*th Riesz transform* R_j, *originally considered as in* (4.9.10) *extends, by density to a linear and bounded operator*

$$R_j : L^2(\mathbb{R}^n) \longrightarrow L^2(\mathbb{R}^n) \tag{4.9.14}$$

and the operator R_j *is skew-symmetric (i.e.,* $R_j^* = -R_j$) *in this context. In addition,*

$$\widehat{R_j f}(\xi) = -\frac{i\omega_n}{2} \frac{\xi_j}{|\xi|} \widehat{f}(\xi) \quad \text{a.e. in} \quad \mathbb{R}^n, \qquad \forall \, f \in L^2(\mathbb{R}^n), \tag{4.9.15}$$

(f) For every $k \in \{1, \ldots, n\}$, *we have*

$$R_j R_k = R_k R_j \quad \text{as operators on} \quad L^2(\mathbb{R}^n), \tag{4.9.16}$$

and

$$\sum_{k=1}^{n} R_k^2 = -\left(\tfrac{\omega_n}{2}\right)^2 I \quad \text{as operators on} \quad L^2(\mathbb{R}^n), \tag{4.9.17}$$

where I denotes the identity operator on $L^2(\mathbb{R}^n)$.

Proof. The claims in *(a)–(b)* are immediate consequence of (4.9.1)–(4.9.2) and Proposition 4.70, upon recalling the discussion in Example 4.69. Also, formula (4.9.12) follows from (4.9.1), (4.9.6), and Proposition 4.84.

Turning our attention to *(d)*, fix an index $j \in \{1, \dots, n\}$ along with two arbitrary functions $\varphi, \psi \in \mathcal{S}(\mathbb{R}^n)$. Based on Exercise 3.27, (4.2.2), (4.9.12), Cauchy–Schwarz's inequality, and (3.2.28), we may then write

$$\left| \langle R_j\varphi, \psi \rangle \right| = (2\pi)^{-n} \left| \langle \widehat{R_j\varphi}, \widehat{\psi}^{\vee} \rangle \right| = (2\pi)^{-n} \frac{\omega_n}{2} \left| \langle \frac{\xi_j}{|\xi|} \widehat{\varphi}, \widehat{\psi}^{\vee} \rangle \right|$$

$$= (2\pi)^{-n} \frac{\omega_n}{2} \left| \int_{\mathbb{R}^n} \frac{\xi_j}{|\xi|} \widehat{\varphi}(\xi) \widehat{\psi}(-\xi) \, d\xi \right|$$

$$\leq (2\pi)^{-n} \frac{\omega_n}{2} \int_{\mathbb{R}^n} \left| \widehat{\varphi}(\xi) \right| \left| \widehat{\psi}(-\xi) \right| d\xi$$

$$\leq (2\pi)^{-n} \frac{\omega_n}{2} \| \widehat{\varphi} \|_{L^2(\mathbb{R}^n)} \| \widehat{\psi} \|_{L^2(\mathbb{R}^n)}$$

$$= \tfrac{\omega_n}{2} \| \varphi \|_{L^2(\mathbb{R}^n)} \| \psi \|_{L^2(\mathbb{R}^n)}. \tag{4.9.18}$$

This computation may be summarized by saying that, for each $\varphi \in \mathcal{S}(\mathbb{R}^n)$ fixed, the linear functional

$$\Lambda_\varphi : \mathcal{S}(\mathbb{R}^n) \to \mathbb{C}, \quad \Lambda_\varphi(\psi) := \langle R_j\varphi, \psi \rangle, \quad \forall \psi \in \mathcal{S}(\mathbb{R}^n), \tag{4.9.19}$$

satisfies $\left| \Lambda_\varphi(\psi) \right| \leq \tfrac{\omega_n}{2} \| \varphi \|_{L^2(\mathbb{R}^n)} \| \psi \|_{L^2(\mathbb{R}^n)}$ for every $\psi \in \mathcal{S}(\mathbb{R}^n)$. Given that $\mathcal{S}(\mathbb{R}^n)$ is a dense subspace of $L^2(\mathbb{R}^n)$, it follows that the mapping Λ_φ from (4.9.19) has a unique extension $\widetilde{\Lambda}_\varphi : L^2(\mathbb{R}^n) \to \mathbb{C}$ satisfying

$$\left| \widetilde{\Lambda}_\varphi(g) \right| \leq \tfrac{\omega_n}{2} \| \varphi \|_{L^2(\mathbb{R}^n)} \| g \|_{L^2(\mathbb{R}^n)} \quad \text{for every} \quad g \in L^2(\mathbb{R}^n). \tag{4.9.20}$$

According to Riesz's representation theorem for such functionals, there exists a unique $f_\varphi \in L^2(\mathbb{R}^n)$ satisfying the following two properties:

$$\widetilde{\Lambda}_\varphi(g) = \int_{\mathbb{R}^n} f_\varphi(x) g(x) \, dx \quad \text{for every} \quad g \in L^2(\mathbb{R}^n), \tag{4.9.21}$$

and

$$\| f_\varphi \|_{L^2(\mathbb{R}^n)} \leq \tfrac{\omega_n}{2} \| \varphi \|_{L^2(\mathbb{R}^n)}. \tag{4.9.22}$$

Since $L^2(\mathbb{R}^n) \subseteq \mathcal{S}'(\mathbb{R}^n)$ (cf. (4.1.9)), it follows that f_φ may be regarded as a tempered distribution. At this stage we make the claim that $R_j\varphi = f_\varphi$ as tempered distributions. Indeed, for every $\psi \in \mathcal{S}(\mathbb{R}^n)$ we have

$$\langle R_j\varphi, \psi \rangle = \Lambda_\varphi(\psi) = \widetilde{\Lambda}_\varphi(\psi) = \int_{\mathbb{R}^n} f_\varphi(x)\psi(x)\,dx = \langle f_\varphi, \psi \rangle, \qquad (4.9.23)$$

from which the claim follows. With this in hand, (4.9.13) now follows from (4.9.22), finishing the proof of part *(d)*. In turn, estimate (4.9.13) and a standard density argument give that R_j extends to a linear and bounded operator in the context of (4.9.14).

The next item in part *(e)* is showing that R_j in (4.9.14) is skew-symmetric. To this end, first note that, given any $\varphi, \psi \in \mathcal{S}(\mathbb{R}^n)$, by reasoning much as in (4.9.18) we obtain

$$\langle R_j\varphi, \overline{\psi} \rangle = (2\pi)^{-n} \langle \widehat{R_j\varphi}, \widehat{\overline{\psi}}^{\vee} \rangle = -\mathrm{i}(2\pi)^{-n}\frac{\omega_n}{2}\Big\langle \frac{\xi_j}{|\xi|}\widehat{\varphi}, \widehat{\overline{\psi}}^{\vee} \Big\rangle$$

$$= -\mathrm{i}(2\pi)^{-n}\frac{\omega_n}{2}\int_{\mathbb{R}^n}\frac{\xi_j}{|\xi|}\widehat{\varphi}(\xi)\overline{\widehat{\psi(-\xi)}}\,d\xi$$

$$= \mathrm{i}(2\pi)^{-n}\frac{\omega_n}{2}\int_{\mathbb{R}^n}\frac{\xi_j}{|\xi|}\overline{\widehat{\psi(\xi)}}\widehat{\varphi}(-\xi)\,d\xi$$

$$= \mathrm{i}(2\pi)^{-n}\frac{\omega_n}{2}\Big\langle \frac{\xi_j}{|\xi|}\widehat{\overline{\psi}}, \widehat{\varphi}^{\vee} \Big\rangle = -(2\pi)^{-n}\langle \widehat{R_j\overline{\psi}}, \widehat{\varphi}^{\vee} \rangle$$

$$= -\langle R_j\overline{\psi}, \varphi \rangle = -\langle \overline{R_j\psi}, \varphi \rangle, \qquad (4.9.24)$$

where the last step uses the fact that, as seen from (4.9.11), $R_j\overline{\psi} = \overline{R_j\psi}$. Since all functions involved are (by part *(d)*) in $L^2(\mathbb{R}^n)$, the distributional pairings may be interpreted as pairings in $L^2(\mathbb{R}^n)$. With this in mind, the equality of the most extreme sides of (4.9.24) reads

$$\int_{\mathbb{R}^n} (R_j\varphi)(x)\overline{\psi(x)}\,dx = -\int_{\mathbb{R}^n} \varphi(x)\overline{(R_j\psi)(x)}\,dx. \qquad (4.9.25)$$

By density and part *(d)*, we deduce from (4.9.25) that

$$\int_{\mathbb{R}^n} (R_jf)(x)\overline{g(x)}\,dx = -\int_{\mathbb{R}^n} f(x)\overline{(R_jg)(x)}\,dx, \qquad \forall f, g \in L^2(\mathbb{R}^n). \qquad (4.9.26)$$

In light of (4.9.9), this shows that $R_j^* = -R_j$ as mappings on $L^2(\mathbb{R}^n)$, as wanted. To complete the treatment of part *(e)*, there remains to observe that, collectively, the boundedness of R_j and of the Fourier transform on $L^2(\mathbb{R}^n)$, (4.9.12), and the density of the Schwartz class in $L^2(\mathbb{R}^n)$, readily yield (4.9.15) via a limiting argument.

In turn, given $k \in \{1, \ldots, n\}$, for each $f \in L^2(\mathbb{R}^n)$ formula (4.9.15) allows us to compute

$$\mathcal{F}(R_j(R_kf)) = -\frac{\mathrm{i}\omega_n}{2}\frac{\xi_j}{|\xi|}\widehat{R_kf} = -\Big(\frac{\omega_n}{2}\Big)^2\frac{\xi_j\xi_k}{|\xi|^2}\widehat{f} \quad \text{in} \quad L^2(\mathbb{R}^n). \qquad (4.9.27)$$

A similar computation also gives that $\mathcal{F}(R_k(R_j f)) = -\left(\frac{\omega_n}{2}\right)^2 \frac{\xi_j \xi_k}{|\xi|^2} \widehat{f}$ in $L^2(\mathbb{R}^n)$. From these, (4.9.16) follows upon recalling (cf. (3.2.31)) that the Fourier transform is an isomorphism of $L^2(\mathbb{R}^n)$.

Finally, as regards (4.9.17), for any function $f \in L^2(\mathbb{R}^n)$, formula (4.9.27) (with $j = k$) permits us to write

$$\mathcal{F}\Big(\sum_{k=1}^n R_k^2 f\Big) = \sum_{k=1}^n \mathcal{F}(R_k(R_k f)) = -\Big(\frac{\omega_n}{2}\Big)^2\Big(\sum_{k=1}^n \frac{\xi_k^2}{|\xi|^2}\Big)\widehat{f}$$

$$= -\Big(\frac{\omega_n}{2}\Big)^2 \widehat{f} \quad \text{in} \quad L^2(\mathbb{R}^n), \tag{4.9.28}$$

where the last equality uses the fact that $\sum_{k=1}^n \frac{\xi_k^2}{|\xi|^2} = 1$ a.e. in $\mathbb{R}^n$. Now identity (4.9.17) follows from (4.9.28) and (3.2.31). $\qquad\square$

Corollary 4.98. *Let $j \in \{1, \ldots, n-1\}$ and $t \in (0, \infty)$, and recall the functions p_t and $(q_j)_t$ from (4.8.8) and (4.8.23), respectively. Then*

$$(q_j)_t = \frac{2}{\omega_{n-1}} R_j p_t \quad \text{in} \quad L^2(\mathbb{R}^{n-1}), \tag{4.9.29}$$

where R_j in (4.9.29) is the j-th Riesz transform from Theorem 4.97 corresponding to n replaced by $n-1$.

Proof. By (3.2.31), it suffices to check identity (4.9.29) on the Fourier transform side. That the latter holds is an immediate consequence of part *(2)* in Proposition 4.92, (4.9.15), and part *(2)* in Proposition 4.90. $\qquad\square$

In the one dimensional setting, Theorem 4.97 yields the following type of information about the Hilbert transform.

Corollary 4.99. *The Hilbert transform defined in (4.9.2) satisfies the following properties:*

(a) The operator

$$H : \mathcal{S}(\mathbb{R}) \to \mathcal{S}'(\mathbb{R}) \quad \text{is well defined,}$$
$$\text{linear, and sequentially continuous.} \tag{4.9.30}$$

(b) For each $\varphi \in \mathcal{S}(\mathbb{R})$ we have $H\varphi \in C^\infty(\mathbb{R})$ and H may be expressed as the singular integral operator

$$(H\varphi)(x) = \lim_{\varepsilon \to 0^+} \int_{|x-y| \geq \varepsilon} \frac{1}{x-y} \varphi(y)\, dy, \qquad \forall\, x \in \mathbb{R}. \tag{4.9.31}$$

(c) The Hilbert transform is a multiplier with symbol $m(\xi) := -i\pi\,(\operatorname{sgn}\xi)$ belonging to $L^\infty(\mathbb{R})$. In other words, for each $\varphi \in \mathcal{S}(\mathbb{R})$,

$$\widehat{H\varphi}(\xi) = -i\pi\,(\operatorname{sgn}\xi)\widehat{\varphi}(\xi) \quad \text{in} \quad \mathcal{S}'(\mathbb{R}). \tag{4.9.32}$$

(d) *For each $\varphi \in S(\mathbb{R})$, we have that $H\varphi$ originally viewed as a tempered distribution belongs to the subspace $L^2(\mathbb{R})$ of $S'(\mathbb{R})$, and*

$$\|H\varphi\|_{L^2(\mathbb{R}^n)} \le \pi\|\varphi\|_{L^2(\mathbb{R}^n)}. \tag{4.9.33}$$

(e) *The Hilbert transform H, originally considered as in (4.9.30) extends, by density to a linear and bounded operator*

$$H : L^2(\mathbb{R}) \longrightarrow L^2(\mathbb{R}) \tag{4.9.34}$$

and the operator H is skew-symmetric (i.e., $H^* = -H$) in this context. In addition,

$$\widehat{Hf}(\xi) = -i\pi\,(\operatorname{sgn}\xi)\,\widehat{f}(\xi) \quad a.e. \text{ in } \mathbb{R}, \qquad \forall f \in L^2(\mathbb{R}). \tag{4.9.35}$$

(f) *In the context of (4.9.34),*

$$H^2 = -\pi^2 I \quad in \quad L^2(\mathbb{R}), \tag{4.9.36}$$

where I denotes the identity operator on $L^2(\mathbb{R})$.

Proof. This corresponds to Theorem 4.97 in the case when $n = 1$. □

The Riesz transforms are prototypes of for the more general class of operators introduced in Definition 4.93 and most of the properties deduced for the Riesz transforms in Theorem 4.97 have natural counterparts in this more general setting.

Theorem 4.100. *Assume that Θ is a function satisfying the conditions in (4.4.1) and let T_Θ be the singular integral operator associated with Θ as in (4.9.3). Then the following statements are true.*

(a) *The operator*

$$T_\Theta : S(\mathbb{R}^n) \to S'(\mathbb{R}^n) \quad is \text{ well defined,}$$
$$\text{linear, and sequentially continuous.} \tag{4.9.37}$$

(b) *For each $\varphi \in S(\mathbb{R}^n)$ we have $T_\Theta\varphi \in C^\infty(\mathbb{R}^n)$ and the singular integral operator T_Θ may be expressed as the convolution*

$$T_\Theta\varphi = (\text{P.V. }\Theta) * \varphi \qquad \forall \varphi \in S(\mathbb{R}^n). \tag{4.9.38}$$

(c) *The function m_Θ, associated with Θ as in (4.5.6), is the symbol of the multiplier T_Θ. This means that for each $\varphi \in S(\mathbb{R}^n)$,*

$$\widehat{T_\Theta\varphi}(\xi) = m_\Theta(\xi)\widehat{\varphi}(\xi) \quad in \quad S'(\mathbb{R}^n). \tag{4.9.39}$$

(d) *For each $\varphi \in S(\mathbb{R}^n)$, we have that $T_\Theta\varphi$ originally viewed as a tempered distribution belongs to the subspace $L^2(\mathbb{R}^n)$ of $S'(\mathbb{R}^n)$, and*

$$\|T_\Theta\varphi\|_{L^2(\mathbb{R}^n)} \le C_n\|\Theta\|_{L^\infty(S^{n-1})}\|\varphi\|_{L^2(\mathbb{R}^n)}, \tag{4.9.40}$$

where the constant $C_n \in (0, \infty)$ is as in (4.5.9).

*(e) The operator T_Θ, originally considered as in (4.9.37) extends, by density to a
linear and bounded operator*

$$T_\Theta : L^2(\mathbb{R}^n) \longrightarrow L^2(\mathbb{R}^n). \tag{4.9.41}$$

Moreover, in this context,

$$\widehat{T_\Theta f}(\xi) = m_\Theta(\xi)\,\widehat{f}(\xi) \quad a.e. \text{ in } \mathbb{R}^n, \qquad \forall f \in L^2(\mathbb{R}^n), \tag{4.9.42}$$

and the operator T_Θ satisfies

$$(T_\Theta)^* = T_{\overline{\Theta}^\vee} \quad in \quad L^2(\mathbb{R}^n). \tag{4.9.43}$$

*In particular, if Θ is odd and real-valued, then the operator T_Θ is skew-symmetric
(i.e., $(T_\Theta)^* = -T_\Theta$), while if Θ is even and real-valued then the operator T_Θ is
self-adjoint (i.e., $(T_\Theta)^* = T_\Theta$).*

Proof. Parts part *(a)*–*(b)* are contained in Proposition 4.94, while part *(c)* follows
from (4.9.6) and Theorem 4.74. Part *(d)* is justified by reasoning as in the proof of
part *(d)* in Theorem 4.97, keeping in mind (4.5.8). As for part *(e)*, in a first stage we
note that, for any $\varphi, \psi \in \mathcal{S}(\mathbb{R}^n)$, by reasoning much as in (4.9.24) while relying on
(4.9.39) and (4.5.10), we obtain

$$\langle T_\Theta \varphi, \overline{\psi} \rangle = (2\pi)^{-n}\langle \widehat{T_\Theta \varphi}, \widehat{\overline{\psi}}^\vee \rangle = (2\pi)^{-n}\Big\langle m_\Theta\,\widehat{\varphi}, \widehat{\overline{\psi}}^\vee \Big\rangle$$

$$= (2\pi)^{-n} \int_{\mathbb{R}^n} m_\Theta(\xi)\,\widehat{\varphi}(\xi)\overline{\widehat{\psi(-\xi)}}\,d\xi$$

$$= (2\pi)^{-n} \int_{\mathbb{R}^n} m_\Theta(-\xi)\,\overline{\widehat{\psi(\xi)}}\widehat{\varphi}(-\xi)\,d\xi$$

$$= (2\pi)^{-n}\Big\langle m_{\Theta^\vee}\,\widehat{\overline{\psi}}, \widehat{\varphi}^\vee \Big\rangle = (2\pi)^{-n}\langle \widehat{T_{\Theta^\vee}\overline{\psi}}, \widehat{\varphi}^\vee \rangle$$

$$= \langle T_{\Theta^\vee}\overline{\psi}, \varphi \rangle = \langle T_{\overline{\Theta}^\vee}\,\psi, \varphi \rangle, \tag{4.9.44}$$

where the last step uses the fact that, as seen from (4.9.3), $T_{\Theta^\vee}\overline{\psi} = \overline{T_{\overline{\Theta}^\vee}\,\psi}$. Passing
from (4.9.44) to (4.9.43) may now be done by arguing as in the proof of part *(e)* of
Theorem 4.97 (a pattern of reasoning which also gives all remaining claims in the
current case). □

We conclude this section with a discussion of the following result of basic importance from Calderón–Zygmund theory (originally proved in [7]; for a more timely
presentation see, e.g., [26], [68], [69]).

Theorem 4.101. *Let Θ be a function satisfying (4.4.1) and, in analogy with (4.9.3),
given some $p \in (1, \infty)$ set*

$$(T_\Theta f)(x) := \lim_{\varepsilon \to 0^+} \int_{|x-y|\geq\varepsilon} \Theta(x-y)f(y)\,dy, \qquad x \in \mathbb{R}^n, \ f \in L^p(\mathbb{R}^n). \tag{4.9.45}$$

Then for every $f \in L^p(\mathbb{R}^n)$ we have that $(T_\Theta f)(x)$ exists for almost every $x \in \mathbb{R}^n$ and

$$\|T_\Theta f\|_{L^p(\mathbb{R}^n)} \leq C_\Theta \|f\|_{L^p(\mathbb{R}^n)}, \tag{4.9.46}$$

where C_Θ is a finite positive constant independent of f.

Compared with Theorem 4.100, a basic achievement of Calderón and Zygmund is allowing functions from $L^p(\mathbb{R}^n)$ in lieu of $\mathcal{S}(\mathbb{R}^n)$. While the estimate in (4.9.46) for $p = 2$ is essentially contained in part *(d)* of Theorem 4.100, the fact that for each $f \in L^2(\mathbb{R}^n)$ the limit defining $(T_\Theta f)(x)$ as in (4.9.45) exists for almost every $x \in \mathbb{R}^n$ is a bit more subtle. To elaborate on this issue, recall that we have already proved that the limit in question exists at every point when f is a Schwartz function (cf. Proposition 4.70). Starting from this, one can pass to arbitrary functions in $L^2(\mathbb{R}^n)$ granted the availability of additional tools from harmonic analysis, including the boundedness in $L^2(\mathbb{R}^n)$ of the so-called maximal operator

$$T_* f(x) := \left| \sup_{\varepsilon > 0} \int_{|x-y| \geq \varepsilon} \Theta(x-y) f(y) \, dy \right| \quad \text{for each } x \in \mathbb{R}^n. \tag{4.9.47}$$

Having dealt with (4.9.46) in the case $p = 2$, its proof for $p \in (1, \infty)$ proceeds as follows. In a first step, a suitable version of (4.9.46) is established for $p = 1$ which, when combined with the case $p = 2$ already treated yields (via a technique called interpolation) (4.9.46) for $p \in (1, 2)$. In a second step, one uses duality to handle the case $p \in (2, \infty)$. There are two important factors at play here: that the dual of $L^p(\mathbb{R}^n)$ with $p \in (1, 2)$ is $L^{p'}(\mathbb{R}^n)$ with $p' = \frac{p}{p-1} \in (2, \infty)$, and the formula for the adjoint of T_Θ we deduced in (4.9.43).

In particular, Theorem 4.101 gives that, given $p \in (1, \infty)$, the j-th Riesz transform R_j, $j \in \{1, \dots, n\}$, originally considered as in (4.9.10) extends by density so that

$$R_j : L^p(\mathbb{R}^n) \longrightarrow L^p(\mathbb{R}^n) \quad \text{is linear and bounded,} \tag{4.9.48}$$

and for each $f \in L^p(\mathbb{R}^n)$ we have

$$(R_j f)(x) = \lim_{\varepsilon \to 0^+} \int_{|x-y| \geq \varepsilon} \frac{x_j - y_j}{|x - y|^{n+1}} f(y) \, dy \quad \text{for a.e. } x \in \mathbb{R}^n. \tag{4.9.49}$$

Of course, the same type of result is true for the Hilbert transform on the real line.

4.10 Derivatives of Volume Potentials

One basic integral operator in analysis is the so-called Newtonian potential, given by

$$(\mathbf{N}_\Omega f)(x) := \frac{-1}{(n-2)\omega_{n-1}} \int_\Omega \frac{1}{|x-y|^{n-2}} f(y) \, dy, \quad \forall x \in \mathbb{R}^n, \tag{4.10.1}$$

where Ω is an open set in $\mathbb{R}^n$, $n \geq 3$, and $f \in L^\infty(\Omega)$ with bounded support. In this regard, an important issue is that of computing $\partial_j \partial_k \mathbf{N}_\Omega f$ in the sense of distributions in $\mathbb{R}^n$, where $j, k \in \{1, \ldots, n\}$. First, we will show (cf. Theorem 4.105) that $\mathbf{N}_\Omega f \in C^1(\mathbb{R}^n)$ and we have

$$\partial_k(\mathbf{N}_\Omega f)(x) = \frac{1}{\omega_{n-1}} \int_\Omega \frac{x_k - y_k}{|x-y|^n} f(y)\, dy, \quad \forall\, x \in \mathbb{R}^n. \tag{4.10.2}$$

Hence, $\partial_k(\mathbf{N}_\Omega f)(x) = \int_\Omega \Phi(x-y)f(y)\, dy$ where $\Phi(z) := \frac{1}{\omega_{n-1}} \frac{z_k}{|z|^n}$ for each point $z \in \mathbb{R}^n \setminus \{0\}$. Note that for x fixed, $\Phi(x - \cdot)$ is locally integrable, though this is no longer the case for $(\partial_j \Phi)(x - \cdot)$. This makes the job of computing $\partial_j(\partial_k(\mathbf{N}_\Omega f))$ considerably more subtle. There is a good reason for this since, as it turns out, the latter distributional derivative involves (as we shall see momentarily) singular integral operators. We note that the function Φ considered above is C^∞ and positive homogeneous of degree $1 - n$ in $\mathbb{R}^n \setminus \{0\}$. These are going to be key features in our subsequent analysis.

Our most general results encompassing the discussion about the Newtonian potential introduced above are contained in Theorem 4.104 and Theorem 4.105. We begin by first considering the case when $\Omega = \mathbb{R}^n$ and f is a Schwartz function.

Proposition 4.102. Let $\Phi \in C^1(\mathbb{R}^n \setminus \{0\})$ be a function that is positive homogeneous of degree $1 - n$ and let $f \in \mathcal{S}(\mathbb{R}^n)$. Then $\int_{\mathbb{R}^n} \Phi(x - y)f(y)\, dy$ is differentiable in $\mathbb{R}^n$ and for each $j \in \{1, \ldots, n\}$ and at every $x \in \mathbb{R}^n$ we have

$$\partial_{x_j}\Big[\int_{\mathbb{R}^n} \Phi(x-y)f(y)\, dy \Big] = \Big(\int_{S^{n-1}} \Phi(\omega)\omega_j\, d\sigma(\omega) \Big) f(x) \tag{4.10.3}$$

$$+ \lim_{\varepsilon \to 0^+} \int_{|y-x| \geq \varepsilon} (\partial_j \Phi)(x-y)f(y)\, dy.$$

Proof. The key ingredient in the proof of (4.10.3) is formula (4.4.19). To see how (4.4.19) applies, first note that by using Exercise 4.53 we have for each $R > 0$,

$$\Big| \int_{\mathbb{R}^n} \Phi(x-y)f(y)\, dy \Big| \tag{4.10.4}$$

$$\leq \|\Phi\|_{L^\infty(S^{n-1})} \int_{\mathbb{R}^n} \frac{|f(y)|}{|x-y|^{n-1}}\, dy$$

$$\leq \|\Phi\|_{L^\infty(S^{n-1})} \|f\|_{L^\infty(\mathbb{R}^n)} \int_{|y-x| \leq R} \frac{dy}{|x-y|^{n-1}}$$

$$+ \|\Phi\|_{L^\infty(S^{n-1})} \|(1+|y|)^2 f\|_{L^\infty(\mathbb{R}^n)} \int_{|y-x| > R} \frac{dy}{(1+|y|^2)|x-y|^{n-1}} < \infty,$$

thus $\int_{\mathbb{R}^n} \Phi(x-y)f(y)\, dy$ is well defined. Second, since

$$\int_{\mathbb{R}^n} \Phi(x-y)f(y)\,\mathrm{d}y = \int_{\mathbb{R}^n} \Phi(y)f(x-y)\,\mathrm{d}y \qquad \forall\, x \in \mathbb{R}^n, \tag{4.10.5}$$

we have that $\int_{\mathbb{R}^n} \Phi(x-y)f(y)\,\mathrm{d}y$ is differentiable and

$$\partial_{x_j} \int_{\mathbb{R}^n} \Phi(x-y)f(y)\,\mathrm{d}y = \int_{\mathbb{R}^n} \Phi(y)(\partial_j f)(x-y)\,\mathrm{d}y \qquad \forall\, x \in \mathbb{R}^n. \tag{4.10.6}$$

Third, for each $x \in \mathbb{R}^n$, with t_x as in (2.8.42) (and recalling (3.2.22)), we may write

$$\int_{\mathbb{R}^n} \Phi(x-y)\varphi(y)\,\mathrm{d}y = \langle t_x(\Phi^\vee), \varphi \rangle, \qquad \forall\, \varphi \in \mathcal{S}(\mathbb{R}^n). \tag{4.10.7}$$

Now fix $j \in \{1, \ldots, n\}$. Then for $x \in \mathbb{R}^n$ we have

$$\partial_{x_j}\Big[\int_{\mathbb{R}^n} \Phi(x-y)f(y)\,\mathrm{d}y \Big] \tag{4.10.8}$$

$$= \langle t_x(\Phi^\vee), \partial_j f \rangle = \langle \Phi^\vee, t_{-x}(\partial_j f) \rangle = \langle \Phi^\vee, \partial_j[t_{-x}f] \rangle = -\langle \partial_j[\Phi^\vee], t_{-x}f \rangle$$

$$= -\Big(\int_{S^{n-1}} \omega_j \Phi^\vee(\omega)\,\mathrm{d}\sigma(\omega) \Big) \langle \delta, t_{-x}f \rangle - \big\langle \mathrm{P.V.}(\partial_j(\Phi^\vee)), t_{-x}f \big\rangle$$

$$= f(x)\Big(\int_{S^{n-1}} \omega_j \Phi(\omega)\,\mathrm{d}\sigma(\omega) \Big) + \lim_{\varepsilon \to 0^+} \int_{|y| \geq \varepsilon} (\partial_j \Phi)(-y)f(x+y)\,\mathrm{d}y$$

$$= f(x)\Big(\int_{S^{n-1}} \omega_j \Phi(\omega)\,\mathrm{d}\sigma(\omega) \Big) + \lim_{\varepsilon \to 0^+} \int_{|y| \geq \varepsilon} (\partial_j \Phi)(x-y)f(y)\,\mathrm{d}y.$$

The first equality in (4.10.8) uses (4.10.6) and (4.10.7), the fifth uses (4.4.19), the sixth the fact that $\partial_j(\Phi^\vee) = -(\partial_j\Phi)^\vee$, and the last one a suitable change of variables. This completes the proof of the corollary. $\qquad \square$

Next, we present a version of Proposition 4.102 when the function f is lacking any type of differentiability properties. Here we make use of the basic Calderón–Zygmund result recorded in Theorem 4.101.

Theorem 4.103. *Let $\Phi \in C^1(\mathbb{R}^n \setminus \{0\})$ be a function that is positive homogeneous of degree $1-n$ and let $f \in L^p(\mathbb{R}^n)$ for some $p \in (1, n)$. Then $\Phi * f \in L^1_{loc}(\mathbb{R}^n)$ and for each $j \in \{1, \ldots, n\}$ we have $T_{\partial_j\Phi}f \in L^p(\mathbb{R}^n)$ and*

$$\partial_j(\Phi * f) = \Big(\int_{S^{n-1}} \Phi(\omega)\omega_j\,\mathrm{d}\sigma(\omega) \Big)f + T_{\partial_j\Phi}f \qquad \text{in } \mathcal{D}'(\mathbb{R}^n), \tag{4.10.9}$$

where $T_{\partial_j\Phi}$ is the operator from (4.9.46) with ψ replaced by $\partial_j\Phi$.

Proof. Fix $p \in (1, n)$, $f \in L^p(\mathbb{R}^n)$, and $R \in (0, \infty)$. Then we write

$$\int_{B(0,R)} \int_{\mathbb{R}^n} \frac{|f(y)|}{|x-y|^{n-1}}\, dy\, dx = \int_{|y|\le 2R} |f(y)| \int_{B(0,R)} \frac{1}{|x-y|^{n-1}}\, dx\, dy$$

$$+ \int_{|y|>2R} |f(y)| \int_{B(0,R)} \frac{1}{|x-y|^{n-1}}\, dx\, dy$$

$$=: I + II. \tag{4.10.10}$$

If $y \in B(0,2R)$ and $x \in B(0,R)$, then $|x-y| \le 3R$, thus

$$I \le \int_{|y|\le 2R} |f(y)| \int_{|x-y|\le 3R} \frac{1}{|x-y|^{n-1}}\, dx\, dy$$

$$\le \Big(\int_{|z|\le 3R} \frac{1}{|z|^{n-1}}\, dz\Big)|B(0,2R)|^{1-\frac{1}{p}} \|f\|_{L^p(B(0,2R))} < \infty, \tag{4.10.11}$$

where the second inequality in (4.10.11) uses Hölder's inequality. Also, whenever $x \in B(0,R)$ and $y \in \mathbb{R}^n \setminus B(0,2R)$ we have

$$|y| \le |y-x| + |x| \le |y-x| + R \le |y-x| + \frac{|y|}{2}, \tag{4.10.12}$$

which implies $|y-x| \ge |y|/2$. Using this, II is estimated as

$$II \le 2^{n-1} \int_{|y|>2R} |f(y)| \int_{|x|\le R} \frac{1}{|y|^{n-1}}\, dx\, dy$$

$$\le 2^{n-1}|B(0,R)| \int_{|y|>2R} \frac{|f(y)|}{|y|^{n-1}}\, dy$$

$$\le 2^{n-1}|B(0,R)|\|f\|_{L^p(\mathbb{R}^n)}\Big(\int_{|y|>2R} \frac{dy}{|y|^{(n-1)p'}}\Big)^{1/p'} < \infty, \tag{4.10.13}$$

where $p' := \frac{p}{p-1}$ is the Hölder conjugate exponent for p. In the third inequality in (4.10.13) the assumption $p \in (1,n)$ has been used to ensure that $(n-1)p' > n$.

A combination of (4.10.10), (4.10.11), and (4.10.13) yields the following conclusion: for every $R \in (0,\infty)$ there exists some finite positive constant C, depending on R, n, and p, such that

$$\int_{B(0,R)} \int_{\mathbb{R}^n} \frac{|f(y)|}{|x-y|^{n-1}}\, dy\, dx \le C\|f\|_{L^p(\mathbb{R}^n)}, \quad \forall f \in L^p(\mathbb{R}^n). \tag{4.10.14}$$

In turn, (4.10.14) entails several useful conclusions. Recall that the assumptions on Φ imply $|\Phi(x)| \le \|\Phi\|_{L^\infty(S^{n-1})}|x|^{1-n}$ for each $x \in \mathbb{R}^n \setminus \{0\}$ (cf. Exercise 4.53). The first conclusion is that for each $f \in L^p(\mathbb{R}^n)$ and each $R \in (0,\infty)$,

$$\int_{\mathbb{R}^n} |\Phi(x-y)||f(y)|\, dy < \infty \quad \text{for a.e.} \quad x \in B(0,R). \tag{4.10.15}$$

This shows that $(\Phi * f)(x)$ is well defined for a.e. $x \in \mathbb{R}^n$. Second, from what we have just proved and (4.10.14) we may conclude that $\Phi * f \in L^1_{loc}(\mathbb{R}^n)$.

Next, recall that $C_0^\infty(\mathbb{R}^n)$ is dense in $L^p(\mathbb{R}^n)$ for each $p \in (1, \infty)$. As such there exists a sequence $\{f_k\}_{k \in \mathbb{N}}$ of function in $C_0^\infty(\mathbb{R}^n)$ with the property that $\lim_{k \to \infty} f_k = f$ in $L^p(\mathbb{R}^n)$. Based on (4.10.14) we may conclude that the sequence $\{\Phi * f_k\}_{k \in \mathbb{N}}$ converges in $L^p(\mathbb{R}^n)$ to $\Phi * f$. Therefore, by Exercise 2.25,

$$\Phi * f_k \xrightarrow[k \to \infty]{\mathcal{D}'(\mathbb{R}^n)} \Phi * f. \tag{4.10.16}$$

Moving on, fix $j \in \{1, \ldots, n\}$ and note that, by Exercise 4.51, $\partial_j \Phi$ is positive homogeneous of degree $-n$. The function $\Phi * f_k$ belongs to $C^\infty(\mathbb{R}^n)$ and Proposition 4.102 applies and implies that

$$\partial_j(\Phi * f_k) = \Big(\int_{S^{n-1}} \Phi(\omega)\omega_j \, d\sigma(\omega) \Big) f_k + T_{\partial_j \Phi} f_k, \quad \forall k \in \mathbb{N}, \tag{4.10.17}$$

pointwise in $\mathbb{R}^n$. In particular, from (4.10.17) we infer that $T_{\partial_j \Phi} f_k \in C^\infty(\mathbb{R}^n)$ for all $k \in \mathbb{N}$, and that the equality in (4.10.17) also holds in $\mathcal{D}'(\mathbb{R}^n)$. Since Theorem 4.101 gives that $T_{\partial_j \Phi} f_k \xrightarrow[k \to \infty]{L^p(\mathbb{R}^n)} T_{\partial_j \Phi} f$, based on Exercise 2.25 it follows that

$$T_{\partial_j \Phi} f_k \xrightarrow[k \to \infty]{\mathcal{D}'(\mathbb{R}^n)} T_{\partial_j \Phi} f \quad \text{and} \quad f_k \xrightarrow[k \to \infty]{\mathcal{D}'(\mathbb{R}^n)} f. \tag{4.10.18}$$

Since an immediate consequence of (4.10.16) is that

$$\partial_j(\Phi * f_k) \xrightarrow[k \to \infty]{\mathcal{D}'(\mathbb{R}^n)} \partial_j(\Phi * f), \tag{4.10.19}$$

combining this, (4.10.18), and the fact that (4.10.17) holds in $\mathcal{D}'(\mathbb{R}^n)$, we obtain (4.10.9). $\qquad\square$

We are prepared to state and prove our most general results regarding distributional derivatives of volume potentials. In particular, the next two theorems contain solutions to the questions formulated at the beginning of this section.

Theorem 4.104. *Let $\Phi \in C^1(\mathbb{R}^n \setminus \{0\})$ be a function that is positive homogeneous of degree $1-n$. Consider a measurable set $\Omega \subseteq \mathbb{R}^n$ and assume that $f \in L^p(\Omega)$ for some $p \in (1, n)$. Then $\int_\Omega \Phi(x-y)f(y) \, dy$ is absolutely convergent for a.e. point $x \in \mathbb{R}^n$ and belongs to $L^1_{loc}(\mathbb{R}^n)$ as a function of the variable x. Moreover, if $\widetilde{f}$ denotes the extension of f by zero to $\mathbb{R}^n$, then for each $j \in \{1, \ldots, n\}$ we have*

$$\partial_{x_j} \Big[\int_\Omega \Phi(x-y)f(y) \, dy \Big] = \Big(\int_{S^{n-1}} \Phi(\omega)\omega_j \, d\sigma(\omega) \Big) \widetilde{f}(x)$$

$$+ \lim_{\varepsilon \to 0^+} \int_{y \in \Omega \setminus B(x,\varepsilon)} (\partial_j \Phi)(x-y)f(y) \, dy \tag{4.10.20}$$

where the derivative in the left-hand side is taken in $\mathcal{D}'(\mathbb{R}^n)$ and the equality is also understood in $\mathcal{D}'(\mathbb{R}^n)$.

In particular, formula (4.10.20) holds for any function f belonging to some $L^q(\Omega)$, with $q \in (1, \infty)$, that vanishes outside of a measurable subset of Ω of finite measure.

Proof. The main claim in the statement follows by applying Theorem 4.103 to the function

$$\widetilde{f} := \begin{cases} f & \text{in } \Omega, \\ 0 & \text{in } \mathbb{R}^n \setminus \Omega, \end{cases} \tag{4.10.21}$$

upon observing that $\widetilde{f} \in L^p(\mathbb{R}^n)$. Moreover, if f is as in the last claim in the statement then Hölder's inequality may be invoked to show that f belongs to $L^p(\Omega)$ for some $p \in (1, n)$. $\qquad\square$

Recall that if $a \in \mathbb{R}$, the integer part of a is denoted by $\lfloor a \rfloor$ and is by definition the largest integer that is less than or equal to a. To state the next theorem we introduce

$$\langle a \rangle := \begin{cases} \lfloor a \rfloor & \text{if } a \notin \mathbb{Z}, \\ a - 1 & \text{if } a \in \mathbb{Z}. \end{cases} \tag{4.10.22}$$

Hence, $\langle a \rangle$ is the largest integer strictly less than a (thus, in particular, $\langle a \rangle < a$ for every $a \in \mathbb{R}$).

Theorem 4.105. *Let $\Phi \in C^\infty(\mathbb{R}^n \setminus \{0\})$ be a function that is positive homogeneous of degree $m \in \mathbb{R}$ where $m > -n$. Define the* generalized volume potential *associated with Φ by setting for each $f \in L^\infty_{comp}(\mathbb{R}^n)$*

$$(\Pi_\Phi f)(x) := \int_{\mathbb{R}^n} \Phi(x - y) f(y) \, dy, \qquad \forall x \in \mathbb{R}^n. \tag{4.10.23}$$

Then for each $f \in L^\infty_{comp}(\mathbb{R}^n)$ one has $\Pi_\Phi f \in C^{\langle m+n \rangle}(\mathbb{R}^n)$ and

$$\partial^\alpha(\Pi_\Phi f) = \Pi_{\partial^\alpha \Phi} f \quad \text{pointwise in } \mathbb{R}^n,$$
$$\text{for each } \alpha \in \mathbb{N}_0^n \text{ with } |\alpha| \le \langle m + n \rangle. \tag{4.10.24}$$

Moreover, if $\alpha \in \mathbb{N}_0^n$ is such that $|\alpha| = m + n$, then for each $f \in L^\infty_{comp}(\mathbb{R}^n)$, the distributional derivative $\partial^\alpha[\Pi_\Phi f]$ is of function type and satisfies

$$\partial^\alpha[\Pi_\Phi f](x) = \Big(\int_{S^{n-1}} (\partial^\beta \Phi)(\omega) \omega_j \, d\sigma(\omega) \Big) f(x) \tag{4.10.25}$$

$$+ \lim_{\varepsilon \to 0^+} \int_{\mathbb{R}^n \setminus B(x,\varepsilon)} (\partial^\alpha \Phi)(x - y) \, f(y) \, dy \qquad \text{in } \mathcal{D}'(\mathbb{R}^n),$$

for any $\beta \in \mathbb{N}_0^n$ and $j \in \{1, \dots, n\}$ such that $\beta + \mathbf{e}_j = \alpha$.

Proof. Fix $f \in L^\infty_{comp}(\mathbb{R}^n)$ and let $K := \operatorname{supp} f$ which is a compact set in $\mathbb{R}^n$. By Exercise 4.51 and Exercise 4.53,

for every $\alpha \in \mathbb{N}_0^n$ we have that $\partial^\alpha \Phi \in C^\infty(\mathbb{R}^n \setminus \{0\})$,

$\partial^\alpha \Phi$ is positive homogeneous of degree $m - |\alpha|$, and $\quad$ (4.10.26)

$$\left| \partial^\alpha \Phi(x - y) \right| \leq \|\partial^\alpha \Phi\|_{L^\infty(S^{n-1})} |x - y|^{m-|\alpha|} \quad \forall x, y \in \mathbb{R}^n, \ x \neq y.$$

Fix now $\alpha \in \mathbb{N}_0^n$ such that $|\alpha| \leq \langle m + n \rangle$. Then $m - |\alpha| \geq \langle m + n \rangle > -n$, hence (4.10.26) further yields

$$\int_{\mathbb{R}^n} |(\partial^\alpha \Phi)(x - y) f(y)| \, dy \leq C \|f\|_{L^\infty(\mathbb{R}^n)} \int_K |x - y|^{m-|\alpha|} \, dy < \infty. \quad (4.10.27)$$

This proves that $\Pi_{\partial^\alpha \Phi} f$ is well defined.

Next, we focus on proving

$$\Pi_{\partial^\alpha \Phi} f \in C^0(\mathbb{R}^n). \quad (4.10.28)$$

To see this, fix $x_0 \in \mathbb{R}^n$ and pick an arbitrary sequence $\{x_k\}_{k \in \mathbb{N}}$ of points in $\mathbb{R}^n$ satisfying $\lim_{k \to \infty} x_k = x_0$. Consider the following functions defined a.e. in $\mathbb{R}^n$:

$$v_k := (\partial^\alpha \Phi)(x_k - \cdot) f, \quad \forall k \in \mathbb{N}, \quad \text{and} \quad v := (\partial^\alpha \Phi)(x_0 - \cdot) f. \quad (4.10.29)$$

To conclude that $\Pi_{\partial^\alpha \Phi} f$ is continuous at x_0, it suffices to show that

$$\lim_{k \to \infty} \int_K v_k(y) \, dy = \int_K v(y) \, dy. \quad (4.10.30)$$

The strategy for proving (4.10.30) is to apply Vitali's theorem (cf. Theorem 14.29) with $X := K$ and μ being the restriction to K of the Lebesgue measure in $\mathbb{R}^n$. Since K is compact we have $\mu(X) < \infty$ and from (4.10.27) we know that $v_k \in L^1(X, \mu)$ for all $k \in \mathbb{N}$. Clearly, $|v(x)| < \infty$ for μ-a.e. $x \in K$. Also, $\lim_{k \to \infty} v_k(y) = v(y)$ for μ-almost every $y \in K$. Hence, in order to obtain (4.10.30), the only hypothesis left to verify in Vitali's theorem is that the sequence $\{v_k\}_{k \in \mathbb{N}}$ is uniformly integrable in (X, μ). With this goal in mind, let $\varepsilon > 0$ be fixed and consider a μ-measurable set $A \subset K$ such that $\mu(A)$ is sufficiently small, to a degree to be specified later. Then for every $k \in \mathbb{N}$, based on (4.10.26), we have

$$\left| \int_A v_k(x) \, d\mu(x) \right| \leq \|\partial^\alpha \Phi\|_{L^\infty(S^{n-1})} \|f\|_{L^\infty(\mathbb{R}^n)} \int_A |x_k - y|^{m-|\alpha|} \, dy$$

$$= C \int_{A - x_k} |x|^{m-|\alpha|} \, dx, \quad (4.10.31)$$

where $C := \|\partial^\alpha \Phi\|_{L^\infty(S^{n-1})} \|f\|_{L^\infty(\mathbb{R}^n)}$. Note that $\mu(A - x_k) = \mu(A)$ for each $k \in \mathbb{N}$. Also, since the sequence $\{x_k\}_{k \in \mathbb{N}}$ and the set K are bounded,

$$\exists R \in (0, \infty) \text{ such that } A - x_k \subset K - x_k \subset B(0, R) \quad \forall k \in \mathbb{N}. \tag{4.10.32}$$

Given that $m - |\alpha| > -n$, we have $|x|^{m-|\alpha|} \in L^1(B(0, R))$ and we may invoke Proposition 14.30 to conclude that there exists $\delta > 0$ such that for every Lebesgue measurable set $E \subset B(0, R)$ with Lebesgue measure less than δ we have $\int_E |x|^{m-|\alpha|} \, dx < \varepsilon/C$. At this point return with the latter estimate in (4.10.31) to conclude that

$$\text{if } \mu(A) < \delta \text{ then } \left| \int_A v_k(x) \, d\mu(x) \right| < \varepsilon \quad \forall k \in \mathbb{N}. \tag{4.10.33}$$

This proves that the sequence $\{v_k\}_{k \in \mathbb{N}}$ is uniformly integrable in (X, μ) and finishes the proof of the fact that $\Pi_{\partial^\alpha \Phi} f$ is continuous at x_0. Since x_0 was arbitrary in $\mathbb{R}^n$ the membership in (4.10.28) follows.

Our next goal is to show (4.10.24). First note that based on what we proved so far we have $\Pi_\Phi f, \Pi_{\partial^\alpha \Phi} f \in L^1_{loc}(\mathbb{R}^n)$, thus they define distributions in $\mathbb{R}^n$. We claim that the distribution $\Pi_{\partial^\alpha \Phi} f$ is equal to the distributional derivative $\partial^\alpha [\Pi_\Phi f]$. To see this, fix $\varphi \in C_0^\infty(\mathbb{R}^n)$ and using the definition of distributional derivatives and the definition of $\Pi_\Phi f$ write

$$\langle \partial^\alpha \Pi_\Phi f, \varphi \rangle = (-1)^{|\alpha|} \langle \Pi_\Phi f, \partial^\alpha \varphi \rangle$$

$$= (-1)^{|\alpha|} \int_{\mathbb{R}^n} \left(\int_{\mathbb{R}^n} \Phi(x - y) f(y) \, dy \right) \partial^\alpha \varphi(x) \, dx$$

$$= (-1)^{|\alpha|} \int_{\mathbb{R}^n} f(y) \left(\int_{\mathbb{R}^n} \Phi(x - y) \partial^\alpha \varphi(x) \, dx \right) dy. \tag{4.10.34}$$

Based on (4.10.26), the assumptions on φ, and the fact that $m - |\alpha| > -n$, we may use Lebesgue's Dominated Convergence Theorem and formula (4.10.26) repeatedly to further write

$$\int_{\mathbb{R}^n} \Phi(x - y) \partial^\alpha \varphi(x) \, dx = \lim_{\varepsilon \to 0^+} \int_{\mathbb{R}^n \setminus B(y, \varepsilon)} \Phi(x - y) \partial^\alpha \varphi(x) \, dx \tag{4.10.35}$$

$$= \lim_{\varepsilon \to 0^+} \left\{ (-1)^{|\alpha|} \int_{\mathbb{R}^n \setminus B(y, \varepsilon)} (\partial^\alpha \Phi)(x - y) \varphi(x) \, dx \right.$$

$$\left. + \sum_{\alpha = \beta + \gamma + e_j} c_{\beta\gamma}^j \int_{\partial B(y, \varepsilon)} (\partial^\beta \Phi)(x - y) \frac{x_j - y_j}{\varepsilon} \partial^\gamma \varphi(x) \, d\sigma(x) \right\},$$

where $c_{\beta\gamma}^j$ are suitable constants independent of ε. Note that, for each β, γ, and j such that $\alpha = \beta + \gamma + \mathbf{e}_j$, in light of (4.10.26) we have

$$\left| \int_{\partial B(y,\varepsilon)} (\partial^\beta \Phi)\,(x-y)\frac{x_j - y_j}{\varepsilon}\,\partial^\gamma \varphi(x)\,d\sigma(x) \right|$$

$$\leq \int_{\partial B(y,\varepsilon)} |(\partial^\beta \Phi)(x-y)|\,|\partial^\gamma \varphi(x)|\,d\sigma(x)$$

$$\leq \|\partial^\gamma \varphi\|_{L^\infty(\mathbb{R}^n)} \|\partial^\beta \Phi\|_{L^\infty(S^{n-1})} \int_{\partial B(y,\varepsilon)} |x-y|^{m-|\beta|}\,d\sigma(x)$$

$$= C\varepsilon^{m-|\beta|+n-1} \xrightarrow[\varepsilon\to 0^+]{} 0, \tag{4.10.36}$$

since $m - |\beta| + n - 1 \geq m + n - |\alpha| > 0$. Returning with (4.10.36) to (4.10.35), and applying one more time Lebesgue's Dominated Convergence Theorem, it follows that

$$\int_{\mathbb{R}^n} \Phi(x-y)\partial^\alpha \varphi(x)\,dx = (-1)^{|\alpha|} \int_{\mathbb{R}^n} (\partial^\alpha \Phi)(x-y)\varphi(x)\,dx. \tag{4.10.37}$$

This, when used in (4.10.34) further yields

$$\langle \partial^\alpha \Pi_\Phi f, \varphi \rangle = \int_{\mathbb{R}^n} f(y)\Big(\int_{\mathbb{R}^n} (\partial^\alpha \Phi)(x-y)\varphi(x)\,dx \Big)\,dy$$

$$= \int_{\mathbb{R}^n} \Big(\int_{\mathbb{R}^n} (\partial^\alpha \Phi)(x-y)f(y)\,dy \Big)\varphi(x)\,dx$$

$$= \langle \Pi_{\partial^\alpha \Phi} f, \varphi \rangle. \tag{4.10.38}$$

Since $\varphi \in C_0^\infty(\mathbb{R}^n)$ is arbitrary, from (4.10.38) we conclude

$$\partial^\alpha \Pi_\Phi f = \Pi_{\partial^\alpha \Phi} f \quad \text{in} \quad \mathcal{D}'(\mathbb{R}^n), \quad \forall\, \alpha \in \mathbb{N}_0^n \text{ with } |\alpha| \leq \langle m+n \rangle. \tag{4.10.39}$$

Upon observing that (4.10.28) and (4.10.39) hold for every $\alpha \in \mathbb{N}_0^n$ with the property that $|\alpha| = \langle m + n \rangle$, we may invoke Theorem 2.112 to infer that $\Pi_\Phi f$ belongs to $C^{\langle m+n \rangle}(\mathbb{R}^n)$ and that (4.10.24) is valid.

We are left with proving the very last statement in the theorem. To this end, suppose there exists $\alpha \in \mathbb{N}_0^n$ satisfying $|\alpha| = m + n$. In particular, we have $m \in \mathbb{Z}$ and $|\alpha| \geq 1$. Hence, there exists $j \in \{1, \ldots, n\}$ with the property that $\alpha_j \geq 1$. Set $\beta := (\alpha_1, \ldots, \alpha_{j-1}, \alpha_j - 1, \alpha_{j+1}, \ldots, \alpha_n)$, so that

$$\alpha = \beta + \mathbf{e}_j \quad \text{and} \quad |\beta| = m + n - 1 = \langle m + n \rangle. \tag{4.10.40}$$

Based on what we proved earlier, we have $\partial^\beta[\Pi_\Phi f] = \Pi_{\partial^\beta \Phi} f$ pointwise in $\mathbb{R}^n$. Also, $\partial^\beta \Phi$ is of class C^∞ and positive homogeneous of degree $1 - n$ in $\mathbb{R}^n \setminus \{0\}$. Thus, we may apply Theorem 4.104 with Φ replaced by $\partial^\beta \Phi$ and Ω replaced by K and obtain

$$\partial^\alpha[\Pi_\Phi f](x) = \partial_j[\partial^\beta[\Pi_\Phi f]](x) = \partial_j[\Pi_{\partial^\beta \Phi} f](x)$$

$$= \partial_{x_j}\Big[\int_K (\partial^\beta \Phi)(x - y) f(y)\,\mathrm{d}y\Big]$$

$$= \Big(\int_{S^{n-1}} (\partial^\beta \Phi)(\omega)\omega_j\,\mathrm{d}\sigma(\omega)\Big) f(x)$$

$$+ \lim_{\varepsilon \to 0^+} \int_{\mathbb{R}^n \setminus B(x,\varepsilon)} (\partial^\alpha \Phi)(x - y)\,f(y)\,\mathrm{d}y, \qquad (4.10.41)$$

where the derivative in the left-hand side of (4.10.41) is taken in $\mathcal{D}'(\mathbb{R}^n)$ and the equality is understood in $\mathcal{D}'(\mathbb{R}^n)$. This proves (4.10.25) and finishes the proof of the theorem. $\qquad\qquad\square$

Exercise 4.106. In the context of Theorem 4.105 prove directly, without relying on Theorem (2.112), that whenever $f \in L^\infty_{comp}(\mathbb{R}^n)$ one has $\Pi_\Phi f \in C^{\langle m+n\rangle}(\mathbb{R}^n)$.

Hint: Show that for each $x \in \mathbb{R}^n$ and each $j \in \{1, \ldots, n\}$ fixed, one has

$$\lim_{h \to 0}\Big[(\Pi_\Phi f)(x + h\mathbf{e}_j) - (\Pi_\Phi f)(x)\Big] = (\Pi_{\partial_j \Phi} f)(x) \qquad (4.10.42)$$

The proof of (4.10.42) may be done by using Vitali's theorem (cf. Theorem 14.29) in a manner analogous to the proof of (4.10.28). Once (4.10.42) is established, iterate to allow higher order partial derivatives.

Further Notes for Chapter 4. The significance of the class of tempered distributions stems from the fact that this class is stable under the action of the Fourier transform. The topics discussed in Sections 4.1–4.6 are classical and a variety of expositions is present in the literature, though they differ in terms of length and depth, and the current presentation is no exception. For example, while the convolution product of distributions is often confined to the case in which one of the distributions in question is compactly supported, i.e., $\mathcal{E}'(\mathbb{R}^n) * \mathcal{D}'(\mathbb{R}^n)$, in Theorem 4.19 we have seen that $\mathcal{S}(\mathbb{R}^n) * \mathcal{S}'(\mathbb{R}^n)$ continues to be meaningfully defined in $\mathcal{S}'(\mathbb{R}^n)$. For us, extending the action of the convolution product in this manner is motivated by the discussion in Section 4.9, indicating how singular integral operators may be interpreted as multipliers.

The main result in Section 4.7, Theorem 4.79, appears to be new at least in the formulation and the degree of generality in which it has been presented. This result may be regarded as a far-reaching generalization of Sokhotsky's formula (2.10.3); cf. Remark 4.82 for details. Theorem 4.79 has a number of remarkable consequences, and we use it to offer a new perspective on the treatment of the classical harmonic Poisson kernel in Section 4.8. Later on, in Section 11.6, Theorem 4.79 resurfaces as the key ingredient in the study of boundary behavior of layer potential operators in the upper-half space.

The treatment of the singular integral operators from in Section 4.9 highlights the interplay between distribution theory, harmonic analysis, and partial differential equations. Specifically, first the action of a singular integral operator of the form T_Θ on a Schwartz function φ is interpreted as $(\mathrm{P.V.}\,\Theta) * \varphi$ (which, as pointed out earlier, is a well-defined object in $\mathcal{S}(\mathbb{R}^n) * \mathcal{S}'(\mathbb{R}^n) \subset \mathcal{S}'(\mathbb{R}^n)$). Second, the Fourier analysis of tempered distributions is invoked to conclude that $\widehat{T_\Theta \varphi} = \widehat{\varphi}\,\mathcal{F}(\mathrm{P.V.}\,\Theta)$ which shifts the focus of understanding the properties of the singular integral operator T_Θ to clarifying the nature of the tempered distribution $m_\Theta := \mathcal{F}(\mathrm{P.V.}\,\Theta)$. Third, it turns out that the distribution

m_Θ is of function type and, in fact, a suitable pointwise formula may be deduced for it. In particular, it is apparent from this formula that $m_\Theta \in L^\infty(\mathbb{R}^n)$. Having established this, in a fourth step we may return to the original focus of our investigation, namely the singular integral operator T_Θ, and use the fact that $\widehat{T_\Theta \varphi} = m_\Theta \widehat{\varphi}$, along with Plancherel's formula and the boundedness of m_Θ to eventually conclude (via a density argument) that T_Θ extends to a linear and bounded operator on $L^2(\mathbb{R}^n)$. In turn, singular integral operators naturally intervene in the derivatives of volume potentials discussed in Section 4.10, and the boundedness result just derived ultimately becomes the key tool in obtaining estimates for the solution of the Poisson problem, treated later.

The discussion in the previous paragraph also sheds light on some of the main aims of the theory of singular integral operators (SIO), as a subbranch of harmonic analysis. For example, one would like to extend the action of SIO to $L^p(\mathbb{R}^n)$ for any $p \in (1, \infty)$, rather than just $L^2(\mathbb{R}^n)$. Also, it is desirable to consider SIO which are not necessarily of convolution type, and/or in a setting in which $\mathbb{R}^n$ is replaced by a more general type of ambient (e.g., some type of surface in $\mathbb{R}^{n+1}$). The execution of this program, essentially originating in the work of A. P. Calderón, A. Zygmund, and S. G. Mikhlin in the 1950s, stretches all the way to the present day. The reader is referred to the exposition in the monographs [8], [9], [19], [26], [52], [68], [69], where references to specific research articles may be found. Here, we only wish to mention that, in turn, such progress in harmonic analysis has led to significant advances in those areas of partial differential equations where SIO play an important role. In this regard, the reader is referred to the discussion in [31], [39], [52], [57].

4.11 Additional Exercises for Chapter 4

Exercise 4.107. Prove that P.V. $\frac{1}{x} \in \mathcal{S}'(\mathbb{R})$.

Exercise 4.108. Prove that $|x|^N \ln |x| \in \mathcal{S}'(\mathbb{R}^n)$ if N is a real number satisfying $N > -n$.

Exercise 4.109. Prove that $(\ln |x|)' = \text{P.V.} \frac{1}{x}$ in $\mathcal{S}'(\mathbb{R})$.

Exercise 4.110. Let $a \in (0, \infty)$ and recall the Heaviside function H from (1.2.9). Prove that the function $e^{ax} H(x)$, $x \in \mathbb{R}$, does not belong to $\mathcal{S}'(\mathbb{R})$. Do any of the functions $e^{-ax} H(-x)$, $e^{ax} H(-x)$, or $e^{-ax} H(x)$, defined for $x \in \mathbb{R}$, belong to $\mathcal{S}'(\mathbb{R})$?

Exercise 4.111. In each case determine if the given sequence of tempered distributions indexed over $j \in \mathbb{N}$ converges in $\mathcal{S}'(\mathbb{R})$. Whenever convergent, determine its limit.

(a) $f_j(x) = \frac{x}{x^2 + j^{-2}}$, $x \in \mathbb{R}$;

(b) $f_j(x) = \frac{1}{j} \cdot \frac{1}{x^2 + j^{-2}}$, $x \in \mathbb{R}$;

(c) $f_j(x) = \frac{1}{\pi} \frac{\sin jx}{x}$, $x \in \mathbb{R} \setminus \{0\}$;

(d) $f_j(x) = e^{j^2} \delta_j$;

Exercise 4.112. For each $j \in \mathbb{N}$ let $f_j(x) := j^n\theta(jx)$ for each $x \in \mathbb{R}^n$, where the function $\theta \in \mathcal{S}(\mathbb{R}^n)$ is fixed and satisfies $\int_{\mathbb{R}^n} \theta(x)\,dx = 1$. Does the sequence $\{f_j\}_{j\in\mathbb{N}}$ converge in $\mathcal{S}'(\mathbb{R}^n)$? If yes, determine its limit.

Exercise 4.113. For each $j \in \mathbb{N}$ consider the function $f_j(x) := e^{x^2}\chi_{[-j,j]}(x)$ defined for $x \in \mathbb{R}$. Prove that the sequence $\{f_j\}_{j\in\mathbb{N}}$ converges in $\mathcal{D}'(\mathbb{R})$ but not in $\mathcal{S}'(\mathbb{R})$.

Exercise 4.114. For each $j \in \mathbb{N}$ let $f_j(x) := \chi_{[j-1,j]}(x)$ for $x \in \mathbb{R}$. Prove that the sequence $\{f_j\}_{j\in\mathbb{N}}$ does not converge in $L^p(\mathbb{R})$ for any $p \geq 1$ but it converges to zero in $\mathcal{S}'(\mathbb{R})$.

Exercise 4.115. For each $j \in \mathbb{N}$ consider the tempered distribution $u_j \in \mathcal{S}'(\mathbb{R})$ defined by $u_j := \chi_{[-j,j]}e^x \sin(e^x)$. Prove that $u_j \xrightarrow[j\to\infty]{\mathcal{D}'(\mathbb{R})} e^x\sin(e^x)$. Does the sequence $\{u_j\}_{j\in\mathbb{N}}$ converge in $\mathcal{S}'(\mathbb{R})$? If yes, what is the limit?

Exercise 4.116. Let $m \in \mathbb{N}$. Prove that any solution u of the equation $x^m u = 0$ in $\mathcal{D}'(\mathbb{R})$ satisfies $u \in \mathcal{S}'(\mathbb{R})$. Use the Fourier transform to show that the general solution to this equation is $u = \sum\limits_{k=0}^{m-1} c_k\delta^{(k)}$ in $\mathcal{S}'(\mathbb{R})$, where $c_k \in \mathbb{C}$, for every $k = 0, 1, \ldots, m-1$.

Exercise 4.117. Prove that for any $f \in \mathcal{S}'(\mathbb{R})$ there exists $u \in \mathcal{S}'(\mathbb{R})$ such that $xu = f$ in $\mathcal{S}'(\mathbb{R})$.

Exercise 4.118. Does the equation $e^{-|x|^2}u = 1$ in $\mathcal{S}'(\mathbb{R}^n)$ have a solution?

Exercise 4.119. Prove that the Heaviside function H belongs to $\mathcal{S}'(\mathbb{R})$ and compute its Fourier transform in $\mathcal{S}'(\mathbb{R})$.

Exercise 4.120. Compute the Fourier transform of P.V. $\frac{1}{x}$ in $\mathcal{S}'(\mathbb{R})$.

Exercise 4.121. Compute the Fourier transform in $\mathcal{S}'(\mathbb{R})$ of each of the following tempered distribution (all are considered in $\mathcal{S}'(\mathbb{R})$ and recall the definition of the sgn function from (1.4.3):

(a) $\operatorname{sgn} x$;

(b) $|x|^k$ for $k \in \mathbb{N}$;

(c) $\frac{\sin(ax)}{x}$ for $a \in \mathbb{R} \setminus \{0\}$;

(d) $\frac{\sin(ax)\sin(bx)}{x}$ for $a, b \in \mathbb{R} \setminus \{0\}$;

(e) $\sin(x^2)$;

(f) $\ln|x|$.

Exercise 4.122. Suppose $a, b \in \mathbb{R}$ are such that $a > 0$. Consider $f(x) := \frac{1}{x^2-(b+ia)^2}$ for every $x \in \mathbb{R}$. Note that $f \in L^1(\mathbb{R})$ hence (4.1.9) implies $f \in \mathcal{S}'(\mathbb{R})$. Compute the Fourier transform in $\mathcal{S}'(\mathbb{R})$ of f.

Exercise 4.123. Let $n = 3$ and $R \in (0, \infty)$. Compute the Fourier transform of the tempered distribution $\delta_{\partial B(0,R)}$ in $S'(\mathbb{R}^3)$, where $\delta_{\partial B(0,R)}$ the distribution defined as in Exercise 2.146 corresponding to $\Sigma := \partial B(0, R)$.

Exercise 4.124. Let $k \in \mathbb{N}_0$ and suppose $f \in L^1(\mathbb{R}^n)$ has the property that $x^\alpha f$ belongs to $L^1(\mathbb{R}^n)$ for every $\alpha \in \mathbb{N}_0^n$ satisfying $|\alpha| \leq k$. Prove that $\widehat{f}$, the Fourier transform of f in $S'(\mathbb{R}^n)$ satisfies $\widehat{f} \in C^k(\mathbb{R}^n)$.

Exercise 4.125. Show that $\chi_{[-1,1]} \in S'(\mathbb{R})$ and compute $\widehat{\chi_{[-1,1]}}$ in $S'(\mathbb{R})$. Do we have $\widehat{\chi_{[-1,1]}} \in L^1(\mathbb{R})$?

Exercise 4.126. Fix $x_0 \in \mathbb{R}^n$ and set

$$t_{x_0} : S'(\mathbb{R}^n) \to S'(\mathbb{R}^n)$$
$$\langle t_{x_0} u, \varphi \rangle := \langle u, t_{-x_0}(\varphi) \rangle, \quad \forall \varphi \in S(\mathbb{R}^n), \tag{4.11.1}$$

where $t_{-x_0}(\varphi)$ is understood as in Exercise 3.44. Prove that the map in (4.11.1) is well defined, linear, and continuous and is an extension of the map from Exercise 3.44 in the sense that, if $f \in S(\mathbb{R}^n)$, then $t_{x_0} u_f = u_{t_{x_0}(f)}$ in $S'(\mathbb{R}^n)$. Also show that

$$\mathcal{F}(t_{x_0} u) = e^{-i x_0 \cdot (\cdot)} \widehat{u} \quad \text{in} \quad S'(\mathbb{R}^n) \quad \text{for every } u \in S'(\mathbb{R}^n). \tag{4.11.2}$$

Exercise 4.127. Let $a \in \mathbb{R}$ and consider the function $g(x) := \cos(ax)$ for every $x \in \mathbb{R}$. Prove that $g \in S'(\mathbb{R})$ and compute $\widehat{g}$ in $S'(\mathbb{R})$.

Exercise 4.128. Let $m \in \mathbb{N}$ and suppose P is a polynomial in $\mathbb{R}^n$ of degree $2m$ that has no real roots. Prove that $\frac{1}{P} \in S'(\mathbb{R}^n)$, and that if $2m > n$ then $\mathcal{F}(\frac{1}{P}) \in C^{2m-n-1}(\mathbb{R}^n)$.

Exercise 4.129. Let P be a polynomial in $\mathbb{R}^n$ and suppose $k \in \mathbb{N}$. Prove that if P is homogeneous of degree $-k$ then $P \equiv 0$.

Exercise 4.130. Let $\zeta, \eta \in \mathbb{R}^n$, $n \geq 2$, be two unit vectors such that $\zeta \cdot \eta \neq -1$, and consider the linear mapping $R : \mathbb{R}^n \to \mathbb{R}^n$ define by

$$R\xi := \xi - \left[\frac{\xi \cdot (\eta + \zeta)}{1 + \eta \cdot \zeta} \right] \zeta + \left[\frac{\xi \cdot [(1 + 2\eta \cdot \zeta)\zeta - \eta]}{1 + \eta \cdot \zeta} \right] \eta, \quad \forall \xi \in \mathbb{R}^n. \tag{4.11.3}$$

Show that this is an orthogonal transformation in $\mathbb{R}^n$ that satisfies

$$R\zeta = \eta \quad and \quad R^\top \eta = \zeta. \tag{4.11.4}$$

Exercise 4.131. Prove that for every $c \in \mathbb{R} \setminus \{0\}$ the following formulas hold:

$$\lim_{\substack{\varepsilon \to 0^+ \\ R \to \infty}} \int_\varepsilon^R \frac{\cos(c\rho) - \cos\rho}{\rho}\, d\rho = -\ln|c|, \tag{4.11.5}$$

$$\lim_{\substack{\varepsilon \to 0^+ \\ R \to \infty}} \int_\varepsilon^R \frac{\sin(c\rho)}{\rho}\, d\rho = \frac{\pi}{2}\operatorname{sgn} c, \tag{4.11.6}$$

$$\sup_{0 < \varepsilon < R} \left| \int_\varepsilon^R \frac{\cos(c\rho) - \cos\rho}{\rho}\, d\rho \right| \le 2\left|\ln|c|\right|, \tag{4.11.7}$$

$$\sup_{0 < \varepsilon < R} \left| \int_\varepsilon^R \frac{\sin(c\rho)}{\rho}\, d\rho \right| \le 4. \tag{4.11.8}$$

Chapter 5
The Concept of Fundamental Solution

Abstract The first explicit encounter with the notion of fundamental solution takes place in this chapter. We consider constant coefficient linear differential operators and discuss the existence of a fundamental solution for such operators based on the classical Malgrange–Ehrenpreis theorem.

5.1 Constant Coefficient Linear Differential Operators

Recall (3.1.10). To fix notation, for $m \in \mathbb{N}_0$, let

$$P(D) := \sum_{|\alpha| \leq m} a_\alpha D^\alpha, \quad a_\alpha \in \mathbb{C}, \quad \forall \alpha \in \mathbb{N}_0^n, \ |\alpha| \leq m. \tag{5.1.1}$$

Also, whenever $P(D)$ is as above, define $P(-D) := \sum_{|\alpha| \leq m} (-1)^{|\alpha|} a_\alpha D^\alpha$, and set

$$P(\xi) := \sum_{|\alpha| \leq m} a_\alpha \xi^\alpha, \quad \forall \xi \in \mathbb{R}^n. \tag{5.1.2}$$

Typically, an object $P(D)$ as in (5.1.1) is called a constant coefficient linear differential operator, while $P(\xi)$ from (5.1.2) is referred to as its `total` (or full) `symbol`. Moreover,

$$P(D) \text{ is said to have order } m \text{ provided}$$

$$\text{there exists } \alpha_* \in \mathbb{N}_0^n \text{ with } |\alpha_*| = m \text{ and } a_{\alpha_*} \neq 0. \tag{5.1.3}$$

Whenever $P(D)$ is a constant coefficient linear differential operator of order m we define its `principal symbol` to be

$$P_m(\xi) := \sum_{|\alpha|=m} a_\alpha \xi^\alpha, \quad \text{for } \xi \in \mathbb{R}^n. \tag{5.1.4}$$

© Springer Nature Switzerland AG 2018
D. Mitrea, *Distributions, Partial Differential Equations, and Harmonic Analysis*,
Universitext, https://doi.org/10.1007/978-3-030-03296-8_5

Remark 5.1. Obviously, any operator of the form

$$Q(\partial) := \sum_{|\alpha| \leq m} b_\alpha \partial^\alpha, \tag{5.1.5}$$

may be expressed as $P(D)$ for the choice of coefficients $a_\alpha := i^{|\alpha|} b_\alpha$ for each $\alpha \in \mathbb{N}_0^n$ with $|\alpha| \leq m$. Hence, we may move freely back and forth between the writings of a differential operator in (5.1.1) and (5.1.5). While we shall naturally define in this case $Q(\xi) := \sum_{|\alpha| \leq m} b_\alpha \xi^\alpha$, the reader is advised that the full symbol of $Q(\partial)$ from (5.1.5) is the expression $\sum_{|\alpha| \leq m} i^{|\alpha|} b_\alpha \xi^\alpha$, for $\xi \in \mathbb{R}^n$, while its principal symbol is $\sum_{|\alpha|=m} i^m b_\alpha \xi^\alpha$, for $\xi \in \mathbb{R}^n$.

The terminology and formalism described above will play a significant role in our subsequent work. In the last part of this section, we prove a regularity result involving constant coefficient linear differential operators with nonvanishing total symbol outside the origin, that is further used to establish a rather general Liouville type theorem for this class of operators.

Proposition 5.2. *Suppose $P(D)$ is a constant coefficient linear differential operator in $\mathbb{R}^n$ with the property that*

$$P(\xi) \neq 0 \quad \text{for every} \quad \xi \in \mathbb{R}^n \setminus \{0\}. \tag{5.1.6}$$

Then if $u \in S'(\mathbb{R}^n)$ is such that $P(D)u = 0$ in $S'(\mathbb{R}^n)$ it follows that u is a polynomial in $\mathbb{R}^n$ and $P(D)u = 0$ pointwise in $\mathbb{R}^n$.

Proof. If $u \in S'(\mathbb{R}^n)$ is such that $P(D)u = 0$ in $S'(\mathbb{R}^n)$, after taking the Fourier transform, we obtain $P(\xi)\widehat{u} = 0$ in $S'(\mathbb{R}^n)$. Granted (5.1.6), the latter identity implies $\operatorname{supp}\widehat{u} \subseteq \{0\}$. By Exercise 4.37, it follows that u is a polynomial in $\mathbb{R}^n$. Since $u \in C^\infty(\mathbb{R}^n)$, the condition $P(D)u = 0$ in $S'(\mathbb{R}^n)$ further yields $P(D)u = 0$ pointwise in $\mathbb{R}^n$. □

In preparation for the general Liouville type theorem advertised earlier, we prove the following result relating the growth of a polynomial to its degree.

Lemma 5.3. *Let $P(x)$ be a polynomial in $\mathbb{R}^n$ with the property that there exist $N \in \mathbb{N}_0$ and $C, R \in [0, \infty)$ such that*

$$|P(x)| \leq C|x|^N \quad \text{whenever} \quad |x| \geq R. \tag{5.1.7}$$

Then $P(x)$ has degree at most N. In particular, the only bounded polynomials in $\mathbb{R}^n$ are constants.

Proof. We begin by noting that the case $n = 1$ is easily handled. In the multidimensional setting, assume that $P(x) = \sum_{|\alpha| \leq m} a_\alpha x^\alpha$ is a polynomial in $\mathbb{R}^n$ satisfying (5.1.7) for some $N \in \mathbb{N}_0$ and $C, R \in [0, \infty)$. Suppose $m \geq N + 1$, since otherwise there is nothing to prove. Fix an arbitrary $\omega \in S^{n-1}$ and consider the one-dimensional polynomial $p_\omega(t) := P(t\omega)$ for $t \in \mathbb{R}$, i.e.,

$$p_\omega(t) = \sum_{j=0}^{m} \Big(\sum_{|\alpha|=j} a_\alpha \omega^\alpha \Big) t^j, \qquad t \in \mathbb{R}. \tag{5.1.8}$$

Given that $|p_\omega(t)| = |P(t\omega)| \le C|t\omega|^N = Ct^N$ whenever $|t| \ge R$, the one-dimensional result applies and gives that the coefficient of t^j in (5.1.8) vanishes if $j \ge N + 1$. Thus,

$$P(t\omega) = p_\omega(t) = \sum_{j=0}^{N} \Big(\sum_{|\alpha|=j} a_\alpha \omega^\alpha \Big) t^j = \sum_{|\alpha| \le N} a_\alpha (t\omega)^\alpha, \qquad \forall\, t \in \mathbb{R}. \tag{5.1.9}$$

Given that $\omega \in S^{n-1}$ was arbitrary, it follows that $P(x) = \sum_{|\alpha| \le N} a_\alpha x^\alpha$ for every $x \in \mathbb{R}^n$. $\qquad \square$

All ingredients are now in place for dealing with the following result.

Theorem 5.4 (A general Liouville type theorem). *Assume $P(D)$ is a constant coefficient linear differential operator in $\mathbb{R}^n$ such that (5.1.6) holds. Also, suppose $u \in L^1_{loc}(\mathbb{R}^n)$ satisfies $P(D)u = 0$ in $S'(\mathbb{R}^n)$ and has the property that there exist $N \in \mathbb{N}_0$ and $C, R \in [0, \infty)$ such that*

$$|u(x)| \le C|x|^N \quad \text{whenever} \quad |x| \ge R. \tag{5.1.10}$$

Then u is a polynomial in $\mathbb{R}^n$ of degree at most N.

In particular, if $u \in L^\infty(\mathbb{R}^n)$ satisfies $P(D)u = 0$ in $S'(\mathbb{R}^n)$ then u is necessarily a constant.

Proof. Since (5.1.10) implies that the locally integrable function u belongs to $S'(\mathbb{R}^n)$ (cf. Example 4.4), Proposition 5.2 implies that u is a polynomial in $\mathbb{R}^n$. Moreover, Lemma 5.3 gives that the degree of this polynomial is at most N. $\qquad \square$

5.2 A First Look at Fundamental Solutions

Definition 5.5. Let $P(D)$ be a constant coefficient linear differential operator in $\mathbb{R}^n$. Then $E \in \mathcal{D}'(\mathbb{R}^n)$ is called a `fundamental solution` for $P(D)$ provided

$$P(D)E = \delta \quad \text{in} \quad \mathcal{D}'(\mathbb{R}^n). \tag{5.2.1}$$

A simple, useful observation is that $E \in \mathcal{D}'(\mathbb{R}^n)$ is a fundamental solution for $P(D)$ if and only if

$$\langle E, P(-D)\varphi \rangle = \varphi(0), \qquad \forall\, \varphi \in C_0^\infty(\mathbb{R}^n). \tag{5.2.2}$$

Indeed, this is immediate from the fact that $\langle P(D)E, \varphi \rangle = \langle E, P(-D)\varphi \rangle$ for each function $\varphi \in C_0^\infty(\mathbb{R}^n)$.

Remark 5.6. Suppose $P(D)$ is a constant coefficient linear differential operator in $\mathbb{R}^n$. Then the equation

$$P(D)u = f \qquad (5.2.3)$$

is called the `Poisson equation for the operator` $P(D)$. If the operator $P(D)$ has a fundamental solution $E \in \mathcal{D}'(\mathbb{R}^n)$, then for any $f \in \mathcal{D}'(\mathbb{R}^n)$ for which $E * f$ exists as an element of $\mathcal{D}'(\mathbb{R}^n)$ (for example, this is the case if $f \in \mathcal{E}'(\mathbb{R}^n)$) the Poisson equation (5.2.3) is solvable in $\mathcal{D}'(\mathbb{R}^n)$. Indeed, if we set $u := E * f$, then

$$P(D)u = (P(D)E) * f = \delta * f = f \quad \text{in} \quad \mathcal{D}'(\mathbb{R}^n). \qquad (5.2.4)$$

This is one of the many reasons for which fundamental solutions are important when solving partial differential equations.

Remark 5.7. Let $P(D)$ and $Q(D)$ be two constant coefficient linear differential operators in $\mathbb{R}^n$ and let $E \in \mathcal{D}'(\mathbb{R}^n)$ be a fundamental solution for the operator $Q(D)P(D)$ in $\mathcal{D}'(\mathbb{R}^n)$. Then $P(D)E$ is a fundamental solution for $Q(D)$ in $\mathcal{D}'(\mathbb{R}^n)$. Moreover, $P(D)E \in \mathcal{S}'(\mathbb{R}^n)$ if $E \in \mathcal{S}'(\mathbb{R}^n)$.

In general, there are two types of questions we are going to be concerned with in this regard, namely whether fundamental solutions exist, and if so whether they can all be cataloged. The proposition below shows that for any differential operator with a nonvanishing total symbol on $\mathbb{R}^n \setminus \{0\}$ a fundamental solution which is a tempered distribution is unique up to polynomials.

Proposition 5.8. *Suppose $P(D)$ is a constant coefficient linear differential operator in $\mathbb{R}^n$ with the property that $P(\xi) \neq 0$ for every $\xi \in \mathbb{R}^n \setminus \{0\}$.*

If $P(D)$ has a fundamental solution E that is a tempered distribution, then any other fundamental solution of $P(D)$ belonging to $\mathcal{S}'(\mathbb{R}^n)$ differs from E by a polynomial Q satisfying $P(D)Q = 0$ pointwise in $\mathbb{R}^n$.

Proof. Suppose $u \in \mathcal{S}'(\mathbb{R}^n)$ is such that $P(D)u = \delta$ in $\mathcal{S}'(\mathbb{R}^n)$. Then $P(D)(u - E) = 0$ in $\mathcal{S}'(\mathbb{R}^n)$ and the desired conclusion is provided by Proposition 5.2. $\square$

Consider now the task of finding a formula for a fundamental solution. In a first stage, suppose $P(D)$ is as in (5.1.1) and make the additional assumption that

$$\text{there exists } c \in (0, \infty) \text{ such that } |P(\xi)| \geq c \text{ for every } \xi \in \mathbb{R}^n. \qquad (5.2.5)$$

In such a scenario, $\frac{1}{P(\xi)} \in \mathcal{L}(\mathbb{R}^n)$, hence $\frac{1}{P(\xi)} \in \mathcal{S}'(\mathbb{R}^n)$ by Exercise 4.7. Thus, if some $E \in \mathcal{S}'(\mathbb{R}^n)$ happens to be a fundamental solution of $P(D)$ (we will see later in Theorem 5.14 that such an E always exists), then $P(D)E = \delta$ in $\mathcal{S}'(\mathbb{R}^n)$. After applying the Fourier transform, this implies $P(\xi)\widehat{E} = 1$ in $\mathcal{S}'(\mathbb{R}^n)$ (recall part *(b)* in Theorem 4.26 and Example 4.22). In particular, since $P(\xi)\frac{1}{P(\xi)} = 1$ in $\mathcal{S}'(\mathbb{R}^n)$, we obtain

$$P(\xi)\Big(\widehat{E}(\xi) - \frac{1}{P(\xi)}\Big) = 0 \quad \text{in} \quad S'(\mathbb{R}^n). \tag{5.2.6}$$

Multiplying by $\frac{1}{P(\xi)} \in \mathcal{L}(\mathbb{R}^n)$ then gives $\widehat{E}(\xi) = \frac{1}{P(\xi)}$ hence, further,

$$\langle E, \varphi \rangle = (2\pi)^{-n} \big\langle \widehat{\widehat{E}}^{\vee}, \varphi \big\rangle = (2\pi)^{-n} \langle \widehat{E}, \widehat{\varphi}^{\vee} \rangle$$

$$= (2\pi)^{-n} \int_{\mathbb{R}^n} \frac{\widehat{\varphi}(-\xi)}{P(\xi)} \, d\xi, \qquad \forall \varphi \in S(\mathbb{R}^n). \tag{5.2.7}$$

Formula (5.2.7), derived under the assumption that (5.2.5) holds, is the departure point for the construction of a fundamental solution for a constant coefficient linear differential operator $P(D)$ in the more general case when (5.2.5) is no longer assumed. Of course, we would have to assume that $P(D) \neq 0$ (since otherwise (5.2.1) clearly does not have any solution).

The following comments are designed to shed some light into how the issues created by the zeros of $P(\xi)$ (in the context of (5.2.7)) may be circumvented when condition (5.2.5) is dropped. The first observation is that if $\varphi \in C_0^\infty(\mathbb{R}^n)$ we may extend its Fourier transform $\widehat{\varphi}$ to $\mathbb{C}^n$ by setting

$$\widehat{\varphi}(\zeta) := \int_{\mathbb{R}^n} e^{-i \sum_{j=1}^n x_j \zeta_j} \varphi(x) \, dx \quad \text{for every} \quad \zeta = (\zeta_1, \ldots, \zeta_n) \in \mathbb{C}^n. \tag{5.2.8}$$

Note that the compact support condition on φ is needed since for each $\zeta \in \mathbb{C}^n \setminus \mathbb{R}^n$ fixed the function $|e^{-i \sum_{j=1}^n x_j \zeta_j}|$ grows rapidly as $|x| \to \infty$.

Second, if (5.2.5) holds then $P(\xi) \neq 0$ for each $\xi \in \mathbb{R}^n$. Consequently, for each $\xi \in \mathbb{R}^n$ there exists a small number $r(\xi) > 0$ such that $P(z) \neq 0$ whenever $z \in \mathbb{C}^n$ satisfies $|z - \xi| < 2r(\xi)$. Hence, for every $\eta \in S^{n-1}$ the mapping

$$\{\tau \in \mathbb{C} : |\tau| < 2r(\xi)\} \ni \tau \mapsto \frac{\widehat{\varphi}(-\xi - \tau\eta)}{P(\xi + \tau\eta)} \in \mathbb{C} \tag{5.2.9}$$

is well defined and analytic. Using Cauchy's formula (and keeping in mind the convention (5.2.8)) we may therefore rewrite (5.2.7) in the form

$$\langle E, \varphi \rangle = (2\pi)^{-n} \int_{\mathbb{R}^n} \frac{\widehat{\varphi}(-\xi)}{P(\xi)} d\xi \tag{5.2.10}$$

$$= (2\pi)^{-n} \int_{\mathbb{R}^n} \Big[\frac{1}{2\pi i} \int_{|\tau| = r(\xi)} \frac{\widehat{\varphi}(-\xi - \tau\eta)}{P(\xi + \tau\eta)} \cdot \frac{d\tau}{\tau}\Big] d\xi, \qquad \forall \varphi \in C_0^\infty(\mathbb{R}^n).$$

The advantage of the expression in the last line of (5.2.10), compared to the expression in the last line of (5.2.7), is the added flexibility in the choice of the set over which $\frac{1}{P}$ is integrated. This discussion suggests that we try to construct a fundamental solution for $P(D)$ in a manner similar to the expression in the last line of (5.2.10), with the integration taking place on a set avoiding the zeros of P, even in the case when (5.2.5) is dropped.

We shall implement this strategy in Section 5.3. For the time being, we discuss a preliminary result that will play a basic role in this endeavor. To state, it recalls the notion of principal symbol from (5.1.4).

Lemma 5.9. *Let $m \in \mathbb{N}$ and let $P(D)$ be a differential operator of order m as in (5.1.1). In addition, suppose that $P_m(\eta) \neq 0$ for some $\eta \in S^{n-1}$. Then there exist $\varepsilon > 0$, $r_0, r_1, \ldots, r_m > 0$ and closed sets $F_0, F_1, \ldots, F_m$ in $\mathbb{R}^n$ such that*

(a) $\mathbb{R}^n = \bigcup\limits_{j=0}^{m} F_j$ *and*

(b) if $\xi \in F_j$, $\tau \in \mathbb{C}$, $|\tau| = r_j$, then $|P(\xi + \tau\eta)| \geq \varepsilon$, for each $j = 0, 1, \ldots, m$.

Proof. Choose $r_j := \frac{j+1}{m}$, for $j \in \{0, 1, \ldots, m\}$ and set $\varepsilon := (2m)^{-m}|P_m(\eta)| > 0$. Also, for each $j \in \{1, \ldots, m\}$ set

$$F_j := \{\xi \in \mathbb{R}^n : |P(\xi + \tau\eta)| \geq \varepsilon, \ \forall \tau \in \mathbb{C}, \ |\tau| = r_j\}. \tag{5.2.11}$$

These sets are closed in $\mathbb{R}^n$ and clearly satisfy *(b)*. There remains to show that *(a)* holds.

Fix $\xi \in \mathbb{R}^n$, and consider the mapping $\mathbb{C} \ni \tau \mapsto P(\xi + \tau\eta) \in \mathbb{C}$. This is a polynomial in τ of order m with the coefficient of τ^m equal to $P_m(\eta)$. Let $\tau_j(\xi)$, $j = 1, \ldots, m$, be the zeros of this polynomial. Then

$$P(\xi + \tau\eta) = P_m(\eta) \prod_{j=1}^{m} (\tau - \tau_j(\xi)), \qquad \forall \tau \in \mathbb{C}. \tag{5.2.12}$$

Consider

$$M_j := \left\{\tau \in \mathbb{C} : \operatorname{dist}(\tau, \partial B(0, r_j)) < \frac{1}{2m}\right\}, \quad \text{for } j \in \{0, 1, \ldots, m\}. \tag{5.2.13}$$

Given that $r_j - r_{j-1} = 1/m$ for $j = 1, \ldots, m$, the sets in the collection $\{M_j\}_{j=0}^{m}$ are pairwise disjoint. Consequently, among these $m+1$ sets there is at least one that does not contain any of the numbers $\tau_j(\xi)$, $j = 1, \ldots, m$ (a simple form of the pigeon hole principle). Hence, there exists $k \in \{0, 1, \ldots, m\}$ such that

$$\operatorname{dist}(\tau_j(\xi), \partial B(0, r_k)) \geq \frac{1}{2m}, \qquad \forall j \in \{1, \ldots, m\}. \tag{5.2.14}$$

Then, $|\tau_j(\xi) - \tau| \geq \frac{1}{2m}$, for every $|\tau| = r_k$ and every $j = 1, \ldots, m$, which when used in (5.2.12) implies

$$|P(\xi + \tau\eta)| = \left|P_m(\eta) \prod_{j=1}^{m} (\tau - \tau_j(\xi))\right| \geq \frac{|P_m(\eta)|}{(2m)^m} = \varepsilon, \qquad \text{if } |\tau| = r_k. \tag{5.2.15}$$

This shows that $\xi \in F_k$, thus *(a)* is satisfied for the sets defined in (5.2.11). $\qquad \square$

5.3 The Malgrange–Ehrenpreis Theorem

The theorem referred to in the title reads as follows.

Theorem 5.10 (Malgrange–Ehrenpreis). *Let $P(D)$ be a linear constant coefficient differential operator in $\mathbb{R}^n$ which is not identically zero. Then there exists some $E \in \mathcal{D}'(\mathbb{R}^n)$ which is a fundamental solution for $P(D)$.*

Proof. Let $m \in \mathbb{N}_0$ and assume that $P(D)$ is as in (5.1.1) of order m. If $m = 0$ then $P(D) = a \in \mathbb{C} \setminus \{0\}$ and the distribution $E := \frac{1}{a}\delta$ is a fundamental solution for $P(D)$. Suppose now that $m \geq 1$. Since the principal symbol P_m does not vanish identically in $\mathbb{R}^n \setminus \{0\}$ and since

$$P_m(\xi) = |\xi|^m P_m\left(\frac{\xi}{|\xi|}\right) \quad \text{for all} \quad \xi \in \mathbb{R}^n \setminus \{0\}, \tag{5.3.1}$$

we conclude that there exists $\eta \in S^{n-1}$ such that $P_m(\eta) \neq 0$. Fix such an η and retain the notation introduced in the proof of Lemma 5.9. Then the functions

$$\chi_j : \mathbb{R}^n \to \mathbb{R}, \quad \chi_j(\xi) := \begin{cases} 1, & \xi \in F_j, \\ 0, & \xi \in \mathbb{R}^n \setminus F_j, \end{cases} \quad j \in \{0, 1, \ldots, m\}, \tag{5.3.2}$$

and

$$f_j : \mathbb{R}^n \to \mathbb{R}, \quad f_j := \frac{\chi_j}{\sum\limits_{k=0}^{m} \chi_k}, \quad j \in \{0, 1, \ldots, m\}, \tag{5.3.3}$$

are well defined, measurable (recall that the F_j's are closed), and $\sum\limits_{j=0}^{m} f_j = 1$ on $\mathbb{R}^n$. In addition, $0 \leq f_j \leq 1$, and $f_j = 0$ on $\mathbb{R}^n \setminus F_j$ for each $j \in \{0, 1, \ldots, m\}$.

Define now the linear functional $E : \mathcal{D}(\mathbb{R}^n) \to \mathbb{C}$ by setting, for each function $\varphi \in C_0^\infty(\mathbb{R}^n)$,

$$E(\varphi) := (2\pi)^{-n} \sum_{j=0}^{m} \int_{\mathbb{R}^n} f_j(\xi) \left[\frac{1}{2\pi i} \int_{|\tau|=r_j} \frac{\widehat{\varphi}(-\xi - \tau\eta)}{P(\xi + \tau\eta)} \cdot \frac{d\tau}{\tau}\right] d\xi. \tag{5.3.4}$$

In order to prove that E is well defined, fix an arbitrary compact set $K \subset \mathbb{R}^n$ and suppose $\varphi \in C_0^\infty(\mathbb{R}^n)$ is such that $\operatorname{supp}\varphi \subseteq K$. First, observe that for each $j = 0, 1, \ldots, m$, if $\xi \in F_j = \operatorname{supp} f_j$, then $|P(\xi + \tau\eta)| \geq \varepsilon$ for every $\tau \in \mathbb{C}$, $|\tau| = r_j$, which makes the integral in (5.3.4) over $|\tau| = r_j$ finite. Second, we claim that the integrals over $\mathbb{R}^n$ in (5.3.4) are convergent. Indeed, since φ is compactly supported and

$$\widehat{\varphi}(-\xi - \tau\eta) = \int_{\mathbb{R}^n} e^{ix\cdot\xi + i\tau x\cdot\eta} \varphi(x)\, dx, \tag{5.3.5}$$

it follows that $\widehat{\varphi}(-\xi - \tau\eta)$ is analytic in τ. Then, if we fix $j \in \{0, 1, \ldots, m\}$, and take $N \in \mathbb{N}_0$, using (5.3.5) and integration by parts, for each $|\tau| = r_j$ we may write

$$(1 + |\xi|^2)^N \widehat{\varphi}(-\xi - \tau\eta) = \int_{\mathbb{R}^n} \left[\left(1 - \sum_{\ell=1}^n \partial_{x_\ell}^2 \right)^N e^{ix\cdot\xi} \right] e^{i\tau x\cdot\eta} \varphi(x) \, dx$$

$$= \int_{\mathbb{R}^n} e^{ix\cdot\xi} \left(-1 - \sum_{\ell=1}^n \partial_{x_\ell}^2 \right)^N [e^{i\tau x\cdot\eta} \varphi(x)] \, dx, \qquad (5.3.6)$$

for each $\xi \in \mathbb{R}^n$. An inspection of the last integral in (5.3.6) then yields

$$|\widehat{\varphi}(-\xi - \tau\eta)| \leq C_j (1 + |\xi|^2)^{-N} \sup_{\substack{x\in K \\ |\alpha|\leq 2N}} |\partial_x^\alpha \varphi(x)| \qquad \text{if } |\tau| = r_j, \qquad (5.3.7)$$

for some constant $C_j = C_j(K, r_j, \eta) \in (0, \infty)$. In turn, we may use (5.3.7) in (5.3.4) to obtain

$$|E(\varphi)| \leq (2\pi)^{-n} \sum_{j=0}^m C_j \int_{\mathbb{R}^n} f_j(\xi) \frac{1}{2\pi} \cdot \frac{2\pi r_j}{\varepsilon r_j} (1 + |\xi|^2)^{-N} \sup_{\substack{x\in K \\ |\alpha|\leq 2N}} |\partial^\alpha \varphi(x)| \, d\xi$$

$$\leq C \sup_{\substack{x\in K \\ |\alpha|\leq 2N}} |\partial^\alpha \varphi(x)| \int_{\mathbb{R}^n} (1 + |\xi|^2)^{-N} \, d\xi$$

$$= C \sup_{\substack{x\in K \\ |\alpha|\leq 2N}} |\partial^\alpha \varphi(x)| \qquad (5.3.8)$$

for some $C \in (0, \infty)$ independent of φ, by choosing $N > \frac{n}{2}$. This proves that E is well defined.

Clearly E is linear which, together with (5.3.8) and Proposition 2.4, implies that $E \in \mathcal{D}'(\mathbb{R}^n)$. To finish the proof of the theorem we are left with showing that $P(D)E = \delta$ in $\mathcal{D}'(\mathbb{R}^n)$, i.e., that (5.2.2) holds. Since $\mathcal{F}(P(-D)\varphi) = P^\vee \widehat{\varphi}$, recalling (5.3.4) we may write

$$\langle E, P(-D)\varphi \rangle = (2\pi)^{-n} \sum_{j=0}^m \int_{\mathbb{R}^n} f_j(\xi) \left[\frac{1}{2\pi i} \int_{|\tau|=r_j} \frac{\widehat{\varphi}(-\xi - \tau\eta)}{\tau} \, d\tau \right] d\xi$$

$$= (2\pi)^{-n} \sum_{j=0}^m \int_{\mathbb{R}^n} f_j(\xi) \widehat{\varphi}(-\xi) \, d\xi$$

$$= (2\pi)^{-n} \int_{\mathbb{R}^n} \widehat{\varphi}(\xi) \, d\xi$$

$$= \varphi(0). \qquad (5.3.9)$$

For the second equality in (5.3.9), we used the fact that $\widehat{\varphi}(-\xi - \tau\eta)$ is analytic in τ, hence by Cauchy's formula we have $\frac{1}{2\pi i} \int_{|\tau|=r_j} \frac{\widehat{\varphi}(-\xi-\tau\eta)}{\tau} \, d\tau = \widehat{\varphi}(-\xi)$. The third equality relies on $\sum_{j=0}^m f_j = 1$ on $\mathbb{R}^n$, while the last one follows from part (4) in Exercise 3.27.

On account of (5.3.9) and (5.2.2) we may therefore conclude that E is a fundamental solution for $P(D)$. □

One of the consequences of Theorem 5.10 is the fact that the Poisson problem for nonzero constant coefficient linear differential operators is locally solvable. Specifically, we have the following result.

Corollary 5.11. *Let $P(D)$ be a constant coefficient linear differential operator in $\mathbb{R}^n$ which is not identically zero. Suppose Ω is a non-empty, open subset of $\mathbb{R}^n$ and that $f \in \mathcal{D}'(\Omega)$ is given. Then for every non-empty, open subset ω of Ω with the property that $\overline{\omega}$ is a compact subset of Ω, there exists $u \in \mathcal{D}'(\Omega)$ such that $P(D)u = f$ in $\mathcal{D}'(\omega)$.*

Proof. Let $E \in \mathcal{D}'(\mathbb{R}^n)$ be a fundamental solution of $P(D)$, which is known to exist by Theorem 5.10. Also, fix $\psi \in C_0^\infty(\Omega)$ such that $\psi \equiv 1$ on ω. Then $\psi f \in \mathcal{E}'(\Omega)$ hence, by Proposition 2.69, it has a unique extension $\widetilde{\psi f} \in \mathcal{E}'(\mathbb{R}^n)$ and $\mathrm{supp}\,(\widetilde{\psi f}) \subseteq \mathrm{supp}\,\psi$. Now let $v := E * (\widetilde{\psi f}) \in \mathcal{D}'(\mathbb{R}^n)$. By Proposition 2.50 we have $v\big|_\Omega \in \mathcal{D}'(\Omega)$ and we claim that $u := v\big|_\Omega$ is a solution of $P(D)u = f$ in $\mathcal{D}'(\omega)$ (in the sense that $P(D)(u\big|_\omega) = f\big|_\omega$ in $\mathcal{D}'(\omega)$). Indeed,

$$P(D)(u\big|_\omega) = [P(D)u]\big|_\omega = [P(D)(v\big|_\Omega)]\big|_\omega = [P(D)(E * \widetilde{\psi f})]\big|_\omega$$

$$= \left[(P(D)E) * \widetilde{\psi f}\right]\big|_\omega = [\delta * \widetilde{\psi f}]\big|_\omega$$

$$= [\widetilde{\psi f}]\big|_\omega = [\psi f]\big|_\omega = f\big|_\omega \quad \text{in } \mathcal{D}'(\omega). \tag{5.3.10}$$

For the first and third equality in (5.3.10) we used (2.5.4), for the fourth equality we used *(e)* in Theorem 2.96, while for the sixth equality we used *(d)* in Theorem 2.96. □

The next example gives an approach for computing fundamental solutions for linear constant coefficient operators on the real line.

Example 5.12. Let $m \in \mathbb{N}$ and let $P := (\frac{\mathrm{d}}{\mathrm{d}x})^m + \sum_{j=0}^{m-1} a_j(\frac{\mathrm{d}}{\mathrm{d}x})^j$ be a differential operator of order m in $\mathbb{R}$ with constant coefficients. Suppose that v is the solution to

$$\begin{cases} v \in C^\infty(\mathbb{R}), \\ Pv = 0 \quad \text{in } \mathbb{R}, \\ v^{(j)}(0) = 0, \qquad 0 \le j \le m-2, \\ v^{(m-1)}(0) = 1, \end{cases} \tag{5.3.11}$$

with the understanding that the third condition is void if $m = 1$. Recall the Heaviside function H from (1.2.9) and define the distributions

$$u_+ := vH \quad \text{and} \quad u_- := -vH^\vee \quad \text{in } \mathcal{D}'(\mathbb{R}). \tag{5.3.12}$$

We claim that u_+ and u_- are fundamental solutions for P in $\mathcal{D}'(\mathbb{R})$. To see that this is the case, express P as $P(D) = \mathrm{i}^m(\frac{1}{\mathrm{i}}\frac{\mathrm{d}}{\mathrm{d}x})^m + \sum_{j=0}^{m-1} \mathrm{i}^j a_j(\frac{1}{\mathrm{i}}\frac{\mathrm{d}}{\mathrm{d}x})^j$, in line with (5.1.1), and note that this forces

$$P(-D) = (-\mathrm{i})^m(\frac{1}{\mathrm{i}}\frac{\mathrm{d}}{\mathrm{d}x})^m + \sum_{j=0}^{m-1}(-\mathrm{i})^j a_j(\frac{1}{\mathrm{i}}\frac{\mathrm{d}}{\mathrm{d}x})^j$$

$$= (-1)^m(\frac{\mathrm{d}}{\mathrm{d}x})^m + \sum_{j=0}^{m-1}(-1)^j a_j(\frac{\mathrm{d}}{\mathrm{d}x})^j. \tag{5.3.13}$$

As such, for every $\varphi \in C_0^\infty(\mathbb{R})$ we have

$$\langle P(D)u_+, \varphi \rangle = \langle u_+, P(-D)\varphi \rangle = \int_0^\infty v(x)(P(-D)\varphi)(x)\,\mathrm{d}x$$

$$= v^{(m-1)}(0)\varphi(0) + \int_0^\infty (Pv)(x)\varphi(x)\,\mathrm{d}x = \langle \delta, \varphi \rangle, \tag{5.3.14}$$

where the third equality uses (5.3.11) and repeated integrations by parts. This proves that $Pu_+ = \delta$ in $\mathcal{D}'(\mathbb{R})$. Similarly, one obtains that $Pu_- = \delta$ in $\mathcal{D}'(\mathbb{R})$. We note that the two fundamental solutions just described satisfy $\operatorname{supp} u_+ \subseteq [0, \infty)$ and $\operatorname{supp} u_- \subseteq (-\infty, 0]$.

Example 5.13. Let $c \in \mathbb{C}$ and consider the differential operator $P := \frac{\mathrm{d}}{\mathrm{d}x} + c$ in $\mathbb{R}$. It is easy to see that $v(x) = \mathrm{e}^{-cx}$, $x \in \mathbb{R}$ is the solution (in the classical sense) of the initial value problem $Pv = 0$ in $\mathbb{R}$, $v(0) = 1$. Hence, by Example 5.12, the distributions $u_+ := vH$ and $u_- := -vH^\vee$ in $\mathcal{D}'(\mathbb{R})$ are fundamental solutions for P. Note that instead of using Example 5.12, one may show that $Pu_+ = \delta$ (and similarly that $Pu_- = \delta$) in $\mathcal{D}'(\mathbb{R})$ by applying *(4)* in Proposition 2.43 and the fact that $H' = \delta$ in $\mathcal{D}'(\mathbb{R})$ to write

$$P(vH) = (vH)' + cvH = v'H + vH' + cvH = v\delta = \delta \quad \text{in } \mathcal{D}'(\mathbb{R}). \tag{5.3.15}$$

We close this section by proving a strengthened version of the Malgrange–Ehrenpreis theorem that addresses the issue of existence of fundamental solutions that are tempered distributions. Such a result is important in the study of partial differential equations as it opens the door for using the Fourier transform.

Theorem 5.14. *Let $P(D)$ be a nonzero constant coefficient linear differential operator in $\mathbb{R}^n$. Then there exists $E \in \mathcal{S}'(\mathbb{R}^n)$ which is a fundamental solution for $P(D)$. In particular, the equality $P(D)E = \delta$ holds in $\mathcal{S}'(\mathbb{R}^n)$.*

We prove this theorem by relying on a deep result due to L. Hörmander and S. Lojasiewicz. For a proof of Theorem 5.15 the interested reader is referred to [32] and [44].

Theorem 5.15 (Hörmander-Lojasiewicz). *Let P be a polynomial in $\mathbb{R}^n$ that does not vanish identically and consider the pointwise multiplication mapping*

$$M_P : S(\mathbb{R}^n) \to S(\mathbb{R}^n), \qquad M_P(\varphi) := P\varphi, \quad \forall\, \varphi \in S(\mathbb{R}^n). \tag{5.3.16}$$

Then M_P is linear, continuous, injective, with closed range $S_P(\mathbb{R}^n)$ in $S(\mathbb{R}^n)$. Moreover, the mapping $M_P : S(\mathbb{R}^n) \to S_P(\mathbb{R}^n)$ has a linear and continuous inverse $T : S_P(\mathbb{R}^n) \to S(\mathbb{R}^n)$.

Granted this, we can now turn to the proof of the theorem stated earlier.

Proof of Theorem 5.14. Let $P(D)$ be as in the statement and consider the polynomial $P(\xi)$ associated to it as in (5.1.2). Denote by $S_P(\mathbb{R}^n)$ the range of the mapping M_P from (5.3.16) corresponding to this polynomial and by T the inverse of $M_P : S(\mathbb{R}^n) \to S_P(\mathbb{R}^n)$. Then Theorem 5.15 combined with Proposition 14.2 give that the transpose of T is linear and continuous as a mapping

$$T^t : S'(\mathbb{R}^n) \to S_P(\mathbb{R}^n)' \tag{5.3.17}$$

where $S_P(\mathbb{R}^n)'$ is the dual of $S_P(\mathbb{R}^n)$ endowed with the weak-$*$ topology.

Next, assume that some $f \in S'(\mathbb{R}^n)$ has been fixed and define $u := T^t f$. Then $u \in S_P(\mathbb{R}^n)'$ and, making use of (14.1.3), (14.1.7), and Theorem 14.5, it is possible to extend u to an element in $S'(\mathbb{R}^n)$, which we continue to denote by u. We claim that $Pu = f$ in $S'(\mathbb{R}^n)$, where Pu is the tempered distribution obtained by multiplying u with the Schwartz function $P(\xi)$. To see this, fix $\varphi \in S(\mathbb{R}^n)$ and write

$$\langle Pu, \varphi \rangle = \langle u, P\varphi \rangle = \langle T^t f, M_P\varphi \rangle = \langle f, T M_P\varphi \rangle = \langle f, \varphi \rangle, \tag{5.3.18}$$

proving the claim. In particular, the above procedure may be implemented for the distribution $f := 1 \in S'(\mathbb{R}^n)$. Corresponding to this choice, we obtain $u \in S'(\mathbb{R}^n)$ with the property that $Pu = 1$ in $S'(\mathbb{R}^n)$. Define $E := \mathcal{F}^{-1}u$. Then E belongs to $S'(\mathbb{R}^n)$ and

$$\widehat{P(D)E} = P(\xi)\widehat{E} = Pu = 1 = \widehat{\delta} \quad \text{in} \quad S'(\mathbb{R}^n). \tag{5.3.19}$$

Taking the Fourier transform we see that, as desired, E is a tempered distribution fundamental solution for $P(D)$. $\square$

Further Notes for Chapter 5. The existence of fundamental solutions for arbitrary nonzero, constant coefficient, linear, partial differential operators were first established in this degree of generality by L. Ehrenpreis and B. Malgrange (cf. [13], [45]), via proofs relying on the Hahn–Banach theorem. This basic result served as a first compelling piece of evidence of the impact of the theory of distributions in the field of partial differential equations. The general discussion initiated here is going to be further augmented by considering in later chapters concrete examples of fundamental solutions associated with basic operators arising in mathematics, physics, and engineering.

5.4 Additional Exercises for Chapter 5

Exercise 5.16. Let $m \in \mathbb{N}$. Prove that

$$\left\{ u \in \mathcal{S}'(\mathbb{R}) : u^{(m)} = \delta \ \text{ in } \ \mathcal{S}'(\mathbb{R}) \right\} \tag{5.4.1}$$

$$= \left\{ \tfrac{1}{(m-1)!} x^{m-1} H + \sum_{k=0}^{m-1} c_k x^k : c_k \in \mathbb{C}, \ k = 1, \dots, m-1 \right\}.$$

Exercise 5.17. Let $n, m \in \mathbb{N}$ and $k_1, k_2 \in \mathbb{N}_0$. Consider two scalar operators

$$P_1(D_x) := \sum_{|\alpha| \le k_1} a_\alpha D_x^\alpha, \quad a_\alpha \in \mathbb{C}, \ \forall \, \alpha \in \mathbb{N}_0^n, \ |\alpha| \le k_1, \tag{5.4.2}$$

$$P_2(D_y) := \sum_{|\beta| \le k_2} b_\beta D_y^\beta, \quad b_\beta \in \mathbb{C}, \ \forall \, \beta \in \mathbb{N}_0^m, \ |\beta| \le k_2, \tag{5.4.3}$$

and assume that $E_1 \in \mathcal{D}'(\mathbb{R}^n)$ and $E_2 \in \mathcal{D}'(\mathbb{R}^m)$ are fundamental solutions for P_1 in $\mathbb{R}^n$ and P_2 in $\mathbb{R}^m$, respectively. Prove that $E_1 \otimes E_2$ is a fundamental solution for $P_1(D_x) \otimes P_2(D_y) := \sum_{|\alpha| \le k_1} \sum_{|\beta| \le k_2} a_\alpha b_\beta D_x^\alpha D_y^\beta$ in $\mathbb{R}^n \times \mathbb{R}^m$.

Chapter 6
Hypoelliptic Operators

Abstract The concept of hypoelliptic operator is introduced and studied. A classical result, due to L. Schwartz, is proved here to the effect that a necessary and sufficient condition for a linear, constant coefficient differential operator to be hypoelliptic in the entire ambient space is that the named operator possesses a fundamental solution with singular support consisting of the origin alone. In this chapter an integral representation formula and interior estimates for a subclass of hypoelliptic operators are proved as well.

6.1 Definition and Properties

As usual, assume that $\Omega \subseteq \mathbb{R}^n$ is an open set, and fix $m \in \mathbb{N}$. From Theorem 2.112 we know that if $u \in \mathcal{D}'(\Omega)$ is such that for each $\alpha \in \mathbb{N}_0^n$ satisfying $|\alpha| = m$ the distributional derivative $\partial^\alpha u$ is continuous on Ω, then $u \in C^m(\Omega)$. To re-phrase this result in a manner which lends itself more naturally to generalizations, define $\mathcal{P}^m u := (\partial^\alpha u)_{|\alpha|=m}$ for each $u \in \mathcal{D}'(\Omega)$. In this notation, the earlier result reads

$$u \in \mathcal{D}'(\Omega) \quad \text{and} \quad \mathcal{P}^m u \in C^0(\Omega) \implies u \in C^m(\Omega). \qquad (6.1.1)$$

We are interested in proving results in the same spirit with the role of mapping $\mathcal{P}^m$ played by (certain) linear differential operators.

Recall the notation from (3.1.10) and consider the linear differential operator

$$P(x, D) = \sum_{|\alpha| \le m} a_\alpha(x) D^\alpha, \quad a_\alpha \in C^\infty(\Omega), \quad \text{for all } \alpha \in \mathbb{N}_0^n \text{ with } |\alpha| \le m. \qquad (6.1.2)$$

By Proposition 2.41, if $f \in C^0(\Omega)$ and $u \in C^m(\Omega)$ is a solution of $P(x, D)u = f$ in $\mathcal{D}'(\Omega)$, then u is a solution of $P(x, D)u = f$ in Ω in the classical sense. However, this reasoning requires knowing in advance that $u \in C^m(\Omega)$. It is desirable to have a more general statement to the effect that if $u \in \mathcal{D}'(\Omega)$ is a solution of the equation

© Springer Nature Switzerland AG 2018
D. Mitrea, *Distributions, Partial Differential Equations, and Harmonic Analysis*,
Universitext, https://doi.org/10.1007/978-3-030-03296-8_6

$P(x, D)u = f$ in $\mathcal{D}'(\Omega)$ and f has certain regularity properties, then the distribution u also belongs to an appropriate smoothness class (e.g., if $f \in C^\infty(\Omega)$ then $u \in C^\infty(\Omega)$). While such a phenomenon is not to be expected in general, there exist classes of operators $P(x, D)$ for which there are a priori regularity results for the solution u based on the regularity of the datum f. Such a class of operators is singled out in this chapter.

Definition 6.1. An operator $P(x, D)$ as in (6.1.2) is called `hypoelliptic` in Ω if for all open subsets ω of Ω the following property holds: whenever $u \in \mathcal{D}'(\omega)$ is such that $P(x, D)u \in C^\infty(\omega)$ then $u \in C^\infty(\omega)$.

Remark 6.2. It is easy to see from Definition 6.1 that if an operator is hypoelliptic in Ω, then it is also hypoelliptic in any other open subset of Ω.

Definition 6.3. Let $u \in \mathcal{D}'(\Omega)$ be arbitrary. Then the `singular support` of u, denoted by sing supp u, is defined by

$$\text{sing supp } u := \{x \in \Omega : \text{there is no open set } \omega$$

$$\text{such that } x \in \omega \subseteq \Omega \text{ and } u\big|_\omega \in C^\infty(\omega)\}. \tag{6.1.3}$$

Example 6.4. Not surprisingly,

$$\text{sing supp } \delta = \{0\}. \tag{6.1.4}$$

To see why this is true, note that clearly sing supp $\delta \subset \{0\}$ and sing supp $\delta \neq \varnothing$ since, by Example 2.13, the distribution δ is not of function type.

Remark 6.5. (1) Since $\Omega \setminus \text{sing supp } u$ is an open set, it follows that sing supp u is relatively closed in Ω.

(2) By Exercise 2.53, we have that $u\big|_{\Omega \setminus \text{sing supp } u} \in C^\infty(\Omega \setminus \text{sing supp } u)$.

(3) The singular support of a distribution $u \in \mathcal{D}'(\Omega)$ is the smallest relatively closed set in Ω with the property that u restricted to its complement in Ω is of class C^∞.

(4) If $u \in \mathcal{D}'(\Omega)$ is such that its singular support is a compact subset of Ω, then u is the sum between a distribution of function type and a compactly supported distribution in Ω. Indeed, if we denote by f the function obtained by extending the restriction $u\big|_{\Omega \setminus \text{sing supp } u}$ by zero to Ω, then $f \in L^1_{loc}(\Omega)$ and $\text{supp}(u - u_f) = \text{sing supp } u$, hence $u - u_f \in \mathcal{E}'(\Omega)$.

(5) If the operator $P(x, D)$ is as in (6.1.2), then

$$\text{sing supp } [P(x, D)u] \subseteq \text{sing supp } u, \qquad \forall u \in \mathcal{D}'(\Omega). \tag{6.1.5}$$

Indeed, if $x_0 \in \Omega \setminus \text{sing supp } u$, then there exists an open neighborhood ω of x_0 contained in Ω and such that $u\big|_\omega \in C^\infty(\omega)$. Hence,

$$[P(x, D)u]\big|_\omega = P(x, D)[u\big|_\omega] \in C^\infty(\omega), \tag{6.1.6}$$

which implies that $x_0 \in \Omega \setminus \text{sing supp } [P(x, D)u]$.

The quality of being a hypoelliptic operator turns out to be equivalent to (6.1.5) being valid with equality.

Proposition 6.6. *Let $P(x, D)$ be as in (6.1.2). Then $P(x, D)$ is hypoelliptic in Ω if and only if*

$$\text{sing supp}\,[P(x, D)u] = \text{sing supp}\,u, \qquad \forall\, u \in \mathcal{D}'(\Omega). \tag{6.1.7}$$

Proof. Suppose $P(x, D)$ is hypoelliptic in Ω. The left-to-right inclusion in (6.1.7) has been proved in part *(5)* of Remark 6.5. To prove the opposite inclusion, fix a point $x_0 \in \Omega \setminus \text{sing supp}\,[P(x, D)u]$. Then there exists an open subset ω of Ω with $x_0 \in \omega$ and $[P(x, D)u]\big|_\omega \in C^\infty(\omega)$. Since $P(x, D)$ is hypoelliptic in Ω it is also hypoelliptic in ω and, as such, $u\big|_\omega \in C^\infty(\omega)$. This proves that $x_0 \in \Omega \setminus \text{sing supp}\,u$, as wanted.

Now suppose (6.1.7) holds and let ω be an open subset of Ω. Fix some $u \in \mathcal{D}'(\omega)$ such that $P(x, D)u \in C^\infty(\omega)$. The desired conclusion will follow as soon as we show that $u \in C^\infty(\omega)$. By Exercise 2.53, it suffices to prove that for each point $x_0 \in \omega$ there exists an open neighborhood $\omega_0 \subseteq \omega$ of x_0 such that $u\big|_{\omega_0}$ belongs to $C^\infty(\omega_0)$. To this end, fix $x_0 \in \omega$ along with some open neighborhood ω_0 of x_0 whose closure is a compact subset of ω. Furthermore, pick a function ψ in $C_0^\infty(\omega)$ with the property that $\psi \equiv 1$ in a neighborhood of ω_0, and note that necessarily $\psi u \in \mathcal{E}'(\omega)$. By Proposition 2.69, every $v \in \mathcal{E}'(\omega)$ has a unique extension to $\mathcal{E}'(\Omega)$ that we will denote by $\text{Ext}(v)$. Since the operation of taking the extension Ext commutes with applying $P(x, D)$ (recall Exercise 2.70) we may write

$$P(x, D)[\text{Ext}(\psi u)] = \text{Ext}(P(x, D)(\psi u)) = \text{Ext}(\psi P(x, D)u + g)$$

$$= \text{Ext}(\psi P(x, D)u) + \text{Ext}(g) \quad \text{in } \mathcal{D}'(\Omega), \tag{6.1.8}$$

where $g := P(x, D)(\psi u) - \psi P(x, D)u \in \mathcal{E}'(\omega)$. On the other hand, from part *(4)* in Proposition 2.43 it follows that $g\big|_{\omega_0} = 0$, thus

$$(\text{sing supp}\,g) \cap \omega_0 = \varnothing. \tag{6.1.9}$$

In concert, (6.1.9), the fact that $\psi(P(x, D)u) \in C_0^\infty(\omega)$ (recall that, by assumption, $P(x, D)u \in C^\infty(\omega)$), and (6.1.8), imply that

$$\text{sing supp}\,(P(x, D)[\text{Ext}(\psi u)]) \cap \omega_0 = \varnothing. \tag{6.1.10}$$

Invoking the hypothesis (6.1.7) then yields $\text{sing supp}\,[\text{Ext}(\psi u)] \cap \omega_0 = \varnothing$ which, in turn, forces $[\text{Ext}(\psi u)]\big|_{\omega_0} \in C^\infty(\omega_0)$. The desired conclusion now follows upon observing that $[\text{Ext}(\psi u)]\big|_{\omega_0} = (\psi u)\big|_{\omega_0} = u\big|_{\omega_0}$. $\qquad \square$

Remark 6.7. A noteworthy consequence of Proposition 6.6 is as follows: any two fundamental solutions of a hypoelliptic operator in $\mathbb{R}^n$ differ by a C^∞ function. More specifically, if an operator $P(x, D)$ as in (6.1.2) is hypoelliptic in $\mathbb{R}^n$, then $E_1 - E_2 \in C^\infty(\mathbb{R}^n)$ for every $E_1, E_2 \in \mathcal{D}'(\mathbb{R}^n)$ satisfying $P(x, D)E_1 = \delta$ and $P(x, D)E_2 = \delta$ in $\mathcal{D}'(\mathbb{R}^n)$.

In particular, if an operator $P(x, D)$ as in (6.1.2) is hypoelliptic in $\mathbb{R}^n$ and has a fundamental solution $E \in \mathcal{D}'(\mathbb{R}^n)$ with the property that sing supp $E = \{0\}$ then the singular support of any other fundamental solution of $P(x, D)$ is also $\{0\}$.

6.2 Hypoelliptic Operators with Constant Coefficients

In this subsection we further analyze hypoelliptic operators with constant coefficients. First we present a result due to L. Schwartz regarding a necessary and sufficient condition for a linear, constant coefficient operator to be hypoelliptic in the entire space.

Theorem 6.8. *Let $P(D)$ be a constant coefficient linear differential operator in $\mathbb{R}^n$. Then $P(D)$ is hypoelliptic in $\mathbb{R}^n$ if and only if there exists $E \in \mathcal{D}'(\mathbb{R}^n)$ fundamental solution of $P(D)$ in $\mathcal{D}'(\mathbb{R}^n)$ satisfying sing supp $E = \{0\}$.*

Proof. Suppose $P(D)$ is hypoelliptic. Then necessarily $P(D)$ is not identically zero (since the zero operator sends the Dirac distribution to the C^∞ function zero). Granted this, Theorem 5.10 applies and gives that the operator $P(D)$ has a fundamental solution $E \in \mathcal{D}'(\mathbb{R}^n)$. In concert with the hypoellipticity of $P(D)$, Proposition 6.6, and (6.1.4), this allows us to write

$$\text{sing supp } E = \text{sing supp } [P(D)E] = \text{sing supp } \delta = \{0\}, \qquad (6.2.1)$$

as wanted.

Consider next the converse implication. Suppose that there exists $E \in \mathcal{D}'(\mathbb{R}^n)$ with $P(D)E = \delta$ and sing supp $E = \{0\}$. Fix $u \in \mathcal{D}'(\mathbb{R}^n)$ arbitrary. Combining (5) in Remark 6.5 with Proposition 6.6, in order to conclude that $P(D)$ is hypoelliptic, it suffices to prove that

$$\text{sing supp } u \subseteq \text{sing supp } [P(D)u]. \qquad (6.2.2)$$

To this end, fix a point $x_0 \in \mathbb{R}^n \setminus \text{sing supp}[P(D)u]$. Then, there exists an open neighborhood ω of x_0 such that $[P(D)u]\big|_\omega \in C^\infty(\omega)$. The fact that x_0 does not belong to sing supp$u$ will follow if we are able to show that there exists $r \in (0, \infty)$ such that $u\big|_{B(x_0,r)} \in C^\infty(B(x_0, r))$.

Pick $r_1 \in (0, \infty)$ with the property that $\overline{B(x_0, r_1)} \subset \omega$. Also, fix $r_0 \in (0, r_1)$ and select a function $\psi \in C_0^\infty(\omega)$ such that $\psi \equiv 1$ on $B(x_0, r_1)$. Set $v := \psi u$. Then $v\big|_{B(x_0,r_1)} = u\big|_{B(x_0,r_1)}$ and $v \in \mathcal{E}'(\omega)$. By virtue of Proposition 2.69 we may extend v to a distribution in $\mathcal{E}'(\mathbb{R}^n)$ (and we continue to denote this extension by v).

Define $g := P(D)v - \psi[P(D)u]$. Since $[P(D)u]\big|_\omega \in C^\infty(\omega)$ and supp $\psi \subset \omega$, we deduce that $\psi[P(D)u] \in C_0^\infty(\omega)$. In particular, $\psi[P(D)u] \in C_0^\infty(\mathbb{R}^n)$. Consequently, $g \in \mathcal{E}'(\mathbb{R}^n)$. In addition, combining (4) in Proposition 2.43 with the fact that $\psi \equiv 1$ on $B(x_0, r_1)$ shows that $g\big|_{B(x_0,r_1)} = 0$. Thus,

$$\text{sing supp}\, g \cap B(x_0, r_1) = \varnothing. \tag{6.2.3}$$

On the other hand, invoking *(d)* and *(e)* in Theorem 2.96 we may write

$$v = \delta * v = (P(D)E) * v = E * (P(D)v)$$

$$= E * g + E * (\psi P(D)u) \quad \text{in } \mathcal{D}'(\mathbb{R}^n). \tag{6.2.4}$$

We claim that

$$\text{sing supp}(E * g) \cap B(x_0, r_0) = \varnothing. \tag{6.2.5}$$

To see why (6.2.5) is true, fix $\varepsilon \in (0, r_1 - r_0)$ and $\phi \in C_0^\infty(B(0, \varepsilon))$ with the property that $\phi \equiv 1$ on $B\left(0, \frac{\varepsilon}{2}\right)$. Then $E * g = [(1 - \phi)E] * g + (\phi E) * g$. Hence, recalling *(a)* in Theorem 2.96, we may write

$$\text{supp}\,[(\phi E) * g] \subseteq \text{supp}\,(\phi E) + \text{supp}\, g \subseteq B(0, \varepsilon) + [\mathbb{R}^n \setminus B(x_0, r_1)]$$

$$\subseteq \mathbb{R}^n \setminus B(x_0, r_0). \tag{6.2.6}$$

This shows that

$$\text{sing supp}[(\phi E) * g] \cap B(x_0, r_0) = \varnothing. \tag{6.2.7}$$

In addition, since by assumption sing supp $E = \{0\}$, it follows that $(1 - \phi)E$ belongs to $C^\infty(\mathbb{R}^n)$ and invoking Exercise 2.103, we obtain that $[(1 - \phi)E] * g$ is of class C^∞ in $\mathbb{R}^n$. The latter combined with (6.2.7) now yields (6.2.5).

After using (6.2.5) back in (6.2.4) and observing that $E * (\psi P(D)u)$ belongs to $C^\infty(\mathbb{R}^n)$ (itself a consequence of Proposition 2.102) we conclude that $v\big|_{B(x_0, r_0)}$ is of class C^∞ in $B(x_0, r_0)$. Upon recalling that $\psi \equiv 1$ on $B(x_0, r_0)$, we necessarily have $u\big|_{B(x_0, r_0)} \in C^\infty(B(x_0, r_0))$. The proof of the theorem is now complete. $\square$

Exercise 6.9. Show that if $P(D)$ is a constant coefficient linear differential operator which is hypoelliptic in $\mathbb{R}^n$, then any fundamental solution $E \in \mathcal{D}'(\mathbb{R}^n)$ of $P(D)$ has the property that sing supp $E = \{0\}$.

Definition 6.10. Let $P(D)$ be a constant coefficient linear differential operator in $\mathbb{R}^n$. Then $F \in \mathcal{D}'(\mathbb{R}^n)$ is called a `parametrix` of $P(D)$ if there exists $w \in C^\infty(\mathbb{R}^n)$ with the property that $P(D)F = \delta + w$ in $\mathcal{D}'(\mathbb{R}^n)$.

It is natural to think of the concept of parametrix as a relaxation of the notion of fundamental solution since, clearly, any fundamental solution is a parametrix. We have seen that Theorem 6.8 is an important tool in determining if a given operator $P(D)$ is hypoelliptic. One apparent drawback of this particular result is the fact that constructing a fundamental solution for a given operator $P(D)$ may be a delicate task. It is therefore desirable to have a more flexible criterion for deciding whether a certain operator is hypoelliptic and, remarkably, Theorem 6.11 shows that one can use a parametrix in place of a fundamental solution to the same effect.

Theorem 6.11. *Let $P(D)$ be a constant coefficient linear differential operator in $\mathbb{R}^n$. Then $P(D)$ is hypoelliptic in $\mathbb{R}^n$ if and only if there exists $F \in \mathcal{D}'(\mathbb{R}^n)$ which is a parametrix of $P(D)$ and satisfies* sing supp $F = \{0\}$.

Proof. In one direction, if $P(D)$ is hypoelliptic in $\mathbb{R}^n$, then by Theorem 6.8 there exists a fundamental solution E of $P(D)$ with $\operatorname{sing\,supp} E = \{0\}$. The desired conclusion follows by taking $F := E$. Conversely, suppose there exists a parametrix $F \in \mathcal{D}'(\mathbb{R}^n)$ of $P(D)$ satisfying $\operatorname{sing\,supp} F = \{0\}$. Let $w \in C^\infty(\mathbb{R}^n)$ be such that $P(D)F - \delta = w$ in $\mathcal{D}'(\mathbb{R}^n)$. To show that $P(D)$ is hypoelliptic in $\mathbb{R}^n$, we adjust the argument used in the proof of Theorem 6.8. Specifically, starting with $u \in \mathcal{D}'(\mathbb{R}^n)$ arbitrary, as before, matters are reduced to proving that for each point x_0 in $\mathbb{R}^n \setminus \operatorname{sing\,supp} [P(D)u]$ there exists $r \in (0, \infty)$ such that $u|_{B(x_0,r)} \in C^\infty(B(x_0, r))$. Retaining the notation and context from the proof of Theorem 6.8, the reasoning there applies and allows us to write (in place of (6.2.4))

$$v = \delta * v = (P(D)F - w) * v = F * (P(D)v) - w * v$$

$$= F * g + F * (\psi P(D)u) - w * v \quad \text{in } \mathcal{D}'(\mathbb{R}^n). \tag{6.2.8}$$

Since $\operatorname{sing\,supp} F = \{0\}$, the same argument as in the proof of (6.2.5) shows that

$$\operatorname{sing\,supp}(F * g) \cap B(x_0, r_0) = \varnothing, \tag{6.2.9}$$

while since $\psi[P(D)u] \in C_0^\infty(\mathbb{R}^n)$, Proposition 2.102 gives that

$$F * (\psi P(D)u) \in C^\infty(\mathbb{R}^n). \tag{6.2.10}$$

The new observation is that from $v \in \mathcal{E}'(\mathbb{R}^n)$, $w \in C^\infty(\mathbb{R}^n)$, and Exercise 2.103, we have

$$w * v \in C^\infty(\mathbb{R}^n). \tag{6.2.11}$$

Combining (6.2.8)–(6.2.11) it follows that $v|_{B(x_0,r_0)} \in C^\infty(B(x_0, r_0))$. The latter forces $u|_{B(x_0,r_0)} \in C^\infty(B(x_0, r_0))$ (recall that $\psi \equiv 1$ on $B(x_0, r_0)$). The proof of the theorem is now complete. $\qquad\square$

In the class of constant coefficient linear differential operators, Theorem 6.8 and Theorem 6.11 provide useful characterizations of hypoellipticity. In order to offer a wider perspective on this matter, we state, without proof, another such characterization[1] due to L. Hörmander.

Theorem 6.12. *Let $P(D)$ be a nonzero constant coefficient linear differential operator in $\mathbb{R}^n$. Then $P(D)$ is hypoelliptic if and only if there exist constants $C, R, c \in (0, \infty)$ with the property that*

$$\left| \frac{\partial^\beta P(\xi)}{P(\xi)} \right| \le C|\xi|^{-c|\beta|}, \qquad \forall \xi \in \mathbb{R}^n \text{ with } |\xi| \ge R, \tag{6.2.12}$$

for every $\beta \in \mathbb{N}_0^n$.

---

[1] This characterization is not going to play a significant role for us here. For a proof, the interested reader is referred to [35, Theorem 11.1.3, p. 62].

An important subclass of linear, constant coefficient operators is that of elliptic operators. Assume that an operator $P(D)$ of order $m \in \mathbb{N}_0$ as in (5.1.1) has been given and recall that its principal symbol is $P_m(\xi) = \sum\limits_{|\alpha|=m} a_\alpha \xi^\alpha, \xi \in \mathbb{R}^n$.

Definition 6.13. Let $P(D)$ be a constant coefficient linear differential operator of order $m \in \mathbb{N}_0$ in $\mathbb{R}^n$. Then $P(D)$ is called elliptic if its principal symbol vanishes only at zero, i.e., if $P_m(\xi) \neq 0$ for every $\xi \in \mathbb{R}^n \setminus \{0\}$.

Remark 6.14. An example of a constant coefficient elliptic operator playing a significant role in partial differential equations is the poly-harmonic operator of order $2m$, where $m \in \mathbb{N}$, defined as

$$\Delta^m := \sum_{j_1,\dots,j_m=1}^n \partial_{j_1}^2 \cdots \partial_{j_m}^2 \quad \text{in} \quad \mathbb{R}^n. \tag{6.2.13}$$

Note that Δ^m is simply $\left(\sum_{j=1}^n \partial_j^2 \right)^m$; hence, its principal symbol is $(-1)^m |\xi|^{2m}$ and it does not vanish for any $\xi \in \mathbb{R}^n \setminus \{0\}$. The operators obtained for $m = 1$ and $m = 2$ are called, respectively, the Laplacian and the bi-Laplacian operator. By the above discussion they are also elliptic.

The relevance of the class of elliptic operators is apparent from the following important consequence of Theorem 6.11.

Theorem 6.15. *Linear, constant coefficient elliptic operators are hypoelliptic in $\mathbb{R}^n$.*

Proof. Let $P(D)$ be a constant coefficient linear elliptic operator of order $m \in \mathbb{N}_0$. By Theorem 6.11, to conclude that $P(D)$ is hypoelliptic, matters reduce to showing that $P(D)$ has a parametrix F with sing supp $F = \{0\}$.

Starting from the ellipticity condition $P_m(\xi) \neq 0$ for every $\xi \in \mathbb{R}^n \setminus \{0\}$, we note that if $C_0 := \inf\limits_{|\xi|=1} |P_m(\xi)|$, then $C_0 > 0$. Keeping in mind that $P_m(\xi)$ is a homogeneous polynomial of degree m we may then estimate

$$|P_m(\xi)| = \left| P_m\left(|\xi| \frac{\xi}{|\xi|} \right) \right| = |\xi|^m \left| P_m\left(\frac{\xi}{|\xi|} \right) \right| \geq C_0 |\xi|^m, \quad \forall \xi \in \mathbb{R}^n \setminus \{0\}. \tag{6.2.14}$$

In addition, there exists $C_1 \in (0, \infty)$ such that

$$|P(\xi) - P_m(\xi)| \leq C_1 (1 + |\xi|)^{m-1} \quad \forall \xi \in \mathbb{R}^n. \tag{6.2.15}$$

Hence, from (6.2.14) and (6.2.15) it follows that for every $\xi \in \mathbb{R}^n \setminus \{0\}$ we have

$$|P(\xi)| \geq |P_m(\xi)| - |P(\xi) - P_m(\xi)| \geq C_0 |\xi|^m - C_1 (1 + |\xi|)^{m-1}. \tag{6.2.16}$$

If $|\xi|$ is sufficiently large, then $C_1(1 + |\xi|)^{m-1} \leq \frac{C_0}{2} |\xi|^m$, which implies that there exists $R \in (0, \infty)$ such that

$$|P(\xi)| \geq \frac{C_0}{2}|\xi|^m \quad \text{whenever} \quad |\xi| \geq R. \tag{6.2.17}$$

Hence, if we take $\psi \in C_0^\infty(\mathbb{R}^n)$ satisfying $\psi \equiv 1$ on $B(0, R)$, then

$$\frac{1 - \psi(\xi)}{P(\xi)} \in C^\infty(\mathbb{R}^n) \cap L^\infty(\mathbb{R}^n). \tag{6.2.18}$$

Recalling (4.1.9) and *(a)* in Theorem 4.26, it follows that there exists some tempered distribution $F \in \mathcal{S}'(\mathbb{R}^n)$ with the property that

$$\widehat{F} = \frac{1 - \psi(\xi)}{P(\xi)} \quad \text{in} \quad \mathcal{S}'(\mathbb{R}^n). \tag{6.2.19}$$

Since $P(\xi) \in \mathcal{L}(\mathbb{R}^n)$, we further have $P(\xi)\widehat{F} = 1 - \psi(\xi)$ in $\mathcal{S}'(\mathbb{R}^n)$. Making use of *(b)* in Theorem 4.26 and (4.2.4), then taking the Fourier transform of the latter identity and invoking (4.2.34), we arrive at

$$P(D)F - \delta = -(2\pi)^{-n}\widehat{\psi}^{\vee} \quad \text{in} \quad \mathcal{S}'(\mathbb{R}^n). \tag{6.2.20}$$

Since $\widehat{\psi} \in \mathcal{S}(\mathbb{R}^n)$, identity (6.2.20) shows that F is a parametrix for $P(D)$.

There remains to prove sing supp $F = \{0\}$. The inclusion $\{0\} \subseteq$ sing supp F is immediate from (6.2.20) given (6.1.4) and (6.1.5). The opposite inclusion will follow once we show

$$F\big|_{\mathbb{R}^n \setminus \{0\}} \in C^\infty(\mathbb{R}^n \setminus \{0\}). \tag{6.2.21}$$

To this end, we claim that if $\beta, \alpha \in \mathbb{N}_0^n$ then

$$\mathcal{F}(D^\beta x^\alpha F) \in L^1(\mathbb{R}^n) \quad \text{whenever} \quad |\beta| - |\alpha| - m < -n. \tag{6.2.22}$$

Assume (6.2.22) for now. Then, recalling (3.1.3), it follows that

$$D^\beta(x^\alpha F) \in C^0(\mathbb{R}^n) \quad \text{for} \quad \beta, \alpha \in \mathbb{N}_0^n, \ |\beta| - |\alpha| - m < -n. \tag{6.2.23}$$

Next, fix $k \in \mathbb{N}_0$ and choose $N \in \mathbb{N}_0$ with the property that $2N > n + k - m$. Since $|x|^{2N} = \sum_{|\alpha|=N} \frac{N!}{\alpha!} x^{2\alpha}$ for every $x \in \mathbb{R}^n$ (cf. (14.2.3)) from (6.2.23) we may conclude that $D^\beta(|x|^{2N}F) \in C^0(\mathbb{R}^n)$ for every $\beta \in \mathbb{N}_0^n$ such that $|\beta| = k$. Hence, $|x|^{2N}F \in C^k(\mathbb{R}^n)$. Because $\frac{1}{|x|^{2N}}\big|_{\mathbb{R}^n \setminus \{0\}} \in C^\infty(\mathbb{R}^n \setminus \{0\})$, the latter further gives $F\big|_{\mathbb{R}^n \setminus \{0\}} \in C^k(\mathbb{R}^n \setminus \{0\})$. The membership in (6.2.21) now follows since $k \in \mathbb{N}_0$ is arbitrary.

Returning to the proof of (6.2.22), fix $\beta, \alpha \in \mathbb{N}_0^n$ with $|\beta| - |\alpha| - m < -n$. In light of *(b)–(c)* in Theorem 4.26, it suffices to show $\xi^\beta D^\alpha \widehat{F} \in L^1(\mathbb{R}^n)$. Note that based on (6.2.19) and (2.4.18) we may write

$$D^\alpha \widehat{F} = \sum_{\gamma \leq \alpha} \frac{\alpha!}{\gamma!(\alpha - \gamma)!} D^\gamma\Big[\frac{1}{P(\xi)}\Big] D^{\alpha-\gamma}[1 - \psi(\xi)] \quad \text{in} \quad \mathcal{S}'(\mathbb{R}^n). \tag{6.2.24}$$

We now claim that for each $\mu, \nu \in \mathbb{N}_0^n$ and with R as in (6.2.17), there exists a finite constant $C > 0$ independent of ξ such that

$$\left| \xi^\mu D^\nu \left[\frac{1}{P(\xi)} \right] \right| \leq C |\xi|^{|\mu| - |\nu| - m} \qquad \text{for} \quad |\xi| > R. \tag{6.2.25}$$

Indeed, since by induction one can see that $D^\nu \left[\frac{1}{P(\xi)} \right] = \frac{Q(\xi)}{P(\xi)^{|\nu|+1}}$ for some polynomial Q of degree at most $(m-1)|\nu|$, by making also use of (6.2.17), for $|\xi| \geq R$ we may write

$$\left| \xi^\mu D^\nu \left[\frac{1}{P(\xi)} \right] \right| = |\xi|^{|\mu|} \frac{|Q(\xi)|}{|P(\xi)|^{|\nu|+1}} \leq C |\xi|^{|\mu|} \frac{|\xi|^{(m-1)|\nu|}}{|\xi|^{(|\nu|+1)m}} = C |\xi|^{|\mu| - |\nu| - m}, \tag{6.2.26}$$

where $C \in (0, \infty)$ is independent of ξ. This proves (6.2.25).

At this point, we combine (6.2.24), (6.2.25), (6.2.17), and the properties of ψ to estimate

$$\left| \xi^\beta D^\alpha \widehat{F} \right| \leq \sum_{\gamma < \alpha} \frac{\alpha!}{\gamma!(\alpha - \gamma)!} \left| \xi^\beta D^\gamma \left[\frac{1}{P(\xi)} \right] D^{\alpha - \gamma} [1 - \psi(\xi)] \right|$$

$$+ \left| \xi^\beta D^\alpha \left[\frac{1}{P(\xi)} \right] [1 - \psi(\xi)] \right|$$

$$\leq \sum_{\gamma < \alpha} C_{\alpha, \gamma} |\xi|^{|\beta| - |\gamma| - m} \chi_{\operatorname{supp} \psi \setminus B(0, R)}$$

$$+ C |\xi|^{|\beta| - |\alpha| - m} \chi_{\mathbb{R}^n \setminus B(0, R)}, \qquad \forall \xi \in \mathbb{R}^n. \tag{6.2.27}$$

Recalling that $|\beta| - |\alpha| - m < -n$, from (6.2.27) we obtain that $\xi^\beta D^\alpha \widehat{F}$ belongs to $L^1(\mathbb{R}^n)$, as desired. This completes the proof of (6.2.22), and with it the proof of the theorem. $\qquad\qquad\square$

Exercise 6.16. Use Theorem 6.12 to give an alternative proof of Theorem 6.15, i.e., that linear, constant coefficient elliptic operators are hypoelliptic in $\mathbb{R}^n$.

Hint: Show that C_1 may be chosen sufficiently large so that, in addition to (6.2.15), one also has $|\partial^\beta P(\xi)| \leq C_1 (1 + |\xi|)^{m - |\beta|}$. Then use the latter and (6.2.17) to show that (6.2.12) is verified by taking $C := \frac{2C_1}{C_0} (1 + \frac{1}{R})^m$ and $c := 1$.

Proposition 6.17. *The poly-harmonic operator is hypoelliptic. In particular, the operators Δ and Δ^2 are hypoelliptic.*

Proof. This is a consequence of Theorem 6.15 and Remark 6.14. $\qquad\qquad\square$

Corollary 6.18. *Let $P(D)$ be a linear, constant coefficient elliptic operator in $\mathbb{R}^n$ and let Ω be an open subset of $\mathbb{R}^n$. If $u \in \mathcal{D}'(\Omega)$ is such that for some open subset ω of Ω we have $P(D)u\big|_\omega \in C^\infty(\omega)$, then $u\big|_\omega \in C^\infty(\omega)$.*

Proof. From Theorem 6.15 and Remark 6.2 we know that the restriction of $P(D)$ to ω is a hypoelliptic operator. Since $P(D)u\big|_\omega \in \mathcal{D}'(\omega)$ has an empty singular support,

Proposition 6.6 gives that the singular support of $u\big|_\omega$ is also empty. Consequently, $u\big|_\omega \in C^\infty(\omega)$. □

An example of a linear, constant coefficient, differential operator that is not hypoelliptic is the operator $P = \partial_1^2 - \partial_2^2$ used in (1.1.1) to describe the equation governing the displacement of a vibrating string in $\mathbb{R}^2$. Indeed, as we shall see later (cf. (9.1.28)), the distribution u_f associated with the locally integrable function

$$f(x_1, x_2) := H(x_2 - |x_1|), \qquad \forall\, (x_1, x_2) \in \mathbb{R}^2, \tag{6.2.28}$$

(where H stands for the Heaviside function) satisfies $(\partial_1^2 - \partial_2^2)u_f = 2\delta$ in $\mathcal{D}'(\mathbb{R}^2)$ and sing supp $u_f = \{(x_1, x_2) \in \mathbb{R}^2 : x_2 = |x_1|\}$. Hence,

$$\text{sing supp}\, u_f \neq \text{sing supp}\, Pu_f \tag{6.2.29}$$

which, in light of Proposition 6.6, shows that P is not hypoelliptic.

6.3 Integral Representation Formulas and Interior Estimates

Consider the constant coefficient, linear, differential operator

$$P(\partial) = \sum_{|\alpha| \leq m} a_\alpha \partial^\alpha, \quad a_\alpha \in \mathbb{C}, \quad \text{for all } \alpha \in \mathbb{N}_0^n \text{ with } |\alpha| \leq m. \tag{6.3.1}$$

Theorem 6.19 (A general integral representation formula). *Assume that the constant coefficient, linear, differential operator $P(\partial)$ is as in (6.3.1) and is hypoelliptic in $\mathbb{R}^n$. Let $E \in \mathcal{D}'(\mathbb{R}^n)$ be a fundamental solution for $P(\partial)$ as provided by Theorem 6.8 (hence, in particular, $E \in C^\infty(\mathbb{R}^n \setminus \{0\})$).*

Let Ω be an open subset of $\mathbb{R}^n$ and suppose $u \in \mathcal{D}'(\Omega)$ satisfies $P(\partial)u = 0$ in $\mathcal{D}'(\Omega)$. Then $u \in C^\infty(\Omega)$ and for each $x_0 \in \Omega$, each $r \in (0, \text{dist}\,(x_0, \partial\Omega))$, and each function $\psi \in C_0^\infty(B(x_0, r))$ such that $\psi \equiv 1$ near $\overline{B(x_0, r/2)}$, we have

$$u(x) = -\sum_{|\alpha| \leq m} \sum_{\gamma < \alpha} (-1)^{|\alpha|+|\gamma|} a_\alpha \frac{\alpha!}{(\alpha-\gamma)!\gamma!} \times \tag{6.3.2}$$

$$\times \int_{B(x_0,r)\setminus\overline{B(x_0,r/2)}} (\partial^\gamma E)(x-y)(\partial^{\alpha-\gamma}\psi)(y)u(y)\, dy,$$

for each $x \in B(x_0, r/2)$. In particular, for every $\mu \in \mathbb{N}_0^n$,

$$(\partial^\mu u)(x) = -\sum_{|\alpha|\le m}\sum_{\gamma<\alpha}(-1)^{|\alpha|+|\gamma|}\,a_\alpha\,\frac{\alpha!}{(\alpha-\gamma)!\gamma!}\times \tag{6.3.3}$$

$$\times \int_{B(x_0,r)\setminus\overline{B(x_0,r/2)}}(\partial^{\gamma+\mu}E)(x-y)(\partial^{\alpha-\gamma}\psi)(y)u(y)\,dy,$$

for each $x \in B(x_0,r/2)$.

Proof. The fact that $u \in C^\infty(\Omega)$ is a consequence of the hypoellipticity of the operator $P(\partial)$ (cf. Definition 6.1 and Remark 6.2). As regards (6.3.2), pick an arbitrary $x_0 \in \Omega$, fix $r \in (0, \mathrm{dist}\,(x_0, \partial\Omega))$, and let $\psi \in C_0^\infty(B(x_0,r))$ be such that $\psi \equiv 1$ near $\overline{B(x_0,r/2)}$. Then $\psi u \in C_0^\infty(\mathbb{R}^n)$ and since $P(\partial)E = \delta$ in $\mathcal{D}'(\mathbb{R}^n)$ we may write, for each point $x \in B(x_0,r/2)$,

$$u(x) = (\psi u)(x) = (\delta * (\psi u))(x)$$

$$= ((P(\partial)E) * (\psi u))(x) = (E * (P(\partial)(\psi u)))(x)$$

$$= \sum_{|\alpha|\le m}\sum_{0<\beta\le\alpha}a_\alpha\frac{\alpha!}{\beta!(\alpha-\beta)!}(E * (\partial^\beta\psi\partial^{\alpha-\beta}u))(x) \tag{6.3.4}$$

where we have used (6.3.1), part *(e)* of Theorem 2.96, and that $P(\partial)u = 0$ in Ω. Note that for each $\beta > 0$ the function $\partial^\beta\psi$ is compactly supported in $B(x_0,r) \setminus \overline{B(x_0,r/2)}$. Keeping this in mind, we may then integrate by parts in order to further write the last expression in (6.3.4) as

$$\sum_{|\alpha|\le m}\sum_{0<\beta\le\alpha}a_\alpha\frac{\alpha!}{\beta!(\alpha-\beta)!}(E * (\partial^\beta\psi\partial^{\alpha-\beta}u))(x) \tag{6.3.5}$$

$$= \sum_{|\alpha|\le m}\sum_{0<\beta\le\alpha}a_\alpha\frac{\alpha!}{\beta!(\alpha-\beta)!}\int_{B(x_0,r)\setminus\overline{B(x_0,r/2)}}E(x-y)\partial^\beta\psi(y)\partial^{\alpha-\beta}u(y)\,dy$$

$$= \sum_{|\alpha|\le m}\sum_{0<\beta\le\alpha}a_\alpha\frac{(-1)^{|\alpha|+|\beta|}\alpha!}{\beta!(\alpha-\beta)!}\int_{B(x_0,r)\setminus\overline{B(x_0,r/2)}}\partial_y^{\alpha-\beta}[E(x-y)\partial^\beta\psi(y)]u(y)\,dy$$

$$= \sum_{|\alpha|\le m}\sum_{0<\beta\le\alpha}\sum_{\gamma\le\alpha-\beta}a_\alpha\frac{(-1)^{|\alpha|+|\beta|+|\gamma|}\alpha!}{\beta!(\alpha-\beta)!}\frac{(\alpha-\beta)!}{\gamma!(\alpha-\beta-\gamma)!}\times$$

$$\times \int_{B(x_0,r)\setminus\overline{B(x_0,r/2)}}(\partial^\gamma E)(x-y)\partial^{\alpha-\gamma}\psi(y)\,u(y)\,dy.$$

To proceed, observe that whenever $\gamma < \alpha$ formula (14.2.5) gives

$$\sum_{0<\beta\le\alpha-\gamma}(-1)^{|\beta|}\frac{(\alpha-\gamma)!}{\beta!(\alpha-\gamma-\beta)!} = \Big(\sum_{\beta\le\alpha-\gamma}(-1)^{|\beta|}\frac{(\alpha-\gamma)!}{\beta!(\alpha-\gamma-\beta)!}\Big) - 1$$

$$= 0 - 1 = -1. \tag{6.3.6}$$

Making use of (6.3.6) back in (6.3.5) yields

$$\sum_{|\alpha|\le m}\sum_{0<\beta\le\alpha}a_\alpha\frac{\alpha!}{\beta!(\alpha-\beta)!}(E*(\partial^\beta\psi\partial^{\alpha-\beta}u))(x) \tag{6.3.7}$$

$$=-\sum_{|\alpha|\le m}\sum_{\gamma<\alpha}a_\alpha\frac{(-1)^{|\alpha|+|\gamma|}\alpha!}{\gamma!(\alpha-\gamma)!}\times$$

$$\times\int_{B(x_0,r)\setminus\overline{B(x_0,r/2)}}(\partial^\gamma E)(x-y)\partial^{\alpha-\gamma}\psi(y)\,u(y)\,dy,$$

and (6.3.2) follows from (6.3.4) and (6.3.7). Finally, (6.3.3) is obtained by differentiating (6.3.2). $\qquad\square$

To state our next result, recall that the operator $P(\partial)$ from (6.3.1) is said to be homogeneous (of degree m) whenever $a_\alpha=0$ for all multi-indices α with $|\alpha|<m$.

Theorem 6.20 (Interior estimates). *Let $P(\partial)$ be a constant coefficient, linear, differential operator, of order $m\in\mathbb{N}_0$, which is hypoelliptic in $\mathbb{R}^n$. Also suppose $E\in\mathcal{D}'(\mathbb{R}^n)$ is a fundamental solution for $P(\partial)$ as provided by Theorem 6.8 (thus, in particular, $E\in C^\infty(\mathbb{R}^n\setminus\{0\})$).*

Then for every $\mu\in\mathbb{N}_0^n$ there exists a constant $C_\mu\in(0,\infty)$ (which also depends on the coefficients of $P(\partial)$ and n) with the following significance. If Ω be an open subset of $\mathbb{R}^n$ and $u\in\mathcal{D}'(\Omega)$ satisfies $P(\partial)u=0$ in $\mathcal{D}'(\Omega)$, then u belongs to $C^\infty(\Omega)$ and for each $x_0\in\Omega$ and each $r\in(0,\operatorname{dist}(x_0,\partial\Omega))$ we have

$$\sup_{x\in B(x_0,r/2)}|(\partial^\mu u)(x)|\le C_\mu\max_{\substack{|\alpha|\le m\\ \gamma<\alpha}}\Big\{r^{|\gamma|-|\alpha|}\cdot\sup_{r/4<|z|<r}|(\partial^{\gamma+\mu}E)(z)|\Big\}\times$$

$$\times\int_{B(x_0,r)\setminus\overline{B(x_0,r/2)}}|u(y)|\,dy. \tag{6.3.8}$$

In the particular case when $P(\partial)$ is homogeneous (of degree m) and one also assumes that there exists $k\in\mathbb{N}$ such that the fundamental solution E satisfies

$$|\partial^\beta E(x)|\le C_\beta|x|^{-n+m-|\beta|},$$
$$\forall\,x\in\mathbb{R}^n\setminus\{0\},\ \forall\beta\in\mathbb{N}_0^n\ \text{with}\ |\beta|\ge k, \tag{6.3.9}$$

then whenever $|\mu|\ge k$ we have

$$|(\partial^\mu u)(x_0)|\le\frac{C_\mu}{r^{|\mu|}}\int_{B(x_0,r)}|u(y)|\,dy. \tag{6.3.10}$$

Proof. Pick a function $\phi\in C_0^\infty(B(0,1))$ such that $\phi\equiv1$ near $\overline{B(0,3/4)}$, and note that if we set

$$\psi(x):=\phi((x-x_0)/r),\qquad\forall\,x\in\mathbb{R}^n, \tag{6.3.11}$$

then $\psi \in C_0^\infty(B(x_0, r))$ and $\psi \equiv 1$ near $\overline{B(x_0, 3r/4)}$. In particular, for every $\gamma \in \mathbb{N}_0^n$ with $|\gamma| > 0$ there exists $c_\gamma \in (0, \infty)$ independent of r such that

$$\operatorname{supp}(\partial^\gamma \psi) \subseteq B(x_0, r) \setminus \overline{B(x_0, 3r/4)} \quad \text{and} \quad \|\partial^\gamma \psi\|_{L^\infty(\mathbb{R}^n)} \le c_\gamma \, r^{-|\gamma|}. \tag{6.3.12}$$

For such a choice of ψ, estimate (6.3.8) then follows in a straightforward manner from the integral representation formula (6.3.3). In turn, (6.3.8) readily implies (6.3.10) in the case when $P(\partial)$ is homogeneous (of degree m), E satisfies (6.3.9) for some $k \in \mathbb{N}$, and $|\mu| \ge k$.   □

The procedure described abstractly in the following lemma is useful for deriving higher order interior estimates, with precise control of the constants involved, for various classes of functions.

Lemma 6.21. *Let $\mathcal{A} \subset C^\infty(\Omega)$ be a family of functions satisfying the following two conditions:*

(1) $\partial^\alpha u \in \mathcal{A}$ for every $u \in \mathcal{A}$ and every $\alpha \in \mathbb{N}_0^n$;
(2) there exists $C \in (0, \infty)$ such that for every $u \in \mathcal{A}$ we have

$$|\partial_j u(x)| \le \frac{C}{r} \max_{y \in B(x,r)} |u(y)|, \qquad \forall \, j \in \{1, \dots, n\}, \tag{6.3.13}$$

whenever $x \in \Omega$ and $r \in (0, \operatorname{dist}(x, \partial\Omega))$.

Then given any $u \in \mathcal{A}$, for every $x \in \Omega$, every $r \in (0, \operatorname{dist}(x, \partial\Omega))$, every $k \in \mathbb{N}$, and every $\lambda \in (0, 1)$, we have (with C as in (6.3.13))

$$\max_{y \in B(x,\lambda r)} |\partial^\alpha u(y)| \le \frac{C^k (1-\lambda)^{-k} e^{k-1} k!}{r^k} \max_{y \in B(x,r)} |u(y)|, \tag{6.3.14}$$

for every multi-index $\alpha \in \mathbb{N}_0^n$ with $|\alpha| = k$.

Proof. In a first stage, we propose to prove by induction over k that, given any $u \in \mathcal{A}$, for every $x \in \Omega$, every $r \in (0, \operatorname{dist}(x, \partial\Omega))$, and every $k \in \mathbb{N}$ we have (with C as in (6.3.13))

$$|\partial^\alpha u(x)| \le \frac{C^k e^{k-1} k!}{r^k} \max_{y \in \overline{B(x,r)}} |u(y)|, \qquad \forall \, \alpha \in \mathbb{N}_0^n \text{ with } |\alpha| = k. \tag{6.3.15}$$

Note that if $k = 1$, this is contained in (6.3.13). Suppose now that (6.3.15) holds for some $k \in \mathbb{N}$ and every $u \in \mathcal{A}$, every $x \in \Omega$, and every $r \in (0, \operatorname{dist}(x, \partial\Omega))$, and pick an arbitrary $\alpha \in \mathbb{N}_0^n$ with $|\alpha| = k + 1$. Then there exist $j \in \{1, \dots, n\}$ and $\beta \in \mathbb{N}_0^n$ for which $\partial^\alpha = \partial_j \partial^\beta$. Fix such j and β and note that, in particular, $|\beta| = k$. Next, take $x \in \Omega$ and $r \in (0, \operatorname{dist}(x, \partial\Omega))$ arbitrary, and pick some $\varepsilon \in (0, 1)$ to be specified shortly. Since $\partial^\beta u \in \mathcal{A}$ we may use (6.3.13) with u replaced by $\partial^\beta u$ and r replaced by $(1 - \varepsilon)r$ to obtain

$$|\partial^\alpha u(x)| \le \frac{C}{(1-\varepsilon)r} \max_{y \in B(x,(1-\varepsilon)r)} |\partial^\beta u(y)|. \tag{6.3.16}$$

Note that if $y \in B(x, (1 - \varepsilon)r)$ then $\overline{B(y, \varepsilon r)} \subset B(x, r) \subset \Omega$. Thus, for every point $y \in B(x, (1 - \varepsilon)r)$, we may use the induction hypothesis to estimate

$$|\partial^\beta u(y)| \le \frac{C^k e^{k-1} k!}{(\varepsilon r)^k} \max_{z \in B(y, \varepsilon r)} |u(z)| \le \frac{C^k e^{k-1} k!}{\varepsilon^k r^k} \max_{z \in B(x, r)} |u(z)|. \tag{6.3.17}$$

Combined, (6.3.16) and (6.3.17) yield

$$|\partial^\alpha u(x)| \le \frac{C^{k+1} e^{k-1} k!}{(1 - \varepsilon) \varepsilon^k r^{k+1}} \max_{z \in B(x, r)} |u(z)|. \tag{6.3.18}$$

Set now $\varepsilon := \frac{k}{k+1} \in (0, 1)$ in (6.3.18) which, given that $\varepsilon^{-k} < e$, further implies

$$|\partial^\alpha u(x)| \le \frac{C^{k+1} e^k (k + 1)!}{r^{k+1}} \max_{z \in B(x, r)} |u(z)|. \tag{6.3.19}$$

Hence, (6.3.15) holds with $k + 1$ in place of k, as desired.

As far as (6.3.14) is concerned, pick $x_0 \in \Omega$, $r \in (0, \operatorname{dist}(x_0, \partial\Omega))$, $\lambda \in (0, 1)$, as well as $k \in \mathbb{N}$. Next, select an arbitrary point $x \in B(x_0, \lambda r)$ and note that this forces $(1 - \lambda)r \in (0, \operatorname{dist}(x, \partial\Omega))$. In concert with the easily seen fact that $B(x, (1 - \lambda)r) \subseteq B(x_0, r)$, this permits us to invoke (6.3.15) in order to estimate

$$|\partial^\alpha u(x)| \le \frac{C^k e^{k-1} k!}{((1 - \lambda)r)^k} \max_{y \in B(x, (1-\lambda)r)} |u(y)|$$

$$\le \frac{C^k (1 - \lambda)^{-k} e^{k-1} k!}{r^k} \max_{y \in B(x_0, r)} |u(y)|, \tag{6.3.20}$$

whenever $\alpha \in \mathbb{N}_0^n$ satisfies $|\alpha| = k$. Taking the supremum over $x \in B(x_0, \lambda r)$ then yields (6.3.14) (written with x_0 in place of x). $\qquad\square$

Definition 6.22. A function $u \in C^\infty(\Omega)$ is called `real-analytic` in Ω provided for every $x \in \Omega$ there exists $r_x \in (0, \infty)$ with the property that $B(x, r_x) \subseteq \Omega$ and the Taylor series for u at x converges uniformly to u on $B(x, r_x)$, i.e.,

$$u(y) = \sum_{\alpha \in \mathbb{N}_0^n} \frac{1}{\alpha!} (y - x)^\alpha (\partial^\alpha u)(x) \quad \text{uniformly for } y \in B(x, r_x). \tag{6.3.21}$$

The following observation justifies the name "real-analytic" used for the class of functions introduced above.

Remark 6.23. In the context of Definition 6.22, it is clear that, for each $x \in \Omega$,

$$\widetilde{u}(z) := \sum_{\alpha \in \mathbb{N}_0^n} \frac{1}{\alpha!} (z - x)^\alpha (\partial^\alpha u)(x) \quad \text{for } z \in \mathbb{C}^n \text{ with } |z - x| < r_x, \tag{6.3.22}$$

is a holomorphic function in $\{z \in \mathbb{C}^n : |z - x| < r_x\}$ which locally extends the original real-analytic function u. Such local extensions may be constructed near each point $x \in \Omega$ and Lemma 7.67 ensures that any two local extensions coincide on their common domain. The conclusion is that the real-analytic function u has a well-defined extension to a holomorphic function defined in a neighborhood of Ω in $\mathbb{C}^n$.

Our next lemma gives a sufficient condition (which is in the nature of best possible) ensuring real-analyticity.

Lemma 6.24. *Suppose $u \in C^\infty(\Omega)$ is a function with the property that for each $x \in \Omega$ there exist $r = r(x) \in (0, \text{dist}\,(x, \partial\Omega))$, $M = M(x) \in (0, \infty)$, and $C = C(x) \in (0, \infty)$, such that for every $k \in \mathbb{N}$ the following estimate holds:*

$$\max_{y \in \overline{B(x,r)}} |\partial^\alpha u(y)| \leq MC^k k! \qquad \forall \alpha \in \mathbb{N}_0^n \text{ with } |\alpha| = k. \tag{6.3.23}$$

Then u is real-analytic in Ω.

Proof. Fix $x \in \Omega$ and let $r \in (0, \text{dist}\,(x, \partial\Omega))$ be such that (6.3.23) holds for every $k \in \mathbb{N}$. Write Taylor's formula (14.2.10) for u at x to obtain that for each $N \in \mathbb{N}$ and each $y \in B(x, r/2)$ there exists $\theta \in (0, 1)$ such that

$$u(y) = \sum_{|\alpha| \leq N-1} \frac{1}{\alpha!}(y - x)^\alpha(\partial^\alpha u)(x) + R_{N,u}(y), \tag{6.3.24}$$

where

$$R_{N,u}(y) := \sum_{|\alpha|=N} \frac{1}{\alpha!}(y - x)^\alpha(\partial^\alpha u)(x + \theta(y - x)). \tag{6.3.25}$$

Using (6.3.23), for each $\alpha \in \mathbb{N}_0^n$ with $|\alpha| = N$ and $y \in B(x, r/2)$, we may estimate

$$\left|(\partial^\alpha u)(x + \theta(y - x))\right| \leq \max_{z \in B(x,r)} |(\partial^\alpha u)(z)| \leq MC^N N! \tag{6.3.26}$$

since $B(x + \theta(y - x), r) \subset B(x, r)$. Together, (6.3.25) and (6.3.26) imply that, for each $y \in B(x, r/2)$,

$$|R_{N,u}(y)| \leq MC^N|y - x|^N \sum_{|\alpha|=N} \frac{N!}{\alpha!} = M\,(nC|y - x|)^N. \tag{6.3.27}$$

For the equality in (6.3.27) we used formula (14.2.3) for $x_1 = \cdots = x_n = 1$. Hence, if say, $|y - x| < \min\{\frac{1}{2nC}, \frac{r}{2}\}$, then $\lim_{N \to \infty} R_{N,u}(y) = 0$ uniformly with respect to y. Consequently, the Taylor series for u converges uniformly to u in a neighborhood of x. Since x is arbitrary in Ω it follows that u is real-analytic in Ω. $\qquad\square$

Theorem 6.25 (Unique continuation). *Suppose $\Omega \subseteq \mathbb{R}^n$ is an open and connected set. Then any real-analytic function u in Ω with the property that there exists $x_0 \in \Omega$*

such that $\partial^\alpha u(x_0) = 0$ for all $\alpha \in \mathbb{N}_0^n$ (which is the case if, e.g., u is zero in a neighborhood of x_0) vanishes identically in Ω.

Proof. Suppose u satisfies the hypotheses of the theorem and define the set

$$U := \{x \in \Omega : \partial^\alpha u(x) = 0 \text{ for all } \alpha \in \mathbb{N}_0^n\}. \tag{6.3.28}$$

Since $x_0 \in U$ we have $U \neq \varnothing$. Also, U is relatively closed in Ω given that it is the intersection of the relatively closed sets $(\partial^\alpha u)^{-1}(\{0\})$, $\alpha \in \mathbb{N}_0^n$. In addition, if $x \in U$ then, on the one hand, the Taylor series of u at x is identically zero, while on the other hand, this series converges to u in an open neighborhood of x. Thus, u is identically zero in that neighborhood of x, which ultimately proves that U is also open. Recalling that Ω is connected, it follows that $U = \Omega$ as desired. □

Theorem 6.25 highlights the much more restrictive nature of real-analyticity compared to indefinite differentiability, since obviously there are plenty of C^∞ functions which vanish in a neighborhood of a point without being identically zero.

Theorem 6.26. *Let $P(\partial)$ be a constant coefficient, linear, differential operator in $\mathbb{R}^n$, which is homogeneous of degree $m \in \mathbb{N}_0$. Assume that $P(\partial)$ has a fundamental solution $E \in \mathcal{D}'(\mathbb{R}^n)$ with the property that $E \in C^\infty(\mathbb{R}^n \setminus \{0\})$ and*

$$|\partial^\beta E(x)| \leq C_\beta |x|^{-n+m-|\beta|},$$
$$\forall x \in \mathbb{R}^n \setminus \{0\}, \quad \forall \beta \in \mathbb{N}_0^n \text{ with } |\beta| > 0. \tag{6.3.29}$$

Finally, suppose that $\Omega \subseteq \mathbb{R}^n$ is open and $u \in \mathcal{D}'(\Omega)$ satisfies $P(\partial)u = 0$ in $\mathcal{D}'(\Omega)$.

Then $u \in C^\infty(\Omega)$ and there exists a constant $C \in (0, \infty)$ such that if $\lambda \in (0, 1)$ we have

$$\max_{y \in B(x,\lambda r)} |\partial^\alpha u(y)| \leq \frac{C^{|\alpha|}(1-\lambda)^{-|\alpha|}|\alpha|!}{r^{|\alpha|}} \max_{y \in B(x,r)} |u(y)|, \quad \forall \alpha \in \mathbb{N}_0^n, \tag{6.3.30}$$

whenever $x \in \Omega$ and $r \in (0, \text{dist}\,(x, \partial\Omega))$. In particular, u is real-analytic in Ω.

Proof. The fact that $P(\partial)$ has a fundamental solution $E \in \mathcal{D}'(\mathbb{R}^n)$ with the property that $E \in C^\infty(\mathbb{R}^n \setminus \{0\})$ implies $\text{sing supp}\, E = \{0\}$; hence, $P(\partial)$ is hypoelliptic in $\mathbb{R}^n$ by Theorem 6.8. Granted the properties enjoyed by $P(\partial)$ and condition (6.3.29), estimate (6.3.30) follows from (6.3.9) (used with $|\mu| = 1$) and Lemma 6.21 (in which we take $\mathcal{A} := \{u \in C^\infty(\Omega) : P(\partial)u = 0\}$). Finally, the last claim in the statement of the theorem is a consequence of (6.3.30) and Lemma 6.24. □

Further Notes for Chapter 6. For more extensive discussion of the notion of hypoellipticity the interested reader is referred to the excellent presentation in [35]. Further information about real-analytic functions may be found in, e.g., [40].

6.4 Additional Exercises for Chapter 6

Exercise 6.27. Let $k \in (0, \infty)$. Prove that

$$\left\{ u \in \mathcal{S}'(\mathbb{R}) : \frac{\mathrm{d}^2 u}{\mathrm{d}x^2} + ku = \delta \ \text{ in } \ \mathcal{S}'(\mathbb{R}) \right\} \tag{6.4.1}$$

$$= \left\{ -\frac{\mathrm{i}}{2k} \mathrm{e}^{\mathrm{i}kt} H + \frac{\mathrm{i}}{2k} \mathrm{e}^{-\mathrm{i}kt} H + c_1 \mathrm{e}^{\mathrm{i}kt} + c_2 \mathrm{e}^{-\mathrm{i}kt} : c_1, c_2 \in \mathbb{C} \right\}.$$

Exercise 6.28. Let $a > 0$ be fixed.

(a) Compute a fundamental solution for the operator $\frac{\mathrm{d}^2}{\mathrm{d}x^2} - a^2$ in $\mathbb{R}$.

(b) Use the result from part (a) to compute the Fourier transform in $\mathcal{S}'(\mathbb{R})$ of the tempered distribution associated with the function $f(x) := \frac{1}{x^2 + a^2}$ for each $x \in \mathbb{R}$ (compare with Example 4.24).

Exercise 6.29. Prove that $\operatorname{sing\,supp} u \subseteq \operatorname{supp} u$ for every $u \in \mathcal{D}'(\Omega)$.

Exercise 6.30. Give an example of a distribution u for which the inclusion from Exercise 6.29 is strict. Give an example of a distribution u for which $\operatorname{sing\,supp} u = \operatorname{supp} u$.

Exercise 6.31. Let $x_0 \in \mathbb{R}^n$. Prove that $\operatorname{sing\,supp} \delta_{x_0}^\alpha = \{x_0\}$ for every $\alpha \in \mathbb{N}_0^n$.

Exercise 6.32. Let $a \in C_0^\infty(\Omega)$ and $u \in \mathcal{D}'(\Omega)$. Prove that

$$\operatorname{sing\,supp}(au) \subseteq \operatorname{sing\,supp} u.$$

Exercise 6.33. Determine $\operatorname{sing\,supp}(\mathrm{P.V.} \frac{1}{x})$, where $\mathrm{P.V.} \frac{1}{x}$ is the distribution defined in (2.1.13).

Exercise 6.34. Let $m \in \mathbb{N}$ and let P be a polynomial in $\mathbb{R}^n$ of degree $2m$ with no real roots. Recall from Exercise 4.128 that $\frac{1}{P} \in \mathcal{S}'(\mathbb{R}^n)$. Prove that in this scenario $\operatorname{sing\,supp}\left(\mathcal{F}(\frac{1}{P})\right)$ is nonempty.

Exercise 6.35. Recall $\mathrm{P.V.} \frac{1}{x} \in \mathcal{D}'(\mathbb{R})$ from (2.1.13) and define the distribution $u \in \mathcal{D}'(\mathbb{R}^2)$ by $u := \mathrm{P.V.} \frac{1}{x} \otimes \delta(y)$. Determine $\operatorname{sing\,supp} u$.

Chapter 7
The Laplacian and Related Operators

Abstract Starting from first principles, all fundamental solutions (that are tempered distributions) for scalar elliptic operators are identified in this chapter. While the natural starting point is the Laplacian, this study encompasses a variety of related operators, such as the bi-Laplacian, the poly-harmonic operator, the Helmholtz operator and its iterations, the Cauchy–Riemann operator, the Dirac operator, the perturbed Dirac operator and its iterations, as well as general second-order constant coefficient strongly elliptic operators. Having accomplished this task then makes it possible to prove the well-posedness of the Poisson problem (equipped with a boundary condition at infinity), and derive qualitative/quantitative properties for the solution. Along the way, Cauchy-like integral operators are also introduced and their connections with Hardy spaces are brought to light in the setting of both complex and Clifford analyses.

7.1 Fundamental Solutions for the Laplace Operator

One of the most important operators in partial differential equations is the `Laplace operator`[1] Δ (also called the `Laplacian`) in $\mathbb{R}^n$ which is defined as $\Delta := \sum_{j=1}^{n} \partial_j^2$.
Functions f satisfying $\Delta f = 0$ pointwise in $\mathbb{R}^n$ are called `harmonic` in $\mathbb{R}^n$. The Laplace operator arises in many applications such as in the modeling of heat conduction, electrical conduction, and chemical concentration, to name a few. The focus for us will be on finding all fundamental solutions for Δ that are tempered distributions. Part of the motivation is that, as explained in Remark 5.6, this greatly facilitates the study of the Poisson equation $\Delta u = f$ in $\mathcal{D}'(\mathbb{R}^n)$, a task we take up in Section 7.2.

The goal in this section is to determine all fundamental solutions of Δ that are also in $\mathcal{S}'(\mathbb{R}^n)$. This is done by employing the properties of the Fourier transform we have proved so far. To get started, fix some $E \in \mathcal{S}'(\mathbb{R}^n)$ that is fundamental

---

[1] Named after the French mathematician and astronomer Pierre Simon de Laplace (1749–1827).

© Springer Nature Switzerland AG 2018

D. Mitrea, *Distributions, Partial Differential Equations, and Harmonic Analysis*,
Universitext, https://doi.org/10.1007/978-3-030-03296-8_7

solution for Δ. Note that the existence of such a tempered distribution is guaranteed by Theorem 5.14. Since $\Delta E = \delta$ in $\mathcal{D}'(\mathbb{R}^n)$ and $\delta \in \mathcal{S}'(\mathbb{R}^n)$, by (4.1.33) it follows that

$$\Delta E = \delta \quad \text{in } \mathcal{S}'(\mathbb{R}^n). \tag{7.1.1}$$

Applying $\mathcal{F}$ to the equation in (7.1.1) (and using that $-\Delta = \sum_{j=1}^{n} D_{x_j}^2$, part *(b)* in Theorem 4.26, and (4.2.3)) we obtain

$$-|\xi|^2 \widehat{E} = 1 \quad \text{in } \mathcal{S}'(\mathbb{R}^n). \tag{7.1.2}$$

To proceed with determining E, we discuss separately the cases $n \geq 3$, $n = 2$, and $n = 1$.

Case $n \geq 3$. From Exercise 4.5 we have that $\frac{1}{|\xi|^2} \in \mathcal{S}'(\mathbb{R}^n)$. In addition, we have $|\xi|^2 \in \mathcal{L}(\mathbb{R}^n)$, thus $|\xi|^2 \cdot \frac{1}{|\xi|^2} \in \mathcal{S}'(\mathbb{R}^n)$ (recall *(b)* in Theorem 4.14), and it is not difficult to check that $|\xi|^2 \cdot \frac{1}{|\xi|^2} = 1$ in $\mathcal{S}'(\mathbb{R}^n)$. Hence, $|\xi|^2(\widehat{E} + \frac{1}{|\xi|^2}) = 0$ in $\mathcal{S}'(\mathbb{R}^n)$. Thus, $\text{supp}\left(\widehat{E} + \frac{1}{|\xi|^2}\right) \subseteq \{0\}$ and by Exercise 4.37, it follows that

$$\widehat{E} + \frac{1}{|\xi|^2} = \widehat{P}(\xi) \quad \text{in } \mathcal{S}'(\mathbb{R}^n), \tag{7.1.3}$$

where P is a polynomial in $\mathbb{R}^n$ satisfying $|\xi|^2 \widehat{P} = 0$ in $\mathcal{S}'(\mathbb{R}^n)$. The latter implies that $\Delta P = 0$ in $\mathcal{S}'(\mathbb{R}^n)$. Since $P \in C^\infty(\mathbb{R}^n)$, we may conclude that $\Delta P = 0$ pointwise in $\mathbb{R}^n$, hence P is a harmonic polynomial in $\mathbb{R}^n$. To compute E (which is equal to $\mathcal{F}^{-1}(\widehat{E})$) apply $\mathcal{F}^{-1}$ to the identity in (7.1.3) and then recall Proposition 4.64 with $\lambda = 2$ to write

$$E(x) = -\mathcal{F}^{-1}\left(\frac{1}{|\xi|^2}\right)(x) + P(x)$$

$$= -2^{-2}\pi^{-\frac{n}{2}} \frac{\Gamma\left(\frac{n-2}{2}\right)}{\Gamma(1)} |x|^{2-n} + P(x)$$

$$= -\frac{1}{(n-2)\omega_{n-1}} \cdot \frac{1}{|x|^{n-2}} + P(x), \qquad \forall\, x \in \mathbb{R}^n \setminus \{0\}, \tag{7.1.4}$$

where the last equality in (7.1.4) is also based on the fact that $\Gamma\left(\frac{n}{2}\right)$ is equal to $\left(\frac{n}{2} - 1\right)\Gamma\left(\frac{n}{2} - 1\right)$ (recall that $n \geq 3$), that $\Gamma(1) = 1$, and on (14.5.6). Moreover,

$$\Delta\left(-\frac{1}{(n-2)\omega_{n-1}} \cdot \frac{1}{|x|^{n-2}}\right) = \delta \quad \text{in } \mathcal{S}'(\mathbb{R}^n), \tag{7.1.5}$$

since (7.1.5) is equivalent (via the Fourier transform, as a consequence of *(a)* and *(b)* in Theorem 4.26, as well as (4.2.3)) with $|\xi|^2 \mathcal{F}\left(\frac{1}{(n-2)\omega_{n-1}} \cdot \frac{1}{|x|^{n-2}}\right) = 1$ in $\mathcal{S}'(\mathbb{R}^n)$, or equivalently, with $|\xi|^2 \cdot \frac{1}{|\xi|^2} = 1$ in $\mathcal{S}'(\mathbb{R}^n)$, which we know to be true.

Case $n = 2$. Since $\frac{1}{|\xi|^2} \notin L^1_{loc}(\mathbb{R}^2)$, we cannot proceed as in the case $n \geq 3$. Instead, we rely on Theorem 4.76. Retaining the notation from Theorem 4.76, fix $\psi \in Q$ and observe that

$$|\xi|^2 w_\psi = 1 \quad \text{in } \mathcal{S}'(\mathbb{R}^2). \tag{7.1.6}$$

To see why this is true, first note that since $|\xi|^2 \in \mathcal{L}(\mathbb{R}^2)$, by *(b)* in Theorem 4.14 it follows that $|\xi|^2 w_\psi \in \mathcal{S}'(\mathbb{R}^2)$ while by *(a)* in Theorem 3.14 we have that $|\xi|^2 \varphi$ belongs to $\mathcal{S}(\mathbb{R}^2)$ for every $\varphi \in \mathcal{S}(\mathbb{R}^2)$. Hence, we may write

$$\langle |\xi|^2 w_\psi, \varphi \rangle = \langle w_\psi, |\xi|^2 \varphi \rangle = \int_{\mathbb{R}^n} \frac{|\xi|^2 \varphi(\xi) - (|\xi|^2 \varphi(\xi))\big|_{\xi=0} \psi(\xi)}{|\xi|^2} \, d\xi$$

$$= \int_{\mathbb{R}^2} \varphi(\xi) \, d\xi = \langle 1, \varphi \rangle, \qquad \forall \, \varphi \in \mathcal{S}(\mathbb{R}^2), \tag{7.1.7}$$

finishing the proof of (7.1.6).

A combination of (7.1.2) and (7.1.6) yields $|\xi|^2(\widehat{E} + w_\psi) = 0$ in $\mathcal{S}'(\mathbb{R}^2)$. Reasoning as in the case $n \geq 3$, the latter implies

$$\widehat{E} = -w_\psi + \widehat{P} \quad \text{in } \mathcal{S}'(\mathbb{R}^n), \tag{7.1.8}$$

for some harmonic polynomial P. After applying the Fourier transform to the equality in (7.1.8) and recalling (4.6.4), we arrive at

$$E(x) = \frac{1}{2\pi} \ln|x| + P(x), \qquad \forall \, x \in \mathbb{R}^2 \setminus \{0\}. \tag{7.1.9}$$

In addition, by making use of the Fourier transform (recall *(a)* and *(c)* in Theorem 4.26, (4.2.3), and Proposition 4.32), equation (7.1.6) is equivalent with

$$-\Delta(\widehat{w_\psi}) = (2\pi)^2 \delta \quad \text{in } \mathcal{S}'(\mathbb{R}^2). \tag{7.1.10}$$

Combining (7.1.10) with (4.6.4), we arrive at

$$\Delta\left(\frac{1}{2\pi} \ln|x|\right) = \delta \quad \text{in } \mathcal{S}'(\mathbb{R}^2). \tag{7.1.11}$$

Case $n = 1$. In this case we use Exercise 5.16 to conclude that $E = xH + c_1 x + c_0$ in $\mathcal{S}'(\mathbb{R})$, for some $c_0, c_1 \in \mathbb{C}$. Hence, E is of function type and $E(x) = xH + c_1 x + c_0$, for every $x \in \mathbb{R}$.

Remark 7.1. From Example 5.12 we know that $-xH^\vee$ is another fundamental solution for Δ in $\mathcal{D}'(\mathbb{R})$. This is in agreement with what we proved above since $H + H^\vee = 1$ in $\mathcal{S}'(\mathbb{R})$, thus $-xH^\vee = xH + x$ in $\mathcal{S}'(\mathbb{R})$.

In summary, we have proved the following result.

Theorem 7.2. *The function $E \in L^1_{loc}(\mathbb{R}^n)$ defined as*

$$E(x) := \begin{cases} \dfrac{-1}{(n-2)\omega_{n-1}} \dfrac{1}{|x|^{n-2}} & \text{if } x \in \mathbb{R}^n \setminus \{0\}, \ n \geq 3, \\[2mm] \dfrac{1}{2\pi} \ln |x| & \text{if } x \in \mathbb{R}^2 \setminus \{0\}, \ n = 2, \\[2mm] xH(x) & \text{if } x \in \mathbb{R}, \ n = 1, \end{cases} \tag{7.1.12}$$

belongs to $S'(\mathbb{R}^n)$ and is a fundamental solution[2] for the Laplace operator Δ in $\mathbb{R}^n$. Moreover,

$$\{u \in S'(\mathbb{R}^n) : \Delta u = \delta \ \text{in} \ S'(\mathbb{R}^n)\} \tag{7.1.13}$$

$$= \{E + P : P \ \text{harmonic polynomial in} \ \mathbb{R}^n\}.$$

Remark 7.3. Note that, as a consequence of Proposition 5.2 and Example 4.4, for a harmonic function in $\mathbb{R}^n$, the respective function is a polynomial if and only if it is a tempered distribution.

The collection of harmonic functions is strictly larger than the collection of harmonic polynomial. For example, the function $u(x, y) := e^x \cos y$ defined for $(x, y) \in \mathbb{R}^2$, is harmonic in $\mathbb{R}^2$ without being a polynomial. In particular, we obtain that $u \notin S'(\mathbb{R}^2)$.

Moving on, we recall that if $\Omega \subset \mathbb{R}^n$ is a domain of class C^1 (see Definition 14.59) with outward unit normal $\nu = (\nu_1, \ldots, \nu_n)$ and u is a function of class C^1 in an open neighborhood of $\partial\Omega$, then the normal derivative of u on $\partial\Omega$, denoted by $\frac{\partial u}{\partial \nu}$, is the directional derivative of the function u along the unit vector ν, i.e.,

$$\frac{\partial u}{\partial \nu}(y) := \sum_{j=1}^{n} \nu_j(y)\partial_j u(y), \qquad \forall\, y \in \partial\Omega. \tag{7.1.14}$$

Remark 7.4. Via a direct computation, one may check that E as in (7.1.12) defines a fundamental solution for the Laplacian. We outline the computation for $n \geq 3$ and leave the cases $n = 2$ and $n = 1$ as an exercise.

First, observe that $E \in C^\infty(\mathbb{R}^n \setminus \{0\})$ and $\Delta E = 0$ in $\mathbb{R}^n \setminus \{0\}$. Next, pick a function $\varphi \in C_0^\infty(\mathbb{R}^n)$ such that $\operatorname{supp}\varphi \subset B(0, R)$, for some $R \in (0, \infty)$. Then, starting with the definition of distributional derivatives and then using Lebesgue's Dominated Convergence Theorem, we write

---

[2] In the case $n = 3$, the expression for E was used in 1789 by Pierre Simon de Laplace to show that for f smooth, compactly supported, $\Delta(E * f) = 0$ outside the support of f (cf. [41]).

$$\langle \Delta E, \varphi \rangle = \langle E, \Delta \varphi \rangle = \lim_{\varepsilon \to 0^+} \int_{\varepsilon \leq |x| \leq R} E(x) \Delta \varphi(x) \, dx$$

$$= \lim_{\varepsilon \to 0^+} \left[\int_{\varepsilon \leq |x| \leq R} \Delta E(x) \varphi(x) \, dx + \int_{\partial B(0,R)} E \frac{\partial \varphi}{\partial \nu} \, d\sigma \right.$$

$$\left. - \int_{\partial B(0,\varepsilon)} E \frac{\partial \varphi}{\partial \nu} \, d\sigma - \int_{\partial B(0,R)} \frac{\partial E}{\partial \nu} \varphi \, d\sigma + \int_{\partial B(0,\varepsilon)} \frac{\partial E}{\partial \nu} \varphi \, d\sigma \right]$$

$$= \lim_{\varepsilon \to 0^+} \left[\int_{\partial B(0,\varepsilon)} \left(\frac{\partial E}{\partial \nu} \varphi - E \frac{\partial \varphi}{\partial \nu} \right) d\sigma \right] = \varphi(0), \tag{7.1.15}$$

where for each integral considered over $\partial B(0, r)$, for some $r \in (0, \infty)$, ν denotes the outward unit normal to $B(0, r)$. For the third equality in (7.1.15) we used (14.8.5) while for the fourth equality in (7.1.15) we used the fact that the support of φ is contained inside $B(0, R)$. In addition, to see why the last equality in (7.1.15) holds, use the fact that $\nabla E(x) = \frac{1}{\omega_{n-1}} \cdot \frac{x}{|x|^n}$ for every $x \in \mathbb{R}^n \setminus \{0\}$ to write

$$\int_{\partial B(0,\varepsilon)} \frac{\partial E}{\partial \nu}(x) \varphi(x) \, d\sigma(x) = \frac{1}{\omega_{n-1} \varepsilon^{n-1}} \int_{\partial B(0,\varepsilon)} \varphi(x) d\sigma(x) \tag{7.1.16}$$

$$= \frac{1}{\omega_{n-1} \varepsilon^{n-1}} \int_{\partial B(0,\varepsilon)} [\varphi(x) - \varphi(0)] d\sigma(x) + \varphi(0)$$

$$= \frac{1}{\omega_{n-1}} \int_{S^{n-1}} [\varphi(\varepsilon y) - \varphi(0)] d\sigma(y) + \varphi(0) \xrightarrow[\varepsilon \to 0^+]{} \varphi(0)$$

where the convergence follows by invoking Lebesgue's Dominated Convergence Theorem. Also,

$$\left| \int_{\partial B(0,\varepsilon)} E(x) \frac{\partial \varphi}{\partial \nu}(x) \, d\sigma(x) \right| \leq \frac{\|\nabla \varphi\|_{L^\infty}}{(n-2)\omega_{n-1}} \int_{\partial B(0,\varepsilon)} \frac{1}{|x|^{n-2}} \, d\sigma(x)$$

$$= \varepsilon \, C(n, \varphi) \xrightarrow[\varepsilon \to 0^+]{} 0. \tag{7.1.17}$$

This computation shows that if $n \geq 3$, the $L^1_{loc}(\mathbb{R}^n)$ function E from (7.1.12) satisfies $\langle \Delta E, \varphi \rangle = \langle \delta, \varphi \rangle$ for every $\varphi \in C_0^\infty(\mathbb{R}^n)$, thus E is a fundamental solution for the Laplacian.

We conclude this section by presenting a result relating the composition of Riesz transforms to singular integral operators whose kernels are two derivatives on the fundamental solution for the Laplacian.

Proposition 7.5. *Let E be the fundamental solution for Δ from (7.1.12) and recall the Riesz transforms in $\mathbb{R}^n$ (cf. Theorem 4.97). Then for every function $\varphi \in S(\mathbb{R}^n)$ and each $j, k \in \{1, \ldots, n\}$ we have*

$$R_j(R_k\varphi)(x) = -\left(\frac{\omega_n}{2}\right)^2 \lim_{\varepsilon \to 0^+} \int_{|x-y|>\varepsilon} (\partial_j \partial_k E)(x-y)\varphi(y)\,dy$$

$$-\left(\frac{\omega_n}{2}\right)^2 \frac{\delta_{jk}}{n}\varphi(x) \quad in \quad \mathcal{S}'(\mathbb{R}^n), \tag{7.1.18}$$

with the left-hand side initially understood in $L^2(\mathbb{R}^n)$ (cf. Theorem 4.97) and the right-hand side considered as a tempered distribution (cf. Proposition 4.70).

Moreover, if $T_{\partial_j \partial_k E}$ is the operator as in part (e) of Theorem 4.100 (for the choice $\Theta := \partial_j \partial_k E$), then for every $f \in L^2(\mathbb{R}^n)$ we have

$$R_j(R_k f) = -\left(\frac{\omega_n}{2}\right)^2 T_{\partial_j \partial_k E} f - \left(\frac{\omega_n}{2}\right)^2 \frac{\delta_{jk}}{n} f \quad in \quad L^2(\mathbb{R}^n). \tag{7.1.19}$$

Proof. Fix $j, k \in \{1, \ldots, n\}$ along with some $\varphi \in \mathcal{S}(\mathbb{R}^n)$. Since the Fourier transform is an isomorphism of $\mathcal{S}'(\mathbb{R}^n)$, it suffices to show that (7.1.18) holds on the Fourier transform side. With this in mind, note that on the one hand,

$$\mathcal{F}(R_j(R_k\varphi)) = -i\frac{\omega_n}{2}\frac{\xi_j}{|\xi|}\mathcal{F}(R_k\varphi) = -\left(\frac{\omega_n}{2}\right)^2 \frac{\xi_j \xi_k}{|\xi|^2}\widehat{\varphi} \quad in \quad \mathcal{S}'(\mathbb{R}^n), \tag{7.1.20}$$

by a twofold application of (4.9.15). On the other hand, since the function $\partial_j \partial_k E$ is as in (4.4.1) (here Example 4.71 is also used), we may invoke Proposition 4.70, (4.9.6), and (4.5.2), to conclude that the Fourier transform of the right-hand side of (7.1.18) is

$$-\left(\frac{\omega_n}{2}\right)^2 \left[\widehat{\varphi}\mathcal{F}(\mathrm{P.V.}\,(\partial_j \partial_k E)) + \frac{\delta_{jk}}{n}\widehat{\varphi}\right]$$

$$= -\left(\frac{\omega_n}{2}\right)^2 \left[\frac{1}{\omega_{n-1}}\widehat{\varphi}\mathcal{F}(\mathrm{P.V.}\,(\partial_j \Phi_k)) + \frac{\delta_{jk}}{n}\widehat{\varphi}\right]$$

$$= -\left(\frac{\omega_n}{2}\right)^2 \left[\left(\frac{\xi_j \xi_k}{|\xi|^2} - \frac{\delta_{jk}}{n}\right)\widehat{\varphi} + \frac{\delta_{jk}}{n}\widehat{\varphi}\right]$$

$$= -\left(\frac{\omega_n}{2}\right)^2 \frac{\xi_j \xi_k}{|\xi|^2}\widehat{\varphi} \quad in \quad \mathcal{S}'(\mathbb{R}^n). \tag{7.1.21}$$

That (7.1.18) holds now follows from (7.1.20) and (7.1.21).

Finally, (7.1.19) is a consequence of what we have proved so far, part (e) in Theorem 4.97, part (e) in Theorem 4.100, and the fact that $\mathcal{S}(\mathbb{R}^n)$ is dense in $L^2(\mathbb{R}^n)$. $\quad\square$

Making use of the full force of Theorem 4.101 we obtain the following L^p version, $1 < p < \infty$, of Proposition 7.5.

Proposition 7.6. Let E be the fundamental solution for Δ as in (7.1.12) and recall the Riesz transforms in $\mathbb{R}^n$ (cf. Theorem 4.97). Also, fix $p \in (1, \infty)$. Then for every $j, k \in \{1, \ldots, n\}$ and $f \in L^p(\mathbb{R}^n)$ we have

$$R_j(R_k f)(x) = -\left(\frac{\omega_n}{2}\right)^2 \lim_{\varepsilon \to 0^+} \int_{|x-y|>\varepsilon} (\partial_j \partial_k E)(x-y) f(y) \, dy$$

$$-\left(\frac{\omega_n}{2}\right)^2 \frac{\delta_{jk}}{n} f(x) \quad \text{for a.e. } x \in \mathbb{R}^n, \qquad (7.1.22)$$

with the Riesz transforms are understood as bounded operators on $L^p(\mathbb{R}^n)$ (cf. (4.9.48)–(4.9.49)) while the singular integral operator in the right-hand side is also considered as a bounded mapping on $L^p(\mathbb{R}^n)$ (cf. Theorem 4.101).

In particular, if $T_{\partial_j \partial_k E}$ is interpreted in the sense of Theorem 4.101 (for the choice $\Theta := \partial_j \partial_k E$), then for every $f \in L^p(\mathbb{R}^n)$ we have

$$R_j(R_k f) = -\left(\frac{\omega_n}{2}\right)^2 T_{\partial_j \partial_k E} f - \left(\frac{\omega_n}{2}\right)^2 \frac{\delta_{jk}}{n} f \quad \text{in} \quad L^p(\mathbb{R}^n). \qquad (7.1.23)$$

Proof. All claims are consequences of Proposition 7.5, Theorem 4.101, and the density of $\mathcal{S}(\mathbb{R}^n)$ in $L^p(\mathbb{R}^n)$. $\qquad \square$

7.2 The Poisson Equation and Layer Potential Representation Formulas

In this section we use the fundamental solutions for the Laplacian determined in the previous section in a number of applications. The first application concerns the Poisson equation[3] for the Laplacian in $\mathbb{R}^n$:

$$\Delta u = f \qquad \text{in } \mathbb{R}^n. \qquad (7.2.1)$$

In (7.2.1) the unknown is u and f is a given function. As a consequence of Remark 5.6, this equation is always solvable in $\mathcal{D}'(\mathbb{R}^n)$ if $f \in C_0^\infty(\mathbb{R}^n)$. We will see that in fact the solution obtained via the convolution of such an f with the fundamental solution for Δ from (7.1.12) is smooth and that we can solve the Poisson equation in $\mathcal{D}'(\mathbb{R}^n)$ for a less restrictive class of functions f. The second application is an integral representation formula, involving layer potentials, for functions of class C^2 on a neighborhood of the closure of a domain of class C^1.

Proposition 7.7. *Let E be the fundamental solution for Δ from (7.1.12). Suppose $f \in C_0^\infty(\mathbb{R}^n)$. Then the function*

$$u(x) := \int_{\mathbb{R}^n} E(x-y) f(y) \, dy, \qquad \forall x \in \mathbb{R}^n, \qquad (7.2.2)$$

---

[3] In 1813, the French mathematician, geometer, and physicist Siméon Denis Poisson (1781–1840) proved that $\Delta(E * f) = f$ in dimension $n = 3$ (cf. [61]), where $E(x) = -\frac{1}{4\pi|x|}$, $x \in \mathbb{R}^3 \setminus \{0\}$, and $f \in C_0^\infty(\mathbb{R}^3)$.

satisfies $u \in C^{\infty}(\mathbb{R}^n)$ and is a classical solution of the Poisson equation (7.2.1) for the Laplacian in $\mathbb{R}^n$.

Proof. Fix $f \in C_0^{\infty}(\mathbb{R}^n)$ and let E be as in (7.1.12). Define $u := E * f$ in $\mathcal{D}'(\mathbb{R}^n)$. Then Remark 5.6 implies that u is a solution of the equation $\Delta u = f$ in $\mathcal{D}'(\mathbb{R}^n)$. In addition, by Proposition 2.102 and the fact that $E \in L^1_{loc}(\mathbb{R}^n)$, we have $u \in C^{\infty}(\mathbb{R}^n)$ and

$$u(x) = \langle E(x-y), f(y) \rangle = \int_{\mathbb{R}^n} E(x-y)f(y)\, dy, \qquad \forall\, x \in \mathbb{R}^n, \tag{7.2.3}$$

or even more explicitly (based on the expressions in (7.1.12)), that

$$u(x) = \begin{cases} \dfrac{-1}{(n-2)\omega_{n-1}} \displaystyle\int_{\mathbb{R}^n} \dfrac{f(y)}{|x-y|^{n-2}}\, dy, & \text{if } n \geq 3, \\[4mm] \dfrac{1}{2\pi} \displaystyle\int_{\mathbb{R}^2} \ln|x-y| f(y)\, dy, & \text{if } n = 2, \qquad \forall\, x \in \mathbb{R}^n. \\[4mm] \displaystyle\int_{\mathbb{R}} (x-y)H(x-y)f(y)\, dy, & \text{if } n = 1, \end{cases} \tag{7.2.4}$$

Furthermore, based on Proposition 2.41, we may conclude that $\Delta u = f$ pointwise in $\mathbb{R}^n$. This proves that u is a solution of (7.7) as desired. $\square$

An inspection of the right-hand side of (7.2.2) reveals that this expression continues to be meaningful under weaker assumptions on f. This observation may be used to prove that (7.7) is solvable in the class of distributions for f belonging to a larger class of functions than $C_0^{\infty}(\mathbb{R}^n)$.

Proposition 7.8. *Let E be the fundamental solution for Δ from (7.1.12). Suppose Ω is an open set in $\mathbb{R}^n$, $n \geq 2$, and $f \in L^{\infty}(\Omega)$ vanishes outside a bounded measurable subset of Ω. Then the function*

$$u(x) := \int_{\Omega} E(x-y)f(y)\, dy, \qquad \forall\, x \in \Omega, \tag{7.2.5}$$

is a distributional solution of the Poisson equation for the Laplacian in Ω, that is, $\Delta u = f$ in $\mathcal{D}'(\Omega)$. In addition, $u \in C^1(\Omega)$ and for each $j \in \{1, \ldots, n\}$,

$$\partial_j u(x) = \int_{\Omega} (\partial_j E)(x-y)f(y)\, dy, \qquad \forall\, x \in \Omega. \tag{7.2.6}$$

Proof. Suppose first that $n \geq 3$. Note that $E \in C^{\infty}(\mathbb{R}^n)$ and is positive homogeneous of degree $2 - n$. Also the extension of f by zero outside Ω, which we continue to denote by f, satisfies $f \in L^{\infty}_{comp}(\mathbb{R}^n)$. From Theorem 4.105 it follows that $u \in C^1(\Omega)$ and (7.2.6) holds for each $j \in \{1, \ldots, n\}$.

If $n = 2$, a reasoning based on Vitali's theorem, similar to the one used in the proof of Theorem 4.105, yields $(\partial_j E) * f \in C^0(\Omega)$ for each $j \in \{1, \ldots, n\}$. Also, the computation in (4.10.35) may be adapted to give that the distributional derivative $\partial_j u$ is equal to $(\partial_j E) * f$ for each $j \in \{1, \ldots, n\}$. Hence, Theorem 2.112 applies and yields $u \in C^1(\Omega)$, as desired.

To finish the proof of the proposition, there remains to show that $\Delta u = f$ in $\mathcal{D}'(\Omega)$. Since $f \in L^1_{loc}(\Omega)$ we have $f \in \mathcal{D}'(\Omega)$. Thus, for each $\varphi \in C_0^\infty(\Omega)$ we may write

$$\langle \Delta u, \varphi \rangle = \langle u, \Delta \varphi \rangle = \int_\Omega u(x) \Delta \varphi(x) \, dx = \int_\Omega \Big(\int_\Omega E(x-y) f(y) \, dy \Big) \Delta \varphi(x) \, dx$$

$$= \int_\Omega f(y) \Big(\int_\Omega E(x-y) \Delta \varphi(x) \, dx \Big) dy$$

$$= \int_\Omega f(y) \Big(\int_{\Omega-y} E(x) \Delta \varphi(x+y) \, dx \Big) dy$$

$$= \int_\Omega f(y) \langle E, \Delta \varphi(\cdot + y) \rangle \, dy = \int_\Omega f(y) \langle \Delta E, \varphi(\cdot + y) \rangle \, dy$$

$$= \int_\Omega f(y) \langle \delta, \varphi(\cdot + y) \rangle \, dy = \int_\Omega f(y) \varphi(y) \, dy = \langle f, \varphi \rangle, \tag{7.2.7}$$

where for each $y \in \Omega$ we set $\Omega - y := \{x - y : x \in \Omega\}$. Note that for each $y \in \Omega$ fixed one has $\varphi(\cdot + y) \in C_0^\infty(\Omega - y)$, $0 \in \Omega - y$, and $E \in \mathcal{D}'(\Omega - y)$, thus the sixth equality in (7.2.7) is justified. The eighth equality in (7.2.7) is a consequence of the fact that E is a fundamental solution for the Laplacian and that $0 \in \Omega - y$. This completes the proof of the proposition. $\qquad\square$

Exercise 7.9. Assume the hypothesis of Proposition 7.8 and recall the fundamental solution E for the Laplacian from (7.1.12). Using Vitali's Theorem 14.29, show that $u(x) = \int_\Omega E(\cdot - y) f(y) \, dy$ is differentiable in Ω (without making use of Theorem 2.112). As a by-product, show that, for every $j \in \{1, \ldots, n\}$,

$$\partial_j u(x) = \int_\Omega \partial_j E(x-y) f(y) \, dy, \qquad \forall \, x \in \Omega. \tag{7.2.8}$$

We next establish mean value formulas for harmonic functions. A clarification of the terminology employed is in order. Traditionally, the name harmonic function in an open set $\Omega \subseteq \mathbb{R}^n$ has been used for functions $u \in C^2(\Omega)$ with the property that $\sum_{j=1}^n \partial_j^2 u = 0$ in a pointwise sense, everywhere in Ω. Such a function is called a classical solution of the equation $\Delta u = 0$ in Ω.

Theorem 7.10. *For an open set $\Omega \subseteq \mathbb{R}^n$ the following are equivalent:*

(i) u is harmonic in Ω;
(ii) $u \in \mathcal{D}'(\Omega)$ and $\Delta u = 0$ in $\mathcal{D}'(\Omega)$;
(iii) $u \in C^\infty(\Omega)$ and $\Delta u = 0$ in a pointwise sense in Ω.

Proof. This is a direct consequence of Theorem 6.8 and Theorem 7.2 (or, alternatively, of Remark 6.14 and Corollary 6.18). $\qquad\square$

Recall the symbol for integral averages from (15).

Theorem 7.11. *Let Ω be an open set in $\mathbb{R}^n$ and u be a harmonic function in Ω. Then for every $x \in \Omega$ and every $r \in (0, \text{dist}(x, \partial\Omega))$ we have*

$$u(x) = \fint_{B(x,r)} u(y)\,dy \quad and \quad u(x) = \fint_{\partial B(x,r)} u(y)\,d\sigma(y). \qquad (7.2.9)$$

Proof. From Theorem 7.10 we know that $u \in C^\infty(\Omega)$ and $\Delta u = 0$ in a pointwise sense in Ω. Fix $x \in \Omega$ and define the function

$$\phi : (0, \text{dist}(x, \partial\Omega)) \to \mathbb{R},$$

$$\phi(r) := \fint_{\partial B(x,r)} u(y)\,d\sigma(y), \quad \forall\, r \in (0, \text{dist}(x, \partial\Omega)). \qquad (7.2.10)$$

A change of variables gives that $\phi(r) = \frac{1}{\omega_{n-1}} \int_{S^{n-1}} u(x+r\omega)\,d\sigma(\omega)$. Taking the derivative of ϕ and then using the integration by parts formula from Theorem 14.60 give that

$$\phi'(r) = \frac{1}{\omega_{n-1}} \int_{S^{n-1}} (\nabla u)(x + r\omega) \cdot \omega\,d\sigma(\omega) \qquad (7.2.11)$$

$$= \frac{r}{\omega_{n-1}} \int_{B(0,1)} (\Delta u)(x + ry)\,dy = \frac{r}{n} \fint_{B(x,r)} \Delta u(z)\,dz.$$

Since $\Delta u = 0$ in Ω we have that $\phi'(r) = 0$, thus $\phi(r) = C$ for some constant C. We claim that $\lim_{r \to 0^+} \phi(r) = u(x)$. Then this claim implies $C = u(x)$ and the first formula in (7.2.9) follows. To see the claim, fix $\varepsilon > 0$ and by the continuity of u at x find $\delta \in (0, \text{dist}(x, \partial\Omega))$ such that $|u(y) - u(x)| < \varepsilon$ if $|y - x| < \delta$. Consequently,

$$r \in (0, \delta) \implies |\phi(r) - u(x)| = \left| \fint_{B(x,r)} [u(y) - u(x)]\,dy \right| \le \varepsilon, \qquad (7.2.12)$$

as desired.

With the help of (14.5.6) and (14.9.5), the first formula in (7.2.9) may be rewritten as

$$\frac{\omega_{n-1} r^n}{n} u(x) = \int_{B(x,r)} u(y)\,d\sigma(y) = \int_0^r \int_{\partial B(x,\rho)} u(\omega)\,d\sigma(\omega)\,d\rho \qquad (7.2.13)$$

for $r \in (0, \text{dist}(x, \partial\Omega))$. Differentiating the left- and right-most terms in (7.2.13) with respect to r and then dividing by $\omega_{n-1} r^{n-1}$ gives the second formula in (7.2.9). $\qquad \square$

Interior estimates for harmonic functions, as in (6.3.30), may be obtained as a particular case of Theorem 6.26 by observing that the fundamental solution for the Laplacian from (7.1.12) satisfies (6.3.29) (with $m = 2$). Below we take a slightly more direct route which also yields explicit constants.

Theorem 7.12 (Interior estimates for the Laplacian). *Suppose u is harmonic in Ω. Then for each $x \in \Omega$, each $r \in (0, \text{dist}\,(x, \partial\Omega))$, and each $k \in \mathbb{N}$, we have*

$$\max_{y \in \overline{B(x, r/2)}} |\partial^{\alpha} u(y)| \leq \frac{(2n)^k e^{k-1} k!}{r^k} \max_{y \in \overline{B(x,r)}} |u(y)|, \qquad \forall\, \alpha \in \mathbb{N}_0^n \text{ with } |\alpha| = k. \quad (7.2.14)$$

Proof. Let $j \in \{1, \ldots, n\}$ be arbitrary and note that, by Theorem 7.10, $u \in C^{\infty}(\Omega)$ and hence, $\partial_j u$ is harmonic in Ω. Thus, by (7.2.9) and the integration-by parts formula (14.8.4), for each $x \in \Omega$ and each $r \in (0, \text{dist}\,(x, \partial\Omega))$, we may write

$$|\partial_j u(x)| = \left| \frac{n}{\omega_{n-1} r^n} \int_{B(x,r)} \partial_j u(y) \, dy \right|$$

$$= \left| \frac{n}{\omega_{n-1} r^n} \int_{\partial B(x,r)} \frac{y_j - x_j}{r} u(y) \, d\sigma(y) \right|$$

$$\leq \frac{n}{r} \max_{y \in \overline{B(x,r)}} |u(y)|. \quad (7.2.15)$$

With this in hand, Lemma 6.21 applies (with $\mathcal{A}$ the class of harmonic functions in Ω and $C := n$) and yields (7.2.14). $\qquad\square$

Theorem 7.13. *Any harmonic function in Ω is real analytic in Ω.*

Proof. This is an immediate consequence of Theorem 7.12 and Lemma 6.24 (or, alternatively, Theorem 6.26). $\qquad\square$

Corollary 7.14. *Suppose u is harmonic in an open, connected set $\Omega \subseteq \mathbb{R}^n$ with the property that there exists $x_0 \in \Omega$ such that $\partial^{\alpha} u(x_0) = 0$ for all $\alpha \in \mathbb{N}_0^n$. Then u vanishes identically in Ω.*

As a consequence, if a harmonic function defined in an open connected set $\Omega \subseteq \mathbb{R}^n$ vanishes on an open subset of Ω, then the respective function vanishes identically in Ω.

Proof. This is an immediate consequence of Theorem 6.25 and Theorem 7.13. $\qquad\square$

Theorem 7.15 (Liouville's Theorem for the Laplacian). *If u is a bounded harmonic function in $\mathbb{R}^n$ then there exists $c \in \mathbb{C}$ such that $u = c$ in $\mathbb{R}^n$.*

Proof. This may be justified in several ways. For example, it suffices to note that this is a particular case of Theorem 5.4. Another proof may be given based on interior estimates. Specifically, let $j \in \{1, \ldots, n\}$ and using (7.2.14), for each $x \in \mathbb{R}^n$, we may write

$$\lim_{r \to \infty} |\partial_j u(x)| \leq \lim_{r \to \infty} \frac{n}{r} \|u\|_{L^{\infty}(\mathbb{R}^n)} = 0. \quad (7.2.16)$$

Hence, $\nabla u = 0$ proving that u is locally constant in $\mathbb{R}^n$. Since $\mathbb{R}^n$ is connected, the desired conclusion follows. $\qquad\square$

All ingredients are now in place for proving the following basic well-posedness result for the Poisson problem for the Laplacian in $\mathbb{R}^n$.

Theorem 7.16. *Assume $n \geq 3$. Then for each $f \in L^\infty_{comp}(\mathbb{R}^n)$ and each $c \in \mathbb{C}$ the Poisson problem for the Laplacian in $\mathbb{R}^n$,*

$$\begin{cases} u \in C^0(\mathbb{R}^n), \\ \Delta u = f \text{ in } \mathcal{D}'(\mathbb{R}^n), \\ \lim_{|x| \to \infty} u(x) = c, \end{cases} \tag{7.2.17}$$

has a unique solution. Moreover, the solution u satisfies the following additional properties.

(1) The function u belongs to $C^1(\mathbb{R}^n)$ and has the integral representation formula

$$u(x) = c + \int_{\mathbb{R}^n} E(x - y) f(y) \, dy, \qquad x \in \mathbb{R}^n, \tag{7.2.18}$$

where E is the fundamental solution for the Laplacian in $\mathbb{R}^n$ from (7.1.12).
(2) If in fact $f \in C_0^\infty(\mathbb{R}^n)$ then actually $u \in C^\infty(\mathbb{R}^n)$.
(3) For every $j, k \in \{1, \ldots, n\}$, we have

$$\partial_j \partial_k u = -\left(\frac{2}{\omega_n}\right)^2 R_j(R_k f) \text{ in } \mathcal{D}'(\mathbb{R}^n), \tag{7.2.19}$$

where R_j and R_k are (j-th and k-th) Riesz transforms in $\mathbb{R}^n$.
(4) For every $p \in (1, \infty)$, the solution u of (7.2.17) satisfies $\partial_j \partial_k u \in L^p(\mathbb{R}^n)$ for each $j, k \in \{1, \ldots, n\}$, where the derivatives are taken in $\mathcal{D}'(\mathbb{R}^n)$, and there exists a constant $C = C(p, n) \in (0, \infty)$ with the property that

$$\sum_{j,k=1}^n \|\partial_j \partial_k u\|_{L^p(\mathbb{R}^n)} \leq C \|f\|_{L^p(\mathbb{R}^n)}. \tag{7.2.20}$$

Proof. The fact that the function u defined as in (7.2.18) is of class $C^1(\mathbb{R}^n)$ and solves $\Delta u = f$ in $\mathcal{D}'(\mathbb{R}^n)$ has been established in Proposition 7.8. To proceed, let $R \in (0, \infty)$ be such that $\operatorname{supp} f \subset B(0, R)$, and note that if $|x| \geq 2R$, then for every $y \in B(0, R)$ we have $|x - y| \geq |x| - |y| \geq R \geq |y|$. Hence, using (7.1.12) (recall that we are assuming $n \geq 3$) and Lebesgue's Dominated Convergence Theorem we obtain

$$\left| \int_{\mathbb{R}^n} E(x - y) f(y) \, dy \right| \leq C \|f\|_{L^\infty(\mathbb{R}^n)} \int_{B(0,R)} \frac{dy}{|x - y|^{n-2}} \to 0 \tag{7.2.21}$$

as $|x| \to \infty$. It is then clear from (7.2.21) that the function u from (7.2.18) also satisfies the limit condition in (7.2.17). That (7.2.18) is the unique solution of (7.2.17) follows from linearity and Theorem 7.15. Next, that $u \in C^\infty(\mathbb{R}^n)$ if $f \in C_0^\infty(\mathbb{R}^n)$, is a consequence of Proposition 7.7 (or, alternatively, Corollary 6.18 and the ellipticity of the Laplacian). This proves *(1)*–*(2)*. As regards *(3)*, start by fixing j, k in $\{1, \ldots, n\}$. Then from (7.2.6), Theorem 4.103, (14.9.45), and Proposition 7.6, we deduce that

$$
(\partial_j \partial_k u)(x) = \partial_j(\partial_k u(x)) = \partial_j\Big[\int_{\mathbb{R}^n} (\partial_k E)(x - y)f(y)\,dy\Big]
$$

$$
= \frac{1}{\omega_{n-1}}\Big(\int_{S^{n-1}} \omega_k \omega_j\, d\sigma(\omega)\Big)f(x)
$$

$$
+ \lim_{\varepsilon \to 0^+} \int_{|x-y|>\varepsilon} (\partial_j \partial_k E)(x - y)f(y)\,dy
$$

$$
= \frac{\delta_{jk}}{n} f(x) + \lim_{\varepsilon \to 0^+} \int_{|x-y|>\varepsilon} (\partial_j \partial_k E)(x - y)f(y)\,dy
$$

$$
= -\Big(\frac{2}{\omega_n}\Big)^2 R_j(R_k f)(x) \quad \text{in } \mathcal{D}'(\mathbb{R}^n). \tag{7.2.22}
$$

Finally, the claim in *(4)* follows from *(3)* and the boundedness of the Riesz transforms on $L^p(\mathbb{R}^n)$ (cf. (4.9.48)). $\qquad\square$

In the last part of this section, we prove an integral representation formula involving layer potentials (a piece of terminology we elaborate on a little later) for arbitrary (as opposed to harmonic) functions $u \in C^2(\overline{\Omega})$, where Ω is a bounded domain of class C^1 in $\mathbb{R}^n$ (as in Definition 14.59). To state this recall the definition of the normal derivative from (7.1.14).

Proposition 7.17. *Suppose $n \geq 2$ and let $\Omega \subset \mathbb{R}^n$ be a bounded domain of class C^1 and let ν denote its outward unit normal. Also let E be the fundamental solution for the Laplacian from (7.1.12). If $u \in C^2(\overline{\Omega})$, then*

$$
\int_\Omega \Delta u(y)E(x - y)\,dy - \int_{\partial\Omega} E(x - y)\frac{\partial u}{\partial \nu}(y)\,d\sigma(y) \tag{7.2.23}
$$

$$
- \int_{\partial\Omega} [\nu(y) \cdot (\nabla E)(x - y)]u(y)\,d\sigma(y) = \begin{cases} u(x), & x \in \Omega, \\ 0, & x \in \mathbb{R}^n \setminus \overline{\Omega}. \end{cases}
$$

Several strategies may be employed to prove Proposition 7.17, and here we choose an approach that highlights the role of distribution theory. Another approach (based on isolating the singularity in the fundamental solution and a limiting argument), more akin to the proof of (7.1.15), is going to be used in the proof of Theorem 7.71 (presented later) where a result of similar flavor to (7.2.23) is established for more general differential operators than the Laplacian.

Proof of Proposition 7.17. In this proof, for a function v defined in Ω, we let $\widetilde{v}$ denote the extension of v with zero outside Ω. Fix some function $u \in C^2(\overline{\Omega})$ and observe that $\widetilde{u} \in L^1_{loc}(\mathbb{R}^n) \subset \mathcal{D}'(\mathbb{R}^n)$. In addition, for each $\varphi \in C_0^\infty(\mathbb{R}^n)$ one has

$$\langle \widetilde{\Delta u}, \varphi \rangle = \langle \widetilde{u}, \Delta\varphi \rangle = \int_\Omega u\Delta\varphi \, dx$$

$$= \int_\Omega \varphi \Delta u \, dx + \int_{\partial\Omega} \left(u\frac{\partial\varphi}{\partial\nu} - \varphi\frac{\partial u}{\partial\nu} \right) d\sigma, \qquad (7.2.24)$$

where the last equality in (7.2.24) is based on (14.8.5).

For $a \in C^0(\partial\Omega)$ define the mappings $a\delta_{\partial\Omega}, \frac{\partial}{\partial\nu}(a\delta_{\partial\Omega}) : C_0^\infty(\mathbb{R}^n) \to \mathbb{C}$ by setting

$$\langle a\delta_{\partial\Omega}, \varphi \rangle := \int_{\partial\Omega} a(x)\varphi(x) \, d\sigma(x), \qquad \forall \, \varphi \in C_0^\infty(\mathbb{R}^n), \qquad (7.2.25)$$

and

$$\left\langle \frac{\partial}{\partial\nu}(a\delta_{\partial\Omega}), \varphi \right\rangle := -\int_{\partial\Omega} a(x)\frac{\partial\varphi}{\partial\nu}(x) \, d\sigma(x), \qquad \forall \, \varphi \in C_0^\infty(\mathbb{R}^n). \qquad (7.2.26)$$

By Exercise 2.146 corresponding to $\Sigma := \partial\Omega$ we have $a\delta_{\partial\Omega} \in \mathcal{D}'(\mathbb{R}^n)$. By a similar reasoning, one also has $\frac{\partial}{\partial\nu}(a\delta_{\partial\Omega}) \in \mathcal{D}'(\mathbb{R}^n)$. In addition, it is easy to see that the supports of the distributions in (7.2.25) and (7.2.26) are contained in $\partial\Omega$, hence $a\delta_{\partial\Omega}, \frac{\partial}{\partial\nu}(a\delta_{\partial\Omega}) \in \mathcal{E}'(\mathbb{R}^n)$. In light of these definitions, (7.2.24) is equivalent with

$$\widetilde{\Delta u} = \widetilde{\Delta u} - \frac{\partial u}{\partial\nu}\delta_{\partial\Omega} - \frac{\partial}{\partial\nu}(u\delta_{\partial\Omega}) \qquad \text{in } \mathcal{D}'(\mathbb{R}^n). \qquad (7.2.27)$$

Note that the supports of the distributions $\widetilde{\Delta u}$ and $\widetilde{\Delta u}$ are contained in the compact set $\overline{\Omega}$. Hence, all distributions in (7.2.27) are compactly supported and their convolutions with E are well defined. Furthermore, since

$$\widetilde{\Delta u} * E = \widetilde{u} * \Delta E = \widetilde{u} * \delta = \widetilde{u} \quad \text{in} \quad \mathcal{D}'(\mathbb{R}^n),$$

after convolving the left- and right-hand sides of (7.2.27) with E we arrive at

$$\widetilde{u} = \widetilde{\Delta u} * E - \left(\frac{\partial u}{\partial\nu}\delta_{\partial\Omega}\right) * E - \left(\frac{\partial}{\partial\nu}(u\delta_{\partial\Omega})\right) * E \quad \text{in } \mathcal{D}'(\mathbb{R}^n). \qquad (7.2.28)$$

Moreover, since $\mathbb{R}^n \setminus \partial\Omega$ is open, it follows that the equality in (7.2.28) also holds in $\mathcal{D}'(\mathbb{R}^n \setminus \partial\Omega)$. The goal is to show that all distributions in (7.2.28) when restricted to $\mathbb{R}^n \setminus \partial\Omega$ are of function type and to determine the respective functions that define them.

The fact that the distribution $\widetilde{u}\big|_{\mathbb{R}^n\setminus\partial\Omega}$ is of function type follows from the definition of $\widetilde{u}$. Also, it is immediate that $\widetilde{u}\big|_{\mathbb{R}^n\setminus\partial\Omega}$ is given by the function equal to u in Ω and equal to zero in $\mathbb{R}^n \setminus \overline{\Omega}$, which is precisely the function in the right-hand side of (7.2.23).

Since $\widetilde{\Delta u}, E \in L_{loc}^1(\mathbb{R}^n)$, with $\widetilde{\Delta u}$ being compactly supported in $\overline{\Omega}$, by Exercise 2.104 we have that $\widetilde{\Delta u} * E$ is of function type, determined by the function $\widetilde{\Delta u} * E \in L_{loc}^1(\mathbb{R}^n)$ defined as

$$(\widetilde{\Delta u} * E)(x) = \int_\Omega E(x - y)\Delta u(y)\,\mathrm{d}y, \qquad \forall\, x \in \mathbb{R}^n. \tag{7.2.29}$$

To proceed with identifying the restrictions to $\mathbb{R}^n \setminus \partial\Omega$ of the other two convolutions in the right-hand side of (7.2.28), let $\varphi \in C_0^\infty(\mathbb{R}^n \setminus \partial\Omega)$ and set $K := \operatorname{supp}\varphi$. Then, the set

$$M_K := \{(x, y) \in \mathbb{R}^n \times \partial\Omega : x + y \in K\}$$

is compact in $\mathbb{R}^{2n}$ and if we take $\psi \in C_0^\infty(\mathbb{R}^{2n})$ such that $\psi \equiv 1$ in a neighborhood of M_K, we have

$$\left\langle \left[\left(\frac{\partial u}{\partial \nu}\right)\delta_{\partial\Omega}\right] * E, \varphi \right\rangle = \left\langle \left(\frac{\partial u}{\partial \nu}\delta_{\partial\Omega}\right)(y), \langle E(x), \psi(x, y)\varphi(x + y)\rangle \right\rangle \tag{7.2.30}$$

$$= \left\langle \left(\frac{\partial u}{\partial \nu}\delta_{\partial\Omega}\right)(y), \int_{\mathbb{R}^n} E(x)\psi(x, y)\varphi(x + y)\,\mathrm{d}x \right\rangle.$$

Note that via a change of variables we have

$$\int_{\mathbb{R}^n} E(x)\psi(x, y)\varphi(x + y)\,\mathrm{d}x = \int_{\mathbb{R}^n} E(z - y)\psi(z - y, y)\varphi(z)\,\mathrm{d}z, \quad \forall\, y \in \mathbb{R}^n. \tag{7.2.31}$$

If we now consider the function

$$h(y) := \int_{\mathbb{R}^n} E(z - y)\psi(z - y, y)\varphi(z)\,\mathrm{d}z, \quad \forall\, y \in \mathbb{R}^n, \tag{7.2.32}$$

then h is well defined. Moreover, since derivatives of order 2 or higher of E are not integrable near the origin, it follows that h does not belong to $C^\infty(\mathbb{R}^n)$, hence (7.2.31) may not be used to rewrite the last term in (7.2.30). To fix this drawback, we impose an additional restriction on ψ. Specifically, since $\partial\Omega \cap K = \varnothing$, there exists $\varepsilon > 0$ sufficiently small such that $(\partial\Omega + B(0, \varepsilon)) \cap (K + B(0, \varepsilon)) = \varnothing$. Then, if U, the neighborhood of M_K where $\psi \equiv 1$, is such that whenever $(x, y) \in U$ we have $y \in \partial\Omega + B(0, \varepsilon/2)$, we may further require that $\psi(\cdot, y) = 0$ for $y \in K + B(0, \varepsilon/2)$. Under this requirement, if $z \in K = \operatorname{supp}\varphi$ and $y \in \mathbb{R}^n$ is such that $|z - y| < \varepsilon/2$, then $\psi(z - y, y) = 0$. This ensures that derivatives in y of any order of the function $E(z - y)\psi(z - y, y)\varphi(z)$ are integrable in z over $\mathbb{R}^n$, thus $h \in C^\infty(\mathbb{R}^n)$. Now (7.2.31) may be used in the last term in (7.2.30) to obtain

$$\left\langle \left[\left(\frac{\partial u}{\partial \nu}\right)\delta_{\partial\Omega}\right] * E, \varphi \right\rangle = \left\langle \left(\frac{\partial u}{\partial \nu}\delta_{\partial\Omega}\right)(y), \int_{\mathbb{R}^n} E(z - y)\psi(z - y, y)\varphi(z)\,\mathrm{d}z \right\rangle. \tag{7.2.33}$$

Since $\frac{\partial u}{\partial \nu} \in C^0(\partial\Omega)$, by (7.2.25) one has

$$\Big\langle \big(\frac{\partial u}{\partial \nu}\delta_{\partial\Omega}\big)(y), \int_{\mathbb{R}^n} E(x-y)\psi(x-y,y)\varphi(x)\,dx\Big\rangle$$

$$= \int_{\partial\Omega} \frac{\partial u}{\partial \nu}(y)\Big[\int_{\mathbb{R}^n} \varphi(x)E(x-y)\psi(x-y,y)\,dx\Big]\,d\sigma(y)$$

$$= \int_{\mathbb{R}^n} \Big[\int_{\partial\Omega} \frac{\partial u}{\partial \nu}(y)E(x-y)\,d\sigma(y)\Big]\varphi(x)\,dx. \qquad (7.2.34)$$

Combining (7.2.33) and (7.2.34) it follows that $\Big(\big[\big(\frac{\partial u}{\partial \nu}\big)\delta_{\partial\Omega}\big] * E\Big)\big|_{\mathbb{R}^n\setminus\partial\Omega}$ is of function type and

$$\Big(\big[\big(\frac{\partial u}{\partial \nu}\big)\delta_{\partial\Omega}\big] * E\Big)(x) = \int_{\partial\Omega} \frac{\partial u}{\partial \nu}(y)E(x-y)\,d\sigma(y), \quad \forall\, x \in \mathbb{R}^n \setminus \partial\Omega. \qquad (7.2.35)$$

Similarly, with φ and ψ as previously specified, since $u \in C^0(\partial\Omega)$, definition (7.2.26) applies and we obtain

$$\Big\langle \big(\frac{\partial}{\partial \nu}(u\delta_{\partial\Omega})\big) * E, \varphi \Big\rangle = \Big\langle \frac{\partial}{\partial \nu}(u\delta_{\partial\Omega})(y), \int_{\mathbb{R}^n} E(x)\psi(x,y)\varphi(x+y)\,dx\Big\rangle$$

$$= \Big\langle \frac{\partial}{\partial \nu}(u\delta_{\partial\Omega})(y), \int_{\mathbb{R}^n} E(x-y)\psi(x-y,y)\varphi(x)\,dx\Big\rangle$$

$$= -\int_{\partial\Omega} u(y)\int_{\mathbb{R}^n} \varphi(x)\Big[\nu(y)\cdot\frac{\partial}{\partial y}(E(x-y))\Big]\,dx\,d\sigma(y)$$

$$= \int_{\mathbb{R}^n} \varphi(x)\int_{\partial\Omega} u(y)[\nu(y)\cdot(\nabla E)(x-y)]\,d\sigma(y)\,dx. \qquad (7.2.36)$$

Hence $\Big(\big(\frac{\partial}{\partial \nu}(u\delta_{\partial\Omega})\big) * E\Big)\big|_{\mathbb{R}^n\setminus\partial\Omega}$ is of function type and

$$\Big(\big(\frac{\partial}{\partial \nu}(u\delta_{\partial\Omega})\big) * E\Big)(x) \qquad (7.2.37)$$

$$= \int_{\partial\Omega} u(y)[\nu(y)\cdot(\nabla E)(x-y)]\,d\sigma(y), \quad \forall\, x \in \mathbb{R}^n \setminus \partial\Omega.$$

Combining (7.2.28), the earlier comments about $\widetilde{u}\big|_{\mathbb{R}^n\setminus\partial\Omega}$, (7.2.29), (7.2.35), and (7.2.37), we may conclude that

$$\widetilde{u}(x) = \int_{\Omega} E(x-y)\Delta u(y)\,dy - \int_{\partial\Omega} E(x-y)\frac{\partial u}{\partial \nu}(y)\,d\sigma(y)$$

$$- \int_{\partial\Omega} [\nu(y)\cdot(\nabla E)(x-y)]u(y)\,d\sigma(y), \quad \forall\, x \in \mathbb{R}^n \setminus \partial\Omega, \qquad (7.2.38)$$

proving (7.2.23).   □

As the observant reader has perhaps noticed, the solid integral appearing in (7.2.23) is simply the volume (or Newtonian) potential defined in (4.10.1) acting on $f := \Delta u$. Starting from this observation, formula (7.2.23) also suggests the consideration of two other types of integral operators associated with a given bounded domain $\Omega \subset \mathbb{R}^n$. Specifically, for each given complex-valued function $\varphi \in C^0(\partial\Omega)$ set

$$
\begin{aligned}
(\mathscr{D}\varphi)(x) &:= -\int_{\partial\Omega} [\nu(y) \cdot (\nabla E)(x - y)]\varphi(y)\, d\sigma(y) \\
&= \frac{1}{\omega_{n-1}} \int_{\partial\Omega} \frac{\nu(y) \cdot (y - x)}{|x - y|^n}\varphi(y)\, d\sigma(y), \qquad x \in \mathbb{R}^n \setminus \partial\Omega, \qquad (7.2.39)
\end{aligned}
$$

and $(\mathscr{S}\varphi)(x) := \int_{\partial\Omega} E(x - y)\varphi(y)\, d\sigma(y)$, for $x \in \mathbb{R}^n \setminus \partial\Omega$. Thus,

$$
(\mathscr{S}\varphi)(x) = \begin{cases} \frac{-1}{(n-2)\omega_{n-1}} \int_{\partial\Omega} \frac{1}{|x-y|^{n-2}}\varphi(y)\, d\sigma(y) & \text{if } n \geq 3, \\ \frac{1}{2\pi} \int_{\partial\Omega} \ln|y - x|\varphi(y)\, d\sigma(y) & \text{if } n = 2. \end{cases} \qquad (7.2.40)
$$

The operators $\mathscr{D}$ and $\mathscr{S}$ are called, respectively, the double and single layer potentials for the Laplacian. In this notation, (7.2.23) reads

$$
(\mathbf{N}_\Omega(\Delta u))(x) - \mathscr{S}\Big(\frac{\partial u}{\partial \nu}\Big)(x) + \mathscr{D}(u|_{\partial\Omega})(x) = \begin{cases} u(x), \ x \in \Omega, \\ 0, \qquad x \in \mathbb{R}^n \setminus \overline{\Omega}, \end{cases} \qquad (7.2.41)
$$

for every $u \in C^2(\overline{\Omega})$. In particular, this formula shows that we may recover any function $u \in C^2(\overline{\Omega})$ knowing Δu in Ω as well as $\frac{\partial u}{\partial \nu}$ and u on $\partial\Omega$.

The above layer potential operators play a crucial role in the treatment of boundary value problems. In this vein, we invite the reader to check that the version of the double layer operator (7.2.39) corresponding to the case when $\Omega = \mathbb{R}^n_+$ is precisely twice the operator $\mathscr{P}$ from (4.8.3) whose relevance in the context of boundary value problems for the Laplacian has been highlighted in (4.8.17)–(4.8.18).

7.3 Fundamental Solutions for the Bi-Laplacian

The bi-Laplacian in $\mathbb{R}^n$ is the operator $\Delta^2 = \Delta\Delta = \big(\sum_{j=1}^{n} \partial_j^2\big)^2$, sometimes also referred to as the biharmonic operator. Functions u of class C^4 in an open set Ω of $\mathbb{R}^n$ satisfying $\Delta^2 u = 0$ pointwise in Ω are called biharmonic in Ω.

Theorem 5.14 guarantees the existence of $E \in \mathcal{S}'(\mathbb{R}^n)$ such that $\Delta^2 E = \delta$ in $\mathcal{S}'(\mathbb{R}^n)$. The goal in this section is to determine all such fundamental solution for Δ^2. In the case when $n \geq 4$, these may be computed by following the approach from Section 7.1 (corresponding to $n \geq 2$ there). Such a line of reasoning will then require treating the cases $n = 1, 2, 3$ via another method. In what follows, we will proceed differently by employing the fact that we have a complete description of all fundamental solutions in $S'(\mathbb{R}^n)$ for the Laplacian and that $\Delta^2 = \Delta\Delta$. This latter approach, will take care of the cases $n \geq 2$, leaving $n = 1$ to be treated separately.

Fix $E \in S'(\mathbb{R}^n)$ satisfying $\Delta^2 E = \delta$ in $\mathcal{D}'(\mathbb{R}^n)$. By (4.1.33) we have $\Delta^2 E = \delta$ in $S'(\mathbb{R}^n)$. Keeping in mind that $\Delta^2 = \Delta\Delta$, it is natural to consider the following equation:

$$\Delta E = E_\Delta \quad \text{in} \quad S'(\mathbb{R}^n), \tag{7.3.1}$$

where E_Δ is the fundamental solution for the Laplacian from (7.1.12). Note that any E as in (7.3.1) is a fundamental solution for Δ^2. We proceed by analyzing three cases.

Case $n \geq 3$. Under the current assumptions, $E_\Delta(x) = -\frac{1}{(n-2)\omega_{n-1}}|x|^{2-n}$ for each $x \in \mathbb{R}^n \setminus \{0\}$. The key observation is that the following identity holds

$$\Delta[|x|^m] = m(m+n-2)|x|^{m-2}, \quad \text{pointwise in } \mathbb{R}^n \setminus \{0\}, \quad \forall m \in \mathbb{R}. \tag{7.3.2}$$

This suggests investigating the validity of a similar identity when derivatives are taken in the distributional sense. As seen in the next lemma, whose proof we postpone for later, a version of identity (7.3.2) holds in $S'(\mathbb{R}^n)$ for a suitable range of exponents that depends on the dimension n.

Lemma 7.18. *Let N be a real number such that $N > 2 - n$. Then*

$$\Delta[|x|^N] = N(N+n-2)|x|^{N-2} \quad in \quad S'(\mathbb{R}^n). \tag{7.3.3}$$

In view of Lemma 7.18 and the definition of E_Δ, it is justified to look for a solution to (7.3.1) of the form $E(x) = c_n|x|^{4-n}$, $x \in \mathbb{R}^n \setminus \{0\}$, where c_n is a constant to be determined. This candidate is in $S'(\mathbb{R}^n)$ (recall Exercise 4.5) and satisfies (7.3.1) if $2c_n(4-n) = -\frac{1}{(n-2)\omega_{n-1}}$. Hence, we obtain $c_n = \frac{1}{2(n-2)(n-4)\omega_{n-1}}$ if $n \neq 4$.

To handle the case $n = 4$, we use another result that we state next (for a proof see the last part of this section).

Lemma 7.19. *Let $n \geq 3$. Then $\Delta[\ln|x|] = (n-2)|x|^{-2}$ in $S'(\mathbb{R}^n)$.*

Lemma 7.19 suggests to take when $n = 4$ the candidate $E(x) = c \ln|x|$ for each $x \in \mathbb{R}^n \setminus \{0\}$, where c is a constant to be determined. From Example 4.9 we know that $E \in S'(\mathbb{R}^n)$. Also, Lemma 7.19 and the expression of E_Δ corresponding to $n = 4$ imply that $E(x) = c \ln|x|$ satisfies (7.3.1) if $2c = -\frac{1}{2\omega_3}$. Since $\omega_3 = 2\pi^2$ (recall (14.5.6) and (14.5.3)), the latter implies $c = -\frac{1}{8\pi^2}$.

In summary, we proved that

$$E(x) = \begin{cases} \dfrac{1}{2(n-2)(n-4)\omega_{n-1}}|x|^{4-n} & \text{for } n \geq 3, \ n \neq 4, \\[2mm] -\frac{1}{8\pi^2}\ln|x| & \text{for } n = 4, \end{cases} \quad x \in \mathbb{R}^n \setminus \{0\}, \tag{7.3.4}$$

satisfies $E \in S'(\mathbb{R}^n)$ and $\Delta^2 E = \delta$ in $S'(\mathbb{R}^n)$, $n \geq 3$.

The treatment of the case $n = 2$ is different from the above considerations since, in such a setting, $E_\Delta(x) = \frac{1}{2\pi} \ln |x|$ for $x \in \mathbb{R}^2 \setminus \{0\}$. Given the format of E_Δ, for some insight on how to choose a candidate E we start with the readily verified identity

$$\Delta[|x|^m \ln |x|] = |x|^{m-2}[m(m + n - 2) \ln |x| + 2m + n - 2]$$

$$(7.3.5)$$

$$\text{pointwise in } \mathbb{R}^n \setminus \{0\}, \quad \forall m \in \mathbb{R}.$$

The function $|x|^a$ belongs to $\mathcal{L}(\mathbb{R}^n)$ only when $a \in \mathbb{N}_0$ and is even, in which case $|x|^a \ln |x| \in S'(\mathbb{R}^n)$ (by Example 4.9 and *(b)* in Theorem 4.14). Hence, these restrictions should be taken into account when considering the analogue of identity (7.3.5) in the distributional sense in $S'(\mathbb{R}^n)$. The latter is stated next and its proof, which is very much in the spirit of the proofs for Lemmas 7.18 and 7.19 (here Exercise 4.108 is also relevant), is left as an exercise.

Lemma 7.20. *Let N be a real number such that $N > 2 - n$. Then*

$$\Delta[|x|^N \ln |x|] = |x|^{N-2}[N(N + n - 2) \ln |x| + 2N + n - 2] \ \text{in} \ S'(\mathbb{R}^n). \quad (7.3.6)$$

We are ready to proceed in earnest with considering:

Case $n = 2$. In view of Lemma 7.20, it is natural to start with a candidate of the form $E(x) := c|x|^2 \ln |x|$, $x \in \mathbb{R}^2 \setminus \{0\}$, with c a constant yet to be determined. Lemma 7.20 applied for $N = 2 = n$ implies $\Delta E = 4c \ln |x| + 4c$ in $S'(\mathbb{R}^2)$, hence (7.3.1) is satisfied for this E provided $4c = \frac{1}{2\pi}$, or equivalently, if $c = \frac{1}{8\pi}$. The bottom line is that, for this value of c, we have $E \in S'(\mathbb{R}^2)$ and $\Delta E = E_\Delta + \frac{1}{2\pi}$ in $S'(\mathbb{R}^2)$. Hence, since $\Delta c = 0$ in $S'(\mathbb{R}^2)$, we conclude that

$$E(x) = \frac{1}{8\pi}|x|^2 \ln |x| \quad \text{for} \ \ x \in \mathbb{R}^2 \setminus \{0\},$$

$$(7.3.7)$$

$$\text{satisfies} \ E \in S'(\mathbb{R}^2) \ \text{and} \ \Delta^2 E = \delta \ \text{in} \ S'(\mathbb{R}^2).$$

Finally, we are left with:

Case $n = 1$. In this situation we use Exercise 5.16 to conclude that any fundamental solution for Δ^2 that is a tempered distribution has the form $\frac{1}{6}x^3 H + P$ for some polynomial P in $\mathbb{R}$ of degree less than or equal to 3.

The main result emerging from this analysis is summarized next.

Theorem 7.21. *The function $E \in L^1_{loc}(\mathbb{R}^n)$ defined as*

$$E(x) := \begin{cases} \dfrac{1}{2(n-2)(n-4)\omega_{n-1}}|x|^{4-n} & \text{if } x \in \mathbb{R}^n \setminus \{0\}, \ n \geq 3, \ n \neq 4, \\[2mm] -\frac{1}{8\pi^2}\ln|x| & \text{if } x \in \mathbb{R}^4 \setminus \{0\}, \ n = 4, \\[2mm] \frac{1}{8\pi}|x|^2\ln|x| & \text{if } x \in \mathbb{R}^2 \setminus \{0\}, \ n = 2, \\[2mm] \frac{1}{6}x^3 H & \text{if } x \in \mathbb{R}, \qquad n = 1, \end{cases} \tag{7.3.8}$$

belongs to $S'(\mathbb{R}^n)$ and is a fundamental solution for the operator Δ^2 in $\mathbb{R}^n$. Moreover,

$$\{u \in S'(\mathbb{R}^n) : \Delta^2 u = \delta \text{ in } S'(\mathbb{R}^n)\} \tag{7.3.9}$$

$$= \{E + P : P \text{ biharmonic polynomial in } \mathbb{R}^n\}.$$

Proof. It is clear from (7.3.8) that $E \in L^1_{loc}(\mathbb{R}^n) \cap S'(\mathbb{R}^n)$. The fact that $\Delta^2 E = \delta$ in $S'(\mathbb{R}^n)$ has been checked in (7.3.4), (7.3.7), and Exercise 5.16. Finally, (7.3.9) is a consequence of Proposition 5.8. $\square$

We now turn to the task of proving Lemmas 7.18 and 7.19.

Proof of Lemma 7.18. Fix $N > 2 - n$. By Exercise 4.5 we have that both $|x|^{N-2}$ and $|x|^N$ belong to $S'(\mathbb{R}^n)$. Thus, invoking (4.1.33), there remains to show that

$$\langle \Delta[|x|^N], \varphi \rangle = N(N + n - 2)\langle |x|^{N-2}, \varphi \rangle, \quad \forall \varphi \in C_0^\infty(\mathbb{R}^n). \tag{7.3.10}$$

To this end, fix $\varphi \in C_0^\infty(\mathbb{R}^n)$ and let $R \in (0, \infty)$ be such that $\operatorname{supp}\varphi \subset B(0, R)$. Then, starting with the definition of distributional derivatives, then using the support condition for φ and Lebesgue's Dominated Convergence Theorem, and then (14.8.5) (keeping in mind that $\operatorname{supp}\varphi \subset B(0, R)$), we have

$$\langle \Delta[|x|^N], \varphi \rangle = \int_{\mathbb{R}^n} |x|^N \Delta\varphi(x)\,dx = \lim_{\varepsilon \to 0^+} \int_{\varepsilon < |x| < R} |x|^N \Delta\varphi(x)\,dx$$

$$= \lim_{\varepsilon \to 0^+} \Big[\int_{\varepsilon < |x| < R} \varphi(x)\Delta|x|^N\,dx - \int_{\partial B(0,\varepsilon)} |x|^N \frac{\partial\varphi}{\partial\nu}(x)\,d\sigma(x)$$

$$+ \int_{\partial B(0,\varepsilon)} \varphi(x)\frac{\partial|x|^N}{\partial\nu}\,d\sigma(x) \Big], \tag{7.3.11}$$

where $\nu(x) = \frac{x}{\varepsilon}$, for $x \in \partial B(0, \varepsilon)$. By (7.3.2) and Lebesgue's Dominated Convergence Theorem 14.15 it follows that

$$\lim_{\varepsilon \to 0^+} \int_{\varepsilon < |x| < R} \varphi(x)\Delta|x|^N\,dx = N(N + n - 2)\int_{\mathbb{R}^n} |x|^{N-2}\varphi(x)\,dx$$

$$= N(N + n - 2)\langle |x|^{N-2}, \varphi \rangle. \tag{7.3.12}$$

Since $\nabla(|x|^N) = N|x|^{N-2}x$ for every $x \neq 0$, we have $\frac{\partial|x|^N}{\partial\nu} = N\varepsilon^{N-1}$ on $\partial B(0, \varepsilon)$. Hence,

$$\left| -\int_{\partial B(0,\varepsilon)} |x|^N \frac{\partial \varphi}{\partial \nu}(x)\, d\sigma(x) + \int_{\partial B(0,\varepsilon)} \varphi(x) \frac{\partial |x|^N}{\partial \nu}\, d\sigma(x) \right| \tag{7.3.13}$$

$$\leq \omega_{n-1} \|\nabla \varphi\|_{L^\infty(\mathbb{R}^n)} \varepsilon^{N+n-1} + N\omega_{n-1} \|\varphi\|_{L^\infty(\mathbb{R}^n)} \varepsilon^{N+n-2} \xrightarrow[\varepsilon \to 0^+]{} 0$$

given that $N > 2 - n$. Now (7.3.10) follows from (7.3.11)–(7.3.13). The proof of the lemma is complete. $\qquad\square$

Proof of Lemma 7.19. From Exercise 4.5 and the assumption $n \geq 3$ it follows that $|x|^{-2} \in \mathcal{S}'(\mathbb{R}^n)$, while from Example 4.9 we have $\ln|x| \in \mathcal{S}'(\mathbb{R}^n)$. Hence, by (4.1.33) matters reduce to showing

$$\langle \Delta[\ln|x|], \varphi \rangle = (n-2)\langle |x|^{-2}, \varphi \rangle, \quad \forall \varphi \in C_0^\infty(\mathbb{R}^n). \tag{7.3.14}$$

Fix $\varphi \in C_0^\infty(\mathbb{R}^n)$ and let $R \in (0, \infty)$ be such that $\operatorname{supp} \varphi \subset B(0, R)$. Using in the current setting a reasoning similar to that applied when deriving (7.3.11) one obtains

$$\langle \Delta[\ln|x|], \varphi \rangle = \int_{\mathbb{R}^n} \ln|x| \Delta \varphi(x)\, dx = \lim_{\varepsilon \to 0^+} \int_{\varepsilon < |x| < R} \ln|x| \Delta \varphi(x)\, dx$$

$$= \lim_{\varepsilon \to 0^+} \Big[\int_{\varepsilon < |x| < R} \varphi(x) \Delta(\ln|x|)\, dx - \int_{\partial B(0,\varepsilon)} \ln|x| \frac{\partial \varphi}{\partial \nu}(x)\, d\sigma(x)$$

$$+ \int_{\partial B(0,\varepsilon)} \varphi(x) \frac{\partial \ln|x|}{\partial \nu}\, d\sigma(x) \Big]. \tag{7.3.15}$$

It is easy to see that $\Delta(\ln|x|) = (n-2)|x|^{-2}$ pointwise in $\mathbb{R}^n \setminus \{0\}$, thus invoking Lebesgue's Dominated Convergence Theorem 14.15 we obtain

$$\lim_{\varepsilon \to 0^+} \int_{\varepsilon < |x| < R} \varphi(x) \Delta(\ln|x|)\, dx = (n-2) \int_{\mathbb{R}^n} |x|^{-2} \varphi(x)\, dx$$

$$= (n-2)\langle |x|^{-2}, \varphi \rangle. \tag{7.3.16}$$

Also, since $\nabla(\ln|x|) = \frac{x}{|x|^2}$ pointwise in $\mathbb{R}^n \setminus \{0\}$, we have $\frac{\partial \ln|x|}{\partial \nu} = \varepsilon^{-1}$ on $\partial B(0, \varepsilon)$. Hence,

$$\left| -\int_{\partial B(0,\varepsilon)} \ln|x| \frac{\partial \varphi}{\partial \nu}(x)\, d\sigma(x) + \int_{\partial B(0,\varepsilon)} \varphi(x) \frac{\partial \ln|x|}{\partial \nu}\, d\sigma(x) \right| \tag{7.3.17}$$

$$\leq \omega_{n-1} \|\nabla \varphi\|_{L^\infty(\mathbb{R}^n)} \varepsilon^{n-1} |\ln\varepsilon| + \omega_{n-1} \|\varphi\|_{L^\infty(\mathbb{R}^n)} \varepsilon^{n-2} \xrightarrow[\varepsilon \to 0^+]{} 0,$$

where for the convergence in (7.3.17) we used the fact that $n \geq 3$. To finish the proof of the lemma we combine (7.3.15)–(7.3.17). $\qquad\square$

Exercise 7.22. Let E be the fundamental solution for the bi-Laplacian operator from (7.3.8). Prove that for every $\alpha \in \mathbb{N}_0^n$ with $|\alpha| = 3$ the function $\partial^\alpha E$ is of class C^∞ and positive homogeneous of degree $1 - n$ in $\mathbb{R}^n \setminus \{0\}$.

7.4 The Poisson Equation for the Bi-Laplacian

Analogously to Theorem 7.10 we have the following regularity result for the bi-Laplacian.

Theorem 7.23. *For an open set $\Omega \subseteq \mathbb{R}^n$ the following are equivalent:*

(i) u is biharmonic *in Ω;*
(ii) $u \in \mathcal{D}'(\Omega)$ and $\Delta^2 u = 0$ in $\mathcal{D}'(\Omega)$;
(iii) $u \in C^\infty(\Omega)$ and $\Delta^2 u = 0$ in a pointwise sense in Ω.

Proof. This is a direct consequence of Theorem 6.8 and Theorem 7.21 (or, alternatively, of Remark 6.14 and Corollary 6.18). □

We begin by establishing mean value formulas for biharmonic functions. Recall the notation from (0.0.15).

Theorem 7.24. *Let Ω be an open set in $\mathbb{R}^n$ and let u be a biharmonic function in Ω. Then for every $x \in \Omega$ and every $r \in (0, \operatorname{dist}(x, \partial\Omega))$ the following formulas hold:*

$$u(x) = \fint_{\partial B(x,r)} u(y)\, d\sigma(y) - \frac{r^2}{2n}(\Delta u)(x), \tag{7.4.1}$$

$$u(x) = \fint_{B(x,r)} u(y)\, dy - \frac{r^2}{2(n+2)}(\Delta u)(x). \tag{7.4.2}$$

Proof. Fix $x \in \Omega$ and recall the function ϕ defined in (7.2.10). Then (7.2.11) holds. Since $\Delta(\Delta u) = 0$ in Ω we have that Δu is harmonic in Ω and we can apply (7.2.9) to obtain that $\phi'(r) = \frac{r}{n}\Delta u(x)$. Consequently, $\phi(r) = \frac{r^2}{2n}\Delta u(x) + C$ for some constant C. To determine C we use the fact that $\lim_{r \to 0^+} \phi(r) = u(x)$ (for the latter see (7.2.15)). Hence formula (7.4.1) follows.

Next, write (7.4.1) as

$$\omega_{n-1} r^{n-1} u(x) = \int_{\partial B(x,r)} u(y)\, d\sigma(y) - \frac{\omega_{n-1} r^{n+1}}{n} \Delta u(x). \tag{7.4.3}$$

Integrating (7.4.3) with respect to r for $r \in (0, R)$ and $R \in (0, \operatorname{dist}(x, \partial\Omega))$, and then dividing by $\omega_{n-1} R^n / n$ gives (7.4.2) with r replaced by R. □

Theorem 7.25 (Liouville's Theorem for Δ^2). *Any bounded biharmonic function in $\mathbb{R}^n$ is constant.*

Proof. One way to justify this result is by observing that it is a particular case of Theorem 5.4. Another proof based on interior estimates goes as follows. Let u be a bounded biharmonic function in $\mathbb{R}^n$. Formula (7.4.2) in the current setting gives that

$$\Delta u(x) = \frac{2(n+2)}{r^2} \fint_{B(x,r)} u(y)\, dy - \frac{2(n+2)}{r^2} u(x) \tag{7.4.4}$$

for every $x \in \mathbb{R}^n$ and every $r \in (0, \infty)$. Since

$$\left| \fint_{B(x,r)} u(y) \, dy \right| \leq \|u\|_{L^\infty(\mathbb{R}^n)}, \tag{7.4.5}$$

if we let $r \to \infty$ in (7.4.4), we see that $\Delta u(x) = 0$, thus u is harmonic. By Liouville's Theorem 7.15 for harmonic functions it follows that there exists $c \in \mathbb{R}$ such that $u(x) = c$ for all $x \in \mathbb{R}^n$. $\qquad\square$

We shall next show that biharmonic functions satisfy interior estimates and are real analytic.

Theorem 7.26. *Suppose $u \in \mathcal{D}'(\Omega)$ satisfies $\Delta^2 u = 0$ in $\mathcal{D}'(\Omega)$. Then u is real analytic in Ω and there exists a dimensional constant $C \in (0, \infty)$ such that*

$$\max_{y \in B(x,r/2)} |\partial^\alpha u(y)| \leq \frac{C^{|\alpha|} |\alpha|!}{r^{|\alpha|}} \max_{y \in B(x,r)} |u(y)|, \qquad \forall \alpha \in \mathbb{N}_0^n, \tag{7.4.6}$$

for each $x \in \Omega$ and each $r \in (0, \operatorname{dist}(x, \partial\Omega))$.

Proof. In the case when $n = 1$ or $n \geq 3$, all claims are direct consequences of Theorem 6.26 and Theorem 7.21. To treat the case $n = 2$ we shall introduce a "dummy" variable. Specifically, in this setting consider the open set $\widetilde{\Omega} := \Omega \times \mathbb{R}$ in $\mathbb{R}^3$ and define the function $\widetilde{u}(x_1, x_2, x_3) := u(x_1, x_2)$ for $(x_1, x_2, x_3) \in \widetilde{\Omega}$. Observe that $\widetilde{u}$ is biharmonic in $\widetilde{\Omega}$. As such, the higher dimensional theory applies and yields that $\widetilde{u}$ satisfies, for some universal constant $C \in (0, \infty)$,

$$\max_{y \in B(x,r/2)} |\partial^\alpha \widetilde{u}(y)| \leq \frac{C^{|\alpha|} |\alpha|!}{r^{|\alpha|}} \max_{y \in B(x,r)} |\widetilde{u}(y)|, \qquad \forall \alpha \in \mathbb{N}_0^3, \tag{7.4.7}$$

for each $x \in \widetilde{\Omega}$ and each $r \in (0, \operatorname{dist}(x, \partial\widetilde{\Omega}))$. Applying (7.4.7) for points of the form $x = (x_1, x_2, 0) \in \widetilde{\Omega}$ and for multi-indices $\alpha = (\alpha_1, \alpha_2, 0)$ then yields (7.4.6) for the original function u. In turn, this and Lemma 6.24 imply that u is real analytic in Ω. $\qquad\square$

We are now prepared to discuss the well-posedness of the Poisson problem for the bi-Laplacian in $\mathbb{R}^n$.

Theorem 7.27. *Assume $n \in \mathbb{N}$ satisfies $n \geq 3$ and $n \neq 4$. Then for each function $f \in L^\infty_{comp}(\mathbb{R}^n)$ and each $c \in \mathbb{C}$ the Poisson problem for the bi-Laplacian in $\mathbb{R}^n$,*

$$\begin{cases} u \in C^0(\mathbb{R}^n), \\ \Delta^2 u = f \text{ in } \mathcal{D}'(\mathbb{R}^n), \\ \lim_{|x| \to \infty} u(x) = c, \end{cases} \tag{7.4.8}$$

has a unique solution. Moreover, the solution u satisfies the following additional properties.

(1) The function u has the integral representation formula

$$u(x) = c + \int_{\mathbb{R}^n} E(x - y) f(y) \, dy, \qquad x \in \mathbb{R}^n, \tag{7.4.9}$$

*where E is the fundamental solution for the bi-Laplacian in $\mathbb{R}^n$ from (7.3.8).
Moreover, u belongs to $C^3(\mathbb{R}^n)$ and for each $\alpha \in \mathbb{N}_0^n$ with $0 < |\alpha| \le 3$ we have*

$$\partial^\alpha u(x) = \int_{\mathbb{R}^n} (\partial^\alpha E)(x - y) f(y) \, dy, \qquad x \in \mathbb{R}^n. \tag{7.4.10}$$

(2) If in fact $f \in C_0^\infty(\mathbb{R}^n)$ then $u \in C^\infty(\mathbb{R}^n)$.
(3) For every $\alpha \in \mathbb{N}_0^n$ with $|\alpha| = 4$ there exists a constant $c_\alpha \in \mathbb{C}$ such that

$$\partial^\alpha u = c_\alpha f + T_{\partial^\alpha E} f \quad \text{in } \mathcal{D}'(\mathbb{R}^n), \tag{7.4.11}$$

*where $T_{\partial^\alpha E}$ is the singular integral operator associated with $\Theta := \partial^\alpha E$ as in
Definition 4.93.*
*(4) For every $p \in (1, \infty)$, the solution u of (7.4.8) satisfies $\partial^\alpha u \in L^p(\mathbb{R}^n)$ for each
$\alpha \in \mathbb{N}_0^n$ with $|\alpha| = 4$, where the derivatives are taken in $\mathcal{D}'(\mathbb{R}^n)$. Moreover, there
exists a constant $C = C(p, n) \in (0, \infty)$ with the property that*

$$\sum_{|\alpha|=4}^n \|\partial^\alpha u\|_{L^p(\mathbb{R}^n)} \le C\|f\|_{L^p(\mathbb{R}^n)}. \tag{7.4.12}$$

Proof. That the function u defined as in (7.4.9) is of class $C^3(\mathbb{R}^n)$ and formula
(7.4.10) holds for each $\alpha \in \mathbb{N}_0^n$ with $|\alpha| \le 3$ can be established much as in the
proof of Proposition 7.8 keeping in mind that $\partial^\alpha E \in L_{loc}^1(\mathbb{R}^n)$ for each $\alpha \in \mathbb{N}_0^n$ with
$|\alpha| \le 3$. Also, this function solves $\Delta^2 u = f$ in $\mathcal{D}'(\mathbb{R}^n)$ by reasoning in a similar
fashion to the computation in (7.2.7). Given the format of E from (7.3.8) under the
current assumptions on n, and the conditions on f, it is clear that the function u from
(7.4.9) satisfies the limit condition in (7.4.8) (see (7.2.21) in the case of the Laplace
operator). The fact that (7.4.9) is the unique solution of (7.4.8) is a consequence of
Theorem 7.25. Moving on, that $u \in C^\infty(\mathbb{R}^n)$ if f belongs to $C_0^\infty(\mathbb{R}^n)$ follows from
Corollary 6.18 and the ellipticity of Δ^2. The above arguments cover the claims in
parts *(1)–(2)*.

Turning to part *(3)*, start by fixing $\alpha \in \mathbb{N}_0^n$ with $|\alpha| = 4$. Then there exists some
$\beta \in \mathbb{N}_0^n$ with $|\beta| = 3$ and $j \in \{1, \dots, n\}$ such that $\partial^\alpha = \partial_j \partial^\beta$. If we define $\Phi := \partial^\beta E$,
it follows that Φ is C^∞ and positive homogeneous of degree $1 - n$ in $\mathbb{R}^n \setminus \{0\}$ (cf.
Exercise 7.22). Consequently, the function $\Theta := \partial^\alpha E$ is as in (4.4.1), thanks to
the discussion in Example 4.71. Then from what we have proved in part *(1)* and
Theorem 4.103 we deduce that the following formula holds in $\mathcal{D}'(\mathbb{R}^n)$:

$$\partial^\alpha u(x) = \partial_j[\partial^\beta u(x)] = \partial_j\Big[\int_{\mathbb{R}^n} (\partial^\beta E)(x-y)f(y)\,dy\Big] \qquad (7.4.13)$$

$$= \Big(\int_{S^{n-1}} (\partial^\beta E)(\omega)\omega_j\,d\sigma(\omega)\Big)f(x) + \lim_{\varepsilon\to 0^+}\int_{|x-y|>\varepsilon}(\partial^\alpha E)(x-y)f(y)\,dy.$$

Choosing $c_\alpha := \int_{S^{n-1}}(\partial^\beta E)(\omega)\omega_j\,d\sigma(\omega)$ the above formula may be written as in (7.4.11), and this finishes the proof of part (3). Finally, the claim in (4) follows from (3) and the boundedness of the singular integral operators $T_{\partial^\alpha E}$ on $L^p(\mathbb{R}^n)$ (cf. Theorem 4.101). □

7.5 Fundamental Solutions for the Poly-harmonic Operator

Let $m \in \mathbb{N}$ and consider the `poly-harmonic` operator

$$\Delta^m := \Big(\sum_{j=1}^n \partial_j^2\Big)^m = \sum_{|\gamma|=m}\frac{m!}{\gamma!}\partial^{2\gamma}, \quad \text{in } \mathbb{R}^n. \qquad (7.5.1)$$

The goal in this section is to identify all fundamental solutions for this operator in $\mathbb{R}^n$ that are tempered distributions. The case $m = 1$ has been treated in Theorem 7.2 and the case $m = 2$ is discussed in Theorem 7.21. We remark that the formulas from Lemma 7.18, Lemma 7.19, and Lemma 7.20, suggest that we should look for a fundamental solution for Δ^m that is a constant multiple of $|x|^{2m-n}$ or a constant multiple of $|x|^{2m-n}\ln|x|$. In precise terms, we will prove the following result (recall that H denotes the Heaviside function).

Theorem 7.28. *Let $m, n \in \mathbb{N}$ and consider the function*

$$F_{m,n}(x) := \begin{cases} \dfrac{(-1)^m\Gamma(n/2-m)}{\pi^{n/2}4^m(m-1)!}\,|x|^{2m-n} & \text{if } n > 2m \text{ or } n \text{ is odd and } n \neq 1, \\[3mm] \dfrac{(-1)^{n/2+1}|x|^{2m-n}}{2\pi^{n/2}4^{m-1}(m-n/2)!(m-1)!}\,\ln|x| & \text{if } n \text{ is even and } \leq 2m, \quad (7.5.2) \\[3mm] \dfrac{1}{(2m-1)!}x^{2m-1}H & \text{if } n = 1, \end{cases}$$

defined for $x \in \mathbb{R}^n \setminus \{0\}$ if $n \geq 2$ and for $x \in \mathbb{R}$ if $n = 1$. Then $F_{m,n}$ belongs to $L^1_{loc}(\mathbb{R}^n) \cap S'(\mathbb{R}^n)$ and is a fundamental solution for Δ^m in $\mathbb{R}^n$. Moreover,

$$\{u \in S'(\mathbb{R}^n) : \Delta^m u = \delta \text{ in } S'(\mathbb{R}^n)\} \qquad (7.5.3)$$

$$= \{F_{m,n} + P : P \text{ poly-harmonic polynomial in } \mathbb{R}^n\}.$$

Proof. The case $n = 1$ is immediate from Exercise 5.16. Assume in what follows that $n \geq 2$. Since $n - 2m < n$, we clearly have $|x|^{2m-n} \in L^1_{loc}(\mathbb{R}^n)$ and furthermore, by Exercise 4.5, that $|x|^{2m-n} \in \mathcal{S}'(\mathbb{R}^n)$. By Exercise 2.124 and Exercise 4.108 we also have $|x|^{2m-n} \ln|x| \in L^1_{loc}(\mathbb{R}^n) \cap \mathcal{S}'(\mathbb{R}^n)$. This proves that $F_{m,n} \in L^1_{loc}(\mathbb{R}^n) \cap \mathcal{S}'(\mathbb{R}^n)$. Once we show that $F_{m,n}$ is a fundamental solution for Δ^m in $\mathbb{R}^n$, Proposition 5.8 readily yields (7.5.3).

There remains to prove $\Delta^m F_{m,n} = \delta$ in $\mathcal{S}'(\mathbb{R}^n)$. If $m = 1$, then it is easy to see that $F_{1,n}$ is the same as the distribution in (7.1.12) (corresponding to $n \geq 2$). Thus, by Theorem 7.2 we have

$$\Delta F_{1,n} = \delta \quad \text{in} \quad \mathcal{S}'(\mathbb{R}^n). \tag{7.5.4}$$

We proceed by breaking up the remaining part of our analysis in a few cases.

The Case $n > 2m$ or n is Odd.
Because of (7.5.4) we may assume $m \geq 2$. Applying Lemma 7.18 and (14.5.2) we obtain

$$\begin{aligned}
\Delta F_{m,n} &= \frac{(-1)^m \Gamma(n/2 - m)}{\pi^{n/2} 4^m (m-1)!} \Delta(|x|^{2m-n}) \\
&= \frac{(-1)^m \Gamma(n/2 - m)}{\pi^{n/2} 4^m (m-1)!} (2m - n)(2m - 2)|x|^{2m-n-2} \\
&= \frac{(-1)^{m-1} \Gamma(n/2 - m)(n/2 - m)(m - 1)}{\pi^{n/2} 4^{m-1}(m-1)!} |x|^{2(m-1)-n} \\
&= F_{m-1,n} \quad \text{in } \mathcal{S}'(\mathbb{R}^n) \text{ and pointwise in } \mathbb{R}^n \setminus \{0\}. \tag{7.5.5}
\end{aligned}$$

Hence, inductively we have $\Delta^{m-1} F_{m,n} = F_{1,n}$ in $\mathcal{S}'(\mathbb{R}^n)$, which when combined with (7.5.4) yields the desired conclusion in the current case.

The Case $n = 2m$.
In this scenario n is even, and the goal is to prove

$$\Delta^{\frac{n}{2}} F_{\frac{n}{2},n} = \delta \quad \text{in} \quad \mathcal{S}'(\mathbb{R}^n). \tag{7.5.6}$$

Note that (7.5.6) holds for $n = 2$ as seen from (7.5.4). Consequently, we may assume that $n \geq 4$. Inductively, using Lemma 7.18, it is not difficult to show that

$$\Delta^k[|x|^{-2}] = \frac{(-1)^k 4^k k!(\frac{n}{2} - 2)!}{(\frac{n}{2} - k - 2)!} |x|^{-2-2k} \quad \text{in} \quad \mathcal{S}'(\mathbb{R}^n) \tag{7.5.7}$$

if n is even, $n \geq 4$ and $k \in \mathbb{N}_0$ is such that $k < \frac{n}{2} - 1$.

In particular, (7.5.7) specialized to $k = \frac{n}{2} - 2$ yields

$$\Delta^{\frac{n}{2}-2}[|x|^{-2}] = (-1)^{\frac{n}{2}} 4^{\frac{n}{2}} \left[\left(\frac{n}{2} - 2\right)!\right]^2 |x|^{2-n} \quad \text{in} \quad \mathcal{S}'(\mathbb{R}^n). \tag{7.5.8}$$

Starting with the expression from (7.5.2) we obtain

$$\Delta^{\frac{n}{2}-1} F_{\frac{n}{2},n} = \frac{(-1)^{\frac{n}{2}+1}}{2\pi^{\frac{n}{2}} 4^{\frac{n}{2}-1}(\frac{n}{2}-1)!} \Delta^{\frac{n}{2}-1}[\ln|x|]$$

$$= \frac{(-1)^{\frac{n}{2}+1}}{2\pi^{\frac{n}{2}} 4^{\frac{n}{2}-1}(\frac{n}{2}-1)!} (n-2)\Delta^{\frac{n}{2}-2}[|x|^{-2}]$$

$$= \frac{(-1)^{\frac{n}{2}+1}}{2\pi^{\frac{n}{2}} 4^{\frac{n}{2}-1}(\frac{n}{2}-1)!} (n-2)(-1)^{\frac{n}{2}} 4^{\frac{n}{2}}\left[\left(\frac{n}{2}-2\right)!\right]^2 |x|^{2-n}$$

$$= \frac{-1}{(n-2)\omega_{n-1}}|x|^{2-n} \quad \text{in} \quad \mathcal{S}'(\mathbb{R}^n). \tag{7.5.9}$$

The second equality in (7.5.9) is based on Lemma 7.19, the third equality uses (7.5.8), while the last equality follows by straightforward calculations and (14.5.6). In summary, we proved that if n is even and $n \geq 4$, then $\Delta^{\frac{n}{2}-1} F_{\frac{n}{2},n} = E_\Delta$ in $\mathcal{S}'(\mathbb{R}^n)$, where E_Δ is the fundamental solution for the Laplacian from (7.1.12). The latter combined with Theorem 7.2 now yields (7.5.6) for $n \geq 4$, and finishes the proof of (7.5.6).

The Case $n \leq 2m$ and n is Even.

Fix an even number $n \in \mathbb{N}$. We will prove that if $m \geq \frac{n}{2}$ then

$$\Delta^m F_{m,n} = \delta \quad \text{in} \quad \mathcal{S}'(\mathbb{R}^n) \tag{7.5.10}$$

by induction on m. The case $m = \frac{n}{2}$ has been treated in the previous case. Suppose (7.5.10) holds for some $m \geq \frac{n}{2}$. Then clearly $2(m+1) > n$ and invoking Lemma 7.20 we obtain

$$\Delta F_{m+1,n} = F_{m,n} + C_{m+1,n}(4m-n+2)|x|^{2m-n} \quad \text{in} \quad \mathcal{S}'(\mathbb{R}^n), \tag{7.5.11}$$

where $C_{m+1,n}$ is the coefficient of $|x|^{2(m+1)-n} \ln|x|$ in the expression for $F_{m+1,n}$. Note that since n is even the expression $|x|^{2m-n}$ is a polynomial of degree $2m - n$. Hence, $\Delta^m[|x|^{2m-n}] = 0$ given that Δ^m is a homogeneous differential operator of degree $2m > 2m - n$. Combined with (7.5.11) and the induction hypothesis on $F_{m,n}$, this gives

$$\Delta^{m+1} F_{m+1,n} = \Delta^m F_{m,n} + C_{m+1,n}(4m-n+2)\Delta^m[|x|^{2m-n}] = \delta \tag{7.5.12}$$

in $\mathcal{S}'(\mathbb{R}^n)$. By induction, it follows that (7.5.10) holds in the current setting. This finishes the proof of the theorem. $\square$

We remark that Theorem 7.28 specialized to $m = 2$ gives another proof of the fact that when $n \geq 2$ the expression from (7.3.8) is a fundamental solution for Δ^2 in $\mathbb{R}^n$.

Proposition 7.29. *Let $m, n \in \mathbb{N}$ be such that $n \geq 2$ and let $F_{m,n}$ be the fundamental solution for Δ^m in $\mathbb{R}^n$ as defined in (7.5.2). Then if $N \in \mathbb{N}_0$ is such that $0 \leq N < m$ the following identity holds*

$$\Delta^N\big[F_{m,n}\big] = F_{m-N,n} + \mathcal{P}_{m,N} \quad in\ \ \mathcal{S}'(\mathbb{R}^n)\ \ and\ pointwise\ in\ \mathbb{R}^n \setminus \{0\}, \qquad (7.5.13)$$

for some homogeneous polynomial $\mathcal{P}_{m,N}$ in $\mathbb{R}^n$ of degree at most $\max\{2m-n-2N, 0\}$ which is identically zero if $n > 2m$ or n is odd, or if n is even such that $n \leq 2m$ and $N > m - \frac{n}{2}$ (in short, $\mathcal{P}_{m,N} \equiv 0$ if either n is odd or $2N + n > 2m$).

Proof. We start by observing that using an inductive reasoning and Lemma 7.20 we obtain that for each $N \in \mathbb{N}_0$

$$\Delta^N[|x|^M \ln|x|] = C_{M,N}^{(1)}|x|^{M-2N} \ln|x| + C_{M,N}^{(2)}|x|^{M-2N} \quad in\ \ \mathcal{S}'(\mathbb{R}^n)$$

$$\text{(7.5.14)}$$

if M is a real number such that $M - 2N > -n$,

for some real constants $C_{M,N}^{(1)}$, $C_{M,N}^{(2)}$, depending on n, M, and N.

Now suppose n is even and $n \leq 2m$. Specializing (7.5.14) to the case $M := 2m-n$ .and taking $N \in \{0, 1, \ldots, m - \frac{n}{2}\}$, we conclude

$$\Delta^N F_{m,n} = c_{m,N}^{(1)} F_{m-N,n} + c_{m,N}^{(2)}|x|^{2m-n-2N} \quad in\ \ \mathcal{S}'(\mathbb{R}^n), \qquad (7.5.15)$$

for some real constants $c_{m,N}^{(1)}$, $c_{m,N}^{(2)}$, depending only on n, m, and N. For each number $N \in \{0, 1, \ldots, m - \frac{n}{2}\}$ set

$$\mathcal{P}_{m,N}(x) := c_{m,N}^{(2)}|x|^{2m-n-2N}, \quad x \in \mathbb{R}^n.$$

Note that since n is even and $n \leq 2m$, each $\mathcal{P}_{m,N}$ is a homogeneous polynomial of degree at most $2m - n - 2N$. Also, for each $N \in \{0, 1, \ldots, m - \frac{n}{2}\}$ the operator Δ^{m-N} is a homogeneous differential operator of degree $2m - 2N > 2m - n - 2N$, thus $\Delta^{m-N}\mathcal{P}_{m,N} = 0$ both in $\mathcal{S}'(\mathbb{R}^n)$ and pointwise in $\mathbb{R}^n$. The latter, (7.5.15), and Theorem 7.28 imply

$$\delta = \Delta^{m-N}[\Delta^N F_{m,n}] = c_{m,N}^{(1)}\Delta^{m-N}[F_{m-N,n}] + \Delta^{m-N}[\mathcal{P}_{m,N}]$$

$$= c_{m,N}^{(1)}\delta \quad in\ \mathcal{S}'(\mathbb{R}^n) \qquad (7.5.16)$$

for each $N \in \{0, 1, \ldots, m - \frac{n}{2}\}$. Thus, $c_{m,N}^{(1)} = 0$ for each $N \in \{0, 1, \ldots, m - \frac{n}{2}\}$. Consequently, (7.5.15) becomes

$$\Delta^N F_{m,n} = F_{m-N,n} + \mathcal{P}_{m,N} \quad in\ \ \mathcal{S}'(\mathbb{R}^n), \ \ \forall N \in \{0, 1, \ldots, m - \frac{n}{2}\}. \qquad (7.5.17)$$

That the above equality also holds pointwise in $\mathbb{R}^n$ is immediate, hence we proved (7.5.13) for n even such that $n \leq 2m$ and $0 \leq N \leq m - \frac{n}{2}$. Note that if $n = 2$ then $m - \frac{n}{2} = m - 1$ so we actually have proved (7.5.13) for all $N \in \{0, 1, \ldots, m - 1\}$.

To finish with the proof of (7.5.13) when n is even and satisfies $n \leq 2m$, there remains the case $N \in \{m - \frac{n}{2} + 1, \ldots, m - 1\}$ under the assumption that $n \geq 4$. For starters, observe that if we write (7.5.17) for $N = m - \frac{n}{2}$, then $\mathcal{P}_{m,N}$ is just a constant and using the expression for $F_{\frac{n}{2},n}$ we have

$$\Delta^{m-\frac{n}{2}} F_{m,n} = F_{\frac{n}{2},n} + C_{m,m-\frac{n}{2}}^{(2)} = c_n \ln |x| + c_{m,m-\frac{n}{2}}^{(2)} \quad \text{in } \mathcal{S}'(\mathbb{R}^n), \tag{7.5.18}$$

for some real constant c_n. Therefore, (7.5.18), Lemma 7.19, and Lemma 7.18 (which we can apply $N - m + \frac{n}{2} - 1$ times since $N < m$) allow us to conclude that for each $N \in \{m - \frac{n}{2} + 1, \ldots, m - 1\}$ there exists some constant $c_{m,N}$ such that

$$\Delta^N F_{m,n} = \Delta^{N-m+\frac{n}{2}} \left[\Delta^{m-\frac{n}{2}} F_{m,n} \right] = c_{m,N} |x|^{-2N+2m-2n} \quad \text{in } \mathcal{S}'(\mathbb{R}^n). \tag{7.5.19}$$

Upon observing that the condition $N > m - \frac{n}{2}$ is equivalent with $n > 2(m - N)$, we may conclude that $c_{m,N} |x|^{-2N+2m-2n} = c_{m,N}^{(1)} F_{m-N,n}$ for some constant $c_{m,N}^{(1)}$. Hence, for each $N \in \{m - \frac{n}{2} + 1, \ldots, m - 1\}$ we have

$$\Delta^N F_{m,n} = c_{m,N}^{(1)} F_{m-N,n} \quad \text{in } \mathcal{S}'(\mathbb{R}^n). \tag{7.5.20}$$

Applying Δ^{m-N} to (7.5.20) and invoking Theorem 7.28 we obtain $c_{m,N}^{(1)} = 0$ for each $N \in \{m - \frac{n}{2} + 1, \ldots, m - 1\}$, so that

$$\Delta^N F_{m,n} = F_{m-N,n} \quad \text{in } \mathcal{S}'(\mathbb{R}^n), \ \forall N \in \{m - \frac{n}{2} + 1, \ldots, m - 1\}. \tag{7.5.21}$$

This proves (7.5.13) when n is even, satisfies $n \leq 2m$, and $m - \frac{n}{2} < N < m$ if we take $\mathcal{P}_{m,N} := 0$.

Suppose next that $n > 2m$ or n is odd. Then as seen in the proof of Theorem 7.28, identity (7.5.12) holds in $\mathcal{S}'(\mathbb{R}^n)$ and pointwise in $\mathbb{R}^n \setminus \{0\}$. Iterating this identity we arrive at

$$\Delta^N [F_{m,n}] = F_{m-N,n} \quad \text{in } \mathcal{S}'(\mathbb{R}^n) \text{ and pointwise in } \mathbb{R}^n \setminus \{0\}. \tag{7.5.22}$$

Consequently, formula (7.5.13) holds in the current setting if we take $\mathcal{P}_{m,N} := 0$ for all $N \in \{0, \ldots, m - 1\}$. This completes the proof of the proposition. $\square$

We next address the issue of interior estimates and real analyticity for poly-harmonic functions.

Theorem 7.30. *Whenever $u \in \mathcal{D}'(\Omega)$ satisfies $\Delta^m u = 0$ in $\mathcal{D}'(\Omega)$, it follows that u is real analytic in Ω and there exists a constant $C = C(n, m) \in (0, \infty)$ with the property that*

$$\max_{y \in B(x,r/2)} |\partial^\alpha u(y)| \le \frac{C^{|\alpha|}|\alpha|!}{r^{|\alpha|}} \max_{y \in B(x,r)} |u(y)|, \qquad \forall \, \alpha \in \mathbb{N}_0^n, \tag{7.5.23}$$

for each $x \in \Omega$ and each $r \in (0, \mathrm{dist}\,(x, \partial\Omega))$.

Proof. In the case when either n is odd, or $n \ge 2m$, all claims are consequences of Theorem 6.26 and Theorem 7.28. The remaining cases may be reduced to the ones just treated by introducing "dummy" variables, as in the proof of Theorem 7.26. $\square$

In the last part of this section we prove an integral representation formula for a fundamental solution for the poly-harmonic operator in $\mathbb{R}^n$ involving a sufficiently large power of the Laplacian (i.e., Δ^m with m positive integer bigger than or equal to $n/2$). The difference between this fundamental solution and that in (7.5.2) turns out to be a homogeneous polynomial that is a null solution of the respective poly-harmonic operator. In what follows log denotes the principal branch of the complex logarithm that is defined for points $z \in \mathbb{C} \setminus (-\infty, 0]$.

Lemma 7.31. *Let $n \in \mathbb{N}$, $n \ge 2$, and suppose $q \in \mathbb{N}_0$ is such that $n + q$ is an even number. Then the function*

$$E_q(x) := -\frac{1}{(2\pi i)^n q!} \int_{S^{n-1}} (x \cdot \xi)^q \log\left(\frac{x \cdot \xi}{i}\right) d\sigma(\xi), \quad \forall \, x \in \mathbb{R}^n \setminus \{0\}, \tag{7.5.24}$$

is a fundamental solution for the poly-harmonic operator $\Delta^{\frac{n+q}{2}}$ in $\mathbb{R}^n$.
Moreover, $E_q \in \mathcal{S}'(\mathbb{R}^n)$ and

$$E_q(x) = F_{\frac{n+q}{2},n}(x) + C(n,q)|x|^q, \quad \forall \, x \in \mathbb{R}^n \setminus \{0\}, \tag{7.5.25}$$

where $F_{\frac{n+q}{2},n}$ is the fundamental solution for $\Delta^{\frac{n+q}{2}}$ in $\mathbb{R}^n$ given in (7.5.2) and

$$C(n,q) := \begin{cases} -\dfrac{2\omega_{n-2}}{(2\pi i)^n q!} \displaystyle\int_0^1 s^q (\ln s)(\sqrt{1-s^2})^{n-3} \, ds & \text{if } n \text{ is even}, \\[4mm] 0 & \text{if } n \text{ is odd}. \end{cases} \tag{7.5.26}$$

Proof. Fix n and q as in the hypotheses of the lemma. Fix $x \in \mathbb{R}^n \setminus \{0\}$ and use the fact that $\log\left(\frac{x \cdot \xi}{i}\right) = \ln |x \cdot \xi| - i\frac{\pi}{2}\mathrm{sgn}\,(x \cdot \xi)$ for every $\xi \in S^{n-1}$ to write

$$\int_{S^{n-1}} (x \cdot \xi)^q \log\left(\frac{x \cdot \xi}{i}\right) d\sigma(\xi)$$

$$= \int_{\xi \in S^{n-1},\, x \cdot \xi > 0} (x \cdot \xi)^q \left[\ln|x \cdot \xi| - i\frac{\pi}{2} \right] d\sigma(\xi)$$

$$+ (-1)^q \int_{\xi \in S^{n-1},\, x \cdot \xi > 0} (x \cdot \xi)^q \left[\ln|x \cdot \xi| + i\frac{\pi}{2} \right] d\sigma(\xi)$$

$$= \frac{1}{2} \int_{S^{n-1}} |x \cdot \xi|^q \left[\ln|x \cdot \xi| - i\frac{\pi}{2} \right] d\sigma(\xi)$$

$$+ \frac{(-1)^q}{2} \int_{S^{n-1}} |x \cdot \xi|^q \left[\ln|x \cdot \xi| + i\frac{\pi}{2} \right] d\sigma(\xi)$$

$$= \frac{1 + (-1)^q}{2} \int_{S^{n-1}} |x \cdot \xi|^q \ln|x \cdot \xi| \, d\sigma(\xi)$$

$$+ \frac{((-1)^q - 1)\pi i}{4} \int_{S^{n-1}} |x \cdot \xi|^q \, d\sigma(\xi) \tag{7.5.27}$$

To compute the integrals in the right-most side of (7.5.27) we use Proposition 14.65. First, applying (14.9.15) with $f(t) := |t|^q$ for $t \in \mathbb{R} \setminus \{0\}$ and $v := x$, we obtain

$$\int_{S^{n-1}} |x \cdot \xi|^q \, d\sigma(\xi) = 2\omega_{n-2}|x|^q \int_0^1 s^q (\sqrt{1 - s^2})^{n-3} \, ds \tag{7.5.28}$$

$$= 2\omega_{n-2}|x|^q \int_0^{\frac{\pi}{2}} (\sin\theta)^q (\cos\theta)^{n-2} \, d\theta$$

$$= \omega_{n-2} \frac{\Gamma(\frac{q+1}{2})\Gamma(\frac{n-1}{2})}{\Gamma(\frac{q+n}{2})} |x|^q = \frac{2\pi^{\frac{n-1}{2}}\Gamma(\frac{q+1}{2})}{\Gamma(\frac{q+n}{2})} |x|^q.$$

For the third equality in (7.5.28) we used (14.5.10), while the last one is based on (14.5.6). Second, formula (14.9.15) with $f(t) := |t|^q \ln|t|$ for $t \in \mathbb{R}$ and $v := x$, in concert with (7.5.28) further yields

$$\int_{S^{n-1}} |x \cdot \xi|^q \ln|x \cdot \xi| \, d\sigma(\xi) = 2\omega_{n-2}|x|^q \int_0^1 s^q [\ln|x| + \ln s](\sqrt{1 - s^2})^{n-3} \, ds$$

$$= \frac{2\pi^{\frac{n-1}{2}}\Gamma(\frac{q+1}{2})}{\Gamma(\frac{q+n}{2})} |x|^q \ln|x| + c(n, q)|x|^q, \tag{7.5.29}$$

where

$$c(n, q) := 2\omega_{n-2} \int_0^1 s^q (\ln s)(\sqrt{1 - s^2})^{n-3} \, ds < \infty. \tag{7.5.30}$$

Hence, a combination of (7.5.27), (7.5.28), (7.5.29), and (7.5.24) implies

$$E_q(x) = \begin{cases} \dfrac{(-1)^{\frac{n}{2}+1}\Gamma(\frac{q+1}{2})}{2^{n-1}\pi^{\frac{n+1}{2}}q!\,\Gamma(\frac{q+n}{2})}\,|x|^q\ln|x| - \dfrac{c(n,q)}{(2\pi i)^n q!}\,|x|^q & \text{if } n \text{ is even,} \\[3mm] \dfrac{(-1)^{\frac{n-1}{2}}\Gamma(\frac{q+1}{2})}{2^n\pi^{\frac{n-1}{2}}q!\,\Gamma(\frac{q+n}{2})}\,|x|^q & \text{if } n \text{ is odd,} \end{cases} \tag{7.5.31}$$

for every $x \in \mathbb{R}^n \setminus \{0\}$. That this expression belongs to $\mathcal{S}'(\mathbb{R}^n)$ follows from Exercise 4.108.

Let $F_{\frac{n+q}{2},n}$ be as in (7.5.2). The goal is to prove that

$$E_q - F_{\frac{n+q}{2},n} = \frac{c(n,q)}{(2\pi i)^n q!}|x|^q \quad \text{pointwise in } \mathbb{R}^n \setminus \{0\}. \tag{7.5.32}$$

Once such an identity is established, Theorem 7.28 may be used to conclude that E_q is a fundamental solution for $\Delta^{\frac{n+q}{2}}$ in $\mathbb{R}^n$ (note that if n is even, then q is even and $|x|^q$ is a homogeneous polynomial of order q that is annihilated by $\Delta^{\frac{n+q}{2}}$). Also, it is easy to see that $C(n,q) = \frac{c(n,q)}{(2\pi i)^n q!}$.

With an eye toward proving (7.5.32), suppose first that n is even. Choosing $m := \frac{n+q}{2}$ in (7.5.2), we have $n \leq 2m$, hence

$$F_{\frac{n+q}{2},n}(x) = \frac{(-1)^{\frac{n}{2}+1}}{\pi^{\frac{n}{2}}2^{n+q-1}(\frac{q}{2})!(\frac{n+q}{2}-1)!}\,|x|^q\ln|x|, \quad \forall x \in \mathbb{R}^n \setminus \{0\}. \tag{7.5.33}$$

Applying (14.5.4) with $q/2$ in place of n, and using (14.5.3) (recall that $n+q$ is even) we have

$$\Gamma\left(\frac{q+1}{2}\right) = \frac{q!\,\pi^{1/2}}{2^q(q/2)!} \quad \text{and} \quad \Gamma\left(\frac{q+n}{2}\right) = \left(\frac{q+n}{2}-1\right)!. \tag{7.5.34}$$

Using now (7.5.34) in (7.5.33) and the expression for E_q, it follows that (7.5.32) holds.

If next we assume that n is odd, then

$$F_{\frac{n+q}{2},n}(x) = \frac{(-1)^{\frac{n+q}{2}}\Gamma(-\frac{q}{2})}{\pi^{\frac{n}{2}}2^{n+q}(\frac{n+q}{2}-1)!}\,|x|^q\ln|x|, \quad \forall x \in \mathbb{R}^n \setminus \{0\}. \tag{7.5.35}$$

Applying (14.5.5) and (14.5.3) with $(q+1)/2$ in place of n we see that

$$\Gamma\left(\frac{q+1}{2}\right) = \frac{q!\,\pi^{1/2}}{2^q(q/2)!} \quad \text{and} \quad \Gamma\left(-\frac{q}{2}\right) = \frac{(-1)^{(q+1)/2}2^{q+1}(\frac{q+1}{2})!\pi^{1/2}}{(q+1)!} \tag{7.5.36}$$

which in concert with the second equality in (7.5.34), the expression in (7.5.35), and the formula for E_q, implies $E_q(x) = F_{\frac{n+q}{2},n}(x)$ in $\mathbb{R}^n \setminus \{0\}$. The proof of the lemma is now complete. $\qquad\square$

7.6 Fundamental Solutions for the Helmholtz Operator

For the purpose of the discussion in this section let $k \in (0, \infty)$ be arbitrary, fixed. The Helmholtz operator in $\mathbb{R}^n$, $n \in \mathbb{N}$, is the operator $\Delta + k^2$. The goal for us is to determine a fundamental solution for the Helmholtz operator which also enjoys a specific decay condition at infinity. As it happens, in dimensions $n \geq 2$, this fundamental solution is related to the Hankel function of the first kind which is a null solution for the Bessel differential operator. As such, we first introduce the latter operator.

The ordinary Bessel differential operator with parameter $\lambda \in \mathbb{R}$ acting on a complex-valued function $v \in C^2((0, \infty))$ is defined as

$$(B_\lambda v)(r) := r^2 v''(r) + r v'(r) + (r^2 - \lambda^2) v(r), \qquad r > 0. \tag{7.6.1}$$

To reveal the connection between the Bessel differential operator and the Helmholtz operator $\Delta + k^2$ in $\mathbb{R}^n$, assume that a smooth radial complex-valued function u defined in $\mathbb{R}^n \setminus \{0\}$, has been given. Then there exists $w \in C^\infty((0, \infty))$ such that

$$u(x) = w(|x|) \quad \text{for each} \quad x \in \mathbb{R}^n \setminus \{0\}. \tag{7.6.2}$$

Via differentiation, formula (7.6.2) implies that for each $j \in \{1, \dots, n\}$ and each $x \in \mathbb{R}^n \setminus \{0\}$, we have

$$(\partial_j u)(x) = w'(|x|) \cdot \frac{x_j}{|x|} \quad \text{and}$$

$$(\partial_j^2 u)(x) = w''(|x|) \cdot \frac{x_j^2}{|x|^2} + w'(|x|) \cdot \frac{1}{|x|} - w'(|x|) \cdot \frac{x_j^2}{|x|^3}, \tag{7.6.3}$$

so that

$$[(\Delta + k^2)u](x) = w''(|x|) + w'(|x|) \cdot \frac{n-1}{|x|} + k^2 w(|x|) \tag{7.6.4}$$

for each $x \in \mathbb{R}^n \setminus \{0\}$. To bring to light the Bessel differential operator define

$$v(r) := r^{(n-2)/2} w(r/k) \quad \text{for each} \quad r \in (0, \infty). \tag{7.6.5}$$

Taking derivatives we obtain

$$v'(r) = \frac{n-2}{2} r^{\frac{n-4}{2}} w(r/k) + \frac{1}{k} r^{\frac{n-2}{2}} w'(r/k),$$

$$v''(r) = \frac{(n-2)(n-4)}{4} r^{\frac{n-6}{2}} w(r/k) + \frac{n-2}{k} r^{\frac{n-4}{2}} w'(r/k) + \frac{1}{k^2} r^{\frac{n-2}{2}} w''(r/k), \tag{7.6.6}$$

for each $r \in (0, \infty)$. This may be then used to check that

$$B_{(n-2)/2} v(r) = \frac{1}{k^2} r^{\frac{n+2}{2}} \left[w''(r/k) + \frac{k}{r}(n-1) w'(r/k) + k^2 w(r/k) \right] \tag{7.6.7}$$

for each $r \in (0, \infty)$. From (7.6.7) and (7.6.4) it follows that

$$[(\Delta + k^2)u](x) = (k|x|)^{-(n+2)/2} k^2 (B_{(n-2)/2} v)(k|x|) \tag{7.6.8}$$

for each $x \in \mathbb{R}^n \setminus \{0\}$. Having obtained (7.6.8), it is natural to consider the Hankel function of the first kind with index $\lambda \in \mathbb{R}$, denoted by $H_\lambda^{(1)}(\cdot)$, which is a null solution of B_λ in the sense that

$$B_\lambda H_\lambda^{(1)} = 0 \quad \text{on } (0, \infty), \quad \text{for each } \lambda \in \mathbb{R}. \tag{7.6.9}$$

(The definition of the Hankel function of the first kind and some of the properties this function enjoys are reviewed in Appendix 14.10.) In particular, corresponding to choosing $\nu := H_{(n-2)/2}^{(1)}$, which via (7.6.5) and (7.6.2) and $r = k|x|$, yields a function $u(x)$ of the form $(k|x|)^{-(n-2)/2} H_{(n-2)/2}^{(1)}(k|x|)$, the identities (7.6.9) and (7.6.8) imply that

$$\text{if } n \geq 2 \text{ and } u(x) := |x|^{-(n-2)/2} H_{(n-2)/2}^{(1)}(k|x|) \text{ for}$$
$$\text{each } x \in \mathbb{R}^n \setminus \{0\}, \text{ then } u \in C^\infty(\mathbb{R}^n \setminus \{0\}) \text{ and } u \tag{7.6.10}$$
$$\text{satisfies the equation } (\Delta + k^2)u = 0 \text{ in } \mathbb{R}^n \setminus \{0\}.$$

In fact, as a function defined a.e. in $\mathbb{R}^n$, it turns out that u is locally integrable and a fundamental solution for the Helmholtz operator in $\mathbb{R}^n$, $n \geq 2$, may be obtained by suitably normalizing u from (7.6.10).

Specifically, consider the function

$$\Phi_k(x) := \begin{cases} c_n k^{(n-2)/2} \dfrac{H_{(n-2)/2}^{(1)}(k|x|)}{|x|^{(n-2)/2}} & \text{if } x \in \mathbb{R}^n \setminus \{0\}, \ n \geq 2, \\[3mm] -\dfrac{\mathrm{i}}{2k} e^{\mathrm{i}k|x|} & \text{if } x \in \mathbb{R}, \ n = 1, \end{cases} \tag{7.6.11}$$

where

$$c_n := \frac{1}{4\mathrm{i}(2\pi)^{(n-2)/2}}. \tag{7.6.12}$$

Corresponding to the case $n = 3$, the Hankel function takes a simple form, namely $H_{1/2}^{(1)}(r) = -\mathrm{i}\sqrt{\frac{2}{\pi r}} \, e^{\mathrm{i}r}$ for $r > 0$ (see, e.g., [74, (6.37), p. 231]), hence formula (7.6.11) becomes

$$\Phi_k(x) = -\frac{e^{\mathrm{i}k|x|}}{4\pi|x|}, \quad \forall x \in \mathbb{R}^3 \setminus \{0\}. \tag{7.6.13}$$

To better understand the behavior of Φ_k near the origin let us introduce, for each $\lambda \in \mathbb{R}$, the function Ψ_λ defined at each $r \in (0, \infty)$ by

$$\Psi_\lambda(r) := \begin{cases} \dfrac{\pi}{2\mathrm{i}} H_0^{(1)}(r)(\ln(r))^{-1} & \text{if } \lambda = 0, \\[3mm] \dfrac{\mathrm{i}\pi}{2^\lambda \Gamma(\lambda)} H_\lambda^{(1)}(r)\, r^\lambda & \text{if } \lambda \in (0, \infty), \\[3mm] \dfrac{\mathrm{i}\pi}{e^{-\mathrm{i}\pi\lambda}\, 2^{-\lambda}\Gamma(-\lambda)} H_\lambda^{(1)}(r)\, r^{-\lambda} & \text{if } \lambda \in (-\infty, 0). \end{cases} \tag{7.6.14}$$

By Lemma 14.72 we have

$$\lim_{r \to 0^+} \Psi_\lambda(r) = 1, \quad \forall \lambda \in \mathbb{R}. \tag{7.6.15}$$

For further reference we also note that item *(3)* in Lemma 14.71 combined with the Chain Rule implies

$$\nabla\left[\frac{H_\lambda^{(1)}(k|x|)}{|x|^\lambda}\right] = k^\lambda\,\nabla\left[\frac{H_\lambda^{(1)}(k|x|)}{(k|x|)^\lambda}\right] = -k^\lambda\,\frac{H_{\lambda+1}^{(1)}(k|x|)}{(k|x|)^\lambda}\,k\,\frac{x}{|x|}$$

$$= -k\frac{H_{\lambda+1}^{(1)}(k|x|)}{|x|^{\lambda+1}}\,x \tag{7.6.16}$$

for each $\lambda \in \mathbb{R}$ and every $x \in \mathbb{R}^n \setminus \{0\}$.

Clearly, Ψ_λ is closely related to the function Φ_k and to the fundamental solution for the Laplacian in $\mathbb{R}^n$ for $n \geq 2$. The latter, which we now denote by E_Δ, is given by (when $n \geq 2$)

$$E_\Delta(x) = \begin{cases} \dfrac{-1}{(n-2)\omega_{n-1}}\dfrac{1}{|x|^{n-2}} & \text{if } x \in \mathbb{R}^n \setminus \{0\},\ n \geq 3, \\[2mm] \frac{1}{2\pi}\ln|x| & \text{if } x \in \mathbb{R}^2 \setminus \{0\},\ n = 2, \end{cases} \tag{7.6.17}$$

See (7.1.12). Here is the result linking Φ_k to E_Δ.

Proposition 7.32. *Let $k \in (0, \infty)$ and let $n \in \mathbb{N}$ be such that $n \geq 2$. Then for each $x \in \mathbb{R}^n \setminus \{0\}$ one has*

$$\Phi_k(x) = \begin{cases} \Psi_{(n-2)/2}(k|x|)\,E_\Delta(x) & \text{if } n \geq 3, \\[2mm] \Psi_0(k|x|)\left(\frac{1}{2\pi}\ln k + E_\Delta(x)\right) & \text{if } n = 2. \end{cases} \tag{7.6.18}$$

In addition, if $n \geq 2$ then for each $x \in \mathbb{R}^n \setminus \{0\}$ one has

$$\nabla\Phi_k(x) = \frac{1}{\omega_{n-1}}\,\Psi_{n/2}(k|x|)\frac{x}{|x|^n} = \Psi_{n/2}(k|x|)\nabla E_\Delta(x). \tag{7.6.19}$$

Finally, if $n \geq 2$ and $j, \ell \in \{1, \ldots, n\}$, then

$$(\partial_\ell\partial_j\Phi_k)(x) = -\frac{n}{k\,\omega_{n-1}}\Psi_{(n+2)/2}(k|x|)\frac{x_jx_\ell}{|x|^{n+3}}$$

$$+ \frac{1}{\omega_{n-1}}\Psi_{n/2}(k|x|)\frac{\delta_{j\ell}}{|x|^n}, \tag{7.6.20}$$

for each $x \in \mathbb{R}^n \setminus \{0\}$.

Proof. Assume for now that $n \geq 3$. Then, starting with the expression in (7.6.11), at each point $x \in \mathbb{R}^n \setminus \{0\}$ we may write

$$\Phi_k(x) = \frac{c_n 2^{(n-2)/2}\Gamma(\frac{n-2}{2})}{i\pi} \cdot \frac{H^{(1)}_{(n-2)/2}(k|x|)}{\underbrace{\frac{2^{(n-2)/2}\Gamma(\frac{n-2}{2})}{i\pi}}} (k|x|)^{(n-2)/2} |x|^{2-n}. \tag{7.6.21}$$

Upon recalling (7.6.12), the fact that $\Gamma(\frac{n}{2}) = \frac{n-2}{2}\Gamma(\frac{n-2}{2})$ (cf. (14.5.2)), and the expression for ω_{n-1} from (14.5.6), a direct computation gives

$$\frac{c_n 2^{(n-2)/2}\Gamma(\frac{n-2}{2})}{i\pi} = -\frac{1}{(n-2)\omega_{n-1}}. \tag{7.6.22}$$

Formula (7.6.18) for $n \geq 3$ now follows by combining (7.6.21), (7.6.22), the definition in (7.6.14) with $\lambda = (n-2)/2 > 0$ and $r = k|x|$, and (7.6.17).

The same circle of ideas gives (7.6.18) in the two-dimensional case. Specifically, if we assume $n = 2$, then starting with (7.6.11) and (7.6.12), then using (7.6.14) with $\lambda = 0$ and $r = k|x|$, and then invoking (7.6.17) (for $n = 2$), we may write

$$\Phi_k(x) = \frac{1}{4i}H^{(1)}_0(k|x|) = \frac{1}{2\pi}\Psi_0(k|x|)\ln(k|x|)$$

$$= \Psi_0(k|x|)\Big(\frac{1}{2\pi}\ln k + E_\Delta(x)\Big), \quad \forall x \in \mathbb{R}^2 \setminus \{0\}, \tag{7.6.23}$$

as wanted.

Next, we prove the statement regarding $\nabla\Phi_k$. Suppose $n \geq 2$. Using formula (7.6.16) with $\lambda = (n-2)/2$ we obtain

$$\nabla\Phi_k(x) = c_n k^{(n-2)/2}\nabla\Big[\frac{H^{(1)}_{(n-2)/2}(k|x|)}{|x|^{(n-2)/2}}\Big]$$

$$= -c_n k^{n/2}\frac{H^{(1)}_{n/2}(k|x|)}{|x|^{n/2}}\,x, \quad \forall x \in \mathbb{R}^n \setminus \{0\}. \tag{7.6.24}$$

Furthermore, for each $x \in \mathbb{R}^n \setminus \{0\}$ we have

$$-c_n k^{n/2}\frac{H^{(1)}_{n/2}(k|x|)}{|x|^{n/2}} = -\frac{c_n 2^{n/2}\Gamma(\frac{n}{2})}{i\pi} \cdot \frac{H^{(1)}_{n/2}(k|x|)}{\frac{2^{n/2}\Gamma(\frac{n}{2})}{i\pi}} (k|x|)^{n/2}|x|^{-n}$$

$$= \frac{1}{\omega_{n-1}}\Psi_{n/2}(k|x|)|x|^{-n} \tag{7.6.25}$$

where in the last equality we have used (7.6.12), (7.6.14) with $\lambda = n/2$, and the expression for ω_{n-1} from (14.5.6). In concert, (7.6.24) and (7.6.25) give the first equality in (7.6.19). The second equality in (7.6.19) is obtained by a direct computation which makes use of (7.6.17).

Moving on to the proof of (7.6.20), fix $j, \ell \in \{1, \ldots, n\}$ and $x \in \mathbb{R}^n \setminus \{0\}$. Then from (7.6.24) we know that

$$(\partial_j \Phi_k)(x) = -c_n k^n \frac{H^{(1)}_{n/2}(k|x|)}{(k|x|)^{n/2}} x_j. \tag{7.6.26}$$

Apply ∂_ℓ. Using the product rule, the Chain Rule, and item *(3)* in Lemma 14.71 (with $\lambda = n/2$) we further obtain

$$(\partial_\ell \partial_j \Phi_k)(x) = c_n k^n \frac{H^{(1)}_{(n+2)/2}(k|x|)}{(k|x|)^{(n+2)/2}} k \frac{x_\ell}{|x|} x_j - c_n k^n \frac{H^{(1)}_{n/2}(k|x|)}{(k|x|)^{n/2}} \delta_{j\ell}$$

$$= \frac{c_n}{k} H^{(1)}_{(n+2)/2}(k|x|)(k|x|)^{(n+2)/2} \frac{x_j x_\ell}{|x|^{n+3}}$$

$$+ \frac{1}{\omega_{n-1}} \Psi_{n/2}(k|x|) \frac{\delta_{j\ell}}{|x|^{-n}} \tag{7.6.27}$$

where the second equality in (7.6.27) uses (7.6.25). Moreover, since

$$\frac{c_n}{k} \cdot \frac{2^{(n+2)/2} \Gamma(\frac{n+2}{2})}{i\pi} = \frac{2^{(n+2)/2} \Gamma(\frac{n+2}{2})}{k \, 4i \, 2^{(n-2)/2} \pi^{(n-2)/2} \, i\pi}$$

$$= -\frac{\frac{n}{2} \Gamma(\frac{n}{2})}{k \pi^{n/2}} = -\frac{n}{k \omega_{n-1}}, \tag{7.6.28}$$

by recalling (7.6.14) (with $\lambda = (n+2)/2$) we may write

$$\frac{c_n}{k} H^{(1)}_{(n+2)/2}(k|x|)(k|x|)^{(n+2)/2} = -\frac{n}{k \omega_{n-1}} \Psi_{(n+2)/2}(k|x|). \tag{7.6.29}$$

At this stage, (7.6.20) follows by combining (7.6.27) and (7.6.29). This finishes the proof of Proposition 7.32. $\qquad \square$

In the next theorem we show that the function Φ_k is a fundamental solution for the Helmholtz operator $\Delta + k^2$.

Theorem 7.33. *Suppose $n \in \mathbb{N}$ and fix $k \in (0, \infty)$. Then the function Φ_k defined in (7.6.11)–(7.6.12) satisfies $\Phi_k \in L^1_{loc}(\mathbb{R}^n) \cap \mathcal{S}'(\mathbb{R}^n)$ and is a fundamental solution for the Helmholtz operator $\Delta + k^2$ in $\mathbb{R}^n$. Moreover,*

$$\{u \in \mathcal{S}'(\mathbb{R}^n) : (\Delta + k^2)u = \delta \ in \ \mathcal{S}'(\mathbb{R}^n)\} \tag{7.6.30}$$

$$= \{\Phi_k + f : f \in C^\infty(\mathbb{R}^n), \ (\Delta + k^2)f = 0 \ in \ \mathbb{R}^n\}.$$

Corresponding to the case $n = 1$,

$$\{u \in \mathcal{S}'(\mathbb{R}) : u'' + ku = \delta \ in \ \mathcal{S}'(\mathbb{R})\} \tag{7.6.31}$$

$$= \left\{ -\frac{i}{2k} e^{ik|x|} + c_1 e^{ikx} + c_2 e^{-ikx} : c_1, c_2 \in \mathbb{C} \right\}.$$

Proof. Consider first the case $n \geq 2$. The fact that $\Phi_k \in L^1_{loc}(\mathbb{R}^n)$ for $n \geq 2$ is a consequence of (7.6.18) in Proposition 7.32, formula (7.6.15), and the membership $E_\Delta \in L^1_{loc}(\mathbb{R}^n)$. Also, item *(9)* in Lemma 14.71 implies

$$\Phi_k(x) = c_n k^{(n-3)/2} \sqrt{\frac{2}{\pi}} e^{i(k|x|-(n-1)\pi/4)} |x|^{-(n-1)/2} + O(|x|^{-(n+1)/2}) \tag{7.6.32}$$

$$\text{as } |x| \to \infty,$$

which further guarantees the existence of some constant $C \in (0, \infty)$ such that $|\Phi_k(x)| \leq C|x|^{-(n-1)/2}$ for every $x \in \mathbb{R}^n \setminus B(0, 1)$. Hence, Example 4.4 applies (condition (4.1.4) is satisfied for any $m > (n+1)/2$) and gives that $\Phi_k \in \mathcal{S}'(\mathbb{R}^n)$.

Next we take up the task of proving that Φ_k in (7.6.11) is a fundamental solution for $\Delta + k^2$ in $\mathbb{R}^n$, $n \geq 2$. By (7.6.10) and (7.6.11) we have

$$\Phi_k \in C^\infty(\mathbb{R}^n \setminus \{0\}) \quad \text{and} \quad (\Delta + k^2)\Phi_k = 0 \quad \text{in } \mathbb{R}^n \setminus \{0\}. \tag{7.6.33}$$

Now let $\varphi \in C_0^\infty(\mathbb{R}^n)$ be arbitrary. To conclude that Φ_k is a fundamental solution for $\Delta + k^2$, since Φ_k is locally integrable, it suffices to establish that

$$\int_{\mathbb{R}^n} \Phi_k(x)[(\Delta + k^2)\varphi](x)\, dx = \varphi(0), \quad \forall\, \varphi \in C_0^\infty(\mathbb{R}^n). \tag{7.6.34}$$

Applying Lebesgue's Dominated Convergence Theorem, then using integrations by parts twice (cf. Theorem 14.60), and then (7.6.33), we may write

$$\int_{\mathbb{R}^n} \Phi_k(x)[(\Delta + k^2)\varphi](x)\, dx = \lim_{\varepsilon \to 0^+} \int_{\mathbb{R}^n \setminus B(0,\varepsilon)} \Phi_k(x)[(\Delta + k^2)\varphi](x)\, dx$$

$$= \lim_{\varepsilon \to 0^+} \left\{ \int_{\mathbb{R}^n \setminus B(0,\varepsilon)} [(\Delta + k^2)\Phi_k](x)\varphi(x)\, dx + I_\varepsilon + II_\varepsilon \right\}$$

$$= \lim_{\varepsilon \to 0^+} I_\varepsilon + \lim_{\varepsilon \to 0^+} II_\varepsilon, \tag{7.6.35}$$

where we have set (recall the definition of the normal derivative from (7.1.14))

$$I_\varepsilon := -\int_{\partial B(0,\varepsilon)} \Phi_k(x) \frac{\partial \varphi}{\partial \nu}(x)\, d\sigma(x), \tag{7.6.36}$$

$$II_\varepsilon := \int_{\partial B(0,\varepsilon)} \frac{\partial \Phi_k}{\partial \nu}(x)\varphi(x)\, d\sigma(x). \tag{7.6.37}$$

Above, ν denotes the outward unit normal to $B(0, \varepsilon)$, that is, $\nu(x) = \frac{x}{\varepsilon}$ for each $x \in \partial B(0, \varepsilon)$, and σ denotes the surface measure on $\partial B(0, \varepsilon)$.

We impose for now the restriction that $n \geq 3$. Note that for each $\varepsilon \in (0, 1)$, formula (7.6.18) permits us to estimate

$$|I_\varepsilon| \leq \int_{\partial B(0,\varepsilon)} |\Phi_k(x)| \left| \frac{\partial \varphi}{\partial \nu}(x) \right| d\sigma(x)$$

$$\leq \|\nabla \varphi\|_{L^\infty(B(0,1))} |\Psi_{(n-2)/2}(k\varepsilon)| \int_{\partial B(0,\varepsilon)} |E_\Delta(x)| \, d\sigma(x)$$

$$= \|\nabla \varphi\|_{L^\infty(B(0,1))} |\Psi_{(n-2)/2}(k\varepsilon)| \frac{\varepsilon}{n-2}. \tag{7.6.38}$$

Taking the limit as $\varepsilon \to 0^+$ and invoking (7.6.15), we obtain

$$\lim_{\varepsilon \to 0^+} I_\varepsilon = 0. \tag{7.6.39}$$

We claim that (7.6.39) is also valid if $n = 2$. Indeed, the estimate for I_ε from (7.6.38) now becomes

$$|I_\varepsilon| \leq \|\nabla \varphi\|_{L^\infty(B(0,1))} |\Psi_0(k\varepsilon)| \int_{\partial B(0,\varepsilon)} \left| \frac{1}{2\pi} \ln k + E_\Delta(x) \right| d\sigma(x)$$

$$= \|\nabla \varphi\|_{L^\infty(B(0,1))} |\Psi_0(k\varepsilon)| |\ln(k\varepsilon)| \varepsilon. \tag{7.6.40}$$

Taking the limit as $\varepsilon \to 0^+$ and invoking (7.6.15) yields (7.6.39). To summarize, we have proved (7.6.39) for $n \geq 2$.

To handle II_ε use (7.6.19) to write, for each $x \in \partial B(0, \varepsilon)$,

$$\frac{\partial \Phi_k}{\partial \nu}(x) = \nabla \Phi_k(x) \cdot \frac{x}{|x|} = \frac{1}{\omega_{n-1}} \Psi_{n/2}(k\varepsilon) \varepsilon^{-(n-1)}. \tag{7.6.41}$$

When used back in (7.6.37), this allows us to recast II_ε as

$$II_\varepsilon = \Psi_{n/2}(k\varepsilon) \cdot \frac{1}{\omega_{n-1}\varepsilon^{n-1}} \int_{\partial B(0,\varepsilon)} \varphi(x) \, d\sigma(x). \tag{7.6.42}$$

Since $\int_{\partial B(0,\varepsilon)} 1 \, d\sigma = \omega_{n-1}\varepsilon^{n-1}$, we may write

$$\frac{1}{\omega_{n-1}\varepsilon^{n-1}} \int_{\partial B(0,\varepsilon)} \varphi(x) \, d\sigma(x)$$

$$= \frac{1}{\omega_{n-1}\varepsilon^{n-1}} \int_{\partial B(0,\varepsilon)} (\varphi(x) - \varphi(0)) \, d\sigma(x) + \varphi(0). \tag{7.6.43}$$

By the Mean Value Theorem we have $|\varphi(x) - \varphi(0)| \leq \varepsilon \|\nabla \varphi\|_{L^\infty(\mathbb{R}^n)}$ for each x in $\partial B(0, \varepsilon)$, thus the integral in the right-hand side of (7.6.43) converges to zero as $\varepsilon \to 0^+$. This proves that

$$\lim_{\varepsilon \to 0^+} \frac{1}{\omega_{n-1}\varepsilon^{n-1}} \int_{\partial B(0,\varepsilon)} \varphi(x) \, d\sigma(x) = \varphi(0). \tag{7.6.44}$$

In concert, (7.6.42), (7.6.44), and (7.6.15) (applied with $\lambda = n/2$) imply

$$\lim_{\varepsilon \to 0^+} II_\varepsilon = \varphi(0). \tag{7.6.45}$$

At this point formula (7.6.34) follows from (7.6.35), (7.6.39), and (7.6.45). This completes the proof of the fact that Φ_k is a fundamental solution for $\Delta + k^2$ in $\mathbb{R}^n$ when $n \geq 2$.

To treat the case when $n = 1$, observe first that

$$\Phi_k(x) = -\frac{i}{2k} e^{ik|x|} = -\frac{i}{2k} e^{ikx} H(x) - \frac{i}{2k} e^{-ikx} H^\vee(x), \tag{7.6.46}$$

for every $x \in \mathbb{R} \setminus \{0\}$. Clearly $\Phi_k \in L^1_{loc}(\mathbb{R})$ and $e^{\pm ikx} \in \mathcal{L}(\mathbb{R})$. Also, Exercise 4.119 guarantees that $H \in \mathcal{S}'(\mathbb{R})$. Granted these properties, item *(b)* in Theorem 4.14 applies and yields $\Phi_k \in \mathcal{S}'(\mathbb{R})$. In addition,

$$\Phi_k(x) + \frac{i}{2k} e^{-ikx} = -\frac{i}{2k} e^{ikx} H(x) + \frac{i}{2k} e^{-ikx} H(x), \tag{7.6.47}$$

for every $x \in \mathbb{R} \setminus \{0\}$. With (7.6.47) in hand, (7.6.31) becomes a consequence of Exercise 6.27.

There remains to check (7.6.30). The right-to-left inclusion is clear from what we have proved already. Also, if $u \in \mathcal{S}'(\mathbb{R}^n)$ is a fundamental solution for $\Delta + k^2$ in $\mathbb{R}^n$, then $(\Delta + k^2)(u - \Phi_k) = 0$ in $\mathcal{D}'(\mathbb{R}^n)$. Observe that the operator $\Delta + k^2$ is elliptic (cf. Definition 6.13), hence hypoelliptic by Theorem 6.15. Consequently, by Remark 6.7, we have $f := u - \Phi_k \in C^\infty(\mathbb{R}^n)$, which in turn implies $(\Delta + k^2)f = 0$ pointwise in $\mathbb{R}^n$. With this the left-to-right inclusion in (7.6.30) also follows. $\square$

Next, we concern ourselves with the behavior of Φ_k at infinity. In this regard, we shall make use of the asymptotic results for Hankel functions from §14.10 in order to establish the following proposition.

Proposition 7.34. *Fix $k \in (0, \infty)$ along with $n \in \mathbb{N}$, $n \geq 2$, and set*

$$b_{n,k} := \frac{k^{(n-3)/2}}{4i(2\pi)^{(n-2)/2}} \left(\frac{2}{\pi}\right)^{1/2} e^{-i\pi(n-1)/4}. \tag{7.6.48}$$

Also, throughout abbreviate

$$\widehat{x} := \frac{x}{|x|} \text{ for each } x \in \mathbb{R}^n \setminus \{0\}. \tag{7.6.49}$$

Then for each multi-index $\alpha \in \mathbb{N}_0^n$ one has

$$(\partial^\alpha \Phi_k)(x - y) = b_{n,k} \frac{e^{ik|x|} e^{-ik\langle y, \widehat{x} \rangle}}{|x|^{(n-1)/2}} (ik\widehat{x})^\alpha + O(|x|^{-(n+1)/2}) \tag{7.6.50}$$

$$= (ik\widehat{x})^\alpha \Phi_k(x - y) + O(|x|^{-(n+1)/2}) \quad as \ |x| \to \infty,$$

uniformly for y in compact subsets of $\mathbb{R}^n$, and

$$(\partial^\alpha \Phi_k)(x-y) = b_{n,k} \frac{e^{ik|y|}e^{-ik\langle x,\widehat{y}\rangle}}{|y|^{(n-1)/2}}(-ik\widehat{y})^\alpha + O(|y|^{-(n+1)/2}) \tag{7.6.51}$$

$$= (-ik\widehat{y})^\alpha \Phi_k(x-y) + O(|y|^{-(n+1)/2}) \quad as \ |y| \to \infty,$$

uniformly for x in compact subsets of $\mathbb{R}^n$.

In particular, for each multi-index $\alpha \in \mathbb{N}_0^n$ *one has*

$$(\partial^\alpha \Phi_k)(x-y) = O(|x|^{-(n-1)/2}) \quad as \ |x| \to \infty,$$

uniformly for y in compact subsets of $\mathbb{R}^n$, $\tag{7.6.52}$

and

$$(\partial^\alpha \Phi_k)(x-y) = O(|y|^{-(n-1)/2}) \quad as \ |y| \to \infty,$$

uniformly for x in compact subsets of $\mathbb{R}^n$. $\tag{7.6.53}$

Proof. Fix $\alpha \in \mathbb{N}_0^n$ arbitrary. As a first step we shall prove that

$$(\partial^\alpha \Phi_k)(x) = b_{n,k} \frac{e^{ik|x|}}{|x|^{(n-1)/2}}(ik\widehat{x})^\alpha + O(|x|^{-(n+1)/2}) \quad as \ |x| \to \infty. \tag{7.6.54}$$

To get started, combine (7.6.12) and item *(9)* in Lemma 14.71 to write

$$\Phi_k(x) = \frac{k^{(n-2)/2}}{4i(2\pi)^{(n-2)/2}}|x|^{-(n-2)/2}\left[\left(\frac{2}{\pi k|x|}\right)^{1/2}e^{i(|x|-(n-2)\pi/4-\pi/4)} + O(|x|^{-3/2})\right]$$

$$= \frac{k^{(n-3)/2}}{4i(2\pi)^{(n-2)/2}}\left(\frac{2}{\pi}\right)^{1/2}e^{-i\pi(n-1)/4}\frac{e^{ik|x|}}{|x|^{(n-1)/2}} + O(|x|^{-(n+1)/2})$$

$$= b_{n,k}\frac{e^{ik|x|}}{|x|^{(n-1)/2}} + O(|x|^{-(n+1)/2}) \quad as \ |x| \to \infty, \tag{7.6.55}$$

which proves (7.6.54) for $\alpha = (0,\dots,0)$.

Next, from (7.6.12) and Leibniz' formula we have

$$(\partial^\alpha \Phi_k)(x) = c_n k^{(n-2)/2}\, \partial^\alpha[H^{(1)}_{(n-2)/2}(k|x|)]|x|^{-(n-2)/2} \tag{7.6.56}$$

$$+ c_n k^{(n-2)/2}\sum_{\beta+\gamma=\alpha,\,|\gamma|>0}\frac{\alpha!}{\beta!\gamma!}\,\partial^\beta(H^{(1)}_{(n-2)/2}(k|x|))\partial^\gamma(|x|^{-(n-2)/2}).$$

By invoking (14.10.17), the first term in the right-hand side of (7.6.56) becomes

$$c_n k^{(n-2)/2} \partial^\alpha [H^{(1)}_{(n-2)/2}(k|x|)]|x|^{-(n-2)/2}$$

$$= c_n k^{(n-2)/2} \Big(H^{(1)}_{(n-2)/2-|\alpha|}(k|x|)(\widehat{kx})^\alpha + O(|x|^{-3/2}) \Big) |x|^{-(n-2)/2}$$

$$= c_n k^{(n-2)/2} \frac{H^{(1)}_{(n-2)/2-|\alpha|}(k|x|)(\widehat{kx})^\alpha}{|x|^{(n-2)/2}} + O(|x|^{-(n+1)/2})$$

$$= c_n k^{(n-2)/2} \frac{\Big((\frac{2}{\pi k|x|})^{1/2} e^{i(k|x|-(n-2)\pi/4+|\alpha|\pi/2-\pi/4)} + O(|x|^{-3/2}) \Big)(\widehat{kx})^\alpha}{|x|^{(n-2)/2}}$$

$$+ O(|x|^{-(n+1)/2})$$

$$= b_{n,k} \frac{e^{ik|x|}}{|x|^{(n-1)/2}} (\widehat{ikx})^\alpha + O(|x|^{-(n+1)/2}) \qquad \text{as } |x| \to \infty, \qquad (7.6.57)$$

where for the third equality in (7.6.57) we used item *(9)* in Lemma 14.71. Regarding the second term in the right-hand side of (7.6.56), it is immediate that $\partial^\gamma(|x|^{-(n-2)/2}) = O(|x|^{-(n-2)/2-|\gamma|})$ as $|x| \to \infty$ for $\gamma \in \mathbb{N}_0^n$, while from (14.10.10) it follows that $\partial^\beta(H^{(1)}_{(n-2)/2}(k|x|)) = O(|x|^{-1/2})$ as $|x| \to \infty$. As such, we may conclude that the sum in (7.6.56) is $O(|x|^{-(n+1)/2})$ as $|x| \to \infty$. This, together with (7.6.57), finishes the proof of (7.6.54).

With an eye on (7.6.50), let K be a compact subset of $\mathbb{R}^n$ and assume for now that $y \in K$ and that $|x|$ is sufficiently large. In this case $|x - y|$ is proportional to $|x|$, hence $O(|x - y|^{-(n+1)/2}) = O(|x|^{-(n+1)/2})$ as $|x| \to \infty$, uniformly for $y \in K$. Moreover, the Mean Value Theorem gives

$$\frac{x-y}{|x-y|} = \frac{x}{|x|} + O\Big(\frac{1}{|x|}\Big), \qquad \text{as } |x| \to \infty, \qquad (7.6.58)$$

uniformly for $y \in K$. Thus, on the one hand,

$$\Big(\frac{x-y}{|x-y|}\Big)^\alpha = \Big(\frac{x}{|x|}\Big)^\alpha + O\Big(\frac{1}{|x|}\Big) \qquad \text{as } |x| \to \infty, \qquad (7.6.59)$$

uniformly for $y \in K$. On the other hand, as a consequence of (7.6.54) we obtain

$$(\partial^\alpha \Phi_k)(x - y) = b_{n,k} \frac{e^{ik|x-y|}}{|x - y|^{(n-1)/2}} (ik\widehat{(x - y)})^\alpha$$

$$+ O(|x - y|^{-(n+1)/2}) \qquad \text{as } |x - y| \to \infty. \qquad (7.6.60)$$

Next we claim that

$$\frac{e^{ik|x-y|}}{|x - y|^{(n-1)/2}} - \frac{e^{ik|x|}e^{-ik\langle y,\widehat{x}\rangle}}{|x|^{(n-1)/2}} = O(|x|^{-(n+1)/2}) \qquad (7.6.61)$$

$$\text{uniformly for } y \in K, \text{ as } |x| \to \infty.$$

Indeed

$$\frac{e^{ik|x-y|}}{|x-y|^{(n-1)/2}} - \frac{e^{ik|x|}e^{-ik\langle y,\widehat{x}\rangle}}{|x|^{(n-1)/2}} \tag{7.6.62}$$

$$= e^{ik|x-y|}\left\{\frac{1}{|x-y|^{(n-1)/2}} - \frac{1}{|x|^{(n-1)/2}}\right\}$$

$$+ \frac{e^{ik|x-y|}}{|x|^{(n-1)/2}}\{1 - e^{-ik|x-y|+ik|x|-ik\langle y,\widehat{x}\rangle}\} := I + II.$$

Based on the Mean Value Theorem we see that

$$I = O(|x|^{-(n+1)/2}) \quad \text{as} \quad |x| \to \infty, \quad \text{uniformly for } y \in K. \tag{7.6.63}$$

Consequently, (7.6.61) will follow once we show that

$$\left|1 - e^{-ik|x-y|+ik|x|-ik\langle y,\widehat{x}\rangle}\right| = O(|x|^{-1})$$

$$\text{as} \quad |x| \to \infty, \quad \text{uniformly for } y \in K. \tag{7.6.64}$$

With this goal in mind, write (for x large)

$$|x-y| = (\langle x-y, x-y\rangle)^{1/2} = (|x|^2 - 2\langle x,y\rangle + |y|^2)^{1/2}$$

$$= |x|\left(1 - 2\frac{\langle \widehat{x}, y\rangle}{|x|} + \frac{|y|^2}{|x|^2}\right)^{1/2}. \tag{7.6.65}$$

Recalling that the Taylor series expansion of the function $t \mapsto (1+t)^{1/2}$ around zero is $(1+t)^{1/2} = 1 + \frac{t}{2} + O(t^2)$, from (7.6.65) we further deduce that

$$|x-y| = |x|\left(1 + \frac{1}{2}\left(\frac{|y|^2}{|x|^2} - 2\frac{\langle \widehat{x}, y\rangle}{|x|}\right)\right) + O\left(\left|\frac{|y|^2}{|x|^2} - 2\frac{\langle \widehat{x}, y\rangle}{|x|}\right|^2\right)$$

$$= |x|\left(1 - \frac{\langle \widehat{x}, y\rangle}{|x|} + O(|x|^{-2})\right) \tag{7.6.66}$$

$$= |x| - \langle \widehat{x}, y\rangle + O(|x|^{-1}) \quad \text{as} \quad |x| \to \infty, \quad \text{uniformly for } y \in K.$$

Now (7.6.64) follows from (7.6.66) and the fact that $\left|1 - e^{ia}\right| \le 2|a|$ for every $a \in \mathbb{R}$. With this, the proof of (7.6.61) is finished.

A combination of (7.6.60), (7.6.59), and (7.6.61) yields

$$(\partial^\alpha \Phi_k)(x-y) = b_{n,k} \frac{e^{ik|x-y|}}{|x-y|^{(n-1)/2}} (ik\widehat{(x-y)})^\alpha + O(|x-y|^{-(n+1)/2})$$

$$= b_{n,k} \left(\frac{e^{ik|x|} e^{-ik\langle y, \widehat{x} \rangle}}{|x|^{(n-1)/2}} + O(|x|^{-(n+1)/2}) \right) \left((ik\widehat{x})^\alpha + O\left(\frac{1}{|x|}\right) \right)$$

$$+ O(|x|^{-(n+1)/2})$$

$$= b_{n,k} \frac{e^{ik|x|} e^{-ik\langle y, \widehat{x} \rangle}}{|x|^{(n-1)/2}} (ik\widehat{x})^\alpha + O(|x|^{-(n+1)/2})$$

$$\text{as } |x| \to \infty, \text{ uniformly for } y \in K, \tag{7.6.67}$$

and proves the first equality in (7.6.50). In particular, corresponding to $|\alpha| = 0$, we have

$$\Phi_k(x-y) = b_{n,k} \frac{e^{ik|x|} e^{-ik\langle y, \widehat{x} \rangle}}{|x|^{(n-1)/2}} + O(|x|^{-(n+1)/2})$$

$$\text{as } |x| \to \infty, \text{ uniformly for } y \in K. \tag{7.6.68}$$

Making use of (7.6.68) back in (7.6.67) implies

$$(\partial^\alpha \Phi_k)(x-y) = (\Phi_k(x-y) + O(|x|^{-(n+1)/2}))(ik\widehat{x})^\alpha + O(|x|^{-(n+1)/2})$$

$$= (ik\widehat{x})^\alpha \Phi_k(x-y) + O(|x|^{-(n+1)/2})$$

$$\text{as } |x| \to \infty, \text{ uniformly for } y \in K. \tag{7.6.69}$$

This finishes the proof of (7.6.50).

Finally, upon noting that Φ_k is even, it follows that

$$(\partial^\alpha \Phi_k)(x-y) = (-1)^{|\alpha|} (\partial^\alpha \Phi_k)(y-x), \qquad x \neq y, \tag{7.6.70}$$

so (7.6.51) becomes a consequence of (7.6.70) and (7.6.50).   □

Exercise 7.35. Prove that Φ_k satisfies Sommerfeld's radiation condition

$$iku(x) - \sum_{j=1}^{n} \frac{x_j}{|x|} (\partial_j u)(x) = o(|x|^{-(n-1)/2}) \text{ as } |x| \to \infty. \tag{7.6.71}$$

Exercise 7.36. Consider $k \in (0, \infty)$ along with some $n \in \mathbb{N}$, $n \geq 2$. Show that for every multi-index $\alpha \in \mathbb{N}_0^n$ one has

$$\langle \widehat{x}, [\nabla(\partial^\alpha \Phi_k)](x-y) \rangle - ik(\partial^\alpha \Phi_k)(x-y)$$

$$= O(|x|^{-(n+1)/2}) \quad \text{as } |x| \to \infty, \tag{7.6.72}$$

uniformly for y in compact subsets of $\mathbb{R}^n$. As a consequence, for each $y \in \mathbb{R}^n$ fixed, conclude that

$(\partial^\alpha \Phi_k)(\cdot - y)$ satisfies Sommerfeld's radiation condition (7.6.71). $\hspace{1em}$ (7.6.73)

Exercise 7.37. Fix $k \in (0, \infty)$ together with $n \in \mathbb{N}$, $n \geq 2$.

(1) Show that for any two multi-indices $\alpha, \beta \in \mathbb{N}_0^n$ one has

$$(\partial^{\alpha+\beta} \Phi_k)(x - y) = (\mathrm{i} k \widehat{x}\,)^\alpha (\partial^\beta \Phi_k)(x - y) + O(|x|^{-(n+1)/2})$$

as $|x| \to \infty$, uniformly for y in compact subsets of $\mathbb{R}^n$,

$$(7.6.74)$$

and

$$(\partial^{\alpha+\beta} \Phi_k)(x - y) = (-\mathrm{i} k \widehat{y}\,)^\alpha (\partial^\beta \Phi_k)(x - y) + O(|y|^{-(n+1)/2})$$

as $|y| \to \infty$, uniformly for x in compact subsets of $\mathbb{R}^n$.

$$(7.6.75)$$

(2) Use the results in part *(1)* to show that for any multi-index $\alpha \in \mathbb{N}_0^n$ and any indexes $j, \ell \in \{1, \ldots, n\}$,

$$\widehat{x}_j (\partial_\ell \partial^\alpha \Phi_k)(x - y) - \widehat{x}_\ell (\partial_j \partial^\alpha \Phi_k)(x - y) = O(|x|^{-(n+1)/2})$$

as $|x| \to \infty$, uniformly for y in compact subsets of $\mathbb{R}^n$,

$$(7.6.76)$$

and

$$\widehat{y}_j (\partial_\ell \partial^\alpha \Phi_k)(x - y) - \widehat{y}_\ell (\partial_j \partial^\alpha \Phi_k)(x - y) = O(|y|^{-(n+1)/2})$$

as $|y| \to \infty$, uniformly for x in compact subsets of $\mathbb{R}^n$.

$$(7.6.77)$$

We close this section by isolating a result which is useful in the computation of fundamental solutions for the perturbed Dirac operator and its iterations (later on, in §7.10–§7.11).

Proposition 7.38. *Let $\lambda \in (-\infty, n/2)$ and consider the function*

$$F_\lambda(x) := \frac{H_\lambda^{(1)}(k|x|)}{|x|^\lambda}, \qquad \forall\, x \in \mathbb{R}^n \setminus \{0\}.$$

$$(7.6.78)$$

Then F_λ belongs to $L_{loc}^1(\mathbb{R}^n)$, hence it defines a distribution of function type. Also, if $\lambda < (n - 1)/2$ then for each $j \in \{1, \ldots, n\}$ it follows that $F_{\lambda+1}(x)\, x_j$ belongs to $L_{loc}^1(\mathbb{R}^n)$ and

$$\partial_j F_\lambda = -k F_{\lambda+1}(x)\, x_j \quad \text{in } \mathcal{D}'(\mathbb{R}^n).$$

$$(7.6.79)$$

Proof. First work under the assumption that $\lambda \in (-\infty, n/2)$. Recall the function Ψ_λ from (7.6.14). Then for each $x \in \mathbb{R}^n \setminus \{0\}$ we have

$$F_\lambda(x) = \begin{cases} \dfrac{2i}{\pi}\, \Psi_0(k|x|)\ln(k|x|) & \text{if } \lambda = 0, \\[2ex] \dfrac{2^\lambda \Gamma(\lambda)}{i\pi\, k^\lambda}\, \Psi_\lambda(k|x|)\, \dfrac{1}{|x|^{2\lambda}} & \text{if } \lambda \in (0, n/2), \\[2ex] \dfrac{e^{-i\pi\lambda}\, 2^{-\lambda}\Gamma(-\lambda)\, k^\lambda}{i\pi}\, \Psi_\lambda(k|x|) & \text{if } \lambda \in (-\infty, 0), \end{cases} \tag{7.6.80}$$

In combination with (7.6.15) this implies that $F_\lambda \in L^1_{loc}(\mathbb{R}^n)$.

Next, suppose $\lambda \in (-\infty, (n-1)/2)$ and fix $j \in \{1, \dots, n\}$. From (7.6.16) we know that when differentiating pointwise we have

$$(\partial_j F_\lambda)(x) = -k F_{\lambda+1}(x)\, x_j \qquad \forall\, x \in \mathbb{R}^n \setminus \{0\}. \tag{7.6.81}$$

Pick an arbitrary $\varphi \in C_0^\infty(\mathbb{R}^n)$. Since $F_\lambda \in L^1_{loc}(\mathbb{R}^n)$ and φ has compact support we may use Lebesgue's Dominated Convergence Theorem, integration by parts, and (7.6.81), to write

$$\begin{aligned} \langle \partial_j F_\lambda, \varphi \rangle &= -\langle F_\lambda, \partial_j \varphi \rangle = -\int_{\mathbb{R}^n} F_\lambda(x) \partial_j \varphi(x)\, dx \\[1ex] &= -\lim_{\varepsilon \to 0^+} \int_{|x| \geq \varepsilon} F_\lambda(x) \partial_j \varphi(x)\, dx \\[1ex] &= \lim_{\varepsilon \to 0^+} \Big[-k \int_{|x| \geq \varepsilon} F_{\lambda+1}(x)\, x_j\, \varphi(x)\, dx - \int_{|x| = \varepsilon} F_\lambda(x) \frac{x_j}{\varepsilon} \varphi(x)\, d\sigma(x) \Big]. \end{aligned} \tag{7.6.82}$$

The reasoning that has produced (7.6.80) also gives that for each $x \in \mathbb{R}^n \setminus \{0\}$ we have

$$F_{\lambda+1}(x) x_j = \begin{cases} C\, \Psi_0(k|x|)\ln(k|x|)\, x_j & \text{if } \lambda = -1, \\[1.5ex] C\, \Psi_{\lambda+1}(k|x|)\, \dfrac{x_j}{|x|^{2(\lambda+1)}} & \text{if } \lambda \in (-1, (n-1)/2), \\[1.5ex] C\, \Psi_{\lambda+1}(k|x|)\, x_j & \text{if } \lambda \in (-\infty, -1), \end{cases} \tag{7.6.83}$$

where C denotes in each case some constant which may depend on n, λ, and k, but not on x. Bearing in mind that we are currently assuming $\lambda < (n-1)/2$, by combining (7.6.83) and (7.6.15) we see that $F_{\lambda+1}(x) x_j$ belongs to $L^1_{loc}(\mathbb{R}^n)$. Also, Lebesgue's Dominated Convergence Theorem applies and gives

$$-k \lim_{\varepsilon \to 0^+} \int_{|x| \geq \varepsilon} F_{\lambda+1}(x)\, x_j \varphi(x)\, dx = -k \int_{\mathbb{R}^n} F_{\lambda+1}(x)\, x_j \varphi(x)\, dx$$

$$= \langle -k F_{\lambda+1}(x) x_j, \varphi \rangle. \tag{7.6.84}$$

Moreover, (7.6.80) used in the range $\lambda < (n-1)/2$ gives

$$
\sup_{|x|=\varepsilon} |F_\lambda(x)| = \begin{cases} O(\ln(k\varepsilon)) & \text{if } \lambda = 0, \\ O(\varepsilon^{-2\lambda}) & \text{if } \lambda \in (0, (n-1)/2), \\ O(1) & \text{if } \lambda \in (-\infty, 0), \end{cases} \quad \text{as } \varepsilon \to 0^+. \tag{7.6.85}
$$

This, together with the fact that φ is bounded, implies

$$
\lim_{\varepsilon \to 0^+} \int_{|x|=\varepsilon} F_\lambda(x) \frac{x_j}{\varepsilon} \varphi(x) \, d\sigma(x) = 0. \tag{7.6.86}
$$

From (7.6.82), (7.6.84), and (7.6.86), we conclude

$$
\langle \partial_j F_\lambda, \varphi \rangle = \langle -k F_{\lambda+1}(x) x_j, \varphi \rangle \tag{7.6.87}
$$

which, in view of the arbitrariness of $\varphi \in C_0^\infty(\mathbb{R}^n)$, proves (7.6.79). $\qquad\square$

7.7 Fundamental Solutions for the Iterated Helmholtz Operator

Suppose $n \in \mathbb{N}$ is such that $n \geq 2$. The goal in this section is to determine a fundamental solution for the iterated Helmholtz operator in $\mathbb{R}^n$, that is, for the operator $(\Delta + k^2)^N$ in $\mathbb{R}^n$, where $k \in (0, \infty)$ and $N \in \mathbb{N}$. The case $N = 1$ has been treated in the previous section; cf. Theorem 7.33. To get started, recall the function Φ_k defined in (7.6.11). Bearing in mind that the Hankel function of the first kind is of class C^∞ in $(0, \infty)$ (cf. Lemma 14.71), this implies that

$$
(0, \infty) \times (\mathbb{R}^n \setminus \{0\}) \ni (k, x) \mapsto \Phi_k(x) \in \mathbb{C}
$$

$$
\text{is a mapping of class } C^\infty. \tag{7.7.1}
$$

This allows us to apply $\frac{d}{dk}$ to both sides of the identity in (7.6.33) and compute (recall that the Laplacian Δ is taken in the variable x)

$$
\begin{aligned}
0 &= \frac{d}{dk}\left[(\Delta + k^2)\Phi_k\right] = \frac{d}{dk}\left[\Delta \Phi_k\right] + \frac{d}{dk}\left[k^2 \Phi_k\right] \\
&= \Delta\left[\frac{d}{dk}\Phi_k\right] + 2k\Phi_k + k^2 \frac{d}{dk}\Phi_k \\
&= (\Delta + k^2)\left[\frac{d}{dk}\Phi_k\right] + 2k\Phi_k \quad \text{pointwise in } \mathbb{R}^n \setminus \{0\}.
\end{aligned} \tag{7.7.2}
$$

Hence,

$$(\Delta + k^2)\left[-\frac{1}{2k} \cdot \frac{d}{dk} \Phi_k \right] = \Phi_k \quad \text{pointwise in } \mathbb{R}^n \setminus \{0\}. \tag{7.7.3}$$

In particular, if we set

$$\Phi_k^{(2)} := -\frac{1}{2k} \cdot \frac{d}{dk} \Phi_k \quad \text{for each } (k, x) \in (0, \infty) \times (\mathbb{R}^n \setminus \{0\}), \tag{7.7.4}$$

then (7.7.3) becomes

$$(\Delta + k^2)\Phi_k^{(2)} = \Phi_k \quad \text{pointwise in } \mathbb{R}^n \setminus \{0\}. \tag{7.7.5}$$

In light of (7.6.33), formula (7.7.5) also implies that for each $k \in (0, \infty)$ there holds $(\Delta + k^2)^2 \Phi_k^{(2)} = (\Delta + k^2)\Phi_k = 0$ pointwise in $\mathbb{R}^n \setminus \{0\}$. Given that from Theorem 7.33 we know that $(\Delta + k^2)\Phi_k = \delta$ in $\mathcal{D}'(\mathbb{R}^n)$ it is natural to ask whether the identity $(\Delta + k^2)^2 \Phi_k^{(2)} = \delta$ holds in $\mathcal{D}'(\mathbb{R}^n)$. The good news is that the answer to this question is yes, and the approach employed to arrive at a formula for $\Phi_k^{(2)}$ is resourceful enough to provide a good candidate $\Phi_k^{(N)}$ for a fundamental solution for the iterated Helmholtz operator $(\Delta + k^2)^N$ with $N \in \mathbb{N}$ arbitrary. This is done in the proposition below via a recurrence relation. Before stating the aforementioned proposition, we wish to note that if $u \in C^\infty((0, \infty) \times (\mathbb{R}^n \setminus \{0\}))$ is arbitrary, then for $k \in (0, \infty)$ and $N \in \mathbb{N}$ the following differentiation formula holds

$$\frac{d}{dk}\left[(\Delta + k^2)^N u \right] = (\Delta + k^2)^N \left[\frac{d}{dk} u \right] + 2Nk(\Delta + k^2)^{N-1} u \tag{7.7.6}$$

pointwise in $\mathbb{R}^n \setminus \{0\}$. This may be easily justified via an induction over N.

Proposition 7.39. *Let $n \in \mathbb{N}$, $n \geq 2$, and take $k \in (0, \infty)$. Recall the function Φ_k from (7.6.11)–(7.6.12). Consider the sequence of functions $\{\Phi_k^{(N)}\}_{N \in \mathbb{N}}$ recurrently defined by setting*

$$\Phi_k^{(1)}(x) := \Phi_k(x), \qquad \forall x \in \mathbb{R}^n \setminus \{0\}, \tag{7.7.7}$$

and, for each $N \in \mathbb{N}$,

$$\Phi_k^{(N+1)}(x) := \frac{-1}{2Nk} \cdot \frac{d}{dk} \Phi_k^{(N)}(x), \qquad \forall x \in \mathbb{R}^n \setminus \{0\}. \tag{7.7.8}$$

Then these functions satisfy the following properties:

(1) *The assignment $(0, \infty) \times (\mathbb{R}^n \setminus \{0\}) \ni (k, x) \mapsto \Phi_k^{(N)}(x) \in \mathbb{C}$ is a function of class C^∞, for each $N \in \mathbb{N}$.*
(2) *For each $N \in \mathbb{N}$ and $k \in (0, \infty)$ one has $(\Delta + k^2)^N \Phi_k^{(N)}(x) = 0$ for every $x \in \mathbb{R}^n \setminus \{0\}$.*
(3) *For each $N \in \mathbb{N}$ and $k \in (0, \infty)$ one has $(\Delta + k^2)\Phi_k^{(N+1)}(x) = \Phi_k^{(N)}(x)$ for every $x \in \mathbb{R}^n \setminus \{0\}$.*
(4) *Having fixed $k \in (0, \infty)$ and $N \in \mathbb{N}$, the following formula holds for each point $x \in \mathbb{R}^n \setminus \{0\}$:*

$$\Phi_k^{(N)}(x) = \frac{(-1)^{N-1} c_n k^{(n-2N)/2} H_{(n-2N)/2}^{(1)}(k|x|)}{2^{N-1}(N-1)! \, |x|^{(n-2N)/2}},\tag{7.7.9}$$

where c_n is as in (7.6.12).

(5) Recall the function Ψ_λ from (7.6.14). Then, for each $x \in \mathbb{R}^n \setminus \{0\}$,

$$\Phi_k^{(N)}(x) = \begin{cases} C_1(n,N) \, \Psi_{(n-2N)/2}(k|x|) \, |x|^{-(n-2N)}, & \text{if } n > 2N, \\ C_2(n,N) \, \Psi_0(k|x|) \, \ln(k|x|), & \text{if } n = 2N, \\ C_3(n,N,k) \, \Psi_{(n-2N)/2}(k|x|), & \text{if } n < 2N, \end{cases}\tag{7.7.10}$$

where

$$C_1(n,N) := \frac{(-1)^N \, 2^{-2N} \, \Gamma\left(\frac{n-2N}{2}\right)}{(N-1)! \, \pi^{n/2}},\tag{7.7.11}$$

$$C_2(n,N) := \frac{(-1)^{N-1}}{2^{(n+2N-2)/2}(N-1)! \, \pi^{n/2}},\tag{7.7.12}$$

$$C_3(n,N,k) := \frac{(-1)^N k^{n-2N} \, e^{-i\pi(n-2N)/2} \, \Gamma\left(\frac{2N-n}{2}\right)}{(N-1)! \, 2^n \pi^{n/2}}.\tag{7.7.13}$$

Consequently,

$$\Phi_k^{(N)}(x) = \begin{cases} O(|x|^{-(n-2N)}), & \text{if } n > 2N, \\ O(\ln(k|x|)), & \text{if } n = 2N, \\ O(1), & \text{if } n < 2N, \end{cases} \quad \text{as } x \to 0.\tag{7.7.14}$$

(6) For each $k \in (0, \infty)$ and each $N \in \mathbb{N}$ one has $\Phi_k^{(N)} \in L_{loc}^1(\mathbb{R}^n)$.

(7) For each $k \in (0, \infty)$ and each $N \in \mathbb{N}$ one has

$$\Phi_k^{(N)}(x) = \frac{(-1)^{N-1} c_n \, k^{(n-2N-1)/2} \, \sqrt{2}}{2^{N-1}(N-1)! \, \sqrt{\pi}} \cdot \frac{e^{i(k|x|-(2n-4N+1)\pi/4)}}{|x|^{(n-2N+1)/2}}$$

$$+ \, O(|x|^{-(n-2N+3)/2}) \text{ as } |x| \to \infty.\tag{7.7.15}$$

(8) For each $k \in (0, \infty)$ and each $N \in \mathbb{N}$ one has $\Phi_k^{(N)} \in \mathcal{S}'(\mathbb{R}^n)$.

(9) For each $k \in (0, \infty)$ and each $N \in \mathbb{N}$ one has

$$\nabla \Phi_k^{(N)}(x) = \begin{cases} C_4(n,N) \, \Psi_{(n-2N+2)/2}(k|x|) \cdot \dfrac{x}{|x|^{n-2N+2}}, & \text{if } n > 2N - 2, \\ C_5(n,N) \, \Psi_0(k|x|) \, \ln(k|x|) \, x, & \text{if } n = 2N - 2, \\ C_6(n,N,k) \, \Psi_{(n-2N+2)/2}(k|x|) \, x, & \text{if } n < 2N - 2, \end{cases}\tag{7.7.16}$$

where

$$C_4(n, N) := \frac{(-1)^{N+1} \, \Gamma(\frac{n-2N+2}{2})}{(N-1)! \, 2^{2N-1} \pi^{n/2}}, \tag{7.7.17}$$

$$C_5(n, N) := \frac{(-1)^N}{2^{(n+2N-2)/2}(N-1)! \, \pi^{n/2}} = -C_2(n, N), \tag{7.7.18}$$

$$C_6(n, N, k) := \frac{(-1)^N k^{n-2N+2} \, e^{-i\pi(n-2N)/2} \, \Gamma(\frac{2N-2-n}{2})}{(N-1)! \, 2^{n+1} \, \pi^{n/2}}. \tag{7.7.19}$$

In particular,

$$\nabla \Phi_k^{(N)}(x) = \begin{cases} O(|x|^{-(n-2N+1)}), & \text{if } n > 2N - 2, \\ O(\ln(k|x|)|x|), & \text{if } n = 2N - 2, \\ O(|x|), & \text{if } n < 2N - 2, \end{cases} \qquad \text{as } x \to 0. \tag{7.7.20}$$

Proof. We shall first prove by induction on $N \in \mathbb{N}$ the properties claimed in items *(1)–(3)* hold. That this is the case when $N = 1$ is seen from (7.7.1), (7.6.33), and (7.7.3). Assume next that the properties claimed in items *(1)–(3)* hold for some $N \in \mathbb{N}$. Then the fact that

$$(0, \infty) \times (\mathbb{R}^n \setminus \{0\}) \ni (k, x) \mapsto \Phi_k^{(N+1)}(x) \in \mathbb{C} \tag{7.7.21}$$

is a function of class C^∞ is a consequence of (7.7.8) and the induction hypothesis. Also, applying $(\Delta + k^2)^N$ to the identity in item *(3)* yields

$$(\Delta + k^2)^{N+1} \Phi_k^{(N+1)} = (\Delta + k^2)^N \Phi_k^{(N)} = 0 \text{ in } \mathbb{R}^n \setminus \{0\}, \tag{7.7.22}$$

where the last equality uses item *(2)* in the current proposition. Finally, to prove the version of item *(3)* written for $N + 1$ in place of N, start by applying $\frac{d}{dk}$ to both sides of the formula $\Phi_k^{(N)} = (\Delta + k^2)\Phi_k^{(N+1)}$ and, reasoning much as in (7.7.2) (while bearing (7.7.8) in mind), obtain

$$\frac{d}{dk} \Phi_k^{(N)} = \frac{d}{dk}\left[(\Delta + k^2)\Phi_k^{(N+1)} \right] = \frac{d}{dk}\left[\Delta \Phi_k^{(N+1)} \right] + \frac{d}{dk}\left[k^2 \Phi_k^{(N+1)} \right]$$

$$= \Delta\left[\frac{d}{dk} \Phi_k^{(N+1)} \right] + 2k\Phi_k^{(N+1)} + k^2 \frac{d}{dk} \Phi_k^{(N+1)}$$

$$= (\Delta + k^2)\left[\frac{d}{dk} \Phi_k^{(N+1)} \right] + 2k\Phi_k^{(N+1)}$$

$$= -2(N + 1)k(\Delta + k^2)\Phi_k^{(N+2)} + 2k\Phi_k^{(N+1)} \tag{7.7.23}$$

pointwise in $\mathbb{R}^n \setminus \{0\}$. Together, (7.7.23) and (7.7.8) give

$$-2Nk\Phi_k^{(N+1)} = -2(N + 1)k(\Delta + k^2)\Phi_k^{(N+2)} + 2k\Phi_k^{(N+1)} \tag{7.7.24}$$

pointwise in $\mathbb{R}^n \setminus \{0\}$, hence $(\Delta + k^2)\Phi_k^{(N+2)} = \Phi_k^{(N+1)}$ pointwise in $\mathbb{R}^n \setminus \{0\}$. This concludes the proof (by induction on N) of the statements in items (1)–(3) of the proposition.

To prove the statement in item (4) fix $k \in (0, \infty)$. We will show by induction on $N \in \mathbb{N}$ that formula (7.7.9) holds. If $N = 1$, formula (7.7.9) is an immediate consequence of (7.6.11) since $\Phi_k^{(1)} = \Phi_k$ and we are assuming $n \geq 2$. Suppose (7.7.9) holds for some $N \in \mathbb{N}$. Then, (7.7.8), the induction hypothesis, item (2) in Lemma 14.71 (applied with $\lambda = (n - 2N)/2$), and the Chain Rule permit us to compute

$$
\begin{aligned}
\Phi_k^{(N+1)}(x) &= \frac{-1}{2Nk} \cdot \frac{d}{dk} \Phi_k^{(N)}(x) \\
&= \frac{(-1)^N c_n}{2^N N! \, k \, |x|^{n-2N}} \cdot \frac{d}{dk} \Big[(k|x|)^{(n-2N)/2} H_{(n-2N)/2}^{(1)}(k|x|) \Big] \\
&= \frac{(-1)^N c_n}{2^N N! \, k \, |x|^{n-2N}} \cdot (k|x|)^{(n-2N)/2} H_{(n-2N-2)/2}^{(1)}(k|x|) \, |x| \\
&= \frac{(-1)^N c_n k^{(n-2N-2)/2} H_{(n-2N-2)/2}^{(1)}(k|x|)}{2^N N! \, |x|^{(n-2N-2)/2}}
\end{aligned}
\tag{7.7.25}
$$

for all $x \in \mathbb{R}^n \setminus \{0\}$. Hence, (7.7.9) holds for $N + 1$ in place of N. This finishes the proof of the statement in item (4).

Moving on to item (5), let $k \in (0, \infty)$, $N \in \mathbb{N}$ be arbitrary and recall the function Ψ_λ from (7.6.14). We separate our discussion in three cases: $n > 2N$, $n = 2N$, and $n < 2N$.

Assume first that $n > 2N$. Then based on (7.7.9) and (7.6.14) corresponding to the case $\lambda = (n - 2N)/2 > 0$ we may write

$$
\Phi_k^{(N)}(x) = \frac{(-1)^{N-1} c_n 2^{(n-2N)/2} \Gamma\left(\frac{n-2N}{2}\right)}{2^{N-1}(N-1)! \, i\pi} \cdot \Psi_{(n-2N)/2}(k|x|) \cdot \frac{1}{|x|^{n-2N}}
\tag{7.7.26}
$$

for every $x \in \mathbb{R}^n \setminus \{0\}$. Using (7.6.12), the notation from (7.7.11), and some elementary algebra, this gives the equality in (7.7.10) if $n > 2N$. Moreover, (7.7.26) and (7.6.15) imply that $\Phi_k^{(N)}(x) = O(|x|^{-(n-2N)})$ as $x \to 0$ whenever $n > 2N$.

Next, suppose $n = 2N$. Then formulas (7.7.9) and (7.6.14) applied with $\lambda = 0$ give

$$
\Phi_k^{(N)}(x) = \frac{(-1)^{N-1} c_n 2i}{2^{N-1}(N-1)! \, \pi} \cdot \Psi_0(k|x|) \ln(k|x|)
\tag{7.7.27}
$$

for every $x \in \mathbb{R}^n \setminus \{0\}$. Making use of (7.6.12) and some algebra yields the equality in (7.7.10) corresponding to $n = 2N$. In addition, (7.7.27) and (7.6.15) ensure that we have $\Phi_k^{(N)}(x) = O(\ln(k|x|))$ as $x \to 0$ in the current case.

Finally, if $n < 2N$, from (7.7.9) and (7.6.14) (used with $\lambda = (n - 2N)/2 < 0$) we obtain

$$\Phi_k^{(N)}(x) = \frac{(-1)^{N-1}c_n k^{n-2N} e^{-i\pi(n-2N)/2} 2^{(2N-n)/2} \Gamma\left(\frac{2N-n}{2}\right)}{2^{N-1}(N-1)! \, i\pi} \cdot \Psi_{(n-2N)/2}(k|x|) \quad (7.7.28)$$

for every $x \in \mathbb{R}^n \setminus \{0\}$. From this and (7.6.12) the equality in (7.7.10) corresponding to $n < 2N$ follows. Invoking (7.6.15) in concert with (7.7.28) we also see that
$\Phi_k^{(N)}(x) = O(1)$ as $x \to 0$ if $n < 2N$. This completes the proof of the statement in
item (5).

The statement in item (6) follows immediately from (7.7.14) since the functions
in the right-hand side of (7.7.14) are locally integrable in $\mathbb{R}^n$. The behavior of $\Phi_k^{(N)}$
at infinity claimed in item (7) is a direct consequence of (7.7.9) and item (9) in
Lemma 14.71.

As regards item (8), in order to show that for each $k \in (0, \infty)$ and each $N \in \mathbb{N}$ the
function $\Phi_k^{(N)}$ defines a tempered distribution in $\mathbb{R}^n$, we note that (7.7.15) implies the
existence of some $C \in (0, \infty)$ with the property that $|\Phi_k^{(N)}(x)| \leq C|x|^{-(n-2N+1)/2}$ for
every $x \in \mathbb{R}^n \setminus B(0, 1)$. From this, the membership $\Phi_k^{(N)} \in L_{loc}^1(\mathbb{R}^n)$, and Example 4.4
(note that condition (4.1.4) holds for any $m > (n+2N-1)/2$) the desired conclusion
follows.

Turning to the proof of the statements in item (9), consider first (7.7.16). As an
application of (7.6.16) (with $\lambda = (n - 2N)/2$) we have

$$\nabla \left[\frac{H_{(n-2N)/2}^{(1)}(k|x|)}{|x|^{(n-2N)/2}} \right] = -k \frac{H_{(n-2N+2)/2}^{(1)}(k|x|)}{|x|^{(n-2N+2)/2}} \, x \quad (7.7.29)$$

for every $x \in \mathbb{R}^n \setminus \{0\}$. Together, (7.7.29) and (7.7.9) then give

$$\nabla \Phi_k^{(N)}(x) = \frac{(-1)^N c_n k^{(n-2N+2)/2} H_{(n-2N+2)/2}^{(1)}(k|x|)}{2^{N-1}(N-1)! \, |x|^{(n-2N+2)/2}} \, x, \quad (7.7.30)$$

for every $x \in \mathbb{R}^n \setminus \{0\}$. This, the definition of the function Ψ_λ from (7.6.14), and
with (7.6.12) then yield (7.7.16). Finally, (7.7.20) is an immediate consequence of
(7.7.16). This finishes the proof of Proposition 7.39. □

We now take up the task of establishing that $\Phi_k^{(N)}$ is a fundamental solution for
the iterated Helmholtz operator $(\Delta + k^2)^N$.

Theorem 7.40. *Let $n \in \mathbb{N}$, $n \geq 2$, and $k \in (0, \infty)$. For $N \in \mathbb{N}$ recall the function
$\Phi_k^{(N)} \in L_{loc}^1(\mathbb{R}^n)$ from (7.7.9), which is explicitly given by*

$$\Phi_k^{(N)}(x) = \frac{(-1)^N i \, k^{(n-2N)/2}}{2^{(n+2N)/2}(N-1)! \pi^{(n-2)/2}} \cdot \frac{H_{(n-2N)/2}^{(1)}(k|x|)}{|x|^{(n-2N)/2}} \quad (7.7.31)$$

*for every $x \in \mathbb{R}^n \setminus \{0\}$. Then, the distribution of function type defined by $\Phi_k^{(N)}$, which
by Proposition 7.39 is known to belong to $\mathcal{S}'(\mathbb{R}^n)$, is a fundamental solution for the
iterated Helmholtz operator $(\Delta + k^2)^N$ in $\mathbb{R}^n$.*

Proof. We shall reason by induction over N. The case $N = 1$ has been dealt with
in Theorem 7.33 (recall that $\Phi_k^{(1)} = \Phi_k$). Suppose next that for some positive inte-

ger N the distribution $\Phi_k^{(N)}$ is known to be a fundamental solution for the operator $(\Delta + k^2)^N$ in $\mathbb{R}^n$. In order to show that $\Phi_k^{(N+1)}$ is a fundamental solution for the operator $(\Delta + k^2)^{N+1}$ in $\mathbb{R}^n$, take $\varphi \in C_0^\infty(\mathbb{R}^n)$. The idea is to reason as in the proof of Theorem 7.33 and employ Proposition 7.39. Starting with the fact that the distribution $\Phi_k^{(N+1)}$ is of function type (cf. *(6)* in Proposition 7.39), then applying Lebesgue's Dominated Convergence Theorem, then integrating by parts twice (cf. Theorem 14.60), then invoking item *(3)* in Proposition 7.39, then item *(6)* in Proposition 7.39 and Lebesgue's Dominated Convergence Theorem, we may write

$$\langle (\Delta + k^2)^{N+1} \Phi_k^{(N+1)}, \varphi \rangle = \int_{\mathbb{R}^n} \Phi_k^{(N+1)}(x)[(\Delta + k^2)^{N+1}\varphi](x)\,dx$$

$$= \lim_{\varepsilon \to 0^+} \int_{\mathbb{R}^n \setminus B(0,\varepsilon)} \Phi_k^{(N+1)}(x)[(\Delta + k^2)^{N+1}\varphi](x)\,dx$$

$$= \lim_{\varepsilon \to 0^+} \left\{ \int_{\mathbb{R}^n \setminus B(0,\varepsilon)} [(\Delta + k^2)\Phi_k^{(N+1)}](x)[(\Delta + k^2)^N\varphi](x)\,dx + I_\varepsilon + II_\varepsilon \right\}$$

$$= \int_{\mathbb{R}^n} \Phi_k^N(x)[(\Delta + k^2)^N\varphi](x)\,dx + \lim_{\varepsilon \to 0^+} I_\varepsilon + \lim_{\varepsilon \to 0^+} II_\varepsilon, \qquad (7.7.32)$$

where we have set (recall the definition of the normal derivative from (7.1.14))

$$I_\varepsilon := -\int_{\partial B(0,\varepsilon)} \Phi_k^{(N+1)}(x)\frac{\partial}{\partial \nu}[(\Delta + k^2)^N\varphi](x)\,d\sigma(x), \qquad (7.7.33)$$

$$II_\varepsilon := \int_{\partial B(0,\varepsilon)} \frac{\partial \Phi_k^{(N+1)}}{\partial \nu}(x)[(\Delta + k^2)^N\varphi](x)\,d\sigma(x). \qquad (7.7.34)$$

Above, $\nu(x) = \frac{x}{\varepsilon}$ for each $x \in \partial B(0,\varepsilon)$, and σ denotes the surface measure on $\partial B(0,\varepsilon)$. By *(6)* in Proposition 7.39 and the induction hypothesis we have

$$\int_{\mathbb{R}^n} \Phi_k^N(x)[(\Delta + k^2)^N\varphi](x)\,dx = \langle \Phi_k^N, (\Delta + k^2)^N\varphi \rangle$$

$$= \langle (\Delta + k^2)^N \Phi_k^N, \varphi \rangle = \langle \delta, \varphi \rangle = \varphi(0). \qquad (7.7.35)$$

Next, let $\varepsilon \in (0, 1)$ be arbitrary and focus on estimating $|I_\varepsilon|$. From (7.7.33) we see that

$$|I_\varepsilon| \le \int_{\partial B(0,\varepsilon)} \left| \Phi_k^{(N+1)}(x) \right| \left| \frac{\partial}{\partial \nu}[(\Delta + k^2)^N\varphi](x) \right| d\sigma(x)$$

$$\le \max_{\alpha \in \mathbb{N}_0^n, |\alpha| \le 2N+1} \left\| \partial^\alpha \varphi \right\|_{L^\infty(B(0,1))} \int_{\partial B(0,\varepsilon)} \left| \Phi_k^{(N+1)}(x) \right| d\sigma(x), \qquad (7.7.36)$$

Bringing in (7.7.14) (applied with N replaced by $N+1$), and denoting by C constants that may depend on φ, N, n, k, but are independent of ε, we may further estimate

$$|I_\varepsilon| \le \begin{cases} C\varepsilon^{2N+1}, & \text{if } n > 2N+2, \\ C|\ln(k\varepsilon)|\varepsilon^{n-1}, & \text{if } n = 2N+2, \\ C\varepsilon^{n-1}, & \text{if } n < 2N+2. \end{cases} \tag{7.7.37}$$

Recalling that we are assuming $n \ge 2$, it is immediate that the terms in the right-hand side of (7.7.37) vanish as $\varepsilon \to 0^+$. Consequently,

$$\lim_{\varepsilon \to 0^+} I_\varepsilon = 0. \tag{7.7.38}$$

Moving on, for each $\varepsilon \in (0, 1)$ starting with (7.7.34) we estimate

$$|II_\varepsilon| \le \int_{\partial B(0,\varepsilon)} \left| \frac{\partial}{\partial \nu} \Phi_k^{(N+1)}(x) \right| \left| [(\Delta + k^2)^N \varphi](x) \right| \, d\sigma(x),$$

$$\le \max_{\alpha \in \mathbb{N}_0^n, |\alpha| \le 2N} \left\| \partial^\alpha \varphi \right\|_{L^\infty(B(0,1))} \int_{\partial B(0,\varepsilon)} \left| \frac{\partial}{\partial \nu} \Phi_k^{(N+1)}(x) \right| \, d\sigma(x). \tag{7.7.39}$$

On account of (7.7.20) (applied with N replaced by $N + 1$),

$$|II_\varepsilon| \le \begin{cases} C\varepsilon^{2N}, & \text{if } n > 2N, \\ C|\ln(k\varepsilon)|\varepsilon^n, & \text{if } n = 2N, \\ C\varepsilon^n, & \text{if } n < 2N, \end{cases} \tag{7.7.40}$$

hence,

$$\lim_{\varepsilon \to 0^+} II_\varepsilon = 0. \tag{7.7.41}$$

Combining (7.7.32), (7.7.35), (7.7.38), and (7.7.41), we arrive at

$$\langle (\Delta + k^2)^{N+1} \Phi_k^{(N+1)}, \varphi \rangle = \varphi(0). \tag{7.7.42}$$

Since $\varphi \in C_0^\infty(\mathbb{R}^n)$ is arbitrary, this implies $(\Delta + k^2)^{N+1} \Phi_k^{(N+1)} = \delta$ in $\mathcal{D}'(\mathbb{R}^n)$, finishing the proof of the theorem. $\qquad \square$

It turns out that the pointwise identity from item *(3)* of Proposition 7.39 is a manifestation of a more general formula, valid in the sense of distributions, described below. This may then be used to give an alternative proof of Theorem 7.40 (see Exercise 7.42).

Proposition 7.41. *Let $n \in \mathbb{N}$, $n \ge 2$, and $k \in (0, \infty)$. For $N \in \mathbb{N}$ recall the function $\Phi_k^{(N)}$ from (7.7.9). Then for each $N \in \mathbb{N}$ one has*

$$(\Delta + k^2)\Phi_k^{(N+1)} = \Phi_k^{(N)} \qquad in \ \mathcal{D}'(\mathbb{R}^n). \tag{7.7.43}$$

Proof. Fix $N \in \mathbb{N}$ and recall the family of functions F_λ, indexed by λ, as in (7.6.78). In terms of this, $\Phi_k^{(N)}$ may be expressed as

$$\Phi_k^{(N)}(x) = a_{N,k} F_{(n-2N)/2}(x), \qquad \forall\, x \in \mathbb{R}^n \setminus \{0\}, \tag{7.7.44}$$

where we have abbreviated

$$a_{N,k} := \frac{(-1)^N \mathrm{i}\, k^{(n-2N)/2}}{2^{(n+2N)/2}(N-1)!\,\pi^{(n-2)/2}}. \tag{7.7.45}$$

Then, for each $j \in \{1, \ldots, n\}$, formula (7.6.79) in Proposition 7.38 (applied with $\lambda := (n - 2N - 2)/2 < (n-1)/2$) implies

$$\partial_j \Phi_k^{(N+1)} = -a_{N+1,k}\, k F_{(n-2N)/2}(x) x_j \quad \text{in } \mathcal{D}'(\mathbb{R}^n). \tag{7.7.46}$$

A second application of (7.6.79) (this time with $\lambda := (n - 2N)/2 < (n-1)/2$) yields that for each $j \in \{1, \ldots, n\}$ we have

$$\partial_j^2 \Phi_k^{(N+1)} = a_{N+1,k}\, k^2 F_{(n-2N+2)/2}(x) x_j^2 - a_{N+1,k}\, k F_{(n-2N)/2} \tag{7.7.47}$$

in $\mathcal{D}'(\mathbb{R}^n)$. Consequently, (7.7.47) and the definition of F_λ from (7.6.78) give

$$(\Delta + k^2)\Phi_k^{(N+1)} = a_{N+1,k}\, k^2 F_{(n-2N+2)/2}(x)|x|^2 - n a_{N+1,k}\, k F_{(n-2N)/2}$$

$$+ k^2 a_{N+1,k} F_{(n-2N-2)/2}$$

$$= k^2 a_{N+1,k}|x|^{-(n-2N-2)/2}\big[H_{(n-2N+2)/2}^{(1)}(k|x|) + H_{(n-2N-2)/2}^{(1)}(k|x|)\big]$$

$$- n\, a_{N+1,k}\, k|x|^{-(n-2N)/2} H_{(n-2N)/2}^{(1)}(k|x|) \tag{7.7.48}$$

in $\mathcal{D}'(\mathbb{R}^n)$. By (6) in Lemma 14.71 we have

$$H_{(n-2N+2)/2}^{(1)}(k|x|) + H_{(n-2N-2)/2}^{(1)}(k|x|) = \frac{n-2N}{k|x|} H_{(n-2N)/2}^{(1)}(k|x|) \tag{7.7.49}$$

for each $x \in \mathbb{R}^n \setminus \{0\}$, hence also in $\mathcal{D}'(\mathbb{R}^n)$. In concert, (7.7.48) and (7.7.49) imply

$$(\Delta + k^2)\Phi_k^{(N+1)} = -2N a_{N+1,k}\, k|x|^{-(n-2N)/2} H_{(n-2N)/2}^{(1)}(k|x|) \tag{7.7.50}$$

in $\mathcal{D}'(\mathbb{R}^n)$. The desired conclusion now follows from (7.7.50) and (7.7.9) upon observing that $-2N a_{N+1,k}\, k = a_{N,k}$. $\square$

Exercise 7.42. Use Proposition 7.41 and induction over N to give another proof of Theorem 7.40, i.e., that $\Phi_k^{(N)}$ as in (7.7.9) is a fundamental solution for the iterated Helmholtz operator $(\Delta + k^2)^N$ in $\mathbb{R}^n$.

7.8 Fundamental Solutions for the Cauchy–Riemann Operator

In this section we determine all fundamental solutions for the `Cauchy-Riemann` operator $\frac{\partial}{\partial \bar{z}} := \frac{1}{2}(\partial_1 + i\partial_2)$ and its conjugate, $\frac{\partial}{\partial z} := \frac{1}{2}(\partial_1 - i\partial_2)$ that belong to $\mathcal{S}'(\mathbb{R}^2)$. It is immediate that $\frac{\partial}{\partial \bar{z}} \frac{\partial}{\partial z} = \frac{1}{4}\Delta$ when acting on distributions from $\mathcal{D}'(\mathbb{R}^2)$. Given this, Remark 5.7 is particularly useful. Specifically, since by Theorem 7.2 we have that $E(x) = \frac{1}{2\pi} \ln |x|$, $x \in \mathbb{R}^2 \setminus \{0\}$, is a fundamental solution for Δ in $\mathbb{R}^2$ and $E \in \mathcal{S}'(\mathbb{R}^2)$, Remark 5.7 implies that

$$\frac{\partial}{\partial \bar{z}}\Big(\frac{2}{\pi} \ln |x|\Big) \in \mathcal{S}'(\mathbb{R}^2) \text{ is a fundamental solution for } \frac{\partial}{\partial z} \text{ in } \mathcal{S}'(\mathbb{R}^2), \qquad (7.8.1)$$

$$\frac{\partial}{\partial z}\Big(\frac{2}{\pi} \ln |x|\Big) \in \mathcal{S}'(\mathbb{R}^2) \text{ is a fundamental solution for } \frac{\partial}{\partial \bar{z}} \text{ in } \mathcal{S}'(\mathbb{R}^2). \qquad (7.8.2)$$

We proceed with computing $\frac{\partial}{\partial \bar{z}} \left(\frac{2}{\pi} \ln |x|\right)$ in the distributional sense in $\mathcal{S}'(\mathbb{R}^2)$. By (c) in Theorem 4.14 it suffices to compute the distributional derivative $\frac{\partial}{\partial \bar{z}} \left(\frac{2}{\pi} \ln |x|\right)$ in $\mathcal{D}'(\mathbb{R}^2)$. We will show that this distributional derivative is equal to the distribution given by the function obtained by taking $\frac{\partial}{\partial \bar{z}} \left(\frac{2}{\pi} \ln |x|\right)$ in the classical sense for $x \neq 0$. To this end, fix a function $\varphi \in C_0^\infty(\mathbb{R}^2)$. Since $\ln |x| \in L^1_{loc}(\mathbb{R}^2)$ (recall Example 2.9) we may write

$$\Big\langle \frac{\partial}{\partial \bar{z}}\Big(\frac{2}{\pi} \ln |x|\Big), \varphi \Big\rangle = -\Big\langle \frac{2}{\pi} \ln |x|, \frac{\partial \varphi}{\partial \bar{z}} \Big\rangle$$

$$= -\frac{1}{\pi} \int_{\mathbb{R}^2} \ln |x| [(\partial_1 + i\partial_2)\varphi(x)] \, dx. \qquad (7.8.3)$$

Let $R > 0$ be such that $\operatorname{supp} \varphi \subset B(0, R)$. By Lebesgue's Dominated Convergence Theorem 14.15 we further have

$$\int_{\mathbb{R}^2} \ln |x| \, [(\partial_1 + i\partial_2)\varphi(x)] \, dx = \lim_{\varepsilon \to 0^+} \int_{\varepsilon \leq |x| \leq R} \ln |x| \, [(\partial_1 + i\partial_2)\varphi(x)] \, dx. \qquad (7.8.4)$$

Fix $\varepsilon \in (0, R/2)$ and with $x = (x_1, x_2)$ use (14.8.4) to write

$$\int_{\varepsilon \leq |x| \leq R} \ln |x| \, (\partial_1 + i\partial_2)\varphi(x) \, dx \qquad (7.8.5)$$

$$= -\int_{|x| \geq \varepsilon} \varphi(x)(\partial_1 + i\partial_2) \ln |x| \, dx - \frac{\ln \varepsilon}{\varepsilon} \int_{|x| = \varepsilon} (x_1 + ix_2)\varphi(x) \, d\sigma(x),$$

where we have used the fact that the outward unit normal to $B(0, \varepsilon)$ at a point x on $\partial B(0, \varepsilon)$ is $\frac{1}{\varepsilon}(x_1, x_2)$ and that φ vanishes on $\partial B(0, R)$. Using polar coordinates, we further write

$$\left| \frac{\ln \varepsilon}{\varepsilon} \int_{|x|=\varepsilon} (x_1 + ix_2)\varphi(x)\, d\sigma(x) \right|$$

$$= \left| \varepsilon \ln \varepsilon \int_0^{2\pi} (\cos \theta + i \sin \theta)\varphi(\varepsilon \cos \theta, \varepsilon \sin \theta)\, d\theta \right|$$

$$\leq C\varepsilon |\ln \varepsilon| \xrightarrow[\varepsilon \to 0^+]{} 0. \tag{7.8.6}$$

Since $(\partial_1 + i\partial_2)[\ln |x|] \in L^1_{loc}(\mathbb{R}^2)$ and φ is compactly supported, by Lebesgue's Dominated Convergence Theorem we have

$$\lim_{\varepsilon \to 0^+} \int_{|x|\geq\varepsilon} \varphi(x)(\partial_1 + i\partial_2)\ln |x|\, dx = \int_{\mathbb{R}^2} \varphi(x)(\partial_1 + i\partial_2)\ln |x|\, dx. \tag{7.8.7}$$

By combining (7.8.3)–(7.8.7) we conclude that

$$\frac{\partial}{\partial \bar{z}}\left(\frac{2}{\pi} \ln |x| \right) = \frac{1}{\pi}(\partial_1 + i\partial_2)[\ln |x|] = \frac{1}{\pi} \cdot \frac{1}{x_1 - ix_2} \quad \text{in } \mathcal{S}'(\mathbb{R}^2). \tag{7.8.8}$$

Similarly,

$$\frac{\partial}{\partial z}\left(\frac{2}{\pi} \ln |x| \right) = \frac{1}{\pi}(\partial_1 - i\partial_2)[\ln |x|] = \frac{1}{\pi} \cdot \frac{1}{x_1 + ix_2} \quad \text{in } \mathcal{S}'(\mathbb{R}^2). \tag{7.8.9}$$

With this at hand, by invoking Proposition 5.8, we have a complete description of all fundamental solutions for $\frac{\partial}{\partial z}$ and $\frac{\partial}{\partial \bar{z}}$ that are also tempered distributions.

In summary, we proved the following theorem.

Theorem 7.43. *Consider the functions*

$$E(x_1, x_2) := \frac{1}{\pi} \cdot \frac{1}{x_1 - ix_2}, \qquad F(x_1, x_2) := \frac{1}{\pi} \cdot \frac{1}{x_1 + ix_2}, \tag{7.8.10}$$

defined for $x = (x_1, x_2) \in \mathbb{R}^2 \setminus \{0\}$. Then $E \in \mathcal{S}'(\mathbb{R}^2)$ and is a fundamental solution for the operator $\frac{\partial}{\partial z}$. Also $F \in \mathcal{S}'(\mathbb{R}^2)$ and is a fundamental solution for the operator $\frac{\partial}{\partial \bar{z}}$.

Moreover,

$$\{ u \in \mathcal{S}'(\mathbb{R}^2) : \tfrac{\partial}{\partial z} u = \delta \text{ in } \mathcal{S}'(\mathbb{R}^2) \} \tag{7.8.11}$$

$$= \{ E + P : P \text{ polynomial in } \mathbb{R}^2 \text{ satisfying } \frac{\partial}{\partial z} P = 0 \text{ in } \mathbb{R}^2 \}$$

and

$$\{ u \in \mathcal{S}'(\mathbb{R}^2) : \tfrac{\partial}{\partial \bar{z}} u = \delta \text{ in } \mathcal{S}'(\mathbb{R}^2) \} \tag{7.8.12}$$

$$= \{ F + P : P \text{ polynomial in } \mathbb{R}^2 \text{ satisfying } \frac{\partial}{\partial \bar{z}} P = 0 \text{ in } \mathbb{R}^2 \}.$$

Having identified the fundamental solutions in $\mathbb{R}^2$ for the Cauchy–Riemann operator $\overline{\partial} = \frac{1}{2}(\partial_x + i\partial_y)$, we shall now compute the Fourier transform of $1/z$. Throughout, we shall repeatedly identify $\mathbb{R}^2$ with $\mathbb{C}$.

Proposition 7.44. *Let* $u(z) := \frac{1}{z}$ *for all* $z \in \mathbb{C} \setminus \{0\}$. *Then* $u \in \mathcal{S}'(\mathbb{R}^2)$ *and*

$$\widehat{u}(\xi) = \frac{2\pi}{i\xi} \quad in \ \ \mathcal{S}'(\mathbb{R}^2). \tag{7.8.13}$$

Proof. That $u \in \mathcal{S}'(\mathbb{R}^2)$ follows by observing that u is locally integrable near the origin, while $|u(z)|$ decays like $|z|^{-1}$ for $|z| > 1$. To prove (7.8.13), first note that, since $\overline{\partial}\left(\frac{1}{\pi z}\right) = \delta$ in $\mathcal{S}'(\mathbb{R}^2)$ by Theorem 7.43, it follows that $\overline{\partial}u = \pi\delta$ in $\mathcal{S}'(\mathbb{R}^2)$. Hence

$$\pi = \widehat{\overline{\partial}u}(\xi) = \frac{i}{2}\xi\widehat{u}(\xi) \Longrightarrow \xi\widehat{u}(\xi) = \frac{2\pi}{i}. \tag{7.8.14}$$

Thus, $\xi\left(\frac{2\pi}{i\xi} - \widehat{u}(\xi)\right) = 0$ in $\mathcal{S}'(\mathbb{R}^2)$ which when combined with Example 2.76 implies

$$\frac{2\pi}{i\xi} - \widehat{u}(\xi) = c\,\delta, \quad \text{for some} \ \ c \in \mathbb{C}. \tag{7.8.15}$$

Thus,

$$\frac{2\pi}{i}u - \widehat{u} = c\,\delta \quad in \ \ \mathcal{S}'(\mathbb{R}^2). \tag{7.8.16}$$

Taking another Fourier transform yields

$$\frac{2\pi}{i}\widehat{u} + (2\pi)^2 u = c \quad in \ \ \mathcal{S}'(\mathbb{R}^2), \tag{7.8.17}$$

given that $\widehat{\widehat{u}} = -(2\pi)^2 u$ (keeping in mind that u is odd). A linear combination of (7.8.16) and (7.8.17) which eliminates $\widehat{u}$ then leads us to the conclusion that $(-2i\pi\delta + 1)c = 0$ in $\mathcal{S}'(\mathbb{R}^2)$. In turn, this forces $c = 0$, concluding the proof of the proposition. $\square$

Exercise 7.45. Consider $u(z) := 1/\overline{z}$ for all $z \in \mathbb{C} \setminus \{0\}$. Show that $u \in \mathcal{S}'(\mathbb{R}^2)$ and

$$\widehat{u}(\xi) = \frac{2\pi}{i\overline{\xi}} \quad in \ \ \mathcal{S}'(\mathbb{R}^2). \tag{7.8.18}$$

Hint: Use Proposition 7.44 and Exercise 3.27.

Proposition 7.46. *Let* Ω *be an open set in* $\mathbb{C}$ *that contains the origin. Suppose* $u \in C^0(\Omega)$ *is such that* $\frac{\partial u}{\partial \overline{z}} \in L^1_{loc}(\Omega)$ *and* $\frac{1}{z}\frac{\partial u}{\partial \overline{z}} \in L^1_{loc}(\Omega)$. *Then* $\frac{u}{z} \in L^1_{loc}(\Omega)$ *and*

$$\frac{\partial}{\partial \overline{z}}\left[\frac{u}{z}\right] = \pi u(0)\delta + \frac{1}{z}\frac{\partial u}{\partial \overline{z}} \quad in \ \ \mathcal{D}'(\Omega). \tag{7.8.19}$$

Proof. Pick a function $\theta \in C^\infty(\mathbb{C})$ with the property that $\theta = 0$ on $B(0,1)$ and $\theta \equiv 1$ on $\mathbb{C} \setminus B(0,2)$. For each $\varepsilon \in (0,1)$ define the function $\theta_\varepsilon : \mathbb{C} \to \mathbb{C}$ by setting

$\theta_\varepsilon(z) := \theta(z/\varepsilon)$ for each $z \in \mathbb{C}$. Then there exists some constant $C \in (0, \infty)$ such that

$$\theta_\varepsilon \in C^\infty(\mathbb{C}), \quad \mathrm{supp}\,(\nabla\theta_\varepsilon) \subseteq \overline{B(0, 2\varepsilon)} \setminus B(0, \varepsilon), \tag{7.8.20}$$

$$\lim_{\varepsilon \to 0^+} \theta_\varepsilon(z) = 1 \quad \text{and} \quad |\theta_\varepsilon(z)| \le C \quad \forall z \in \mathbb{C}. \tag{7.8.21}$$

Next, fix $\varphi \in C_0^\infty(\Omega)$ and, with $\mathcal{L}^2$ denoting the Lebesgue measure in $\mathbb{R}^2$, write

$$\left\langle \frac{\partial}{\partial \bar{z}}\left[\frac{u}{z}\right], \varphi \right\rangle = -\left\langle \frac{u}{z}, \frac{\partial \varphi}{\partial \bar{z}} \right\rangle = -\int_\Omega \frac{u}{z} \frac{\partial \varphi}{\partial \bar{z}} \, d\mathcal{L}^2$$

$$= -\lim_{\varepsilon \to 0^+} \int_\Omega \frac{u}{z} \theta_\varepsilon \frac{\partial \varphi}{\partial \bar{z}} \, d\mathcal{L}^2, \tag{7.8.22}$$

where for the last equality in (7.8.22) we used (7.8.21) and Lebesgue's Dominated Convergence Theorem. Note that

$$\frac{1}{z}\varphi\theta_\varepsilon \in C_0^\infty(\Omega) \quad \text{and} \quad \frac{\partial}{\partial \bar{z}}\left[\frac{1}{z}\right] = 0 \quad \text{in} \quad \mathbb{C} \setminus \{0\}. \tag{7.8.23}$$

Therefore,

$$\frac{\partial}{\partial \bar{z}}\left[\frac{1}{z}\varphi\theta_\varepsilon\right] = \frac{1}{z}\theta_\varepsilon \frac{\partial \varphi}{\partial \bar{z}} + \frac{1}{z}\varphi \frac{\partial \theta_\varepsilon}{\partial \bar{z}} \quad \text{in} \quad \Omega. \tag{7.8.24}$$

Combining (7.8.22) and (7.8.24) we obtain

$$\left\langle \frac{\partial}{\partial \bar{z}}\left[\frac{u}{z}\right], \varphi \right\rangle = -\lim_{\varepsilon \to 0^+} \int_\Omega u \frac{\partial}{\partial \bar{z}}\left[\frac{1}{z}\varphi\theta_\varepsilon\right] d\mathcal{L}^2 + \lim_{\varepsilon \to 0^+} \int_\Omega \frac{u}{z}\varphi \frac{\partial \theta_\varepsilon}{\partial \bar{z}} \, d\mathcal{L}^2$$

$$= \lim_{\varepsilon \to 0^+} \left\langle \frac{\partial u}{\partial \bar{z}}, \frac{1}{z}\varphi\theta_\varepsilon \right\rangle + \lim_{\varepsilon \to 0^+} \int_\Omega \frac{u}{z}\varphi \frac{\partial \theta_\varepsilon}{\partial \bar{z}} \, d\mathcal{L}^2 =: I + II. \tag{7.8.25}$$

Using the hypotheses on u, (7.8.21), and Lebesgue's Dominated Convergence Theorem, we obtain

$$I = \lim_{\varepsilon \to 0^+} \int_\Omega \frac{\partial u}{\partial \bar{z}} \frac{1}{z}\varphi\theta_\varepsilon \, d\mathcal{L}^2 = \int_\Omega \frac{\partial u}{\partial \bar{z}} \frac{1}{z}\varphi \, d\mathcal{L}^2 = \left\langle \frac{1}{z}\frac{\partial u}{\partial \bar{z}}, \varphi \right\rangle. \tag{7.8.26}$$

To compute II, first write

$$\int_\Omega \frac{u}{z}\varphi \frac{\partial \theta_\varepsilon}{\partial \bar{z}} \, d\mathcal{L}^2 = \int_\Omega \frac{u - u(0)}{z}\varphi \frac{1}{\varepsilon}\frac{\partial \theta}{\partial \bar{z}}(\cdot/\varepsilon)\, d\mathcal{L}^2 + u(0)\int_\Omega \frac{1}{z}\varphi \frac{\partial \theta_\varepsilon}{\partial \bar{z}} \, d\mathcal{L}^2$$

$$=: III + IV. \tag{7.8.27}$$

Using the support condition from (7.8.20), the continuity of u and the properties of φ, term III may be estimated by

$$|III| \le C\|\varphi\|_{L^\infty(\Omega)}\|\nabla\theta\|_{L^\infty(\mathbb{R}^n)}\Big(\sup_{B(0,2\varepsilon)} |u - u(0)|\Big) \xrightarrow[\varepsilon \to 0^+]{} 0. \tag{7.8.28}$$

As for IV, we have

$$\lim_{\varepsilon \to 0^+} IV = u(0) \lim_{\varepsilon \to 0^+} \int_\Omega \frac{1}{z} \varphi \frac{\partial \theta_\varepsilon}{\partial \bar{z}} \, d\mathcal{L}^2 = -u(0) \lim_{\varepsilon \to 0^+} \int_\Omega \frac{1}{z} \frac{\partial \varphi}{\partial \bar{z}} \theta_\varepsilon \, d\mathcal{L}^2$$

$$= -u(0) \int_\Omega \frac{1}{z} \frac{\partial \varphi}{\partial \bar{z}} \, d\mathcal{L}^2 = -u(0)\Big\langle \frac{1}{z}, \frac{\partial \varphi}{\partial \bar{z}} \Big\rangle = u(0)\Big\langle \frac{\partial}{\partial \bar{z}}\Big[\frac{1}{z}\Big], \varphi \Big\rangle$$

$$= \pi u(0)\langle \delta, \varphi \rangle. \tag{7.8.29}$$

For the second equality in (7.8.29), for each $\varepsilon \in (0,1)$ fixed, we used integration by parts (cf. (14.8.4)) on $D \setminus \overline{B(0,\varepsilon/2)}$, where D is a bounded open subset of Ω with the property that supp $\varphi \subset D$. At this step is also useful to recall (7.8.23). For the third equality in (7.8.29), we applied Lebesgue's Dominated Convergence Theorem, while the fifth is based on Theorem 7.43. Now (7.8.19) follows by combining (7.8.25)–(7.8.29), since φ is arbitrary in $C_0^\infty(\Omega)$. $\square$

Exercise 7.47. Let Ω be an open set in $\mathbb{C}$ and let $z_0 \in \Omega$. Suppose $u \in C^0(\Omega)$ is such that $\frac{\partial u}{\partial \bar{z}} \in L^1_{loc}(\Omega)$ and $\frac{1}{z-z_0} \frac{\partial u}{\partial \bar{z}} \in L^1_{loc}(\Omega)$. Then $\frac{u}{z-z_0} \in L^1_{loc}(\Omega)$ and

$$\frac{\partial}{\partial \bar{z}}\Big[\frac{u}{z-z_0}\Big] = \pi u(z_0)\delta_{z_0} + \frac{1}{z-z_0} \frac{\partial u}{\partial \bar{z}} \quad \text{in} \quad \mathcal{D}'(\Omega). \tag{7.8.30}$$

Proposition 7.48. *If $\varphi \in \mathcal{S}(\mathbb{R})$ is a complex-valued function, define the* Cauchy operator *by*

$$(\mathscr{C}\varphi)(z) := \frac{1}{2\pi i} \int_\mathbb{R} \frac{\varphi(x)}{x-z} \, dx, \qquad \forall z \in \mathbb{C} \setminus \mathbb{R}. \tag{7.8.31}$$

Then the following Plemelj jump-formula holds at every $x \in \mathbb{R}$:

$$\lim_{y \to 0^\pm} (\mathscr{C}\varphi)(x+iy) = \pm\frac{1}{2}\varphi(x) + \lim_{\varepsilon \to 0^+} \frac{1}{2\pi i} \int_{\substack{y \in \mathbb{R} \\ |x-y|>\varepsilon}} \frac{\varphi(y)}{y-x} \, dy. \tag{7.8.32}$$

Proof. Apply Corollary 4.81 to the function $\Phi : \mathbb{R}^2 \setminus \{(0,0)\} \to \mathbb{C}$ defined by

$$\Phi(x,y) := \frac{-1}{2\pi i(x+iy)} \quad \text{for all} \quad (x,y) \in \mathbb{R}^2 \setminus \{(0,0)\}. \tag{7.8.33}$$

Note that Φ is C^∞, odd, and homogeneous of degree -1 and, under the canonical identification $\mathbb{R}^2 \equiv \mathbb{C}$, takes the form $\Phi(z) = \frac{-1}{2\pi i z}$ for $z \in \mathbb{C} \setminus \{0\}$. Proposition 7.44 then gives $\widehat{\Phi}(\xi) = \frac{1}{\xi}$ for all $\xi \in \mathbb{C} \setminus \{0\}$ which, in particular, yields $\widehat{\Phi}(0,1) = -i$. Having established this, (7.8.32) follows directly from (4.7.46). $\square$

Remark 7.49. Upon recalling formula (4.9.31) for the Hilbert transform H on the real line, we may recast the version of Plemelj jump-formula (7.8.32) corresponding to considering the Cauchy operator in the upper half-plane in the form

$$\left[\mathscr{C}\varphi\right]\Big|_{\partial\mathbb{R}^2_+}^{\text{ver}} = \frac{1}{2}\varphi + \frac{i}{2\pi}H\varphi \quad \text{in} \quad \mathbb{R}, \qquad \forall\,\varphi \in S(\mathbb{R}), \qquad (7.8.34)$$

where the "vertical limit" of $\mathscr{C}\varphi$ to the boundary of the upper half-plane is understood as in (4.8.20). In turn, formula (7.8.34) suggests the consideration of the operator (with I denoting the identity)

$$P := \frac{1}{2}I + \frac{i}{2\pi}H. \qquad (7.8.35)$$

From Corollary 4.99 it follows that P is a well-defined, linear, and bounded operator on $L^2(\mathbb{R})$. Using the fact that $H^2 = -\pi^2 I$ and $H^* = -H$ on $L^2(\mathbb{R})$ (again, see Corollary 4.99), we may then compute

$$P^2 = \left(\frac{1}{2}I + \frac{i}{2\pi}H\right)^2 = \frac{1}{4}I + \frac{i}{2\pi}H + \left(\frac{i}{2\pi}\right)^2 H^2$$

$$= \frac{1}{4}I + \frac{i}{2\pi}H + \frac{1}{4}I = \frac{1}{2}I + \frac{i}{2\pi}H$$

$$= P, \qquad (7.8.36)$$

and

$$P^* = \frac{1}{2}I - \frac{i}{2\pi}H^* = \frac{1}{2}I + \frac{i}{2\pi}H = P. \qquad (7.8.37)$$

Any linear and bounded operator on $L^2(\mathbb{R})$ satisfying these two properties (i.e., $P^2 = P$ and $P^* = P$) is called a projection. Then one may readily verify that $I - P = \frac{1}{2}I - \frac{i}{2\pi}H$ is also a projection and if we introduce (what are commonly referred to as Hardy spaces on the real line)

$$\mathscr{H}^2_\pm(\mathbb{R}) := \left\{\left(\frac{1}{2}I \pm \frac{i}{2\pi}H\right)f : f \in L^2(\mathbb{R})\right\}, \qquad (7.8.38)$$

then any complex-valued function $f \in L^2(\mathbb{R})$ may be uniquely decomposed as $f = f_+ + f_-$ with $f_\pm \in \mathscr{H}^2_\pm(\mathbb{R})$ and, moreover, any two functions $f_\pm \in \mathscr{H}^2_\pm(\mathbb{R})$ are orthogonal, in the sense that

$$\int_{\mathbb{R}} f_+(x)\overline{f_-(x)}\,dx = 0. \qquad (7.8.39)$$

7.9 Fundamental Solutions for the Dirac Operator

In a nutshell, Dirac operators are first-order differential operators factoring the Laplacian. When $n = 1$, $\Delta = d^2/dx^2$, hence if we set $D := i(d/dx)$, then $D^2 = -\Delta$. We seek a higher dimensional generalization of the latter factorization formula. The natural context in which such a generalization may be carried out is the Clifford algebra setting.

The Clifford algebra with n generators $\mathcal{C}\!\ell_n$ is the associative algebra with unit $(\mathcal{C}\!\ell_n, \odot, +, 1)$ freely generated over $\mathbb{R}$ by the family $\{\mathbf{e}_j\}_{1 \le j \le n}$, of the standard orthonormal base in $\mathbb{R}^n$, now called imaginary units, subject to the following axioms:

$$\mathbf{e}_j \odot \mathbf{e}_k + \mathbf{e}_k \odot \mathbf{e}_j = -2\delta_{jk}, \qquad \forall\, j,k \in \{1, ..., n\}. \tag{7.9.1}$$

Hence,

$$\mathbf{e}_j \odot \mathbf{e}_k = -\mathbf{e}_k \odot \mathbf{e}_j \quad \text{if } 1 \le j \ne k \le n,$$
$$\text{and } \mathbf{e}_j \odot \mathbf{e}_j = -1 \quad \text{for } j \in \{1, ..., n\}. \tag{7.9.2}$$

The first condition above indicates that $\mathcal{C}\!\ell_n$ is noncommutative if $n > 1$, while the second condition justifies calling $\{\mathbf{e}_j\}_{1 \le j \le n}$ imaginary units. Elements in the Clifford algebra $\mathcal{C}\!\ell_n$ can be uniquely written in the form

$$a = \sum_{l=0}^{n} {\sum_{|I|=l}}' a_I \, \mathbf{e}_I \tag{7.9.3}$$

with $a_I \in \mathbb{C}$, where $\mathbf{e}_I$ stands for the product $\mathbf{e}_{i_1} \odot \mathbf{e}_{i_2} \odot \cdots \odot \mathbf{e}_{i_l}$ if $I = (i_1, i_2, \ldots, i_l)$ with $1 \le i_1 < i_2 < \cdots < i_l \le n$, $\mathbf{e}_\varnothing := 1 \in \mathbb{R}$ (that plays the role of the multiplicative unit in $\mathcal{C}\!\ell_n$) and $\sum'_{|I|=l}$ indicates that the sum is performed over strictly increasingly ordered indexes I with l components (selected from the set $\{1, ..., n\}$). In the writing (7.9.3) we shall refer to the numbers $a_I \in \mathbb{C}$ as the scalar components of a.

Exercise 7.50. Given any $a, b \in \mathcal{C}\!\ell_n$ with scalar components $a_I, b_I \in \mathbb{C}$ define

$$(a,b) := \sum_{l=0}^{n} {\sum_{|I|=l}}' a_I b_I, \qquad \overline{a} := \sum_{l=0}^{n} {\sum_{|I|=l}}' \overline{a_I} \, \mathbf{e}_I, \tag{7.9.4}$$

and abbreviate $M_a b := a \odot b$. Prove that

$$\overline{M_a b} = M_{\overline{a}} \, \overline{b} \quad \text{and} \quad (M_a b, c) = (b, M_a c) \quad \text{for every} \quad a, b, c \in \mathcal{C}\!\ell_n. \tag{7.9.5}$$

Clifford algebra-valued functions defined in an open set $\Omega \subseteq \mathbb{R}^n$ may be defined naturally. Specifically, any function $f : \Omega \to \mathcal{C}\!\ell_n$ is an object of the form

$$f = \sum_{l=0}^{n} {\sum_{|I|=l}}' f_I \, \mathbf{e}_I, \tag{7.9.6}$$

where each component f_I is a complex-valued function defined in Ω. Given any $k \in \mathbb{N}_0 \cup \{\infty\}$, we shall denote by $C^k(\Omega, \mathcal{C}\!\ell_n)$ the collection of all Clifford algebra-valued functions f whose scalar components f_I are of class C^k in Ω. In a similar manner we may define $C_0^\infty(\Omega, \mathcal{C}\!\ell_n)$, $L^1_{loc}(\Omega, \mathcal{C}\!\ell_n)$, $L^p(\Omega, \mathcal{C}\!\ell_n)$, etc.

In fact, we may also consider Clifford algebra-valued distributions in an open set $\Omega \subseteq \mathbb{R}^n$. Specifically, write $u \in \mathcal{D}'(\Omega, \mathcal{C}\!\ell_n)$ provided

$$u = \sum_{l=0}^{n} \sideset{}{'}\sum_{|I|=l} u_I \, \mathbf{e}_I \quad \text{with} \quad u_I \in \mathcal{D}'(\Omega) \quad \text{for each } I. \tag{7.9.7}$$

In particular we have that the Dirac distribution $\delta = \delta \mathbf{e}_\varnothing$, and the action of a Clifford algebra-valued distribution $u \in \mathcal{D}'(\Omega, \mathcal{Cl}_m)$ as in (7.9.7) on a test function $\varphi \in C_0^\infty(\Omega)$ is naturally defined by

$$u(\varphi) := \sum_{\ell=0}^{m} \sideset{}{'}\sum_{|I|=\ell} \langle u_I, \varphi \rangle \, \mathbf{e}_I. \tag{7.9.8}$$

Much of the theory originally developed for scalar-valued distributions extends in a natural fashion to the current setting. For example, if $u \in \mathcal{D}'(\Omega, \mathcal{Cl}_n)$ is as in (7.9.7), we may define

$$\partial^\alpha u := \sum_{l=0}^{n} \sideset{}{'}\sum_{|I|=l} \partial^\alpha u_I \, \mathbf{e}_I, \qquad \forall \, \alpha \in \mathbb{N}_0^n, \tag{7.9.9}$$

and

$$fu := \sum_{\ell=0}^{n} \sideset{}{'}\sum_{|J|=\ell} \sum_{k=0}^{n} \sideset{}{'}\sum_{|I|=k} f_J u_I \, \mathbf{e}_J \odot \mathbf{e}_I, \tag{7.9.10}$$

for any $f = \sum_{k=0}^{n} \sideset{}{'}\sum_{|J|=k} f_J \, \mathbf{e}_J \in C^\infty(\Omega, \mathcal{Cl}_n)$.

We are ready to introduce the Dirac operator D associated with $\mathcal{Cl}_n$. Specifically, given a Clifford algebra-valued distribution $u \in \mathcal{D}'(\Omega, \mathcal{Cl}_n)$ as in (7.9.7), we define $Du \in \mathcal{D}'(\Omega, \mathcal{Cl}_n)$ by setting

$$Du := \sum_{j=1}^{n} \sum_{l=0}^{n} \sideset{}{'}\sum_{|I|=l} \partial_j u_I \, \mathbf{e}_j \odot \mathbf{e}_I. \tag{7.9.11}$$

In other words,

$$D := \sum_{j=1}^{n} M_{\mathbf{e}_j} \partial_j, \tag{7.9.12}$$

where $M_{\mathbf{e}_j}$ denotes the operator of Clifford algebra multiplication by $\mathbf{e}_j$ from the left.

Proposition 7.51. *Let $\Omega \subseteq \mathbb{R}^n$ be an open set. Then the Dirac operator D satisfies*

$$D^2 = -\Delta \quad \text{in} \quad \mathcal{D}'(\Omega, \mathcal{Cl}_n), \tag{7.9.13}$$

where Δ is the Laplacian.

Proof. Pick an arbitrary $u \in \mathcal{D}'(\Omega, \mathcal{Cl}_n)$, say $u = \sum_{l=0}^{n} \sideset{}{'}\sum_{|I|=l} u_I \, \mathbf{e}_I$ with $u_I \in \mathcal{D}'(\Omega)$ for each I. Then using (7.9.11) twice yields

$$D^2 u = D(Du) = \sum_{j,k=1}^{n} \sum_{l=0}^{n} \sideset{}{'}\sum_{|I|=l} \partial_k \partial_j u_I \, \mathbf{e}_k \odot \mathbf{e}_j \odot \mathbf{e}_I. \tag{7.9.14}$$

Observe that, on the one hand,

$$\sum_{1\leq j\neq k\leq n} \sum_{l=0}^{n} \sideset{}{'}\sum_{|I|=l} \partial_k\partial_j u_I \, \mathbf{e}_k \odot \mathbf{e}_j \odot \mathbf{e}_I = 0, \tag{7.9.15}$$

since for each I, we have $\partial_k\partial_j u_I = \partial_j\partial_k u_I$ and $\mathbf{e}_k \odot \mathbf{e}_j \odot \mathbf{e}_I = -\mathbf{e}_j \odot \mathbf{e}_k \odot \mathbf{e}_I$ whenever $j \neq k$ by the first formula in (7.9.2). On the other hand, corresponding to the case when $j = k$

$$\sum_{k=1}^{n}\sum_{l=0}^{n} \sideset{}{'}\sum_{|I|=l} \partial_k^2 u_I \, \mathbf{e}_k \odot \mathbf{e}_k \odot \mathbf{e}_I = -\sum_{k=1}^{n}\sum_{l=0}^{n}\sideset{}{'}\sum_{|I|=l} \partial_k^2 u_I \, \mathbf{e}_I = -\Delta u, \tag{7.9.16}$$

since $\mathbf{e}_k \odot \mathbf{e}_k = -1$ by the second formula in (7.9.2). $\square$

Exercise 7.52. Consider the embedding

$$\mathbb{R}^n \hookrightarrow \mathcal{Cl}_n, \qquad \mathbb{R}^n \ni x = (x_j)_{1\leq j\leq n} \equiv \sum_{j=1}^{n} x_j \mathbf{e}_j \in \mathcal{Cl}_n, \tag{7.9.17}$$

which identifies vectors from $\mathbb{R}^n$ with elements in the Clifford algebra $\mathcal{Cl}_n$. With this identification in mind, show that

$$x \odot x = -|x|^2 \quad \text{for any } x \in \mathbb{R}^n, \tag{7.9.18}$$

$$x \odot y + y \odot x = -2x \cdot y \quad \text{for any } x, y \in \mathbb{R}^n. \tag{7.9.19}$$

In light of the embedding described in (7.9.17) we may regard the assignment $\mathbb{R}^n \setminus \{0\} \ni x \mapsto \frac{x}{|x|^n}$ as a Clifford algebra-valued function.

Proposition 7.51 combined with Remark 5.7 yields the following result.

Theorem 7.53. *The Clifford algebra-valued function*

$$E(x) := -\frac{1}{\omega_{n-1}} \frac{x}{|x|^n} \in L_{loc}^1(\mathbb{R}^n, \mathcal{Cl}_n) \cap S'(\mathbb{R}^n, \mathcal{Cl}_n) \tag{7.9.20}$$

is a fundamental solution for the Dirac operator D in $\mathbb{R}^n$,, i.e.,

$$DE = \delta \quad \text{in} \ \ S'(\mathbb{R}^n, \mathcal{Cl}_n). \tag{7.9.21}$$

Moreover, any $u \in S'(\mathbb{R}^n, \mathcal{Cl}_n)$ satisfying $Du = \delta$ in $S'(\mathbb{R}^n, \mathcal{Cl}_n)$ is of the form $E + P$ where E is as in (7.9.20) and P is a Clifford algebra-valued function whose components are polynomials and such that $DP = 0$ in $\mathbb{R}^n$.

Proof. Let E_Δ be the fundamental solution for Δ as described in (7.1.12) for $n \geq 2$. From (7.9.13) and Remark 5.7, we may infer that $-DE_\Delta$ computed in $\mathcal{D}'(\mathbb{R}^n)$ is a fundamental solution for the Dirac operator D in $\mathbb{R}^n$. From Exercise 2.128 (when $n = 2$) and Exercise 2.129 (when $n \geq 3$) we obtain

$$-DE_\Delta = -\sum_{j=1}^{n} \partial_j(E_\Delta)\, \mathbf{e}_j = -\sum_{j=1}^{n} \frac{1}{\omega_{n-1}} \cdot \frac{x_j}{|x|^n}\, \mathbf{e}_j$$

$$= -\frac{1}{\omega_{n-1}} \cdot \frac{x}{|x|^n}, \qquad \text{in } \mathcal{D}'(\mathbb{R}^n). \tag{7.9.22}$$

This proves that the distribution in (7.9.20) is indeed a fundamental solution for the Dirac operator D in $\mathbb{R}^n$. To justify the last claim in the statement of the theorem, let $F \in \mathcal{S}'(\mathbb{R}^n, \mathcal{C}\ell_n)$ be an arbitrary fundamental solution for the Dirac operator D in $\mathbb{R}^n$. Then the tempered distribution $P := F - E$ satisfies $DP = 0$ in $\mathcal{S}'(\mathbb{R}^n, \mathcal{C}\ell_n)$ and, on the Fourier transform side, we have $\xi \odot \widehat{P} = 0$ in $\mathcal{S}'(\mathbb{R}^n, \mathcal{C}\ell_n)$. Multiplying (in the Clifford algebra sense) this equality with the Clifford algebra-valued function with polynomial growth ξ then yields $-|\xi|^2 \widehat{P} = 0$ in $\mathcal{S}'(\mathbb{R}^n, \mathcal{C}\ell_n)$ (cf. (7.9.18)). In turn, the latter implies $\operatorname{supp} \widehat{P} \subseteq \{0\}$, hence the components of P are polynomials in $\mathbb{R}^n$ (cf. Exercise 4.37). □

There are other versions of the Dirac operator D from (7.9.12) that are more in line with the classical Cauchy–Riemann operator $\frac{\partial}{\partial \bar{z}} := \frac{1}{2}(\partial_1 + i\partial_2)$. Specifically, in $\mathbb{R}^{n+1}$ set $x = (x_0, x_1, x_2, \dots, x_n) \in \mathbb{R}^{n+1}$ and consider

$$D^\pm := \partial_0 \pm D = \partial_0 \pm \sum_{j=1}^{n} M_{\mathbf{e}_j} \partial_j, \tag{7.9.23}$$

Note that in the case when $n = 1$ the Dirac operator D^- corresponds to a constant multiple of the Cauchy–Riemann operator $\frac{\partial}{\partial \bar{z}}$. A reasoning similar to that used above for D also yields fundamental solutions for $D^\pm$. We leave this as an exercise for the interested reader.

Exercise 7.54. Let $\Omega \subseteq \mathbb{R}^{n+1}$ be an open set and let $x = (x_0, x_1, x_2, \dots, x_n) \in \mathbb{R}^{n+1}$. Then the Dirac operators $D^\pm$ satisfy

$$D^+ D^- = D^- D^+ = \Delta_{n+1} \quad \text{in } \mathcal{D}'(\Omega, \mathcal{C}\ell_n), \tag{7.9.24}$$

where Δ_{n+1} is the Laplacian in $\mathbb{R}^{n+1}$.

Moreover, use a reasoning similar to the one used in the proof of Theorem 7.53 to show that the functions

$$E^+(x) := \frac{1}{\omega_n} \frac{x_0 + \sum_{j=1}^{n} x_j \mathbf{e}_j}{|x|^{n+1}} \in L^1_{loc}(\mathbb{R}^{n+1}, \mathcal{C}\ell_n) \hookrightarrow \mathcal{D}'(\mathbb{R}^{n+1}, \mathcal{C}\ell_n) \tag{7.9.25}$$

$$E^-(x) := \frac{1}{\omega_n} \frac{x_0 - \sum_{j=1}^{n} x_j \mathbf{e}_j}{|x|^{n+1}} \in L^1_{loc}(\mathbb{R}^{n+1}, \mathcal{C}\ell_n) \hookrightarrow \mathcal{D}'(\mathbb{R}^{n+1}, \mathcal{C}\ell_n) \tag{7.9.26}$$

are fundamental solution for the Dirac operators D^- and D^+, respectively, in $\mathbb{R}^{n+1}$.

We next introduce the `Cauchy-Clifford` operator and discuss its jump-formulas in the upper and lower half-spaces.

Theorem 7.55. *For each $\mathcal{Cl}_n$-valued function $\varphi \in \mathcal{S}(\mathbb{R}^{n-1})$ define the Cauchy–Clifford operator*

$$(\mathscr{C}\varphi)(x) := -\frac{1}{\omega_{n-1}} \int_{\mathbb{R}^{n-1}} \frac{\sum_{j=1}^{n-1}(x_j - y_j)\mathbf{e}_j + x_n\mathbf{e}_n}{|x - (y', 0)|^n} \odot \mathbf{e}_n \odot \varphi(y')\,dy', \qquad (7.9.27)$$

for each $x = (x_1, \ldots, x_n) \in \mathbb{R}^n$ with $x_n \neq 0$. Then for each $x' \in \mathbb{R}^{n-1}$ one has

$$\lim_{x_n \to 0^{\pm}} (\mathscr{C}\varphi)(x', x_n) \qquad\qquad\qquad\qquad\qquad\qquad\qquad (7.9.28)$$

$$= \pm\frac{1}{2}\varphi(x') + \lim_{\varepsilon \to 0^+} \frac{1}{\omega_{n-1}} \int_{\substack{y' \in \mathbb{R}^{n-1} \\ |x' - y'| > \varepsilon}} \frac{\sum_{j=1}^{n-1}(y_j - x_j)\mathbf{e}_j}{|x' - y'|^n} \odot \mathbf{e}_n \odot \varphi(y')\,dy'.$$

Proof. Consider the Clifford algebra-value function $\Phi : \mathbb{R}^n \setminus \{0\} \to \mathcal{Cl}_n$ given by

$$\Phi(x) := -\frac{\sum_{j=1}^n x_j\mathbf{e}_j}{\omega_{n-1}|x|^n} \odot \mathbf{e}_n \quad \text{for each} \quad x = (x_1, \ldots, x_n) \in \mathbb{R}^n \setminus \{0\}. \qquad (7.9.29)$$

Then Φ is C^∞, odd, and positive homogeneous of degree $1 - n$ in $\mathbb{R}^n \setminus \{0\}$. As such, Corollary 4.81 may be applied (to each component of Φ). In this regard note from Corollary 4.65 that

$$\widehat{\Phi}(\xi) = -\frac{1}{\omega_{n-1}}\left[\sum_{j=1}^n \mathcal{F}\left(\frac{x_j}{|x|^n}\right)(\xi)\,\mathbf{e}_j\right] \odot \mathbf{e}_n = i\left[\sum_{j=1}^n \frac{\xi_j}{|\xi|^2}\,\mathbf{e}_j\right] \odot \mathbf{e}_n \qquad (7.9.30)$$

in $\mathcal{S}'(\mathbb{R}^n)$. In particular, since $\mathbf{e}_n \odot \mathbf{e}_n = -1$, we obtain

$$\widehat{\Phi}(0', 1) = i\,\mathbf{e}_n \odot \mathbf{e}_n = -i. \qquad (7.9.31)$$

Given that, as seen from (7.9.27) and (7.9.29), we have

$$(\mathscr{C}\varphi)(x) = \int_{\mathbb{R}^{n-1}} \Phi(x' - y', x_n) \odot \varphi(y')\,dy', \qquad (7.9.32)$$

for each $x = (x_1, \ldots, x_n) \in \mathbb{R}^n$ with $x_n \neq 0$, the jump-formulas for the Cauchy–Clifford operator in (7.9.28) follow from (4.7.46) and (7.9.31). $\qquad\square$

Paralleling the discussion in Remark 7.49, in the higher dimensional setting we have the following connection between the Cauchy–Clifford operator and Riesz transforms.

Remark 7.56. Let R_j, $j \in \{1, \ldots, n-1\}$, be the Riesz transforms in $\mathbb{R}^{n-1}$ (i.e., singular integral operators defined as in (4.9.11) with $n - 1$ in place of n). Also, recall the definition of the "vertical limit" of a function defined in $\mathbb{R}^n_+$ to the boundary of the upper half-space from (4.8.20). Then we may express the version of the jump-formula (7.9.28) corresponding to considering the Cauchy–Clifford operator in the upper half-space as

$$\left[\mathscr{C}\varphi\right]\Big|^{ver}_{\partial\mathbb{R}^n_+} = \frac{1}{2}\,\varphi - \frac{1}{\omega_{n-1}}\sum_{j=1}^{n-1}\mathbf{e}_j\odot\mathbf{e}_n\odot(R_j\varphi) \quad \text{in} \quad \mathbb{R}^{n-1} \tag{7.9.33}$$

for each $\mathcal{C}\!l_n$-valued function $\varphi \in \mathcal{S}(\mathbb{R}^{n-1})$, where the Riesz transforms act on φ componentwise.

The format of the jump-formula displayed in (7.9.33) suggests considering the operator acting on $\mathcal{C}\!l_n$-valued functions according to

$$P := \frac{1}{2}I - \frac{1}{\omega_{n-1}}\sum_{j=1}^{n-1}\mathbf{e}_j\odot\mathbf{e}_n\odot R_j$$

$$= \frac{1}{2}I + \frac{1}{\omega_{n-1}}\,\mathbf{e}_n\odot\Big(\sum_{j=1}^{n-1}\mathbf{e}_j\odot R_j\Big), \tag{7.9.34}$$

where I stands for the identity operator. In the second line of (7.9.34) the change in sign is due to the formula (cf. (7.9.2))

$$\mathbf{e}_j\odot\mathbf{e}_n = -\mathbf{e}_n\odot\mathbf{e}_j \quad \text{for each} \quad j\in\{1,\dots,n-1\}. \tag{7.9.35}$$

Theorem 4.97 then gives that P is a well-defined, linear, and bounded operator on $L^2(\mathbb{R}^{n-1},\mathcal{C}\!l_n)$. Moreover, since each R_j has a real-valued kernel, its action commutes with multiplication by elements from $\mathcal{C}\!l_n$ (i.e., for every $a \in \mathcal{C}\!l_n$ we have $R_jM_a = M_aR_j$, in the notation from Exercise 7.50). Keeping this in mind and relying on (7.9.35) and the fact that $\mathbf{e}_n^2 = -1$ (cf. (7.9.2)), we may then write

$$P^2 = \Big(\frac{1}{2}I + \frac{1}{\omega_{n-1}}\,\mathbf{e}_n\odot\Big(\sum_{j=1}^{n-1}\mathbf{e}_j\odot R_j\Big)\Big)^2$$

$$= \frac{1}{4}I + \frac{1}{\omega_{n-1}}\,\mathbf{e}_n\odot\Big(\sum_{j=1}^{n-1}\mathbf{e}_j\odot R_j\Big) + \Big(\frac{1}{\omega_{n-1}}\Big)^2\Big(\sum_{j=1}^{n-1}\mathbf{e}_j\odot R_j\Big)^2. \tag{7.9.36}$$

Furthermore, using the fact that, as proved in Theorem 4.97, the Riesz transforms commute with one another and satisfy $\sum_{j=1}^{n-1}R_j^2 = -(\frac{\omega_{n-1}}{2})^2 I$ on $L^2(\mathbb{R}^{n-1})$, we may expand

$$\Big(\sum_{j=1}^{n-1} \mathbf{e}_j \odot R_j\Big)^2 = \sum_{j,k=1}^{n-1} \mathbf{e}_j \odot \mathbf{e}_k \odot R_j R_k$$

$$= \sum_{j=1}^{n-1} \mathbf{e}_j^2 R_j^2 + \sum_{1 \le j \ne k \le n-1} \mathbf{e}_j \odot \mathbf{e}_k \odot R_j R_k$$

$$= -\sum_{j=1}^{n-1} R_j^2 = \big(\frac{\omega_{n-1}}{2}\big)^2 I, \tag{7.9.37}$$

where the source of the cancelation taking place in the third equality above is the observation that $\mathbf{e}_j \odot \mathbf{e}_k \odot R_j R_k = -\mathbf{e}_k \odot \mathbf{e}_j \odot R_k R_j$ whenever $1 \le j \ne k \le n-1$. Combining (7.9.36)–(7.9.37) and recalling (7.9.34) then yields

$$P^2 = P \quad \text{on} \quad L^2(\mathbb{R}^{n-1}, \mathcal{C}\!\ell_n). \tag{7.9.38}$$

Note that this is in agreement with the result obtained in (7.8.36) in the case of the two-dimensional setting. Let us also consider the higher dimensional analogue of (7.8.37). In this regard, we first observe that based on Exercise 7.50 for any $f, g \in L^2(\mathbb{R}^{n-1}, \mathcal{C}\!\ell_n)$ we may write

$$\int_{\mathbb{R}^{n-1}} \Big(M_{\mathbf{e}_n} \sum_{j=1}^{n-1} M_{\mathbf{e}_j}(R_j f)(x'), \overline{g(x')}\Big) dx'$$

$$= \int_{\mathbb{R}^{n-1}} \sum_{j=1}^{n-1} \Big((R_j f)(x'), \overline{M_{\mathbf{e}_j} M_{\mathbf{e}_n} g(x')}\Big) dx'$$

$$= -\int_{\mathbb{R}^{n-1}} \sum_{j=1}^{n-1} \Big(f(x'), \overline{M_{\mathbf{e}_j} M_{\mathbf{e}_n}(R_j g)(x')}\Big) dx'$$

$$= \int_{\mathbb{R}^{n-1}} \Big(f(x'), \overline{M_{\mathbf{e}_n} \sum_{j=1}^{n-1} M_{\mathbf{e}_j}(R_j g)(x')}\Big) dx'. \tag{7.9.39}$$

From this and (7.9.34) it follows that for every $f, g \in L^2(\mathbb{R}^{n-1}, \mathcal{C}\!\ell_n)$ we have

$$\int_{\mathbb{R}^{n-1}} ((Pf)(x'), \overline{g(x')}) dx' = \int_{\mathbb{R}^{n-1}} (f(x'), \overline{(Pg)(x')}) dx', \tag{7.9.40}$$

a condition that we shall interpret simply as

$$P^* = P \quad \text{on} \quad L^2(\mathbb{R}^{n-1}, \mathcal{C}\!\ell_n). \tag{7.9.41}$$

In summary, the above analysis shows that the operator P defined as in (7.9.34) is a projection on $L^2(\mathbb{R}^{n-1}, \mathcal{C}\!\ell_n)$. Starting from this result, a corresponding higher dimensional Hardy space theory may be developed in the Clifford algebra setting as well.

7.10 Fundamental Solutions for the Perturbed Dirac Operator

In this section we will work within the framework of $\mathcal{C}\ell_{n+1}$, the Clifford algebra with $n + 1$ imaginary units $\{e_j\}_{1 \leq j \leq n+1}$. This is the Clifford algebra introduced in Section 7.9 with n replaced by $n + 1$. We continue to denote by Ω an arbitrary open set in $\mathbb{R}^n$ and will be working with $\mathcal{C}\ell_{n+1}$-valued functions defined in Ω, which are functions $f : \Omega \to \mathcal{C}\ell_{n+1}$ of the form

$$f = \sum_{l=0}^{n+1} \sideset{}{'}\sum_{|I|=l} f_I \, e_I, \tag{7.10.1}$$

whose components f_I's are complex-valued functions defined in Ω. As before, we will also use the notation $C^m(\Omega, \mathcal{C}\ell_{n+1})$, $L^1_{loc}(\Omega, \mathcal{C}\ell_{n+1})$, $L^p(\Omega, \mathcal{C}\ell_{n+1})$, etc., for the collection of $\mathcal{C}\ell_{n+1}$-valued functions defined in Ω with scalar components in $C^m(\Omega)$, $L^1_{loc}(\Omega)$, $L^p(\Omega)$, etc., respectively (here $m \in \mathbb{N} \cup \{\infty\}$). Similarly, $\mathcal{D}'(\Omega, \mathcal{C}\ell_{n+1})$ denotes $\mathcal{C}\ell_{n+1}$-valued distributions in Ω. Hence, any $u \in \mathcal{D}'(\Omega, \mathcal{C}\ell_{n+1})$ is of the form

$$u = \sum_{l=0}^{n+1} \sideset{}{'}\sum_{|I|=l} u_I \, e_I \qquad \text{with } u_I \in \mathcal{D}'(\Omega) \text{ for each } I, \tag{7.10.2}$$

and the action of the Dirac operator D from (7.9.12) on u becomes

$$Du := \sum_{j=1}^{n} \sum_{l=0}^{n+1} \sideset{}{'}\sum_{|I|=l} \partial_j u_I \, e_j \odot e_I. \tag{7.10.3}$$

Next, fix $k \in \mathbb{C}$ arbitrary. Then the `perturbed Dirac operator` is denoted by D_k and is defined as

$$D_k := D + k e_{n+1} \odot . \tag{7.10.4}$$

Hence, if $u \in \mathcal{D}'(\Omega, \mathcal{C}\ell_{n+1})$ is as in (7.10.2), then the action of D_k on u is given as

$$D_k u = \sum_{j=1}^{n} \sum_{l=0}^{n+1} \sideset{}{'}\sum_{|I|=l} \partial_j u_I \, e_j \odot e_I + \sum_{l=0}^{n+1} \sideset{}{'}\sum_{|I|=l} k u_I \, e_{n+1} \odot e_I. \tag{7.10.5}$$

It is easy to see by an inspection of the proof of Proposition 7.51 that

$$D^2 = -\Delta \quad \text{in} \quad \mathcal{D}'(\Omega, \mathcal{C}\ell_{n+1}), \tag{7.10.6}$$

and that

$$D_k^2 = -(\Delta + k^2) \quad \text{in} \quad \mathcal{D}'(\Omega, \mathcal{C}\ell_{n+1}), \tag{7.10.7}$$

where $\Delta := \sum_{j=1}^{n} \partial_j^2$ is the Laplacian in $\mathbb{R}^n$. Moreover, since $\mathcal{Cl}_n \hookrightarrow \mathcal{Cl}_{n+1}$, the embedding from (7.9.17) now becomes

$$\mathbb{R}^n \hookrightarrow \mathcal{Cl}_{n+1}, \qquad \mathbb{R}^n \ni x = (x_j)_{1 \leq j \leq n} \equiv \sum_{j=1}^{n} x_j \mathbf{e}_j \in \mathcal{Cl}_{n+1}, \qquad (7.10.8)$$

the identities in (7.9.18) and (7.9.19) continue to hold in $\mathcal{Cl}_{n+1}$, and the assignment $\mathbb{R}^n \setminus \{0\} \ni x \mapsto \frac{x}{|x|^m}$ is a $\mathcal{Cl}_{n+1}$-valued function for each $m \in \mathbb{R}$.

Identity (7.10.7), Remark 5.7, and Theorem 7.33 are the main ingredients in the proof of the following result.

Theorem 7.57. *Let $k \in (0, \infty)$ and suppose $n \in \mathbb{N}$, $n \geq 2$. Then the Clifford algebra-valued function*

$$E_k(x) := \frac{c_n k^{n/2}}{|x|^{(n-2)/2}} \Big[H_{n/2}^{(1)}(k|x|) \frac{x}{|x|} - H_{(n-2)/2}^{(1)}(k|x|) \mathbf{e}_{n+1} \Big], \qquad (7.10.9)$$

for all $x \in \mathbb{R}^n \setminus \{0\}$, belongs to $L_{loc}^1(\mathbb{R}^n, \mathcal{Cl}_{n+1}) \cap \mathcal{S}'(\mathbb{R}^n, \mathcal{Cl}_{n+1})$ and satisfies

$$D_k E_k = \delta \quad in \ \ \mathcal{S}'(\mathbb{R}^n, \mathcal{Cl}_{n+1}), \qquad (7.10.10)$$

thus is a fundamental solution for the perturbed Dirac operator D_k in $\mathbb{R}^n$.

Similarly, the Clifford algebra-valued function

$$E_{-k}(x) := \frac{c_n k^{n/2}}{|x|^{(n-2)/2}} \Big[H_{n/2}^{(1)}(k|x|) \frac{x}{|x|} + H_{(n-2)/2}^{(1)}(k|x|) \mathbf{e}_{n+1} \Big], \qquad (7.10.11)$$

for all $x \in \mathbb{R}^n \setminus \{0\}$, belongs to $L_{loc}^1(\mathbb{R}^n, \mathcal{Cl}_{n+1}) \cap \mathcal{S}'(\mathbb{R}^n, \mathcal{Cl}_{n+1})$ and is a fundamental solution for the perturbed Dirac operator D_{-k} in $\mathbb{R}^n$.

Proof. Recall F_λ from (7.6.78). Then Φ_k from (7.6.11)–(7.6.12) may be written in terms of F_λ corresponding to $\lambda := (n-2)/2$ as

$$\Phi_k(x) = c_n k^{(n-2)/2} F_{(n-2)/2}(x), \qquad \forall x \in \mathbb{R}^n \setminus \{0\}. \qquad (7.10.12)$$

Also, the function E_k from (7.10.9) satisfies

$$E_k(x) = c_n k^{n/2} F_{n/2}(x) \, x - c_n k^{n/2} F_{(n-2)/2}(x) \mathbf{e}_{n+1}, \qquad \forall x \in \mathbb{R}^n \setminus \{0\}. \qquad (7.10.13)$$

Proposition 7.38 (applied with $\lambda = (n-2)/2$) ensures that $E_k \in L_{loc}^1(\mathbb{R}^n)$. In addition, (7.10.4) and Proposition 7.38 imply

$$-D_k \Phi_k = -\sum_{j=1}^{n} \partial_j \Phi_k \, \mathbf{e}_j - k \Phi_k \, \mathbf{e}_{n+1}$$

$$= -c_n k^{(n-2)/2} \sum_{j=1}^{n} \partial_j F_{(n-2)/2} \, \mathbf{e}_j - c_n k^{n/2} F_{(n-2)/2} \, \mathbf{e}_{n+1}$$

$$= c_n k^{n/2} \sum_{j=1}^{n} F_{n/2} \, x_j \, \mathbf{e}_j - c_n k^{n/2} F_{(n-2)/2} \, \mathbf{e}_{n+1}$$

$$= E_k \quad \text{in} \quad \mathcal{D}'(\mathbb{R}^n, \mathcal{C}\!\ell_{n+1}). \tag{7.10.14}$$

This, (7.10.7), and Theorem 7.33, imply that $D_k E_k = \delta$ in $\mathcal{D}'(\mathbb{R}^n, \mathcal{C}\!\ell_{n+1})$. Hence, E_k is a fundamental solution for the perturbed Dirac operator D_k in $\mathbb{R}^n$. Also, item *(9)* in Lemma 14.71 implies that

$$E_k(x) + c_n k^{(n-3)/2} \sqrt{\frac{2}{\pi}} e^{i(k|x| - (n-1)\pi/4)} |x|^{-(n-1)/2} \, \mathbf{e}_{n+1}$$
$$= O(|x|^{-(n+1)/2}) \qquad \text{as} \quad |x| \to \infty. \tag{7.10.15}$$

Consequently, there exists some constant $C \in (0, \infty)$ such that the coefficients of the Clifford algebra-valued function $E_k(x)$ are bounded by $C|x|^{-(n-1)/2}$ for every point $x \in \mathbb{R}^n \setminus B(0,1)$. Hence, Example 4.4 applies (condition (4.1.4) is satisfied for any $m > (n+1)/2$) and gives that E_k belongs to $\mathcal{S}'(\mathbb{R}^n, \mathcal{C}\!\ell_{n+1})$.

The same reasoning also gives that E_{-k} belongs to $L^1_{loc}(\mathbb{R}^n, \mathcal{C}\!\ell_{n+1}) \cap \mathcal{S}'(\mathbb{R}^n, \mathcal{C}\!\ell_{n+1})$ and that $-D_{-k} \Phi_k = E_{-k}$ in $\mathcal{D}'(\mathbb{R}^n, \mathcal{C}\!\ell_{n+1})$, which ultimately, in concert with (7.10.7) and Theorem 7.33, implies that E_{-k} is a fundamental solution for D_{-k} in $\mathbb{R}^n$. □

Recall that the complex number $b_{n,k}$ has been defined in (7.6.48).

Exercise 7.58. Pick $k \in (0, \infty)$ and $n \in \mathbb{N}$, $n \geq 2$. Prove that

$$(D_k \Phi_k)(x - y) = \Phi_k(x - y)(ik\widehat{x} + k e_{n+1}) + O(|x|^{-(n+1)/2})$$

$$= b_{n,k} \frac{e^{ik|x|} e^{-ik\langle y, \widehat{x} \rangle}}{|x|^{(n-1)/2}} k(i\,\widehat{x} + e_{n+1}) + O(|x|^{-(n+1)/2})$$

$$\text{as} \quad |x| \to \infty, \tag{7.10.16}$$

uniformly for y in compact subsets of $\mathbb{R}^n$, and

$$(D_k \Phi_k)(x - y) = \Phi_k(x - y)(-ik\widehat{y} + k e_{n+1}) + O(|y|^{-(n+1)/2})$$

$$= b_{n,k} \frac{e^{ik|y|} e^{-ik\langle x, \widehat{y} \rangle}}{|y|^{(n-1)/2}} k(-i\,\widehat{y} + e_{n+1}) + O(|y|^{-(n+1)/2})$$

$$\text{as} \quad |y| \to \infty, \tag{7.10.17}$$

uniformly for x in compact subsets of $\mathbb{R}^n$.

7.11 Fundamental Solutions for the Iterated Perturbed Dirac Operator

Suppose $n \in \mathbb{N}$ is such that $n \geq 2$ and let $k \in (0, \infty)$. In this section we focus on determining a fundamental solution for the iterated Dirac operator D_k^N in $\mathbb{R}^n$, where $N \in \mathbb{N}$. In Section 7.10 we have treated the case $N = 1$; cf. Theorem 7.57. As seen in (7.10.7), the identity $D_k^2 = -(\Delta + k^2)$ holds in $\mathcal{D}'(\Omega, \mathcal{C}\!\ell_{n+1})$. Since in Section 7.7 we have determined a fundamental solution $\Phi_k^{(N)}$ for the iterated Helmholtz operator $(\Delta + k^2)^N$ (cf. Theorem 7.40), we now can obtain a fundamental solution for D_k^N as follows.

Case I: N is even. Since $D_k^N = (-1)^{N/2}(\Delta + k^2)^{N/2}$ in $\mathcal{D}'(\mathbb{R}^n, \mathcal{C}\!\ell_{n+1})$ and $\Phi_k^{(N/2)}$ is a fundamental solution for $(\Delta + k^2)^{N/2}$ in $\mathbb{R}^n$, it follows that $(-1)^{N/2}\Phi_k^{(N/2)}$ is a fundamental solution for D_k^N in $\mathbb{R}^n$.

Case II: N is odd. We write

$$(-1)^{(N+1)/2}(\Delta + k^2)^{(N+1)/2} = D_k^N D_k \quad \text{in} \ \ \mathcal{D}'(\Omega, \mathcal{C}\!\ell_{n+1}), \tag{7.11.1}$$

which in light of Remark 5.7 gives that $(-1)^{(N+1)/2}D_k\Phi_k^{((N+1)/2)}$, computed in the sense of distributions, is a fundamental solution for D_k^N in $\mathbb{R}^n$. To determine an explicit expression for this fundamental solution, based on (7.7.9), (7.10.4), and an application of Proposition 7.38 with $\lambda = (n - N - 1)/2 < (n - 1)/2$, we obtain

$$
\begin{aligned}
D_k\Phi_k^{((N+1)/2)}(x) &= \frac{(-1)^{(N-1)/2}c_n k^{(n-N-1)/2}}{2^{(N-1)/2}(\frac{N-1}{2})!} D_k\left[\frac{H^{(1)}_{(n-N-1)/2}(k|x|)}{|x|^{(n-N-1)/2}}\right] \\
&= \frac{(-1)^{(N-1)/2}c_n k^{(n-N-1)/2}}{2^{(N-1)/2}(\frac{N-1}{2})!} \sum_{j=1}^{n} \partial_j\left[\frac{H^{(1)}_{(n-N-1)/2}(k|x|)}{|x|^{(n-N-1)/2}}\right]\mathbf{e}_j \\
&\quad + \frac{(-1)^{(N-1)/2}c_n k^{(n-N+1)/2}}{2^{(N-1)/2}(\frac{N-1}{2})!} \cdot \frac{H^{(1)}_{(n-N-1)/2}(k|x|)}{|x|^{(n-N-1)/2}}\mathbf{e}_{n+1} \\
&= \frac{(-1)^{(N+1)/2}c_n k^{(n-N+1)/2}}{2^{(N-1)/2}(\frac{N-1}{2})!} \cdot \frac{H^{(1)}_{(n-N+1)/2}(k|x|)}{|x|^{(n-N+1)/2}}x_j\,\mathbf{e}_j \\
&\quad + \frac{(-1)^{(N-1)/2}c_n k^{(n-N+1)/2}}{2^{(N-1)/2}(\frac{N-1}{2})!} \cdot \frac{H^{(1)}_{(n-N-1)/2}(k|x|)}{|x|^{(n-N-1)/2}}\mathbf{e}_{n+1} \\
&= \frac{(-1)^{(N+1)/2}c_n k^{(n-N+1)/2}}{2^{(N-1)/2}(\frac{N-1}{2})!|x|^{(n-N-1)/2}} \times
\end{aligned}
$$

$$\times\left[H^{(1)}_{(n-N+1)/2}(k|x|)\frac{x}{|x|} - H^{(1)}_{(n-N-1)/2}(k|x|)\,\mathbf{e}_{n+1}\right] \tag{7.11.2}$$

in $\mathcal{D}'(\mathbb{R}^n, \mathcal{C}\!\ell_{n+1})$, where c_n is as in (7.6.12). In addition, Proposition 7.38 also gives that the resulting expression in the variable $x \in \mathbb{R}^n \setminus \{0\}$ defines a function in $L^1_{loc}(\mathbb{R}^n, \mathcal{C}\!\ell_{n+1})$ which is a Clifford algebra-valued tempered distribution in $\mathbb{R}^n$.

The above analysis combined with the definition of c_n from (7.6.12) justifies the following result.

Theorem 7.59. *Let* $n \in \mathbb{N}$, $n \geq 2$, *and let* $k \in (0, \infty)$. *Recall the functions* $\Phi_k^{(N)}$, $N \in \mathbb{N}$, *from Theorem 7.40 which belong to* $L^1_{loc}(\mathbb{R}^n) \cap S'(\mathbb{R}^n)$. *For each* $N \in \mathbb{N}$ *define, with derivatives taken in the sense of distributions in* $\mathbb{R}^n$,

$$\Theta_k^{(N)} := \begin{cases} (-1)^{N/2}\, \Phi_k^{(N/2)} & \text{if } N \text{ even,} \\ (-1)^{(N+1)/2}\, D_k\, \Phi_k^{((N+1)/2)} & \text{if } N \text{ odd.} \end{cases} \tag{7.11.3}$$

Then $\Theta_k^{(N)}$ *belongs to* $L^1_{loc}(\mathbb{R}^n, \mathcal{C}\!\ell_{n+1}) \cap S'(\mathbb{R}^n, \mathcal{C}\!\ell_{n+1})$ *and is a fundamental solution for the iterated perturbed Dirac operator* D_k^N *in* $\mathbb{R}^n$. *Moreover,*

$$\Theta_k^{(N)}(x) = \frac{ik^{(n-N)/2}}{2^{(n+N)/2}\, \left(\frac{N-2}{2}\right)!\, \pi^{(n-2)/2}\, |x|^{(n-N)/2}}\, H^{(1)}_{(n-N)/2}(k|x|), \tag{7.11.4}$$

for all $x \in \mathbb{R}^n \setminus \{0\}$, *if* N *is even,*

and

$$\Theta_k^{(N)}(x) = \frac{-ik^{(n-N+1)/2}}{2^{(n+N+1)/2}\, \left(\frac{N-1}{2}\right)!\, \pi^{(n-2)/2}\, |x|^{(n-N-1)/2}} \times$$
$$\times \left[H^{(1)}_{(n-N+1)/2}(k|x|)\, \frac{x}{|x|} - H^{(1)}_{(n-N-1)/2}(k|x|)\, e_{n+1} \right] \tag{7.11.5}$$

for all $x \in \mathbb{R}^n \setminus \{0\}$, *if* N *is odd.*

7.12 Fundamental Solutions for General Second-Order Operators

Consider a constant, complex coefficient, homogeneous, second-order differential operator $L = \sum\limits_{j,k=1}^{n} a_{jk}\partial_j\partial_k$ in $\mathbb{R}^n$. In a first stage, our goal is to find necessary and sufficient conditions, that can be expressed without reference to the theory of distributions, guaranteeing that a function $E \in L^1_{loc}(\mathbb{R}^n)$ is a fundamental solution for L. Several necessary conditions readily present themselves. First, it is clear that LE is the zero distribution in $\mathbb{R}^n \setminus \{0\}$. In addition, if L is elliptic then this necessarily implies that $E \in C^\infty(\mathbb{R}^n \setminus \{0\})$. In the absence of ellipticity as a hypothesis for L, we may wish to assume that E is reasonably regular, say $E \in C^2(\mathbb{R}^n \setminus \{0\})$. Second, if the fundamental solution E is a priori known to be a tempered distribution then $-L(\xi)\widehat{E}(\xi) = 1$ in $S'(\mathbb{R}^n)$. If L is elliptic and $n \geq 3$, this forces $E = E_0 + P$ where

$E_0 := -\mathcal{F}^{-1}(L(\xi)^{-1})$ is homogeneous of degree $2 - n$ and P is a polynomial that is annihilated by L. Hence, working with E_0 in place of E, there is no loss of generality in assuming that E is homogeneous of degree $2 - n$. The case $n = 2$ may also be included in this discussion by demanding that ∇E is homogeneous of degree $1 - n$.

In summary, it is reasonable to restrict our search for a fundamental solution for L in the class of functions satisfying $E \in C^2(\mathbb{R}^n \setminus \{0\}) \cap L^1_{loc}(\mathbb{R}^n)$ with the property that ∇E is positive homogeneous of degree $1 - n$ in $\mathbb{R}^n \setminus \{0\}$. However, these conditions do not rule out such trivial candidates as the zero distribution. In Theorem 7.60 we identify the key nondegeneracy property (7.12.2) guaranteeing that E is in fact a fundamental solution. Theorem 7.60 is then later used to find an explicit formula for such a fundamental solution, under a strong ellipticity assumption on L (cf. Theorem 7.68).

Theorem 7.60. *Assume that $n \geq 2$ and consider*

$$L = \sum_{j,k=1}^n a_{jk} \partial_j \partial_k, \qquad a_{jk} \in \mathbb{C}. \tag{7.12.1}$$

Then for a function $E \in C^2(\mathbb{R}^n \setminus \{0\}) \cap L^1_{loc}(\mathbb{R}^n)$ with the property that ∇E is positive homogeneous of degree $1 - n$ in $\mathbb{R}^n \setminus \{0\}$, the following statements are equivalent:

(1) When viewed in $L^1_{loc}(\mathbb{R}^n)$, the function E is a fundamental solution for L in $\mathbb{R}^n$;
(2) One has $LE = 0$ pointwise in $\mathbb{R}^n \setminus \{0\}$ and

$$\sum_{j,k=1}^n \int_{S^{n-1}} a_{jk} \omega_j \partial_k E(\omega) \, d\sigma(\omega) = 1. \tag{7.12.2}$$

Remark 7.61. (i) In the partial differential equation parlance, the integrand in (7.12.2), i.e., the expression $\sum_{j,k=1}^n a_{jk} \omega_j \partial_k E(\omega)$, is referred to as the conormal derivative of E on S^{n-1}.

(ii) One remarkable aspect of Theorem 7.60 is that the description of a fundamental solution from part *(2)* is purely in terms of ordinary calculus (i.e., without any reference to the theory of distributions).

Proof of Theorem 7.60. Let $E \in C^2(\mathbb{R}^n \setminus \{0\}) \cap L^1_{loc}(\mathbb{R}^n)$ with the property that ∇E is positive homogeneous of degree $1 - n$ in $\mathbb{R}^n \setminus \{0\}$. Exercise 4.53 then implies that $\nabla E \in L^1_{loc}(\mathbb{R}^n)$. Fix an arbitrary $f \in C_0^\infty(\mathbb{R}^n)$. Then making use of (4.4.19) and Proposition 4.70 (applied to each $\partial_k E$) we obtain

$$(LE) * f = \sum_{j,k=1}^{n} a_{jk} (\partial_j (\partial_k E)) * f$$

$$= \sum_{j,k=1}^{n} a_{jk} \Big(\int_{S^{n-1}} \partial_k E(\omega) \omega_j \, d\sigma(\omega) \Big) (\delta * f) + \sum_{j,k=1}^{n} a_{jk} \big(P.V.(\partial_j \partial_k E) \big) * f$$

$$= \Big(\int_{S^{n-1}} \sum_{j,k=1}^{n} a_{jk} \partial_k E(\omega) \omega_j \, d\sigma(\omega) \Big) f \tag{7.12.3}$$

$$+ \lim_{\varepsilon \to 0^+} \int_{|y-\cdot|>\varepsilon} \sum_{j,k=1}^{n} a_{jk} (\partial_j \partial_k E)(\cdot - y) f(y) \, dy \quad \text{in} \quad \mathcal{D}'(\mathbb{R}^n),$$

where we have also used the fact that $\delta * f = f$. Having proved this, we now turn in earnest to the proof of the equivalence in the statement of the theorem.

First, assume that E is a fundamental solution for L in $\mathbb{R}^n$. Then $LE = \delta$ in $\mathcal{D}'(\mathbb{R}^n)$ implies that $L(E|_{\mathbb{R}^n \setminus \{0\}}) = 0$ in $\mathcal{D}'(\mathbb{R}^n)$. Since by assumption $E|_{\mathbb{R}^n \setminus \{0\}}$ belongs to $C^2(\mathbb{R}^n \setminus \{0\})$, we arrive at the conclusion that $LE = 0$ pointwise in $\mathbb{R}^n \setminus \{0\}$. Explicitly,

$$\sum_{j,k=1}^{n} a_{jk} \partial_j \partial_k E(x) = 0, \quad \forall x \in \mathbb{R}^n \setminus \{0\}. \tag{7.12.4}$$

This proves the first claim in *(2)*. Next, for each function $f \in C_0^\infty(\mathbb{R}^n)$ we may write $f = \delta * f = (LE) * f$ in $\mathcal{D}'(\mathbb{R}^n)$ which, in light of (7.12.3) and (7.12.4), forces

$$f = \Big(\int_{S^{n-1}} \sum_{j,k=1}^{n} a_{jk} \partial_k E(\omega) \omega_j \, d\sigma(\omega) \Big) f. \tag{7.12.5}$$

Since $f \in C_0^\infty(\mathbb{R}^n)$ was arbitrary, (7.12.2) follows. This finishes the proof of *(1)* $\Rightarrow$ *(2)*.

Conversely, suppose that $E \in C^2(\mathbb{R}^n \setminus \{0\}) \cap L_{loc}^1(\mathbb{R}^n)$ with the property that ∇E is positive homogeneous of degree $1 - n$ in $\mathbb{R}^n \setminus \{0\}$, such that $LE = 0$ pointwise in $\mathbb{R}^n \setminus \{0\}$, and (7.12.2) holds. Then for every $f \in C_0^\infty(\mathbb{R}^n)$ formula (7.12.3) simply reduces to $(LE) * f = f$. Now Exercise 2.97 may be invoked to conclude that $LE = \delta$ in $\mathcal{D}'(\mathbb{R}^n)$, as wanted. $\qquad\square$

Next, we turn to the task of finding all fundamental solutions that are tempered distributions for general homogeneous, second-order, constant coefficient operators that are strongly elliptic. We begin by defining this stronger (than originally introduced in Definition 6.13) notion of ellipticity.

Let $A = (a_{jk})_{1 \le j,k \le n} \in \mathcal{M}_{n\times n}(\mathbb{C})$ and associated to such a matrix A, consider the operator

$$L_A := L_A(\partial) := \sum_{j,k=1}^{n} a_{jk} \partial_j \partial_k. \tag{7.12.6}$$

This is a homogeneous, second-order, constant coefficient operator for which the ellipticity condition reads

$$L_A(\xi) := \sum_{j,k=1}^{n} a_{jk}\xi_j\xi_k \neq 0, \qquad \forall \xi = (\xi_1,\ldots,\xi_n) \in \mathbb{R}^n \setminus \{0\}. \tag{7.12.7}$$

As a trivial consequence of the Malgrange–Ehrenpreis theorem (cf. Theorem 5.10), any elliptic operator has a fundamental solution. The goal is to obtain explicit formulas for such fundamental solutions for a subclass of homogeneous, elliptic, second-order, constant coefficient operators satisfying a stronger condition than (7.12.7).

Definition 7.62. Call an operator L_A as in (7.12.6) `strongly elliptic`, if there exists a constant $C \in (0,\infty)$ such that

$$\text{Re}\Big[\sum_{j,k=1}^{n} a_{jk}\xi_j\xi_k \Big] \geq C|\xi|^2, \quad \forall \xi = (\xi_1,\ldots,\xi_n) \in \mathbb{R}^n. \tag{7.12.8}$$

By extension, call a matrix $A = (a_{jk})_{1\leq j,k\leq n} \in \mathcal{M}_{n\times n}(\mathbb{C})$ `strongly elliptic` provided there exists some $C \in (0,\infty)$ with the property that (7.12.8) holds.

Remark 7.63.

(1) It is obvious that any operator L_A as in (7.12.6) that is strongly elliptic is elliptic.

(2) Up to changing L to $-L$, any elliptic, homogeneous, second- order, constant coefficient differential operator L with real coefficients is strongly elliptic. To see why this is the case let $A \in \mathcal{M}_{n\times n}(\mathbb{R})$ and suppose that L_A is elliptic. Consider the function $f : S^{n-1} \to \mathbb{R}$ defined by $f(\xi) := \sum_{j,k=1}^{n} a_{jk}\xi_j\xi_k$ for $\xi \in S^{n-1}$. Then f is continuous and since L_A is elliptic the number 0 is not in the image of f. The unit sphere S^{n-1} being compact and connected, it is mapped by f in a compact, connected, subset of $\mathbb{R}$, not containing 0. This forces the image of f to be a compact interval that does not contain 0. Hence, there exists $c \in (0,\infty)$ with the property that either $f(\xi) \geq c$ for every $\xi \in S^{n-1}$ or $-f(\xi) \geq c$ for every $\xi \in S^{n-1}$. This implies that either $f(\xi/|\xi|) \geq c$ for every $\xi \in \mathbb{R}^n \setminus \{0\}$ or $-f(\xi/|\xi|) \geq c$ for every $\xi \in \mathbb{R}^n \setminus \{0\}$, or equivalently, that either L_A or $-L_A$ is strongly elliptic.

(3) Consider the operator $L = \partial_1^2 + i\partial_2^2$ in $\mathbb{R}^2$. If we take $A := \begin{pmatrix} 1 & 0 \\ 0 & i \end{pmatrix}$ then $L = L_A$ is a homogeneous, second-order, constant complex coefficient differential operator, and $L(\xi) = \xi_1^2 + i\xi_2^2$, $\xi = (\xi_1,\xi_2) \in \mathbb{R}^2$. Clearly $L(\xi) \neq 0$ if $\xi \neq 0$, so L is elliptic. However, L is not strongly elliptic since $\text{Re}\,[L(\xi)] = \xi_1^2$ which cannot be bounded from below by a constant multiple of $|\xi|^2$ since the latter blows up if $|\xi_2| \to \infty$.

Fix $A = (a_{jk})_{1\leq j,k\leq n} \in \mathcal{M}_{n\times n}(\mathbb{C})$ and consider the operator L_A as in (7.12.6). Due to the symmetry of mixed partial derivatives in the sense of distributions, it is immediate that

$$L_A = L_{A_{sym}}, \quad \text{where} \quad A_{sym} := \frac{A + A^\top}{2}. \tag{7.12.9}$$

As such, any fundamental solution for $L_{A_{sym}}$ is also a fundamental solution for L_A. Also, since $(A_{sym}\xi) \cdot \xi = (A\xi) \cdot \xi$ for every $\xi \in \mathbb{R}^n$, we have that

$$L_A \quad \text{is strongly elliptic if and only if} \quad L_{A_{sym}} \quad \text{is strongly elliptic.} \tag{7.12.10}$$

Consequently,

> when computing the fundamental solution for L_A
> we may assume without loss of generality that A (7.12.11)
> is symmetric, i.e., $A = A^\top$.

For further reference we summarize a few basic properties of symmetric matrices (throughout, the symbol "$\cdot$" denotes the real inner product of vectors with complex components).

$$A \in \mathcal{M}_{n \times n}(\mathbb{R}), \;\; A = A^\top \implies (A\zeta) \cdot \overline{\zeta} \in \mathbb{R}, \quad \forall \zeta \in \mathbb{C}^n, \tag{7.12.12}$$

$$A \in \mathcal{M}_{n \times n}(\mathbb{C}), \;\; A = A^\top \implies \operatorname{Re} A = (\operatorname{Re} A)^\top \text{ and } \operatorname{Im} A = (\operatorname{Im} A)^\top, \tag{7.12.13}$$

$$A \in \mathcal{M}_{n \times n}(\mathbb{C}), \;\; A = A^\top \implies \operatorname{Re}[(A\zeta) \cdot \overline{\zeta}] = ((\operatorname{Re} A)\zeta) \cdot \overline{\zeta}, \quad \forall \zeta \in \mathbb{C}^n. \tag{7.12.14}$$

It is easy to see that (7.12.12)–(7.12.13) hold, while (7.12.14) follows from (7.12.12)–(7.12.13) after writing $A = \operatorname{Re} A + i \operatorname{Im} A$.

Also, recall that a matrix $A \in \mathcal{M}_{n \times n}(\mathbb{C})$ is said to be positive definite provided

$$(A\zeta) \cdot \overline{\zeta} \quad \text{is real and strictly positive for each} \quad \zeta \in \mathbb{C}^n \setminus \{0\}. \tag{7.12.15}$$

It is easy to see that any positive-definite matrix $A \in \mathcal{M}_{n \times n}(\mathbb{C})$ satisfies $(\overline{A})^\top = A$, and there exists $c \in (0, \infty)$ such that

$$(A\zeta) \cdot \overline{\zeta} \geq c|\zeta|^2, \quad \forall \zeta \in \mathbb{C}^n. \tag{7.12.16}$$

Remark 7.64. Fix $A \in \mathcal{M}_{n \times n}(\mathbb{C})$ that is symmetric and satisfies (7.12.8). Then, for each $\zeta \in \mathbb{C}^n$ we have (with $C \in (0, \infty)$ as in (7.12.8))

$$\operatorname{Re}[(A\zeta) \cdot \overline{\zeta}] = \operatorname{Re}[(A(\operatorname{Re}\zeta + i\operatorname{Im}\zeta)) \cdot (\operatorname{Re}\zeta - i\operatorname{Im}\zeta)]$$

$$= \operatorname{Re}[(A\operatorname{Re}\zeta) \cdot \operatorname{Re}\zeta + (A\operatorname{Im}\zeta) \cdot \operatorname{Im}\zeta]$$

$$\geq C|\operatorname{Re}\zeta|^2 + C|\operatorname{Im}\zeta|^2 = C|\zeta|^2. \tag{7.12.17}$$

The second equality in (7.12.17) uses the fact that A is symmetric, while (7.12.8) is used for the inequality in (7.12.17). Thus, combining (7.12.17) with the Cauchy–Schwarz inequality, we obtain

$$|A\zeta| \geq C|\zeta| \quad \text{for every} \quad \zeta \in \mathbb{C}^n, \tag{7.12.18}$$

proving that the linear map $A : \mathbb{C}^n \to \mathbb{C}^n$ is injective, thus invertible. In particular, $\det A \neq 0$. Also, thanks to (7.12.14) and (7.12.17) we have that $\operatorname{Re} A$ is a positive-definite matrix. More precisely, with C as in (7.12.8), we have

$$((\operatorname{Re} A)\zeta) \cdot \overline{\zeta} \geq C|\zeta|^2 \qquad \forall\, \zeta \in \mathbb{C}^n. \tag{7.12.19}$$

From Remark 7.64 and definitions we see that, given any $A \in \mathcal{M}_{n\times n}(\mathbb{C})$, the following implications hold:

$$\operatorname{Re} A \text{ positive definite} \implies A \text{ strongly elliptic}, \tag{7.12.20}$$

$$A \text{ strongly elliptic} \iff A_{sym} \text{ strongly elliptic}, \tag{7.12.21}$$

$$A \text{ strongly elliptic and symmetric} \implies \operatorname{Re} A \text{ positive definite}. \tag{7.12.22}$$

Remark 7.65. Assume that $A \in \mathcal{M}_{n\times n}(\mathbb{C})$ is symmetric and satisfies (7.12.8). From Remark 7.64 it follows that A is invertible. Moreover, if we define

$$\|A\| := \sup\{|A\zeta| : \zeta \in \mathbb{C}^n,\ |\zeta| = 1\}, \tag{7.12.23}$$

then (7.12.18) ensures that $\|A\| > 0$. We claim that

$$\operatorname{Re}[(A^{-1}\zeta) \cdot \overline{\zeta}] \geq \frac{C}{\|A\|^2}|\zeta|^2 \qquad \forall\, \zeta \in \mathbb{C}^n, \tag{7.12.24}$$

where C is as in (7.12.8). To justify this, first note that

$$|\zeta|^2 = |AA^{-1}\zeta|^2 \leq \|A\|^2|A^{-1}\zeta|^2 \quad \text{for each } \zeta \in \mathbb{C}^n. \tag{7.12.25}$$

In turn, (7.12.25) and (7.12.17) permit us to estimate

$$\operatorname{Re}[(A^{-1}\zeta) \cdot \overline{\zeta}] = \operatorname{Re}[\overline{(A^{-1}\zeta)} \cdot \zeta] = \operatorname{Re}[\overline{(A^{-1}\zeta)} \cdot (AA^{-1}\zeta)] \geq C|A^{-1}\zeta|^2$$

$$\geq \frac{C}{\|A\|^2}|\zeta|^2 \qquad \forall\, \zeta \in \mathbb{C}^n. \tag{7.12.26}$$

This proves (7.12.24). In particular, (7.12.24) yields

$$\|A^{-1}\|\|\xi|^2 \geq |(A^{-1}\xi) \cdot \xi| \geq \frac{C}{\|A\|^2}|\xi|^2 \qquad \forall\, \xi \in \mathbb{R}^n. \tag{7.12.27}$$

Remark 7.66. Consider the set

$$\mathfrak{M} := \{A \in \mathcal{M}_{n\times n}(\mathbb{C}) : A = A^\top,\ \operatorname{Re} A \text{ is positive definite}\}. \tag{7.12.28}$$

Since the $n\times n$ symmetric matrices $A = (a_{jk})_{1\leq j,k\leq n}$ with complex entries are uniquely determined by the elements a_{jk} with $1 \leq j \leq k \leq n$, we may naturally identify $\mathfrak{M}$ with an open convex subset of $\mathbb{C}^{n(n+1)/2}$. Throughout, this identification is implicitly assumed. Moreover, every $A \in \mathfrak{M}$ satisfies $\det A \neq 0$, since if $A\zeta = 0$ for some

$\zeta \in \mathbb{C}^n$, then (7.12.14) and the fact that $\mathrm{Re}\, A$ is positive definite force $\zeta = 0$. The fact that $\mathfrak{M}$ is convex, implies that there is a unique analytic branch of the mapping $\mathfrak{M} \ni A \mapsto (\det A)^{\frac{1}{2}} \in \mathbb{C}$ such that $(\det A)^{\frac{1}{2}} > 0$ when A is real. Thus $(\det A)^{\frac{1}{2}}$ is unambiguously defined for $A \in \mathfrak{M}$.

To proceed with the discussion regarding determining a fundamental solution for a strongly elliptic operator L_A we first analyze the case when A has real entries.

The Case When A is Real, Symmetric and Satisfies (7.12.8).
Since A is real, symmetric, and positive definite, A is diagonalizable and has the form $A = U^{-1}DU$ for some orthogonal matrix $U \in M_{n \times n}(\mathbb{R})$ and some diagonal $n \times n$ matrix D whose entries on the main diagonal are strictly positive real numbers. Hence, $D^{\frac{1}{2}}$ is meaningfully defined as the $n \times n$ diagonal matrix with the entries on the diagonal being equal to the square roots of the entries on the main diagonal in D. In addition, $A^{\frac{1}{2}} := U^{-1}D^{\frac{1}{2}}U$ is well-defined, symmetric, invertible, $A^{\frac{1}{2}}A^{\frac{1}{2}} = A$ and $\det(A^{\frac{1}{2}}) = \sqrt{\det A}$. The idea now is to apply Exercise 7.77 with $A_1 := A$ and $B := A^{\frac{1}{2}}$. In this vein, observe that $A_2 = A^{-\frac{1}{2}}A(A^{-\frac{1}{2}})^{\top} = A^{-\frac{1}{2}}AA^{-\frac{1}{2}} = I$. Thus, $L_{A_2} = \varDelta$ and formula (7.14.3) becomes

$$(L_A u) \circ A^{\frac{1}{2}} = \varDelta(u \circ A^{\frac{1}{2}}) \quad \text{for every} \ \ u \in \mathcal{D}'(\mathbb{R}^n). \tag{7.12.29}$$

Our goal is to find $E_A \in \mathcal{D}'(\mathbb{R}^n)$ such that $L_A E_A = \delta$ in $\mathcal{D}'(\mathbb{R}^n)$. The latter equality is equivalent with $(L_A E_A) \circ A^{\frac{1}{2}} = \delta \circ A^{\frac{1}{2}}$ in $\mathcal{D}'(\mathbb{R}^n)$. Using (2.2.7) and the definition of δ we see that

$$\langle \delta \circ A^{\frac{1}{2}}, \varphi \rangle = \frac{1}{\sqrt{\det A}} \langle \delta, \varphi \circ A^{-\frac{1}{2}} \rangle = \frac{1}{\sqrt{\det A}} \varphi(0)$$

$$= \Big\langle \frac{1}{\sqrt{\det A}} \delta, \varphi \Big\rangle, \qquad \forall \varphi \in C_0^\infty(\mathbb{R}^n). \tag{7.12.30}$$

Hence, in light of (7.12.29) and (7.12.30), it suffices to find $E_A \in \mathcal{D}'(\mathbb{R}^n)$ such that

$$\varDelta(E_A \circ A^{\frac{1}{2}}) = \frac{1}{\sqrt{\det A}} \delta \quad \text{in} \ \ \mathcal{D}'(\mathbb{R}^n). \tag{7.12.31}$$

In particular, if we now choose

$$E_A(y) := \frac{1}{\sqrt{\det A}}(E_\varDelta \circ A^{-\frac{1}{2}})(y) \quad \text{for} \ \ y \in \mathbb{R}^n \setminus \{0\}, \tag{7.12.32}$$

where $E_\varDelta$ is the fundamental solution for the Laplacian from (7.1.12), then E_A satisfies (7.12.31), hence is a fundamental solution for L_A in $\mathbb{R}^n$. The additional properties of $E_\varDelta$, (7.12.27), and Exercise 4.6, imply that $E_A \in L^1_{loc}(\mathbb{R}^n)$ and E_A is a tempered distribution in $\mathbb{R}^n$. Keeping in mind that

$$|A^{-\frac{1}{2}}x|^2 = (A^{-\frac{1}{2}}x) \cdot (A^{-\frac{1}{2}}x) = (A^{-1}x) \cdot x, \qquad \forall x \in \mathbb{R}^n, \tag{7.12.33}$$

formulas (7.1.12) and (7.12.32) give

$$
E_A(x) = \begin{cases} \dfrac{-1}{(n-2)\omega_{n-1}\sqrt{\det A}} \cdot \dfrac{1}{[(A^{-1}x)\cdot x]^{\frac{n-2}{2}}} & \text{if } n \ge 3, \\[2ex] \dfrac{1}{4\pi\sqrt{\det A}} \ln\left[(A^{-1}x)\cdot x\right] & \text{if } n = 2, \end{cases} \tag{7.12.34}
$$

for every $x \in \mathbb{R}^n \setminus \{0\}$. This completes the discussion for determining a fundamental solution for L_A in the case when A is real, symmetric, and satisfies (7.12.8).

In preparation to dealing with the case of matrices with complex entries, we state and prove the following useful complex analysis result.

Lemma 7.67. *Let $N \in \mathbb{N}$ and assume O is an open and convex subset of $\mathbb{C}^N$ with the property that $O \cap \mathbb{R}^N \neq \varnothing$ (where $\mathbb{R}^N$ is canonically embedded into $\mathbb{C}^N$). Also, suppose $f, g : O \to \mathbb{C}$ are two functions which are separately holomorphic (i.e., in each scalar complex component in $\mathbb{C}^N$) such that*

$$
f\big|_{O\cap\mathbb{R}^N} = g\big|_{O\cap\mathbb{R}^N}. \tag{7.12.35}
$$

Then $f = g$ in O.

Proof. Fix an arbitrary point $(x_1, \ldots, x_N) \in O \cap \mathbb{R}^N$ and consider

$$
O^1 := \{z_1 \in \mathbb{C} : (z_1, x_2, \ldots, x_N) \in O\}. \tag{7.12.36}
$$

Then O^1 is an open convex subset of $\mathbb{C}$, which contains x_1, hence $O^1 \cap \mathbb{R} \neq \varnothing$. Define the functions $f_1, g_1 : O^1 \to \mathbb{C}$ by

$$
f_1(z) := f(z, x_2, \ldots, x_N) \text{ and } g_1(z) := g(z, x_2, \ldots, x_N), \quad \forall z \in O^1. \tag{7.12.37}
$$

Then f_1 and g_1 are holomorphic functions in O^1 which coincide on $O^1 \cap \mathbb{R}$. Since the latter contains an accumulation point in the convex (hence connected) set O^1, it follows that $f_1 = g_1$ on O^1 by the coincidence theorem for holomorphic functions of one complex variable. Since $(x_1, \ldots, x_N) \in O \cap \mathbb{R}^N$ was arbitrary, we may conclude that

$$
f\big|_{O\cap(\mathbb{C}\times\mathbb{R}^{N-1})} = g\big|_{O\cap(\mathbb{C}\times\mathbb{R}^{N-1})}. \tag{7.12.38}
$$

Next, fix $(z_1, x_2, \ldots, x_N) \in O \cap (\mathbb{C} \times \mathbb{R}^{N-1})$ and define

$$
O^2 := \{z_2 \in \mathbb{C} : (z_1, z_2, x_3, \ldots, x_N) \in O\}. \tag{7.12.39}
$$

Once again, O^2 is an open convex subset of $\mathbb{C}$, which contains x_2, hence $O^2 \cap \mathbb{R} \neq \varnothing$. If we now define the functions $f_2, g_2 : O^2 \to \mathbb{C}$ by

$$
f_2(z) := f(z_1, z, x_3, \ldots, x_N) \text{ and } g_2(z) := g(z_1, z, x_3, \ldots, x_N) \tag{7.12.40}
$$

for each point z belonging to the set O^2,

it follows that f_2, g_2 are holomorphic in O^2 which, by (7.12.38), coincide on $O^2 \cap \mathbb{R}$. Given that the latter set contains an accumulation point in the convex set O^1, we deduce that $f_2 = g_2$ on O^2 by once again invoking the coincidence theorem for holomorphic functions of one complex variable. Upon recalling that $(z_1, x_2, \ldots, x_N) \in O \cap (\mathbb{C} \times \mathbb{R}^{N-1})$ was arbitrary, we conclude that

$$f\big|_{O \cap (\mathbb{C}^2 \times \mathbb{R}^{N-2})} = g\big|_{O \cap (\mathbb{C}^2 \times \mathbb{R}^{N-2})}. \tag{7.12.41}$$

Continuing this process inductively, we arrive at the conclusion that $f = g$ in O. $\square$

After this preamble, we are ready to consider the general case.

The Case When A Has Complex Entries, is Symmetric and Satisfies (7.12.8).
As observed in Remark 7.64, under the current assumptions, A continues to be invertible. Also, (7.12.27) holds. In addition, under the current assumptions we have that $A \in \mathfrak{M}$ and $(\det A)^{\frac{1}{2}}$ is unambiguously defined (see in Remark 7.66).

These comments show that the function E_A from (7.12.34) continues to be well defined under the current assumption on A if ln is replaced by the principal branch of the complex log (defined for points $z \in \mathbb{C} \setminus (-\infty, 0]$ so that $z^a = e^{a \log z}$ for each $a \in \mathbb{R}$). In addition, E_A continues to belong to $L^1_{loc}(\mathbb{R}^n)$ and E_A is of class C^∞ in $\mathbb{R}^n \setminus \{0\}$. Furthermore, from (7.12.27) and Exercise 4.6 it follows that the function E_A continues to be a tempered distribution in $\mathbb{R}^n$. The goal is to prove that this expression is a fundamental solution for L_A in the current case.

First, observe that since A^{-1} is symmetric, for each $j, k \in \{1, \ldots, n\}$ we have

$$\partial_k[(A^{-1}x) \cdot x] = 2(A^{-1}x)_k \quad \text{and} \quad \partial_j[(A^{-1}x)_k] = (A^{-1})_{kj}. \tag{7.12.42}$$

Hence, for every $x \in \mathbb{R}^n \setminus \{0\}$, differentiating pointwise we obtain

$$L_A((A^{-1}x) \cdot x)^{\frac{2-n}{2}} = \sum_{j,k=1}^n a_{jk} \partial_j \left[(2-n)((A^{-1}x) \cdot x)^{\frac{-n}{2}} (A^{-1}x)_k \right]$$

$$= -n(2-n) \sum_{j,k=1}^n a_{jk}((A^{-1}x) \cdot x)^{\frac{-n-2}{2}} (A^{-1}x)_j (A^{-1}x)_k$$

$$+ (2-n) \sum_{j,k=1}^n a_{jk}((A^{-1}x) \cdot x)^{\frac{-n}{2}} (A^{-1})_{kj} = 0. \tag{7.12.43}$$

Similarly, $L_A \log [(A^{-1}x) \cdot x] = 0$ for $x \in \mathbb{R}^2 \setminus \{0\}$. Thus, we may conclude that

$$L_A E_A(x) = 0 \qquad \forall x \in \mathbb{R}^n \setminus \{0\}. \tag{7.12.44}$$

Second, by making use of (7.12.42) and the expression for E_A we obtain

$$\nabla E_A(x) = \frac{1}{\omega_{n-1} \sqrt{\det A}} \cdot \frac{A^{-1}x}{[(A^{-1}x) \cdot x]^{\frac{n}{2}}} \qquad \forall x \in \mathbb{R}^n \setminus \{0\} \tag{7.12.45}$$

which, in particular, shows that ∇E_A is positive homogeneous of degree $1 - n$ in $\mathbb{R}^n \setminus \{0\}$. Furthermore, for each $x \in S^{n-1}$ we have

$$\sum_{j,k=1}^{n} a_{jk} x_j \partial_k E_A(x) = \frac{1}{\omega_{n-1} \sqrt{\det A}} \cdot \frac{\sum\limits_{j,k=1}^{n} a_{jk} x_j (A^{-1}x)_k}{[(A^{-1}x) \cdot x]^{\frac{n}{2}}}$$

$$= \frac{1}{\omega_{n-1} \sqrt{\det A}} \cdot \frac{(A^\top x) \cdot (A^{-1}x)}{[(A^{-1}x) \cdot x]^{\frac{n}{2}}}$$

$$= \frac{1}{\omega_{n-1} \sqrt{\det A}} \cdot \frac{1}{[(A^{-1}x) \cdot x]^{\frac{n}{2}}}, \tag{7.12.46}$$

where the last equality uses the fact that $|x| = 1$.

Invoking Theorem 7.60 we conclude that E_A is a fundamental solution for L_A in $\mathbb{R}^n$ if and only if

$$\omega_{n-1} \sqrt{\det A} = \int_{S^{n-1}} \frac{d\sigma(x)}{[(A^{-1}x) \cdot x]^{\frac{n}{2}}}. \tag{7.12.47}$$

The fact that we already know that E_A is a fundamental solution for L_A in $\mathbb{R}^n$ in the case when $A \in M_{n \times n}(\mathbb{R})$ satisfies $A = A^\top$ and condition (7.12.8), implies that formula (7.12.47) holds for this class of matrices.

We make the claim that in fact (7.12.47) actually holds for the larger class of matrices $A \in M_{n \times n}(\mathbb{C})$ satisfying $A = A^\top$ and condition (7.12.8). To see why this is true, recall the open subset $\mathfrak{M}$ of $\mathbb{C}^{n(n+1)/2}$ from (7.12.28) and consider the functions $f, g : \mathfrak{M} \to \mathbb{C}$ defined for every $A = (a_{jk})_{1 \le j,k \le n} \in \mathfrak{M}$ by

$$f((a_{jk})_{1 \le j \le k \le n}) := \omega_{n-1} \sqrt{\det A}, \tag{7.12.48}$$

$$g((a_{jk})_{1 \le j \le k \le n}) := \int_{S^{n-1}} \frac{d\sigma(x)}{[(A^{-1}x) \cdot x]^{\frac{n}{2}}}. \tag{7.12.49}$$

Then f and g are analytic (as functions of several complex variables) on $\mathfrak{M}$, which is an open convex set in $\mathbb{C}^{n(n+1)/2}$. If $A \in \mathfrak{M}$ has real entries, then A satisfies (7.12.8), and (7.12.47) holds for such A, hence $f = g$ on $\mathfrak{M} \cap M_{n \times n}(\mathbb{R})$. Invoking Lemma 7.67 we may therefore conclude that $f = g$ on $\mathfrak{M}$. Thus, (7.12.47) holds for every $A \in M_{n \times n}(\mathbb{C})$ satisfying $A = A^\top$ and condition (7.12.8).

Finally, we note that thanks to Proposition 5.8 and the current strong ellipticity assumption, any other fundamental solution of L_A belonging to $S'(\mathbb{R}^n)$ differs from E_A by a polynomial that L_A annihilates.

In summary, the above analysis proves the following result.

Theorem 7.68. *Suppose $A = (a_{jk})_{1 \le j,k \le n} \in M_{n \times n}(\mathbb{C})$ and consider the operator L_A associated to A as in (7.12.6). If L_A is strongly elliptic, then the function defined by*

$$E_A(x) := \begin{cases} -\dfrac{1}{(n-2)\omega_{n-1}\sqrt{\det A_{sym}}} \cdot \dfrac{1}{[((A_{sym})^{-1}x)\cdot x]^{\frac{n-2}{2}}} & \text{if } n \geq 3, \\[4mm] \dfrac{1}{4\pi\sqrt{\det A_{sym}}}\log\left[((A_{sym})^{-1}x)\cdot x\right] & \text{if } n = 2, \end{cases} \qquad (7.12.50)$$

for $x \in \mathbb{R}^n \setminus \{0\}$ belongs to $L^1_{loc}(\mathbb{R}^n) \cap \mathcal{S}'(\mathbb{R}^n) \cap C^\infty(\mathbb{R}^n \setminus \{0\})$ and is a fundamental solution for L_A in $\mathbb{R}^n$. Above, $A_{sym} := \frac{1}{2}(A + A^\top)$, $\sqrt{\det A_{sym}}$ is defined as in Remark 7.66, and $\log$ denotes the principal branch of the complex logarithm (defined for complex numbers $z \in \mathbb{C} \setminus (-\infty, 0]$ so that $z^a = e^{a\log z}$ for each $a \in \mathbb{R}$). Moreover,

$$\{u \in \mathcal{S}'(\mathbb{R}^n) : L_A u = \delta \text{ in } \mathcal{S}'(\mathbb{R}^n)\} \qquad (7.12.51)$$
$$= \{E_A + P : P \text{ polynomial such that } L_A P = 0 \text{ in } \mathbb{R}^n\}.$$

We conclude this section with a couple of related exercises about fundamental solutions for second-order, constant coefficient, differential operators.

Exercise 7.69. Let $n \geq 2$ and consider a differential operator $L = \sum\limits_{j,k=1}^{n} a_{jk}\partial_j\partial_k$ with complex coefficients. Assume that $E \in C^2(\mathbb{R}^n \setminus \{0\}) \cap L^1_{loc}(\mathbb{R}^n)$ is a function with the property that ∇E is positive homogeneous of degree $1 - n$ in $\mathbb{R}^n \setminus \{0\}$. In addition, suppose that

$$\lambda := \sum_{j,k=1}^{n}\int_{S^{n-1}} a_{jk}\omega_j\partial_k E(\omega)\,d\sigma(\omega) \neq 0. \qquad (7.12.52)$$

Prove that $\lambda^{-1}E$ is a fundamental solution for L in $\mathbb{R}^n$. Use this result to find the proper normalization for the standard fundamental solution for the Laplacian in $\mathbb{R}^n$, starting with $E(x) := |x|^{2-n}$ when $n \geq 3$, and with $E(x) := \ln|x|$ for $n = 2$.

Exercise 7.70. Let $n \geq 2$ and suppose that $E \in C^2(\mathbb{R}^n \setminus \{0\}) \cap L^1_{loc}(\mathbb{R}^n)$ has the property that ∇E is odd and positive homogeneous of degree $1 - n$ in $\mathbb{R}^n \setminus \{0\}$. In addition, assume that the function E is a fundamental solution for the complex constant coefficient differential operator $L = \sum\limits_{j,k=1}^{n} a_{jk}\partial_j\partial_k$ in $\mathbb{R}^n$. Prove that for every $\xi \in S^{n-1}$ one has

$$\int_{\substack{\omega \in S^{n-1} \\ \omega\cdot\xi > 0}} \sum_{j,k=1}^{n} a_{jk}\omega_j\partial_k E(\omega)\,d\sigma(\omega) = \frac{1}{2}. \qquad (7.12.53)$$

Hint: Let $f(\xi) := \displaystyle\int_{\substack{\omega \in S^{n-1} \\ \omega\cdot\xi > 0}} \sum_{j,k=1}^{n} a_{jk}\omega_j\partial_k E(\omega)\,d\sigma(\omega)$ for each $\xi \in S^{n-1}$. Show that f is even and make use of Theorem 7.60.

7.13 Layer Potential Representation Formulas Revisited

The goal here is to derive a layer potential representation formula generalizing the identity from Proposition 7.17 for the Laplacian. We begin by describing the setting in which we intend to work.

Given a matrix $A = (a_{jk})_{1 \leq j,k \leq n} \in \mathcal{M}_{n \times n}(\mathbb{C})$, we associate the homogeneous second-order differential operator

$$L_A = \sum_{j,k=1}^{n} a_{jk} \partial_j \partial_k \quad \text{in} \quad \mathbb{R}^n, \tag{7.13.1}$$

and for every unit vector $\nu = (\nu_1, \ldots, \nu_n)$ and any complex-valued function u of class C^1, define the conormal derivative of u associated with the matrix A (along ν) as

$$\partial_\nu^A u := \sum_{j,k=1}^{n} \nu_j a_{jk} \partial_k u. \tag{7.13.2}$$

Theorem 7.71. *Suppose $n \geq 2$ and let $\Omega \subset \mathbb{R}^n$ be a bounded domain of class C^1, with outward unit normal ν and surface measure σ. In addition, assume that the matrix $A = (a_{jk})_{1 \leq j,k \leq n} \in \mathcal{M}_{n \times n}(\mathbb{C})$ is such that the operator L_A associated with A as in (7.13.1) is strongly elliptic, and recall the fundamental solution E_A for L_A defined in (7.12.50).*

Then for every complex-valued function $u \in C^2(\overline{\Omega})$ one has

$$\int_\Omega (L_A u)(y) E_A(x - y) \, dy - \int_{\partial \Omega} E_A(x - y)(\partial_\nu^A u)(y) \, d\sigma(y)$$

$$- \int_{\partial \Omega} u(y)(\partial_\nu^{A^\top} E)(x - y) \, d\sigma(y) = \begin{cases} u(x), & x \in \Omega, \\ 0, & x \in \mathbb{R}^n \setminus \overline{\Omega}, \end{cases} \tag{7.13.3}$$

where is the conormal derivative associated with the matrix $A^\top$ (along ν).

Proof. When $x \in \mathbb{R}^n \setminus \overline{\Omega}$, it clear from (7.12.50) that $E_A(x - \cdot) \in C^\infty(\overline{\Omega})$. Also, since E_A is a fundamental solution for L_A in $\mathbb{R}^n$ we have that $L_A[E_A(x - \cdot)] = 0$ in $\mathbb{R}^n \setminus \{x\}$, hence $(L_A E_A)(x - \cdot) = 0$ in Ω. Based on these and repeated use of (14.8.4) we may then write

$$\int_{\Omega} (L_A u)(y) E_A(x-y)\,dy = \sum_{j,k=1}^{n} \int_{\Omega} a_{jk}(\partial_j \partial_k u)(y) E_A(x-y)\,dy$$

$$= \sum_{j,k=1}^{n} \int_{\Omega} a_{jk}(\partial_k u)(y)(\partial_j E_A)(x-y)\,dy$$

$$+ \sum_{j,k=1}^{n} \int_{\partial\Omega} a_{jk}\nu_j(y)(\partial_k u)(y) E_A(x-y)\,d\sigma(y)$$

$$= \int_{\Omega} u(y)(L_A E_A)(x-y)\,dy + \int_{\partial\Omega} u(y)(\partial_\nu^{A^\top} E_A)(x-y)\,d\sigma(y)$$

$$+ \int_{\partial\Omega} E_A(x-y)(\partial_\nu^{A} u)(y)\,d\sigma(y). \tag{7.13.4}$$

Upon recalling that $(L_A E_A)(x - \cdot) = 0$ in Ω, the last solid integral drops out and the resulting identity is in agreement with (7.13.3).

Consider now the case when $x \in \Omega$. Since Ω is open, there exists $r > 0$ such that $B(x, r) \subseteq \Omega$. For each $\varepsilon \in (0, r)$ define $\Omega_\varepsilon := \Omega \setminus \overline{B(x, \varepsilon)}$ which is a bounded domain of class C^1. Since $E_A(x - \cdot) \in C^\infty(\overline{\Omega_\varepsilon})$ and $(L_A E_A)(x - \cdot) = 0$ in Ω_ε, the same type of reasoning as above gives (keeping in mind that $\partial\Omega_\varepsilon = \partial\Omega \cup \partial B(x, \varepsilon)$)

$$\int_{\Omega_\varepsilon} (L_A u)(y) E_A(x-y)\,dy = \int_{\partial\Omega} u(y)(\partial_\nu^{A^\top} E_A)(x-y)\,d\sigma(y)$$

$$- \int_{\partial B(x,\varepsilon)} u(y)(\partial_\nu^{A^\top} E_A)(x-y)\,d\sigma(y) + \int_{\partial\Omega} E_A(x-y)(\partial_\nu^{A} u)(y)\,d\sigma(y)$$

$$- \int_{\partial B(x,\varepsilon)} E_A(x-y)(\partial_\nu^{A} u)(y)\,d\sigma(y) =: I + II + III + IV. \tag{7.13.5}$$

As seen from (7.12.50), we have $|IV| \le C_A \|\nabla u\|_{L^\infty(\Omega)}\, \varepsilon \max\{1, |\ln|\varepsilon||\}$, from which we deduce that $\lim_{\varepsilon \to 0^+} IV = 0$. Next, split $II = II' + u(x)II''$ where

$$II' := - \int_{\partial B(x,\varepsilon)} [u(y) - u(x)](\partial_\nu^{A^\top} E_A)(x-y)\,d\sigma(y), \tag{7.13.6}$$

and observe that

$$II'' := -\int_{\partial B(x,\varepsilon)} (\partial_\nu^{A^\top} E_A)(x-y)\, d\sigma(y)$$

$$= -\int_{\partial B(x,\varepsilon)} \sum_{j,k=1}^n a_{jk} \frac{y_k - x_k}{\varepsilon} (\partial_j E_A)(x-y)\, d\sigma(y)$$

$$= \int_{S^{n-1}} \sum_{j,k=1}^n a_{jk}\omega_k(\partial_j E_A)(\omega)\, d\sigma(\omega)$$

$$= \int_{S^{n-1}} \sum_{j,k=1}^n (A^\top)_{jk}\, \omega_j (\partial_k E_{A^\top})(\omega)\, d\sigma(\omega) = 1. \tag{7.13.7}$$

Above, the first equality defines II'', the second equality uses the definition of the conormal derivative and the outward unit normal to the ball, the third equality is based on the change of variables $\omega = (x-y)/\varepsilon$ and the fact that ∇E_A is positive homogeneous of degree $1-n$. Finally, in the fourth equality we have interchanged j and k in the summation and used the identities $E_A = E_{A^\top}$, $L_A = L_{A^\top}$, while the last equality is due to (7.12.2).

In addition, $|II'| \le C_A \|\nabla u\|_{L^\infty(\Omega)}\, \varepsilon$, hence $\lim_{\varepsilon \to 0^+} II' = 0$, and

$$\lim_{\varepsilon \to 0^+} \int_{\Omega_\varepsilon} (L_A u)(y) E_A(x-y)\, dy = \int_\Omega (L_A u)(y) E_A(x-y)\, dy$$

by Lebesgue's Dominated Convergence Theorem (recall that $E_A(x-\cdot) \in L^1(\Omega)$ since Ω is bounded). Collectively, the results deduced in the above analysis yield (7.13.3) in the case when $x \in \Omega$, finishing the proof of the theorem. $\square$

Further Notes for Chapter 7. As evidenced by the treatment of the Poisson problem for the Laplacian and bi-Laplacian (from Section 7.2 and Section 7.4, respectively), fundamental solutions play a key role both for establishing integral representation formulas and for deriving estimates for the solution. This type of application to partial differential equations amply substantiate the utility of the tools from distribution theory and harmonic analysis derived in Section 4.10 (dealing with derivatives of volume potentials) and Section 4.9 (dealing with singular integral operators). The aforementioned Poisson problems serve as a prototype for other types of boundary value problems formulated for other differential operators and with the entire space $\mathbb{R}^n$ replaced by an open set $\Omega \subset \mathbb{R}^n$. In the latter scenario one specifies boundary conditions on $\partial\Omega$ in place of ∞ as in the case of $\mathbb{R}^n$ (note that ∞ plays the role of the topological boundary of $\mathbb{R}^n$ regarded as an open subset of its compactification $\mathbb{R}^n \cup \{\infty\}$).

In Section 7.9 the Dirac operator has been considered in the natural setting of Clifford algebras. For more information pertaining to this topic, the interested reader is referred to the monographs [6], [25], [58]. The last two references also contain a discussion of Hardy spaces in the context of Clifford algebras (a topic touched upon in Section 7.9). A classical reference to Hardy spaces in the ordinary context of $\mathbb{C}$ (which appeared at the end of Section 7.8) is the book [30].

7.14 Additional Exercises for Chapter 7

Exercise 7.72. Prove that there exists a unique $E \in S'(\mathbb{R}^n)$ such that $\Delta E - E = \delta$ in $S'(\mathbb{R}^n)$.

Exercise 7.73. Does there exist $E \in L^1(\mathbb{R}^n)$ such that $\Delta E = \delta$ in $\mathcal{D}'(\mathbb{R}^n)$?

Exercise 7.74. Prove that for every $u \in \mathcal{D}'(\Omega)$ we have

$$\text{div}(\nabla u) = \Delta u \quad \text{in} \quad \mathcal{D}'(\Omega). \tag{7.14.1}$$

Exercise 7.75. Prove that if $f \in C^\infty(\Omega)$ and $u \in \mathcal{D}'(\Omega)$, then

$$\Delta(fu) = (\Delta f)u + 2(\nabla f) \cdot (\nabla u) + f\Delta u \quad \text{in} \quad \mathcal{D}'(\Omega). \tag{7.14.2}$$

Exercise 7.76. Let Ω be an open set in $\mathbb{R}^2$ and suppose K is a compact set with $K \subset \Omega$. Prove that if $u \in C^2(\Omega)$ is such that its restriction to $\Omega \setminus K$ is holomorphic, then Δu is absolutely integrable on Ω and $\int_\Omega \Delta u \, dx = 0$.

Exercise 7.77. Suppose $A_1 \in M_{n\times n}(\mathbb{C})$ and $B \in M_{n\times n}(\mathbb{R})$ are given and B is invertible. Define the matrix $A_2 \in M_{n\times n}(\mathbb{C})$ by $A_2 := B^{-1}A_1(B^{-1})^\top$. Recall (7.12.6). Prove that

$$(L_{A_1}u) \circ B = L_{A_2}(u \circ B) \quad \text{for every} \ \ u \in \mathcal{D}'(\mathbb{R}^n). \tag{7.14.3}$$

Exercise 7.78. Suppose $n \geq 2$ and denote by E_Δ the fundamental solution for the Laplacian operator Δ given in (7.1.12). Without making use of Corollary 4.65, prove that for each $j \in \{1, \ldots, n\}$ one has

$$\mathcal{F}(\partial_j E_\Delta) = -\mathrm{i}\frac{\xi_j}{|\xi|^2} \quad \text{in} \quad S'(\mathbb{R}^n). \tag{7.14.4}$$

In turn, use (7.14.4) to show that

$$\mathcal{F}\left(\frac{x_j}{|x|^n}\right) = -\mathrm{i}\omega_{n-1}\frac{\xi_j}{|\xi|^2} \quad \text{in} \quad S'(\mathbb{R}^n), \tag{7.14.5}$$

and

$$\mathcal{F}^{-1}\left(\frac{\xi_j}{|\xi|^2}\right) = \frac{\mathrm{i}}{\omega_{n-1}} \cdot \frac{x_j}{|x|^n} \quad \text{in} \quad S'(\mathbb{R}^n). \tag{7.14.6}$$

Exercise 7.79. Suppose $n \geq 3$ and denote by E_{Δ^2} the fundamental solution for the bi-Laplacian operator Δ^2 given in (7.3.8). Prove that for each $j, k \in \{1, \ldots, n\}$ one has

$$\mathcal{F}(\partial_j\partial_k E_{\Delta^2}) = -\frac{\xi_j\xi_k}{|\xi|^4} \quad \text{in} \quad S'(\mathbb{R}^n). \tag{7.14.7}$$

Consequently,

$$\mathcal{F}^{-1}\left(\frac{\xi_j\xi_k}{|\xi|^4}\right) = \frac{1}{2(n-2)\omega_{n-1}} \cdot \frac{\delta_{jk}}{|x|^{n-2}} - \frac{1}{2\omega_{n-1}} \cdot \frac{x_jx_k}{|x|^n} \quad \text{in} \quad S'(\mathbb{R}^n), \tag{7.14.8}$$

and

$$\mathcal{F}\left(\frac{x_j x_k}{|x|^n}\right) = \omega_{n-1}\frac{\delta_{jk}}{|\xi|^2} - 2\omega_{n-1}\frac{\xi_j \xi_k}{|\xi|^4} \quad \text{in} \quad \mathcal{S}'(\mathbb{R}^n). \tag{7.14.9}$$

Exercise 7.80. Let $P(D)$ be a nonzero linear constant coefficient operator of order $m \in \mathbb{N}_0$. Prove that $P(D)$ is elliptic if and only if there exist $C, R \in (0, \infty)$ such that $|P(\xi)| \geq C|\xi|^m$ for every $\xi \in \mathbb{R}^n \setminus B(0, R)$.

Exercise 7.81. Give a second proof to Theorem 7.68 without making any appeal to Theorem 7.60.

Chapter 8
The Heat Operator and Related Versions

Abstract This chapter has a twofold aim: determine all fundamental solutions that are tempered distributions for the heat operator and related versions (including the Schrödinger operator), then use this as a tool in obtaining the solution of the generalized Cauchy problem for the heat operator.

8.1 Fundamental Solutions for the Heat Operator

Throughout this chapter we use the notation $(x, t) := (x_1, \ldots, x_n, t) \in \mathbb{R}^{n+1}$. The heat operator[1] is then defined as $L := \partial_t - \Delta_x = \partial_t - \sum_{j=1}^{n} \partial_{x_j}^2$. The starting point in determining all fundamental solutions for the heat operator L that are tempered distributions is Theorem 5.14 which guarantees that such fundamental solutions do exist. As we have seen in the case of the Laplace operator, the Fourier transform is an important tool in determining explicit expressions for fundamental solutions that are tempered distributions. We will continue to make use of this tool in the case of the heat operator with the adjustment that, this time, we work with the partial Fourier transform $\mathcal{F}_x$ discussed at the end of Section 4.2.

Let $E \in \mathcal{S}'(\mathbb{R}^{n+1})$ be a fundamental solution for L. Thus, in view of Exercise 2.90 and (4.1.33), we have

$$\partial_t E - \Delta_x E = \delta(x) \otimes \delta(t) \quad \text{in} \quad \mathcal{S}'(\mathbb{R}^{n+1}). \tag{8.1.1}$$

Applying $\mathcal{F}_x$ to (8.1.1), denoting $\mathcal{F}_x(E)$ by $\widehat{E}_x$, and using Exercise 4.42, it follows that

$$\partial_t \widehat{E}_x + |\xi|^2 \widehat{E}_x = 1(\xi) \otimes \delta(t) \quad \text{in} \quad \mathcal{S}'(\mathbb{R}^{n+1}). \tag{8.1.2}$$

In particular, for each fixed $\xi \in \mathbb{R}^n$, we have

---

[1] First considered in 1809 for $n = 1$ by Laplace (cf. [42]) and then for higher dimensions by Poisson (cf. [62] for $n = 2$).

© Springer Nature Switzerland AG 2018
D. Mitrea, *Distributions, Partial Differential Equations, and Harmonic Analysis*,
Universitext, https://doi.org/10.1007/978-3-030-03296-8_8

$$[\partial_t + |\xi|^2]\widehat{E}_x = \delta(t) \quad \text{in } \mathcal{D}'(\mathbb{R}). \tag{8.1.3}$$

Applying Example 5.13, we obtain that $\widehat{E}_x(\xi, t) := H(t)e^{-|\xi|^2 t}$ is a solution of (8.1.3), where as before, H denotes the Heaviside function from (1.2.9). Consequently, we have $\widehat{E}_x(\xi, t) \in \mathcal{S}'(\mathbb{R}^{n+1})$. Also, if $\varphi \in C_0^\infty(\mathbb{R}^{n+1})$ then, integration by parts yields

$$\langle \partial_t \widehat{E}_x + |\xi|^2 \widehat{E}_x, \varphi \rangle = -\langle H(t)e^{-|\xi|^2 t}, \partial_t \varphi(\xi, t) \rangle + \langle H(t)e^{-|\xi|^2 t}, |\xi|^2 \varphi(\xi, t) \rangle$$

$$= -\int_{\mathbb{R}^n} \int_0^\infty e^{-|\xi|^2 t} \partial_t \varphi(\xi, t)\, dt\, d\xi$$

$$+ \int_0^\infty \int_{\mathbb{R}^n} |\xi|^2 e^{-|\xi|^2 t} \varphi(\xi, t)\, d\xi\, dt$$

$$= \int_{\mathbb{R}^n} \varphi(\xi, 0)\, d\xi = \langle 1(\xi) \otimes \delta(t), \varphi \rangle, \tag{8.1.4}$$

hence $\widehat{E}_x$ verifies $\partial_t \widehat{E}_x + |\xi|^2 \widehat{E}_x = 1(\xi) \otimes \delta(t)$ in $\mathcal{D}'(\mathbb{R}^{n+1})$. Invoking (4.1.33), it follows that $\widehat{E}_x(\xi, t) = H(t)e^{-|\xi|^2 t}$ verifies (8.1.2). We remark here that while the distribution $-H(-t)e^{-|\xi|^2 t}$ satisfies (8.1.3), based on our earlier discussion pertaining to the nature of the function in (4.1.35), it does not belong to $\mathcal{S}'(\mathbb{R}^{n+1})$, thus we cannot apply $\mathcal{F}_x^{-1}$ to it.

Starting from the identity $\mathcal{F}_x(\mathcal{F}_x E(x, t)) = (2\pi)^n E(-x, t)$ (which is easy to check), we may write

$$E(-x, t) = (2\pi)^{-n} \mathcal{F}_\xi(\widehat{E}_x(\xi))(x) = (2\pi)^{-n} H(t) \mathcal{F}_\xi(e^{-t|\xi|^2})(x)$$

$$= (2\pi)^{-n} H(t) \left(\frac{\pi}{t}\right)^{\frac{n}{2}} e^{-\frac{|x|^2}{4t}} = H(t)(4\pi t)^{-\frac{n}{2}} e^{-\frac{|x|^2}{4t}} \quad \text{in } \mathcal{S}'(\mathbb{R}^{n+1}), \tag{8.1.5}$$

where for the third equality in (8.1.5) we used Remark 4.23 and (3.2.6). Hence, we may conclude that the tempered distribution from (8.1.5) is a fundamental solution for the heat operator.

Note that, with the notation from (3.1.10), we have

$$\partial_t - \Delta_x = iD_{n+1} + \sum_{j=1}^n D_j^2, \tag{8.1.6}$$

hence the heat operator satisfies the hypothesis of Proposition 5.8. Consequently, if $u \in \mathcal{S}'(\mathbb{R}^{n+1})$ is an arbitrary fundamental solution for the heat operator in $\mathbb{R}^{n+1}$, then $u - E = P(x, t)$ in $\mathcal{S}'(\mathbb{R}^{n+1})$, for some polynomial $P(x, t)$ in $\mathbb{R}^{n+1}$ satisfying $(\partial_t - \Delta_x)P(x, t) = 0$ in $\mathbb{R}^{n+1}$.

As a final remark, we claim that E as in (8.1.5) satisfies $E \in L^1_{loc}(\mathbb{R}^{n+1})$. Indeed, if K is a compact subset of $\mathbb{R}^n \times [-R, R]$ for some $R \in (0, \infty)$, then

$$0 \leq \int_K E(x,t)\,dx\,dt \leq \int_0^R \int_{\mathbb{R}^n} (4\pi t)^{-\frac{n}{2}} e^{-\frac{|x|^2}{4t}}\,dx\,dt$$

$$= \pi^{-\frac{n}{2}} \int_0^R \int_{\mathbb{R}^n} e^{-|y|^2}\,dy\,dt = R < \infty. \tag{8.1.7}$$

In summary, we proved the following theorem.

Theorem 8.1. *The function defined as*

$$E(x,t) := H(t)(4\pi t)^{-\frac{n}{2}} e^{-\frac{|x|^2}{4t}}, \qquad \forall\, (x,t) \in \mathbb{R}^{n+1}, \tag{8.1.8}$$

belongs to $S'(\mathbb{R}^{n+1}) \cap L^1_{loc}(\mathbb{R}^{n+1}) \cap C^\infty(\mathbb{R}^{n+1} \setminus \{0\})$ *and is a fundamental solution for the heat operator* $\partial_t - \Delta_x$ *in* $\mathbb{R}^{n+1}$. *Moreover,*

$$\{u \in S'(\mathbb{R}^{n+1}) : (\partial_t - \Delta_x)u = \delta(x) \otimes \delta(t) \text{ in } S'(\mathbb{R}^{n+1})\} \tag{8.1.9}$$

$$= \{E + P : P \text{ polynomial in } \mathbb{R}^{n+1} \text{ satisfying } (\partial_t - \Delta_x)P(x,t) = 0\}.$$

Corollary 8.2. *The heat operator* $\partial_t - \Delta_x$ *is hypoelliptic in* $\mathbb{R}^{n+1}$.

Proof. This is a consequence of Theorem 6.8 and Theorem 8.1. $\qquad\square$

Exercise 8.3. Let E be the function defined in (8.1.9) and for each $(x,t) \in \mathbb{R}^{n+1}$ set $F(x,t) := -E(x,-t)$. Prove that F is a fundamental solution for the operator $\partial_t + \Delta$ in $\mathbb{R}^{n+1}$.

Remark 8.4. Given the expression $E(x,t) = H(t)(4\pi t)^{-\frac{n}{2}} e^{-\frac{|x|^2}{4t}}$ for $(x,t) \in \mathbb{R}^{n+1}$, then one may check via a direct computation that this is a fundamental solution for the heat operator $L = \partial_t - \Delta_x$. First, the computation in (8.1.7) gives that $E \in L^1_{loc}(\mathbb{R}^{n+1})$, which in turn implies $E \in \mathcal{D}'(\mathbb{R}^{n+1})$.

Second, if $\varphi \in C_0^\infty(\mathbb{R}^{n+1})$ is arbitrary, then using integration by parts we may write

$$\langle LE, \varphi \rangle = -\langle E, \partial_t \varphi + \Delta_x \varphi \rangle$$

$$= -\lim_{\varepsilon \to 0^+} \left[\int_\varepsilon^\infty \int_{\mathbb{R}^n} E(x,t)(\partial_t \varphi(x,t) + \Delta_x \varphi(x,t))\,dx\,dt \right]$$

$$= \lim_{\varepsilon \to 0^+} \left[\int_{\mathbb{R}^n} E(x,\varepsilon)\varphi(x,\varepsilon)\,dx + \int_\varepsilon^\infty \int_{\mathbb{R}^n} LE(x,t)\varphi(x,t)\,dx\,dt \right]$$

$$= \lim_{\varepsilon \to 0^+} \int_{\mathbb{R}^n} (4\pi\varepsilon)^{-\frac{n}{2}} e^{-\frac{|x|^2}{4\varepsilon}} \varphi(x,\varepsilon)\,dx$$

$$= \lim_{\varepsilon \to 0^+} \int_{\mathbb{R}^n} \pi^{-\frac{n}{2}} e^{-|y|^2} \varphi(2\sqrt{\varepsilon}y, \varepsilon)\,dy$$

$$= \varphi(0) = \langle \delta, \varphi \rangle. \tag{8.1.10}$$

For the fourth equality in (8.1.10) we have used the fact that $LE = 0$ pointwise in $\mathbb{R}^n \times (0, \infty)$, for the fifth a suitable change of variables, while for the sixth equality we applied Lebesgue's Dominated Convergence Theorem. This proves that $LE = \delta$ in $\mathcal{D}'(\mathbb{R}^{n+1})$, thus E is a fundamental solution for L.

In closing we record a Liouville type theorem for the operator $\partial_t - \Delta$, which is a particular case of Theorem 5.4.

Theorem 8.5 (Liouville's Theorem for the heat operator). *Any bounded function in $\mathbb{R}^{n+1}$ that is a null solution of the heat operator is constant.*

8.2 The Generalized Cauchy Problem for the Heat Operator

Let $F \in C^0(\mathbb{R}^n \times [0, \infty))$, $f \in C^0(\mathbb{R}^n)$ and suppose u is a solution of the Cauchy problem for the heat operator:

$$
\begin{cases}
u \in C^0(\mathbb{R}^n \times [0, \infty)), \\
\partial_t u, \partial_{x_j}^2 u \in C^0(\mathbb{R}^n \times (0, \infty)), \quad j = 1, \ldots, n, \\
\partial_t u - \Delta_x u = F \text{ in } \mathbb{R}^n \times (0, \infty), \\
u(\cdot, 0) = f \text{ in } \mathbb{R}^n.
\end{cases}
\tag{8.2.1}
$$

Denote by $\widetilde{u}$ and $\widetilde{F}$ the extensions by zero of u and F to the entire space $\mathbb{R}^{n+1}$. Then, if $\varphi \in C_0^\infty(\mathbb{R}^{n+1})$, integrating by parts and using Lebesgue's Dominated Convergence Theorem we obtain

$$
\begin{aligned}
\langle (\partial_t - \Delta_x)\widetilde{u}, \varphi \rangle &= -\langle \widetilde{u}, \partial_t u + \Delta_x \varphi \rangle = -\lim_{\varepsilon \to 0^+} \int_\varepsilon^\infty \int_{\mathbb{R}^n} u(\partial_t \varphi + \Delta_x \varphi) \, dx \, dt \\
&= \lim_{\varepsilon \to 0^+} \left[\int_\varepsilon^\infty \int_{\mathbb{R}^n} (\partial_t u - \Delta_x u)\varphi \, dx \, dt + \int_{\mathbb{R}^n} u(x, \varepsilon)\varphi(x, \varepsilon) \, dx \right] \\
&= \int_0^\infty \int_{\mathbb{R}^n} F(x, t)\varphi(x, t) \, dx \, dt + \int_{\mathbb{R}^n} f(x)\varphi(x, 0) \, dx \\
&= \langle \widetilde{F}, \varphi \rangle + \langle f(x) \otimes \delta(t), \varphi(x, t) \rangle.
\end{aligned}
\tag{8.2.2}
$$

This proves that $(\partial_t - \Delta_x)\widetilde{u} = \widetilde{F} + f(x) \otimes \delta(t)$ in $\mathcal{D}'(\mathbb{R}^{n+1})$ and suggests the definition made below. As a preamble, we introduce the notation

$$
\overline{\mathbb{R}_+^{n+1}} := \{(x, t) \in \mathbb{R}^{n+1} : t \geq 0\}
\tag{8.2.3}
$$

and the space

$$
\mathcal{D}_+'(\mathbb{R}^{n+1}) := \{u \in \mathcal{D}'(\mathbb{R}^{n+1}) : \operatorname{supp} u \subseteq \overline{\mathbb{R}_+^{n+1}}\}.
\tag{8.2.4}
$$

Definition 8.6. Let $F_0 \in \mathcal{D}'_+(\mathbb{R}^{n+1})$ and $f \in \mathcal{D}'(\mathbb{R}^n)$ be given. Then a distribution $u \in \mathcal{D}'_+(\mathbb{R}^{n+1})$ is called a solution of the `generalized Cauchy problem` for the heat operator for the data F_0 and f, if u verifies

$$(\partial_t - \Delta_x)u = F_0 + f(x) \otimes \delta(t) \quad \text{in } \mathcal{D}'(\mathbb{R}^{n+1}). \tag{8.2.5}$$

The issue of solvability of equation (8.2.5) fits into the framework presented in Remark 5.6. More precisely, let E be the fundamental solution for the heat operator as given in (8.1.8). Then, if $F_0 \in \mathcal{D}'_+(\mathbb{R}^{n+1})$ and $f \in \mathcal{D}'(\mathbb{R}^n)$ are such that

$$u := E * [F_0 + f \otimes \delta] \quad \text{exists in } \mathcal{D}'(\mathbb{R}^{n+1}) \tag{8.2.6}$$

the distribution u satisfies (8.2.5). For this u to be a solution of the generalized Cauchy problem for the heat operator, we would also need $\operatorname{supp} u \subseteq \overline{\mathbb{R}^{n+1}_+}$. Since it is not difficult to check that $\operatorname{supp} E \subseteq \overline{\mathbb{R}^{n+1}_+}$ and $\operatorname{supp}(f \otimes \delta) \subseteq \overline{\mathbb{R}^{n+1}_+}$, and since by assumption $\operatorname{supp} F_0 \subseteq \overline{\mathbb{R}^{n+1}_+}$, it follows that whenever condition (8.2.6) is verified, the distribution u defined in (8.2.6) also satisfies $\operatorname{supp} u \subseteq \overline{\mathbb{R}^{n+1}_+}$ (by (a) in Theorem 2.96). While the convolution of two arbitrary distributions in $\mathcal{D}'_+(\mathbb{R}^n)$ does not always exist (you might want to check that an exception is the case $n = 1$), under the additional assumptions $f \in \mathcal{E}'(\mathbb{R}^n)$ and $F_0 \in \mathcal{E}'(\mathbb{R}^{n+1})$ condition (8.2.6) is verified.

The above discussion is the reason why we analyze in detail the following setting. Retain the notation introduced at the beginning of the section, and assume that

$$\widetilde{F} \in C_0^\infty(\mathbb{R}^{n+1}) \quad \text{and} \quad f \in C_0^\infty(\mathbb{R}^n). \tag{8.2.7}$$

Then

$$u := E * \widetilde{F} + E * (f(x) \otimes \delta(t)) \quad \text{exists, belongs to } \mathcal{D}'_+(\mathbb{R}^{n+1}), \tag{8.2.8}$$

and is a solution of (8.2.5). We proceed by rewriting the expression for u in a more explicit form. First, by Proposition 2.102 and the fact that $E \in L^1_{loc}(\mathbb{R}^{n+1})$, we have $E * \widetilde{F} \in C^\infty(\mathbb{R}^{n+1})$ and

$$(E * \widetilde{F})(x,t) = \langle E(y,\tau), \widetilde{F}(x-y, t-\tau) \rangle = \langle E(x-y, t-\tau), \widetilde{F}(y,\tau) \rangle$$

$$= \int_0^\infty \int_{\mathbb{R}^n} H(t-\tau)[4\pi(t-\tau)]^{-\frac{n}{2}} e^{-\frac{(x-y)^2}{4(t-\tau)}} F(y,\tau) \, dy \, d\tau$$

$$= \int_0^t \int_{\mathbb{R}^n} E(x-y, t-\tau) F(y,\tau) \, dy \, d\tau. \tag{8.2.9}$$

To compute $E * (f \otimes \delta)$, fix an arbitrary compact set $K \subset \mathbb{R}^{n+1}$, consider a function $\varphi \in C_0^\infty(\mathbb{R}^{n+1})$ such that $\operatorname{supp} \varphi \subseteq K$, and pick some $\psi \in C_0^\infty(\mathbb{R}^{2n+2})$ with the property that $\psi \equiv 1$ in a neighborhood of the set

$$\{(x,t,y,0) \in \mathbb{R}^n \times \mathbb{R} \times \mathbb{R}^n \times \mathbb{R} : y \in \operatorname{supp} f \text{ and } (x+y,t) \in K\}. \tag{8.2.10}$$

Relying on the definition of convolution we have

$$\langle E * (f \otimes \delta), \varphi \rangle = \langle E(x,t), \langle f(y) \otimes \delta(\tau), \psi(x,t,y,\tau)\varphi(x+y,t+\tau) \rangle \rangle$$

$$= \left\langle E(x,t), \int_{\mathbb{R}^n} f(y)\varphi(x+y,t)\,\mathrm{d}y \right\rangle$$

$$= \int_{\mathbb{R}} \int_{\mathbb{R}^n} \int_{\mathbb{R}^n} E(x,t)f(y)\varphi(x+y,t)\,\mathrm{d}x\,\mathrm{d}y\,\mathrm{d}t$$

$$= \int_{\mathbb{R}} \int_{\mathbb{R}^n} \int_{\mathbb{R}^n} E(z-y,t)f(y)\varphi(z,t)\,\mathrm{d}z\,\mathrm{d}y\,\mathrm{d}t. \qquad (8.2.11)$$

Hence, $E * (f(x) \otimes \delta(t))$ is given by the function

$$\mathbb{R}^n \times \mathbb{R} \ni (x,t) \mapsto \int_{\mathbb{R}^n} E(x-y,t)f(y)\,\mathrm{d}y \in \mathbb{C}, \qquad (8.2.12)$$

whose restriction to $\mathbb{R}^n \times (0,\infty)$ is of class C^∞.

In summary, this analysis proves the following result.

Proposition 8.7. *Let $f \in C_0^\infty(\mathbb{R}^n)$ and assume that $F \in C^0(\mathbb{R}^n \times [0,\infty))$ is such that its extension $\widetilde{F}$ by zero to $\mathbb{R}^{n+1}$ satisfies $\widetilde{F} \in C_0^\infty(\mathbb{R}^{n+1})$. Also let E be the fundamental solution for the heat operator $\partial_t - \Delta_x$ as given in (8.1.8).*

Then the generalized Cauchy problem for the heat operator for the data $\widetilde{F}$ and f has a solution $u \in \mathcal{D}'_+(\mathbb{R}^{n+1})$ that is of function type, whose restriction to the set $\mathbb{R}^n \times (0,\infty)$ is of class C^∞, and has the expression

$$u(x,t) = \int_{\mathbb{R}^n} E(x-y,t)f(y)\,\mathrm{d}y + \int_0^t \int_{\mathbb{R}^n} E(x-y,t-\tau)\widetilde{F}(y,\tau)\,\mathrm{d}y\,\mathrm{d}\tau \qquad (8.2.13)$$

for every $x \in \mathbb{R}^n$ and $t \in (0,\infty)$.

Note that the integrals in (8.2.13) are meaningfully defined under weaker assumptions on F and f. In fact, starting with u as in (8.2.13) one may prove that this is a solution to a version of (8.2.1) (corresponding to finite time, i.e., $t \in (0,T)$, for some $T > 0$) under suitable yet less stringent conditions on F and f.

8.3 Fundamental Solutions for General Second-Order Parabolic Operators

We continue to work in $\mathbb{R}^{n+1}$ and use the notation $(x,t) := (x_1, \ldots, x_n, t) \in \mathbb{R}^{n+1}$. We look at a more general parabolic operator than the heat operator $\partial_t - \Delta_x$ discussed so far in this chapter. Specifically, if $A = (a_{jk})_{1 \leq j,k \leq n} \in \mathcal{M}_{n \times n}(\mathbb{C})$, associate to such a matrix A the parabolic operator

$$\mathcal{L}_A := \mathcal{L}_A(\partial) := \partial_t - \sum_{j,k=1}^{n} a_{jk}\partial_j\partial_k, \tag{8.3.1}$$

where $\partial_j = \partial_{x_j}$, $j = 1,\ldots,n$. The goal is to obtain explicit formulas for all tempered distributions in $\mathbb{R}^{n+1}$ that are fundamental solutions for $\mathcal{L}_A$ in $\mathbb{R}^{n+1}$ under the additional assumption that there exists a constant $C \in (0, \infty)$ such that the matrix A satisfies the strict positiveness condition

$$\text{Re}\Big[\sum_{j,k=1}^{n} a_{jk}\xi_j\xi_k \Big] \geq C|\xi|^2, \quad \forall \xi = (\xi_1,\ldots,\xi_n) \in \mathbb{R}^n. \tag{8.3.2}$$

The approach is an adaptation to the parabolic setting of the ideas used in Section 7.12 for the derivation of (7.12.50). In a first stage, we note that, via the same reasoning as in Section 7.12, when computing the fundamental solution for $\mathcal{L}_A$ we may assume without loss of generality that A is symmetric, i.e., $A = A^\top$. Also, we treat first the case when A has real entries.

The Case When A is Real, Symmetric and Satisfies (8.3.2).
Since A is real, symmetric and positive definite, $A^{\frac{1}{2}}$ is well defined, real, symmetric, invertible, $A^{\frac{1}{2}}A^{\frac{1}{2}} = A$, and $\det(A^{\frac{1}{2}}) = \sqrt{\det A}$. Consider next the matrix $\widetilde{B}$ in $\mathcal{M}_{(n+1)\times(n+1)}(\mathbb{R})$ whose entries on the positions (j,k), for $j,k \in \{1,\ldots,n\}$, coincide with the entries of $A^{\frac{1}{2}}$, has 1 on the entry $(n+1, n+1)$ and zeros on the rest of the entries. Hence, using matrix block notation, $\widetilde{B} = \begin{pmatrix} A^{\frac{1}{2}} & 0 \\ 0 & 1 \end{pmatrix}$. Then $\widetilde{B}$ is symmetric, invertible, $\widetilde{B}^{-1} = \begin{pmatrix} A^{-\frac{1}{2}} & 0 \\ 0 & 1 \end{pmatrix}$ and $\det\widetilde{B} = \sqrt{\det A}$. A direct computation based on the Chain Rule gives that

$$\Big(-\partial_t - \sum_{j,k=1}^{n} a_{jk}\partial_j\partial_k\Big)(\varphi \circ \widetilde{B}^{-1}) = [(-\partial_t - \Delta_x)\varphi] \circ \widetilde{B}^{-1} \tag{8.3.3}$$
$$\text{pointwise in } \mathbb{R}^{n+1} \text{ for every } \varphi \in C_0^\infty(\mathbb{R}^{n+1}).$$

This and (2.2.7) then allow us to write, for each $u \in \mathcal{D}'(\mathbb{R}^{n+1})$,

$$\langle (\mathcal{L}_A u) \circ \widetilde{B}, \varphi \rangle = \frac{1}{|\det\widetilde{B}|}\langle \mathcal{L}_A u, \varphi \circ \widetilde{B}^{-1} \rangle$$

$$= \frac{1}{|\det\widetilde{B}|}\Big\langle u, \big(-\partial_t - \sum_{j,k=1}^{n} a_{jk}\partial_j\partial_k\big)(\varphi \circ \widetilde{B}^{-1})\Big\rangle$$

$$= \frac{1}{|\det\widetilde{B}|}\langle u, [(-\partial_t - \Delta_x)\varphi] \circ \widetilde{B}^{-1} \rangle$$

$$= \langle u \circ \widetilde{B}, (-\partial_t - \Delta_x)\varphi \rangle$$

$$= \langle (\partial_t - \Delta_x)(u \circ \widetilde{B}), \varphi \rangle, \quad \forall \varphi \in C_0^\infty(\mathbb{R}^{n+1}). \tag{8.3.4}$$

In addition, with δ denoting the Dirac distribution in $\mathbb{R}^{n+1}$, we also have

$$\langle \delta \circ \widetilde{B}, \varphi \rangle = \frac{1}{\det \widetilde{B}} \langle \delta, \varphi \circ \widetilde{B}^{-1} \rangle = \frac{1}{\sqrt{\det A}} \varphi(0)$$

$$= \Big\langle \frac{1}{\sqrt{\det A}} \delta, \varphi \Big\rangle, \quad \forall \varphi \in C_0^\infty(\mathbb{R}^{n+1}). \tag{8.3.5}$$

Our goal is to find $E_A \in \mathcal{D}'(\mathbb{R}^{n+1})$ such that $\mathcal{L}_A E_A = \delta$ in $\mathcal{D}'(\mathbb{R}^{n+1})$. The latter equality is equivalent with $(\mathcal{L}_A E_A) \circ \widetilde{B} = \delta \circ \widetilde{B}$ in $\mathcal{D}'(\mathbb{R}^{n+1})$, which furthermore, in view of (8.3.4)–(8.3.5), becomes

$$(\partial_t - \Delta_x)(E_A \circ \widetilde{B}) = \frac{1}{\sqrt{\det A}} \delta \quad \text{in } \mathcal{D}'(\mathbb{R}^{n+1}). \tag{8.3.6}$$

If now $E_{\partial_t - \Delta}$ denotes the fundamental solution for the heat operator from (8.1.8), the properties of $E_{\partial_t - \Delta}$, (8.3.6), and the properties of $\widetilde{B}$, ensure that the function

$$E_A(x,t) := \frac{1}{\sqrt{\det A}} E_{\partial_t - \Delta}(A^{-\frac{1}{2}} x, t) \quad \text{for each } (x,t) \in \mathbb{R}^{n+1}, \tag{8.3.7}$$

is a fundamental solution for $\mathcal{L}_A$ in $\mathbb{R}^{n+1}$. Keeping in mind that for every $x \in \mathbb{R}^n$ we have $|A^{-\frac{1}{2}} x|^2 = (A^{-\frac{1}{2}} x) \cdot (A^{-\frac{1}{2}} x) = (A^{-1} x) \cdot x$, the function E_A may be re-written as

$$E_A(x,t) = \frac{1}{\sqrt{\det A}} H(t)(4\pi t)^{-\frac{n}{2}} e^{-\frac{(A^{-1}x) \cdot x}{4t}} \quad \text{for each } (x,t) \in \mathbb{R}^{n+1}. \tag{8.3.8}$$

Moreover, from (8.3.7), Theorem 8.1, and Proposition 4.43, it follows that E_A belongs to $\mathcal{S}'(\mathbb{R}^{n+1}) \cap L_{loc}^1(\mathbb{R}^{n+1}) \cap C^\infty(\mathbb{R}^{n+1} \setminus \{0\})$.

The Case When A Has Complex Entries, is Symmetric and Satisfies (8.3.2).
As observed in Remark 7.64, under the current assumptions, A continues to be invertible. Also, (7.12.27) holds. In addition, under the current assumptions $(\det A)^{\frac{1}{2}}$ is unambiguously defined (see in Remark 7.66).

These comments show that the function E_A from (8.3.8) continues to be well-defined under the current assumption on A if ln is replaced by the principal branch of the complex log (defined for points $z \in \mathbb{C} \setminus (-\infty, 0]$ so that $z^a = e^{a \log z}$ for each $a \in \mathbb{R}$). In addition, E_A continues to belong to $L_{loc}^1(\mathbb{R}^{n+1})$ (this can be seen by a computation similar to that in (8.1.7), keeping in mind (7.12.24)), and $E_A \in C^\infty(\mathbb{R}^{n+1} \setminus \{0\})$. Moreover, from (7.12.42) and Exercise 4.6 it follows that E_A belongs to $\mathcal{S}'(\mathbb{R}^{n+1})$.

The goal is to prove that this expression is a fundamental solution for $\mathcal{L}_A$ in the current case. Making use of (7.12.42) for every $(x,t) \in \mathbb{R}^{n+1}$ with $t \neq 0$ differentiating pointwise we obtain

$$\sum_{j,k=1}^n a_{jk} \partial_j \partial_k \Big[e^{-\frac{(A^{-1}x) \cdot x}{4t}} \Big] = e^{-\frac{(A^{-1}x) \cdot x}{4t}} \Big[\frac{(A^{-1}x) \cdot x}{4t^2} - \frac{n}{2t} \Big], \tag{8.3.9}$$

while

$$\partial_t\Big[t^{-n/2}e^{-\frac{(A^{-1}x)\cdot x}{4t}}\Big] = t^{-n/2}e^{-\frac{(A^{-1}x)\cdot x}{4t}}\Big[\frac{(A^{-1}x)\cdot x}{4t^2} - \frac{n}{2t}\Big]. \qquad (8.3.10)$$

From (8.3.9), (8.3.10), and the expression for E_A we may conclude that

$$\mathcal{L}_A E_A(x,t) = 0 \qquad \forall\, x \in \mathbb{R}^n,\ \forall\, t \in \mathbb{R}\setminus\{0\}. \qquad (8.3.11)$$

In addition, for each $\varphi \in C_0^\infty(\mathbb{R}^{n+1})$ we may compute

$$\langle \mathcal{L}_A E_A, \varphi \rangle = -\Big\langle E_A, \partial_t\varphi + \sum_{j,k=1}^n a_{jk}\partial_j\partial_k\varphi \Big\rangle$$

$$= -\lim_{\varepsilon\to 0^+}\Big[\int_\varepsilon^\infty \int_{\mathbb{R}^n} E_A(x,t)\Big(\partial_t\varphi(x,t) + \sum_{j,k=1}^n a_{jk}\partial_j\partial_k\varphi(x,t)\Big)\,dx\,dt\Big]$$

$$= \lim_{\varepsilon\to 0^+}\Big[\int_{\mathbb{R}^n} E_A(x,\varepsilon)\varphi(x,\varepsilon)\,dx + \int_\varepsilon^\infty \int_{\mathbb{R}^n} (\mathcal{L}_A E_A)(x,t)\varphi(x,t)\,dx\,dt\Big]$$

$$= \lim_{\varepsilon\to 0^+} \frac{1}{\sqrt{\det A}}\int_{\mathbb{R}^n} (4\pi\varepsilon)^{-\frac{n}{2}}e^{-\frac{(A^{-1}x)\cdot x}{4\varepsilon}}\,\varphi(x,\varepsilon)\,dx$$

$$= \lim_{\varepsilon\to 0^+} \frac{\pi^{-\frac{n}{2}}}{\sqrt{\det A}}\int_{\mathbb{R}^n} e^{-(A^{-1}y)\cdot y}\,\varphi(2\sqrt{\varepsilon}y,\varepsilon)\,dy$$

$$= \varphi(0)\frac{\pi^{-\frac{n}{2}}}{\sqrt{\det A}}\int_{\mathbb{R}^n} e^{-(A^{-1}y)\cdot y}\,dy. \qquad (8.3.12)$$

For the fourth equality in (8.3.12) we have used (8.3.11), for the fifth the change of variables $x = 2\sqrt{\varepsilon}y$, while for the sixth equality we applied Lebesgue's Dominated Convergence Theorem. From (8.3.12) we then conclude that E_A is a fundamental solution for $\mathcal{L}_A$ in $\mathbb{R}^{n+1}$ if and only if

$$\int_{\mathbb{R}^n} e^{-(A^{-1}y)\cdot y}\,dy = \pi^{\frac{n}{2}}\sqrt{\det A}. \qquad (8.3.13)$$

The fact that we already know that E_A is a fundamental solution for $\mathcal{L}_A$ in $\mathbb{R}^{n+1}$ in the case when $A \in \mathcal{M}_{n\times n}(\mathbb{R})$ satisfies $A = A^\top$ and condition (8.3.2), implies that formula (8.3.13) holds for this class of matrices. By using the same circle of ideas as the ones employed in proving (7.12.47) (based on Lemma 7.67), we conclude that (8.3.13) holds for the larger class of matrices $A \in \mathcal{M}_{n\times n}(\mathbb{C})$ satisfying $A = A^\top$ and condition (8.3.2). Hence, E_A is indeed a fundamental solution for $\mathcal{L}_A$ under the current assumptions on A.

Next, we claim that the hypotheses of Proposition 5.8 are satisfied in the case when $P(D) := \mathcal{L}_A$. To justify this, note that if $\xi = (\xi_1,\dots,\xi_{n+1}) \in \mathbb{R}^{n+1}$ is such that

$P(\xi) = 0$ then

$$i\xi_{n+1} + \sum_{j,k=1}^{n} a_{jk}\xi_j\xi_k = 0. \tag{8.3.14}$$

Taking reals parts, condition (8.3.2) implies $\xi_1 = \cdots = \xi_n = 0$ which, in combination with (8.3.14), also forces $\xi_{n+1} = 0$. Hence, $\xi = 0$ as wanted. Applying now Proposition 5.8 gives that if $u \in S'(\mathbb{R}^{n+1})$ is an arbitrary fundamental solution for $\mathcal{L}_A$ in $\mathbb{R}^{n+1}$, then $u = E_A + P$ in $S'(\mathbb{R}^{n+1})$, for some polynomial P in $\mathbb{R}^{n+1}$ satisfying $\mathcal{L}_A P = 0$ pointwise in $\mathbb{R}^{n+1}$.

In summary, the above analysis proves the following result.

Theorem 8.8. *Suppose $A = (a_{jk})_{1 \le j,k \le n} \in \mathcal{M}_{n \times n}(\mathbb{C})$ satisfies (8.3.2) and consider the operator $\mathcal{L}_A$ associated to A as in (8.3.1). Then the function defined by*

$$E_A(x) := \frac{1}{\sqrt{\det A_{sym}}} \, H(t)(4\pi t)^{-\frac{n}{2}} e^{-\dfrac{(A_{sym}^{-1}x)\cdot x}{4t}} \qquad \text{for all } (x,t) \in \mathbb{R}^{n+1}, \tag{8.3.15}$$

belongs to $S'(\mathbb{R}^{n+1}) \cap L^1_{loc}(\mathbb{R}^{n+1}) \cap C^\infty(\mathbb{R}^{n+1} \setminus \{0\})$ and is a fundamental solution for $\mathcal{L}_A$ in $\mathbb{R}^{n+1}$. Above, $A_{sym} := \frac{1}{2}(A + A^\top)$, $\sqrt{\det A_{sym}}$ is defined as in Remark 7.64, and $\log$ denotes the principal branch of the complex logarithm (defined for points $z \in \mathbb{C} \setminus (-\infty, 0]$ so that $z^a = e^{a \log z}$ for each $a \in \mathbb{R}$).

Moreover,

$$\{u \in S'(\mathbb{R}^{n+1}) : \mathcal{L}_A u = \delta \ \text{in} \ S'(\mathbb{R}^{n+1})\} \tag{8.3.16}$$

$$= \{E_A + P : P \ \text{polynomial in} \ \mathbb{R}^{n+1} \ \text{satisfying} \ \mathcal{L}_A P = 0\}.$$

8.4 Fundamental Solution for the Schrödinger Operator

Let $x \in \mathbb{R}^n$ and $t \in \mathbb{R}$. The operator $\frac{1}{i}\partial_t - \Delta_x$ is called the (time dependent) Schrödinger operator in $\mathbb{R}^{n+1}$ (with zero potential). In this section we determine a fundamental solution for this operator. By Theorem 5.14, there exists some $E \in S'(\mathbb{R}^{n+1})$ such that

$$\left(\frac{1}{i}\partial_t - \Delta_x\right)E = \delta(x) \otimes \delta(t) \quad \text{in} \quad S'(\mathbb{R}^{n+1}). \tag{8.4.1}$$

Fix such a distribution E and take the partial Fourier transform $\mathcal{F}_x$ of (8.4.1) (recall the discussion on partial Fourier transform at the end of Section 4.2) to obtain

$$\frac{1}{i}\partial_t \widehat{E}_x + |\xi|^2 \widehat{E}_x = 1(\xi) \otimes \delta(t) \quad \text{in} \ S'(\mathbb{R}^{n+1}). \tag{8.4.2}$$

Fix $\xi \in \mathbb{R}^n$ and consider the equation

$$\frac{1}{i}\partial_t u + |\xi|^2 u = \delta \quad \text{in} \quad \mathcal{D}'(\mathbb{R}). \tag{8.4.3}$$

Using Example 5.12, we obtain that $iH(t)e^{-i|\xi|^2 t}$ and $-iH(-t)e^{-i|\xi|^2 t}$ are solution of this equation. This suggests considering $F := iH(t)e^{-i|\xi|^2 t}$ that belongs to $\mathcal{S}'(\mathbb{R}^{n+1})$ and satisfies (based on a computation similar to that from (8.1.4))

$$\frac{1}{i}\partial_t F + |\xi|^2 F = 1(\xi) \otimes \delta(t) \quad \text{in} \quad \mathcal{S}'(\mathbb{R}^{n+1}). \tag{8.4.4}$$

Then $\mathcal{F}_x^{-1}(F) \in \mathcal{S}'(\mathbb{R}^{n+1})$ and

$$\left(\frac{1}{i}\partial_t - \Delta_x\right)\mathcal{F}_x^{-1}(F) = \delta(x) \otimes \delta(t) \quad \text{in} \quad \mathcal{S}'(\mathbb{R}^{n+1}). \tag{8.4.5}$$

To compute $\mathcal{F}_x^{-1}(F)$, pick $\varphi \in \mathcal{S}(\mathbb{R}^{n+1})$ and, based on the properties of $\mathcal{F}_x$ as well as the expression for F, write

$$\langle \mathcal{F}_x^{-1}(F), \varphi \rangle = (2\pi)^{-n}\langle \mathcal{F}_x(F^\vee), \varphi \rangle = (2\pi)^{-n}\langle F^\vee, \mathcal{F}_x\varphi \rangle$$

$$= (2\pi)^{-n}i \int_{\mathbb{R}^{n+1}} H(t)e^{-i|\xi|^2 t}(\mathcal{F}_x\varphi)(\xi,t)\,d\xi\,dt$$

$$= (2\pi)^{-n}i \int_0^\infty H(t)\left[\int_{\mathbb{R}^n} e^{-i|\xi|^2 t}(\mathcal{F}_x\varphi)(\xi,t)\,d\xi\right] dt. \tag{8.4.6}$$

Observe that, for each $t > 0$, we have

$$\int_{\mathbb{R}^n} e^{-i|\xi|^2 t}(\mathcal{F}_x\varphi)(\xi,t)\,d\xi = \langle \mathcal{F}_x(e^{-it|x|^2}), \varphi(\cdot,t) \rangle$$

$$= \left(\frac{\pi}{it}\right)^{\frac{n}{2}} \int_{\mathbb{R}^n} e^{-\frac{|\xi|^2}{4it}}\varphi(\xi,t)\,d\xi, \tag{8.4.7}$$

where the last equality is a consequence of Example 4.25 (used with $a = t$). Hence, for every $\varphi \in \mathcal{S}(\mathbb{R}^{n+1})$ we have

$$\langle \mathcal{F}_x^{-1}(F), \varphi \rangle = (2\pi)^{-n}i \int_0^\infty H(t)\left(\frac{\pi}{it}\right)^{\frac{n}{2}}\left[\int_{\mathbb{R}^n} e^{-\frac{|\xi|^2}{4it}}\varphi(\xi,t)\,d\xi\right] dt$$

$$= i^{1-\frac{n}{2}} \int_0^\infty H(t)(4\pi t)^{-\frac{n}{2}}\left[\int_{\mathbb{R}^n} e^{-\frac{|\xi|^2}{4it}}\varphi(\xi,t)\,d\xi\right] dt. \tag{8.4.8}$$

We remark here that $H(t)(4\pi t)^{-\frac{n}{2}}e^{-\frac{|x|^2}{4it}} \in L_{loc}^1(\mathbb{R}^{n+1})$ only if $n = 1$. In particular, $\mathcal{F}_x^{-1}(F)$ is of function type only if $n = 1$. In general, this distribution belongs to $\mathcal{S}'(\mathbb{R}^{n+1})$ and its action on $\varphi \in \mathcal{S}(\mathbb{R}^{n+1})$ is given as in (8.4.8).

In summary we proved the following result.

Theorem 8.9. *The distribution $E \in \mathcal{S}'(\mathbb{R}^{n+1})$ defined by*

$$\langle E, \varphi \rangle := i^{1-\frac{n}{2}} \int_0^\infty H(t)(4\pi t)^{-\frac{n}{2}} \left[\int_{\mathbb{R}^n} e^{-\frac{|\xi|^2}{4it}} \varphi(\xi, t) \, d\xi \right] dt \qquad (8.4.9)$$

for each $\varphi \in \mathcal{S}(\mathbb{R}^{n+1})$, is a fundamental solution for the Schrödinger operator $\frac{1}{i}\partial_t - \Delta_x$ in $\mathbb{R}^{n+1}$. In particular, if $n = 1$ then E is of function type and is given by the $L^1_{loc}(\mathbb{R}^2)$

$$E(x, t) = H(t)(4\pi t)^{-\frac{1}{2}} e^{-\frac{|x|^2}{4it}} \quad \text{for each } x \in \mathbb{R} \text{ and } t \in \mathbb{R}. \qquad (8.4.10)$$

Further Notes for Chapter 8. The heat equation is one example of what is commonly referred to as linear evolution equations. Originally derived in physics from Fourier's law and conservation of energy (see, e.g., [77] for details), the heat equation has come to play a role of fundamental importance in mathematics and applied sciences. In mathematics, the heat operator is the prototype for a larger class, called parabolic partial differential operators, that includes the operators studied in Section 8.3. The Schrödinger operator is named after the Austrian physicist Erwin Schrödinger who first introduced it in 1926. It plays a fundamental role in quantum mechanics, where it describes how the quantum state of certain physical systems changes in time.

Chapter 9
The Wave Operator

Abstract Here all fundamental solutions that are tempered distributions for the wave operator are determined and then used as a tool in the solution of the generalized Cauchy problem for this operator.

9.1 Fundamental Solution for the Wave Operator

The operator $\square := \partial_t^2 - \Delta_x$, $x \in \mathbb{R}^n$, $t \in \mathbb{R}$, is called the wave operator in $\mathbb{R}^{n+1}$. The wave operator arises from modeling vibrations in a string, membrane, or elastic solid. The goal is to determine fundamental solutions for this operator that are tempered distributions. By Theorem 5.14, we know that the wave operator $\square$ admits a fundamental solution $E \in \mathcal{S}'(\mathbb{R}^{n+1})$. Hence, by (4.1.33) and Exercise 2.90, we have

$$\partial_t^2 E - \Delta_x E = \delta(x) \otimes \delta(t) \quad \text{in} \quad \mathcal{S}'(\mathbb{R}^{n+1}). \tag{9.1.1}$$

In this section we determine an explicit expression for two such fundamental solutions.

Fix $E \in \mathcal{S}'(\mathbb{R}^{n+1})$ that satisfies (9.1.1) and apply the partial Fourier transform $\mathcal{F}_x$ to this equation (recall the discussion about partial Fourier transforms from the last part of Section 4.2) to obtain

$$\partial_t^2 \widehat{E}_x + |\xi|^2 \widehat{E}_x = 1(\xi) \otimes \delta(t) \quad \text{in} \quad \mathcal{S}'(\mathbb{R}^{n+1}), \tag{9.1.2}$$

where $\widehat{E}_x := \mathcal{F}_x(E) \in \mathcal{S}'(\mathbb{R}^{n+1})$. For $\xi \in \mathbb{R}^n \setminus \{0\}$ fixed consider the initial value problem (in the variable t)

$$\begin{cases} \frac{d^2}{dt^2} v + |\xi|^2 v = 0 \text{ in } \mathbb{R}, \\ (\partial_t v)(0) = 1, \quad v(0) = 0, \end{cases} \tag{9.1.3}$$

© Springer Nature Switzerland AG 2018
D. Mitrea, *Distributions, Partial Differential Equations, and Harmonic Analysis*,
Universitext, https://doi.org/10.1007/978-3-030-03296-8_9

which admits the solution $v(t) = \frac{\sin(t|\xi|)}{|\xi|}$ for $t \in \mathbb{R}$. By Example 5.12, it follows that vH and $-vH^\vee$ are fundamental solutions for the operator $\frac{d^2}{dt^2} + |\xi|^2$ in $\mathcal{D}'(\mathbb{R})$. In addition, vH and $-vH^\vee$ belong to $\mathcal{S}'(\mathbb{R})$ (based on *(b)* in Theorem 4.14 and Exercise 4.119), thus

$$\left(\frac{d^2}{dt^2} + |\xi|^2\right)(vH) = \delta \quad \text{and} \quad \left(\frac{d^2}{dt^2} + |\xi|^2\right)(-vH^\vee) = \delta \quad \text{in} \quad \mathcal{S}'(\mathbb{R}). \tag{9.1.4}$$

Moreover, there exists $c \in (0, \infty)$ such that

$$\left| H(\pm t) \frac{\sin(t|y|)}{|y|} \right| \leq c|t| \quad \text{for} \ (y, t) \in B(0, 1) \setminus \{0\} \subset \mathbb{R}^{n+1}. \tag{9.1.5}$$

In particular, if we define the functions

$$f_\pm(y, t) := H(\pm t) \frac{\sin(t|y|)}{|y|} \quad \text{for} \ (y, t) \in \mathbb{R}^{n+1} \text{with } t \neq 0, \tag{9.1.6}$$

then (9.1.5) implies $f_\pm \in L^1_{loc}(\mathbb{R}^{n+1})$. Furthermore,

$$(1 + |y|^2 + t^2)^{-n-2} f_\pm \in L^1(\mathbb{R}^{n+1}), \tag{9.1.7}$$

thus by Example 4.4, we obtain $H(\pm t) \frac{\sin(t|y|)}{|y|} \in \mathcal{S}'(\mathbb{R}^{n+1})$. Based on all these facts, if we introduce

$$F^+ := H(t) \frac{\sin(t|\xi|)}{|\xi|}, \qquad F^- := -H(-t) \frac{\sin(t|\xi|)}{|\xi|}, \tag{9.1.8}$$

then

$$F^+, F^- \in \mathcal{S}'(\mathbb{R}^{n+1}) \quad \text{and} \tag{9.1.9}$$

$$\partial_t^2 F^\pm + |\xi|^2 F^\pm = 1(\xi) \otimes \delta(t) \quad \text{in} \quad \mathcal{S}'(\mathbb{R}^{n+1}). \tag{9.1.10}$$

In particular, $\mathcal{F}_x^{-1}(F^+)$ and $\mathcal{F}_x^{-1}(F^-)$ are meaningfully defined in $\mathcal{S}'(\mathbb{R}^{n+1})$. Thus, if we set

$$E_+ := \mathcal{F}_x^{-1}\left(H(t) \frac{\sin(t|\xi|)}{|\xi|}\right), \tag{9.1.11}$$

$$E_- := \mathcal{F}_x^{-1}\left(-H(-t) \frac{\sin(t|\xi|)}{|\xi|}\right), \tag{9.1.12}$$

we have

$$E_\pm \in \mathcal{S}'(\mathbb{R}^{n+1}), \quad (\partial_t^2 - \Delta_x)E_\pm = \delta \quad \text{in} \quad \mathcal{S}'(\mathbb{R}^{n+1}), \tag{9.1.13}$$

$$\text{supp } E_+ \subseteq \mathbb{R}^n \times [0, \infty), \quad \text{supp } E_- \subseteq \mathbb{R}^n \times (-\infty, 0]. \tag{9.1.14}$$

The next task is to find explicit expressions for $\mathcal{F}_x^{-1}(F^{\pm})$. To this end, fix a function $\varphi \in \mathcal{S}(\mathbb{R}^{n+1})$ and, with the operation $\cdot^{\vee}$ considered only in the variable x, write

$$\langle E_+, \varphi \rangle = (2\pi)^{-n} \langle (\mathcal{F}_x \mathcal{F}_x(E_+))^{\vee}, \varphi \rangle = (2\pi)^{-n} \langle \mathcal{F}_x(E_+), (\mathcal{F}_x\varphi)^{\vee} \rangle$$

$$= (2\pi)^{-n} \int_{\mathbb{R}^{n+1}} H(t) \frac{\sin t|\xi|}{|\xi|} \widehat{\varphi}_x(-\xi, t) \, d\xi \, dt. \tag{9.1.15}$$

Note that if one replaces $\widehat{\varphi}_x(-\xi, t)$ above with $\int_{\mathbb{R}^n} e^{ix \cdot \xi} \varphi(x, t) \, dx$ then the order of integration in the resulting iterated integral may not be switched since $\frac{1}{|\xi|}$ is not integrable at infinity, thus Fubini's theorem does not apply. This is why we should proceed with more care and, based on Lebesgue's Dominated Convergence Theorem, we introduce a convergence factor which enables us to eventually make the use of Fubini's theorem. More precisely, we have

$$\langle E_+, \varphi \rangle = (2\pi)^{-n} \int_{\mathbb{R}^{n+1}} F^+(\xi, t) \widehat{\varphi}_x(-\xi, t) \, d\xi \, dt$$

$$= \lim_{\varepsilon \to 0^+} (2\pi)^{-n} \int_{\mathbb{R}^{n+1}} F^+(\xi, t) e^{-\varepsilon|\xi|} \widehat{\varphi}_x(-\xi, t) \, d\xi \, dt$$

$$= \lim_{\varepsilon \to 0^+} (2\pi)^{-n} \int_{\mathbb{R}^{n+1}} \varphi(x, t) \Big(\int_{\mathbb{R}^n} e^{ix \cdot \xi - \varepsilon|\xi|} F^+(\xi, t) \, d\xi \Big) dx \, dt$$

$$= \lim_{\varepsilon \to 0^+} \langle E_{\varepsilon}, \varphi \rangle, \tag{9.1.16}$$

where

$$E_{\varepsilon}(x, t) := (2\pi)^{-n} H(t) \int_{\mathbb{R}^n} e^{ix \cdot \xi - \varepsilon|\xi|} \frac{\sin(t|\xi|)}{|\xi|} \, d\xi, \quad \forall x \in \mathbb{R}^n, \ \forall t \in \mathbb{R}. \tag{9.1.17}$$

To compute the last limit in (9.1.16), we separate our analysis into three cases: $n = 1$, $n = 2p + 1$ with $p \geq 1$, and $n = 2p$ with $p \geq 1$.

9.1.1 The Case $n = 1$

Fix $\varepsilon > 0$ and $x \in \mathbb{R}$ and define the function

$$f_{\varepsilon}(t) := \int_{-\infty}^{\infty} e^{ix\xi - \varepsilon|\xi|} \frac{\sin(t|\xi|)}{|\xi|} \, d\xi \qquad \text{for} \quad t \in \mathbb{R}. \tag{9.1.18}$$

Then

$$\partial_t f_\varepsilon(t) = \int_{-\infty}^{\infty} e^{ix\xi - \varepsilon|\xi|} \cos(t|\xi|) \, d\xi = \int_{-\infty}^{\infty} e^{ix\xi - \varepsilon|\xi|} \frac{e^{it|\xi|} + e^{-it|\xi|}}{2} \, d\xi$$

$$= \frac{1}{2} \Big[\int_0^\infty e^{\xi(ix+it-\varepsilon)} \, d\xi + \int_0^\infty e^{\xi(ix-it-\varepsilon)} \, d\xi \Big]$$

$$+ \frac{1}{2} \Big[\int_{-\infty}^0 e^{\xi(ix-it+\varepsilon)} \, d\xi + \int_{-\infty}^0 e^{\xi(ix+it+\varepsilon)} \, d\xi \Big]$$

$$= \frac{1}{2} \Big[-\frac{1}{ix+it-\varepsilon} - \frac{1}{ix-it-\varepsilon} + \frac{1}{ix-it+\varepsilon} + \frac{1}{ix+it+\varepsilon} \Big]$$

$$= \frac{\varepsilon}{(x+t)^2 + \varepsilon^2} + \frac{\varepsilon}{(x-t)^2 + \varepsilon^2}, \qquad \forall\, t \in \mathbb{R}. \tag{9.1.19}$$

Consequently, (9.1.19) and the fact that $f_\varepsilon(0) = 0$ imply

$$f_\varepsilon(t) = \arctan\Big(\frac{x+t}{\varepsilon}\Big) - \arctan\Big(\frac{x-t}{\varepsilon}\Big) \qquad \text{for} \quad t \in \mathbb{R}. \tag{9.1.20}$$

Making use of (9.1.20) back in (9.1.17) (written for $n = 1$) then gives

$$E_\varepsilon(x,t) = \frac{1}{2\pi} H(t) \Big(\arctan\Big(\frac{x+t}{\varepsilon}\Big) - \arctan\Big(\frac{x-t}{\varepsilon}\Big) \Big), \qquad \forall\, x, t \in \mathbb{R}. \tag{9.1.21}$$

Hence, for $\varphi \in \mathcal{S}(\mathbb{R}^2)$ fixed, we may write

$$\lim_{\varepsilon \to 0^+} \langle E_\varepsilon(x,t), \varphi(x,t) \rangle \tag{9.1.22}$$

$$= \lim_{\varepsilon \to 0^+} \frac{1}{2\pi} \int_{\mathbb{R}} \int_0^\infty \Big[\arctan\Big(\frac{x+t}{\varepsilon}\Big) - \arctan\Big(\frac{x-t}{\varepsilon}\Big) \Big] \varphi(x,t) \, dt \, dx.$$

To continue with our calculation, we further decompose

$$\int_0^\infty \int_{\mathbb{R}} \arctan\Big(\frac{x+t}{\varepsilon}\Big) \varphi(x,t) \, dx \, dt$$

$$= \int_0^\infty \int_{-\infty}^{-t} \arctan\Big(\frac{x+t}{\varepsilon}\Big) \varphi(x,t) \, dx \, dt + \int_0^\infty \int_{-t}^\infty \arctan\Big(\frac{x+t}{\varepsilon}\Big) \varphi(x,t) \, dx \, dt$$

$$=: I_\varepsilon + II_\varepsilon. \tag{9.1.23}$$

By Lebesgue's Dominated Convergence Theorem,

$$\lim_{\varepsilon \to 0^+} I_\varepsilon = -\frac{\pi}{2} \int_0^\infty \int_{-\infty}^{-t} \varphi(x,t) \, dx \, dt \tag{9.1.24}$$

and

$$\lim_{\varepsilon \to 0^+} II_\varepsilon = \frac{\pi}{2} \int_0^\infty \int_{-t}^\infty \varphi(x,t) \, dx \, dt. \tag{9.1.25}$$

Similarly,

$$\lim_{\varepsilon \to 0^+} \int_0^\infty \int_{\mathbb{R}} \arctan\left(\frac{x-t}{\varepsilon}\right)\varphi(x,t)\,dx\,dt$$

$$= -\frac{\pi}{2}\int_0^\infty \int_{-\infty}^t \varphi(x,t)\,dx\,dt + \frac{\pi}{2}\int_0^\infty \int_t^\infty \varphi(x,t)\,dx\,dt. \tag{9.1.26}$$

Combining (9.1.22)–(9.1.26) permits us to write

$$\lim_{\varepsilon \to 0^+} \langle E_\varepsilon(x,t), \varphi(x,t)\rangle \tag{9.1.27}$$

$$= \frac{1}{4}\int_0^\infty \left[-\int_{-\infty}^{-t} \varphi(x,t)\,dx + \int_{-t}^\infty \varphi(x,t)\,dx\right]dt$$

$$+ \frac{1}{4}\int_0^\infty \left[\int_{-\infty}^t \varphi(x,t)\,dx - \int_t^\infty \varphi(x,t)\,dx\right]dt$$

$$= \frac{1}{2}\int_0^\infty \int_{-t}^t \varphi(x,t)\,dx\,dt = \frac{1}{2}\int_{\mathbb{R}^2} H(t-|x|)\varphi(x,t)\,dx\,dt.$$

From (9.1.27) and (9.1.16) we then conclude that[1] $E_+(x,t) = \frac{H(t-|x|)}{2}$ for each point $(x,t) \in \mathbb{R}^2$. In summary, we proved that

$$\begin{cases} E_+(x,t) = \dfrac{H(t-|x|)}{2}, \quad \forall\,(x,t) \in \mathbb{R}^2, \quad \text{is a tempered distribution} \\ \text{and satisfies } (\partial_t^2 - \partial_x^2)E_+ = \delta \text{ in } \mathcal{S}'(\mathbb{R}^2). \end{cases} \tag{9.1.28}$$

Remark 9.1.

(1) The reasoning used to obtain E_+ also yields that

$$\begin{cases} E_-(x,t) := \dfrac{H(-t-|x|)}{2}, \quad \forall\,(x,t) \in \mathbb{R}^2, \quad \text{is a tempered distribution} \\ \text{and satisfies } (\partial_t^2 - \partial_x^2)E_- = \delta \text{ in } \mathcal{S}'(\mathbb{R}^2). \end{cases} \tag{9.1.29}$$

(2) An inspection of (9.1.28) and (9.1.29) reveals that

$$\operatorname{supp} E_+ = \{(x,t) \in \mathbb{R}^2 : |x| \le t\}$$

$$\text{and } \operatorname{supp} E_- = \{(x,t) \in \mathbb{R}^2 : |x| \le -t\}. \tag{9.1.30}$$

---

[1] this expression for a fundamental solution for the wave operator when $n = 1$ was first used by Jean d'Alembert in 1747 in connection with a vibrating string

9.1.2 The Case $n = 2p + 1$, $p \geq 1$

It is immediate from (9.1.17) that $E_\varepsilon(x, t)$ is invariant under orthogonal transformations in the variable x. Also, if $y \in \mathbb{R}^n \setminus \{0\}$, we may set $v_1 := \frac{1}{|y|} y$, then complete to an orthonormal basis $\{v_1, v_2, \ldots, v_n\}$ in $\mathbb{R}^n$, and consider the orthogonal transformation $A : \mathbb{R}^n \to \mathbb{R}^n$ satisfying $A(v_j) = \mathbf{e}_j$, for $j \in \{1, \ldots, n\}$. Then $A(y) = |y| \mathbf{e}_1$ which when combined with the invariance of $E_\varepsilon(x, t)$ under orthogonal transformations in the variable x, implies that $E_\varepsilon(y, t) = E_\varepsilon(|y|, 0, \ldots, 0, t)$ for every y in $\mathbb{R}^n$ and $t \in \mathbb{R}$. As such, in what follows we may assume $x = (|x|, 0, \ldots, 0)$. Fix such an x and set $r := |x|$. With $t \in \mathbb{R}$ fixed, using polar coordinates (see (14.9.1), (14.9.2), and (14.9.3)), rewrite the expression from (9.1.17) as

$$E_\varepsilon(x, t) = (2\pi)^{-n} H(t) \int_0^\infty \int_{(0,\pi)^{n-2}} \int_0^{2\pi} e^{i\rho r \cos\theta_1 - \varepsilon\rho} \frac{\sin t\rho}{\rho} \rho^{n-1} \times$$

$$\times (\sin\theta_1)^{n-2} \cdots \sin(\theta_{n-2}) \, d\theta_{n-1} \cdots d\theta_1 \, d\rho$$

$$= (2\pi)^{-n} H(t) \int_0^\infty \rho^{n-2} \sin(t\rho) \int_0^\pi e^{i\rho r \cos\theta_1 - \varepsilon\rho} (\sin\theta_1)^{n-2} \times$$

$$\times \left[\int_{(0,\pi)^{n-3}} \int_0^{2\pi} (\sin\theta_2)^{n-3} \ldots \sin(\theta_{n-2}) \, d\theta_{n-1} \cdots d\theta_2 \right] d\theta_1 \, d\rho$$

$$= (2\pi)^{-n} \omega_{n-2} H(t) \int_0^\infty \int_0^\pi e^{i\rho r \cos\theta - \varepsilon\rho} \rho^{n-2} (\sin\theta)^{n-2} \sin(t\rho) \, d\theta \, d\rho \quad (9.1.31)$$

where ω_{n-2} denotes the surface area of the unit ball in $\mathbb{R}^{n-1}$ (see also (14.6.6) for more details on why the expression inside the right brackets in (9.1.31) is equal to ω_{n-2}).

To proceed with the computation of the integrals in the rightmost term in (9.1.31), recall that $n = 2p + 1$ and, for $\rho > 0$ fixed, set

$$I_p := \rho^{2p-1} \int_0^\pi e^{i\rho r \cos\theta} (\sin\theta)^{2p-1} \, d\theta, \qquad \forall \, p \in \mathbb{N}. \quad (9.1.32)$$

We claim that

$$I_{p+1} = -\frac{2p}{r} \partial_r I_p \quad \text{for all } p \geq 1. \quad (9.1.33)$$

In order to prove (9.1.33) note that, for each $p \geq 1$, integration by parts yields

$$I_{p+1} = \rho^{2p+1} \int_0^\pi e^{i\rho r \cos\theta} (\sin\theta)^{2p+1} \, d\theta$$

$$= \rho^{2p+1} \frac{-1}{i\rho r} \int_0^\pi \partial_\theta (e^{i\rho r \cos\theta})(\sin\theta)^{2p} \, d\theta$$

$$= \frac{i}{r} \rho^{2p} \left[e^{i\rho r \cos\theta} (\sin\theta)^{2p} \Big|_{\theta=0}^{\theta=\pi} - 2p \int_0^\pi e^{i\rho r \cos\theta} (\sin\theta)^{2p-1} \cos\theta \, d\theta \right]$$

$$= -\frac{2pi}{r} \rho^{2p} \int_0^\pi e^{i\rho r \cos\theta} (\sin\theta)^{2p-1} \cos\theta \, d\theta$$

$$= -\frac{2pi}{r} \rho^{2p} \frac{1}{i\rho} \partial_r \int_0^\pi e^{i\rho r \cos\theta} (\sin\theta)^{2p-1} \, d\theta = -\frac{2p}{r} (\partial_r I_p), \tag{9.1.34}$$

as wanted. By induction, from the recurrence relation in (9.1.33) it follows that

$$I_p = (-2)^{p-1}(p-1)! \left(\frac{1}{r} \partial_r \right)^{(p-1)} I_1 \quad \text{for each } p \in \mathbb{N}. \tag{9.1.35}$$

As for I_1, we have

$$I_1 = \rho \int_0^\pi e^{i\rho r \cos\theta} \sin\theta \, d\theta = \frac{-1}{ir} e^{i\rho r \cos\theta} \Big|_{\theta=0}^{\theta=\pi} = \frac{2\sin(\rho r)}{r}. \tag{9.1.36}$$

Recalling (9.1.31) and the fact that $n = 2p + 1$ for some $p \in \mathbb{N}$, formulas (9.1.35) and (9.1.36) yield

$$E_\varepsilon(x, t) = (2\pi)^{-n} \omega_{n-2} H(t) \int_0^\infty I_p e^{-\varepsilon\rho} \sin(t\rho) \, d\rho \tag{9.1.37}$$

$$= (2\pi)^{-n} \omega_{n-2} H(t) (-2)^{p-1}(p-1)! \, 2 \times$$

$$\times \int_0^\infty e^{-\varepsilon\rho} \sin(t\rho) \left(\frac{1}{r} \partial_r \right)^{p-1} \left(\frac{\sin(\rho r)}{r} \right) d\rho.$$

Furthermore, using (14.5.6) we have

$$\omega_{n-2} = \frac{2\pi^{\frac{n-1}{2}}}{\Gamma\left(\frac{n-1}{2}\right)} = \frac{2\pi^p}{\Gamma(p)} = \frac{2\pi^p}{(p-1)!} \tag{9.1.38}$$

which further simplifies the expression in (9.1.37) to

$$E_\varepsilon(x, t) \tag{9.1.39}$$

$$= 2(-2\pi)^{-p-1} H(t) \left(\frac{1}{r} \partial_r \right)^{p-1} \left[\frac{1}{r} \int_0^\infty e^{-\varepsilon\rho} \sin(t\rho) \sin(\rho r) \, d\rho \right].$$

Our next claim is that

$$\int_0^\infty e^{-\varepsilon\rho} \sin(t\rho) \sin(\rho r)\, d\rho = \frac{1}{2}\left[\frac{\varepsilon}{\varepsilon^2 + (t-r)^2} - \frac{\varepsilon}{\varepsilon^2 + (t+r)^2}\right]. \qquad (9.1.40)$$

Indeed,

$$\int_0^\infty e^{-\varepsilon\rho} \sin(t\rho) \sin(\rho r)\, d\rho \qquad\qquad\qquad (9.1.41)$$

$$= \frac{1}{2}\int_0^\infty e^{-\varepsilon\rho}[\cos(t-r)\rho - \cos(t+r)\rho]\, d\rho$$

$$= \frac{1}{2}\mathrm{Re}\int_0^\infty [e^{-\varepsilon\rho + \mathrm{i}(t-r)\rho} - e^{-\varepsilon\rho + \mathrm{i}(t+r)\rho}]\, d\rho$$

$$= \frac{1}{2}\mathrm{Re}\left[-\frac{1}{-\varepsilon + \mathrm{i}(t-r)} + \frac{1}{-\varepsilon + \mathrm{i}(t+r)}\right]$$

$$= \frac{1}{2}\mathrm{Re}\left[\frac{\varepsilon + \mathrm{i}(t-r)}{\varepsilon^2 + (t-r)^2} + \frac{-\varepsilon - \mathrm{i}(t+r)}{\varepsilon^2 + (t+r)^2}\right]$$

$$= \frac{1}{2}\left[\frac{\varepsilon}{\varepsilon^2 + (t-r)^2} - \frac{\varepsilon}{\varepsilon^2 + (t+r)^2}\right],$$

proving (9.1.40). The identity resulting from (9.1.41) further simplifies the expression in (9.1.39) as

$$E_\varepsilon(x,t) \qquad\qquad\qquad (9.1.42)$$

$$= (-2\pi)^{-p-1} H(t)\left(\frac{1}{r}\partial_r\right)^{(p-1)}\left[\frac{1}{r}\left(\frac{\varepsilon}{\varepsilon^2 + (t-r)^2} - \frac{\varepsilon}{\varepsilon^2 + (t+r)^2}\right)\right]$$

for every $x \in \mathbb{R}^n$ and every $t \in \mathbb{R}$, where $r = |x|$.

Recall from (9.1.16) that in order to determine E_+ we further need to compute $\lim_{\varepsilon \to 0^+} E_\varepsilon(x,t)$ in $\mathcal{S}'(\mathbb{R}^{n+1})$. With this goal in mind, fix $\varphi \in \mathcal{S}(\mathbb{R}^{n+1})$ and use (9.1.42) in concert with 14.9.8 to write

$$\langle E_\varepsilon, \varphi \rangle \tag{9.1.43}$$

$$= (-2\pi)^{-p-1} \int_0^\infty \int_0^\infty \int_{\partial B(0,r)} \left(\frac{1}{r}\partial_r\right)^{p-1} \left[\frac{\varepsilon}{r}\left(\frac{1}{\varepsilon^2 + (t-r)^2} - \frac{1}{\varepsilon^2 + (t+r)^2}\right)\right] \times$$

$$\times \varphi(\omega, t)\, d\sigma(\omega)\, dr\, dt$$

$$= (-2\pi)^{-p-1}\varepsilon \int_0^\infty \int_0^\infty \left(\frac{1}{r}\partial_r\right)^{p-1} \left[\frac{1}{r}\left(\frac{1}{\varepsilon^2 + (t-r)^2} - \frac{1}{\varepsilon^2 + (t+r)^2}\right)\right] \times$$

$$\times \int_{\partial B(0,r)} \varphi(\omega, t)\, d\sigma(\omega)\, dr\, dt.$$

A natural question to ask is whether, if $p \geq 2$, we may integrate by parts $(p-1)$ times in the r variable in the rightmost expression in (9.1.43). Observe that, at least formally,

$$\int_0^\infty \left[\left(\frac{1}{r}\frac{d}{dr}\right)f(r)\right]g(r)\, dr = -\int_0^\infty f(r)\frac{d}{dr}\left[\frac{1}{r}g(r)\right] dr \tag{9.1.44}$$

if

$$\lim_{r\to\infty} \frac{f(r)g(r)}{r} = 0 = \lim_{r\to 0^+} \frac{f(r)g(r)}{r}. \tag{9.1.45}$$

In the setting of the last expression in (9.1.43), for each $\varepsilon, t > 0$ fixed, and assuming $p \geq 2$, if we set, for each $r > 0$,

$$f(r) := \left(\frac{1}{r}\partial_r\right)^{p-2} \left[\frac{1}{r}\left(\frac{1}{\varepsilon^2 + (t-r)^2} - \frac{1}{\varepsilon^2 + (t+r)^2}\right)\right], \tag{9.1.46}$$

$$g(r) := \int_{\partial B(0,r)} \varphi(\omega, t)\, d\sigma(\omega) = r^{n-1} \int_{S^{n-1}} \varphi(r\omega, t)\, d\sigma(\omega), \tag{9.1.47}$$

then these functions satisfy (9.1.45) and (9.1.44). Proceeding by induction (with $p \geq 2$), we apply $(p-1)$ times formula (9.1.44), pick up $(-1)^{p-1}$ in the process (the last factor $\frac{1}{r}$ bundled up with the derivative) and write

$$\langle E_\varepsilon, \varphi \rangle = (2\pi)^{-p-1} \int_0^\infty \int_0^\infty \left(\frac{\varepsilon}{\varepsilon^2 + (t-r)^2} - \frac{\varepsilon}{\varepsilon^2 + (t+r)^2}\right) \times \tag{9.1.48}$$

$$\times \left(\frac{1}{r}\partial_r\right)^{p-1} \left[\frac{1}{r}\int_{\partial B(0,r)} \varphi(\omega, t)\, d\sigma(\omega)\right] dr\, dt.$$

Note that (9.1.48) is also valid if $p = 1$ without any need of integration by parts.

We are left with taking the limit as $\varepsilon \to 0^+$ in (9.1.48) a task we complete by using Lemma 9.3 (which is stated and proved at the end of this subsection). Specifically, we apply Lemma 9.3 with

$$h(r) := \left(\frac{1}{r}\partial_r\right)^{p-1}\left[\frac{1}{r}\int_{\partial B(0,r)}\varphi(\omega,t)\,d\sigma(\omega)\right]. \tag{9.1.49}$$

Note that the second equality in (9.1.47) and the fact that $\varphi \in S(\mathbb{R}^{n+1})$ guarantee that h in (9.1.49) satisfies the hypothesis of Lemma 9.3. These facts combined with Lebesgue's Dominated Convergence Theorem yield

$$\lim_{\varepsilon\to0^+}\langle E_\varepsilon,\varphi\rangle \tag{9.1.50}$$

$$= (2\pi)^{-p-1}\pi\int_0^\infty\left[\left(\frac{1}{r}\partial_r\right)^{p-1}\left(\frac{1}{r}\int_{\partial B(0,r)}\varphi(\omega,t)\,d\sigma(\omega)\right)\right]\Bigg|_{r=t}dt.$$

In summary, we proved the following result:

If $n = 2p + 1$, for some $p \in \mathbb{N}$, then $E_+ \in S'(\mathbb{R}^{n+1})$ defined by

$$\langle E_+,\varphi\rangle = (2\pi)^{-p-1}\pi\int_0^\infty\left[\left(\frac{1}{r}\partial_r\right)^{p-1}\left(\frac{1}{r}\int_{\partial B(0,r)}\varphi(\omega,t)\,d\sigma(\omega)\right)\right]\Bigg|_{r=t}dt \tag{9.1.51}$$

for $\varphi \in S(\mathbb{R}^{n+1})$ is a fundamental solution for the wave operator in $\mathbb{R}^{n+1}$.

The reasoning used to obtain (9.1.51) also yields an expression for E_-. More precisely, similar to (9.1.16), we arrive at

$$\langle E_-,\varphi\rangle = \lim_{\varepsilon\to0^+}\langle E_\varepsilon,\varphi\rangle \quad\text{for each}\quad \varphi \in S(\mathbb{R}^{n+1}), \tag{9.1.52}$$

where, this time,

$$\langle E_\varepsilon,\varphi\rangle = -(2\pi)^{-p-1}\int_{-\infty}^0\int_0^\infty\left(\frac{\varepsilon}{\varepsilon^2+(t-r)^2}-\frac{\varepsilon}{\varepsilon^2+(t+r)^2}\right)\times$$

$$\times\left(\frac{1}{r}\partial_r\right)^{p-1}\left[\frac{1}{r}\int_{\partial B(0,r)}\varphi(\omega,t)\,d\sigma(\omega)\right]dr\,dt$$

$$= (2\pi)^{-p-1}\int_0^\infty\int_0^\infty\left(\frac{\varepsilon}{\varepsilon^2+(t-r)^2}-\frac{\varepsilon}{\varepsilon^2+(t+r)^2}\right)\times \tag{9.1.53}$$

$$\times\left(\frac{1}{r}\partial_r\right)^{p-1}\left[\frac{1}{r}\int_{\partial B(0,r)}\varphi(\omega,-t)\,d\sigma(\omega)\right]dr\,dt.$$

Invoking Lemma 9.3 we obtain:

If $n = 2p + 1$, for some $p \in \mathbb{N}$, then $E_- \in \mathcal{S}'(\mathbb{R}^{n+1})$, defined by

$$\langle E_-, \varphi \rangle = (2\pi)^{-p-1} \pi \int_0^\infty \left[\left(\frac{1}{r} \partial_r \right)^{p-1} \left(\frac{1}{r} \int_{\partial B(0,r)} \varphi(\omega, -t) \, d\sigma(\omega) \right) \right]_{r=t} dt, \qquad (9.1.54)$$

for $\varphi \in \mathcal{S}(\mathbb{R}^{n+1})$, is a fundamental solution for the wave operator in $\mathbb{R}^{n+1}$.

Remark 9.2. (1) The distributions E_+ and E_- from (9.1.51) and (9.1.54), respectively, satisfy

$$\text{supp}\, E_+ = \{(x, t) \in \mathbb{R}^{2p+1} \times [0, \infty) : |x| = t\}, \qquad (9.1.55)$$

$$\text{supp}\, E_+ = \{(x, t) \in \mathbb{R}^{2p+1} \times (-\infty, 0] : |x| = -t\}. \qquad (9.1.56)$$

(2) In the case when $n = 3$ (thus, for $p = 1$), formulas (9.1.51) and (9.1.54) become

$$\langle E_+, \varphi \rangle = \frac{1}{4\pi} \int_0^\infty \frac{1}{t} \int_{\partial B(0,t)} \varphi(\omega, t) \, d\sigma(\omega) \, dt$$

$$= \left\langle \frac{H(t)}{4\pi t} \delta_{\partial B(0,t)}, \varphi \right\rangle, \qquad (9.1.57)$$

and

$$\langle E_-, \varphi \rangle = \frac{1}{4\pi} \int_0^\infty \frac{1}{t} \int_{\partial B(0,t)} \varphi(\omega, -t) \, d\sigma(\omega) \, dt$$

$$= \left\langle -\frac{H(-t)}{4\pi t} \delta_{\partial B(0,-t)}, \varphi \right\rangle, \qquad (9.1.58)$$

for every $\varphi \in \mathcal{S}(\mathbb{R}^4)$ where, for each $R \in (0, \infty)$, the symbol $\delta_{\partial B(0,R)}$ stands for the distribution defined as in Exercise 2.146 with $\Sigma := \partial B(0, R)$.

(3) If $n = 2p$, $p \in \mathbb{N}$, the approach used to obtain (9.1.51) works up to the point where the general formula for I_p was obtained. More precisely, with $\rho > 0$ fixed, if we define

$$J_n := \rho^n \int_0^\pi e^{i\rho r \cos\theta} (\sin\theta)^n \, d\theta, \qquad \forall n \geq 2, \qquad (9.1.59)$$

then the recurrence formula $J_n = -\frac{n-1}{r} \partial_r J_{n-2}$ is valid for all $n \geq 2$ (observe that $I_p = J_{2p-1}$). However, the integral $J_0 = \int_0^\pi e^{i\rho r \cos\theta} \, d\theta$ cannot be computed explicitly (as opposed to the computation of I_1), hence the recurrence formula for J_n may not be used inductively to obtain an explicit expression for J_n. This is why, in order to obtain explicit expressions for $E_\pm$ when n is even we resort to a proof different than the one used when n is odd.

Lemma 9.3. *If $h : [0, \infty) \to \mathbb{R}$ is continuous and bounded, then for each t in $(0, \infty)$ we have*

$$\lim_{\varepsilon \to 0^+} \int_0^\infty \left(\frac{\varepsilon}{\varepsilon^2 + (t-r)^2} - \frac{\varepsilon}{\varepsilon^2 + (t+r)^2} \right) h(r)\, dr = \pi h(t). \tag{9.1.60}$$

Proof. If $t \geq 0$ is fixed, then via suitable changes of variables we obtain

$$\lim_{\varepsilon \to 0^+} \int_0^\infty \left(\frac{\varepsilon}{\varepsilon^2 + (t-r)^2} - \frac{\varepsilon}{\varepsilon^2 + (t+r)^2} \right) h(r)\, dr$$

$$= \lim_{\varepsilon \to 0^+} \left[\int_{-\frac{t}{\varepsilon}}^\infty \frac{h(t+\varepsilon\lambda)}{1+\lambda^2}\, d\lambda - \int_{\frac{t}{\varepsilon}}^\infty \frac{h(-t+\varepsilon\lambda)}{1+\lambda^2}\, d\lambda \right]$$

$$= \int_{-\infty}^\infty \frac{h(t)}{1+\lambda^2}\, d\lambda = \pi h(t), \tag{9.1.61}$$

where for the second to the last equality in (9.1.61) we applied Lebesgue's Dominated Convergence Theorem.   □

9.1.3 The Method of Descent

In order to treat the case when n is even, we use a procedure called the method of descent. The ultimate goal is to use the method of descent to deduce from a fundamental solution for the wave operator in dimension $n+1$ with n even, a fundamental solution for the wave operator in dimension n. To set the stage we make the following definition.

Definition 9.4. A sequence $\{\psi_j\}_{j \in \mathbb{N}}$ of functions in $C_0^\infty(\mathbb{R})$ is said to `converge in a dominated fashion` to 1 if the following two conditions are satisfied:

(i) for every compact subset K of $\mathbb{R}$ there exists $j_0 = j_0(K) \in \mathbb{N}$ such that $\psi_j(x) = 1$ for all $x \in K$ if $j \geq j_0$;
(ii) for every $q \in \mathbb{N}_0$, the sequence $\{\psi_j^{(q)}\}_{j \in \mathbb{N}}^\infty$ is uniformly bounded on $\mathbb{R}$.

An example of a sequence $\{\psi_j\}_{j \in \mathbb{N}}$ converging in a dominated fashion to 1 is given by

$$\psi_j(x) := \psi\left(\frac{x}{j}\right), \quad \forall\, x \in \mathbb{R}, \qquad \forall\, j \in \mathbb{N}, \tag{9.1.62}$$

where

$$\psi \in C_0^\infty(\mathbb{R}) \quad \text{satisfies} \quad \psi(x) \equiv 1 \text{ whenever } |x| \leq 1. \tag{9.1.63}$$

In what follows we will use the notation

$$x = (x', x_n) \in \mathbb{R}^n,\ x' \in \mathbb{R}^{n-1},\ x_n \in \mathbb{R},$$
$$\partial = (\partial', \partial_n),\ \partial' = (\partial_1, \dots, \partial_{n-1}). \tag{9.1.64}$$

Definition 9.5. Call a distribution $u \in \mathcal{D}'(\mathbb{R}^n)$ `integrable with respect to` x_n if for any $\varphi \in C_0^\infty(\mathbb{R}^{n-1})$ and any sequence $\{\psi_j\}_{j\in\mathbb{N}} \subset C_0^\infty(\mathbb{R})$ converging in a dominated fashion to 1, the sequence $\{\langle u, \varphi \otimes \psi_j \rangle\}_{j\in\mathbb{N}}$ is convergent and its limit does not depend on the selection of the sequence $\{\psi_j\}_{j\in\mathbb{N}}$.

Suppose $u \in \mathcal{D}'(\mathbb{R}^n)$ is integrable with respect to x_n and $\{\psi_j\}_{j\in\mathbb{N}} \subset C_0^\infty(\mathbb{R})$ is a sequence converging in a dominated fashion to 1. For each $j \in \mathbb{N}$ define the linear mapping

$$u_j : \mathcal{D}(\mathbb{R}^{n-1}) \to \mathbb{C}, \quad u_j(\varphi) := \langle u, \varphi \otimes \psi_j \rangle, \quad \forall \varphi \in C_0^\infty(\mathbb{R}^{n-1}). \qquad (9.1.65)$$

Then $u_j \in \mathcal{D}'(\mathbb{R}^{n-1})$ for each $j \in \mathbb{N}$ and $\lim_{j,k\to\infty} \langle u_j - u_k, \varphi \rangle = 0$ for every test function $\varphi \in C_0^\infty(\mathbb{R}^{n-1})$. Thus, the sequence $\{u_j\}_{j\in\mathbb{N}}$ is Cauchy in $\mathcal{D}'(\mathbb{R}^{n-1})$ (see Section 14.1.0.5). Since $\mathcal{D}'(\mathbb{R}^{n-1})$ is sequentially complete (recall Fact 2.22), it follows that

there exists some $u_0 \in \mathcal{D}'(\mathbb{R}^{n-1})$ with the property that

$$u_j \xrightarrow[j\to\infty]{\mathcal{D}'(\mathbb{R}^{n-1})} u_0 \quad \text{and} \quad \langle u_0, \varphi \rangle = \lim_{j\to\infty} \langle u_j, \varphi \rangle \quad \forall \varphi \in C_0^\infty(\mathbb{R}^{n-1}). \qquad (9.1.66)$$

Moreover, u_0 is independent of the choice of the sequence $\{\psi_j\}_{j\in\mathbb{N}}$ converging in a dominated fashion to 1 (prove this as an exercise). The distribution u_0 will be called the `integral of` u `with respect to` x_n and will be denoted by $\int_{-\infty}^\infty u \, dx_n$. The reason for using this terminology and notation for u_0 is evident from the next proposition. We denote by $\mathcal{L}^{n-1}$ the Lebesgue measure in $\mathbb{R}^{n-1}$.

Proposition 9.6. *If $f \in L_{loc}^1(\mathbb{R}^n)$ is a function with the property that*

$$\int_{\mathbb{R}} |f(x', x_n)| \, dx_n < \infty \quad \text{for } \mathcal{L}^{n-1}\text{-almost every } x' \in \mathbb{R}^{n-1} \qquad (9.1.67)$$

and $f \in L^1(K \times \mathbb{R})$ *for every compact set $K \subset \mathbb{R}^{n-1}$,*

then the distribution $u_f \in \mathcal{D}'(\mathbb{R}^n)$ determined by f (recall (2.1.6)) is integrable with respect to x_n. In addition, the distribution $u_0 := \int_{-\infty}^\infty u_f \, dx_n$ is of function type and is given by the function

$$g(x') := \int_{\mathbb{R}} f(x', x_n) \, dx_n \quad \text{defined for } \mathcal{L}^{n-1}\text{-almost every } x' \in \mathbb{R}^{n-1}. \qquad (9.1.68)$$

Proof. Since $f \in L_{loc}^1(\mathbb{R}^n)$ we have that $u_f \in \mathcal{D}'(\mathbb{R}^n)$ as in (2.1.6) is well defined. Fix $\varphi \in C_0^\infty(\mathbb{R}^{n-1})$. Since f is absolutely integrable on $\mathrm{supp}\,\varphi \times \mathbb{R}$, whenever $\{\psi_j\}_{j\in\mathbb{N}} \subset C_0^\infty(\mathbb{R})$ is a sequence converging in a dominated fashion to 1, we may apply Fubini's theorem and then Lebesgue's Dominated Convergence Theorem to write

$$\lim_{j\to\infty} \langle u_f, \varphi \otimes \psi_j \rangle = \lim_{j\to\infty} \int_{\mathbb{R}^{n-1}} \varphi(x') \int_{\mathbb{R}} f(x', x_n) \psi_j(x_n)\, dx_n\, dx'$$

$$= \int_{\mathbb{R}^{n-1}} \varphi(x') \int_{\mathbb{R}} f(x', x_n)\, dx_n\, dx', \qquad (9.1.69)$$

and the desired conclusion follows. $\square$

The proposition that is the engine in the method of descent is proved next.

Proposition 9.7. *Let $m \in \mathbb{N}_0$ and let $P(\partial) = P(\partial', \partial_n)$ be a constant coefficient, linear operator of order m in $\mathbb{R}^n$. Define the differential operator $P_0(\partial') := P(\partial', 0)$ in $\mathbb{R}^{n-1}$. If $f \in \mathcal{D}'(\mathbb{R}^{n-1})$ and $u \in \mathcal{D}'(\mathbb{R}^n)$ are such that*

$$P(\partial)u = f(x') \otimes \delta(x_n) \quad in \quad \mathcal{D}'(\mathbb{R}^n) \qquad (9.1.70)$$

and u is integrable with respect to x_n, then $u_0 := \int_{-\infty}^{\infty} u\, dx_n$ is a solution of the equation $P_0(\partial')u_0 = f$ in $\mathcal{D}'(\mathbb{R}^{n-1})$.

Proof. Fix a sequence $\{\psi_j\}_{j\in\mathbb{N}} \subset C_0^\infty(\mathbb{R})$ that converges in a dominated fashion to 1 and let $\varphi \in C_0^\infty(\mathbb{R}^{n-1})$. Using the definition of u_0 we may write

$$\langle P_0(\partial')u_0, \varphi \rangle \qquad (9.1.71)$$

$$= \langle u_0, P_0(-\partial')\varphi \rangle = \lim_{j\to\infty} \langle u, (P_0(-\partial')\varphi) \otimes \psi_j \rangle$$

$$= \lim_{j\to\infty} \Big[\big\langle u, P(-\partial)(\varphi \otimes \psi_j) \big\rangle + \big\langle u, (P_0(-\partial')\varphi) \otimes \psi_j - P(-\partial)(\varphi \otimes \psi_j) \big\rangle \Big].$$

We claim that

$$\lim_{j\to\infty} \big\langle u, (P_0(-\partial')\varphi) \otimes \psi_j - P(-\partial)(\varphi \otimes \psi_j) \big\rangle = 0. \qquad (9.1.72)$$

Assume the claim for now. Then returning to (9.1.71) we have

$$\langle P_0(\partial')u_0, \varphi \rangle = \lim_{j\to\infty} \langle u, P(-\partial)(\varphi \otimes \psi_j) \rangle = \lim_{j\to\infty} \langle P(\partial)u, \varphi \otimes \psi_j \rangle$$

$$= \lim_{j\to\infty} \langle f \otimes \delta, \varphi \otimes \psi_j \rangle = \lim_{j\to\infty} [\langle f, \varphi \rangle \psi_j(0)] = \langle f, \varphi \rangle, \qquad (9.1.73)$$

where for the last equality in (9.1.73) we used property (*i*) in Definition 9.4 with $K = \{0\}$. Hence, the desired conclusion follows.

To prove (9.1.72), observe that

$$(P_0(-\partial')\varphi) \otimes \psi_j - P(-\partial)(\varphi \otimes \psi_j) = \sum_{q=1}^{m} (P_q(\partial')\varphi) \otimes \psi_j^{(q)}$$

where P_q is a differential operator of order $\leq m - q$. Then for each $q \in \{1, \dots m\}$, the sequence $\{\psi_j + \psi_j^{(q)}\}_{j\in\mathbb{N}}$ also converges in a dominated fashion to 1, which combined with the fact that u is integrable with respect to x_n, further yields

$$\lim_{j \to} \langle u, (P_q(\partial')\varphi) \otimes \psi_j^{(q)} \rangle$$

$$= \lim_{j \to \infty} \langle u, (P_q(\partial')\varphi) \otimes (\psi_j + \psi_j^{(q)}) \rangle - \lim_{j \to \infty} \langle u, (P_q(\partial')\varphi) \otimes \psi_j \rangle$$

$$= \langle u_0, P_q(\partial')\varphi \rangle - \langle u_0, P_q(\partial')\varphi \rangle = 0, \tag{9.1.74}$$

proving (9.1.72). The proof of the proposition is now complete. $\qquad\square$

9.1.4 The Case $n = 2p$, $p \geq 1$

We are now ready to proceed with determining a fundamental solution for the wave operator in the case when $n = 2p$, $p \geq 1$. The main idea is to use Proposition 9.7 corresponding to $P := \partial_t^2 - \sum_{j=1}^{2p+1} \partial_j^2$ being the wave operator in $\mathbb{R}^{n+2}$, $f := \delta(x_1, \ldots, x_{2p}) \otimes \delta(t)$, and u equal to E_{2p+1}, the fundamental solution from (9.1.51). Note that under these assumptions we have $P_0 := \partial_t^2 - \sum_{j=1}^{2p} \partial_j^2$ which is the wave operator in $\mathbb{R}^{n+1}$. Thus, if E_{2p+1} is integrable with respect to x_{2p+1}, then by Proposition 9.7 it follows that $u := \int_{-\infty}^{\infty} E_{2p+1} \, dx_{2p+1}$ satisfies

$$\left(\partial_t^2 - \sum_{j=1}^{2p} \partial_j^2\right)u = \delta(x_1, \ldots, x_{2p}) \otimes \delta(t) \quad \text{in} \quad \mathcal{D}'(\mathbb{R}^{2p+1}). \tag{9.1.75}$$

Therefore, u is a fundamental solution for the wave operator corresponding to $n = 2p$.

Let us first show that the distribution E_{2p+1} given by the formula in (9.1.51) is integrable with respect to x_{2p+1}. Fix an arbitrary function $\varphi \in C_0^\infty(\mathbb{R}^{2p+1})$ and let $\{\psi_j\}_{j \in \mathbb{N}} \subset C_0^\infty(\mathbb{R})$ be a sequence that converges in a dominated fashion to 1. Then, using the notation $x = (x', x_{2p+1}) \in \mathbb{R}^{2p} \times \mathbb{R}$, we have

$$\lim_{j \to \infty} \langle E_{2p+1}, \varphi \otimes \psi_j \rangle$$

$$= \lim_{j \to \infty} 2^{-(p+1)} \pi^{-p} \int_0^\infty \left[\left(\frac{1}{r}\partial_r\right)^{p-1}\left(\frac{1}{r} \int_{x \in \mathbb{R}^{2p+1}, |x|=r} \varphi(x', t)\psi_j(x_{2p+1}) \, d\sigma(x)\right)\right]_{r=t} dt$$

$$= 2^{-(p+1)} \pi^{-p} \int_0^\infty \left[\left(\frac{1}{r}\partial_r\right)^{p-1}\left(\frac{1}{r} \int_{x \in \mathbb{R}^{2p+1}, |x|=r} \varphi(x', t) \, d\sigma(x)\right)\right]_{r=t} dt, \tag{9.1.76}$$

where for the last equality in (9.1.76) we used Lebesgue's Dominated Convergence Theorem (here we note that the second equality in (9.1.47) and the properties satisfied by $\{\psi_j\}_{j\in\mathbb{N}}$ play an important role). With (9.1.76) in hand, we may conclude that E_{2p+1} is integrable with respect to x_{2p+1}. Consequently, if one sets $E_{2p} := \int_{-\infty}^{\infty} E_{2p+1}\, dx_{2p+1}$, then $E_{2p} \in \mathcal{D}'(\mathbb{R}^{2p+1})$ and

$$\langle E_{2p}, \varphi \rangle = 2^{-(p+1)}\pi^{-p} \int_0^{\infty} \Big[\Big(\frac{1}{r}\partial_r\Big)^{p-1} \Big(\frac{1}{r} \int_{x\in\mathbb{R}^{2p+1},\,|(x',x_{2p+1})|=r} \varphi(x',t)\, d\sigma(x)\Big) \Big]_{r=t} dt \quad (9.1.77)$$

for every $\varphi \in C_0^{\infty}(\mathbb{R}^{2p+1})$. In addition, by Proposition 9.7, it follows that E_{2p} is a fundamental solution for the wave operator in $\mathbb{R}^{n+1}$.

Note that the function φ appearing under the second integral in (9.1.77) does not depend on the variable x_{2p+1}. Hence, it is natural to proceed further in order to rewrite (9.1.77) in a form that does not involve the variable x_{2p+1}. Fix $r > 0$ and denote by $B(0,r)$ the ball in $\mathbb{R}^n$ of radius r and centered at $0 \in \mathbb{R}^n$. Define the mappings $P_{\pm} : B(0,r) \to \mathbb{R}^{2p+1}$ by setting

$$P_{\pm}(x') := (x_1, x_2, \ldots, x_{2p}, \pm\sqrt{r^2 - |x'|^2})$$
$$\text{for each } \ x' = (x_1, \ldots, x_{2p}) \in B(0,r). \quad (9.1.78)$$

Then P_+ and P_- are parametrizations of the (open) upper and lower, respectively, hemispheres of the surface ball in $\mathbb{R}^{2p+1}$ of radius r and centered at 0. Hence, (keeping in mind Definition 14.47 and Definition 14.48) we may write

$$\int_{x\in\mathbb{R}^{2p+1},\,|(x',x_{2p+1})|=r} \varphi(x',t)\, d\sigma(x) \quad (9.1.79)$$

$$= \int_{B(0,r)} \varphi(x',t)\,|\partial_1 P_+ \times \cdots \times \partial_{2p}P_+|\, dx'$$

$$+ \int_{B(0,r)} \varphi(x',t)\,|\partial_1 P_- \times \cdots \times \partial_{2p}P_-|\, dx'.$$

A direct computation based on (9.1.78) and 14.6.4 further yields

$$\partial_1 P_{\pm} \times \cdots \times \partial_{2p}P_{\pm} = \begin{vmatrix} 1 & 0 & \ldots & 0 & \mp\frac{x_1}{\sqrt{r^2-|x'|^2}} \\ 0 & 1 & \ldots & 0 & \mp\frac{x_2}{\sqrt{r^2-|x'|^2}} \\ \ldots & \ldots & \ldots & \ldots & \ldots \\ 0 & 0 & \ldots & 1 & \mp\frac{x_{2p}}{\sqrt{r^2-|x'|^2}} \\ \mathbf{e}_1 & \mathbf{e}_2 & \ldots & \mathbf{e}_{2p} & \mathbf{e}_{2p+1} \end{vmatrix} =: \det A^{\pm}, \quad (9.1.80)$$

where $\mathbf{e}_j$ is the unit vector in $\mathbb{R}^{2p+1}$ with 1 on the j-th position, for each j in $\{1, \ldots, 2p+1\}$. Hence, the components of the vector $\partial_1 P_{\pm} \times \cdots \times \partial_{2p}P_{\pm}$ are

$$(-1)^{k+1} \det A_k^{\pm}, \quad k \in \{1, \ldots, 2p, 2p+1\}, \tag{9.1.81}$$

where $A_k^{\pm}$ is the $2p \times 2p$ matrix obtained from $A^{\pm}$ by deleting column k and row $2p+1$. It is easy to see from (9.1.80) that

$$\det A_{2p+1}^{\pm} = 1, \quad \det A_k^{\pm} = (-1)^k \frac{\mp x_k}{\sqrt{r^2 - |x'|^2}}, \quad \forall k \in \{1, \ldots, 2p\}. \tag{9.1.82}$$

Consequently,

$$|\partial_1 P_{\pm} \times \cdots \times \partial_{2p} P_{\pm}| = \frac{r}{\sqrt{r^2 - |x'|^2}}. \tag{9.1.83}$$

Formula (9.1.83) combined with (9.1.79) and (9.1.77) gives

$$\langle E_{2p}, \varphi \rangle = (2\pi)^{-p} \int_0^{\infty} \left[\left(\frac{1}{r} \partial_r \right)^{p-1} \left(\frac{1}{r} \int_{x' \in B(0,r)} \varphi(x', r) \frac{r}{\sqrt{r^2 - |x'|^2}} \, dx' \right) \right]_{r=t} dt$$

$$= (2\pi)^{-p} \int_0^{\infty} \left[\left(\frac{1}{r} \partial_r \right)^{p-1} \left(\int_{B(0,r)} \frac{\varphi(x', t)}{\sqrt{r^2 - |x'|^2}} \, dx' \right) \right]_{r=t} dt. \tag{9.1.84}$$

In summary, we proved:

If $n = 2p$, for some $p \in \mathbb{N}$, then $E_+ \in \mathcal{S}'(\mathbb{R}^{n+1})$, defined by

$$\langle E_+, \varphi \rangle = (2\pi)^{-p} \int_0^{\infty} \left[\left(\frac{1}{r} \partial_r \right)^{p-1} \left(\int_{x \in \mathbb{R}^n, |x| < r} \frac{\varphi(x, t)}{\sqrt{r^2 - |x|^2}} \, dx \right) \right]_{r=t} dt, \tag{9.1.85}$$

for $\varphi \in \mathcal{S}(\mathbb{R}^{n+1})$, is a fundamental solution for the wave operator in $\mathbb{R}^{n+1}$.

The reasoning used to obtain E_+ also applies if one starts with E_{2p+1} being the distribution from (9.1.54). Under this scenario the conclusion is that

If $n = 2p$, for some $p \in \mathbb{N}$, then $E_- \in \mathcal{S}'(\mathbb{R}^{n+1})$, defined by

$$\langle E_-, \varphi \rangle = (2\pi)^{-p} \int_0^{\infty} \left[\left(\frac{1}{r} \partial_r \right)^{p-1} \left(\int_{x \in \mathbb{R}^n, |x| < r} \frac{\varphi(x, -t)}{\sqrt{r^2 - |x|^2}} \, dx \right) \right]_{r=t} dt, \tag{9.1.86}$$

for $\varphi \in \mathcal{S}(\mathbb{R}^{n+1})$, is a fundamental solution for the wave operator in $\mathbb{R}^{n+1}$.

Remark 9.8.(1) If E_+ and E_- are as in (9.1.85) and (9.1.86), respectively, then

$$\operatorname{supp} E_+ = \{(x, t) \in \mathbb{R}^{2p} \times [0, \infty) : |x| \leq t\}, \quad \text{and} \tag{9.1.87}$$

$$\operatorname{supp} E_- = \{(x, t) \in \mathbb{R}^{2p} \times (-\infty, 0] : |x| \leq -t\}. \tag{9.1.88}$$

(2) If $p = 1$, then for each $\varphi \in \mathcal{S}(\mathbb{R}^3)$ the expression[2] in (9.1.85) becomes

$$\langle E_+, \varphi \rangle = \frac{1}{2\pi} \int_0^\infty \int_{|x|<t} \frac{\varphi(x,t)}{\sqrt{t^2 - |x|^2}} \, dx \, dt = \left\langle \frac{H(t - |x|)}{2\pi \sqrt{t^2 - |x|^2}}, \varphi \right\rangle.$$

Hence, if $n = 2$ then E_+ is of function type and

$$E_+(x, t) = \frac{H(t - |x|)}{2\pi \sqrt{t^2 - |x|^2}} \quad \text{for every} \quad x \in \mathbb{R}^2 \quad \text{and every} \quad t \in \mathbb{R}.$$

Similarly, when $n = 2$ it follows that E_- is of function type and

$$E_-(x, t) = \frac{H(-t - |x|)}{2\pi \sqrt{t^2 - |x|^2}} \quad \text{for every} \quad x \in \mathbb{R}^2 \quad \text{and every} \quad t \in \mathbb{R}.$$

9.1.5 Summary for Arbitrary n

Here we combine the results obtained in Section 9.1.1, Section 9.1.2, and Section 9.1.4 regarding fundamental solutions for the heat operator. These results have been summarized in (9.1.28), Remark 9.1, (9.1.51), (9.1.54), Remark 9.2, (9.1.85), (9.1.86), and Remark 9.8.

Theorem 9.9. *Consider the wave operator $\square = \partial_t^2 - \Delta_x$ in $\mathbb{R}^{n+1}$, where $x \in \mathbb{R}^n$ and $t \in \mathbb{R}$. Then the following are true.*

(1) Suppose $n = 1$. Then the $L^1_{loc}(\mathbb{R}^2)$ functions

$$E_+(x, t) := \frac{H(t - |x|)}{2}, \qquad \forall\, (x, t) \in \mathbb{R}^2, \tag{9.1.89}$$

$$E_-(x, t) := \frac{H(-t - |x|)}{2}, \qquad \forall\, (x, t) \in \mathbb{R}^2, \tag{9.1.90}$$

satisfy $E_+, E_- \in \mathcal{S}'(\mathbb{R}^2)$ and are fundamental solutions for the wave operator in $\mathbb{R}^2$. Moreover,

$$\operatorname{supp} E_+ = \{(x, t) \in \mathbb{R}^2 : |x| \leq t\}, \tag{9.1.91}$$

$$\operatorname{supp} E_- = \{(x, t) \in \mathbb{R}^2 : |x| \leq -t\}. \tag{9.1.92}$$

(2) Suppose $n = 2p + 1$, for some $p \in \mathbb{N}$. Then the distributions $E_+ \in \mathcal{S}'(\mathbb{R}^{n+1})$ and $E_- \in \mathcal{S}'(\mathbb{R}^{n+1})$ defined by

---

[2] this expression was first found by Vito Volterra (cf. [79])

$$\langle E_+, \varphi \rangle := (2\pi)^{-p-1} \pi \int_0^\infty \left[\left(\frac{1}{r} \partial_r \right)^{p-1} \left(\frac{1}{r} \int_{\partial B(0,r)} \varphi(\omega, t) \, d\sigma(\omega) \right) \right]_{r=t} dt, \quad (9.1.93)$$

$$\langle E_-, \varphi \rangle := (2\pi)^{-p-1} \pi \int_0^\infty \left[\left(\frac{1}{r} \partial_r \right)^{p-1} \left(\frac{1}{r} \int_{\partial B(0,r)} \varphi(\omega, -t) \, d\sigma(\omega) \right) \right]_{r=t} dt, \quad (9.1.94)$$

for every $\varphi \in \mathcal{S}(\mathbb{R}^{n+1})$, are fundamental solutions for the wave operator in $\mathbb{R}^{n+1}$. Moreover,

$$\operatorname{supp} E_+ = \{ (x, t) \in \mathbb{R}^{2p+1} \times [0, \infty) : |x| = t \}, \quad (9.1.95)$$

$$\operatorname{supp} E_+ = \{ (x, t) \in \mathbb{R}^{2p+1} \times (-\infty, 0] : |x| = -t \}. \quad (9.1.96)$$

Corresponding to the case $n = 3$ (thus, for $p = 1$), formulas (9.1.93) and (9.1.94) become

$$\langle E_+, \varphi \rangle = \frac{1}{4\pi} \int_0^\infty \frac{1}{t} \int_{\partial B(0,t)} \varphi(\omega, t) \, d\sigma(\omega) \, dt = \left\langle \frac{H(t)}{4\pi t} \delta_{\partial B(0,t)}, \varphi \right\rangle, \quad (9.1.97)$$

$$\langle E_-, \varphi \rangle = \frac{1}{4\pi} \int_0^\infty \frac{1}{t} \int_{\partial B(0,t)} \varphi(\omega, -t) \, d\sigma(\omega) \, dt = \left\langle -\frac{H(-t)}{4\pi t} \delta_{\partial B(0,-t)}, \varphi \right\rangle, \quad (9.1.98)$$

for every $\psi \in \mathcal{S}(\mathbb{R}^4)$.

(3) Suppose $n = 2p$, for some $p \in \mathbb{N}$. Then the distributions $E_+ \in \mathcal{S}'(\mathbb{R}^{n+1})$ and $E_- \in \mathcal{S}'(\mathbb{R}^{n+1})$ defined by

$$\langle E_+, \varphi \rangle := (2\pi)^{-p} \int_0^\infty \left[\left(\frac{1}{r} \partial_r \right)^{p-1} \left(\int_{x \in \mathbb{R}^n, |x| < r} \frac{\varphi(x, t)}{\sqrt{r^2 - |x|^2}} \, dx \right) \right]_{r=t} dt, \quad (9.1.99)$$

$$\langle E_-, \varphi \rangle := (2\pi)^{-p} \int_0^\infty \left[\left(\frac{1}{r} \partial_r \right)^{p-1} \left(\int_{x \in \mathbb{R}^n, |x| < r} \frac{\varphi(x, -t)}{\sqrt{r^2 - |x|^2}} \, dx \right) \right]_{r=t} dt, \quad (9.1.100)$$

for every $\varphi \in \mathcal{S}(\mathbb{R}^{n+1})$, are fundamental solutions for the wave operator in $\mathbb{R}^{n+1}$. Moreover,

$$\operatorname{supp} E_+ = \{ (x, t) \in \mathbb{R}^{2p} \times [0, \infty) : |x| \le t \}, \quad \text{and} \quad (9.1.101)$$

$$\operatorname{supp} E_- = \{ (x, t) \in \mathbb{R}^{2p} \times (-\infty, 0] : |x| \le -t \}. \quad (9.1.102)$$

Corresponding to the case $n = 2$, thus $p = 1$, the distributions E_+ and E_- defined in (9.1.99) and (9.1.100) are of function type and given by the functions

$$E_+(x,t) = \frac{H(t - |x|)}{2\pi\sqrt{t^2 - |x|^2}}, \qquad \forall\, x \in \mathbb{R}^2, \forall\, t \in \mathbb{R}, \tag{9.1.103}$$

$$E_-(x,t) = \frac{H(-t - |x|)}{2\pi\sqrt{t^2 - |x|^2}}, \qquad \forall\, x \in \mathbb{R}^2, \forall\, t \in \mathbb{R}. \tag{9.1.104}$$

9.2 The Generalized Cauchy Problem for the Wave Operator

In this section we discuss the generalized Cauchy problem for the wave operator. Let $F \in C^0(\mathbb{R}^{n+1})$, $f, g \in C^0(\mathbb{R}^n)$ and, as before, use the notation $x \in \mathbb{R}^n$, $t \in \mathbb{R}$. Suppose $u \in C^2(\mathbb{R}^{n+1})$ solves in the classical sense the Cauchy problem

$$\begin{cases} (\partial_t^2 - \Delta_x)u = F & \text{in } \mathbb{R}^{n+1}, \\ u\big|_{t=0} = f, \ \partial_t u\big|_{t=0} = g & \text{in } \mathbb{R}^n. \end{cases} \tag{9.2.1}$$

Define $\widetilde{u}(x,t) := H(t)u(x,t)$ and $\widetilde{F}(x,t) := H(t)F(x,t)$ for every $x \in \mathbb{R}^n$ and every $t \in \mathbb{R}$. Then $\widetilde{u}, \widetilde{F} \in L^1_{loc}(\mathbb{R}^{n+1})$ and for each $\varphi \in C_0^\infty(\mathbb{R}^{n+1})$ we have

$$\langle (\partial_t^2 - \Delta_x)\widetilde{u}, \varphi \rangle = \langle \widetilde{u}, (\partial_t^2 - \Delta_x)\varphi \rangle = \lim_{\varepsilon \to 0^+} \int_\varepsilon^\infty \int_{\mathbb{R}^n} u\,(\partial_t^2 - \Delta_x)\varphi \,dx\,dt$$

$$= \lim_{\varepsilon \to 0^+} \int_\varepsilon^\infty \int_{\mathbb{R}^n} (\partial_t^2 - \Delta_x)u\,\varphi \,dx\,dt \tag{9.2.2}$$

$$+ \lim_{\varepsilon \to 0^+} \int_{\mathbb{R}^n} \big(\partial_t u(x,\varepsilon)\varphi(x,\varepsilon) - u(x,\varepsilon)\partial_t\varphi(x,\varepsilon) \big)\,dx$$

$$= \langle \widetilde{F}, \varphi \rangle + \langle f(x) \otimes \delta'(t), \varphi(x,t) \rangle + \langle g(x) \otimes \delta(x), \varphi(x,t) \rangle.$$

This suggests the following definition (recall (8.2.4)).

Definition 9.10. Suppose $\widetilde{F} \in \mathcal{D}'_+(\mathbb{R}^{n+1})$ and $f, g \in \mathcal{D}(\mathbb{R}^n)$ are given. Then a distribution $\widetilde{u} \in \mathcal{D}'_+(\mathbb{R}^{n+1})$ is called a solution of the generalized Cauchy problem for the wave operator with data $\widetilde{F}$, f, and g, if

$$(\partial_t^2 - \Delta_x)\widetilde{u} = \widetilde{F} + f(x) \otimes \delta'(t) + g(x) \otimes \delta(t) \quad \text{in } \mathcal{D}'(\mathbb{R}^{n+1}). \tag{9.2.3}$$

Theorem 9.11. *Let $\widetilde{F} \in \mathcal{D}'_+(\mathbb{R}^{n+1})$ and $f, g \in \mathcal{D}(\mathbb{R}^n)$ be given. Then the generalized Cauchy problem (9.2.3) has the unique solution*

$$\widetilde{u} = E * \widetilde{F} + E * (f \otimes \delta') + E * (g \otimes \delta), \tag{9.2.4}$$

where E is the fundamental solution for the wave operator in $\mathbb{R}^{n+1}$ as specified in (9.1.89) if $n = 1$, in (9.1.93) if $n \geq 3$ is odd, and in (9.1.99) if n is even.

Proof. Let E be as in the statement of the theorem. Then, by (9.1.91), by (9.1.95), and by (9.1.101), for any $G \in \mathcal{D}'_+(\mathbb{R}^{n+1})$ we have that, whenever K is a compact subset of $\mathbb{R}^{n+1}$, the set

$$M_K := \{((x,t),(y,s)) : (x,t) \in \operatorname{supp} E,$$
$$(y,s) \in \operatorname{supp} G, (x+y, t+s) \in K\} \qquad (9.2.5)$$

is compact in $\mathbb{R}^{n+1}$. Hence, by Theorem 2.94, the convolution $E * G$ exists. This proves that $\widetilde{u}$ as in (9.2.4) is a well-defined element of $\mathcal{D}'(\mathbb{R}^{n+1})$. Moreover, since $\widetilde{F}, f \otimes \delta', g \otimes \delta, E \in \mathcal{D}'_+(\mathbb{R}^{n+1})$, by *(a)* in Theorem 2.96, we may conclude that $\widetilde{u}$ belongs to $\mathcal{D}'_+(\mathbb{R}^{n+1})$. In addition, by Remark 5.6, we obtain that $\widetilde{u}$ as in (9.2.4) is a solution of the generalized Cauchy problem (9.2.3).

To prove uniqueness, observe that if $\widetilde{u} \in \mathcal{D}'_+(\mathbb{R}^{n+1})$ is such that $(\partial_t^2 - \Delta_x)\widetilde{u} = 0$ in $\mathcal{D}'(\mathbb{R}^{n+1})$, then

$$\widetilde{u} = \widetilde{u} * \delta = \widetilde{u} * ((\partial_t^2 - \Delta_x)E) = ((\partial_t^2 - \Delta_x)\widetilde{u}) * E = 0. \qquad (9.2.6)$$

This completes the proof of the theorem. $\qquad\qquad\square$

Further Notes for Chapter 9. The wave operator $\square := \partial_t^2 - \Delta_x$, where $x \in \mathbb{R}^n$ and $t \in \mathbb{R}$, was originally discovered by the French mathematician and physicist Jean le Rond d'Alembert in 1747 in the case $n = 1$ in his studies of vibration of strings. For this reason, $\square$ is also called the d'Alembert operator or, simply, the d'Alembertian. Like the heat operator discussed in Chapter 8, the wave operator is another basic example of a partial differential operator governing a linear evolution equation (though, unlike the heat operator, the wave operator belongs to a class of operators called hyperbolic operators).

9.3 Additional Exercises for Chapter 9

Exercise 9.12. Use the Method of Descent to compute a fundamental solution for the Laplace operator $\Delta = \sum_{j=1}^{n} \partial_j^2$ in $\mathbb{R}^n$, $n \geq 3$, by starting from a fundamental solution for the heat operator $L = \partial_t - \sum_{j=1}^{n} \partial_j^2$ in $\mathbb{R}^{n+1}$.

Chapter 10
The Lamé and Stokes Operators

Abstract The material here is centered around two basic systems: the Lamé operator arising in the theory of elasticity, and the Stokes operator arising in hydrodynamics. Among other things, all of their fundamental solutions that are tempered distributions are identified, and the well-posedness of the Poisson problem for the Lamé system is established.

Throughout this chapter, it is assumed that $n \in \mathbb{N}$ satisfies $n \geq 2$.

10.1 General Remarks About Vector and Matrix Distributions

The material developed up to this point may be regarded as a theory for scalar distributions. Nonetheless, practical considerations dictate the necessity of considering vectors/matrices whose components/entries are themselves distributions. It is therefore natural to refer to such objects as vector and matrix distributions. A significant portion of the theory of scalar distributions then readily extends to this more general setting. The philosophy in the vector/matrix case is that we perform the same type of analysis as in the scalar case, at the level of individual components, while at the same time obeying the natural algebraic rules that are now in effect (e.g., keeping in mind the algebraic mechanism according to which two matrices are multiplies, etc.).

To offer some examples, fix an open set $\Omega \subseteq \mathbb{R}^n$ and consider a matrix distribution

$$\mathbb{U} = (u_{k\ell})_{\substack{1 \leq k \leq N \\ 1 \leq \ell \leq K}} \in \mathcal{M}_{N \times K}(\mathcal{D}'(\Omega)), \tag{10.1.1}$$

i.e., $\mathbb{U}$ is an $N \times K$ matrix whose entries are from $\mathcal{D}'(\Omega)$. Naturally, equality of elements in $\mathcal{M}_{N \times K}(\mathcal{D}'(\Omega))$ is understood entry by entry in $\mathcal{D}'(\Omega)$. We agree to define the support of $\mathbb{U}$ as

$$\operatorname{supp} \mathbb{U} := \bigcup_{k=1}^{K} \bigcup_{\ell=1}^{N} \operatorname{supp} u_{k\ell}. \tag{10.1.2}$$

© Springer Nature Switzerland AG 2018
D. Mitrea, *Distributions, Partial Differential Equations, and Harmonic Analysis*,
Universitext, https://doi.org/10.1007/978-3-030-03296-8_10

Note that supp $\mathbb{U}$ is the smallest relatively closed subset of Ω outside of which all entries of $\mathbb{U}$ vanish. Similarly, we define the singular support of $\mathbb{U}$ as

$$\text{sing supp}\,\mathbb{U} := \bigcup_{k=1}^{K}\bigcup_{\ell=1}^{N}\text{sing supp}\,u_{k\ell}. \tag{10.1.3}$$

Hence, sing supp $\mathbb{U}$ is the smallest relatively closed subset of Ω outside of which all entries of $\mathbb{U}$ are C^{∞}.

Next, if $A = (a^{jk})_{\substack{1 \le j \le M \\ 1 \le k \le N}} \in \mathcal{M}_{M \times N}(C^{\infty}(\Omega))$, then we define (with $\mathbb{U}$ as in (10.1.1))

$$A\mathbb{U} := \Big(\sum_{k=1}^{N} a^{jk} u_{k\ell}\Big)_{\substack{1 \le j \le M \\ 1 \le \ell \le K}} \in \mathcal{M}_{M \times K}(\mathcal{D}'(\Omega)). \tag{10.1.4}$$

Some trivial, yet useful properties include

$$\text{if} \quad \mathbb{U}, \mathbb{V} \in \mathcal{M}_{N \times K}(\mathcal{D}'(\Omega)), \quad A \in \mathcal{M}_{J \times M}(C^{\infty}(\Omega)), \quad B \in \mathcal{M}_{M \times N}(C^{\infty}(\Omega)),$$
$$\text{then} \quad A(B\mathbb{U}) = (AB)\mathbb{U} \quad \text{and}$$
$$\mathbb{U} = \mathbb{V} \implies A\mathbb{U} = A\mathbb{V}. \tag{10.1.5}$$

Given an operator of the form

$$L = \sum_{|\alpha| \le m} A_\alpha \partial^\alpha, \qquad A_\alpha = (a_\alpha^{jk})_{\substack{1 \le j \le M \\ 1 \le k \le N}} \in \mathcal{M}_{M \times N}(C^{\infty}(\Omega)), \tag{10.1.6}$$

referred to as an $M \times N$ system (of differential operators) of order $m \in \mathbb{N}$, the natural way in which L acts on $\mathbb{U}$ as in (10.1.1) is to regard $L\mathbb{U}$ as the matrix distribution from $\mathcal{M}_{M \times K}(\mathcal{D}'(\Omega))$ whose entries are given by

$$(L\mathbb{U})_{j\ell} := \sum_{|\alpha| \le m} \sum_{k=1}^{N} a_\alpha^{jk} \partial^\alpha u_{k\ell} \quad \text{for all} \quad 1 \le j \le M, \ 1 \le \ell \le K. \tag{10.1.7}$$

A matrix distribution with a single column is referred to a vector distribution and simply denoted by $\mathbf{u} = (u_k)_{1 \le k \le N}$ with components $u_k \in \mathcal{D}'(\Omega), 1 \le k \le N$. The collection of all such vector distributions is denoted by $[\mathcal{D}'(\Omega)]^N$.

Regarding the convolution product for matrix distributions, if

$$\mathbb{U} = (u_{jk})_{\substack{1 \le k \le N \\ 1 \le \ell \le K}} \in \mathcal{M}_{M \times N}(\mathcal{D}'(\mathbb{R}^n)) \quad \text{and}$$
$$\mathbb{V} = (v_{k\ell})_{\substack{1 \le k \le N \\ 1 \le \ell \le K}} \in \mathcal{M}_{N \times K}(\mathcal{E}'(\mathbb{R}^n)) \tag{10.1.8}$$

(hence, $\mathbb{V}$ is an $N \times K$ matrix whose entries are from $\mathcal{E}'(\mathbb{R}^n)$), we define $\mathbb{U} * \mathbb{V}$ as being the $M \times K$ matrix whose entries are the distributions from $\mathcal{D}'(\mathbb{R}^n)$ given by

$$(\mathbb{U} * \mathbb{V})_{j\ell} := \sum_{k=1}^{N} u_{jk} * v_{k\ell}, \qquad 1 \le j \le M, \ 1 \le \ell \le K. \tag{10.1.9}$$

Formula (10.1.9) is also taken as definition for $\mathbb{U} * \mathbb{V}$ in the case when

$$\mathbb{U} = (u_{jk})_{\substack{1 \le k \le N \\ 1 \le \ell \le K}} \in \mathcal{M}_{M \times N}(\mathcal{E}'(\mathbb{R}^n)) \tag{10.1.10}$$

and

$$\mathbb{V} = (v_{k\ell})_{\substack{1 \le k \le N \\ 1 \le \ell \le K}} \in \mathcal{M}_{N \times K}(\mathcal{D}'(\mathbb{R}^n)). \tag{10.1.11}$$

The reader is advised that, as opposed to the scalar case, the commutativity property for the convolution product is lost in the setting of (say, square) matrix distributions.

For each natural number M we agree to denote by $\delta I_{M \times M}$ the matrix distribution from $\mathcal{M}_{M \times M}(\mathcal{D}'(\mathbb{R}^n))$ with entries $\delta_{jk}\delta$, for all $j, k \in \{1, \dots, M\}$, where $\delta_{jk} = 1$ if $j = k$ and 0 otherwise, is the Kronecker symbol. Then, one can check using (10.1.9) and part (d) in Theorem (2.96) that

$$(\delta I_{M \times M}) * \mathbb{U} = \mathbb{U}, \qquad \forall\, \mathbb{U} \in \mathcal{M}_{M \times N}(\mathcal{D}'(\mathbb{R}^n)), \tag{10.1.12}$$

$$\mathbb{V} * (\delta I_{M \times M}) = \mathbb{V}, \qquad \forall\, \mathbb{V} \in \mathcal{M}_{N \times M}(\mathcal{D}'(\mathbb{R}^n)). \tag{10.1.13}$$

Two important first-order systems of differential operators are the gradient ∇ and the divergence div. Specifically, the gradient $\nabla := (\partial_1, \dots, \partial_n)$ acts on a scalar distribution according to $\nabla u := (\partial_1 u, \dots, \partial_n u) \in [\mathcal{D}'(\Omega)]^n$ for every $u \in \mathcal{D}'(\Omega)$, while the divergence operator div acts on a vector distribution

$$\mathbf{v} := (v_1, \dots, v_n) \in [\mathcal{D}'(\Omega)]^n \tag{10.1.14}$$

according to $\operatorname{div}\mathbf{v} := \sum_{j=1}^{n} \partial_j v_j$.

To give an example of a specific result from the theory of scalar distributions and scalar differential operators that naturally carries over to the vector/matrix case, we note here that if L is an $J \times M$ system with constant coefficients, then for every matrix distributions $\mathbb{U} \in \mathcal{M}_{M \times N}(\mathcal{D}'(\mathbb{R}^n))$ and $\mathbb{V} \in \mathcal{M}_{N \times K}(\mathcal{E}'(\mathbb{R}^n))$ then

$$L(\mathbb{U} * \mathbb{V}) = (L\mathbb{U}) * \mathbb{V} = \mathbb{U} * (L\mathbb{V}) \quad \text{in} \quad \mathcal{M}_{J \times K}(\mathcal{D}'(\mathbb{R}^n)). \tag{10.1.15}$$

Of course, similar equalities hold when

$$\mathbb{U} \in \mathcal{M}_{M \times N}(\mathcal{E}'(\mathbb{R}^n)) \ \text{ and } \ \mathbb{V} \in \mathcal{M}_{N \times K}(\mathcal{D}'(\mathbb{R}^n)). \tag{10.1.16}$$

Definition 10.1. Given an $M \times N$ system of order $m \in \mathbb{N}$,

$$L = L(\partial) = \sum_{|\alpha| \le m} A_\alpha \partial^\alpha, \qquad A_\alpha = (a_\alpha^{jk})_{\substack{1 \le j \le M \\ 1 \le k \le N}} \in \mathcal{M}_{M \times N}(C^\infty(\mathbb{R}^n)), \tag{10.1.17}$$

define

$$L(\zeta) := \sum_{|\alpha| \leq m} \zeta^\alpha A_\alpha \in M_{M \times N}(C^\infty(\mathbb{R}^n)), \qquad \forall\, \zeta \in \mathbb{C}^n. \tag{10.1.18}$$

Also, call a matrix distribution $\mathbb{E} \in M_{N \times M}(\mathcal{D}'(\mathbb{R}^n))$ a fundamental solution for L in $\mathbb{R}^n$ provided

$$L\mathbb{E} = \delta I_{M \times M} \quad \text{in} \quad M_{M \times M}(\mathcal{D}'(\mathbb{R}^n)). \tag{10.1.19}$$

We record next the analogue of Proposition 5.2 and Proposition 5.8 in the current setting. The reader is advised to recall Remark 5.1.

Proposition 10.2. *Suppose L is a constant coefficient $M \times M$ system of order $m \in \mathbb{N}$,*

$$L = L(\partial) := \sum_{|\alpha| \leq m} A_\alpha \partial^\alpha, \quad A_\alpha = (a_\alpha^{jk})_{1 \leq j,k \leq M} \in M_{M \times M}(\mathbb{C}), \tag{10.1.20}$$

with the property that

$$\det(L(i\xi)) \neq 0 \;\; \text{for every} \;\; \xi \in \mathbb{R}^n \setminus \{0\}. \tag{10.1.21}$$

Then the following hold.

(1) If $\mathbb{U} \in M_{M \times K}(\mathcal{S}'(\mathbb{R}^n))$ is such that $L\mathbb{U} = 0$ in $M_{M \times K}(\mathcal{S}'(\mathbb{R}^n))$, then the entries of $\mathbb{U}$ are polynomials in $\mathbb{R}^n$ and $L\mathbb{U} = 0 \in M_{M \times K}(\mathbb{C})$ pointwise in $\mathbb{R}^n$.

(2) If L has a fundamental solution $\mathbb{E} \in M_{M \times M}(\mathcal{S}'(\mathbb{R}^n))$, then any other fundamental solution of L belonging to $M_{M \times M}(\mathcal{S}'(\mathbb{R}^n))$ differs from $\mathbb{E}$ by an $M \times M$ matrix $\mathbb{P}$ whose entries are polynomials in $\mathbb{R}^n$ satisfying

$$L\mathbb{P} = 0 \in M_{M \times M}(\mathbb{C}) \;\; \text{pointwise in} \;\; \mathbb{R}^n. \tag{10.1.22}$$

Proof. If $\mathbb{U} \in M_{M \times K}(\mathcal{S}'(\mathbb{R}^n))$ is such that $L\mathbb{U} = 0$ in $M_{M \times K}(\mathcal{S}'(\mathbb{R}^n))$, taking the Fourier transform, we obtain $L(i\xi)\widehat{\mathbb{U}} = 0$ in $M_{M \times K}(\mathcal{S}'(\mathbb{R}^n))$. In particular,

$$L(i\xi)\widehat{\mathbb{U}} = 0 \quad \text{in} \quad M_{M \times K}(\mathcal{D}'(\mathbb{R}^n \setminus \{0\})). \tag{10.1.23}$$

Since (10.1.21) holds, the matrix $L(i\xi)$ is invertible for every $\xi \in \mathbb{R}^n \setminus \{0\}$, hence $(L(i\xi))^{-1} \in M_{M \times M}(C^\infty(\mathbb{R}^n \setminus \{0\}))$. Based on this, (10.1.23), and (10.1.5), we may write

$$\widehat{\mathbb{U}} = (L(i\xi))^{-1} L(i\xi)\widehat{\mathbb{U}} = (L(i\xi))^{-1} 0 = 0 \quad \text{in} \quad M_{M \times K}(\mathcal{D}'(\mathbb{R}^n \setminus \{0\})). \tag{10.1.24}$$

Hence, $\widehat{\mathbb{U}} = 0$ in $M_{M \times K}(\mathcal{D}'(\mathbb{R}^n \setminus \{0\}))$ which implies $\operatorname{supp} \widehat{\mathbb{U}} \subseteq \{0\}$. Exercise 4.37 (applied to each entry of $\mathbb{U}$) then gives that each component of $\mathbb{U}$ is a polynomial in $\mathbb{R}^n$. In addition, $L\mathbb{U} = 0$ pointwise in $\mathbb{R}^n$. This proves *(1)*.

Suppose $\mathbb{U} \in M_{M \times M}(\mathcal{S}'(\mathbb{R}^n))$ is such that $L\mathbb{U} = \delta I_{M \times M}$ in $M_{M \times M}(\mathcal{S}'(\mathbb{R}^n))$. Then $L(\mathbb{U} - \mathbb{E}) = 0$ in $M_{M \times M}(\mathcal{S}'(\mathbb{R}^n))$ and the desired conclusion follows by applying part *(1)*. $\qquad\square$

We also have a the general Liouville type theorem for system with total symbol invertible outside the origin that we prove next.

Theorem 10.3 (A general Liouville type theorem for systems). *Assume L is an $M \times M$ system of order $m \in \mathbb{N}$ of the form (10.1.20) which satisfies (10.1.21). Also, suppose $\mathbf{u} = (u_1, \ldots, u_M) \in [L^1_{loc}(\mathbb{R}^n)]^M$ satisfies $L\mathbf{u} = 0$ in $[S'(\mathbb{R}^n)]^M$ and has the property that there exist $N \in \mathbb{N}_0$ and $C, R \in [0, \infty)$ such that*

$$|u_j(x)| \leq C|x|^N \quad \text{whenever} \quad |x| \geq R \quad \text{and} \quad j \in \{1, \ldots, M\}. \qquad (10.1.25)$$

Then u_j is a polynomial in $\mathbb{R}^n$ of degree at most N for each $j \in \{1, \ldots, M\}$.

In particular, if $\mathbf{u} \in [L^\infty(\mathbb{R}^n)]^M$ satisfies $L\mathbf{u} = 0$ in $[S'(\mathbb{R}^n)]^M$ then the components of $\mathbf{u}$ are necessarily constants.

Proof. Since (5.1.10) implies that the locally integrable function u_j belongs to $S'(\mathbb{R}^n)$ (cf. Example 4.4) for each $j \in \{1, \ldots, M\}$, Proposition 10.2 implies that all the entries of $\mathbf{u}$ are polynomials in $\mathbb{R}^n$. Moreover, Lemma 5.3 gives that the degree of these polynomials are at most N. $\qquad \square$

We conclude this section by noting a couple of useful results.

Proposition 10.4. *Assume that*

$$L = \sum_{|\alpha| \leq m} A_\alpha \partial^\alpha, \qquad A_\alpha = (a_\alpha^{jk})_{1 \leq j,k \leq M} \in M_{M \times M}(\mathbb{C})), \qquad (10.1.26)$$

is an $M \times M$ system of order $m \in \mathbb{N}$ with constant complex coefficients and suppose that $\mathbb{E} \in M_{M \times M}(\mathcal{D}'(\mathbb{R}^n))$ is a fundamental solution for the system L in $\mathbb{R}^n$. Then for any $\mathbb{U} \in M_{M \times M}(\mathcal{E}'(\mathbb{R}^n))$ one has

$$L(\mathbb{E} * \mathbb{U}) = \mathbb{U} \quad \text{in} \quad M_{M \times M}(\mathcal{D}'(\mathbb{R}^n)). \qquad (10.1.27)$$

Proof. This follows from (10.1.15), (10.1.19), and (10.1.12). $\qquad \square$

Our last result in this section is the following counterpart of Theorem 7.60 at the level of systems.

Theorem 10.5. *Assume that $n \geq 2$, $M \in \mathbb{N}$, and consider the $M \times M$ system*

$$L = \sum_{j,k=1}^n A_{jk} \partial_j \partial_k, \qquad A_{jk} \in M_{M \times M}(\mathbb{C}). \qquad (10.1.28)$$

Then for an $M \times M$ matrix-valued function $\mathbb{E}$ whose components are contained in $C^2(\mathbb{R}^n \setminus \{0\}) \cap L^1_{loc}(\mathbb{R}^n)$ and have the property that their gradients are positive homogeneous of degree $1 - n$ in $\mathbb{R}^n \setminus \{0\}$, the following statements are equivalent:

(1) When viewed in $L^1_{loc}(\mathbb{R}^n)$, the matrix-valued function $\mathbb{E}$ is a fundamental solution for the system L in $\mathbb{R}^n$;

(2) One has $L\mathbb{E} = 0$ pointwise in $\mathbb{R}^n \setminus \{0\}$ and

$$\sum_{j,k=1}^n \int_{S^{n-1}} \omega_j A_{jk} \partial_k \mathbb{E}(\omega) \, d\sigma(\omega) = I_{M \times M}. \qquad (10.1.29)$$

Proof. This is established much as in the scalar case, following the argument in the proof of Theorem 7.60, keeping in mind the matrix algebra formalism.   □

10.2 Fundamental Solutions and Regularity for General Systems

Definition 10.6. Let $\mathbb{K}$ be a field, and let $(\mathfrak{R}, +, 0)$ be a vector space over $\mathbb{K}$ equipped with an additional binary operation from $\mathfrak{R} \times \mathfrak{R}$ to $\mathfrak{R}$, called product (or multiplication). Then $\mathfrak{R}$ is said to be a `commutative associative algebra` over $\mathbb{K}$ if the following identities hold for any three elements $a, b, c \in \mathfrak{R}$, and any two scalars $x, y \in \mathbb{K}$:

$$\text{commutativity}: \quad ab = ba, \tag{10.2.1}$$

$$\text{associativity}: \quad (ab)c = a(bc), \tag{10.2.2}$$

$$\text{distributivity}: \quad (a+b)c = ac + bc, \tag{10.2.3}$$

$$\text{compatibility with scalars}: \quad (xa)(yb) = (xy)(ab). \tag{10.2.4}$$

Finally, an element $e \in \mathfrak{R}$ is said to be `multiplicative unit` provided $ea = ae = a$ for each $a \in \mathfrak{R}$.

Let $\mathfrak{R}$ be a commutative associative algebra over a field $\mathbb{K}$, with multiplicative unit e and additive neutral element 0. We make the convention that $a - b := a + (-b)$ if $a, b \in \mathfrak{R}$ and $-b$ is the inverse of b with respect to the addition operation in the algebra $\mathfrak{R}$. Also, for $j \in \mathbb{N}$, we define $(-1)^j a$ simply a when j is even, and as $-a$ when j is odd.

Given $M \in \mathbb{N}$, we shall let $\mathcal{M}_{M \times M}(\mathfrak{R})$ stand for the collection of $M \times M$ matrices with entries from $\mathfrak{R}$. Then, if $A = (a_{jk})_{1 \leq j,k \leq M} \in \mathcal{M}_{M \times M}(\mathfrak{R})$ and $c \in \mathfrak{R}$, one defines in the usual fashion the operations

$$A^\top := (a_{kj})_{1 \leq j,k \leq M} \in \mathcal{M}_{M \times M}(\mathfrak{R}), \tag{10.2.5}$$

$$cA := (c\, a_{jk})_{1 \leq j,k \leq M} \in \mathcal{M}_{M \times M}(\mathfrak{R}), \tag{10.2.6}$$

$$\det A := \sum_{\sigma \in \mathcal{P}_M} (-1)^{\operatorname{sgn}\sigma} a_{1\sigma(1)} a_{2\sigma(2)} \cdots a_{M\sigma(M)} \in \mathfrak{R}, \tag{10.2.7}$$

where $\mathcal{P}_M$ is the collection of all permutations of the set $\{1, 2, \ldots, M\}$, and the matrix of cofactors

$$\operatorname{adj}(A) := \left((-1)^{j+k} \det(A_{jk})\right)_{1 \leq j,k \leq M}, \tag{10.2.8}$$

where, for any given $j_0, k_0 \in \{1, \ldots, M\}$, the minor $A_{j_0 k_0}$ is defined by

$$A_{j_0 k_0} := (a_{jk})_{1 \leq j,k \leq M,\, j \neq j_0,\, k \neq k_0} \in \mathcal{M}_{(M-1) \times (M-1)}(\mathfrak{R}). \tag{10.2.9}$$

Furthermore, given another matrix $B = (b_{jk})_{1 \le j,k \le M} \in \mathcal{M}_{M \times M}(\mathfrak{R})$, we set

$$A \pm B := (a_{jk} \pm b_{jk})_{1 \le j,k \le M} \quad \text{and} \quad A \cdot B := \Big(\sum_{r=1}^{M} a_{jr} b_{rk} \Big)_{1 \le j,k \le M}. \quad (10.2.10)$$

A number of basic properties satisfied by these operations are collected in the next proposition.

Proposition 10.7. *Let $\mathfrak{R}$ be a commutative associative algebra over a field $\mathbb{K}$ with multiplicative unit e and additive neutral element 0. Also, let $M \in \mathbb{N}$ be arbitrary. Then the following statements are true:*

(i) $(A \cdot B)^{\top} = B^{\top} \cdot A^{\top}$ for all $A, B \in \mathcal{M}_{M \times M}(\mathfrak{R})$;

(ii) the identity matrix $I_{M \times M} := \begin{pmatrix} e & 0 & \dots & 0 \\ 0 & e & \dots & 0 \\ \dots\dots\dots\dots \\ 0 & 0 & \dots & e \end{pmatrix} \in \mathcal{M}_{M \times M}(\mathfrak{R})$ has the property

that $I_{M \times M} \cdot A = A \cdot I_{M \times M} = A$ for each $A \in \mathcal{M}_{M \times M}(\mathfrak{R})$;

(iii) $\det (A^{\top}) = \det A$ for each $A \in \mathcal{M}_{M \times M}(\mathfrak{R})$;

(iv) $\det (c\,A) = c^{M} \det A$ for each $A \in \mathcal{M}_{M \times M}(\mathfrak{R})$ and each $c \in \mathbb{K}$;

(v) given any $A \in \mathcal{M}_{M \times M}(\mathfrak{R})$, for each $j,k \in \{1, \dots, M\}$ we have

$$\det A = \sum_{\ell=1}^{M} a_{j\ell}(-1)^{j+\ell} \det (A_{j\ell}) = \sum_{\ell=1}^{M} a_{\ell k}(-1)^{\ell+k} \det (A_{\ell k}); \quad (10.2.11)$$

(vi) $\det A = 0$ whenever $A \in \mathcal{M}_{M \times M}(\mathfrak{R})$ has either two identical columns, or two identical rows;

(vii) $\operatorname{adj} (A^{\top}) \cdot A = A \cdot \operatorname{adj} (A^{\top}) = [\det A] I_{M \times M}$ for each $A \in \mathcal{M}_{M \times M}(\mathfrak{R})$;

(viii) $\det (A \cdot B) = (\det A)(\det B)$ for all $A, B \in \mathcal{M}_{M \times M}(\mathfrak{R})$;

(ix) $\operatorname{adj} (A \cdot B) = \operatorname{adj} (A) \cdot \operatorname{adj} (B)$ for all $A, B \in \mathcal{M}_{M \times M}(\mathfrak{R})$;

(x) $\mathcal{M}_{M \times M}(\mathfrak{R})$ is a (in general, noncommutative) ring, with multiplicative unit $I_{M \times M}$, and $A \in \mathcal{M}_{M \times M}(\mathfrak{R})$ has a multiplicative inverse in $\mathcal{M}_{M \times M}(\mathfrak{R})$ if and only if $\det (A)$ has a multiplicative inverse in $\mathfrak{R}$.

Proof. All properties are established much as in the standard case $\mathfrak{R} \equiv \mathbb{C}$. Here we only wish to mention that *(vii)* is a direct consequence of *(v)-(vi)* (complete proofs may be found in, e.g., [36]). □

An example that is relevant for our future considerations pertaining to the Lamé system is as follows. Suppose $\mathfrak{R}$ is a commutative associative algebra over a field $\mathbb{K}$ and that $n \in \mathbb{N}$, $\lambda, \mu \in \mathbb{K}$, with $\mu \neq 0$. Also, fix $a \in \mathfrak{R}^n$ (i.e., $a = (a_1, \dots, a_n)$ with $a_j \in \mathfrak{R}$ for each $j \in \{1, \dots, n\}$), and consider $A \in \mathcal{M}_{n \times n}(\mathfrak{R})$ given by

$$A := \mu \Big(\sum_{j=1}^{n} a_j a_j \Big) I_{n \times n} + (\lambda + \mu) a \otimes a, \quad (10.2.12)$$

where $a \otimes a \in M_{n \times n}(\mathfrak{R})$ is defined as $a \otimes a := (a_j a_k)_{1 \leq j, k \leq n}$. Then a direct calculation (compare with Exercise 10.13) shows that

$$\det A = \mu^{n-1}(\lambda + 2\mu)\Big(\sum_{j=1}^{n} a_j a_j\Big)^n, \qquad (10.2.13)$$

and

$$\mathrm{adj}\,(A) \qquad\qquad\qquad\qquad\qquad\qquad\qquad\qquad\qquad (10.2.14)$$

$$= \mu^{n-2}\Big(\sum_{j=1}^{n} a_j a_j\Big)^{n-2}\Big[(\lambda + 2\mu)\Big(\sum_{j=1}^{n} a_j a_j\Big)I_{n \times n} - (\lambda + \mu)a \otimes a\Big].$$

Our main interest in the algebraic framework developed so far in this section lies with the particular case when, for some fixed $n \in \mathbb{N}$,

$$\mathfrak{R} := \Big\{ \sum_{\alpha \in \mathbb{N}_0^n, |\alpha| \leq m} a_\alpha \partial^\alpha : a_\alpha \in \mathbb{C}, \quad m \in \mathbb{N}_0 \Big\}, \qquad (10.2.15)$$

with the natural operations of addition and multiplication, and with the convention that $\partial^{(0,\dots,0)} = 1$. Then clearly $\mathfrak{R}$ is a commutative associative algebra over $\mathbb{C}$ with multiplicative unit 1 and, for each $M \in \mathbb{N}$, the set $M_{M \times M}(\mathfrak{R})$ consists of all $M \times M$ systems of constant, complex coefficient differential operators. Thus, if $M \in \mathbb{N}$, $m \in \mathbb{N}_0$, are fixed an $M \times M$ system of constant, complex coefficient differential operators of degree m has the form $L(\partial) \in M_{M \times M}(\mathfrak{R})$ with

$$L(\partial) := \Big(\sum_{|\alpha| \leq m} a_\alpha^{jk} \partial^\alpha\Big)_{1 \leq j, k \leq M}. \qquad (10.2.16)$$

We shall call $L(\partial)$ a homogeneous (linear) system (of order m) if $a_\alpha^{jk} = 0$ whenever $|\alpha| < m$ and $j, k \in \{1, \dots, M\}$.

According to (10.2.5), the *algebraic* transpose of $L(\partial)$ from (10.2.16) in the space $M_{M \times M}(\mathfrak{R})$ is given by

$$L(\partial)^\top = \Big(\sum_{|\alpha| \leq m} a_\alpha^{kj} \partial^\alpha\Big)_{1 \leq j, k \leq M}. \qquad (10.2.17)$$

Next, for each $L(\partial)$ as in (10.2.16) set

$$D_L(\partial) := \det(L(\partial)) \qquad\qquad\qquad\qquad\qquad (10.2.18)$$

and notice that $D_L(\partial)$ is a scalar, constant (complex) coefficient, differential operator of order $\leq Mm$.

Consequently, statement *(vii)* in Proposition 10.7, (10.2.17), and (10.2.18) readily give that

$$L(\partial) \cdot \mathrm{adj}\,[L(\partial)^\top] = D_L(\partial)I_{M \times M}, \qquad \forall\, L(\partial) \in M_{M \times M}(\mathfrak{R}). \qquad (10.2.19)$$

Let us also observe that, at the level of symbols,

$$D_L(\xi) = \det(L(\xi)) \quad \text{for each} \quad \xi \in \mathbb{R}^n \tag{10.2.20}$$

and that

$$D_L(\partial) \text{ is a homogeneous scalar operator} \tag{10.2.21}$$
$$\text{whenever } L(\partial) \text{ is a homogeneous system.}$$

In particular, from (10.2.20) and (10.2.21) we deduce that

$$\left. \begin{array}{l} \text{if } L(\partial) \text{ is a homogeneous system with} \\ \det(L(\xi)) \neq 0 \quad \text{for each} \quad \xi \in \mathbb{R}^n \setminus \{0\} \end{array} \right\} \implies D_L(\partial) \text{ is elliptic.} \tag{10.2.22}$$

Our goal is to use (10.2.19) as a link between systems and scalar partial differential operators. We shall prove two results based on this scheme, the first of which is a procedure to reduce the task of finding a fundamental solution for a given system to the case of scalar operators. Specifically, we have the following proposition.

Theorem 10.8. *Let $L(\partial)$ be an $M \times M$ system with constant coefficients as in (10.2.16), and let $D_L(\partial)$ be the scalar differential operator associated with L as in (10.2.18). Then if $E \in \mathcal{D}'(\mathbb{R}^n)$ is a fundamental solution for $D_L(\partial)$ in $\mathbb{R}^n$, it follows that*

$$\mathbb{E} := \text{adj}\,[L(\partial)^\top](EI_{M \times M}) \in \mathcal{M}_{M \times M}(\mathcal{D}'(\mathbb{R}^n)) \tag{10.2.23}$$

is a fundamental solution for the system $L(\partial)$ in $\mathbb{R}^n$. Moreover,

$$E \in \mathcal{S}'(\mathbb{R}^n) \implies \mathbb{E} \in \mathcal{M}_{M \times M}(\mathcal{S}'(\mathbb{R}^n)). \tag{10.2.24}$$

In particular, if $D_L(\partial)$ is not identically zero, then $L(\partial)$ has a fundamental solution that is a tempered distribution.

Proof. Let E be a fundamental solution for $D_L(\partial)$ and define $\mathbb{E}$ as in (10.2.23). That $\mathbb{E}$ is a fundamental solution for $L(\partial)$ follows from (10.2.19). Also it is clear that $E \in \mathcal{S}'(\mathbb{R}^n)$ forces $\mathbb{E} \in \mathcal{M}_{M \times M}(\mathcal{S}'(\mathbb{R}^n))$. As for the last claim in the statement of the theorem, note that if $D_L(\partial)$ is not identically zero, then Theorem 5.14 ensures the existence of a fundamental solution $E \in \mathcal{S}'(\mathbb{R}^n)$ for $D_L(\partial)$ which, in turn, yields a solution for $L(\partial)$ via the recipe (10.2.23). $\qquad\square$

Next we record the following consequence of (10.2.19) and (10.2.22). As a preamble, recall the notion of singular support from (10.1.3).

Theorem 10.9. *Assume that $L = L(\partial)$ is an $M \times M$ homogeneous system in $\mathbb{R}^n$, with constant complex coefficients, such that*

$$\det[L(\xi)] \neq 0, \quad \forall \xi \in \mathbb{R}^n \setminus \{0\}. \tag{10.2.25}$$

Then, for each open set $\Omega \subseteq \mathbb{R}^n$ and for each $\mathbf{u} \in [\mathcal{D}'(\Omega)]^M$ one has

$$\text{sing supp } \mathbf{u} = \text{sing supp}(L\mathbf{u}). \tag{10.2.26}$$

In particular, if $\mathbf{u} \in [\mathcal{D}'(\Omega)]^M$ *is such that* $L\mathbf{u} \in [C^\infty(\Omega)]^M$ *then* $\mathbf{u} \in [C^\infty(\Omega)]^M$.

Proof. Let L be a system as in the statement of the theorem, and define the operator $D_L(\partial) := \det[L(\partial)]$. Then (10.2.22) ensures that $D_L(\partial)$ is an elliptic, scalar differential operator, with constant coefficients. Next, fix a vector-valued distribution $\mathbf{u} = (u_1, \ldots, u_M) \in [\mathcal{D}'(\Omega)]^M$ and let $\omega \subseteq \Omega$ be an open set such that $(L\mathbf{u})\big|_\omega \in C^\infty(\omega)$. Then, for each $j \in \{1, \ldots, M\}$ there holds

$$(D_L(\partial)u_j)\big|_\omega = \Big(\text{adj}\,[L(\partial)^\top]L(\partial)\mathbf{u}\Big)_j\Big|_\omega = \Big(\text{adj}\,[L(\partial)^\top]L(\partial)\mathbf{u}\big|_\omega\Big)_j. \qquad (10.2.27)$$

From (10.2.27) it follows that $D_L(\partial)u_j \in C^\infty(\omega)$ for each $j \in \{1, \ldots, M\}$. This and Corollary 6.18 imply that $u_j\big|_\omega \in C^\infty(\omega)$ for each $j \in \{1, \ldots, M\}$. Consequently $\omega \subseteq \Omega \setminus \text{sing supp}\,\mathbf{u}$, from which the left-to-right inclusion in (10.2.26) immediately follows. The opposite inclusion is readily seen from definitions. $\qquad \square$

Corollary 10.10. *Let* $L = L(\partial)$ *be an* $M \times M$ *homogeneous differential system in* $\mathbb{R}^n$, *with constant complex coefficients, with the property that* $\det[L(\xi)] \neq 0$ *for all* $\xi \in \mathbb{R}^n \setminus \{0\}$. *Then* L *has a fundamental solution* $\mathbb{E} \in \mathcal{M}_{M \times M}(\mathcal{S}'(\mathbb{R}^n))$ *which also satisfies* $\text{sing supp}\,\mathbb{E} = \{0\}$.

Proof. This is an immediate consequence of Theorem 10.8 and Theorem 10.9. $\qquad \square$

Exercise 10.11. Suppose $M \in \mathbb{N}$ and consider an $M \times M$ homogeneous second-order system with constant complex coefficients $L(\partial) = \Big(\sum_{r,s=1}^n a_{rs}^{jk} \partial_r \partial_s\Big)_{1 \le j,k \le M}$ in $\mathbb{R}^n$. Assume that there exists $c \in (0, \infty)$ such that this system satisfies the Legendre–Hadamard ellipticity condition

$$\text{Re}[(L(\xi)\eta) \cdot \overline{\eta}] \ge c|\xi|^2|\eta|^2, \qquad \forall\,\xi \in \mathbb{R}^n, \ \forall\,\eta \in \mathbb{C}^M. \qquad (10.2.28)$$

Then, $|\det[L(\xi)]| \ge c^M|\xi|^{2M}$ for every $\xi \in \mathbb{R}^n$. In particular, $\det[L(\xi)] \neq 0$ for every $\xi \in \mathbb{R}^n \setminus \{0\}$.

Sketch of proof: Show that

$$A \in \mathcal{M}_{M \times M}(\mathbb{C}), \ |A\eta| \ge c|\eta|, \ \forall\,\eta \in \mathbb{C}^M \implies |\det A| \ge c^M. \qquad (10.2.29)$$

This may be proved by considering the self-adjoint matrix $B := A^*A$ (where the matrix $A^* := (\overline{A})^\top$ is the adjoint of A) that satisfies $(B\eta) \cdot \overline{\eta} = |A\eta|^2 \ge c^2|\eta|^2$ for every $\eta \in \mathbb{C}^M$. Use the latter and the fact that B is diagonalizable to conclude that $|\det A|^2 = \det B \in [c^{2M}, \infty)$. To finish the proof of the exercise, apply (10.2.29) with $A := L(\xi), \xi \in \mathbb{R}^n$.

10.3 Fundamental Solutions for the Lamé Operator

The Lamé operator L in $\mathbb{R}^n$ is a differential operator that acts on vector distributions. Specifically, if $\mathbf{u} \in [\mathcal{D}'(\mathbb{R}^n)]^n$, i.e., $\mathbf{u} = (u_1, \ldots, u_n)$ where $u_k \in \mathcal{D}'(\mathbb{R}^n)$, $k = 1, \ldots, n$, then the action of the Lamé operator on $\mathbf{u}$ is defined by

$$L\mathbf{u} := \mu \Delta \mathbf{u} + (\lambda + \mu)\nabla \operatorname{div} \mathbf{u}$$

$$= \Big(\mu \Delta u_j + (\lambda + \mu) \sum_{\ell=1}^{n} \partial_j \partial_\ell u_\ell\Big)_{1 \le j \le n} \in [\mathcal{D}'(\mathbb{R}^n)]^n, \tag{10.3.1}$$

where the constants $\lambda, \mu \in \mathbb{C}$ (typically called Lamé moduli) are assumed to satisfy

$$\mu \neq 0 \quad \text{and} \quad 2\mu + \lambda \neq 0. \tag{10.3.2}$$

To study in greater detail the structure of the Lamé operator, we need to discuss some useful algebraic formalism.

Definition 10.12. Given, $a = (a_1, ..., a_n) \in \mathbb{C}^n$, $b = (b_1, ..., b_n) \in \mathbb{C}^n$, define $a \otimes b$ to be the matrix

$$a \otimes b := (a_j b_k)_{1 \le j, k \le n} \in \mathcal{M}_{n \times n}(\mathbb{C}). \tag{10.3.3}$$

Exercise 10.13. Prove that:

(1) $(a \otimes b)^\top = b \otimes a$ for all $a, b \in \mathbb{C}^n$;

(2) $\operatorname{Tr}(a \otimes b) = a \cdot b$ for all $a, b \in \mathbb{C}^n$;

(3) $(a \otimes b)c = (b \cdot c)a$ for all $a, b, c \in \mathbb{C}^n$;

(4) $\det(I_{n \times n} + a \otimes b) = 1 + a \cdot b$ for all $a, b \in \mathbb{C}^n$;

(5) $(a \otimes b)(c \otimes d) = (b \cdot c)\, a \otimes d$ for all $a, b, c, d \in \mathbb{C}^n$;

(6) for every $a \in \mathbb{R}^n$ and every numbers $\mu, \lambda \in \mathbb{C}$ the matrix $\mu I_{n \times n} + \lambda\, a \otimes a$ is invertible if and only if $\mu \neq 0$ and $\mu \neq -\lambda|a|^2$; moreover, whenever $\mu \neq 0$ and $\mu \neq -\lambda|a|^2$,

$$(\mu I_{n \times n} + \lambda\, a \otimes a)^{-1} = \frac{1}{\mu}\Big[I_{n \times n} - \Big(\frac{\lambda}{\mu + \lambda|a|^2}\Big)a \otimes a\Big]. \tag{10.3.4}$$

Hint for (4): Fix $a = (a_1, ..., a_n) \in \mathbb{C}^n$ and $b = (b_1, ..., b_n) \in \mathbb{C}^n$. By continuity, it suffices to prove the formula in *(4)* when $a_j \neq 0$ and $b_j \neq 0$ for every $j \in \{1, ..., n\}$. Assuming that this is the case, we may write

$$\det\left(I_{n\times n} + a\otimes b\right) = \det \begin{vmatrix} 1+a_1b_1 & a_1b_2 & \cdots & a_1b_n \\ a_2b_1 & 1+a_2b_2 & \cdots & a_2b_n \\ \cdots & \cdots & \cdots & \cdots \\ a_nb_1 & a_nb_2 & \cdots & 1+a_nb_n \end{vmatrix}$$

$$= \left(\prod_{j=1}^{n} a_j\right)\det \begin{vmatrix} \frac{1}{a_1}+b_1 & b_2 & \cdots & b_n \\ b_1 & \frac{1}{a_2}+b_2 & \cdots & b_n \\ \cdots & \cdots & \cdots & \cdots \\ b_1 & b_2 & \cdots & \frac{1}{a_n}+b_n \end{vmatrix}$$

$$= \left(\prod_{j=1}^{n} a_j\right)\left(\prod_{j=1}^{n} b_j\right)\det \begin{vmatrix} \frac{1}{a_1b_1}+1 & 1 & \cdots & 1 \\ 1 & \frac{1}{a_2b_2}+1 & \cdots & 1 \\ \cdots & \cdots & \cdots & \cdots \\ 1 & 1 & \cdots & \frac{1}{a_nb_n}+1 \end{vmatrix}$$

$$= \left(\prod_{j=1}^{n} a_j\right)\left(\prod_{j=1}^{n} b_j\right)\left(\prod_{j=1}^{n} \frac{1}{a_jb_j}\right)\left(1+\sum_{j=1}^{n} a_jb_j\right)$$

$$= 1 + \sum_{j=1}^{n} a_jb_j = 1 + a\cdot b. \tag{10.3.5}$$

The format of the Lamé system (10.3.1) suggests the following definition and result, shedding light on the conditions imposed on the Lamé moduli in (10.3.2).

Proposition 10.14. *Given any Lamé moduli $\lambda, \mu \in \mathbb{C}$, define the characteristic matrix for the Lamé system (10.3.1) as*

$$L(\xi) := \mu|\xi|^2 I_{n\times n} + (\lambda+\mu)\xi\otimes\xi, \qquad \forall\,\xi\in\mathbb{R}^n. \tag{10.3.6}$$

Then the following statements are equivalent:

(1) $L(\xi)$ is invertible for every $\xi \in \mathbb{R}^n \setminus \{0\}$;
(2) $L(\xi)$ is invertible for some $\xi \in \mathbb{R}^n \setminus \{0\}$;
(3) $\mu \neq 0$ and $\lambda + 2\mu \neq 0$.

Moreover, if $\mu \neq 0$ and $\lambda + 2\mu \neq 0$, then for each $\xi \in \mathbb{R}^n \setminus \{0\}$ one has

$$[L(\xi)]^{-1} = \frac{1}{\mu|\xi|^2}\Big[I_{n\times n} - \frac{\lambda+\mu}{\lambda+2\mu}\frac{\xi}{|\xi|}\otimes\frac{\xi}{|\xi|}\Big]. \tag{10.3.7}$$

Proof. This is a direct consequence of Exercise 10.13.   □

Now we are ready to tackle the issue of finding all fundamental solutions for the Lamé operator (10.3.1) when the Lamé moduli satisfy (10.3.2). In this context, according to Definition 10.1, a fundamental solution for the Lamé operator is a matrix distribution $\mathbb{E} \in \mathcal{M}_{n\times n}(\mathcal{D}'(\mathbb{R}^n))$ with the property that $L\mathbb{E} = \delta I_{n\times n}$ in

$\mathcal{M}_{M\times M}(\mathcal{D}'(\mathbb{R}^n))$. It follows that the columns $\mathbf{E}_k$, $k = 1, \ldots, n$, of $\mathbb{E}$ satisfy the equations

$$L\mathbf{E}_k = \delta\,\mathbf{e}_k \quad \text{in} \quad [\mathcal{D}'(\mathbb{R}^n)]^n \quad \text{for} \quad k = 1, \ldots, n, \tag{10.3.8}$$

where $\mathbf{e}_k$ denotes the unit vector in $\mathbb{R}^n$ with 1 on the k-th entry, for each $k \in \{1, \ldots, n\}$.

Our goal is to determine, under the standing assumption $n \geq 3$, all the fundamental solutions for the Lamé operator with entries in $\mathcal{S}'(\mathbb{R}^n)$. We do so by relying on the tools developed for scalar operators. To get started, suppose that there exists some matrix $\mathbb{E} = (E_{jk})_{1 \leq j,k \leq n} \in \mathcal{M}_{n\times n}(\mathcal{S}'(\mathbb{R}^n))$ whose columns satisfy (10.3.8). Since $\delta \in \mathcal{S}'(\mathbb{R}^n)$ and (4.1.33) is true, using (10.3.1), the latter is equivalent with

$$\mu\Delta E_{jk} + (\lambda + \mu) \sum_{\ell=1}^{n} \partial_j \partial_\ell E_{\ell k} = \delta_{jk}\delta \quad \text{in} \quad \mathcal{S}'(\mathbb{R}^n), \quad j, k \in \{1, \ldots, n\}. \tag{10.3.9}$$

Applying the Fourier transform to each equation in (10.3.9), we further write

$$-\mu|\xi|^2\,\widehat{E_{jk}}(\xi) - (\lambda + \mu)\xi_j \sum_{\ell=1}^{n} \xi_\ell\widehat{E_{\ell k}}(\xi) = \delta_{jk} \quad \text{in} \quad \mathcal{S}'(\mathbb{R}^n), \tag{10.3.10}$$

for each $j, k \in \{1, \ldots, n\}$. For k arbitrary, fixed, multiply (10.3.10) with ξ_j and then sum up the resulting identities over $j \in \{1, \ldots, n\}$ to obtain that, for each $k \in \{1, \ldots, n\}$,

$$-\mu|\xi|^2 \sum_{j=1}^{n} \xi_j\widehat{E_{jk}}(\xi) - (\lambda + \mu)\Big(\sum_{j=1}^{n} \xi_j^2\Big) \sum_{\ell=1}^{n} \xi_\ell\widehat{E_{\ell k}}(\xi) = \xi_k \quad \text{in} \quad \mathcal{S}'(\mathbb{R}^n), \tag{10.3.11}$$

or equivalently, that

$$-(\lambda + 2\mu)|\xi|^2 \sum_{j=1}^{n} \xi_j\widehat{E_{jk}}(\xi) = \xi_k \quad \text{in} \quad \mathcal{S}'(\mathbb{R}^n), \quad k \in \{1, \ldots, n\}. \tag{10.3.12}$$

If we now multiply (10.3.10) by $(\lambda + 2\mu)|\xi|^2$ and make use of (10.3.12), we may conclude that, for each $j, k \in \{1, \ldots, n\}$,

$$-\mu(\lambda + 2\mu)|\xi|^4\,\widehat{E_{jk}}(\xi) = (\lambda + 2\mu)|\xi|^2\delta_{jk} - (\lambda + \mu)\xi_j\xi_k \quad \text{in} \quad \mathcal{S}'(\mathbb{R}^n). \tag{10.3.13}$$

To proceed fix $j, k \in \{1, \ldots, n\}$ and note that since $n \geq 3$, by Exercise 4.5 we have $\frac{1}{|\xi|^2} \in \mathcal{S}'(\mathbb{R}^n)$. Also, $\frac{\xi_j\xi_k}{|\xi|^4} \in L_{loc}^1(\mathbb{R}^n)$ and in view of Example 4.4 one may infer that $\frac{\xi_j\xi_k}{|\xi|^4} \in \mathcal{S}'(\mathbb{R}^n)$. In addition, $|\xi|^4 \in \mathcal{L}(\mathbb{R}^n)$, thus $|\xi|^4 \cdot \frac{\xi_j\xi_k}{|\xi|^4}$, $|\xi|^4 \cdot \frac{1}{|\xi|^2} \in \mathcal{S}'(\mathbb{R}^n)$ (recall (b) in Theorem 4.14), and it is not difficult to check that $|\xi|^4 \cdot \frac{\xi_j\xi_k}{|\xi|^4} = \xi_j\xi_k$ and $|\xi|^4 \cdot \frac{1}{|\xi|^2} = |\xi|^2$ in $\mathcal{S}'(\mathbb{R}^n)$. These conclusions combined with (10.3.13) imply (recall also (10.3.2))

$$\mu(\lambda + 2\mu)|\xi|^4\left[\widehat{E_{jk}}(\xi) + \frac{\delta_{jk}}{\mu} \cdot \frac{1}{|\xi|^2} - \frac{(\lambda + \mu)}{\mu(\lambda + 2\mu)} \cdot \frac{\xi_j\xi_k}{|\xi|^4}\right] = 0 \tag{10.3.14}$$

in $S'(\mathbb{R}^n)$. Thus, by Proposition 5.2 applied with $P(D) := \Delta^2$, it follows that

$$\widehat{E_{jk}}(\xi) + \frac{\delta_{jk}}{\mu} \cdot \frac{1}{|\xi|^2} - \frac{(\lambda + \mu)}{\mu(\lambda + 2\mu)} \cdot \frac{\xi_j \xi_k}{|\xi|^4} = \widehat{P_{jk}}(\xi) \quad \text{in } S'(\mathbb{R}^n), \tag{10.3.15}$$

where P_{jk} is a polynomial in $\mathbb{R}^n$ satisfying $\Delta^2 P_{jk} = 0$ pointwise in $\mathbb{R}^n$. To continue with the computation of E_{jk} we apply the inverse Fourier transform to (10.3.15) and use Proposition 4.64 with $\lambda = 2$ as well as (7.14.8) to write

$$E_{jk} = -\frac{\delta_{jk}}{\mu} \mathcal{F}^{-1}\left(\frac{1}{|\xi|^2}\right) + \frac{(\lambda + \mu)}{\mu(\lambda + 2\mu)} \mathcal{F}^{-1}\left(\frac{\xi_j \xi_k}{|\xi|^4}\right) + P_{jk}$$

$$= -\frac{1}{\mu(n-2)\omega_{n-1}} \cdot \frac{\delta_{jk}}{|x|^{n-2}} + P_{jk}(x)$$

$$+ \frac{(\lambda + \mu)}{\mu(\lambda + 2\mu)}\left[\frac{1}{2(n-2)\omega_{n-1}} \cdot \frac{\delta_{jk}}{|x|^{n-2}} - \frac{1}{2\omega_{n-1}} \cdot \frac{x_j x_k}{|x|^n}\right]$$

$$= \left[\frac{(\lambda + \mu)}{\mu(\lambda + 2\mu)} \cdot \frac{1}{2(n-2)\omega_{n-1}} - \frac{1}{\mu(n-2)\omega_{n-1}}\right] \cdot \frac{\delta_{jk}}{|x|^{n-2}}$$

$$- \frac{(\lambda + \mu)}{2\omega_{n-1}\mu(\lambda + 2\mu)} \cdot \frac{x_j x_k}{|x|^n} + P_{jk}(x)$$

$$= \frac{-1}{2\mu(2\mu + \lambda)\omega_{n-1}}\left[\frac{3\mu + \lambda}{n-2} \frac{\delta_{jk}}{|x|^{n-2}} + \frac{(\mu + \lambda)x_j x_k}{|x|^n}\right] + P_{jk}(x) \tag{10.3.16}$$

in $S'(\mathbb{R}^n)$ (hence, in particular, for all $x \in \mathbb{R}^n \setminus \{0\}$).

Next, we claim that the matrix $\mathbb{F} = (F_{jk})_{1 \leq j,k \leq n} \in \mathcal{M}_{n \times n}(S'(\mathbb{R}^n))$ with entries defined by

$$F_{jk} := \frac{-1}{2\mu(2\mu + \lambda)\omega_{n-1}}\left[\frac{3\mu + \lambda}{n-2} \frac{\delta_{jk}}{|x|^{n-2}} + \frac{(\mu + \lambda)x_j x_k}{|x|^n}\right], \tag{10.3.17}$$

for $j, k \in \{1, \ldots, n\}$, is a fundamental solution for the Lamé operator. Note that, based on the properties of the Fourier transform, this claim is equivalent with having the entries of $\mathbb{F}$ satisfy

$$-\mu|\xi|^2 \widehat{F_{jk}}(\xi) - (\lambda + \mu)\xi_j \sum_{\ell=1}^{n} \xi_\ell \widehat{F_{\ell k}}(\xi) = \delta_{jk} \quad \text{in } S'(\mathbb{R}^n) \tag{10.3.18}$$

for each $j, k \in \{1, \ldots, n\}$. To check (10.3.18) we use (10.3.17), Proposition 4.64 (with $\lambda = n - 2$) and (7.14.9) to first write

$$\widehat{F_{jk}}(\xi) = \frac{-1}{2\mu(2\mu+\lambda)\omega_{n-1}}\left[\frac{(3\mu+\lambda)\delta_{jk}}{n-2}\mathcal{F}\left(\frac{1}{|x|^{n-2}}\right) + (\mu+\lambda)\mathcal{F}\left(\frac{x_j x_k}{|x|^n}\right)\right]$$

$$= \frac{-(3\mu+\lambda)\delta_{jk}}{2\mu(2\mu+\lambda)\omega_{n-1}(n-2)}\cdot\frac{(n-2)\omega_{n-1}}{|\xi|^2}$$

$$-\frac{\mu+\lambda}{2\mu(2\mu+\lambda)\omega_{n-1}}\left[\omega_{n-1}\frac{\delta_{jk}}{|\xi|^2} - 2\omega_{n-1}\frac{\xi_j\xi_k}{|\xi|^4}\right]$$

$$= -\frac{\delta_{jk}}{\mu}\cdot\frac{1}{|\xi|^2} + \frac{(\lambda+\mu)}{\mu(\lambda+2\mu)}\cdot\frac{\xi_j\xi_k}{|\xi|^4} \quad \text{in} \ \ S'(\mathbb{R}^n). \tag{10.3.19}$$

Next, we use (10.3.19) to rewrite the term in the left-hand side of (10.3.18) as

$$-\mu|\xi|^2\,\widehat{F_{jk}}(\xi) - (\lambda+\mu)\xi_j\sum_{\ell=1}^{n}\xi_\ell\widehat{F_{\ell k}}(\xi)$$

$$= \mu|\xi|^2\left[\frac{\delta_{jk}}{\mu|\xi|^2} - \frac{(\lambda+\mu)\xi_j\xi_k}{\mu(\lambda+2\mu)|\xi|^4}\right] + (\lambda+\mu)\xi_j\sum_{\ell=1}^{n}\xi_\ell\left[\frac{\delta_{\ell k}}{\mu|\xi|^2} - \frac{(\lambda+\mu)\xi_\ell\xi_k}{\mu(\lambda+2\mu)|\xi|^4}\right]$$

$$= \delta_{jk} - \frac{(\lambda+\mu)\xi_j\xi_k}{(\lambda+2\mu)|\xi|^2} + (\lambda+\mu)\xi_j\left[\frac{\xi_k}{\mu|\xi|^2} - \frac{(\lambda+\mu)\xi_k}{\mu(\lambda+2\mu)|\xi|^2}\right]$$

$$= \delta_{jk} \quad \text{in} \ \ S'(\mathbb{R}^n), \tag{10.3.20}$$

proving that the matrix $\mathbb{F}$ is a fundamental solution for the Lamé operator.

The main result emerging from this analysis is summarized next.

Theorem 10.15. *Assume $n \geq 3$ and let L be the Lamé operator from (10.3.1) such that the constants the $\lambda, \mu \in \mathbb{C}$ satisfy (10.3.2). Define the matrix $\mathbb{E} = (E_{jk})_{1\leq j,k\leq n}$ with entries given by the $L^1_{loc}(\mathbb{R}^n)$ functions*

$$E_{jk}(x) := \frac{-1}{2\mu(2\mu+\lambda)\omega_{n-1}}\left[\frac{3\mu+\lambda}{n-2}\frac{\delta_{jk}}{|x|^{n-2}} + \frac{(\mu+\lambda)x_j x_k}{|x|^n}\right] \tag{10.3.21}$$

for each $x \in \mathbb{R}^n \setminus \{0\}$ and $j, k \in \{1,\dots,n\}$. Then $\mathbb{E}$ belongs to $M_{n\times n}(S'(\mathbb{R}^n))$ and is a fundamental solution for the Lamé operator in $\mathbb{R}^n$.

In addition, any fundamental solution $\mathbb{U} \in M_{n\times n}(S'(\mathbb{R}^n))$ of the Lamé operator in $\mathbb{R}^n$ is of the form $\mathbb{U} = \mathbb{E} + \mathbb{P}$, for some matrix $\mathbb{P} := (P_{jk})_{1\leq j,k\leq n}$ whose entries are polynomials in $\mathbb{R}^n$ and whose columns, $\mathbf{P}_k, k = 1,\dots,n$, satisfy the pointwise equations $L\mathbf{P}_k = (0,\dots,0) \in \mathbb{C}^n$ in $\mathbb{R}^n$ for $k = 1,\dots,n$.

Proof. Since the entries of $\mathbb{E}$ as given in (10.3.21) are the same as the expressions from (10.3.17), the earlier analysis shows that $\mathbb{E}$ belongs to $M_{n\times n}(S'(\mathbb{R}^n))$ and is a fundamental solution for the Lamé operator in $\mathbb{R}^n$. To justify the claim in the last paragraph of the statement of the theorem we shall invoke Proposition 10.2. Concretely, since condition (10.3.2) and Proposition 10.14 imply $\det(L(\xi)) \neq 0$ for $\xi \in \mathbb{R}^n \setminus \{0\}$, it follows that $\det(L(i\xi)) = -\det(L(\xi)) \neq 0$ for $\xi \in \mathbb{R}^n \setminus \{0\}$.

This shows that (10.1.21) is satisfied, hence Proposition 10.2 applies and yields the desired conclusion.   □

Note that, as Lemma 10.17 shows, if $\mathbb{P} = (P_{jk})_{1 \leq j,k \leq n}$ is as in Theorem 10.15, then $\Delta^2 P_{jk} = 0$ pointwise in $\mathbb{R}^n$ for every $j, k \in \{1, \ldots, n\}$.

One may check without using of the Fourier transform that the matrix from Theorem 10.15 is a fundamental solution for the Lamé operator (much as in the spirit of Remark 7.4).

Exercise 10.16. Follow the outline below to check, without the use of the Fourier transform, that the matrix $\mathbb{E} = (E_{jk})_{1 \leq j,k \leq n} \in \mathcal{M}_{n \times n}(\mathcal{S}'(\mathbb{R}^n))$, with entries of function type defined by the functions from (10.3.21), is a fundamental solution for the Lamé operator in $\mathbb{R}^n$, $n \geq 3$.

Step 1. Show that $\mu \Delta E_{jk} + (\lambda + \mu) \sum_{\ell=1}^{n} \partial_j \partial_\ell E_{\ell k} = 0$ pointwise in $\mathbb{R}^n \setminus \{0\}$ for each $j, k \in \{1, \ldots, n\}$.

Step 2. Show that the desired conclusion (i.e., that the given E is a fundamental solution for the Lamé operator in $\mathbb{R}^n$, $n \geq 3$) is equivalent with the condition that

$$\lim_{\varepsilon \to 0^+} \left[\mu \int_{|x| \geq \varepsilon} E_{jk}(x) \Delta \varphi(x) \, dx + (\lambda + \mu) \int_{|x| \geq \varepsilon} \sum_{\ell=1}^{n} E_{\ell k}(x) \partial_j \partial_\ell \varphi(x) \, dx \right]$$

$$= \varphi(0) \delta_{jk} \qquad (10.3.22)$$

for every $j, k \in \{1, \ldots, n\}$ and every $\varphi \in C_0^\infty(\mathbb{R}^n)$.

Step 3. Fix $j, k \in \{1, \ldots, n\}$, $\varphi \in C_0^\infty(\mathbb{R}^n)$, and let $R \in (0, \infty)$ be such that $\operatorname{supp} \varphi \subseteq B(0, R)$, so that one may replace the domain of integration for the integrals in the left-hand side of (10.3.22) with $\{x \in \mathbb{R}^n : \varepsilon < |x| < R\}$. Use this domain of integration, (14.8.5), (14.8.4), and the result from Step 1, to prove that (10.3.22) is equivalent with

$$\lim_{\varepsilon \to 0^+} \left[-\mu \int_{\partial B(0,\varepsilon)} E_{jk}(x) \frac{\partial \varphi}{\partial \nu}(x) \, d\sigma(x) + \mu \int_{\partial B(0,\varepsilon)} \varphi(x) \frac{\partial E_{jk}}{\partial \nu}(x) \, d\sigma(x) \right.$$

$$- (\lambda + \mu) \int_{\partial B(0,\varepsilon)} \left[\sum_{\ell=1}^{n} E_{\ell k}(x) \partial_\ell \varphi(x) \right] \nu_j(x) \, d\sigma(x) \qquad (10.3.23)$$

$$\left. + (\lambda + \mu) \int_{\partial B(0,\varepsilon)} \left[\sum_{\ell=1}^{n} \nu_\ell(x) \partial_j E_{\ell k}(x) \right] \varphi(x) \, d\sigma(x) \right] = \varphi(0) \delta_{jk},$$

where $\nu(x) = \frac{x}{\varepsilon}$ for each $x \in \partial B(0, \varepsilon)$.

Step 4. Prove that there exists a constant $C \in (0, \infty)$ independent of ε such that each of the quantities:

$$\left|\int_{\partial B(0,\varepsilon)} E_{jk}(x)\frac{\partial\varphi}{\partial\nu}(x)\,\mathrm{d}\sigma(x)\right|,\qquad \left|\int_{\partial B(0,\varepsilon)}\frac{\partial E_{jk}}{\partial\nu}(x)[\varphi(x)-\varphi(0)]\,\mathrm{d}\sigma(x)\right|,$$

$$\left|\int_{\partial B(0,\varepsilon)}\left[\sum_{\ell=1}^{n}E_{\ell k}(x)\partial_\ell\varphi(x)\right]\nu_j(x)\,\mathrm{d}\sigma(x)\right|,\quad\text{and}\tag{10.3.24}$$

$$\left|\int_{\partial B(0,\varepsilon)}\left[\sum_{\ell=1}^{n}\nu_\ell(x)\partial_j E_{\ell k}(x)\right][\varphi(x)-\varphi(0)]\,\mathrm{d}\sigma(x)\right|,$$

is bounded by $C\|\nabla\varphi\|_{L^\infty(\mathbb{R}^n)}\,\varepsilon$, thus convergent to zero as $\varepsilon\to 0^+$.

Step 5. Combine Steps 2-4 to reduce matters to proving that

$$\lim_{\varepsilon\to 0^+}\left[\mu\int_{\partial B(0,\varepsilon)}\sum_{s=1}^{n}\frac{x_s}{\varepsilon}(\partial_s E_{jk})(x)\,\mathrm{d}\sigma(x)\right.\tag{10.3.25}$$

$$\left.+(\lambda+\mu)\int_{\partial B(0,\varepsilon)}\sum_{\ell=1}^{n}\frac{x_\ell}{\varepsilon}(\partial_j E_{\ell k})(x)\,\mathrm{d}\sigma(x)\right]=\delta_{jk}.$$

Step 6. Prove that for every $x\in\partial B(0,\varepsilon)$ we have

$$\sum_{s=1}^{n}\frac{x_s}{\varepsilon}(\partial_s E_{jk})(x)=\frac{(3\mu+\lambda)\delta_{jk}}{2\mu(2\mu+\lambda)\,\omega_{n-1}\varepsilon^{n-1}}-\frac{(\mu+\lambda)(2-n)x_j x_k}{2\mu(2\mu+\lambda)\,\omega_{n-1}\varepsilon^{n+1}}\tag{10.3.26}$$

and

$$\sum_{\ell=1}^{n}\frac{x_\ell}{\varepsilon}(\partial_j E_{\ell k})(x)=\frac{(3\mu+\lambda)-(\mu+\lambda)(1-n)}{2\mu(2\mu+\lambda)\,\omega_{n-1}\varepsilon^{n+1}}x_j x_k$$

$$-\frac{(\lambda+\mu)\delta_{jk}}{2\mu(2\mu+\lambda)\,\omega_{n-1}\varepsilon^{n-1}}.\tag{10.3.27}$$

Step 7. Using the fact that (cf. (14.9.45))

$$\int_{\partial B(0,\varepsilon)}x_j x_k\,\mathrm{d}\sigma(x)=\frac{\varepsilon^{n+1}\omega_{n-1}}{n}\delta_{jk},\tag{10.3.28}$$

integrate the expressions in (10.3.26)–(10.3.27) to conclude that

$$\int_{\partial B(0,\varepsilon)}\sum_{s=1}^{n}\frac{x_s}{\varepsilon}(\partial_s E_{jk})(x)\,\mathrm{d}\sigma(x)\tag{10.3.29}$$

$$=\left[3\mu+\lambda-\frac{(\mu+\lambda)(2-n)}{n}\right]\frac{\delta_{jk}}{2\mu(2\mu+\lambda)}\tag{10.3.30}$$

and

$$\int\limits_{\partial B(x,\varepsilon)} \sum_{\ell=1}^{n} \frac{x_\ell}{\varepsilon}(\partial_j E_{\ell k})(x)\,d\sigma(x) = \frac{\delta_{jk}}{n(2\mu+\lambda)}, \qquad (10.3.31)$$

then finish the proof of (10.3.25).

10.4 Mean Value Formulas and Interior Estimates for the Lamé Operator

The goal here is to prove that solutions of the Lamé system satisfy certain mean value formulas similar in spirit to those for harmonic functions. We start by establishing a few properties of elastic vector fields.

Lemma 10.17. *Let $\lambda, \mu \in \mathbb{C}$ be such that $\mu \neq 0$ and $\lambda + 2\mu \neq 0$. Assume that the vector distribution $\mathbf{u} = (u_1, u_2, \ldots, u_n) \in [\mathcal{D}'(\Omega)]^n$ satisfies the Lamé system*

$$\mu\Delta\mathbf{u} + (\lambda+\mu)\nabla div\,\mathbf{u} = \mathbf{0} \quad in\ [\mathcal{D}'(\Omega)]^n. \qquad (10.4.1)$$

Then $\mathbf{u} \in [C^\infty(\Omega)]^n$ and the following statements are true.

(i) The function $div\,\mathbf{u}$ is harmonic in Ω.
(ii) The function u_j is biharmonic in Ω for each $j = 1, \ldots, n$.
(iii) The function $\partial_j(div\,\mathbf{u})$ is harmonic in Ω for each $j = 1, \ldots, n$.

Proof. The fact that $\mathbf{u} \in [C^\infty(\Omega)]^n$ is a consequence of Theorem 10.9, Proposition 10.14, and our assumptions on λ, μ. To show *(i)*, we apply div to (10.4.1). Since $div\nabla = \Delta$ and $div\Delta = \Delta div$ we obtain $(\lambda + 2\mu)\Delta div\,\mathbf{u} = 0$ in Ω, thus *(i)* follows since $\lambda + 2\mu \neq 0$. Now if we return to (10.4.1) and apply Δ to it, the second term will be zero because of the harmonicity of $div\,\mathbf{u}$. Consequently, $\mu\Delta^2\mathbf{u} = 0$ in Ω which proves *(ii)* since $\mu \neq 0$. Finally, applying Δ to (10.4.1) and using *(ii)* yields $(\lambda+\mu)\Delta(\nabla div\,\mathbf{u}) = 0$, and *(iii)* follows from this in the case $\lambda + \mu \neq 0$. If the latter condition fails, then from (10.4.1) we have that u is harmonic, so *(iii)* also holds in this case. $\qquad\qquad\Box$

As an auxiliary step in the direction of Theorem 10.19 that contains the mean value formulas alluded to earlier, we prove the following useful result.

Proposition 10.18. *Let $\lambda, \mu \in \mathbb{C}$ be such that $\mu \neq 0$ and $\lambda + 2\mu \neq 0$. Assume $\mathbf{u} = (u_1, u_2, \ldots, u_n) \in [\mathcal{D}'(\Omega)]^n$ satisfies the Lamé system (10.4.1). Then the components of $\mathbf{u}$ belong to $C^\infty(\Omega)$ and the following formulas hold*

$$u_j(x) + \frac{(\mu-\lambda)r^2}{2\mu(n+2)}\partial_j(div\,\mathbf{u})(x) \qquad (10.4.2)$$

$$= \frac{n}{r^2}\int_{\partial B(x,r)} (y_j - x_j)(y - x) \cdot \mathbf{u}(y)\,d\sigma(y),$$

and

$$u_j(x) - \frac{[(n+3)\mu + (n+1)\lambda]r^2}{2\mu(n+2)(n-1)} \, \partial_j(\operatorname{div} \mathbf{u})(x) \tag{10.4.3}$$

$$= -\frac{n}{(n-1)r^2} \int_{\partial B(x,r)} [(y_j - x_j)(y - x) \cdot \mathbf{u}(y) - r^2 u_j(y)] \, d\sigma(y),$$

for every $x \in \Omega$, every $r \in (0, \operatorname{dist}(x, \partial\Omega))$, and every $j \in \{1, \dots, n\}$.

Proof. The fact that $\mathbf{u} \in [C^\infty(\Omega)]^n$ follows from Lemma 10.17. To proceed, fix some $x \in \Omega$, $r \in (0, \operatorname{dist}(x, \partial\Omega))$, and $j \in \{1, \dots, n\}$. By *(iii)* in Lemma 10.17, we have that $\partial_j(\operatorname{div} \mathbf{u})(x)$ is harmonic in Ω. Thus, we may apply the mean value formula on solid balls for harmonic function (c.f. the first formula in (7.2.9)), followed by an application of the integration by parts formula (14.8.4), to write

$$\partial_j(\operatorname{div} \mathbf{u})(x) = \frac{n}{\omega_{n-1} r^n} \int_{B(x,r)} \partial_j(\operatorname{div} \mathbf{u})(y) \, dy$$

$$= \frac{n}{\omega_{n-1} r^n} \int_{\partial B(x,r)} \frac{y_j - x_j}{r} \operatorname{div} \mathbf{u}(y) \, d\sigma(y)$$

$$= \frac{n}{\omega_{n-1} r^{n+1}} \int_{\partial B(x,r)} \operatorname{div}[(y_j - x_j)\mathbf{u}(y)] \, d\sigma(y)$$

$$- \frac{n}{\omega_{n-1} r^{n+1}} \int_{\partial B(x,r)} u_j(y) \, d\sigma(y) =: I + II. \tag{10.4.4}$$

Since u_j is biharmonic (recall *(ii)* in Lemma 10.17), we may write (7.4.1) for u_j, then use the latter formula to simplify II, and then replace Δu_j by $-\frac{\lambda+\mu}{\mu} \partial_j(\operatorname{div} \mathbf{u})$ (recall that $\mathbf{u}$ satisfies the Lamé system) to obtain

$$II = -\frac{n}{r^2} u_j(x) - \frac{1}{2} \Delta u_j(x) = -\frac{n}{r^2} u_j(x) + \frac{\lambda+\mu}{2\mu} \partial_j(\operatorname{div} \mathbf{u})(x). \tag{10.4.5}$$

Combining (10.4.4) and (10.4.5) we see that

$$\frac{\mu - \lambda}{2\mu} r^{n+1} \partial_j(\operatorname{div} \mathbf{u})(x) + nr^{n-1} u_j(x) \tag{10.4.6}$$

$$= \frac{n}{\omega_{n-1}} \int_{\partial B(x,r)} \operatorname{div}[(y_j - x_j)\mathbf{u}(y)] \, d\sigma(y).$$

Fix $R \in (0, \operatorname{dist}(x, \partial\Omega))$. Integrating (10.4.6) with respect to r for $r \in (0, R)$, applying (14.9.5) and then (14.8.4), we arrive at

$$R^n u_j(x) + \frac{\mu - \lambda}{2\mu(n+2)} R^{n+2} \, \partial_j(\mathrm{div}\,\mathbf{u})(x) \tag{10.4.7}$$

$$= \frac{n}{\omega_{n-1}} \int_{B(x,R)} \mathrm{div}\big((y_j - x_j)\mathbf{u}(y)\big)\,\mathrm{d}y$$

$$= \frac{n}{\omega_{n-1}} \int_{\partial B(x,R)} (y_j - x_j)\frac{y - x}{R} \cdot \mathbf{u}(y)\,\mathrm{d}\sigma(y)$$

which, in turn, yields (10.4.2) (with R in place of r) after dividing by R^n.

To prove (10.4.3) we start with the term in the right-hand side of formula (10.4.2) in which we add and subtract $r^2 u_j(y)$ under the integral sign and then split it in two integrals. One integral, call it I_1, corresponds to the expression

$$(y_j - x_j)(y - x) \cdot \mathbf{u}(y) - r^2 u_j(y). \tag{10.4.8}$$

The other integral, call it I_2, corresponds to $u_j(y)$. For I_2 we recall that u_j is biharmonic and use (7.4.1) after which we replace $\Delta u_j(x)$ by $-\frac{\lambda+\mu}{\mu} \, \partial_j(\mathrm{div}\,\mathbf{u})(x)$. Hence,

$$\frac{n}{\omega_{n-1} r^{n+1}} \int_{\partial B(x,r)} (y_j - x_j)(y - x) \cdot \mathbf{u}(y)\,\mathrm{d}\sigma(y) \tag{10.4.9}$$

$$= \frac{n}{\omega_{n-1} r^{n+1}} \int_{\partial B(x,r)} [(y_j - x_j)(y - x) \cdot \mathbf{u}(y) - r^2 u_j(y)]\,\mathrm{d}\sigma(y)$$

$$+ n u_j(x) - \frac{r^2}{2} \cdot \frac{\lambda + \mu}{\mu} \partial_j(\mathrm{div}\,\mathbf{u})(x).$$

Finally, (10.4.3) follows by adding (10.4.2) and (10.4.9). □

As mentioned before, Proposition 10.18 is an important ingredient in proving the following mean value formulas for the Lamé system.

Theorem 10.19. *Let $\lambda, \mu \in \mathbb{C}$ be such that $\mu \neq 0$, $\lambda + 2\mu \neq 0$, and $(n+1)\mu + \lambda \neq 0$. Assume $\mathbf{u} \in [\mathcal{D}'(\Omega)]^n$ satisfies the Lamé system (10.4.1). Then $\mathbf{u} \in [C^\infty(\Omega)]^n$, and for every $x \in \Omega$ and every $r \in (0, \mathrm{dist}(x, \partial\Omega))$ the following formulas hold:*

$$\mathbf{u}(x) = \frac{n(\lambda + \mu)(n+2)}{2[(n+1)\mu + \lambda]r^2} \fint_{\partial B(x,r)} [(y - x) \cdot \mathbf{u}(y)](y - x)\,\mathrm{d}\sigma(y)$$

$$+ \frac{n(\mu - \lambda)}{2[(n+1)\mu + \lambda]} \fint_{\partial B(x,r)} \mathbf{u}(y)\,\mathrm{d}\sigma(y) \tag{10.4.10}$$

and

$$\mathbf{u}(x) = \frac{n(\lambda + \mu)(n+2)}{2[(n+1)\mu + \lambda]} \fint_{B(x,r)} \left[\frac{y - x}{|y - x|} \cdot \mathbf{u}(y)\right] \frac{y - x}{|y - x|}\,\mathrm{d}y \tag{10.4.11}$$

$$+ \frac{n(\mu - \lambda)}{2[(n+1)\mu + \lambda]} \fint_{B(x,r)} \mathbf{u}(y)\,\mathrm{d}y.$$

Proof. Once again, that $\mathbf{u} \in [C^\infty(\Omega)]^n$ is contained in Lemma 10.17. Formula (10.4.10) follows by taking a suitable linear combination of (10.4.2) and (10.4.3) so that $\partial_j(\operatorname{div}\mathbf{u})$ cancels (here $(n+1)\mu + \lambda \neq 0$ is used). To prove (10.4.11), multiply (10.4.10) by $\omega_{n-1}r^{n-1}$, then move $\frac{1}{r^2}$ from the front of the first integral in the right-hand side of the equality obtained inside of that integral as $\frac{1}{|y-x|^2}$, then integrate with respect to $r \in (0, R)$, where $R \in (0, \operatorname{dist}(x, \partial\Omega))$ and use (14.9.5). Finally divide the very last expression by $\omega_{n-1}R^n/n$ which is precisely the volume of $B(x, R)$. This gives (10.4.11) written with R in place of r. □

We shall now discuss how the mean value formulas from Theorem 10.19 can be used to obtain interior estimates for solutions of the Lamé system. Two such versions are proved in Theorem 10.20 and Theorem 10.22 (cf. also Exercise 10.21).

Theorem 10.20 (L^1-Interior estimates for the Lamé operator). *Let $\lambda, \mu \in \mathbb{C}$ be such that $\mu \neq 0$, $\lambda + 2\mu \neq 0$, and $(n+1)\mu + \lambda \neq 0$. Assume $\mathbf{u} \in [\mathcal{D}'(\Omega)]^n$ is a vector distribution satisfying the Lamé system (10.4.1). Then $\mathbf{u} \in [C^\infty(\Omega)]^n$ and there exists $C \in (0, \infty)$ depending only on n, λ, and μ, such that for every $x \in \Omega$, every $r \in (0, \operatorname{dist}(x, \partial\Omega))$ and each $k \in \{1, \ldots, n\}$, one has*

$$|\partial_k \mathbf{u}(x)| \leq \frac{C}{r} \fint_{B(x,r)} |\mathbf{u}(y)| \, dy. \tag{10.4.12}$$

Proof. The fact that $\mathbf{u} \in [C^\infty(\Omega)]^n$ has been established in Lemma 10.17. Fix $k \in \{1, \ldots, n\}$ and observe that since $\mathbf{u} = (u_1, \ldots, u_n)$ satisfies the Lamé system in Ω, then so does $\partial_k \mathbf{u}$. Fix $x^* \in \Omega$ arbitrary and select $R \in (0, \operatorname{dist}(x^*, \partial\Omega))$. Writing formula (10.4.11) for $\mathbf{u}$ replaced by $\partial_k \mathbf{u}$ and r replaced by $R/2$ shows that for each $j \in \{1, \ldots, n\}$,

$$\partial_k u_j(x^*) = \frac{c_1 2^n n}{\omega_{n-1} R^n} \int_{B(x^*, R/2)} \left(\frac{y - x^*}{|y - x^*|} \cdot \partial_k \mathbf{u}(y) \right) \frac{y_j - x_j^*}{|y - x^*|} \, dy \tag{10.4.13}$$

$$+ \frac{c_2 2^n n}{\omega_{n-1} R^n} \int_{B(x^*, R/2)} \partial_k u_j(y) \, dy,$$

where

$$c_1 := \frac{n(\lambda + \mu)(n + 2)}{2[(n+1)\mu + \lambda]} \quad \text{and} \quad c_2 := \frac{n(\mu - \lambda)}{2[(n+1)\mu + \lambda]}. \tag{10.4.14}$$

Integrating by parts (using (14.8.4)) in both integrals in (10.4.13) yields

$$\partial_k u_j(x^*) = -\frac{c_1 2^n n}{\omega_{n-1} R^n} \int\limits_{B(x^*, R/2)} \sum_{\ell=1}^n \partial_{y_k}\Big[\frac{(y_j - x_j^*)(y_\ell - x_\ell^*)}{|y - x^*|^2}\Big] u_\ell(y)\, dy \qquad (10.4.15)$$

$$+ \frac{c_1 2^n n}{\omega_{n-1} R^n} \int\limits_{\partial B(x^*, R/2)} \frac{y_k - x_k^*}{|y - x^*|} \sum_{\ell=1}^n \frac{(y_j - x_j^*)(y_\ell - x_\ell^*)}{|y - x^*|^2} u_\ell(y)\, d\sigma(y)$$

$$+ \frac{c_2 2^n n}{\omega_{n-1} R^n} \int\limits_{\partial B(x^*, R/2)} \frac{y_k - x_k^*}{|y - x^*|} u_j(y)\, d\sigma(y).$$

Thus,

$$|\partial_k \mathbf{u}(x^*)| \le CR^{-n} \int\limits_{B(x^*, R/2)} \frac{|\mathbf{u}(y)|}{|y - x^*|}\, dy + CR^{-n} \int\limits_{\partial B(x^*, R/2)} |\mathbf{u}(y)|\, d\sigma(y), \qquad (10.4.16)$$

where C stands for a finite positive constant depending only on n, λ, and μ. Before continuing let us note a useful consequence of (10.4.11). Specifically, for every $x \in \Omega$ and every $r \in (0, \operatorname{dist}(x, \partial\Omega))$ one has

$$|\mathbf{u}(x)| \le C \fint_{B(x,r)} |\mathbf{u}(z)|\, dz. \qquad (10.4.17)$$

Taking $y \in \overline{B(x^*, R/2)}$ forces $B(y, R/2) \subset B(x^*, R)$ which, when used in estimate (10.4.17) written for x replaced by y and r by $R/2$, gives

$$|\mathbf{u}(y)| \le C \fint_{B(y, R/2)} |\mathbf{u}(z)|\, dz \le C \fint_{B(x^*, R)} |\mathbf{u}(z)|\, dz, \quad \forall\, y \in \overline{B(x^*, R/2)}. \qquad (10.4.18)$$

This, in turn, allows us to estimate the integrals in (10.4.16) as follows. For the boundary integral, (10.4.18) yields

$$\int\limits_{\partial B(x^*, R/2)} |\mathbf{u}(y)|\, d\sigma(y) \le C\Big(\fint_{B(x^*, R)} |\mathbf{u}(z)|\, dz\Big) R^{n-1}$$

$$= \frac{C}{R} \int_{B(x^*, R)} |\mathbf{u}(z)|\, dz, \qquad (10.4.19)$$

while for the solid integral

$$\int\limits_{B(x^*, R/2)} \frac{|\mathbf{u}(y)|}{|y - x^*|}\, dy \le C\Big(\fint_{B(x^*, R)} |\mathbf{u}(z)|\, dz\Big)\Big(\int_{B(x^*, R/2)} \frac{dy}{|y - x^*|}\Big)$$

$$\le \frac{C}{R} \int_{B(x^*, R)} |\mathbf{u}(z)|\, dz, \qquad (10.4.20)$$

since $\int_{B(x^*,R/2)} \frac{dy}{|y-x^*|} = C \int_0^{\frac{R}{2}} \rho^{n-2} \, d\rho = CR^{n-1}$. Now (10.4.12) (with r replaced by R) follows from (10.4.16), (10.4.19), and (10.4.20). $\qquad\square$

Exercise 10.21. Under the same background assumptions as in Theorem 10.20, prove that for every multi-index $\alpha \in \mathbb{N}_0^n$ there exists a constant $C_\alpha \in (0, \infty)$, that depends only on n, α, λ, μ, and with the property that for every $x \in \Omega$ and every $r \in (0, \text{dist}(x, \partial\Omega))$

$$|\partial^\alpha \mathbf{u}(x)| \le \frac{C_\alpha}{r^{|\alpha|}} \int_{B(x,r)} |\mathbf{u}(y)| \, dy. \qquad (10.4.21)$$

Hint: Use induction on $|\alpha|$, (10.4.12) written for $r/2$, and the fact that $\partial^\alpha \mathbf{u}$ continues to be a solution of the Lamé system (10.4.1)

Theorem 10.22 (L^∞-**Interior estimates for the Lamé operator**). *Let* $\lambda, \mu \in \mathbb{C}$ *be such that* $\mu \ne 0$, $\lambda + 2\mu \ne 0$, *and* $(n+1)\mu + \lambda \ne 0$ *and suppose* $\mathbf{u} \in [\mathcal{D}'(\Omega)]^n$ *satisfies the Lamé system* (10.4.1). *Then* $\mathbf{u} \in [C^\infty(\Omega)]^n$ *and, with* C *as in Theorem 10.20, for each* $x \in \Omega$, *each* $r \in (0, \text{dist}(x, \partial\Omega))$, *and each* $k \in \mathbb{N}$, *we have*

$$|\partial^\alpha \mathbf{u}(x)| \le \frac{C^k e^{k-1} k!}{r^k} \max_{y \in \overline{B(x,r)}} |\mathbf{u}(y)|, \qquad \forall \, \alpha \in \mathbb{N}_0^n \text{ with } |\alpha| = k. \qquad (10.4.22)$$

Proof. The fact that $\mathbf{u} \in [C^\infty(\Omega)]^n$ follows from Lemma 10.17. In particular, this implies that $\partial^\alpha \mathbf{u}$ satisfies (10.4.1) for every $\alpha \in \mathbb{N}_0^n$. The case $k = 1$ is an immediate consequence of (10.4.12) since, clearly,

$$\int_{B(x,r)} |\mathbf{u}(y)| \, dy \le \max_{y \in \overline{B(x,r)}} |\mathbf{u}(y)|. \qquad (10.4.23)$$

Having established this, the desired conclusion follows by invoking Lemma 6.21 (with $\mathcal{A}$ the class of null solutions for the Lamé system in Ω). $\qquad\square$

Recall Definition 6.22.

Theorem 10.23. *Suppose* $\lambda, \mu \in \mathbb{C}$ *are such that*

$$\mu \ne 0, \quad \lambda + 2\mu \ne 0, \quad \text{and} \quad (n+1)\mu + \lambda \ne 0. \qquad (10.4.24)$$

Then any null solution of the Lamé system (10.4.1) *has components that are real-analytic in* Ω.

Proof. This is an immediate consequence of Theorem 10.22 and Lemma 6.24. $\qquad\square$

Theorem 10.24. *Suppose* $\lambda, \mu \in \mathbb{C}$ *satisfy the conditions in* (10.4.24) *and* $\Omega \subseteq \mathbb{R}^n$ *is open and connected. Assume* $\mathbf{u}$ *is a null solution of the Lamé system* (10.4.1) *with the property that* $\partial^\alpha \mathbf{u}(x_0) = 0$ *for some* $x_0 \in \Omega$ *and for all* $\alpha \in \mathbb{N}_0^n$. *Then* $\mathbf{u} = 0$ *in* Ω.

Proof. This follows from Theorem 10.23 and Theorem 6.25. $\qquad\square$

Next we record the analogue of the classical Liouville's theorem for the Laplacian (cf. Theorem 7.15) in the case of the Lamé system.

Theorem 10.25 (Liouville's Theorem for the Lamé system). *Let $\lambda, \mu \in \mathbb{C}$ be such that $\mu \neq 0$, $\lambda + 2\mu \neq 0$ and suppose $\mathbf{u} \in [L^\infty(\mathbb{R}^n)]^n$ satisfies the Lamé system*

$$\mu \Delta \mathbf{u} + (\lambda + \mu) \nabla div\, \mathbf{u} = \mathbf{0} \quad in\ [\mathcal{D}'(\mathbb{R}^n)]^n. \tag{10.4.25}$$

Then there exists a constant vector $\mathbf{c} \in \mathbb{C}^n$ such that $\mathbf{u}(x) = \mathbf{c}$ for all $x \in \mathbb{R}^n$.

Proof. This is a particular case of Theorem 10.3 since based on the current assumptions on λ and μ, Proposition 10.14 ensures that condition (10.1.21) is satisfied. $\square$

Exercise 10.26. Assuming that $\lambda, \mu \in \mathbb{C}$ satisfy the conditions in (10.4.24), give an alternative proof of Theorem 10.25 by relying on the interior estimates from (10.4.12).

10.5 The Poisson Equation for the Lamé Operator

Let L be the operator from (10.3.1) with coefficients $\mu, \lambda \in \mathbb{C}$ satisfying (10.3.2). Then the Poisson equation for the Lamé operator in an open subset Ω of $\mathbb{R}^n$ reads

$$L\mathbf{u} = \mathbf{f}, \tag{10.5.1}$$

where the vector $\mathbf{f}$ is given and the vector $\mathbf{u}$ is the unknown. If $\mathbf{u}$ is a priori known to be of class C^2, then the equality in (10.5.1) is considered in the pointwise sense, everywhere in $\mathbb{R}^n$. Such a solution is called classical. Often, one starts with $\mathbf{u}$ simply a vector distribution in which scenario (10.5.1) is interpreted in $[\mathcal{D}'(\Omega)]^n$. In this case, we shall refer to $\mathbf{u}$ as a distributional solution. In this vein, it is worth pointing out that if the datum $\mathbf{f}$ is of class C^∞ in Ω then any distributional solution of (10.5.1) is also of class C^∞ in Ω, as seen from Theorem 10.9, Proposition 10.14, and (10.3.2).

A key ingredient in solving the Poisson equation (10.5.1) is going to be the fundamental solution for the Lamé system derived in (10.3.21).

Proposition 10.27. *Let L be the Lamé operator from (10.3.1) such that (10.3.2) is satisfied. Assume $n \geq 2$ and let $\mathbb{E} = (E_{jk})_{1 \leq j,k \leq n}$ be the fundamental solution for L in $\mathbb{R}^n$ with entries as in (10.3.21) for $n \geq 3$ and as in (10.7.2) for $n = 2$. Suppose that Ω is an open set in $\mathbb{R}^n$ and that $\mathbf{f} \in [L^\infty(\Omega)]^n$ vanishes outside a bounded subset of Ω. Then*

$$\mathbf{u}(x) = \int_\Omega \mathbb{E}(x - y)\mathbf{f}(y)\,dy, \qquad \forall\, x \in \Omega, \tag{10.5.2}$$

is a distributional solution of the Poisson equation for the Lamé system in Ω, i.e., $L\mathbf{u} = \mathbf{f}$ in $[\mathcal{D}'(\Omega)]^n$. In addition, $\mathbf{u} \in [C^1(\Omega)]^n$.

Proof. This is established by arguing along the lines of the proof of Proposition 7.8.
$\square$

The main result in this section is the following well-posedness result for the Poisson problem for the Lamé operator in $\mathbb{R}^n$.

Theorem 10.28. *Assume $n \geq 3$, and let L be the Lamé operator from (10.3.1) with $\lambda, \mu \in \mathbb{C}$ satisfying $\mu \neq 0$ and $\lambda + 2\mu \neq 0$. Also, suppose a vector-valued function $\mathbf{f} \in [L_{comp}^\infty(\mathbb{R}^n)]^n$ and $\mathbf{c} \in \mathbb{C}^n$ are given. Then the Poisson problem for the Lamé operator in $\mathbb{R}^n$,*

$$\begin{cases} \mathbf{u} \in [C^0(\mathbb{R}^n)]^n, \\ L\mathbf{u} = \mathbf{f} \quad in \quad [\mathscr{D}'(\mathbb{R}^n)]^n, \\ \lim_{|x| \to \infty} \mathbf{u}(x) = \mathbf{c}, \end{cases} \tag{10.5.3}$$

has a unique solution. Moreover, the solution $\mathbf{u}$ satisfies the following additional properties.

(1) The function $\mathbf{u}$ is of class C^1 in $\mathbb{R}^n$ and admits the integral representation formula

$$\mathbf{u}(x) = \mathbf{c} + \int_{\mathbb{R}^n} \mathbb{E}(x - y)\mathbf{f}(y) \, dy, \quad \forall x \in \mathbb{R}^n, \tag{10.5.4}$$

where $\mathbb{E} = (E_{jk})_{1 \leq j,k \leq n}$ is the fundamental solution for L in $\mathbb{R}^n$ with entries as in (10.3.21). Moreover, for each $j \in \{1, \ldots, n\}$ we have

$$\partial_j \mathbf{u}(x) = \int_{\mathbb{R}^n} (\partial_j \mathbb{E})(x - y)\mathbf{f}(y) \, dy, \quad \forall x \in \mathbb{R}^n. \tag{10.5.5}$$

(2) If in fact $\mathbf{f} \in [C_0^\infty(\mathbb{R}^n)]^n$ then $\mathbf{u} \in [C^\infty(\mathbb{R}^n)]^n$.
(3) For every $j, k \in \{1, \ldots, n\}$ there exists a matrix $C_{jk} \in M_{n \times n}(\mathbb{C})$ such that

$$\partial_j \partial_k \mathbf{u} = C_{jk} \mathbf{f} + T_{\partial_j \partial_k \mathbb{E}} \mathbf{f} \quad in \quad [\mathscr{D}'(\mathbb{R}^n)]^n, \tag{10.5.6}$$

where $T_{\partial_j \partial_k \mathbb{E}}$ is the singular integral operator associated with the matrix-valued function $\Theta := \partial_j \partial_k \mathbb{E}$ (cf. Definition 4.93).
(4) For every integrability exponent $p \in (1, \infty)$, the solution $\mathbf{u}$ of (10.5.3) satisfies $\partial_j \partial_k \mathbf{u} \in [L^p(\mathbb{R}^n)]^n$ for each $j, k \in \{1, \ldots, n\}$, where the derivatives are taken in $[\mathscr{D}'(\mathbb{R}^n)]^n$. Moreover, there exists a constant $C = C(p, n) \in (0, \infty)$ with the property that

$$\sum_{j,k=1}^n \|\partial_j \partial_k \mathbf{u}\|_{[L^p(\mathbb{R}^n)]^n} \leq C \|\mathbf{f}\|_{[L^p(\mathbb{R}^n)]^n}. \tag{10.5.7}$$

Proof. From Proposition 10.27 we have that $\mathbf{u}$ defined as in (10.5.4) is of class C^1 in $\mathbb{R}^n$ and satisfies $L\mathbf{u} = \mathbf{f}$ in $[\mathscr{D}'(\mathbb{R}^n)]^n$. In addition, by reasoning as in the proof of Proposition 7.8 we also see that formula (10.5.5) holds for each $j \in \{1, \ldots, n\}$. Furthermore, the same type of estimate as in (7.2.21) proves that the function $\mathbf{u}$ from (10.5.4) also satisfies the limit condition in (10.5.3). This concludes the treatment of the existence. As for uniqueness, suppose $\mathbf{u}$ is a solution of (10.5.3) for $\mathbf{f} = \mathbf{0} \in \mathbb{C}^n$. Given that $\mathbf{u}$ satisfies (10.5.3), it follows that $\mathbf{u}$ is bounded in $\mathbb{R}^n$. As such, Liouville's theorem (cf. Theorem 10.25) applies and gives that $\mathbf{u} = \mathbf{0}$. This proves that $\mathbf{u}$

defined as in (10.5.4) is the unique solution of (10.5.3). Next, the regularity result in part *(2)* may be seen either directly from (10.5.4), or by relying on Theorem 10.9, Proposition 10.14, and the assumptions on the Lamé moduli λ, μ. This concludes the proof of the claims made in parts *(1)–(2)*.

Consider now the claim made in part *(3)*. Fix $j, k \in \{1, \ldots, n\}$ arbitrary. Then, as seen from (10.3.21), the function $\Phi := \partial_k \mathbb{E}$ is C^∞ and positive homogeneous of degree $1 - n$ in $\mathbb{R}^n \setminus \{0\}$. In turn, this implies that the matrix-valued function $\Theta := \partial_j \partial_k \mathbb{E}$ has entries satisfying the conditions in (4.4.1) (here the discussion in Example (4.71) is relevant). From what we have proved in part *(1)* and Theorem 4.103 we then conclude that, in the sense of $[\mathcal{D}'(\mathbb{R}^n)]^n$,

$$\partial_j \partial_k \mathbf{u}(x) = \partial_j [\partial_k \mathbf{u}(x)] = \partial_j \Big[\int_{\mathbb{R}^n} (\partial_k \mathbb{E})(x - y) \mathbf{f}(y) \, dy \Big] \qquad (10.5.8)$$

$$= \Big(\int_{S^{n-1}} (\partial_k \mathbb{E})(\omega) \omega_j \, d\sigma(\omega) \Big) \mathbf{f}(x)$$

$$+ \lim_{\varepsilon \to 0^+} \int_{|x-y| > \varepsilon} (\partial_j \partial_k \mathbb{E})(x - y) \mathbf{f}(y) \, dy.$$

Upon taking $C_{jk} := \int_{S^{n-1}} (\partial_k \mathbb{E})(\omega) \omega_j \, d\sigma(\omega) \in M_{n \times n}(\mathbb{C})$, formula (10.5.6) follows, finishing the proof of part *(3)*. Lastly, the claim in *(4)* is a consequence of *(3)* and the boundedness of the singular integral operators $T_{\partial_j \partial_k \mathbb{E}}$ on $[L^p(\mathbb{R}^n)]^n$ (as seen by applying Theorem 4.101 componentwise). $\qquad\qquad\square$

10.6 Fundamental Solutions for the Stokes Operator

Let $\mathbf{u} = (u_1, \ldots, u_n) \in [\mathcal{D}'(\mathbb{R}^n)]^n$ and $p \in \mathcal{D}'(\mathbb{R}^n)$. Then the Stokes operator L_S acting on $(\mathbf{u}, p) = (u_1, \ldots, u_n, p)$ is defined by

$$L_S(\mathbf{u}, p) := \Big(\Delta u_1 - \partial_1 p, \ldots, \Delta u_n - \partial_n p, \sum_{s=1}^{n} \partial_s u_s \Big) \in [\mathcal{D}'(\mathbb{R}^n)]^{n+1}. \qquad (10.6.1)$$

In practice, $\mathbf{u}$ and p are referred to as the velocity field and the pressure function, respectively.

A fundamental solution for the Stokes operator is given by a pair $(\mathbb{E}, \mathbf{p})$, where $\mathbb{E} = (E_{jk})_{1 \le j, k \le n} \in M_{n \times n}(\mathcal{D}'(\mathbb{R}^n))$, $\mathbf{p} = (p_1, \ldots, p_n) \in [\mathcal{D}'(\mathbb{R}^n)]^n$, satisfy the following conditions. If for each $k \in \{1, \ldots, n\}$ we set $\Gamma_k := (E_{1k}, \ldots, E_{nk}, p_k)$ and $\mathbf{e}_k^{n+1}$ denotes the unit vector in $\mathbb{R}^{n+1}$ with 1 on the k-th entry, then

$$L_S \Gamma_k = \delta \mathbf{e}_k^{n+1} \quad \text{in} \quad [\mathcal{D}'(\mathbb{R}^n)]^{n+1} \quad \text{for} \quad k = 1, \ldots, n. \qquad (10.6.2)$$

We propose to determine all the fundamental solutions for the Stokes operator with entries in $\mathcal{S}'(\mathbb{R}^n)$ in the case when $n \ge 3$, a condition assumed in this section unless otherwise specified. Suppose $(\mathbb{E}, \mathbf{p})$ is a fundamental solution for L_S with the property that $\mathbb{E} = (E_{jk})_{1 \le j,k \le n} \in M_{n \times n}(\mathcal{S}'(\mathbb{R}^n))$ and $\mathbf{p} = (p_1, \ldots, p_n)$ belongs to $[\mathcal{S}'(\mathbb{R}^n)]^n$. Then (10.6.2) implies

$$\Delta E_{jk} - \partial_j p_k = \delta_{jk}\delta \quad \text{in} \quad \mathcal{S}'(\mathbb{R}^n), \quad \forall j,k \in \{1, \ldots, n\}, \tag{10.6.3}$$

$$\sum_{s=1}^{n} \partial_s E_{sk} = 0 \quad \text{in} \quad \mathcal{S}'(\mathbb{R}^n), \quad \forall k \in \{1, \ldots, n\}. \tag{10.6.4}$$

Apply the Fourier transform to each of the equalities in (10.6.3) and (10.6.4) to obtain

$$-|\xi|^2 \widehat{E_{jk}} - i\xi_j \widehat{p_k} = \delta_{jk} \quad \text{in} \quad \mathcal{S}'(\mathbb{R}^n), \quad \forall j,k \in \{1, \ldots, n\}, \tag{10.6.5}$$

$$\sum_{s=1}^{n} \xi_s \widehat{E_{sk}} = 0 \quad \text{in} \quad \mathcal{S}'(\mathbb{R}^n), \quad \forall k \in \{1, \ldots, n\}. \tag{10.6.6}$$

Fix $k \in \{1, \ldots, n\}$ and, for each $j = 1, \ldots, n$, multiply the identity in (10.6.5) corresponding to this j with ξ_j, then sum up over j and use (10.6.6) to arrive at

$$|\xi|^2 \widehat{p_k} = i\xi_k \quad \text{in} \quad \mathcal{S}'(\mathbb{R}^n). \tag{10.6.7}$$

Reasoning as in the derivation of (10.3.15) from (10.3.13), the last identity implies

$$\widehat{p_k} = i\frac{\xi_k}{|\xi|^2} + \widehat{r_k}(\xi) \quad \text{in} \quad \mathcal{S}'(\mathbb{R}^n), \tag{10.6.8}$$

for some harmonic polynomial r_k in $\mathbb{R}^n$. An application of the inverse Fourier transform to (10.6.8) combined with (7.14.6) then yields

$$p_k = i\mathcal{F}^{-1}\left(\frac{\xi_k}{|\xi|^2}\right) + r_k = -\frac{1}{\omega_{n-1}} \cdot \frac{x_k}{|x|^n} + r_k \quad \text{in} \quad \mathcal{S}'(\mathbb{R}^n). \tag{10.6.9}$$

In particular, if we take $p_k = -\frac{1}{\omega_{n-1}} \cdot \frac{x_k}{|x|^n}$ and use it to substitute for $\widehat{p_k}$ back in (10.6.5), we arrive at the condition

$$|\xi|^2 \widehat{E_{jk}} = \frac{\xi_j \xi_k}{|\xi|^2} - \delta_{jk} \quad \text{in} \quad \mathcal{S}'(\mathbb{R}^n), \quad \forall j \in \{1, \ldots, n\}. \tag{10.6.10}$$

Consequently, since $n \ge 3$, by reasoning as in the derivation of (10.3.15) from (10.3.13), identity (10.6.10) implies

$$\widehat{E_{jk}} = \frac{\xi_j \xi_k}{|\xi|^4} - \frac{\delta_{jk}}{|\xi|^2} + \widehat{R_{jk}} \quad \text{in} \quad \mathcal{S}'(\mathbb{R}^n), \quad \forall j \in \{1, \ldots, n\}, \tag{10.6.11}$$

for some polynomials R_{jk} in $\mathbb{R}^n$ satisfying $\Delta R_{jk} = 0$ in $\mathbb{R}^n$, $j = 1, \ldots, n$. Taking the Fourier transform in (10.3.13), then using (7.14.8) and Proposition 4.64 with $\lambda = n - 2$, we obtain

$$E_{jk} = \mathcal{F}^{-1}\left(\frac{\xi_j \xi_k}{|\xi|^4}\right) - \mathcal{F}^{-1}\left(\frac{1}{|\xi|^2}\right)\delta_{jk} + R_{jk}$$

$$= -\frac{1}{2(n-2)\omega_{n-1}}\frac{\delta_{jk}}{|x|^{n-2}} - \frac{1}{2\omega_{n-1}}\frac{x_j x_k}{|x|^n} + R_{jk} \quad \text{in } \mathcal{S}'(\mathbb{R}^n). \qquad (10.6.12)$$

We are now ready to state our main result regarding fundamental solutions for the Stokes operator that are tempered distributions.

Theorem 10.29. *Let $n \geq 3$ and let L_S be the Stokes operator from* (10.6.1). *Consider the following functions in $L^1_{loc}(\mathbb{R}^n)$:*

$$E_{jk}(x) := -\frac{1}{2(n-2)\omega_{n-1}}\frac{\delta_{jk}}{|x|^{n-2}} - \frac{1}{2\omega_{n-1}}\frac{x_j x_k}{|x|^n}, \quad \forall\, j,k \in \{1,\dots,n\}, \qquad (10.6.13)$$

$$p_k(x) := -\frac{1}{\omega_{n-1}}\frac{x_k}{|x|^n}, \quad \forall\, k \in \{1,\dots,n\}, \qquad (10.6.14)$$

defined for $x \in \mathbb{R}^n \setminus \{0\}$. Then if we set $\mathbb{E} := (E_{jk})_{1 \leq j,k \leq n}$ and $\mathbf{p} := (p_1,\dots,p_n)$, we have $\mathbb{E} \in \mathcal{M}_{n \times n}(\mathcal{S}'(\mathbb{R}^n))$, $\mathbf{p} \in [\mathcal{S}'(\mathbb{R}^n)]^n$, and the pair $(\mathbb{E}, \mathbf{p})$ is a fundamental solution for the Stokes operator in $\mathbb{R}^n$.

Moreover, any fundamental solution $(\mathbb{U}, \mathbf{q})$ for the Stokes operator with the property that $\mathbb{U} \in \mathcal{M}_{n \times n}(\mathcal{S}'(\mathbb{R}^n))$, $\mathbf{q} \in [\mathcal{S}'(\mathbb{R}^n)]^n$ is of the form $\mathbb{U} = \mathbb{E} + \mathbb{P}$, $\mathbf{q} = \mathbf{p} + \mathbf{r}$, for some matrix $\mathbb{P} := (P_{jk})_{1 \leq j,k \leq n}$ whose entries are polynomials in $\mathbb{R}^n$ satisfying $\Delta^2 P_{jk} = 0$ pointwise in $\mathbb{R}^n$ for $j,k = 1,\dots,n$, and some vector $\mathbf{r} := (r_1,\dots,r_n)$ whose entries are polynomials in $\mathbb{R}^n$ satisfying $\Delta r_k = 0$ pointwise in $\mathbb{R}^n$ for $k = 1,\dots,n$.

Proof. From (10.6.12) and (10.6.9) we have $\mathbb{E} \in \mathcal{M}_{n \times n}(\mathcal{S}'(\mathbb{R}^n))$, $\mathbf{p} \in [\mathcal{S}'(\mathbb{R}^n)]^n$, and

$$\widehat{E_{jk}} = \frac{\xi_j \xi_k}{|\xi|^4} - \frac{\delta_{jk}}{|\xi|^2}, \quad \widehat{p_k} = \mathrm{i}\frac{\xi_k}{|\xi|^2}, \quad \text{in } \mathcal{S}'(\mathbb{R}^n), \quad \forall\, j,k \in \{1,\dots,n\}. \qquad (10.6.15)$$

Based on the properties of the Fourier transform, we have that $(\mathbb{E}, \mathbf{p})$ is a fundamental solution for the Stokes operator if and only if its components satisfy (10.6.5)–(10.6.6). By making use of (10.6.15) we may write

$$-|\xi|^2 \widehat{E_{jk}} - \mathrm{i}\xi_j \widehat{p_k} = -\frac{\xi_j \xi_k}{|\xi|^2} + \delta_{jk} + \frac{\xi_j \xi_k}{|\xi|^2} = \delta_{jk} \text{ in } \mathcal{S}'(\mathbb{R}^n), \quad \forall\, j,k \in \{1,\dots,n\}$$

$$\qquad\qquad (10.6.16)$$

and

$$\sum_{j=1}^n \xi_j \widehat{E_{jk}} = \sum_{j=1}^n \frac{\xi_j^2 \xi_k}{|\xi|^4} - \sum_{j=1}^n \frac{\delta_{jk}\xi_j}{|\xi|^2} = 0 \text{ in } \mathcal{S}'(\mathbb{R}^n), \qquad (10.6.17)$$

proving that $(\mathbb{E}, \mathbf{p})$ is a fundamental solution for the Stokes operator.

Suppose $(\mathbb{U}, \mathbf{q})$ is another fundamental solution for the Stokes operator such that $\mathbb{U} = (U_{jk})_{1 \leq j,k \leq n} \in \mathcal{M}_{n \times n}(\mathcal{S}'(\mathbb{R}^n))$ and $\mathbf{q} = (q_1,\dots,q_n) \in [\mathcal{S}'(\mathbb{R}^n)]^n$. Then for each $k \in \{1,\dots,n\}$ we have $q_k = p_k + r_k$ for some harmonic polynomial r_k in $\mathbb{R}^n$ (recall the computation that lead to (10.6.9)). From this fact and the equations corresponding to (10.6.3) written for the components of $(\mathbb{E}, \mathbf{p})$ and $(\mathbb{U}, \mathbf{q})$, we further obtain that

$$\Delta(U_{jk} - E_{jk}) = \partial_j r_k \text{ in } \mathcal{S}'(\mathbb{R}^n) \text{ for each } j, k \in \{1, \ldots, n\}. \quad (10.6.18)$$

Fix $j, k \in \{1, \ldots, n\}$ arbitrary. Then by applying Δ to both sides of the equation in (10.6.18) it follows that $\Delta^2(U_{jk} - E_{jk}) = 0$ in $\mathcal{S}'(\mathbb{R}^n)$. Consequently, after taking the Fourier transform of the latter identity we arrive at $|\xi|^4(\widehat{U_{jk}} - \widehat{E_{jk}}) = 0$ in $\mathcal{S}'(\mathbb{R}^n)$. In turn, this implies $U_{jk} - E_{jk} = R_{jk}$ for some polynomials R_{jk} in $\mathbb{R}^n$ satisfying $\Delta^2 R_{jk} = 0$ pointwise in $\mathbb{R}^n$. The proof of the theorem is now complete. $\qquad \square$

Exercise 10.30. Follow the outline below to check, without the use of the Fourier transform, that $(\mathbb{E}, \mathbf{p})$ with entries $\mathbb{E} = (E_{jk})_{1 \le j,k \le n}$ and $\mathbf{p} = (p_1, \ldots, p_n)$ of function type defined by the functions from (10.6.13)–(10.6.14) is a fundamental solution for the Stokes operator in $\mathbb{R}^n$, $n \ge 3$.

Step 1. Show that $\Delta E_{jk} - \partial_j p_k = 0$ and $\sum_{\ell=1}^{n} \partial_\ell E_{\ell k} = 0$ pointwise in $\mathbb{R}^n \setminus \{0\}$ for every $j, k \in \{1, \ldots, n\}$.

Step 2. Show that the desired conclusion (i.e., that the given $(\mathbb{E}, \mathbf{p})$ is a fundamental solution for the Stokes operator in $\mathbb{R}^n$, $n \ge 3$) is equivalent with the conditions that

$$\lim_{\varepsilon \to 0^+} \left[\int_{|x| \ge \varepsilon} E_{jk}(x) \Delta \varphi(x) \, dx + \int_{|x| \ge \varepsilon} p_k(x) \partial_j \varphi(x) \, dx \right] = \varphi(0) \delta_{jk} \quad (10.6.19)$$

and

$$\lim_{\varepsilon \to 0^+} \left[\int_{|x| \ge \varepsilon} \sum_{\ell=1}^{n} E_{\ell k}(x) \partial_\ell \varphi(x) \, dx \right] = 0 \quad (10.6.20)$$

for every $j, k \in \{1, \ldots, n\}$ and every $\varphi \in C_0^\infty(\mathbb{R}^n)$.

Step 3. Fix $j, k \in \{1, \ldots, n\}$, $\varphi \in C_0^\infty(\mathbb{R}^n)$, and let $R \in (0, \infty)$ be such that $\text{supp}\, \varphi \subseteq B(0, R)$, so that one may replace the domain of integration for the integrals in (10.6.19) and (10.6.20) with $\{x \in \mathbb{R}^n : \varepsilon < |x| < R\}$. Use this domain of integration, (14.8.5), (14.8.4), and the result from Step 1, to prove that (10.6.19) and (10.6.20) are equivalent with

$$\lim_{\varepsilon \to 0^+} \left[-\int_{\partial B(0,\varepsilon)} E_{jk}(x) \frac{\partial \varphi}{\partial \nu}(x) \, d\sigma(x) + \int_{\partial B(0,\varepsilon)} \varphi(x) \frac{\partial E_{jk}}{\partial \nu}(x) \, d\sigma(x) \right.$$
$$\left. - \int_{\partial B(0,\varepsilon)} p_k(x) \varphi(x) \nu_j(x) \, d\sigma(x) \right] = \varphi(0) \delta_{jk} \quad (10.6.21)$$

and

$$\lim_{\varepsilon \to 0^+} \left[\int_{\partial B(0,\varepsilon)} \sum_{\ell=1}^{n} E_{\ell k}(x) \varphi(x) \nu_l(x) \, d\sigma(x) \right] = 0, \quad (10.6.22)$$

where $\nu(x) = \frac{x}{\varepsilon}$ for each $x \in \partial B(0, \varepsilon)$.

Step 4. Prove that there exists a constant $C \in (0, \infty)$ independent of ε such that each of the quantities:

$$\left| \int_{\partial B(0,\varepsilon)} E_{jk}(x) \frac{\partial \varphi}{\partial \nu}(x) \, d\sigma(x) \right|,$$

$$\left| \int_{\partial B(0,\varepsilon)} \frac{\partial E_{jk}}{\partial \nu}(x) [\varphi(x) - \varphi(0)] \, d\sigma(x) \right|, \tag{10.6.23}$$

$$\left| \int_{\partial B(0,\varepsilon)} p_k(x)[\varphi(x) - \varphi(0)] \nu_j(x) \, d\sigma(x) \right|,$$

is bounded by $C \|\nabla \varphi\|_{L^\infty(\mathbb{R}^n)} \, \varepsilon$, thus convergent to zero as $\varepsilon \to 0^+$, and such that

$$\left| \int_{\partial B(0,\varepsilon)} \sum_{\ell=1}^n E_{\ell k}(x) \varphi(x) \nu_l(x) \, dx \right| \le C \|\varphi\|_{L^\infty(\mathbb{R}^n)} \, \varepsilon \xrightarrow[\varepsilon \to 0^+]{} 0. \tag{10.6.24}$$

Step 5. Combine Steps 2, 3, and 4 to reduce matters to proving that

$$\lim_{\varepsilon \to 0^+} \left[\int_{\partial B(0,\varepsilon)} \sum_{s=1}^n \frac{x_s}{\varepsilon} (\partial_s E_{jk})(x) \, d\sigma(x) - \int_{\partial B(0,\varepsilon)} \frac{x_j}{\varepsilon} p_k(x) \, d\sigma(x) \right] = \delta_{jk}. \tag{10.6.25}$$

Step 6. Prove that for every $x \in \partial B(0, \varepsilon)$ we have

$$\sum_{s=1}^n \frac{x_s}{\varepsilon} (\partial_s E_{jk})(x) - \frac{x_j}{\varepsilon} p_k(x) = \frac{\delta_{jk}}{2\,\omega_{n-1}\varepsilon^{n-1}} + \frac{n x_j x_k}{2\,\omega_{n-1}\varepsilon^{n+1}}. \tag{10.6.26}$$

Step 7. Integrate the expression in (10.6.26) and use (10.3.28) to finish the proof of (10.6.25).

Further Notes for Chapter 10. The discussion in Chapters 5–6 about the existence and nature of fundamental solutions has been limited to the case of scalar differential operators, and in Section 10.2 these scalar results have been extended to generic constant coefficient systems (of arbitrary order) via an approach that appears to be new. Moreover, this approach offers further options for finding an explicit form for a fundamental solution for a given system, such as the Lamé system. While we shall pursue this idea later, in Section 11.1, we felt it is natural and beneficial to first deal with the issue of computing a fundamental solution for the Lamé and Stokes systems via Fourier analysis (as done in Section 10.3 and Section 10.6, respectively).

The inclusion of a section on mean value formulas for the Lamé operator is justified by the fact that such formulas directly yield interior estimates, without resorting to quantitative elliptic regularity, which is typically formulated in the language of Sobolev spaces. In turn, these interior estimates play a pivotal role in establishing uniqueness for the corresponding Poisson problem. A more systematic treatment of the issue of mean value formulas for the Lamé operator may be found in [56].

10.7 Additional Exercises for Chapter 10

Exercise 10.31. Fix $m, N, M \in \mathbb{N}$ and assume

$$L = L(\partial) = \sum_{|\alpha| \leq m} A_\alpha \partial^\alpha, \quad A_\alpha = (a_\alpha^{jk})_{\substack{1 \leq j \leq M \\ 1 \leq k \leq N}} \in M_{M \times N}(\mathbb{C}), \tag{10.7.1}$$

is a constant, complex coefficient, homogeneous system of order m. Make the assumption that $L(\xi) \in M_{M \times N}(\mathbb{C})$ is a surjective linear map from $\mathbb{C}^N \to \mathbb{C}^M$ for each $\xi \in \mathbb{R}^n \setminus \{0\}$. Follow the outline below to construct a fundamental solution for L.

(1) Consider $\mathbb{L} := LL^*$. Show that $\mathbb{L}$ is a homogeneous constant coefficient $M \times M$ system of order $2m$ and $\mathbb{L}(\xi) = L(\xi)L(\xi)^*$ in $M_{M \times N}(\mathbb{C})$ for each $\xi \in \mathbb{R}^n$.
(2) Show that $\mathbb{L}(\xi)$ is injective for each $\xi \in \mathbb{R}^n \setminus \{0\}$ and use this to conclude that $\det[\mathbb{L}(\xi)] \neq 0$ for each $\xi \in \mathbb{R}^n \setminus \{0\}$.
(3) Corollary 10.10 guarantees the existence of a matrix-valued fundamental solution $\mathbb{E} \in M_{M \times M}(\mathcal{S}'(\mathbb{R}^n))$ for $\mathbb{L}$. Show that $E := L^*\mathbb{E}$ is a fundamental solution for L.

Exercise 10.32. Let L_S be the Stokes operator from (10.6.1). Also, suppose p in $\mathcal{D}'(\Omega)$ and $\mathbf{u} = (u_1, \dots, u_n) \in [\mathcal{D}'(\Omega)]^n$ satisfy $L_S(\mathbf{u}, p) = 0$ in $[\mathcal{D}'(\Omega)]^{n+1}$. Under these conditions prove that $p, u_j \in C^\infty(\Omega)$ for $j \in \{1, \dots, n\}$, and $\Delta p = 0$ pointwise in Ω while $\Delta^2 u_j = 0$ pointwise in Ω for $j \in \{1, \dots, n\}$.

Exercise 10.33. For each $x \in \mathbb{R}^2 \setminus \{0\}$ and $j, k \in \{1, 2\}$ consider the functions

$$E_{jk}(x) := \frac{1}{4\pi\mu(2\mu + \lambda)} \left[(3\mu + \lambda)\delta_{jk}\ln|x| - \frac{(\mu + \lambda)x_j x_k}{|x|^2} \right]. \tag{10.7.2}$$

Prove that $\mathbb{E} := (E_{j,k})_{1 \leq j,k \leq 2}$ is a fundamental solution for the Lamé operator in $\mathbb{R}^2$.

Exercise 10.34. Consider the functions

$$E_{jk}(x) := \frac{1}{4\pi}\delta_{jk}\ln|x| - \frac{1}{4\pi} \cdot \frac{x_j x_k}{|x|^2}, \quad p_k(x) := -\frac{1}{2\pi} \cdot \frac{x_k}{|x|^2}, \tag{10.7.3}$$

defined for $x \in \mathbb{R}^2 \setminus \{0\}$ and $j, k \in \{1, 2\}$. Prove that if $\mathbb{E} := (E_{j,k})_{1 \leq j,k \leq 2}$ and $\mathbf{p} = (p_1, p_2)$, then $(\mathbb{E}, \mathbf{p})$ is a fundamental solution for the Stokes operator in $\mathbb{R}^2$.

Exercise 10.35. Assume $n \geq 3$, let L be the Lamé operator from (10.3.1) such that (10.3.2) is satisfied and consider $\mathbf{f} = (f_1, \dots, f_n) \in [C_0^\infty(\mathbb{R}^n)]^n$. Prove that for each $A \in M_{n \times n}(\mathbb{R})$ and each $b \in \mathbb{R}^n$, there exists a unique solution for the problem

$$\begin{cases} \mathbf{u} = (u_1, \dots, u_n) \in [C^\infty(\mathbb{R}^n)]^n, \\ L\mathbf{u} = \mathbf{f} \quad \text{pointwise in } \mathbb{R}^n, \\ \lim_{|x| \to \infty} |\mathbf{u}(x) - Ax - B| = 0. \end{cases} \tag{10.7.4}$$

Exercise 10.36. Assume $n \geq 3$, let L be the Lamé operator from (10.3.1) such that (10.3.2) is satisfied. Prove that if $m \in \mathbb{N}_0$ and $\mathbf{u}$ is a solution of the problem

$$\begin{cases} \mathbf{u} \in [C^\infty(\mathbb{R}^n)]^n, \\ L\mathbf{u} = 0 \quad \text{pointwise in } \mathbb{R}^n, \\ |\mathbf{u}(x)| \leq C(|x|^m + 1), \end{cases} \tag{10.7.5}$$

for some $C \in (0, \infty)$, then the components of $\mathbf{u}$ are polynomials of degree $\leq m$.

Chapter 11
More on Fundamental Solutions for Systems

Abstract The issue of identifying fundamental solutions for homogeneous constant coefficient systems of arbitrary order is a central topic here. As particular cases of the approach is developed, fundamental solutions that are tempered distributions for the Lamé and Stokes operators are derived. The fact that integral representation formulas and interior estimates hold for null-solutions of homogeneous systems with nonvanishing full symbol is also proved. As a consequence, null-solutions are real-analytic and shown to satisfy reverse Hölder estimates. Finally, layer potentials associated with arbitrary constant coefficient second- order systems in the upper-half space, and the relevance of these operators vis-a-vis to the solvability of boundary value problems for such systems in this setting, are discussed.

11.1 Computing a Fundamental Solution for the Lamé Operator

In this section, we use Theorem 10.8 in order to find a fundamental solution for the Lamé operator of elastostatics in $\mathbb{R}^n$, with $n \geq 2$. This is possible since in the case when $L(\partial)$ is the Lamé operator we have that $\det(L(\partial))$ is a constant factor of Δ^n.

Concretely, fix $n \in \mathbb{N}$, $n \geq 2$, and consider the algebra $\mathfrak{R}$ as in (10.2.15). Fix constants $\lambda, \mu \in \mathbb{C}$ satisfying $\mu \neq 0$, $\lambda + 2\mu \neq 0$, and set

$$L(\partial) := \mu\Delta + (\lambda + \mu)\nabla\text{div} \in \mathcal{M}_{n \times n}(\mathfrak{R}), \qquad (11.1.1)$$

(recall that $\nabla = (\partial_1, \ldots, \partial_n)$ and the operator div was defined in Exercise 7.74). This is precisely the Lamé operator from (10.3.1) and $L(\partial)$ has the form (10.2.12), with $a := \nabla$. Thus, for this choice of a, formula (10.2.13) gives

$$P(\partial) := \det L(\partial) = \mu^{n-1}(\lambda + 2\mu)\Delta^n, \qquad (11.1.2)$$

while (10.2.14) implies

$$\text{adj}\,[L(\partial)] = \mu^{n-2}\big[(\lambda + 2\mu)\Delta I_{n \times n} - (\lambda + \mu)\nabla\text{div}\big]\Delta^{n-2}. \qquad (11.1.3)$$

© Springer Nature Switzerland AG 2018

D. Mitrea, *Distributions, Partial Differential Equations, and Harmonic Analysis*, Universitext, https://doi.org/10.1007/978-3-030-03296-8_11

Going further, for each $m \in \mathbb{N}$, denoted by E_{Δ^m}, the fundamental solution for the poly-harmonic differential operator Δ^m in $\mathbb{R}^n$ from (7.5.2) and apply Proposition 7.29 to conclude that

$$\Delta^{n-2}\big[E_{\Delta^n}\big] = E_{\Delta^2} + c(n) \quad \text{in } \mathcal{S}'(\mathbb{R}^n) \quad \text{and pointwise in } \mathbb{R}^n \setminus \{0\}, \qquad (11.1.4)$$

where $c(n) = 0$ if $n \neq 4$ and $c(4) \in \mathbb{R}$. A direct computation starting with (7.5.2) corresponding to $n = 4 = m$ and using Lemma 7.20 yields $c(4) = -\frac{5}{24\pi^2}$. Thus, a combination of (11.1.4) and (7.3.8) yields

$$\Delta^{n-2}E_{\Delta^n}(x) = \begin{cases} \dfrac{|x|^{4-n}}{2(n-2)(n-4)\omega_{n-1}} & \text{if } n \geq 3,\ n \neq 4, \\[2ex] -\dfrac{1}{8\pi^2}\cdot\ln|x| - \dfrac{5}{24\pi^2} & \text{if } n = 4, \\[2ex] \dfrac{1}{8\pi}\cdot|x|^2\ln|x| & \text{if } n = 2, \end{cases} \qquad (11.1.5)$$

both for each $x \in \mathbb{R}^n \setminus \{0\}$ as well as in $\mathcal{S}'(\mathbb{R}^n)$. Also, by applying one more Δ to the identity in (11.1.4) and recalling (7.1.12), we have

$$\Delta^{n-1}E_{\Delta^n}(x) = E_{\Delta}(x) = \begin{cases} -\dfrac{1}{(n-2)\omega_{n-1}}\cdot|x|^{2-n} & \text{if } n \geq 3, \\[2ex] \dfrac{1}{2\pi}\cdot\ln|x| & \text{if } n = 2, \end{cases} \qquad (11.1.6)$$

for every $x \in \mathbb{R}^n \setminus \{0\}$ and in $\mathcal{S}'(\mathbb{R}^n)$.

Hence, based on (10.2.23) and (11.1.2), we obtain that a fundamental (matrix) solution $\mathbb{E}^L$ for the operator $L(\partial)$ as in (11.1.1) is given by

$$\mathbb{E}^L(x) = \text{adj}\,[L(\partial)]\big(E_{\mu^{n-1}(\lambda+2\mu)\Delta^n}(x)I_{n\times n}\big), \qquad x \in \mathbb{R}^n \setminus \{0\}, \qquad (11.1.7)$$

where $E_{\mu^{n-1}(\lambda+2\mu)\Delta^n}$ is a fundamental solution for $P(\partial) := \mu^{n-1}(\lambda + 2\mu)\Delta^n$. Using (11.1.3) and (11.1.4) in the right-most expression in (11.1.7) further gives that, for each $x \in \mathbb{R}^n \setminus \{0\}$,

$$\mathbb{E}^L(x) = \frac{\mu^{n-2}}{\mu^{n-1}(\lambda + 2\mu)}\big[(\lambda + 2\mu)\Delta I_{n\times n} - (\lambda + \mu)\nabla\text{div}\big]\Delta^{n-2}\big(E_{\Delta^n}(x)I_{n\times n}\big)$$

$$= \left(\frac{1}{\mu}\delta_{jk}E_{\Delta}(x) - \frac{\lambda + \mu}{\mu(\lambda + 2\mu)}(\partial_j\partial_k E_{\Delta^2})(x)\right)_{1\leq j,k\leq n}. \qquad (11.1.8)$$

A quick inspection of (7.3.8) and the last equality in (11.1.6) shows that both for each $x = (x_1,\ldots,x_n) \in \mathbb{R}^n \setminus \{0\}$ as well as in $\mathcal{S}'(\mathbb{R}^n)$, for each j,k in $\{1,\ldots,n\}$ we have

$$\partial_j \partial_k E_{\Delta^2}(x) = \begin{cases} \dfrac{\delta_{jk}}{2} E_\Delta(x) + \dfrac{1}{2\omega_{n-1}} \cdot \dfrac{x_j x_k}{|x|^n} & \text{if } n \geq 3, \\[4mm] \dfrac{\delta_{jk}}{2} E_\Delta(x) + \dfrac{1}{2\omega_{n-1}} \cdot \dfrac{x_j x_k}{|x|^n} + \dfrac{\delta_{jk}}{8}, & \text{if } n = 2. \end{cases} \qquad (11.1.9)$$

Note that, for the purpose of computing E^L, the additive constant $\frac{\delta_{jk}}{8}$ from (11.1.9) may be dropped. Thus, based on (11.1.8) and (11.1.9) (with $\delta_{jk}/8$ dropped), it follows that

$$\mathbb{E}^L(x) = \frac{3\mu + \lambda}{2\mu(2\mu + \lambda)} E_\Delta(x) I_{n \times n} \qquad (11.1.10)$$

$$- \frac{\lambda + \mu}{2\mu(2\mu + \lambda)\omega_{n-1}} \cdot \frac{x}{|x|^{n/2}} \otimes \frac{x}{|x|^{n/2}}, \quad \forall x \in \mathbb{R}^n \setminus \{0\},$$

is a tempered distribution that is a fundamental solution for the operator Lamé operator (11.1.1) (with $\mu \neq 0$, $\lambda + 2\mu \neq 0$) in $\mathbb{R}^n$. Note that this formula coincides with the one from (10.2.21) if $n \geq 3$ and with the one from (10.7.2) if $n = 2$.

11.2 Computing a Fundamental Solution for the Stokes Operator

In this section, the goal is to compute a fundamental solution for the Stokes operator in $\mathbb{R}^n$ starting from the explicit expression for the fundamental solution for the operator Lamé operator $\mathbb{E}^L$ from (11.1.10). To this end, fix $n \in \mathbb{N}$, $n \geq 2$ and $\lambda, \mu \in \mathbb{C}$ satisfying $\mu \neq 0$, $\lambda + 2\mu \neq 0$, and let $x = (x_1, \ldots, x_n)$ for $x \in \mathbb{R}^n$.

For each $k \in \{1, \ldots, n\}$, denote by $\mathbf{E}_k^L$ the k-th column of the fundamental solution $\mathbb{E}^L$ from (11.1.10). Then a straightforward calculation gives that for each $x \in \mathbb{R}^n \setminus \{0\}$ and in $\mathcal{S}'(\mathbb{R}^n)$, there holds

$$\text{div} [\mathbf{E}_k^L(x)] = \frac{1}{(\lambda + 2\mu)\omega_{n-1}} \cdot \frac{x_k}{|x|^n}, \qquad \forall k \in \{1, \ldots, n\}. \qquad (11.2.11)$$

Define

$$p_k(x) := -\frac{1}{\omega_{n-1}} \cdot \frac{x_k}{|x|^n}, \qquad \forall x \in \mathbb{R}^n \setminus \{0\}, \quad \forall k \in \{1, \ldots, n\}. \qquad (11.2.12)$$

Then, from (11.2.11) it follows that for each $k \in \{1, \ldots, n\}$ we have

$$\lim_{\lambda \to \infty} \text{div} [\mathbf{E}_k^L] = 0 \text{ pointwise in } \mathbb{R}^n \setminus \{0\} \text{ and in } \mathcal{S}'(\mathbb{R}^n), \qquad (11.2.13)$$

and

$$\lim_{\lambda \to \infty} \left[-(\lambda + \mu)\text{div}[\mathbf{E}_k^L] \right] = p_k \quad \text{pointwise in } \mathbb{R}^n \setminus \{0\} \text{ and in } \mathcal{S}'(\mathbb{R}^n). \qquad (11.2.14)$$

In addition, from (11.1.10) we obtain

$$\lim_{\lambda \to \infty} \mathbb{E}^L = \frac{1}{\mu} \mathbb{E}^S \quad \text{pointwise in } \mathbb{R}^n \setminus \{0\} \text{ and in } \mathcal{S}'(\mathbb{R}^n), \tag{11.2.15}$$

where, we have set

$$\mathbb{E}^S(x) := \Big(\frac{\delta_{jk}}{2} \cdot E_\Delta(x) - \frac{1}{2\omega_{n-1}} \cdot \frac{x_j x_k}{|x|^n} \Big)_{1 \le j,k \le n}, \quad \forall x \in \mathbb{R}^n \setminus \{0\}, \tag{11.2.16}$$

with E_Δ as in (11.1.6). Consequently, a combination of (11.2.14) and (11.2.16) (in view of the definition (11.1.1) of the Lamé operator), implies

$$\Delta E^S_{jk} - \partial_j p_k = \lim_{\lambda \to \infty} \Big[\mu E^L_{jk} + (\lambda + \mu) \sum_{s=1}^n \partial_j \partial_s E^L_{sk} \Big]$$

$$= \delta_{jk} \delta \quad \text{in } \mathcal{S}'(\mathbb{R}^n), \quad \forall j,k \in \{1,\dots,n\}. \tag{11.2.17}$$

Also, (11.2.15) and (11.2.13), with the convention that $\mathbf{E}^S_k$ stands for the k-th column in the matrix $\mathbb{E}^S$ from (11.2.16), give

$$\text{div}\,[\mathbf{E}^S_k] = \lim_{\lambda \to \infty} \Big(\text{div}\,[\mathbf{E}^L_k] \Big) = 0 \quad \text{in } \mathcal{S}'(\mathbb{R}^n), \quad \forall k \in \{1,\dots,n\}. \tag{11.2.18}$$

A quick inspection of (11.2.16) and (11.2.12), shows that the entries in the matrix $\mathbb{E}^S$ and the components of the vector $\mathbf{p} := (p_1, \dots, p_n)$ belong to $L^1_{loc}(\mathbb{R}^n)$, while (11.2.17) guarantees that

$$\begin{cases} \Delta \mathbb{E}^S - \nabla \mathbf{p} = \delta_0 I_{n \times n}, & \text{in } \mathcal{M}_{n \times n}(\mathcal{S}'(\mathbb{R}^n)), \\ \text{div}\,\mathbb{E}^S = 0, & \text{in } [\mathcal{S}'(\mathbb{R}^n)]^n. \end{cases} \tag{11.2.19}$$

Recalling now the definition of the Stokes operator from (10.6.1), we may conclude that

$$(\mathbb{E}^S, \mathbf{p}) \text{ is a fundamental solution for the Stokes operator in } \mathbb{R}^n. \tag{11.2.20}$$

Note that the expressions we obtained for $(\mathbb{E}^S, \mathbf{p})$ as given in (11.2.16) and (11.2.12) are the same as the ones from (10.6.13)–(10.6.14) if $n \ge 3$ and as the one from (10.7.3) if $n = 2$.

11.3 Fundamental Solutions for Higher Order Systems

In this section, we discuss an approach for computing all fundamental solutions that are tempered distributions for a certain subclass of systems of the form 10.2.16. To specify this class, let us before $n \in \mathbb{N}$ denote the Euclidean dimension, fix $M, m \in \mathbb{N}$, and consider $M \times M$ systems of homogeneous differential operators of order $2m$ with complex constant coefficients that have the form

$$L = L(\partial) := \Big(\sum_{|\alpha|=2m} a_\alpha^{jk} \partial^\alpha \Big)_{1 \le j,k \le M} =: (L_{jk}(\partial))_{1 \le j,k \le M} \qquad (11.3.1)$$

where $a_\alpha^{jk} \in \mathbb{C}$, $j, k \in \{1, ..., M\}$. Define the characteristic matrix of L to be

$$L(\xi) := \Big(\sum_{|\alpha|=2m} a_\alpha^{jk} \xi^\alpha \Big)_{1 \le j,k \le M}, \qquad \forall \xi \in \mathbb{R}^n. \qquad (11.3.2)$$

The standing assumption under which we will identify all fundamental solutions for L that are tempered distributions is

$$\det [L(\xi)] \neq 0, \qquad \forall \xi \in \mathbb{R}^n \setminus \{0\}. \qquad (11.3.3)$$

Several types of operators discussed earlier fall under the scope of these specifications, including the class of strictly elliptic, homogeneous, second order, and constant coefficient operators from Section 7.12, the polyharmonic operator from Section 7.5, and the Lamé operator from Section 10.3. Also recall that, as seen in Corollary 10.10, assumption (11.3.3) guarantees that L has a fundamental solution that is a tempered distribution.

For any operator L as in (11.3.1), its transpose and complex conjugate are, respectively, given by

$$L^\top := \Big(\sum_{|\alpha|=2m} a_\alpha^{jk} \partial^\alpha \Big)_{1 \le k,j \le M}, \qquad \overline{L} = \Big(\sum_{|\alpha|=2m} \overline{a_\alpha^{jk}} \partial^\alpha \Big)_{1 \le j,k \le M}, \qquad (11.3.4)$$

while the formal adjoint of L is defined as

$$L^* := \overline{L}^\top. \qquad (11.3.5)$$

The main result in this section is Theorem 11.1 below, describing the nature and properties of all fundamental solutions for a higher order system L as in (11.3.1)–(11.3.3) that are tempered distributions. Before presenting this basic result, we first describe a strategy that points to a natural candidate for a fundamental solution for L.

Suppose that Q is a scalar constant coefficient differential operator (of an auxiliary nature) which, from other sources of information, is known to have a fundamental solution in $\mathbb{R}^n$ of the form

$$E_Q(x) = \int_{S^{n-1}} F(x \cdot \xi) \, d\sigma(\xi), \qquad (11.3.6)$$

where F is a sufficiently regular scalar-valued function on the real line. Granted this, we then proceed to define

$$E_L(x) := Q\Big[\int_{S^{n-1}} G(x \cdot \xi)[L(\xi)]^{-1} \, d\sigma(\xi)\Big], \tag{11.3.7}$$

where G is a scalar-valued function on the real line, chosen in such a way that

$$p(t) := G^{(2m)}(t) - F(t) \text{ is a polynomial in } t \in \mathbb{R} \text{ of degree} < \text{order } Q. \tag{11.3.8}$$

Then

$$P(x) := \int_{S^{n-1}} p(x \cdot \xi) \, d\sigma(\xi), \qquad x \in \mathbb{R}^n, \tag{11.3.9}$$

is a polynomial in $\mathbb{R}^n$ of degree $<$ order Q. Keeping this in mind and observing that

$$LQ = QL \quad \text{and} \quad L[G(x \cdot \xi)] = G^{(2m)}(x \cdot \xi)L(\xi), \tag{11.3.10}$$

we may compute

$$\begin{aligned}
L[E_L(x)] &= Q\Big[\int_{S^{n-1}} G^{(2m)}(x \cdot \xi)L(\xi)[L(\xi)]^{-1} \, d\sigma(\xi)\Big] \\
&= Q\Big[\int_{S^{n-1}} F(x \cdot \xi) \, d\sigma(\xi) + P(x)\Big] I_{M\times M} \\
&= Q[E_Q(x)]I_{M\times M} + (QP)(x)I_{M\times M} = \delta I_{M\times M} + 0\,I_{M\times M} \\
&= \delta I_{M\times M}, \tag{11.3.11}
\end{aligned}$$

which shows that E_L is a fundamental solution for the system L in $\mathbb{R}^n$.

To implement the procedure just described, it is natural to take the polyharmonic operator Δ^N to play the role of the auxiliary scalar differential operator Q. This is not just a matter of convenience, but from Lemma 7.31 we already know that if $N \in \mathbb{N}$ satisfies $N \geq n/2$ then there exists a fundamental solution for this operator that has the form requested in (11.3.6). Specifically, in this case we have (compare with (7.5.24))

$$F(t) := -\frac{1}{(2\pi i)^n(2N - n)!}\, t^{2N-n} \log(t/i). \tag{11.3.12}$$

To find a function G on the real line that satisfies (11.3.8), we may try

$$G(t) := a\, t^A \log(t/i), \tag{11.3.13}$$

where $a \in \mathbb{R}$, $A \in \mathbb{N}$ with $A \geq 2m$, and note there exists a constant $C_{A,a,m}$ such that

$$\Big(\frac{d}{dt}\Big)^{2m}[G(t)] = a\frac{A!}{(A - 2m)!}\, t^{A-2m} \log(t/i) + C_{A,a,m}\, t^{A-2m}, \tag{11.3.14}$$

which means that (11.3.8) is going to hold if we take

$$A := 2N - n + 2m \quad \text{and} \quad a := -\frac{1}{(2\pi i)^n(2N - n + 2m)!}, \tag{11.3.15}$$

for a polynomial p on $\mathbb{R}$ of degree $2N - n$, which is strictly less than the order of $Q = \Delta^N$.

To bring matters more in line with notation employed in the proof of Theorem 11.1, pick now a number $q \in \mathbb{N}_0$ with the same parity as n, and choose $N := (n + q)/2$ (which satisfies $N \in \mathbb{N}$ and $N \geq n/2$). For this choice of N,

$$\text{the operator } Q \text{ becomes } \Delta^{\frac{n+q}{2}}, \tag{11.3.16}$$

and if A, a are as in (11.3.15) the function G from (11.3.13) takes the form

$$G(t) = -\frac{1}{(2\pi i)^n (2m + q)!} \, t^{2m+q} \log(t/i). \tag{11.3.17}$$

For these specifications, the function E_L from (11.3.7) then becomes precisely the function

$$\mathbb{E}(x) := \frac{-\Delta_x^{(n+q)/2}}{(2\pi i)^n (2m + q)!} \int_{S^{n-1}} (x \cdot \xi)^{2m+q} \log\left(\frac{x \cdot \xi}{i}\right) [L(\xi)]^{-1} \, d\sigma(\xi). \tag{11.3.18}$$

Having explained the genesis of this formula, we now proceed to rigorously show that not only is this candidate for a fundamental solution natural but it also does job.

Theorem 11.1. *Fix $n, m, M \in \mathbb{N}$ with $n \geq 2$, and assume L is an $M \times M$ system in $\mathbb{R}^n$ of order $2m$ as in (11.3.1)–(11.3.3). Then the $M \times M$ matrix $\mathbb{E}$ defined at each $x \in \mathbb{R}^n \setminus \{0\}$ by*

$$\mathbb{E}(x) := \frac{\Delta_x^{(n-1)/2}}{4(2\pi i)^{n-1}(2m - 1)!} \int_{S^{n-1}} (x \cdot \xi)^{2m-1} \operatorname{sgn}(x \cdot \xi) [L(\xi)]^{-1} \, d\sigma(\xi) \tag{11.3.19}$$

if n is odd, and

$$\mathbb{E}(x) := \frac{-\Delta_x^{n/2}}{(2\pi i)^n (2m)!} \int_{S^{n-1}} (x \cdot \xi)^{2m} \ln|x \cdot \xi| [L(\xi)]^{-1} \, d\sigma(\xi) \tag{11.3.20}$$

if n is even, satisfies the following properties.

(1) One has

$$\mathbb{E} \in \mathcal{M}_{M \times M}(\mathcal{S}'(\mathbb{R}^n)) \cap \mathcal{M}_{M \times M}(C^\infty(\mathbb{R}^n \setminus \{0\}))$$

$$\cap \, \mathcal{M}_{M \times M}(L^1_{loc}(\mathbb{R}^n)) \tag{11.3.21}$$

and

$$\mathbb{E}(-x) = \mathbb{E}(x) \quad \text{for all} \quad x \in \mathbb{R}^n \setminus \{0\}. \tag{11.3.22}$$

Moreover, the entries in $\mathbb{E}$ are real-analytic functions in $\mathbb{R}^n \setminus \{0\}$.
(2) If $I_{M \times M}$ is the $M \times M$ identity matrix, then for each $y \in \mathbb{R}^n$ one has

$$L^x[\mathbb{E}(x - y)] = \delta_y(x) I_{M \times M} \quad \text{in} \quad \mathcal{M}_{M \times M}(\mathcal{S}'(\mathbb{R}^n)), \tag{11.3.23}$$

*where the superscript x denotes the fact that the operator L in (11.3.23) is applied
to each column of $\mathbb{E}$ in the variable x.*

(3) Define the $M \times M$ matrix-valued function

$$\mathbb{P}(x) := \frac{-1}{(2\pi i)^n (2m-n)!} \int_{S^{n-1}} (x \cdot \xi)^{2m-n} [L(\xi)]^{-1} \, d\sigma(\xi), \quad \forall x \in \mathbb{R}^n. \quad (11.3.24)$$

*Then, the entries of $\mathbb{P}$ are identically zero when either n is odd or $n > 2m$, and are
homogeneous polynomials of degree $2m - n$ when $n \leq 2m$. Moreover, there exists
a function $\Phi \in M_{M \times M}(C^\infty(\mathbb{R}^n \setminus \{0\}))$ that is positive homogeneous of degree
$2m - n$ such that*

$$\mathbb{E}(x) = \Phi(x) + (\ln|x|)\mathbb{P}(x), \qquad \forall x \in \mathbb{R}^n \setminus \{0\}. \quad (11.3.25)$$

As a consequence,

> *if either n is odd or $n > 2m$ then $\mathbb{E}$ is positive homoge-
> neous of degree $2m - n$ in $\mathbb{R}^n \setminus \{0\}$.* $\quad (11.3.26)$

*(4) For each $\beta \in \mathbb{N}_0^n$ with $|\beta| \geq 2m - 1$, the restriction to $\mathbb{R}^n \setminus \{0\}$ of the matrix
distribution $\partial^\beta \mathbb{E}$ is of class C^∞ and positive homogeneous of degree $2m - n - |\beta|$.*
(5) For each $\beta \in \mathbb{N}_0^n$ there exists $C_\beta \in (0, \infty)$ such that the estimate

$$|\partial^\beta \mathbb{E}(x)| \leq \begin{cases} \dfrac{C_\beta}{|x|^{n-2m+|\beta|}} & \text{if either n is odd, or } n > 2m, \text{ or } |\beta| > 2m - n, \\[4mm] \dfrac{C_\beta(1 + |\ln|x||)}{|x|^{n-2m+|\beta|}} & \text{if } 0 \leq |\beta| \leq 2m - n, \end{cases}$$

$$(11.3.27)$$

holds for each $x \in \mathbb{R}^n \setminus \{0\}$.
*(6) When restricted to $\mathbb{R}^n \setminus \{0\}$, the entries of $\widehat{\mathbb{E}}$ (with "hat" denoting the Fourier
transform) are C^∞ functions and, moreover,*

$$\widehat{\mathbb{E}}(\xi) = (-1)^m [L(\xi)]^{-1} \quad \text{for each} \quad \xi \in \mathbb{R}^n \setminus \{0\}. \quad (11.3.28)$$

In addition,

> *if $n > 2m$ then $\widehat{\mathbb{E}} = (-1)^m [L(\cdot)]^{-1}$*
> *as tempered distributions in $\mathbb{R}^n$,* $\quad (11.3.29)$

which also implies

> *if $n > 2m$ then $\mathbb{E} = (-1)^m (2\pi)^{-n} \widehat{[L(\cdot)]^{-1}}$*
> *as tempered distributions in $\mathbb{R}^n$.* $\quad (11.3.30)$

*A more general version of (11.3.28) is as follows: given any $\gamma \in \mathbb{N}_0^n$, the tempered
distribution $\widehat{\partial^\gamma \mathbb{E}}$ is of class $\mathscr{C}^\infty$ when restricted to $\mathbb{R}^n \setminus \{0\}$ and, regarded as such,
satisfies*

$$\widehat{\partial^\gamma \mathbb{E}}(\xi) = (-1)^m i^{|\gamma|} \xi^\gamma [L(\xi)]^{-1} \ \text{for every} \ \xi \in \mathbb{R}^n \setminus \{0\}. \tag{11.3.31}$$

(7) Writing $\mathbb{E}_L$ in place of $\mathbb{E}$ to emphasize the dependence on L, the fundamental solution $\mathbb{E}_L$ with entries as in (11.3.19)–(11.3.20) satisfies

$$(\mathbb{E}_L)^\top = \mathbb{E}_{L^\top}, \quad \overline{\mathbb{E}_L} = \mathbb{E}_{\overline{L}}, \quad (\mathbb{E}_L)^* = \mathbb{E}_{L^*},$$
$$\text{and} \quad \mathbb{E}_{\lambda L} = \lambda^{-1} \mathbb{E}_L \quad \forall \lambda \in \mathbb{C} \setminus \{0\}. \tag{11.3.32}$$

As a consequence of this and (11.3.23), for each $y \in \mathbb{R}^n$ one also has

$$(L^x)^\top [\mathbb{E}^\top(x - y)] = \delta_y(x) I_{M \times M} \quad \text{in} \ \mathcal{M}_{M \times M}(\mathcal{S}'(\mathbb{R}^n)). \tag{11.3.33}$$

(8) Any fundamental solution $\mathbb{U} \in \mathcal{M}_{M \times M}(\mathcal{S}'(\mathbb{R}^n))$ of the system L in $\mathbb{R}^n$ is of the form $\mathbb{U} = \mathbb{E} + \mathbb{Q}$ where $\mathbb{E}$ is as in (11.3.19)–(11.3.20) and $\mathbb{Q}$ is an $M \times M$ matrix whose entries are polynomials in $\mathbb{R}^n$ and whose columns, $\mathbf{Q}_k$, $k \in \{1, \ldots, M\}$, satisfy the pointwise equations $L\mathbf{Q}_k = \mathbf{0} \in \mathbb{C}^M$ in $\mathbb{R}^n$ for each $k \in \{1, \ldots, M\}$.

Remark 11.2.(1) Let $m \in \mathbb{N}$ and $a \in \mathbb{R} \setminus \{0\}$. Since $a = |a| \operatorname{sgn} a$, it follows that $a^{2m-1} \operatorname{sgn} a = |a|^{2m-1} (\operatorname{sgn} a)^{2m} = |a|^{2m-1}$. This shows that $\mathbb{E}$ in (11.3.19) corresponding to $n \in \mathbb{N}$, n odd, may be written as

$$\mathbb{E}(x) = \frac{\Delta_x^{(n-1)/2}}{4(2\pi i)^{n-1}(2m-1)!} \int_{S^{n-1}} |x \cdot \xi|^{2m-1} [L(\xi)]^{-1} \, d\sigma(\xi) \tag{11.3.34}$$

for each $x \in \mathbb{R}^n \setminus \{0\}$.

(2) For further reference, we single out the expressions of $\mathbb{E}$ from Theorem 11.1 corresponding to the case $m = 1$. Specifically, if $n, M \in \mathbb{N}$ with $n \geq 2$, and L is the $M \times M$ system of order 2 with complex constant coefficients in $\mathbb{R}^n$ of the form

$$L = L(\partial) := \left(\sum_{r,s \in \{1,\ldots,n\}} a_{rs}^{jk} \partial_r \partial_s \right)_{1 \leq j,k \leq M}, \tag{11.3.35}$$

where $a_{rs}^{jk} \in \mathbb{C}$, $j, k \in \{1, \ldots, M\}$, $r, s \in \{1, \ldots, n\}$, and L satisfies (11.3.3), then for each $x \in \mathbb{R}^n \setminus \{0\}$, the $M \times M$ matrix $\mathbb{E}$ from (11.3.19)–(11.3.20) has the expression

$$\mathbb{E}(x) = \frac{\Delta_x^{(n-1)/2}}{4(2\pi i)^{n-1}} \int_{S^{n-1}} |x \cdot \xi| [L(\xi)]^{-1} \, d\sigma(\xi) \tag{11.3.36}$$

if n is odd, and

$$\mathbb{E}(x) = \frac{-\Delta_x^{n/2}}{2(2\pi i)^n} \int_{S^{n-1}} (x \cdot \xi)^2 \ln |x \cdot \xi| [L(\xi)]^{-1} \, d\sigma(\xi) \tag{11.3.37}$$

if n is even.
Moreover, by making use of the observation that for all $x, \xi \in \mathbb{R}^n \setminus \{0\}$ we have

$$\Delta_x [(x \cdot \xi)^2 \ln [(x \cdot \xi)^2]] = 8|\xi|^2 + 4|\xi|^2 \ln |x \cdot \xi|, \tag{11.3.38}$$

one may absorb Δ_x inside the integral in formula (11.3.37) and obtain that if n is even then

$$\mathbb{E}(x) = c_n - \frac{\Delta_x^{(n-2)/2}}{(2\pi i)^n} \int_{S^{n-1}} \ln|x \cdot \xi| [L(\xi)]^{-1} \, d\sigma(\xi) \tag{11.3.39}$$

for each $x \in \mathbb{R}^n \setminus \{0\}$, where

$$c_n := \begin{cases} 0, & \text{if } n > 2, \\ \dfrac{1}{2\pi^2} \displaystyle\int_{S^1} [L(\xi)]^{-1} \, d\sigma(\xi), & \text{if } n = 2. \end{cases} \tag{11.3.40}$$

Proof of Theorem 11.1. To facilitate the subsequent discussion, we agree to denote by $(P^{jk}(\xi))_{1 \leq j,k \leq M}$ the inverse of the characteristic matrix $L(\xi)$, i.e.,

$$\left(P^{jk}(\xi)\right)_{1 \leq j,k \leq M} := [L(\xi)]^{-1}, \qquad \forall \, \xi \in \mathbb{R}^n \setminus \{0\}. \tag{11.3.41}$$

Then the entries $(E_{jk})_{1 \leq j,k \leq M}$ of the matrix $\mathbb{E}$ are given at each $x \in \mathbb{R}^n \setminus \{0\}$ by

$$E_{jk}(x) = \frac{\Delta_x^{(n-1)/2}}{4(2\pi i)^{n-1}(2m-1)!} \int_{S^{n-1}} (x \cdot \xi)^{2m-1} \operatorname{sgn}(x \cdot \xi) P^{jk}(\xi) \, d\sigma(\xi) \tag{11.3.42}$$

if n is odd, and

$$E_{jk}(x) = \frac{-\Delta_x^{n/2}}{(2\pi i)^n(2m)!} \int_{S^{n-1}} (x \cdot \xi)^{2m} \ln|x \cdot \xi| P^{jk}(\xi) \, d\sigma(\xi) \tag{11.3.43}$$

if n is even. The proof of the theorem is completed in eight steps.

Step I. *We claim that if we set* $q := 0$ *if* n *is even and* $q := 1$ *if* n *is odd, then*

$$E_{jk}(x) = \frac{-\Delta_x^{(n+q)/2}}{(2\pi i)^n(2m+q)!} \int_{S^{n-1}} (x \cdot \xi)^{2m+q} \log\left(\frac{x \cdot \xi}{i}\right) P^{jk}(\xi) \, d\sigma(\xi) \tag{11.3.44}$$

for $x \in \mathbb{R}^n \setminus \{0\}$ *and* $j, k \in \{1, \ldots, M\}$*, where* log *denotes the principal branch of the complex logarithm defined for points* $z \in \mathbb{C} \setminus \{x : x \in \mathbb{R}, \, x \leq 0\}$.

Suppose first that n is even, thus $q = 0$, and start from (11.3.43), then use the formula $\log\left(\frac{x \cdot \xi}{i}\right) = \ln|x \cdot \xi| - \frac{\pi i}{2} \operatorname{sgn}(x \cdot \xi)$ for the term under the integral sign and the fact that the integral over S^{n-1} of the function $(x \cdot \xi)^{2m} \operatorname{sgn}(x \cdot \xi) P^{jk}(\xi)$ (which is odd in ξ) is zero, to obtain that (11.3.44) holds.

Moving on to the case when n is odd, consider the function

$$F(t) := \begin{cases} t^{2m+q} \log(t/i) & \text{if } t \in \mathbb{R} \setminus \{0\}, \\ 0 & \text{if } t = 0. \end{cases} \tag{11.3.45}$$

It is not difficult to see that

$$F \in C^\infty(\mathbb{R} \setminus \{0\}) \cap C^{2m+q-1}(\mathbb{R}), \quad F^{(k)}(0) = 0 \quad \text{if } k = 1, \ldots, 2m+q-1, \quad (11.3.46)$$

and that for each $k \in \{1, \ldots, 2m\}$ there exists a constant $C_{m,q,k}$ such that

$$\Big(\frac{d}{dt}\Big)^k F(t) = \frac{(2m+q)!}{(2m+q-k)!} t^{2m+q-k} \log\frac{t}{i} + C_{m,q,k} \, t^{2m+q-k} \quad (11.3.47)$$

in $\mathbb{R} \setminus \{0\}$. In particular, by the Chain Rule, if $\beta \in \mathbb{N}_0^n$ is such that $|\beta| \leq 2m+q-1$, then

$$\partial_x^\beta [F(x \cdot \xi)] = (F^{(|\beta|)})(x \cdot \xi)\xi^\beta, \quad \forall x \in \mathbb{R}^n, \ \forall \xi \in S^{n-1}. \quad (11.3.48)$$

Based on (11.3.48) and (11.3.47) for each $x \in \mathbb{R}^n \setminus \{0\}$ we may write

$$\Delta_x \int_{S^{n-1}} (x \cdot \xi)^{2m+1} \log\Big(\frac{x \cdot \xi}{i}\Big) P^{jk}(\xi) \, d\sigma(\xi)$$

$$= \Delta_x \int_{S^{n-1}} F(x \cdot \xi) P^{jk}(\xi) \, d\sigma(\xi) = \int_{S^{n-1}} F''(x \cdot \xi) P^{jk}(\xi) \, d\sigma(\xi)$$

$$= 2m(2m+1) \int_{S^{n-1}} (x \cdot \xi)^{2m-1} \log\Big(\frac{x \cdot \xi}{i}\Big) P^{jk}(\xi) \, d\sigma(\xi)$$

$$+ C_{m,q} \int_{S^{n-1}} (x \cdot \xi)^{2m-1} P^{jk}(\xi) \, d\sigma(\xi)$$

$$= 2m(2m+1) \int_{S^{n-1}} (x \cdot \xi)^{2m-1} \Big[\ln|x \cdot \xi| - \tfrac{\pi i}{2}\mathrm{sgn}(x \cdot \xi)\Big] P^{jk}(\xi) \, d\sigma(\xi)$$

$$= -\pi i m(2m+1) \int_{S^{n-1}} (x \cdot \xi)^{2m-1} \mathrm{sgn}(x \cdot \xi) P^{jk}(\xi) \, d\sigma(\xi). \quad (11.3.49)$$

Hence, if one starts with the expression in the right-hand side of (11.3.44), then one transfers Δ under the integral sign using from (11.3.49), one arrives at the expression of E_{jk} from (11.3.42). This completes the proof of Step I.

Step II. *Proof of the fact that the entries of $\mathbb{E}$ are C^∞ and even in $\mathbb{R}^n \setminus \{0\}$.*
That the functions in (11.3.42) and (11.3.43) are even is immediate from their respective expressions. To show that the components of $\mathbb{E}$ belong to the space $C^\infty(\mathbb{R}^n \setminus \{0\})$, for every $\ell \in \{1, \ldots, n\}$ let $\mathbf{e}_\ell$ be the unit vector in $\mathbb{R}^n$ with one on the ℓ-th component, and consider the open set

$$O_\ell := \mathbb{R}^n \setminus \{\lambda \mathbf{e}_\ell : \lambda \leq 0\} = \{x = (x_1, \ldots, x_n) \in \mathbb{R}^n : x_\ell > 0\}. \quad (11.3.50)$$

Fix an arbitrary index $\ell \in \{1, \ldots, n\}$ and for each $x = (x_1, \ldots, x_n) \in O_\ell$ define the linear map $R_{\ell,x} : \mathbb{R}^n \to \mathbb{R}^n$ by

$$R_{\ell,x}(\xi) := \xi - \frac{\xi \cdot x + \xi_\ell |x|}{|x| + x_\ell} \mathbf{e}_\ell + \frac{\xi_\ell(|x| + 2x_\ell) - \xi \cdot x}{|x|(|x| + x_\ell)} x, \quad \forall \xi \in \mathbb{R}^n. \quad (11.3.51)$$

By Exercise 4.130 (presently used with $\zeta := \mathbf{e}_\ell$ and $\eta := \frac{x}{|x|}$) we have that this is a unitary transformation and

$$x \cdot R_{\ell,x}(\xi) = |x|\xi_\ell, \qquad \forall\, x \in O_\ell, \quad \forall\, \xi \in \mathbb{R}^n. \tag{11.3.52}$$

Also,

$$R_{\ell,\lambda x} = R_{\ell,x} \quad \text{whenever} \quad x \in O_\ell \quad \text{and} \quad \lambda > 0, \tag{11.3.53}$$

and the joint application

$$O_\ell \times \mathbb{R}^n \ni (x,\xi) \mapsto R_{\ell,x}(\xi) \in \mathbb{R}^n \qquad \text{is of class } C^\infty. \tag{11.3.54}$$

Fix $j, k \in \{1, \ldots, M\}$. Using the invariance under unitary transformations of the operation of integration over S^{n-1}, for each $x \in O_\ell$ we may then write

$$\int_{S^{n-1}} \left[(x \cdot \xi)^{2m+q} \cdot \log\left(\frac{x \cdot \xi}{i}\right) \right] P^{jk}(\xi)\, d\sigma(\xi)$$

$$= \int_{S^{n-1}} \left[(x \cdot R_{\ell,x}(\xi))^{2m+q} \cdot \log\left(\frac{x \cdot R_{\ell,x}(\xi)}{i}\right) \right] P^{jk}(R_{\ell,x}(\xi))\, d\sigma(\xi)$$

$$= \int_{S^{n-1}} \left[(|x|\xi_\ell)^{2m+q} \cdot \log\left(\frac{|x|\xi_\ell}{i}\right) \right] P^{jk}(R_{\ell,x}(\xi))\, d\sigma(\xi)$$

$$= |x|^{2m+q} \int_{S^{n-1}} \xi_\ell^{2m+q} \left\{ \ln|x| + \log\left(\frac{\xi_\ell}{i}\right) \right\} P^{jk}(R_{\ell,x}(\xi))\, d\sigma(\xi). \tag{11.3.55}$$

From this representation, (11.3.54), and (11.3.44), it is clear that $E_{jk} \in C^\infty(O_\ell)$.

Next, for each $\ell \in \{1, \ldots, n\}$ define the open set

$$O_\ell^- := \mathbb{R}^n \setminus \{\lambda\, \mathbf{e}_\ell : \lambda \geq 0\} = \{x = (x_1, \ldots, x_n) \in \mathbb{R}^n : x_\ell < 0\} \tag{11.3.56}$$

and for each given $x \in O_\ell^-$ define the linear map $R_{\ell,x}^- : \mathbb{R}^n \to \mathbb{R}^n$ by

$$R_{\ell,x}^-(\xi) := \xi + \frac{\xi \cdot x - \xi_\ell|x|}{|x| - x_\ell}\, \mathbf{e}_\ell - \frac{\xi_\ell(|x| - 2x_\ell) + \xi \cdot x}{|x|(|x| - x_\ell)}\, x, \qquad \forall\, \xi \in \mathbb{R}^n. \tag{11.3.57}$$

Running the reasoning above with O_ℓ replaced by O_ℓ^- and $R_{\ell,x}$ replaced by $R_{\ell,x}^-$ (this time invoking Exercise 4.130 with $\zeta := \mathbf{e}_\ell$ and $\eta := -\frac{x}{|x|}$, and with identity (11.3.52) replaced by $x \cdot R_{\ell,x}^-(\xi) = -|x|\xi_\ell$ for all $x \in O_\ell^-$ and all $\xi \in \mathbb{R}^n$) ultimately implies $E_{jk} \in C^\infty(O_\ell^-)$ for each $j, k \in \{1, \ldots, M\}$ and each $\ell \in \{1, \ldots, n\}$.

To complete Step II, there remains to observe that $\mathbb{R}^n \setminus \{0\} = \bigcup_{\ell=1}^n (O_\ell \cup O_\ell^-)$.

Step III. *Proof of part (3) in the statement of the theorem.*
To facilitate the discussion, fix $j, k \in \{1, \ldots, M\}$, introduce

$$Q_{jk}(x) := \int_{S^{n-1}} (x \cdot \xi)^{2m+q} P^{jk}(\xi)\, d\sigma(\xi), \qquad \forall\, x \in \mathbb{R}^n, \tag{11.3.58}$$

and define $\Psi_{jk} : \mathbb{R}^n \setminus \{0\} \to \mathbb{C}$ by setting

$$\Psi_{jk}(x) := \int_{S^{n-1}} \left[(x \cdot \xi)^{2m+q} \cdot \log\left(\frac{x \cdot \xi}{i}\right) \right] P^{jk}(\xi) \, d\sigma(\xi) - \left(\ln|x| \right) Q_{jk}(x) \quad (11.3.59)$$

for each $\mathbb{R}^n \setminus \{0\}$. Observe that Q_{jk} is a polynomial of degree $2m + q$ that vanishes when n is odd (since in that case the integrand in (11.3.58) is odd). Also, from our earlier analysis in (11.3.55), we know that the integral in the right-hand side of (11.3.59) depends in a C^∞ fashion on the variable $x \in \mathbb{R}^n \setminus \{0\}$. These comments and (11.3.59) imply that Ψ_{jk} is of class C^∞ in $\mathbb{R}^n \setminus \{0\}$.

Our next goal is to prove that

$$\Psi_{jk} \text{ is positive homogeneous of degree } 2m + q \text{ in } \mathbb{R}^n \setminus \{0\}. \quad (11.3.60)$$

To this end, first note that for each $\ell \in \{1, \ldots, n\}$ we may write

$$Q_{jk}(x) = |x|^{2m+q} \int_{S^{n-1}} \xi_\ell^{2m+q} P^{jk}(R_{\ell,x}(\xi)) \, d\sigma(\xi), \qquad \forall \, x \in O_\ell, \quad (11.3.61)$$

by (11.3.58) and (11.3.52). Consequently, from (11.3.61), (11.3.55), and (11.3.59), we deduce that for each $\ell \in \{1, \ldots, n\}$,

$$\Psi_{jk}(x) := |x|^{2m+q} \int_{S^{n-1}} \xi_\ell^{2m+q} \log\left(\frac{\xi_\ell}{i}\right) P^{jk}(R_{\ell,x}(\xi)) \, d\sigma(\xi), \quad (11.3.62)$$

for all $x \in O_\ell$. In turn, (11.3.62) and (11.3.53) readily show that Ψ_{jk} is positive homogeneous of degree $2m + q$ when restricted to the cone-like region O_ℓ. Since $\mathbb{R}^n \setminus \{0\} = \bigcup_{\ell=1}^n O_\ell$, the claim in (11.3.60) follows.

Next, an induction argument shows that for each $\xi \in \mathbb{R}^n \setminus \{0\}$ and $k, N \in \mathbb{N}$ satisfying $N \geq 2k$, the following formulas hold:

$$\Delta_x^k[(x \cdot \xi)^N] = \frac{N!}{(N-2k)!}(x \cdot \xi)^{N-2k}|\xi|^{2k} \quad (11.3.63)$$

and

$$\Delta_x^k[(x \cdot \xi)^N \ln|x|] = (\ln|x|)\Delta_x^k[(x \cdot \xi)^N] \quad (11.3.64)$$

$$+ \sum_{r=1}^k c(r, k, N, n)|x|^{-2r}(x \cdot \xi)^{N-2k+2r}.$$

We now observe that if P_{jk} is the (j, k)-entry in the matrix $\mathbb{P}$ from (11.3.24), then

$$P_{jk}(x) = \frac{-1}{(2\pi i)^n (2m + q)!} \Delta_x^{(n+q)/2} \int_{S^{n-1}} (x \cdot \xi)^{2m+q} P^{jk}(\xi) \, d\sigma(\xi), \quad (11.3.65)$$

as seen from identity (11.3.63) with $N := 2m + q$ and $k := (n + q)/2$. It is also immediate from (11.3.65) that P_{jk} is identically zero when either n is odd (due to parity considerations) or $n > 2m$ (due to degree considerations). In addition, formula (11.3.65) shows that P_{jk} is a homogeneous polynomial of degree $2m - n$ when $n \leq 2m$. Finally, we note that combining (11.3.58), (11.3.65), and (11.3.64) used with $N := 2m + q$ and $k := (n + q)/2$, yields

$$\frac{-1}{(2\pi i)^n (2m + q)!} \Delta_x^{(n+q)/2} [(\ln |x|) Q_{jk}(x)] = (\ln |x|) P_{jk}(x) \tag{11.3.66}$$

$$+ \sum_{r=1}^{(n+q)/2} C_r |x|^{-2r} \int_{S^{n-1}} (x \cdot \xi)^{2m-n+2r} P^{jk}(\xi) \, d\sigma(\xi), \quad \forall x \in \mathbb{R}^n \setminus \{0\},$$

for some constants C_r depending only on r, n, q, and m. It is easy to see that the sum in the right-hand side of (11.3.66) gives rise to a function that belongs to $C^\infty(\mathbb{R}^n \setminus \{0\})$ and is positive homogeneous of degree $2m - n$. At this point, if we define

$$\Phi_{jk}(x) := \frac{-1}{(2\pi i)^n (2m + q)!} \Delta_x^{(n+q)/2} \Psi_{jk}(x) \tag{11.3.67}$$

$$+ \sum_{r=1}^{(n+q)/2} C_r |x|^{-2r} \int_{S^{n-1}} (x \cdot \xi)^{2m-n+2r} P^{jk}(\xi) \, d\sigma(\xi)$$

for each $x \in \mathbb{R}^n \setminus \{0\}$, then (11.3.25) follows from (11.3.44), (11.3.59), (11.3.66), and (11.3.67). Moreover, from (11.3.60) and (11.3.67), it is clear that Φ_{jk} is positive homogeneous of degree $2m - n$, while the regularity of Ψ_{jk} established earlier entails $\Phi_{jk} \in C^\infty(\mathbb{R}^n \setminus \{0\})$.

Step IV. *Proof of the fact that* $\mathbb{E} \in M_{M \times M}(\mathcal{S}'(\mathbb{R}^n)) \cap M_{M \times M}(L^1_{loc}(\mathbb{R}^n \setminus \{0\}))$.
Fix $j, k \in \{1, \ldots, M\}$ and recall (11.3.25). By what we proved in Step III, Φ_{jk} is positive homogeneous of degree $2m - n$ in $\mathbb{R}^n \setminus \{0\}$, and since $m \geq 1$ we may invoke Exercise 4.53 to conclude that $\Phi_{jk} \in L^1_{loc}(\mathbb{R}^n)$. Now the estimate from Exercise 4.53 and Example 4.4 give $\Phi_{jk} \in \mathcal{S}'(\mathbb{R}^n)$. In addition, by Exercise 4.18, it follows that $(\ln |x|) P_{jk} \in \mathcal{S}'(\mathbb{R}^n) \cap L^1_{loc}(\mathbb{R}^n \setminus \{0\})$. In summary, we conclude that $E_{jk} \in \mathcal{S}'(\mathbb{R}^n) \cap L^1_{loc}(\mathbb{R}^n \setminus \{0\})$.

Step V. *Proof of* (11.3.23).
Componentwise, (11.3.23) reads as follows: for each $j, k \in \{1, \ldots, M\}$,

$$\sum_{r=1}^M L_{jr}^x [E_{rk}(x - y)] = \begin{cases} 0 & \text{if } j \neq k, \\ \delta_y(x) & \text{if } j = k, \end{cases} \quad \text{in } \mathcal{S}'(\mathbb{R}^n), \tag{11.3.68}$$

where the superscript x denotes the fact that the operator L_{jr} (defined as in (11.3.1)) is applied in the variable x. To justify this, fix $j, k, r \in \{1, \ldots, M\}$. By (11.3.47) and (11.3.48) we have

$$L_{jr}^x F(x \cdot \xi) = \Big[\frac{(2m+q)!}{q!} (x \cdot \xi)^q \log \frac{x \cdot \xi}{i} + C_{m,q} (x \cdot \xi)^q \Big] L_{jr}(\xi) \qquad (11.3.69)$$

for every $x \in \mathbb{R}^n \setminus \{0\}$ and every $\xi \in S^{n-1}$ such that $x \cdot \xi \neq 0$. To continue, fix $\varphi \in C_0^\infty(\mathbb{R}^n)$ and write

$$\Big\langle L_{jr}^x \Big[\int_{S^{n-1}} F(x \cdot \xi) P^{rk}(\xi) \, d\sigma(\xi) \Big], \varphi(x) \Big\rangle$$

$$= \Big\langle \int_{S^{n-1}} F(x \cdot \xi) P^{rk}(\xi) \, d\sigma(\xi), L_{jr}\varphi(x) \Big\rangle$$

$$= \int_{\mathbb{R}^n} \int_{S^{n-1}} F(x \cdot \xi) P^{rk}(\xi) \, d\sigma(\xi) L_{jr}\varphi(x) \, dx$$

$$= \int_{S^{n-1}} P^{rk}(\xi) \int_{x \in \mathbb{R}^n, \, x \cdot \xi > 0} F(x \cdot \xi) L_{jr}\varphi(x) \, dx \, d\sigma(\xi)$$

$$+ \int_{S^{n-1}} P^{rk}(\xi) \int_{x \in \mathbb{R}^n, \, x \cdot \xi < 0} F(x \cdot \xi) L_{jr}\varphi(x) \, dx \, d\sigma(\xi). \qquad (11.3.70)$$

At this point, we integrate by parts repeatedly with respect to x (according to formula (14.8.4)) in the inner most integrals in (11.3.70), until we transfer all the derivatives from φ onto $F(x \cdot \xi)$. This is justified by (11.3.46), the fact that

$$\partial_x^\alpha [F(x \cdot \xi)] \in L_{loc}^1(\mathbb{R}^n) \quad \text{whenever } \alpha \in \mathbb{N}_0^n \text{ has } |\alpha| \leq 2m \qquad (11.3.71)$$

(cf. (11.3.48) and Exercise 2.125), and the fact that φ has compact support. Note that in the process, the terms corresponding to boundary integrals (i.e., integrals over the set $\{x \in \mathbb{R}^n : x \cdot \xi = 0\} \cap \mathrm{supp}\, \varphi$) are zero thanks to the formulas in (11.3.46). Hence, summing up over r the resulting identity in (11.3.70), then using (11.3.69) and the fact that $\sum_{r=1}^M L_{jr}(\xi) P^{rk}(\xi) = \delta_{jk}$ for every $\xi \in S^{n-1}$, we arrive at

$$\Big\langle \sum_{r=1}^M L_{jr}^x \Big[\int_{S^{n-1}} F(x \cdot \xi) P^{rk}(\xi) \, d\sigma(\xi) \Big], \varphi(x) \Big\rangle \qquad (11.3.72)$$

$$= \sum_{r=1}^M \int_{S^{n-1}} P^{rk}(\xi) \int_{x \in \mathbb{R}^n} [L_{jr}^x F(x \cdot \xi)] \varphi(x) \, dx \, d\sigma(\xi)$$

$$= \int_{\mathbb{R}^n} \int_{S^{n-1}} \Big[\frac{(2m+q)!}{q!} (x \cdot \xi)^q \log \frac{x \cdot \xi}{i} + C_{m,q} (x \cdot \xi)^q \Big] \delta_{jk} \, d\sigma(\xi) \varphi(x) \, dx.$$

Consequently, since $\varphi \in C_0^\infty(\mathbb{R}^n)$ is arbitrary, a combination of (11.3.72), (11.3.45), and (11.3.44), yields

$$\sum_{r=1}^{M} L_{jr} E_{rk} = \Delta^{(n+q)/2} \big(E_q + P_q \big) \delta_{jk} \quad \text{in} \quad \mathcal{D}'(\mathbb{R}^n), \tag{11.3.73}$$

where

$$E_q(x) := \frac{-1}{(2\pi i)^n q!} \int_{S^{n-1}} (x \cdot \xi)^q \cdot \log\Big(\frac{x \cdot \xi}{i}\Big) \, d\sigma(\xi), \tag{11.3.74}$$

for each $x \in \mathbb{R}^n \setminus \{0\}$, and

$$P_q(x) := \frac{-C_{m,q}}{(2\pi i)^n (2m+q)!} \int_{S^{n-1}} (x \cdot \xi)^q \, d\sigma(\xi), \quad \forall x \in \mathbb{R}^n. \tag{11.3.75}$$

Given our choice of q, from Lemma 7.31 we have that E_q is a fundamental solution for $\Delta^{(n+q)/2}$, so $\Delta^{(n+q)/2} E_q = \delta$ in $\mathcal{D}'(\mathbb{R}^n)$. Also, P_q is a homogeneous polynomial of degree q in $\mathbb{R}^n$, thus $\Delta^{(n+q)/2} P_q = 0$ pointwise and in $\mathcal{D}'(\mathbb{R}^n)$. Therefore, (11.3.73) becomes

$$\sum_{r=1}^{M} L_{jr} E_{rk} = \delta_{jk} \delta \quad \text{in} \quad \mathcal{D}'(\mathbb{R}^n), \quad \forall j, k \in \{1, \dots, M\}. \tag{11.3.76}$$

The statement in part (2) of the theorem follows from (11.3.76), (4.1.33), and the result from Step IV.

Step VI. *Proof of claims in parts (4)–(7) in the statement of the theorem.*
The claim in part (4) follows from part (1), (11.3.25), the fact that each P_{jk} is a homogeneous polynomial of degree at most $2m - n$, and the observation that when computing $\partial^\beta E_{jk}$ with $|\beta| \geq 2m - 1$ at least one derivative falls on ln. The estimates in (11.3.27) are a direct consequence of (11.3.25). This takes care of parts (4) and (5) in the statement of the theorem.

Moving on to the proof of part (6), fix $j, k \in \{1, \dots, M\}$ and recall (11.3.25). From the proof in Step IV, we have that Φ_{jk} and $(\ln|x|)P_{jk}$ are tempered distributions in $\mathbb{R}^n$. Since $\Phi_{jk} \in C^\infty(\mathbb{R}^n \setminus \{0\})$ and is positive homogeneous of degree $2m - n$ in $\mathbb{R}^n \setminus \{0\}$, Proposition 4.60 implies that its Fourier transform coincides with a C^∞ function on $\mathbb{R}^n \setminus \{0\}$. To analyze the effect of taking the Fourier transform of $(\ln|x|)P_{jk}$ pick some $\theta \in C_0^\infty(\mathbb{R}^n)$ such that $\operatorname{supp}\theta \subset B(0, 2)$ and $\theta \equiv 1$ on $B(0, 1)$ and write

$$(\ln|x|)P_{jk} = (1 - \theta)(\ln|x|)P_{jk} + \theta(\ln|x|)P_{jk} \quad \text{in} \quad \mathbb{R}^n \setminus \{0\}. \tag{11.3.77}$$

The two terms in the right-hand side of (11.3.77) continue to belong to $\mathcal{S}'(\mathbb{R}^n)$. From Example 2.9 and the fact that θ is compactly supported, we obtain that the function $\theta(\ln|x|)P_{jk}$ belongs to $L^1(\mathbb{R}^n)$ and has compact support. Consequently, $\mathcal{F}(\theta(\ln|x|)P_{jk}) \in C^\infty(\mathbb{R}^n)$ (recall Exercise 3.31). Regarding $(1 - \theta)(\ln|x|)P_{jk}$, note that this function is of class C^∞ in $\mathbb{R}^n$. Also, for every $\beta \in \mathbb{N}_0^n$, the function $x^\beta \partial^\alpha [(1 - \theta)(\ln|x|)P_{jk}]$ belongs to $L^1(\mathbb{R}^n)$ provided $\alpha \in \mathbb{N}_0^n$ and $|\alpha|$ is sufficiently large. Since the Fourier transform of any L^1 function is continuous, this readily implies that for any $r \in \mathbb{N}$ there exists $\alpha \in \mathbb{N}_0^n$ such that

$$\mathcal{F}(\partial^\alpha[(1-\theta)(\ln|x|)P_{jk}]) = (\mathrm{i}\xi)^\alpha \mathcal{F}((1-\theta)(\ln|x|)P_{jk}) \in C^r(\mathbb{R}^n). \qquad (11.3.78)$$

Thus, we necessarily have $\mathcal{F}((1-\theta)(\ln|x|)P_{jk}) \in C^\infty(\mathbb{R}^n \setminus \{0\})$.

The reasoning above shows that the Fourier transform of (the matrix-valued tempered distribution) $\mathbb{E}$ when restricted to $\mathbb{R}^n \setminus \{0\}$ is a function of class C^∞. Taking the Fourier transforms of both sides of (11.3.76) gives

$$(-1)^m \sum_{r=1}^M L_{jr}(\xi)\widehat{E_{rk}}(\xi) = \delta_{jk} \quad \text{in} \quad \mathcal{S}'(\mathbb{R}^n), \quad \text{for all} \ \ j,k \in \{1,\dots,M\}. \qquad (11.3.79)$$

Restricting (11.3.79) to $\mathbb{R}^n \setminus \{0\}$ then readily implies (11.3.28). Also, for each $\gamma \in \mathbb{N}_0^n$ we have that $\partial^\gamma \mathbb{E}$ is a tempered distribution and item (b) in Theorem 4.26 implies $\widehat{\partial^\gamma \mathbb{E}} = (-1)^m \mathrm{i}^{|\gamma|} \xi^\gamma \widehat{\mathbb{E}}$ in $\mathcal{S}'(\mathbb{R}^n)$. The latter combined with (11.3.28) then yields (11.3.31).

Consider next the task of justifying (11.3.29). The assumption that $n > 2m$ guarantees that the matrix-valued function $[L(\cdot)]^{-1}$ has entries that are locally integrable in $\mathbb{R}^n$ and satisfy (4.1.4). As remarked in Example 4.4, the distribution of function type defined by $[L(\cdot)]^{-1}$ is a well-defined tempered distribution in $\mathbb{R}^n$. Moreover, Exercise 4.57 implies that this tempered distribution is positive homogeneous of degree $-2m$. To proceed, define

$$u := \widehat{\mathbb{E}} - (-1)^m[L(\cdot)]^{-1}. \qquad (11.3.80)$$

From the above discussion, it follows that $u \in \mathcal{M}_{M\times M}(\mathcal{S}'(\mathbb{R}^n))$. In addition, since from (11.3.26) and Proposition 4.59 we know that $\widehat{\mathbb{E}} \in \mathcal{M}_{M\times M}(\mathcal{S}'(\mathbb{R}^n))$ is positive homogeneous of degree $-2m$, we conclude that u is also positive homogeneous of degree $-2m$. Observe next that $L(\xi) \in \mathcal{M}_{M\times M}(\mathcal{L}(\mathbb{R}^n))$ and

$$L(\xi)u = L(\xi)\widehat{\mathbb{E}} - (-1)^m L(\xi)[L(\cdot)]^{-1}$$

$$= (-1)^m\widehat{L\mathbb{E}} - (-1)^m I_{M\times M} = (-1)^m\widehat{\delta I_{M\times M}} - (-1)^m I_{M\times M}$$

$$= (-1)^m I_{M\times M} - (-1)^m I_{M\times M} = 0 \quad \text{in} \quad \mathcal{M}_{M\times M}(\mathcal{S}'(\mathbb{R}^n)); \qquad (11.3.81)$$

cf. item (b) in Theorem 4.26 and (4.2.4). As a consequence of (11.3.81) and (11.3.3), we deduce that $\operatorname{supp} u \subseteq \{0\}$. Granted this, we may invoke Exercise 2.75 to conclude that each entry in u is of the form $\sum_{|\alpha|\leq N} c_\alpha \partial^\alpha \delta$. Hence, on the one hand, $\widehat{u}$ has polynomial entries (again, see item (b) in Theorem 4.26 and (4.2.4)). On the other hand, Proposition 4.59 ensures that $\widehat{u} \in \mathcal{M}_{M\times M}(\mathcal{S}'(\mathbb{R}^n))$ is positive homogeneous of degree $-n+2m$. Since the current assumption forces $-n+2m < 0$, this implies $\widehat{u} = 0$ hence, ultimately, $u = 0$. Now (11.3.29) follows from (11.3.80). In turn, (11.3.30) is a consequence of (11.3.29), (4.2.34), and (11.3.22). This finishes the proof of part (6) in the statement of the theorem.

Finally, the identities in (11.3.32) can be seen directly from (11.3.42)–(11.3.43).

Step VII. *Proof of the claim in part (8) in the statement of the theorem.*
Let $\mathbb{U} \in M_{M \times M}(\mathcal{S}'(\mathbb{R}^n))$ be an arbitrary fundamental solution of the system L in $\mathbb{R}^n$
and set $\mathbb{Q} := \mathbb{U} - \mathbb{E}$. Then $L\mathbb{Q} = 0$ in $M_{M \times M}(\mathcal{S}'(\mathbb{R}^n))$ and, on the Fourier transform
side, $\mathbb{Q}$ satisfies $L(\xi)\widehat{\mathbb{Q}} = 0$ in $M_{M \times M}(\mathcal{S}'(\mathbb{R}^n))$. In light of (11.3.3), this implies
$\operatorname{supp} \widehat{\mathbb{Q}} \subseteq \{0\}$, hence $\mathbb{Q}$ is an $M \times M$ matrix whose entries are polynomials in $\mathbb{R}^n$ by
Exercise 4.37 (applied to each entry).

Step VIII. *Proof of the fact that the entries of $\mathbb{E}$ are real-analytic in $\mathbb{R}^n \setminus \{0\}$.*
This is a direct consequence of the fact that (cf. (11.3.23))

$$L\mathbb{E} = \delta I_{M \times M} \quad \text{in} \quad M_{M \times M}(\mathcal{S}'(\mathbb{R}^n)), \tag{11.3.82}$$

where the operator L is applied to each column of $\mathbb{E}$, which implies that $L\mathbb{E} = 0$ in
$M_{M \times M}(\mathcal{D}'(\mathbb{R}^n \setminus \{0\}))$, and Theorem 11.7, established in the next section.
    The proof of Theorem 11.1 is now complete. $\square$

Theorem 11.1 describes all fundamental solutions, which are tempered distributions, for any homogeneous constant coefficient system with an invertible characteristic matrix, and elaborates on the properties of such fundamental solutions.
In specific cases, it is possible to use formulas (11.3.19)–(11.3.20) to find an explicit expression for a specific fundamental solution. The case of the polyharmonic
operator is discussed in Exercise 11.24. Here we study in detail the case of the three-
dimensional Lamé operator (cf. also Exercise 11.20 and Exercise 11.21). We remark
that the argument in the proof of Proposition 11.3 is different from the one used to
derive the fundamental solution for the Lamé operator in Section 11.1.

Proposition 11.3. *Let $\lambda, \mu \in \mathbb{C}$ be such that $\mu \neq 0$, $\lambda + 2\mu \neq 0$. A fundamental
solution for the Lamé operator (10.3.1) in $\mathbb{R}^3$ is*

$$E(x) = -\frac{1}{8\pi} \frac{\lambda + 3\mu}{\mu(\lambda + 2\mu)} \frac{1}{|x|} I_{3 \times 3} - \frac{1}{8\pi} \frac{\lambda + \mu}{\mu(\lambda + 2\mu)} \frac{x \otimes x}{|x|^3}, \quad x \in \mathbb{R}^3 \setminus \{0\}. \tag{11.3.83}$$

Proof. We start by recalling formula (11.3.42) that gives an expression for the fundamental solution of a homogeneous differential operator for n odd. In our case,

$$L = (L_{jk})_{1 \leq j,k \leq 3}, \quad L_{jk} = \mu \delta_{jk} \Delta + (\lambda + \mu) \partial_j \partial_k,$$
$$\text{and} \quad L(\xi) = \mu I_{3 \times 3} + (\lambda + \mu)\xi \otimes \xi \quad \text{for each} \quad \xi \in S^2. \tag{11.3.84}$$

Observe that $(x \cdot \xi) \operatorname{sgn}(x \cdot \xi) = |x \cdot \xi|$. As such, for any $x \in \mathbb{R}^3 \setminus \{0\}$ we may compute

$$E(x) = -\frac{1}{16\pi^2}\Delta_x \int_{S^2} |x \cdot \xi|(L(\xi))^{-1}\, d\sigma(\xi) \tag{11.3.85}$$

$$= -\frac{1}{16\pi^2}\Delta_x \int_{S^2} |x \cdot \xi|\Big(\frac{1}{\mu}I_{3\times 3} - \frac{\lambda+\mu}{\mu(\lambda+2\mu)}\xi \otimes \xi\Big) d\sigma(\xi)$$

$$= -\frac{1}{16\pi^2}\Delta_x\Big[\frac{\omega_1}{\mu}|x|I_{3\times 3} - \frac{\lambda+\mu}{\mu(\lambda+2\mu)}\frac{\omega_1}{4}\Big(|x|I_{3\times 3} + \frac{x \otimes x}{|x|}\Big)\Big]$$

$$= -\frac{1}{16\pi}\Delta_x\Big[\frac{4\lambda+8\mu-\lambda-\mu}{2\mu(\lambda+2\mu)}|x|I_{3\times 3} - \frac{\lambda+\mu}{2\mu(\lambda+2\mu)}\frac{x \otimes x}{|x|}\Big]$$

$$= -\frac{1}{16\pi}\Big[\frac{3\lambda+7\mu}{2\mu(\lambda+2\mu)}\frac{2}{|x|}I_{3\times 3} - \frac{\lambda+\mu}{2\mu(\lambda+2\mu)}\Big(\frac{2}{|x|}I_{3\times 3} - \frac{4x \otimes x}{|x|^3}\Big)\Big]$$

$$= -\frac{1}{8\pi}\frac{\lambda+3\mu}{\mu(\lambda+2\mu)}\frac{1}{|x|}I_{3\times 3} - \frac{1}{8\pi}\frac{\lambda+\mu}{\mu(\lambda+2\mu)}\frac{x \otimes x}{|x|^3}.$$

For the second equality in (11.3.85), we have used Proposition 10.14, for the third we have used Proposition 14.67 and Proposition 14.68, while for the fifth we have used (7.3.2) and the readily verified fact that if N is a nonzero integer then

$$\Delta_x\Big[\frac{x \otimes x}{|x|^N}\Big] = \frac{2}{|x|^N}I_{n\times n} + \frac{N(N-n-2)}{|x|^{N+2}}\,x \otimes x, \tag{11.3.86}$$

for each $x \in \mathbb{R}^n \setminus \{0\}$. $\qquad\square$

Exercise 11.4. Fix $n \in \mathbb{N}$ with $n \geq 2$ along with $M \in \mathbb{N}$ and consider a second order $M \times M$ system

$$L = \Big(\sum_{j,k=1}^{n} a_{jk}^{\alpha\beta}\partial_j\partial_k\Big)_{1\leq\alpha,\beta\leq M}, \qquad a_{jk}^{\alpha\beta} \in \mathbb{C}. \tag{11.3.87}$$

Also, suppose $E = (E_{\alpha\beta})_{1\leq\alpha,\beta\leq M} : \mathbb{R}^n \setminus \{0\} \to \mathcal{M}_{M\times M}(\mathbb{C})$ is a matrix-valued function whose entries belong to $C^2(\mathbb{R}^n \setminus \{0\}) \cap L^1_{loc}(\mathbb{R}^n)$ and have gradients positive homogeneous of degree $1 - n$ in $\mathbb{R}^n \setminus \{0\}$.

Prove that the following statements are equivalent:

(1) When viewed in $\mathcal{M}_{M\times M}(L^1_{loc}(\mathbb{R}^n))$, the matrix-valued function E is a fundamental solution for L in $\mathbb{R}^n$;

(2) With the system L acting on the columns of E one has $LE = 0 \in \mathcal{M}_{M\times M}(\mathbb{C})$ pointwise in $\mathbb{R}^n \setminus \{0\}$ and for each $\alpha, \gamma \in \{1,\dots,M\}$ there holds

$$\sum_{j,k=1}^{n}\sum_{\beta=1}^{M} \int_{S^{n-1}} a_{jk}^{\alpha\beta}\omega_j\partial_k E_{\beta\gamma}(\omega)\, d\sigma(\omega) = \delta_{\alpha\gamma}. \tag{11.3.88}$$

Hint: Adapt the proof of Theorem 7.60 (this time taking the test function to be vector valued, i.e., $f \in [C_0^\infty(\mathbb{R}^n)]^M$).

Exercise 11.5. Suppose $n \in \mathbb{N}$ satisfies $n \geq 2$, and fix some $M \in \mathbb{N}$. Consider a second order $M \times M$ system L as in (11.3.87) with the property that (11.3.3) holds, and suppose $E = (E_{\alpha\beta})_{1 \leq \alpha,\beta \leq M}$ is the fundamental solution associated with L as in Theorem 11.1 (with $m := 1$). Finally, fix two indexes $\alpha, \gamma \in \{1, \ldots, M\}$. Prove that

$$\sum_{j,k=1}^{n} \sum_{\beta=1}^{M} \int_{S^{n-1}} a_{jk}^{\alpha\beta} \omega_j (\partial_k E_{\beta\gamma})(\omega) \, d\sigma(\omega) = \delta_{\alpha\gamma}$$

$$= \sum_{j,k=1}^{n} \sum_{\beta=1}^{M} \int_{S^{n-1}} a_{kj}^{\beta\alpha} \omega_j (\partial_k E_{\gamma\beta})(\omega) \, d\sigma(\omega), \tag{11.3.89}$$

and that, given any vector $\xi \in S^{n-1}$,

$$\sum_{j,k=1}^{n} \sum_{\beta=1}^{M} \int_{\substack{\omega \in S^{n-1} \\ \langle \omega, \xi \rangle > 0}} a_{jk}^{\alpha\beta} \omega_j (\partial_k E_{\beta\gamma})(\omega) \, d\sigma(\omega) = \frac{1}{2} \delta_{\alpha\gamma}$$

$$= \sum_{j,k=1}^{n} \sum_{\beta=1}^{M} \int_{\substack{\omega \in S^{n-1} \\ \langle \omega, \xi \rangle > 0}} a_{kj}^{\beta\alpha} \omega_j (\partial_k E_{\gamma\beta})(\omega) \, d\sigma(\omega). \tag{11.3.90}$$

Hint: For the first equality in (11.3.89) use Exercise 11.4 and Theorem 11.1. Then, the last equality in (11.3.89) follows from the first (written for $L^\top$ in place of L, bearing in mind (11.3.32)). Formulas (11.3.90) are consequences of (11.3.89) and the fact that the integrands are even (cf. Theorem 11.1).

11.4 Interior Estimates and Real-Analyticity for Null-Solutions of Systems

The aim in this section is to explore the extent to which results such as integral representation formula and interior estimates, proved earlier in §6.3 in the scalar context, continue to hold for vector-valued functions which are null-solutions of a certain class of systems of differential operators.

Proposition 11.6. *Fix $n, m, M \in \mathbb{N}$ with $n \geq 2$, and assume that L is an $M \times M$ system in $\mathbb{R}^n$ of order $2m$ of the form*

$$L = \sum_{|\alpha|=2m} A_\alpha \partial^\alpha, \qquad A_\alpha = (a_\alpha^{jk})_{1 \leq j,k \leq M} \in \mathcal{M}_{M \times M}(\mathbb{C}), \tag{11.4.1}$$

with the property that $\det [L(\xi)] \neq 0$ for each $\xi \in \mathbb{R}^n \setminus \{0\}$. In addition, suppose $\Omega \subseteq \mathbb{R}^n$ is an open set and $\mathbf{u} = (u_1, \dots, u_M) \in [\mathcal{D}'(\Omega)]^M$ is such that $L\mathbf{u} = 0$ in $[\mathcal{D}'(\Omega)]^M$.

Then $\mathbf{u} \in [C^\infty(\Omega)]^M$ and for each $x_0 \in \Omega$, each $r \in (0, \operatorname{dist}(x_0, \partial\Omega))$, and each function $\psi \in C_0^\infty(B(x_0, r))$ such that $\psi \equiv 1$ near $\overline{B(x_0, r/2)}$, we have

$$u_\ell(x) = - \sum_{j,k=1}^M \sum_{|\alpha|=2m} \sum_{\gamma < \alpha} (-1)^{|\gamma|} a_\alpha^{jk} \frac{\alpha!}{\gamma!(\alpha - \gamma)!} \times$$

$$\times \int_{B(x_0,r) \setminus \overline{B(x_0,r/2)}} (\partial^\gamma E_{j\ell})(x - y)\partial^{\alpha - \gamma}\psi(y)\, u_k(y)\, dy \qquad (11.4.2)$$

for each $\ell \in \{1, \dots, M\}$ and each $x \in B(x_0, r/2)$, where

$$\mathbb{E} = (E_{jk})_{1 \leq j,k \leq M} \in \mathcal{M}_{M \times M}(\mathcal{S}'(\mathbb{R}^n)) \cap \mathcal{M}_{M \times M}(C^\infty(\mathbb{R}^n \setminus \{0\})) \qquad (11.4.3)$$

is the fundamental matrix for $L^\top$ (the transpose of L) as given by Theorem 11.1.

In particular, for every $\mu \in \mathbb{N}_0^n$,

$$\partial^\mu u_\ell(x) = - \sum_{j,k=1}^M \sum_{|\alpha|=2m} \sum_{\gamma < \alpha} (-1)^{|\gamma|} a_\alpha^{jk} \frac{\alpha!}{\gamma!(\alpha - \gamma)!} \times$$

$$\times \int_{B(x_0,r) \setminus \overline{B(x_0,r/2)}} (\partial^{\gamma + \mu} E_{j\ell})(x - y)\partial^{\alpha - \gamma}\psi(y)\, u_k(y)\, dy \qquad (11.4.4)$$

for each $\ell \in \{1, \dots, M\}$ and each $x \in B(x_0, r/2)$. Also, if either n is odd, or $n > 2m$, or if $|\mu| > 2m - n$, we have

$$|(\partial^\mu \mathbf{u})(x_0)| \leq \frac{C_\mu}{r^{|\mu|}} \int_{B(x_0,r)} |\mathbf{u}(y)|\, dy, \qquad (11.4.5)$$

where $C_\mu \in (0, \infty)$ is independent of $\mathbf{u}$, x_0, r, and Ω.

Proof. Let $\mathbf{u} = (u_\ell)_{1 \leq \ell \leq M}$ and $\mathbb{E} = (E_{jk})_{1 \leq j,k \leq M}$ be as specified in the statement of the proposition. In particular,

$$\mathbb{E} \in \mathcal{M}_{M \times M}(\mathcal{S}'(\mathbb{R}^n)) \cap \mathcal{M}_{M \times M}(C^\infty(\mathbb{R}^n \setminus \{0\})) \qquad (11.4.6)$$

by Theorem 11.1. Also, pick $x_0 \in \Omega$, a number $r \in (0, \operatorname{dist}(x_0, \partial\Omega))$, and fix some function $\psi \in C_0^\infty(B(x_0, r))$ such that $\psi \equiv 1$ near $\overline{B(x_0, r/2)}$. Then, from Theorem 10.9 it follows that $\mathbf{u} \in [C^\infty(\Omega)]^M$, hence also $\psi\mathbf{u} \in [C_0^\infty(\Omega)]^M$. Granted these, for each $x \in B(x_0, r/2)$ we may then write (keeping in mind that $\mathbb{E}$ is a fundamental solution for $L^\top$, the transposed of L),

$$\mathbf{u}(x) = (\psi \mathbf{u})(x) = ((\psi u_\ell)(x))_{1 \le \ell \le M}$$

$$= \Big(\sum_{k=1}^{M} \langle \delta_{k\ell} \, \delta \, , \, (\psi u_k)(x - \cdot) \rangle \Big)_{1 \le \ell \le M}$$

$$= \Big(\sum_{k=1}^{M} \langle (L^\top \mathbb{E})_{k\ell} \, , \, (\psi u_k)(x - \cdot) \rangle \Big)_{1 \le \ell \le M}. \qquad (11.4.7)$$

Note that

$$L^\top = \sum_{|\alpha|=2m} (-1)^{|\alpha|} A_\alpha^\top \partial^\alpha = \sum_{|\alpha|=2m} A_\alpha^\top \partial^\alpha, \qquad (11.4.8)$$

where $A_\alpha^\top$ is the transposed of the matrix A_α, for each α. Thus, (11.4.7) implies

$$\mathbf{u}(x) = \Big(\sum_{k=1}^{M} \Big\langle \sum_{|\alpha|=2m} \sum_{j=1}^{M} (A_\alpha^\top)_{kj} \partial^\alpha E_{j\ell}, (\psi u_k)(x - \cdot) \Big\rangle \Big)_{1 \le \ell \le M} \qquad (11.4.9)$$

$$= \Big(\sum_{k=1}^{M} \Big\langle \sum_{|\alpha|=2m} \sum_{j=1}^{M} a_\alpha^{jk} E_{j\ell}, \partial^\alpha [(\psi u_k)(x - \cdot)] \Big\rangle \Big)_{1 \le \ell \le M}$$

$$= \Big(\sum_{j=1}^{M} \Big\langle E_{j\ell}, \sum_{|\alpha|=2m} \sum_{k=1}^{M} a_\alpha^{jk} (\partial^\alpha (\psi u_k))(x - \cdot) \Big\rangle \Big)_{1 \le \ell \le M}$$

$$= \Big(\sum_{j=1}^{M} \Big\langle E_{j\ell}(x - \cdot), \sum_{|\alpha|=2m} \sum_{k=1}^{M} a_\alpha^{jk} \sum_{0 < \beta \le \alpha} \frac{\alpha!}{\beta!(\alpha - \beta)!} \partial^\beta \psi \partial^{\alpha - \beta} u_k \Big\rangle \Big)_{1 \le \ell \le M}$$

$$+ \Big(\sum_{j=1}^{M} \Big\langle E_{j\ell}(x - \cdot), \psi \sum_{|\alpha|=2m} \sum_{k=1}^{M} a_\alpha^{jk} \partial^\alpha u_k \Big\rangle \Big)_{1 \le \ell \le M}$$

$$= \Big(\sum_{j,k=1}^{M} \sum_{|\alpha|=2m} \sum_{0 < \beta \le \alpha} a_\alpha^{jk} \frac{\alpha!}{\beta!(\alpha - \beta)!} \Big\langle E_{j\ell}(x - \cdot), \partial^\beta \psi \, \partial^{\alpha - \beta} u_k \Big\rangle \Big)_{1 \le \ell \le M}$$

since for each $j \in \{1, \ldots, M\}$,

$$\psi \sum_{|\alpha|=2m} \sum_{k=1}^{M} a_\alpha^{jk} \partial^\alpha u_k = \psi (L\mathbf{u})_j = 0 \quad \text{in } \mathbb{R}^n. \qquad (11.4.10)$$

Next, for each $\ell, j, k \in \{1, \ldots, M\}$ and each $\alpha, \beta \in \mathbb{N}_0^n$ satisfying $|\alpha| = 2m$ and $0 < \beta \le \alpha$, we observe that

$$\langle E_{j\ell}(x - \cdot), \partial^\beta \psi \, \partial^{\alpha-\beta} u_k \rangle \tag{11.4.11}$$

$$= \int_{B(x_0,r) \setminus \overline{B(x_0,r/2)}} E_{j\ell}(x - y) \partial^\beta \psi(y) \partial^{\alpha-\beta} u_k(y) \, dy$$

$$= (-1)^{|\beta|} \int_{B(x_0,r) \setminus \overline{B(x_0,r/2)}} \partial_y^{\alpha-\beta} [E_{j\ell}(x - y) \partial^\beta \psi(y)] u_k(y) \, dy$$

$$= \sum_{\gamma \leq \alpha-\beta} (-1)^{|\beta|+|\gamma|} \frac{(\alpha - \beta)!}{\gamma!(\alpha - \beta - \gamma)!} \times$$

$$\times \int_{B(x_0,r) \setminus \overline{B(x_0,r/2)}} (\partial^\gamma E_{j\ell})(x - y) \partial^{\alpha-\gamma} \psi(y) \, u_k(y) \, dy,$$

and that, whenever $\gamma < \alpha$,

$$\sum_{0 < \beta \leq \alpha-\gamma} (-1)^{|\beta|} \frac{(\alpha - \gamma)!}{\beta!(\alpha - \beta - \gamma)!} = \Big(\sum_{\beta \leq \alpha-\gamma} (-1)^{|\beta|} \frac{(\alpha - \gamma)!}{\beta!(\alpha - \beta - \gamma)!} \Big) - 1$$

$$= 0 - 1 = -1. \tag{11.4.12}$$

At this stage, (11.4.2) follows from (11.4.9), (11.4.11), and (11.4.12). In turn, (11.4.2) readily implies (11.4.4). Finally, (11.4.5) is a consequence of (11.4.4), the assumptions on n and μ, and (11.3.27), by choosing ψ as in (6.3.11)–(6.3.12). □

An L^∞-version of the interior estimate (11.4.5), valid in all space dimensions, is established in the next theorem. In particular, this version allows us to prove the real-analyticity of null-solutions of the systems considered here.

Theorem 11.7. *Let $n, m, M \in \mathbb{N}$ and suppose L is an $M \times M$ constant (complex) coefficient system in $\mathbb{R}^n$, homogeneous of order $2m$, and with the property that $\det [L(\xi)] \neq 0$ for each $\xi \in \mathbb{R}^n \setminus \{0\}$. Assume also that $\Omega \subseteq \mathbb{R}^n$ is an open set and $\mathbf{u} \in [\mathcal{D}'(\Omega)]^M$ is such that $L\mathbf{u} = 0$ in $[\mathcal{D}'(\Omega)]^M$.*

Then $\mathbf{u} \in [C^\infty(\Omega)]^M$ and there exists a constant $C \in (0, \infty)$ such that

$$\max_{y \in B(x,\lambda r)} |\partial^\alpha \mathbf{u}(y)| \leq \frac{C^{|\alpha|}(1 - \lambda)^{-|\alpha|}|\alpha|!}{r^{|\alpha|}} \max_{y \in B(x,r)} |\mathbf{u}(y)|, \quad \forall \alpha \in \mathbb{N}_0^n, \tag{11.4.13}$$

for each $x \in \Omega$, each $r \in (0, \text{dist}(x, \partial\Omega))$, and each $\lambda \in (0, 1)$. In particular, each component of $\mathbf{u}$ is real-analytic in Ω.

Proof. Theorem 10.9 gives that $\mathbf{u} \in [C^\infty(\Omega)]^M$. As far as (11.4.13) is concerned, consider first the case when either n is odd, or $n > 2m$. In this scenario, (11.4.5) gives that there exists some $C \in (0, \infty)$, independent of $\mathbf{u}$ and Ω, such that

$$|\partial_j \mathbf{u}(x)| \leq \frac{C}{r} \fint_{B(x,r)} |\mathbf{u}(y)| \, dy, \quad \forall j \in \{1, \ldots, n\}, \tag{11.4.14}$$

whenever $x \in \Omega$ and $r \in (0, \text{dist}(x, \partial\Omega))$. A quick inspection reveals that the proof of Lemma 6.21 carries through in the case when the functions involved are vector-valued. When applied with

$$\mathcal{A} := \{\mathbf{u} \in [C^\infty(\Omega)]^M : L\mathbf{u} = 0 \text{ in } \Omega\}, \tag{11.4.15}$$

this lemma gives (thanks to (11.4.14)) that for every point $x \in \Omega$, every radius $r \in (0, \text{dist}(x, \partial\Omega))$, every $k \in \mathbb{N}$, and every $\lambda \in (0, 1)$, we have (with C as in (11.4.14))

$$\max_{y \in B(x, \lambda r)} |\partial^\alpha \mathbf{u}(y)| \leq \frac{C^k (1 - \lambda)^{-k} e^{k-1} k!}{r^k} \max_{y \in B(x,r)} |\mathbf{u}(y)|, \tag{11.4.16}$$

for every multi-index $\alpha \in \mathbb{N}_0^n$ with $|\alpha| = k$.

This proves (11.4.13) under the current assumptions on n.

Fix now an arbitrary $n \in \mathbb{N}$ and pick some $k \in \mathbb{N}$. With L as in (11.4.1) consider the constant coefficient, homogeneous, $M \times M$ system in $\mathbb{R}^n$, of order $4m$, given by

$$\mathcal{L}(\partial) := L^* L = \sum_{|\alpha|=|\beta|=2m} A_\alpha^* A_\beta \partial^{\alpha+\beta}, \tag{11.4.17}$$

and note that for each $\xi \in \mathbb{R}^n$ we have

$$\mathcal{L}(\xi) = \sum_{|\alpha|=|\beta|=2m} \xi^{\alpha+\beta} A_\alpha^* A_\beta = \sum_{|\alpha|=|\beta|=2m} (\xi^\alpha A_\alpha)^* (\xi^\beta A_\beta)$$

$$= (L(\xi))^* L(\xi). \tag{11.4.18}$$

In particular,

$$\mathcal{L}(\xi)\eta \cdot \overline{\eta} = |L(\xi)\eta|^2, \qquad \forall \xi \in \mathbb{R}^n, \quad \forall \eta \in \mathbb{C}^M. \tag{11.4.19}$$

Finally, define the constant coefficient, homogeneous, $M \times M$ system in $\mathbb{R}^{n+k}$, of order $4m$,

$$\widetilde{\mathcal{L}}(\partial, \partial_{n+1}, \ldots, \partial_{n+k}) := \mathcal{L}(\partial) + (\partial_{n+1}^{4m} + \cdots + \partial_{n+k}^{4m}) I_{M \times M}. \tag{11.4.20}$$

We claim that

$$\det[\widetilde{\mathcal{L}}(\xi, \xi_{n+1}, \ldots, \xi_{n+k})] \neq 0, \quad \forall (\xi, \xi_{n+1}, \ldots, \xi_{n+k}) \in \mathbb{R}^{n+k} \setminus \{0\}. \tag{11.4.21}$$

To see that this is the case, fix some $(\xi, \xi_{n+1}, \ldots, \xi_{n+k}) \in \mathbb{R}^{n+k} \setminus \{0\}$ and note that it suffices to show that the $M \times M$ complex matrix $\widetilde{\mathcal{L}}(\xi, \xi_{n+1}, \ldots, \xi_{n+k})$ acts injectively on $\mathbb{C}^M$. With this in mind, pick some $\eta \in \mathbb{C}^M$ with the property that $\widetilde{\mathcal{L}}(\xi, \xi_{n+1}, \ldots, \xi_{n+k})\eta = 0 \in \mathbb{C}^M$. Then

$$0 = \widetilde{\mathcal{L}}(\xi, \xi_{n+1}, \ldots, \xi_{n+k})\eta \cdot \overline{\eta} = |L(\xi)\eta|^2 + (\xi_{n+1}^{4m} + \cdots + \xi_{n+k}^{4m})|\eta|^2 \tag{11.4.22}$$

which forces

$$L(\xi)\eta = 0 \quad \text{and} \quad (\xi_{n+1}^{4m} + \cdots + \xi_{n+k}^{4m})|\eta|^2 = 0. \tag{11.4.23}$$

If $\xi \in \mathbb{R}^n \setminus \{0\}$ then the first condition in (11.4.23) implies $\eta = 0$, given the assumptions on L. If $\xi = 0$ then necessarily there exists $j \in \{1, \ldots, k\}$ such that $\xi_{n+j} \neq 0$, and the second condition in (11.4.23) now implies $\eta = 0$. Thus, $\eta = 0$ in all alternatives which finishes the proof of (11.4.21).

To proceed, assume $\mathbf{u} \in [C^\infty(\Omega)]^M$ satisfies $L\mathbf{u} = 0$ in Ω, and define

$$\widetilde{\mathbf{u}}(x, x_{n+1}, \ldots, x_{n+k}) := \mathbf{u}(x) \quad \text{for each}$$

$$(x, x_{n+1}, \ldots, x_{n+k}) \in \widetilde{\Omega} := \Omega \times \mathbb{R}^k. \tag{11.4.24}$$

Then $\widetilde{\Omega}$ is an open subset of $\mathbb{R}^{n+k}$ and $\widetilde{\mathbf{u}} \in [C^\infty(\widetilde{\Omega})]^M$. Moreover, as is apparent from (11.4.17), (11.4.20), and (11.4.24), we have

$$\widetilde{\mathscr{L}}(\partial, \partial_{n+1}, \ldots, \partial_{n+k})\widetilde{\mathbf{u}} = 0 \quad \text{in} \quad \widetilde{\Omega}. \tag{11.4.25}$$

Assume now that $n \in \mathbb{N}$ is an arbitrary given number and pick $k \in \mathbb{N}$ such that $n + k$ is either odd or $n + k > 4m$. Then from the first part in the proof, applied to $\widetilde{\mathscr{L}}$ and $\widetilde{\mathbf{u}}$, we know that there exists a constant $C \in (0, \infty)$ such that

$$\max_{B(\overline{x}, \lambda r)} |\partial^\alpha \widetilde{\mathbf{u}}| \leq \frac{C^{|\alpha|}(1-\lambda)^{-|\alpha|}|\alpha|!}{r^{|\alpha|}} \max_{B(\overline{x}, r)} |\widetilde{\mathbf{u}}|, \quad \forall \alpha \in \mathbb{N}_0^n, \tag{11.4.26}$$

for each $\widetilde{x} \in \widetilde{\Omega}$, each $r \in (0, \text{dist}(\widetilde{x}, \partial\widetilde{\Omega}))$, and each $\lambda \in (0, 1)$, where the balls appearing in (11.4.26) are considered in $\mathbb{R}^{n+k}$. Now, given $x \in \Omega$ and some radius $r \in (0, \text{dist}(x, \partial\Omega))$, specializing (11.4.26) to the case when $\widetilde{x} := (x, 0, \ldots, 0)$ yields (11.4.13) for the current n. Thus, (11.4.13) holds for any n. Finally, the last claim in the statement of the theorem is a consequence of (11.4.13) and Lemma 6.24. $\qquad \square$

11.5 Reverse Hölder Estimates for Null-Solutions of Systems

The principal result in this section is the version of interior estimates for null-solutions of certain higher order systems stated in Theorem 11.12. In particular, this contains the fact that such null-solutions satisfy reverse Hölder estimates (a topic worthy of investigation in its own right). We begin by making the following definition.

Definition 11.8. A continuous (complex) vector-valued function $\mathbf{u}$ defined in Ω is said to be p-subaveraging for some $p \in (0, \infty)$ if there exists a finite constant $C > 0$ with the property that

$$|\mathbf{u}(x)| \leq C \left(\fint_{B(x,r)} |\mathbf{u}(y)|^p \, dy \right)^{\frac{1}{p}} \tag{11.5.1}$$

for every $x \in \Omega$ and every $r \in (0, \operatorname{dist}(x, \partial\Omega))$.

The class of p-subaveraging functions exhibits a number of self-improvement properties, the first of which is discussed below.

Lemma 11.9. *Assume that* $\mathbf{u}$ *is a continuous (complex) vector-valued function defined in* Ω, *and suppose that* $0 < p < \infty$. *Then* $\mathbf{u}$ *is* p-*subaveraging if and only if there exists a finite constant* $C > 0$ *such that*

$$\sup_{z \in B(x, \lambda r)} |\mathbf{u}(z)| \leq C(1 - \lambda)^{-n/p} \left(\fint_{B(x,r)} |\mathbf{u}(y)|^p \, dy \right)^{\frac{1}{p}} \tag{11.5.2}$$

for every $x \in \Omega$, *every* $r \in (0, \operatorname{dist}(x, \partial\Omega))$, *and every* $\lambda \in (0, 1)$.

Proof. The fact that (11.5.2) implies (11.5.1) is obvious. Conversely, suppose that u is p-subaveraging. Pick $x \in \Omega$, $r \in (0, \operatorname{dist}(x, \partial\Omega))$, and $z \in B(x, \lambda r)$. Then, if $R := (1 - \lambda)r$, it follows that $z \in \Omega$ and $0 < R < \operatorname{dist}(z, \partial\Omega)$. Furthermore, $B(z, R) \subseteq B(x, r)$. Consequently, with C as in (11.5.1),

$$|\mathbf{u}(z)| \leq C \left(\fint_{B(z,R)} |\mathbf{u}(y)|^p \, dy \right)^{\frac{1}{p}} = C \left(\frac{(1-\lambda)^{-n}}{|B(z,r)|} \int_{B(z,R)} |\mathbf{u}(y)|^p \, dy \right)^{\frac{1}{p}}$$

$$\leq C(1-\lambda)^{-n/p} \left(\fint_{B(x,r)} |\mathbf{u}(y)|^p \, dy \right)^{\frac{1}{p}}, \tag{11.5.3}$$

which readily implies (11.5.2) by taking the supremum over $z \in B(x, \lambda r)$.   □

The second self-improvement within the class of p-subaveraging functions is the fact that the value of the parameter p is immaterial.

Lemma 11.10. *If there exists* $p_0 > 0$ *such that* $\mathbf{u}$ *is* p_0-*subaveraging function, then* $\mathbf{u}$ *is* p-*subaveraging for every* $p \in (0, \infty)$.

In light of Lemma 11.10, it is unequivocal to refer to a function $\mathbf{u}$ as simply being subaveraging if it is p-subaveraging for some $p \in (0, \infty)$. The optimal constant which can be used in (11.5.1) is referred to as the p-subaveraging constant of the function $\mathbf{u}$.

Proof of Lemma 11.10. The proof is based on ideas used in the work of G. Hardy and J. Littlewood [29] (cf. also [16, Lemma 2, pp. 172–173]). The case $p > p_0$ can be handled directly utilizing Hölder's inequality with $q = \frac{p}{p_0} > 1$. Henceforth, we shall focus on the case when $p < p_0$. Replacing $\mathbf{u}$ by a suitable power of $|\mathbf{u}|$, there is no loss of generality in assuming that, in fact, $p_0 = 1$ and $p < 1$. We may also assume (by rescaling and making a translation) that $\overline{B(0, 1)} \subseteq \Omega$, $x = 0$, and $\int_{B(0,1)} |\mathbf{u}(y)|^p \, dy = 1$. The goal is then to prove the estimate $|\mathbf{u}(0)| \leq C$ with a finite constant $C > 0$ independent of $\mathbf{u}$. Continuing our series of reductions, we may assume that $|\mathbf{u}(0)| > 1$. Next, for each $r \in (0, 1]$ and $q \in (0, \infty]$, introduce

$$m_q(r) := \begin{cases} \left(\int_{B(0,r)} |\mathbf{u}(y)|^p \, dy \right)^{\frac{1}{p}} & \text{if } q < \infty, \\ \sup_{B(0,r)} |\mathbf{u}(y)| & \text{if } q = \infty. \end{cases} \tag{11.5.4}$$

Observe that for each $r \in (0,1)$,

$$m_1(r) \le (m_p(r))^p \, (m_\infty(r))^{1-p} \le (m_\infty(r))^{1-p}, \quad \forall\, p \in (0,1), \tag{11.5.5}$$

where the last inequality holds by virtue of the trivial estimate $m_p(r) \le m_p(1)$, valid for every $r \in (0,1)$, and the assumption $m_p(1) = 1$. On the other hand, for every $x \in \Omega$ and every $r \in (0, \text{dist}\,(x, \partial\Omega))$

$$|\mathbf{u}(x)| \le \frac{C}{r^n} \int_{B(x,r)} |\mathbf{u}(y)| \, dy, \tag{11.5.6}$$

and, consequently,

$$|\mathbf{u}(z)| \le \frac{C}{(r-\rho)^n} \int_{B(z,r-\rho)} |\mathbf{u}(y)| \, dy$$

$$\le \frac{C}{(r-\rho)^n} \int_{B(0,r)} |\mathbf{u}(y)| \, dy \tag{11.5.7}$$

whenever $|z| = \rho \in (0,r)$. Then for any $z^* \in B(0,\rho)$ such that $|z^*| = \rho^* < \rho$ we obtain

$$|\mathbf{u}(z^*)| \le \frac{C}{(r-\rho^*)^n} \int_{B(0,r)} |\mathbf{u}(y)| \, dy$$

$$\le \frac{C}{(r-\rho)^n} \int_{B(0,r)} |\mathbf{u}(y)| \, dy, \tag{11.5.8}$$

which, in concert with (11.5.5), yields the estimate

$$m_\infty(\rho) \le \frac{C}{(r-\rho)^n} m_\infty(r)^{1-p}. \tag{11.5.9}$$

To continue, set $\rho := r^a$ for some $a \in (1, \infty)$ to be specified momentarily. Then (11.5.9) entails

$$\int_{1/2}^1 \ln m_\infty(r^a) \, \frac{dr}{r} \le C + n \int_{1/2}^1 \ln \frac{1}{(r - r^a)} \frac{dr}{r}$$

$$+ (1-p) \int_{1/2}^1 \ln m_\infty(r) \, \frac{dr}{r}, \tag{11.5.10}$$

and for the first integral above the change of variables $t := r^a$ gives

$$\int_{1/2}^{1} \ln m_\infty(r^a) \, \frac{dr}{r} = \frac{1}{a} \int_{(1/2)^a}^{1} \ln m_\infty(t) \, \frac{dt}{t}. \tag{11.5.11}$$

Since our assumption $|\mathbf{u}(0)| > 1$ implies $m_\infty(t) \geq 1$, the right-hand side of (11.5.11) is bounded from below by

$$\frac{1}{a} \int_{1/2}^{1} \ln m_\infty(r) \, \frac{dr}{r}. \tag{11.5.12}$$

Therefore, (11.5.10)–(11.5.12) imply

$$\left(\frac{1}{a} - 1 + p \right) \int_{1/2}^{1} \ln m_\infty(r) \, \frac{dr}{r} \leq C + C \int_{1/2}^{1} \ln \frac{1}{(r - r^a)} \, \frac{dr}{r}$$

$$\leq C < \infty. \tag{11.5.13}$$

Choose now $a > 1$ such that $\frac{1}{a} - 1 + p > 0$. Then (11.5.13) forces

$$\int_{1/2}^{1} \ln m_\infty(r) \, dr \leq C, \tag{11.5.14}$$

and hence, $\ln m_\infty(1/2) \leq C$ for some finite constant $C > 0$ independent of initial function $\mathbf{u}$. In concert with the inequality $|\mathbf{u}(0)| \leq m_\infty(1/2)$, this finishes the proof of the lemma.   □

There are certain connections between the subaveraging property and reverse Hölder estimates, brought to light by our next two results.

Lemma 11.11. *Let $\mathbf{u}$ be a subaveraging function. Then for every $p, q \in (0, \infty)$ and $\lambda \in (0, 1)$ the following reverse Hölder estimate holds*

$$\left(\fint_{B(x,\lambda r)} |\mathbf{u}(y)|^q \, dy \right)^{\frac{1}{q}} \leq C \left(\fint_{B(x,r)} |\mathbf{u}(y)|^p \, dy \right)^{\frac{1}{p}}, \tag{11.5.15}$$

for $x \in \Omega$ and $0 < r < \mathrm{dist}\,(x, \partial\Omega)$, where $C > 0$ is a finite constant depending only on p, q, λ, n and the p-subaveraging constant of $\mathbf{u}$.

Proof. If $x \in \Omega$ and $0 < r < \mathrm{dist}\,(x, \partial\Omega)$, we write

$$\left(\fint_{B(x,\lambda r)} |\mathbf{u}(y)|^q \, dy \right)^{\frac{1}{q}} \leq \sup_{z \in B(x,\lambda r)} |\mathbf{u}(z)|$$

$$\leq C \left(\fint_{B(x,r)} |\mathbf{u}(y)|^p \, dy \right)^{\frac{1}{p}}, \tag{11.5.16}$$

by Lemma 11.9.   □

The main result in this section is the combination of interior estimates and reverse Hölder estimates contained in the following theorem.

Theorem 11.12. *Let $n, m, M \in \mathbb{N}$ and suppose L is an $M \times M$ constant (complex) coefficient system in $\mathbb{R}^n$, homogeneous of order $2m$, and with the property that $\det[L(\xi)] \neq 0$ for each $\xi \in \mathbb{R}^n \setminus \{0\}$. Assume also that $\Omega \subseteq \mathbb{R}^n$ is an open set and $\mathbf{u} \in [\mathcal{D}'(\Omega)]^M$ is such that $L\mathbf{u} = 0$ in $[\mathcal{D}'(\Omega)]^M$.*

Then $\mathbf{u} \in [C^{\infty}(\Omega)]^M$ and $\mathbf{u}$ is subaveraging. Moreover, there exists a constant $C = C(L, n) \in (0, \infty)$ with the property that given $p \in (0, \infty)$ there exists some $c = c(L, n, p) \in (0, \infty)$ satisfying

$$\max_{y \in \overline{B(x, \lambda r)}} |\partial^{\alpha} \mathbf{u}(y)| \leq c \, (1 - \lambda)^{-|\alpha| - n/p} \frac{C^{|\alpha|} |\alpha|!}{r^{|\alpha|}} \Big(\fint_{B(x, r)} |\mathbf{u}(y)|^p \, dy \Big)^{1/p} \qquad (11.5.17)$$

whenever $x \in \Omega$, $0 < r < \operatorname{dist}(x, \partial\Omega)$, $\lambda \in (0, 1)$, and $\alpha \in \mathbb{N}_0^n$. In particular, the components of $\mathbf{u}$ are real-analytic in Ω.

Proof. As in the past, Theorem 10.9 gives that $\mathbf{u} \in [C^{\infty}(\Omega)]^M$. By working with the system $\mathscr{L}$ defined as in (11.4.20) and the function $\widetilde{\mathbf{u}}$ defined as in (11.4.24), the same type of reasoning as in the proof of Theorem 11.7 shows that estimate (11.4.5) actually holds without any restrictions on the dimension n. Keeping this in mind, the version of this estimate corresponding to $\mu = (0, \dots, 0)$ may be interpreted as saying that $\mathbf{u}$ is 1-subaveraging. Hence, by Lemma 11.10 and the ensuing remark, it follows that $\mathbf{u}$ is subaveraging. Based on this and (11.4.13), for each $x \in \Omega$, each $r \in (0, \operatorname{dist}(x, \partial\Omega))$, each $\theta, \eta \in (0, 1)$ and each $\alpha \in \mathbb{N}_0^n$ we may write

$$\max_{y \in \overline{B(x, \theta\eta r)}} |\partial^{\alpha} \mathbf{u}(y)| \leq \frac{C^{|\alpha|} (1 - \theta)^{-|\alpha|} |\alpha|!}{(\eta r)^{|\alpha|}} \max_{y \in \overline{B(x, \eta r)}} |\mathbf{u}(y)| \qquad (11.5.18)$$

$$\leq c(1 - \eta)^{-n/p} \frac{C^{|\alpha|} (1 - \theta)^{-|\alpha|} |\alpha|!}{(\eta r)^{|\alpha|}} \Big(\fint_{B(x, r)} |\mathbf{u}(y)|^p \, dy \Big)^{1/p},$$

where $c = c(L, n, p) \in (0, \infty)$. Next, given any $\lambda \in (0, 1)$, specialize (11.5.18) to the case when $\theta := 2\lambda/(\lambda + 1)$ and $\eta := (\lambda + 1)/2$. Note that $\theta, \eta \in (0, 1)$, $\theta\eta = \lambda$, $1 - \eta = (1 - \lambda)/2$, and $1 - \theta = (1 - \lambda)/(1 + \lambda)$. Based on these, we may transform (11.5.18) into

$$\max_{y \in \overline{B(x, \lambda r)}} |\partial^{\alpha} \mathbf{u}(y)| \qquad (11.5.19)$$

$$\leq c \, 2^{n/p} (1 - \lambda)^{-|\alpha| - n/p} \frac{(2C)^{|\alpha|} |\alpha|!}{r^{|\alpha|}} \Big(\fint_{B(x, r)} |\mathbf{u}(y)|^p \, dy \Big)^{1/p},$$

from which (11.5.17) follows after adjusting notation. Also, the last claim in the statement of the theorem is a consequence of (11.5.17) and Lemma 6.24. $\square$

11.6 Layer Potentials and Jump Relations for Systems

The goal of this section is to introduce and study certain types of integral operators, typically called layer potentials, that are particularly useful in the treatment of
boundary value problems for systems. This is also an excellent opportunity to il-
lustrate how a good understanding of the nature of fundamental solutions coupled
with a versatile command of distribution theory yield powerful tools in the study of
partial differential equations.

To set the stage, we introduce some notation and make some background assumptions. Throughout this section we let

$$L = \sum_{j,k=1}^{n} A_{jk} \partial_j \partial_k, \qquad A_{jk} = (a_{jk}^{\alpha\beta})_{1 \le \alpha,\beta \le M} \in \mathcal{M}_{M \times M}(\mathbb{C}), \qquad (11.6.1)$$

be a second order, homogeneous, complex constant coefficient, $M \times M$ system,
where $M \in \mathbb{N}$, with the property that its characteristic matrix, $L(\xi)$, defined for
each vector $\xi = (\xi_1, \ldots, \xi_n) \in \mathbb{R}^n$ by the formula

$$L(\xi) := \sum_{j,k=1}^{n} \xi_j \xi_k A_{jk} = \Big(\sum_{j,k=1}^{n} a_{jk}^{\alpha\beta} \xi_j \xi_k \Big)_{1 \le \alpha,\beta \le M} \in \mathcal{M}_{M \times M}(\mathbb{C}), \qquad (11.6.2)$$

satisfies the ellipticity condition

$$\det[L(\xi)] \ne 0, \quad \forall \xi \in \mathbb{R}^n \setminus \{0\}. \qquad (11.6.3)$$

In particular, since $L(\mathbf{e}_j) = A_{jj}$ for each $j \in \{1, \ldots, n\}$, the ellipticity condition
(11.6.3) entails

$$A_{jj} \in \mathcal{M}_{M \times M}(\mathbb{C}) \text{ is an invertible matrix for each } j \in \{1, \ldots, n\}. \qquad (11.6.4)$$

Thanks to (11.6.3), Theorem 11.1 is applicable (with $m = 1$) and we denote by

$$\mathbb{E} = (E_{\alpha\beta})_{1 \le \alpha,\beta \le M} \in \mathcal{M}_{M \times M}(\mathcal{S}'(\mathbb{R}^n)) \qquad (11.6.5)$$

the matrix-valued fundamental solution of L in $\mathbb{R}^n$ whose entries are constructed
according to the recipe devised there. Thus, among other things, for each $\alpha, \beta \in
\{1, \ldots, M\}$ we have

$$E_{\alpha\beta}\big|_{\mathbb{R}^n \setminus \{0\}} \text{ is even and belongs to } C^\infty(\mathbb{R}^n \setminus \{0\}), \qquad (11.6.6)$$

$$\nabla E_{\alpha\beta} \text{ is positive homogeneous of degree } 1 - n. \qquad (11.6.7)$$

First, by specializing Theorem 4.79 to the case when Φ is a first-order partial
derivative of $\mathbb{E}$, we obtain the following basic result.

Theorem 11.13. *Assume that the system L is as in* (11.6.1)–(11.6.3) *and suppose
that $\mathbb{E}$ is the fundamental solution for L in $\mathbb{R}^n$ constructed in Theorem 11.1. Then*

for each $j \in \{1, \ldots, n\}$ one has

$$\lim_{\varepsilon \to 0^{\pm}} (\partial_j \mathbb{E})(x', \varepsilon) = \pm \tfrac{1}{2} \delta_{jn} \delta(x')(A_{nn})^{-1} + \text{P.V.}\,(\partial_j \mathbb{E})(x', 0) \qquad (11.6.8)$$

in $\mathcal{M}_{M \times M}(\mathcal{S}'(\mathbb{R}^{n-1}))$.

Proof. Fix $j \in \{1, \ldots, n\}$. Then $\Phi := \partial_j \mathbb{E}$ is of class C^{∞}, odd, and positive homogeneous of degree $1 - n$ in $\mathbb{R}^n \setminus \{0\}$ (cf. (11.6.6)–(11.6.7)). Moreover, by virtue of (11.3.28) used with $m = 1$, we have

$$\widehat{\Phi}(\xi) = \widehat{\partial_j \mathbb{E}}(\xi) = i\,\xi_j \widehat{\mathbb{E}}(\xi) = -i\,\xi_j [L(\xi)]^{-1} \quad \text{in} \quad \mathcal{M}_{M \times M}(\mathcal{S}'(\mathbb{R}^{n-1})). \qquad (11.6.9)$$

This further implies that

$$\widehat{\Phi}(0', 1) = \widehat{\Phi}(\mathbf{e}_n) = -i\,\delta_{jn}[L(\mathbf{e}_n)]^{-1} = -i\,\delta_{jn}(A_{nn})^{-1}. \qquad (11.6.10)$$

With this in hand, (11.6.8) follows with the help of (4.7.1). □

In the next result, we introduce the so-called single layer potential operator associated with the system L and study the boundary behavior of its first-order derivatives. Here and elsewhere, the integral of a vector-valued function is applied to each individual component.

Theorem 11.14. *Suppose that the system L is as in (11.6.1)–(11.6.3) and assume that $\mathbb{E}$ is the fundamental solution for L in $\mathbb{R}^n$ constructed in Theorem 11.1. Given any $\mathbb{C}^M$-valued function $\varphi \in \mathcal{S}(\mathbb{R}^{n-1})$, define the $\mathbb{C}^M$-valued function*

$$(\mathscr{S}\varphi)(x) := \int_{\mathbb{R}^{n-1}} \mathbb{E}(x' - y', t)\varphi(y')\,dy' \qquad (11.6.11)$$

for each $x = (x', t) \in \mathbb{R}^n$ with $t \neq 0$.

Then for any $\mathbb{C}^M$-valued function $\varphi \in \mathcal{S}(\mathbb{R}^{n-1})$,

$$\mathscr{S}\varphi \in C^{\infty}(\mathbb{R}^n_{\pm}), \quad L(\mathscr{S}\varphi) = 0 \text{ pointwise in } \mathbb{R}^n_{\pm}, \qquad (11.6.12)$$

and for each $j \in \{1, \ldots, n\}$ and each $x' \in \mathbb{R}^{n-1}$ one has

$$\lim_{t \to 0^{\pm}} \partial_j (\mathscr{S}\varphi)(x', t) = \pm \tfrac{1}{2} \delta_{jn}(A_{nn})^{-1} \varphi(x')$$

$$+ \lim_{\varepsilon \to 0^+} \int_{\substack{y' \in \mathbb{R}^{n-1} \\ |x'-y'|>\varepsilon}} (\partial_j \mathbb{E})(x' - y', 0)\varphi(y')\,dy'. \qquad (11.6.13)$$

Proof. Let $\varphi \in \mathcal{S}(\mathbb{R}^{n-1})$ be an arbitrary $\mathbb{C}^M$-valued function. That $\mathscr{S}\varphi$ is of class C^{∞} in $\mathbb{R}^n \setminus \{x_n = 0\}$ is clear from (11.6.11), (11.6.6), and estimates for the derivatives of $\mathbb{E}$ (see (11.3.27)). In addition, for each $x = (x', x_n) \in \mathbb{R}^n$ with $x_n \neq 0$,

$$L(\mathscr{S}\varphi)(x) = \int_{\mathbb{R}^{n-1}} (L\mathbb{E})(x' - y', x_n)\varphi(y')\,dy' = 0 \qquad (11.6.14)$$

since $L\mathbb{E} = \delta I_{M \times M}$ in $\mathcal{D}'(\mathbb{R}^n)$ which (upon recalling that $\mathbb{E} \in C^\infty(\mathbb{R}^n \setminus \{0\})$) forces

$$L\mathbb{E} = 0 \quad \text{pointwise in} \quad \mathbb{R}^n_\pm. \tag{11.6.15}$$

Finally, (11.6.13) follows from Corollary 4.81 applied to the same function Φ used in the proof of Theorem 11.13. $\qquad\square$

Remark 11.15. The integral operator $\mathscr{S}$ from (11.6.11) is called the single layer potential (or single layer operator) associated to L in the upper and lower half-spaces in $\mathbb{R}^n$, while (11.6.13) may be naturally regarded as the jump formula for the gradient of the single layer potential in this setting.

Our next result brings to the forefront another basic integral operator, typically referred to as the double layer potential.

Theorem 11.16. *Let the system L be as in* (11.6.1)–(11.6.3) *and assume that $\mathbb{E}$ is the fundamental solution for L in $\mathbb{R}^n$ constructed in Theorem 11.1. Given any $\mathbb{C}^M$-valued function $\varphi \in \mathcal{S}(\mathbb{R}^{n-1})$, define for each $x = (x', t) \in \mathbb{R}^n$ with $t \neq 0$*

$$(\mathscr{D}\varphi)(x) := \int_{\mathbb{R}^{n-1}} \sum_{k=1}^n (\partial_k \mathbb{E})(x' - y', t) A_{kn} \varphi(y') \, dy'. \tag{11.6.16}$$

Then, for any $\mathbb{C}^M$-valued function $\varphi \in \mathcal{S}(\mathbb{R}^{n-1})$,

$$\mathscr{D}\varphi \in C^\infty(\mathbb{R}^n_\pm), \quad L(\mathscr{D}\varphi) = 0 \text{ pointwise in } \mathbb{R}^n_\pm, \tag{11.6.17}$$

and for every $x' \in \mathbb{R}^{n-1}$ one has

$$\lim_{t \to 0^\pm} (\mathscr{D}\varphi)(x', t) = \pm \tfrac{1}{2} \varphi(x')$$

$$+ \lim_{\varepsilon \to 0^+} \int_{\substack{y' \in \mathbb{R}^{n-1} \\ |x' - y'| > \varepsilon}} \sum_{k=1}^n (\partial_k \mathbb{E})(x' - y', 0) A_{kn} \varphi(y') \, dy'. \tag{11.6.18}$$

Proof. Fix some $\mathbb{C}^M$-valued function $\varphi \in \mathcal{S}(\mathbb{R}^{n-1})$. The fact that $\mathscr{D}\varphi$ is of class C^∞ in $\mathbb{R}^n \setminus \{x_n = 0\}$ is seen from (11.6.16), (11.6.6), and estimates for the derivatives of $\mathbb{E}$ (cf. (11.3.27)). Moreover, for each $x = (x', x_n) \in \mathbb{R}^n$ with $x_n \neq 0$,

$$L(\mathscr{D}\varphi)(x) = \int_{\mathbb{R}^{n-1}} \sum_{k=1}^n \partial_{x_k}[(L\mathbb{E})(x' - y', x_n)] A_{kn} \varphi(y') \, dy' = 0 \tag{11.6.19}$$

where the last equality uses (11.6.15).

At this stage, there remains to prove (11.6.18). In this regard, if $x' \in \mathbb{R}^{n-1}$ has been fixed, making use of Corollary 11.14 we may write

$$\lim_{t \to 0^\pm} (\mathscr{D}\varphi)(x', t) = \sum_{k=1}^{n} \lim_{t \to 0^\pm} \int_{\mathbb{R}^{n-1}} (\partial_k \mathbb{E})(x' - y', t) A_{kn}\varphi(y')\, dy'$$

$$= \sum_{k=1}^{n} \lim_{t \to 0^\pm} \partial_k (\mathscr{S} A_{kn}\varphi)(x', t)$$

$$= \pm \frac{1}{2} \sum_{k=1}^{n} \delta_{kn} (A_{nn})^{-1} A_{kn}\varphi(x') \qquad (11.6.20)$$

$$+ \lim_{\varepsilon \to 0^+} \int_{\substack{y' \in \mathbb{R}^{n-1} \\ |x'-y'| > \varepsilon}} \sum_{k=1}^{n} (\partial_k \mathbb{E})(x' - y', 0) A_{kn}\varphi(y')\, dy'$$

$$= \pm \frac{1}{2}\, \varphi(x') + \lim_{\varepsilon \to 0^+} \int_{\substack{y' \in \mathbb{R}^{n-1} \\ |x'-y'| > \varepsilon}} \sum_{k=1}^{n} (\partial_k \mathbb{E})(x' - y', 0) A_{kn}\varphi(y')\, dy',$$

as wanted. $\qquad\qquad\qquad\qquad\qquad\qquad\qquad\qquad\qquad\qquad\qquad\quad$ □

Remark 11.17. The mapping $\mathscr{D}$ from (11.6.16) is called the double layer potential (or double layer operator) associated to the system L in $\mathbb{R}_\pm^n$ and (11.6.18) may be naturally regarded as the jump formula for the double layer potential in this setting.

When dealing with the so-called Neumann boundary value problem for the system L (discussed later in (11.6.34)), as boundary condition one prescribes a certain combination of first-order derivatives of the unknown function, namely $-\sum_{k=1}^{n} A_{nk}\partial_k$, amounting to what is called the conormal derivative associated with the system L. Our next result elaborates on the nature of boundary behavior of this conormal derivative of the single layer potential $\mathscr{S}$ introduced earlier in (11.6.11).

Corollary 11.18. *Assume that the system L is as in* (11.6.1)–(11.6.3) *and let $\mathbb{E}$ be the fundamental solution for L in $\mathbb{R}^n$ constructed in Theorem* 11.1. *Then for every $\mathbb{C}^M$-valued function $\varphi \in S(\mathbb{R}^{n-1})$,*

$$\lim_{t \to 0^\pm} \sum_{k=1}^{n} A_{nk}\partial_k (\mathscr{S}\varphi)(x', t) = \pm \frac{1}{2}\, \varphi(x') \qquad (11.6.21)$$

$$+ \lim_{\varepsilon \to 0^+} \int_{\substack{y' \in \mathbb{R}^{n-1} \\ |x'-y'| > \varepsilon}} \sum_{k=1}^{n} A_{nk}(\partial_k \mathbb{E})(x' - y', 0)\varphi(y')\, dy'.$$

Proof. For each $\mathbb{C}^M$-valued function $\varphi \in S(\mathbb{R}^{n-1})$, formula (11.6.21) follows using Corollary 11.14 by writing

$$\lim_{t \to 0^{\pm}} \sum_{k=1}^{n} A_{nk} \, \partial_k (\mathscr{S}\varphi)(x', t) = \sum_{k=1}^{n} A_{nk} \lim_{t \to 0^{\pm}} \partial_k (\mathscr{S}\varphi)(x', t)$$

$$= \pm \frac{1}{2} \sum_{k=1}^{n} A_{nk} \delta_{kn} (A_{nn})^{-1} \varphi(x') \tag{11.6.22}$$

$$+ \lim_{\varepsilon \to 0^+} \int_{\substack{y' \in \mathbb{R}^{n-1} \\ |x'-y'| > \varepsilon}} \sum_{k=1}^{n} A_{nk} (\partial_k \mathbb{E})(x' - y', 0) \varphi(y') \, dy'$$

$$= \pm \tfrac{1}{2} \, \varphi(x') + \lim_{\varepsilon \to 0^+} \int_{\substack{y' \in \mathbb{R}^{n-1} \\ |x'-y'| > \varepsilon}} \sum_{k=1}^{n} A_{nk} (\partial_k \mathbb{E})(x' - y', 0) \varphi(y') \, dy',$$

at every point $x' \in \mathbb{R}^{n-1}$. $\square$

Exercise 11.19. State and prove the analogue of Theorem 4.100 in the scenario when $\Theta : \mathbb{R}^n \setminus \{0\} \to M_{M \times M}(\mathbb{C})$, for some $M \in \mathbb{N}$, is a matrix-valued function whose scalar entries satisfy the conditions in (4.4.1). In particular, pay attention to the fact that the format of (4.9.43) now becomes

$$(T_{\Theta})^* = T_{\overline{\Theta^\top}^\vee} \quad \text{in} \quad [L^2(\mathbb{R}^n)]^M \tag{11.6.23}$$

where, as before, $\Theta^\top$ denotes the transposed of the matrix Θ.

Hint: Prove and use the fact that $[m_\Theta]^\top = m_{\Theta^\top}$.

The fact that for any $\varphi \in [S(\mathbb{R}^{n-1})]^M$ we have $L(\mathscr{D}\varphi) = 0$ in $\mathbb{R}_{\pm}^n$ (cf. (11.6.17)) means that we may regard the double layer potential operator $\mathscr{D}$ defined in (11.6.16) as a mechanism for generating an abundance of solutions of the partial differential equation $Lu = 0$ in $\mathbb{R}_{\pm}^n$. In addition, the jump formula (11.6.18) fully clarifies the boundary behavior of the function

$$\mathbf{u} := \mathscr{D}\varphi. \tag{11.6.24}$$

This is particularly relevant in the context of the Dirichlet problem for the $M \times M$ system L (assumed to be as in (11.6.1)–(11.6.3)), whose formulation in, say, the upper-half space reads (compare with (4.8.17) in the case $L := \Delta$)

$$\begin{cases} \mathbf{u} \in [C^\infty(\mathbb{R}_+^n)]^M, \\ L\mathbf{u} = 0 & \text{in } \mathbb{R}_+^n, \\ \mathbf{u}\Big|_{\partial \mathbb{R}_+^n}^{\text{ver}} = \psi & \text{on } \mathbb{R}^{n-1} \equiv \partial \mathbb{R}_+^n. \end{cases} \tag{11.6.25}$$

Above, $\psi \in [S(\mathbb{R}^{n-1})]^M$ is a given function (called the boundary datum) and, much as in the case of (4.8.17), the symbol $\mathbf{u}\Big|_{\partial \mathbb{R}_+^n}^{\text{ver}}$ stands for the "vertical limit" of $\mathbf{u}$

to the boundary of the upper-half space, understood at each point $x' \in \mathbb{R}^{n-1}$ as $\lim_{x_n \to 0^+} \mathbf{u}(x', x_n)$.

Indeed, focusing the search for a solution of (11.6.25) in the class of functions $\mathbf{u}$ defined as in (11.6.24) (with $\varphi \in [S(\mathbb{R}^{n-1})]^M$ yet to be determined) has the distinct advantage that the first two conditions in (11.6.25) are automatically satisfied (thanks to (11.6.17)), irrespective of the choice of φ. Keeping in mind (11.6.18), matters have been therefore reduced to solving the boundary integral equation

$$\tfrac{1}{2}\varphi(x') + \lim_{\varepsilon \to 0^+} \int_{\substack{y' \in \mathbb{R}^{n-1} \\ |x'-y'|>\varepsilon}} \sum_{k=1}^{n}(\partial_k \mathbb{E})(x' - y', 0)A_{kn}\varphi(y')\,\mathrm{d}y' = \psi(x'), \qquad (11.6.26)$$

with $x' \in \mathbb{R}^{n-1}$, so named since $\mathbb{R}^{n-1}$ may be regarded as the boundary of $\mathbb{R}^n_+$. To streamline notation, introduce the singular integral operator

$$(K\varphi)(x') := \lim_{\varepsilon \to 0^+} \int_{\substack{y' \in \mathbb{R}^{n-1} \\ |x'-y'|>\varepsilon}} \sum_{k=1}^{n}(\partial_k \mathbb{E})(x' - y', 0)A_{kn}\varphi(y')\,\mathrm{d}y', \qquad (11.6.27)$$

for each $x' \in \mathbb{R}^{n-1}$, typically referred to as the principal value (or boundary version) of the double layer associated with the system L in the upper-half space. Employing (11.6.27), we may write in place of (11.6.26)

$$(\tfrac{1}{2}I_{M\times M} + K)\varphi = \psi \quad \text{in} \quad \mathbb{R}^{n-1}. \qquad (11.6.28)$$

Reducing the entire boundary value problem (11.6.25) to solving the boundary integral equation (11.6.28) is perhaps the most distinguished feature of the technology described thus far for dealing with (11.6.25), called the method of boundary layer potentials (in standard partial differential equations parlance).

Regarding the boundary integral equation (11.6.28), involving the singular integral operator K introduced in (11.6.27), we wish to note that by taking the Fourier transform (in $\mathbb{R}^{n-1}$), then invoking Theorem 4.74 (whose applicability in the present setting is justified by virtue of the properties of $\mathbb{E}$ from (11.6.6)–(11.6.7)), we arrive at

$$(\tfrac{1}{2}I_{M\times M} + a_L)\widehat{\varphi} = \widehat{\psi} \quad \text{in} \quad S'(\mathbb{R}^{n-1}), \qquad (11.6.29)$$

where the $M \times M$ matrix-valued function a_L is defined by the formula

$$a_L(\xi') := -\sum_{k=1}^{n}\left(\int_{S^{n-2}}(\partial_k \mathbb{E})(\omega', 0)\log(\mathrm{i}(\xi' \cdot \omega'))\,\mathrm{d}\sigma(\omega')\right)A_{kn} \qquad (11.6.30)$$

for each $\xi' \in \mathbb{R}^{n-1} \setminus \{0'\}$. Thus, in the case when the matrix $\tfrac{1}{2}I_{M\times M} + a_L$ is invertible, at least at the formal level we may express the solution of (11.6.29) in the form

$$\varphi = \mathcal{F}^{-1}\left((\tfrac{1}{2}I_{M\times M} + a_L)^{-1}\widehat{\psi}\right). \qquad (11.6.31)$$

then ultimately conclude from (11.6.24) and (11.6.31) that

$$\mathbf{u} = \mathscr{D}\Big[\mathcal{F}^{-1}\big((\tfrac{1}{2}I_{M\times M} + a_L)^{-1}\widehat{\psi}\big)\Big] \tag{11.6.32}$$

solves the Dirichlet problem (11.6.25) for the system L in $\mathbb{R}_+^n$ with ψ as boundary datum.

In certain concrete cases of practical importance, the "symbol" function a_L from (11.6.30) is simple enough in order for us to make sense of the expression appearing in the right-hand side of (11.6.31). For example, it is visible from (11.6.30) that $a_L = 0$ in $\mathbb{R}^{n-1} \setminus \{0'\}$ whenever

$$\sum_{k=1}^{n} (\partial_k \mathbb{E})(x', 0) A_{kn} = 0 \quad \text{for all} \quad x' \in \mathbb{R}^{n-1} \setminus \{0'\}. \tag{11.6.33}$$

This is indeed the case when $L = \Delta$, the Laplacian in $\mathbb{R}^n$. To see that this happens, note that in this scalar ($M = 1$) case, $A_{jk} = \delta_{jk}$ for each $j, k \in \{1, \ldots, n\}$, and if E_Δ is the fundamental solution for the Laplacian from (7.1.12) then $(\partial_n E_\Delta)(x', x_n) = \frac{1}{\omega_{n-1}} \frac{x_n}{|x|^n}$ for each $x = (x', x_n) \in \mathbb{R}^n \setminus \{0\}$, hence $(\partial_n E_\Delta)(x', 0) = 0$ for each $x' \in \mathbb{R}^{n-1} \setminus \{0'\}$. The bottom line is that the symbol function a_Δ vanishes identically. In light of (11.6.31), this shows that for the Laplacian we must chose the (originally unspecified) function φ to be equal to the (given) boundary datum ψ. For this choice, the general formula (11.6.32) then reduces precisely to the classical expression (4.8.19, as the reader may verify without difficulty.

However, in general, the assignment $\psi \mapsto \mathcal{F}^{-1}\big((\tfrac{1}{2}I_{M\times M} + a_L)^{-1}\widehat{\psi}\big)$ can be of an intricate nature, leading to operators that are well beyond the class of singular integral operators introduced in Definition 4.93. This leads to the consideration of more exotic classes of operators, such as pseudodifferential operators, Fourier integral operators, etc., which are outside of the scope of the present monograph. This being said, the material developed here serves both as preparation and motivation for the reader interested in further pursuing such matters.

Similar considerations apply in the case of the Neumann problem for a system L as in (11.6.1)–(11.6.3), whose formulation reads

$$\begin{cases} \mathbf{u} \in [C^\infty(\mathbb{R}_+^n)]^M, \\[4pt] L\mathbf{u} = 0 & \text{in } \mathbb{R}_+^n, \\[4pt] -\sum_{k=1}^{n} A_{nk}[\partial_k \mathbf{u}]\Big|_{\partial\mathbb{R}_+^n}^{ver} = \psi & \text{on } \mathbb{R}^{n-1} \equiv \partial\mathbb{R}_+^n, \end{cases} \tag{11.6.34}$$

where $\psi \in [\mathcal{S}(\mathbb{R}^{n-1})]^M$ is the boundary datum. Granted the results proved in Theorem 11.14 and Corollary 11.18, this time it is natural to seek a solution in the form (compare with (11.6.24))

$$\mathbf{u} := \mathscr{S}\varphi \quad \text{in } \mathbb{R}_+^n, \tag{11.6.35}$$

where the function $\varphi \in [\mathcal{S}(\mathbb{R}^{n-1})]^M$ is subject to the boundary integral equation (implied by (11.6.21))

$$-\tfrac{1}{2}\varphi(x') - \lim_{\varepsilon \to 0^+} \int\limits_{\substack{y' \in \mathbb{R}^{n-1} \\ |x'-y'|>\varepsilon}} \sum_{k=1}^{n} A_{nk}(\partial_k \mathbb{E})(x'-y',0)\varphi(y')\,dy' = \psi(x') \qquad (11.6.36)$$

for each $x' \in \mathbb{R}^{n-1}$. Once again, Theorem 4.74 may be used in order to rewrite (11.6.36) on the Fourier transform side as

$$\left(-\tfrac{1}{2}I_{M\times M} + \widetilde{a}_L\right)\widehat{\varphi} = \widehat{\psi} \quad \text{in} \quad \mathcal{S}'(\mathbb{R}^{n-1}), \qquad (11.6.37)$$

where the $M \times M$ matrix-valued function $\widetilde{a}_L$ is now defined by the formula

$$\widetilde{a}_L(\xi') := \sum_{k=1}^{n} A_{nk}\left(\int_{S^{n-2}} (\partial_k \mathbb{E})(\omega',0)\log(\mathrm{i}(\xi' \cdot \omega'))\,d\sigma(\omega')\right) \qquad (11.6.38)$$

for each $\xi' \in \mathbb{R}^{n-1} \setminus \{0'\}$. As regards the symbol function $\widetilde{a}_L$ for the Neumann problem, one can see from (11.6.38), (11.6.30), and (11.3.32) that

$$\widetilde{a}_L = -[a_{L^\top}]^\top \qquad (11.6.39)$$

which shows that $\widetilde{a}_L$ has, up to transpositions, the same nature as the symbol function for the Dirichlet problem. Hence, the same type of considerations as in the latter case apply. For example, corresponding to the case when $L = \Delta$ we have $\widetilde{a}_\Delta = 0$ which means that a solution to the Neumann problem for the Laplacian in the upper-half space, i.e.,

$$\begin{cases} u \in C^\infty(\mathbb{R}^n_+), \\ \Delta u = 0 & \text{in } \mathbb{R}^n_+, \\ -[\partial_n u]\big|_{\partial\mathbb{R}^n_+}^{ver} = \psi & \text{on } \mathbb{R}^{n-1} \equiv \partial\mathbb{R}^n_+, \end{cases} \qquad (11.6.40)$$

where $\psi \in \mathcal{S}(\mathbb{R}^{n-1})$ is the boundary datum, is given by

$$u(x) := \int_{\mathbb{R}^{n-1}} E_\Delta(x'-y',x_n)\psi(y')\,dy' \quad \text{for} \quad x = (x',x_n) \in \mathbb{R}^n_+, \qquad (11.6.41)$$

with E_Δ denoting the fundamental solution for the Laplacian (cf. (7.1.12)).

Further Notes for Chapter 11. As already mentioned, the approach developed in Section 10.2 contains a constructive procedure for reducing the task of finding a fundamental solution for the Lamé operator to the scalar case. This scheme has been implemented in Section 11.1 based on the knowledge of a fundamental solution for the polyharmonic operator from Section 7.5. Subsequently, in Section 11.2 a fundamental solution for the Stokes operator is computed indirectly by rigorously making sense of the informal observation that this operator is a limiting case of the Lamé operator (taking $\mu = 1$ and sending $\lambda \to \infty$).

From the considerations in Section 10.2 it was also already known that the higher order homogeneous elliptic constant coefficient linear systems considered in Section 11.3 posses fundamental solutions that are tempered distributions with singular support at the origin. The new issue addressed in Theorem 11.1 is to find an explicit formula and to study other properties of such fundamental solutions. This theorem refines and further builds on the results proved in [34], [38], [57], [59], and [67], in various degrees of generality. Fundamental solutions for variable coefficient elliptic operators on manifolds have been studied in [55].

11.7 Additional Exercises for Chapter 11

Exercise 11.20. Use Theorem 11.1 in a similar manner as in the proof of Proposition 11.3 in order to derive a formula for a fundamental solution for the Lamé operator in $\mathbb{R}^n$ when $n \geq 3$ is arbitrary and odd.

Exercise 11.21. Similarly to Exercise 11.20, derive a formula for a fundamental solution for the Lamé operator in $\mathbb{R}^n$ when n is even.

Exercise 11.22. Let L be a homogeneous differential operator in $\mathbb{R}^n$, $n \geq 2$, of order $2m$, $m \in \mathbb{N}$, with complex constant coefficients as in (11.3.1) that satisfies (11.3.3), and recall (11.3.41). Also, let $q \in \mathbb{N}_0$ be such that $n + q$ is even. Consider the matrix $\mathbb{E} := (E_{jk})_{1 \leq j,k \leq M}$ with entries given by

$$E_{jk}(x) = \frac{-1}{(2\pi i)^n (2m + q)!} \int_{S^{n-1}} (x \cdot \xi)^{2m+q} \log\left(\frac{x \cdot \xi}{i}\right) P^{jk}(\xi) \, d\sigma(\xi) \qquad (11.7.42)$$

for $x \in \mathbb{R}^n \setminus \{0\}$ and $j, k \in \{1, \ldots, M\}$, where log denotes the principal branch of the complex logarithm defined for points $z \in \mathbb{C} \setminus \{x : x \in \mathbb{R}, x \leq 0\}$. Prove that $\mathbb{E} \in \mathcal{M}_{M \times M}(\mathcal{S}'(\mathbb{R}^n))$ is a fundamental solution for the system $\Delta_x^{(n+q)/2} L$ in $\mathbb{R}^n$.

Exercise 11.23. Let L_1, L_2 be two homogeneous $M \times M$ systems of differential operators in $\mathbb{R}^n$, $n \geq 2$, of orders $2m_1$ and $2m_2$, respectively, where $m_1, m_2 \in \mathbb{N}$, with complex constant coefficients, satisfying $\det L_1(\xi) \neq 0$ and $\det L_2(\xi) \neq 0$ for every $\xi \in \mathbb{R}^n \setminus \{0\}$. Prove that, with the notational convention employed in part (7) of Theorem 11.1, one has

$$L_2 \mathbb{E}_{L_1 L_2} = \mathbb{E}_{L_1} + P \quad \text{in} \quad \mathcal{S}'(\mathbb{R}^n), \qquad (11.7.43)$$

where P is zero if either n is odd or $2m_1 < n$, and P is an $M \times M$ matrix whose entries are homogeneous polynomials of degree $2m_1 - n$ if $2m_1 \geq n$.

Exercise 11.24. Recall the notational convention employed in part (7) of Theorem 11.1 as well as $F_{m,n}$ from (7.5.2). Prove that for every $m \in \mathbb{N}$ one has

$$\mathbb{E}_{\Delta^m} = F_{m,n} + P \quad \text{in} \quad \mathcal{S}'(\mathbb{R}^n), \qquad (11.7.44)$$

where P is zero if either n is odd or $2m < n$, and P is a homogeneous polynomial of degree $2m - n$ if $2m \geq n$.

Chapter 12
Sobolev Spaces

Abstract While Lebesgue spaces play a most basic role in analysis, it is highly desirable to consider a scale of spaces which contains provisions for quantifying smoothness (measured in a suitable sense). This is the key feature of the so-called Sobolev spaces, introduced and studied at some length in this chapter in a completely self-contained manner. The starting point is the treatment of global L^2-based Sobolev spaces of arbitrary smoothness in the entire Euclidean space, using the Fourier transform as the main tool. We then proceed to define Sobolev spaces in arbitrary open sets, both via restriction (which permits the consideration of arbitrary amounts of smoothness) and in an intrinsic fashion (for integer amounts of smoothness, demanding that distributional derivatives up to a certain order are square-integrable in the respective open set). When the underlying set is a bounded Lipschitz domain, both these brands of Sobolev spaces (defined intrinsically and via restriction) coincide for an integer amount of smoothness. A key role in the proof of this result is played by Calderón's extension operator, mapping functions originally defined in the said Lipschitz domain to the entire Euclidean ambient with preservation of Sobolev class. Finally, the fractional Sobolev space of order $1/2$ is defined on the boundary of a Lipschitz domain as the space of square-integrable functions satisfying a finiteness condition involving a suitable Gagliardo–Slobodeckij semi-norm. This is then linked to Sobolev spaces in Lipschitz domains via trace and extension results.

12.1 Global Sobolev Spaces $H^s(\mathbb{R}^n)$, $s \in \mathbb{R}^n$

Throughout this section we will use the notation

$$\langle \xi \rangle := (1 + |\xi|^2)^{1/2}, \qquad \forall \xi \in \mathbb{R}^n. \tag{12.1.1}$$

Then, for $s \in \mathbb{R}$, the Sobolev space $H^s(\mathbb{R}^n)$ is defined by

© Springer Nature Switzerland AG 2018

D. Mitrea, *Distributions, Partial Differential Equations, and Harmonic Analysis*,
Universitext, https://doi.org/10.1007/978-3-030-03296-8_12

$$H^s(\mathbb{R}^n) := \{u \in S'(\mathbb{R}^n) : \langle \xi \rangle^s \widehat{u} \in L^2(\mathbb{R}^n)\} \tag{12.1.2}$$

where $\widehat{u}$ is the Fourier transform of u (cf. Proposition 4.21). Note that for each distribution $u \in S'(\mathbb{R}^n)$ we have $\widehat{u} \in S'(\mathbb{R}^n)$ and $\langle \xi \rangle^s \widehat{u}$ is understood as the multiplication between the slowly increasing function $\langle \xi \rangle^s$ (cf. Exercise 3.13) and the tempered distribution $\widehat{u}$ (see *(b)* in Theorem 4.14). Hence, we are demanding that the tempered distribution $\langle \xi \rangle^s \widehat{u}$ is given by a square-integrable function in $\mathbb{R}^n$ (cf. (4.1.9) in this regard). In particular, this implies that $\widehat{u} \in L^1_{loc}(\mathbb{R}^n)$.

The space $H^s(\mathbb{R}^n)$ introduced in (12.1.2) should be thought of as the collection of all tempered distributions which, in a suitable sense, have "derivatives up to order s" in $L^2(\mathbb{R}^n)$. This heuristic principle is substantiated by Theorem 12.24 which shows this is indeed the case when s is a natural number. In particular, it is natural to think of the parameter s as being a smoothness index.

We begin by systematically studying the basic properties of the scale of global Sobolev spaces just introduced. For starters, it is easy to check that $H^s(\mathbb{R}^n)$ is a complex vector space. When endowed with an appropriate inner product, $H^s(\mathbb{R}^n)$ becomes a Hilbert space. This issue is discussed in the next theorem which also contains an important density result.

Theorem 12.1. *Let $s \in \mathbb{R}$. Then the following are true.*

(1) Define

$$(u, v)_{H^s(\mathbb{R}^n)} := \int_{\mathbb{R}^n} \langle \xi \rangle^{2s} \widehat{u}(\xi) \overline{\widehat{v}(\xi)} \, d\xi, \qquad \forall \, u, v \in H^s(\mathbb{R}^n). \tag{12.1.3}$$

Then this is an inner product on $H^s(\mathbb{R}^n)$ with respect to which $H^s(\mathbb{R}^n)$ is a Hilbert space, with norm

$$\|u\|_{H^s(\mathbb{R}^n)} := \sqrt{(u, u)_{H^s(\mathbb{R}^n)}} = \left\| \langle \xi \rangle^s \widehat{u} \right\|_{L^2(\mathbb{R}^n)}, \qquad \forall \, u \in H^s(\mathbb{R}^n). \tag{12.1.4}$$

(2) The following topological inclusions hold: $S(\mathbb{R}^n) \hookrightarrow H^s(\mathbb{R}^n) \hookrightarrow S'(\mathbb{R}^n)$.
(3) The space $C_0^\infty(\mathbb{R}^n)$ is dense in $H^s(\mathbb{R}^n)$.

Remark 12.2. From items *(1)–(2)* in Theorem 12.1 and part *(b)* of Theorem 14.1 it follows that whenever $\{u_j\}_{j \in \mathbb{N}}$ is a sequence that converges both in $H^{s_1}(\mathbb{R}^n)$ and in $H^{s_2}(\mathbb{R}^n)$, for some real numbers s_1, s_2, then necessarily the limits coincide (as tempered distributions).

Before presenting the proof of Theorem 12.1 we isolate a number of technical results in the next three lemmas.

Lemma 12.3. *The following are true.*

(1) If $s_1, s_2 \in \mathbb{R}$ then $\langle \xi \rangle^{s_1 + s_2} = \langle \xi \rangle^{s_1} \langle \xi \rangle^{s_2}$ for every $\xi \in \mathbb{R}^n$.
(2) If $s \in \mathbb{R}$ then $\langle \xi + \eta \rangle^s \leq 2^{|s|/2} \langle \xi \rangle^s \langle \eta \rangle^{|s|}$ for every $\xi, \eta \in \mathbb{R}^n$.

Proof. The identity in item *(1)* is immediate. To prove the inequality in item *(2)*, note that for each $\xi, \eta \in \mathbb{R}^n$ we have $1 + |\xi + \eta|^2 \leq 2(1 + |\xi|^2)(1 + |\eta|^2)$. If $s \geq 0$ the latter inequality raised to power $s/2$ gives the inequality in *(2)*. Also, if $s < 0$, from what we just proved (written for $-s > 0$) it follows that $\langle \xi + \eta \rangle^{-s} \leq 2^{-s/2} \langle \xi \rangle^{-s} \langle \eta \rangle^{-s}$ for every $\xi, \eta \in \mathbb{R}^n$. In particular, replacing ξ by $\xi + \eta$ and η by $-\eta$, we obtain $\langle \xi \rangle^{-s} \leq 2^{-s/2} \langle \xi + \eta \rangle^{-s} \langle \eta \rangle^{-s}$ for every $\xi, \eta \in \mathbb{R}^n$. The desired conclusion follows since $-s = |s|$. $\qquad\square$

Lemma 12.4. *Let $s \in \mathbb{R}$. Define the space*

$$L_s^2(\mathbb{R}^n) := \{ v : \mathbb{R}^n \to \mathbb{C} : \langle \xi \rangle^s v \in L^2(\mathbb{R}^n) \}, \qquad (12.1.5)$$

and consider

$$\|v\|_{L_s^2(\mathbb{R}^n)} := \left\| \langle \xi \rangle^s v \right\|_{L^2(\mathbb{R}^n)}, \qquad \forall\, v \in L_s^2(\mathbb{R}^n). \qquad (12.1.6)$$

Then the following are true.

(1) The expression in (12.1.6) defines a norm on $L_s^2(\mathbb{R}^n)$, which is actually induced by the inner product

$$(u, v)_{L_s^2(\mathbb{R}^n)} := \int_{\mathbb{R}^n} \langle \xi \rangle^{2s} u(\xi) \overline{v}(\xi) \, \mathrm{d}\xi, \qquad \forall\, u, v \in L_s^2(\mathbb{R}^n) \qquad (12.1.7)$$

with respect to which $L_s^2(\mathbb{R}^n)$ becomes a Hilbert space.

(2) The space of Schwartz functions $\mathcal{S}(\mathbb{R}^n)$ is continuously imbedded in $L_s^2(\mathbb{R}^n)$.

(3) The identification of functions from $L_s^2(\mathbb{R}^n)$ with tempered distributions via integration against Schwartz functions induces a continuous embedding of the space $L_s^2(\mathbb{R}^n)$ into $\mathcal{S}'(\mathbb{R}^n)$.

(4) Let $\varphi \in \mathcal{S}(\mathbb{R}^n)$, $v \in L_s^2(\mathbb{R}^n)$, and define the function

$$f(\xi) := \int_{\mathbb{R}^n} \varphi(\xi - \eta) v(\eta) \, \mathrm{d}\eta, \qquad \forall\, \xi \in \mathbb{R}^n. \qquad (12.1.8)$$

Then f is well-defined (via an absolutely convergent integral for each $\xi \in \mathbb{R}^n$), and may be alternatively expressed as

$$f(\xi) = \int_{\mathbb{R}^n} \varphi(\eta) v(\xi - \eta) \, \mathrm{d}\eta, \qquad \forall\, \xi \in \mathbb{R}^n. \qquad (12.1.9)$$

Moreover, f belongs to the space of slowly increasing functions $\mathcal{L}(\mathbb{R}^n)$ (in particular, f induces a tempered distribution via integration against Schwartz functions), and for each multi-index $\alpha \in \mathbb{N}_0^n$ the following formula holds

$$(\partial^\alpha f)(\xi) = \int_{\mathbb{R}^n} (\partial^\alpha \varphi)(\xi - \eta) v(\eta) \, \mathrm{d}\eta, \qquad \forall\, \xi \in \mathbb{R}^n. \qquad (12.1.10)$$

(5) Fix $\varphi \in \mathcal{S}(\mathbb{R}^n)$ and $v \in L_s^2(\mathbb{R}^n)$. Regarding v as a tempered distribution (cf. item (3) above) define $\varphi * v \in \mathcal{S}'(\mathbb{R}^n)$ in the sense of item (e) in Theorem 4.19. Then, with f as in item (4) above, the following is true

$$\varphi * v = f \quad in \quad \mathcal{S}'(\mathbb{R}^n). \qquad (12.1.11)$$

Proof. That (12.1.6) is a norm on $L_s^2(\mathbb{R}^n)$ is easily seen from the properties of the norm on $L^2(\mathbb{R}^n)$. Also, it is immediate that the norm (12.1.6) is induced by the inner product (12.1.7). To prove that $L_s^2(\mathbb{R}^n)$ is complete with respect to the norm (12.1.6), consider the mapping

$$L_s^2(\mathbb{R}^n) \ni v \mapsto \langle \xi \rangle^s v \in L^2(\mathbb{R}^n). \tag{12.1.12}$$

Clearly, this mapping is well-defined, injective, and an isometry. We claim that it is also surjective. Indeed, if $w \in L^2(\mathbb{R}^n)$ then the function $v := \langle \xi \rangle^{-s} w$ belongs to $L_s^2(\mathbb{R}^n)$ and satisfies $\langle \xi \rangle^s v = w$. Hence, the mapping in (12.1.12) is a bijective isometry. Consequently, since $L^2(\mathbb{R}^n)$ is complete, we have that $L_s^2(\mathbb{R}^n)$ is also complete.

As regards the statement in item *(2)*, let $\varphi \in \mathcal{S}(\mathbb{R}^n)$. Choose a positive number m satisfying $m > n/2 + s$ and estimate

$$\Big(\int_{\mathbb{R}^n} \langle \xi \rangle^{2s} |\varphi(\xi)|^2 \, d\xi \Big)^{1/2} \leq \sup_{\xi \in \mathbb{R}^n} [(1 + |\xi|)^m |\varphi(\xi)|] \times \tag{12.1.13}$$

$$\times \Big(\int_{\mathbb{R}^n} \langle \xi \rangle^{2(s-m)} \, d\xi \Big)^{1/2} < \infty,$$

where for the last inequality in (12.1.13) we have used Remark 3.4. This shows that $\mathcal{S}(\mathbb{R}^n) \subseteq L_s^2(\mathbb{R}^n)$. In addition, the estimate in (12.1.13) and Exercise 3.8 also give that the latter embedding is sequentially continuous, thus continuous (recall that $\mathcal{S}(\mathbb{R}^n)$ is metrizable; cf. Fact 3.6). This completes the proof of the statement in item *(2)*.

Moving on to item *(3)*, let $v \in L_s^2(\mathbb{R}^n)$ and $\varphi \in \mathcal{S}(\mathbb{R}^n)$. By item *(2)* (used with $-s$ in place of s) we know that $\varphi \in L_{-s}^2(\mathbb{R}^n)$. Based on Hölder's inequality, (12.1.13) (with $-s$ in place of s, and $m > n/2 - s$), and Remark 3.4 we may estimate

$$\int_{\mathbb{R}^n} |v(\xi)\varphi(\xi)| \, d\xi \leq \|v\|_{L_s^2(\mathbb{R}^n)} \|\varphi\|_{L_{-s}^2(\mathbb{R}^n)}$$

$$\leq C\|v\|_{L_s^2(\mathbb{R}^n)} \sup_{\xi \in \mathbb{R}^n} [(1 + |\xi|)^m |\varphi(\xi)|] < \infty, \tag{12.1.14}$$

for a finite constant $C = C(s, n) > 0$ independent of φ. This proves that the measurable function $v\varphi$ is absolutely integrable in $\mathbb{R}^n$ and that

$$\Big| \int_{\mathbb{R}^n} v(\xi)\varphi(\xi) \, d\xi \Big| \leq C\|v\|_{L_s^2(\mathbb{R}^n)} \sup_{\xi \in \mathbb{R}^n} [(1 + |\xi|)^m |\varphi(\xi)|]. \tag{12.1.15}$$

By invoking Exercise 4.2 we conclude that v induces a tempered distribution defined via integration against Schwartz functions. Finally, the assignment taking any function $v \in L_s^2(\mathbb{R}^n)$ into the tempered distribution

$$\mathcal{S}(\mathbb{R}^n) \ni \varphi \mapsto \int_{\mathbb{R}^n} v(\xi)\varphi(\xi) \, d\xi \in \mathbb{C} \tag{12.1.16}$$

is linear and injective (thanks to Theorem 1.3). In addition, estimate (12.1.15) and Fact 4.11 imply that the respective embedding is sequentially continuous, thus continuous, as wanted. This shows that we have the embedding claimed in item *(3)*.

To prove the statement in item *(4)*, fix $\varphi \in \mathcal{S}(\mathbb{R}^n)$ and $v \in L_s^2(\mathbb{R}^n)$. Also, pick $\xi \in \mathbb{R}^n$. Based on the Cauchy–Schwarz inequality, item *(2)* in Lemma 12.3, and the current item *(2)* we may estimate

$$\int_{\mathbb{R}^n} |\varphi(\xi - \eta)v(\eta)| \, d\eta \leq \|v\|_{L_s^2(\mathbb{R}^n)} \Big(\int_{\mathbb{R}^n} \langle \eta \rangle^{-2s} |\varphi(\xi - \eta)|^2 \, d\eta \Big)^{1/2}$$

$$\leq 2^{|s|} \langle \xi \rangle^{-2s} \|v\|_{L_s^2(\mathbb{R}^n)} \|\varphi\|_{L_{|s|}^2(\mathbb{R}^n)} < \infty. \tag{12.1.17}$$

Thus, the integral $\int_{\mathbb{R}^n} \varphi(\xi - \eta)v(\eta) \, d\eta$ is absolutely convergent for each $\xi \in \mathbb{R}^n$. Hence, f is well defined and a simple change of variables shows that (12.1.9) holds. In addition, (12.1.17) gives that, if $C := 2^{|s|} \|v\|_{L_s^2(\mathbb{R}^n)} \|\varphi\|_{L_{|s|}^2(\mathbb{R}^n)} \in [0, \infty)$, then

$$|f(\xi)| \leq C \langle \xi \rangle^{-2s}, \qquad \forall \xi \in \mathbb{R}^n. \tag{12.1.18}$$

To conclude that f belongs to $\mathcal{L}(\mathbb{R}^n)$ there remains to prove that f is smooth and its derivatives satisfy estimates similar in spirit to (12.1.18). To this end, based on what we have proved so far and induction over $|\alpha| \in \mathbb{N}$, it suffices to show that for each $j \in \{1, \ldots, n\}$ the following formula holds

$$(\partial_j f)(\xi) = \int_{\mathbb{R}^n} (\partial_j \varphi)(\xi - \eta)v(\eta) \, d\eta, \qquad \forall \xi \in \mathbb{R}^n. \tag{12.1.19}$$

With the goal of proving (12.1.19), fix $j \in \{1, \ldots, n\}$ and $\xi \in \mathbb{R}^n$. From (12.1.17) (used with $\partial_j \varphi$ in place of φ) we know that the integral in (12.1.19) is absolutely convergent. Next, for each $h \in [-1, 0) \cup (0, 1]$ define

$$F_h(\eta) := \frac{\varphi(\xi + h\mathbf{e}_j - \eta) - \varphi(\xi - \eta)}{h}, \qquad \forall \eta \in \mathbb{R}^n. \tag{12.1.20}$$

Since $\varphi \in C^\infty(\mathbb{R}^n)$, we have that

$$\lim_{h \to 0} F_h(\eta) = (\partial_j \varphi)(\xi - \eta) \qquad \text{for every } \eta \in \mathbb{R}^n, \tag{12.1.21}$$

while the Mean Value Theorem gives

$$|F_h(\eta)| \leq \sup_{|t| \leq 1} \left| (\partial_j \varphi)(\xi + t\mathbf{e}_j - \eta) \right| \qquad \text{for each } \eta \in \mathbb{R}^n. \tag{12.1.22}$$

Note that if we set $c(\xi) := |\xi| + 2$ then

$$1 + |\eta| \leq c(\xi)(1 + |\xi + t\mathbf{e}_j - \eta|), \qquad \forall t \in [-1, 1], \ \forall \eta \in \mathbb{R}^n. \tag{12.1.23}$$

This, combined with Remark 3.4 applied for $m > n/2 + |s|$, gives that there exists some constant $C \in (0, \infty)$ depending only on φ, m, and ξ, such that

$$\sup_{|t|\leq 1}\left|(\partial_j\varphi)(\xi + te_j - \eta)\right| \leq \frac{C}{(1 + |\eta|)^m}, \qquad \forall\, \eta \in \mathbb{R}^n. \tag{12.1.24}$$

In concert, (12.1.22) and (12.1.24) imply

$$\sup_{|h|\leq 1} |F_h(\eta)||v(\eta)| \leq C\frac{\langle\eta\rangle^{-s}}{(1 + |\eta|)^m}\,\langle\eta\rangle^s|v(\eta)| \leq C\frac{(1 + |\eta|)^{|s|}}{(1 + |\eta|)^m}\,\langle\eta\rangle^s|v(\eta)|$$

$$\leq C\frac{1}{(1 + |\eta|)^{m-|s|}}\,\langle\eta\rangle^s|v(\eta)|, \qquad \forall\, \eta \in \mathbb{R}^n. \tag{12.1.25}$$

Recalling that $\langle\eta\rangle^s|v(\eta)| \in L^2(\mathbb{R}^n)$ and $m - |s| > n/2$, from (12.1.25) and an application of Hölder's inequality we obtain $\sup_{|h|\leq 1} |F_h||v| \leq G$ for some function $G \in L^1(\mathbb{R}^n)$. The latter, (12.1.21), and Lebesgue's Dominated Convergence Theorem allow us to write

$$\lim_{h\to 0}\frac{f(\xi + he_j) - f(\xi)}{h} = \lim_{h\to 0}\int_{\mathbb{R}^n} F_h(\eta)v(\eta)\,d\eta$$

$$= \int_{\mathbb{R}^n}(\partial_j\varphi)(\xi - \eta)v(\eta)\,d\eta. \tag{12.1.26}$$

This completes the proof of (12.1.19), finishing the treatment of item *(4)*.

On to item *(5)*, given any $\psi \in \mathcal{S}(\mathbb{R}^n)$ we may use (4.1.43), item *(3)* in the current lemma, the definition of the convolution at the level of functions, Fubini's theorem, and (12.1.8) to write

$$\langle\varphi * v, \psi\rangle = \langle v, \varphi^\vee * \psi\rangle = \int_{\mathbb{R}^n} v(\xi)(\varphi^\vee * \psi)(\xi)\,d\xi$$

$$= \int_{\mathbb{R}^n} v(\xi)\Big(\int_{\mathbb{R}^n}\varphi(\eta - \xi)\psi(\eta)\,d\eta\Big)d\xi$$

$$= \int_{\mathbb{R}^n}\psi(\eta)\Big(\int_{\mathbb{R}^n}\varphi(\eta - \xi)v(\xi)\,d\xi\Big)d\eta$$

$$= \int_{\mathbb{R}^n}\psi(\eta)f(\eta)\,d\eta. \tag{12.1.27}$$

To justify the applicability of Fubini's theorem in the fourth equality, note that Hölder's inequality, item *(2)* in Lemma 12.3 which gives

$$\langle\xi\rangle^{-2s} \leq 2^{|s|}\langle\eta - \xi\rangle^{-2s}\langle\eta\rangle^{2|s|} \quad \text{for}\ \ \xi, \eta \in \mathbb{R}^n, \tag{12.1.28}$$

item *(2)* in the current lemma, and Remark 3.4, imply

$$\int_{\mathbb{R}^n} \int_{\mathbb{R}^n} |v(\xi)||\varphi(\eta - \xi)||\psi(\eta)| \, d\xi \, d\eta$$

$$\leq \|v\|_{L_s^2(\mathbb{R}^n)} \int_{\mathbb{R}^n} |\psi(\eta)| \Big(\int_{\mathbb{R}^n} \langle \xi \rangle^{-2s} |\varphi(\eta - \xi)|^2 \, d\xi \Big)^{1/2} \, d\eta$$

$$\leq 2^{|s|} \|v\|_{L_s^2(\mathbb{R}^n)} \int_{\mathbb{R}^n} |\psi(\eta)| \langle \eta \rangle^{2|s|} \Big(\int_{\mathbb{R}^n} \langle \eta - \xi \rangle^{-2s} |\varphi(\eta - \xi)|^2 \, d\xi \Big)^{1/2} \, d\eta$$

$$\leq 2^{|s|} \|v\|_{L_s^2(\mathbb{R}^n)} \|\varphi\|_{L_{-s}^2(\mathbb{R}^n)} \times$$

$$\times \sup_{\eta \in \mathbb{R}^n} \big[(1 + |\eta|)^{2|s|+n+1} |\psi(\eta)| \big] \int_{\mathbb{R}^n} \frac{d\eta}{(1 + |\eta|)^{n+1}} < \infty. \quad (12.1.29)$$

Thus $\int_{\mathbb{R}^n} \int_{\mathbb{R}^n} v(\xi) \varphi(\eta - \xi) \psi(\eta) \, d\xi \, d\eta$ is absolutely convergent, so we may indeed reverse the order of integration. Having established (12.1.27), we conclude that (12.1.11) holds. This finishes the proof of the lemma. □

Lemma 12.5. *If $u \in H^s(\mathbb{R}^n)$, with $s \in \mathbb{R}$ arbitrary, then*

$$\langle u, \varphi \rangle = (2\pi)^{-n} \int_{\mathbb{R}^n} \widehat{u}(\xi) \widehat{\varphi}(-\xi) \, d\xi, \quad \forall \varphi \in \mathcal{S}(\mathbb{R}^n). \quad (12.1.30)$$

Consequently, for each $s \in \mathbb{R}$ the following estimate holds:

$$|\langle u, \varphi \rangle| \leq (2\pi)^{-n} \|\widehat{u}\|_{L_s^2(\mathbb{R}^n)} \|\widehat{\varphi}\|_{L_{-s}^2(\mathbb{R}^n)}$$
$$\forall u \in H^s(\mathbb{R}^n), \ \forall \varphi \in \mathcal{S}(\mathbb{R}^n). \quad (12.1.31)$$

Proof. For $\varphi \in \mathcal{S}(\mathbb{R}^n)$ write

$$\langle u, \varphi \rangle = (2\pi)^{-n} \langle \widehat{u}^{\vee}, \varphi \rangle = (2\pi)^{-n} \langle \widehat{u}, \varphi^{\vee} \rangle = (2\pi)^{-n} \langle \widehat{u}, \langle \xi \rangle^s \langle \xi \rangle^{-s} \widehat{\varphi^{\vee}} \rangle$$

$$= (2\pi)^{-n} \big\langle \langle \xi \rangle^s \widehat{u}, \langle \xi \rangle^{-s} \widehat{\varphi^{\vee}} \big\rangle$$

$$= (2\pi)^{-n} \int_{\mathbb{R}^n} \langle \xi \rangle^s \widehat{u}(\xi) \langle \xi \rangle^{-s} \widehat{\varphi}(-\xi) \, d\xi$$

$$= (2\pi)^{-n} \int_{\mathbb{R}^n} \widehat{u}(\xi) \widehat{\varphi}(-\xi) \, d\xi. \quad (12.1.32)$$

The first and second equality in (12.1.32) use Proposition 4.32 and Proposition 4.21. The fourth equality is based on the definition of the multiplication of the tempered distribution $\widehat{u}$ with the slowly increasing function $\langle \xi \rangle^s$ (keeping in mind that $\langle \xi \rangle^{-s} \widehat{\varphi^{\vee}}$ belongs to $\mathcal{S}(\mathbb{R}^n)$ by *(a)* in Theorem 3.14 and *(c)* in Theorem 3.21). Also, the fifth equality uses the fact that $\langle \xi \rangle^s \widehat{u} \in L^2(\mathbb{R}^n)$ and Example 4.8. This proves (12.1.30).

Finally, the estimate in (12.1.31) becomes a consequence of (12.1.30), item *(2)* in Lemma 12.4, and Hölder's inequality. □

After this preamble we are ready to proceed with the proof of Theorem 12.1.

Proof of Theorem 12.1. The fact that (12.1.3) is an inner product on $H^s(\mathbb{R}^n)$ follows from the properties of the Fourier transform, those of the inner product in $L^2(\mathbb{R}^n)$, and (12.1.31).

We will show that $H^s(\mathbb{R}^n)$ is a Hilbert space by proving that it is isometrically isomorphic to the Hilbert space $L^2_s(\mathbb{R}^n)$. The starting point is the observation that

$$H^s(\mathbb{R}^n) = \{u \in \mathcal{S}'(\mathbb{R}^n) : \widehat{u} \in L^2_s(\mathbb{R}^n)\}. \tag{12.1.33}$$

Then a combination of (12.1.33), item *(3)* in Lemma 12.4, and part *(a)* of Theorem 4.26, give that (the restriction of) the Fourier transform $\mathcal{F}$ at the level of tempered distributions is a well-defined and injective map from $H^s(\mathbb{R}^n)$ into $L^2_s(\mathbb{R}^n)$. This map is also surjective since for $v \in L^2_s(\mathbb{R}^n)$, if we take $u := \mathcal{F}^{-1}v$ (where $\mathcal{F}^{-1}$ is the inverse of the map in part *(a)* of Theorem 4.26) then $u \in \mathcal{S}'(\mathbb{R}^n)$ and $\widehat{u} = v \in L^2_s(\mathbb{R}^n)$, thus $u \in H^s(\mathbb{R}^n)$. In addition, (12.1.4) and (12.1.6) imply that the Fourier transform is an isometry from $H^s(\mathbb{R}^n)$ into $L^2_s(\mathbb{R}^n)$. In summary,

$$\mathcal{F} : H^s(\mathbb{R}^n) \to L^2_s(\mathbb{R}^n) \quad \text{is a bijective isometry.} \tag{12.1.34}$$

Having established that $L^2_s(\mathbb{R}^n)$ is a Hilbert space (cf. item *(1)* in Lemma 12.4), from (12.1.34) we deduce that $H^s(\mathbb{R}^n)$ is also a Hilbert space. This proves the statement in item *(1)*.

The statement in item *(2)* is a consequence of items *(2)*–*(3)* in Lemma 12.4 and (12.1.34). We are left with proving the statement in item *(3)*. In a first stage, we observe that as a consequence of the fact that $C_0^\infty(\mathbb{R}^n)$ is dense in $L^2(\mathbb{R}^n)$, we have

$$C_0^\infty(\mathbb{R}^n) \subseteq L^2_s(\mathbb{R}^n) \quad \text{densely.} \tag{12.1.35}$$

Specifically, if $v \in L^2_s(\mathbb{R}^n)$, then $\langle \xi \rangle^s v \in L^2(\mathbb{R}^n)$, so there exists a sequence $\{\varphi_j\}_{j \in \mathbb{N}}$ in $C_0^\infty(\mathbb{R}^n)$ that converges to $\langle \xi \rangle^s v$ in $L^2(\mathbb{R}^n)$. Taking $\psi_j := \langle \xi \rangle^{-s} \varphi_j$ for each $j \in \mathbb{N}$, it follows that $\{\psi_j\}_{j \in \mathbb{N}}$ is a sequence of functions in $C_0^\infty(\mathbb{R}^n)$ which converges to v in $L^2_s(\mathbb{R}^n)$.

If we now take $u \in H^s(\mathbb{R}^n)$, then there exists a sequence $\{\psi_j\}_{j \in \mathbb{N}} \subseteq C_0^\infty(\mathbb{R}^n)$ which converges to $\widehat{u}$ in $L^2_s(\mathbb{R}^n)$. Hence, by (12.1.34) and Remark 4.23, we have that $\{\mathcal{F}^{-1}\psi_j\}_{j \in \mathbb{N}}$ is a sequence of Schwartz functions which converges to $\mathcal{F}^{-1}(\widehat{u}) = u$ in $H^s(\mathbb{R}^n)$. This shows that $\mathcal{S}(\mathbb{R}^n)$ is dense in $H^s(\mathbb{R}^n)$. To finish the proof of the statement in item *(3)* we now invoke the first inclusion in item *(2)* of the current theorem and the density result from part *(d)* of Theorem 3.14. $\square$

Example 12.6. Corresponding to $s = 0$, from (12.1.2) and Remark 3.29 we have

$$H^0(\mathbb{R}^n) = \{u \in \mathcal{S}'(\mathbb{R}^n) : \widehat{u} \in L^2(\mathbb{R}^n)\}$$

$$= \{u \in \mathcal{S}'(\mathbb{R}^n) : u \in L^2(\mathbb{R}^n)\} = L^2(\mathbb{R}^n) \tag{12.1.36}$$

and $\|u\|_{H^0(\mathbb{R}^n)} = (2\pi)^{n/2}\|u\|_{L^2(\mathbb{R}^n)}$. Hence, $H^0(\mathbb{R}^n) = L^2(\mathbb{R}^n)$ with equivalent norms.

The Sobolev spaces $H^s(\mathbb{R}^n)$ are stable under multiplication with Schwartz functions and are nested with respect to s. These and other useful properties are proved next. Let us also note that, as seen from Exercise 12.54, the scale $\{H^s(\mathbb{R}^n)\}_{s \in \mathbb{R}}$ consists of distinct spaces.

Theorem 12.7. *The following are true.*

(1) Let $s \in \mathbb{R}$ and $\varphi \in \mathcal{S}(\mathbb{R}^n)$. Then for every $u \in H^s(\mathbb{R}^n)$ one has $\varphi u \in H^s(\mathbb{R}^n)$ and

$$\|\varphi u\|_{H^s(\mathbb{R}^n)} \leq C(\varphi, s) \|u\|_{H^s(\mathbb{R}^n)}, \tag{12.1.37}$$

for some constant $C(\varphi, s) \in (0, \infty)$ independent of u.

(2) If $s_1, s_2 \in \mathbb{R}$ are two numbers satisfying $s_1 \leq s_2$ then $H^{s_2}(\mathbb{R}^n) \subseteq H^{s_1}(\mathbb{R}^n)$ and $\|u\|_{H^{s_1}(\mathbb{R}^n)} \leq \|u\|_{H^{s_2}(\mathbb{R}^n)}$ for every $u \in H^{s_2}(\mathbb{R}^n)$. Consequently, the continuous embedding $H^{s_2}(\mathbb{R}^n) \hookrightarrow H^{s_1}(\mathbb{R}^n)$ holds. In particular, the sequence of spaces $\{H^s(\mathbb{R}^n)\}_{s \in \mathbb{R}}$ is nested with respect to the smoothness index s.

(3) If $P(D)$ is a constant coefficient operator of order $m \in \mathbb{N}_0$ (recall (5.1.1)) then $P(D) : H^s(\mathbb{R}^n) \to H^{s-m}(\mathbb{R}^n)$ is well-defined and continuous for each $s \in \mathbb{R}$.

(4) For each compactly supported distribution u in $\mathbb{R}^n$ there exists $s \in (-\infty, 0)$ with the property that u belongs to the Sobolev space $H^s(\mathbb{R}^n)$. Consequently, we have $\mathcal{E}'(\mathbb{R}^n) \subseteq \bigcup_{s<0} H^s(\mathbb{R}^n)$.

Proof. Let $\varphi \in \mathcal{S}(\mathbb{R}^n)$ and $u \in H^s(\mathbb{R}^n)$. Then $\widehat{\varphi} \in \mathcal{S}(\mathbb{R}^n)$ and $\widehat{u} \in \mathcal{S}'(\mathbb{R}^n)$ and $\widehat{\varphi} * \widehat{u} \in \mathcal{S}'(\mathbb{R}^n)$ (see *(e)* in Theorem 4.19). Hence, by item *(a)* in Theorem 4.35 and (4.2.34) we have $\widehat{\varphi} * \widehat{u} = \widehat{\widehat{\varphi u}} = (2\pi)^{2n}(\varphi u)^\vee$ in $\mathcal{S}'(\mathbb{R}^n)$. Taking the Fourier transform of the latter identity (combined with (4.2.34)) yields $\widehat{\varphi} * \widehat{u} = (2\pi)^n \widehat{\varphi u}$ in $\mathcal{S}'(\mathbb{R}^n)$. This, item *(5)* in Lemma 12.4, item *(2)* in Lemma 12.3, and the Cauchy–Schwarz inequality give

$$\left| \langle \xi \rangle^s \widehat{\varphi u}(\xi) \right| = (2\pi)^{-n} \left| \langle \xi \rangle^s (\widehat{\varphi} * \widehat{u})(\xi) \right|$$

$$\leq (2\pi)^{-n} 2^{|s|/2} \int_{\mathbb{R}^n} \langle \eta \rangle^{|s|} |\widehat{\varphi}(\eta)| \langle \xi - \eta \rangle^s |\widehat{u}(\xi - \eta)| \, d\eta$$

$$\leq (2\pi)^{-n} 2^{|s|/2} \left(\int_{\mathbb{R}^n} \langle \eta \rangle^{|s|} |\widehat{\varphi}(\eta)| \, d\eta \right)^{1/2} \times$$

$$\times \left(\int_{\mathbb{R}^n} \langle \eta \rangle^{|s|} |\widehat{\varphi}(\eta)| \langle \xi - \eta \rangle^{2s} |\widehat{u}(\xi - \eta)|^2 \, d\eta \right)^{1/2} \tag{12.1.38}$$

for each $\xi \in \mathbb{R}^n$. Squaring the left-most terms in (12.1.38), then integrating them with respect to ξ over $\mathbb{R}^n$, and finally using Fubini's theorem and a simple change of variables further yields

$$\int_{\mathbb{R}^n} \left| \langle \xi \rangle^s \widehat{\varphi u}(\xi) \right|^2 \, d\xi \leq (2\pi)^{-2n} 2^{|s|} \int_{\mathbb{R}^n} \langle \eta \rangle^{|s|} |\widehat{\varphi}(\eta)| \, d\eta \times$$

$$\times \int_{\mathbb{R}^n} \int_{\mathbb{R}^n} \langle \eta \rangle^{|s|} |\widehat{\varphi}(\eta)| \langle \xi - \eta \rangle^{2s} |\widehat{u}(\xi - \eta)|^2 \, d\eta \, d\xi$$

$$= (2\pi)^{-2n} 2^{|s|} \left(\int_{\mathbb{R}^n} \langle \eta \rangle^{|s|} |\widehat{\varphi}(\eta)| \, d\eta \right)^2 \int_{\mathbb{R}^n} \left| \langle z \rangle^s \widehat{u}(z) \right|^2 \, dz. \tag{12.1.39}$$

This now implies (12.1.37) with

$$C(\varphi, s) := (2\pi)^{-n} 2^{|s|/2} \int_{\mathbb{R}^n} |\langle\eta\rangle^{|s|} \widehat{\varphi}(\eta)| \, d\eta.$$ (12.1.40)

Next, let $s_1, s_2 \in \mathbb{R}$ be such that $s_1 \le s_2$. Then $\langle\xi\rangle^{s_1} \le \langle\xi\rangle^{s_2}$ for every $\xi \in \mathbb{R}^n$. Hence, if $u \in H^{s_2}(\mathbb{R}^n)$ then $\langle\xi\rangle^{s_2}\widehat{u} \in L^2(\mathbb{R}^n)$ which further forces

$$\langle\xi\rangle^{s_1}\widehat{u} = \langle\xi\rangle^{s_1-s_2}(\langle\xi\rangle^{s_2}\widehat{u}) \in L^\infty(\mathbb{R}^n) \cdot L^2(\mathbb{R}^n) \subseteq L^2(\mathbb{R}^n).$$ (12.1.41)

This proves that $u \in H^{s_1}(\mathbb{R}^n)$ and also gives that $\|u\|_{H^{s_1}(\mathbb{R}^n)} \le \|u\|_{H^{s_2}(\mathbb{R}^n)}$. Ultimately, $H^{s_2}(\mathbb{R}^n) \subseteq H^{s_1}(\mathbb{R}^n)$ and the embedding $H^{s_2}(\mathbb{R}^n) \hookrightarrow H^{s_1}(\mathbb{R}^n)$ is continuous.

Moving on, fix $m \in \mathbb{N}_0$ and suppose $P(D) = \sum_{|\alpha| \le m} a_\alpha D^\alpha$ is a constant (complex) coefficient differential operator of order m. Then there exists $C \in (0, \infty)$, depending only on P, such that

$$|P(\xi)| \le C\langle\xi\rangle^m \qquad \forall \xi \in \mathbb{R}^n.$$ (12.1.42)

Indeed, making use of (3.1.18) we may estimate

$$|P(\xi)| \le \big(\max_{|\alpha| \le m} |a_\alpha|\big) \sum_{|\alpha| \le m} |\xi|^{|\alpha|} \le \big(\max_{|\alpha| \le m} |a_\alpha|\big) \sum_{|\alpha| \le m} (1 + |\xi|^2)^{|\alpha|/2}$$

$$\le C(1 + |\xi|^2)^{m/2}, \qquad \forall \xi \in \mathbb{R}^n,$$ (12.1.43)

for some finite constant $C = C(P, n, m) > 0$, and (12.1.42) follows. Going further, if $s \in \mathbb{R}$ and $u \in H^s(\mathbb{R}^n)$, then $P(D)u \in \mathcal{S}'(\mathbb{R}^n)$ and $\widehat{P(D)u}(\xi) = P(\xi)\widehat{u}(\xi)$ in $\mathcal{S}'(\mathbb{R}^n)$ (cf. part (b) in Theorem 4.26). As such, we may compute

$$\langle\xi\rangle^{s-m}\widehat{P(D)u}(\xi) = \langle\xi\rangle^{s-m}P(\xi)\widehat{u}(\xi)$$ (12.1.44)

$$= \Big(\frac{P(\xi)}{\langle\xi\rangle^m}\Big)(\langle\xi\rangle^s\widehat{u}(\xi)) \in L^\infty(\mathbb{R}^n) \cdot L^2(\mathbb{R}^n) \subseteq L^2(\mathbb{R}^n).$$

This implies

$$P(D)u \in H^{s-m}(\mathbb{R}^n)$$ (12.1.45)

and

$$\|P(D)u\|_{H^{s-m}(\mathbb{R}^n)} = \big\|\langle\xi\rangle^{s-m}\widehat{P(D)u}(\xi)\big\|_{L^2(\mathbb{R}^n)}$$ (12.1.46)

$$\le \Big(\sup_{\xi \in \mathbb{R}^n} \frac{|P(\xi)|}{\langle\xi\rangle^m}\Big)\|\langle\xi\rangle^s\widehat{u}\|_{L^2(\mathbb{R}^n)} = C\|u\|_{H^s(\mathbb{R}^n)}$$

for some constant $C = C(P) \in (0, \infty)$ independent of u. This establishes the statement in item (3) of the current theorem.

Finally, if $u \in \mathcal{E}'(\mathbb{R}^n)$, then $\widehat{u} \in \mathcal{L}(\mathbb{R}^n)$ (cf. item (b) in Theorem 4.35) thus $\widehat{u}$ is a smooth function and there exists $k \in \mathbb{N}_0$ such that $\sup_{\xi \in \mathbb{R}^n} [(1 + |\xi|)^{-k}|\widehat{u}(\xi)|] < \infty$

(recall Definition 3.11). Consequently, there exists a constant $C \in (0, \infty)$ with the property that $|\widehat{u}(\xi)| \leq C \langle \xi \rangle^k$ for all $\xi \in \mathbb{R}^n$. In turn, this implies

$$\int_{\mathbb{R}^n} \langle \xi \rangle^{2s} |\widehat{u}(\xi)|^2 \, d\xi \leq C \int_{\mathbb{R}^n} \langle \xi \rangle^{2s+2k} \, d\xi < \infty \qquad (12.1.47)$$

whenever $2s + 2k < -n$. Thus, $u \in H^s(\mathbb{R}^n)$ for $s < -k - n/2$. $\qquad\square$

In the next theorem we show that under some additional assumptions the embedding from item (2) in Theorem 12.7 is compact.

Theorem 12.8. *Let $s_1, s_2 \in \mathbb{R}$ satisfy $s_1 < s_2$ and suppose K is a compact set in $\mathbb{R}^n$. Then the embedding $H^{s_2}(\mathbb{R}^n) \cap \mathcal{E}'_K(\mathbb{R}^n) \hookrightarrow H^{s_1}(\mathbb{R}^n)$ is compact, where $\mathcal{E}'_K(\mathbb{R}^n)$ denotes the collection of distributions in $\mathbb{R}^n$ with support contained in K.*

We remark that, as seen from Exercise 12.52, the aforementioned embedding is no longer compact if $s_1 = s_2$.

Proof of Theorem 12.8. Let $\{u_j\}_{j \in \mathbb{N}}$ be a sequence in $H^{s_2}(\mathbb{R}^n) \cap \mathcal{E}'_K(\mathbb{R}^n)$ such that there exists $C_0 \in (0, \infty)$ with

$$\|u_j\|_{H^{s_2}(\mathbb{R}^n)} \leq C_0, \qquad \forall j \in \mathbb{N}. \qquad (12.1.48)$$

The goal is to prove that this sequence contains a subsequence that is convergent in $H^{s_1}(\mathbb{R}^n)$. To proceed, pick $\varphi \in C_0^\infty(\mathbb{R}^n)$ satisfying $\varphi \equiv 1$ in a neighborhood of K. Fix $j \in \mathbb{N}$. Then $u_j = \varphi u_j$ in $\mathcal{S}'(\mathbb{R}^n)$ and, reasoning as in the first part of the proof of Theorem 12.7, we obtain

$$\widehat{u_j} = \widehat{\varphi u_j} = (2\pi)^{-n} \widehat{\varphi} * \widehat{u_j} \quad \text{in } \mathcal{S}'(\mathbb{R}^n). \qquad (12.1.49)$$

This and item (5) in Lemma 12.4 imply that the action of $\widehat{u_j}$ on Schwartz functions is given by integration against the function

$$f_j(\xi) := (2\pi)^{-n} \int_{\mathbb{R}^n} \widehat{\varphi}(\xi - \eta) \, \widehat{u_j}(\eta) \, d\eta, \qquad \forall \xi \in \mathbb{R}^n. \qquad (12.1.50)$$

Fix now $\alpha \in \mathbb{N}_0^n$ and invoke (4) in Lemma 12.4 together with (b) in Theorem 3.21 to compute

$$(\partial^\alpha f_j)(\xi) = (2\pi)^{-n} \int_{\mathbb{R}^n} (\partial^\alpha \widehat{\varphi})(\xi - \eta) \, \widehat{u_j}(\eta) \, d\eta$$

$$= (2\pi)^{-n} (-i)^{|\alpha|} \int_{\mathbb{R}^n} \widehat{\zeta^\alpha \varphi}(\xi - \eta) \, \widehat{u_j}(\eta) \, d\eta, \qquad \forall \xi \in \mathbb{R}^n. \qquad (12.1.51)$$

This identity, item (2) in Lemma 12.3, and the Cauchy–Schwarz inequality, allow us to further estimate

$$\left|\langle\xi\rangle^{s_2}(\partial^\alpha f_j)(\xi)\right| \le (2\pi)^{-n} 2^{|s_2|/2} \int_{\mathbb{R}^n} \langle\xi-\eta\rangle^{|s_2|} \left|\widehat{\zeta^\alpha\varphi}(\xi-\eta)\right| \langle\eta\rangle^{s_2} \left|\widehat{u}_j(\eta)\right| d\eta$$

$$\le (2\pi)^{-n} 2^{|s_2|/2} \|\zeta^\alpha\varphi\|_{H^{|s_2|}(\mathbb{R}^n)} \|u_j\|_{H^{s_2}(\mathbb{R}^n)}$$

$$\le C\|\zeta^\alpha\varphi\|_{H^{|s_2|}(\mathbb{R}^n)}, \qquad \forall\,\xi\in\mathbb{R}^n, \tag{12.1.52}$$

where $C := (2\pi)^{-n} 2^{|s_2|/2} C_0$. This goes to show that the sequence $\{\partial^\alpha f_j\}_{j\in\mathbb{N}}$ is uniformly bounded on compact sets in $\mathbb{R}^n$. In particular, the sequence $\{f_j\}_{j\in\mathbb{N}}$ is equicontinuous at each point in $\mathbb{R}^n$. Hence, Arzelà–Ascoli's theorem applies (see Theorem 14.24) and proves the existence of a subsequence $\{f_{j_k}\}_{k\in\mathbb{N}}$ that converges uniformly on compact subsets of $\mathbb{R}^n$ to a continuous function. Upon recalling that $f_j = \widehat{u}_j$ for each $j \in \mathbb{N}$, this uniform convergence on compacts implies

> for each $R > 0$ and $\delta > 0$ there exists $N \in \mathbb{N}$ such that $|\widehat{u}_{j_k}(\xi) - \widehat{u}_{j_\ell}(\xi)| < \delta$ for all $\xi \in B(0,R)$ if $k,\ell \ge N$. $\tag{12.1.53}$

The claim we make at this stage is that

$$\text{the sequence } \{u_{j_k}\}_{k\in\mathbb{N}} \text{ is Cauchy in } H^{s_1}(\mathbb{R}^n). \tag{12.1.54}$$

To prove (12.1.54), fix a threshold $\varepsilon > 0$ and observe that since we have $s_1 - s_2 < 0$ there exists $R_\varepsilon > 0$ such that $\langle\xi\rangle^{s_1-s_2} < \varepsilon/(4C_0)$ if $|\xi| \ge R_\varepsilon$. In particular,

$$\langle\xi\rangle^{s_1} = \langle\xi\rangle^{s_1-s_2}\langle\xi\rangle^{s_2} < \frac{\varepsilon}{4C_0}\langle\xi\rangle^{s_2}, \qquad \forall\,\xi\in\mathbb{R}^n\setminus B(0,R_\varepsilon). \tag{12.1.55}$$

Then, using (12.1.55) and (12.1.48), for any $k,\ell \in \mathbb{N}$ we may write

$$\|u_{j_k} - u_{j_\ell}\|_{H^{s_1}(\mathbb{R}^n)} = \left(\int_{\mathbb{R}^n} \langle\xi\rangle^{2s_1} |\widehat{u}_{j_k}(\xi) - \widehat{u}_{j_\ell}(\xi)|^2 \, d\xi\right)^{1/2}$$

$$\le \left(\int_{B(0,R_\varepsilon)} \langle\xi\rangle^{2s_1} |\widehat{u}_{j_k}(\xi) - \widehat{u}_{j_\ell}(\xi)|^2 \, d\xi\right)^{1/2}$$

$$+ \frac{\varepsilon}{4C_0}\left(\int_{\mathbb{R}^n\setminus B(0,R_\varepsilon)} \langle\xi\rangle^{2s_2} |\widehat{u}_{j_k}(\xi) - \widehat{u}_{j_\ell}(\xi)|^2 \, d\xi\right)^{1/2}$$

$$\le \left(\int_{B(0,R_\varepsilon)} \langle\xi\rangle^{2s_1} |\widehat{u}_{j_k}(\xi) - \widehat{u}_{j_\ell}(\xi)|^2 \, d\xi\right)^{1/2}$$

$$+ \frac{\varepsilon}{4C_0}\|u_{j_k} - u_{j_\ell}\|_{H^{s_2}(\mathbb{R}^n)}$$

$$\le \left(\int_{B(0,R_\varepsilon)} \langle\xi\rangle^{2s_1} |\widehat{u}_{j_k}(\xi) - \widehat{u}_{j_\ell}(\xi)|^2 \, d\xi\right)^{1/2} + \frac{\varepsilon}{2}. \tag{12.1.56}$$

By applying now (12.1.53) with $\delta := \frac{\varepsilon}{2}\|\langle\xi\rangle^{s_1}\|^{-1}_{L^2(B(0,R_\varepsilon))}$ and $R := R_\varepsilon$, we conclude that there exists $N \in \mathbb{N}$ such that the first term in the right-most side of (12.1.56) is $\le \varepsilon/2$ if $k,\ell \ge N$. When used in concert with (12.1.56), this shows that

$$\|u_{j_k} - u_{j_\ell}\|_{H^{s_1}(\mathbb{R}^n)} < \varepsilon \text{ for } k, \ell \geq N. \tag{12.1.57}$$

The proof of the claim in (12.1.54) is therefore finished. Now we may invoke the fact that $H^{s_1}(\mathbb{R}^n)$ is complete (cf. item *(1)* in Theorem 12.1) in order to conclude that the subsequence $\{u_{j_k}\}_{k \in \mathbb{N}}$ of $\{u_j\}_{j \in \mathbb{N}}$ is convergent in $H^{s_1}(\mathbb{R}^n)$. $\square$

While for arbitrary $s \in \mathbb{R}$ one cannot talk about pointwise differentiability for distributions belonging to the Sobolev space $H^s(\mathbb{R}^n)$, if s is sufficiently large these distributions are actually given by differentiable functions. This fact is made precise in the next theorem.

Theorem 12.9. *If $s \in \mathbb{R}$ and $k \in \mathbb{N}_0$ are such that $k < s - n/2$ then $H^s(\mathbb{R}^n)$ is contained in $C^k(\mathbb{R}^n)$.*

Proof. Suppose s, k satisfy the hypotheses of the theorem and let $\alpha \in \mathbb{N}_0^n$ such that $|\alpha| \leq k$. Then $s - |\alpha| > n/2$, thus $\xi^\alpha \langle \xi \rangle^{-s} \in L^2(\mathbb{R}^n)$. Let now $u \in H^s(\mathbb{R}^n)$. Then

$$\xi^\alpha \widehat{u}(\xi) = (\xi^\alpha \langle \xi \rangle^{-s})(\langle \xi \rangle^s \widehat{u}(\xi)) \in L^2(\mathbb{R}^n) \cdot L^2(\mathbb{R}^n) \subseteq L^1(\mathbb{R}^n). \tag{12.1.58}$$

By combining (4.2.34), (4.2.33), (4.2.2), and *(b)* in Theorem 4.26, we obtain

$$\langle \partial^\alpha u, \varphi \rangle = (2\pi)^{-n} \langle \widehat{\widetilde{\partial^\alpha u}}^\vee, \varphi \rangle = (2\pi)^{-n} \langle \widehat{\partial^\alpha u}, \varphi^\vee \rangle$$

$$= (2\pi)^{-n} i^{|\alpha|} \langle \widehat{\xi^\alpha u}, \varphi^\vee \rangle, \qquad \forall \varphi \in \mathcal{S}(\mathbb{R}^n). \tag{12.1.59}$$

From (12.1.58) and (4.1.9) (see also Example 4.8), we know that the last pairing in (12.1.59) is given by an integral, thus

$$\langle \partial^\alpha u, \varphi \rangle = (2\pi)^{-n} i^{|\alpha|} \int_{\mathbb{R}^n} \xi^\alpha \widehat{u}(\xi) \widehat{\varphi^\vee}(\xi) \, d\xi$$

$$= (2\pi)^{-n} i^{|\alpha|} \int_{\mathbb{R}^n} \int_{\mathbb{R}^n} \xi^\alpha \widehat{u}(\xi) e^{ix \cdot \xi} \varphi(x) \, dx \, d\xi, \qquad \forall \varphi \in \mathcal{S}(\mathbb{R}^n). \tag{12.1.60}$$

Combined, (12.1.58) and the fact that $\varphi \in L^1(\mathbb{R}^n)$ (cf. Exercise 3.5) ensure that the double integral in (12.1.60) is absolutely convergent and we may apply Fubini's theorem to further write

$$\langle \partial^\alpha u, \varphi \rangle = (2\pi)^{-n} i^{|\alpha|} \int_{\mathbb{R}^n} \varphi(x) \int_{\mathbb{R}^n} \xi^\alpha \widehat{u}(\xi) e^{ix \cdot \xi} \, d\xi \, dx$$

$$= (2\pi)^{-n} i^{|\alpha|} \int_{\mathbb{R}^n} \varphi(x) \mathcal{F}(\xi^\alpha \widehat{u}(\xi))(-x) \, dx$$

$$= \langle (2\pi)^{-n} i^{|\alpha|} \mathcal{F}(\xi^\alpha \widehat{u}(\xi))^\vee, \varphi \rangle, \qquad \forall \varphi \in \mathcal{S}(\mathbb{R}^n). \tag{12.1.61}$$

Hence,

$$\partial^\alpha u = (2\pi)^{-n} i^{|\alpha|} \mathcal{F}(\xi^\alpha \widehat{u}(\xi))^\vee \text{ in } \mathcal{S}'(\mathbb{R}^n). \tag{12.1.62}$$

In turn, from (12.1.62), (12.1.58), and (3.1.3) we see that the distributional derivative $\partial^\alpha u$ belongs to $C^0(\mathbb{R}^n)$. Since this conclusion holds for all $\alpha \in \mathbb{N}_0^n$ with $|\alpha| \leq k$, we may now invoke Proposition 2.109 to obtain that $u \in C^k(\mathbb{R}^n)$. $\square$

A corollary of Theorem 12.9 worth singling out is the fact that

$$\bigcap_{s\in\mathbb{R}} H^s(\mathbb{R}^n) \subset C^\infty(\mathbb{R}^n). \tag{12.1.63}$$

As seen from Exercise 12.55 this is a strict inclusion.

12.2 Restriction Sobolev Spaces $H^s(\Omega)$, $s \in \mathbb{R}$

Let Ω be an open subset of $\mathbb{R}^n$ and let $s \in \mathbb{R}$. Then the Sobolev space of order s over Ω, denoted by $H^s(\Omega)$, is defined via restriction, as

$$H^s(\Omega) := \Big\{ u \in \mathcal{D}'(\Omega) : \text{ there exists some } U \in H^s(\mathbb{R}^n) \tag{12.2.1}$$
$$\text{with the property that } U\big|_\Omega = u \text{ in } \mathcal{D}'(\Omega)\Big\}.$$

It is immediate that $H^s(\Omega)$ is a vector space over $\mathbb{C}$. This may be endowed with a natural norm, as explained in the next lemma.

Lemma 12.10. *For each open subset Ω of $\mathbb{R}^n$ and each $s \in \mathbb{R}$ the mapping*

$$H^s(\Omega) \ni u \longmapsto \|u\|_{H^s(\Omega)} := \inf_{U\in H^s(\mathbb{R}^n),\, U|_\Omega=u} \|U\|_{H^s(\mathbb{R}^n)} \tag{12.2.2}$$

defines a norm on $H^s(\Omega)$. Moreover, when regarding $H^s(\Omega)$ equipped with this norm,

$$\text{the inclusion } H^s(\Omega) \hookrightarrow \mathcal{D}'(\Omega) \text{ is continuous.} \tag{12.2.3}$$

Proof. The homogeneity, subadditivity, and positivity are consequences of the fact that $\| \cdot \|_{H^s(\mathbb{R}^n)}$ is a norm on $H^s(\mathbb{R}^n)$. There remains to prove that if $u \in H^s(\Omega)$ is such that $\|u\|_{H^s(\Omega)} = 0$ then $u = 0$ in $\mathcal{D}'(\Omega)$. Consider such a u. Then by (12.2.2) there exists a sequence $\{U_j\}_{j\in\mathbb{N}}$ contained in $H^s(\mathbb{R}^n)$ such that $U_j\big|_\Omega = u$ in $\mathcal{D}'(\Omega)$ for each $j \in \mathbb{N}$ and $\lim_{j\to\infty} \|U_j\|_{H^s(\mathbb{R}^n)} = 0$. For each $\varphi \in C_0^\infty(\Omega)$ we have $\widetilde{\varphi} \in C_0^\infty(\mathbb{R}^n)$ (considering φ extended by zero outside Ω) and

$$|\langle u,\varphi\rangle| = |\langle U_j,\widetilde{\varphi}\rangle| \leq \|U_j\|_{H^s(\mathbb{R}^n)} \big\|\widehat{\varphi}\big\|_{L^2_{s}(\mathbb{R}^n)}, \quad \forall j \in \mathbb{N}. \tag{12.2.4}$$

The equality in (12.2.4) uses Proposition 2.50, while the inequality is based on (12.1.31). Passing to the limit with $j \to \infty$ in (12.2.4) we arrive at $\langle u,\varphi\rangle = 0$. Since φ is arbitrary in $C_0^\infty(\Omega)$, we conclude that $u = 0$ in $\mathcal{D}'(\Omega)$, as wanted.

Finally, the claim in (12.2.3) may be justified based on (12.2.1), (12.2.2), item *(2)* in Theorem 12.1, and part *(b)* of Theorem 14.1. $\square$

Example 12.11. Corresponding to $s = 0$ we have $H^0(\Omega) = L^2(\Omega)$ as vector spaces, and

$$\|u\|_{H^0(\Omega)} = (2\pi)^{n/2}\|u\|_{L^2(\Omega)}, \qquad \forall\, u \in H^0(\Omega). \tag{12.2.5}$$

This is a consequence of Example 12.6 and Remark 3.29. Specifically, if $u \in H^0(\Omega)$ then there exists $U \in H^0(\mathbb{R}^n) = L^2(\mathbb{R}^n)$ satisfying $U\big|_\Omega = u$, hence u belongs to $L^2(\Omega)$. For the opposite inclusion, if $u \in L^2(\Omega)$ and we denote by $\widetilde{u}$ the extension by zero of u to $\mathbb{R}^n$, then $\widetilde{u} \in L^2(\mathbb{R}^n) = H^0(\mathbb{R}^n)$, $\widetilde{u}\big|_\Omega = u$, thus $u \in H^0(\Omega)$. In addition, for every $u \in L^2(\Omega)$ we have $\|\widetilde{u}\|_{H^0(\mathbb{R}^n)} = (2\pi)^{n/2}\|\widetilde{u}\|_{L^2(\mathbb{R}^n)} = (2\pi)^{n/2}\|u\|_{L^2(\Omega)}$, and any function $U \in L^2(\mathbb{R}^n)$ with the property that $U\big|_\Omega = u$ necessarily satisfies $\|U\|_{L^2(\mathbb{R}^n)} \geq \|U|_\Omega\|_{L^2(\Omega)} = \|u\|_{L^2(\Omega)}$. Hence, (12.2.5) holds.

Remark 12.12. As regards the scale of restriction Sobolev spaces defined in this section, we wish to remark that if $\Omega \subseteq \mathbb{R}^n$ is an open set and $s \in \mathbb{R}$, then the restriction operator $R_\Omega : \mathcal{D}'(\mathbb{R}^n) \to \mathcal{D}'(\Omega)$, defined by $R_\Omega(U) := U\big|_\Omega$ for every $U \in \mathcal{D}'(\mathbb{R}^n)$, induces a well-defined, linear, continuous, and surjective map

$$R_\Omega : H^s(\mathbb{R}^n) \to H^s(\Omega). \tag{12.2.6}$$

That (12.2.6) is well-defined, linear, and surjective follows from the definition of $H^s(\Omega)$, while its continuity is a consequence of (12.2.2).

Remark 12.13. As a consequence of (12.2.3), if $s_1, s_2 \in \mathbb{R}$ and $\{u_j\}_{j \in \mathbb{N}}$ is a sequence that converges both in $H^{s_1}(\Omega)$ and in $H^{s_2}(\Omega)$, then the limits coincide (as distributions in Ω).

Theorem 12.14. *Let Ω be an open subset of $\mathbb{R}^n$ and let $s \in \mathbb{R}$. Then the following are true.*

(1) The space $H^s(\Omega)$ endowed with the norm $\| \cdot \|_{H^s(\Omega)}$ is complete, hence Banach.

(2) The set $C_0^\infty(\overline{\Omega}) := \{\varphi\big|_\Omega : \varphi \in C_0^\infty(\mathbb{R}^n)\}$ is dense in $H^s(\Omega)$ with respect to the norm $\| \cdot \|_{H^s(\Omega)}$.

Proof. Let $A := \{v \in H^s(\mathbb{R}^n) : v\big|_\Omega = 0 \text{ in } \mathcal{D}'(\Omega)\}$. Then A is a closed subspace of the Banach space $(H^s(\mathbb{R}^n), \| \cdot \|_{H^s(\mathbb{R}^n)})$. Hence the quotient space $H^s(\mathbb{R}^n)/A$ is a Banach space when equipped with the natural quotient norm

$$\big\|[U]\big\|_{H^s(\mathbb{R}^n)/A} := \inf_{v \in A}\|U - v\|_{H^s(\mathbb{R}^n)}, \qquad \forall\, [U] \in H^s(\mathbb{R}^n)/A, \tag{12.2.7}$$

where $[U]$ denotes the class of $U \in H^s(\mathbb{R}^n)$ in the space $H^s(\mathbb{R}^n)/A$. Define the mapping

$$H^s(\Omega) \ni u \longmapsto [U_u] \in H^s(\mathbb{R}^n)/A, \tag{12.2.8}$$

where for $u \in H^s(\Omega)$ we denote by U_u an element in $H^s(\mathbb{R}^n)$ with $U_u\big|_\Omega = u$ in $\mathcal{D}'(\Omega)$. This map is well defined since if $U, V \in H^s(\mathbb{R}^n)$ satisfy $U\big|_\Omega = u = V\big|_\Omega$ in $\mathcal{D}'(\Omega)$ then $U - V \in A$, so $[U] = [V]$. The map (12.2.8) is clearly linear and injective. To see that it is also surjective, note that if $[U] \in H^s(\mathbb{R}^n)/A$ and $V \in H^s(\mathbb{R}^n)$ satisfies

$U - V \in A$, then by setting $u := U\big|_{\Omega} = V\big|_{\Omega}$ we have $u \in H^s(\Omega)$ and u is mapped by (12.2.8) into $[U]$. In addition, for each $u \in H^s(\Omega)$, and each $U_u \in H^s(\mathbb{R}^n)$ with $U_u\big|_{\Omega} = u$ in $\mathcal{D}'(\Omega)$, we have

$$\{U \in H^s(\mathbb{R}^n) : U\big|_{\Omega} = u\} = \{U_u - v : v \in A\}. \tag{12.2.9}$$

In concert with (12.2.7) and (12.2.2), this ultimately implies that the map (12.2.8) is an isometry. In summary, the space $(H^s(\Omega), \|\cdot\|_{H^s(\Omega)})$ is isomorphically isometric to a Banach space, hence it is Banach.

Moving on to the density statement in (2), let $u \in H^s(\Omega)$ and let $U \in H^s(\mathbb{R}^n)$ be such that $U\big|_{\Omega} = u$ in $\mathcal{D}'(\Omega)$. By (3) in Theorem 12.1, there exists a sequence $\{\varphi_j\}_{j\in\mathbb{N}} \subset C_0^\infty(\mathbb{R}^n)$ that converges to U in $H^s(\mathbb{R}^n)$. If we set $\psi_j := \varphi_j\big|_{\Omega}$ for each $j \in \mathbb{N}$, then $\{\psi_j\}_{j\in\mathbb{N}} \subset C_0^\infty(\overline{\Omega})$ and we claim that $\lim_{j\to\infty} \psi_j = u$ in $H^s(\Omega)$. Indeed, for each $j \in \mathbb{N}$ we have $U - \varphi_j \in H^s(\mathbb{R}^n)$ and $(U - \varphi_j)\big|_{\Omega} = u - \psi_j$ in $\mathcal{D}'(\Omega)$. Hence,

$$\|u - \psi_j\|_{H^s(\Omega)} \le \|U - \varphi_j\|_{H^s(\mathbb{R}^n)}, \qquad \forall\, j \in \mathbb{N}. \tag{12.2.10}$$

The desired conclusion follows by passing to the limit in (12.2.10) with $j \to \infty$. $\quad\square$

The analogues of the results in Theorem 12.7 and Theorem 12.9 for spaces $H^s(\Omega)$ are discussed next.

Theorem 12.15. *The following are true.*

(1) Let $s \in \mathbb{R}$ and $\varphi \in C_0^\infty(\overline{\Omega})$. Then for every $u \in H^s(\Omega)$ we have $\varphi u \in H^s(\Omega)$ and

$$\|\varphi u\|_{H^s(\Omega)} \le C\|u\|_{H^s(\Omega)}, \tag{12.2.11}$$

for some constant $C \in (0, \infty)$ independent of u.

(2) If $s_1, s_2 \in \mathbb{R}$ satisfying $s_1 \le s_2$ then $H^{s_2}(\Omega) \subseteq H^{s_1}(\Omega)$ and

$$\|u\|_{H^{s_1}(\Omega)} \le \|u\|_{H^{s_2}(\Omega)} \quad \text{for every } u \in H^{s_2}(\Omega). \tag{12.2.12}$$

As such, one has the continuous embedding $H^{s_2}(\Omega) \hookrightarrow H^{s_1}(\Omega)$. Consequently, the sequence of spaces $\{H^s(\Omega)\}_{s\in\mathbb{R}}$ is nested with respect to s.

(3) If $s_1, s_2 \in \mathbb{R}$ are such that $s_1 < s_2$ and Ω is bounded then the embedding $H^{s_2}(\Omega) \hookrightarrow H^{s_1}(\Omega)$ is in fact compact.

(4) If $m \in \mathbb{N}_0$ and $P(D) = \sum\limits_{|\alpha|\le m} a_\alpha D^\alpha$ with $a_\alpha \in C_0^\infty(\overline{\Omega})$ for all $\alpha \in \mathbb{N}_0^n$, $|\alpha| \le m$, then $P(D) : H^s(\Omega) \to H^{s-m}(\Omega)$ is a well-defined and continuous operator for each $s \in \mathbb{R}$.

(5) If $s \in \mathbb{R}$ and $k \in \mathbb{N}_0$ are such that $k < s - n/2$ then $H^s(\Omega) \subset C^k(\overline{\Omega})$.

Proof. If $u \in H^s(\Omega)$ and $\varphi \in C_0^\infty(\overline{\Omega})$ then there exist $U \in H^s(\mathbb{R}^n)$ and $\Phi \in C_0^\infty(\mathbb{R}^n)$ such that $U\big|_{\Omega} = u$ in $\mathcal{D}'(\Omega)$ and $\Phi\big|_{\Omega} = \varphi$ in Ω. By (1) in Theorem 12.7 we have $\Phi U \in H^s(\mathbb{R}^n)$ and

$$\|\Phi U\|_{H^s(\mathbb{R}^n)} \le C(\Phi, s)\|U\|_{H^s(\mathbb{R}^n)}. \tag{12.2.13}$$

Also, (2.5.5) ensures that $(\Phi U)\big|_\Omega = \varphi u$ in $\mathcal{D}'(\Omega)$, hence $\varphi u \in H^s(\Omega)$. In addition, by making use of (12.2.13) we have

$$\|\varphi u\|_{H^s(\Omega)} \le \|\Phi U\|_{H^s(\mathbb{R}^n)} \le C\|U\|_{H^s(\mathbb{R}^n)}. \tag{12.2.14}$$

Taking now the infimum over all $U \in H^s(\mathbb{R}^n)$ with $U\big|_\Omega = u$ in $\mathcal{D}'(\Omega)$, from (12.2.14) we obtain

$$\|\varphi u\|_{H^s(\Omega)} \le C\|u\|_{H^s(\Omega)}. \tag{12.2.15}$$

This proves the statement in item *(1)*.

Next, let $s_1, s_2 \in \mathbb{R}$ be such that $s_1 \le s_2$ and consider $u \in H^{s_2}(\Omega)$. Then there exists $U \in H^{s_2}(\mathbb{R}^n)$ satisfying $U\big|_\Omega = u$ in $\mathcal{D}'(\Omega)$ and *(2)* in Theorem 12.7 implies $U \in H^{s_1}(\mathbb{R}^n)$ and $\|U\|_{H^{s_1}(\mathbb{R}^n)} \le \|U\|_{H^{s_2}(\mathbb{R}^n)}$. Hence, $u \in H^{s_1}(\Omega)$ and since we have $\|u\|_{H^{s_1}(\Omega)} \le \|U\|_{H^{s_1}(\mathbb{R}^n)}$ it follows that $\|u\|_{H^{s_1}(\Omega)} \le \|U\|_{H^{s_2}(\mathbb{R}^n)}$. Taking the infimum over all such U's we obtain $\|u\|_{H^{s_1}(\Omega)} \le \|u\|_{H^{s_2}(\Omega)}$, proving the statement in item *(2)*.

In order to prove the statement in item *(3)* make the additional assumption that Ω is bounded. Let $\{u_j\}_{j\in\mathbb{N}}$ be a bounded sequence in $H^{s_2}(\Omega)$, say $\|u_j\|_{H^{s_2}(\Omega)} \le C_0$, for some $C_0 \in (0, \infty)$. Then for each $j \in \mathbb{N}$ there exists $U_j \in H^{s_2}(\mathbb{R}^n)$ with the property that $U_j\big|_\Omega = u_j$ in $\mathcal{D}'(\Omega)$ and such that $\|U_j\|_{H^{s_2}(\mathbb{R}^n)} \le 2C_0$ (recall the definition from (12.2.2)). Pick now $\varphi \in C_0^\infty(\mathbb{R}^n)$ with the property that $\varphi \equiv 1$ in a neighborhood of $\overline{\Omega}$. Then *(1)* in Theorem 12.7 implies $\varphi U_j \in H^{s_2}(\mathbb{R}^n)$ and the existence of a constant $C(\varphi, s_2) \in (0, \infty)$ such that $\|\varphi U_j\|_{H^{s_2}(\mathbb{R}^n)} \le 2C_0 C(\varphi, s_2)$ for all $j \in \mathbb{N}$. Thus, the sequence $\{\varphi U_j\}_{j\in\mathbb{N}}$ is bounded in $H^{s_2}(\mathbb{R}^n)$. Being also a sequence of distributions supported in $\mathrm{supp}\,\varphi$, we may invoke Theorem 12.8 to extract a subsequence $\{\varphi U_{j_k}\}_{k\in\mathbb{N}}$ which converges in $H^{s_1}(\mathbb{R}^n)$. Since for each $k \in \mathbb{N}$ we have $(\varphi U_{j_k})\big|_\Omega = u_{j_k}$ in $\mathcal{D}'(\Omega)$, Remark 12.12 implies that the sequence $\{u_{j_k}\}_{k\in\mathbb{N}}$ is convergent in $H^{s_1}(\Omega)$ and finishes the proof of the statement in item *(3)*.

Moving on, observe that having proved the statements in items *(1)* and *(2)*, to show that the statement in item *(4)* is true it suffices to prove that ∂_j is a continuous operator from $H^s(\Omega)$ into $H^{s-1}(\Omega)$. To this end, start with $u \in H^s(\Omega)$ and pick some $U \in H^s(\mathbb{R}^n)$ such that $U\big|_\Omega = u$ in $\mathcal{D}'(\Omega)$. We may then apply *(3)* in Theorem 12.7 to obtain that $\partial_j U \in H^{s-1}(\mathbb{R}^n)$ and

$$\|\partial_j U\|_{H^{s-1}(\mathbb{R}^n)} \le \|U\|_{H^s(\mathbb{R}^n)}. \tag{12.2.16}$$

(The fact that the norm inequality in (12.2.16) holds as stated is seen from (12.1.46).) In addition, since distributional derivatives commute with restrictions to open sets (cf. item *(2)* in Exercise 2.51), we have

$$(\partial_j U)\big|_\Omega = \partial_j u \quad \text{in } \mathcal{D}'(\Omega). \tag{12.2.17}$$

In concert, (12.2.2), (12.2.17), and (12.2.16) imply

$$\|\partial_j u\|_{H^{s-1}(\Omega)} \le \|U\|_{H^s(\mathbb{R}^n)}, \qquad \forall\, U \in H^s(\mathbb{R}^n),\ U\big|_\Omega = u. \tag{12.2.18}$$

Taking now the infimum over all $U \in H^s(\mathbb{R}^n)$ such that $U\big|_\Omega = u$ in $\mathcal{D}'(\Omega)$, we obtain

$$\|\partial_j u\|_{H^{s-1}(\Omega)} \leq \|u\|_{H^s(\Omega)}. \tag{12.2.19}$$

This proves that ∂_j is a continuous operator from $H^s(\Omega)$ into $H^{s-1}(\Omega)$ and completes the proof of the statement in item *(4)*.

Finally, the statement in item *(5)* is an immediate consequence of (12.2.1) and Theorem 12.9. $\square$

12.3 Intrinsic Sobolev Spaces $\mathcal{H}^m(\Omega)$, $m \in \mathbb{N}$

Recall the restriction Sobolev spaces considered in §12.2. At least for (positive) integer amounts of smoothness there is yet another venue, which is intrinsic in nature, of introducing a natural brand of Sobolev spaces on any given open set. Specifically, let Ω be an open set in $\mathbb{R}^n$ and fix some $m \in \mathbb{N}_0$. We then define the intrinsic Sobolev space $\mathcal{H}^m(\Omega)$ by setting

$$\mathcal{H}^m(\Omega) := \{u \in L^2(\Omega) : \partial^\alpha u \in L^2(\Omega), \ \forall \alpha \in \mathbb{N}_0^n, |\alpha| \leq m\}, \tag{12.3.1}$$

where the partial derivatives are considered in the sense of distributions. As is apparent from the above definition,

$$\mathcal{H}^0(\Omega) = L^2(\Omega). \tag{12.3.2}$$

Also, $\mathcal{H}^m(\Omega)$ is a vector space over $\mathbb{C}$, and it is easy to check that

$$(u, v)_{\mathcal{H}^m(\Omega)} := \sum_{\alpha \in \mathbb{N}_0^n, |\alpha| \leq m} \int_\Omega \partial^\alpha u \, \overline{\partial^\alpha v} \, \mathrm{d}x, \qquad \forall u, v \in \mathcal{H}^m(\Omega), \tag{12.3.3}$$

is an inner product on $\mathcal{H}^m(\Omega)$. The norm induced on $\mathcal{H}^m(\Omega)$ by this inner product is

$$\|u\|_{\mathcal{H}^m(\Omega)} := \sqrt{(u, u)_{\mathcal{H}^m(\Omega)}}$$

$$= \Big(\sum_{\alpha \in \mathbb{N}_0^n, |\alpha| \leq m} \|\partial^\alpha u\|_{L^2(\Omega)}^2 \Big)^{1/2}, \qquad \forall u \in \mathcal{H}^m(\Omega). \tag{12.3.4}$$

In particular,

$$\| \cdot \|_{\mathcal{H}^0(\Omega)} = \| \cdot \|_{L^2(\Omega)}, \tag{12.3.5}$$

and for each $u \in \mathcal{H}^m(\Omega)$ we have

$$\|\partial^\alpha u\|_{L^2(\Omega)} \leq \|u\|_{\mathcal{H}^m(\Omega)}, \qquad \forall \alpha \in \mathbb{N}_0^n, |\alpha| \leq m. \tag{12.3.6}$$

Theorem 12.16. *For each open subset Ω of $\mathbb{R}^n$ and each $m \in \mathbb{N}_0$, the space $\mathcal{H}^m(\Omega)$ endowed with the inner product (12.3.3) is a Hilbert space.*

Proof. In order to prove that $\mathcal{H}^m(\Omega)$ is complete with respect to the norm (12.3.4), consider a Cauchy sequence $\{u_j\}_{j \in \mathbb{N}} \subseteq \mathcal{H}^m(\Omega)$. In view of (12.3.6), this implies that for each $\alpha \in \mathbb{N}_0^n$ with $|\alpha| \leq m$ the sequence $\{\partial^\alpha u_j\}_{j \in \mathbb{N}}$ is Cauchy in $L^2(\Omega)$. Since $L^2(\Omega)$ is complete, for each $\alpha \in \mathbb{N}_0^n$ with $|\alpha| \leq m$ there exists $u_\alpha \in L^2(\Omega)$ such that

$$\lim_{j \to \infty} \partial^\alpha u_j = u_\alpha \qquad \text{in } L^2(\Omega). \tag{12.3.7}$$

Abbreviate $u := u_{(0,\dots,0)}$. Then, in order to conclude that $\lim_{j \to \infty} u_j = u$ in $\mathcal{H}^m(\Omega)$, there remains to show that for each $\alpha \in \mathbb{N}_0^n$ with $|\alpha| \leq m$ the weak derivative $\partial^\alpha u$ exists and equals u_α. Fix such an α and pick $\varphi \in C_0^\infty(\Omega)$. We may then compute

$$\langle \partial^\alpha u, \varphi \rangle = (-1)^\alpha \langle u, \partial^\alpha \varphi \rangle = (-1)^\alpha \int_\Omega u\, \partial^\alpha \varphi \, dx$$

$$= \lim_{j \to \infty} (-1)^\alpha \int_\Omega u_j\, \partial^\alpha \varphi \, dx = \lim_{j \to \infty} \int_\Omega (\partial^\alpha u_j) \varphi \, dx$$

$$= \int_\Omega u_\alpha \varphi \, dx = \langle u_\alpha, \varphi \rangle. \tag{12.3.8}$$

Above, the first equality is based on the definition of the weak derivative ∂^α, the second equality uses the fact that $u \in L^2(\Omega) \subseteq L^1_{loc}(\Omega)$, the third equality is justified by observing that

$$\left| \int_\Omega u\, \partial^\alpha \varphi \, dx - \int_\Omega u_j\, \partial^\alpha \varphi \, dx \right| \leq \int_\Omega |u - u_j| |\partial^\alpha \varphi| \, dx \tag{12.3.9}$$

$$\leq \|u - u_j\|_{L^2(\Omega)} \|\partial^\alpha \varphi\|_{L^2(\Omega)} \to 0 \quad \text{as } j \to \infty,$$

(thanks to (12.3.7)), the fourth equality is integration by parts (recall that φ has compact support in Ω), and the fifth equality has a justification similar to (12.3.9). Hence, $\partial^\alpha u = u_\alpha$, and the proof of the theorem is finished. $\qquad\square$

Remark 12.17. Let $\Omega_1 \subseteq \Omega_2$ be open sets in $\mathbb{R}^n$ and let $m \in \mathbb{N}$. Then the restriction operator $R_{\Omega_2} : \mathcal{D}'(\Omega_1) \to \mathcal{D}'(\Omega_2)$ defined by $R_{\Omega_2}(w) := w|_{\Omega_2}$ in $\mathcal{D}'(\Omega_2)$ for every $w \in \mathcal{D}'(\Omega_1)$, induces a well-defined, linear, and continuous map

$$R_{\Omega_2} : \mathcal{H}^m(\Omega_1) \to \mathcal{H}^m(\Omega_2). \tag{12.3.10}$$

Indeed this is a consequence of (12.3.1), (12.3.4), the fact that distributional derivatives commute with restrictions to open sets, and the continuity of the restriction to Ω_2 as an operator from $L^2(\Omega_1)$ into $L^2(\Omega_2)$.

Exercise 12.18. If Ω is an open set in $\mathbb{R}^n$ and $m \in \mathbb{N}_0$ then the following are true.

(1) For each $\varphi \in C^\infty(\mathbb{R}^n)$ with $\partial^\alpha \varphi$ bounded for all $\alpha \in \mathbb{N}_0^n$ with $|\alpha| \leq m$ (hence, in particular, for each $\varphi \in C_0^\infty(\overline{\Omega})$) the mapping

$$\mathcal{H}^m(\Omega) \ni u \longmapsto \varphi u \in \mathcal{H}^m(\Omega) \tag{12.3.11}$$

is well-defined, linear, and continuous.

(2) If Ω is also bounded then $C^m(\overline{\Omega}) \subset \mathcal{H}^m(\Omega)$.

Hint: To prove *(1)* use the Generalized Leibniz Formula from Proposition 2.49 while for the claim in *(2)* recall Exercise 2.40.

Example 12.19. Suppose Ω is an open set in $\mathbb{R}^n$ and $u \in \mathcal{H}^m(\Omega)$ for some $m \in \mathbb{N}_0$. If $\psi \in C_0^\infty(\Omega)$ then $\widetilde{\psi u}$, which is the extension by zero of ψu outside Ω belongs to $\mathcal{H}^m(\mathbb{R}^n)$ and $\|\widetilde{\psi u}\|_{\mathcal{H}^m(\mathbb{R}^n)} \leq C\|u\|_{\mathcal{H}^m(\Omega)}$ for some $C \in (0, \infty)$ depending on ψ and independent of u.

Indeed, if $\varphi \in C_0^\infty(\mathbb{R}^n)$ and $\xi \in C_0^\infty(\Omega)$ satisfies $\xi \equiv 1$ near $\operatorname{supp}\psi$, then for each $\alpha \in \mathbb{N}_0^n$ we have that $\partial^\alpha\varphi - \partial^\alpha(\varphi\xi)$ vanishes in a neighborhood of the support of $\widetilde{\psi u}$, thus

$$(-1)^{|\alpha|} \int_{\mathbb{R}^n} \widetilde{\psi u}\, \partial^\alpha\varphi\, dx = (-1)^{|\alpha|} \int_{\mathbb{R}^n} \widetilde{\psi u}\, \partial^\alpha(\xi\varphi)\, dx$$

$$= (-1)^{|\alpha|} \int_\Omega \psi u\, \partial^\alpha(\xi\varphi)\, dx$$

$$= \int_\Omega \partial^\alpha(\psi u)\, \xi\varphi\, dx = \int_\Omega \partial^\alpha(\psi u)\, \varphi\, dx. \qquad (12.3.12)$$

This implies that $\partial^\alpha(\widetilde{\psi u}) = \widetilde{\partial^\alpha(\psi u)}$ in $\mathcal{D}'(\mathbb{R}^n)$ which further entails

$$\|\widetilde{\psi u}\|_{\mathcal{H}^m(\mathbb{R}^n)}^2 = \sum_{|\alpha| \leq m} \|\partial^\alpha(\widetilde{\psi u})\|_{L^2(\mathbb{R}^n)}^2 = \sum_{|\alpha| \leq m} \|\partial^\alpha(\psi u)\|_{L^2(\Omega)}^2$$

$$= \|\psi u\|_{\mathcal{H}^m(\Omega)}^2 \leq C\|u\|_{\mathcal{H}^m(\Omega)}^2 \qquad (12.3.13)$$

for some $C \in (0, \infty)$ dependent on ψ and independent of u (for the inequality in (12.3.13) we used item *(1)* in Exercise 12.18).

Example 12.20. Suppose Ω_0, Ω_1, and C are open sets in $\mathbb{R}^n$ such that $\Omega_0 \cap C = \Omega_1 \cap C$. Fix $m \in \mathbb{N}$ and suppose $u \in \mathcal{H}^m(\Omega_0)$ is such that $u = 0$ a.e. in $\Omega_0 \setminus K$ for some compact set $K \subset C$. If we define $v : \Omega_1 \to \mathbb{C}$ by

$$v := \begin{cases} u & \text{in } \Omega_1 \cap C, \\ 0 & \text{in } \Omega_1 \setminus C, \end{cases} \qquad (12.3.14)$$

then $v \in \mathcal{H}^m(\Omega_1)$, for each $\alpha \in \mathbb{N}_0^n$ with $|\alpha| \leq m$ we have

$$\partial^\alpha v = \begin{cases} \partial^\alpha u & \text{in } \Omega_1 \cap C, \\ 0 & \text{in } \Omega_1 \setminus C, \end{cases} \qquad (12.3.15)$$

and $\|v\|_{\mathcal{H}^m(\Omega_1)} = \|u\|_{\mathcal{H}^m(\Omega_2)}$.

To see why v has the aforementioned properties fix some $\xi \in C_0^\infty(C)$ with the property that $\xi \equiv 1$ near K. Then for each $\varphi \in C_0^\infty(\Omega_1)$ and each $\alpha \in \mathbb{N}_0^n$ we may write

$$(-1)^{|\alpha|} \int_{\Omega_1} v \, \partial^\alpha \varphi \, dx = (-1)^{|\alpha|} \int_{\Omega_1 \cap C} u \, \partial^\alpha \varphi \, dx = (-1)^{|\alpha|} \int_{\Omega_0 \cap C} u \, \partial^\alpha (\varphi \xi) \, dx$$

$$= (-1)^{|\alpha|} \int_{\Omega_0} u \, \partial^\alpha (\varphi \xi) \, dx = \int_{\Omega_0} (\partial^\alpha u) \varphi \xi \, dx$$

$$= \int_{\Omega_0 \cap C} (\partial^\alpha u) \varphi \, dx = \int_{\Omega_1 \cap C} (\partial^\alpha u) \varphi \, dx. \qquad (12.3.16)$$

The first equality in (12.3.16) is due to (12.3.14), the second uses the fact that $\Omega_0 \cap C = \Omega_1 \cap C$ and the observation that $\partial^\alpha \varphi - \partial^\alpha (\varphi \xi)$ is identically zero near K, thus has support disjoint from the support of u. The third equality uses the support condition on u. The fourth equality uses the fact that $u \in \mathcal{H}^m(\Omega_0)$ and $\varphi \xi$ is a smooth function compactly supported in $\Omega_1 \cap C = \Omega_0 \cap C$, hence compactly supported in Ω_0. The fifth equality is based on the inclusions $\operatorname{supp}(\partial^\alpha u) \subseteq \operatorname{supp} u \subseteq K \cap C$ and the fact that $\xi \equiv 1$ near K, while the last equality is a consequence of the fact that $\Omega_0 \cap C = \Omega_1 \cap C$. The resulting equality in (12.3.16) proves (12.3.15). Also, from (12.3.15) we see that

$$\|v\|_{\mathcal{H}^m(\Omega_1)}^2 = \sum_{\alpha \in \mathbb{N}_0^n, |\alpha| \le m} \|\partial^\alpha u\|_{L^2(\Omega_1 \cap C)}^2$$

$$= \sum_{\alpha \in \mathbb{N}_0^n, |\alpha| \le m} \|\partial^\alpha u\|_{L^2(\Omega_0 \cap C)}^2 = \|u\|_{\mathcal{H}^m(\Omega_0)}^2, \qquad (12.3.17)$$

hence the rest of the claims about v follow.

Moving on, we elaborate on the manner on which the approximation by convolution with a smooth mollifier behaves on the scale of intrinsic Sobolev spaces in the Euclidean ambient.

Lemma 12.21. *Let ϕ be a function as in (1.2.3) and for each $\varepsilon > 0$ define ϕ_ε as in (1.2.4). Also, let $u \in \mathcal{H}^m(\mathbb{R}^n)$ for some $m \in \mathbb{N}$ and set $u_\varepsilon := u * \phi_\varepsilon$ for every $\varepsilon > 0$. Then $u_\varepsilon \in C^\infty(\mathbb{R}^n) \cap \mathcal{H}^m(\mathbb{R}^n)$ for each $\varepsilon > 0$ and $u_\varepsilon \xrightarrow[\varepsilon \to 0^+]{\mathcal{H}^m(\mathbb{R}^n)} u$.*

Proof. That $u_\varepsilon \in C^\infty(\mathbb{R}^n)$ for each $\varepsilon > 0$ is a consequence of Proposition 2.102. Next, pick a multi-index $\alpha \in \mathbb{N}_0^n$ such that $|\alpha| \le m$. By part *(e)* in Theorem 2.96 we have $\partial^\alpha u_\varepsilon = (\partial^\alpha u) * \phi_\varepsilon$ for each $\varepsilon > 0$. Since by assumption $\partial^\alpha u \in L^2(\mathbb{R}^n)$, we may invoke Lemma 14.19 to conclude that $\partial^\alpha u_\varepsilon \in L^2(\mathbb{R}^n)$ for each $\varepsilon > 0$, and that $\partial^\alpha u_\varepsilon \xrightarrow[\varepsilon \to 0^+]{L^2(\mathbb{R}^n)} \partial^\alpha u$. Granted these, the desired conclusion follows from (12.3.1)–(12.3.4). $\qquad \square$

Our focus next is proving a couple of density results for intrinsic Sobolev spaces defined in arbitrary open sets. Here is the first such result.

Theorem 12.22. *Let Ω be an open set in $\mathbb{R}^n$ and suppose $m \in \mathbb{N}$. Then the set $C^\infty(\Omega) \cap \mathcal{H}^m(\Omega)$ is dense in $\mathcal{H}^m(\Omega)$.*

Proof. To get started fix $u \in \mathcal{H}^m(\Omega)$ along with some threshold $\varepsilon_* > 0$. The desired conclusion then will follow once we show that

$$\begin{array}{c} \text{there exists a function } w \in C^\infty(\Omega) \cap \mathcal{H}^m(\Omega) \\ \text{with the property that } \|u - w\|_{\mathcal{H}^m(\Omega)} < \varepsilon_*. \end{array} \qquad (12.3.18)$$

To this end, for each $k \in \mathbb{N}$, define the open set

$$O_k := \left\{ x \in \Omega : \tfrac{1}{k+1} < \operatorname{dist}(x, \partial\Omega) < \tfrac{1}{k-1} \text{ and } |x| < k \right\}. \qquad (12.3.19)$$

This definition ensures that

$$\begin{array}{c} O_k \text{ is open, bounded, } \overline{O_k} \subset \Omega, \text{ for each } k \in \mathbb{N}, \\ \text{and } \bigcup_{k \in \mathbb{N}} O_k = \Omega. \end{array} \qquad (12.3.20)$$

By Theorem 14.42, there exists a partition of unity subordinate to the family $(O_k)_{k \in \mathbb{N}}$. Specifically, there exists an at most countable collection $(\varphi_j)_{j \in J}$ of C^∞ functions $\varphi_j : \Omega \to \mathbb{R}$, neither of which is identically zero, and satisfying the following properties:

(a) For every $j \in J$ one has $0 \leq \varphi_j \leq 1$ in Ω and there exists $k \in I$ such that φ_j is compactly supported in O_k;
(b) The family of sets $(\{x \in \Omega : \varphi_j(x) \neq 0\})_{j \in J}$ is locally finite in Ω;
(c) $\sum_{j \in J} \varphi_j(x) = 1$ for every $x \in \Omega$.

Next, for each $k \in \mathbb{N}$ define

$$J_k := \left\{ j \in J : \operatorname{supp} \varphi_j \subseteq O_k \setminus \bigcup_{i=1}^{k-1} O_i \right\}. \qquad (12.3.21)$$

We claim that

$$J_k \text{ is finite for each } k \in \mathbb{N}. \qquad (12.3.22)$$

Suppose the claim in (12.3.22) is false. Then there exists $k_0 \in \mathbb{N}$ and an infinite sequence $\{j_i\}_{i \in \mathbb{N}} \subseteq J_{k_0}$. Upon recalling that neither function in the partition of unity is identically zero, for each $i \in \mathbb{N}$ we may pick $x_i \in \Omega$ such that $\varphi_{j_i}(x_i) \neq 0$. By definition (12.3.21), it follows that the sequence $\{x_i\}_{i \in \mathbb{N}}$ is contained in the compact set $\overline{O_{k_0}}$, thus it contains a subsequence $\{x_{i_\ell}\}_{\ell \in \mathbb{N}}$ convergent to some point $x_* \in \overline{O_{k_0}}$. In particular,

$$\begin{array}{c} \text{for each } r \in (0, \infty) \text{ there exists } \ell_r \in \mathbb{N} \text{ such that} \\ x_{i_\ell} \in B(x_*, r) \cap \{x \in \Omega : \varphi_{j_{i_\ell}}(x) \neq 0\} \text{ for all } \ell \geq \ell_r. \end{array} \qquad (12.3.23)$$

However, by property *(b)* above, there exists some $r_0 > 0$ such that $B(x_*, r_0)$ intersects only finitely many of the sets $\{x \in \Omega : \varphi_{j_{i_\ell}}(x) \neq 0\}$, $\ell \in \mathbb{N}$. The latter is in contradiction with (12.3.23). This shows that the claim in (12.3.22) must be true.

Having proved (12.3.22), we may define the functions

$$\psi_k := \sum_{j \in J_k} \varphi_j, \quad \text{for each } k \in \mathbb{N}. \tag{12.3.24}$$

Based on (12.3.22), the observation that $\bigcup_{k \in \mathbb{N}} J_k = J$ (where the union is disjoint), and the properties of the partition of unity recalled earlier we conclude that

$$\psi_k \in C_0^\infty(O_k), \ \forall k \in \mathbb{N}, \quad \text{and} \quad \sum_{k \in \mathbb{N}} \psi_k = \sum_{j \in J} \varphi_j = 1 \text{ in } \Omega. \tag{12.3.25}$$

Next, fix $k \in \mathbb{N}$ and proceed as follows. First, by Example 12.19, we have $\widetilde{\psi_k u} \in \mathcal{H}^m(\mathbb{R}^n)$, where tilde denotes extension by zero outside Ω. Second, invoking Lemma 12.21 (with $u := \widetilde{\psi_k u}$), we construct the functions $(\widetilde{\psi_k u})_\varepsilon \in C^\infty(\mathbb{R}^n)$, for each $\varepsilon > 0$, with the property that

$$(\widetilde{\psi_k u})_\varepsilon \xrightarrow[\varepsilon \to 0^+]{\mathcal{H}^m(\mathbb{R}^n)} \widetilde{\psi_k u}. \tag{12.3.26}$$

In addition, since $\widetilde{\psi_k u}$ is compactly supported in O_k, we also have that

$$(\widetilde{\psi_k u})_\varepsilon \text{ has compact supported contained in } O_k + B(0, \varepsilon), \tag{12.3.27}$$

which is a consequence of the fact that the support of $(\widetilde{\psi_k u})_\varepsilon$ is contained in $\text{supp}\,(\widetilde{\psi_k u}) + \text{supp}\,\phi_\varepsilon$, where ϕ_ε is as in Lemma 12.21. Third, we define the bounded open set

$$\mathcal{U}_k := \left\{ x \in \Omega : \tfrac{1}{2(k+1)} < \text{dist}\,(x, \partial\Omega) < \tfrac{2}{k-1} \text{ and } |x| < 2k \right\}. \tag{12.3.28}$$

This is a slight enlargement of O_k. Invoking (12.3.26) and (12.3.27) we may pick some $\varepsilon_k > 0$ sufficiently small to ensure that

$$\text{supp}\,(\widetilde{\psi_k u})_{\varepsilon_k} \subseteq \mathcal{U}_k \quad \text{and} \quad \left\| (\widetilde{\psi_k u})_{\varepsilon_k} - \widetilde{\psi_k u} \right\|_{\mathcal{H}^m(\mathbb{R}^n)} < \frac{\varepsilon_*}{2^k}. \tag{12.3.29}$$

Now we are ready to finish the proof of the theorem. Specifically, set

$$w := \sum_{k=1}^\infty (\widetilde{\psi_k u})_{\varepsilon_k}\big|_\Omega \quad \text{in } \Omega. \tag{12.3.30}$$

Note that as a consequence of the definition in (12.3.28) and the support property in (12.3.29), the sum in (12.3.30) is locally finite, which further implies that w is well defined and belongs to $C^\infty(\Omega)$. In fact, if we consider the open subsets of Ω given by

$$\Omega_j := \left\{ x \in \Omega : \text{dist}\,(x, \partial\Omega) > \tfrac{1}{j} \right\}, \quad \text{for each } j \in \mathbb{N}, \tag{12.3.31}$$

then, for each $j \in \mathbb{N}$, it follows (from the support property in (12.3.29) and the definition in (12.3.28)) that whenever $x \in \Omega_j$ we have $\quad w(x) = \sum_{k=1}^{2j} (\widetilde{\psi_k u})_{\varepsilon_k}(x).$ \hfill (12.3.32)

Also, from (12.3.25) and (12.3.19), we may conclude that, for each $j \in \mathbb{N}$,

whenever $x \in \Omega_j$ we have $\quad u(x) = \sum_{k=1}^{j} (\widetilde{\psi_k u})(x) = \sum_{k=1}^{2j} (\widetilde{\psi_k u})(x)$ \hfill (12.3.33)

(note that the terms in the last sum corresponding to $k \in \{j+1, \ldots, 2j\}$ are zero). In concert, (12.3.32), (12.3.33), Minkowski's inequality (cf. Theorem 14.22), and (12.3.29) imply

$$\Big(\sum_{|\alpha| \leq m} \|\partial^\alpha w - \partial^\alpha u\|_{L^2(\Omega_j)}^2 \Big)^{1/2}$$

$$\leq \Big(\sum_{|\alpha| \leq m} \Big(\sum_{k=1}^{2j} \big\| \partial^\alpha (\widetilde{\psi_k u})_{\varepsilon_k} - \partial^\alpha (\widetilde{\psi_k u}) \big\|_{L^2(\Omega_j)} \Big)^2 \Big)^{1/2}$$

$$\leq \sum_{k=1}^{2j} \Big(\sum_{|\alpha| \leq m} \big\| \partial^\alpha (\widetilde{\psi_k u})_{\varepsilon_k} - \partial^\alpha (\widetilde{\psi_k u}) \big\|_{L^2(\Omega_j)}^2 \Big)^{1/2}$$

$$\leq \sum_{k=1}^{2j} \big\| (\widetilde{\psi_k u})_{\varepsilon_k} - \widetilde{\psi_k u} \big\|_{\mathcal{H}^m(\mathbb{R}^n)}$$

$$\leq \sum_{k=1}^{2j} \frac{\varepsilon_*}{2^k} \leq \varepsilon_*, \quad \text{for all} \quad j \in \mathbb{N}. \hfill (12.3.34)$$

Observing that $\Omega_j \nearrow \Omega$ as $j \to \infty$ and relying on Lebesgue's Monotone Convergence Theorem, from (12.3.34) we deduce that

$$\Big(\sum_{|\alpha| \leq m} \|\partial^\alpha w - \partial^\alpha u\|_{L^2(\Omega)}^2 \Big)^{1/2} \leq \varepsilon_*. \hfill (12.3.35)$$

In particular,

$$\Big(\sum_{|\alpha| \leq m} \|\partial^\alpha w\|_{L^2(\Omega)}^2 \Big)^{1/2}$$

$$\leq \Big(\sum_{|\alpha| \leq m} \|\partial^\alpha w - \partial^\alpha u\|_{L^2(\Omega)}^2 \Big)^{1/2} + \Big(\sum_{|\alpha| \leq m} \|\partial^\alpha u\|_{L^2(\Omega)}^2 \Big)^{1/2}$$

$$\leq \varepsilon_* + \|u\|_{\mathcal{H}^m(\Omega)} < \infty, \hfill (12.3.36)$$

which goes to show that $w \in \mathcal{H}^m(\Omega)$. Thus, ultimately, $w \in C^\infty(\Omega) \cap \mathcal{H}^m(\Omega)$ and we may recast (12.3.35) simply as $\|w - u\|_{\mathcal{H}^m(\Omega)} \leq \varepsilon_*$. This establishes (12.3.18) which completes the proof of the theorem. $\qquad \square$

Here is our second density result for intrinsic Sobolev spaces defined in open sets, advertised earlier.

Theorem 12.23. *Suppose $\Omega \subseteq \mathbb{R}^n$ is an open set and fix $m \in \mathbb{N}$. Then the set $\{u \in \mathcal{H}^m(\Omega): u$ vanishes outside of a bounded subset of $\Omega\}$ is dense in the intrinsic Sobolev space $\mathcal{H}^m(\Omega)$.*

Proof. Pick some $\theta \in C_0^\infty(\mathbb{R}^n)$ with the property that $\theta \equiv 1$ on $B(0, 1)$ and, for each $R > 0$, define $\theta_R(x) := \theta(x/R)$, $x \in \mathbb{R}^n$. Having fixed a function $u \in \mathcal{H}^m(\Omega)$, we may invoke part *(1)* of Exercise 12.18 to conclude that for each $R > 0$ the function $u_R := \theta_R u$ belongs to $\mathcal{H}^m(\Omega)$. In addition, it is clear that u_R vanishes outside of a bounded subset of Ω. As such, the proof is complete as soon as we establish that

$$u_R \xrightarrow[R \to \infty]{\mathcal{H}^m(\Omega)} u. \tag{12.3.37}$$

To justify this, observe that for each $\alpha \in \mathbb{N}_0^n$ with $|\alpha| \leq m$ the Generalized Leibniz Formula from Proposition 2.49 gives

$$\partial^\alpha u_R = \theta_R \partial^\alpha u + \sum_{\substack{\beta + \gamma = \alpha \\ |\beta| > 0}} \frac{\alpha!}{\beta! \gamma!} \partial^\beta \theta_R \partial^\gamma u \quad \text{in } \mathcal{D}'(\Omega). \tag{12.3.38}$$

Lebesgue's Dominated Convergence Theorem (cf. Theorem 14.16 for $p = 2$) ensures that $\theta_R \partial^\alpha u \xrightarrow[R \to \infty]{L^2(\Omega)} \partial^\alpha u$ and that $\partial^\beta \theta_R \partial^\gamma u \xrightarrow[R \to \infty]{L^2(\Omega)} 0$ for each $\beta, \gamma \in \mathbb{N}_0^n$ with $0 < |\beta| \leq m$. On account of these observations, (12.3.37) follows from (12.3.38). $\qquad \square$

We next take up the task of determining how the space $H^m(\Omega)$ relates to $\mathcal{H}^m(\Omega)$ for $m \in \mathbb{N}_0$. First we consider the case $\Omega = \mathbb{R}^n$.

Theorem 12.24. *For each $m \in \mathbb{N}_0$ one has $H^m(\mathbb{R}^n) = \mathcal{H}^m(\mathbb{R}^n)$ as vector spaces, with equivalent norms.*

Proof. If $u \in H^m(\mathbb{R}^n)$, for each multi-index $\alpha \in \mathbb{N}_0^n$ with $|\alpha| \leq m$ we may write (keeping in mind *(b)* in Theorem 4.26)

$$\widehat{\partial^\alpha u}(\xi) = \xi^\alpha \widehat{u}(\xi) = \left(\frac{\xi^\alpha}{\langle \xi \rangle^m} \right) (\langle \xi \rangle^m \widehat{u}(\xi))$$

$$\in L^\infty(\mathbb{R}^n) \cdot L^2(\mathbb{R}^n) \subseteq L^2(\mathbb{R}^n). \tag{12.3.39}$$

Hence, for each multi-index $\alpha \in \mathbb{N}_0^n$ with $|\alpha| \leq m$ we have $\widehat{\partial^\alpha u} \in L^2(\mathbb{R}^n)$ and $\|\widehat{\partial^\alpha u}\|_{L^2(\mathbb{R}^n)} \leq C\|u\|_{H^m(\mathbb{R}^n)}$ for some constant $C \in (0, \infty)$ independent of u. Via Plancherel's theorem (cf. part *(2)* in Remark 3.29), this ultimately shows that $\partial^\alpha u$

belongs to $L^2(\mathbb{R}^n)$ and $\|\partial^\alpha u\|_{L^2(\mathbb{R}^n)} \leq C\|u\|_{H^m(\mathbb{R}^n)}$ for each $\alpha \in \mathbb{N}_0^n$ with $|\alpha| \leq m$. Thus, $u \in \mathcal{H}^m(\mathbb{R}^n)$ and $\|u\|_{\mathcal{H}^m(\mathbb{R}^n)} \leq C\|u\|_{H^m(\mathbb{R}^n)}$. This proves that we have a continuous embedding $H^m(\mathbb{R}^n) \hookrightarrow \mathcal{H}^m(\mathbb{R}^n)$.

To deal with the converse embedding, we first note that (3.1.7) implies that there exist constants $C_1, C_2 \in (0, \infty)$ depending only on n and m such that

$$C_1 \langle \xi \rangle^{2m} \leq \sum_{|\alpha| \leq m} |\xi^{2\alpha}| \leq C_2 \langle \xi \rangle^{2m}, \qquad \forall \xi \in \mathbb{R}^n, \tag{12.3.40}$$

(recall that we adopted the convention that $\xi^{(0,\dots,0)} := 1$). If $u \in \mathcal{H}^m(\mathbb{R}^n)$ then $\partial^\alpha u \in L^2(\mathbb{R}^n)$ for all $\alpha \in \mathbb{N}_0^n$ with $|\alpha| \leq m$. Hence, invoking (3.2.31) and (b) in Theorem 4.26 we have $i^{|\alpha|} \xi^\alpha \widehat{u} = \widehat{\partial^\alpha u} \in L^2(\mathbb{R}^n)$ for all $\alpha \in \mathbb{N}_0^n$ with $|\alpha| \leq m$. The latter combined with the first inequality in (12.3.40) yields $\langle \xi \rangle^m \widehat{u} \in L^2(\mathbb{R}^n)$, thus $u \in H^m(\mathbb{R}^n)$, and the estimate

$$\|u\|_{H^m(\mathbb{R}^n)} \leq (2\pi)^{n/2} C_1^{-1/2} \|u\|_{\mathcal{H}^m(\mathbb{R}^n)} \tag{12.3.41}$$

follows after one more use of (3.2.29). $\square$

In general, for an open set Ω in $\mathbb{R}^n$ and $m \in \mathbb{N}_0$, one cannot expect to have $H^m(\Omega) = \mathcal{H}^m(\Omega)$. Nonetheless, the left-to-right inclusion always holds.

Proposition 12.25. *If Ω is an open set in $\mathbb{R}^n$ and $m \in \mathbb{N}_0$ then $H^m(\Omega)$ embeds continuously into $\mathcal{H}^m(\Omega)$.*

Proof. Let $u \in H^m(\Omega)$ and pick $U \in H^m(\mathbb{R}^n)$ with $U|_\Omega = u$ in $\mathcal{D}'(\Omega)$. By Theorem 12.24 we have $U \in \mathcal{H}^m(\mathbb{R}^n)$. Given that taking restrictions of distributions to open sets commutes with taking distributional derivatives, the latter implies that $u \in \mathcal{H}^m(\Omega)$ and

$$\|u\|_{\mathcal{H}^m(\Omega)} \leq \|U\|_{\mathcal{H}^m(\mathbb{R}^n)}. \tag{12.3.42}$$

Since by Theorem 12.24 there exists a constant $C \in (0, \infty)$ independent of U, such that $\|U\|_{\mathcal{H}^m(\mathbb{R}^n)} \leq C\|U\|_{H^m(\mathbb{R}^n)}$, we may conclude

$$\|u\|_{\mathcal{H}^m(\Omega)} \leq C\|U\|_{H^m(\mathbb{R}^n)}. \tag{12.3.43}$$

Recalling (12.2.2), if we take in (12.3.43) the infimum over all $U \in H^m(\mathbb{R}^n)$ such that $U|_\Omega = u$ in $\mathcal{D}'(\Omega)$, we arrive at $\|u\|_{\mathcal{H}^m(\Omega)} \leq C\|u\|_{H^m(\Omega)}$, as wanted. $\square$

There are open sets Ω in $\mathbb{R}^n$ for which the embedding from Proposition 12.25 is strict. Here is an example.

Example 12.26. Let $\Omega := \{(x, y) \in \mathbb{R}^n : 0 < x < 1, \, 0 < y < x^4\}$. Note that if $\lambda < 5/2$ then

$$\|x^{-\lambda}\|_{L^2(\Omega)} = \int_0^1 \int_0^{x^4} x^{-2\lambda} \, dy \, dx = \int_0^1 x^{4-2\lambda} \, dx = \frac{1}{5 - 2\lambda}. \tag{12.3.44}$$

Hence, if we define $u(x, y) := x^{-1/4}$ for all $(x, y) \in \Omega$ then we have $u \in C^\infty(\Omega)$, $\partial_1 u(x, y) = -\frac{1}{4}x^{-5/4}$, $\partial_1^2 u(x, y) = \frac{5}{16}x^{-9/4}$, and $\partial_2 u(x, y) = \partial_2^2 u(x, y) = 0$, for all $(x, y) \in \Omega$. Invoking (12.3.44) it follows that $\partial^\alpha u \in L^2(\Omega)$ for all $\alpha \in \mathbb{N}_0^2$, $|\alpha| \leq 2$. This implies $u \in \mathcal{H}^2(\Omega)$. On the other hand, if we assume that $u \in H^2(\Omega)$, then by (5) in Theorem 12.15 there exists a function $U \in C^0(\overline{\Omega})$ such that $U|_\Omega = u$. The latter is not possible since u cannot be extended continuously near the origin. Hence, necessarily $u \notin H^2(\Omega)$.

Example 12.26 points to the fact that the equality between the space $\mathcal{H}^m(\Omega)$ and $H^m(\Omega)$ requires additional assumptions on the open set Ω. In Theorem 12.30 we will show that this equality holds in the case of bounded Lipschitz domains (recall Definition 14.49). Among other things, the proof of Theorem 12.30 relies on a density result. In part (2) of Theorem 12.14 we have established an important density result for $(H^s(\Omega), \|\cdot\|_{H^s(\Omega)})$ with $s \in \mathbb{R}$ and Ω an open set in $\mathbb{R}^n$. A natural question is whether a similar result is available for the intrinsic Sobolev spaces $(\mathcal{H}^m(\Omega), \|\cdot\|_{\mathcal{H}^m(\Omega)})$ with $m \in \mathbb{N}$. The answer turns out to be more delicate, and the smoothness of the domain Ω plays an important role as is seen in the next theorem.

Theorem 12.27. *Fix $m \in \mathbb{N}$ and let Ω be either an upper-graph Lipschitz domain, or a bounded Lipschitz domain in $\mathbb{R}^n$. Then the set $C_0^\infty(\overline{\Omega})$ is dense in $\mathcal{H}^m(\Omega)$ with respect to the norm $\|\cdot\|_{\mathcal{H}^m(\Omega)}$.*

Proof. We divide the proof into two steps.

Step I. *The case of an upper-graph Lipschitz domain.* Fix an upper-graph Lipschitz domain $\Omega \subseteq \mathbb{R}^n$ and a function $u_o \in \mathcal{H}^m(\Omega)$. Also, pick $\varepsilon > 0$. According to Theorem 12.23 there exists some $u \in \mathcal{H}^m(\Omega)$ which vanishes a.e. outside of a bounded subset of Ω and such that

$$\|u_o - u\|_{\mathcal{H}^m(\Omega)} \leq \varepsilon/2. \tag{12.3.45}$$

Suppose we are able to show that there exists a sequence

$$\{\varphi_k\}_{k\in\mathbb{N}} \subset C_0^\infty(\mathbb{R}^n) \text{ satisfying } \varphi_k\big|_\Omega \xrightarrow[k\to\infty]{\mathcal{H}^m(\Omega)} u. \tag{12.3.46}$$

Then if $k \in \mathbb{N}$ is sufficiently large we have $\left\|\varphi_k\big|_\Omega - u\right\|_{\mathcal{H}^m(\Omega)} \leq \varepsilon/2$ which, in concert with (12.3.45) proves that $\left\|\varphi_k\big|_\Omega - u_o\right\|_{\mathcal{H}^m(\Omega)} \leq \varepsilon$. Since $\varphi_k\big|_\Omega \in C_0^\infty(\overline{\Omega})$ and $\varepsilon > 0$ is arbitrary, we conclude that $C_0^\infty(\overline{\Omega})$ is dense in $\mathcal{H}^m(\Omega)$ in this case.

Back to the issue of constructing a sequence as in (12.3.46), first recall the notation for a circular cone from (14.7.23). Lemma 14.55 guarantees that there exists an angle $\theta \in (0, \pi/2)$ (which is determined by the Lipschitz constant of the function whose upper-graph is Ω) such that

$$\Gamma_\theta(x, -\mathbf{e}_n) \subseteq \mathbb{R}^n \setminus \Omega \quad \text{for each } x \in \partial\Omega. \tag{12.3.47}$$

To simplify notation, abbreviate $\Gamma := \Gamma_\theta(0, -\mathbf{e}_n)$.

Next, pick a function $\Theta \in C_0^\infty(\Gamma)$ with the property that $\int_{\mathbb{R}^n} \Theta \, dx = 1$ and for each $\varepsilon \in (0, \infty)$ define $\Theta_\varepsilon(x) := \varepsilon^{-n}\Theta(x/\varepsilon)$, for all $x \in \mathbb{R}^n$. It follows that for each $\varepsilon \in (0, \infty)$ we have

$$\Theta_\varepsilon \in C_0^\infty(\Gamma), \quad \text{supp}\,\Theta_\varepsilon \subseteq \Gamma, \quad \text{and} \quad \int_{\mathbb{R}^n} \Theta_\varepsilon \, dx = 1, \tag{12.3.48}$$

Also, denote by tilde the extension by zero outside Ω to $\mathbb{R}^n$, and pick some $\alpha \in \mathbb{N}_0^n$ with $|\alpha| \leq m$. Since u belongs to $\mathcal{H}^m(\Omega)$ and vanishes a.e. outside of a bounded subset of Ω, it follows that the distribution $\widetilde{\partial^\alpha u}$ belongs to $L^2(\mathbb{R}^n)$ and has compact support. Moreover, $\widetilde{u}$ is a compactly supported distribution in $\mathbb{R}^n$ and so is $\partial^\alpha \widetilde{u}$. Hence, if we define $\omega_\alpha := \partial^\alpha \widetilde{u} - \widetilde{\partial^\alpha u}$, then $\omega_\alpha \in \mathcal{E}'(\mathbb{R}^n)$ and, in fact,

$$\text{supp}\,\omega_\alpha \subseteq \partial\Omega. \tag{12.3.49}$$

Now fix $\varepsilon \in (0, \infty)$. Proposition 2.102 ensures that $\widetilde{u} * \Theta_\varepsilon \in C_0^\infty(\mathbb{R}^n)$, hence

$$u_\varepsilon := (\widetilde{u} * \Theta_\varepsilon)\big|_\Omega \in C_0^\infty(\overline{\Omega}). \tag{12.3.50}$$

Given that differentiation of distributions commutes with restrictions to open sets, part *(e)* in Theorem 2.96 permits us to write

$$\partial^\alpha u_\varepsilon = ((\partial^\alpha \widetilde{u}) * \Theta_\varepsilon)\big|_\Omega = (\widetilde{\partial^\alpha u} * \Theta_\varepsilon)\big|_\Omega + (\omega_\alpha * \Theta_\varepsilon)\big|_\Omega \tag{12.3.51}$$

in $\mathcal{D}'(\Omega)$. In light of (12.3.49), the support inclusion in (12.3.48), item *(a)* in Theorem 2.96, and (12.3.47), we obtain that

$$\text{supp}\,(\omega_\alpha * \Theta_\varepsilon) \subseteq \partial\Omega + \Gamma \subseteq \mathbb{R}^n \setminus \Omega, \tag{12.3.52}$$

which implies $(\omega_\alpha * \Theta_\varepsilon)\big|_\Omega = 0$ in $\mathcal{D}'(\Omega)$. In concert with (12.3.51) this ultimately yields

$$\partial^\alpha u_\varepsilon = (\widetilde{\partial^\alpha u} * \Theta_\varepsilon)\big|_\Omega \qquad \text{for all } \varepsilon > 0. \tag{12.3.53}$$

In addition, from Exercise 4.30 we know that $\widetilde{\partial^\alpha u} * \Theta_\varepsilon \xrightarrow[\varepsilon \to 0^+]{L^2(\mathbb{R}^n)} \widetilde{\partial^\alpha u}$ which, in turn, implies

$$(\widetilde{\partial^\alpha u} * \Theta_\varepsilon)\big|_\Omega \xrightarrow[\varepsilon \to 0^+]{L^2(\Omega)} \partial^\alpha u \tag{12.3.54}$$

since the operation of restriction to Ω is continuous from $L^2(\mathbb{R}^n)$ into $L^2(\Omega)$. From (12.3.53) and (12.3.54) it follows that

$$\partial^\alpha u_\varepsilon \xrightarrow[\varepsilon \to 0^+]{L^2(\Omega)} \partial^\alpha u, \quad \text{for each } \alpha \in \mathbb{N}_0^n, \ |\alpha| \leq m. \tag{12.3.55}$$

Finally, if we now define $\varphi_k := \widetilde{u} * \Theta_{\frac{1}{k}}$ for every $k \in \mathbb{N}$, then from (12.3.55) and (12.3.50) we conclude that $\{\varphi_k\}_{k \in \mathbb{N}}$ is as in (12.3.46). This finishes the proof of Step I.

Step II. *The case of a bounded Lipschitz domain.* Suppose Ω is a bounded Lipschitz domain in $\mathbb{R}^n$ and fix $u \in \mathcal{H}^m(\Omega)$. By Remark 14.54 there exist $N \in \mathbb{N}$, points

$x_1^*, \ldots, x_N^* \in \partial\Omega$, open cylinders $\{C_{x_j^*}\}_{j=1}^N$ satisfying (14.7.19), upper-graph Lipschitz domains $\{\Omega_j\}_{j=1}^N$ such that

$$\Omega_j \cap C_{x_j^*} = \Omega \cap C_{x_j^*}, \quad \forall\, j \in \{1, \ldots, N\}, \tag{12.3.56}$$

a set $O \subset \Omega$ with $\overline{O} \subset \Omega$ and such that $O \cup \bigcup_{j=1}^N C_{x_j^*}$ is an open cover of $\overline{\Omega}$, and a partition of unity $\{\psi_j\}_{j=0}^N$ subordinate to this cover (cf. (14.7.22)).

Corresponding to $j = 0$, let $\widetilde{\psi_0 u}$ denote the extension of $\psi_0 u$ by zero outside Ω to $\mathbb{R}^n$. By Example 12.19 we have $\widetilde{\psi_0 u} \in \mathcal{H}^m(\mathbb{R}^n)$ which when combined with Theorem 12.24 gives $\widetilde{\psi_0 u} \in H^m(\mathbb{R}^n)$. Thus, we may apply *(3)* in Theorem 12.1 to find a sequence

$$\{\varphi_k^0\}_{k \in \mathbb{N}} \subset C_0^\infty(\mathbb{R}^n) \quad \text{such that} \quad \varphi_k^0 \xrightarrow[k \to \infty]{H^m(\mathbb{R}^n)} \widetilde{\psi_0 u}. \tag{12.3.57}$$

By Remark 12.17 (applied with $\Omega_1 := \mathbb{R}^n$ and $\Omega_2 := \Omega$) it follows that $R_\Omega(\varphi_k^0)$ converges to $R_\Omega(\widetilde{\psi_0 u})$ in $\mathcal{H}^m(\Omega)$ as $k \to \infty$. Upon observing that $R_\Omega(\varphi_k^0) = \varphi_k^0\big|_\Omega$ for each $k \in \mathbb{N}$, and that $R_\Omega(\widetilde{\psi_0 u}) = \psi_0 u$, the latter becomes

$$\varphi_k^0\big|_\Omega \xrightarrow[k \to \infty]{\mathcal{H}^m(\Omega)} \psi_0 u. \tag{12.3.58}$$

Next, fix $j \in \{1, \ldots, N\}$. By Exercise 12.18 we have $\psi_j u \in \mathcal{H}^m(\Omega)$. If we now define the function

$$u_j := \begin{cases} \psi_j u & \text{in } \Omega_j \cap C_{x_j^*}, \\ 0 & \text{in } \Omega_j \setminus C_{x_j^*}, \end{cases} \tag{12.3.59}$$

then Example 12.20, applied with $\Omega_0 := \Omega$, $\Omega_1 := \Omega_j$, $u := \psi_j u$, and $v := u_j$ (whose applicability is ensured by (12.3.59), (12.3.56), and the properties of $\psi_j u$) gives

$$u_j \in \mathcal{H}^m(\Omega_j), \quad \|u_j\|_{\mathcal{H}^m(\Omega_j)} = \|\psi_j u\|_{\mathcal{H}^m(\Omega)}, \quad \text{and}$$

$$\partial^\alpha u_j = \begin{cases} \partial^\alpha(\psi_j u) & \text{in } \Omega_j \cap C_{x_j^*}, \\ 0 & \text{in } \Omega_j \setminus C_{x_j^*}, \end{cases} \quad \forall\, \alpha \in \mathbb{N}_0^n, \ |\alpha| \le m. \tag{12.3.60}$$

At this point we may apply Step I corresponding to the upper-graph Lipschitz domain Ω_j and the function $u_j \in \mathcal{H}^m(\Omega_j)$ to obtain

$$\begin{array}{l} \text{a sequence } \{\varphi_k^j\}_{k \in \mathbb{N}} \text{ of functions in } C_0^\infty(\mathbb{R}^n) \text{ with} \\ \text{the property that } \lim_{k \to \infty} \varphi_k^j\big|_{\Omega_j} = u_j \text{ in } \mathcal{H}^m(\Omega_j). \end{array} \tag{12.3.61}$$

Also, let $\eta_j \in C_0^\infty(C_{x_j^*})$ satisfying $\eta_j \equiv 1$ near $\operatorname{supp}\psi_j$. In particular, $\eta_j \equiv 1$ near the support of u_j, hence $\eta_j u_j = u_j$ in $\mathcal{D}'(\Omega_j)$. Moreover, if we set

$$\Phi_k^j := \eta_j \varphi_k^j \quad \text{for each } k \in \mathbb{N}, \tag{12.3.62}$$

then

$$\Phi_k^j \in C_0^\infty(C_{x_j}) \quad \text{for each } k \in \mathbb{N}. \tag{12.3.63}$$

A combination of (12.3.62), (12.3.61), and part *(1)* in Exercise 12.18 gives

$$\lim_{k \to \infty} \Phi_k^j\big|_{\Omega_j} = \eta_j u_j = u_j \quad \text{in } \mathcal{H}^m(\Omega_j). \tag{12.3.64}$$

In addition, for each $\alpha \in \mathbb{N}_0^n$ with $|\alpha| \le m$, we may write

$$\begin{aligned}
\left\| \partial^\alpha(\Phi_k^j\big|_\Omega) - \partial^\alpha(\psi_j u) \right\|_{L^2(\Omega)} &= \left\| \partial^\alpha(\Phi_k^j\big|_{\Omega \cap C_{x_j^*}}) - \partial^\alpha(\psi_j u) \right\|_{L^2(\Omega \cap C_{x_j^*})} \\
&= \left\| \partial^\alpha(\Phi_k^j\big|_{\Omega_j \cap C_{x_j^*}}) - \partial^\alpha u_j \right\|_{L^2(\Omega_j \cap C_{x_j^*})} \\
&= \left\| \partial^\alpha(\Phi_k^j\big|_{\Omega_j}) - \partial^\alpha u_j \right\|_{L^2(\Omega_j)}.
\end{aligned} \tag{12.3.65}$$

The first equality in (12.3.65) follows from (12.3.63) and the fact that the support of ψ_j is contained in $C_{x_j^*}$, the second equality is due to (12.3.56) and (12.3.60), while the last equality uses (12.3.63) and (12.3.60). Recalling the definition of the $\mathcal{H}^m$-norm (cf. (12.3.4)), from (12.3.65) and (12.3.64) we ultimately conclude that

$$\lim_{k \to \infty} \Phi_k^j\big|_\Omega = \psi_j u \quad \text{in } \mathcal{H}^m(\Omega), \quad \text{for each } j \in \{1, \ldots, N\}. \tag{12.3.66}$$

At this stage, for each $k \in \mathbb{N}$ define

$$\Phi_k := \varphi_k^0 + \sum_{j=1}^N \Phi_k^j. \tag{12.3.67}$$

Then (12.3.57) and (12.3.63) guarantee that $\Phi_k \in C_0^\infty(\mathbb{R}^n)$ for each $k \in \mathbb{N}$. Also, based on (12.3.67), (12.3.58), and (12.3.66), and the last line in (14.7.22) we obtain

$$\lim_{k \to \infty} \Phi_k\big|_\Omega = \sum_{j=0}^N \psi_j u = u \quad \text{in } \mathcal{H}^m(\Omega). \tag{12.3.68}$$

This finishes the proof of Theorem 12.27. $\square$

Extending functions from Sobolev spaces defined intrinsically on open sets to the entire Euclidean space with preservation of Sobolev class is a delicate issue. As is apparent from Example 12.26, such an extension is not possible for arbitrary bounded open sets. We will prove that an extension exists in the case of bounded Lipschitz domains. This will be done via the so-called Calderón extension operator. To set the stage, in the next lemma we prove a useful integral representation formula for smooth functions.

Lemma 12.28. *Let Γ be an infinite open upright circular cone with vertex at the origin and aperture $\theta \in (0, \pi)$, that is $\Gamma := \{x \in \mathbb{R}^n : \cos(\theta/2)|x| < x \cdot \mathbf{e}_n\}$. Fix $m \in \mathbb{N}$ and suppose*

$$\phi \in C^\infty(S^{n-1}) \quad \text{is supported in} \quad \Gamma \cap S^{n-1} \tag{12.3.69}$$

and satisfies

$$\int_{S^{n-1}} \phi(\omega)\,d\omega = \frac{(-1)^m}{(m-1)!}. \tag{12.3.70}$$

Then for every $\Theta \in C_0^\infty(\mathbb{R}^n)$ there holds

$$\Theta(0) = \sum_{|\alpha|=m} \int_\Gamma \frac{m!}{\alpha!}\, \phi\Big(\frac{y}{|y|}\Big)\, \frac{y^\alpha}{|y|^n}\, (\partial^\alpha \Theta)(y)\,dy. \tag{12.3.71}$$

Proof. To get started, fix $\omega \in S^{n-1}$ and apply formula (14.2.8) to $f(t) := \Theta(t\omega)$, $t \in \mathbb{R}$, to obtain

$$\Theta(0) = \frac{(-1)^m}{(m-1)!} \int_0^\infty t^{m-1} \frac{d^m}{dt^m}[\Theta(t\omega)]\,dt. \tag{12.3.72}$$

We claim that

$$\frac{d^m}{dt^m}[\Theta(t\omega)] = \sum_{|\alpha|=m} \frac{m!}{\alpha!}\, \omega^\alpha (\partial^\alpha \Theta)(t\omega), \quad t \in \mathbb{R}. \tag{12.3.73}$$

This is proved by induction over m. If $m = 1$, formula (12.3.73) is just the Chain Rule. Suppose (12.3.73) holds for some $m \in \mathbb{N}$. Then taking one more derivative in t and using the Chain Rule from (12.3.73) we obtain

$$\frac{d^{m+1}}{dt^{m+1}}[\Theta(t\omega)] = \sum_{j=1}^n \sum_{|\alpha|=m} \frac{m!}{\alpha!}\, \omega^{\alpha+\mathbf{e}_j}(\partial^{\alpha+\mathbf{e}_j}\Theta)(t\omega), \quad t \in \mathbb{R}. \tag{12.3.74}$$

Now fix $j \in \{1, \dots, n\}$ and invoke Lemma 14.7 in the following setting:

$$D := \{\alpha \in \mathbb{N}_0^n : |\alpha| = m\}, \quad R := \{\beta \in \mathbb{N}_0^n : |\beta| = m+1\},$$

$$\mathcal{F}(\alpha) := \alpha + \mathbf{e}_j, \quad \text{for each } \alpha \in D, \quad G := \mathbb{C}, \tag{12.3.75}$$

$$G(\beta) := \frac{m!}{(\beta - \mathbf{e}_j)!}\, \omega^\beta (\partial^\beta \Theta)(t\omega), \quad \text{for each } \beta \in R,$$

In particular, if $\operatorname{supp}\beta := \{i \in \{1, \dots, n\} : \beta_i \neq 0\}$ for each $\beta = (\beta_1, \dots, \beta_n) \in \mathbb{N}_0^n$, we have

$$\#\mathcal{F}^{-1}(\{\beta\}) = \begin{cases} 0 & \text{if } j \notin \operatorname{supp}\beta, \\ 1 & \text{if } j \in \operatorname{supp}\beta, \end{cases} \quad \forall \beta \in R, \tag{12.3.76}$$

and (14.2.1) (applied for each j) in the current yields

$$\sum_{j=1}^{n} \sum_{|\alpha|=m} \frac{m!}{\alpha!} \, \omega^{\alpha+\mathbf{e}_j} (\partial^{\alpha+\mathbf{e}_j} \Theta)(t\omega)$$

$$= \sum_{j=1}^{n} \sum_{|\beta|=m+1} \mathbf{1}_{\mathrm{supp}\beta}(j) \frac{m!}{(\beta - \mathbf{e}_j)!} \, \omega^{\beta} (\partial^{\beta} \Theta)(t\omega)$$

$$= \sum_{|\beta|=m+1} m! \Big(\sum_{j \in \mathrm{supp}\beta} \frac{1}{(\beta - \mathbf{e}_j)!} \Big) \omega^{\beta} (\partial^{\beta} \Theta)(t\omega) \qquad (12.3.77)$$

for each $t \in \mathbb{R}$. In addition, since

$$\sum_{j \in \mathrm{supp}\beta} \frac{1}{(\beta - \mathbf{e}_j)!} = \sum_{j \in \mathrm{supp}\beta} \frac{\beta_j}{\beta!} = \frac{1}{\beta!} \sum_{j \in \mathrm{supp}\beta} \beta_j = \frac{m+1}{\beta!}, \qquad (12.3.78)$$

formula (12.3.77) further gives, for each $t \in \mathbb{R}$,

$$\sum_{j=1}^{n} \sum_{|\alpha|=m} \frac{m!}{\alpha!} \, \omega^{\alpha+\mathbf{e}_j} (\partial^{\alpha+\mathbf{e}_j} \Theta)(t\omega) = \sum_{|\beta|=m+1} \frac{(m+1)}{\beta!} \, \omega^{\beta} (\partial^{\beta} \Theta)(t\omega). \qquad (12.3.79)$$

This and (12.3.74) completes the proof by induction of (12.3.73).

A combination of (12.3.72) and (12.3.73) then gives

$$\Theta(0) = \frac{(-1)^m}{(m-1)!} \sum_{|\alpha|=m} \frac{m!}{\alpha!} \int_0^{\infty} t^{m-1} \omega^{\alpha} (\partial^{\alpha} \Theta)(t\omega) \, \mathrm{d}t. \qquad (12.3.80)$$

Multiplying (12.3.80) by $\frac{(m-1)!}{(-1)^m} \phi(\omega)$, then integrating in $\omega \in S^{n-1}$ and using (12.3.70), permits us to write

$$\Theta(0) = \sum_{|\alpha|=m} \frac{m!}{\alpha!} \int_{S^{n-1}} \int_0^{\infty} t^{m-1} \omega^{\alpha} \phi(\omega) (\partial^{\alpha} \Theta)(t\omega) \, \mathrm{d}t \, \mathrm{d}\omega$$

$$= \sum_{|\alpha|=m} \frac{m!}{\alpha!} \int_{\Gamma} |y|^{m-1} \Big(\frac{y}{|y|} \Big)^{\alpha} \phi(y/|y|) (\partial^{\alpha} \Theta)(y) \frac{1}{|y|^{n-1}} \, \mathrm{d}y$$

$$= \sum_{|\alpha|=m} \frac{m!}{\alpha!} \int_{\Gamma} \frac{y^{\alpha}}{|y|^n} \phi(y/|y|) (\partial^{\alpha} \Theta)(y) \, \mathrm{d}y. \qquad (12.3.81)$$

For the second equality in (12.3.81) we changed variables $y = t\omega$ and observed that if $\omega \in \mathrm{supp}\,\phi \subseteq \Gamma \cap S^{n-1}$ then $y \in \Gamma$. The proof of (12.3.71) is now finished. $\qquad \square$

The construction of Calderón's extension operator makes the object of the next theorem.

Theorem 12.29 (Calderón's Extension Operator). *Let Ω be a bounded Lipschitz domain in $\mathbb{R}^n$ and let $m \in \mathbb{N}$. Then there exists a linear and bounded operator*

$$\mathbb{E}_m : \mathcal{H}^m(\Omega) \to \mathcal{H}^m(\mathbb{R}^n) = H^m(\mathbb{R}^n) \tag{12.3.82}$$

with the property that

$$(\mathbb{E}_m u)\big|_\Omega = u \ \ for \ every \ \ u \in \mathcal{H}^m(\Omega). \tag{12.3.83}$$

Proof. We split the proof of the theorem in a number of steps.

Step I. *The Extension operator for upper-graph Lipschitz domains and functions that vanish outside of a fixed bounded set.*
Suppose Ω is the upper-graph of a Lipschitz function $\varphi : \mathbb{R}^{n-1} \to \mathbb{R}$, with Lipschitz constant M, i.e., $\Omega = \{(y', y_n) \in \mathbb{R}^{n-1} \times \mathbb{R} : y_n > \varphi(y')\}$. In this setting, the goal is to show that

> for each compact set $K \subset \mathbb{R}^n$ and each $\eta \in C_0^\infty(\mathbb{R}^n)$
> such that $\eta \equiv 1$ near K, there exists a linear and
> bounded operator $E_m : \mathcal{H}^m(\Omega) \to H^m(\mathbb{R}^n)$ satisfy-
> ing $\mathrm{supp}\,(E_m v) \subseteq \mathrm{supp}\,\eta$ and $(E_m v)\big|_\Omega = \eta^2 v$ a.e. in Ω,
> for every $v \in \mathcal{H}^m(\Omega)$. $\tag{12.3.84}$

To this end, fix a compact set K and a function η as in (12.3.84). Also, pick an angle $\theta \in (0, 2\arctan(\frac{1}{M})]$ and apply Lemma 14.55 to conclude that (14.7.24) holds. To lighten notation, abbreviate $\Gamma := \Gamma_\theta(0, \mathbf{e}_n)$ (recall (14.7.23)). Then (14.7.24) implies

$$x + \Gamma \subseteq \Omega \quad \text{for every } x \in \overline{\Omega}. \tag{12.3.85}$$

Let ϕ be as in Lemma 12.28 associated with this cone Γ. For each $\alpha \in \mathbb{N}_0^n$ with $|\alpha| = m$ define the function

$$\phi_\alpha(x) := (-1)^m \frac{m!}{\alpha!} \phi\left(-\frac{x}{|x|}\right) \frac{x^\alpha}{|x|^n}, \qquad \forall x \in \mathbb{R}^n \setminus \{0\}. \tag{12.3.86}$$

From (12.3.69) and (12.3.86) we see that

$$\mathrm{supp}\,\phi_\alpha \subset -\Gamma, \quad \phi_\alpha \in C^\infty(\mathbb{R}^n \setminus \{0\}),$$
$$\phi_\alpha(\lambda x) = \lambda^{m-n}\phi_\alpha(x), \ \ \forall x \in \mathbb{R}^n \setminus \{0\}, \ \ \forall \lambda \in (0, \infty). \tag{12.3.87}$$

The properties listed in (12.3.87) allow us to define the generalized volume potential associated with ϕ_α (cf. (4.10.23)). This acts on each $f \in L_{comp}^\infty(\mathbb{R}^n)$ according to

$$(\Pi_{\phi_\alpha} f)(x) := \int_{\mathbb{R}^n} \phi_\alpha(x - y) f(y)\,\mathrm{d}y, \quad \forall x \in \mathbb{R}^n. \tag{12.3.88}$$

From Theorem 4.105 we know that $\Pi_{\phi_\alpha} f \in C^{m-1}(\mathbb{R}^n)$ for each $f \in L_{comp}^\infty(\mathbb{R}^n)$ (note that ϕ_α is positive homogeneous of degree $m - n \in \mathbb{Z}$) and

$$\partial^\gamma[\Pi_{\phi_\alpha} f] = \Pi_{\partial^\gamma \phi_\alpha} f \quad \text{pointwise in } \mathbb{R}^n$$
$$\text{for every } \gamma \in \mathbb{N}_0^n \text{ with } |\gamma| \le m - 1. \tag{12.3.89}$$

Moreover, Theorem 4.105 gives that if $\gamma = (\gamma_1, \ldots, \gamma_n) \in \mathbb{N}_0^n$ has $|\gamma| = m$ then for each $f \in L_{comp}^\infty(\mathbb{R}^n)$ the distributional derivative $\partial^\gamma[\Pi_{\phi_\alpha} f]$ is of function type and for each $j \in \{1, \ldots, n\}$ such that $\gamma_j \neq 0$ we have

$$\partial^\gamma[\Pi_{\phi_\alpha} f](x) = \Big(\int_{S^{n-1}} (\partial^{\gamma-e_j}\phi_\alpha)(\omega)\omega_j \, d\sigma(\omega)\Big)f(x) \tag{12.3.90}$$

$$+ \lim_{\varepsilon \to 0^+} \int_{|y-x|\geq\varepsilon} (\partial^\gamma\phi_\alpha)(x-y)f(y)\,dy \qquad \text{for a.e. } x \in \mathbb{R}^n,$$

where the derivative in the left-hand side is taken in $\mathcal{D}'(\mathbb{R}^n)$.

Next, consider the operator $\mathcal{E}_m : C_0^\infty(\overline{\Omega}) \to \mathbb{C}$ acting on each $w \in C_0^\infty(\overline{\Omega})$ according to

$$(\mathcal{E}_m w)(x) := \sum_{|\alpha|=m} \eta(x)\big(\Pi_{\phi_\alpha}\widetilde{\partial^\alpha(\eta w)}\big)(x), \quad \forall\, x \in \mathbb{R}^n, \tag{12.3.91}$$

where tilde denotes extension by zero from Ω to $\mathbb{R}^n$. Note that since $w \in C_0^\infty(\overline{\Omega})$ then $\widetilde{\partial^\alpha(\eta w)} \in L_{comp}^\infty(\mathbb{R}^n)$. In light of the fact that the operator in (12.3.88) is well defined, this ensures that the operator in (12.3.91) is also well defined. We make two claims pertaining to the nature of $\mathcal{E}_m$.

Claim 1. $(\mathcal{E}_m w)\big|_\Omega = \eta^2 w$ pointwise in Ω, for each $w \in C_0^\infty(\overline{\Omega})$.

Claim 2. There exists a constant $C \in (0, \infty)$ such that

$$\|\mathcal{E}_m w\|_{H^m(\mathbb{R}^n)} \leq C\|w\|_{\mathcal{H}^m(\Omega)}, \quad \forall\, w \in C_0^\infty(\overline{\Omega}). \tag{12.3.92}$$

Let us assume *Claim 1* and *Claim 2* for the moment and see how they may be used to finish the construction of the desired extension operator.

The operator $\mathcal{E}_m$ is obviously linear which, when combined with the estimate in (12.3.92), implies that $\mathcal{E}_m : (C_0^\infty(\overline{\Omega}), \|\cdot\|_{\mathcal{H}^m(\Omega)}) \to H^m(\mathbb{R}^n)$ is linear and bounded. Since $C_0^\infty(\overline{\Omega})$ is dense in $\mathcal{H}^m(\Omega)$ (cf. Theorem 12.27) it follows that $\mathcal{E}_m$ extends continuously to $\mathcal{H}^m(\Omega)$, thus

$$\begin{array}{l}\text{there exists } E_m : \mathcal{H}^m(\Omega) \to H^m(\mathbb{R}^n) \text{ linear and} \\ \text{bounded and such that } E_m\big|_{C_0^\infty(\overline{\Omega})} = \mathcal{E}_m.\end{array} \tag{12.3.93}$$

This extension by density means that

$$\begin{array}{l}E_m v = \lim_{j\to\infty} \mathcal{E}_m w_j \text{ for each given } v \in \mathcal{H}^m(\Omega) \text{ and each} \\ \text{sequence } w_j \in C_0^\infty(\overline{\Omega}),\, j \in \mathbb{N}, \text{ such that } w_j \xrightarrow[j\to\infty]{\mathcal{H}^m(\Omega)} v.\end{array} \tag{12.3.94}$$

In particular, since $\operatorname{supp}(\mathcal{E}_m w) \subseteq \operatorname{supp}\eta$ for every $w \in C_0^\infty(\overline{\Omega})$ (as seen from an inspection of formula (12.3.91)), we also have that

$$\operatorname{supp}(E_m v) \subseteq \operatorname{supp} \eta \quad \text{for all} \quad v \in \mathcal{H}^m(\Omega). \tag{12.3.95}$$

Hence, to show that the operator E_m from (12.3.93) satisfies all the conditions listed in (12.3.84), there remains to prove (modulo the justifications of the two earlier claims) that $(E_m v)\big|_\Omega = \eta^2 v$, a.e. in Ω, for each $v \in \mathcal{H}^m(\Omega)$. Fix such a function v and apply Theorem 12.27 to obtain a sequence $\{w_j\}_{j \in \mathbb{N}}$ of functions in $C_0^\infty(\overline{\Omega})$ such that $w_j \xrightarrow[j \to \infty]{\mathcal{H}^m(\Omega)} v$. In particular, $w_j \xrightarrow[j \to \infty]{L^2(\Omega)} v$ and, by passing to a subsequence (which we still denote by $\{w_j\}_{j \in \mathbb{N}}$) and multiplying by η^2, we have

$$\eta^2 w_j \to \eta^2 v \text{ pointwise a.e. in } \Omega \text{ as } j \to \infty. \tag{12.3.96}$$

In addition, $E_m v \in H^m(\mathbb{R}^n)$ and the continuity of E_m further yields

$$\mathcal{E}_m w_j = E_m w_j \xrightarrow[j \to \infty]{H^m(\mathbb{R}^n)} E_m v. \tag{12.3.97}$$

In particular, (12.3.97) implies $\mathcal{E}_m w_j \xrightarrow[j \to \infty]{L^2(\mathbb{R}^n)} E_m v$ and, passing to a subsequence (which is a subsequence of the subsequence in (12.3.96)) we obtain (again, keeping the same notation for this sub-subsequence)

$$\mathcal{E}_m w_j \to E_m v \text{ pointwise a.e. in } \mathbb{R}^n \text{ as } j \to \infty. \tag{12.3.98}$$

By restricting to Ω and invoking *Claim 1*, to the effect that $\eta^2 w_j = (\mathcal{E}_m w_j)\big|_\Omega$, the convergence in (12.3.98) further gives

$$\eta^2 w_j \to (E_m v)\big|_\Omega \text{ pointwise a.e. in } \Omega \text{ as } j \to \infty. \tag{12.3.99}$$

In concert, (12.3.96) and (12.3.99) imply

$$(E_m v)\big|_\Omega = \eta^2 v \quad \text{a.e. in } \Omega. \tag{12.3.100}$$

In summary, we have established (12.3.84), modulo the proofs of *Claim 1* and *Claim 2* which we shall deal with in the next two steps.

Step II. *Proof of Claim 1 from Step I.*
Let $w \in C_0^\infty(\overline{\Omega})$ be given. Pick $W \in C_0^\infty(\mathbb{R}^n)$ such that $W\big|_\Omega = w$, and fix $x \in \Omega$. Starting with (12.3.91) and recalling (12.3.88) we may write

$$(\mathcal{E}_m w)(x) = \sum_{|\alpha|=m} \eta(x) \int_\Omega \phi_\alpha(x-y) \partial^\alpha(\eta W)(y) \, dy$$

$$= \sum_{|\alpha|=m} \eta(x) \int_{-\Gamma} \phi_\alpha(z) \partial^\alpha(\eta W)(x-z) \, dz, \tag{12.3.101}$$

where for the second equality in (12.3.101) we made the change of variables $y = x-z$ and used the support condition $\operatorname{supp} \phi_\alpha \subset -\Gamma$ (cf. (12.3.87)) and the fact that if $y \in \Omega$

then $z \in (x - \Omega) \cap (-\Gamma) = -\Gamma$ (due to (12.3.85)). Next, we bring in formula (12.3.86) and make the change of variables $z = -y$ under the integral to further obtain

$$(\mathcal{E}_m w)(x) = \eta(x) \sum_{|\alpha|=m} \frac{m!}{\alpha!} \int_\Gamma \phi\Big(\frac{y}{|y|}\Big) \frac{y^\alpha}{|y|^n} \partial^\alpha(\eta W)(x+y)\, dy. \qquad (12.3.102)$$

This last expression in the right-hand side of (12.3.102) allows us to apply (12.3.71) in Lemma 12.28 with $\Theta(y) := \partial^\alpha(\eta W)(x+y)$ for $y \in \mathbb{R}^n$ (recall that $x \in \Omega$ is fixed) and conclude that

$$(\mathcal{E}_m w)(x) = \eta(x)(\eta W)(x) = \eta^2(x) w(x), \qquad (12.3.103)$$

as desired.

Step III. *Proof of Claim 2 from Step I.*
Fix again $w \in C_0^\infty(\overline{\Omega})$ and pick some $\beta \in \mathbb{N}_0^n$ with $|\beta| \leq m$. Then (12.3.91) and the Generalized Leibniz Formula from Proposition 2.49 give

$$\partial^\beta(\mathcal{E}_m w) = \sum_{|\alpha|=m} \sum_{\gamma \leq \beta} \frac{\beta!}{\gamma!(\beta - \gamma)!} (\partial^{\beta-\gamma}\eta)\partial^\gamma\Big(\Pi_{\phi_\alpha}\widetilde{\partial^\alpha(\eta w)}\Big) \qquad (12.3.104)$$

in $\mathcal{D}'(\mathbb{R}^n)$. To estimate the $L^2(\mathbb{R}^n)$ norm of the terms in the right-hand side we distinguish two cases.

First, suppose $\gamma \in \mathbb{N}_0^n$ is such that $\gamma \leq \beta$ and $|\gamma| \leq m - 1$. Observe that since η is compactly supported, the set

$$K_\eta := \{x - y : x, y \in \operatorname{supp}\eta\} \quad \text{is compact in } \mathbb{R}^n. \qquad (12.3.105)$$

From (12.3.87) it follows that the function $\partial^\gamma\phi_\alpha$ is smooth outside the origin and is positive homogeneous of degree $m - n - |\gamma| \geq 1 - n$ (recall Exercise 4.51) so it belongs to $L^1_{loc}(\mathbb{R}^n)$. The latter and (12.3.105) then ensure $(\partial^\gamma\phi_\alpha)\mathbf{1}_{K_\eta} \in L^1(\mathbb{R}^n)$, so for each $x \in \operatorname{supp}\eta$ we may write

$$\Big(\Pi_{\partial^\gamma\phi_\alpha}\widetilde{\partial^\alpha(\eta w)}\Big)(x) = \int_{\mathbb{R}^n} ((\partial^\gamma\phi_\alpha)\mathbf{1}_{K_\eta})(x-y)\widetilde{\partial^\alpha(\eta w)}(y)\, dy$$

$$= \Big(((\partial^\gamma\phi_\alpha)\mathbf{1}_{K_\eta}) * \widetilde{\partial^\alpha(\eta w)}\Big)(x). \qquad (12.3.106)$$

Since $\partial^\alpha(\eta w) \in L^2(\mathbb{R}^n)$ we may invoke Young's Inequality and Example 12.19 to estimate

$$\left\|\left(\partial^{\beta-\gamma}\eta\right)\left(\Pi_{\partial^\gamma\phi_\alpha}\widetilde{\partial^\alpha(\eta w)}\right)\right\|_{L^2(\mathbb{R}^n)}$$

$$\leq \left\|\partial^{\beta-\gamma}\eta\right\|_{L^\infty(\mathbb{R}^n)}\left\|(\partial^\gamma\phi_\alpha)\mathbf{1}_{K_\eta}\right\|_{L^1(\mathbb{R}^n)}\left\|\widetilde{\partial^\alpha(\eta w)}\right\|_{L^2(\mathbb{R}^n)}$$

$$\leq C\left\|\partial^{\beta-\gamma}\eta\right\|_{L^\infty(\mathbb{R}^n)}\left\|(\partial^\gamma\phi_\alpha)\mathbf{1}_{K_\eta}\right\|_{L^1(\mathbb{R}^n)}\|w\|_{\mathcal{H}^m(\Omega)}$$

$$\leq C\|w\|_{\mathcal{H}^m(\Omega)}, \tag{12.3.107}$$

for some $C = C(\Omega, \eta, \phi, \alpha, \gamma) \in (0, \infty)$ independent of w.

Second, suppose $\gamma \in \mathbb{N}_0^n$ is such that $\gamma \leq \beta$ and $|\gamma| = m$ (a scenario in which we necessarily have $|\beta| = m$ and, in fact, $\gamma = \beta$). If $\gamma = (\gamma_1, \ldots, \gamma_n)$ and $j \in \{1, \ldots, n\}$ is such that $\gamma_j \neq 0$, then according to (12.3.90) we have

$$\partial^\gamma[\Pi_{\phi_\alpha}\widetilde{\partial^\alpha(\eta w)}] = C(\gamma, \phi_\alpha)\widetilde{\partial^\alpha(\eta w)} + T_{\partial^\gamma\phi_\alpha}\widetilde{\partial^\alpha(\eta w)} \tag{12.3.108}$$

where the constant $C(\gamma, \phi_\alpha)$ is defined as

$$C(\gamma, \phi_\alpha) := \int_{S^{n-1}} (\partial^{\gamma-\mathbf{e}_j}\phi_\alpha)(\omega)\omega_j\,d\sigma(\omega) \tag{12.3.109}$$

and, for a.e. $x \in \mathbb{R}^n$,

$$T_{\partial^\gamma\phi_\alpha}\widetilde{\partial^\alpha(\eta w)}(x) := \lim_{\varepsilon\to 0^+} \int_{|y-x|\geq\varepsilon} (\partial^\gamma\phi_\alpha)(x-y)\widetilde{\partial^\alpha(\eta w)}(y)\,dy. \tag{12.3.110}$$

From (12.3.87), the current assumptions on γ, Exercise 4.51, and Example 4.71 (applied to $\partial^{\gamma-\mathbf{e}_j}\phi_\alpha$), it follows that the function $\partial^\gamma\phi_\alpha$ satisfies the conditions in (4.4.1). Consequently, part *(e)* in Theorem 4.100 ensures that the operator $T_{\partial^\gamma\phi_\alpha}$ is bounded from $L^2(\mathbb{R}^n)$ into $L^2(\mathbb{R}^n)$, thus

$$\left\|T_{\partial^\gamma\phi_\alpha}\widetilde{\partial^\alpha(\eta w)}\right\|_{L^2(\mathbb{R}^n)} \leq C\left\|\widetilde{\partial^\alpha(\eta w)}\right\|_{L^2(\mathbb{R}^n)}. \tag{12.3.111}$$

In concert, (12.3.108), (12.3.111), and Example 12.19 give

$$\left\|\eta\left(\Pi_{\partial^\gamma\phi_\alpha}\widetilde{\partial^\alpha(\eta w)}\right)\right\|_{L^2(\mathbb{R}^n)} \leq C\|\eta\|_{L^\infty(\mathbb{R}^n)}\left\|\widetilde{\partial^\alpha(\eta w)}\right\|_{L^2(\mathbb{R}^n)}$$

$$\leq C\|w\|_{\mathcal{H}^m(\Omega)}, \tag{12.3.112}$$

for some $C = C(\Omega, \eta, \phi, \alpha, \gamma) \in (0, \infty)$ independent of w.

At this stage, combining (12.3.104), (12.3.107), (12.3.112) proves that there exists a constant $C \in (0, \infty)$ with the property that

$$\|\partial^\beta(\mathcal{E}_m w)\|_{L^2(\mathbb{R}^n)} \leq C\|w\|_{\mathcal{H}^m(\Omega)}, \tag{12.3.113}$$

for all $w \in C_0^\infty(\overline{\Omega})$ and $\beta \in \mathbb{N}_0^n$ with $|\beta| \leq m$. The desired estimate in *Claim 2* now follows by invoking Theorem 12.24 and (12.3.4).

Step IV. *The extension operator for a bounded Lipschitz domain.*
Work in the case when Ω is a bounded Lipschitz domain. By Remark 14.54 there
exist a natural number $N \in \mathbb{N}$, points $x_1^*, \ldots, x_N^* \in \partial\Omega$, upper-graph Lipschitz domains $\{\Omega_j\}_{j=1}^N$, open cylinders $\{C_{x_j^*}\}_{j=1}^N$ satisfying (14.7.19), a set $O \subset \Omega$ with $\overline{O} \subset \Omega$
and such that $O \cup \bigcup_{j=1}^N C_{x_j^*}$ is an open cover of $\overline{\Omega}$, and a partition of unity $\{\psi_j\}_{j=0}^N$
subordinate to this cover (cf. (14.7.22)). In addition, for each $j \in \{1, \ldots, N\}$, consider

$$\eta_j \in C_0^\infty(C_{x_j^*}), \quad \eta_j \equiv 1 \text{ near } \operatorname{supp}\psi_j. \tag{12.3.114}$$

To proceed, select $u \in \mathcal{H}^m(\Omega)$ and fix $j \in \{1, \ldots, n\}$. Making use of Exercise 12.18 we obtain

$$\psi_j u \in \mathcal{H}^m(\Omega) \quad \text{and} \quad \|\psi_j u\|_{\mathcal{H}^m(\Omega)} \le C\|u\|_{\mathcal{H}^m(\Omega)}, \tag{12.3.115}$$

for some constant $C = C(\Omega) \in (0, \infty)$ (given that the partition of unity ultimately
depends only on Ω). Since by (14.7.19) we have

$$\Omega_j \cap C_{x_j^*} = \Omega \cap C_{x_j^*}, \tag{12.3.116}$$

we may define the function

$$v_j := \begin{cases} \psi_j u & \text{in } \Omega_j \cap C_{x_j^*}, \\ 0 & \text{in } \Omega_j \setminus C_{x_j^*}. \end{cases} \tag{12.3.117}$$

Granted (12.3.115) and (12.3.117), Example 12.20 may be invoked to conclude that
there exists some $C = C(\Omega) \in (0, \infty)$ such that

v_j belongs to $\mathcal{H}^m(\Omega_j)$, vanishes outside a bounded subset
of Ω_j, and obeys the estimate $\|v_j\|_{\mathcal{H}^m(\Omega_j)} \le C\|u\|_{\mathcal{H}^m(\Omega)}$. $\tag{12.3.118}$

Next, let E_m^j denote the extension operator from Step I associated with the upper-
graph Lipschitz domain Ω_j, the compact set $K := \operatorname{supp}\psi_j$, and the function $\eta := \eta_j$.
In particular, we have

$$E_m^j v_j \in H^m(\mathbb{R}^n), \quad \left\|E_m^j v_j\right\|_{H^m(\mathbb{R}^n)} \le C\|v_j\|_{\mathcal{H}^m(\Omega_j)},$$

$$\operatorname{supp}(E_m^j v_j) \subseteq \operatorname{supp}\eta_j \subset C_{x_j^*}, \quad \text{and} \tag{12.3.119}$$

$$(E_m^j v_j)\big|_{\Omega_j} = \eta_j^2 v_j \quad \text{a.e. in } \Omega_j.$$

Taking into account (12.3.114) and (12.3.116), the last condition in (12.3.119) further yields

$$\left(E_m^j v_j\right)\big|_{\Omega \cap C_{x_j^*}} = \left(E_m^j v_j\right)\big|_{\Omega_j \cap C_{x_j^*}} = \left(\eta_j^2 v_j\right)\big|_{\Omega_j \cap C_{x_j^*}}$$

$$= \left(\eta_j^2 \psi_j u\right)\big|_{\Omega_j \cap C_{x_j^*}} = \left(\psi_j u\right)\big|_{\Omega_j \cap C_{x_j^*}}$$

$$= \left(\psi_j u\right)\big|_{\Omega \cap C_{x_j^*}}. \tag{12.3.120}$$

In concert, (12.3.120), the support condition in (12.3.119), and the fact that ψ_j is supported in $C_{x_j^*}$ imply

$$\left(E_m^j v_j\right)\big|_\Omega = \widetilde{\left(E_m^j v_j\right)\big|_{\Omega \cap C_{x_j^*}}} = \widetilde{\left(\psi_j u\right)\big|_{\Omega \cap C_{x_j^*}}} = \psi_j u, \tag{12.3.121}$$

where tilde denotes the extension by zero to Ω. Also, by combining the norm estimate in (12.3.119) with the norm estimate in (12.3.118) we obtain

$$\left\| E_m^j v_j \right\|_{H^m(\mathbb{R}^n)} \le C \| u \|_{\mathcal{H}^m(\Omega)}. \tag{12.3.122}$$

At this point, we are ready to define an extension of u to $H^m(\mathbb{R}^n)$. Specifically, set

$$\mathbb{E}_m u := \sum_{j=1}^N E_m^j v_j + \widetilde{\psi_0 u}, \tag{12.3.123}$$

where $\widetilde{\psi_0 u}$ denotes the extension of $\psi_0 u$ by zero outside Ω to $\mathbb{R}^n$. From Example 12.19 we know that

$$\widetilde{\psi_0 u} \in H^m(\mathbb{R}^n) \text{ and } \| \widetilde{\psi_0 u} \|_{H^m(\mathbb{R}^n)} \le C \| u \|_{\mathcal{H}^m(\Omega)}, \tag{12.3.124}$$

for some $C = C(\psi_0) = C(\Omega) \in (0, \infty)$. From (12.3.123), (12.3.122), and (12.3.124), it follows that

$$\mathbb{E}_m u \in H^m(\mathbb{R}^n) \text{ and } \left\| \mathbb{E}_m u \right\|_{H^m(\mathbb{R}^n)} \le C \| u \|_{\mathcal{H}^m(\Omega)}, \tag{12.3.125}$$

for some finite constant $C = C(\Omega)$ independent of u. Moreover, (12.3.123), (12.3.121), the fact that $\widetilde{\psi_0 u}\big|_\Omega = \psi_0 u$, and that $\sum_{j=0}^N \psi_j = 1$ in Ω, imply

$$\left(\mathbb{E}_m u\right)\big|_\Omega = \sum_{j=1}^N \left(E_m^j v_j\right)\big|_\Omega + \widetilde{\psi_0 u}\big|_\Omega = \sum_{j=0}^N \psi_j u = u. \tag{12.3.126}$$

Since $u \in \mathcal{H}^m(\Omega)$ is arbitrary, in light of (12.3.125) and (12.3.126), we have that the operator defined in (12.3.123) is linear and bounded from $\mathcal{H}^m(\Omega)$ into $H^m(\mathbb{R}^n)$ and is an extension operator, in the sense that $\left(\mathbb{E}_m u\right)\big|_\Omega = u$ for every $u \in \mathcal{H}^m(\Omega)$. This finishes the proof of Theorem 12.29. $\square$

An important consequence of the existence of Calderón's extension operator as proved in the last theorem is the fact that for a bounded Lipschitz domain Ω the spaces $\mathcal{H}^m(\Omega)$ and $H^m(\Omega)$ are equal.

Theorem 12.30. *Let Ω be a bounded Lipschitz domain in $\mathbb{R}^n$ and fix $m \in \mathbb{N}$. Then $\mathcal{H}^m(\Omega) = H^m(\Omega)$ as vector spaces, with equivalent norms. In particular,*

$$\|u\|_{H^m(\Omega)} \approx \Big(\sum_{\alpha \in \mathbb{N}_0^n, |\alpha| \leq m} \|\partial^\alpha u\|_{L^2(\Omega)}^2 \Big)^{1/2}, \qquad \forall\, u \in H^m(\Omega). \tag{12.3.127}$$

Proof. Since the inclusion $H^m(\Omega) \subseteq \mathcal{H}^m(\Omega)$ and the corresponding norm inequality hold for arbitrary open sets Ω (cf. Proposition 12.25) there remains to prove the opposite inclusion and the naturally accompanying norm inequality. To see this, pick an arbitrary $u \in \mathcal{H}^m(\Omega)$ and recall from Theorem 12.29 that $u = (\mathbb{E}_m u)\big|_\Omega$, where $\mathbb{E}_m : \mathcal{H}^m(\Omega) \to \mathcal{H}^m(\mathbb{R}^n)$ is Calderón's extension operator. Since the latter is bounded, the desired conclusion follows. $\qquad\square$

Collectively, Theorem 12.30 and Theorem 12.15 yield the following result.

Corollary 12.31. *Let Ω be a bounded Lipschitz domain in $\mathbb{R}^n$. Then the following statements are true.*

(1) For every $m \in \mathbb{N}$ the embedding $\mathcal{H}^m(\Omega) \hookrightarrow \mathcal{H}^{m-1}(\Omega)$ is compact.
(2) If $m \in \mathbb{N}$ and $k \in \mathbb{N}_0$ are such that $k < m - n/2$ then $\mathcal{H}^m(\Omega) \subset C^k(\overline{\Omega})$.

We close this section by proving a Chain Rule formula for functions in intrinsic Sobolev spaces of order one. To set the stage, the reader is reminded that a function $\Psi : E \to F$ is said to be bi-Lipschitz (where $E, F \subseteq \mathbb{R}^n$) provided there exists $c \in (0, 1)$ such that $c|x - y| \leq |\Psi(x) - \Psi(y)| \leq c^{-1}|x - y|$ for each $x, y \in E$.

Theorem 12.32. *Suppose $\Psi : \mathbb{R}^n \to \mathbb{R}^n$ is a bijective and bi-Lipschitz function. Let O be an open set in $\mathbb{R}^n$ and consider $\mathcal{U} := \Psi(O)$. Then $\mathcal{U}$ is an open subset of $\mathbb{R}^n$, for every $u \in \mathcal{H}^1(\mathcal{U})$ we have $u \circ \Psi \in \mathcal{H}^1(O)$, and for each $j \in \{1, \ldots, n\}$ the following Chain Rule formula holds*

$$\partial_j(u \circ \Psi) = \sum_{k=1}^n ((\partial_k u) \circ \Psi) \cdot \partial_j \Psi_k \quad \textit{a.e. in } O. \tag{12.3.128}$$

Moreover, there exists a constant $C = C(n, \Psi) \in (0, \infty)$ such that

$$\|u \circ \Psi\|_{\mathcal{H}^1(O)} \leq C\|u\|_{\mathcal{H}^1(\mathcal{U})}, \quad \textit{for all } u \in \mathcal{H}^1(\mathcal{U}). \tag{12.3.129}$$

Proof. Since Ψ is bijective and bi-Lipschitz, it follows that both Ψ and its inverse Ψ^{-1} are Lipschitz functions (in particular, continuous). The fact that $\mathcal{U}$ is the preimage of the open set O under the continuous function Ψ^{-1} then implies that $\mathcal{U}$ is itself open.

To proceed, we first observe that for any measurable function $w : \mathcal{U} \to \mathbb{C}$, the change of variables $y = \Psi(x)$ (cf. Theorem 14.58) yields

$$\int_O \left|w(\Psi(x))\right|^2 dx = \int_{\mathcal{U}} |w(y)|^2 \left|\det(D\Psi^{-1})(y)\right| dy$$

$$\leq C \int_{\mathcal{U}} |w(y)|^2 \, dy, \tag{12.3.130}$$

where the inequality in (12.3.130) is based on the fact that pointwise partial derivatives of Lipschitz functions are bounded (cf. Theorem 2.115 and Exercise 2.116). In particular, from (12.3.130) we obtain the implication

$$w \in L^2(\mathcal{U}) \implies w \circ \Psi \in L^2(O) \text{ and } \|w \circ \Psi\|_{L^2(O)} \leq C\|w\|_{L^2(\mathcal{U})}, \tag{12.3.131}$$

for some finite constant $C = C(n, \Psi) > 0$ independent of w.

Now fix $u \in \mathcal{H}^1(\mathcal{U})$. Since $u \in L^2(\mathcal{U})$ and $\partial_k u \in L^2(\mathcal{U})$ for $k \in \{1, \ldots, n\}$, implication (12.3.131) gives that $u \circ \Psi \in L^2(O)$ and $(\partial_k u) \circ \Psi \in L^2(O)$ for all integers $k \in \{1, \ldots, n\}$. Recalling that all the entries in $D\Psi$ are uniformly bounded, to complete the proof of the theorem there remains to show

$$\partial_j(u \circ \Psi) = \sum_{k=1}^n ((\partial_k u) \circ \Psi) \cdot \partial_j \Psi_k \quad \text{in } \mathcal{D}'(O). \tag{12.3.132}$$

To this end, apply Theorem 12.22 (with $m := 1$) to obtain a sequence of functions $\{u_m\}_{m \in \mathbb{N}} \subseteq C^\infty(\mathcal{U}) \cap \mathcal{H}^1(\mathcal{U})$ with the property that

$$u_m \xrightarrow[m \to \infty]{\mathcal{H}^1(\mathcal{U})} u. \tag{12.3.133}$$

Next, pick $j \in \{1, \ldots, n\}$. Since the Lipschitz function $\Psi : O \to \mathcal{U}$ is differentiable a.e. on O, for every $m \in \mathbb{N}$ we may rely on the classical Chain Rule to write that

$$\partial_j(u_m \circ \Psi) = \sum_{k=1}^n ((\partial_k u_m) \circ \Psi) \cdot \partial_j \Psi_k \quad \text{pointwise a.e. in } O. \tag{12.3.134}$$

Pick $k \in \{1, \ldots, n\}$ and write the inequality resulting when applying (12.3.130) to the function $w := \partial_k u_m - \partial_k u$. Since $\partial_k u_m \xrightarrow[m \to \infty]{L^2(\mathcal{U})} \partial_k u$ (a consequence of (12.3.133)) we may further conclude that

$$(\partial_k u_m) \circ \Psi \xrightarrow[m \to \infty]{L^2(O)} (\partial_k u) \circ \Psi, \quad \forall k \in \{1, \ldots, n\}. \tag{12.3.135}$$

A similar reasoning (working with u_m in place of $\partial_k u_m$) yields

$$u_m \circ \Psi \xrightarrow[m \to \infty]{L^2(O)} u \circ \Psi. \tag{12.3.136}$$

At this point we take $\varphi \in C_0^\infty(O)$ and write

$$\langle \partial_j(u \circ \Psi), \varphi \rangle = -\langle u \circ \Psi, \partial_j \varphi \rangle = - \int_O (u \circ \Psi)(x)(\partial_j \varphi)(x) \, dx$$

$$= - \lim_{m \to \infty} \int_O (u_m \circ \Psi)(x)(\partial_j \varphi)(x) \, dx$$

$$= \lim_{m \to \infty} \int_O \partial_j(u_m \circ \Psi)(x)\varphi(x) \, dx$$

$$= \lim_{m \to \infty} \sum_{k=1}^n \int_O ((\partial_k u_m) \circ \Psi)(x)(\partial_j \Psi)(x)\varphi(x) \, dx$$

$$= \sum_{k=1}^n \int_O ((\partial_k u) \circ \Psi)(x)(\partial_j \Psi)(x)\varphi(x) \, dx$$

$$= \Big\langle \sum_{k=1}^n ((\partial_k u) \circ \Psi) \cdot \partial_j \Psi_k , \varphi \Big\rangle. \qquad (12.3.137)$$

The second and last equality in (12.3.137) are consequences of the fact that the distributions $u \circ \Psi$ and $((\partial_k u) \circ \Psi) \cdot \partial_j \Psi_k$, for each $k \in \{1, \ldots, n\}$, are given by locally integrable functions in O. The third equality in (12.3.137) is based on (12.3.136), Cauchy–Schwarz's inequality, and the (obvious) membership of $\partial_j \varphi$ to $L^2(O)$. The fourth equality in (12.3.137) uses the integration by parts formula (2.9.39) (bearing in mind that $u_m \circ \Psi$ is a locally Lipschitz function in O; cf. Exercise 2.120 and Exercise 2.119). The fifth equality in (12.3.137) is a consequence of (12.3.134), while for the sixth equality we used (12.3.135), Hölder's inequality, and the membership $\varphi \in L^2(O)$. This proves (12.3.132) and finishes the proof of the theorem. $\qquad \square$

Corollary 12.33. *Let Ω be an upper-graph Lipschitz domain in $\mathbb{R}^n$ and let Φ be the bijective bi-Lipschitz flattening map associated with Ω as in Remark 14.52. Then the following are true.*

(i) For a measurable function u in Ω we have

$$u \in L^2(\Omega) \iff u \circ \Phi \in L^2(\mathbb{R}^n_+) \qquad (12.3.138)$$

and

$$u \in \mathcal{H}^1(\Omega) \iff u \circ \Phi \in \mathcal{H}^1(\mathbb{R}^n_+), \qquad (12.3.139)$$

and the equivalences hold with naturally accompanying estimates.
(ii) For a measurable function v in $\mathbb{R}^n_+$ we have

$$v \in L^2(\mathbb{R}^n_+) \iff v \circ \Phi^{-1} \in L^2(\Omega) \qquad (12.3.140)$$

and

$$v \in \mathcal{H}^1(\mathbb{R}^n_+) \iff v \circ \Phi^{-1} \in \mathcal{H}^1(\Omega), \qquad (12.3.141)$$

and the equivalences hold with naturally accompanying estimates.

Proof. All equivalences are consequences of Theorem 12.32. For example, for the left-to-right implications in *(i)* apply Theorem 12.32 with $u := u$, $\Psi := \Phi$, $O := \mathbb{R}^n_+$ and $\mathcal{U} := \Omega$. □

In the last portion of this section we present yet another extension result, which is a companion to Theorem 12.29. While the latter theorem dealt with bounded Lipschitz domains and intrinsic Sobolev spaces of any order (via the Calderón extension operator), the new result treats Sobolev spaces of order one in upper-graph Lipschitz domains (via flattening and extension by reflection).

Theorem 12.34. *Let Ω be an upper-graph Lipschitz domain in $\mathbb{R}^n$. Then there exists a linear and bounded operator*

$$\mathcal{E}_\Omega : \mathcal{H}^1(\Omega) \to \mathcal{H}^1(\mathbb{R}^n) \tag{12.3.142}$$

with the property that

$$(\mathcal{E}_\Omega u)\big|_\Omega = u \ \text{ for every } \ u \in \mathcal{H}^1(\Omega). \tag{12.3.143}$$

Proof. The proof is organized into two steps.

Step I. *The case of the upper-half space.* Fix $u \in C_0^\infty(\overline{\mathbb{R}^n_+})$ and define the function $\mathcal{E}u \in C^0(\mathbb{R}^n)$ by setting

$$(\mathcal{E}u)(x) := \begin{cases} u(x) & \text{if } x \in \overline{\mathbb{R}^n_+}, \\ u(x', -x_n) & \text{if } x = (x', x_n) \in \mathbb{R}^n_-, \end{cases} \tag{12.3.144}$$

for each $x \in \mathbb{R}^n$. Note that

$$\|\mathcal{E}u\|_{L^2(\mathbb{R}^n)} \leq \sqrt{2}\|u\|_{L^2(\mathbb{R}^n_+)}. \tag{12.3.145}$$

For each $j \in \{1, \ldots, n\}$ and each $x \in \mathbb{R}^n$ let us also introduce

$$f_j(x) := \begin{cases} (\partial_j u)(x) & \text{if } x \in \overline{\mathbb{R}^n_+}, \\ (\partial_j u)(x', -x_n) & \text{if } x = (x', x_n) \in \mathbb{R}^n_-, \end{cases} \tag{12.3.146}$$

if $1 \leq j \leq n - 1$ and, corresponding to $j = n$,

$$f_n(x) := \begin{cases} (\partial_n u)(x) & \text{if } x \in \overline{\mathbb{R}^n_+}, \\ -(\partial_n u)(x', -x_n) & \text{if } x = (x', x_n) \in \mathbb{R}^n_-. \end{cases} \tag{12.3.147}$$

It is then clear that $f_j \in L^2(\mathbb{R}^n)$ for each $j \in \{1, \ldots, n\}$ and there exists a purely dimensional constant $C_n \in (0, \infty)$ such that

$$\sum_{j=1}^n \|f_j\|_{L^2(\mathbb{R}^n)} \leq C_n\|\nabla u\|_{L^2(\mathbb{R}^n_+)}. \tag{12.3.148}$$

Fix $\varphi \in C_0^\infty(\mathbb{R}^n)$ and assume $j \in \{1, \ldots, n-1\}$. Then based on (12.3.144), (12.3.146), and straightforward integrations by parts we may compute

$$\int_{\mathbb{R}^n} (\mathcal{E}u)(x)(\partial_j\varphi)(x)\,dx = \int_{\mathbb{R}^n_+} u(x)(\partial_j\varphi)(x)\,dx + \int_{\mathbb{R}^n_-} u(x',-x_n)(\partial_j\varphi)(x)\,dx$$

$$= -\int_{\mathbb{R}^n_+} (\partial_j u)(x)\varphi(x)\,dx - \int_{\mathbb{R}^n_-} (\partial_j u)(x',-x_n)\varphi(x)\,dx$$

$$= -\int_{\mathbb{R}^n} f_j(x)\varphi(x)\,dx. \tag{12.3.149}$$

Also, corresponding to $j = n$,

$$\int_{\mathbb{R}^n} (\mathcal{E}u)(x)(\partial_n\varphi)(x)\,dx = \int_{\mathbb{R}^n_+} u(x)(\partial_n\varphi)(x)\,dx + \int_{\mathbb{R}^n_-} u(x',-x_n)(\partial_n\varphi)(x)\,dx$$

$$= -\int_{\mathbb{R}^n_+} (\partial_n u)(x)\varphi(x)\,dx - \int_{\mathbb{R}^{n-1}} u(x',0)\varphi(x',0)\,dx'$$

$$+ \int_{\mathbb{R}^n_-} (\partial_n u)(x',-x_n)\varphi(x)\,dx + \int_{\mathbb{R}^{n-1}} u(x',0)\varphi(x',0)\,dx'$$

$$= -\int_{\mathbb{R}^n} f_n(x)\varphi(x)\,dx. \tag{12.3.150}$$

All together, (12.3.149)–(12.3.150) prove that

$$\partial_j(\mathcal{E}u) = f_j \text{ in } \mathcal{D}'(\mathbb{R}^n) \text{ for each } j \in \{1, \ldots, n\}. \tag{12.3.151}$$

In particular, from (12.3.145), (12.3.151), and (12.3.148) we conclude that

$$\text{for each } u \in C_0^\infty(\overline{\mathbb{R}^n_+}) \text{ we have } \mathcal{E}u \in \mathcal{H}^1(\mathbb{R}^n)$$
$$\text{and } \|\mathcal{E}u\|_{\mathcal{H}^1(\mathbb{R}^n)} \leq C_n \|u\|_{\mathcal{H}^1(\mathbb{R}^n_+)}. \tag{12.3.152}$$

Hence the assignment $C_0^\infty(\overline{\mathbb{R}^n_+}) \ni u \mapsto \mathcal{E}u \in \mathcal{H}^1(\mathbb{R}^n)$ is linear and bounded. In concert with the density result from Theorem 12.27 (used for $\Omega := \mathbb{R}^n_+$), this allows us to conclude that there exists a (unique) linear and bounded mapping

$$\mathcal{E}_{\mathbb{R}^n_+} : \mathcal{H}^1(\mathbb{R}^n_+) \to \mathcal{H}^1(\mathbb{R}^n) \tag{12.3.153}$$

with the property that

$$\mathcal{E}_{\mathbb{R}^n_+} u = \mathcal{E}u \text{ for every } u \in C_0^\infty(\overline{\mathbb{R}^n_+}). \tag{12.3.154}$$

In particular, if $R_{\mathbb{R}^n_+} : \mathcal{H}^1(\mathbb{R}^n) \to \mathcal{H}^1(\mathbb{R}^n_+)$ denotes the operator of restriction to $\mathbb{R}^n_+$ which is linear and bounded in this context (cf. Remark 12.17), from (12.3.154) and

(12.3.144) we conclude that

$$(R_{\mathbb{R}^n_+} \circ \mathcal{E}_{\mathbb{R}^n_+})u = u \text{ for every } u \in C_0^\infty(\overline{\mathbb{R}^n_+}). \tag{12.3.155}$$

Thus, $R_{\mathbb{R}^n_+} \circ \mathcal{E}_{\mathbb{R}^n_+}$ and I, the identity operator, are two continuous operators from $\mathcal{H}^1(\mathbb{R}^n_+)$ into itself which agree on $C_0^\infty(\overline{\mathbb{R}^n_+})$, a dense subset of the latter space (cf. Theorem 12.27). As such, the said operators agree on the entire space $\mathcal{H}^1(\mathbb{R}^n_+)$, i.e.,

$$\left.\left(\mathcal{E}_{\mathbb{R}^n_+}u\right)\right|_{\mathbb{R}^n_+} = u \text{ for every } u \in \mathcal{H}^1(\mathbb{R}^n_+). \tag{12.3.156}$$

This concludes the treatment of the case when the underlying domain is the upper-half space.

Step II. *The case of an upper-graph Lipschitz domain.* Assume $\Omega \subseteq \mathbb{R}^n$ is an upper-graph Lipschitz domain and bring in the bijective bi-Lipschitz map Φ associated with Ω as in Remark 14.52. In this context, we define

$$\mathcal{E}_\Omega : \mathcal{H}^1(\Omega) \to \mathcal{H}^1(\mathbb{R}^n)$$

$$\mathcal{E}_\Omega u := \left[\mathcal{E}_{\mathbb{R}^n_+}(u \circ \Phi)\right] \circ \Phi^{-1} \text{ for each } u \in \mathcal{H}^1(\Omega). \tag{12.3.157}$$

From Corollary 12.33 and Step I we see that this is a well-defined, linear, and bounded operator. Moreover, for $x \in \Omega$ we have $\Phi^{-1}(x) \in \mathbb{R}^n_+$ hence given any $u \in \mathcal{H}^1(\Omega)$ we may write

$$(\mathcal{E}_\Omega u)(x) = \left[\mathcal{E}_{\mathbb{R}^n_+}(u \circ \Phi)\right](\Phi^{-1}(x)) = (u \circ \Phi)(\Phi^{-1}(x)) = u(x), \tag{12.3.158}$$

where the first equality is the definition in (12.3.157), and the second equality is a consequence of (12.3.156) (bearing in mind that $u \circ \Phi \in \mathcal{H}^1(\mathbb{R}^n_+)$; cf. Corollary 12.33). This establishes (12.3.143) and finishes the proof of Theorem 12.34. □

Here is a companion result to Theorem 12.30.

Corollary 12.35. *Suppose Ω is an upper-graph Lipschitz domain in $\mathbb{R}^n$. Then $\mathcal{H}^1(\Omega) = H^1(\Omega)$ as vector spaces, with equivalent norms.*

Proof. This is an immediate consequence of Proposition 12.25, Theorem 12.34, Theorem 12.24, and definitions. □

12.4 The Space $H^{1/2}(\partial\Omega)$ on Boundaries of Lipschitz Domains

In this section we define the Sobolev space of order $s = 1/2$ on the boundary of a set $\Omega \subset \mathbb{R}^n$ which is either an upper-graph Lipschitz domain, or a bounded Lipschitz domain. To set the stage for the subsequent discussion, we first revisit the space $H^{1/2}(\mathbb{R}^n)$ and define an equivalent norm on it which may be adapted to more general geometries. Recall from (12.1.2) that $u \in H^{1/2}(\mathbb{R}^n)$ provided $u \in \mathcal{S}'(\mathbb{R}^n)$ and $\langle \xi \rangle^{1/2}\widehat{u} \in L^2(\mathbb{R}^n)$. Since

$$\frac{1}{\sqrt{2}}(1 + |\xi|) \le (1 + |\xi|^2)^{1/2} \le 1 + |\xi| \quad \text{for all } \xi \in \mathbb{R}^n, \tag{12.4.1}$$

it follows that for any $u \in \mathcal{S}'(\mathbb{R}^n)$ we have

$$\langle \xi \rangle^{1/2} \widehat{u} \in L^2(\mathbb{R}^n) \iff u \in L^2(\mathbb{R}^n) \text{ and } \int_{\mathbb{R}^n} |\xi| |\widehat{u}(\xi)|^2 \, d\xi < \infty, \tag{12.4.2}$$

plus naturally accompanying estimates.

A useful fact about the integral in (12.4.2) is proved in the next lemma. In what follows, we employ the notation $\omega = (\omega_1, \ldots, \omega_n)$ for $\omega \in S^{n-1}$.

Lemma 12.36. *Let $u \in L^2(\mathbb{R}^n)$. Then*

$$c_0 \int_{\mathbb{R}^n} |\xi| |\widehat{u}(\xi)|^2 \, d\xi = \int_{\mathbb{R}^n} \int_{\mathbb{R}^n} \frac{|u(x) - u(y)|^2}{|x - y|^{n+1}} \, dx \, dy \tag{12.4.3}$$

where $c_0 := (2\pi)^{-n} \int_0^\infty \int_{S^{n-1}} \frac{4 \sin^2(t\omega_1/2)}{t^2} \, d\sigma(\omega) \, dt \in (0, \infty)$.

Proof. Starting with the substitution $x = y + h$, then reversing the order of integration and applying Plancherel's identity (cf. formula (3.2.29)), using the fact that $\mathcal{F}(u(\cdot + h) - u)(\xi) = (e^{ih \cdot \xi} - 1)\widehat{u}(\xi)$ for each $\xi \in \mathbb{R}^n$, and again reversing the order of integration, we obtain

$$\int_{\mathbb{R}^n} \int_{\mathbb{R}^n} \frac{|u(x) - u(y)|^2}{|x - y|^{n+1}} \, dx \, dy$$

$$= (2\pi)^{-n} \int_{\mathbb{R}^n} \frac{\|\mathcal{F}(u(\cdot + h) - u)\|^2_{L^2(\mathbb{R}^n)}}{|h|^{n+1}} \, dh$$

$$= (2\pi)^{-n} \int_{\mathbb{R}^n} |\widehat{u}(\xi)|^2 \int_{\mathbb{R}^n} \frac{|e^{ih \cdot \xi} - 1|^2}{|h|^{n+1}} \, dh \, d\xi. \tag{12.4.4}$$

Making use of polar coordinates the inner integral becomes

$$\int_{\mathbb{R}^n} \frac{|e^{ih \cdot \xi} - 1|^2}{|h|^{n+1}} \, dh = \int_0^\infty \frac{1}{\rho^2} \int_{S^{n-1}} |e^{i\rho\omega \cdot \xi} - 1|^2 \, d\sigma(\omega) \, d\rho$$

$$= |\xi| \int_0^\infty \frac{1}{t^2} \int_{S^{n-1}} |e^{it\omega \cdot \frac{\xi}{|\xi|}} - 1|^2 \, d\sigma(\omega) \, dt$$

$$= |\xi| \int_0^\infty \frac{1}{t^2} \int_{S^{n-1}} |e^{it\omega_1} - 1|^2 \, d\sigma(\omega) \, dt, \tag{12.4.5}$$

where in the second equality we made the change of variables $\rho = t/|\xi|$, and the last equality is based on (14.9.11) applied with $f(\omega) := |e^{it\omega \cdot \frac{\xi}{|\xi|}} - 1|^2$ for each $\omega \in S^{n-1}$, and $\mathcal{R}$ being the unitary transformation mapping $\frac{\xi}{|\xi|}$ into the vector $(1, 0, \ldots, 0)$. Note that $|e^{it\omega_1} - 1|^2 = 4 \sin^2(t\omega_1/2)$ which, together with the fact that $|\sin(x)| \le |x|$ for every $x \in \mathbb{R}$, implies

$$\int_0^\infty \frac{1}{t^2} \int_{S^{n-1}} |e^{it\omega_1} - 1|^2 \, d\sigma(\omega) \, dt$$

$$\leq \int_0^1 \frac{1}{t^2} \int_{S^{n-1}} t^2 \omega_1^2 \, d\sigma(\omega) \, dt + \int_1^\infty \frac{1}{t^2} \int_{S^{n-1}} 4 \, d\sigma(\omega) \, dt < \infty. \qquad (12.4.6)$$

Now (12.4.3) follows from (12.4.4)–(12.4.6). $\qquad\square$

An important consequence of this discussion is the following description of the vector space $H^{1/2}(\mathbb{R}^n)$.

Proposition 12.37. *The vector space $H^{1/2}(\mathbb{R}^n)$ is equal to the collection of all functions $u \in L^2(\mathbb{R}^n)$ with the property that*

$$\int_{\mathbb{R}^n} \int_{\mathbb{R}^n} \frac{|u(x) - u(y)|^2}{|x - y|^{n+1}} \, dx \, dy < \infty. \qquad (12.4.7)$$

In addition, if for each $u \in H^{1/2}(\mathbb{R}^n)$ one defines

$$|||u|||_{\frac{1}{2}, \mathbb{R}^n} := ||u||_{L^2(\mathbb{R}^n)} + \left(\int_{\mathbb{R}^n} \int_{\mathbb{R}^n} \frac{|u(x) - u(y)|^2}{|x - y|^{n+1}} \, dx \, dy \right)^{1/2}, \qquad (12.4.8)$$

then $||| \cdot |||_{\frac{1}{2}, \mathbb{R}^n}$ is a norm on $H^{1/2}(\mathbb{R}^n)$ which is equivalent with the norm $|| \cdot ||_{H^{1/2}(\mathbb{R}^n)}$.

Proof. Let us check that $||| \cdot |||_{\frac{1}{2}, \mathbb{R}^n}$ is indeed a norm on the vector space $H^{1/2}(\mathbb{R}^n)$. Start by considering the ambient set $X := \mathbb{R}^n \times \mathbb{R}^n$ endowed with the measure

$$\mu(x, y) := |x - y|^{-n-1} \mathcal{L}^n(x) \otimes \mathcal{L}^n(y) \quad \text{for } (x, y) \in X, \qquad (12.4.9)$$

where $\mathcal{L}^n$ is the Lebesgue measure in $\mathbb{R}^n$. For $u, v \in H^{1/2}(\mathbb{R}^n)$, define the functions $F(x, y) := u(x) - u(y)$ and $G(x, y) := v(x) - v(y)$, for every $(x, y) \in X$. Using the properties of a generic L^2-norm we may then write

$$|||u + v|||_{\frac{1}{2}, \mathbb{R}^n} = ||u + v||_{L^2(\mathbb{R}^n)} + ||F + G||_{L^2(X, \mu)}$$

$$\leq ||u||_{L^2(\mathbb{R}^n)} + ||v||_{L^2(\mathbb{R}^n)} + ||F||_{L^2(X, \mu)} + ||G||_{L^2(X, \mu)}$$

$$= |||u|||_{\frac{1}{2}, \mathbb{R}^n} + |||v|||_{\frac{1}{2}, \mathbb{R}^n} \qquad (12.4.10)$$

hence $|||\cdot|||_{\frac{1}{2}, \mathbb{R}^n}$ satisfies the triangle inequality. The homogeneity is immediate, while the nondegeneracy of $||| \cdot |||_{\frac{1}{2}, \mathbb{R}^n}$ is inherited from that of $|| \cdot ||_{L^2(\mathbb{R}^n)}$.

Moving on, the first statement in the proposition is a consequence of (12.1.2), (12.4.2), and (12.4.3). These and the definition of $|| \cdot ||_{H^{1/2}(\mathbb{R}^n)}$ also imply the equivalence of the latter norm with $||| \cdot |||_{\frac{1}{2}, \mathbb{R}^n}$. $\qquad\square$

An immediate consequence of Proposition 12.37 and part *(3)* in Theorem 12.1 is the next density result.

Corollary 12.38. *The space $C_0^\infty(\mathbb{R}^n)$ is dense in $(H^{1/2}(\mathbb{R}^n), ||| \cdot |||_{\frac{1}{2}, \mathbb{R}^n})$.*

Proposition 12.37 (used with n replaced by $n-1$) motivates the following definition for Sobolev spaces of order $1/2$ on Lipschitz surfaces in $\mathbb{R}^n$.

Definition 12.39. Let Ω be either an upper-graph Lipschitz domain in $\mathbb{R}^n$, or a bounded Lipschitz domain in $\mathbb{R}^n$. Then the Sobolev space $H^{1/2}(\partial\Omega)$ is defined as

$$H^{1/2}(\partial\Omega) := \Big\{ u \in L^2(\partial\Omega) : \Big(\int_{\partial\Omega} \int_{\partial\Omega} \frac{|u(x)-u(y)|^2}{|x-y|^n} \, d\sigma(x) \, d\sigma(y) \Big)^{1/2} < \infty \Big\}.$$

(12.4.11)

Proposition 12.40. *Let Ω be an upper-graph Lipschitz domain in $\mathbb{R}^n$, or a bounded Lipschitz domain in $\mathbb{R}^n$. For each function $u \in H^{1/2}(\partial\Omega)$ set*

$$\|u\|_{H^{1/2}(\partial\Omega)} := \|u\|_{L^2(\partial\Omega)}$$

(12.4.12)

$$+ \Big(\int_{\partial\Omega} \int_{\partial\Omega} \frac{|u(x)-u(y)|^2}{|x-y|^n} \, d\sigma(x) \, d\sigma(y) \Big)^{1/2}.$$

Then (12.4.12) is a norm on $H^{1/2}(\partial\Omega)$ and $(H^{1/2}(\partial\Omega), \|\cdot\|_{H^{1/2}(\partial\Omega)})$ is a Banach space.

Proof. The proof of the fact that (12.4.12) is a norm on $H^{1/2}(\partial\Omega)$ is similar to the proof used in Proposition 12.37, this time considering the set $X := \partial\Omega \times \partial\Omega$ endowed with the measure $\mu(x,y) := |x-y|^{-n}\sigma(x) \otimes \sigma(y)$, for $(x,y) \in X$.

To show that $H^{1/2}(\partial\Omega)$ is complete with respect to the norm $\|\cdot\|_{H^{1/2}(\partial\Omega)}$, let $\{u_j\}_{j \in \mathbb{N}} \subseteq H^{1/2}(\partial\Omega)$ be a Cauchy sequence. Recalling (12.4.12) it follows that $\{u_j\}_{j \in \mathbb{N}}$ is Cauchy in $L^2(\partial\Omega)$ and $\{F_j\}_{j \in \mathbb{N}}$ is Cauchy in $L^2(X,\mu)$, where we have set $F_j(x,y) := u_j(x) - u_j(y)$ for each point $(x,y) \in X$. In particular, there exists some $u \in L^2(\partial\Omega)$ such that $\{u_j\}_{j \in \mathbb{N}}$ converges to u both pointwise σ-a.e. on $\partial\Omega$ and in $L^2(\partial\Omega)$ as $j \to \infty$, as well as some $G \in L^2(X,\mu)$ such that $\{F_j\}_{j \in \mathbb{N}}$ converges to G both pointwise μ-a.e. on X and in $L^2(X,\mu)$ as $j \to \infty$. The pointwise convergence of these sequences forces $G(x,y)$ to be equal to $F(x,y) := u(x) - u(y)$ for σ-a.e. $x, y \in \partial\Omega$. Consequently, $u \in H^{1/2}(\partial\Omega)$ and

$$\lim_{j \to \infty} \|u_j - u\|_{H^{1/2}(\partial\Omega)}$$

$$= \lim_{j \to \infty} \|u_j - u\|_{L^2(\partial\Omega)} + \lim_{j \to \infty} \|F_j - F\|_{L^2(X,\mu)} = 0.$$

(12.4.13)

This proves that $(H^{1/2}(\partial\Omega), \|\cdot\|_{H^{1/2}(\partial\Omega)})$ is indeed a Banach space. $\qquad\square$

The next lemma ensures that the Sobolev space in Definition 12.39 is rather rich. For a set $E \subset \mathbb{R}^n$ we denote by $Lip_{comp}(E)$ the collection of functions in $Lip(E)$ which are compactly supported.

Lemma 12.41. *Let Ω be either an upper-graph Lipschitz domain in $\mathbb{R}^n$, or a bounded Lipschitz domain in $\mathbb{R}^n$. Then*

$$Lip_{comp}(\partial\Omega) \subseteq H^{1/2}(\partial\Omega).$$

(12.4.14)

Proof. Since $Lip_{comp}(\partial\Omega) \subseteq L^2(\partial\Omega)$, in view of (12.4.11) there remains to prove that

$$\int_{\partial\Omega}\int_{\partial\Omega} \frac{|u(x) - u(y)|^2}{|x - y|^n} \, d\sigma(x) \, d\sigma(y) < \infty \quad \text{for each } u \in Lip_{comp}(\partial\Omega). \quad (12.4.15)$$

Pick $u \in Lip_{comp}(\partial\Omega)$. Then we may estimate

$$\int_{\partial\Omega}\int_{\partial\Omega} \frac{|u(x) - u(y)|^2}{|x - y|^n} \mathbf{1}_{|x-y|>1} \, d\sigma(x) \, d\sigma(y)$$

$$\leq 4 \int_{\partial\Omega} |u(y)|^2 \Big(\int_{\partial\Omega} \frac{\mathbf{1}_{|x-y|>1}}{|x - y|^n} \, d\sigma(x) \Big) d\sigma(y)$$

$$\leq C \|u\|^2_{L^2(\partial\Omega)} < \infty, \quad (12.4.16)$$

where the second inequality is due to Lemma 14.57 (applied with $\alpha := 1$ and $r := 1$).

Fix a reference point $x_0 \in \partial\Omega$ and pick some radius $R > 0$ large enough so that $\text{supp}\, u \subseteq B(x_0, R)$. Then for every choice of $y \in \partial\Omega \setminus B(x_0, R + 2)$ and $x \in \partial\Omega$ with $|x - y| \leq 1$ we necessarily have $u(x) = 0 = u(y)$. Similarly, $u(x) = 0 = u(y)$ whenever $x \in \partial\Omega \setminus B(x_0, R + 2)$ and $y \in \partial\Omega$ satisfy $|x - y| \leq 1$. Hence, if M denotes the Lipschitz constant of u, i.e., $M := \sup_{x,y \in \partial\Omega,\ x \neq y} \frac{|u(x)-u(y)|}{|x-y|}$, we may write for some $C = C(\Omega) \in (0, \infty)$,

$$\int_{\partial\Omega}\int_{\partial\Omega} \frac{|u(x) - u(y)|^2}{|x - y|^n} \mathbf{1}_{|x-y|\leq 1} \, d\sigma(x) \, d\sigma(y)$$

$$= \int_{\partial\Omega\cap B(x_0,R+2)}\int_{\partial\Omega\cap B(x_0,R+2)} \frac{|u(x) - u(y)|^2}{|x - y|^n} \mathbf{1}_{|x-y|\leq 1} \, d\sigma(x) \, d\sigma(y)$$

$$\leq M^2 \int_{\partial\Omega\cap B(x_0,R+2)} \Big(\int_{\partial\Omega} \frac{\mathbf{1}_{|x-y|\leq 1}}{|x - y|^{n-2}} \, d\sigma(x) \Big) d\sigma(y)$$

$$\leq CM^2 \sigma(\partial\Omega \cap B(x_0, R + 2)) < \infty. \quad (12.4.17)$$

Above, the inner-most integral in the penultimate line has been estimated using (14.7.33) with $\alpha := 1$ and $r := 1$, and the last inequality uses (14.7.34). Now (12.4.15) follows from (12.4.16) and (12.4.17). $\qquad\qquad\square$

Exercise 12.42. Let Ω be an upper-graph Lipschitz domain in $\mathbb{R}^n$ and let Φ be the bijective bi-Lipschitz flattening map associated with Ω as in Remark 14.52. Throughout, identify $\partial\mathbb{R}^n_+$ canonically with $\mathbb{R}^{n-1}$. Prove that the following are true.

(i) For a measurable function u on $\partial\Omega$ we have

$$u \in L^2(\partial\Omega) \iff u \circ \Phi \in L^2(\partial\mathbb{R}^n_+) \quad (12.4.18)$$

and

$$u \in H^{1/2}(\partial\Omega) \iff u \circ \Phi \in H^{1/2}(\partial\mathbb{R}^n_+),$$ (12.4.19)

and the equivalences hold with naturally accompanying estimates.

(ii) For a measurable function v on $\partial\mathbb{R}^n_+$ we have

$$v \in L^2(\partial\mathbb{R}^n_+) \iff v \circ \Phi^{-1} \in L^2(\partial\Omega)$$ (12.4.20)

and

$$v \in H^{1/2}(\partial\mathbb{R}^n_+) \iff v \circ \Phi^{-1} \in H^{1/2}(\partial\Omega),$$ (12.4.21)

and the equivalences hold with naturally accompanying estimates.

Hint: If $\Omega \subseteq \mathbb{R}^n$ is the upper-graph of a Lipschitz function $\varphi : \mathbb{R}^{n-1} \to \mathbb{R}$ with Lipschitz constant M, then for all $x', y' \in \mathbb{R}^{n-1}$ we have

$$1 \leq \sqrt{1 + |\nabla\varphi(x')|^2} \leq \sqrt{1 + M^2} \quad \text{and}$$
$$|x' - y'|^2 \leq \left|(x', \varphi(x')) - (y', \varphi(y'))\right|^2 \leq (1 + M^2)|x' - y'|^2.$$ (12.4.22)

Use (12.4.12), (14.7.31), and (12.4.22) to prove the equivalences in *(i)*. The statements in *(ii)* are implied by *(i)* used with $u := v \circ \Phi^{-1}$.

Lemma 12.43. *Let Ω be either a bounded Lipschitz domain, or an upper-graph Lipschitz domain in $\mathbb{R}^n$. Also, suppose Ω_1 is either a bounded Lipschitz domain, or an upper-graph Lipschitz domain in $\mathbb{R}^n$ and assume that there exist $x_0 \in \partial\Omega \cap \partial\Omega_1$ and an open neighborhood C of x_0 with the property that $\partial\Omega \cap C = \partial\Omega_1 \cap C$. Let K be a compact set in $\mathbb{R}^n$ such that $K \subset C$. For $u \in H^{1/2}(\partial\Omega)$ which vanishes outside $\partial\Omega \setminus K$ define the function*

$$v := \begin{cases} u & \text{on } \partial\Omega_1 \cap C = \partial\Omega \cap C, \\ 0 & \text{on } \partial\Omega_1 \setminus C. \end{cases}$$ (12.4.23)

Then $v \in H^{1/2}(\partial\Omega_1)$ and there exist two constants $C_1, C_2 \in (0, \infty)$ depending only Ω, Ω_1, C, and K, such that

$$C_1\|v\|_{H^{1/2}(\partial\Omega_1)} \leq \|u\|_{H^{1/2}(\partial\Omega)} \leq C_2\|v\|_{H^{1/2}(\partial\Omega_1)}.$$ (12.4.24)

Proof. It is immediate that

$$\|v\|_{L^2(\partial\Omega_1)} = \|u\|_{L^2(\partial\Omega_1 \cap C)} = \|u\|_{L^2(\partial\Omega \cap C)} = \|u\|_{L^2(\partial\Omega)}.$$ (12.4.25)

Next, pick $\varepsilon \in (0, \text{dist}(K, \partial C))$ and break up the double integral in the $H^{1/2}(\partial\Omega_1)$ norm of v as follows (with σ_1 denoting the surface measure on $\partial\Omega_1$)

$$\int_{\partial\Omega_1} \int_{\partial\Omega_1} \frac{|v(x) - v(y)|^2}{|x - y|^n} \, d\sigma_1(x) \, d\sigma_1(y) =: I + II,$$ (12.4.26)

where

$$I := \int_{\partial\Omega_1} \int_{\partial\Omega_1} \frac{|v(x) - v(y)|^2}{|x - y|^n} \mathbf{1}_{|x-y|>\varepsilon}(y) \, d\sigma_1(x) \, d\sigma_1(y), \tag{12.4.27}$$

$$II := \int_{\partial\Omega_1} \int_{\partial\Omega_1} \frac{|v(x) - v(y)|^2}{|x - y|^n} \mathbf{1}_{|x-y|\le\varepsilon}(y) \, d\sigma_1(x) \, d\sigma_1(y). \tag{12.4.28}$$

To estimate I write

$$I \le 2 \int_{\partial\Omega_1} \int_{\partial\Omega_1} \frac{|v(x)|^2 + |v(y)|^2}{|x - y|^n} \mathbf{1}_{|x-y|>\varepsilon} \, d\sigma_1(x) \, d\sigma_1(y)$$

$$= 4 \int_{\partial\Omega_1} |v(x)|^2 \int_{\partial\Omega_1} \frac{\mathbf{1}_{|x-y|>\varepsilon}}{|x - y|^n} \, d\sigma_1(y) \, d\sigma_1(x)$$

$$\le C\varepsilon^{-1} \|v\|^2_{L^2(\partial\Omega_1)} = C\varepsilon^{-1} \|u\|^2_{L^2(\partial\Omega)}, \tag{12.4.29}$$

where for the first equality in (12.4.29) we used Fubini's theorem, the next inequality is due to (14.7.39) in Lemma 14.57 (with $\alpha := 1$ and $r := \varepsilon$), and the last equality uses (12.4.25).

In order to estimate II observe that if $x \in \partial\Omega_1 \setminus C$, $y \in \partial\Omega_1$, and $|x - y| \le \varepsilon$, then the choice of ε and the fact that $\operatorname{supp} v \subseteq K$ imply $v(x) = 0 = v(y)$. Similarly, $v(x) = 0 = v(y)$ whenever $x \in \partial\Omega_1$, $y \in \partial\Omega_1 \setminus C$, and $|x - y| \le \varepsilon$. Consequently,

$$II = \int_{\partial\Omega_1 \cap C} \int_{\partial\Omega_1 \cap C} \frac{|v(x) - v(y)|^2}{|x - y|^n} \mathbf{1}_{|x-y|\le\varepsilon} \, d\sigma_1(x) \, d\sigma_1(y)$$

$$= \int_{\partial\Omega \cap C} \int_{\partial\Omega \cap C} \frac{|u(x) - u(y)|^2}{|x - y|^n} \mathbf{1}_{|x-y|\le\varepsilon} \, d\sigma(x) \, d\sigma(y)$$

$$\le \int_{\partial\Omega} \int_{\partial\Omega} \frac{|u(x) - u(y)|^2}{|x - y|^n} \, d\sigma(x) \, d\sigma(y). \tag{12.4.30}$$

In concert, (12.4.25)–(12.4.30) and (12.4.12) yield the first inequality in (12.4.24). The second inequality in (12.4.24) is proved similarly. $\qquad\square$

Theorem 12.44. *Let Ω be either an upper-graph Lipschitz domain in $\mathbb{R}^n$, or a bounded Lipschitz domain in $\mathbb{R}^n$. Then the following are true.*

(1) If $\varphi \in Lip_{comp}(\partial\Omega)$ then $\varphi u \in H^{1/2}(\partial\Omega)$ for every $u \in H^{1/2}(\partial\Omega)$ and

$$\|\varphi u\|_{H^{1/2}(\partial\Omega)} \le C(\varphi, \Omega)\|u\|_{H^{1/2}(\partial\Omega)}, \tag{12.4.31}$$

for some constant $C(\varphi, \Omega) \in (0, \infty)$ independent of u.
(2) The set $Lip_{comp}(\partial\Omega)$ is dense in $H^{1/2}(\partial\Omega)$.

Proof. Fix $\varphi \in Lip_{comp}(\partial\Omega)$ and $u \in H^{1/2}(\partial\Omega)$. Then

$$\|\varphi u\|_{L^2(\partial\Omega)} \le \|\varphi\|_{L^\infty(\partial\Omega)}\|u\|_{L^2(\partial\Omega)}. \tag{12.4.32}$$

Also, adding and subtracting $\varphi(x)u(y)$ we may write

$$\int_{\partial\Omega}\int_{\partial\Omega}\frac{|(\varphi u)(x)-(\varphi u)(y)|^2}{|x-y|^n}\,d\sigma(x)\,d\sigma(y) \tag{12.4.33}$$

$$\leq 2\int_{\partial\Omega}\int_{\partial\Omega}\frac{|\varphi(x)|^2|u(x)-u(y)|^2}{|x-y|^n}\,d\sigma(x)\,d\sigma(y)$$

$$+ 2\int_{\partial\Omega}|u(y)|^2\int_{\partial\Omega}\frac{|\varphi(x)-\varphi(y)|^2}{|x-y|^n}\,d\sigma(x)\,d\sigma(y) =: I + II.$$

It is immediate that

$$I \leq 2\|\varphi\|_{L^\infty(\partial\Omega)}\int_{\partial\Omega}\int_{\partial\Omega}\frac{|u(x)-u(y)|^2}{|x-y|^n}\,d\sigma(x)\,d\sigma(y). \tag{12.4.34}$$

Moreover, if M denotes the Lipschitz constant of φ, for each $y \in \partial\Omega$ we may apply Lemma 14.56 and Lemma 14.57 with $\alpha := 1$ and $r := 1$ to estimate

$$\int_{\partial\Omega}\frac{|\varphi(x)-\varphi(y)|^2}{|x-y|^n}\,d\sigma(x)\,d\sigma(y) \tag{12.4.35}$$

$$\leq M^2\int_{\partial\Omega}\frac{\mathbf{1}_{|x-y|\leq 1}}{|x-y|^{n-2}}\,d\sigma(x) + 4\|\varphi\|_{L^\infty(\partial\Omega)}\int_{\partial\Omega}\frac{\mathbf{1}_{|x-y|>1}}{|x-y|^n}\,d\sigma(x) \leq C,$$

for some constant $C = C(\varphi,\Omega) \in (0,\infty)$ independent of y. In turn, (12.4.35) may be used to estimate II as

$$II \leq C\int_{\partial\Omega}|u(y)|^2\,d\sigma(x). \tag{12.4.36}$$

Now (12.4.31) follows by combining (12.4.12), (12.4.32), (12.4.33), (12.4.34), and (12.4.36).

Moving on to the proof of the density result, consider first the case when Ω is an upper-graph Lipschitz domain and let Φ be the function associated with Ω as in Remark 14.52. We pick some $u \in H^{1/2}(\partial\Omega)$ and make use of (12.4.19) to observe that $u \circ \Phi \in H^{1/2}(\partial\mathbb{R}^n_+) \equiv H^{1/2}(\mathbb{R}^{n-1})$. Apply Corollary 12.38 to obtain a sequence

$$\{g_k\}_{k\in\mathbb{N}} \subseteq C_0^\infty(\mathbb{R}^{n-1}) \subseteq Lip_{comp}(\mathbb{R}^{n-1}) \tag{12.4.37}$$

with the property that $g_k \xrightarrow[k\to\infty]{H^{1/2}(\mathbb{R}^{n-1})} u \circ \Phi$. Since Φ is a bijective bi-Lipschitz function we have that $g_k \circ \Phi^{-1} \in Lip_{comp}(\partial\Omega)$ for each $k \in \mathbb{N}$. Also, (12.4.21) and its accompanying estimate further imply $g_k \circ \Phi^{-1} \xrightarrow[k\to\infty]{H^{1/2}(\partial\Omega)} u$. This proves the density result stated in (2) when Ω is an upper-graph Lipschitz domain.

Next, consider the case when Ω is a bounded Lipschitz domain. Let $\{C_{x_j^*}\}_{j=1}^N$, $\{\Omega_j\}_{j=1}^N$, and $\{\psi_j\}_{j=1}^N$ be the families of cylinders, of upper-graph Lipschitz domains, and the partition of unity associated with Ω as in Remark 14.54. Also, for each j in $\{1,\ldots,N\}$ fix $\xi_j \in C_0^\infty(C_{x_j^*})$ with the property that $\xi_j \equiv 1$ near $\mathrm{supp}\,\psi_j$. Then, if

$u \in H^{1/2}(\partial\Omega)$ we have $u = \sum_{j=1}^{N} \psi_j u$. Theorem 12.44 implies $\psi_j u \in H^{1/2}(\partial\Omega)$ for all $j \in \{1, \dots, N\}$. Fix $j \in \{1, \dots, N\}$ and define the function

$$u_j := \begin{cases} \psi_j u & \text{on } \partial\Omega_j \cap C_{x_j^*} = \partial\Omega \cap C_{x_j^*}, \\ 0 & \text{on } \partial\Omega_j \setminus C_{x_j^*}. \end{cases} \tag{12.4.38}$$

By Lemma 12.43 applied with $\Omega := \Omega$, $\Omega_1 := \Omega_j$, $u := \psi_j u$, and $v := u_j$ we have $u_j \in H^{1/2}(\partial\Omega_j)$. Since Ω_j is an upper-graph Lipschitz domain, based on what we proved earlier, there exists a sequence $\{g_k^j\}_{k\in\mathbb{N}} \subseteq Lip_{comp}(\partial\Omega_j)$ with the property that $g_k^j \xrightarrow[k\to\infty]{H^{1/2}(\partial\Omega_j)} u_j$. In addition, by relying on item (1) in Theorem 12.44, the fact that $\operatorname{supp} u_j \subseteq (\operatorname{supp}\psi_j) \cap \partial\Omega_j$, and that $\xi_j \equiv 1$ near $\operatorname{supp}\psi_j$, we obtain

$$\xi_j g_k^j \xrightarrow[k\to\infty]{H^{1/2}(\partial\Omega_j)} \xi_j u_j = u_j. \tag{12.4.39}$$

If we now define

$$\Psi_k^j := \begin{cases} \xi_j g_k^j & \text{on } \partial\Omega \cap C_{x_j^*} = \partial\Omega_j \cap C_{x_j^*}, \\ 0 & \text{on } \partial\Omega \setminus C_{x_j^*}, \end{cases} \quad \forall k \in \mathbb{N}, \tag{12.4.40}$$

then $\{\Psi_k^j\}_{k\in\mathbb{N}} \subseteq Lip_{comp}(\partial\Omega)$ and we may invoke Lemma 12.43 to conclude that $\Psi_k^j \xrightarrow[k\to\infty]{H^{1/2}(\partial\Omega)} \psi_j u$. Since the latter holds for each $j \in \{1, \dots, N\}$ and $u = \sum_{j=1}^{N} \psi_j u$, the desired conclusion follows. $\qquad\square$

12.5 Traces and Extensions

In the first part of this section we address the issue of existence of traces of functions in the Sobolev space $H^1(\Omega)$ considered in a bounded Lipschitz domain Ω on the boundary of the respective domain. The reader is reminded that in such a setting we have $H^1(\Omega) = \mathcal{H}^1(\Omega)$, with equivalent norms (cf. Theorem 12.30).

Theorem 12.45. *Let Ω be a bounded Lipschitz domain in $\mathbb{R}^n$. Then the following are true.*

(1) *The vector space $\mathcal{V}(\Omega) := C^0(\overline{\Omega}) \cap Lip_{loc}(\Omega) \cap H^1(\Omega)$ is a dense subspace of $H^1(\Omega)$.*

(2) *For each $u \in \mathcal{V}(\Omega)$ its restriction to the boundary $u|_{\partial\Omega} \in C^0(\partial\Omega)$ also belongs to $H^{1/2}(\partial\Omega)$ and there exists $C = C(\Omega) \in (0, \infty)$ such that*

$$\left\|u|_{\partial\Omega}\right\|_{H^{1/2}(\partial\Omega)} \leq C\|u\|_{H^1(\Omega)}. \tag{12.5.1}$$

(3) *The map $\mathcal{V}(\Omega) \ni u \mapsto u|_{\partial\Omega} \in C^0(\partial\Omega) \cap H^{1/2}(\partial\Omega)$ has a unique extension to a linear and bounded operator*

$$\text{Tr} : H^1(\Omega) \longrightarrow H^{1/2}(\partial\Omega). \tag{12.5.2}$$

Proof. The density statement in *(1)* follows by observing that $C_0^\infty(\overline{\Omega}) \subseteq \mathcal{V}(\Omega)$ and recalling that the set $C_0^\infty(\overline{\Omega})$ is dense in $H^1(\Omega)$ (cf. Theorem 12.14). Also, the statement in *(3)* is an immediate consequence of items *(1)-(2)*. Hence, we are left with showing (12.5.1) whose proof we divide in three steps.

Step I. *The case when $\Omega = \mathbb{R}^n_+$.* In this scenario fix $R > 0$ and assume that

$$u \in C^0(\overline{\mathbb{R}^n_+}) \cap Lip_{loc}(\mathbb{R}^n_+) \cap \mathcal{H}^1(\mathbb{R}^n_+)$$
$$\text{and } \operatorname{supp} u \subseteq [-R, R]^{n-1} \times [0, R]. \tag{12.5.3}$$

The goal is to show that

$$u\big|_{\partial\mathbb{R}^n_+} \text{ belongs to the space } H^{1/2}(\partial\mathbb{R}^n_+) \equiv H^{1/2}(\mathbb{R}^{n-1})$$
$$\text{and } \big\|u\big|_{\partial\mathbb{R}^n_+}\big\|_{H^{1/2}(\partial\mathbb{R}^n_+)} \leq C\|u\|_{\mathcal{H}^1(\mathbb{R}^n_+)}, \tag{12.5.4}$$

for some constant $C = C(R, n) \in (0, \infty)$ independent of u. To get started, introduce the notation $x = (x', 0)$ for points $x \in \partial\mathbb{R}^n_+$. In view of (12.4.12) the aim is to estimate

$$\Big(\int_{\mathbb{R}^{n-1}} |u(x', 0)|^2 \, dx' \Big)^{1/2} \tag{12.5.5}$$
$$+ \Big(\int_{\mathbb{R}^{n-1}} \int_{\mathbb{R}^{n-1}} \frac{|u(x', 0) - u(y', 0)|^2}{|x' - y'|^n} \, dy' \, dx' \Big)^{1/2}$$

by a fixed multiple of $\|u\|_{\mathcal{H}^1(\mathbb{R}^n_+)}$.

Using the assumptions on u, for each $x' \in [-R, R]^{n-1}$, we may estimate

$$|u(x', 0)|^2 = |u(x', 0) - u(x', R)|^2 \leq \Big(\int_0^R |(\partial_n u)(x', t)| \, dt \Big)^2$$
$$\leq R \int_0^R |\nabla u(x', t)|^2 \, dt, \tag{12.5.6}$$

where for the first inequality we used Lemma 14.20 with $f(t) := u(x', t)$, $t \in [0, R]$, and for the second inequality we used Hölder. Consequently,

$$\int_{\mathbb{R}^{n-1}} |u(x', 0)|^2 \, dx' = \int_{|x'| \leq R} |u(x', 0)|^2 \, dx'$$
$$\leq R \int_{|x'| \leq R} \int_0^R |\nabla u(x', t)|^2 \, dt \, dx'$$
$$= R \int_{\mathbb{R}^n_+} |(\nabla u)(x)|^2 \, dx. \tag{12.5.7}$$

To treat the double integral in (12.5.5), apply the triangle inequality to write

$$\int_{\mathbb{R}^{n-1}} \int_{\mathbb{R}^{n-1}} \frac{|u(x',0) - u(y',0)|^2}{|x' - y'|^n} \, dy' \, dx' \le 3(I_1 + I_2 + I_3), \tag{12.5.8}$$

where

$$I_1 := \int_{\mathbb{R}^{n-1}} \int_{\mathbb{R}^{n-1}} \frac{|u(x',0) - u(x',|x' - y'|)|^2}{|x' - y'|^n} \, dy' \, dx', \tag{12.5.9}$$

$$I_2 := \int_{\mathbb{R}^{n-1}} \int_{\mathbb{R}^{n-1}} \frac{|u(x',|x' - y'|) - u(y',|x' - y'|)|^2}{|x' - y'|^n} \, dy' \, dx', \tag{12.5.10}$$

$$I_3 := \int_{\mathbb{R}^{n-1}} \int_{\mathbb{R}^{n-1}} \frac{|u(y',|x' - y'|) - u(y',0)|^2}{|x' - y'|^n} \, dy' \, dx'. \tag{12.5.11}$$

We proceed to estimate the integrals in (12.5.9)–(12.5.11), starting with integral I_1. For each $x', y' \in \mathbb{R}^{n-1}$ we apply Lemma 14.20 to the real-valued function defined by $f(t) := u(x', t|x' - y'|)$ for $t \in [0, 1]$ and write

$$I_1 \le \int_{\mathbb{R}^{n-1}} \int_{\mathbb{R}^{n-1}} \frac{1}{|x' - y'|^n} \left| \int_0^1 |x' - y'| \, (\partial_n u)(x', (1-t)|x' - y'|) \, dt \right|^2 \, dy' \, dx'$$

$$\le \int_{\mathbb{R}^{n-1}} \int_{\mathbb{R}^{n-1}} \frac{1}{|x' - y'|^{n-2}} \left(\int_0^1 |(\partial_n u)(x', t|x' - y'|)| \, dt \right)^2 \, dy' \, dx'. \tag{12.5.12}$$

Next, for each fixed $x' \in \mathbb{R}^{n-1}$, we use polar coordinates to write $y' = x' + \rho \omega$, where $\omega \in S^{n-2}$ and $\rho \in (0, \infty)$. Then (12.5.12) further yields

$$I_1 \le \int_{\mathbb{R}^{n-1}} \int_{S^{n-2}} \int_0^\infty \frac{\rho^{n-2}}{\rho^{n-2}} \left(\int_0^1 |(\partial_n u)(x', \rho t)| \, dt \right)^2 \, d\rho \, d\omega \, dx'$$

$$= \omega_{n-2} \int_{\mathbb{R}^{n-1}} \int_0^\infty \left(\int_0^1 |(\partial_n u)(x', \rho t)| \, dt \right)^2 \, d\rho \, dx'$$

$$= \omega_{n-2} \int_{\mathbb{R}^{n-1}} \int_0^\infty \rho^{-2} \left(\int_0^\rho |(\partial_n u)(x', \tau)| \, d\tau \right)^2 \, d\rho \, dx' \tag{12.5.13}$$

where ω_{n-2} denotes the area of the unit sphere in $\mathbb{R}^{n-1}$ and the last equality in (12.5.13) is based on the change of variables $\tau := \rho t$. Moreover, for each $x' \in \mathbb{R}^{n-1}$ we apply Hardy's inequality (cf. (14.2.31) in Theorem 14.21, presently used with $p := 2$, $r := 1$, and $f := |(\partial_n u)(x', \cdot)|$) to obtain

$$\int_0^\infty \rho^{-2} \left(\int_0^\rho |(\partial_n u)(x', \tau)| \, d\tau \right)^2 \, d\rho \le 4 \int_0^\infty |(\partial_n u)(x', \tau)|^2 \, d\tau. \tag{12.5.14}$$

Combining (12.5.13), (12.5.14), and Fubini's theorem we arrive at

$$I_1 \leq 4\,\omega_{n-2} \int_{\mathbb{R}_+^n} |(\partial_n u)(x)|^2 \, dx. \tag{12.5.15}$$

By interchanging the roles of x' and y', a similar argument yields

$$I_3 \leq 4\,\omega_{n-2} \int_{\mathbb{R}_+^n} |(\partial_n u)(x)|^2 \, dx. \tag{12.5.16}$$

Next, we estimate I_2. By the integral version of the Mean Value Theorem in $\mathbb{R}^{n-1}$ and the Cauchy–Schwarz inequality we may write

$$I_2 = \int_{\mathbb{R}^{n-1}} \int_{\mathbb{R}^{n-1}} \frac{1}{|x'-y'|^n} \Big| \int_0^1 ((x',|x'-y'|) - (y',|x'-y'|)) \cdot$$

$$\cdot (\nabla u)(t(x',|x'-y'|) + (1-t)(y',|x'-y'|)) \, dt \Big|^2 \, dy' \, dx'$$

$$= \int_{\mathbb{R}^{n-1}} \int_{\mathbb{R}^{n-1}} \frac{1}{|x'-y'|^n} \times$$

$$\times \Big| \int_0^1 (x'-y',0) \cdot (\nabla u)(tx' + (1-t)y', |x'-y'|) \, dt \Big|^2 \, dy' \, dx'$$

$$\leq \int_{\mathbb{R}^{n-1}} \int_{\mathbb{R}^{n-1}} \Big[\int_0^1 \Big(\frac{1}{|x'-y'|^{n-2}} \Big)^{1/2} \times$$

$$\times |(\nabla_{n-1} u)(tx' + (1-t)y', |x'-y'|)| \, dt \Big]^2 \, dy' \, dx', \tag{12.5.17}$$

where $\nabla_{n-1} u$ denotes the first $n-1$ components of ∇u. To further estimate the last expression in (12.5.17) we invoke the generalized Minkowski inequality (cf. Theorem 14.22) and obtain

$$I_2 \leq \Big[\int_0^1 \Big(\int_{\mathbb{R}^{n-1}} \int_{\mathbb{R}^{n-1}} \frac{1}{|x'-y'|^{n-2}} \times$$

$$\times |(\nabla_{n-1} u)(y' + t(x'-y'), |x'-y'|)|^2 \, dx' \, dy' \Big)^{1/2} \, dt \Big]^2. \tag{12.5.18}$$

Next, we make a series of changes of variables. First, for each fixed $y' \in \mathbb{R}^{n-1}$ we set $z' := x' - y'$ and then we use Fubini's theorem to write

$$I_2 \leq \Big[\int_0^1 \Big(\int_{\mathbb{R}^{n-1}} \int_{\mathbb{R}^{n-1}} \frac{1}{|z'|^{n-2}} |(\nabla_{n-1} u)(y' + tz', |z'|)|^2 \, dy' \, dz' \Big)^{1/2} \, dt \Big]^2. \tag{12.5.19}$$

Second, for each $t \in [0,1]$ and each $z' \in \mathbb{R}^{n-1}$ fixed we let $\xi' := y' + tz'$ and the estimate in (12.5.19) becomes

$$I_2 \leq \Big[\int_0^1 \Big(\int_{\mathbb{R}^{n-1}} \int_{\mathbb{R}^{n-1}} \frac{1}{|z'|^{n-2}} |(\nabla_{n-1} u)(\xi', |z'|)|^2 \, d\xi' \, dz' \Big)^{1/2} dt \Big]^2$$

$$= \int_{\mathbb{R}^{n-1}} \int_{\mathbb{R}^{n-1}} \frac{1}{|z'|^{n-2}} |(\nabla_{n-1} u)(\xi', |z'|)|^2 \, d\xi' \, dz'. \tag{12.5.20}$$

Third, we pass to polar coordinates in z' by setting $z' := \rho \omega$ with $\rho \in (0, \infty)$ and $\omega \in S^{n-2}$, which allows us to write the last double integral in (12.5.20) as

$$\int_{\mathbb{R}^{n-1}} \int_{\mathbb{R}^{n-1}} \frac{1}{|z'|^{n-2}} |(\nabla_{n-1} u)(\xi', |z'|)|^2 \, d\xi' \, dz'$$

$$= \int_0^\infty \int_{S^{n-2}} \int_{\mathbb{R}^{n-1}} \frac{\rho^{n-2}}{\rho^{n-2}} |(\nabla_{n-1} u)(\xi', \rho)|^2 \, d\xi' \, d\omega \, d\rho$$

$$= \omega_{n-2} \int_0^\infty \int_{\mathbb{R}^{n-1}} |(\nabla_{n-1} u)(\xi', \rho)|^2 \, d\xi' \, d\rho$$

$$= \omega_{n-2} \int_{\mathbb{R}^n_+} |(\nabla_{n-1} u)(x)|^p \, dx, \tag{12.5.21}$$

where the last equality uses Fubini's theorem. From (12.5.21) and (12.5.20) it follows that

$$I_2 \leq \omega_{n-2} \int_{\mathbb{R}^n_+} |(\nabla_{n-1} u)(x)|^2 \, dx. \tag{12.5.22}$$

In concert, (12.5.8), (12.5.22), (12.5.16), and (12.5.15), then yield

$$\int_{\mathbb{R}^{n-1}} \int_{\mathbb{R}^{n-1}} \frac{|u(x', 0) - u(y', 0)|^2}{|x' - y'|^n} \, dy' \, dx' \tag{12.5.23}$$

$$\leq 27 \, \omega_{n-2} \int_{\mathbb{R}^n_+} |(\nabla u)(x)|^2 \, dx.$$

Finally, by combining (12.5.7), (12.5.23), and (12.3.127), it follows that if u is as in (12.5.3) then u satisfies (12.5.4). This finishes the proof of Step I.

Step II. *The upper-graph Lipschitz domain case.* The goal is to prove that

if $\Omega \subseteq \mathbb{R}^n$ is an upper-graph Lipschitz domain, K is some compact subset of $\mathbb{R}^n$, and $u \in C^0(\overline{\Omega}) \cap Lip_{loc}(\Omega) \cap \mathcal{H}^1(\Omega)$ vanishes outside $\Omega \setminus K$, then there exists a finite constant $C = C(\Omega, K) > 0$ such that $\big\| u\big|_{\partial\Omega} \big\|_{H^{1/2}(\partial\Omega)} \leq C \|u\|_{\mathcal{H}^1(\Omega)}$. $\tag{12.5.24}$

To this end, let $\varphi : \mathbb{R}^{n-1} \to \mathbb{R}$ be the Lipschitz function with the property that

$$\Omega = \{(y', y_n) \in \mathbb{R}^{n-1} \times \mathbb{R} : y_n > \varphi(y')\} \tag{12.5.25}$$

and denote by M its Lipschitz constant. In particular, we have

$$\partial\Omega = \{(y', \varphi(y')) : y' \in \mathbb{R}^{n-1}\}. \tag{12.5.26}$$

Recall the function Φ from Remark 14.52. Having fixed some compact set $K \subset \mathbb{R}^n$, pick $u \in C^0(\overline{\Omega}) \cap Lip_{loc}(\Omega) \cap \mathcal{H}^1(\Omega)$ which vanishes identically on $\Omega \setminus K$, and define the function $w(x', t) := u(x', \varphi(x') + t)$ for each $x' \in \mathbb{R}^{n-1}$ and each $t \in [0, \infty)$. Since $w = u \circ \Phi$, with Φ as in (14.7.6), we have that

$$w \in C^0(\overline{\mathbb{R}^n_+}) \cap Lip_{loc}(\mathbb{R}^n_+) \cap \mathcal{H}^1(\mathbb{R}^n_+) \quad \text{and has bounded support.} \tag{12.5.27}$$

Also, from the chain rule formula (12.3.128) proved in Theorem 12.32 and the fact that $D\Phi$ and $D\Phi^{-1}$ are bounded, we see that there exists a constant $C \in (0, \infty)$ depending only on Ω such that

$$|(\nabla w)(x', t)| \leq C |(\nabla u)(\Phi(x', t))| \quad \text{for a.e. } (x', t) \in \mathbb{R}^n_+. \tag{12.5.28}$$

After this preamble we are ready to proceed with the estimate in (12.5.24). Based on (14.7.30) we obtain

$$\int_{\partial\Omega} |u(x)|^2 \, d\sigma(x) = \int_{\mathbb{R}^{n-1}} |u(x', \varphi(x'))|^2 \sqrt{1 + |\nabla\varphi(x')|^2} \, dx'$$

$$\leq \sqrt{1 + M^2} \int_{\mathbb{R}^{n-1}} |w(x', 0)|^2 \, dx' \tag{12.5.29}$$

$$\leq C_K \sqrt{1 + M^2} \int_{\mathbb{R}^n_+} |(\nabla w)(x)|^2 \, dx$$

$$\leq C \int_{\mathbb{R}^n_+} |(\nabla u)(\Phi(x))|^2 \, dx \leq C \int_{\Omega} |(\nabla u)(y)|^2 \, dy.$$

For the first inequality in (12.5.29) we used the fact that φ is Lipschitz and the definition of w, for the second one we applied (12.5.7) (bearing in mind (12.5.27)), for the third inequality we invoked (12.5.28), while for the last inequality we have used Corollary 12.33.

A similar circle of ideas also gives

$$\int_{\partial\Omega}\int_{\partial\Omega}\frac{|u(x)-u(y)|^2}{|x-y|^n}\,d\sigma(x)\,d\sigma(y) \tag{12.5.30}$$

$$=\int_{\mathbb{R}^{n-1}}\int_{\mathbb{R}^{n-1}}\frac{\left|u(x',\varphi(x'))-u(y',\varphi(y'))\right|^2}{\left|(x'-y',\varphi(x')-\varphi(y'))\right|^n}\times$$

$$\times\sqrt{1+|\nabla\varphi(x')|^2}\sqrt{1+|\nabla\varphi(y')|^2}\,dx'\,dy'$$

$$\le(1+M^2)\int_{\mathbb{R}^{n-1}}\int_{\mathbb{R}^{n-1}}\frac{\left|w(x',0)-w(y',0)\right|^2}{|x'-y'|^n}\,dx'\,dy'$$

$$\le(1+M^2)\int_{\mathbb{R}^n_+}\left|(\nabla w)(x)\right|^2\,dx\le C\int_{\mathbb{R}^n_+}\left|(\nabla u)(\Phi(x))\right|^2\,dx$$

$$\le C\int_\Omega|(\nabla u)(y)|^2\,dy.$$

The equality in (12.5.30) comes from (14.7.30). The first inequality in (12.5.30) uses the fact that φ is Lipschitz and the definition of w, the second inequality is based on (12.5.23) (whose applicability is ensured by (12.5.27)), the third inequality uses (12.5.28), while the last inequality is justified by Corollary 12.33.

Finally, combining (12.5.29), (12.5.30), and (12.4.12) at this stage we may conclude that $u\big|_{\partial\Omega}\in H^{1/2}(\partial\Omega)$ and $\big\|u\big|_{\partial\Omega}\big\|_{H^{1/2}(\partial\Omega)}\le C\|u\|_{\mathcal{H}^1(\Omega)}$, finishing the proof of (12.5.24).

Step III. *Localization.* Assume Ω is a bounded Lipschitz domain and recall that in such a setting we have $H^1(\Omega)=\mathcal{H}^1(\Omega)$, with equivalent norms (cf. Theorem 12.30). Remark 14.54 ensures the existence of a family of cylinders $\{C_{x_j^*}\}_{j=1}^N$ satisfying (14.7.19), a family of upper-graph Lipschitz domains $\{\Omega_j\}_{j=1}^N$, and a partition of unity $\{\psi_j\}_{j=0}^N$ (cf. (14.7.22)). In view of (14.7.19) and Lemma 14.53, we have

$$\Omega_j\cap C_{x_j^*}=\Omega\cap C_{x_j^*}, \tag{12.5.31}$$

$$\overline{\Omega_j}\cap C_{x_j^*}=\overline{\Omega}\cap C_{x_j^*}, \tag{12.5.32}$$

$$\partial\Omega_j\cap C_{x_j^*}=\partial\Omega\cap C_{x_j^*}, \tag{12.5.33}$$

for each $j\in\{0,1,\dots,N\}$. Pick $u\in\mathcal{V}(\Omega)$. Item *(1)* in Theorem 12.15 then implies that $\psi_j u\in H^1(\Omega)$ for each $j\in\{0,1,\dots,N\}$.

Next, fix $j\in\{1,\dots,N\}$ and define the function

$$u_j:=\begin{cases}\psi_j u & \text{in }\overline{\Omega_j}\cap C_{x_j^*}=\overline{\Omega}\cap C_{x_j^*},\\ 0 & \text{in }\overline{\Omega_j}\setminus C_{x_j^*}.\end{cases} \tag{12.5.34}$$

From (12.5.34), the properties of u, and Example 12.20 (also mindful of Theorem 12.30 and the generalized Leibniz formula (2.4.18)) it follows that

> $u_j \in C^0(\overline{\Omega_j}) \cap Lip_{loc}(\Omega_j) \cap \mathcal{H}^1(\Omega_j)$, it vanishes outside a
> bounded subset of Ω_j, and has $\|u_j\|_{\mathcal{H}^1(\Omega_j)} \leq C\|u\|_{H^1(\Omega)}$ for   (12.5.35)
> some constant $C \in (0, \infty)$ depending only on Ω.

Granted these properties, we may invoke (12.5.24) to conclude that

$$u_j\big|_{\partial\Omega_j} \text{ belongs to } H^{1/2}(\partial\Omega_j) \text{ and}$$
$$\left\|u_j\big|_{\partial\Omega_j}\right\|_{H^{1/2}(\partial\Omega_j)} \leq C\|u_j\|_{\mathcal{H}^1(\Omega_j)} \leq C\|u\|_{H^1(\Omega)}. \qquad (12.5.36)$$

Let us also observe that (12.5.34) implies that

$$(\psi_j u)\big|_{\partial\Omega} = \begin{cases} u_j\big|_{\partial\Omega_j} & \text{on } \partial\Omega \cap C_{x_j^*} = \partial\Omega_j \cap C_{x_j^*}, \\ 0 & \text{on } \partial\Omega \setminus C_{x_j^*}. \end{cases} \qquad (12.5.37)$$

With (12.5.36)–(12.5.37) in hand, Lemma 12.43 applies (with $\Omega := \Omega_j$, $\Omega_1 := \Omega$, $u := u_j\big|_{\partial\Omega_j}$, and $v := (\psi_j u)\big|_{\partial\Omega}$) and gives that $(\psi_j u)\big|_{\partial\Omega} \in H^{1/2}(\partial\Omega)$ and

$$\left\|(\psi_j u)\big|_{\partial\Omega}\right\|_{H^{1/2}(\partial\Omega)} \leq C\left\|u_j\big|_{\partial\Omega_j}\right\|_{H^{1/2}(\partial\Omega_j)} \leq C\|u\|_{H^1(\Omega)}, \qquad (12.5.38)$$

for some constant $C \in (0, \infty)$ depending only on Ω. Upon recalling that we may express $u\big|_{\partial\Omega} = \sum_{j=1}^{N} (\psi_j u)\big|_{\partial\Omega}$ on $\partial\Omega$, and employing (12.5.38), we obtain

$$u\big|_{\partial\Omega} \in H^{1/2}(\partial\Omega) \text{ and } \left\|u\big|_{\partial\Omega}\right\|_{H^{1/2}(\partial\Omega)} \leq C\|u\|_{H^1(\Omega)} \qquad (12.5.39)$$

for some constant $C \in (0, \infty)$ depending only on Ω. This finishes the proof of Theorem 12.45. $\square$

It turns out that the trace operator from Theorem 12.45 is surjective in the context of (12.5.2). Remarkably, the said trace operator has an inverse from the right, which is the extension mapping described in the theorem below.

Theorem 12.46. *Let $\Omega \subseteq \mathbb{R}^n$ be a bounded Lipschitz domain. Then there exists a mapping*

$$\mathrm{Ex} : H^{1/2}(\partial\Omega) \longrightarrow H^1(\Omega) \qquad (12.5.40)$$

that is linear, bounded, and satisfies

$$\mathrm{Tr}(\mathrm{Ex}(f)) = f, \qquad \forall f \in H^{1/2}(\partial\Omega). \qquad (12.5.41)$$

where Tr *is the trace operator from (12.5.2).*

Proof. We divide the proof in a number of steps.

Step I. *The case when* $\Omega = \mathbb{R}^n_+$. The goal in this step is to prove that there exists a linear operator $\mathrm{Ex}_{\mathbb{R}^n_+} : Lip_{comp}(\mathbb{R}^{n-1}) \to C^0(\overline{\mathbb{R}^n_+}) \cap C^\infty(\mathbb{R}^n_+)$ satisfying

$$(\mathrm{Ex}_{\mathbb{R}_+^n} f)\big|_{\partial \mathbb{R}_+^n} = f \quad \text{for each } f \in Lip_{comp}(\mathbb{R}^{n-1}), \tag{12.5.42}$$

and with the property that there exists a finite positive constant C_n such that

$$\big\|\nabla(\mathrm{Ex}_{\mathbb{R}_+^n} f)\big\|_{L^2(\mathbb{R}_+^n)} \leq C_n \Big(\int_{\mathbb{R}^{n-1}} \int_{\mathbb{R}^{n-1}} \frac{|f(y') - f(z')|^2}{|y' - z'|^n} \, dy' \, dz' \Big)^{1/2} \tag{12.5.43}$$

for every $f \in Lip_{comp}(\mathbb{R}^{n-1})$, and for each compact set $K \subset \mathbb{R}^n$ there exists a finite positive constant C_K such that

$$\big\|\mathrm{Ex}_{\mathbb{R}_+^n} f\big\|_{L^2(\mathbb{R}_+^n \cap K)} \leq C_K \|f\|_{L^2(\mathbb{R}^{n-1})} \tag{12.5.44}$$

for every $f \in Lip_{comp}(\mathbb{R}^{n-1})$.

To this end, we begin by picking a function

$$\begin{gathered} \eta \in C^\infty(\mathbb{R}^n), \quad 0 \leq \eta \leq 1 \text{ on } \mathbb{R}^n, \\ \mathrm{supp}\,\eta \subseteq B(0,4), \quad \eta \equiv 1 \cdot \text{ on } B(0,2). \end{gathered} \tag{12.5.45}$$

Next, define the kernel

$$k : \mathbb{R}_+^n \times \overline{\mathbb{R}_+^n} \longrightarrow \mathbb{R} \tag{12.5.46}$$

by setting

$$k(x,y) := \eta\Big(\frac{x-y}{x_n}\Big) \Big[\int_{\mathbb{R}^{n-1}} \eta\Big(\frac{x-(z',0)}{x_n}\Big) dz' \Big]^{-1},$$
$$\text{for each } x = (x_1, \ldots, x_n) \in \mathbb{R}_+^n \text{ and each } y \in \overline{\mathbb{R}_+^n}. \tag{12.5.47}$$

Note that k is well-defined. Indeed,

$$\int_{\mathbb{R}^{n-1}} \eta\Big(\frac{x-(z',0)}{x_n}\Big) dz' = \int_{\mathbb{R}^{n-1}} \eta\Big(\frac{x'-z'}{x_n}, 1\Big) dz'$$

$$= x_n^{n-1} \int_{\mathbb{R}^{n-1}} \eta(w', 1) \, dw' = c_o \, x_n^{n-1} > 0, \tag{12.5.48}$$

where $c_o := \int_{\mathbb{R}^{n-1}} \eta(w', 1) \, dw'$ is a real nonzero constant (as seen from (12.5.45)). In particular, we have

$$k(x,y) = c_o\, \eta\Big(\frac{x-y}{x_n}\Big) x_n^{1-n},$$
$$\text{for each } x = (x', x_n) \in \mathbb{R}^{n-1} \times (0, \infty) \text{ and each } y \in \overline{\mathbb{R}_+^n}. \tag{12.5.49}$$

Clearly k is a nonnegative function and, by design,

$$\int_{\mathbb{R}^{n-1}} k(x, (y', 0)) \, dy' = 1, \qquad \forall\, x \in \mathbb{R}_+^n. \tag{12.5.50}$$

Moreover, (12.5.49) and the regularity of η, imply $k \in C^\infty(\mathbb{R}^n_+ \times \overline{\mathbb{R}^n_+})$.

In what follows we will need estimates for derivatives in the first variable of k (see (12.5.57) below), a task we take up next. To get started, fix $\alpha \in \mathbb{N}^n_0$ and use Leibniz's rule to compute (starting from (12.5.49))

$$\partial^\alpha_x k(x, y) = c_o \sum_{\beta + \gamma = \alpha} \frac{\alpha!}{\beta! \gamma!} \partial^\beta_x \left\{ \eta \left(\frac{x - y}{x_n} \right) \right\} \partial^\gamma_x [x_n^{1-n}]. \tag{12.5.51}$$

Furthermore,

$$\partial^\gamma_x (x_n^{1-n}) = \begin{cases} \left(\prod_{j=0}^{|\gamma|-1} (1 - n - j) \right) x_n^{1-n-|\gamma|}, & \text{if } \gamma = (0, ..., 0, \gamma_n), \\ 0, & \text{otherwise,} \end{cases} \tag{12.5.52}$$

while by the Chain Rule we have

$$\partial^\beta_x \left[\eta \left(\frac{x - y}{x_n} \right) \right] \tag{12.5.53}$$

$$= \sum_{|\delta| \le |\beta|} (\partial^\delta \eta) \left(\frac{x - y}{x_n} \right) \frac{P^{\beta,\delta}_{2|\beta|-|\beta|}(x_1 - y_1, ..., x_n - y_n, x_n)}{x_n^{2|\beta|}}.$$

Above, generally speaking, $P^{\beta,\delta}_r(t_1, ..., t_n, t_{n+1})$ is a homogeneous polynomial of degree r in the variables $t_1, ..., t_{n+1}$, that is,

$$P^{\beta,\delta}_r(t) = \sum_{|\gamma|=r} d^{\beta,\delta}_\gamma t^\gamma, \qquad t = (t_1, ..., t_n, t_{n+1}) \in \mathbb{R}^{n+1}, \tag{12.5.54}$$

where the $d^{\beta,\delta}_\gamma$'s are real-coefficients. Formula (12.5.53) may be justified by starting from the observation that, for each $j \in \{1, ..., n\}$ and for each differentiable function F, there holds

$$\partial_{x_j} \left[F \left(\frac{x - y}{x_n} \right) \right] = \sum_{k=1}^n (\partial_k F) \left(\frac{x - y}{x_n} \right) \frac{\delta_{jk} x_n - (x_k - y_k) \delta_{jn}}{x_n^2}, \tag{12.5.55}$$

and then using induction on the length of the multi-index $\beta \in \mathbb{N}^n_0$. In particular, from (12.5.53) we see that for each $x = (x', x_n) \in \mathbb{R}^n_+$ and each $y \in \overline{\mathbb{R}^n_+}$ we have

$$\frac{x - y}{x_n} \in \text{supp}\,(\partial^\delta \eta) \implies |x - y| \le 4x_n$$

$$\implies \left| P^{\beta,\delta}_{2|\beta|-|\beta|}(x_1 - y_1, ..., x_n - y_n, x_n) \right| \le C_{n,\beta,\delta}\, x_n^{2|\beta|-|\beta|}$$

$$\implies \left| \partial^\beta_x \left[\eta \left(\frac{x - y}{x_n} \right) \right] \right| \le C\, x_n^{-|\beta|} \chi_{|x-y|<4x_n}. \tag{12.5.56}$$

Collectively, (12.5.51), (12.5.52), and (12.5.56) imply that the function k satisfies

$$\left|(\partial_x^\alpha k)(x,y)\right| \leq C_{n,\alpha}\, x_n^{1-n-|\alpha|}\chi_{|x-y|<4x_n},$$

$$\forall x = (x', x_n) \in \mathbb{R}_+^n, \quad \forall y \in \overline{\mathbb{R}_+^n}, \quad \forall \alpha \in \mathbb{N}_0^n. \tag{12.5.57}$$

We are now ready to define the extension operator

$$(\mathrm{Ex}_{\mathbb{R}_+^n} f)(x) := \int_{\mathbb{R}^{n-1}} k(x, (y', 0))\, f(y')\, dy',$$

$$\text{for each } f \in Lip_{comp}(\mathbb{R}^{n-1}) \text{ and all } x \in \mathbb{R}_+^n. \tag{12.5.58}$$

Note that (12.5.57) (applied with $\alpha = (0,\ldots,0)$) ensures that $\mathrm{Ex}_{\mathbb{R}_+^n} f$ is well-defined whenever $f \in Lip_{comp}(\mathbb{R}^{n-1})$. Also, thanks to (12.5.57), we have that $\mathrm{Ex}_{\mathbb{R}_+^n} f$ inherits the regularity of k, i.e., $\mathrm{Ex}_{\mathbb{R}_+^n} f \in C^\infty(\mathbb{R}_+^n)$.

We claim that there exists some $C_n \in (0, \infty)$ such that (12.5.43) holds for each $f \in Lip_{comp}(\mathbb{R}^{n-1})$. To justify this claim, fix $f \in Lip_{comp}(\mathbb{R}^{n-1})$. Also fix $x \in \mathbb{R}_+^n$ and $z' \in \mathbb{R}^{n-1}$. Then (12.5.50) gives

$$f(z') = \int_{\mathbb{R}^{n-1}} k(x, (y', 0)) f(z')\, dy', \tag{12.5.59}$$

which further implies

$$0 = \int_{\mathbb{R}^{n-1}} \partial_{x_j} k(x, (y', 0)) f(z')\, dy', \quad \forall j \in \{1, \ldots, n\}. \tag{12.5.60}$$

Combined, (12.5.60) and (12.5.58) imply

$$\left|[\nabla(\mathrm{Ex}_{\mathbb{R}_+^n} f)](x)\right| \leq \int_{\mathbb{R}^{n-1}} \left|(\nabla_x k)(x, (y', 0))\right| |f(y') - f(z')|\, dy'. \tag{12.5.61}$$

In turn, from (12.5.61), (12.5.57), and Hölder's inequality we obtain

$$\left|[\nabla(\mathrm{Ex}_{\mathbb{R}_+^n} f)](x)\right|^2 \leq C\left(x_n^{-n} \int_{|x-(y',0)|<4x_n} |f(y') - f(z')|\, dy'\right)^2$$

$$\leq C x_n^{-2n} \cdot x_n^{n-1} \int_{|x-(y',0)|<4x_n} |f(y') - f(z')|^2\, dy'. \tag{12.5.62}$$

At this stage, average the most extreme sides of (12.5.62) in $z' \in \mathbb{R}^{n-1}$ such that $|x - (z', 0)| < 4x_n$ in order to obtain

$$\left|[\nabla(\mathrm{Ex}_{\mathbb{R}_+^n} f)](x)\right|^2 \tag{12.5.63}$$

$$\leq C x_n^{-2n} \int_{|x-(z',0)|<4x_n} \int_{|x-(y',0)|<4x_n} |f(y') - f(z')|^2\, dy'\, dz'$$

for each $x = (x', x_n) \in \mathbb{R}^n_+$. Consequently, Fubini's theorem allows us to write

$$
\int_{\mathbb{R}^n_+} \left| [\nabla(\mathrm{Ex}_{\mathbb{R}^n_+} f)](x) \right|^2 \, dx
$$

$$
\leq C \int_{\mathbb{R}^{n-1}} \int_{\mathbb{R}^{n-1}} |f(y') - f(z')|^2 \left[\int_{\substack{|x-(z',0)|<4x_n \\ |x-(y',0)|<4x_n}} x_n^{-2n} \, dx \right] dy' \, dz'. \qquad (12.5.64)
$$

Observe that on the domain of integration of the inner-most integral $|x' - z'| < \sqrt{15}\, x_n$ and $|x' - y'| < \sqrt{15}\, x_n$, hence $|y' - z'| < 2\sqrt{15}\, x_n$ by the triangle inequality. Bearing this in mind and using Fubini's theorem, we may write

$$
\int_{\substack{|x-(z',0)|<4x_n \\ |x-(y',0)|<4x_n}} x_n^{-2n} \, dx \leq \int_{|y'-z'|/(2\sqrt{15})}^{\infty} \left(\int_{|x'-z'| < \sqrt{15}\, x_n} 1 \, dx' \right) x_n^{-2n} \, dx_n
$$

$$
= C \int_{|y'-z'|/(2\sqrt{15})}^{\infty} x_n^{-1-n} \, dx_n = C|y' - z'|^{-n}, \qquad (12.5.65)
$$

where $C = C(n) > 0$ is a finite constant given that $-1 - n < -1$. At this stage, (12.5.43) follows from (12.5.64) and (12.5.65).

Moving on, we claim that for each compact set $K \subset \mathbb{R}^n$ there exists a constant $C_K \in (0, \infty)$ with the property that the estimate recorded in (12.5.44) holds for each function $f \in Lip_{comp}(\mathbb{R}^{n-1})$. To see why this is the case, given such a compact K choose some radius $R \in (0, \infty)$ with the property that $K \subset B(0, R)$. Also, fix $f \in Lip_{comp}(\mathbb{R}^{n-1})$ and let $x \in \mathbb{R}^n_+$. Then formula (12.5.58), estimate (12.5.57), and Hölder's inequality give

$$
|(\mathrm{Ex}_{\mathbb{R}^n_+} f)(x)|^2 \leq \left(\int_{\mathbb{R}^{n-1}} |k(x, (y', 0))| |f(y')| \, dy' \right)^2
$$

$$
\leq C \left(\int_{|x-(y',0)|<4x_n} x_n^{1-n} |f(y')| \, dy' \right)^2
$$

$$
\leq C x_n^{2-2n} x_n^{n-1} \int_{|x-(y',0)|<4x_n} |f(y')|^2 \, dy'
$$

$$
= C x_n^{1-n} \int_{|x-(y',0)|<4x_n} |f(y')|^2 \, dy' \qquad (12.5.66)
$$

for each $x = (x', x_n) \in \mathbb{R}^n_+$. Hence,

$$\int_{\mathbb{R}^n_+ \cap K} |(\mathrm{Ex}_{\mathbb{R}^n_+} f)(x)|^2 \, dx \leq \int_{\mathbb{R}^n_+ \cap B(0,R)} |(\mathrm{Ex}_{\mathbb{R}^n_+} f)(x)|^2 \, dx$$

$$\leq C \int_{\mathbb{R}^n_+ \cap B(0,R)} \int_{|x-(y',0)|<4x_n} x_n^{1-n} |f(y')|^2 \, dy' \, dx$$

$$\leq C \int_{\mathbb{R}^{n-1}} |f(y')|^2 \Big[\int_0^R \Big(\int_{|x'-y'|<\sqrt{15}\,x_n} 1 \, dx' \Big) x_n^{1-n} \, dx_n \Big] \, dy'$$

$$\leq C R \int_{\mathbb{R}^{n-1}} |f(y')|^2 \, dy', \tag{12.5.67}$$

where for the second inequality we applied Fubini's theorem. This proves (12.5.44).

To complete the proof of the goal for Step I we are left with the task of establishing that for each $f \in Lip_{comp}(\mathbb{R}^{n-1})$

$$\mathrm{Ex}_{\mathbb{R}^n_+} f \text{ extends continuously to } \overline{\mathbb{R}^n_+} \text{ and}$$
$$[(\mathrm{Ex}_{\mathbb{R}^n_+} f)\big|_{\partial \mathbb{R}^n_+}](x') = f(x'), \quad \forall\, x' \in \mathbb{R}^{n-1} \equiv \partial \mathbb{R}^n_+. \tag{12.5.68}$$

To this end, fix $f \in Lip_{comp}(\mathbb{R}^{n-1})$ along with some $x'_* \in \mathbb{R}^{n-1}$. Also, let $\varepsilon > 0$ be fixed. Since f is continuous at x'_*, there exists $\delta > 0$ such that

$$y' \in \mathbb{R}^{n-1}, \quad |y' - x'_*| < \delta \implies |f(y') - f(x'_*)| < \varepsilon. \tag{12.5.69}$$

Then for each $x = (x', x_n) \in \mathbb{R}^n_+$ we may estimate

$$|(\mathrm{Ex}_{\mathbb{R}^n_+} f)(x) - f(x'_*)| = \Big| \int_{\mathbb{R}^{n-1}} k(x,(y',0))(f(y') - f(x'_*)) \, dy' \Big|$$

$$\leq \int_{\mathbb{R}^{n-1}} |k(x,(y',0))| \, |f(y') - f(x'_*)| \, dy'$$

$$\leq C_n \fint_{|x'-y'|<\sqrt{15}\,x_n} |f(y') - f(x'_*)| \, dy', \tag{12.5.70}$$

where the equality uses (12.5.50), while for the last inequality we have used (12.5.57) and the fact that the set $\{y' \in \mathbb{R}^{n-1} : |x-(y',0)| < 4x_n\}$ is contained in the set $\{y' \in \mathbb{R}^{n-1} : |x'-y'| < \sqrt{15}\,x_n\}$. Thus, (12.5.69) and (12.5.70) yield

$$|(\mathrm{Ex}_{\mathbb{R}^n_+} f)(x) - f(x'_*)| \leq \varepsilon \text{ if } |x' - x'_*| < \delta/2 \text{ and } x_n < \delta/(2\sqrt{15}), \tag{12.5.71}$$

and the claims in (12.5.68) readily follow from this. In particular, (12.5.71) implies $(\mathrm{Ex}_{\mathbb{R}^n_+} f)\big|_{\partial \mathbb{R}^n_+} = f$. This completes the proof of the stated goal for Step I.

Step II. *The case when Ω is an upper-graph Lipschitz domain.* The claim we make is that there exists a linear map $\mathrm{Ex}_\Omega : Lip_{comp}(\partial\Omega) \to C^0(\overline{\Omega}) \cap Lip_{loc}(\Omega)$ satisfying

$$(\mathrm{Ex}_\Omega f)\big|_{\partial\Omega} = f \quad \text{for each } f \in Lip_{comp}(\partial\Omega), \tag{12.5.72}$$

and with the property that there exists a finite positive constant C_n such that

$$\left\|\nabla(\mathrm{Ex}_\Omega f)\right\|_{L^2(\Omega)} \le C_n\Big(\int_{\partial\Omega} \int_{\partial\Omega} \frac{|f(y) - f(z)|^2}{|y - z|^n} \, d\sigma(y) \, d\sigma(z)\Big)^{1/2} \qquad (12.5.73)$$

for every $f \in Lip_{comp}(\partial\Omega)$, and for each compact set $K \subset \mathbb{R}^n$, there exists a finite positive constant C_K such that

$$\left\|\mathrm{Ex}_\Omega f\right\|_{L^2(\Omega \cap K)} \le C_K\|f\|_{L^2(\partial\Omega)} \qquad (12.5.74)$$

for every $f \in Lip_{comp}(\partial\Omega)$.

In order to prove this claim suppose Ω is the upper-graph of some Lipschitz function $\varphi : \mathbb{R}^{n-1} \to \mathbb{R}$ and recall the bijective bi-Lipschitz flattening map Φ from Remark 14.52. Pick $f \in Lip_{comp}(\partial\Omega)$. Identifying $\mathbb{R}^{n-1} \times \{0\}$ canonically with $\mathbb{R}^{n-1}$, it follows that the function $f \circ \Phi : \mathbb{R}^{n-1} \times \{0\} \to \mathbb{R}$ belongs to $Lip_{comp}(\mathbb{R}^{n-1})$. Then, according to Step I, the function $\mathrm{Ex}_{\mathbb{R}_+^n}(f \circ \Phi)$ belongs to $C^0(\overline{\mathbb{R}_+^n}) \cap C^\infty(\mathbb{R}_+^n)$, satisfies the trace identity

$$\left(\mathrm{Ex}_{\mathbb{R}_+^n}(f \circ \Phi)\right)\big|_{\partial\mathbb{R}_+^n} = f \circ \Phi \quad \text{on} \quad \mathbb{R}^{n-1}, \qquad (12.5.75)$$

the estimate

$$\left\|\nabla(\mathrm{Ex}_{\mathbb{R}_+^n}(f \circ \Phi))\right\|_{L^2(\mathbb{R}_+^n)} \qquad (12.5.76)$$

$$\le C_n\Big(\int_{\mathbb{R}^{n-1}} \int_{\mathbb{R}^{n-1}} \frac{|(f \circ \Phi)(y') - (f \circ \Phi)(z')|^2}{|y' - z'|^n} \, dy' \, dz'\Big)^{1/2}$$

and, for each compact $K \subset \mathbb{R}^n$, the estimate

$$\left\|\mathrm{Ex}_{\mathbb{R}_+^n}(f \circ \Phi)\right\|_{L^2(\mathbb{R}_+^n \cap K)} \le C_K\|f \circ \Phi\|_{L^2(\mathbb{R}^{n-1})}, \qquad (12.5.77)$$

where $C_n, C_K \in (0, \infty)$ are independent of f. Hence, if we further compose this extension of $f \circ \Phi$ to $\mathbb{R}_+^n$ with Φ^{-1}, we obtain the function $[\mathrm{Ex}_{\mathbb{R}_+^n}(f \circ \Phi)] \circ \Phi^{-1}$ which belongs to $C^0(\overline{\Omega}) \cap Lip_{loc}(\Omega)$. This suggests defining the mapping

$$\mathrm{Ex}_\Omega : Lip_{comp}(\partial\Omega) \to C^0(\overline{\Omega}) \cap Lip_{loc}(\Omega)$$
$$\mathrm{Ex}_\Omega f := [\mathrm{Ex}_{\mathbb{R}_+^n}(f \circ \Phi)] \circ \Phi^{-1}, \quad \forall f \in Lip_{comp}(\partial\Omega). \qquad (12.5.78)$$

To see why this map satisfies (12.5.72)–(12.5.74), fix some $f \in Lip_{comp}(\partial\Omega)$. If $x' \in \mathbb{R}^{n-1}$, then $(x', \varphi(x')) \in \partial\Omega$, we have $\Phi^{-1}(x', \varphi(x') = (x', 0)$, and we may write

$$(\mathrm{Ex}_\Omega f)(x', \varphi(x')) = (\mathrm{Ex}_{\mathbb{R}_+^n}(f \circ \Phi))(\Phi^{-1}(x', \varphi(x')))$$

$$= (\mathrm{Ex}_{\mathbb{R}_+^n}(f \circ \Phi))(x', 0)$$

$$= (f \circ \Phi)(x', 0) = f(x', \varphi(x')), \qquad (12.5.79)$$

where the third equality in (12.5.79) is just (12.5.75). This proves (12.5.72).

Moving on, for each index $j \in \{1, \ldots, n\}$ apply the standard Chain Rule (recall that $\mathrm{Ex}_{\mathbb{R}^n_+}(f \circ \Phi)$ is smooth in $\mathbb{R}^n_+$ and Φ is bijective and bi-Lipschitz, hence the components of Φ and Φ^{-1} are differentiable a.e. by Rademacher's theorem) to obtain that

$$\partial_j\big([\mathrm{Ex}_{\mathbb{R}^n_+}(f \circ \Phi)] \circ \Phi^{-1}\big) = \sum_{k=1}^{n} \partial_k[\mathrm{Ex}_{\mathbb{R}^n_+}(f \circ \Phi)] \circ \Phi^{-1} \cdot \partial_j \Phi^{-1}_k \qquad (12.5.80)$$

at a.e. point in Ω. On account of this, (12.5.78), and Proposition 2.121 (guaranteeing that the distributional derivatives of any locally Lipschitz function agree with the pointwise partial derivatives) we therefore conclude that, for each index $j \in \{1, \ldots, n\}$,

$$\partial_j(\mathrm{Ex}_\Omega f) = \sum_{k=1}^{n} \partial_k[\mathrm{Ex}_{\mathbb{R}^n_+}(f \circ \Phi)] \circ \Phi^{-1} \cdot \partial_j \Phi^{-1}_k \quad \text{in } \mathcal{D}'(\Omega). \qquad (12.5.81)$$

Hence, the distributional partial derivatives of $\mathrm{Ex}_\Omega f$ in Ω are of function type and

$$\big\|\nabla(\mathrm{Ex}_\Omega f)\big\|_{L^2(\Omega)} \le C\big\|\nabla(\mathrm{Ex}_{\mathbb{R}^n_+}(f \circ \Phi)) \circ \Phi^{-1}\big\|_{L^2(\Omega)}$$

$$\le C\big\|\nabla(\mathrm{Ex}_{\mathbb{R}^n_+}(f \circ \Phi))\big\|_{L^2(\mathbb{R}^n_+)}$$

$$\le C\Big(\int_{\mathbb{R}^{n-1}} \int_{\mathbb{R}^{n-1}} \frac{|f(y', \varphi(y')) - f(z', \varphi(z'))|^2}{|y' - z'|^n} \, dy' \, dz' \Big)^{1/2}$$

$$\le C\Big(\int_{\partial\Omega} \int_{\partial\Omega} \frac{|f(y) - f(z)|^2}{|y - z|^n} \, d\sigma(y) \, d\sigma(z) \Big)^{1/2}, \qquad (12.5.82)$$

for some constant $C \in (0, \infty)$ depending only on n and Φ. The first inequality in (12.5.82) is a consequence of (12.5.80) and the fact that $D\Phi^{-1}$ has bounded components. The second inequality is due to (12.3.138) and the corresponding norm estimate, the third inequality follows from (12.5.76), while the last inequality is based on (12.4.22) and (14.7.31). This shows that (12.5.73) is also satisfied.

In order to check (12.5.74), let K be a fixed compact subset of $\mathbb{R}^n$. Then, since Φ^{-1} is continuous, $\widetilde{K} := \Phi^{-1}(K)$ is a compact set. Thus, we may write

$$\big\|\mathrm{Ex}_\Omega f\big\|_{L^2(\Omega \cap K)} = \big\|[\mathrm{Ex}_{\mathbb{R}^n_+}(f \circ \Phi)] \circ \Phi^{-1}\big\|_{L^2(\Omega \cap K)}$$

$$\le C(\Omega)\big\|\mathrm{Ex}_{\mathbb{R}^n_+}(f \circ \Phi)\big\|_{L^2(\mathbb{R}^n_+ \cap \widetilde{K})}$$

$$\le C(\widetilde{K}, \Omega)\|f \circ \Phi\|_{L^2(\mathbb{R}^{n-1})} \le C(K, \Omega)\|f\|_{L^2(\partial\Omega)}, \qquad (12.5.83)$$

where the first inequality uses (12.3.140) (for $v := [\mathrm{Ex}_{\mathbb{R}^n_+}(f \circ \Phi)]\mathbf{1}_{\widetilde{K}}$), the second inequality is (12.5.77) (applied with $\widetilde{K}$ in place of K), while the third inequality is obtained by applying (12.3.138) and the accompanying norm estimate. This shows that (12.5.74) holds for the choice of the extension (12.5.78). This finishes the proof of the claim in Step II.

Step III. *The case when Ω is a bounded Lipschitz domain and the extension operator acts on Lipschitz functions defined on $\partial\Omega$.* The goal here is to show that there exists

> Ex : $Lip(\partial\Omega) \to C^0(\overline{\Omega}) \cap Lip_{loc}(\Omega) \cap H^1(\Omega)$ linear map such
> that $(\mathrm{Ex}f)\big|_{\partial\Omega} = f$ and $\|\mathrm{Ex}f\|_{H^1(\Omega)} \leq C\|f\|_{H^{1/2}(\partial\Omega)}$ for every
> $f \in Lip(\partial\Omega)$, where C is a constant depending only on the
> set Ω and is independent of f. $\qquad(12.5.84)$

Recall that in the present setting we have (cf. Theorem 12.30)

$$H^1(\Omega) = \mathcal{H}^1(\Omega), \quad \text{with equivalent norms.} \qquad (12.5.85)$$

Start by applying Remark 14.54 to obtain a family of cylinders $\{C_{x_j^*}\}_{j=1}^N$ satisfying (14.7.19), a family of upper-graph Lipschitz domains $\{\Omega_j\}_{j=1}^N$, and a partition of unity $\{\psi_j\}_{j=0}^N$ as in (14.7.22). In view of (14.7.19) and Lemma 14.53, we have

$$\overline{\Omega_j} \cap C_{x_j^*} = \overline{\Omega} \cap C_{x_j^*}, \qquad (12.5.86)$$

$$\partial\Omega_j \cap C_{x_j^*} = \partial\Omega \cap C_{x_j^*}, \qquad (12.5.87)$$

for each $j \in \{0, 1, \ldots, N\}$. In addition, consider functions $\xi_j \in C_0^\infty(C_{x_j^*})$ with $\xi_j \equiv 1$ near $\operatorname{supp}\psi_j$, for every $j \in \{0, 1, \ldots, N\}$.

Fix $f \in Lip(\partial\Omega)$. Then we may decompose $f = \sum_{j=1}^N \psi_j f$ with each term satisfying

$$\psi_j f \in Lip(\partial\Omega) \quad \text{and} \quad \operatorname{supp}(\psi_j f) \subseteq \partial\Omega \cap C_{x_j^*} = \partial\Omega_j \cap C_{x_j^*}. \qquad (12.5.88)$$

Pick $j \in \{1, \ldots, N\}$ and define

$$f_j := \begin{cases} \psi_j f & \text{in } \partial\Omega_j \cap C_{x_j^*}, \\ 0 & \text{in } \partial\Omega_j \setminus C_{x_j^*}. \end{cases} \qquad (12.5.89)$$

Then $f_j \in Lip_{comp}(\partial\Omega_j)$. As such, we may apply Step II to construct an extension of f_j. Specifically, if Ex_{Ω_j} denotes the extension operator from Step II associated with the upper-graph Lipschitz domain Ω_j, then what we have proved in Step II implies $\mathrm{Ex}_{\Omega_j} f_j \in C^0(\overline{\Omega_j}) \cap Lip_{loc}(\Omega_j)$, hence

$$\xi_j(\mathrm{Ex}_{\Omega_j} f_j) \in C^0(\overline{\Omega_j}) \cap Lip_{loc}(\Omega_j)$$
$$\text{and } \operatorname{supp}[\xi_j(\mathrm{Ex}_{\Omega_j} f_j)] \subseteq C_{x_j^*}. \qquad (12.5.90)$$

Also, by Step II (cf. (12.5.72)–(12.5.73) with Ω replaced by Ω_j and f replaced by f_j) we have the trace property

$$(\mathrm{Ex}_{\Omega_j} f_j)\big|_{\partial\Omega_j} = f_j, \tag{12.5.91}$$

and the estimate (with σ_j denoting the surface measure on $\partial\Omega_j$)

$$\left\|\nabla(\mathrm{Ex}_{\Omega_j} f_j)\right\|_{L^2(\Omega_j)} \le C\Big(\int_{\partial\Omega_j}\int_{\partial\Omega_j} \frac{|f_j(y)-f_j(z)|^2}{|y-z|^n}\,d\sigma_j(y)\,d\sigma_j(z)\Big)^{1/2}$$

$$\le C\|f_j\|_{H^{1/2}(\partial\Omega_j)}$$

$$\le C\|\psi_j f\|_{H^{1/2}(\partial\Omega)} \le C\|f\|_{H^{1/2}(\partial\Omega)}, \tag{12.5.92}$$

for some constant $C = C(n,\Omega) > 0$. The first inequality in (12.5.92) is a rewriting of (12.5.73) in the current setting, the second inequality is immediate from the definition of the $H^{1/2}$-norm, the third is a consequence of Lemma 12.43, and the last one follows from item (1) in Theorem 12.44.

Going further, use estimate (12.5.74) from Step II with $\Omega := \Omega_j$, $f := f_j$, and $K := \mathrm{supp}\,\xi_j$ to write

$$\left\|\xi_j \mathrm{Ex}_{\Omega_j} f_j\right\|_{L^2(\Omega_j)} = \left\|\xi_j \mathrm{Ex}_{\Omega_j} f_j\right\|_{L^2(\Omega_j \cap K)}$$

$$\le \|\xi_j\|_{L^\infty(\mathbb{R}^n)}\left\|\mathrm{Ex}_{\Omega_j} f_j\right\|_{L^2(\Omega_j \cap K)}$$

$$\le C\|f_j\|_{L^2(\partial\Omega_j)} = C\|\psi_j f\|_{L^2(\partial\Omega_j \cap C_{x_j^*})}$$

$$\le C\|f\|_{L^2(\partial\Omega \cap C_{x_j^*})} \le C\|f\|_{L^2(\partial\Omega)}, \tag{12.5.93}$$

for some finite constant $C = C(n,\Omega) > 0$.

Let us now set

$$u_j := \begin{cases} \xi_j(\mathrm{Ex}_{\Omega_j} f_j) & \text{in } \Omega \cap C_{x_j^*} = \Omega_j \cap C_{x_j^*}, \\ 0 & \text{in } \Omega \setminus C_{x_j^*}, \end{cases} \tag{12.5.94}$$

and note that from (12.5.90) and (12.5.93) we have

$$u_j \in C^0(\overline{\Omega}) \cap Lip_{loc}(\Omega) \quad\text{and}\quad \|u_j\|_{L^2(\Omega)} \le C\|f\|_{L^2(\partial\Omega)}. \tag{12.5.95}$$

Also, by the Chain Rule, (12.5.93) (applied with ξ_j replaced by $\nabla\xi_j$) and (12.5.92), we obtain

$$\|\nabla u_j\|_{L^2(\Omega)} = \left\|\nabla[\xi_j(\mathrm{Ex}_{\Omega_j} f_j)]\right\|_{L^2(\Omega_j)}$$

$$\le \left\|(\nabla\xi_j)(\mathrm{Ex}_{\Omega_j} f_j)\right\|_{L^2(\Omega_j)} + \left\|\xi_j\nabla(\mathrm{Ex}_{\Omega_j} f_j)\right\|_{L^2(\Omega_j)}$$

$$\le C\|f\|_{H^{1/2}(\partial\Omega)} + \|\xi_j\|_{L^\infty(\mathbb{R}^n)}\left\|\nabla(\mathrm{Ex}_{\Omega_j} f_j)\right\|_{L^2(\Omega_j)}$$

$$\le C\|f\|_{H^{1/2}(\partial\Omega)}, \tag{12.5.96}$$

for some finite constant $C = C(n, \Omega) > 0$. From (12.5.95), (12.5.96), and (12.5.85) we conclude that there exists some $C = C(n, \Omega) \in (0, \infty)$ such that

$$u_j \in H^1(\Omega) \quad \text{and} \quad \|u_j\|_{H^1(\Omega)} \leq C\|f\|_{H^{1/2}(\partial\Omega)}. \tag{12.5.97}$$

We are now ready to define the extension operator by setting

$$\text{Ex} f := \sum_{j=1}^{N} u_j. \tag{12.5.98}$$

From (12.5.95), (12.5.97), and (12.5.98) it follows that

$$\text{Ex} f \in C^0(\overline{\Omega}) \cap Lip_{loc}(\Omega) \cap H^1(\Omega) \quad \text{and}$$
$$\|\text{Ex} f\|_{H^1(\Omega)} \leq C\|f\|_{H^{1/2}(\partial\Omega)}. \tag{12.5.99}$$

Moreover, for each $j \in \{1, \ldots, N\}$ we have

$$u_j\big|_{\partial\Omega \cap C_{x_j^*}} = [\xi_j(\text{Ex}_{\Omega_j} f_j)]\big|_{\partial\Omega_j \cap C_{x_j^*}}$$

$$= (\xi_j\big|_{\partial\Omega_j \cap C_{x_j^*}})[(\text{Ex}_{\Omega_j} f_j)]\big|_{\partial\Omega_j \cap C_{x_j^*}}$$

$$= (\xi_j\big|_{\partial\Omega_j \cap C_{x_j^*}}) f_j\big|_{\partial\Omega_j \cap C_{x_j^*}}$$

$$= (\xi_j\big|_{\partial\Omega_j \cap C_{x_j^*}})(\psi_j f)\big|_{\partial\Omega_j \cap C_{x_j^*}}$$

$$= (\psi_j f)\big|_{\partial\Omega \cap C_{x_j^*}}, \tag{12.5.100}$$

taking into account (12.5.94), (12.5.91), (12.5.88)–(12.5.89), and the fact that ξ_j is identically one on the support of ψ_j. Hence, if tilde denotes the extension by zero outside support (of functions defined on subsets of $\partial\Omega$) to the entire set $\partial\Omega$, from (12.5.98) and (12.5.100) we obtain

$$(\text{Ex} f)\big|_{\partial\Omega} = \sum_{j=1}^{N} u_j\big|_{\partial\Omega} = \sum_{j=1}^{N} \widetilde{u_j\big|_{\partial\Omega \cap C_{x_j^*}}}$$

$$= \sum_{j=1}^{N} \widetilde{(\psi_j f)\big|_{\partial\Omega \cap C_{x_j^*}}} = \sum_{j=1}^{N} \psi_j f = f. \tag{12.5.101}$$

Thus, the operator Ex defined in (12.5.98) satisfies all the properties listed in (12.5.84). The proof of Step III is therefore complete.

Step IV. *The proof of the existence of an extension operator as in the statement of the theorem.* Having proved the existence of the extension operator Ex as in Step III (cf. (12.5.84)), since for bounded Lipschitz domains Ω in $\mathbb{R}^n$ we know that $Lip(\partial\Omega)$ is a dense subspace of $H^{1/2}(\partial\Omega)$ (cf. item *(2)* in Theorem 12.44) and $H^1(\Omega)$ is a Banach space (cf. Theorem 12.14) it follows that the operator Ex from (12.5.84) has a unique extension to a linear and bounded operator

$$\widetilde{\mathrm{Ex}} : H^{1/2}(\partial\Omega) \to H^1(\Omega). \tag{12.5.102}$$

In addition, if $f \in Lip(\partial\Omega)$ then

$$\widetilde{\mathrm{Ex}}f = \mathrm{Ex}f \in C^0(\overline{\Omega}) \cap Lip_{loc}(\Omega) \cap H^1(\Omega) = \mathcal{V}(\Omega), \tag{12.5.103}$$

(where the last equality is a definition; cf. item *(1)* in Theorem 12.45), hence if Tr denotes the trace operator from Theorem 12.45 we have

$$(\mathrm{Tr} \circ \widetilde{\mathrm{Ex}})f = \mathrm{Tr}(\widetilde{\mathrm{Ex}}f) = \mathrm{Tr}(\mathrm{Ex}f) = (\mathrm{Ex}f)\big|_{\partial\Omega} = f, \tag{12.5.104}$$

where for the third equality in (12.5.104) we used the fact that Tr acts as the pointwise restriction to $\partial\Omega$ for functions in the space $\mathcal{V}(\Omega)$. This shows that the operator $\mathrm{Tr} \circ \widetilde{\mathrm{Ex}}$ acting from $H^{1/2}(\partial\Omega)$ into $H^{1/2}(\partial\Omega)$ is well-defined, linear, bounded, and coincides with I, the identity operator, on $Lip(\partial\Omega)$. By density, it follows that $\mathrm{Tr} \circ \widetilde{\mathrm{Ex}} = I$ on $H^{1/2}(\partial\Omega)$. The proof of Theorem 12.46 is now finished. $\square$

Further Notes for Chapter 12. Sobolev spaces constitute a classical topic at the confluence of harmonic analysis, functional analysis, and partial differential equations and there is a vast literature dedicated to this subject. In this regard, the interested reader may consult [2], [43]. See also [15], [47], [48], [49], for more specialized, in-depth treatments.

12.6 Additional Exercises for Chapter 12

Exercise 12.47. From Theorem 12.1 we know that $\mathcal{S}(\mathbb{R}^n) \subset \bigcap_{s>0} H^s(\mathbb{R}^n)$. Prove that this is a strict inclusion by considering the function $u(x) := \prod_{j=1}^n f(x_j)$ for all $x = (x_1, \ldots, x_n) \in \mathbb{R}^n$, where

$$f(\xi) := \begin{cases} \dfrac{\sin\xi}{\xi}, & \text{if } \xi \neq 0, \\ 1, & \text{if } \xi = 0, \end{cases} \qquad \text{for all } \xi \in \mathbb{R}. \tag{12.6.1}$$

Exercise 12.48. By Theorem 12.7 we know that $\mathcal{E}'(\mathbb{R}^n) \subset \bigcup_{s<0} H^s(\mathbb{R}^n)$. Prove that this is a strict inclusion.

Exercise 12.49. Show that for every $\varepsilon > 0$ we have $\delta \in H^{-n/2-\varepsilon}(\mathbb{R}^n)$ while, at the same time, $\delta \notin H^{-n/2}(\mathbb{R}^n)$.

Exercise 12.50. Suppose $u \in H^{-n/2}(\mathbb{R}^n) \cap \mathcal{E}'(\mathbb{R}^n)$ and $\mathrm{supp}\, u \subset \{a_1, \ldots, a_N\}$ for some distinct points $a_1, \ldots, a_N \in \mathbb{R}^n$. Show that $u \equiv 0$.

Exercise 12.51. Let $t \in \mathbb{R}$ and $\varphi \in \mathcal{S}(\mathbb{R}^n)$ and define

$$\mathcal{B}_t\varphi(x) := (2\pi)^{-n} \int_{\mathbb{R}^n} e^{ix\cdot\xi}\langle\xi\rangle^{-t}\widehat{\varphi}(\xi)\,d\xi, \qquad \forall\, x \in \mathbb{R}^n. \tag{12.6.2}$$

Prove the following statements.

(1) $\mathcal{B}_t\varphi \in \mathcal{S}(\mathbb{R}^n)$ for each $t \in \mathbb{R}$ and $\varphi \in \mathcal{S}(\mathbb{R}^n)$.
(2) $\|\mathcal{B}_t\varphi\|_{H^{s+t}(\mathbb{R}^n)} = \|\varphi\|_{H^s(\mathbb{R}^n)}$ for each $s, t \in \mathbb{R}$ and $\varphi \in \mathcal{S}(\mathbb{R}^n)$.
(3) $\mathcal{B}_t \circ \mathcal{B}_r = \mathcal{B}_{t+r}$ for all $t, r \in \mathbb{R}$.
(4) $\mathcal{B}_0\varphi = \varphi$ for all $\varphi \in \mathcal{S}(\mathbb{R}^n)$.
(5) For each $s, t \in \mathbb{R}$ the mapping $\mathcal{B}_t$ extends uniquely as a linear and continuous map from $H^s(\mathbb{R}^n)$ into $H^{s+t}(\mathbb{R}^n)$.

Exercise 12.52. Consider a nonzero function $\theta \in C_0^\infty(\mathbb{R})$ with $\operatorname{supp}\theta \subseteq [-1, 1]$ and define

$$u_j(x) := \theta(x)\frac{\sin(jx)}{j}, \qquad \forall\, x \in \mathbb{R}, \ \forall\, j \in \mathbb{N}. \tag{12.6.3}$$

Prove that the sequence $\{u_j\}_{j\in\mathbb{N}}$ is bounded in $H^1(\mathbb{R})$, it is convergent in $H^s(\mathbb{R})$ provided $s < 1$, but it has no convergent subsequence in $H^1(\mathbb{R})$.

Exercise 12.53. Consider a function $\theta \in C_0^\infty(\mathbb{R})$ with $\operatorname{supp}\theta \subseteq [0, 1]$ and define

$$u_j(x) := \theta(x)\Big[1 + (-1)^j + \frac{\cos(jx)}{j}\Big], \qquad \forall\, x \in \mathbb{R}, \ \forall\, j \in \mathbb{N}. \tag{12.6.4}$$

Prove that the sequence $\{u_j\}_{j\in\mathbb{N}}$ has a convergent subsequence in $H^{1/2}(\mathbb{R})$.

Exercise 12.54. Let $s_1, s_2 \in \mathbb{R}$ be such that $s_1 < s_2$. Prove that the embedding $H^{s_2}(\mathbb{R}^n) \hookrightarrow H^{s_1}(\mathbb{R}^n)$ is strict.

Exercise 12.55. Prove that the inclusion $\bigcap_{s\in\mathbb{R}} H^s(\mathbb{R}^n) \subset C^\infty(\mathbb{R}^n)$ (cf. (12.1.63)) is strict.

Exercise 12.56. Prove that $|x| \in H^1((-1, 1))$ and $|x| \notin H^2((-1, 1))$.

Exercise 12.57. Let Ω be a bounded open set in $\mathbb{R}^n$. Prove that $Lip(\Omega) \subset H^1(\Omega)$.

Exercise 12.58. Show that the membership $|x|^{-\lambda} \in H^1(B(0, 1))$, when considered in $\mathbb{R}^n$, holds for all $\lambda < 1$ if and only if $n \geq 4$.

Chapter 13
Solutions to Selected Exercises

Abstract In this chapter, solutions to various problems listed at the end of each of the Chapters 1–12 are provided.

13.1 Solutions to Exercises from Section 1.4

Exercise 1.27 $a = 0$; see Examples 1.4 and 1.5.

Exercise 1.28 No; see Examples 1.4 and 1.5.

Exercise 1.29 Let $\varphi \in C_0^\infty(\mathbb{R}^2)$. Based on integration by parts and the fact that φ is compactly supported we have

$$\langle \partial_1 \partial_2 f, \varphi \rangle = \int_{\mathbb{R}^2} f(x, y)(\partial_1 \partial_2 \varphi)(x, y) \, dx \, dy$$

$$= \int_0^\infty \int_{\mathbb{R}} \partial_2(\partial_1 \varphi)(x, y) \, dy \, dx + \int_0^\infty \int_{\mathbb{R}} \partial_1(\partial_2 \varphi)(x, y) \, dx \, dy = 0. \quad (13.1.1)$$

This shows that $\partial_1 \partial_2 f = 0$ in the weak sense. Similarly, the other weak derivatives are $\partial_2(\partial_1 \partial_2 f) = 0$ and $\partial_1^2 \partial_2 f = 0$. Suppose now that $\partial_1 f$ exists in the weak sense, i.e., there exists $g \in L^1_{loc}(\mathbb{R}^2)$ such that

$$\int_{\mathbb{R}^2} f(x, y)(\partial_1 \varphi)(x, y) \, dx \, dy = -\int_{\mathbb{R}^2} g(x, y)\varphi(x, y) \, dx \, dy, \quad (13.1.2)$$

for all $\varphi \in C_0^\infty(\mathbb{R}^2)$. Since $\int_{\mathbb{R}^2} f(x, y)(\partial_1 \varphi)(x, y) \, dx \, dy = -\int_{\mathbb{R}} \varphi(0, y) \, dy$ (as seen using the definition of f and integration by parts), (13.1.2) becomes

$$\int_{\mathbb{R}} \varphi(0, y) \, dy = \int_{\mathbb{R}^2} g(x, y)\varphi(x, y) \, dx \, dy, \quad \forall \varphi \in C_0^\infty(\mathbb{R}^2). \quad (13.1.3)$$

© Springer Nature Switzerland AG 2018
D. Mitrea, *Distributions, Partial Differential Equations, and Harmonic Analysis*,
Universitext, https://doi.org/10.1007/978-3-030-03296-8_13

Taking now φ in (13.1.3) to be supported in $\{(x, y) \in \mathbb{R}^2 : x \neq 0\}$ and recalling Theorem 1.3, it follows that $g = 0$ almost everywhere on $\{(x, y) \in \mathbb{R}^2 : x \neq 0\}$. Thus, $g = 0$ almost everywhere in $\mathbb{R}^2$. The latter, when used in (13.1.3), implies $\int_{\mathbb{R}} \varphi(0, y) \, dy = 0$ for every $\varphi \in C_0^\infty(\mathbb{R}^2)$. However this leads to a contradiction by taking $\varphi(x, y) = \varphi_1(x)\varphi_2(y)$ for $(x, y) \in \mathbb{R}^2$, where $\varphi_1, \varphi_2 \in C_0^\infty(\mathbb{R})$ are such that $\varphi_1(0) = 1$ and $\int_{\mathbb{R}} \varphi_2(y) \, dy = 1$.

Exercise 1.30 Let $\varphi \in C_0^\infty((-1, 1))$ be arbitrary. Then Lebesgue's Dominated Convergence Theorem and integration by parts give

$$
-\int_{-1}^1 f(x)\varphi'(x) \, dx = \lim_{\varepsilon \to 0^+} \int_{-1}^{-\varepsilon} \sqrt{-x}\, \varphi'(x) \, dx - \lim_{\varepsilon \to 0^+} \int_\varepsilon^1 \sqrt{x}\, \varphi'(x) \, dx
$$

$$
= \lim_{\varepsilon \to 0^+} \left[\sqrt{-x}\, \varphi(x)\big|_{-1}^{-\varepsilon} + \int_{-1}^{-\varepsilon} \frac{1}{2\sqrt{-x}}\, \varphi(x) \, dx \right]
$$

$$
- \lim_{\varepsilon \to 0^+} \left[\sqrt{x}\varphi(x)\big|_{-1}^{\varepsilon} - \int_\varepsilon^1 \frac{1}{2\sqrt{x}}\, \varphi(x) \, dx \right]
$$

$$
= \int_{-1}^1 \frac{1}{2\sqrt{|x|}}\, \varphi(x) \, dx, \tag{13.1.4}
$$

where we have used the fact that $\frac{1}{2\sqrt{|x|}} \in L^1((-1, 1))$. Hence the weak derivative of order one of f is the function given by $g(x) := \frac{1}{2\sqrt{|x|}}$ for $x \in (-1, 1) \setminus \{0\}$.

Exercise 1.31 Consider a function $\varphi \in C_0^\infty(\mathbb{R}^2)$ and pick a number $R > 0$ such that $\operatorname{supp}\varphi \subset (-R, R) \times (-R, R)$. Then

$$
-\int_{\mathbb{R}^2} x|y|(\partial_1^2 \partial_2 \varphi)(x, y) \, dx \, dy = -\int_{\mathbb{R}} |y| \int_{\mathbb{R}} x(\partial_1^2 \partial_2 \varphi)(x, y) \, dx \, dy = 0, \tag{13.1.5}
$$

where for the last equality is obtained integrating by parts twice with respect to x, keeping in mind that φ has compact support. This proves that $\partial_1^2 \partial_2 f = 0$ in the weak sense. On the other hand,

$$
-\int_{\mathbb{R}^2} x|y|(\partial_1 \partial_2^2 \varphi)(x, y) \, dx \, dy = \int_{\mathbb{R}^2} |y|(\partial_2^2 \varphi)(x, y) \, dx \, dy
$$

$$
= \int_{\mathbb{R}} \left[-\int_{-R}^0 y(\partial_2^2 \varphi)(x, y) \, dy + \int_0^R y(\partial_2^2 \varphi)(x, y) \, dy \right] dx
$$

$$
= -\int_{\mathbb{R}^2} \operatorname{sgn}(y)(\partial_2 \varphi)(x, y) \, dy \, dx = 2 \int_{\mathbb{R}} \varphi(x, 0) \, dx \tag{13.1.6}
$$

after integrating by parts, first with respect to x, then twice with respect to y. Now reasoning as in the proof of Exercise 1.29 one can show that if g belongs to $L_{loc}^1(\mathbb{R}^2)$ is such that

$$2 \int_{\mathbb{R}} \varphi(x, 0) \, dx = \int_{\mathbb{R}^2} g(x, y) \varphi(x, y) \, dx \, dy, \qquad \forall \varphi \in C_0^\infty(\mathbb{R}^2),$$

leads to a contradiction. Hence, $\partial_1 \partial_2^2 f$ does not exist in the weak sense.

Exercise 1.32 Reasoning as in the proofs of Exercise 1.29 and Exercise 1.31, one can show that $\partial_1^k f$ and $\partial_2^k f$ do not exist in the weak sense for any $k \in \mathbb{N}$, and that $\partial^\alpha f = 0$ in the weak sense whenever $\alpha = (\alpha_1, \alpha_2) \in \mathbb{N}_0^2$ with $\alpha_1 \neq 0$ and $\alpha_2 \neq 0$.

Exercise 1.33 Let $\varphi \in C^\infty(\mathbb{R})$ with $\operatorname{supp} \varphi \subset (-R, R)$ for some $R \in (0, \infty)$. Then,

$$-\int_{\mathbb{R}} f(x) \varphi'(x) \, dx = -\int_{-R}^0 \sin(-x) \varphi'(x) \, dx - \int_0^R \sin(x) \varphi'(x) \, dx$$

$$= -\int_{-R}^0 \cos(x) \varphi(x) \, dx + \int_0^R \cos(x) \varphi(x) \, dx$$

$$= -\int_{\mathbb{R}} \operatorname{sgn}(x) \cos(x) \varphi(x) \, dx. \tag{13.1.7}$$

Since $\operatorname{sgn}(x) \cos(x) \in L_{loc}^1(\mathbb{R})$, we deduce from (13.1.7) that f' exists in the weak sense and that $f'(x) = \operatorname{sgn}(x) \cos(x)$ for $x \in \mathbb{R}$. Also,

$$\int_{\mathbb{R}} f(x) \varphi''(x) \, dx = \int_{-R}^0 \cos(x) \varphi'(x) \, dx - \int_0^R \cos(x) \varphi'(x) \, dx$$

$$= 2\varphi(0) - \int_{\mathbb{R}} \operatorname{sgn}(x) \sin(x) \varphi(x) \, dx. \tag{13.1.8}$$

If the weak derivative f'' were to exist we could find $g \in L_{loc}^1(\mathbb{R})$ such that

$$\int_{\mathbb{R}} g(x) \varphi(x) \, dx = 2\varphi(0) - \int_{\mathbb{R}} \operatorname{sgn}(x) \sin(x) \varphi(x) \, dx, \qquad \forall \varphi \in C_0^\infty(\mathbb{R}). \tag{13.1.9}$$

In particular, (13.1.9) would also hold for all $\varphi \in C_0^\infty(\mathbb{R})$ supported in $(0, \infty)$, which would give $g|_{(0,\infty)} = -\sin(x)$. Similarly, we would conclude that $g|_{(-\infty,0)} = \sin(x)$ and, hence, returning to (13.1.9), we would obtain $\varphi(0) = 0$ for all $\varphi \in C_0^\infty(\mathbb{R})$. The latter is false, thus the weak derivative f'' does not exist.

Exercise 1.35 Suppose $\partial_j f$, $j = 1, \ldots, n$, exist in the weak sense. Fix j and note that by Exercise 1.34 we have that the weak derivative $(\partial_j f)|_{\mathbb{R}^n \setminus \{0\}}$ equals the weak derivative $\partial_j(f|_{\mathbb{R}^n \setminus \{0\}})$. The function $f|_{\mathbb{R}^n \setminus \{0\}} = \frac{1}{|x|^a}$ is smooth in $\mathbb{R}^n \setminus \{0\}$ so the latter weak derivative equals the classical pointwise derivative which is $-a \frac{x_j}{|x|^{a+2}}$ for each $x \in \mathbb{R}^n \setminus \{0\}$. Since the Lebesgue measure of the set $\{0\}$ is zero, we obtain that the weak derivative $\partial_j f(x) = -a \frac{x_j}{|x|^{a+2}}$ for almost every $x \in \mathbb{R}^n$. However, by assumption, this weak derivative is in $L_{loc}^1(\mathbb{R}^n)$ and since j is arbitrary, we obtain that $\frac{a}{|x|^{a+1}} = \left[\sum_{j=1}^n (\partial_j f)^2 \right]^{1/2} \in L_{loc}^1(\mathbb{R}^n)$. This necessarily implies $a < n - 1$.

Conversely, suppose $a < n - 1$ and consider the functions $g_j(x) := (-a)\frac{x_j}{|x|^{a+2}}$ defined for almost every $x \in \mathbb{R}^n$ and each $j \in \{1, \ldots, n\}$. Then $g_j \in L^1_{loc}(\mathbb{R}^n)$ for each $j \in \{1, \ldots, n\}$. To complete the proof it suffices to show that

$$\int_{\mathbb{R}^n} \frac{1}{|x|^a}(\partial_j \varphi)(x)\,\mathrm{d}x = a \int_{\mathbb{R}^n} \frac{x_j}{|x|^{a+2}} \varphi(x)\,\mathrm{d}x \tag{13.1.10}$$

for every $\varphi \in C_0^\infty(\mathbb{R}^n)$ and each $j \in \{1, \ldots, n\}$. For φ and j arbitrary, fixed, using Lebesgue's Dominated Convergence Theorem and integration by parts (cf. formula (14.8.4)) we write

$$\int_{\mathbb{R}^n} \frac{1}{|x|^a}(\partial_j \varphi)(x)\,\mathrm{d}x = \lim_{\varepsilon \to 0^+} \int_{\mathbb{R}^n \setminus B(0,\varepsilon)} \frac{1}{|x|^a}(\partial_j \varphi)(x)\,\mathrm{d}x$$

$$= \lim_{\varepsilon \to 0^+} a \int_{\mathbb{R}^n \setminus B(0,\varepsilon)} \frac{x_j}{|x|^{a+2}} \varphi(x)\,\mathrm{d}x - \lim_{\varepsilon \to 0^+} \int_{|x|=\varepsilon} \frac{x_j \varphi(x)}{|x|^{a+1}}\,\mathrm{d}\sigma(x)$$

$$= a \int_{\mathbb{R}^n} \frac{x_j}{|x|^{a+2}} \varphi(x)\,\mathrm{d}x. \tag{13.1.11}$$

For the last equality in (13.1.11) we have used the fact that

$$\left| \int_{|x|=\varepsilon} \frac{x_j \varphi(x)}{|x|^{a+1}}\,\mathrm{d}\sigma(x) \right| \leq \|\varphi\|_{L^\infty(\mathbb{R}^n} \frac{\omega_{n-1}}{n} \varepsilon^{n-1-a} \to 0 \text{ as } \varepsilon \to 0^+ \tag{13.1.12}$$

since $n - 1 - a > 0$. This proves (13.1.10) as wanted.

Exercise 1.36 Let $\varphi \in C_0^\infty(\Omega)$ be arbitrary. Then $\partial^{\alpha+\beta}\varphi = \partial^\alpha(\partial^\beta \varphi)$ and since the weak derivative $\partial^\alpha f$ exists we have

$$\int_\Omega f \partial^{\alpha+\beta}\varphi\,\mathrm{d}x = (-1)^{|\alpha|} \int_\Omega (\partial^\alpha f)(\partial^\beta \varphi)\,\mathrm{d}x. \tag{13.1.13}$$

Also, since the weak derivative $\partial^\beta(\partial^\alpha f)$ exists we have

$$\int_\Omega (\partial^\alpha f)(\partial^\beta \varphi)\,\mathrm{d}x = (-1)^{|\beta|} \int_\Omega \partial^\beta(\partial^\alpha f)\varphi\,\mathrm{d}x. \tag{13.1.14}$$

The desired conclusion follows from (13.1.13), (13.1.14), and the definition of weak derivative.

Exercise 1.37 Note that $f \in L^1_{loc}(\mathbb{R}^n)$ for every $n \in \mathbb{N}$. Suppose $n \geq 2$, and fix some index $j \in \{1, \ldots, n\}$ along with a function $\varphi \in C_0^\infty(\mathbb{R}^n)$. If $R > 0$ is such that $\operatorname{supp}\varphi \subset B(0, R)$, then using Lebesgue's Dominated Convergence Theorem and (14.8.4) we obtain

$$-\int_{\mathbb{R}^n} f(x)\partial_j\varphi(x)\,dx = -\lim_{r\to 0^+}\int_{r<|x|<R} f(x)\partial_j\varphi(x)\,dx$$

$$= \lim_{r\to 0^+}\Big[\int_{\partial B(0,r)}\frac{x_j}{r^{\varepsilon+1}}\varphi(x)\,d\sigma(x)$$

$$-\varepsilon\int_{r<|x|<R}\frac{x_j}{|x|^{\varepsilon+2}}\varphi(x)\,dx\Big]$$

$$= -\varepsilon\int_{\mathbb{R}^n}\frac{x_j}{|x|^{\varepsilon+2}}\varphi(x)\,dx, \tag{13.1.15}$$

since, given that under the current assumptions $n-1-\varepsilon > 0$, we have

$$\Big|\int_{\partial B(0,r)}\frac{x_j}{r^{\varepsilon+1}}\varphi(x)\,d\sigma(x)\Big| \le \omega_{n-1}\|\varphi\|_{L^\infty(\mathbb{R}^n)}r^{n-1-\varepsilon} \to 0 \quad\text{as } r\to 0^+. \tag{13.1.16}$$

This proves that if $n \ge 2$ then the weak derivative $\partial_j f$ exists and is equal to the function $-\varepsilon\frac{x_j}{|x|^{\varepsilon+2}} \in L^1_{loc}(\mathbb{R}^n)$.

Assume next that $n = 1$ and suppose that there exists some $g \in L^1_{loc}(\mathbb{R})$ with the property that $-\int_\mathbb{R} f\varphi'\,dx = \int_\mathbb{R} g\varphi\,dx$ for every $\varphi \in C_0^\infty(\mathbb{R})$. Then, with $R > 0$ such that $\mathrm{supp}\,\varphi \subset (-R, R)$, using Lebesgue's Dominated Convergence Theorem we may write

$$\int_\mathbb{R} f\varphi'\,dx = \lim_{r\to 0^+}\Big[\int_{-R}^{-r}\frac{\varphi'(x)}{(-x)^\varepsilon}\,dx + \int_r^R\frac{\varphi'(x)}{x^\varepsilon}\,dx\Big] \tag{13.1.17}$$

$$= \lim_{r\to 0^+}\Big[\frac{\varphi(-r)}{r^\varepsilon} - \varepsilon\int_{-R}^{-r}\frac{\varphi(x)}{(-x)^{\varepsilon+1}}\,dx - \frac{\varphi(r)}{r^\varepsilon} + \varepsilon\int_r^R\frac{\varphi(x)}{x^{\varepsilon+1}}\,dx\Big].$$

In particular, identity (13.1.17) holds if $\mathrm{supp}\,\varphi \subset (0,\infty)$, in which case (13.1.17) implies $g(x) = \varepsilon x^{-\varepsilon-1}$ for $x > 0$, and if $\mathrm{supp}\,\varphi \subset (-\infty, 0)$, when we obtain $g(x) = -\varepsilon(-x)^{-\varepsilon-1}$ for $x < 0$ (recall Theorem 1.3). However, the function g thus obtained does not belong to $L^1_{loc}(\mathbb{R})$. Consequently, f does not have a weak derivative of order one in the case when $n = 1$.

Exercise 1.38 (a) Let $f \in L^1_{loc}((a, b))$ be such that the weak derivative f' exists and equals 0. That is,

$$\int_a^b f(x)\varphi'(x)\,dx = 0, \qquad \forall\,\varphi \in C_0^\infty((a, b)). \tag{13.1.18}$$

Select $\varphi_0 \in C_0^\infty((a, b))$ such that $\int_a^b \varphi_0(s)\,ds = 1$. Then, each $\varphi \in C_0^\infty((a, b))$ may be written as

$$\varphi(t) = \varphi_0(t)\int_a^b \varphi(s)\,ds + \psi'(t), \qquad \forall\,t \in (a, b), \tag{13.1.19}$$

where

$$\psi(x) := \int_a^x \left[\varphi(t) - \varphi_0(t) \int_a^b \varphi(s) \, ds \right] dt, \qquad \forall \, x \in (a, b). \tag{13.1.20}$$

An inspection of (13.1.20) reveals that $\psi \in C^\infty((a, b))$, and if $[a_0, b_0] \subseteq (a, b)$ is an interval such that $\operatorname{supp} \varphi$, $\operatorname{supp} \varphi_0 \subseteq [a_0, b_0]$, then we also have $\operatorname{supp} \psi \subseteq [a_0, b_0]$ (as seen by analyzing the cases $x \in (a, a_0)$ and $x \in (b_0, b)$). Now define the constant $c := \int_a^b f(x)\varphi_0(x) \, dx$ and use (13.1.18) to obtain

$$\int_a^b f(x)\varphi(x) \, dx = \int_a^b c\,\varphi(s) \, ds, \qquad \forall \, \varphi \in C_0^\infty((a, b)). \tag{13.1.21}$$

Hence, by invoking Theorem 1.3 we obtain $f = c$ almost everywhere in (a, b).

(b) Let $g \in L^1_{loc}((a, b))$, $x_0 \in (a, b)$, and set $f(x) := \int_{x_0}^x g(t) \, dt$ for every $x \in (a, b)$. By Lebesgue's Dominated Convergence Theorem, it follows that f is continuous on (a, b), hence $f \in L^1_{loc}((a, b))$. There remains to prove that g is the weak derivative of f.

Parenthetically, we note that under the current assumptions, one may not expect f to necessarily be pointwise differentiable in (a, b). As an example, take the function $f : \mathbb{R} \to \mathbb{R}$ defined by $f(x) := x$ for $x \geq 0$ and $f(x) := 0$ for $x < 0$. Then f is not differentiable at zero but its weak derivative f' exists and is equal to $\chi_{(0,\infty)} \in L^1_{loc}(\mathbb{R})$. Also, observe that the fundamental theorem of calculus does not apply in this case since while g is in $L^1_{loc}(\mathbb{R})$ it is not continuous.

Let $\varphi \in C_0^\infty((a, b))$ and $c \in (a, b)$ such that $\operatorname{supp} \varphi \subset (c, b)$. Consider two cases.

Case 1. Assume $c > x_0$. Then $\operatorname{supp} \varphi \subset (x_0, b)$ and using the expression for f and Fubini's theorem we may write

$$\int_a^b f(x)\varphi'(x) \, dx = \int_{x_0}^b \left(\int_{x_0}^x g(t) \, dt \right) \varphi'(x) \, dx = \int_{x_0}^b \left(\int_t^b \varphi'(x) \, dx \right) g(t) \, dt$$

$$= -\int_{x_0}^b \varphi(t)g(t) \, dt = -\int_a^b g(t)\varphi(t) \, dt. \tag{13.1.22}$$

Case 2. Assume $c < x_0$. Again, using the expression for f and Fubini's theorem it follows that

$$\int_a^b f(x)\varphi'(x)\,dx = \int_a^{x_0} f(x)\varphi'(x)\,dx + \int_{x_0}^b f(x)\varphi'(x)\,dx$$

$$= \int_a^{x_0} \left(\int_{x_0}^x g(t)\,dt\right)\varphi'(x)\,dx + \int_{x_0}^b \left(\int_{x_0}^x g(t)\,dt\right)\varphi'(x)\,dx$$

$$= -\int_a^{x_0} g(t)\int_a^t \varphi'(x)\,dx\,dt + \int_{x_0}^b g(t)\int_t^b \varphi'(x)\,dx\,dt$$

$$= -\int_a^b g(t)\varphi(t)\,dt. \tag{13.1.23}$$

From (13.1.22) and (13.1.23) the desired conclusion follows.

(c) Denote by $h \in L^1_{loc}((a,b))$ the weak derivative $f^{(k)}$ and fix an arbitrary point $x_0 \in (a,b)$. If one sets $g(x) := \int_{x_0}^x h(s)\,ds$ for $x \in (a,b)$, then by Lebesgue's Dominated Convergence Theorem, it follows that g is continuous. Thus, in particular, $g \in L^1_{loc}((a,b))$. Fix $\varphi_0 \in C_0^\infty((a,b))$ with $\int_a^b \varphi_0(t)\,dt = 1$. Using induction and what we have proved in part *(b)*, the desired conclusion will follow if we show that

$$f^{(k-1)} = g - \int_a^b g(t)\varphi_0(t)\,dt + (-1)^{k-1}\int_a^b f(t)\varphi_0^{(k-1)}(t)\,dt \quad \text{a.e. on } (a,b). \tag{13.1.24}$$

Let $\varphi \in C_0^\infty((a,b))$ and write it as in (13.1.19)–(13.1.20). Then

$$\int_a^b g(x)\varphi(x)\,dx = \left(\int_a^b g(x)\varphi_0(x)\,dx\right)\left(\int_a^b \varphi(t)\,dt\right) + \int_a^b g(x)\psi'(x)\,dx$$

$$= \left(\int_a^b g(x)\varphi_0(x)\,dx\right)\left(\int_a^b \varphi(t)\,dt\right) - \int_a^b h(x)\psi(x)\,dx$$

$$= \left(\int_a^b g(x)\varphi_0(x)\,dx\right)\left(\int_a^b \varphi(t)\,dt\right) + (-1)^{k-1}\int_a^b f(x)\psi^{(k)}(x)\,dx$$

$$= \left(\int_a^b g(x)\varphi_0(x)\,dx\right)\left(\int_a^b \varphi(t)\,dt\right) + (-1)^{k-1}\int_a^b f(x)\varphi^{(k-1)}(x)\,dx$$

$$+ (-1)^k\left(\int_a^b f(x)\varphi_0^{(k-1)}(x)\,dx\right)\left(\int_a^b \varphi(t)\,dt\right). \tag{13.1.25}$$

Now (13.1.24) follows from (13.1.25) and Theorem 1.3.

(d) Using what we have proved in part *(c)*, we obtain that the weak derivative $f^{(j)}$ exists for each $j \in \mathbb{N}$ satisfying $j \le k$. From what we proved in part *(a)* we know that $f^{(k-1)} = a_{k-1} \in \mathbb{C}$, while from part *(c)* we know that $f^{(k-2)} = a_{k-1}x - a_{k-1}x_0 - c$. The proof may be now completed by induction.

(e) No; see Exercise 1.29.

Exercise 1.39 Let $R > 0$ be such that $\operatorname{supp}\theta \subseteq \overline{B(0,R)}$. Then $\varphi_j \in C_0^\infty(\mathbb{R}^n)$ and $\operatorname{supp}\varphi_j \subseteq \overline{B(0,R)}$ for each $j \in \mathbb{N}$. Also, given any $\alpha \in \mathbb{N}_0^n$ we have

$$\partial^\alpha \varphi_j = e^{-j} j^{m+|\alpha|}(\partial^\alpha \theta)(j\cdot) \qquad \text{for every } j \in \mathbb{N}. \tag{13.1.26}$$

Hence,

$$\sup_{x \in \overline{B(0,R)}} |\partial^\alpha \varphi_j(x)| \le e^{-j} j^{m+|\alpha|} \|\theta\|_{L^\infty(\mathbb{R}^n)} \xrightarrow[j\to\infty]{} 0.$$

This shows that $\varphi_j \xrightarrow[j\to\infty]{\mathcal{D}(\mathbb{R}^n)} 0$.

Exercise 1.40 Clearly, for every $j \in \mathbb{N}$, $\varphi_j \in C_0^\infty(\mathbb{R}^n)$ and $\operatorname{supp}\varphi_j = \operatorname{supp}\theta - jh$. Hence, for every $R > 0$ there exists $j_0 \in \mathbb{N}$ such that

$$(\operatorname{supp}\theta - jh) \cap B(0,R) = \varnothing \quad \text{for} \quad j \ge j_0. \tag{13.1.27}$$

This shows that there is no compact set $K \subset \mathbb{R}^n$ such that $\operatorname{supp}\varphi_j \subseteq K$ for all $j \in \mathbb{N}$, which implies that $\{\varphi_j\}_{j\in\mathbb{N}}$ does not converge in $\mathcal{D}(\mathbb{R}^n)$. In addition, if $\alpha \in \mathbb{N}_0^n$, K is a compact set in $\mathbb{R}^n$, and $R > 0$ is such that $K \subseteq B(0,R)$, then $\sup_{x\in K}|\partial^\alpha \varphi_j(x)| = \sup_{x\in K}|(\partial^\alpha \theta)(x+jh)| = 0$ for all $j \ge j_0$, where $j_0 \in \mathbb{N}$ is such that $(\operatorname{supp}\theta - jh) \cap B(0,R) = \varnothing$. This proves that $\varphi_j \xrightarrow[j\to\infty]{\mathcal{E}(\mathbb{R}^n)} 0$.

Exercise 1.41 Suppose there exists $\varphi \in C_0^\infty(\mathbb{R}^n)$ such that $\varphi_j \xrightarrow[j\to\infty]{\mathcal{E}(\mathbb{R}^n)} \varphi$. Then necessarily $\varphi_j \xrightarrow[j\to\infty]{} \varphi$ pointwise (explain why). Since $\lim_{j\to\infty} \varphi_j(x) = 0$ for every point $x \in \mathbb{R}^n$, this forces $\varphi = 0$. However, if $\alpha \in \mathbb{N}_0$ is such that $|\alpha| = 1$, then we have $\partial^\alpha \varphi_j = (\partial^\alpha \theta)(j\cdot)$ for every $j \in \mathbb{N}$, hence for each compact K we have $\sup_{x\in K}|\partial^\alpha \varphi_j| = \|\partial^\alpha \theta\|_{L^\infty(K)}$ which does not converge to zero as $j \to \infty$, leading to a contradiction.

Exercise 1.42 One implication is easy. If $\theta \ne 0$, then $\operatorname{supp}\varphi_j = jh + \frac{1}{j}\operatorname{supp}\theta$ for each $j \in \mathbb{N}$, and since $h \ne 0$, there is no compact $K \subset \mathbb{R}^n$ such that $\operatorname{supp}\varphi_j \subseteq K$ for all $j \in \mathbb{N}$. This shows that $\{\varphi_j\}_{j\in\mathbb{N}}$ does not converge in $\mathcal{D}(\mathbb{R}^n)$.

Exercise 1.43 $\{\varphi_j\}_{j\in\mathbb{N}}$ does not converges in $\mathcal{D}(\mathbb{R}^n)$ since $\operatorname{supp}\varphi_j = j\operatorname{supp}\theta$ for every $j \in \mathbb{N}$, thus there is no compact $K \subset \mathbb{R}^n$ such that $\operatorname{supp}\varphi_j \subseteq K$ for all $j \in \mathbb{N}$. On the other hand, if $\alpha \in \mathbb{N}_0^n$ then $|\partial^\alpha \varphi_j(x)| \le j^{-1-|\alpha|}\|\partial^\alpha \theta\|_{L^\infty(\mathbb{R}^n)}$ for every $x \in \mathbb{R}^n$. Thus, $\partial^\alpha \varphi_j$ converges to zero uniformly on $\mathbb{R}^n$, proving that $\varphi_j \xrightarrow[j\to\infty]{\mathcal{E}(\mathbb{R}^n)} 0$.

Exercise 1.44 Let $x_0 \in \Omega$ and $R > 0$ be such that $B(x_0,R) \subset \Omega$. Consider a nonzero function $\varphi \in C_0^\infty(\mathbb{R})$ with the property that $\operatorname{supp}\varphi \subset B(0,R)$ and define the sequence $\varphi_j(x) := \frac{1}{j}\varphi(j(x-x_0))$, $x \in \Omega$, $j \in \mathbb{N}$. Show that the sequence $\{\varphi_j\}_{j\in\mathbb{N}}$ satisfies the hypotheses in the problem but it does not converge in $\mathcal{D}(\mathbb{R}^n)$.

13.2 Solutions to Exercises from Section 2.10

Exercise 2.122 Let K be a compact set in $\mathbb{R}$ and fix an arbitrary $\varphi \in C_0^\infty(\mathbb{R})$ with $\operatorname{supp} \varphi \subseteq K$. By the Mean Value Theorem, $|\varphi(\frac{1}{j^2}) - \varphi(0)| \leq \frac{1}{j^2}\|\varphi'\|_{L^\infty(K)}$. This shows that the series in (2.10.1) is absolutely convergent, hence u is well defined. Clearly, u is linear and since $|u(\varphi)| \leq \left(\sum_{j=1}^\infty j^{-2}\right)\sup_{x\in K}|\varphi'(x)|$ it follows that $u \in \mathcal{D}'(\mathbb{R})$, it has finite order, and its order is at most 1. To see that u is not of order zero, consider the sequence of functions $\{\varphi_k\}_{k\in\mathbb{N}}$ satisfying, for each $k \in \mathbb{N}$, the conditions

$$\varphi_k \in C_0^\infty((0,2)), \ \ 0 \leq \varphi_k \leq \tfrac{1}{k},$$
$$\varphi_k(x) = 0 \text{ if } x \leq \tfrac{1}{(k+1)^2}, \ \ \varphi_k(x) = \tfrac{1}{k} \text{ if } x \in \left[\tfrac{1}{k^2}, 1\right]. \tag{13.2.1}$$

Then $\langle u, \varphi_k \rangle = \sum_{j=1}^k \varphi_k(\frac{1}{j^2}) = 1$. If we assume that u has order zero, then there exists a finite positive number C such that $|\langle u, \varphi \rangle| \leq C \sup_{x\in[0,2]}|\varphi(x)|$ for every $\varphi \in C_0^\infty(\mathbb{R})$ with $\operatorname{supp}\varphi \subseteq [0,2]$. This implies $1 = |\langle u, \varphi_k \rangle| \leq \frac{C}{k}$ for every $k \in \mathbb{N}$, which leads to a contradiction. Hence the order of u is 1.

Exercise 2.123 Take u to be the distribution given by the constant function 1. Then this u does not satisfy (2.10.2). Indeed, for any compact $K \subset \Omega$ we can choose $\varphi \in C_0^\infty(\Omega)$ with $\operatorname{supp}\varphi \subseteq \Omega\setminus K$ and $\int_\Omega \varphi\,dx \neq 0$ which would lead to a contradiction if (2.10.2) were true for $u = 1$.

Exercise 2.124 Use a reasoning similar in spirit to the one from Example 2.9.

Exercise 2.125 Estimate $\int_{B(0,R)} |f(x)|\,dx$ by working in polar coordinates (cf. (14.9.7)) and using Proposition 14.65, as well as (2.1.9).

Exercise 2.126 Let $\varphi \in C_0^\infty(\mathbb{R})$ with $\operatorname{supp}\varphi \subset (-R, R)$. Then using integration by parts we have

$$\langle (\ln|x|)', \varphi \rangle = -\langle \ln|x|, \varphi' \rangle$$

$$= -\int_\mathbb{R} \varphi'(x)\ln|x|\,dx = -\lim_{\varepsilon\to 0^+}\int_{\varepsilon<|x|<R}\varphi'(x)\ln|x|\,dx$$

$$= \lim_{\varepsilon\to 0^+}\left[\int_{\varepsilon<|x|<R}\frac{\varphi(x)}{x}\,dx - \varphi(x)\ln|x|\Big|_{-R}^{-\varepsilon} - \varphi(x)\ln|x|\Big|_{\varepsilon}^{R}\right]$$

$$= \left\langle \mathrm{P.V.}\,\frac{1}{x}, \varphi \right\rangle + \lim_{\varepsilon\to 0^+}\left[(\varphi(\varepsilon) - \varphi(-\varepsilon))\ln\varepsilon\right]$$

$$= \left\langle \mathrm{P.V.}\,\frac{1}{x}, \varphi \right\rangle. \tag{13.2.2}$$

For the last equality in (13.2.2) note that

$$\left|(\varphi(\varepsilon) - \varphi(-\varepsilon))\ln\varepsilon\right| \le 2\|\varphi'\|_{L^\infty(\mathbb{R})}\,\varepsilon|\ln\varepsilon| \xrightarrow[\varepsilon\to0^+]{} 0. \qquad (13.2.3)$$

Exercise 2.127 It is immediate that f is continuous on $\mathbb{R} \setminus \{0\}$, while its continuity at $x = 0$ follows from the fact that

$$\lim_{x\to0} x\ln|x| = 0. \qquad (13.2.4)$$

Fix $\varphi \in C_0^\infty(\mathbb{R})$. Starting with the definition of distributional derivative, then applying Lebesgue's Dominated Convergence Theorem (in concert with the properties of φ), then integrating by parts and using (13.2.4), we obtain

$$\begin{aligned}
\langle f', \varphi \rangle &= -\langle f, \varphi' \rangle \\
&= -\lim_{\varepsilon\to0^+}\Big[\int_{-\infty}^{-\varepsilon}(x\ln(-x)-x)\varphi'(x)\,dx + \int_\varepsilon^\infty(x\ln x - x)\varphi'(x)\,dx\Big] \\
&= \lim_{\varepsilon\to0^+}\Big[\int_{-\infty}^{-\varepsilon}\ln(-x)\varphi(x)\,dx + \int_\varepsilon^\infty(\ln x)\varphi(x)\,dx\Big] \\
&= \langle \ln|x|, \varphi \rangle. \qquad (13.2.5)
\end{aligned}$$

The last equality in (13.2.5) uses the fact that $\ln|x| \in L^1_{loc}(\mathbb{R})$.

Alternatively, using (4) in Proposition 2.43, Exercise 2.126, the fact that $x' = 1$ in $\mathcal{D}'(\mathbb{R})$, and (2.3.7), we have

$$(x\ln|x| - x)' = \ln|x| - x\Big(\text{P.V.}\,\frac{1}{x}\Big) - 1 = \ln|x| \quad \text{in} \quad \mathcal{D}'(\mathbb{R}). \qquad (13.2.6)$$

Exercise 2.128 Fix $j \in \{1, \ldots, n\}$ and $\varphi \in C_0^\infty(\mathbb{R}^n)$. Then

$$\begin{aligned}
\langle \partial_j(\ln|x|), \varphi \rangle &= -\langle \ln|x|, \partial_j\varphi \rangle = -\int_{\mathbb{R}^n}(\ln|x|)\partial_j\varphi(x)\,dx \\
&= -\lim_{\varepsilon\to0^+}\int_{|x|\ge\varepsilon}(\ln|x|)\partial_j\varphi(x)\,dx \\
&= \lim_{\varepsilon\to0^+}\int_{|x|\ge\varepsilon}\frac{x_j}{|x|^2}\varphi(x)\,dx + \lim_{\varepsilon\to0^+}\int_{|x|=\varepsilon}(\ln\varepsilon)\frac{x_j}{\varepsilon}\varphi(x)\,d\sigma(x) \\
&= \int_{\mathbb{R}^n}\frac{x_j}{|x|^2}\varphi(x)\,dx + \lim_{\varepsilon\to0^+}\varepsilon^{n-1}(\ln\varepsilon)\int_{S^{n-1}}\omega_j\varphi(\varepsilon\omega)\,d\sigma(\omega) \quad (13.2.7)
\end{aligned}$$

For the third and last equality in (13.2.7) we used Lebesgue's Dominated Convergence Theorem (note that $\frac{x_j}{|x|^2}$, $\ln|x| \in L^1_{loc}(\mathbb{R}^n)$ and φ is compactly supported). The fourth equality uses the integration by parts formula (14.8.4). Also, for the last equality, in the integral on $\partial B(0, \varepsilon)$ we made the change of variables $x = \varepsilon\omega$ with $\omega \in S^{n-1}$. One more application of Lebesgue's Dominated Convergence Theorem yields

$$\lim_{\varepsilon \to 0^+} \int_{S^{n-1}} \omega_j \varphi(\varepsilon\omega) \, d\sigma(\omega) = \varphi(0) \int_{S^{n-1}} \omega_j \, d\sigma(\omega) = 0, \tag{13.2.8}$$

where the last equality is due to the fact that the integral of an odd function over the unit sphere is zero. Moreover, (2.1.9) implies $\lim_{\varepsilon \to 0^+} \varepsilon^{n-1}(\ln \varepsilon) = 0$ (recall that we are assuming $n \geq 2$). Returning with all these comments to (13.2.7) we arrive at the conclusion that

$$\langle \partial_j(\ln|x|), \varphi \rangle = \left\langle \frac{x_j}{|x|^2}, \varphi \right\rangle \quad \text{for every} \quad \varphi \in C_0^\infty(\mathbb{R}^n).$$

Hence, if $n \geq 2$ then $\partial_j(\ln|x|) = \frac{x_j}{|x|^2}$ in $\mathcal{D}'(\mathbb{R}^n)$.

Exercise 2.129 Fix $j \in \{1, \ldots, n\}$ and $\varphi \in C_0^\infty(\mathbb{R}^n)$. Then

$$\left\langle \partial_j\left(\frac{1}{|x|^{n-2}}\right), \varphi \right\rangle = -\left\langle \frac{1}{|x|^{n-2}}, \partial_j \varphi \right\rangle = -\int_{\mathbb{R}^n} \frac{1}{|x|^{n-2}} \partial_j \varphi(x) \, dx$$

$$= -\lim_{\varepsilon \to 0^+} \int_{|x| \geq \varepsilon} \frac{1}{|x|^{n-2}} \partial_j \varphi(x) \, dx$$

$$= \lim_{\varepsilon \to 0^+} \int_{|x| \geq \varepsilon} \frac{(2-n)x_j}{|x|^n} \varphi(x) \, dx + \lim_{\varepsilon \to 0^+} \int_{|x| = \varepsilon} \frac{x_j}{\varepsilon^{n-1}} \varphi(x) \, d\sigma(x)$$

$$= (2-n) \int_{\mathbb{R}^n} \frac{x_j}{|x|^n} \varphi(x) \, dx + \lim_{\varepsilon \to 0^+} \int_{S^{n-1}} \omega_j \varphi(\varepsilon\omega) \, d\sigma(\omega) \tag{13.2.9}$$

For the third and last equality in (13.2.7) we used Lebesgue's Dominated Convergence Theorem (note that $\frac{x_j}{|x|^n}, \frac{1}{|x|^{n-2}} \in L^1_{loc}(\mathbb{R}^n)$ and φ is compactly supported). The fourth equality uses the integration by parts formula (14.8.4). Also, for the last equality, in the integral on $\partial B(0, \varepsilon)$ we made the change of variables $x = \varepsilon\omega$ with $x \in S^{n-1}$. One more application of Lebesgue's Dominated Convergence Theorem yields

$$\lim_{\varepsilon \to 0^+} \int_{S^{n-1}} \omega_j \varphi(\varepsilon\omega) \, d\sigma(\omega) = \varphi(0) \int_{S^{n-1}} \omega_j \, d\sigma(\omega) = 0, \tag{13.2.10}$$

where the last equality is due to the fact that the integral of an odd function over the unit sphere is zero. Returning with all these comments to (13.2.9) we arrive at the conclusion that

$$\left\langle \partial_j\left(\frac{1}{|x|^{n-2}}\right), \varphi \right\rangle = \left\langle (2-n)\frac{x_j}{|x|^n}, \varphi \right\rangle \quad \text{for every} \quad \varphi \in C_0^\infty(\mathbb{R}^n).$$

Hence, $\partial_j\left(\frac{1}{|x|^{n-2}}\right) = (2-n)\frac{x_j}{|x|^n}$ in $\mathcal{D}'(\mathbb{R}^n)$.

Exercise 2.130 Fix $j \in \mathbb{N}$. Using the change of variables $y = jx - jt$ the expression for ψ_j becomes

$$\psi_j(x) = \int_{jx-j^2}^{jx-1} \theta(y)\,dy \qquad \text{for} \quad x \in \mathbb{R}.$$

Hence $\psi_j(x) = 0$ if $x \leq 0$, while for $x > 0$ we have $\psi_j(x) = \int_{-1}^{1} \theta(y)\,dy = 1$ if $j \geq j_0$, where $j_0 \in \mathbb{N}$ is such that $j_0 x - j_0^2 \leq -1$ and $j_0 x - 1 \geq 1$. This shows that ψ_j converges pointwise to H as $j \to \infty$. In addition, $|\psi_j(x)| \leq \int_{-1}^{1} |\theta(y)|\,dy < \infty$ for every $j \in \mathbb{N}$. Hence, by applying Lebesgue's Dominated Convergence Theorem,

$$\lim_{j \to \infty} \int_{\mathbb{R}} \psi_j(x)\varphi(x)\,dx = \int_{\mathbb{R}} H(x)\varphi(x)\,dx \qquad \text{for each} \quad \varphi \in C_0^\infty(\mathbb{R}).$$

This proves that $\psi_j \xrightarrow[j \to \infty]{\mathcal{D}'(\mathbb{R})} H$ as wanted.

Exercise 2.131 If $\varphi \in C_0^\infty(\mathbb{R})$ then the support condition for φ guarantees that only finitely many terms in the sum $\sum_{j=1}^{\infty} \varphi^{(j)}(j)$ are nonzero, hence u is well defined. Clearly u is linear. If K is a compact in $\mathbb{R}$ and $k \in \mathbb{N}$ is such that $K \subseteq [-k, k]$, then

$$|u(\varphi)| \leq \sum_{j=1}^{k} \left|\varphi^{(j)}(j)\right| \leq k \sup_{x \in K,\, j \leq k} \left|\varphi^{(j)}(x)\right|$$

for $\varphi \in C_0^\infty(\mathbb{R})$ with $\operatorname{supp} \varphi \subseteq K$. This proves that $u \in \mathcal{D}'(\mathbb{R})$.

Suppose the distribution u has finite order. Then there exists a nonnegative integer k_0 with the property that for each compact set $K \subset \mathbb{R}$ there is a finite constant $C_K \geq 0$ such that $|\langle u, \varphi \rangle| \leq C_K \sup_{x \in K,\, j \leq k_0} |\varphi^{(j)}(x)|$ for every $\varphi \in C_0^\infty(\mathbb{R})$ with $\operatorname{supp} \varphi \subseteq K$. In particular, from this and the definition of u it follows that there exists $C \in [0, \infty)$ such that whenever $\varphi \in C_0^\infty(\mathbb{R})$ satisfies the support condition $\operatorname{supp} \varphi \subseteq [k_0 + \frac{1}{2}, k_0 + \frac{3}{2}]$ we have

$$|\varphi^{(k_0+1)}(k_0 + 1)| = |\langle u, \varphi \rangle| \leq C \sup_{x \in [k_0 + \frac{1}{2}, k_0 + \frac{3}{2}],\, \ell \leq k_0} |\varphi^{(\ell)}(x)|. \qquad (13.2.11)$$

Now consider $\theta \in C_0^\infty((-1/2, 1/2))$ satisfying $\theta(0) = 1$ and construct the sequence of smooth functions $\varphi_j(x) := \theta(jx - j(k_0 + 1))$ for $x \in \mathbb{R}$ and $j \in \mathbb{N}$. Then, for each $j \in \mathbb{N}$, we have

$$\operatorname{supp} \varphi_j \subseteq [k_0 + \frac{1}{2}, k_0 + \frac{3}{2}],$$

$$\varphi_j^{(\ell)} = j^\ell \theta^{(\ell)}(j \cdot - j(k_0 + 1)), \quad \forall \ell \in \mathbb{N}_0,$$

$$\text{and} \quad \varphi_j^{(k_0+1)}(k_0 + 1) = j^{k_0+1}.$$

Combining all these facts with (13.2.11) it follows that for each $j \in \mathbb{N}$,

$$j^{k_0+1} \leq C \sup_{x \in \left[-\frac{1}{2},\frac{1}{2}\right], \ell \leq k_0} j^{\ell} |\theta^{(\ell)}(x)|$$

$$\leq C \max \left\{ \|\theta^{(\ell)}\|_{L^{\infty}([-1/2,1/2])} : \ell \leq k_0 \right\} j^{k_0}. \tag{13.2.12}$$

Choosing now j sufficiently large in (13.2.12) leads to a contradiction. Hence u does not have finite order.

Exercise 2.132 Note that for each $j \in \mathbb{N}$ we have $f_j \in L^1_{loc}(\mathbb{R}^n)$. Pick some $\varphi \in C_0^{\infty}(\mathbb{R}^n)$ and suppose $R > 0$ is such that $\operatorname{supp} \varphi \subset B(0, R)$. Then

$$\langle f_j, \varphi \rangle = \frac{1}{j} \int_{B(0,R)} \frac{\varphi(x) - \varphi(0)}{|x|^{n-\frac{1}{j}}} \, dx + \frac{\varphi(0)}{j} \int_{B(0,R)} \frac{1}{|x|^{n-\frac{1}{j}}} \, dx. \tag{13.2.13}$$

Using the Mean Value Theorem and then (14.9.6) we may further write

$$\left| \frac{1}{j} \int_{B(0,R)} \frac{\varphi(x) - \varphi(0)}{|x|^{n-\frac{1}{j}}} \, dx \right| \leq \frac{\|\nabla \varphi\|_{L^{\infty}(\mathbb{R}^n)}}{j} \int_{B(0,R)} \frac{1}{|x|^{n-1-\frac{1}{j}}} \, dx$$

$$= \frac{\omega_{n-1} \|\nabla \varphi\|_{L^{\infty}(\mathbb{R}^n)}}{j} \int_0^R \rho^{\frac{1}{j}} \, d\rho = \frac{\omega_{n-1} \|\nabla \varphi\|_{L^{\infty}(\mathbb{R}^n)} R^{\frac{1}{j}+1}}{j+1} \xrightarrow[j \to \infty]{} 0. \tag{13.2.14}$$

One more use of (14.9.6) implies

$$\frac{1}{j} \int_{B(0,R)} \frac{1}{|x|^{n-\frac{1}{j}}} \, dx = \frac{\omega_{n-1}}{j} \int_0^R \rho^{\frac{1}{j}-1} \, d\rho$$

$$= \omega_{n-1} R^{\frac{1}{j}} \xrightarrow[j \to \infty]{} \omega_{n-1}. \tag{13.2.15}$$

By combining (13.2.13)–(13.2.15) we obtain

$$\lim_{j \to \infty} \langle f_j, \varphi \rangle = \omega_{n-1} \varphi(0) = \langle \omega_{n-1} \delta, \varphi \rangle. \tag{13.2.16}$$

The desired conclusion now follows.

Exercise 2.133 Note that $f_{\varepsilon} \in L^1_{loc}(\mathbb{R})$ for every $\varepsilon > 0$, hence $f_{\varepsilon} \in \mathcal{D}'(\mathbb{R})$. Also, $\lim_{\varepsilon \to 0^+} f_{\varepsilon}(x) = 0$ for every $x \in \mathbb{R} \setminus \{0\}$. For a given function $\varphi \in C_0^{\infty}(\mathbb{R})$ let $R > 0$ be such that $\operatorname{supp} \varphi \subset (-R, R)$ and for $\varepsilon > 0$ write

$$\langle f_{\varepsilon}, \varphi \rangle = \frac{1}{\pi} \int_{-R}^R \frac{x\varepsilon}{x^2 + \varepsilon^2} \cdot \frac{\varphi(x) - \varphi(0)}{x} \, dx + \frac{\varphi(0)\varepsilon}{\pi} \int_{-R}^R \frac{dx}{x^2 + \varepsilon^2}$$

$$=: I + II. \tag{13.2.17}$$

Then $\left| \frac{x\varepsilon}{x^2+\varepsilon^2} \right| \leq 1/2$ for each $x \in \mathbb{R}$ if $\varepsilon > 0$, so we may apply Lebesgue's Dominated Convergence Theorem to conclude that $\lim_{\varepsilon \to 0^+} I = 0$. Also, integrating the second term and then taking the limit yields $\lim_{\varepsilon \to 0^+} II = \frac{2\varphi(0)}{\pi} \lim_{\varepsilon \to 0^+} \arctan(R/\varepsilon) = \varphi(0)$. In conclusion, $\lim_{\varepsilon \to 0^+} \langle f_\varepsilon, \varphi \rangle = \varphi(0) = \langle \delta, \varphi \rangle$ as desired.

Exercise 2.134 Let $\varphi \in C_0^\infty(\mathbb{R}^n)$ and $\varepsilon > 0$. Using first the change of variables $x = 2\sqrt{\varepsilon}y$ and then Lebesgue's Dominated Convergence Theorem, we have

$$\langle f_\varepsilon, \varphi \rangle = (4\pi\varepsilon)^{-\frac{n}{2}} \int_{\mathbb{R}^n} e^{-\frac{|x|^2}{4\varepsilon}} \varphi(x)\, dx \tag{13.2.18}$$

$$= \pi^{-\frac{n}{2}} \int_{\mathbb{R}^n} e^{-|y|^2} \varphi(2\sqrt{\varepsilon}y)\, dy \xrightarrow[\varepsilon \to 0^+]{} \pi^{-\frac{n}{2}} \int_{\mathbb{R}^n} e^{-|y|^2} \varphi(0)\, dy = \varphi(0).$$

Exercise 2.135 Note that for every $\varepsilon > 0$ we have $|f_\varepsilon^\pm| \le \frac{1}{\varepsilon}$, hence these are functions in $L_{loc}^1(\mathbb{R}) \subset \mathcal{D}'(\mathbb{R})$. Let $\varphi \in C_0^\infty(\mathbb{R})$ and let $R > 0$ be such that $\operatorname{supp}\varphi \subset (-R, R)$. Then

$$\langle f_\varepsilon^\pm(x), \varphi \rangle = \varphi(0) \int_{-R}^R \frac{x \mp i\varepsilon}{x^2 + \varepsilon^2}\, dx + \int_{-R}^R \frac{x \mp i\varepsilon}{x^2 + \varepsilon^2} [\varphi(x) - \varphi(0)]\, dx =: I + II. \tag{13.2.19}$$

Since $\frac{x}{x^2+\varepsilon^2}$ is odd, we further obtain

$$I = \mp 2i\varphi(0) \arctan(R/\varepsilon) \xrightarrow[\varepsilon \to 0^+]{} \mp i\pi\varphi(0) = \langle \mp i\pi\delta, \varphi \rangle. \tag{13.2.20}$$

As for II, since $\lim_{\varepsilon \to 0^+} \frac{x \mp i\varepsilon}{x^2 + \varepsilon^2} = \frac{1}{x}$ for every $x \ne 0$ and

$$\left| \frac{x \mp i\varepsilon}{x^2 + \varepsilon^2} [\varphi(x) - \varphi(0)] \right| \le \frac{|\varphi(x) - \varphi(0)|}{|x|} \in L^1((-R, R)),$$

we may apply Lebesgue's Dominated Convergence Theorem to obtain

$$\lim_{\varepsilon \to 0^+} II = \int_{-R}^R \frac{\varphi(x) - \varphi(0)}{x}\, dx = \left\langle \mathrm{P.V.}\,\frac{1}{x}, \varphi \right\rangle. \tag{13.2.21}$$

Exercise 2.136 You may use the following outline:
(a) Show that for $f \in L^1(\mathbb{R})$ one has $\lim_{j \to \infty} \int_{\mathbb{R}} f(x) \sin(jx)\, dx = 0$ by reducing to the case $f \in C_0^\infty(\mathbb{R})$ based on density arguments.
(b) For $\varphi \in C_0^\infty(\mathbb{R})$ and $R > 0$, write the expression $\langle \frac{1}{\pi} \frac{\sin jx}{x}, \varphi(x) \rangle$ as the sum of two integrals, one over the region $\{x \in \mathbb{R} : |x| \le R\}$, the other over $\{x \in \mathbb{R} : |x| > R\}$, and use (a) to obtain the desired conclusion. Recall that $\int_{\mathbb{R}} \frac{\sin x}{x}\, dx = \pi$.

Exercise 2.137 Let $\varphi \in C_0^\infty(\mathbb{R})$.
(a) In the expression for $\langle f_j, \varphi \rangle$ use the change of variables $x = y/\sqrt{j}$ and then Lebesgue's Dominated Convergence Theorem to conclude $f_j \xrightarrow[j \to \infty]{\mathcal{D}'(\mathbb{R})} \delta$.
(b) Integrate by parts $m + 1$ times to conclude that

$$|\langle f_j, \varphi \rangle| = j^m \left| \int_{\mathbb{R}} \cos(jx)\varphi(x)\,dx \right| \le j^{-1} \int_{\mathbb{R}} |\varphi^{(m+1)}(x)|\,dx \xrightarrow[j\to\infty]{} 0,$$

hence $f_j \xrightarrow[j\to\infty]{\mathscr{D}'(\mathbb{R})} 0$.

(c) Use a reasoning similar to one in the proof of (a) to conclude that $f_j \xrightarrow[j\to\infty]{\mathscr{D}'(\mathbb{R})} \delta$, this time via the change of variables $x = y/j$.

(d) Not convergent since if the function $\varphi \in C_0^\infty(\mathbb{R})$ is such that $\varphi(0) \ne 0$, then $\langle u_j, \varphi \rangle = (-1)^j \varphi(1/j)$ and the sequence $\{(-1)^j \varphi(1/j)\}_{j\in\mathbb{N}}$ is not convergent.

(e) Use the Mean Value Theorem to obtain that $\langle u_j, \varphi \rangle \xrightarrow[j\to\infty]{} \varphi'(0) = \langle -\delta', \varphi \rangle$.

(f) $f_j \xrightarrow[j\to\infty]{\mathscr{D}'(\mathbb{R})} \text{P.V.}\,\frac{1}{x}$.

(g) Use a reasoning similar to the one in the proof of (a) to conclude that $f_j \xrightarrow[j\to\infty]{\mathscr{D}'(\mathbb{R})} \delta$, this time via the change of variables $x = y/j$ and recalling that $\int_{\mathbb{R}} \frac{(\sin y)^2}{y^2}\,dy = \pi$.

(h) Integrate by parts $m + 1$ times and then use Lebesgue's Dominated Convergence Theorem to conclude that $f_j \xrightarrow[j\to\infty]{\mathscr{D}'(\mathbb{R})} 0$.

(j) Integrate by parts twice to obtain

$$\langle u_j, \varphi \rangle = i\varphi(0) + i \int_0^\infty e^{ijx}\varphi'(x)\,dx$$

$$= i\varphi(0) - \frac{1}{j}\varphi'(0) - \frac{1}{j}\int_0^\infty e^{ijx}\varphi''(x)\,dx \xrightarrow[j\to\infty]{} i\varphi(0) = \langle i\delta, \varphi \rangle. \qquad (13.2.22)$$

Exercise 2.138 $(H(\cdot - a))' = \delta_a$ in $\mathscr{D}'(\mathbb{R})$.

Exercise 2.139 $(u_f)' = a\delta_a + H(\cdot - a)$ in $\mathscr{D}'(\mathbb{R})$.

Exercise 2.140 Let $\varphi \in C_0^\infty(\mathbb{R})$. Then integration by parts yields

$$-\int_{\mathbb{R}} f(x)\varphi'(x)\,dx = \int_{-\infty}^0 \sin(x)\varphi'(x)\,dx - \int_0^\infty \sin(x)\varphi'(x)\,dx$$

$$= -\int_{-\infty}^0 \cos(x)\varphi(x)\,dx + \int_0^\infty \cos(x)\varphi(x)\,dx, \qquad (13.2.23)$$

hence $(u_f)' = \cos(x)H(x) - \cos(x)H(-x)$ in $\mathscr{D}'(\mathbb{R})$. Similarly,

$$\int_{\mathbb{R}} f(x)\varphi''(x)\,dx = 2\varphi(0) + \int_{-\infty}^0 \sin(x)\varphi(x)\,dx - \int_0^\infty \sin(x)\varphi(x)\,dx, \qquad (13.2.24)$$

hence $(u_f)'' = 2\delta - \sin(x)H(x) + \sin(x)H(-x)$ in $\mathscr{D}'(\mathbb{R})$.

Exercise 2.141 Let $a, b \in I$ be such that $a < x_0 < b$. Since for every $x \in [a, x_0)$ we have $f(x) = f(a) + \int_a^x f'(t)\,dt$, Lebesgue's Dominated Convergence Theorem gives that $\lim_{x \to x_0^-} f(x)$ exists and equals $f(a) + \int_a^{x_0} f'(t)\,dt$. A similar argument proves that

$\lim_{x \to x_0^+} f(x) = f(b) - \int_{x_0}^b f'(t)\,dt$. Note that $f \in L^1([a,b])$. Suppose now that $\varphi \in C_0^\infty(I)$

satisfies $\operatorname{supp}\varphi \subset [c,d]$ for some $c < x_0 < d$. Then for $\varepsilon > 0$ small enough we have

$$\langle (u_f)', \varphi \rangle = -\langle u_f, \varphi' \rangle = -\int_c^d f(t)\varphi'(t)\,dt$$

$$= -f(x_0 - \varepsilon)\varphi(x_0 - \varepsilon) + \int_c^{x_0-\varepsilon} f'(t)\varphi(t)\,dt - \int_{x_0-\varepsilon}^{x_0+\varepsilon} f(t)\varphi'(t)\,dt$$

$$+ f(x_0 + \varepsilon)\varphi(x_0 + \varepsilon) + \int_{x_0+\varepsilon}^d f'(t)\varphi(t)\,dt. \tag{13.2.25}$$

Send $\varepsilon \to 0^+$ in (13.2.25) and observe that $\lim_{\varepsilon \to 0^+} \int_{x_0-\varepsilon}^{x_0+\varepsilon} f(t)\varphi'(t)\,dt = 0$ by Lebesgue's Dominated Convergence Theorem. The case when x_0 is not in the interior of the support of φ is simpler, and can be handled via a direct integration by parts.

Exercise 2.143 Use Exercise 2.141 and induction.

Exercise 2.144 Use Exercise 2.141 and the fact that since $\{x_k\}_{k \in \mathbb{N}}$ has no accumulation point in I, for each $R > 0$ only finitely many terms of the sequence $\{x_k\}_{k \in \mathbb{N}}$ are contained in $(-R, R)$.

Exercise 2.145 Use Exercise 2.144.

Exercise 2.146 Clearly δ_Σ is well defined and linear. Also, for each compact set $K \subset \mathbb{R}^n$ and every $\varphi \in C_0^\infty(\mathbb{R}^n)$ satisfying $\operatorname{supp}\varphi \subseteq K$ we have

$$|\delta_\Sigma(\varphi)| \leq \sigma(\Sigma \cap K) \sup_{x \in K} |\varphi(x)|. \tag{13.2.26}$$

This shows that $\delta_\Sigma \in \mathcal{D}'(\mathbb{R}^n)$ and has order zero. Also, if $\operatorname{supp}\varphi \cap \Sigma = \varnothing$ then $\langle \delta_\Sigma, \varphi \rangle = 0$, thus $\operatorname{supp}\delta_\Sigma \subseteq \Sigma$.

To prove that $\Sigma \subseteq \operatorname{supp}\delta_\Sigma$, note that if $x^* \in \Sigma$, then there exists a neighborhood $U(x^*)$ of x^* and a local parametrization P of class C^1 near x^* as in (14.6.2)–(14.6.3). In particular, if $u_0 \in O$ is such that $P(u_0) = x^*$, then the vectors $\partial_1 P(u_0)$, ..., $\partial_{n-1} P(u_0)$, are linearly independent. Upon recalling the cross product from (14.6.4), this ensures that

$$c_0 := \|\partial_1 P(u_0) \times \cdots \times \partial_{n-1} P(u_0)\| \neq 0. \tag{13.2.27}$$

Since P is of class C^1, it follows that $\|\partial_1 P(u) \times \cdots \times \partial_{n-1} P(u)\| \geq c_0/2$ for every u belonging to some small open neighborhood $\widetilde{O} \subseteq O$ of u_0 in $\mathbb{R}^{n-1}$. Using the fact that $P : O \to P(O)$ is a homeomorphism (see Proposition 14.46), it follows that there exists some open set $V(x^*)$ in $\mathbb{R}^n$ contained in $U(x^*)$ and containing x^* with the property that $P(\widetilde{O}) = V(x^*) \cap \Sigma$. By further shrinking $\widetilde{O}$ if necessary, there is no loss of generality in assuming that $V(x^*)$ is bounded.

Now consider $0 < r_1 < r_2 < \infty$ such that $V(x^*) \subseteq \overline{B(x^*, r_1)} \subseteq B(x^*, r_2)$. Pick a function $\varphi \in C_0^\infty(B(x^*, r_2))$ with $\varphi \geq 0$, $\varphi \equiv 1$ in a neighborhood of $\overline{B(x^*, r_1)}$ (see Proposition 14.34). Then

$$\langle \delta_\Sigma, \varphi \rangle = \int_\Sigma \varphi(x)\, d\sigma(x) \geq \int_{\Sigma \cap V(x^*)} \varphi(x)\, d\sigma(x) \tag{13.2.28}$$

$$= \int_{\widetilde{O}} \|\partial_1 P(u) \times \cdots \times \partial_{n-1} P(u)\|\, du \geq c_0 |\widetilde{O}| > 0. \tag{13.2.29}$$

In a similar manner, for each function g satisfying $g \in L^\infty(K \cap \Sigma)$ for any compact set $K \subseteq \mathbb{R}^n$, we have $g\delta_\Sigma \in \mathcal{D}'(\mathbb{R}^n)$ and for each compact K one has

$$|(g\delta_\Sigma)(\varphi)| \leq \|g\|_{L^\infty(K)} \sigma(\Sigma \cap K) \sup_{x \in K} |\varphi(x)|, \quad \forall \varphi \in C_0^\infty(\mathbb{R}^n), \ \ \mathrm{supp}\, \varphi \subseteq K.$$

Exercise 2.147 Use integration by parts (see Theorem 14.60) and Exercise 2.146.

Exercise 2.148 Use the definition of distributional derivative, integration by parts (cf. Theorem 14.60), and Exercise 2.146.

Exercise 2.149 Let $\varphi \in C_0^\infty(\mathbb{R}^n)$ be such that $\mathrm{supp}\, \varphi \cap \partial B(0, R) = \varnothing$. In this scenario, we have $\frac{\varphi}{|x|^2 - R^2} \in C_0^\infty(\mathbb{R}^n)$, hence we may write

$$\langle u, \varphi \rangle = \left\langle (|x|^2 - R^2)u, \frac{\varphi}{|x|^2 - R^2} \right\rangle = 0.$$

This proves that $\mathrm{supp}\, u \subseteq \partial B(0, R)$, thus u is compactly supported. Examples of distributions satisfying the given equation include $\delta_{\partial B(0, R)}$ and δ_{x_0}, for any $x_0 \in \mathbb{R}^n$ with $|x_0| = R$.

Exercise 2.150 Use Example 2.56.

Exercise 2.151 The derivative of order m is equal to:
(a) $\mathrm{sgn}(x)$ if $m = 1$ and $2\delta^{(m-2)}$ if $m \geq 2$;
(b) $2\delta^{(m-1)}$;
(c) $\cos x\, H + \sum_{j=1}^n [\delta^{(4j-1)} - \delta^{(4j-3)}]$ if $m = 4j$, $j \in \mathbb{N}_0$; $-\sin x\, H + \delta + \sum_{j=1}^n [\delta^{(4j)} - \delta^{(4j-2)}]$ if $m = 4j + 1$, $j \in \mathbb{N}_0$; $-\cos x\, H + \delta' + \sum_{j=1}^n [\delta^{(4j+1)} - \delta^{(4j-1)}]$ if $m = 4j$, $j \in \mathbb{N}_0$;
$\sin x\, H + \sum_{j=0}^n [\delta^{(4j+2)} - \delta^{(4j)}]$ if $m = 4j$, $j \in \mathbb{N}_0$; the convention is that a sum is void if the upper bound is lower than the lower bound in the summation sign; ·
(d) $(\sin x\, H)' = \cos x\, H$ and use (c);
(e) $-\delta_1 + \delta_{-1} + 2x\chi_{[-1,1]}$ if $m = 1$, $-(\delta_1)' + (\delta_{-1})' - 2\delta_1 - 2\delta_{-1} + 2\chi_{[-1,1]}$ if $m = 2$, and $\delta_{-1}^{(m-1)} - \delta_1^{(m-1)} - 2\delta_1^{(m-2)} - 2\delta_{-1}^{(m-2)} - 2\delta_1^{(m-3)} + 2\delta_{-1}^{(m-3)}$ if $m \geq 3$.

Exercise 2.152 For $\varphi \in C_0^\infty(\mathbb{R}^2)$ use the change of variables $u = x + y$, $v = x - y$ and the reasoning in (1.1.6)–(1.1.8) with $\psi(u, v) := \varphi\left(\frac{u+v}{2}, \frac{u-v}{2}\right)$ for $u \in [2, 4]$, $v \in [0, 2]$;

to write

$$\langle (\partial_1^2 - \partial_2^2)\chi_A, \varphi \rangle = \int \int_A [(\partial_1^2 \varphi)(x, y) - (\partial_2^2 \varphi)(x, y)] \, dx \, dy$$

$$= 2 \int_2^4 \int_0^2 \partial_u \partial_v \psi(u, v) \, dv \, du$$

$$= 2[\psi(4, 2) - \psi(2, 2) - \psi(4, 0) + \psi(2, 0)]$$

$$= 2[\varphi(3, 1) - \varphi(2, 0) - \varphi(2, 2) + \varphi(1, 1)]$$

$$= \langle 2[\delta_{(3,1)} - \delta_{(2,0)} - \delta_{(2,2)} + \delta_{(1,1)}], \varphi \rangle. \qquad (13.2.30)$$

Exercise 2.153 Fix $\varphi \in C_0^\infty(\mathbb{R}^2)$. Then, using the change of variables $x = t + y$ we obtain

$$\langle \partial_1(u_f), \varphi \rangle = - \int \int_{\mathbb{R}^2} \chi_{[0,1]}(x - y)\partial_1 \varphi(x, y) \, dx \, dy$$

$$= - \int \int_{\mathbb{R}^2} \chi_{[0,1]}(t)(\partial_1 \varphi)(t + y, y) \, dt \, dy$$

$$= \int_{\mathbb{R}} [\varphi(y, y) - \varphi(1 + y, y)] \, dy. \qquad (13.2.31)$$

Similarly,

$$\langle \partial_2(u_f), \varphi \rangle = \int_{\mathbb{R}} [\varphi(x, x - 1) - \varphi(x, x)] \, dx. \qquad (13.2.32)$$

Combining (13.2.31)–(13.2.32) it follows that $\partial_1(u_f) - \partial_2(u_f) = 0$ in $\mathcal{D}'(\mathbb{R}^2)$ and, in turn, that $\partial_1^2(u_f) - \partial_2^2(u_f) = (\partial_1 + \partial_2)(\partial_1(u_f) - \partial_2(u_f)) = 0$ in $\mathcal{D}'(\mathbb{R}^2)$.

Exercise 2.156 The uniqueness statement is a consequence of Proposition 2.52. Let K be a compact set in $\mathbb{R}^n$ such that $K \subset \Omega$. Refine $\{\Omega_j\}_{j \in I}$ to a finite subcover $\{\Omega_{\ell_k}\}_{k=1}^N$, with $\ell_k \in I$ for $k = 1, \dots, N$, of K. Consider a partition of unity $\{\varphi_k\}_{1 \le k \le N}$ subordinate to the cover $\{\Omega_{\ell_k}\}_{k=1}^N$ of K, as given by Theorem 14.37. Hence, $\varphi_k \in C_0^\infty(\Omega)$ with $\operatorname{supp} \varphi_k \subset \Omega_{\ell_k}$ for each k, and $\sum_{k=1}^N \varphi_k = 1$ in a neighborhood of K. Next, for each function $\varphi \in C_0^\infty(\Omega)$ with $\operatorname{supp} \varphi \subseteq K$ define $u_K(\varphi) := \sum_{k=1}^N \langle u_{\ell_k}, \varphi_k \varphi \rangle$. Show that $\sum_{k=1}^N \langle u_{\ell_k}, \varphi_k \varphi \rangle$ is independent of the cover of K chosen and of the partition of unity, thus $u_K : \mathcal{D}_K(\Omega) \to \mathbb{C}$ is well defined. The map u_K is clearly linear. Now set $u : \mathcal{D}(\Omega) \to \mathbb{C}$ to be the map given by $u(\varphi) := u_K(\varphi)$ for each $\varphi \in C_0^\infty(\Omega)$ such that $\operatorname{supp} \varphi \subseteq K$. Show that this map is well defined, satisfies $u \in \mathcal{D}'(\Omega)$, and $u|_{\Omega_j} = u_j$ in $\mathcal{D}'(\Omega_j)$ for every $j \in I$.

Exercise 2.157 Fix $\varphi_0 \in C_0^\infty(\mathbb{R}^n)$ with the property that $\int_{\mathbb{R}^n} \varphi_0(x) \, dx = 1$. Next, let $\varphi \in C_0^\infty(\mathbb{R}^n)$ be arbitrary and set $\lambda := \int_{\mathbb{R}^n} \varphi(x) \, dx$. Hence, if $\psi := \varphi - \lambda \varphi_0$, then $\psi \in C_0^\infty(\mathbb{R}^n)$ and $\int_{\mathbb{R}^n} \psi(x) \, dx = 0$, so

$$0 = \langle u, \psi \rangle = \langle u, \varphi \rangle - \lambda \langle u, \varphi_0 \rangle. \tag{13.2.33}$$

Consequently, $\langle u, \varphi \rangle = \lambda \langle u, \varphi_0 \rangle = \langle c, \varphi \rangle$, where $c := \langle u, \varphi_0 \rangle \in \mathbb{C}$.

Exercise 2.158 You may proceed by completing the following steps. Let $\varphi \in C_0^\infty(\mathbb{R}^n)$.

Step I. Prove that $\int_{\mathbb{R}^n} \varphi(x) \, dx = 0$ if and only if there exist $\varphi_1, \ldots, \varphi_n \in C_0^\infty(\mathbb{R}^n)$ such that $\varphi = \sum_{j=1}^n \partial_j \varphi_j$. Do so by induction over n. Corresponding to $n = 1$ show that if $a, b \in \mathbb{R}$ satisfy $a < b$ and for $\varphi \in C_0^\infty((a, b))$ one defines $\phi_1(x) := \int_a^x \varphi(t) \, dt$, $x \in (a, b)$, then $\phi_1 \in C_0^\infty((a, b))$ if and only if $\int_a^b \varphi(x) \, dx = 0$.

Step II. Show that the statement from Step I continues to hold if $\mathbb{R}^n$ is replaced by $(a_1, b_1) \times \cdots \times (a_n, b_n)$, where $a_j, b_j \in \mathbb{R}$, $a_j < b_j$ for each $j = 1, \ldots, n$.

Step III. Fix $a_j, b_j \in \mathbb{R}$, $a_j < b_j$ for $j = 1, \ldots, n$ and consider the n-dimensional rectangle $Q := (a_1, b_1) \times \cdots \times (a_n, b_n)$. Let $\varphi_0 \in C_0^\infty(Q)$ be such that $\int_Q \varphi_0 \, dx = 1$. Then, if $\varphi \in C_0^\infty(Q)$ is arbitrary, the function $\psi := \varphi - [\int_Q \varphi \, dx] \varphi_0$ belongs to $C_0^\infty(Q)$ and satisfies $\int_Q \psi \, dx = 0$. As such, Step II applies and shows that there exist $\varphi_1, \ldots, \varphi_n \in C_0^\infty(Q)$ such that $\varphi = [\int_Q \varphi \, dx] \varphi_0 + \sum_{j=1}^n \partial_j \varphi_j$. In turn, this permits us to write

$$\langle u, \varphi \rangle = \langle u, \varphi_0 \rangle \int_Q \varphi \, dx - \sum_{j=1}^n \langle \partial_j u, \varphi_j \rangle = \langle \langle u, \varphi_0 \rangle, \varphi \rangle, \tag{13.2.34}$$

which shows if $c_Q := \langle u, \varphi_0 \rangle \in \mathbb{C}$, then $u\big|_Q = c_Q$ in $\mathcal{D}'(Q)$.

Step IV. Since Ω is connected and open, it is path connected. Now combine this with the fact that u is locally constant (as proved in Step III) to finish the proof.

Exercise 2.159 By Proposition 2.81, it suffices to prove that there exists $v \in \mathcal{D}'(\mathbb{R}^{n-1})$ such that $\langle u, \varphi \otimes \psi \rangle = \langle v, \varphi \rangle \langle \delta, \psi \rangle$ for every $\varphi \in C_0^\infty(\mathbb{R}^{n-1})$ and every $\psi \in C_0^\infty(\mathbb{R})$. Fix $\varphi \in C_0^\infty(\mathbb{R}^{n-1})$, $\psi \in C_0^\infty(\mathbb{R})$, and consider some $\psi_0 \in C_0^\infty(\mathbb{R})$ with the property that $\psi_0(0) = 1$. Then there exists $h \in C_0^\infty(\mathbb{R})$ satisfying $\psi(x_n) - \psi(0)\psi_0(x_n) = x_n h(x_n)$ for every $x_n \in \mathbb{R}$. This and the fact that $x_n u = 0$ allows us to write

$$\langle u, \varphi \psi \rangle = \langle u, \varphi(\psi - \psi(0)\psi_0) \rangle + \langle u, \varphi \psi_0 \rangle \psi(0)$$

$$= \langle x_n u, \varphi h \rangle + \langle u, \varphi \psi_0 \rangle \psi(0) = \langle u, \varphi \psi_0 \rangle \langle \delta, \psi \rangle. \tag{13.2.35}$$

Now define $v : \mathcal{D}(\mathbb{R}^{n-1}) \to \mathbb{C}$ by $v(\varphi) := \langle u, \varphi \psi_0 \rangle$ for $\varphi \in C_0^\infty(\mathbb{R}^{n-1})$, and show that $v \in \mathcal{D}(\mathbb{R}^{n-1})$. By (13.2.35) this v does the job.

Exercise 2.160 Fix $\psi \in C_0^\infty(\mathbb{R})$ with the property that $\psi(0) = 1$. Use Exercise 2.159 and induction to show that $u = c \, \delta(x_1) \otimes \cdots \otimes \delta(x_n)$, where $c := \langle u, \psi \otimes \cdots \otimes \psi \rangle \in \mathbb{C}$.

Exercise 2.161 Fix $\psi \in C_0^\infty(\mathbb{R})$ with the property that $\int_\mathbb{R} \psi(s)\,\mathrm{d}s = 1$. Given any function $\varphi \in C_0^\infty(\mathbb{R}^n)$, at each point $x = (x', x_n) \in \mathbb{R}^{n-1} \times \mathbb{R}$ we may write

$$\varphi(x) = \varphi(x) - \psi(x_n) \int_\mathbb{R} \varphi(x', s)\,\mathrm{d}s + \psi(x_n) \int_\mathbb{R} \varphi(x', s)\,\mathrm{d}s$$

$$= \partial_{x_n}\Big[\int_{-\infty}^{x_n} [\varphi(x', t) - \psi(t) \int_\mathbb{R} \varphi(x', s)\,\mathrm{d}s]\,\mathrm{d}t \Big] + \psi(x_n) \int_\mathbb{R} \varphi(x', s)\,\mathrm{d}s.$$

Since $\partial_n u = 0$ in $\mathcal{D}'(\mathbb{R}^n)$, this yields

$$\langle u, \varphi \rangle = \Big\langle u, \psi(x_n) \int_\mathbb{R} \varphi(x', s)\,\mathrm{d}s \Big\rangle. \tag{13.2.36}$$

In particular, if $\varphi = \varphi_1 \otimes \varphi_2$ for some $\varphi_1 \in C_0^\infty(\mathbb{R}^{n-1})$ and some $\varphi_2 \in C_0^\infty(\mathbb{R})$, then

$$\langle u, \varphi \rangle = \Big\langle u, \varphi_1 \otimes \psi\Big(\int_\mathbb{R} \varphi_2(s)\,\mathrm{d}s \Big) \Big\rangle = \langle u, \varphi_1 \otimes \psi \rangle \langle 1, \varphi_2 \rangle. \tag{13.2.37}$$

Define $v : C_0^\infty(\mathbb{R}^{n-1}) \to \mathbb{C}$ by $v(\theta) := \langle u, \theta \otimes \psi \rangle$ for every $\theta \in C_0^\infty(\mathbb{R}^{n-1})$. Then $v \in \mathcal{D}'(\mathbb{R}^{n-1})$ and $u(x', x_n) = v(x') \otimes 1$ when restricted to $C_0^\infty(\mathbb{R}^{n-1}) \otimes C_0^\infty(\mathbb{R})$. The desired conclusion follows by recalling Proposition 2.81.

Exercise 2.164 Fix $j \in \mathbb{N}$ and note that for each $x = (x_1, \ldots, x_n) \in \mathbb{R}^n$ we may write

$$f_j(x) = \frac{1}{2\pi} \int_{-j}^j e^{ix_1\xi_1}\,\mathrm{d}\xi_1 \otimes \cdots \otimes \frac{1}{2\pi} \int_{-j}^j e^{ix_n\xi_n}\,\mathrm{d}\xi_n. \tag{13.2.38}$$

Also, for each $j \in \mathbb{N}$ and each $k \in \{1, \ldots, n\}$, the fundamental theorem of calculus gives

$$\int_{-j}^j e^{ix_k\xi_k}\,\mathrm{d}\xi_k = \int_{-j}^j \cos(x_k\xi_k)\,\mathrm{d}\xi_k = 2\frac{\sin(jx_k)}{x_k}, \quad \text{assuming} \quad x_k \neq 0. \tag{13.2.39}$$

Now use (13.2.38), (13.2.39), Exercise 2.136, and part *(d)* in Theorem 2.89 to finish the proof.

Exercise 2.165 Note that $u = -\delta$ is a solution for the equation in (1). Hence, if u is any other solution of the equation $(x - 1)u = \delta$, then setting $v := u + \delta$ it follows that $(x - 1)v = 0$ in $\mathcal{D}'(\mathbb{R})$. Next use this and the reasoning from Example 2.76 to conclude that the general solution for the equation in (1) is $-\delta + c\,\delta_1$, with $\dot{c} \in \mathbb{C}$.

Fix $\psi \in C_0^\infty(\mathbb{R})$ with the property that $\psi(0) = 1$. Show that any solution u of the equation in (2) satisfies

$$\langle u, \varphi \rangle = \int_\mathbb{R} a(x)\frac{\varphi(x) - \varphi(0)\psi(x)}{x}\,\mathrm{d}x + \langle\langle u, \psi \rangle \delta, \varphi \rangle, \quad \forall \varphi \in C_0^\infty(\mathbb{R}) \tag{13.2.40}$$

and use this to obtain that the general solution of the equation in (2) is $v_a + c\delta$, $c \in \mathbb{C}$, where v_a is the distribution given by

$$\langle v_a, \varphi \rangle := \int_{\mathbb{R}} a(x) \frac{\varphi(x) - \varphi(0)\psi(x)}{x} \, dx, \quad \forall \varphi \in C_0^\infty(\mathbb{R}).$$

Similarly, any solution u of the equation in (3) satisfies

$$\langle u, \varphi \rangle = \left\langle v, \frac{\varphi - \varphi(0)\psi}{x} \right\rangle + \langle \langle u, \psi \rangle \delta, \varphi \rangle, \quad \forall \varphi \in C_0^\infty(\mathbb{R}), \tag{13.2.41}$$

so the general solution for (3) is $w + c\delta$, where $c \in \mathbb{C}$ and w is the distribution given by

$$\langle w, \varphi \rangle := \left\langle v, \frac{\varphi - \varphi(0)\psi}{x} \right\rangle, \quad \forall \varphi \in C_0^\infty(\mathbb{R}).$$

Exercise 2.166 (a) Since $H \in L_{loc}^1(\mathbb{R})$ and

$$M_R := \{(x, y) \in [0, \infty) \times [0, \infty) : |x + y| \le R\}$$

is a compact set in $\mathbb{R}^2$ for each $R > 0$, by Remark 2.93 and Theorem 2.94, it follows that $H * H$ is well defined and belongs to $\mathcal{D}'(\mathbb{R})$. Fix a compact set K in $\mathbb{R}$ and let $R > 0$ be such that $K \subset (-R, R)$. Pick now $\varphi \in C_0^\infty(\mathbb{R})$ with supp $\varphi \subseteq K$, and suppose that $\psi \in C_0^\infty(\mathbb{R}^2)$ satisfies $\psi \equiv 1$ on the set

$$M_K := \{(x, y) \in [0, \infty) \times [0, \infty) : x + y \in K\}.$$

Then

$$\langle H * H, \varphi \rangle = \int_{\mathbb{R}} \int_{\mathbb{R}} H(x)H(y)\varphi(x + y)\psi(x, y) \, dy \, dx$$

$$= \int_0^\infty \int_0^\infty \varphi(x + y)\psi(x, y) \, dy \, dx \tag{13.2.42}$$

Note that

$$\{(x, y) \in [0, \infty) \times [0, \infty) : |x + y| \le R\} \tag{13.2.43}$$
$$= \{(x, y) : 0 \le x \le R, \, 0 \le y \le R - x\},$$

hence

$$\langle H * H, \varphi \rangle = \int_0^R \int_0^{R-x} \varphi(x + y)\psi(x, y) \, dy \, dx = \int_0^R \int_0^{R-x} \varphi(x + y) \, dy \, dx$$

$$= \int_0^R \int_x^R \varphi(z) \, dz \, dx = \int_0^R \int_0^z \varphi(z) \, dx \, dz = \int_0^R z\varphi(z) \, dz$$

$$= \langle xH, \varphi \rangle. \tag{13.2.44}$$

Alternatively, one may use Remark 2.95, to observe that $H * H$ in the distributional sense is the distribution given by the function obtained by taking the convolution, in the sense of (2.8.2), of the function H with itself. Hence, $(H * H)(x) = \int_0^\infty \chi_{[0,\infty)}(x - y)\,dy = xH(x)$ for every $x \in \mathbb{R}$.

(b) $-xH(-x)$ (c) $(x^2 - 2 + 2\cos x)H(x)$ (d) $\frac{x^2}{2}H(x) - \frac{(x-1)^2}{2}H(x - 1)$

(e) Exercise 2.146 tells us that $\delta_{\partial B(0,r)}$ is compactly supported, so the given convolution is well defined. Also, by Exercise 2.103, $|x|^2 * \delta_{\partial B(0,r)} \in C^\infty(\mathbb{R}^n)$ and equals

$$\langle \delta_{\partial B(0,r)}, |x - y|^2 \rangle = \int_{\partial B(0,r)} |x - y|^2\,d\sigma(y) = r^{n-1} \int_{S^{n-1}} |x - r\omega|^2\,d\sigma(\omega)$$

$$= r^{n-1} \int_{S^{n-1}} [r^2 + |x|^2 - 2rx \cdot \omega]\,d\sigma(\omega)$$

$$= (r^{n+1} + |x|^2 r^{n-1})\omega_{n-1}, \qquad \forall\, x \in \mathbb{R}^n. \tag{13.2.45}$$

For the second equality in (13.2.45) we used the change of variables $y = r\omega$, $\omega \in S^{n-1}$, while for the last equality we used the fact that since $x \cdot \omega$ as a function in ω is odd, its integral over S^{n-1} is zero.

Exercise 2.167 $u_j \xrightarrow[j\to\infty]{\mathcal{D}'(\mathbb{R}^n)} 0$, $v_j \xrightarrow[j\to\infty]{\mathcal{D}'(\mathbb{R}^n)} 0$. Given that $u_j, v_j \in \mathcal{E}'(\mathbb{R}^n)$, we deduce that $u_j * v_j \in \mathcal{E}'(\mathbb{R}^n)$ for each $j \in \mathbb{N}$. Also, $u_j * v_j \xrightarrow[j\to\infty]{\mathcal{D}'(\mathbb{R}^n)} \delta$.

Exercise 2.168 The limits in (a) and (b) do not exist. Since for each $j \in \mathbb{N}$ we have $f_j \in \mathcal{E}'(\mathbb{R})$ and $g_j \in C^\infty(\mathbb{R})$, Exercise 2.103 may be used to conclude that $f_j * g_j \in C^\infty(\mathbb{R})$ and that $f_j * g_j = 1$ for every j. Thus, $f_j * g_j \xrightarrow[j\to\infty]{\mathcal{D}'(\mathbb{R}^n)} 1$.

Exercise 2.169 Note that Λ is well defined based on Proposition 2.102. You may want to use Theorem 14.6 to prove the continuity of Λ.

Exercise 2.170 For $f : \mathbb{R}^n \to \mathbb{C}$ set $f^\vee(x) := f(-x)$, $x \in \mathbb{R}^n$. If $u \in \mathcal{D}'(\mathbb{R}^n)$ is such that $u * \varphi = 0$ for every $\varphi \in C_0^\infty(\mathbb{R}^n)$, then $0 = (u * \varphi^\vee)(0) = \langle u, \varphi \rangle$ for every $\varphi \in C_0^\infty(\mathbb{R}^n)$, thus $u = 0$. This proves the uniqueness part in the statement. As for existence, given Λ as specified, define $u_0 : \mathcal{D}(\mathbb{R}^n) \to \mathbb{C}$ by $u_0(\varphi) := \langle \delta, \Lambda(\varphi^\vee) \rangle$ for $\varphi \in C_0^\infty(\mathbb{R}^n)$. Being a composition of linear and continuous maps, u_0 is linear and continuous, thus $u_0 \in \mathcal{D}'(\mathbb{R}^n)$. Also, if $\varphi \in C_0^\infty(\mathbb{R}^n)$ is fixed, we have

$$(u_0 * \varphi)(x) = \langle u_0, \varphi(x - \cdot) \rangle = \langle u_0, (\varphi^\vee)(\cdot - x) \rangle = \langle \delta, \Lambda(\varphi(\cdot + x)) \rangle$$

$$= \langle \delta, (\Lambda\varphi)(\cdot + x) \rangle = \Lambda(\varphi)(x), \qquad \forall\, x \in \mathbb{R}^n. \tag{13.2.46}$$

For the first equality in (13.2.46) we used Proposition 2.102, the third equality is based on the definition of u_0, while the forth equality uses the fact that Λ is commutes with translations.

Exercise 2.171 From hypotheses we obtain $\langle u, P \rangle = 0$ for every polynomial P in $\mathbb{R}^n$. Now use Lemma 2.83 to conclude that $\langle u, \varphi \rangle = 0$ for every $\varphi \in C_0^\infty(\mathbb{R}^n)$. Let $\psi \in C_0^\infty(\mathbb{R}^n)$ be such that $\psi \equiv 1$ in a neighborhood of supp u. Then for every function $\varphi \in C^\infty(\mathbb{R}^n)$ we have $0 = \langle u, \psi\varphi \rangle = \langle u, \varphi \rangle$ since the support condition on u implies $\langle u, (1 - \psi)\varphi \rangle = 0$.

Exercise 2.172 Fix $\varphi \in C^\infty(\mathbb{R})$ and write

$$\sum_{j=1}^k \varphi(\tfrac{1}{j}) - k\varphi(0) - \varphi'(0)\ln k$$

$$= \sum_{j=1}^k \left[\varphi(\tfrac{1}{j}) - \varphi(0) - \tfrac{1}{j}\varphi'(0) \right] + \varphi'(0)\left[\sum_{j=1}^k \tfrac{1}{j} - \ln k \right]. \qquad (13.2.47)$$

Since

$$\left| \varphi(\tfrac{1}{j}) - \varphi(0) - \tfrac{1}{j}\varphi'(0) \right| \le \tfrac{1}{j^2}\|\varphi''\|_{L^\infty([0,1])} \quad \forall j \in \mathbb{N},$$

taking the limit as $k \to \infty$ in (13.2.47) (also recall Euler's constant γ from (4.6.24)) we obtain

$$u(\varphi) = \sum_{j=1}^\infty \left[\varphi(\tfrac{1}{j}) - \varphi(0) - \tfrac{1}{j}\varphi'(0) \right] + \gamma\,\varphi'(0).$$

Now apply Fact 2.63 with $K := [0, 1]$, $m := 2$ and $C := \sum_{j=1}^\infty \tfrac{1}{j^2} + \gamma$ to conclude $u \in \mathcal{E}'(\mathbb{R})$. Also, show that supp $u = \{0\} \cup \left\{ \tfrac{1}{j} : j \in \mathbb{N} \right\}$.

Exercise 2.173 Note that $f_j \in L_{comp}^1(\mathbb{R})$ for each $j \in \mathbb{N}$, hence $\{f_j\}_{j\in\mathbb{N}} \subset \mathcal{E}'(\mathbb{R})$. If $\varphi \in C^\infty(\mathbb{R})$ then we may write

$$\left| \langle f_j, \varphi \rangle - \langle \delta, \varphi \rangle \right| = \left| \int_{-1/j}^{1/j} \frac{j}{2}(\varphi(x) - \varphi(0))\, dx \right| \le \frac{j}{2} \int_{-1/j}^{1/j} |\varphi(x) - \varphi(0)|\, dx$$

$$\le \frac{1}{j}\|\varphi'\|_{L^\infty([-1,1])} \to 0 \quad \text{as} \quad j \to \infty. \qquad (13.2.48)$$

Exercise 2.174 Since $f_j \in L_{comp}^1(\mathbb{R})$ for each $j \in \mathbb{N}$, we have $\{f_j\}_{j\in\mathbb{N}} \subset \mathcal{E}'(\mathbb{R})$. Let $\varphi \in C_0^\infty(\mathbb{R})$ and suppose $R \in (0, \infty)$ is such that supp $\varphi \subset (-R, R)$. Then for $j \in \mathbb{N}$ with $j \ge R$ we have

$$|\langle f_j, \varphi \rangle| = \left| \int_{-j}^{j} \frac{\varphi(x)}{j}\, dx \right| = \left| \int_{-R}^{R} \frac{\varphi(x)}{j}\, dx \right| \le \frac{2R\|\varphi\|_{L^\infty(\mathbb{R})}}{j} \qquad (13.2.49)$$

which proves that $f_j \xrightarrow[j\to\infty]{\mathcal{D}'(\mathbb{R}^n)} 0$. Suppose next that there exists a distribution $u \in \mathcal{E}'(\mathbb{R}^n)$ such that $f_j \xrightarrow[j\to\infty]{\mathcal{E}'(\mathbb{R}^n)} u$. In particular, we have $\lim_{j\to\infty}\langle f_j, \varphi \rangle = \langle u, \varphi \rangle$ for every

$\varphi \in C_0^\infty(\mathbb{R})$. Together with what we have proved before, this forces $u = 0$. However, $\langle f_j, 1 \rangle = 2$ for every $j \in \mathbb{N}$, which contradicts the fact that $u = 0$. Thus, $\{f_j\}_{j \in \mathbb{N}}$ does not converge in $\mathcal{E}'(\mathbb{R}^n)$.

Exercise 2.175 Since $f_j \in C_0^\infty(\mathbb{R})$ for each $j \in \mathbb{N}$, we have $\{f_j\}_{j \in \mathbb{N}} \subset \mathcal{E}'(\mathbb{R})$. Let $\varphi \in C^\infty(\mathbb{R}^n)$. Then using the change of variables $jx = y$ we may write

$$|\langle f_j, \varphi \rangle - \langle \delta, \varphi \rangle| = \left| \int_{\mathbb{R}^n} j^n \psi(jx) \varphi(x) \, dx - \varphi(0) \int_{\mathbb{R}^n} \psi(x) \, dx \right|$$

$$= \left| \int_{\mathbb{R}^n} \psi(y) \varphi(y/j) \, dy - \varphi(0) \int_{\mathbb{R}^n} \psi(x) \, dx \right|$$

$$\leq \int_{\operatorname{supp} \psi} |\varphi(y/j) - \varphi(0)| |\psi(y)| \, dy \to 0 \quad \text{as} \quad j \to \infty, \qquad (13.2.50)$$

where the convergence is based on Lebesgue's Dominated Convergence Theorem.

Exercise 2.176 For each $k \in \{1, \ldots, n\}$ consider the sequence $\{\langle \delta_{x_j}, \varphi_k \rangle\}_{j \in \mathbb{N}}$ where the function φ_k is defined by $\varphi_k(x) := x_k$ for each $x = (x_1 \ldots, x_n) \in \mathbb{R}^n$.

13.3 Solutions to Exercises from Section 3.3

Exercise 3.33 Let $f \in \mathcal{S}(\mathbb{R}^n)$ and $\alpha, \beta \in \mathbb{N}_0^n$. Let $N := |\alpha| + 1$. Then Exercise 3.5 implies the existence of a constant $C \in (0, \infty)$ such that $|\partial^\beta f(x)| \leq C(1 + |x|)^{-N}$ for all $x \in \mathbb{R}^n$. Hence,

$$\sup_{|x| \geq R} |x^\alpha \partial^\beta f(x)| \leq \sup_{|x| \geq R} \frac{C|x^\alpha|}{(1 + |x|)^{-N}} \leq CR^{-1}. \qquad (13.3.1)$$

The desired conclusion follows after taking the limit as $R \to \infty$.

Exercise 3.34 Let $R > 0$ be such that $\operatorname{supp} \varphi \subset B(0, R)$. Then $\varphi_j \in C_0^\infty(\mathbb{R}^n)$ and $\operatorname{supp} \varphi_j \subset B(0, jR)$ for each $j \in \mathbb{N}$. Since there is no compact $K \subset \mathbb{R}^n$ such that $\operatorname{supp} \varphi_j \subset K$ for all $j \in \mathbb{N}$, the sequence $\{\varphi_j\}_{j \in \mathbb{N}}$ does not converge in $\mathcal{D}(\mathbb{R}^n)$. Also, for each $\alpha \in \mathbb{N}_0^n$ we have $\partial^\alpha \varphi_j(x) = \frac{1}{j^{|\alpha|}} e^{-j} (\partial^\alpha \varphi)(x/j)$. Hence if we also take $\beta \in \mathbb{N}_0^n$ arbitrary, then

$$\sup_{x \in \mathbb{R}^n} |x^\beta \partial^\alpha \varphi_j(x)| \leq e^{-j} j^{-|\alpha|} \sup_{x \in B(0, jR)} |x^\beta (\partial^\alpha \varphi)(x/j)|$$

$$\leq e^{-j} j^{-|\alpha|} \sup_{x \in B(0, jR)} |x^\beta| \|\partial^\alpha \varphi\|_{L^\infty(B(0,R))}$$

$$\leq e^{-j} j^{|\beta| - |\alpha|} R^{|\beta|} \|\partial^\alpha \varphi\|_{L^\infty(B(0,R))} \xrightarrow[j \to \infty]{} 0, \qquad (13.3.2)$$

which implies that $\{\varphi_j\}_{j \in \mathbb{N}}$ converges to zero in $\mathcal{S}(\mathbb{R}^n)$.

Exercise 3.35 For each $\alpha \in \mathbb{N}_0^n$ we have $\partial^\alpha \varphi_j(x) = \frac{1}{j^{|\alpha|+1}}(\partial^\alpha \varphi)(x/j)$, for each $x \in \mathbb{R}^n$. Consequently, for every compact subset K of $\mathbb{R}^n$ we may write

$$\sup_{x \in K} |\partial^\alpha \varphi_j(x)| \le j^{-|\alpha|-1} \sup_{x \in K} |(\partial^\alpha \varphi)(x/j)|$$

$$\le j^{-|\alpha|-1} \|\partial^\alpha \varphi\|_{L^\infty(\mathbb{R}^n)} \xrightarrow[j \to \infty]{} 0, \qquad (13.3.3)$$

which proves that $\varphi_j \xrightarrow[j \to \infty]{\mathcal{E}(\mathbb{R}^n)} 0$. Moreover, if

$$O := \{x = (x_1, \ldots, x_n) \in \mathbb{R}^n : x_j \ne 0 \text{ for } j = 1, \ldots, n\}$$

we claim that there exists $x_* \in O$ with the property that $\varphi(x_*) \ne 0$. Indeed, if this were not to be the case, we would have $\varphi = 0$ in O, hence $\varphi = 0$ in $\mathbb{R}^n$ since φ is continuous and O is dense in $\mathbb{R}^n$. Having fixed such a point x_* we then proceed to estimate

$$\sup_{x \in \mathbb{R}^n} |x^\beta \varphi_j(x)| = j^{-1} \sup_{x \in \mathbb{R}^n} |x^\beta \varphi(x/j)| = j^{|\beta|-1} \sup_{x \in \mathbb{R}^n} |x^\beta \varphi(x)|$$

$$\ge j^{|\beta|-1} |x_*^\beta| |\varphi(x_*)| \xrightarrow[j \to \infty]{} \infty, \qquad (13.3.4)$$

for $\beta \in \mathbb{N}_0^n$ satisfying $|\beta| > 1$. Thus, $\{\varphi_j\}_{j \in \mathbb{N}}$ does not converge in $\mathcal{S}(\mathbb{R}^n)$.

Exercise 3.36 Clearly $\{\varphi_j\}_{j \in \mathbb{N}} \subset C^\infty(\mathbb{R})$. If $m \in \mathbb{N}$ then for each $j \in \mathbb{N}_0$ we have

$$\varphi_j^{(m)}(x) = \frac{1}{j} \sum_{k=0}^m \frac{m!}{k!(m-k)!} \psi^{(k)}(x) \theta^{(m-k)}(x/j) \frac{1}{j^{m-k}}, \qquad \forall x \in \mathbb{R}. \qquad (13.3.5)$$

Hence, if $\ell \in \mathbb{N}_0$, then using the properties of θ and ψ we may write

$$\sup_{x \in \mathbb{R}} |x^\ell \varphi_j^{(m)}(x)| = \frac{1}{j} \sup_{x \in \mathbb{R}} \left| \sum_{k=0}^m \frac{m!}{k!(m-k)!} \psi^{(k)}(x) \theta^{(m-k)}(x/j) \frac{x^\ell}{j^{m-k}} \right|$$

$$\le \frac{1}{j} \sup_{|x| \le 1} |\psi^{(m)}(x)| + \frac{1}{j} \sup_{x > 1} \left| \sum_{k=0}^m \frac{m!}{k!(m-k)!} e^{-x} \theta^{(m-k)}(x/j) \frac{x^\ell}{j^{m-k}} \right|$$

$$\le \frac{C}{j} + \frac{C}{j} \sup_{x > 1} |e^{-x} x^\ell| \le \frac{C}{j} \xrightarrow[j \to \infty]{} 0. \qquad (13.3.6)$$

In conclusion, $\varphi_j \xrightarrow[j \to \infty]{\mathcal{S}(\mathbb{R}^n)} 0$.

Exercise 3.37 (a) Not in $\mathcal{S}(\mathbb{R}^n)$ since it is not bounded since $\lim\limits_{x_1 \to -\infty} e^{-x_1} = \infty$.

(b) Since $e^{-|x|^2} \in \mathcal{S}(\mathbb{R}^n)$ and $|x|^{2n!} \in \mathcal{L}(\mathbb{R}^n)$, by (a) in Theorem 3.14, their product is in $\mathcal{S}(\mathbb{R}^n)$.

(c) $(1 + |x|^2)^{2^n}$ is a polynomial function and if $(1 + |x|^2)^{-2^n} \in S(\mathbb{R}^n)$ then their product, which is equal to 1, would belong to $S(\mathbb{R}^n)$, which is not true.

(d) Show first that $\sin(e^{-|x|^2}) \in S(\mathbb{R}^n)$. Then, given that $\frac{1}{1+|x|^2} \in \mathcal{L}(\mathbb{R}^n)$, it follows that $\frac{\sin(e^{-|x|^2})}{1+|x|^2} \in S(\mathbb{R}^n)$.

(e) Not in $S(\mathbb{R}^n)$ since $\lim_{|x|\to\infty} (1 + |x|^2)^{n+1} \frac{\cos(e^{-|x|^2})}{(1+|x|^2)^n} = \infty$.

(f) Set $\varphi(x) := e^{-|x|^2} \sin(e^{x_1^2})$, $x \in \mathbb{R}^n$. If φ were to belong to $S(\mathbb{R}^n)$ then $\partial_1 \varphi$ and $x_1 \partial \varphi$ would be bounded. However, for every $x = (x_1, ..., x_n) \in \mathbb{R}^n$,

$$\partial_1 \varphi(x) = 2e^{-x_2^2 - x_3^2 - \cdots - x_n^2} x_1 [\cos(e^{x_1^2}) - e^{-x_1^2} \sin(e^{x_1^2})]. \qquad (13.3.7)$$

In particular, $(\partial_1 \varphi)(x_1, 0, ..., 0) + 2x_1 \varphi(x_1, 0, ..., 0) = 2x_1 \cos(e^{x_1^2})$ which is not bounded.

(g) Since A is positive definite, there exists a real, symmetric, positive definite $n \times n$ matrix B such that $B^2 = A$. Hence, $\varphi(x) = e^{-|Bx|^2}$ for every $x \in \mathbb{R}^n$, which means that φ is the composition between the function $e^{-|x|^2} \in S(\mathbb{R}^n)$ (recall Exercise 3.3) and the linear transformation B that maps $S(\mathbb{R}^n)$ into itself. Recalling Exercise 3.17 we may conclude that $\varphi \in S(\mathbb{R}^n)$.

Exercise 3.38 From part *(g)* in Exercise 3.37 we have $f \in S(\mathbb{R}^n)$. Pick some $B \in M_{n \times n}(\mathbb{R})$ with the property that $B^2 = A$, so that $f(x) = e^{-|Bx|^2}$ for every $x \in \mathbb{R}^n$. Now use Exercise 3.24 and Example 3.22.

Exercise 3.39 Let $A := \begin{pmatrix} 1 & 1/2 \\ 1/2 & 1 \end{pmatrix}$. Then $f(x) = e^{-(Ax)\cdot x}$ for every $x \in \mathbb{R}^2$ and Exercise 3.38 applies and yields

$$\widehat{f}(\xi) = \frac{2\pi}{\sqrt{3}} e^{-(\xi_1^2 - \xi_1 \xi_2 + \xi_2^2)/3} \qquad \text{for } \xi = (\xi_1, \xi_2) \in \mathbb{R}^2. \qquad (13.3.8)$$

Exercise 3.40 Use (b) in Theorem 3.21 and Example 3.22.

Exercise 3.41 Since $\sin(x \cdot x_0) = \frac{1}{2i}[e^{ix \cdot x_0} - e^{-ix \cdot x_0}]$ matters reduced to the case $n = 1$ after which we may apply Exercise 3.23. Hence, if $x = (x_1, \ldots, x_n)$, $x_0 = (x_{01}, \ldots, x_{0n})$, and $\xi = (\xi_1, \ldots, \xi_n)$, we may write

$$\mathcal{F}(e^{-a|x|^2} \sin(x \cdot x_0))(\xi) = \frac{1}{2i} \Big[\prod_{j=1}^n \mathcal{F}(e^{-ax_j^2 + ix_j x_{0j}}) - \prod_{j=1}^n \mathcal{F}(e^{-ax_j^2 - ix_j x_{0j}}) \Big]$$

$$= \frac{1}{2i} \Big(\frac{\pi}{a}\Big)^{\frac{n}{2}} \Big[\prod_{j=1}^n e^{-\frac{(\xi_j - x_{0j})^2}{4a}} - \prod_{j=1}^n e^{-\frac{(\xi_j + x_{0j})^2}{4a}} \Big]$$

$$= \frac{1}{2i} \Big(\frac{\pi}{a}\Big)^{\frac{n}{2}} \Big[e^{-\frac{|\xi - x_0|^2}{4a}} - e^{-\frac{|\xi + x_0|^2}{4a}} \Big]. \qquad (13.3.9)$$

Exercise 3.42 Fix $\varphi \in S(\mathbb{R})$. Suppose $\psi \in S(\mathbb{R})$ is such that $\psi' = \varphi$. Then

$$\int_{\mathbb{R}} \varphi(x)\,dx = \lim_{R_1, R_2 \to \infty} \int_{R_1}^{R_2} \psi'(x)\,dx = \lim_{R_1, R_2 \to \infty} [\psi(R_2) - \psi(R_1)] = 0.$$

Conversely, if $\int_{\mathbb{R}} \varphi(x)\,dx = 0$, then the function $\psi(x) := \int_0^x \varphi(t)\,dt$, for $x \in \mathbb{R}$, belongs to $\mathcal{S}(\mathbb{R})$ and $\psi' = \varphi$.

Exercise 3.43 Use Exercise 3.42.

13.4 Solutions to Exercises from Section 4.11

Exercise 4.107 Using the change of variables $y = -x$ on the region corresponding to $x < 0$, we have

$$\int_{|x| \geq \varepsilon} \frac{\varphi(x)}{x}\,dx = \int_{\varepsilon}^{\infty} \frac{\varphi(x) - \varphi(-x)}{x}\,dx, \qquad \forall\, \varphi \in \mathcal{S}(\mathbb{R}^n). \tag{13.4.1}$$

Moreover, for each $\varepsilon > 0$ and each $\varphi \in \mathcal{S}(\mathbb{R})$ we may write

$$\left| \int_{|x| \geq \varepsilon} \frac{\varphi(x)}{x}\,dx \right| \leq \int_{\varepsilon}^{1} \frac{|\varphi(x) - \varphi(-x)|}{x}\,dx + \int_{1}^{\infty} \frac{|\varphi(x) - \varphi(-x)|x}{x^2}\,dx$$

$$\leq \sup_{x \in \mathbb{R}} |\varphi'(x)| + 2\left(\int_{1}^{\infty} \frac{dx}{x^2} \right) \sup_{x \in \mathbb{R}} |x\varphi(x)| \tag{13.4.2}$$

so that P.V. $\frac{1}{x}$ is well defined and

$$\left| \langle \text{P.V.}\,\frac{1}{x}, \varphi \rangle \right| \leq \left(1 + \int_{|x| > 1} \frac{dx}{|x|^2} \right) \sup_{x \in \mathbb{R},\, k=0,1,\, j=0,1} |x^k \varphi^{(j)}(x)| \tag{13.4.3}$$

for all $\varphi \in \mathcal{S}(\mathbb{R})$ Hence, (4.1.2) holds for $m = k = 1$. Since P.V. $\frac{1}{x}$ is also linear, we obtain P.V. $\frac{1}{x} \in \mathcal{S}'(\mathbb{R}^n)$.

Exercise 4.108 Use Exercise 2.124, (2.1.9), and Exercise 4.6.

Exercise 4.109 Use Exercise 2.126 and (4.1.33).

Exercise 4.110 Use a reasoning similar to the proof of the fact that the function in (4.1.35) is not a tempered distribution to conclude that $e^{ax}H(x)$ and $e^{-ax}H(-x)$ do not belong to $\mathcal{S}'(\mathbb{R})$ while $e^{-ax}H(x)$, $e^{ax}H(-x) \in \mathcal{S}'(\mathbb{R})$.

Exercise 4.111 Let $\varphi \in \mathcal{S}(\mathbb{R})$ and $j \in \mathbb{N}$.
  (a) We may write

$$\left\langle \frac{x}{x^2 + j^{-2}}, \varphi \right\rangle = \int_{|x|>1} \frac{x}{x^2 + j^{-2}} \, \varphi(x) \, dx + \int_{|x|\le 1} \frac{x^2}{x^2 + j^{-2}} \cdot \frac{\varphi(x) - \varphi(0)}{x} \, dx$$

$$+ \varphi(0) \int_{|x|\le 1} \frac{x}{x^2 + j^{-2}} \, dx. \qquad (13.4.4)$$

The last integral in (13.4.4) is equal to zero (the function integrated is odd) while for the other two integrals in (13.4.4) apply Lebesgue's Dominated Convergence Theorem to obtain that the first integral converges to $\int_{|x|>1} \frac{\varphi(x)}{x} \, dx$ and the second integral converges to $\int_{|x|\le 1} \frac{\varphi(x)-\varphi(0)}{x} \, dx$. In conclusion, $\frac{x}{x^2+j^{-2}} \xrightarrow[j\to\infty]{S'(\mathbb{R})} \text{P.V.} \frac{1}{x}$.

(b) Making the change of variables $x = y/j$ and then applying Lebesgue's Dominated Convergence Theorem write

$$\left\langle \frac{1}{j(x^2 + j^{-2})}, \varphi \right\rangle = \int_{\mathbb{R}} \frac{\varphi(x)}{j(x^2 + j^{-2})} \, dx = \int_{\mathbb{R}} \frac{\varphi(y/j)}{y^2 + 1} \, dy \xrightarrow[j\to\infty]{} \pi\varphi(0), \qquad (13.4.5)$$

hence $\frac{1}{j(x^2+j^{-2})} \xrightarrow[j\to\infty]{S'(\mathbb{R})} \pi\delta$.

(c) See Exercise 2.136. $\frac{\sin(jx)}{\pi x} \xrightarrow[j\to\infty]{S'(\mathbb{R})} \delta$.

(d) Prove first that $e^{j^2}\delta_j \xrightarrow[j\to\infty]{\mathcal{D}'(\mathbb{R})} 0$. If $\{e^{j^2}\delta_j\}_{j\in\mathbb{N}}$ would converge in $S'(\mathbb{R})$ it would have to converge to 0. However, for the test function $e^{-\frac{x^2}{2}} \in S(\mathbb{R})$ one has

$$\left\langle e^{j^2}\delta_j, e^{-\frac{x^2}{2}} \right\rangle = e^{\frac{j^2}{2}} \xrightarrow[j\to\infty]{} \infty.$$

Exercise 4.112 Use the change of variables $jx = y$ and Lebesgue's Dominated Convergence Theorem to show that $j^n \theta(jx) \xrightarrow[j\to\infty]{S'(\mathbb{R})} \delta$.

Exercise 4.113 Use Lebesgue's Dominated Convergence Theorem to prove that $f_j \xrightarrow[j\to\infty]{\mathcal{D}'(\mathbb{R})} e^{x^2}$. On the other hand, the sequence $\{f_j\}_{j\in\mathbb{N}}$ does not convergence in $S'(\mathbb{R})$ since

$$\langle f_j, e^{-x^2/2} \rangle = \int_{-j}^{j} e^{x^2/2} \, dx \xrightarrow[j\to\infty]{} \int_{\mathbb{R}} e^{x^2/2} \, dx = \infty.$$

Exercise 4.114 If there were some $f \in L^p(\mathbb{R})$ such that $\lim\limits_{j\to\infty} \|f_j - f\|_{L^p(\mathbb{R})} = 0$, then $\lim\limits_{j\to\infty} f_j(x) = f(x)$ for almost every $x \in \mathbb{R}$. Since $\lim\limits_{j\to\infty} f_j(x) = 0$ for every $x \in \mathbb{R}$, we have that $f = 0$ almost everywhere on $\mathbb{R}$. This leads to a contradiction given that we would have $0 = \lim\limits_{j\to\infty} \|f_j\|_{L^p(\mathbb{R})}^p = \int_{j-1}^{j} 1 \, dx = 1$. Hence, the sequence $\{f_j\}_{j\in\mathbb{N}}$ does not converge in $L^p(\mathbb{R})$. If $\varphi \in S(\mathbb{R})$ then

$$\left| \langle f_j, \varphi \rangle \right| \le \sup_{x \in \mathbb{R}} |x^2 \varphi(x)| \int_{j-1}^{j} \frac{dx}{x^2} \le \frac{C}{j(j-1)} \xrightarrow[j \to \infty]{} 0, \tag{13.4.6}$$

which shows that $f_j \xrightarrow[j \to \infty]{\mathcal{S}'(\mathbb{R})} 0$.

Exercise 4.115 Using Lebesgue's Dominated Convergence Theorem it is not difficult to check that $\langle u_j, \varphi \rangle \xrightarrow[j \to \infty]{} \langle e^x \sin(e^x), \varphi \rangle$ for every $\varphi \in C_0^\infty(\mathbb{R})$.

If $\varphi \in \mathcal{S}(\mathbb{R})$, integration by parts yields

$$\int_{-j}^{j} e^x \sin(e^x) \varphi(x) \, dx = - \cos(e^x) \varphi(x) \Big|_{-j}^{j} + \int_{-j}^{j} \cos(e^x) \varphi'(x) \, dx. \tag{13.4.7}$$

Since $\varphi \in \mathcal{S}(\mathbb{R})$ one has $\cos(e^x) \varphi' \in L^1(\mathbb{R})$ and

$$\left| \cos(e^x) \varphi(x) \Big|_{-j}^{j} \right| = \left| \frac{\cos(e^x)}{x} x \varphi(x) \Big|_{-j}^{j} \right| \le \frac{C}{j} \xrightarrow[j \to \infty]{} 0.$$

Thus, $\lim_{j \to \infty} \langle u_j, \varphi \rangle = \int_{\mathbb{R}} \cos(e^x) \varphi'(x) \, dx$ and if $u : \mathcal{S}(\mathbb{R}) \to \mathbb{C}$ is such that

$$\langle u, \varphi \rangle := \int_{\mathbb{R}} \cos(e^x) \varphi'(x) \, dx \quad \text{for every} \quad \varphi \in \mathcal{S}(\mathbb{R}), \tag{13.4.8}$$

then u is a well defined, linear mapping and

$$|\langle u, \varphi \rangle| \le \left(\int_{\mathbb{R}} \frac{dx}{1 + x^2} \right) \sup_{x \in \mathbb{R}} |(1 + x^2) \varphi'(x)| \quad \text{for every} \quad \varphi \in \mathcal{S}(\mathbb{R}). \tag{13.4.9}$$

Hence $u \in \mathcal{S}'(\mathbb{R})$ and $u_j \xrightarrow[j \to \infty]{\mathcal{S}'(\mathbb{R})} u$.

Exercise 4.117 Fix $f \in \mathcal{S}'(\mathbb{R})$ and let $\psi \in \mathcal{S}(\mathbb{R})$ be such that $\psi(0) = 1$. Then, if u is a solution of the equation $xu = f$ in $\mathcal{S}'(\mathbb{R})$, for each $\varphi \in \mathcal{S}(\mathbb{R})$ we have

$$\langle u, \varphi \rangle = \left\langle u, x \frac{\varphi - \varphi(0)\psi}{x} \right\rangle + \langle u, \varphi(0)\psi \rangle$$

$$= \left\langle f, \frac{\varphi - \varphi(0)\psi}{x} \right\rangle + \langle u, \psi \rangle \langle \delta, \varphi \rangle. \tag{13.4.10}$$

Set

$$g(x) := \begin{cases} \frac{\varphi(x) - \varphi(0)\psi(x)}{x} & \text{if } x \in \mathbb{R} \setminus \{0\}, \\ \varphi'(0) - \varphi(0)\psi'(0) & \text{if } x = 0. \end{cases} \tag{13.4.11}$$

Note that $g \in \mathcal{S}(\mathbb{R})$, since $g(x) = \int_0^1 [\varphi'(tx) - \varphi(0)\psi'(tx)] \, dt$ for $x \in \mathbb{R}$. Thus, if we define

$$\widetilde{\varphi}(x) := \begin{cases} \frac{\varphi(x)-\varphi(0)\psi(x)}{x} & \text{if } |x| > 1, \\ g(x) & \text{if } |x| \le 1, \end{cases} \qquad (13.4.12)$$

then $\widetilde{\varphi} \in S(\mathbb{R})$ and $\widetilde{x\varphi} = \varphi$. There remains to observe that the mapping

$$w_f : S(\mathbb{R}) \to \mathbb{C}, \qquad \langle w_f, \varphi \rangle := \langle f, \widetilde{\varphi} \rangle \qquad \forall \varphi \in S(\mathbb{R}), \qquad (13.4.13)$$

where $\widetilde{\varphi}$ is as in (13.4.12), satisfies $w_f \in S'(\mathbb{R})$ in order to conclude that $u = w_f + c\delta$, some $c \in \mathbb{R}$.

Exercise 4.118 Suppose $u \in S'(\mathbb{R}^n)$ is a solution of the equation $e^{-|x|^2} u = 1$ in $S'(\mathbb{R}^n)$. Then for $\varphi \in C_0^\infty(\mathbb{R}^n)$ we have

$$\langle u, \varphi \rangle = \langle u, e^{-|x|^2} e^{|x|^2} \varphi \rangle = \langle 1, e^{|x|^2} \varphi \rangle = \langle e^{|x|^2}, \varphi \rangle, \qquad (13.4.14)$$

which shows that $u\big|_{C_0^\infty(\mathbb{R}^n)} = e^{|x|^2}$. Since $C_0^\infty(\mathbb{R}^n)$ is sequentially dense in $S(\mathbb{R}^n)$, this would imply that $u = e^{|x|^2}$. However, as proved following Remark 4.16, $e^{|x|^2}$ does not belong to $S'(\mathbb{R}^n)$.

Exercise 4.119 Since H is Lebesgue measurable and $(1 + x^2)^{-1} H \in L^1(\mathbb{R})$, by Exercise 4.6 it follows that H defines a tempered distribution. We have seen that $H \in D'(\mathbb{R})$ and $H' = \delta$ in $D'(\mathbb{R})$ (recall Example 2.44). Since $\delta \in S'(\mathbb{R})$, by (4.1.33) it follows that $H' = \delta$ in $S'(\mathbb{R})$. Taking the Fourier transform of the last equation we arrive at $i\xi\widehat{H} = 1$ in $S'(\mathbb{R})$. On the other hand, $\xi\left(\text{P.V.}\frac{1}{\xi}\right) = 1$ in $S'(\mathbb{R})$ by Exercise 4.107 and (2.3.6). Consequently, $\xi(i\widehat{H} - \text{P.V.}\frac{1}{\xi}) = 0$ in $S'(\mathbb{R})$. Now Example 2.76 may be used to conclude that $i\widehat{H} - \text{P.V.}\frac{1}{\xi} = c\delta$, for some $c \in \mathbb{C}$. Hence

$$\widehat{H} = -i\text{P.V.}\frac{1}{\xi} - c\delta \qquad \text{in } S'(\mathbb{R}). \qquad (13.4.15)$$

To determine c apply $\widehat{H}$ to the function $\varphi(x) := e^{-x^2} \in S(\mathbb{R})$. First, by Example 3.22 write

$$\langle \widehat{H}, e^{-x^2} \rangle = \langle H, \widehat{e^{-x^2}} \rangle = \langle H, \sqrt{\pi}e^{-x^2/4} \rangle = \sqrt{\pi} \int_0^\infty e^{-x^2/4} \, dx$$

$$= \frac{\sqrt{\pi}}{2} \int_{-\infty}^\infty e^{-x^2/4} \, dx = \widehat{e^{-x^2/4}}(0) = \pi. \qquad (13.4.16)$$

Second, from (13.4.15) and the fact that $\frac{e^{-x^2}}{x}$ is an odd function obtain that

$$\left\langle -i\text{P.V.}\frac{1}{\xi} - c\delta, e^{-x^2} \right\rangle = -i \lim_{\varepsilon \to 0^+} \int_{|x|>\varepsilon} \frac{e^{-x^2}}{x} \, dx - c = -c. \qquad (13.4.17)$$

Combining (13.4.15), (13.4.16), and (13.4.17) arrive at $\widehat{H} = -i\text{P.V.}\frac{1}{\xi} + \pi\delta$.

Exercise 4.120 You may use Exercise 4.119 to show that

$$\mathcal{F}\left(\text{P.V.}\,\frac{1}{x}\right)(\xi) = 2\pi i H(\xi) - i\pi = i\pi \operatorname{sgn}(\xi) \quad \text{in} \quad \mathcal{S}'(\mathbb{R}).$$

Exercise 4.121 (a) sgn $= H - H^{\vee}$, hence using Exercise 4.119 one may write

$$\widehat{\operatorname{sgn}} = \widehat{H} - \widehat{H^{\vee}} = \pi\delta - i\,\text{P.V.}\,\frac{1}{\xi} - \pi\delta - i\,\text{P.V.}\,\frac{1}{\xi} = -2i\,\text{P.V.}\,\frac{1}{\xi} \quad (13.4.18)$$

in $\mathcal{S}'(\mathbb{R})$. Alternatively, you may use Exercise 4.120 and take another Fourier transform.

(b) If k is even, then $|x|^k = x^k$, so that $\widehat{|x|^k} = \widehat{x^k\delta} = (-D)^k\widehat{\delta} = 2\pi i^k\delta^{(k)}$ in $\mathcal{S}'(\mathbb{R})$. If k is odd, then $|x|^k = x^k\operatorname{sgn} x$, thus

$$\widehat{|x|^k} = \widehat{x^k\operatorname{sgn} x} = (-D)^k\widehat{\operatorname{sgn} x} = -2i^{k+1}\left(\text{P.V.}\,\frac{1}{\xi}\right)^{(k)} \quad \text{in} \quad \mathcal{S}'(\mathbb{R}). \quad (13.4.19)$$

(c) You may use Example 4.36. Alternatively, take the Fourier transform of the identity $x \cdot \frac{\sin(ax)}{x} = \sin(ax)$ in $\mathcal{S}'(\mathbb{R})$, then use (c) in Theorem 4.26 and (4.2.48) to conclude that

$$\left[\mathcal{F}\left(\frac{\sin(ax)}{x}\right)\right]' = \pi\delta_{-a} - \pi\delta_a \quad \text{in} \quad \mathcal{S}'(\mathbb{R}).$$

Now use Exercise 2.138 and Proposition 2.47 to obtain

$$\mathcal{F}\left(\frac{\sin(ax)}{x}\right) = \pi \operatorname{sgn}(a)\chi_{[-|a|,|a|]} + c \quad \text{in} \quad \mathcal{S}'(\mathbb{R}), \quad (13.4.20)$$

and then show that $c = 0$ by applying the Fourier transform to the last identity and recalling (4.2.45).

(d) Take the Fourier transform of the identity

$$x \cdot \frac{\sin(ax)\sin(bx)}{x} = \sin(ax)\sin(bx) \quad \text{in} \quad \mathcal{S}'(\mathbb{R}),$$

then use (c) in Theorem 4.26 and (4.2.48) to conclude that

$$\left[\mathcal{F}\left(\frac{\sin(ax)\sin(bx)}{x}\right)\right]' = \frac{i\pi}{2}\left[\delta_{a+b} - \delta_{a-b} - \delta_{b-a} + \delta_{-a-b}\right] \quad \text{in} \quad \mathcal{S}'(\mathbb{R}).$$

Now use Exercise 2.138 and Proposition 2.47 to obtain

$$\mathcal{F}\left(\frac{\sin(ax)\sin(bx)}{x}\right)$$

$$= \frac{i\pi}{2}\left[H(x-a-b) - H(x-a+b) - H(x-b+a) + H(x+a+b)\right] + c_0$$

$$= \frac{i\pi}{2}\operatorname{sgn}(b)\left[\chi_{[-a-|b|,-a+|b|]} - \chi_{[a-|b|,a+|b|]}\right] + c_0 \quad (13.4.21)$$

in $\mathcal{S}'(\mathbb{R})$, and then show that $c_0 = 0$ by applying the Fourier transform to the last identity.

(e) Using the identity $\sin(x^2) = \frac{1}{2i}\left[e^{ix^2} - e^{-ix^2}\right]$, (4.2.22) and (4.2.56), we may write

$$\widehat{\sin(x^2)} = \frac{1}{2i}\left[\widehat{e^{ix^2}} - \widehat{e^{-ix^2}}\right]$$

$$= \frac{\sqrt{\pi}}{2i}\left[e^{i\frac{\pi}{4}}e^{-i\frac{\xi^2}{4}} - e^{-i\frac{\pi}{4}}e^{i\frac{\xi^2}{4}}\right]$$

$$= \frac{\sqrt{\pi}}{2i}\left[e^{i\frac{\pi-\xi^2}{4}} - e^{-i\frac{\pi-\xi^2}{4}}\right]. \qquad (13.4.22)$$

(f) Use (4.6.26), (4.2.3), and Proposition 4.32 to show

$$\widehat{\ln|x|} = -2\pi\gamma\delta - \pi w_{\chi_{(-1,1)}} \qquad \text{in } \mathcal{S}'(\mathbb{R}).$$

Exercise 4.122 The computation of the Fourier transform in Example 4.24 naturally adapts to the current setting and yields that

$$\mathcal{F}\left(\frac{1}{x^2 - (b+ia)^2}\right)(\xi) = \frac{\pi i e^{ib|\xi|}}{b+ia}e^{-a|\xi|}, \qquad \forall\,\xi \in \mathbb{R}. \qquad (13.4.23)$$

Exercise 4.123 From Exercise 2.146 we know that $\delta_{\partial B(0,R)} \in \mathcal{E}'(\mathbb{R}^3)$, thus by *(b)* in Theorem 4.35 it follows that

$$\widehat{\delta_{\partial B(0,R)}}(\xi) = \langle \delta_{\partial B(0,R)}, e^{-ix\cdot\xi}\rangle = \int_{\partial B(0,R)} e^{-ix\cdot\xi}\,d\sigma(x), \qquad \forall\,\xi \in \mathbb{R}^3. \qquad (13.4.24)$$

Check that $\delta_{\partial B(0,R)}$ is invariant under orthogonal transformations, and conclude that $\widehat{\delta_{\partial B(0,R)}}$ is invariant under orthogonal transformations. Fix $\xi \in \mathbb{R}^3 \setminus \{0\}$ and show that there exists an orthogonal transformation $A \in \mathcal{M}_{3\times 3}(\mathbb{R})$ such that $A\xi = (|\xi|, 0, 0)$, and furthermore that $\widehat{\delta_{\partial B(0,R)}}(\xi) = \int_{\partial B(0,R)} e^{-ix_1|\xi|}\,d\sigma(x)$. Now use spherical coordinates to compute the latter integral and conclude that $\widehat{\delta_{\partial B(0,R)}}(\xi) = 4\pi R\frac{\sin(R|\xi|)}{|\xi|}$. Treat separately the case $\xi = (0,0,0)$.

Exercise 4.124 Use Lemma 4.28.

Exercise 4.125 Since $\chi_{[-1,1]} \in L^1(\mathbb{R})$, from (4.1.9) it follows that $\chi_{[-1,1]} \in \mathcal{S}'(\mathbb{R})$. Also, by Exercise 4.27, we have that $\widehat{\chi_{[-1,1]}}$ in $\mathcal{S}'(\mathbb{R})$ is the tempered distribution given by the function

$$\int_{-1}^{1} e^{-ix\xi}\,dx = \frac{2\sin\xi}{\xi}, \qquad \xi \in \mathbb{R} \setminus \{0\}.$$

In particular, since $\frac{\sin\xi}{\xi} \notin L^1(\mathbb{R})$ we have $\widehat{\chi_{[-1,1]}} \notin L^1(\mathbb{R})$.

Exercise 4.127 See Example 4.38 and deduce that $\widehat{g} = \pi\delta_a + \pi\delta_{-a}$ in $\mathcal{S}'(\mathbb{R})$.

Exercise 4.128 Use Exercise 4.124.

Exercise 4.129 If $P(x) = \sum_{|\alpha|\leq m} a_\alpha x^\alpha$, then the condition $P(tx) = t^{-k}P(x)$ in $\mathbb{R}^n$ for each $t > 0$ implies $\sum_{|\alpha|\leq m} t^{|\alpha|}a_\alpha x^\alpha = t^{-k}\sum_{|\alpha|\leq m} a_\alpha x^\alpha$ in $\mathbb{R}^n$ for each $t > 0$. Hence, for each $\alpha \in \mathbb{N}_0^n$ we have $t^{|\alpha|}a_\alpha = t^{-k}a_\alpha$ for every $t > 0$, or equivalently that $t^{|\alpha|+k}a_\alpha = a_\alpha$ for every $t > 0$. Now take $t \to 0^+$ in the last equality to obtain $a_\alpha = 0$.

Exercise 4.130 The first identity in (4.11.4) is easy to verify (keeping in mind that $|\eta| = |\zeta| = 1$). Check, via a direct calculation, that

$$R^\top w := w - \Big[\frac{w \cdot [\zeta - (1 + 2\eta \cdot \zeta)\eta]}{1 + \eta \cdot \zeta}\Big]\zeta - \Big[\frac{w \cdot (\zeta + \eta)}{1 + \eta \cdot \zeta}\Big]\eta, \quad \forall\, w \in \mathbb{R}^n. \quad (13.4.25)$$

To show that $R^\top R = I_{n\times n}$, let $\xi \in \mathbb{R}^n$ be arbitrary and set $a := \zeta \cdot \eta \in \mathbb{R} \setminus \{-1\}$. From (13.4.25) and (4.11.3) we have

$$R^\top(R\xi) = R\xi - \Big[\frac{(R\xi) \cdot [\zeta - (1 + 2a)\eta]}{1 + a}\Big]\zeta - \Big[\frac{(R\xi) \cdot (\zeta + \eta)}{1 + a}\Big]\eta$$

$$= \xi - \Big[\frac{\xi \cdot (\eta + \zeta)}{1 + a}\Big]\zeta + \Big[\frac{\xi \cdot [(1 + 2a)\zeta - \eta]}{1 + a}\Big]\eta$$

$$- \frac{1}{1 + a}\Big[\xi \cdot [\zeta - (1 + 2a)\eta] - \frac{\xi \cdot (\eta + \zeta)}{1 + a}(1 - (1 + 2a)a)\Big]\zeta$$

$$- \frac{\xi \cdot [(1 + 2a)\zeta - \eta]}{(1 + a)^2}(a - (1 + 2a))\zeta$$

$$- \frac{1}{1 + a}\Big[\xi \cdot (\eta + \zeta) - \frac{\xi \cdot (\eta + \zeta)}{1 + a}(1 + a)\Big]\eta$$

$$- \frac{\xi \cdot [(1 + 2a)\zeta - \eta]}{(1 + a)^2}(1 + a)\eta = \xi. \quad (13.4.26)$$

Hence, $R^\top R = I_{n\times n}$ as wanted. The latter identity now implies the second identity in (4.11.4) and the fact that $RR^\top = I_{n\times n}$.

Exercise 4.131 Fix $c \in \mathbb{R} \setminus \{0\}$ and use the fact that cosine is an odd function, the fundamental theorem of calculus, and Fubini's theorem to write

$$\int_\varepsilon^R \frac{\cos(c\rho) - \cos\rho}{\rho}\, d\rho = \int_\varepsilon^R \frac{\cos(|c|\rho) - \cos\rho}{\rho}\, d\rho$$

$$= -\int_\varepsilon^R \int_1^{|c|} \sin(r\rho)\, dr\, d\rho$$

$$= -\int_1^{|c|} \int_\varepsilon^R \sin(r\rho)\, d\rho\, dr$$

$$= \int_1^{|c|} \Big[\frac{\cos(Rr)}{r} - \frac{\cos(\varepsilon r)}{r}\Big]\, dr. \qquad (13.4.27)$$

In particular, estimate (4.11.7) readily follows from this formula. Going further, we note that an integration by parts gives

$$\int_1^{|c|} \frac{\cos(Rr)}{r}\, dr = \Big[\frac{\sin(Rr)}{Rr}\Big]_{r=1}^{r=|c|} + \frac{1}{R}\int_1^{|c|} \frac{\sin(Rr)}{r^2}\, dr. \qquad (13.4.28)$$

By combining (13.4.27) with (13.4.28) we arrive at

$$\int_\varepsilon^R \frac{\cos(c\rho) - \cos\rho}{\rho}\, d\rho = \frac{\sin(|c|R)}{|c|R} - \frac{\sin R}{R} + \frac{1}{R}\int_1^{|c|} \frac{\sin(Rr)}{r^2}\, dr$$

$$- \int_1^{|c|} \frac{\cos(\varepsilon r)}{r}\, dr. \qquad (13.4.29)$$

Passing to limit $R \to \infty$ and $\varepsilon \to 0^+$ in (13.4.29) yields (4.11.5) after observing (e.g., by Lebesgue's Dominated Convergence Theorem) that

$$\lim_{\varepsilon \to 0^+} \int_1^{|c|} \frac{\cos(\varepsilon r)}{r}\, dr = \int_1^{|c|} \frac{dr}{r} = \ln|c|. \qquad (13.4.30)$$

Next,

$$\lim_{\substack{\varepsilon \to 0^+ \\ R \to \infty}} \int_\varepsilon^R \frac{\sin(c\rho)}{\rho}\, d\rho = \lim_{\substack{\varepsilon \to 0^+ \\ R \to \infty}} \int_{\varepsilon c}^{Rc} \frac{\sin t}{t}\, dt$$

$$= (\operatorname{sgn} c) \lim_{\substack{\varepsilon \to 0^+ \\ R \to \infty}} \int_\varepsilon^R \frac{\sin t}{t}\, dt = \frac{\pi}{2} \operatorname{sgn} c, \qquad (13.4.31)$$

by a suitable change of variables and the well-known fact (based on a residue calculation) that $\lim_{\substack{\varepsilon \to 0^+ \\ R \to \infty}} \int_\varepsilon^R \frac{\sin t}{t}\, dt = \frac{\pi}{2}$. This proves (4.11.6). Finally, to justify (4.11.8) use the fact that whenever $0 < a < b < \infty$ an integration by parts gives

$$\int_a^b \frac{\sin t}{t}\, dt = -\Big[\frac{\cos t}{t}\Big]_{t=a}^{t=b} - \int_a^b \frac{\cos t}{t^2}\, dt. \qquad (13.4.32)$$

13.5 Solutions to Exercises from Section 5.4

Exercise 5.16 The ordinary differential equation $(\frac{d}{dx})^m v = 0$, with initial conditions $v(0) = v'(0) = \cdots = v^{(m-2)}(0) = 0$, and $v^{(m-1)}(0) = 1$, has the unique solution $v(x) = \frac{1}{(m-1)!} x^{m-1}$ for $x \in \mathbb{R}$. Hence, by Example 5.12 the function $u := \frac{1}{(m-1)!} x^{m-1} H$ is a fundamental solution for $(\frac{d}{dx})^m$ in $\mathbb{R}$. In addition, by Exercise 4.119 we have $H \in \mathcal{S}'(\mathbb{R})$ which, when combined with the fact that $\frac{1}{(m-1)!} x^{m-1} \in \mathcal{L}(\mathbb{R})$, implies (as a consequence of *(b)* in Theorem 4.14) that

$$\frac{1}{(m-1)!} x^{m-1} H \in \mathcal{S}'(\mathbb{R}).$$

If $E \in \mathcal{S}'(\mathbb{R})$ is an arbitrary fundamental solution for $(\frac{d}{dx})^m$, then by Proposition 5.8 we have that $E - u = P$ for some polynomial in $\mathbb{R}$ satisfying $(\frac{d}{dx})^m P = 0$ in $\mathbb{R}$. This forces P to be a polynomial of degree less than or equal to $m - 1$.

Exercise 5.17 First prove that

$$\langle P_1(D_x) \otimes P_2(D_y), \varphi_1(x) \otimes \varphi_2(y) \rangle = \langle \delta_x \otimes \delta_y, \varphi_1(x) \otimes \varphi_2(y) \rangle$$

for every $\varphi_1 \in C_0^\infty(\mathbb{R}^n)$ and every $\varphi_2 \in C_0^\infty(\mathbb{R}^m)$, and then use Proposition 2.81.

13.6 Solutions to Exercises from Section 6.4

Exercise 6.27 The ordinary differential equation $v \in C^\infty(\mathbb{R})$, $v'' + k^2 v = 0$ in $\mathbb{R}$, with initial conditions $v(0) = 0$ and $v'(0) = 1$, has the unique solution

$$v(x) = -\frac{i}{2k} e^{ikt} + \frac{i}{2k} e^{-ikt} \quad \text{for all} \ \ x \in \mathbb{R}. \tag{13.6.1}$$

Hence, by Example 5.12 the function $u := vH = -\frac{i}{2k} e^{ikt} H + \frac{i}{2k} e^{-ikt} H$ is a fundamental solution for $(\frac{d}{dx})^2 + k^2$ in $\mathbb{R}$. In addition, by Exercise 4.119 we have $H \in \mathcal{S}'(\mathbb{R})$ which, when combined with the fact that $v \in \mathcal{L}(\mathbb{R})$, implies (as a consequence of *(b)* in Theorem 4.14) that $u \in \mathcal{S}'(\mathbb{R})$. If $E \in \mathcal{S}'(\mathbb{R})$ is an arbitrary fundamental solution for $(\frac{d}{dx})^2 + k^2$ in $\mathbb{R}$, then

$$\left[\left(\frac{d}{dx} \right)^2 + k^2 \right] (E - u) = 0 \quad \text{in} \ \ \mathcal{S}'(\mathbb{R}).$$

Since $(\frac{d}{dx})^2 + k^2$ is elliptic (cf. Definition 6.13), it is hypoelliptic (cf. Theorem 6.15) in $\mathbb{R}$, hence $E - u \in C^\infty(\mathbb{R})$. Thus, $E - u$ is a solution in the classical sense of the ordinary differential equation $w'' + k^2 w = 0$ in $\mathbb{R}$. The general solution of the latter equation is $c_1 e^{ikt} + c_2 e^{-ikt}$ for any $c_1, c_2 \in \mathbb{C}$. Consequently, $E = u + c_1 e^{ikt} + c_2 e^{-ikt}$ for $c_1, c_2 \in \mathbb{C}$.

Exercise 6.28 (a) The general solution for the ordinary differential equation $v'' - a^2 v = 0$ in $\mathbb{R}$ is $v(x) := c_1 e^{ax} + c_2 e^{-ax}$ for $x \in \mathbb{R}$, where $c_1, c_2 \in \mathbb{C}$. Hence, if we further impose the conditions $v(0) = 0$ and $v'(0) = 1$ then $c_1 = \frac{1}{2a} = -c_2$, and by Example 5.12, we have that $u := \frac{1}{2a} e^{ax} H - \frac{1}{2a} e^{-ax} H$ satisfies $u'' - a^2 u = \delta$ in $\mathcal{D}'(\mathbb{R})$.

(b) By Theorem 5.14 there exists $E \in \mathcal{S}'(\mathbb{R})$ such that $(\frac{d^2}{dx^2} - a^2)E = \delta$ in $\mathcal{S}'(\mathbb{R})$. Fix such an E and apply the Fourier transform to the last equation. Then $-(\xi^2 + a^2)\widehat{E} = 1$ in $\mathcal{S}'(\mathbb{R})$. Since $\frac{1}{\xi^2 + a^2} \in \mathcal{L}(\mathbb{R})$ we may multiply the last equality by $\frac{1}{\xi^2 + a^2}$ to conclude that $-\widehat{E} = \frac{1}{\xi^2 + a^2} = f$, and, furthermore, that $\widehat{f} = -2\pi E^{\vee}$. Therefore we can determine $\widehat{f}$ as soon as we find a fundamental solution $E \in \mathcal{S}'(\mathbb{R})$ for the operator $\frac{d^2}{dx^2} - a^2$.

By Theorem 6.15, $\frac{d^2}{dx^2} - a^2$ is a hypoelliptic operator in $\mathbb{R}$. As a consequence of Remark 6.7 we have (with u as in (a)) that $u - E \in C^{\infty}(\mathbb{R})$ and is a classical solution of the ordinary differential equation $v'' - a^2 v = 0$ in $\mathbb{R}$. Thus,

$$
\begin{aligned}
E &= u + c_1 e^{ax} + c_2 e^{-ax} \\[2mm]
&= \tfrac{1}{2a} e^{ax} H - \tfrac{1}{2a} e^{-ax} H + c_1 e^{ax} H + c_1 e^{ax} H^{\vee} + c_2 e^{-ax} H + c_2 e^{-ax} H^{\vee} \\[2mm]
&= \left[\tfrac{1}{2a} + c_1\right] e^{ax} H + \left[-\tfrac{1}{2a} + c_2\right] e^{-ax} H + c_1 e^{ax} H^{\vee} + c_2 e^{-ax} H^{\vee}.
\end{aligned}
\tag{13.6.2}
$$

The condition $E \in \mathcal{S}'(\mathbb{R})$ implies (in view of Exercise 4.110) that $\frac{1}{2a} + c_1 = 0$ and $c_2 = 0$, which when used in (13.6.2) give $E = -\frac{1}{2a} e^{-a|x|}$. Consequently, $\widehat{f}(x) = \frac{\pi}{a} e^{-a|x|}$ for $x \in \mathbb{R}$.

Exercise 6.30 Recall Exercise 6.4. Consider $u := 1 + \delta$ in $\mathcal{D}'(\mathbb{R}^n)$. It follows that $\operatorname{sing\,supp} u = \{0\}$ and $\operatorname{supp} u = \mathbb{R}^n$, thus $\operatorname{sing\,supp} u \subset \operatorname{supp} u$.

Exercise 6.31 Since $\partial^{\alpha} \delta_{x_0}\big|_{\mathbb{R}^n \setminus \{x_0\}} = 0$ we have $\operatorname{sing\,supp} \partial^{\alpha} \delta_{x_0} \subseteq \{x_0\}$. To see that the latter inclusion is in fact equality, use Example 2.13 if $\alpha = (0, \ldots, 0)$ and if $|\alpha| > 0$ use the fact that the order of the distribution $\partial^{\alpha} \delta_{x_0}$ equals $|\alpha|$ while any distribution of function type is of order zero.

Exercise 6.32 If $x \in \Omega$ is such that there exists an open set $\omega \subseteq \Omega$ with $x \in \omega$ and $u\big|_{\omega} \in C^{\infty}(\omega)$, then $(au)\big|_{\omega} \in C^{\infty}(\omega)$, which gives

$$
\Omega \setminus \operatorname{sing\,supp} u \subseteq \Omega \setminus \operatorname{sing\,supp} (au).
$$

Exercise 6.33 Since $\left(\text{P.V.} \frac{1}{x}\right)\big|_{\mathbb{R}^n \setminus \{0\}} = \frac{1}{x}$ and $\frac{1}{x}$ belongs to $C^{\infty}(\mathbb{R}^n \setminus \{0\})$, it follows that $\operatorname{sing\,supp}\left(\text{P.V.} \frac{1}{x}\right) \subseteq \{0\}$. Now using Example 2.11 we have that the distribution $\text{P.V.} \frac{1}{x}$ is not of function type.

Exercise 6.34 First prove that $P(x) \cdot \frac{1}{P(x)} = 1$ in $\mathcal{S}'(\mathbb{R}^n)$. Then take the Fourier transform to arrive at

$$P(-D)\mathcal{F}\Big(\frac{1}{P}\Big)(\xi) = (2\pi)^n\delta \quad \text{in} \quad S'(\mathbb{R}^n). \tag{13.6.3}$$

Use (13.6.3), (6.1.5), and Example 6.4, to show that it is not possible to have

$$\text{sing supp}\Big(\mathcal{F}(\frac{1}{P})\Big) = \varnothing.$$

Exercise 6.35 sing supp $u = \mathbb{R} \times \{0\}$. To prove this you may want to use the fact that

$$\langle u,\varphi\rangle = \Big\langle \text{P.V.}\,\frac{1}{x},\varphi(x,0)\Big\rangle \tag{13.6.4}$$

$$= \int_{|x|\geq 1} \frac{\varphi(x,0)}{x}\,dx + \int_{|x|\leq 1} \frac{\varphi(x,0) - \varphi(0,0)}{x}\,dx$$

for every $\varphi \in C_0^\infty(\mathbb{R}^2)$.

13.7 Solutions to Exercises from Section 7.14

Exercise 7.72 If the equation has two solutions $E_1, E_2 \in S'(\mathbb{R}^n)$, set $E := E_1 - E_2$. Then $\Delta E - E = 0$ in $S'(\mathbb{R}^n)$ implies (after an application of the Fourier transform) $-(|\xi|^2 + 1)\widehat{E} = 0$ in $S'(\mathbb{R}^n)$. Since $\frac{1}{1+|\xi|^2} \in \mathcal{L}(\mathbb{R}^n)$ (recall Exercise 3.13), by part (a) in Theorem 3.14 we have $\frac{1}{1+|\xi|^2}\varphi \in S(\mathbb{R}^n)$ for every $\varphi \in S(\mathbb{R}^n)$. Thus,

$$\langle \widehat{E},\varphi\rangle = \Big\langle (1+|\xi|^2)\widehat{E},\frac{1}{1+|\xi|^2}\varphi\Big\rangle = 0, \qquad \forall\,\varphi \in S(\mathbb{R}^n).$$

This proves that $\widehat{E} = 0$ in $S'(\mathbb{R}^n)$, thus after applying the Fourier transform we obtain $E = 0$ in $S'(\mathbb{R}^n)$. In turn, the latter implies $E_1 = E_2$ in $S'(\mathbb{R}^n)$ and the proof of the uniqueness statement is complete. Regarding the existence statement, suppose $E \in S'(\mathbb{R}^n)$ satisfies $\Delta E - E = \delta$ in $S'(\mathbb{R}^n)$. This equation, via the Fourier transform, is equivalent with $-(|\xi|^2 + 1)\widehat{E} = 1$ in $S'(\mathbb{R}^n)$. Since $\frac{1}{1+|\xi|^2} \in S'(\mathbb{R}^n)$ (use Exercise 4.7) and $(|\xi|^2 + 1)\frac{1}{1+|\xi|^2} = 1$ in $S'(\mathbb{R}^n)$, it follows that the tempered distribution $E := -\mathcal{F}^{-1}\Big(\frac{1}{1+|\xi|^2}\Big)$ is a solution of $\Delta E - E = \delta$ in $S'(\mathbb{R}^n)$, which must be unique based on the earlier reasoning.

Exercise 7.73 Suppose there exists $E \in L^1(\mathbb{R}^n)$ with $\Delta E = \delta$. Then $-|\xi|^2\widehat{E} = 1$. We know that $\widehat{E} \in \widehat{L^1} \subset C^0(\mathbb{R}^n)$ (recall (3.1.3)), hence $[|\xi|^2\widehat{E}]\big|_{\xi=(0,...,0)} = 0$. This leads to a contradiction.

Exercise 7.76 Since $\frac{\partial}{\partial z}\frac{\partial}{\partial \bar{z}} = \frac{1}{4}\Delta$ and u is holomorphic on $\Omega \setminus K$, it follows that $\text{supp}\,(\Delta u) \subseteq K$. It follows that $\Delta u \in C_0^0(\Omega)$, hence Δu is absolutely integrable on Ω. Take $\varphi \in C_0^\infty(\Omega)$ satisfying $\varphi \equiv 1$ in a neighborhood of K. Then integration by parts, the fact that $\frac{\partial}{\partial \bar{z}} u = 0$ pointwise on $\Omega \setminus K$, and that $\frac{\partial}{\partial z} \varphi = 0$ near K, imply

$$\int_\Omega \Delta u \, dx = \int_\Omega \varphi \Delta u \, dx = -4 \int_\Omega \frac{\partial}{\partial z} \varphi \, \frac{\partial}{\partial \bar{z}} u \, dx = 0. \tag{13.7.1}$$

Exercise 7.77 Let $\varphi \in C_0^\infty(\mathbb{R}^n)$. A direct computation based on the Chain Rule gives that $L_{A_1}(\varphi \circ B^{-1}) = (L_{A_2}\varphi) \circ B^{-1}$ pointwise in $\mathbb{R}^n$. This and (2.2.7) then allow us to write

$$\langle (L_{A_1}u) \circ B, \varphi \rangle = \frac{1}{|\det B|} \langle L_{A_1}u, \varphi \circ B^{-1} \rangle = \frac{1}{|\det B|} \langle u, L_{A_1}(\varphi \circ B^{-1}) \rangle$$

$$= \frac{1}{|\det B|} \langle u, (L_{A_2}\varphi) \circ B^{-1} \rangle = \langle u \circ B, L_{A_2}\varphi \rangle$$

$$= \langle L_{A_2}(u \circ B), \varphi \rangle, \qquad \forall u \in \mathcal{D}'(\mathbb{R}^n). \tag{13.7.2}$$

Exercise 7.78 Since $E_\Delta \in \mathcal{S}'(\mathbb{R}^n)$ is a fundamental solution for Δ, we have $\Delta E_\Delta = \delta$ in $\mathcal{S}'(\mathbb{R}^n)$. Fix $j \in \{1, \ldots, n\}$. Then, $\Delta(\partial_j E_\Delta) = \partial_j \delta$ in $\mathcal{S}'(\mathbb{R}^n)$. Take the Fourier transform of the last equation to arrive at

$$|\xi|^2 \mathcal{F}(\partial_j E_\Delta) = -i\xi_j \quad \text{in} \quad \mathcal{S}'(\mathbb{R}^n). \tag{13.7.3}$$

Since $n \geq 2$, by Example 4.4 we have $\frac{\xi_j}{|\xi|^2} \in \mathcal{S}'(\mathbb{R}^n)$. Also, $|\xi|^2 \in \mathcal{L}(\mathbb{R}^n)$, thus $|\xi|^2 \cdot \frac{\xi_j}{|\xi|^2} \in \mathcal{S}'(\mathbb{R}^n)$ (recall (b) in Theorem 4.14), and it is not difficult to check that $|\xi|^2 \cdot \frac{\xi_j}{|\xi|^2} = \xi_j$ in $\mathcal{S}'(\mathbb{R}^n)$. These facts combined with (13.7.3) imply

$$|\xi|^2 \left(\mathcal{F}(\partial_j E_\Delta) + i\frac{\xi_j}{|\xi|^2} \right) = 0 \quad \text{in} \quad \mathcal{S}'(\mathbb{R}^n). \tag{13.7.4}$$

Thus, $\text{supp}\left(\mathcal{F}(\partial_j E_\Delta) + i\frac{\xi_j}{|\xi|^2} \right) \subseteq \{0\}$ and by Exercise 4.37, it follows that

$$\mathcal{F}(\partial_j E_\Delta) = -i\frac{\xi_j}{|\xi|^2} + \widehat{P}_j(\xi) \quad \text{in } \mathcal{S}'(\mathbb{R}^n), \tag{13.7.5}$$

where P_j is a polynomial in $\mathbb{R}^n$. Now a direct computation gives that if $n \geq 2$, then

$$\partial_j E_\Delta = \frac{1}{\omega_{n-1}} \cdot \frac{x_j}{|x|^n} \quad \text{in} \quad \mathcal{S}'(\mathbb{R}^n). \tag{13.7.6}$$

Hence, $\partial_j E_\Delta$ is positive homogeneous of degree $1 - n$, which in turn when combined with Proposition 4.59 implies that $\mathcal{F}(\partial_j E_\Delta)$ is positive homogeneous of degree -1. Thus, the term in the right-hand side of (13.7.5) is positive homogeneous of degree -1. Since $\frac{\xi_j}{|\xi|^2}$ is positive homogeneous of degree -1, we conclude that $\widehat{P}_j(\xi)$ is positive homogeneous of degree -1, and furthermore, by Proposition 4.59 and Exercise 4.57, that the polynomial P_j is positive homogeneous of degree $1 - n \leq -1$.

Now invoking Exercise 4.129 we obtain $P_j \equiv 0$. The latter when used in (13.7.5) proves (7.14.4). Identities (7.14.5) and (7.14.6) follow from (7.14.4) and (13.7.6).

Exercise 7.79 Since $E_{\Delta^2} \in S'(\mathbb{R}^n)$ is a fundamental solution for Δ^2, we have $\Delta^2 E_{\Delta^2} = \delta$ in $S'(\mathbb{R}^n)$. Fix $j, k \in \{1, \ldots, n\}$. Then, $\Delta^2(\partial_j \partial_k E_{\Delta^2}) = \partial_j \partial_k \delta$ in $S'(\mathbb{R}^n)$. Take the Fourier transform of the last equation to arrive at

$$|\xi|^4 \mathcal{F}(\partial_j \partial_k E_{\Delta^2}) = -\xi_j \xi_k \quad \text{in} \quad S'(\mathbb{R}^n). \tag{13.7.7}$$

Under the current assumption on n, by Exercise 4.5 we have $\frac{1}{|\xi|^2} \in S'(\mathbb{R}^n)$. Also, $\frac{\xi_j \xi_k}{|\xi|^4} \in L_{loc}^1(\mathbb{R}^n)$ and in view of Example 4.4 one may infer that $\frac{\xi_j \xi_k}{|\xi|^4} \in S'(\mathbb{R}^n)$. In addition, $|\xi|^4 \in \mathcal{L}(\mathbb{R}^n)$, thus $|\xi|^4 \cdot \frac{\xi_j \xi_k}{|\xi|^4}, |\xi|^4 \cdot \frac{1}{|\xi|^2} \in S'(\mathbb{R}^n)$ (recall *(b)* in Theorem 4.14), and $|\xi|^4 \cdot \frac{\xi_j \xi_k}{|\xi|^4} = \xi_j \xi_k, |\xi|^4 \cdot \frac{1}{|\xi|^2} = |\xi|^2$ in $S'(\mathbb{R}^n)$. These facts combined with (13.7.7) imply

$$|\xi|^4 \left(\mathcal{F}(\partial_j \partial_k E_{\Delta^2}) + \frac{\xi_j \xi_k}{|\xi|^4} \right) = 0 \quad \text{in} \quad S'(\mathbb{R}^n). \tag{13.7.8}$$

Thus, $\mathrm{supp}\left(\mathcal{F}(\partial_j \partial_k E_{\Delta^2}) + \frac{\xi_j \xi_k}{|\xi|^4} \right) \subseteq \{0\}$ and by Exercise 4.37, it follows that

$$\mathcal{F}(\partial_j \partial_k E_{\Delta^2}) = -\frac{\xi_j \xi_k}{|\xi|^4} + \widehat{R_{jk}}(\xi) \quad \text{in } S'(\mathbb{R}^n), \tag{13.7.9}$$

where R_{jk} is a polynomial in $\mathbb{R}^n$.

It is easy to check that

$$\partial_j \partial_k E_{\Delta^2} = -\frac{1}{2(n-2)\omega_{n-1}} \cdot \frac{\delta_{jk}}{|x|^{n-2}} + \frac{1}{2\omega_{n-1}} \cdot \frac{x_j x_k}{|x|^n} \quad \text{in} \ S'(\mathbb{R}^n). \tag{13.7.10}$$

Hence, $\partial_j \partial_k E_{\Delta^2}$ is positive homogeneous of degree $2 - n$ which, in turn, when combined with Proposition 4.59 implies that $\mathcal{F}(\partial_j \partial_k E_{\Delta^2})$ is positive homogeneous of degree -2. Thus, the term in the right-hand side of (13.7.9) is positive homogeneous of degree -2. Since $\frac{\xi_j \xi_k}{|\xi|^4}$ is positive homogeneous of degree -2, we may conclude that $\widehat{R_{jk}}(\xi)$ is positive homogeneous of degree -2 and furthermore (by Proposition 4.59 and Exercise 4.57) that the polynomial R_{jk} is positive homogeneous of degree $2 - n \leq -1$. Now invoking Exercise 4.129 we obtain $R_{jk} \equiv 0$. The latter when used in (13.7.9) proves (7.14.7). Identity (7.14.8) follows from (7.14.7) and (13.7.10). As for identity (7.14.9), apply the Fourier transform to (7.14.8) and then use Proposition 4.64 with $\lambda = n - 2$.

Exercise 7.80 If $P(D)$ is elliptic then first show that there exists $C \in (0, \infty)$ such that $P_m(\xi) \geq C$ for $\xi \in S^{n-1}$, thus conclude $P_m(\xi) \geq C|\xi|^m$ for every $\xi \in \mathbb{R}^n \setminus \{0\}$. Use the latter and the fact that

$$|P(\xi)| \geq |P_m(\xi)| - \left(\sum_{|\alpha| \leq m-1} a_\alpha \xi^\alpha \right)$$

to obtain the desired conclusion. Conversely, suppose that there exist finite, positive numbers C, R such that $|P(\xi)| \geq C|\xi|^m$ for every $\xi \in \mathbb{R}^n \setminus B(0, R)$ and that $P_m(\xi_*) = 0$ for some $\xi_* \in \mathbb{R}^n \setminus \{0\}$. Then for every $\lambda > R/|\xi_*|$ we have

$$0 < C\lambda^m |\xi_*|^m \leq |P(\lambda\xi_*)| = \left| \sum_{|\alpha| < m} a_\alpha \lambda^{|\alpha|} \xi_*^\alpha \right| \leq \sum_{j=1}^{m-1} c_j \lambda^j$$

and obtain a contradiction by dividing with λ^m and then letting $\lambda \to \infty$.

Exercise 7.81 Fix $\varphi \in C_0^\infty(\mathbb{R}^n)$. Using the fact that φ is compactly supported, Lebesgue's Dominated Convergence Theorem, formula (14.8.4) twice, and the fact that A is symmetric, we may write

$$\int_{\mathbb{R}^n} E_A(x) L_A \varphi(x) \, dx = \lim_{\varepsilon \to 0^+} \int_{\mathbb{R}^n \setminus \overline{B(0,\varepsilon)}} E_A(x) L_A \varphi(x) \, dx$$

$$= \lim_{\varepsilon \to 0^+} \left[\int_{\mathbb{R}^n \setminus \overline{B(0,\varepsilon)}} (L_A E_A)(x) \varphi(x) \, dx - \frac{1}{\varepsilon} \int_{\partial B(0,\varepsilon)} E_A(x) (A\nabla\varphi(x)) \cdot x \, d\sigma(x) \right.$$

$$+ \frac{1}{\varepsilon} \int_{\partial B(0,\varepsilon)} [\varphi(x) - \varphi(0)](A\nabla E_A(x)) \cdot x \, d\sigma(x)$$

$$\left. + \frac{\varphi(0)}{\varepsilon} \int_{\partial B(0,\varepsilon)} (A\nabla E_A(x)) \cdot x \, d\sigma(x) \right]. \tag{13.7.11}$$

For the second equality in (13.7.11) we also used the fact that the outward unit normal to $\partial B(0, \varepsilon)$ is $\nu(x) = \frac{1}{\varepsilon}x$, for $x \in \partial B(0, \varepsilon)$.

Analyze each of the integrals in the rightmost side of (13.7.11). First, a direct computation shows that $L_A E_A = 0$ pointwise in $\mathbb{R}^n \setminus \{0\}$. Second, (7.12.45), (7.12.27), and the Mean Value Theorem for φ, imply

$$\left| \frac{1}{\varepsilon} \int_{\partial B(0,\varepsilon)} [\varphi(x) - \varphi(0)](A\nabla E_A(x)) \cdot x \, d\sigma(x) \right| \leq C\varepsilon \xrightarrow[\varepsilon \to 0^+]{} 0, \tag{13.7.12}$$

where the constant C in (13.7.12) depends on A, $\|\nabla\varphi\|_{L^\infty(\mathbb{R}^n)}$ and n but not on ε. Third, thanks to (7.12.27) we have

$$\left| \frac{1}{\varepsilon} \int_{\partial B(0,\varepsilon)} E_A(x) (A\nabla\varphi(x)) \cdot x \, d\sigma(x) \right|$$

$$\leq \begin{cases} C\varepsilon & \text{if } n \geq 3 \\ C(|\ln\varepsilon| + 1)\varepsilon & \text{if } n = 2 \end{cases} \xrightarrow[\varepsilon \to 0^+]{} 0. \tag{13.7.13}$$

Fourth, (7.12.45) and a natural change of variables allow us to write

$$\frac{1}{\varepsilon} \int_{\partial B(0,\varepsilon)} (A\nabla E_A(x)) \cdot x \, d\sigma(x) = \frac{\varepsilon}{\omega_{n-1} \sqrt{\det A}} \int_{\partial B(0,\varepsilon)} \frac{d\sigma(x)}{[(A^{-1}x) \cdot x]^{\frac{n}{2}}}$$

$$= \frac{1}{\omega_{n-1} \sqrt{\det A}} \int_{S^{n-1}} \frac{d\sigma(x)}{[(A^{-1}x) \cdot x]^{\frac{n}{2}}}. \qquad (13.7.14)$$

Combining (13.7.11), (7.12.44), (13.7.12), (13.7.13), and (13.7.14), we conclude that, under the current assumptions on A, for every $\varphi \in C_0^\infty(\mathbb{R}^n)$ we have

$$\int_{\mathbb{R}^n} E_A(x) L_A \varphi(x) \, dx = \frac{\varphi(0)}{\omega_{n-1} \sqrt{\det A}} \int_{S^{n-1}} \frac{d\sigma(x)}{[(A^{-1}x) \cdot x]^{\frac{n}{2}}}. \qquad (13.7.15)$$

In particular, (13.7.15) holds if A has real entries, is symmetric and satisfies (7.12.8). In this latter situation, we already proved that E_A is a fundamental solution for L_A, thus the left-hand side in (13.7.15) is equal to $\langle E_A, L_A \varphi \rangle = \langle L_A E_A, \varphi \rangle = \langle \delta, \varphi \rangle = \varphi(0)$, which means that (7.12.47) holds in the case when A has real entries, is symmetric and satisfies (7.12.8). As in the proof of Theorem 7.68, the identity (7.12.47) may be extended to the class of complex symmetric matrices satisfying condition (7.12.8).

Based on what we have just proved and (13.7.15) we deduce that

$$\int_{\mathbb{R}^n} E_A(x) L_A \varphi(x) \, dx = \varphi(0) \qquad (13.7.16)$$

holds for every $\varphi \in C_0^\infty(\mathbb{R}^n)$ and every matrix $A \in M_{n \times n}(\mathbb{C}^n)$ satisfying $A = A^\top$ as well as condition (7.12.8). The latter, combined with the fact that $E_A \in L_{loc}^1(\mathbb{R}^n)$ further implies that E_A is a fundamental solution for L_A whenever $A \in M_{n \times n}(\mathbb{C})$ is symmetric and satisfies (7.12.8).

13.8 Solutions to Exercises from Section 9.3

Exercise 9.12 Let E be the fundamental solution for the operator $\partial_t - \Delta_x$ in $\mathbb{R}^{n+1}$ as in (8.1.8). Then $E \in L_{loc}^1(\mathbb{R}^{n+1})$ and we claim that for each $x \in \mathbb{R}^n \setminus \{0\}$, the function $E(x,t)$ is absolutely integrable with respect to $t \in \mathbb{R}$. Indeed, using the change of variables $\tau = \frac{|x|^2}{4t}$, (14.5.1), the fact that $\Gamma\left(\frac{n}{2}\right) = \left(\frac{n}{2} - 1\right)\Gamma\left(\frac{n}{2} - 1\right)$, and (14.5.6), we obtain

$$\int_{-\infty}^\infty E(x,t) \, dt = -\frac{1}{4\pi^{n/2}|x|^{n-2}} \int_0^\infty e^{-\tau} \tau^{\frac{n}{2}-2} \, d\tau = -\frac{\Gamma(n/2 - 1)}{4\pi^{n/2}|x|^{n-2}}$$

$$= -\frac{1}{(n-2)\omega_{n-1}} \cdot \frac{1}{|x|^{n-2}} \quad \text{if } x \in \mathbb{R}^n \setminus \{0\}. \qquad (13.8.1)$$

Hence, we may apply Proposition 9.6 and Proposition 9.7, to conclude that the distribution

$$-\int_{-\infty}^{\infty} E(x,t)\,dt = \frac{1}{(n-2)\omega_{n-1}} \cdot \frac{1}{|x|^{n-2}}, \quad x \in \mathbb{R}^n \setminus \{0\},$$

is a fundamental solution for Δ in $\mathbb{R}^n$, when $n \geq 3$.

13.9 Solutions to Exercises from Section 10.7

Exercise 10.32 We have $\Delta \mathbf{u} + \nabla p = 0$ in $[\mathcal{D}'(\Omega)]^n$ and $\operatorname{div} u = 0$ in $\mathcal{D}'(\Omega)$. Apply div to the first equation and conclude that $\Delta p = 0$ in $\mathcal{D}'(\Omega)$. Next apply Δ to the first equation. Finally, use Theorem 7.10 and Theorem 7.23.

Exercise 10.33 Follow the outline from Exercise 10.16 with the adjustment that the integrals in (10.3.24) this time will be bounded by $C\|\nabla\varphi\|_{L^\infty(\mathbb{R}^n)}\,\varepsilon(1 + |\ln\varepsilon|)$.

Exercise 10.34 Follow the outline from Exercise 10.30 with the adjustment that the integrals in (10.6.23) this time will be bounded by $C\|\nabla\varphi\|_{L^\infty(\mathbb{R}^n)}\,\varepsilon(1 + |\ln\varepsilon|)$.

Exercise 10.35 Note that $\mathbf{u}$ is a solution for (10.7.4) if and only if $\mathbf{u} - Ax - b$ is a solution for (10.5.3).

Exercise 10.36 Show that Theorem 10.3 applies.

13.10 Solutions to Exercises from Section 11.7

Exercise 11.22 Follow the outline of the proof of Theorem 11.1.

Exercise 11.23 Compute $L_2 \mathbb{E}_{L_1 L_2}$ using the analogue of (11.3.44) written for the operator $L := L_1 L_2$, and make use of a suitable version of (11.3.69). Deduce that

$$P(x) = c_{m_1,m_2,n,q}\Delta^{(n+q)/2}\int_{S^{n-1}} (x \cdot \xi)^{2m_1+q}[L_2(\xi)]^{-1}\,d\sigma(\xi) \qquad (13.10.1)$$

for every $x \in \mathbb{R}^n$, from which all the desired properties of P follow.

Exercise 11.24 Start with (11.3.44) written for $L = \Delta^m$ and use (7.5.24), (7.5.25), and (7.5.13), in order to deduce that (11.7.3) holds with

$$P(x) = \mathcal{P}_{m+\frac{n+q}{2},\frac{n+q}{2}}(x) + C(n,q)\Delta^{(n+q)/2}[|x|^q], \quad x \in \mathbb{R}^n, \qquad (13.10.2)$$

where $C(n,q)$ is as in (7.5.26).

13.11 Solutions to Exercises from Section 12.6

Exercise 12.47 Example 4.36 implies $\widehat{u}(\xi) = \pi^n \prod_{j=1}^{n} \chi_{[-1,1]}(\xi_j)$ for all points $\xi = (\xi_1, \ldots, \xi_n) \in \mathbb{R}^n$. Use this to show that $u \in H^s(\mathbb{R}^n)$ for every $s \in \mathbb{R}$. In particular, by (12.1.63), we have $u \in C^\infty(\mathbb{R}^n)$. In addition, Exercise 3.33 applied with $\alpha = (1, \ldots, 1)$ and $\beta = (0, \ldots, 0)$, gives that if we assume $u \in \mathcal{S}(\mathbb{R}^n)$ then necessarily $\lim_{R \to \infty} \left[\sup_{|x| \geq R} |x_1 \cdots x_n \cdot u(x)| \right] = 0$. However, for each $R > 0$ we have $\sup_{|x| \geq R} |x_1 \cdots x_n u(x)| = 1$ so the latter limits cannot be zero. This proves that $u \notin \mathcal{S}(\mathbb{R}^n)$.

Exercise 12.48 Consider the function $u(x) := e^{-|x|^2}$ for $x \in \mathbb{R}^n$ and use Exercise 3.3 and item *(2)* in Theorem 12.1.

Exercise 12.49 Since $\widehat{\delta} = 1$, we have $\delta \in H^s(\mathbb{R}^n)$ if and only if $\langle \xi \rangle^s \in L^2(\mathbb{R}^n)$. The latter membership is true if and only if $s < -n/2$.

Exercise 12.50 Let $r > 0$ be such that the sets $\overline{B(a_j, r)}$, $j \in \{1, \ldots, N\}$, are pairwise disjoint and for each $j \in \{1, \ldots, N\}$ pick a function $\psi \in C_0^\infty(B(a_j, r))$ with $\psi \equiv 1$ on $\overline{B(a_j, r/2)}$. Then $u = \sum_{j=1}^{N} \psi_j u$ and it suffices to show that $\psi_j u \equiv 0$ for each $j \in \{1, \ldots, N\}$. Fix an arbitrary $j \in \{1, \ldots, N\}$ and note that $\psi_j u \in H^{-n/2}(\mathbb{R}^n)$ (cf. item *(1)* in Theorem 12.7) and $\text{supp}(\psi_j u) \subseteq \{a_j\}$. By Exercise 2.75, we have $\psi_j u = \sum_{|\alpha| \leq k_j} c_\alpha^j \partial^\alpha \delta_{a_j}$, for some $k_j \in \mathbb{N}_0$ and coefficients $c_\alpha^j \in \mathbb{C}$. Taking the Fourier transform, applying item *(2)* in Theorem 4.26 and (4.2.6), we obtain $\widehat{\psi_j u}(\xi) = \left(\sum_{|\alpha| \leq k_j} c_\alpha^j i^{|\alpha|} \xi^\alpha \right) e^{-ia_j \cdot \xi}$ for all $\xi \in \mathbb{R}^n$. The membership $\psi_j u \in H^{-n/2}(\mathbb{R}^n)$ implies

$$\left| \sum_{|\alpha| \leq k_j} c_\alpha^j i^{|\alpha|} \xi^\alpha \right|^2 (1 + |\xi|^2)^{-n/2} \in L^1(\mathbb{R}^n). \tag{13.11.1}$$

However, (13.11.1) holds if and only if $\sum_{|\alpha| \leq k_j} c_\alpha^j i^{|\alpha|} \xi^\alpha \equiv 0$, thus $c_\alpha^j = 0$ for all $|\alpha| \leq k_j$. This proves $\psi_j u \equiv 0$.

Exercise 12.51 Pick $\varphi \in \mathcal{S}(\mathbb{R}^n)$. Then $\widehat{\varphi} \in \mathcal{S}(\mathbb{R}^n)$. By Exercise 3.13 we know that $\langle \xi \rangle^{-t} \in \mathcal{L}(\mathbb{R}^n)$, hence item *(a)* in Theorem 3.14 implies $\langle \xi \rangle^{-t} \widehat{\varphi} \in \mathcal{S}(\mathbb{R}^n)$. The expression for $\mathcal{B}_t$ may now be written as

$$\mathcal{B}_t \varphi(x) = \mathcal{F}^{-1}(\langle \xi \rangle^{-t} \widehat{\varphi})(x), \qquad \forall x \in \mathbb{R}^n. \tag{13.11.2}$$

The claim in *(1)* now follows by invoking Theorem 3.25. In addition, identity (13.11.2) further yields

$$\mathcal{F}(\mathcal{B}_t \varphi)(\xi) = \langle \xi \rangle^{-t} \widehat{\varphi}(\xi), \qquad \forall \xi \in \mathbb{R}^n. \tag{13.11.3}$$

This and (12.1.4) then give

$$\|\mathcal{B}_t \varphi\|_{H^{s+t}(\mathbb{R}^n)}^2 = \int_{\mathbb{R}^n} \langle \xi \rangle^{2s+2t} \langle \xi \rangle^{-2t} |\widehat{\varphi}(\xi)|^2 \, d\xi = \|\varphi\|_{H^s(\mathbb{R}^n)}^2, \tag{13.11.4}$$

proving *(2)*.

To show the identity in *(3)*, let $t, r \in \mathbb{R}$ and write (making use of (13.11.3))

$$(\mathcal{B}_t \circ \mathcal{B}_r)(\varphi)(x) = (2\pi)^{-n} \int_{\mathbb{R}^n} e^{ix \cdot \xi} \langle \xi \rangle^{-t} \mathcal{F}(\mathcal{B}_r \varphi)(\xi) \, d\xi$$

$$= (2\pi)^{-n} \int_{\mathbb{R}^n} e^{ix \cdot \xi} \langle \xi \rangle^{-t} \langle \xi \rangle^{-r} \widehat{\varphi}(\xi) \, d\xi$$

$$= (\mathcal{B}_{t+r} \varphi)(x), \qquad \forall x \in \mathbb{R}^n. \tag{13.11.5}$$

Also, the identity in *(4)* is immediate from (13.11.2). Finally, since $\mathcal{B}_t$ is linear, given the norm estimate in (13.11.4), and the fact that $\mathcal{S}(\mathbb{R}^n)$ is dense in $H^s(\mathbb{R}^n)$ (cf. *(3)* in Theorem 12.1), the extension of $\mathcal{B}_t$ by continuity to a linear and continuous map from $H^s(\mathbb{R}^n)$ into $H^{s+t}(\mathbb{R}^n)$ follows.

Exercise 12.52 For each $j \in \mathbb{N}$ we have $u_j \in C_0^\infty(\mathbb{R})$, hence $u_j \in H^s(\mathbb{R})$ for all $s \in \mathbb{R}$. Also, since

$$u_j'(x) = \theta'(x) \frac{\sin(jx)}{j} + \theta(x) \cos(jx), \qquad \forall x \in \mathbb{R}, \ \forall j \in \mathbb{N}, \tag{13.11.6}$$

we have

$$\|u_j\|_{H^1(\mathbb{R})}^2 = \int_{\mathbb{R}} (1 + \xi^2) |\widehat{u_j}(\xi)|^2 \, d\xi = \int_{\mathbb{R}} (|\widehat{u_j}(\xi)|^2 + |\widehat{u_j'}(\xi)|^2) \, d\xi$$

$$= 2\pi \int_{\mathbb{R}} (|u_j(x)|^2 + |u_j'(x)|^2) \, dx$$

$$\leq 2\pi \int_{\mathbb{R}} (3|\theta(x)|^2 + 2|\theta'(x)|^2) \, dx \leq 6\pi \|\theta\|_{H^1(\mathbb{R})}^2 < \infty. \tag{13.11.7}$$

This proves that the sequence $\{u_j\}_{j \in \mathbb{N}}$ is bounded in $H^1(\mathbb{R})$.

Observe that $|u_j| \leq |\theta|/j$ for all $j \in \mathbb{N}$, hence

$$u_j \xrightarrow[j \to \infty]{H^0(\mathbb{R})} 0. \tag{13.11.8}$$

Fix now $s < 1$. Then any subsequence $\{u_{j_k}\}_{k \in \mathbb{N}}$ is bounded in $H^s(\mathbb{R})$ (since it is bounded in $H^1(\mathbb{R})$). By Theorem 12.8 we know that, with $K := [-1, 1]$, the embedding $H^1(\mathbb{R}) \cap \mathcal{E}_K'(\mathbb{R}) \hookrightarrow H^s(\mathbb{R})$ is compact for all $s < 1$. Thus, any subsequence $\{u_{j_k}\}_{k \in \mathbb{N}}$ contains a subsequence convergent in $H^s(\mathbb{R})$, which by Remark 12.2 and (13.11.8) should in fact converge to zero. This ultimately implies that $\{u_j\}_{j \in \mathbb{N}}$ converges to zero in $H^s(\mathbb{R})$.

Assume now that there exists a subsequence $\{u_{j_k}\}_{k \in \mathbb{N}}$ which is convergent in $H^1(\mathbb{R})$. Then, by Remark 12.2 and (13.11.8) the respective limit must be zero. Recalling the computation in the first two equalities in (13.11.7), the latter and (13.11.8) imply $\lim_{k \to \infty} \int_{\mathbb{R}} |u_{j_k}'(x)|^2 \, dx = 0$, and furthermore (in view of (13.11.6)), that

$$0 = \lim_{k \to \infty} \int_{\mathbb{R}} |\theta(x) \cos(j_k x)|^2 \, dx$$

$$= \frac{1}{2} \int_{\mathbb{R}} \theta(x)^2 \, dx + \lim_{k \to \infty} \frac{1}{2} \int_{\mathbb{R}} \theta(x)^2 \cos(2 j_k x) \, dx. \qquad (13.11.9)$$

Note that $\int_{\mathbb{R}} \theta(x)^2 \cos(2 j_k x) \, dx$ is equal to the real part of $\mathcal{F}(\theta^2)(-2 j_k)$ and, according to Proposition 3.1, it converges to zero as $k \to \infty$. Hence, $\int_{\mathbb{R}} \theta(x)^2 \, dx = 0$ which forces $\theta \equiv 0$. The latter is in contradiction with the properties of θ. Ultimately, this proves that $\{u_j\}_{j \in \mathbb{N}}$ has no convergent subsequence.

Exercise 12.53 If $\theta \equiv 0$ the conclusion is obvious. If θ is nonzero, consider the subsequence $\{u_{2k+1}\}_{k \in \mathbb{N}}$ and see Exercise 12.52.

Exercise 12.54 By Exercise 12.49 we have $\delta \in H^s(\mathbb{R}^n)$ for all $s < -n/2$ and $\delta \notin H^{-n/2}(\mathbb{R}^n)$. Applying Exercise 12.51, it follows that $u := \mathcal{B}_{n/2+s_2}\delta$ belongs to $H^{s+n/2+s_2}(\mathbb{R}^n)$ for all $s < -n/2$. Since $s + n/2 + s_2 < s_2$, we have that $u \in H^s(\mathbb{R}^n)$ for all $s < s_2$, hence $u \in H^{s_1}(\mathbb{R}^n)$. Also, $u \notin H^{s_2}(\mathbb{R}^n)$. Otherwise, by Exercise 12.51 we would have $\delta = \mathcal{B}_{-n/2-s_2}u \in H^{-n/2-s_2+s_2}(\mathbb{R}^n) = H^{-n/2}(\mathbb{R}^n)$, a contradiction.

Exercise 12.55 Consider $u(x) = 1$, for all $x \in \mathbb{R}^n$.

Exercise 12.56 We have $|x| \in L^2((-1, 1))$ and by *(a)* in Exercise 2.151, we have $|x|' = \operatorname{sgn} x \in L^2((-1, 1))$. Hence, $|x| \in H^1((-1, 1))$. Also, by Exercise 2.151 we have $|x|'' = 2\delta \notin L^2((-1, 1))$, thus $|x| \notin H^2((-1, 1))$.

Exercise 12.57 Pick a function $\psi \in C_0^\infty(\mathbb{R}^n)$ with the property that $\psi \equiv 1$ near $\overline{\Omega}$. Given any $f \in Lip(\Omega)$, use (2.4.5) to find $F \in Lip(\mathbb{R}^n)$ such that $f = F|_\Omega$. Use the distributional characterization of Lipschitzianity from Theorem 2.114 to show that $u := \psi F$ belongs to $H^1(\mathbb{R}^n)$. Finally, observe that $f = u|_\Omega$, hence $f \in H^1(\Omega)$.

Chapter 14
Appendix

Abstract The appendix contains a summary of topological and functional analysis results in reference to the description of the topology and equivalent characterizations of convergence in spaces of test functions and in spaces of distributions. In addition, a variety of foundational results from calculus, measure theory, and special functions originating outside the scope of this book are included here.

14.1 Summary of Topological and Functional Analytic Results

A `topology` on a set X is a family τ of subsets of X satisfying the following properties:

(1) $X, \varnothing \in \tau$;
(2) if $\{A_\alpha\}_{\alpha \in I}$ is a family of sets contained in τ, then $\bigcup_{\alpha \in I} A_\alpha \in \tau$;
(3) if $A_1, A_2 \in \tau$, then $A_1 \cap A_2 \in \tau$.

If X and τ are as above, the pair (X, τ) is called a `topological` space and the elements of τ are called `open` sets. If x is an element of X (sometimes referred to as point), then any open set containing x is called an `open neighborhood` of x. Any set that contains an open neighborhood of x is called a neighborhood of x. A sequence $\{x_j\}_{j \in \mathbb{N}}$ of elements of a topological space (X, τ) are said to converge to $x \in X$, and we write $\lim_{j \to \infty} x_j = x$, if every neighborhood of x contains all but a finite number of the elements of the sequence.

A family $\mathcal{B}$ of subsets of X is a `base for a topology` τ on X if $\mathcal{B}$ is a subfamily of τ and for each $x \in X$, and each neighborhood U of x, there is $V \in \mathcal{B}$ such that $x \in V \subset U$. We say that the base $\mathcal{B}$ generates the topology τ. An equivalent characterization of bases that is useful in applications is as follows. A base for a topological space (X, τ) is a collection $\mathcal{B}$ of open sets in τ such that every open set in τ can be written as a union of elements of $\mathcal{B}$.

It is important to note that not any family of subsets of a given set is a base for some topology on the set. Given a set X and a family $\mathcal{B}$ of subsets of X, this family

© Springer Nature Switzerland AG 2018

D. Mitrea, *Distributions, Partial Differential Equations, and Harmonic Analysis*,
Universitext, https://doi.org/10.1007/978-3-030-03296-8_14

$\mathcal{B}$ is a base for some topology on X if and only if

$$\begin{cases} \bigcup_{B \in \mathcal{B}} B = X, \\ \forall B_1, B_2 \in \mathcal{B} \text{ and } \forall x \in B_1 \cap B_2, \ \exists B \in \mathcal{B} \text{ such that } x \in B \subseteq B_1 \cap B_2. \end{cases} \quad (14.1.1)$$

If τ_1 and τ_2 are two topologies on a set X, such that every member of τ_2 is also a member of τ_1, then we say that τ_1 is finer (or, larger) than τ_2, and that τ_2 is coarser (or, smaller) than τ_1. If (X, τ) is a topological space and $E \subseteq X$, then the topology induced by τ on E is the topology for which all open sets are intersections of open sets in τ with E.

Let (X, τ) and (Y, τ') be two topological spaces. The topology on $X \times Y$ with base the collection of products of open sets in X and open sets in Y (which satisfies (14.1.1)) is called the product topology. A function $f : X \to Y$ is called continuous at $x \in X$ if the inverse image under f of every open neighborhood of $f(x)$ is an open neighborhood of x. It is called continuous on X if it is continuous at every $x \in X$. In particular, it is easy to see that if $f : X \to Y$ is continuous then $\lim_{j \to \infty} f(x_j) = f(x)$ for every convergent sequence $\{x_j\}_{j \in \mathbb{N}}$ of X converging to $x \in X$, i.e., f is sequentially continuous. While in general the converse is false, in the case when X is in fact a metric space (more on this shortly), if $f : X \to Y$ is sequential continuous, then f is continuous (see Theorem 14.1 below).

A topological space (X, τ) is said to be separated, or Hausdorff, if for any two distinct elements x and y of X, there exist U, a neighborhood of x and V, a neighborhood of y, such that $U \cap V = \varnothing$. In a Hausdorff space the limit of a convergent sequence is unique.

A metric space is a set X equipped with a distance function (also called metric). This is a function $d : X \times X \to [0, \infty)$ satisfying the following properties:

(i) if $x, y \in X$, then $d(x, y) = 0$ if and only if $x = y$ (nondegeneracy);
(ii) $d(x, y) = d(y, x)$ for every $x, y \in X$ (symmetry);
(iii) $d(x, y) \leq d(x, z) + d(z, y)$ for every $x, y, z \in X$ (triangle inequality).

Let d be a metric on X. The open balls $B(x, r) := \{y \in X : d(x, y) < r\}$, $x \in X$, $r \in (0, \infty)$, are then the base of a Hausdorff topology on X (since it satisfies (14.1.1)), denoted τ_d. For a sequence $\{x_j\}_{j \in \mathbb{N}}$ of points in X and $x \in X$ one has $\lim_{j \to \infty} x_j = x$ in the topology τ_d, if and only if the sequence of numbers $\{d(x_j, x)\}_{j \in \mathbb{N}}$ converges to 0 as $j \to \infty$. A sequence $\{x_j\}_{j \in \mathbb{N}}$ of points in (X, τ_d) is called Cauchy if $d(x_j, x_k) \to 0$ as $j, k \to \infty$. A metric space is called complete if every Cauchy sequence is convergent.

A topological space X is called metrizable if there exists a distance function for which the open balls form a base for the topology (i.e., the topology on X coincides with τ_d). For a proof of the theorem below see [65, Theorem, p. 395].

Theorem 14.1. *Let X and Y be two Hausdorff topological spaces.*

(a) If $\Lambda : X \to Y$ is continuous, then Λ is sequentially continuous.
(b) If $\Lambda : X \to Y$ is sequentially continuous and X is metrizable, then Λ is continuous.

Moving on, recall that a vector space X over $\mathbb{C}$ becomes a `topological vector space` if equipped with a topology that is compatible with its vector structure, that is, the operation of vector addition and of multiplication by a complex number are continuous maps $X \times X \to X$ and $\mathbb{C} \times X \to X$, respectively (where $X \times X$ and $\mathbb{C} \times X$ are endowed each with the corresponding product topology). Observe that in order to specify a topology on a vector space, it suffices to give the system of neighborhoods of the zero element $0 \in X$, since the system of neighborhoods of any other element of X is obtained from this via translation. In fact, it suffices to give a base of neighborhoods of 0. This is a family $\mathcal{B}$ of neighborhoods of 0 such that every neighborhood of 0 contains a member of $\mathcal{B}$. A set $E \subseteq X$ is then open if and only if, for every $x \in E$, there exists $U \in \mathcal{B}$ such that $x + U \subseteq E$, where $x + U := \{x + y : y \in U\}$.

A topological vector space is called `locally convex` if it has a base of neighborhoods of 0 consisting of sets that are convex, balanced, and absorbing. A set U is called `convex` provided $tx + (1 - t)y \in U$ whenever $x, y \in U$ and $t \in [0, 1]$. A set U is called `balanced` if $cx \in U$ whenever $x \in U$, $c \in \mathbb{C}$ and $|c| \leq 1$. A set U is called `absorbing` provided for each $x \in X$ there exists $t > 0$ such that $x \in tU := \{ty : y \in U\}$.

A `seminorm` on a vector space X is a function $p : X \to \mathbb{R}$ satisfying the properties

(1) $p(cx) = |c|p(x)$, for every $c \in \mathbb{C}$ and every $x \in X$ (positively homogeneous);
(2) $p(x + y) \leq p(x) + p(y)$, for every $x, y \in X$ (sub-additive).

In particular, a seminorm p on X satisfies $p(0) = 0$ and $p(x) \geq 0$ for every $x \in X$.

A family $\mathcal{P}$ of seminorms on X is called `separating` if, for each $x \in X$, $x \neq 0$, there exists $p \in \mathcal{P}$ such that $p(x) \neq 0$. Given a separating family of seminorms $\mathcal{P}$ on X, let B be the collection of sets of the form

$$\{x : p(x) < \varepsilon \ \forall p \in \mathcal{P}_0\}, \quad \mathcal{P}_0 \subseteq \mathcal{P}, \ \mathcal{P}_0 \text{ finite}, \ \varepsilon > 0. \tag{14.1.2}$$

Then B is a base of neighborhoods of 0 of a locally convex vector space topology $\tau_\mathcal{P}$ on X called the `topology generated by the family of seminorms` $\mathcal{P}$. In this context, it may be readily verified that if Y is a linear subspace of X and $\mathcal{P}|_Y := \{p|_Y : p \in \mathcal{P}\}$, then

$$\text{the topology induced on } Y \text{ by } \tau_\mathcal{P} \text{ coincides with } \tau_{\mathcal{P}|_Y}. \tag{14.1.3}$$

Conversely, if X is a locally convex topological vector space, for each U convex, balanced, and absorbing neighborhood of 0, the mapping $p_U : X \to \mathbb{R}$ defined by $p_U(x) := \inf\{t > 0 : t^{-1}x \in U\}$, $x \in X$ (called the Minkowski functional associated with U) is a seminorm. It is then not hard to see that the topology on X is generated by this family of seminorms (in the manner described above).

Let $\mathcal{P} = \{p_j\}_{j \in \mathbb{N}}$ be a countable family of seminorms that is also separating (thus, if $x \in X$ and $p_j(x) = 0$ for all $j \in \mathbb{N}$, then $x = 0$). The topology generated by this family is metrizable. Indeed, the function $d : X \times X \to \mathbb{R}$ defined by

$$d(x,y) := \sum_{j=1}^{\infty} 2^{-j} \frac{p_j(x-y)}{1 + p_j(x-y)}, \qquad \text{for each } x, y \in X, \qquad (14.1.4)$$

is a distance on X and the topology τ_d induced by the metric d coincides with the topology generated by $\mathcal{P}$. In the converse direction, it can be shown that the topology of a locally convex space that is metrizable, endowed with a translation invariant metric, can be generated by a countable family of seminorms.

A locally convex topological vector space that is metrizable and complete is called a Fréchet space. Thus, if a family $\mathcal{P} = \{p_j\}_{j \in \mathbb{N}}$ of seminorms generates the topology of a Fréchet space X, then whenever $p_j(x_k - x_l) \to 0$ as $k, l \to \infty$ for every $j \in \mathbb{N}$, there exists $x \in X$ such that $p_j(x_k - x) \to 0$ as $k \to \infty$ for every $j \in \mathbb{N}$.

Let X be a vector space, $\{X_j\}_{j \in J}$ be a family of vector subspaces of X such that $X = \bigcup_{j \in J} X_j$ and if $j_1, j_2 \in J$ then there exists $j_3 \in J$ with $X_{j_1} \subset X_{j_3}$ and $X_{j_2} \subset X_{j_3}$. Also assume that there exist topologies τ_j on X_j such that if $X_{j_1} \subset X_{j_2}$ then the topology τ_{j_1} is finer than the topology $\tau_{j_2 j_1}$ induced on X_{j_1} by X_{j_2}. Let

$$\mathcal{W} := \{W \subset X : W \text{ balanced, convex and such that } W \cap X_j$$

$$\text{is a neighborhood of } 0 \text{ in } X_j, \ \forall \, j \in \mathbb{N}\}. \qquad (14.1.5)$$

Then $\mathcal{W}$ is a base of neighborhoods of 0 in a locally convex topology τ on X. Call this topology the inductive limit topology on X.

If in addition whenever $X_{j_1} \subset X_{j_2}$ we also have $\tau_{j_2 j_1} = \tau_{j_1}$, we call the inductive limit strict. If the topology τ on X is the strict inductive limit of the topologies of an increasing sequence $\{X_n\}_{n \in \mathbb{N}}$, then the topology induced on X_n by the topology τ on X coincides with the initial topology τ_n on X_n, for every $n \in \mathbb{N}$.

In general, if X is a topological vector space, its dual space is the collection of all linear mappings $f : X \to \mathbb{C}$ (also referred to as functionals) that are continuous with respect to the topology on X. In the case when X is a locally convex topological vector space and $\mathcal{P}$ is a family of seminorms generating the topology of X, then a seminorm q on X is continuous if and only if there exist $N \in \mathbb{N}$, seminorms $p_1, ..., p_N \in \mathcal{P}$, and a constant $C \in (0, \infty)$, such that

$$|q(x)| \le C \max \{p_1(x), \ldots, p_n(x)\}, \qquad \forall \, x \in X. \qquad (14.1.6)$$

This fact then gives a criterion for continuity for functionals on X, since if $f : X \to \mathbb{C}$ is a linear mapping, then $q(x) := |f(x)|$, for $x \in X$, is a seminorm on X. In addition, if the family of seminorms $\mathcal{P}$ has the property that for any $p_1, p_2 \in \mathcal{P}$ there exists $p_3 \in \mathcal{P}$ with the property that $\max\{p_1(x), p_2(x)\} \le p_3(x)$ for every $x \in X$, then a linear functional $f : X \to \mathbb{C}$ is continuous if and only if there exist a seminorm $p \in \mathcal{P}$ and a constant $C \in (0, \infty)$, such that

$$|f(x)| \le C p(x), \qquad \forall \, x \in X. \qquad (14.1.7)$$

Given a nonempty set X and a family $\mathcal{F}$ of mappings $f : X \to \mathbb{C}$, denote by $\tau_{\mathcal{F}}$ the collection of all unions of finite intersections of sets $f^{-1}(V)$, for $f \in \mathcal{F}$ and V open set in $\mathbb{C}$. Then $\tau_{\mathcal{F}}$ is a topology on X, and is the weakest topology on X that makes every $f \in \mathcal{F}$ continuous. We will refer to it as the $\mathcal{F}$-topology on X.

Let X be a vector space, $\mathcal{F}$ be a separating vector space of linear functionals on X, and $\tau_{\mathcal{F}}$ be the $\mathcal{F}$-topology on X. Then $(X, \tau_{\mathcal{F}})$ is a locally convex topological space and its dual is $\mathcal{F}$. In particular, a sequence $\{x_j\}_j$ in X satisfies $x_j \to 0$ in $\tau_{\mathcal{F}}$ as $j \to \infty$ if and only if $f(x_j) \to 0$ as $j \to \infty$ for every $f \in \mathcal{F}$.

If X is a topological vector space and X' denotes its dual space, then every $x \in X$ induces a linear functional F_x on X' defined by $F_x(\Lambda) := \Lambda(x)$ for every $\Lambda \in X'$ and if we set $\mathcal{F} := \{F_x : x \in X\}$, then $\mathcal{F}$ separates points in X'. In particular, the $\mathcal{F}$-topology on X', called the weak∗-topology on X', is locally convex and every linear functional on X' that is continuous with respect to the weak∗-topology is precisely of the form F_x, for some $x \in X$. Moreover, a sequence $\{\Lambda_j\}_j \subset X'$ converges to some $\Lambda \in X'$, in the weak∗-topology on X', if and only if $\Lambda_j(x) \to \Lambda(x)$ as $j \to \infty$ for every $x \in X$.

An inspection of the definition of the weak∗-topology yields a description of the open sets in this topology. More precisely, if X is a topological vector space and for $A \subseteq X$ and $\varepsilon \in (0, \infty)$ we set

$$O_{A,\varepsilon} := \{f \in X' : |f(x)| < \varepsilon, \ \forall x \in A\}, \tag{14.1.8}$$

then the following equivalence holds:

$$O \subseteq X' \ \text{ is a weak∗-open neighborhood of } 0 \in X' \iff \text{ there exist a set } I,$$

$$A_j \subseteq X \text{ finite and } \varepsilon_j > 0 \text{ for each } j \in I, \text{ such that } O = \bigcup_{j \in I} O_{A_j, \varepsilon_j}. \tag{14.1.9}$$

The transpose of any linear and continuous operator between two topological vector spaces is always continuous at the level of dual spaces equipped with weak∗-topologies.

Proposition 14.2. *Assume X and Y are two given topological vector spaces, and denote by X', Y' their duals, each endowed with the corresponding weak∗-topology. Also, suppose $T : X \to Y$ is a linear and continuous operator, and define its transpose T^t as the mapping*

$$T^t : Y' \to X', \quad T^t(y') := y' \circ T \quad \text{for each} \quad y' \in Y'. \tag{14.1.10}$$

Then T^t is well-defined, linear, and continuous.

Proof. For each $y' \in Y'$ it follows that $y' \circ T$ is a composition of two linear and continuous mappings. Hence, $y' \circ T \in X'$ which proves that $T^t : Y' \to X'$ is well-defined. It is also clear from (14.1.10) that T^t is linear. There remains to prove that T^t is continuous. By linearity it suffices to check that T^t is continuous at 0. With this goal in mind, fix an arbitrary finite subset A of X along with an arbitrary number $\varepsilon \in (0, \infty)$, and define

$$O_{A,\varepsilon} := \{x' \in X' : |x'(x)| < \varepsilon, \ \forall x \in A\}. \tag{14.1.11}$$

Furthermore, introduce $\widetilde{A} := \{Tx : x \in A\}$ and set

$$\widetilde{O}_{\widetilde{A},\varepsilon} := \{y' \in Y' : |y'(y)| < \varepsilon, \ \forall y \in \widetilde{A}\}. \tag{14.1.12}$$

Then $T^t(\widetilde{O}_{\widetilde{A},\varepsilon}) \subseteq O_{A,\varepsilon}$. Invoking the description from (14.1.9) we may conclude that T^t is continuous at 0, and this finishes the proof of the proposition. $\qquad\square$

Proposition 14.3. *Suppose that* X, Y, Z *are topological vector spaces, and denote by* X', Y', Z' *their duals, each endowed with the corresponding weak∗-topology. In addition, assume that* $T : X \to Y$ *and* $R : Y \to Z$ *are two linear and continuous operators. Then*

$$(R \circ T)^t = T^t \circ R^t. \tag{14.1.13}$$

In particular, if $T : X \to Y$ *is a linear, continuous, bijective map, with continuous inverse* $T^{-1} : Y \to X$, *then* $T^t : Y' \to X'$ *is also bijective and has a continuous inverse* $(T^t)^{-1} : X' \to Y'$ *that satisfies* $(T^t)^{-1} = (T^{-1})^t$.

Proof. Formula (14.1.13) is immediate from definitions, while the claims in the last part of the statement are direct consequences of (14.1.13) and the fact that the transpose of the identity is also the identity. $\qquad\square$

We also state and prove an embedding result at the level of dual spaces endowed with the weak∗-topology.

Proposition 14.4. *Suppose* X *and* Y *are topological vector spaces such that* $X \subseteq Y$ *densely and the inclusion map* $\iota : X \to Y$, $\iota(x) := x$ *for each* $x \in X$, *is continuous. Then* Y' *endowed with the weak∗-topology embeds continuously in the space* X' *endowed with the weak∗-topology, in the sense that the mapping*

$$\iota^t : Y' \longrightarrow X' \tag{14.1.14}$$

is well-defined, linear, injective, and continuous. Under the identification of Y' *with* $\iota^t(Y') \subseteq X'$, *we therefore have .*

$$Y' \hookrightarrow X'. \tag{14.1.15}$$

Proof. The fact that ι^t is well-defined, linear, and continuous follows directly from Proposition 14.2 and assumptions. Assume now that $y' \in Y'$ is such that $\iota^t(y') = 0$. Then from the fact that $y' \circ \iota : X \to \mathbb{C}$ is zero we deduce that $y'\big|_X = 0$. Since $y' : Y \to \mathbb{C}$ is continuous and X is dense in Y, we necessarily have that y' vanishes on Y, forcing $y' = 0$ in Y'. Keeping in mind that ι^t is linear, this implies that ι^t is injective. $\qquad\square$

Theorem 14.5 (Hahn–Banach Theorem). *Let* X *be a vector space (over complex numbers) and suppose* $p : X \to [0, \infty)$ *is a seminorm on* X. *Also assume that* Y *is a linear subspace of* X *and that* $\phi : Y \to \mathbb{C}$ *is a linear functional which is dominated by* p *on* Y, *i.e.,*

$$|\phi(y)| \le p(y), \qquad \forall\, y \in Y. \tag{14.1.16}$$

Then there exists a linear functional $\Phi : X \to \mathbb{C}$ satisfying

$$\Phi(y) = \phi(y) \qquad \forall\, y \in Y, \qquad |\Phi(x)| \le p(x) \qquad \forall\, x \in X. \tag{14.1.17}$$

In the next installment we shall specialize the above considerations to various specific settings used in the book. The reader is reminded that $\Omega \subseteq \mathbb{R}^n$ denotes a fixed, nonempty, arbitrary open set.

14.1.0.1 The Topological Vector Space $\mathcal{E}(\Omega)$

By τ we denote the topology on $C^\infty(\Omega)$ generated by the following family of seminorms:

$$p_{K,m} : C^\infty(\Omega) \to \mathbb{R}, \quad p_{K,m}(\varphi) := \sup_{\substack{x \in K \\ \alpha \in \mathbb{N}_0^n,\, |\alpha| \le m}} |\partial^\alpha \varphi(x)|, \quad \forall\, \varphi \in C^\infty(\Omega), \tag{14.1.18}$$

where $K \subset \Omega$ is a compact set and $m \in \mathbb{N}_0$. Consequently, a sequence $\varphi_j \in C^\infty(\Omega)$, $j \in \mathbb{N}$, converges in τ to a function $\varphi \in C^\infty(\Omega)$ as $j \to \infty$, if and only if for any compact set $K \subset \Omega$ and any $m \in \mathbb{N}_0$ one has

$$\lim_{j \to \infty} \sup_{\alpha \in \mathbb{N}_0^n,\, |\alpha| \le m} \sup_{x \in K} |\partial^\alpha(\varphi_j - \varphi)(x)| = 0. \tag{14.1.19}$$

We will use the notation

$$\mathcal{E}(\Omega) = (C^\infty(\Omega), \tau). \tag{14.1.20}$$

The space $\mathcal{E}(\Omega)$ is locally convex and metrizable since its topology is defined by the family of countable seminorms $\{p_{K_m,m}\}_{m \in \mathbb{N}_0}$ where

$$K_m \subset \mathring{K}_{m+1} \subset \Omega, \quad K_m \text{ is compact for each } m \in \mathbb{N}_0 \text{ and } \Omega = \bigcup_{m=0}^{\infty} K_m. \tag{14.1.21}$$

In addition, τ is independent of the family $\{K_m\}_{m \in \mathbb{N}_0}$ with the above properties and $\mathcal{E}(\Omega)$ is complete. Thus, $\mathcal{E}(\Omega)$ is a Frechét space.

14.1.0.2 The Topological Vector Space $\mathcal{E}'(\Omega)$

Based on the discussion on dual spaces for locally convex topological vector spaces (see (14.1.7) and the remarks preceding it), it follows that the dual space of $\mathcal{E}(\Omega)$ is the collection of all linear functionals $u : \mathcal{E}(\Omega) \to \mathbb{C}$ for which there exist $m \in \mathbb{N}$, a compact set $K \subset \mathbb{R}^n$ with $K \subset \Omega$, and a constant $C > 0$ such that

$$|u(\varphi)| \le C \sup_{\alpha \in \mathbb{N}_0^n,\, |\alpha| \le m} \sup_{x \in K} |\partial^\alpha(\varphi)(x)|, \quad \forall\, \varphi \in C^\infty(\Omega). \tag{14.1.22}$$

This dual space will be endowed with the weak∗-topology induced by $\mathcal{E}(\Omega)$ and we denote this topological space by $\mathcal{E}'(\Omega)$. Hence, if for each $\varphi \in C^\infty(\Omega)$ we consider the evaluation mapping F_φ taking any functional u from the dual of $\mathcal{E}(\Omega)$ into the number $F_\varphi(u) := u(\varphi) \in \mathbb{C}$, then the family $\mathcal{F} := \{F_\varphi : \varphi \in \mathcal{E}(\Omega)\}$ separates points in the dual of $\mathcal{E}(\Omega)$, and the weak∗-topology on this dual is the $\mathcal{F}$-topology on it. In particular, if $\{u_j\}_{j\in\mathbb{N}}$ is a sequence in $\mathcal{E}'(\Omega)$ and $u \in \mathcal{E}'(\Omega)$, then

$$
\begin{aligned}
u_j \to u \text{ in } \mathcal{E}'(\Omega) \text{ as } j \to \infty \iff \\
u_j(\varphi) \to u(\varphi) \text{ as } j \to \infty, \ \forall \varphi \in C^\infty(\Omega).
\end{aligned}
\tag{14.1.23}
$$

Moreover, a sequence $\{u_j\}_{j\in\mathbb{N}}$ in $\mathcal{E}'(\Omega)$ is Cauchy provided $\lim\limits_{j,k\to\infty} (u_j - u_k)(\varphi) = 0$ for every $\varphi \in C^\infty(\Omega)$ and the weak∗-topology on the dual of $\mathcal{E}(\Omega)$ is locally convex and complete.

14.1.0.3 The Topological Vector Space $\mathcal{D}_K(\Omega)$

Let $K \subseteq \Omega$ be a compact set in $\mathbb{R}^n$. Denote by $\mathcal{D}_K(\Omega)$ the topological vector space of functions $\{f \in C^\infty(\Omega) : \operatorname{supp} f \subseteq K\}$ with the topology induced by τ, the topology in $\mathcal{E}(\Omega)$. Then the topology on $\mathcal{D}_K(\Omega)$ is generated by the family of seminorms

$$
\begin{cases}
p_m : \mathcal{D}_K(\Omega) \to \mathbb{R}, \\
p_m(\varphi) := \sup\limits_{\substack{x\in K \\ \alpha\in\mathbb{N}_0^n, |\alpha|\le m}} |\partial^\alpha \varphi(x)|, \quad \forall \varphi \in C^\infty(\Omega), \ \operatorname{supp} \varphi \subseteq K,
\end{cases}
\tag{14.1.24}
$$

where $m \in \mathbb{N}_0$. Hence, $\mathcal{D}_K(\Omega)$ is a Fréchet space. In addition, a linear map $u : \mathcal{D}_K(\Omega) \to \mathbb{C}$ is continuous if and only if there exist $m \in \mathbb{N}$ and a constant $C > 0$, both depending on K, such that

$$
|u(\varphi)| \le C \sup\limits_{\alpha\in\mathbb{N}_0^n, |\alpha|\le m} \sup\limits_{x\in K} |\partial^\alpha(\varphi)(x)|, \quad \forall \varphi \in C^\infty(\Omega), \ \operatorname{supp} \varphi \subseteq K.
\tag{14.1.25}
$$

14.1.0.4 The Topological Vector Space $\mathcal{D}(\Omega)$

The topological vector space on $C_0^\infty(\Omega)$ endowed with the inductive limit topology of the Frechét spaces $\mathcal{D}_K(\Omega)$ will be denoted by $\mathcal{D}(\Omega)$. In this setting, we have

$$
\varphi_j \to \varphi \text{ in } \mathcal{D}(\Omega) \text{ as } j \to \infty
$$

$$
\iff
\begin{cases}
\exists K \subseteq \Omega \text{ compact set such that } \varphi_j \in \mathcal{D}_K(\Omega), \ \forall j, \text{ and} \\
\varphi_j \to \varphi \text{ in } \mathcal{D}_K(\Omega) \text{ as } j \to \infty.
\end{cases}
\tag{14.1.26}
$$

The topology $\mathcal{D}(\Omega)$ is locally convex and complete but not metrizable (thus not Fréchet) and is the strict inductive limit of the topologies $\{\mathcal{D}_{K_j}(\Omega)\}_{j\in\mathbb{N}}$, where the family $\{K_j\}_{j\in\mathbb{N}}$ is as in (14.1.21).

We also record an important result that is proved in [65, Theorem 6.6, p. 155].

Theorem 14.6. *Let X be a locally convex topological vector space and suppose the map $\Lambda : \mathcal{D}(\Omega) \to X$ is linear. Then Λ is continuous if and only if for every sequence $\{\varphi_j\}_{j\in\mathbb{N}}$ in $C_0^\infty(\Omega)$ satisfying $\varphi_j \xrightarrow[j\to\infty]{\mathcal{D}(\Omega)} 0$ we have $\lim_{j\to\infty} \Lambda(\varphi_j) = 0$ in X.*

14.1.0.5 The Topological Vector Space $\mathcal{D}'(\Omega)$

The dual space of $\mathcal{D}(\Omega)$ endowed with the weak∗-topology is denoted by $\mathcal{D}'(\Omega)$. Hence, if $\{u_j\}_{j\in\mathbb{N}}$ is a sequence in $\mathcal{D}'(\Omega)$ and $u \in \mathcal{D}'(\Omega)$, then

$$u_j \to u \text{ in } \mathcal{D}'(\Omega) \text{ as } j \to \infty \iff$$
$$u_j(\varphi) \to u(\varphi) \text{ as } j \to \infty, \ \forall\, \varphi \in C_0^\infty(\Omega). \tag{14.1.27}$$

In addition, a sequence $\{u_j\}_{j\in\mathbb{N}}$ in $\mathcal{D}'(\Omega)$ is called Cauchy provided

$$\lim_{j,k\to\infty} (u_j - u_k)(\varphi) = 0 \text{ for every } \varphi \in C_0^\infty(\Omega). \tag{14.1.28}$$

The weak∗-topology on the dual of $\mathcal{D}(\Omega)$ is locally convex and complete and an inspection of this topology reveals that it coincides with the topology defined by the family of seminorms

$$\mathcal{D}'(\Omega) \ni u \mapsto \max_{1\le j\le m} |u(\varphi_j)|, \quad m \in \mathbb{N}, \ \varphi_1,\dots,\varphi_m \in \mathcal{D}(\Omega). \tag{14.1.29}$$

14.1.0.6 The Topological Vector Space $\mathcal{S}(\mathbb{R}^n)$

The Schwartz class of rapidly decreasing functions is the vector space

$$\mathcal{S}(\mathbb{R}^n) := \Big\{\varphi \in C^\infty(\mathbb{R}^n) : \ \forall\, \alpha,\beta \in \mathbb{N}_0^n,\ \sup_{x\in\mathbb{R}^n} |x^\beta \partial^\alpha \varphi(x)| < \infty\Big\}, \tag{14.1.30}$$

endowed with the topology generated by the family of seminorms $\{p_{k,m}\}_{k,m\in\mathbb{N}_0}$ defined by

$$p_{k,m} : \mathcal{S}(\mathbb{R}^n) \to \mathbb{R},$$
$$p_{k,m}(\varphi) := \sup_{\substack{x\in\mathbb{R}^n \\ |\alpha|\le m,\, |\beta|\le k}} |x^\beta \partial^\alpha \varphi(x)|, \quad \forall\, \varphi \in \mathcal{S}(\mathbb{R}^n). \tag{14.1.31}$$

Hence, the topology generated by the family of seminorms $\{p_{k,m}\}_{k,m\in\mathbb{N}_0}$ on $\mathcal{S}(\mathbb{R}^n)$ is locally convex, metrizable, and since it is also complete, the space $\mathcal{S}(\mathbb{R}^n)$ is Frechét. Moreover, a sequence $\varphi_j \in \mathcal{S}(\mathbb{R}^n)$, $j \in \mathbb{N}$, converges in $\mathcal{S}(\mathbb{R}^n)$ to a function $\varphi \in$

$\mathcal{S}(\mathbb{R}^n)$ as $j \to \infty$, if and only if for every $m, k \in \mathbb{N}_0$ one has

$$\lim_{j\to\infty} \sup_{\substack{x\in\mathbb{R}^n \\ |\alpha|\le m, |\beta|\le k}} \left| x^\beta \partial^\alpha (\varphi_j - \varphi)(x) \right| = 0. \tag{14.1.32}$$

14.1.0.7 The Topological Vector Space $\mathcal{S}'(\mathbb{R}^n)$

By the discussion about dual spaces for locally convex topological vector spaces, it follows that the dual space of $\mathcal{S}(\mathbb{R}^n)$ is the collection of all linear functions $u :$ $\mathcal{S}(\mathbb{R}^n) \to \mathbb{C}$ for which there exist $m, k \in \mathbb{N}_0$, and a finite constant $C > 0$, such that

$$|u(\varphi)| \le C \sup_{\alpha,\beta\in\mathbb{N}_0^n, |\alpha|\le m, |\beta|\le k} \sup_{x\in\mathbb{R}^n} \left| x^\beta \partial^\alpha \varphi(x) \right|, \quad \forall \varphi \in \mathcal{S}(\mathbb{R}^n). \tag{14.1.33}$$

We endow the dual of $\mathcal{S}(\mathbb{R}^n)$ with the weak*-topology and denote the resulting locally convex topological vector space by $\mathcal{S}'(\mathbb{R}^n)$. Hence, if $\{u_j\}_{j\in\mathbb{N}}$ is a sequence in $\mathcal{S}'(\mathbb{R}^n)$ and $u \in \mathcal{S}'(\mathbb{R}^n)$, then

$$\begin{aligned} u_j \to u \text{ in } \mathcal{S}'(\mathbb{R}^n) \text{ as } j \to \infty &\Longleftrightarrow \\ u_j(\varphi) \to u(\varphi) \text{ as } j \to \infty, &\ \forall \varphi \in \mathcal{S}(\mathbb{R}^n). \end{aligned} \tag{14.1.34}$$

Also, a sequence $\{u_j\}_{j\in\mathbb{N}}$ in $\mathcal{S}'(\mathbb{R}^n)$ is called Cauchy provided

$$\lim_{j,k\to\infty} (u_j - u_k)(\varphi) = 0 \text{ for every } \varphi \in \mathcal{S}(\mathbb{R}^n). \tag{14.1.35}$$

14.2 Basic Results from Calculus, Measure Theory, and Topology

The next lemma contains a discrete change of variable formula (for sums). To state it, given a set E we denote by $\#E$ its cardinality.

Lemma 14.7. *Let $\mathcal{F} : D \to R$ be a function between finite sets. If $(G, +)$ is an Abelian group, then for every function $\mathcal{G} : R \to G$ one has*

$$\sum_{a\in D} \mathcal{G}(\mathcal{F}(a)) = \sum_{b\in R} \#\mathcal{F}^{-1}(\{b\}) \mathcal{G}(b), \tag{14.2.1}$$

where $\mathcal{F}^{-1}(\{b\}) := \{a \in D : \mathcal{F}(a) = b\}$ for each $b \in R$.

As corollary, under the additional assumption that $\mathcal{F} : D \to R$ is a bijection,

$$\sum_{a\in D} \mathcal{G}(\mathcal{F}(a)) = \sum_{b\in R} \mathcal{G}(b). \tag{14.2.2}$$

Proposition 14.8 (Multinomial Theorem). *If $x = (x_1, \ldots, x_n) \in \mathbb{R}^n$ and $N \in \mathbb{N}$ are arbitrary, then*

$$\Big(\sum_{j=1}^{n} x_j\Big)^N = \sum_{|\alpha|=N} \frac{N!}{\alpha!} x^\alpha. \tag{14.2.3}$$

Theorem 14.9 (Binomial Theorem). *For any* $x, y \in \mathbb{C}^n$ *and any* $\gamma \in \mathbb{N}_0^n$ *we have*

$$(x + y)^\gamma = \sum_{\alpha+\beta=\gamma} \frac{\gamma!}{\alpha!\beta!} x^\alpha y^\beta, \tag{14.2.4}$$

(with the convention that $z^0 := 1$ *for each* $z \in \mathbb{C}$*).*

In the particular case when $x = (1, \ldots, 1) \in \mathbb{C}^n$ and $y = (-1, \ldots, -1) \in \mathbb{C}^n$, formula (14.2.4) yields

$$0 = \sum_{\alpha+\beta=\gamma} \frac{\gamma!}{\alpha!\beta!} (-1)^{|\beta|}, \qquad \forall \gamma \in \mathbb{N}_0^n \text{ with } |\gamma| > 0. \tag{14.2.5}$$

Proposition 14.10 (Leibniz's Formula). *Suppose that* $U \subseteq \mathbb{R}^n$ *is an open set,* $N \in \mathbb{N}$, *and* $f, g : U \to \mathbb{C}$ *are two functions of class* C^N *in* U. *Then*

$$\partial^\alpha(fg) = \sum_{\beta\leq\alpha} \frac{\alpha!}{\beta!(\alpha-\beta)!} (\partial^\beta f)(\partial^{\alpha-\beta} g) \qquad in \ U, \tag{14.2.6}$$

for every multi-index $\alpha \in \mathbb{N}_0^n$ *of length* $\leq N$.

It is useful to note that for each $\alpha, \beta \in \mathbb{N}_0^n$ and $x \in \mathbb{R}^n$,

$$\partial^\beta(x^\alpha) = \begin{cases} \dfrac{\alpha!}{(\alpha-\beta)!} x^{\alpha-\beta} & \text{if } \beta \leq \alpha, \\ 0 & \text{otherwise.} \end{cases} \tag{14.2.7}$$

Theorem 14.11 (Taylor's Formula in dimension one). *For every function* f *belonging to* $C_0^\infty(\mathbb{R})$ *and every* $m \in \mathbb{N}$ *the following formula holds:*

$$f(0) = \frac{(-1)^m}{(m-1)!} \int_0^\infty t^{m-1} \frac{d^m}{dt^m} [f(t)] \, dt. \tag{14.2.8}$$

Theorem 14.12 (Taylor's Formula). *Assume* $U \subseteq \mathbb{R}^n$ *is an open convex set, and that* $N \in \mathbb{N}$. *Also, suppose that* $f : U \to \mathbb{C}$ *is a function of class* C^{N+1} *on* U. *Then for every* $x, y \in U$ *one has*

$$f(x) = \sum_{|\alpha|\leq N} \frac{1}{\alpha!}(x - y)^\alpha (\partial^\alpha f)(y) \tag{14.2.9}$$

$$+ \sum_{|\alpha|=N+1} \frac{N+1}{\alpha!} \int_0^1 (1 - t)^N (x - y)^\alpha (\partial^\alpha f)(tx + (1 - t)y) \, dt.$$

In particular, for each $x, y \in U$ *there exists* $\theta \in (0, 1)$ *with the property that*

$$f(x) = \sum_{|\alpha| \leq N} \frac{1}{\alpha!} (x - y)^\alpha (\partial^\alpha f)(y) \tag{14.2.10}$$

$$+ \sum_{|\alpha| = N+1} \frac{1}{\alpha!} (x - y)^\alpha (\partial^\alpha f)(\theta x + (1 - \theta) y).$$

Theorem 14.13 (Rademacher's Theorem). *If $f : \mathbb{R}^n \to \mathbb{R}$ is a Lipschitz function with Lipschitz constant less than or equal to M, for some $M \in (0, \infty)$, then f is differentiable almost everywhere and $\|\partial_k f\|_{L^\infty(\mathbb{R}^n)} \leq M$ for each $k = 1, \ldots, n$.*

Theorem 14.14 (Lebesgue's Differentiation Theorem). *If $f \in L^1_{loc}(\mathbb{R}^n)$, then*

$$\lim_{\varepsilon \to 0^+} \frac{1}{|B(x, \varepsilon)|} \int_{B(x, \varepsilon)} |f(y) - f(x)| \, dy = 0 \quad \text{for almost every } x \in \mathbb{R}^n. \tag{14.2.11}$$

In particular,

$$\lim_{\varepsilon \to 0^+} \frac{1}{|B(x, \varepsilon)|} \int_{B(x, \varepsilon)} f(y) \, dy = f(x) \quad \text{for almost every } x \in \mathbb{R}^n. \tag{14.2.12}$$

Theorem 14.15 (Lebesgue's Dominated Convergence Theorem). *Let (X, μ) be a positive measure space and assume that $g \in L^1(X, \mu)$ is a nonnegative function. If $\{f_j\}_{j \in \mathbb{N}}$ is a sequence of μ-measurable, complex-valued functions on X, such that $|f_j(x)| \leq g(x)$ for μ-almost every $x \in X$ and $f(x) := \lim_{j \to \infty} f_j(x)$ exists (in $\mathbb{C}$) for μ-almost every $x \in X$, then $f \in L^1(X, \mu)$ and $\lim_{j \to \infty} \int_X |f_j - f| \, d\mu = 0$. In particular, $\lim_{j \to \infty} \int_X f_j \, d\mu = \int_X f \, d\mu$.*

The same circle of ideas that are used to prove Theorem 14.15 also yields a corresponding L^p-version which is stated next.

Theorem 14.16 (L^p version of Lebesgue's Dominated Convergence Theorem). *Let (X, μ) be a positive measure space and assume that $g \in L^p(X, \mu)$ for some $p \in [1, \infty)$ is a nonnegative function. If $\{f_j\}_{j \in \mathbb{N}}$ is a sequence of μ-measurable, complex-valued functions on X, such that $|f_j(x)| \leq g(x)$ for μ-almost every $x \in X$ and $f(x) := \lim_{j \to \infty} f_j(x)$ exists (in $\mathbb{C}$) for μ-almost every $x \in X$, then $f \in L^p(X, \mu)$ and $\lim_{j \to \infty} \int_X |f_j - f|^p \, d\mu = 0$. In particular, $\lim_{j \to \infty} \int_X |f_j|^p \, d\mu = \int_X |f|^p \, d\mu$.*

One can prove Theorem 14.16 by applying Fatou's lemma to the sequence of functions $\{2^{p-1} g^p - |f_j - f|^p\}_{j \in \mathbb{N}}$.

Theorem 14.17 (Young's Inequality). *Assume that $1 \leq p, q, r \leq \infty$ are such that $\frac{1}{p} + \frac{1}{q} = \frac{1}{r} + 1$. Then for every $f \in L^p(\mathbb{R}^n)$, $g \in L^q(\mathbb{R}^n)$ it follows that*

$$\int_{\mathbb{R}^n} |f(x - y)||g(y)| \, dy < \infty \quad \text{for a.e. } x \in \mathbb{R}^n, \tag{14.2.13}$$

the function defined at a.e. $x \in \mathbb{R}^n$ *by*

$$(f * g)(x) := \int_{\mathbb{R}^n} f(x - y)g(y)\,dy \qquad (14.2.14)$$

belongs to $L^r(\mathbb{R}^n)$, *and* $\|f * g\|_{L^r(\mathbb{R}^n)} \leq \|f\|_{L^p(\mathbb{R}^n)}\|g\|_{L^q(\mathbb{R}^n)}$.

We single out two corollaries of Theorem 14.17 obtained by taking particular values for q and r. Specifically, if we take $q = 1 = r$, then Theorem 14.17 implies

$$L^p(\mathbb{R}^n) * L^1(\mathbb{R}^n) \hookrightarrow L^1(\mathbb{R}^n) \quad \text{for all } p \in [1, \infty], \qquad (14.2.15)$$

and if $q = \frac{p}{p-1}$, $r = \infty$, we obtain

$$L^p(\mathbb{R}^n) * L^{p'}(\mathbb{R}^n) \hookrightarrow L^\infty(\mathbb{R}^n) \quad \text{if } p, p' \in [1, \infty], \ \frac{1}{p} + \frac{1}{p'} = 1. \qquad (14.2.16)$$

In turn, (14.2.15) readily yields the following result.

Lemma 14.18. *Let* $\theta \in L^1(\mathbb{R}^n)$ *and for each* $\varepsilon > 0$ *define* $\theta_\varepsilon(x) := \varepsilon^{-n}\theta(x/\varepsilon)$ *for* $x \in \mathbb{R}^n$. *Then for each* $p \in [1, \infty]$ *and each* $f \in L^p(\mathbb{R}^n)$, *the functions* $f_\varepsilon := f * \theta_\varepsilon$, $\varepsilon > 0$, *satisfy* $f_\varepsilon \in L^p(\mathbb{R}^n)$ *for each* $\varepsilon > 0$ *and*

$$\sup_{\varepsilon > 0} \|f_\varepsilon\|_{L^p(\mathbb{R}^n)} \leq \|\theta\|_{L^1(\mathbb{R}^n)}\|f\|_{L^p(\mathbb{R}^n)}. \qquad (14.2.17)$$

Lemma 14.18 is an ingredient in the proof of the following basic approximation result.

Lemma 14.19. *Let* $\theta \in L^1_{comp}(\mathbb{R}^n)$ *be such that* $\int_{\mathbb{R}^n} \theta\,dx = 1$. *For each* $\varepsilon > 0$ *set* $\theta_\varepsilon(x) := \varepsilon^{-n}\theta(x/\varepsilon)$ *for* $x \in \mathbb{R}^n$. *Consider* $f \in L^p(\mathbb{R}^n)$ *with* $p \in [1, \infty)$ *and define the sequence of functions* $f_\varepsilon := f * \theta_\varepsilon$, $\varepsilon > 0$. *Then* $f_\varepsilon \in L^p(\mathbb{R}^n)$ *for each* $\varepsilon > 0$ *and* $f_\varepsilon \to f$ *in* $L^p(\mathbb{R}^n)$ *as* $\varepsilon \to 0^+$.

Proof. First we show that the conclusion of the lemma holds if $f \in C^0_0(\mathbb{R}^n)$. To this end pick $R > 0$ with $\operatorname{supp} \theta \subseteq B(0, R)$. Then

$$\operatorname{supp} \theta_\varepsilon \subseteq B(0, \varepsilon R) \subseteq \overline{B(0, R)} \qquad (14.2.18)$$

for every $\varepsilon \in (0, 1)$ and if we set $K := \operatorname{supp} f + \overline{B(0, R)}$ then K is compact and

$$\operatorname{supp} f \subseteq K, \quad \operatorname{supp} f_\varepsilon \subseteq K \quad \text{for all } \varepsilon \in (0, 1). \qquad (14.2.19)$$

Let $\delta > 0$ be arbitrary. From the uniform continuity of f it follows that there exists $\eta > 0$ such that if $x, y \in \mathbb{R}^n$ are such that $|x - y| < \eta$ then $|f(x) - f(y)| < \delta$. Hence, if $0 < \varepsilon < \frac{\eta}{R}$ and we use the fact that $\int_{\mathbb{R}^n} \theta_\varepsilon\,dx = 1$, for each $x \in \mathbb{R}^n$ we may write

$$|f_\varepsilon(x) - f(x)| \le \int_{B(x,\varepsilon R)} |f(y) - f(x)||\theta_\varepsilon(x - y)| \, dy$$

$$\le \delta \|\theta\|_{L^1(\mathbb{R}^n)}. \tag{14.2.20}$$

This proves that $f_\varepsilon \to f$ in $L^\infty(\mathbb{R}^n)$ as $\varepsilon \to 0^+$. The latter combined with (14.2.19) readily implies $f_\varepsilon \to f$ in $L^p(\mathbb{R}^n)$ as $\varepsilon \to 0^+$, as wanted.

Suppose now that $f \in L^p(\mathbb{R}^n)$ for some $p \in [1, \infty)$ and let $\delta > 0$ be arbitrary. Since $C_0^0(\mathbb{R}^n)$ is dense in $L^p(\mathbb{R}^n)$ we can find $g \in C_0^0(\mathbb{R}^n)$ with the property that $\|f - g\|_{L^p(\mathbb{R}^n)} < \delta/3$. Moreover, based on what we proved earlier, there exists $\varepsilon_0 > 0$ such that $\|g - g_\varepsilon\|_{L^p(\mathbb{R}^n)} < \delta/3$ if $\varepsilon \in (0, \varepsilon_0)$. Hence, if $\varepsilon \in (0, \varepsilon_0)$ we may estimate

$$\|f - f_\varepsilon\|_{L^p(\mathbb{R}^n)} \le \|f - g\|_{L^p(\mathbb{R}^n)} + \|g - g_\varepsilon\|_{L^p(\mathbb{R}^n)} + \|(g - f)_\varepsilon\|_{L^p(\mathbb{R}^n)}$$

$$< \frac{\delta}{3} + \frac{\delta}{3} + \|g - f\|_{L^p(\mathbb{R}^n)} < \delta, \tag{14.2.21}$$

where for the second inequality in (14.2.21) we have also used (14.2.17). This finishes the proof of the lemma. $\qquad\square$

Lemma 14.20. *Let $a, b \in \mathbb{R}$ satisfying $a < b$. Then for every real-valued function $f \in C^0([a, b]) \cap Lip_{loc}((a, b))$ one has*

$$|f(b) - f(a)| \le \int_a^b |f'(t)| \, dt. \tag{14.2.22}$$

Proof. Assume first that (14.2.22) holds for functions in $Lip([a, b])$ and pick some $f \in C^0([a, b]) \cap Lip_{loc}((a, b))$. Then for each $\varepsilon \in (0, (b-a)/4)$ the function f belongs to $Lip([a + \varepsilon, b - \varepsilon])$ and, by the current assumption,

$$|f(b - \varepsilon) - f(a + \varepsilon)| \le \int_{a+\varepsilon}^{b-\varepsilon} |f'(t)| \, dt. \tag{14.2.23}$$

That f satisfies estimate (14.2.22) is now seen by passing to the limit in (14.2.23) as $\varepsilon \to 0^+$ and using the continuity of f combined with the Lebesgue Monotone Convergence Theorem.

We are left with proving that (14.2.22) holds for each function $f \in Lip([a, b])$. Fix such an f and observe that without loss of generality we may assume that actually $f \in Lip_{comp}(\mathbb{R})$. Indeed, a classical result (due to E.J. McShane [51]) states that

any real-valued Lipschitz function defined on a subset of the
Euclidean space may be extended to a Lipschitz function on the $\qquad$ (14.2.24)
entire Euclidean space with preservation of Lipschitz constant.

Multiplying the extension of f provided by (14.2.24) with a smooth compactly supported function that is identically one near $[a, b]$ then yields an extension of f in the space $Lip_{comp}(\mathbb{R})$. In particular, we have that $f' \in L^\infty_{comp}(\mathbb{R}^n)$. Let $\theta \in C_0^\infty(\mathbb{R})$

be even, positive, supported in $[-1, 1]$ with $\int_{\mathbb{R}} \theta \, dx = 1$ (cf. Lemma 14.32) and set $\theta_\varepsilon(x) := \varepsilon^{-1}\theta(x/\varepsilon)$, $x \in \mathbb{R}$ for each $\varepsilon > 0$. Then for each $\varepsilon > 0$ the function $f_\varepsilon := f * \theta_\varepsilon$ belongs to $C_0^\infty(\mathbb{R})$, satisfies $(f_\varepsilon)' := f' * \theta_\varepsilon$ and

$$\left|(f_\varepsilon)'(t)\right| \leq \int_{\mathbb{R}} |f'(s)| \theta_\varepsilon(t - s) \, ds, \quad \forall t \in \mathbb{R}. \tag{14.2.25}$$

Consequently

$$\int_a^b \left|(f_\varepsilon)'(t)\right| dt \leq \int_a^b \int_{\mathbb{R}} |f'(s)| \theta_\varepsilon(t - s) \, ds \, dt \tag{14.2.26}$$

$$= \int_{\mathbb{R}} |f'(s)| \int_a^b \theta_\varepsilon(t - s) \, dt \, ds$$

$$= \int_{\mathbb{R}} |f'(s)| (\mathbf{1}_{[a,b]} * \theta_\varepsilon)(s) \, ds.$$

Lemma 14.19 applied with $n := 1$, $f := \mathbf{1}_{[a,b]}$, and $p := 1$ implies

$$\mathbf{1}_{[a,b]} * \theta_\varepsilon \to \mathbf{1}_{[a,b]} \text{ in } L^1(\mathbb{R}) \text{ as } \varepsilon \to 0^+. \tag{14.2.27}$$

This and the fact that $f' \in L^\infty_{comp}(\mathbb{R}^n)$ allow us to take the limit in (14.2.26) as $\varepsilon \to 0^+$ to obtain

$$\lim_{\varepsilon \to 0^+} \int_a^b \left|(f_\varepsilon)'(t)\right| dt \leq \int_a^b |f'(s)| \, ds. \tag{14.2.28}$$

In addition, the Fundamental Theorem of Calculus implies $f_\varepsilon(b) - f_\varepsilon(a) = \int_a^b (f_\varepsilon)'(t) \, dt$ for each $\varepsilon > 0$ hence

$$\left|f_\varepsilon(b) - f_\varepsilon(a)\right| \leq \int_a^b \left|(f_\varepsilon)'(t)\right| dt, \tag{14.2.29}$$

for each $\varepsilon > 0$. Since f_ε converges to f uniformly on compact subsets of $\mathbb{R}$ as $\varepsilon \to 0^+$ (this may be seen as in the proof of (2.1.32)) taking the limit in (14.2.29) as $\varepsilon \to 0^+$ gives

$$|f(b) - f(a)| \leq \lim_{\varepsilon \to 0^+} \int_a^b \left|(f_\varepsilon)'(t)\right| dt. \tag{14.2.30}$$

Combining (14.2.30) and (14.2.28) it is immediate that f satisfies (14.2.22) as wanted. $\qquad \square$

Theorem 14.21 (Hardy's Inequality). *Suppose* $p \in [1, \infty)$, $r \in (0, \infty)$, *and consider a measurable function* $f : [0, \infty] \longrightarrow [0, \infty]$. *Then*

$$\int_0^\infty \rho^{-1-r}\left(\int_0^\rho f(\theta) \, d\theta\right)^p d\rho \leq \left(\frac{p}{r}\right)^p \int_0^\infty f(\theta)^p \, \theta^{p-r-1} \, d\theta. \tag{14.2.31}$$

See e.g., [68, p. 272, A.4].

Theorem 14.22 (Minkowski's Inequality). *Suppose* $(X, \mathcal{M}, \mu)$ *and* $(Y, \mathcal{N}, \nu)$ *are* σ-*finite measure spaces, and let* f *be an* $(\mathcal{M} \otimes \mathcal{N})$-*measurable function on* $X \times Y$. *If* $f \geq 0$ *and* $p \in [1, \infty)$, *then*

$$\left[\int_X \left(\int_Y f(x, y) \, d\nu(y) \right)^p d\mu(x) \right]^{1/p} \leq \int_Y \left[\int_X f(x, y)^p \, d\mu(x) \right]^{1/p} d\nu(y). \quad (14.2.32)$$

For a proof of Minkowski's Inequality see [18, 6.19, p. 194].

Definition 14.23. A rigid transformation, or isometry, of the Euclidean space $\mathbb{R}^n$ is any distance preserving mapping of $\mathbb{R}^n$, i.e., any function $T : \mathbb{R}^n \to \mathbb{R}^n$ satisfying

$$|T(x) - T(y)| = |x - y|, \quad \forall \, x, y \in \mathbb{R}^n. \quad (14.2.33)$$

A rigid transformation of $\mathbb{R}^n$ is any distance preserving mapping of $\mathbb{R}^n$, i.e., any function $T : \mathbb{R}^n \to \mathbb{R}^n$ satisfying $|Tx - Ty| = |x - y|$ for every $x, y \in \mathbb{R}^n$. The rigid transformations of the Euclidean space $\mathbb{R}^n$ are precisely those obtained by composing a translation with a mapping in $\mathbb{R}^n$ given by an orthogonal matrix. In other words, a mapping $T : \mathbb{R}^n \to \mathbb{R}^n$ is a rigid transformation of $\mathbb{R}^n$ if and only if there exist $x_0 \in \mathbb{R}^n$ and an orthogonal matrix $A : \mathbb{R}^n \to \mathbb{R}^n$ with the property that

$$T(x) = x_0 + Ax, \quad \forall \, x \in \mathbb{R}^n. \quad (14.2.34)$$

For a proof of the following version of the Arzelà–Ascoli theorem see [63, Corollary 34, p. 179].

Theorem 14.24 (Arzelà–Ascoli's Theorem). *Let* $\mathcal{F}$ *be an equicontinuous family of real-valued functions on a sepa0rable space* X. *Then each sequence* $\{f_j\}_{j \in \mathbb{N}}$ *in* $\mathcal{F}$ *which is bounded at each point has a subsequence* $\{f_{j_k}\}_{k \in \mathbb{N}}$ *that converges pointwise to a continuous function, the converges being uniform on each compact subset of* X.

Theorem 14.25 (Riesz's Representation Theorem for Positive Functionals). *Let* X *be a locally compact Hausdorff topological space and* Λ *a positive linear functional on the space of continuous, compactly supported functions on* X *(denoted by* $C_0^0(X)$). *Then there exists a unique* σ-*algebra* $\mathfrak{M}$ *on* X, *which contains all Borel sets on* X, *and a unique measure* $\mu : \mathfrak{M} \to [0, \infty]$ *that represents* Λ, *that is, the following hold:*

(i) $\Lambda f = \int_X f \, d\mu$ *for every continuous, compactly supported function* f *on* X;
(ii) $\mu(K) < \infty$ *for every compact* $K \subset X$;
(iii) for every $E \in \mathfrak{M}$ *we have* $\mu(E) = \inf\{\mu(V) : E \subset V, V \text{ open}\}$;
(iv) $\mu(E) = \sup\{\mu(K) : K \subset E, K \text{ compact}\}$ *for every open set* E *and every* $E \in \mathfrak{M}$ *with* $\mu(E) < \infty$;
(v) if $E \in \mathfrak{M}$, $A \subset E$, *and* $\mu(E) = 0$, *then* $\mu(A) = 0$.

Theorem 14.26 (Riesz's Representation Theorem for Complex Functionals). *Let X be a locally compact Hausdorff topological space and consider the space of continuous functions on X vanishing at infinity, i.e.,*

$$C_{oo}(X) := \{f \in C^0(X) : \forall \, \varepsilon > 0, \, \exists \text{ compact } K \subset X \text{ such that}$$

$$|f(x)| < \varepsilon \text{ for } x \in X \setminus K\}. \tag{14.2.35}$$

Then $C_{oo}(X)$ is the closure in the uniform norm of $C_0^0(X)$ and for every bounded linear functional $\Lambda : C_{oo}(X) \to \mathbb{C}$ there exists a unique regular complex Borel measure μ on X such that $\Lambda f = \int_X f \, d\mu$ for every $f \in C_{oo}(X)$ and $\|\Lambda\| = |\mu|(X)$.

Theorem 14.27 (Riesz's Representation Theorem for Locally Bounded Functionals). *Let X be a locally compact Hausdorff topological space and assume that $\Lambda : C_0^0(X) \to \mathbb{R}$ is a linear functional that is locally bounded, in sense that for each compact set $K \subset X$ there exists a constant $C_K \in (0, \infty)$ such that*

$$|\Lambda f| \leq C_K \sup_{x \in K} |f(x)|, \quad \forall \, f \in C_0^0(X) \text{ with } \operatorname{supp} f \subseteq K. \tag{14.2.36}$$

Then there exist two measures μ_1, μ_2, taking Borel sets from X into $[0, \infty]$, and satisfying properties (ii)-(iv) in Theorem 14.25, such that

$$\Lambda f = \int_X f \, d\mu_1 - \int_X f \, d\mu_2 \quad \text{for every} \quad f \in C_0^0(X). \tag{14.2.37}$$

The reader is warned that since both μ_1 and μ_2 are allowed to take the value ∞, their difference $\mu_1 - \mu_2$ is not well-defined in general. This being said, $\mu_1 - \mu_2$ is a well-defined finite signed measure on each compact subset of X.

Proposition 14.28 (Urysohn's Lemma). *If X is a locally compact Hausdorff space and $K \subset U \subset X$ are such that K is compact and U is open, then there exists a function $f \in C_0^0(U)$ that satisfies $f = 1$ on K and $0 \leq f \leq 1$.*

Theorem 14.29 (Vitali's Convergence Theorem). *Let (X, μ) be a positive measure space with $\mu(X) < \infty$. Suppose $\{f_k\}_{k \in \mathbb{N}}$ is a sequence of functions in $L^1(X, \mu)$ and that f is a function on X (all complex-valued) satisfying:*

(i) $f_k(x) \to f(x)$ for μ-almost every $x \in X$ as $k \to \infty$;

(ii) $|f| < \infty$ μ-almost everywhere in X;

(iii) $\{f_k\}_{k \in \mathbb{N}}$ is uniformly integrable, in the sense that for every $\varepsilon > 0$ there exists $\delta > 0$ such that for every $k \in \mathbb{N}$ we have $\left| \int_E f_k \, d\mu \right| < \varepsilon$ whenever $E \subseteq X$ is a μ-measurable set with $\mu(E) < \delta$.

Then $f \in L^1(X, \mu)$ and

$$\lim_{k \to \infty} \int_X |f_k - f| \, d\mu = 0. \tag{14.2.38}$$

In particular, $\lim_{k \to \infty} \int_X f_k \, d\mu = \int_X f \, d\mu$.

See, e.g., [64, p. 133].

Proposition 14.30. *Let* (X, μ) *be a positive measure space and suppose* $f \in L^1(X, \mu)$. *Then for every* $\varepsilon > 0$ *there exists* $\delta > 0$ *such that for every* μ-*measurable set* $A \subseteq X$ *satisfying* $\mu(A) < \delta$ *we have* $\int_A |f| \, d\mu < \varepsilon$.

Proof. Consider the measure $\lambda := |f|\mu$ on X. Then λ is absolutely continuous with respect to μ and the (ε, δ) characterization of absolute continuity of measures (see, e.g., [64, Theorem 6.11, p. 124]) yields the desired conclusion. $\qquad\square$

14.3 Custom-Designing Smooth Cut-off Functions

Lemma 14.31. *Let* $f : \mathbb{R} \to \mathbb{R}$ *be the function defined by*

$$f(x) := \begin{cases} e^{-1/x}, & \text{if } x > 0, \\ 0, & \text{if } x \le 0, \end{cases} \qquad \forall x \in \mathbb{R}. \qquad (14.3.1)$$

Then f *is of class* C^∞ *on* $\mathbb{R}$.

Proof. Denote by C the collection of functions $g : \mathbb{R} \to \mathbb{R}$ for which there exists a polynomial P such that

$$g(x) := \begin{cases} e^{-1/x} P(1/x), & \text{if } x > 0, \\ 0, & \text{if } x \le 0, \end{cases} \qquad \forall x \in \mathbb{R}. \qquad (14.3.2)$$

Recall that if $h : \mathbb{R} \to \mathbb{R}$ is a continuous function that is differentiable on $\mathbb{R} \setminus \{0\}$ and for which there exists $L \in \mathbb{R}$ such that $\lim_{x \to 0^-} h'(x) = L = \lim_{x \to 0^+} h'(x)$, then h is also differentiable at the origin and $h'(0) = L$. An immediate consequence of this fact is that any $g \in C$ is differentiable and $g' \in C$. In turn, this readily gives that any $g \in C$ is of class C^∞ on $\mathbb{R}$. Since f defined in (14.3.1) clearly belongs to C, it follows that f is of class C^∞ on $\mathbb{R}$. $\qquad\square$

Lemma 14.32. *The function* $\phi : \mathbb{R}^n \to \mathbb{R}$ *defined by*

$$\phi(x) := \begin{cases} C e^{\frac{1}{|x|^2 - 1}} & \text{if } x \in B(0, 1), \\ 0 & \text{if } x \in \mathbb{R}^n \setminus B(0, 1), \end{cases} \qquad (14.3.3)$$

where $C := \left(\omega_{n-1} \int_0^1 e^{1/(\rho^2 - 1)} \rho^{n-1} \, d\rho \right)^{-1} \in (0, \infty)$, *satisfies the following properties:*

$$\phi \in C^\infty(\mathbb{R}^n), \quad \phi \ge 0, \quad \phi \text{ is even,}$$
$$\text{supp } \phi \subseteq \overline{B(0, 1)}, \quad \text{and} \quad \int_{\mathbb{R}^n} \phi(x) \, dx = 1. \qquad (14.3.4)$$

Proof. That $\phi \ge 0$ and $\text{supp } \phi \subseteq \overline{B(0, 1)}$ is immediate from its definition. Also, since $\phi(x) = C f(1 - |x|^2)$ for $x \in \mathbb{R}^n$ where f is as in (14.3.1), invoking Lemma 14.31

it follows that $\phi \in C^\infty(\mathbb{R}^n)$. Finally, the condition $\int_{\mathbb{R}^n} \phi(x)\,dx = 1$ follows upon observing that based on (14.9.9) we have $\int_{B(0,1)} e^{\frac{1}{|x|^2-1}}\,dx = 1/C$. $\qquad\square$

Proposition 14.33. *Let $F_0, F_1 \subset \mathbb{R}^n$ be two nonempty sets with the property that* $\mathrm{dist}(F_0, F_1) > 0$. *Then there exists a function* $\psi : \mathbb{R}^n \to \mathbb{R}$ *with the following properties:*

$$\psi \in C^\infty(\mathbb{R}^n), \ \ 0 \leq \psi \leq 1, \ \ \psi = 0 \text{ on } F_0, \ \psi \equiv 1 \text{ on } F_1, \text{ and}$$

$$\forall \alpha \in \mathbb{N}_0^n \ \exists\, C_\alpha \in (0,\infty) \text{ such that } |\partial^\alpha \psi(x)| \leq \frac{C_\alpha}{\mathrm{dist}\,(F_0,F_1)^{|\alpha|}} \ \ \forall\, x \in \mathbb{R}^n. \tag{14.3.5}$$

Proof. Let $r := \mathrm{dist}(F_0, F_1) > 0$ and set $\widetilde{F_1} := \{x \in \mathbb{R}^n : \mathrm{dist}\,(x, F_1) \leq r/4\}$. Also, with ϕ as in Lemma 14.32, define the function $\theta(x) := (\frac{4}{r})^n \phi(4x/r)$ for $x \in \mathbb{R}^n$. Then

$$\theta \in C^\infty(\mathbb{R}^n), \ \ \theta \geq 0, \ \ \mathrm{supp}\,\theta \subseteq \overline{B(0, r/4)}, \text{ and} \int_{\mathbb{R}^n} \theta(x)\,dx = 1. \tag{14.3.6}$$

We claim that the function $\psi : \mathbb{R}^n \to \mathbb{R}$ defined by $\psi := \chi_{\widetilde{F_1}} * \theta$ has the desired properties. To see why this is true note that since

$$\psi(x) = \int_{\widetilde{F_1}} \theta(x-y)\,dy \qquad \forall\, x \in \mathbb{R}^n, \tag{14.3.7}$$

from the properties of θ it is immediate that $\psi \in C^\infty(\mathbb{R}^n)$ and $0 \leq \psi \leq 1$. Furthermore, for $\alpha \in \mathbb{N}_0^n$ and $x \in \mathbb{R}^n$ we may write

$$|\partial^\alpha \psi(x)| \leq (\tfrac{4}{r})^{|\alpha|} \int_{\widetilde{F_1}} (\tfrac{4}{r})^n \big|(\partial^\alpha \phi)(4(x-y)/r)\big|\,dy$$

$$\leq (\tfrac{4}{r})^{|\alpha|} \int_{\mathbb{R}^n} |\partial^\alpha \phi(y)|\,dy = C_\alpha r^{-|\alpha|} = \frac{C_\alpha}{\mathrm{dist}\,(F_0,F_1)^{|\alpha|}}, \tag{14.3.8}$$

where $C_\alpha := 4^{|\alpha|} \int_{\mathbb{R}^n} |\partial^\alpha \phi(y)|\,dy$ is a positive, finite number independent of r.

We are left with checking the fact that $\psi = j$ on F_j, $j = 0, 1$. First, if $x \in F_0$, then $|x - y| \geq 3r/4$ for every $y \in \widetilde{F_1}$, hence $\theta(x - y) = 0$ for every $y \in \widetilde{F_1}$, which when combined with (14.3.7) implies $\psi(x) = 0$. Second, if $x \in F_1$, then $\overline{B(x, r/4)} \subseteq \widetilde{F_1}$. Since the support of $\theta(x - \cdot) \subset \overline{B(x, r/4)}$ the latter implies $\psi(x) = \int_{\mathbb{R}^n} \theta(x-y)\,dy = 1$. The proof of the proposition is complete. $\qquad\square$

A trivial yet useful consequence of the above result is as follows.

Proposition 14.34. *If $U \subseteq \mathbb{R}^n$ is open and $K \subset U$ is compact, then there exists a function $\psi : \mathbb{R}^n \to \mathbb{R}$ that is of class C^∞, satisfies $0 \leq \psi(x) \leq 1$ for every $x \in \mathbb{R}^n$ and $\psi(x) = 1$ for every $x \in K$, and which has compact support, contained in U.*

Proof. Since K is a compact set contained in U we may define $r := \mathrm{dist}\,(K, \mathbb{R}^n \setminus U)$ which is a positive number. Now Proposition 14.33 applied with $F_1 := K$ and $F_0 := \{x \in \mathbb{R}^n : \mathrm{dist}\,(x, K) > r/2\}$ yields the desired function ψ. $\qquad\square$

14.4 Partition of Unity

Lemma 14.35. *If $C \subset \mathbb{R}^n$ is compact, and $U \subseteq \mathbb{R}^n$ is an open set such that $C \subset U$, then there exists a compact set $D \subseteq \mathbb{R}^n$ such that $C \subset \mathring{D} \subset D \subset U$.*

Proof. Let $V = U^c \cap \overline{B(0,R)}$, where $R > 0$ is large enough so that $\overline{U} \subset B(0,R)$. Then V is compact and disjoint from C so $r := \mathrm{dist}(V,C) = \inf\limits_{\substack{x \in C \\ y \in V}} \|x - y\| > 0$. Hence, if we set

$$D := \bigcup_{x \in C} \overline{B(x, r/4)}, \qquad (14.4.1)$$

then D is compact and $C \subset \mathring{D} \subset D \subset U$. □

Lemma 14.36. *Suppose that $K \subseteq \mathbb{R}^n$ is a compact set and that $\{O_j\}_{1 \leq j \leq k}$ is a finite open cover of K. Then there are compact sets $D_j \subset O_j$, $1 \leq j \leq k$, with the property that*

$$K \subset \bigcup_{j=1}^{k} \mathring{D}_j. \qquad (14.4.2)$$

Proof. Set $C_1 := K \setminus \bigcup_{j=2}^{k} O_j \subseteq O_1$. Since C_1 is compact, Lemma 14.35 shows that there exists a compact set D_1 with the property that $C_1 \subset \mathring{D}_1 \subset D_1 \subset O_1$. Next, proceeding inductively, suppose that $m \in \mathbb{N}$ is such that $1 \leq m < k$ and that m compact sets $D_1, \ldots, D_m$ have been constructed with the property that $K \subset (\bigcup_{j=1}^{m} \mathring{D}_j) \cup (\bigcup_{j=m+1}^{k} O_j)$. Introduce

$$C_{m+1} := K \setminus \left(\bigcup_{j=1}^{m} \mathring{D}_j \cup \bigcup_{j=m+2}^{k} O_j \right). \qquad (14.4.3)$$

Clearly, C_{m+1} is a compact subset of O_{m+1}. By once again invoking Lemma 14.35, there exists a compact set $D_{m+1} \subset O_{m+1}$ such that $C_{m+1} \subset \mathring{D}_{m+1}$. After k iterations, this procedure yields a family of sets $C_1, \ldots, C_k$ which have all the desired properties. □

Theorem 14.37 (Partition of Unity for Compact Sets). *Let $K \subset \mathbb{R}^n$ be a compact set, and let $\{O_j\}_{1 \leq j \leq N}$ be a finite open cover of K. Then there exists a finite collection of C^∞ functions $\varphi_j : \mathbb{R}^n \to \mathbb{R}$, $1 \leq j \leq N$, satisfying the following properties:*

(i) For every $1 \leq j \leq N$, the set $\mathrm{supp}(\varphi_j)$ is compact and contained in O_j;

(ii) For every $1 \leq j \leq N$, one has $0 \leq \varphi_j \leq 1$;

(iii) $\sum\limits_{j=1}^{N} \varphi_j(x) = 1$ for every $x \in K$.

The family $\{\varphi_j : 1 \leq j \leq N\}$ is called a `partition of unity subordinate to the cover` $\{O_j\}_{1 \leq j \leq N}$ *of K.*

Proof. Let $\{O_j\}_{1 \leq j \leq N}$ be any finite open cover for K. From Lemma 14.36 we know that there exist compact sets $D_j \subseteq O_j$, $1 \leq j \leq N$, such that $K \subset \bigcup\limits_{j=1}^{N} \mathring{D}_j$. By

Proposition 14.34, for each $1 \le j \le N$, choose a C^∞ function $\eta_j : \mathbb{R}^n \to [0, \infty)$ that is positive on D_j and has compact support in O_j. It follows that $\sum_{j=1}^{N} \eta_j(x) > 0$ for all $x \in \bigcup_{j=1}^{N} \mathring{D}_j$, so we can define for each $j \in \{1, ..., N\}$ the function

$$\psi_j : \bigcup_{i=1}^{N} \mathring{D}_j \to \mathbb{R}, \qquad \psi_j(x) := \frac{\eta_j(x)}{\sum_{k=1}^{N} \eta_k(x)}, \qquad \forall\, x \in \bigcup_{i=1}^{N} \mathring{D}_j. \tag{14.4.4}$$

By Lemma 14.36, there exists a compact set $U \subseteq \mathbb{R}^n$ with $K \subseteq \mathring{U} \subseteq U \subseteq \bigcup_{j=1}^{N} \mathring{D}_j$. We apply Proposition 14.34 to obtain a C^∞ function $f : \mathbb{R}^n \to [0, 1]$ that satisfies $f(x) = 1$ for $x \in K$ and having compact support contained in $\mathring{U}$. Then for each $j \in \{1, ..., N\}$ we define the function $\varphi_j := f\psi_j$ acting from $\mathbb{R}^n$ into $\mathbb{R}$. It is not hard to see that each φ_j is C^∞ in $\mathbb{R}^n$, has compact support, contained in O_j, $0 \le \varphi_j \le 1$, and $\sum_{1 \le j \le N} \varphi_j(x) = 1$ for all $x \in K$. $\qquad\square$

Definition 14.38. (i) A family $(F_j)_{j \in I}$ of subsets of $\mathbb{R}^n$ is said to be `locally finite` in $E \subseteq \mathbb{R}^n$ provided every $x \in E$ has a neighborhood $O \subseteq \mathbb{R}^n$ with the property that the set $\{j \in I : F_j \cap O \ne \varnothing\}$ is finite.

(ii) Given a collection of functions $f_j : \Omega \to \mathbb{R}$, $j \in I$, defined in some fixed subset Ω of $\mathbb{R}^n$, the sum $\sum_{j \in I} f_j$ is called `locally finite in` $E \subseteq \mathbb{R}^n$ provided the family of sets $\{x \in \Omega : f_j(x) \ne 0\}$, indexed by $j \in I$, is locally finite in E.

Exercise 14.39. Show that a family $(F_j)_{j \in I}$ of subsets of $\mathbb{R}^n$ is locally finite in the open set $E \subseteq \mathbb{R}^n$ if and only if for every compact $K \subseteq E$ the set $\{j \in I : F_j \cap K \ne \varnothing\}$ is finite.

Exercise 14.40. Show that if the family $(F_j)_{j \in I}$ of subsets of $\mathbb{R}^n$ is locally finite in $E \subseteq \mathbb{R}^n$, then $(\overline{F_j})_{j \in I}$ is also locally finite in E.

Exercise 14.41. Show that if the family $(F_j)_{j \in I}$ of closed subsets of $\mathbb{R}^n$ is locally finite in $E \subseteq \mathbb{R}^n$, then $\bigcup_{j \in I}(E \cap F_j)$ is a relatively closed subset of E.

Proof. Let $x^* \in E$ be such that $x^* \notin F_j$ for every $j \in I$. Then there exists $r > 0$ with the property that $B(x^*, r) \cap F_j = \varnothing$ for every $j \in I \setminus I_*$, where I_* is a finite subset of I. Hence, by eventually further decreasing r, it can be assumed that $B(x^*, r) \cap F_j = \varnothing$ for every I_* as well. Thus, $B(x^*, r) \cap E$ is a relative neighborhood of x^* in E that is disjoint from $\bigcup_{j \in I} F_j$. This proves that $E \setminus \left(\bigcup_{j \in I} F_j \right)$ is relatively open in E; hence, $\bigcup_{j \in I}(E \cap F_j)$ is a relatively closed subset of E. $\qquad\square$

Theorem 14.42 (Partition of Unity for Arbitrary Open Covers). *Let $(O_k)_{k \in I}$ be an arbitrary family of open sets in $\mathbb{R}^n$ and set $\Omega := \bigcup_{k \in I} O_k$. Then there exists an at* most countable collection $(\varphi_j)_{j\in J}$ of C^∞ nonzero functions $\varphi_j : \Omega \to \mathbb{R}$ satisfying the following properties:

(i) For every $j \in J$ there exists $k \in I$ such that φ_j is compactly supported in O_k;
(ii) For every $j \in J$, one has $0 \le \varphi_j \le 1$ in Ω;
(iii) The family of sets $\{x \in \Omega : \varphi_j(x) \ne 0\}$, indexed by $j \in J$, is locally finite in Ω;
(iv) $\sum_{j\in J} \varphi_j(x) = 1$ for every $x \in \Omega$.

The family $(\varphi_j)_{j\in J}$ is called a `partition of unity subordinate to the family` $(O_k)_{k\in I}$.

Proof. Start by defining

$$\Omega_j := \{x \in \Omega : \|x\| \le j \text{ and } \mathrm{dist}(x, \partial\Omega) \ge \tfrac{1}{j}\}, \qquad j \in \mathbb{N}, \tag{14.4.5}$$

Then $\Omega = \bigcup_{j=1}^\infty \Omega_j$ and

$$\Omega_j \subseteq \mathbb{R}^n \text{ is compact}, \quad \Omega_j \subset \mathring{\Omega}_{j+1} \quad \text{for every } j \in \mathbb{N}. \tag{14.4.6}$$

Proceed now to define the compact sets

$$K_2 := \Omega_2, \qquad K_j := \Omega_j \setminus \mathring{\Omega}_{j-1} \qquad \text{for every } j \ge 3. \tag{14.4.7}$$

As such, from (14.4.7) and (14.4.6) we have

$$K_j = \Omega_j \cap (\mathring{\Omega}_{j-1})^c \subseteq \mathring{\Omega}_{j+1} \cap (\Omega_{j-2})^c \qquad \text{for every } j \ge 3. \tag{14.4.8}$$

Finally, we define the following families of open sets:

$$O_2 := \{O_k \cap \mathring{\Omega}_3 : k \in I\}, \quad O_j := \{O_k \cap (\Omega_{j+1} \setminus \mathring{\Omega}_{j-2}) : k \in I\} \quad \forall j \ge 3. \tag{14.4.9}$$

Making use of (14.4.6), for every $j \ge 3$ we have that

$$(\Omega_{j+1} \setminus \Omega_{j-2})^\circ = (\Omega_{j+1} \cap (\Omega_{j-2})^c)^\circ = \mathring{\Omega}_{j+1} \cap ((\Omega_{j-2})^c)^\circ$$
$$= \mathring{\Omega}_{j+1} \cap (\overline{\Omega_{j-2}})^c = \mathring{\Omega}_{j+1} \cap (\Omega_{j-2})^c. \tag{14.4.10}$$

Hence, (14.4.7), (14.4.9), and (14.4.10) imply that O_j is an open cover for K_j for every $j \ge 3$. Also, from the definitions of K_2 and O_2 and (14.4.6), we obtain that O_2 is an open cover for K_2. Since the K_j's are compact, these open covers can be refined to finite subcovers in each case. As such, for each $j = 2, 3, \ldots$, we can apply Theorem 14.37 to obtain a finite partition of unity $\{\varphi : \varphi \in \Phi_j\}$ for K_j subordinate to O_j. Also note that due to (14.4.10) and (14.4.9), we necessarily have that, for each $j \in \{2, 3, \ldots\}$, $\Omega_j \cap O = \varnothing$ for every $O \in O_k$, for every $k \in \mathbb{N}$ satisfying $k \ge j + 2$. This ensures that the family $\{\mathrm{supp}\,\varphi : \varphi \in \Phi_j, j \ge 2\}$ is locally finite in Ω, so we can define

$$s(x) := \sum_{j\ge 2} \sum_{\varphi\in\Phi_j} \varphi(x), \qquad \text{for every } x \in \Omega. \tag{14.4.11}$$

Given that differentiability is a local property, it follows that s is of class C^∞ in Ω. Moreover, note that $s(x) > 0$ for every $x \in \Omega$, since $0 \le \varphi \le 1$ for all $\varphi \in \Phi_j$, $j = 2, 3, \ldots$, and if $x \in \Omega_j$, for some $j \in \{2, 3, \ldots\}$, then $\sum_{\varphi \in \Phi_j} \varphi(x) = 1$. Consequently, $1/s$ is also a C^∞ function in Ω. It is then clear that the collection $\Phi := \{\varphi/s : \varphi \in \Phi_j, j = 2, 3, \ldots\}$ is a partition of unity subordinate to the family of open sets $(O_k)_{k \in I}$ (in the sense described in the statement of the theorem).   □

Theorem 14.43 (Partition of Unity with Preservation of Indexes). *Let $(O_k)_{k \in I}$ be an arbitrary family of open sets in $\mathbb{R}^n$ and set $\Omega := \bigcup_{k \in I} O_k$. Then there exists a collection $(\psi_k)_{k \in I}$ of C^∞ functions $\psi_k : \Omega \to \mathbb{R}$ satisfying the following properties:*

(i) For every $k \in I$ the function ψ_k vanishes outside of a relatively closed subset of O_k;
(ii) For every $k \in I$, one has $0 \le \psi_k \le 1$ in Ω;
(iii) The family of sets $\{x \in \Omega : \psi_k(x) \ne 0\}$, indexed by $k \in I$, is locally finite in Ω;
(iv) $\sum_{k \in I} \psi_k(x) = 1$ for every $x \in \Omega$.

Proof. Let $(\varphi_j)_{j \in J}$ be a partition of unity subordinate to the family $(O_k)_{k \in I}$, and denote by $f : J \to I$ a function with the property that, for every $j \in J$, the function φ_j is compactly supported in $O_{f(j)}$. That this exists is guaranteed by Theorem 14.42. For every $k \in I$ then define

$$\psi_k(x) := \sum_{j \in f^{-1}(\{k\})} \varphi_j(x), \qquad \forall x \in \Omega. \tag{14.4.12}$$

Note that the sum is locally finite in Ω, hence $\psi_k : \Omega \to \mathbb{R}$ is a well-defined non-negative function, of class C^∞ in Ω, for every $k \in I$. In addition, the result from Exercise 14.41 shows that, for every $k \in I$, φ_k vanishes outside a relatively closed subset of O_k. Furthermore, since $(f^{-1}(\{k\}))_{k \in I}$ is a partition of J into mutually disjoint subsets, we may compute

$$\sum_{k \in I} \psi_k(x) = \sum_{k \in I} \sum_{j \in f^{-1}(\{k\})} \varphi_j(x) = \sum_{j \in J} \varphi_j(x) = 1 \qquad \forall x \in \Omega. \tag{14.4.13}$$

Incidentally, this also shows that, necessarily, $0 \le \psi_k(x) \le 1$ for every $k \in I$ and $x \in \Omega$. Finally, the fact that the family of sets $\{x \in \Omega : \psi_k(x) \ne 0\}$, $k \in I$, is locally finite in Ω is inherited from the corresponding property of the φ_j's.   □

Theorem 14.44 (Lebesgue Number Theorem). *If (X, d) is a compact metric space and $\{C_j\}_{j=1}^N$ is an open cover of X, then there exists a number $r_* > 0$ such that every subset of X having diameter less than r_* is contained in some member of the respective cover.*

14.5 The Gamma and Beta Functions

The `Gamma function` is defined as

$$\Gamma(z) := \int_0^\infty t^{z-1} e^{-t} \, dt, \qquad \text{for } z \in \mathbb{C}, \ \operatorname{Re} z > 0. \tag{14.5.1}$$

It is easy to check that $\Gamma(1) = 1$, $\Gamma(1/2) = \sqrt{\pi}$ and via integration by parts that $\Gamma(z + 1) = z\Gamma(z)$ for $z \in \mathbb{C}$, $\operatorname{Re} z > 0$. By analytic continuation, the function $\Gamma(z)$ is extended to a meromorphic function defined for all complex numbers z except for $z = -n$, $n \in \mathbb{N}_0$, where the extended function has simple poles with residue $(-1)^n/n!$ and this extension satisfies

$$\Gamma(z + 1) = z\Gamma(z) \qquad \text{for } z \in \mathbb{C} \setminus \{-n : n \in \mathbb{N}_0\}. \tag{14.5.2}$$

By induction it follows that for every $n \in \mathbb{N}$ we have

$$\Gamma(n) = (n - 1)!, \tag{14.5.3}$$

$$\Gamma\left(\frac{1}{2} + n\right) = \frac{1 \cdot 3 \cdot 5 \cdots (2n - 1)}{2^n} \sqrt{\pi} = \frac{(2n)!}{2^{2n} n!} \sqrt{\pi}, \tag{14.5.4}$$

$$\Gamma\left(\frac{1}{2} - n\right) = \frac{(-1)^n 2^n}{1 \cdot 3 \cdot 5 \cdots (2n - 1)} \sqrt{\pi} = \frac{(-1)^n 2^{2n} n!}{(2n)!} \sqrt{\pi}. \tag{14.5.5}$$

The volume of the unit ball $B(0, 1)$ in $\mathbb{R}^n$, which we denote by $|B(0, 1)|$, and the surface area of the unit sphere in $\mathbb{R}^n$, denoted here by ω_{n-1}, have the following formulas:

$$|B(0, 1)| = \frac{\pi^{\frac{n}{2}}}{\Gamma\left(\frac{n}{2} + 1\right)}, \qquad \omega_{n-1} = n|B(0, 1)| = \frac{2\pi^{\frac{n}{2}}}{\Gamma\left(\frac{n}{2}\right)}. \tag{14.5.6}$$

Next, consider the so-called `beta function`

$$B(z, w) := \int_0^1 t^{z-1}(1 - t)^{w-1} \, dt, \qquad \operatorname{Re} z, \operatorname{Re} w > 0. \tag{14.5.7}$$

Clearly $B(z, w) = B(w, z)$. Making the change of variables $t = u/(u + 1)$ for each $u \in (0, \infty)$, it follows that

$$B(z, w) = \int_0^\infty u^{w-1} \left(\frac{1}{u + 1}\right)^{z+w} du \tag{14.5.8}$$

whenever $\operatorname{Re} z, \operatorname{Re} w > 0$.

The basic identity relating the Gamma and Beta functions reads

$$B(z, w) = \frac{\Gamma(z)\Gamma(w)}{\Gamma(z + w)}, \qquad \operatorname{Re} z, \operatorname{Re} w > 0. \tag{14.5.9}$$

This is easily proved starting with (14.5.8), writing $\Gamma(z + w) = \int_0^\infty t^{z+w-1} e^{-t}\, dt$ and expressing $B(z, w)\Gamma(z + w)$ as a double integral, then making the change of variables $s := t/(u + 1)$. A useful consequence of identity (14.5.9) is the following formula:

$$\int_0^{\pi/2} (\sin\theta)^a (\cos\theta)^b\, d\theta = \frac{1}{2} B\left(\tfrac{a+1}{2}, \tfrac{b+1}{2}\right)$$

$$= \frac{1}{2} \frac{\Gamma(\tfrac{a+1}{2})\Gamma(\tfrac{b+1}{2})}{\Gamma(\tfrac{a+b+2}{2})} \quad \text{if } a, b > -1. \tag{14.5.10}$$

Indeed, making the change of variables $u := (\sin\theta)^2$, the integral in the leftmost side of (14.5.10) becomes $\frac{1}{2} \int_0^1 u^{(a-1)/2}(1 - u)^{(b-1)/2}\, du$.

For further reference, let us also note here that

$$\int_0^\pi (\sin\theta)^a (\cos\theta)^b\, d\theta = \frac{1 + (-1)^b}{2} \cdot B\left(\tfrac{a+1}{2}, \tfrac{b+1}{2}\right)$$

$$= \frac{1 + (-1)^b}{2} \cdot \frac{\Gamma(\tfrac{a+1}{2})\Gamma(\tfrac{b+1}{2})}{\Gamma(\tfrac{a+b+2}{2})}, \tag{14.5.11}$$

whenever $a, b > -1$. This is proved by splitting the domain of integration into $(0, \pi/2) \cup (\pi/2, \pi)$, making a change of variables $\theta \mapsto \theta - \pi/2$ in the second integral, and invoking (14.5.10).

14.6 Surfaces in $\mathbb{R}^n$ and Surface Integrals

Definition 14.45. Given $n \geq 2$ and $k \in \mathbb{N} \cup \{\infty\}$, a C^k surface (or, surface of class C^k) in $\mathbb{R}^n$ is a subset Σ of $\mathbb{R}^n$ with the property that for every $x^* \in \Sigma$ there exists an open neighborhood $U(x^*)$ such that

$$\Sigma \cap U(x^*) = P(O) \tag{14.6.1}$$

where O is an open subset of $\mathbb{R}^{n-1}$ and

$$P : O \longrightarrow \mathbb{R}^n \quad \text{is an injective function of class } C^k \text{ satisfying}$$
$$\text{rank}\,[DP(u)] = n - 1, \text{ for all } u = (u_1, \ldots, u_{n-1}) \in O, \tag{14.6.2}$$

where DP is the Jacobian matrix of P, i.e.,

$$DP(u) = \left(\frac{D(P_1, \ldots, P_n)}{D(u_1, \ldots, u_{n-1})}\right)(u), \qquad u \in O. \tag{14.6.3}$$

The function P in (14.6.2) is called a local parametrization of class C^k near x^* and $\Sigma \cap U(x^*)$ a parametrizable patch.

In the case when (14.6.1) holds when we formally take $r = +\infty$, i.e., in the case when $\Sigma = P(O)$, we call P a global parametrization of the surface Σ.

Proposition 14.46. *If O is an open subset of $\mathbb{R}^{n-1}$ and $P : O \to \mathbb{R}^n$ satisfies (14.6.2)–(14.6.3) for $k = 1$, then $P : O \to P(O)$ is a homeomorphism.*

Definition 14.47. Assume $n \geq 3$. If $v_1 = (v_{11}, ..., v_{1n}), ..., v_{n-1} = (v_{n-1\,1}, ..., v_{n-1\,n})$ are $n - 1$ vectors in $\mathbb{R}^n$, their cross product is defined as

$$v_1 \times v_2 \times \cdots \times v_{n-1} := \det \begin{pmatrix} v_{11} & v_{12} & \cdots & v_{1n} \\ v_{21} & v_{22} & \cdots & v_{2n} \\ \vdots & \vdots & \cdots & \vdots \\ v_{n-1\,1} & v_{n-1\,2} & \cdots & v_{n-1\,n} \\ \mathbf{e}_1 & \mathbf{e}_2 & \cdots & \mathbf{e}_n \end{pmatrix}, \tag{14.6.4}$$

where the determinant is understood as computed by formally expanding it with respect to the last row, the result being a vector in $\mathbb{R}^n$. More precisely,

$$v_1 \times \ldots \times v_{n-1} \tag{14.6.5}$$

$$:= \sum_{j=1}^{n} (-1)^{j+1} \det \begin{pmatrix} v_{11} & \cdots & v_{j-1} & v_{j+1} & \cdots & v_{1n} \\ \vdots & \cdots \vdots & \vdots & \cdots & \vdots \\ v_{n-1\,1} & \cdots & v_{n-1\,j-1} & v_{n-1\,j+1} & \cdots & v_{n-1\,n} \end{pmatrix} \mathbf{e}_j.$$

Definition 14.48. Let $\Sigma \subset \mathbb{R}^n$, $n \geq 2$, be a C^1 surface and assume that $P : O \to \mathbb{R}^n$, with O open subset of $\mathbb{R}^{n-1}$, is a local parametrization of Σ of class C^1 near some point $X^* \in \Sigma$. Also, suppose that $f : \Sigma \to \mathbb{R}$ is a continuous function on Σ that vanishes outside of a compact subset of $P(O)$. We then define

$$\int_{\Sigma} f(x) \, d\sigma(x) \tag{14.6.6}$$

$$:= \int_{O} (f \circ P)(u) \, |\partial_1 P(u) \times \ldots \times \partial_{n-1} P(u)| \, du_1 \ldots du_{n-1} \quad \text{if } n \geq 3,$$

and

$$\int_{\Sigma} f(x) \, d\sigma(x) := \int_{O} (f \circ P)(u) |P'(u)| \, du \quad \text{if } n = 2. \tag{14.6.7}$$

In (14.6.6), $d\sigma$ stands for the surface measure (or, surface area element), whereas in (14.6.7), $d\sigma$ stands for the arc-length measure.

14.7 Lipschitz Domains

Definition 14.49. Call $\Omega \subseteq \mathbb{R}^n$ a bounded Lipschitz domain if Ω is a nonempty, proper, bounded open subset of $\mathbb{R}^n$ for which there exist $r, c > 0$ with the following significance. For each point $x^* \in \partial\Omega$ there exist an $(n-1)$-dimensional plane $H \subseteq \mathbb{R}^n$

passing through the point x^* and a choice v of the unit normal to H such that if one defines the open cylinder

$$C_{x^*} := C(x^*, H, v, r, c) := \{x' + tv : x' \in H, |x' - x^*| < r, |t| < c\} \tag{14.7.1}$$

then

$$C_{x^*} \cap \Omega = C_{x^*} \cap \{x' + tv : x' \in H \text{ and } t > \varphi(x')\}, \tag{14.7.2}$$

for some Lipschitz function $\varphi : H \to \mathbb{R}$ satisfying

$$\varphi(x^*) = 0 \text{ and } |\varphi(x')| < c \text{ if } |x' - x^*| \leq r. \tag{14.7.3}$$

Definition 14.50. Call $\Omega \subseteq \mathbb{R}^n$ an upper-graph Lipschitz domain if there exists a Lipschitz function $\varphi : \mathbb{R}^{n-1} \to \mathbb{R}$ with the property that

$$\Omega = \{(y', y_n) \in \mathbb{R}^{n-1} \times \mathbb{R} : y_n > \varphi(y')\}. \tag{14.7.4}$$

Remark 14.51. We note that if $\Omega \subseteq \mathbb{R}^n$ is an upper-graph Lipschitz domain and φ is a Lipschitz function so that (14.7.4) holds, the outward unit normal to Ω is the vector valued function $v = (v_j)_{1 \leq j \leq n}$ with $v_j : \partial\Omega \to \mathbb{R}$, $j \in \{1, \dots, n\}$, and where

$$v_j(x', \varphi(x')) := \frac{\partial_j \varphi(x')}{\sqrt{1 + |\nabla\varphi(x')|^2}}, \quad \text{for } 1 \leq j \leq n-1,$$

$$\text{and } v_n(x') := \frac{-1}{\sqrt{1 + |\nabla\varphi(x')|^2}}, \quad \text{for a.e. } x' \in \mathbb{R}^{n-1}. \tag{14.7.5}$$

Remark 14.52. A very useful feature of upper-graph Lipschitz domains is that they are homeomorphic, via bijective bi-Lipschitz maps, to the upper-half space. To see this, suppose Ω is the upper-graph of a Lipschitz function $\varphi : \mathbb{R}^{n-1} \to \mathbb{R}$, i.e., that (14.7.4) holds. Introduce the function

$$\Phi : \mathbb{R}^n \to \mathbb{R}^n, \quad \Phi(x', t) := (x', \varphi(x') + t),$$

$$\text{for each } x' \in \mathbb{R}^{n-1} \text{ and each } t \in \mathbb{R}. \tag{14.7.6}$$

Clearly Φ is a Lipschitz function and Φ is bijective with inverse $\Phi^{-1} : \mathbb{R}^n \to \mathbb{R}^n$ given by $\Phi^{-1}(x', t) := (x', t - \varphi(x'))$ for each $x' \in \mathbb{R}^{n-1}$ and each $t \in \mathbb{R}$. In particular, Φ is a bijective bi-Lipschitz map. Also, Rademacher's Theorem (cf. Theorem 2.115) ensures that Φ is differentiable (in a classical sense) at almost every point. A direct computation shows that the Jacobian matrix of Φ is

$$D\Phi = \begin{pmatrix} 1 & 0 & \dots & 0 & 0 \\ 0 & 1 & \dots & 0 & 0 \\ \vdots & \vdots & \dots & \vdots & \vdots \\ 0 & 0 & \dots & 1 & 0 \\ \partial_1\varphi & \partial_2\varphi & \dots & \partial_{n-1}\varphi & 1 \end{pmatrix} \quad \text{a.e. in } \mathbb{R}^n. \tag{14.7.7}$$

It is apparent from this that $\det(D\Phi) = \det(D\Phi^{-1}) = 1$ a.e. in $\mathbb{R}^n$. Also, Exercise 2.116 guarantees that

the entries in $D\Phi$ and $D\Phi^{-1}$ belong to $L^\infty(\mathbb{R}^n)$. $\hspace{2cm}$ (14.7.8)

In addition, based on (14.7.4) it is immediate that the following restrictions are bi-Lipschitz bijections:

$$\Phi : \mathbb{R}^n_+ \to \Omega, \hspace{3cm} (14.7.9)$$

$$\Phi : \overline{\mathbb{R}^n_+} \to \overline{\Omega}, \hspace{3cm} (14.7.10)$$

$$\Phi : \partial\mathbb{R}^n_+ \to \partial\Omega, \hspace{3cm} (14.7.11)$$

A few useful properties of bounded Lipschitz domains are collected in the lemma below. For a proof see [4, Proposition 2.8].

Lemma 14.53. *Assume that Ω is a bounded Lipschitz domain in $\mathbb{R}^n$. Let $x^* \in \partial\Omega$ and let C_{x^*} and φ be associated with x^* as in Definition 14.49. Then, in addition to (14.7.2), one also has*

$$C_{x^*} \cap \partial\Omega = C_{x^*} \cap \{x' + tv : x' \in H, \ t = \varphi(x')\}, \hspace{1cm} (14.7.12)$$

$$C_{x^*} \setminus \overline{\Omega} = C_{x^*} \cap \{x' + tv : x' \in H, \ t < \varphi(x')\}. \hspace{1cm} (14.7.13)$$

Furthermore,

$$C_{x^*} \cap \overline{\Omega} = C_{x^*} \cap \{x' + tv : x' \in H, \ t \geq \varphi(x')\}, \hspace{1cm} (14.7.14)$$

$$C_{x^*} \cap \overset{\circ}{\overline{\Omega}} = C_{x^*} \cap \{x' + tv : x' \in H, \ t > \varphi(x')\}, \hspace{1cm} (14.7.15)$$

and, consequently,

$$E \cap \partial\Omega = E \cap \partial(\overline{\Omega}), \hspace{1cm} \forall E \subseteq C_{x^*}. \hspace{1cm} (14.7.16)$$

In the proof of a number of results stated for bounded Lipschitz domains one first reduces matters to an upper-graph Lipschitz domain. This is done via a localization procedure which we describe next.

Remark 14.54. Let Ω be a bounded Lipschitz domain in $\mathbb{R}^n$. Then from Definition 14.49 it follows that for each $x^* \in \partial\Omega$ there exist an $(n-1)$-dimensional plane $H \subseteq \mathbb{R}^n$ passing through x^*, a choice $v \in S^{n-1}$ of the unit normal to H, an open cylinder C_{x^*}, and a Lipschitz function $\varphi : H \to \mathbb{R}$ with $\varphi(x^*) = 0$ and such that

$$C_{x^*} \cap \Omega = C_{x^*} \cap \{x' + tv : x' \in H \text{ and } t > \varphi(x')\}. \hspace{1cm} (14.7.17)$$

Tautologically, $\partial\Omega \subseteq \bigcup_{x^* \in \partial\Omega} C_{x^*}$. Since Ω is assumed to be bounded, it follows that $\partial\Omega$ is a compact set. As such, there exist $N \in \mathbb{N}$ and $x_1^*, \ldots, x_N^* \in \partial\Omega$ with the property that $\partial\Omega \subseteq \bigcup_{j=1}^{N} C_{x_j^*}$. For each $j = 1, \ldots, N$, denote by H_j, ν_j, φ_j the respective hyperplane, unit normal, and Lipschitz function associated with x_j^*. In particular, if for each $j \in \{1, \ldots, N\}$ we set

$$\Omega_j := \{x' + t\nu_j : x' \in H_j \text{ and } t > \varphi_j(x')\}, \tag{14.7.18}$$

then Ω_j is an upper-graph Lipschitz domain in $\mathbb{R}^n$ (when described in a suitable system of coordinates corresponding to H_j being $\partial\mathbb{R}_+^n$) and

$$C_{x_j^*} \cap \Omega = C_{x_j^*} \cap \Omega_j, \tag{14.7.19}$$

$$C_{x_j^*} \cap \partial\Omega = C_{x_j^*} \cap \partial\Omega_j, \tag{14.7.20}$$

for each $j \in \{1, \ldots, N\}$. Also, we complete the collection $\{C_{x_j^*}\}_{j=1}^{N}$ of cylinder to an open cover for $\overline{\Omega}$ by picking an open set $O \subset \Omega$ with the property that $\overline{O} \subset \Omega$ and

$$\overline{\Omega} \subseteq O \cup \bigcup_{j=1}^{N} C_{x_j^*}. \tag{14.7.21}$$

Then, we invoke Theorem 14.37 to further obtain a partition of unity subordinate to this open cover of $\overline{\Omega}$, i.e., a collection $\{\psi_j\}_{j=0}^{N}$ of functions satisfying:

$$\psi_j \in C_0^\infty(\mathbb{R}^n), \quad 0 \leq \psi_j \leq 1, \quad \forall j \in \{0, 1, \ldots, N\},$$

$$\text{supp}(\psi_0) \subset O, \quad \text{supp}(\psi_j) \subset C_{x_j^*}, \quad \forall j \in \{1, \ldots, N\}, \tag{14.7.22}$$

$$\text{and } \sum_{j=0}^{N} \psi_j \equiv 1 \text{ near } \overline{\Omega}.$$

This machinery is used, for example, in the proof of Theorem 12.27, Theorem 12.29, Theorem 12.44, Theorem 12.45, and Theorem 12.46.

To state the next lemma we introduce the (open, infinite) one-component circular cone in $\mathbb{R}^n$ with vertex $x^* \in \mathbb{R}^n$, axis $h \in S^{n-1}$, and (full) aperture $\theta \in (0, \pi)$, denoted by $\Gamma_\theta(x^*, h)$, as the set

$$\Gamma_\theta(x^*, h) := \{x \in \mathbb{R}^n : \cos(\theta/2)\,|x - x^*| < (x - x^*) \cdot h\}. \tag{14.7.23}$$

Lemma 14.55. *Assume $\Omega \subseteq \mathbb{R}^n$ is an upper-graph Lipschitz domain in $\mathbb{R}^n$ and let $\varphi : \mathbb{R}^{n-1} \to \mathbb{R}$ be the Lipschitz function such that (14.7.4) holds. Denote by M the Lipschitz constant of φ and fix $\theta \in (0, 2\arctan(\frac{1}{M})]$. Then*

$$\Gamma_\theta(x, \mathbf{e}_n) \subseteq \Omega \quad \text{and} \quad \Gamma_\theta(x, -\mathbf{e}_n) \subseteq \mathbb{R}^n \setminus \Omega \text{ for each } x \in \partial\Omega. \tag{14.7.24}$$

Proof. As far as the first inclusion in (14.7.24) is concerned, it suffices to show that

$$\text{if } x', y' \in \mathbb{R}^{n-1}, \quad s \in \mathbb{R} \text{ are such that}$$
$$(y', s) \in \Gamma_\theta((x', \varphi(x')), \mathbf{e}_n) \text{ then } s > \varphi(y'). \tag{14.7.25}$$

Fix x', y', s such that $(y', s) \in \Gamma_\theta((x', \varphi(x')), \mathbf{e}_n)$. Then based on (14.7.23) we have

$$\cos(\theta/2)\sqrt{|y' - x'|^2 + (s - \varphi(x'))^2} + \varphi(x') < s. \tag{14.7.26}$$

To conclude that $s > \varphi(y')$ it suffices to show

$$\cos(\theta/2)(|y' - x'|^2 + (s - \varphi(x'))^2)^{\frac{1}{2}} + \varphi(x') \geq \varphi(y'). \tag{14.7.27}$$

This inequality is trivially true if $y' = x'$. Suppose $x' \neq y'$. Then (14.7.27) is equivalent with

$$\cos(\theta/2)\sqrt{1 + \frac{(s - \varphi(x'))^2}{|y' - x'|^2}} + \frac{\varphi(x') - \varphi(y')}{|y' - x'|} \geq 0. \tag{14.7.28}$$

Here is how (14.7.26) implies (14.7.28). Start by setting $A := \frac{|s - \varphi(x')|^2}{|y' - x'|^2}$. Then (14.7.26) becomes $\cos(\frac{\theta}{2})(1 + A)^{\frac{1}{2}} < A^{\frac{1}{2}}$. Since $\theta \neq 0$, squaring both sides this further implies $A > \cot^2(\frac{\theta}{2})$, and furthermore that

$$\cos\left(\frac{\theta}{2}\right)\sqrt{1 + \frac{(s - \varphi(x'))^2}{|y' - x'|^2}} > \cos\left(\frac{\theta}{2}\right)\sqrt{1 + \cot^2\left(\frac{\theta}{2}\right)}$$

$$= \cot\left(\frac{\theta}{2}\right) > M, \tag{14.7.29}$$

where for the last inequality in (14.7.29) we have used our assumption on θ. Now (14.7.28) follows from (14.7.29) upon recalling that φ is a Lipschitz function with Lipschitz constant M.   □

Next we discuss integration on boundaries of Lipschitz domains. In the case when $\Omega \subseteq \mathbb{R}^n$ is an upper-graph Lipschitz domain and $\varphi : \mathbb{R}^{n-1} \to \mathbb{R}$ is the Lipschitz function with the property that (14.7.4) holds, for each σ-measurable nonnegative function $f : \partial\Omega \to \mathbb{R}$ we define its integral over $\partial\Omega$ as

$$\int_{\partial\Omega} f \, d\sigma := \int_{\mathbb{R}^{n-1}} f(x', \varphi(x'))\sqrt{1 + |\nabla\varphi(x')|} \, dx'. \tag{14.7.30}$$

The space $L^1(\partial\Omega)$ is then defined as the collection (of equivalence classes, under the identification of a.e. coincidence) of σ-measurable functions $f : \partial\Omega \to \mathbb{R}$ with the property that $\int_{\partial\Omega} |f| \, d\sigma < \infty$. For each $f \in L^1(\partial\Omega)$ we may then define $\int_{\partial\Omega} f \, d\sigma$ as $\int_{\partial\Omega} f_+ \, d\sigma - \int_{\partial\Omega} f_- \, d\sigma$ where $f_\pm := \max\{\pm f, 0\} \in L^1(\partial\Omega)$. This ultimately implies that

$$\int_{\partial\Omega} f \, d\sigma = \int_{\mathbb{R}^{n-1}} f(x', \varphi(x')) \sqrt{1 + |\nabla\varphi(x')|} \, dx' \quad \text{for all } f \in L^1(\partial\Omega). \tag{14.7.31}$$

In the case when $\Omega \subseteq \mathbb{R}^n$ is a bounded Lipschitz domain the integral over $\partial\Omega$ of a σ-measurable nonnegative function $f : \partial\Omega \to \mathbb{R}$ is defined as

$$\int_{\partial\Omega} f \, d\sigma := \sum_{j=1}^{N} \int_{\partial\Omega_j} \psi_j f \, d\sigma_j, \tag{14.7.32}$$

where $\{\Omega_j\}_{j=1}^N$ are upper-graph Lipschitz domains and $\{\psi_j\}_{j=0}^N$ is a partition of unity associated with Ω as in Remark 14.54. We note that the value of the sum in (14.7.32) is independent of the latter choice of families of upper-graph domains and partition of unity. As before, this definition leads to a naturally defined Lebesgue space $L^1(\partial\Omega)$, within which the integration on $\partial\Omega$ is meaningful.

Lemma 14.56. *Let $\alpha > 0$. Suppose Ω is either an upper-graph Lipschitz domain in $\mathbb{R}^n$, or a bounded Lipschitz domain in $\mathbb{R}^n$. Then there exists some constant $C = C(\Omega, \alpha) \in (0, \infty)$ such that for each $x \in \partial\Omega$ and each $r > 0$ we have*

$$\int_{\partial\Omega} \frac{\mathbf{1}_{|x-y|<r}}{|x-y|^{n-1-\alpha}} \, d\sigma(y) \leq Cr^\alpha. \tag{14.7.33}$$

In particular, corresponding to the special case when $\alpha = n - 1$, estimate (14.7.33) yields

$$\sigma(\partial\Omega \cap B(x, r)) \leq Cr^{n-1}, \quad \forall x \in \partial\Omega, \quad \forall r \in (0, \infty). \tag{14.7.34}$$

Proof. Suppose first that Ω is the upper-graph of a Lipschitz function $\varphi : \mathbb{R}^{n-1} \to \mathbb{R}$ with Lipschitz constant M. Pick arbitrary $x \in \partial\Omega$ and $r > 0$. Then, by (14.7.31) we have

$$\int_{\partial\Omega} \frac{\mathbf{1}_{|x-y|<r}}{|x-y|^{n-1-\alpha}} \, d\sigma(y)$$

$$= \int_{\mathbb{R}^{n-1}} \frac{\mathbf{1}_{|(x'-y', \varphi(x')-\varphi(y'))|<r}}{\left|(x'-y', \varphi(x') - \varphi(y'))\right|^{n-1-\alpha}} \sqrt{1 + |\nabla\varphi(y')|^2} \, dy'$$

$$\leq \sqrt{1 + M^2} \int_{y' \in \mathbb{R}^{n-1}, \, |y'-x'|<r} \frac{dy'}{|x'-y'|^{n-1-\alpha}} \, dy'$$

$$= \omega_{n-2} \sqrt{1 + M^2} \int_0^r \frac{\rho^{n-2}}{\rho^{n-1-\alpha}} \, d\rho = \frac{\omega_{n-2}}{\alpha} \sqrt{1 + M^2} \, r^\alpha. \tag{14.7.35}$$

In the second to the last equality in (14.7.35) we used polar coordinates to write $y' = x' + \rho\,\omega$, for $\omega \in S^{n-2}$ and $\rho \in [0, r]$. This proves (14.7.33). Also, (14.7.34) in the case of an upper-graph Lipschitz domain now follows from (14.7.33) by taking $\alpha = n - 1$.

Next, assume Ω is a bounded Lipschitz domain. Let the integer $N \in \mathbb{N}$, the points $x_1^*, \ldots, x_N^* \in \partial\Omega$, the upper-graph Lipschitz domains $\{\Omega_j\}_{j=1}^N$, and the cylinders $\{C_{x_j^*}\}_{j=1}^N$ be as in Remark 14.54. These cylinders have finite diameters, hence $R := \max_{1 \le j \le N} \operatorname{diam}(C_{x_j^*})$ is finite. In particular, $\partial\Omega \cap C_{x_j^*} \subseteq \partial\Omega_j \cap B(x_j^*, R)$, for each $j \in \{1, \ldots, N\}$, and we may use (14.7.34) to obtain

$$\sigma(\partial\Omega) \le \sum_{j=1}^N \sigma(\partial\Omega \cap C_{x_j^*}) \le \sum_{j=1}^N C(\Omega_j) R^{n-1} < \infty. \tag{14.7.36}$$

Next, apply the Lebesgue Number Theorem 14.44 to the metric space $\partial\Omega$ endowed with the Euclidean distance and cover $\{C_{x_j^*}\}_{j=1}^N$ to find $r_* > 0$ such that for every $x \in \partial\Omega$ the ball $B(x, r_*)$ is contained in one of the cylinders in the respective cover. Pick an arbitrary $x \in \partial\Omega$ and $r > 0$. Let $j \in \{1, \ldots, N\}$ be such that $B(x, r_*) \subseteq C_{x_j^*}$. Invoking (14.7.20) we have $\partial\Omega \cap B(x, r_*) = \partial\Omega_j \cap B(x, r_*)$. We have two cases. First, if $r \le r_*$ then $\partial\Omega \cap B(x, r) = \partial\Omega_j \cap B(x, r)$. Hence, with $d\sigma_j$ denoting the surface measure on $\partial\Omega_j$, we obtain

$$\int_{\partial\Omega} \frac{\mathbf{1}_{|x-y|<r}}{|x-y|^{n-1-\alpha}} \, d\sigma(y) = \int_{\partial\Omega_j} \frac{\mathbf{1}_{|x-y|<r}}{|x-y|^{n-1-\alpha}} \, d\sigma_j(y)$$

$$\le Cr^\alpha, \quad \forall r \in (0, r_*], \tag{14.7.37}$$

where for the inequality in (14.7.37) we used (14.7.33) for the upper-graph Lipschitz domain Ω_j. Second, if $r_* \le r$, then we may use (14.7.37) with $r = r_*$ and (14.7.36) to write

$$\int_{\partial\Omega} \frac{\mathbf{1}_{|x-y|<r}}{|x-y|^{n-1-\alpha}} \, d\sigma(y) = \int_{\partial\Omega} \frac{\mathbf{1}_{|x-y|<r_*}}{|x-y|^{n-1-\alpha}} \, d\sigma(y) + \int_{\partial\Omega} \frac{\mathbf{1}_{r_*<|x-y|<r}}{|x-y|^{n-1-\alpha}} \, d\sigma(y)$$

$$\le Cr_*^\alpha + r_*^{-n+1+\alpha} \sigma(B(x, r) \cap \partial\Omega)$$

$$\le r_*^\alpha(C + r_*^{-n+1} \sigma(\partial\Omega)) = Cr_*^\alpha \le Cr^\alpha, \tag{14.7.38}$$

for some $C \in (0, \infty)$ depending on Ω (including the cover $\{C_{x_j^*}\}_{j=1}^N$) and α. A combination of (14.7.37) and (14.7.38) yields (14.7.33) in the case when Ω is a bounded Lipschitz domain. $\qquad\square$

Lemma 14.57. *Let $\alpha > 0$. Assume that $\Omega \subseteq \mathbb{R}^n$ is either an upper-graph Lipschitz domain, or a bounded Lipschitz domain. Then for each $x \in \partial\Omega$ and each $r > 0$ there exists a constant $C \in (0, \infty)$ depending on Ω, n, and α such that*

$$\int_{\partial\Omega} \frac{\mathbf{1}_{|x-y|>r}}{|x-y|^{n-1+\alpha}} \, d\sigma(y) \le Cr^{-\alpha}. \tag{14.7.39}$$

Proof. Given $x \in \partial\Omega$ and $r > 0$ arbitrary we have

$$\int_{\partial\Omega} \frac{\mathbf{1}_{|x-y|>r}}{|x-y|^{n-1+\alpha}}\, d\sigma(y) = \sum_{j=0}^{\infty} \int_{\partial\Omega} \frac{\mathbf{1}_{2^{j+1}r \geq |x-y| > 2^j r}}{|x-y|^{n-1+\alpha}}\, d\sigma(y)$$

$$\leq \sum_{j=0}^{\infty} (2^j r)^{-n+1-\alpha} \sigma(B(x, 2^{j+1}r) \cap \partial\Omega)$$

$$\leq \sum_{j=0}^{\infty} (2^j r)^{-n+1-\alpha} C(2^{j+1}r)^{n-1}$$

$$= Cr^{-\alpha} \sum_{j=0}^{\infty} 2^{-j\alpha} = Cr^{-\alpha}, \tag{14.7.40}$$

where for the last inequality we have used (14.7.34). $\qquad\square$

We record here a change of variable formula via bijective bi-Lipschitz functions.

Theorem 14.58. *Suppose* $\Psi : \mathbb{R}^n \to \mathbb{R}^n$ *is a bijective and bi-Lipschitz function. Let* O *and* $\mathcal{U}$ *be open sets in* $\mathbb{R}^n$ *such that* $\mathcal{U} = \Psi(O)$. *Then for every function* $u \in L^1(\mathcal{U})$ *one has*

$$\int_{\mathcal{U}} u(x) \big|\det(D(\Psi^{-1})(x))\big|\, dx = \int_{O} u(\Psi(y))\, dy. \tag{14.7.41}$$

Proof. This is a particular case of [15, Theorem 2, p. 99] applied with $n := m$, $f := \Psi^{-1}$, and $g := \widetilde{u}$, the extension by zero of u outside $\mathcal{U}$. $\qquad\square$

14.8 Integration by Parts and Green's Formula

Definition 14.59. We say that a nonempty open set $\Omega \subseteq \mathbb{R}^n$, where $n \geq 2$, is a C^k domain (or, a domain of class C^k), for some $k \in \mathbb{N}_0 \cup \{\infty\}$, provided the following holds. For every point $x^* \in \partial\Omega$ there exist $R > 0$, an open interval $I \subset \mathbb{R}$ with $0 \in I$, a rigid transformation $T : \mathbb{R}^n \to \mathbb{R}^n$ with $T(x^*) = 0$, along with a function ϕ of class C^k that maps $B(0, R) \subseteq \mathbb{R}^{n-1}$ into I with the property that $\phi(0) = 0$ and, if C denotes the (open) cylinder $B(0, R) \times I \subseteq \mathbb{R}^{n-1} \times \mathbb{R} = \mathbb{R}^n$, then

$$C \cap T(\Omega) = \{x = (x', x_n) \in C : x_n > \phi(x')\}, \tag{14.8.1}$$

$$C \cap \partial T(\Omega) = \{x = (x', x_n) \in C : x_n = \phi(x')\}, \tag{14.8.2}$$

$$C \cap (\overline{T(\Omega)})^c = \{x = (x', x_n) \in C : x_n < \phi(x')\}. \tag{14.8.3}$$

If $\Omega \subseteq \mathbb{R}^n$ is a C^k domain for some $k \in \mathbb{N}_0 \cup \{\infty\}$, then it may be easily verified that $\partial\Omega$ is a C^k surface. The fact that boundaries of bounded domains of class C^k, $k \in \mathbb{N} \cup \{\infty\}$, may be locally flattened via diffeomorphisms of class C^k is stated next.

Theorem 14.60 (Integration by Parts Formula). *Suppose $\Omega \subset \mathbb{R}^n$ is a domain of class C^1 and $\nu = (\nu_1, \ldots, \nu_n)$ denotes its outward unit normal. Let $k \in \{1, \ldots, n\}$ and assume $f, g \in C^1(\Omega) \cap C^0(\overline{\Omega})$ are such that $\partial_k f, \partial_k g \in L^1(\Omega)$ and there exists a compact set $K \subset \mathbb{R}^n$ with the property that $f = 0$ on $\Omega \setminus K$. Then,*

$$\int_\Omega (\partial_k f) g \, \mathrm{d}x = -\int_\Omega f(\partial_k g) \, \mathrm{d}x + \int_{\partial\Omega} f g \nu_k \, \mathrm{d}\sigma, \tag{14.8.4}$$

where σ is the surface measure on $\partial\Omega$.

For the sense in which the last integral in (14.8.4) should be understood see Definition 14.48.

An immediate corollary of Theorem 14.60 is Green's formula that is stated next (recall 7.1.14).

Theorem 14.61 (Green's formula). *Suppose $\Omega \subset \mathbb{R}^n$ is a bounded domain of class C^1 and ν denotes its outward unit normal. If $f, g \in C^2(\overline{\Omega})$, then*

$$\int_\Omega f \Delta g \, \mathrm{d}x = \int_\Omega g \Delta f \, \mathrm{d}x + \int_{\partial\Omega} \left(f \frac{\partial g}{\partial \nu} - g \frac{\partial f}{\partial \nu} \right) \mathrm{d}\sigma. \tag{14.8.5}$$

14.9 Polar Coordinates and Integrals on Spheres

Assume that $n \geq 3$ and $R > 0$ are fixed. For $\rho \in (0, R)$, $\theta_j \in (0, \pi)$, $j \in \{1, \ldots, n-2\}$, and $\theta_{n-1} \in (0, 2\pi)$, set

$$x_1 := \rho \cos \theta_1,$$

$$x_2 := \rho \sin \theta_1 \cos \theta_2,$$

$$x_3 := \rho \sin \theta_1 \sin \theta_2 \cos \theta_3, \tag{14.9.1}$$

$$\vdots$$

$$x_{n-1} := \rho \sin \theta_1 \sin \theta_2 \ldots \sin \theta_{n-2} \cos \theta_{n-1},$$

$$x_n := \rho \sin \theta_1 \sin \theta_2 \ldots \sin \theta_{n-2} \sin \theta_{n-1}.$$

The variables $\theta_1, \ldots, \theta_{n-1}, \rho$ are called `polar coordinates`.

Definition 14.62. Assume that $x_* \in \mathbb{R}^n$, $n \geq 3$, is a fixed point, and $R > 0$ is arbitrary. The `standard parametrization of the ball` $B(x_*, R)$ is defined as

$$\mathcal{P} : (0, \pi)^{n-2} \times (0, 2\pi) \times (0, R) \longrightarrow \mathbb{R}^n,$$

$$\mathcal{P}(\theta_1, \theta_2, \ldots, \theta_{n-1}, \rho) := x_* + (x_1, x_2, \ldots, x_n), \tag{14.9.2}$$

where $x_1, \ldots, x_n$ are as in (14.9.1).

The function $\mathcal{P}$ in (14.9.2) is injective, of class C^∞, takes values in $B(x_*, R)$, its image differs from $B(x_*, R)$ by a subset of measure zero and

$$\det(D\mathcal{P})(\theta_1, \theta_2, \ldots, \theta_{n-1}, \rho) = \rho^{n-1}(\sin \theta_1)^{n-2}(\sin \theta_2)^{n-3} \ldots (\sin \theta_{n-2}), \quad (14.9.3)$$

at every point in its domain, where $D\mathcal{P}$ denotes the Jacobian of $\mathcal{P}$. Using this standard parametrization for the unit sphere in $\mathbb{R}^n$, we see that

$$\omega_{n-1} = \int_0^\pi \cdots \int_0^\pi \int_0^{2\pi} (\sin \varphi_1)^{n-2}(\sin \varphi_2)^{n-3} \cdots (\sin \varphi_{n-3})^2 \times$$

$$\times (\sin \varphi_{n-2}) \, d\varphi_{n-1} \, d\varphi_{n-2} \ldots d\varphi_1. \quad (14.9.4)$$

This parametrization of the sphere $B(x_*, R)$ may also be used to prove the following theorem.

Theorem 14.63 (Spherical Fubini and Polar Coordinates). *Let* $f \in L^1_{loc}(\mathbb{R}^n)$, $n \geq 2$. *Then for each* $x_* \in \mathbb{R}^n$ *and each* $R > 0$ *the following formulas hold:*

$$\int_{B(x_*, R)} f \, dx = \int_0^R \left(\int_{\partial B(x_*, \rho)} f \, d\sigma \right) d\rho, \quad (14.9.5)$$

$$\int_{B(x_*, R)} f \, dx = \int_0^R \int_{S^{n-1}} f(x_* + \rho\omega)\rho^{n-1} \, d\sigma(\omega) \, d\rho \quad (14.9.6)$$

$$= \int_0^R \int_{\partial B(x_*, 1)} f(\rho\omega)\rho^{n-1} \, d\sigma(\omega) \, d\rho. \quad (14.9.7)$$

Moreover, if $f \in L^1(\mathbb{R}^n)$, *then*

$$\int_{\mathbb{R}^n} f \, dx = \int_0^\infty \left(\int_{\partial B(0, \rho)} f \, d\sigma \right) d\rho, \quad (14.9.8)$$

$$\int_{\mathbb{R}^n} f \, dx = \int_0^\infty \int_{S^{n-1}} f(\rho\omega)\rho^{n-1} \, d\sigma(\omega) \, d\rho. \quad (14.9.9)$$

Note that if $P : O \to S^{n-1}$ is a parametrization of the unit sphere in $\mathbb{R}^n$ and if $\mathcal{R}$ is a unitary transformation in $\mathbb{R}^n$, then $\mathcal{R} \circ P : O \to S^{n-1}$ is also a parametrization of the unit sphere in $\mathbb{R}^n$. Indeed, this function is injective, has C^1 components, its image is S^{n-1} (up to a negligible set) and

$$\text{Rank}\left[D(\mathcal{R} \circ P)\right] = \dim\left(\text{Im}\left[\mathcal{R}(DP)\right]\right)$$

$$= \dim\left(\text{Im}(DP)\right) = \text{Rank}(DP) = n - 1. \quad (14.9.10)$$

Hence,

$$\int_{S^{n-1}} f \circ \mathcal{R} \, d\sigma = \int_O f(\mathcal{R} \circ P) |\partial_{u_1} P \times \cdots \times \partial_{u_{n-1}} P| \, du_1 \ldots du_{n-1}$$

$$= \int_O f(\mathcal{R} \circ P) |\partial_{u_1} (\mathcal{R} \circ P) \times \cdots \times \partial_{u_{n-1}} (\mathcal{R} \circ P)| \, du_1 \ldots du_{n-1}$$

$$= \int_{S^{n-1}} f \, d\sigma. \tag{14.9.11}$$

The same type of reasoning also yields the following result.

Proposition 14.64. *For each* $x = (x_1, ..., x_n) \in \mathbb{R}^n$, *and each* $j, k \in \{1, \ldots, n\}$ *with* $j \leq k$, *define*

$$R_j(x) := (x_1, ..., x_{j-1}, -x_j, x_{j+1}, ..., x_n),$$
$$R_{jk}(x) := (x_1, ..., x_{j-1}, x_k, x_{j+1}, ..., x_{k-1}, x_j, x_{k+1}, ..., x_n). \tag{14.9.12}$$

Then

$$\int_{S^{n-1}} f \circ R_j \, d\sigma = \int_{S^{n-1}} f \, d\sigma, \qquad 1 \leq j \leq n, \tag{14.9.13}$$

$$\int_{S^{n-1}} f \circ R_{jk} \, d\sigma = \int_{S^{n-1}} f \, d\sigma, \qquad 1 \leq j \leq k \leq n. \tag{14.9.14}$$

Proposition 14.65. *Let* $v \in \mathbb{R}^n \setminus \{0\}$, $n \geq 2$, *be fixed. Then for any measurable and nonnegative function* f *defined on the real line, there holds*

$$\int_{S^{n-1}} f(v \cdot \theta) \, d\sigma(\theta) = \omega_{n-2} \int_{-1}^1 f(s|v|)(\sqrt{1 - s^2})^{n-3} \, ds. \tag{14.9.15}$$

Proof. Since integrals over the unit sphere are invariant under orthogonal transformations, we may assume that $v/|v| = e_1$ and, hence, using polar coordinates and (14.9.3), we have

$$\int_{S^{n-1}} f(v \cdot \theta) \, d\sigma(\theta) = \int_{S^{n-1}} f(|v|\theta_1) \, d\sigma(\theta)$$

$$= \int_0^{2\pi} \left(\int_0^\pi \cdots \int_0^\pi f(|v| \cos \varphi_1) \prod_{j=1}^{n-2} (\sin \varphi_j)^{n-1-j} \, d\varphi_1 \ldots d\varphi_{n-2} \right) d\varphi_{n-1}$$

$$= \omega_{n-2} \int_0^\pi f(|v| \cos \varphi_1)(\sin \varphi_1)^{n-2} \, d\varphi_1. \tag{14.9.16}$$

Making the change of variables $s := \cos \varphi_1$ in the last integral above shows that this matches the right-hand side of (14.9.15). $\square$

Remark 14.66. Of course (14.9.15) remains valid for measurable and non-positive functions. In general, if f is merely measurable and real-valued, then one can write

(14.9.15) for f_+ and f_- and subtract these identities in order to obtain (14.9.15) for f, as long as one does not run into $\infty - \infty$.

Proposition 14.67. *Let $f \in C^0(\mathbb{R})$ be positive homogeneous of degree $m \in \mathbb{R}$ and fix $\eta = (\eta_1, \ldots, \eta_n) \in \mathbb{R}^n$. Then if $n \in \mathbb{N}$ with $n \geq 2$, for $j, k \in \{1, \ldots, n\}$ one has*

$$\int_{S^{n-1}} f(\eta \cdot \xi) \xi_j \xi_k \, d\sigma(\xi) = \alpha |\eta|^m \delta_{jk} + \beta |\eta|^{m-2} \eta_j \eta_k \tag{14.9.17}$$

where

$$\begin{aligned}
\alpha &= \frac{1}{n-1} \int_{S^{n-1}} f(\xi_1)(1 - \xi_1^2) \, d\sigma(\xi), \\
\beta &= \frac{1}{n-1} \int_{S^{n-1}} f(\xi_1)(n\xi_1^2 - 1) \, d\sigma(\xi).
\end{aligned} \tag{14.9.18}$$

Proof. For $j, k \in \{1, \ldots, n\}$ set

$$q_{jk}(\eta) := \int_{S^{n-1}} f(\eta \cdot \xi) \xi_j \xi_k \, d\sigma(\xi), \qquad \forall \eta \in \mathbb{R}^n, \tag{14.9.19}$$

and define the quadratic form

$$Q(\zeta, \eta) := \sum_{j,k=1}^n q_{jk}(\eta) \zeta_j \zeta_k, \qquad \forall \zeta, \eta \in \mathbb{R}^n. \tag{14.9.20}$$

Observe that we can write $Q(\zeta, \eta) = \int_{S^{n-1}} f(\eta \cdot \xi)(\zeta \cdot \xi)^2 \, d\sigma(\xi)$. By the invariance under rotations of integrals over S^{n-1} (see (14.9.11)), we have that for any rotation R in $\mathbb{R}^n$

$$Q(\zeta, \eta) = \int_{S^{n-1}} f(\eta \cdot R^\top \xi)(\zeta \cdot R^\top \xi)^2 \, d\sigma(\xi) = Q(R\zeta, R\eta), \quad \zeta, \eta \in \mathbb{R}^n. \tag{14.9.21}$$

A direct computation also gives that

$$Q(\lambda_1 \zeta, \lambda_2 \eta) = \lambda_1^2 \lambda_2^m Q(\zeta, \eta) \quad \text{for all } \lambda_1, \lambda_2 > 0, \, \zeta, \eta \in \mathbb{R}^n. \tag{14.9.22}$$

Next we claim that

$$Q(\zeta, \eta) = \alpha + \beta (\eta \cdot \zeta)^2, \qquad \forall \zeta, \eta \in S^{n-1}. \tag{14.9.23}$$

To show that (14.9.23) holds, we first observe that it suffices to prove (14.9.23) when $\eta = e_1$. Indeed, if we assume that (14.9.23) is true for $\eta = e_1$, then for an arbitrary η let R be the rotation such that $R\eta = e_1$. Then if we also take into account (14.9.20), we have

$$Q(\zeta, \eta) = Q(R\zeta, R\eta) = Q(R\zeta, e_1) = \alpha + \beta(R\zeta \cdot e_1)^2 \tag{14.9.24}$$

$$= \alpha + \beta(\zeta \cdot R^\top e_1)^2 = \alpha + \beta(\eta \cdot \zeta)^2.$$

Hence, (14.9.23) will follow if we prove that

$$\int_{S^{n-1}} f(\xi_1)(\zeta \cdot \xi)^2 \, d\sigma(\xi) = \alpha + \beta \zeta_1^2, \qquad \forall \zeta \in S^{n-1}. \tag{14.9.25}$$

To see the later let $A_{jk} := \int_{S^{n-1}} f(\xi_1) \xi_j \xi_k \, d\sigma(\xi)$ for $j, k \in \{1, \dots, n\}$. Assume $j \neq k$. Then either $j \neq 1$ or $k \neq 1$. If, say, $j \neq 1$ we use (14.9.13) to conclude that $A_{jk} = 0$ in this case. A similar reasoning applies in the case $k \neq 1$. Clearly, we have $A_{11} = \int_{S^{n-1}} f(\xi_1) \xi_1^2 \, d\sigma(\xi)$. Corresponding to $j = k \neq 1$, we first observe that $A_{22} = \cdots = A_{nn}$ since A_{jj} is independent of $j \in \{2, \dots, n\}$ due to (14.9.14). Moreover, $\sum_{j=1}^{n} A_{jj} = \int_{S^{n-1}} f(\xi_1) \, d\sigma(\xi)$, which in turn implies that

$$A_{jj} = \frac{1}{n-1} \Big(\int_{S^{n-1}} f(\xi_1) \, d\sigma(\xi) - A_{11} \Big) = \alpha \quad \text{for each} \quad j \in \{2, \dots, n\}. \tag{14.9.26}$$

Combining all these we have that for $\zeta \in S^{n-1}$,

$$\int_{S^{n-1}} f(\xi_1)(\zeta \cdot \xi)^2 \, d\sigma(\xi) = \sum_{j,k=1}^{n} \zeta_j \zeta_k A_{jk}$$

$$= \zeta_1^2 \int_{S^{n-1}} f(\xi_1) \xi_1^2 \, d\sigma(\xi) + \alpha \sum_{j=2}^{n} \zeta_j^2$$

$$= \alpha + \beta \zeta_1^2. \tag{14.9.27}$$

This concludes the proof of (14.9.25) which, in turn, implies (14.9.23). Now if ζ, $\eta \in \mathbb{R}^n \setminus \{0\}$, we make use of (14.9.22) and (14.9.23) to write

$$Q(\zeta, \eta) = |\zeta|^2 |\eta|^m Q\Big(\frac{\zeta}{|\zeta|}, \frac{\eta}{|\eta|} \Big) = \alpha |\zeta|^2 |\eta|^m + \beta |\eta|^{m-2} (\zeta \cdot \eta)^2 \tag{14.9.28}$$

$$= \sum_{j,k=1}^{n} (\alpha |\eta|^m \delta_{jk} + \beta |\eta|^{m-2} \eta_j \eta_k) \zeta_j \zeta_k$$

which proves (14.9.17).   □

Proposition 14.68. *Consider $f(t) := |t|$ for $t \in \mathbb{R}$, and let α, β be as in (14.9.17) for this choice of f. Then*

$$\alpha = \beta = \frac{2\omega_{n-2}}{n^2 - 1}, \tag{14.9.29}$$

where ω_{n-2} denotes the surface measure of the unit ball in $\mathbb{R}^{n-1}$.

Proof. Using the standard parametrization of S^{n-1} (see (14.9.1) with $R = 1$) we have

$$\alpha = \frac{1}{n-1} \int_0^\pi \int_0^\pi \cdots \int_0^\pi \int_0^{2\pi} |\cos\varphi_1|(1 - (\cos\varphi_1)^2)(\sin\varphi_1)^{n-2}(\sin\varphi_2)^{n-3} \times$$

$$\times \cdots \times (\sin\varphi_{n-2}) \, d\varphi_{n-1} \, d\varphi_{n-2} \ldots d\varphi_1$$

$$= \frac{\omega_{n-2}}{n-1} \int_0^\pi |\cos\varphi_1|(\sin\varphi_1)^n \, d\varphi_1. \tag{14.9.30}$$

The change of variables $\theta = \pi - y$ yields $-\int_{\frac{\pi}{2}}^\pi \cos\theta(\sin\theta)^n \, d\theta = \int_0^{\frac{\pi}{2}} \cos y(\sin y)^n \, dy$. Using this back in (14.9.30) then gives

$$\alpha = \frac{2\omega_{n-2}}{n-1} \int_0^{\frac{\pi}{2}} \cos\theta\,(\sin\theta)^n \, d\theta = \frac{2\omega_{n-2}}{n-1} \frac{(\sin\theta)^{n+1}}{n+1}\Big|_0^{\frac{\pi}{2}} = \frac{2\omega_{n-2}}{n^2-1}. \tag{14.9.31}$$

Similar arguments in computing β give that

$$\beta = \frac{\omega_{n-2}}{n-1} \int_0^\pi |\cos\theta|[n(\cos\theta)^2 - 1](\sin\theta)^{n-2} \, d\theta \tag{14.9.32}$$

$$= \omega_{n-2} \int_0^\pi |\cos\theta|(\sin\theta)^{n-2} \, d\theta - \frac{n\,\omega_{n-2}}{n-1} \int_0^\pi |\cos\theta|(\sin\theta)^n \, d\theta$$

$$= \frac{2\omega_{n-2}}{n-1}\Big[(n-1) \int_0^{\frac{\pi}{2}} \cos\theta\,(\sin\theta)^{n-2} \, d\theta - n \int_0^{\frac{\pi}{2}} \cos\theta\,(\sin\theta)^n \, d\theta\Big]$$

$$= \frac{2\omega_{n-2}}{n-1}\Big[(\sin\theta)^{n-1}\Big|_0^{\frac{\pi}{2}} - n\frac{(\sin\theta)^{n+1}}{n+1}\Big|_0^{\frac{\pi}{2}}\Big] = \frac{2\omega_{n-2}}{n^2-1}.$$

This finishes the proof. $\qquad\square$

Recall from (0.0.8) that $z^\alpha = z_1^{\alpha_1} z_2^{\alpha_2} \cdots z_n^{\alpha_n}$ whenever $z = (z_1, ..., z_n) \in \mathbb{R}^n$ and $\alpha = (\alpha_1, \alpha_2, \ldots, \alpha_n) \in \mathbb{N}_0^n$. Let us also introduce

$$2\mathbb{N}_0^n := \{(2\alpha_1, 2\alpha_2, \ldots, 2\alpha_n) : \alpha = (\alpha_1, \alpha_2, \ldots, \alpha_n) \in \mathbb{N}_0^n\}. \tag{14.9.33}$$

The next proposition deals with the issue of integrating arbitrary monomials on the unit sphere centered at the origin.

Proposition 14.69. *For each multi-index $\alpha = (\alpha_1, \ldots, \alpha_n) \in \mathbb{N}_0^n$,*

$$\int_{S^{n-1}} z^\alpha \, d\sigma(z) = \begin{cases} 0 & \text{if } \alpha \notin 2\mathbb{N}_0^n, \\[2mm] \dfrac{\left(\frac{|\alpha|}{2}\right)!}{|\alpha|!} \cdot \dfrac{\alpha!}{\left(\frac{\alpha}{2}\right)!} \dfrac{2\pi^{\frac{n-1}{2}}\Gamma(\frac{1+|\alpha|}{2})}{\Gamma(\frac{|\alpha|+n}{2})} & \text{if } \alpha \in 2\mathbb{N}_0^n, \end{cases} \tag{14.9.34}$$

where Γ is the Gamma function introduced in (14.5.1).

Proof. Fix an arbitrary $k \in \mathbb{N}$ and set

$$q_\alpha := \int_{S^{n-1}} z^\alpha \, d\sigma(z), \quad \forall \, \alpha \in \mathbb{N}_0^n \text{ with } |\alpha| = k. \tag{14.9.35}$$

Also, with "dot" standing for the standard inner product in $\mathbb{R}^n$, introduce

$$Q_k(x) := \sum_{\substack{\alpha \in \mathbb{N}_0^n \\ |\alpha|=k}} \frac{k!}{\alpha!} q_\alpha x^\alpha, \quad \forall \, x = (x_1, x_2, \dots, x_n) \in \mathbb{R}^n. \tag{14.9.36}$$

Then (14.2.3) implies that

$$Q_k(x) = \int_{S^{n-1}} (z \cdot x)^k \, d\sigma(z), \quad \forall \, x \in \mathbb{R}^n. \tag{14.9.37}$$

Let us also observe here that, if $x \in S^{n-1}$ is arbitrary but fixed and if $\mathcal{R}$ is a rotation about the origin in $\mathbb{R}^n$ such that $\mathcal{R}^{-1}x = e_n := (0, \dots, 0, 1) \in \mathbb{R}^n$, then by (14.9.36) and the rotation invariance of integrals on S^{n-1} (cf. (14.9.11)), we have

$$Q_k(x) = \int_{S^{n-1}} (\mathcal{R}z \cdot x)^k \, d\sigma(z) = Q_k(e_n). \tag{14.9.38}$$

By the homogeneity of Q_k, (14.9.38) implies that $Q_k(x) = |x|^k Q_k(e_n)$ for all $x \in \mathbb{R}^n$ and, hence,

$$\sum_{|\alpha|=k} \frac{k!}{\alpha!} q_\alpha x^\alpha = |x|^k Q_k(e_n) \quad \text{for all } x \in \mathbb{R}^n. \tag{14.9.39}$$

We now consider two cases, starting with:

Case I: k is odd. In this scenario, the mapping $S^{n-1} \ni z \mapsto z_n^k \in \mathbb{R}$ is odd. In particular, $Q_k(x) = \int_{S^{n-1}} z_n^k \, d\sigma(z) = 0$. This, in turn, along with (14.9.39) then force $\sum_{|\alpha|=k} \frac{k!}{\alpha!} q_\alpha x^\alpha = 0$ for every $x \in \mathbb{R}^n$. From (14.2.7) it is easy to deduce that for each $\beta \in \mathbb{N}_0^n$ we have

$$\partial_x^\gamma [x^\alpha] \Big|_{x=0} = \begin{cases} 0 & \text{if } \alpha \neq \gamma, \\ \alpha! & \text{if } \alpha = \gamma. \end{cases} \tag{14.9.40}$$

We may therefore conclude that $q_\gamma = \partial_x^\gamma \left[\sum_{|\alpha|=k} \frac{1}{\alpha!} q_\alpha x^\alpha \right] = 0$ for each $\gamma \in \mathbb{N}_0^n$ with $|\gamma| = k$, in agreement with (14.9.34).

Case II: k is even. Suppose $k = 2m$ for some $m \in \mathbb{N}$. Then, $|x|^k = \sum_{|\beta|=m} \frac{m!}{\beta!} x^{2\beta}$ and (14.9.39) becomes

$$\sum_{|\beta|=m} \frac{m!}{\beta!} x^{2\beta} Q_k(e_n) = \sum_{|\alpha|=k} \frac{k!}{\alpha!} q_\alpha x^\alpha \quad \text{for all } x \in \mathbb{R}^n. \tag{14.9.41}$$

Fix $\gamma \in \mathbb{N}_0^n$ such that $|\gamma| = k$ and observe that

$$\partial_x^\gamma \left[\frac{m!}{\beta!} x^{2\beta} Q_k(e_n) \right]\Bigg|_{x=0} = \begin{cases} \dfrac{m!}{\beta!} (2\beta)! Q_k(e_n) & \text{if } \gamma = 2\beta, \text{ some } \beta \in \mathbb{N}_0^n, \ |\beta| = m, \\ 0 & \text{otherwise.} \end{cases}$$

$$(14.9.42)$$

This, (14.9.41), and (14.9.40) then imply that

$$q_\gamma = \begin{cases} 0 & \text{if } \gamma \notin 2\mathbb{N}_0^n, \\ \dfrac{\left(\frac{|\gamma|}{2}\right)!}{\left(\frac{\gamma}{2}\right)!} \cdot \dfrac{\gamma!}{|\gamma|!} Q_{|\gamma|}(e_n) & \text{if } \gamma \in 2\mathbb{N}_0^n. \end{cases}$$

$$(14.9.43)$$

We are now left with computing $Q_{2m}(e_n)$ when $m \in \mathbb{N}$. Using spherical coordinates, a direct computation gives that

$$Q_{2m}(e_n) = \int_{S^{n-1}} (z \cdot e_n)^{2m} \, d\sigma(z)$$

$$= \int_0^{2\pi} \left(\int_0^\pi \cdots \int_0^\pi (\cos \theta_1)^{2m} \prod_{j=1}^{n-2} (\sin \theta_j)^{n-1-j} \, d\theta_1 \ldots d\theta_{n-2} \right) d\theta_{n-1}$$

$$= \omega_{n-2} \int_0^\pi (\cos \theta)^{2m} (\sin \theta)^{n-2} \, d\theta = \frac{2\pi^{\frac{n-1}{2}} \Gamma(\frac{1}{2} + m)}{\Gamma(m + \frac{n}{2})}, \qquad (14.9.44)$$

by (14.9.4) and (14.5.6) (considered with $n-1$ in place of n), and (14.5.11). This once again agrees with (14.9.34), and the proof of Proposition 14.69 is finished. $\square$

A simple useful consequence of the general formula (14.9.34) is the fact that

$$\int_{S^{n-1}} z_j z_k \, d\sigma(z) = \frac{\omega_{n-1}}{n} \delta_{jk} \quad \text{whenever } 1 \le j, k \le n. \qquad (14.9.45)$$

14.10 Hankel Functions

Let $H_\lambda^{(1)}(\cdot)$ denote the Hankel function of the first kind with index $\lambda \in \mathbb{R}$. Its definition and some of its basic properties are reviewed next. For more on this topic see e.g., [1], [46], [60], [74].

Definition 14.70. Fix $\lambda \in \mathbb{R}$. The regular Bessel function J_λ is defined by

$$J_\lambda(z) = \sum_{j=0}^\infty \left(\frac{z}{2}\right)^{\lambda+2j} \frac{(-1)^j}{\Gamma(\lambda + j + 1) j!}, \qquad z \in \mathbb{C}, \qquad (14.10.1)$$

where Γ is the gamma function from (14.5.1). Also, by N_λ we denote the irregular Bessel function given by

$$N_\lambda(z) := \frac{J_\lambda(z)\cos(\pi\lambda) - J_{-\lambda}(z)}{\sin(\pi\lambda)}, \qquad z \in \mathbb{C}, \tag{14.10.2}$$

for $\lambda \notin \mathbb{N}_0$ and, corresponding to the case when $\lambda \in \mathbb{N}_0$,

$$N_\lambda(z) := \frac{2}{\pi} J_\lambda(z) \log\left(\frac{z}{2}\right) - \frac{1}{\pi} \sum_{j=0}^{\lambda-1} \frac{(\lambda - j - 1)!}{j!} \left(\frac{z}{2}\right)^{2j-\lambda}$$

$$- \frac{1}{\pi} \sum_{j=0}^{\infty} \left(\frac{z}{2}\right)^{2j+\lambda} \frac{(-1)^j}{j!(j+\lambda)!} \left[\frac{\Gamma'(j+1)}{\Gamma(j+1)} + \frac{\Gamma'(j+\lambda+1)}{\Gamma(j+\lambda+1)}\right], \tag{14.10.3}$$

for $z \in \mathbb{C}$.

The Hankel function of the first kind with index λ is then defined by

$$H_\lambda^{(1)}(\cdot) := J_\lambda(\cdot) + i\,N_\lambda(\cdot). \tag{14.10.4}$$

The following lemma summarizes some well-known standard properties of Hankel functions of the first kind (cf. [1], [60]).

Lemma 14.71. *Let $\lambda \in \mathbb{R}$. Then the Hankel function of the first kind $H_\lambda^{(1)}$ is of class C^∞ in $(0, \infty)$ and the following properties hold for each $r > 0$:*

(1) $H_{-\lambda}^{(1)}(r) = e^{i\pi\lambda} H_\lambda^{(1)}(r)$

(2) $\dfrac{d}{dr}\left[r^\lambda H_\lambda^{(1)}(r)\right] = r^\lambda H_{\lambda-1}^{(1)}(r)$

(3) $\dfrac{d}{dr}\left[r^{-\lambda} H_\lambda^{(1)}(r)\right] = -r^{-\lambda} H_{\lambda+1}^{(1)}(r)$

(4) $\dfrac{d}{dr} H_\lambda^{(1)}(r) = H_{\lambda-1}^{(1)}(r) - \dfrac{\lambda}{r} H_\lambda^{(1)}(r)$

(5) $\dfrac{d}{dr} H_\lambda^{(1)}(r) = -H_{\lambda+1}^{(1)}(r) + \dfrac{\lambda}{r} H_\lambda^{(1)}(r)$

(6) $\dfrac{2\lambda}{r} H_\lambda^{(1)}(r) = H_{\lambda-1}^{(1)}(r) + H_{\lambda+1}^{(1)}(r)$

(7) $\dfrac{d}{dr} H_\lambda^{(1)}(r) = \dfrac{1}{2}\left[H_{\lambda-1}^{(1)}(r) - H_{\lambda+1}^{(1)}(r)\right]$

(8) for each $N \in \mathbb{N}$ we have $\left(\dfrac{d}{dr}\right)^N H_\lambda^{(1)}(r) = \dfrac{1}{2^N} \sum_{j=0}^{N} (-1)^j \binom{N}{j} H_{\lambda-N+2j}^{(1)}(r)$

(9) $H_\lambda^{(1)}(r) = \left(\dfrac{2}{\pi r}\right)^{1/2} e^{i(r - \lambda\pi/2 - \pi/4)} + O(r^{-3/2})$ *as $r \to \infty$.*

The nature of the singularity of Hankel functions of the first kind at the origin is studied next.

Lemma 14.72. *The following formulas are valid:*

$$\lim_{r \to 0^+} \frac{H_0^{(1)}(r)}{\dfrac{2i}{\pi}\ln(r)} = 1 \tag{14.10.5}$$

and

$$\lim_{r \to 0^+} \frac{H_\lambda^{(1)}(r)}{\dfrac{2^\lambda}{i\pi}\Gamma(\lambda)\, r^{-\lambda}} = 1, \qquad \forall \lambda \in (0, \infty). \tag{14.10.6}$$

Moreover,

$$\lim_{r \to 0^+} \frac{H_\lambda^{(1)}(r)}{\dfrac{e^{-i\pi\lambda}\, 2^{-\lambda}}{i\pi}\Gamma(-\lambda)\, r^{\lambda}} = 1, \qquad \forall \lambda \in (-\infty, 0). \tag{14.10.7}$$

Proof. The limits in (14.10.5) and (14.10.6) may be found in [60, 10.7.2 and 10.7.7]. If $\lambda \in (-\infty, 0)$, then item *(1)* in Lemma 14.71 and (14.10.6) imply (14.10.7). $\qquad\square$

We continue by collecting useful asymptotic expansions at infinity for Hankel functions of the first kind in the next lemma.

Lemma 14.73. *Let* $\lambda \in \mathbb{R}$, $N \in \mathbb{N}$, *and suppose* $r > 0$. *Then the following asymptotic expansions of Hankel functions of the first kind and their derivatives hold:*

$$H_\lambda^{(1)}(r) = O(r^{-1/2}) \quad as \ \ r \to \infty, \tag{14.10.8}$$

$$\frac{d}{dr} H_\lambda^{(1)}(r) = H_{\lambda-1}^{(1)}(r) + O(r^{-3/2}) \quad as \ \ r \to \infty, \tag{14.10.9}$$

$$\left(\frac{d}{dr}\right)^N H_\lambda^{(1)}(r) = O(r^{-1/2}) \quad as \ \ r \to \infty, \tag{14.10.10}$$

$$\left(\frac{d}{dr}\right)^N H_\lambda^{(1)}(r) = H_{\lambda-N}^{(1)}(r) + O(r^{-3/2}) \quad as \ \ r \to \infty. \tag{14.10.11}$$

Proof. Property (14.10.8) follows directly from item *(9)* in Lemma 14.71, while (14.10.8) combined with *(4)* in Lemma 14.71 yields (14.10.9). Also, (14.10.8) together with *(8)* in Lemma 14.71 gives (14.10.10). We are left with proving (14.10.11). First, we claim that for each $N \in \mathbb{N}$,

$$\left(\frac{d}{dr}\right)^N H_\lambda^{(1)}(r) = \sum_{j=0}^{N} C_{N,j}^\lambda \frac{1}{r^{N-j}} H_{\lambda-j}^{(1)}(r) \tag{14.10.12}$$

where $C_{N,j}^\lambda \in \mathbb{C}$ are constants depending only on N, j, and λ, defined as follows. Corresponding to $N = 1$ we take

$$C_{1,0}^\lambda := -\lambda \quad and \quad C_{1,1}^\lambda := 1, \tag{14.10.13}$$

then for each $N \in \mathbb{N}$ we recursively define

$$C_{N+1,j}^\lambda := \begin{cases} 1 & \text{if } j = N+1, \\ C_{N,j}^\lambda(2j - N - \lambda) + C_{N,j-1}^\lambda & \text{if } 1 \le j \le N, \\ C_{N,0}^\lambda(-N - \lambda) & \text{if } j = 0. \end{cases} \qquad (14.10.14)$$

We shall now prove that formula (14.10.12) holds for the choice of coefficients as in (14.10.13)–(14.10.14) via an induction argument over N. That the corresponding statement for $N = 1$ is true is seen directly from *(4)* in Lemma 14.71 and (14.10.13). Suppose next that (14.10.12) holds for some $N \in \mathbb{N}$. By differentiating (14.10.12) one more time and using *(4)* in Lemma 14.71 we arrive at

$$\left(\frac{d}{dr}\right)^{N+1} H_\lambda^{(1)}(r) = \sum_{j=0}^{N} C_{N,j}^\lambda \frac{j-N}{r^{N+1-j}} H_{\lambda-j}^{(1)}(r)$$

$$+ \sum_{j=0}^{N} C_{N,j}^\lambda \frac{1}{r^{N-j}} \left[H_{\lambda-j-1}^{(1)}(r) - \frac{\lambda-j}{r} H_{\lambda-j}^{(1)}(r) \right]$$

$$= \sum_{j=0}^{N} C_{N,j}^\lambda \frac{2j-N-\lambda}{r^{N+1-j}} H_{\lambda-j}^{(1)}(r) + \sum_{j=1}^{N+1} C_{N,j-1}^\lambda \frac{1}{r^{N+1-j}} H_{\lambda-j}^{(1)}(r)$$

$$= \sum_{j=0}^{N+1} C_{N+1,j}^\lambda \frac{1}{r^{N+1-j}} H_{\lambda-j}^{(1)}(r), \qquad (14.10.15)$$

where the last step uses the recurrence formula (14.10.14). This completes the proof of (14.10.12).

Moving on, observe that formula (14.10.12) may be written as

$$\left(\frac{d}{dr}\right)^{N} H_\lambda^{(1)}(r) = H_{\lambda-N}^{(1)}(r) + \sum_{j=0}^{N-1} C_{N,j}^\lambda \frac{1}{r^{N-j}} H_{\lambda-j}^{(1)}(r). \qquad (14.10.16)$$

When used in concert with (14.10.8), this now readily yields (14.10.11). $\square$

A combination of Lemma 14.73 and the Chain Rule yields asymptotic expansions for derivatives with respect to x of $H_\lambda^{(1)}(k|x|)$. These are collected in the next proposition. The reader is reminded that for each $x \in \mathbb{R}^n \setminus \{0\}$ we abbreviate $\widehat{x} := x/|x|$.

Proposition 14.74. *Let $\lambda \in \mathbb{R}$, $k \in (0, \infty)$, and fix a multi-index $\beta \in \mathbb{N}_0^n$ with $|\beta| > 0$. Then the following asymptotic expansions hold:*

$$\partial^\beta [H_\lambda^{(1)}(k|x|)] = H_{\lambda-|\beta|}^{(1)}(k|x|)(k\widehat{x})^\beta + O(|x|^{-3/2}) \quad as \ |x| \to \infty, \tag{14.10.17}$$

$$\partial^\beta [H_\lambda^{(1)}(k|x|)] = \left(\left(\frac{2}{\pi k|x|} \right)^{1/2} e^{i(k|x|-\lambda\pi/2-\pi/4)} \right)(ik\widehat{x})^\beta$$

$$+ O(|x|^{-3/2}) \quad as \ |x| \to \infty. \tag{14.10.18}$$

Proof. Fix a multi-index $\beta = (\beta_1, \ldots, \beta_n) \in \mathbb{N}_0^n$ of positive length. For starters observe that repeated applications of the Chain Rule give that, for $x \in \mathbb{R}^n \setminus \{0\}$,

$$\partial^\beta [H_\lambda^{(1)}(k|x|)] = k^{|\beta|} \left(\left(\frac{d}{dr} \right)^{|\beta|} H_\lambda^{(1)} \right)(k|x|) \times$$

$$\times (\partial_1(|x|))^{\beta_1} \cdot (\partial_2(|x|))^{\beta_2} \cdots (\partial_n(|x|))^{\beta_n}$$

$$+ \sum_{\ell=1}^{|\beta|-1} k^\ell \left(\left(\frac{d}{dr} \right)^\ell H_\lambda^{(1)} \right)(k|x|) \times$$

$$\times \sum_{\alpha_1+\cdots+\alpha_\ell=\beta} C_{\alpha_1,\ldots,\alpha_\ell} \partial^{\alpha_1}(|x|) \cdots \partial^{\alpha_\ell}(|x|), \tag{14.10.19}$$

with the convention that the sum over ℓ is void if $|\beta| = 1$. Above, $C_{\alpha_1,\ldots,\alpha_\ell}$ are constants depending only on the multi-indices $\alpha_1,\ldots,\alpha_\ell \in \mathbb{N}_0^n$. Since for each $j \in \{1,\ldots,n\}$ we have $\partial_j(|x|) = \frac{x_j}{|x|} = \widehat{x}_j$, we may write (14.10.19) as

$$\partial^\beta [H_\lambda^{(1)}(k|x|)] = \left(\left(\frac{d}{dr} \right)^{|\beta|} H_\lambda^{(1)} \right)(k|x|) (k\widehat{x})^\beta \tag{14.10.20}$$

$$+ \sum_{\ell=1}^{|\beta|-1} k^\ell \left(\left(\frac{d}{dr} \right)^\ell H_\lambda^{(1)} \right)(k|x|) \times$$

$$\times \sum_{\alpha_1+\cdots+\alpha_\ell=\beta} C_{\alpha_1,\ldots,\alpha_\ell} \partial^{\alpha_1}(|x|) \cdots \partial^{\alpha_\ell}(|x|),$$

again, with the convention that the sum over ℓ is void in the case when $|\beta| = 1$. Invoking (14.10.11), we further transform

$$\left(\left(\frac{d}{dr} \right)^{|\beta|} H_\lambda^{(1)} \right)(k|x|) (k\widehat{x})^\beta = (H_{\lambda-|\beta|}^{(1)}(k|x|) + O(|x|^{-3/2}))(k\widehat{x})^\beta$$

$$= H_{\lambda-|\beta|}^{(1)}(k|x|)(k\widehat{x})^\beta + O(|x|^{-3/2})$$

$$as \ |x| \to \infty. \tag{14.10.21}$$

On the other hand, note that $\partial^\gamma(|x|) = O(|x|^{1-|\gamma|})$ as $|x| \to \infty$, for any $\gamma \in \mathbb{N}_0^n$. On account of this observation we then conclude that, if $|\beta| > 1$, then for each index $\ell \in \{1, \ldots, |\beta| - 1\}$ we have

$$\sum_{\alpha_1 + \cdots + \alpha_\ell = \beta} C_{\alpha_1, \ldots, \alpha_\ell} \partial^{\alpha_1}(|x|) \cdots \partial^{\alpha_\ell}(|x|)$$

$$= O(|x|^{\ell - |\beta|}) = O(|x|^{-1}) \quad \text{as } |x| \to \infty. \tag{14.10.22}$$

Now (14.10.17) follows by combining (14.10.20), (14.10.21), (14.10.22), and (14.10.10). Finally, formula (14.10.18) is a direct consequence of (14.10.17) and item *(9)* in Lemma 14.71. $\square$

14.11 Tables of Fourier Transforms

Fourier Transforms of Schwartz Functions

f	$\mathcal{F}(f)$	Location
$e^{-\lambda|x|^2}$, $\lambda \in \mathbb{C}$, $\mathrm{Re}(\lambda) > 0$	$\left(\frac{\pi}{\lambda}\right)^{\frac{n}{2}} e^{-\frac{|\xi|^2}{4\lambda}}$	Example 3.22
$e^{-ax^2 + ibx}$, if $a > 0$ and $b \in \mathbb{R}$ are fixed	$\left(\frac{\pi}{a}\right)^{\frac{1}{2}} e^{-\frac{(\xi-b)^2}{4a}}$	Exercise 3.23
$e^{-(Ax) \cdot x}$, $A \in \mathcal{M}_{n\times n}(\mathbb{R})$, $A = A^\top$, $(Ax) \cdot x > 0$ for $x \in \mathbb{R}^n \setminus \{0\}$	$\dfrac{\pi^{\frac{n}{2}}}{\sqrt{\det A}} e^{-\frac{(A^{-1}\xi) \cdot \xi}{4}}$	Exercise 3.38
$e^{-(x_1^2 + x_1 x_2 + x_2^2)}$ for $(x_1, x_2) \in \mathbb{R}^2$	$\dfrac{2\pi}{\sqrt{3}} e^{-(\xi_1^2 - \xi_1 \xi_2 + \xi_2^2)/3}$	Exercise 3.39
$e^{-a|x|^2} \sin(x \cdot x_0)$, for $x \in \mathbb{R}^n$, if $a > 0$ and $x_0 \in \mathbb{R}^n$ are fixed	$\frac{1}{2i}\left(\frac{\pi}{a}\right)^{\frac{n}{2}}\left[e^{-\frac{|\xi-x_0|^2}{4a}} - e^{-\frac{|\xi+x_0|^2}{4a}}\right]$	Exercise 3.41

Fourier Transforms of Tempered Distributions in $\mathbb{R}$

Below $x, \xi \in \mathbb{R}$.

u	$\mathcal{F}(u)$	Location
$\frac{1}{x^2+a^2}$, $a \in (0,\infty)$ given	$\frac{\pi}{a}e^{-a|\xi|}$	Example 4.24
$\frac{1}{x^2-(b+ia)^2}$, $a \in (0,\infty)$ given	$\frac{\pi i e^{ib|\xi|}}{b+ia}e^{-a|\xi|}$	Exercise 4.122
$\frac{x}{x^2+a^2}$, $a \in (0,\infty)$ given	$-\pi i(\operatorname{sgn}\xi)e^{-a|\xi|}$	Example 4.31
$\chi_{[-a,a]}$, $a \in (0,\infty)$ given	$\begin{cases} 2\,\frac{\sin(a\xi)}{\xi} & \text{for } \xi \in \mathbb{R}\setminus\{0\} \\[2mm] 2a & \text{for } \xi = 0 \end{cases}$	Example 4.36
$\sin(b\,x)$, $b \in \mathbb{R}$ fixed	$-i\pi\delta_b + i\pi\delta_{-b}$	Example 4.38
$\cos(b\,x)$, $b \in \mathbb{R}$	$\pi\delta_b + \pi\delta_{-b}$	Exercise 4.127
$\sin(b\,x)\sin(c\,x)$, $b,c \in \mathbb{R}$ fixed	$-\frac{\pi}{2}(\delta_{b+c} - \delta_{b-c} - \delta_{c-b} + \delta_{-b-c})$	Example 4.39
H	$-i\text{P.V.}\,\frac{1}{\xi} - c\,\delta$	Exercise 4.119
$\text{P.V.}\,\frac{1}{x}$	$i\pi\operatorname{sgn}(\xi)$	Exercise 4.120
$\operatorname{sgn} x$	$-2i\,\text{P.V.}\,\frac{1}{\xi}$	Exercise 4.121
$|x|^k$, $k \in \mathbb{N}$ even	$2\pi i^k \delta^{(k)}$	Exercise 4.121
$|x|^k$, $k \in \mathbb{N}$ odd	$-2i^{k+1}(\text{P.V.}\,\frac{1}{\xi})^{(k)}$	Exercise 4.121
$\frac{\sin(b\,x)}{x}$, $b \in \mathbb{R}$ fixed	$\pi\operatorname{sgn}(b)\chi_{[-|b|,|b|]}$	Exercise 4.121
$\frac{\sin(b\,x)\sin(c\,x)}{x}$, $b,c \in \mathbb{R}$ fixed	$\frac{i\pi}{2}\operatorname{sgn}(c)\left[\chi_{[-b-|c|,-b+|c|]} - \chi_{[b-|c|,b+|c|]}\right]$	Exercise 4.121
$\sin(x^2)$	$\frac{\sqrt{\pi}}{2i}\left[e^{i\frac{\pi-\xi^2}{4}} - e^{-i\frac{\pi-\xi^2}{4}}\right]$	Exercise 4.121
$\ln|x|$	$-2\pi\gamma\delta - \pi w_{\chi_{(-1,1)}}$	Exercise 4.121

$$w_{\chi_{(-1,1)}} \text{ from } (4.6.22),\ \gamma \text{ from } (4.6.24)$$

Fourier Transforms of Tempered Distributions

Below $x, \xi \in \mathbb{R}^n$ and $x', \xi' \in \mathbb{R}^{n-1}$.

u	$\mathcal{F}(u)$	Location
δ	1	Example 4.22
1	$(2\pi)^n \delta$	Exercise 4.33
$e^{-ia|x|^2}$, $a \in (0,\infty)$ given	$\left(\frac{\pi}{a}\right)^{\frac{n}{2}} e^{-i\frac{n\pi}{4}} e^{i\frac{|\xi|^2}{4a}}$	Example 4.25
$e^{ia|x|^2}$, $a \in (0,\infty)$ given	$\left(\frac{\pi}{a}\right)^{\frac{n}{2}} e^{i\frac{n\pi}{4}} e^{-i\frac{|\xi|^2}{4a}}$	Example 4.40
$|x|^{-\lambda}$, $\lambda \in (0,n)$ fixed	$2^{n-\lambda} \pi^{\frac{n}{2}} \frac{\Gamma\left(\frac{n-\lambda}{2}\right)}{\Gamma\left(\frac{\lambda}{2}\right)} |\xi|^{\lambda-n}$	Proposition 4.64
$\frac{x_j}{|x|^n}$, $n \geq 2$	$-i\,\omega_{n-1} \frac{\xi_j}{|\xi|^2}$	Corollary 4.65
$\frac{x_j}{|x|^{\lambda+2}}$, $n \geq 2$, $\lambda \in [0, n-1)$, $j \in \{1,\dots,n\}$	$-i\,2^{n-\lambda-1} \pi^{\frac{n}{2}} \frac{\Gamma\left(\frac{n-\lambda}{2}\right)}{\Gamma\left(\frac{\lambda}{2}+1\right)} \frac{\xi_j}{|\xi|^{n-\lambda}} \cdot$	Corollary 4.65
$\frac{x_j x_k}{|x|^n}$, $n \geq 3$, $j,k \in \{1,\dots,n\}$	$\omega_{n-1} \frac{\delta_{jk}}{|\xi|^2} - 2\omega_{n-1} \frac{\xi_j \xi_k}{|\xi|^4}$	Exercise 7.79
$\mathrm{P.V.}\left(\partial_j\left[\frac{x_k}{|x|^n}\right]\right)$, $j,k \in \{1,\dots,n\}$	$\omega_{n-1} \frac{\xi_j \xi_k}{|\xi|^2} - \frac{\omega_{n-1}}{n} \delta_{jk}$	Proposition 4.73
$\mathrm{P.V.}\,\Theta$, with Θ as in (4.4.1)	$-\int_{S^{n-1}} \Theta(\omega) \log(i(\xi \cdot \omega))\, d\sigma(\omega)$	Theorem 4.74
$\mathrm{P.V.}\left(\frac{x_j}{|x|^{n+1}}\right)$, $j \in \{1,\dots,n\}$	$-\frac{i\omega_n}{2} \frac{\xi_j}{|\xi|}$	Proposition 4.84
$\frac{2}{\omega_{n-1}} \frac{t}{(t^2+|x'|^2)^{\frac{n}{2}}}$ with $t > 0$ fixed	$e^{-t|\xi'|}$	Proposition 4.90
$\frac{2}{\omega_{n-1}} \frac{x_j}{(t^2+|x'|^2)^{\frac{n}{2}}}$ $t > 0$ fixed, $j \in \{1,\dots,n-1\}$	$-i\frac{\xi_j}{|\xi'|} e^{-t|\xi'|}$	Proposition 4.92

References

1. M. Abramowitz and I. A. Stegun, *Handbook of Mathematical Functions*, Dover, New York, 1972.
2. R. A. Adams and J. J. F. Fournier, *Sobolev Spaces*, Pure and Applied Mathematics, 140, 2nd edition, Academic Press, 2003.
3. D. R. Adams and L. I. Hedberg, *Function Spaces and Potential Theory*, Grundlehren der Mathematischen Wissenschaften, Vol. 314, Springer, Berlin, 1996.
4. R. Alvarado, D. Brigham, V. Maz'ya, M. Mitrea, and E. Ziadé, *On the regularity of domains satisfying a uniform hour-glass condition and a sharp version of the Hopf-Oleinik Boundary Point Principle*, Journal of Mathematical Sciences, 176 (2011), no. 3, 281–360.
5. S. Axler, P. Bourdon, and W. Ramey, *Harmonic Function Theory*, 2nd edition, Graduate Texts in Mathematics, 137, Springer, New York, 2001.
6. F. Brackx, R. Delanghe, and F. Sommen, *Clifford Analysis*, Research Notes in Mathematics, 76, Pitman, Advanced Publishing Program, Boston, MA, 1982.
7. A. P. Calderón and A. Zygmund, *On the existence of certain singular integrals*, Acta Math., 88 (1952), 85–139.
8. M. Christ, *Lectures on Singular Integral Operators*, CBMS Regional Conference Series in Mathematics, No. 77, AMS, 1990.
9. G. David and S. Semmes, *Singular Integrals and Rectifiable Sets in $\mathbb{R}^n$: Beyond Lipschitz Graphs*, Astérisque, No. 193, 1991.
10. J. J. Duistermaat and J. A. C. Kolk, *Distributions. Theory and applications*, Translated from the Dutch by J. P. van Braam Houckgeest, Cornerstones, Birkhäuser Boston, Inc., Boston, MA, 2010.
11. J. Duoandikoetxea. *Fourier Analysis*, Graduate Studies in Mathematics, AMS, 2000.
12. R. E. Edwards, *Functional Analysis. Theory and Applications*, Dover Publications, Inc., New York, 1995.
13. L. Ehrenpreis, *Solution of some problems of division. I. Division by a polynomial of derivation*, Amer. J. Math. 76, (1954).
14. L. Evans, *Partial Differential Equations*, 2nd edition, Graduate Studies in Mathematics, 19, American Mathematical Society, Providence, RI, 2010.
15. L. C. Evans and R. F. Gariepy, *Measure Theory and Fine Properties of Functions*, Studies in Advanced Mathematics, CRC Press, Boca Raton, FL, 1992.
16. C. Fefferman and E. M. Stein, *H^p spaces of several variables*, Acta Math., 129 (1972), no. 3-4, 137–193.
17. G. Friedlander and M. Joshi, *Introduction to the Theory of Distributions*, Cambridge University Press, 2nd edition, 1998.
18. G. B. Folland, *Real Analysis, Modern Techniques and Their Applications*, 2-nd edition, John Wiley and Sons, Inc., New York, 1999.

© Springer Nature Switzerland AG 2018

D. Mitrea, *Distributions, Partial Differential Equations, and Harmonic Analysis*, Universitext, https://doi.org/10.1007/978-3-030-03296-8

19. J. Garcia-Cuerva and J. Rubio de Francia, *Weighted Norm Inequalities and Related Topics*, North Holland, Amsterdam, 1985.

20. I. M. Gel'fand and G. E. Šilov, *Generalized Functions, Vol. 1: Properties and Operations*, Academic Press, New York and London, 1964.

21. I. M. Gel'fand and N. Y. Vilenkin, *Generalized Functions, Vol. 4: Applications of Harmonic Analysis*, Academic Press, New York and London, 1964.

22. I. M. Gel'fand, M. I. Graev, and N. Y. Vilenkin, *Generalized Functions, Vol. 5: Integral Geometry and Representation theory*, Academic Press, 1966.

23. I. M. Gel'fand and G. E. Šilov, *Generalized Functions, Vol. 2: Spaces of Fundamental and Generalized Functions*, Academic Press, New York and London, 1968.

24. D. Gilbarg and N.S. Trudinger, *Elliptic Partial Differential Equations of Second Order*, Springer, 2001.

25. J. Gilbert and M. A. M. Murray, *Clifford Algebras and Dirac Operators in Harmonic Analysis*, Cambridge Studies in Advanced Mathematics, 26, Cambridge University Press, Cambridge, 1991.

26. L. Grafakos, *Classical and modern Fourier analysis*, Pearson Education, Inc., Upper Saddle River, NJ, 2004.

27. G. Grubb, *Distributions and Operators*, Springer, 2009.

28. M. A. Al-Gwaiz, *Theory of Distributions*, Pure and Applied Mathematics, CRC Press, 1992.

29. G. Hardy and J. Littlewood, *Some properties of conjugate functions*, J. Reine Angew. Math., 167 (1932), 405–423.

30. K. Hoffman, *Banach Spaces of Analytic Functions*, Dover Publications, 2007.

31. S. Hofmann, M. Mitrea and M. Taylor, *Singular integrals and elliptic boundary problems on regular Semmes-Kenig-Toro domains*, International Math. Research Notices, 2010 (2010), 2567–2865.

32. L. Hörmander, *On the division of distributions by polynomials*, Ark. Mat., 3 (1958), 555–568.

33. L. Hörmander, *Linear Partial Differential Operators*, Die Grundlehren der Mathematischen Wissenschaften in Einzeldarstellungen, Vol. 116, Springer, 1969.

34. L. Hörmander, *The Analysis of Linear Partial Differential Operators I*, Distribution Theory and Fourier Analysis, Springer-Verlag, 2003.

35. L. Hörmander, *The Analysis of Linear Partial Differential Operators II*, Differential Operators with Constant Coefficients, Springer, 2003.

36. R. Howard, *Rings, determinants, and the Smith normal form*, preprint (2013). http://www.math.sc.edu/ howard/Classes/700c/notes2.pdf

37. G.C. Hsiao and W.L. Wendland, *Boundary integral equations*, Applied Mathematical Sciences, 164, Springer, Berlin, 2008.

38. F. John, *Plane Waves and Spherical Means Applied to Partial Differential Equations*, Interscience Publishers, New York-London, 1955.

39. C. E. Kenig, *Harmonic Analysis Techniques for Second Order Elliptic Boundary Value Problems*, CBMS Regional Conference Series in Mathematics, 83, American Mathematical Society, Providence, RI, 1994.

40. S. G. Krantz and H. R. Parks, *A Primer of Real Analytic Functions*, 2-nd edition, Birkhäuser, Boston, 2002.

41. P. S. Laplace, *Mémoire sur la théorie de l'anneau de Saturne*, Mém. Acad. Roy. Sci. Paris (1787/1789), 201–234.

42. P. S. Laplace, J. École Polytéch. cah., 15 (1809), p. 240 (quoted in Enc. Math. Wiss. Band II, 1. Teil, 2. Hälfte, p. 1198).

43. G. Leoni, *A First Course in Sobolev Spaces*, Vol. 105, Graduate Studies in Mathematics, American Mathematical Soc., 2009.

44. S. Lojasiewicz, *Division d'une distribution par une fonction analytique de variables réelles*, C. R. Acad. Sci. Paris, 246 (1958), 683–686.

45. B. Malgrange, *Existence et approximation des solutions des équations aux dérivées partielles et des équations de convolution*, (French) Ann. Inst. Fourier, Grenoble 6 (1955–1956), p. 271–355.

46. E. Marmolejo-Olea, D. Mitrea, I. Mitrea, and M. Mitrea, *Radiation Conditions and Integral Representations for Clifford Algebra-Valued Null-Solutions of the Helmholtz Operator*, Journal of Mathematical Sciences, Vol. 231, no. 3 (2018), 367–472; https://doi.org/10.1007/s10958-018-3826-9.

47. V. G. Maz'ya, *Sobolev Spaces*, Springer, Berlin-New York, 1985.

48. V. G. Maz'ya, *Sobolev spaces with applications to elliptic partial differential equations*, Second, revised and augmented edition, Grundlehren der Mathematischen Wissenschaften [Fundamental Principles of Mathematical Sciences], 342, Springer, Heidelberg, 2011.

49. V. G. Maz'ya and T. Shaposhnikova, *Theory of Sobolev multipliers. With applications to differential and integral operators*, Grundlehren der Mathematischen Wissenschaften [Fundamental Principles of Mathematical Sciences], 337, Springer, Berlin, 2009.

50. W. McLean, *Strongly Elliptic Systems and Boundary Integral Equations*, Cambridge University Press, Cambridge, 2000.

51. E.J. McShane, *Extension of range of functions*, Bull. Amer. Math. Soc., 40 (1934), 837–842.

52. Y. Meyer, *Ondelettes et Opérateurs II. Opérateurs de Calderón-Zygmund*, Actualités Mathématiques, Hermann, Paris, 1990.

53. Y. Meyer and R. R. Coifman, *Ondelettes et Opérateurs III. Opérateurs Multilinéaires*, Actualités Mathématiques, Hermann, Paris, 1991.

54. D. Mitrea, I. Mitrea, M. Mitrea, and S. Monniaux, *Groupoid Metrization Theory With Applications to Analysis on Quasi-Metric Spaces and Functional Analysis*, Birkhäuser, Springer New York, Heidelberg, Dordrecht, London, 2013.

55. D. Mitrea, M. Mitrea and M. Taylor, *Layer potentials, the Hodge Laplacian and global boundary problems in nonsmooth Riemannian manifolds*, Memoirs of American Mathematical Society, Vol. 150, No. 713, 2001.

56. D. Mitrea and H. Rosenblatt, *A General Converse Theorem for Mean Value Theorems in Linear Elasticity*, Mathematical Methods for the Applied Sciences, Vol. 29, No. 12 (2006), 1349–1361.

57. I. Mitrea and M. Mitrea, *Multi-Layer Potentials and Boundary Problems for Higher-Order Elliptic Systems in Lipschitz Domains*, Lecture Notes in Mathematics 2063, Springer, 2012.

58. M. Mitrea, *Clifford Wavelets, Singular Integrals, and Hardy Spaces*, Lecture Notes in Mathematics, 1575, Springer, Berlin, 1994.

59. C. B. Morrey, *Second order elliptic systems of differential equations. Contributions to the theory of partial differential equations*, Ann. Math. Studies, 33 (1954), 101–159.

60. F. Olver, *NIST Digital Library of Mathematical Functions*, Chapter 10, http://dlmf.nist.gov/10, [edited by] Frank W. J. Olver and L. C. Maximon, New York: Cambridge University Press: NIST, 2010.

61. S. D. Poisson, *Remarques sur une équation qui se présente dans la théorie de l'attraction des sphéroides*, Bulletin de la Société Philomathique de Paris, 3 (1813), 388–392.

62. S. D. Poisson, *Sur l'intégrale de l'équation relative aux vibrations des surfaces élastiques et au mouvement des ondes*, Bulletin de la Société Philomathique de Paris, (1818), 125–128.

63. H. L. Royden, *Real Analysis*, Macmillian Publishing Co., Inc., New York, 1968.

64. W. Rudin, *Real and Complex Analysis*, 2nd edition, International Series in Pure and Applied Mathematics, McGraw-Hill, Inc., 1986.

65. W. Rudin, *Functional Analysis*, 2nd edition, International Series in Pure and Applied Mathematics, McGraw-Hill, Inc., 1991.

66. L. Schwartz, *Théorie des Distributions, I, II*, Hermann, Paris, 1950-51.

67. Z. Shapiro, *On elliptical systems of partial differential equations*, C. R. (Doklady) Acad. Sci. URSS (N. S.), 46 (1945).

68. E.M. Stein, *Singular Integrals and Differentiability Properties of Functions*, Princeton Mathematical Series, 30, Princeton University Press, Princeton, N.J., 1970.

69. E.M. Stein, *Harmonic Analysis: Real-Variable Methods, Orthogonality, and Oscillatory Integrals*, Princeton Mathematical Series, 43, Monographs in Harmonic Analysis, III, Princeton University Press, Princeton, NJ, 1993.

70. E. M. Stein and G. Weiss, *Introduction to Fourier Analysis on Euclidean Spaces*, Princeton University Press, Princeton, New Jersey, 1971.

71. R. Strichartz, *A guide to Distribution Theory and Fourier Transforms*, World Scientific Publishing Co., Inc., River Edge, NJ, 2003.

72. H. Tanabe, *Functional Analytic Methods for Partial Differential Equations*, Monographs and Textbooks in Pure and Applied Mathematics, 204, Marcel Dekker, Inc., New York, 1997.

73. M. E. Taylor, *Pseudodifferential Operators*, Princeton Mathematical Series, 34, Princeton University Press, Princeton, N.J., 1981.

74. M. E. Taylor, *Partial Differential Equations. I. Basic Theory*, Applied Mathematical Sciences, 115, Springer, New York, 1996.

75. M. E. Taylor, *Tools for PDE. Pseudodifferential Operators, Paradifferential Operators, and Layer Potentials*, Mathematical Surveys and Monographs, 81, American Mathematical Society, Providence, RI, 2000.

76. F. Tréves, *Topological Vector Spaces, Distributions and Kernels*, Dover Publications, 2006.

77. V. S. Vladimirov, *Equations of Mathematical Physics*, Marcel Dekker Inc., 1971.

78. V. S. Vladimirov, *Methods of the Theory of Generalized Functions*, Analytical Methods and Special Functions, 6, Taylor & Francis, London, 2002.

79. V. Voltera, *Sur les vibrations des corps élastiques isotropes*, Acta Math., 18 (1894), 161–232.

80. W. Wendland, *Integral Equation Methods for Boundary Value Problems*, Springer, 2002.

81. J. T. Wloka, B. Rowley, and B. Lawruk, *Boundary Value Problems for Elliptic Systems*, Cambridge University Press, 1995.

82. W. P. Ziemer, *Weakly Differentiable Functions: Sobolev Spaces and Functions of Bounded Variation*, Graduate Text in Mathematics, 120, Springer, New York, 1989.

Subject Index

© Springer Nature Switzerland AG 2018

D. Mitrea, *Distributions, Partial Differential Equations, and Harmonic Analysis*,
Universitext, https://doi.org/10.1007/978-3-030-03296-8

Symbol Index

© Springer Nature Switzerland AG 2018
D. Mitrea, *Distributions, Partial Differential Equations, and Harmonic Analysis*,
Universitext, https://doi.org/10.1007/978-3-030-03296-8

Printed in the United States
By Bookmasters